Collected Works

Eugenio Calabi
© Holly Chen, Philadelphia, 2011

Eugenio Calabi

Collected Works

Edited by Jean-Pierre Bourguignon, Xiuxiong Chen
and Simon Donaldson

With Contributions by Shing-Tung Yau, Blaine Lawson,
Marcel Berger and Claude LeBrun

Springer

Eugenio Calabi
University of Pennsylvania
Philadelphia, PA, USA

Editors
Jean-Pierre Bourguignon
Le Bois-Marie
Inst Hautes Etudes Scientifiques
Bures-sur-Yvette, France

Xiuxiong Chen
Mathematics Department
Stony Brook University
Stony Brook, NY, USA

Simon Donaldson
Simons Center for Geometry
and Physics
State University of New York
Stony Brook, NY, USA

ISBN 978-3-662-62133-2

Mathematics Subject Classification: 53C55, 32F07, 35J96, 53A10, 53A15, 53C21, 53C24, 53C26

This Springer imprint is published by the registered company Springer-Verlag GmbH, DE part of Springer Nature.
The registered company address is: Heidelberger Platz 3, 14197 Berlin, Germany

Preface

It is with great pleasure that we accept the privilege to present this volume of collected works of the great differential geometer Eugenio Calabi. Nearly all of Calabi's papers are reproduced here, and in addition we have been most fortunate to obtain invaluable contributions from Blaine Lawson, Claude LeBrun and Shing-Tung Yau discussing aspects of Calabi's work and their interactions with Calabi. We also include a testimony by the late Marcel Berger, an article on Calabi's papers on affine geometry and results of Bernstein type, written by two of us, and a review on the impact of Calabi's work in Kählerian geometry going back to the celebration of Calabi's 70th birthday written by the first editor.

We will not attempt to give a further comprehensive account of Calabi's work and its huge impact and influence in this preface, instead referring the reader to the articles below. But it is impossible not to mention here the extraordinary vision of his contributions, particularly in the case of complex differential geometry where his conjecture on realising any form in the first Chern class as the Ricci curvature of a Kähler metric from the 1950s, known as the "Calabi conjecture", dominated the subject for half a century. This led to exceptional developments in the search for Kähler–Einstein metrics. While many of the main questions about Kähler–Einstein metrics, and more generally metrics with constant scalar curvature, are now settled, the "extremal metrics" which Calabi introduced in the 1980s (although the roots go back to his 1954 International Congress of Mathematicians paper), present another circle of questions which are largely open and furnish a rich and active area of current research. Calabi's visionary contributions will certainly continue to shape the course of this subject far into the future. (We refer to the article of Claude LeBrun below for a detailed discussion of these developments in complex differential geometry.)

We also want to emphasise that the scope of Calabi's work takes in many other topics not discussed in these survey articles. One instance is Calabi's 1970 paper *On the group of automorphisms of a symplectic manifold.* In 1970, symplectic geometry scarcely existed as a mathematical subject, but exploded into a huge field in the 1980s. Among other things, Calabi's paper introduced the fundamental "Calabi homomorphism", which is central in this field. For example, it is required for the statement of the Arnold conjecture which drove much of this explosive development. Another instance is his 1958 paper *Construction and properties of some 6-dimensional almost-complex manifolds*, which introduces constructions highly relevant to the modern theory of exceptional holonomy. In the same vein, in his 1978 paper *Métriques kählériennes et fibrés holomorphes*, Calabi was also the first to discuss and give examples of hyperkähler metrics, a topic which has grown into a very active subject, in particular because of its somewhat unexpected relevance in theoretical physics. A third instance occurs in his 1967 paper *On Ricci curvature and geodesics*, which establishes an integral bound on the Hessian of the distance function along a geodesic which has been a foundation of recent work by Tobias Colding and Aaron Naber on the convexity of regular sets.

There can be no question that Calabi was one of the most important differential geometers of the twentieth century. These collected papers help the reader to appreciate the full scope of his work and to gain fascinating insights into the modern history of the subject, going back more than 60 years. In the 1950s there were few global results in Riemannian geometry, machinery such as sheaf theory was just entering complex geometry, differential topology was in its infancy and the interaction with PDEs (especially nonlinear PDEs) was very limited. Now the picture is very different. In addition there are vast and rich new research fields deriving from physics—the remarkable properties of Calabi–Yau manifolds being a prime example. All of these different strands can be traced through these collected papers.

We are very grateful to the staff of Springer for their excellent and efficient work on this volume and to all the publishers of the journals where the papers first appeared for permission to reproduce them here. Many thanks are due to Élisabeth Jasserand and Sylvie Sallé from the I.H.É.S. for their tireless work in gathering the papers and contacting publishers.

Bures-sur-Yvette — Jean-Pierre Bourguignon
Stony Brook, NY — Xiuxiong Chen
Stony Brook, NY — Simon Donaldson

Contents

Part II: Collected Works

Eugenio Calabi's Short Biography

Jean-Pierre Bourguignon[1]

Eugenio Calabi, born 11 May 1923 in Milan, Italy, is an American mathematician who contributed to many areas of geometry, mainly differential geometry, and partial differential equations and their geometric applications. He grew up in Italy until his family moved to France, and later to the United States. He became an American citizen in 1943.

He studied at MIT in 1946 as a Putnam Fellow for a Bachelor of Science in chemical engineering. He got a masters in mathematics from the University of Illinois at Urbana in 1947, and received a PhD from Princeton University under Solomon Bochner in 1950.

His first position was an assistant professorship at the Louisiana State University that he held from 1951 to 1954. He visited the California Institute of Technology during the academic year 1954–1955. He came to the University of Minnesota in 1955 as an assistant professor, a position which he left in 1964 to join the mathematics faculty of the University of Pennsylvania in Philadelphia. In 1967 he was appointed there to the Thomas A. Scott Professor of Mathematics Chair. After 30 years of service at the University of Pennsylvania, he assumed emeritus status in 1992.

He also held visiting positions at the Institute for Advanced Study in Princeton in the academic years 1958–1959 and 1979–1980, at the Università degli Studi in Pisa in the academic year 1962–1963 and at the Institut des Hautes Études Scientifiques in the academic year 1968–1969.

He was elected a member of the United States National Academy of Sciences in 1982. He received the Steele Prize from the American Mathematical Society (AMS) in 1991 for his contributions to differential geometry. In 2012 he was named a Fellow of the AMS. He received an honorary degree from the University of Pennsylvania in 2013.

[1] Jean-Pierre Bourguignon
Institut des Hautes Études Scientifiques, Université Paris-Saclay, Le Bois-Marie, Bures-sur-Yvette, France
e-mail: JPB@ihes.fr

Bibliographic List of Works

Volume Editors

[1] Calabi, E. (1950). *Isometric, Complex Analytic Imbedding of Kähler Manifolds.* Princeton Thesis. (Published in revised form as [7].)

[2] Calabi, E. & Dvoretzky, A. (1951). Convergence- and sum-factors for series of complex numbers. *Trans. Amer. Math. Soc.* 70, 177–194.

[3] Calabi, E. & Spencer, D. C. (1951). Completely integrable almost complex manifolds. *Bull. Amer. Math. Soc.* 57, 254–255.

[4] Calabi, E. (1953). Metric Riemann surfaces. In: Ahlfors, L. V., Calabi, E., Morse, M., Sario, L., Spencer, D. C. (eds.) *Contributions to the Theory of Riemann Surfaces*, 77–85. Annals of Mathematics Studies, No. 30. Princeton University Press, Princeton, NJ.

[5] Calabi, E. & Rosenlicht, M. (1953). Complex analytic manifolds without countable base. *Proc. Amer. Math. Soc.* 4, 335–340.

[6] Calabi, E. & Eckmann, B. (1953). A class of compact, complex manifolds which are not algebraic. *Ann. of Math.* (2) 58, 494–500.

[7] Calabi, E. (1953). Isometric imbedding of complex manifolds. *Ann. of Math.* (2) 58, 1–23.

[8] Calabi, E. (1954). The space of Kähler metrics. In: Proceedings of the International Congress of Mathematicians, 1954: Amsterdam, September 2–9. Vol. II: Short lectures, 206–207. E. P. Noordhoff, Groningen. (Republished in [53].)

[9] Calabi, E. (1954). The variation of Kähler metrics. I. The structure of the space. *Bull. Amer. Math. Soc.*, 60, 167–168.

[10] Calabi, E. (1954). The variation of Kähler metrics. II. A minimum problem. *Bull. Amer. Math. Soc.*, 60, 168.

[11] Calabi, E. (1957). On Kähler manifolds with vanishing canonical class. In: Fox, R. H. (ed.) *Algebraic Geometry and Topology. A Symposium in Honor of S. Lefschetz*, 78–89. Princeton University Press, Princeton, NJ. (Republished in [54].)

[12] Calabi, E. (1957). Closed, locally euclidean, 4-dimensional manifolds. *Bull. Amer. Math. Soc.* 63, 135.

[13] Calabi, E. et al. (1957). Problems and questions. *Mathematics Magazine* 30(3), 167–173.

[14] Calabi, E. et al. (1957). Problems and questions. *Mathematics Magazine* 31(1), 51–57.

[15] Calabi, E. (1958). Construction and properties of some 6-dimensional almost complex manifolds. *Trans. Amer. Math. Soc.* 87, 407–438.

[16] Calabi, E. (1958). Improper affine hyperspheres of convex type and a generalization of a theorem by K. Jörgens. *Michigan Math. J.* 5, 105–126.
[17] Calabi, E. (1958). An extension of E. Hopf's maximum principle with an application to Riemannian geometry. *Duke Math. J.* 25, 45–56.
[18] Calabi, E. (1959). Errata: An extension of E. Hopf's maximum principle with an application to Riemannian geometry. *Duke Math. J.* 26, 706.
[19] Calabi, E. & Vesentini, E. (1959). Sur les variétés complexes compactes localement symétriques. *Bull. Soc. Math. France* 87, 311–317.
[20] Calabi, E. & Vesentini, E. (1960). On compact, locally symmetric Kähler manifolds. *Ann. of Math.* (2) 71, 472–507.
[21] Calabi, E. (1961). On compact, Riemannian manifolds with constant curvature. I. In: C. B. Allendoerfer (Ed.) *Differential Geometry.* Proc. Sympos. Pure Math., Vol. III, 155–180. American Mathematical Society, Providence, RI.
[22] Calabi, E. & Markus, L. (1962). Relativistic space forms. *Ann. of Math.* (2) 75, 63–76.
[23] Calabi, E. (1964). Linear systems of real quadratic forms. *Proc. Amer. Math. Soc.* 15, 844–846.
[24] Calabi, E. (1966). Quasi-surjective mappings and a generalization of Morse theory. In: Proc. U.S.-Japan Seminar in Differential Geometry (Kyoto, 1965), 13–16. Nippon Hyoronsha, Tokyo.
[25] Calabi, E. (1967). Minimal immersions of surfaces in Euclidean spheres. *J. Differential Geometry* 1, 111–125.
[26] Calabi, E. (1967). On Ricci curvature and geodesics. *Duke Math. J.* 34, 667–676.
[27] Calabi, E. (1968). On differentiable actions of compact Lie groups on compact manifolds. In: *Proc. Conf. on Transformation Groups* (New Orleans, La., 1967), 210–213. Springer, New York.
[28] Calabi, E. (1968). Quelques applications de l'analyse complexe aux surfaces d'aire minima. In: Rossi, H. *Topics in Complex Manifolds.* Séminaire de Mathématiques Supérieures, N° 30 (Été, 1967). Les Presses de l'Université de Montréal, Montréal, Que.
[29] Calabi, E. (1969). An intrinsic characterization of harmonic one-forms. In: *Global Analysis* (Papers in honor of K. Kodaira), 101–117. Univ. Tokyo Press, Tokyo.
[30] Calabi, E. (1970). On the group of automorphisms of a symplectic manifold. In: *Problems in Analysis* (Lectures at the Sympos. in honor of Salomon Bochner, Princeton Univ., Princeton, NJ, 1969), 1–26. Princeton Univ. Press, Princeton, NJ.
[31] Calabi, E. & Hartman, P. (1970). On the smoothness of isometries. *Duke Math. J.* 37, 741–750.
[32] Calabi, E. (1970). Examples of Bernstein problems for some nonlinear equations. In: *Global Analysis* (Proc. Sympos. Pure Math., Vol. XV, Berkeley, Calif., 1968), 223–230. Amer. Math. Soc., Providence, RI.
[33] Calabi, E. (1971). Über singuläre symplektische Strukturen. In: Tagungsbericht 23/1969, *Differentialgeometrie im Großen* 13.7. bis 19.7.1969, Mathematisches Forschungsinstitut Oberwolfach, p. 4.
[34] Calabi, E. (1972). Complete affine hyperspheres. I. In: *Symposia Mathematica,* Vol. X (Convegno di Geometria Differenziale, INDAM, Rome, 1971/1972), 19–38. Academic Press, London.
[35] Calabi, E. (1975). A construction of nonhomogeneous Einstein metrics. In: *Differential Geometry* (Proc. Sympos. Pure Math., Vol. XXVII, Stanford Univ., Stanford, Calif., 1973), Part 2, 17–24. Amer. Math. Soc., Providence, RI.
[36] Calabi, E. & Wilf, H.S. (1977). On the sequential and random selection of subspaces over a finite field. *J. Combinatorial Theory* Ser. A 22, No. 1, 107–109.
[37] Calabi, E. (1979). Métriques kählériennes et fibrés holomorphes. *Ann. Sci. École Norm. Sup.* (4) 12, n° 2, 269–294.

[38] Calabi, E. (1980). Isometric families of Kähler structures. In: *The Chern Symposium 1979* (Proc. Internat. Sympos., Berkeley, Calif., 1979), 23–39. Springer, New York-Berlin.
[39] Calabi, E. (1981). Géométrie différentielle affine des hypersurfaces. In: *Bourbaki Seminar*, Vol. 1980/81, 189–204. Lecture Notes in Math., 901. Springer, Berlin-New York.
[40] Calabi, E. (1982). Linear systems of real quadratic forms. II. *Proc. Amer. Math. Soc.* 84, No. 3, 331–334.
[41] Calabi, E. (1982). Extremal Kähler metrics. In: *Seminar on Differential Geometry*, 259–290. Ann. of Math. Stud., 102. Princeton Univ. Press, Princeton, NJ.
[42] Calabi, E. (1982). Hypersurfaces with maximal affinely invariant area. *Amer. J. Math.* 104, No. 1, 91–126.
[43] Calabi, E. (1985). Extremal Kähler metrics. II. In: *Differential Geometry and Complex Analysis*, Vol. dedic. H. E. Rauch, 95–114. Springer, Berlin.
[44] Calabi, E. (1988). Convex affine maximal surfaces. *Result. Math.* 13, No. 3-4, 199–223.
[45] Calabi, E. (1990). Affine differential geometry and holomorphic curves. In: *Complex Geometry and Analysis*, Proc. Int. Symp. in Honour of E. Vesentini (Pisa, Italy, 1988), 15–21. Lecture Notes in Math., 1422. Springer, Berlin.
[46] Calabi, E. & Cao, J. (1992). Simple closed geodesics on convex surfaces. *J. Differential Geom.* 36, no. 3, 517–549.
[47] Beukers, F., Kolk, J. A. C., Calabi, E. (1993). Sums of generalized harmonic series and volumes. *Nieuw Arch. Wisk.* (4) 11, nr. 3, 217–224.
[48] Calabi, E. & Gluck, H. (1993). What are the best almost-complex structures on the 6-sphere? In: Greene, Robert (ed.) et al., *Differential Geometry: Part 2: Geometry in Mathematical Physics and Related Topics.* Proceedings of a summer research institute, held at the University of California (Los Angeles, CA, USA, July 8–28, 1990), 99–106, Proc. Sympos. Pure Math., 54, Part 2. Amer. Math. Soc., Providence, RI.
[49] Calabi, E. (1996). Extremal isosystolic metrics for compact surfaces. In: Arthur L. (ed.) *Actes de la table ronde de géométrie différentielle en l'honneur de Marcel Berger* (Luminy, France, 12–18 juillet, 1992), 167–204. Sémin. Congr., 1, Soc. Math. France, Paris.
[50] Calabi, E., Olver, P. J., Tannenbaum, A. (1996). Affine geometry, curve flows, and invariant numerical approximations. *Adv. Math.* 124, No. 1, 154–196.
[51] Bourguignon, J.-P., Calabi, E., Eells, J., García-Prada, O., Gromov, M. (2001). Where does geometry go? A research and education perspective. In: Fernández, M. (ed.) et al. *Global Differential Geometry: The Mathematical Legacy of Alfred Gray* (Bilbao, Spain, September 18–23, 2000), 442–457. Contemp. Math., 288. Amer. Math. Soc., Providence, RI.
[52] Calabi, E. & Chen, X. X. (2002). The space of Kähler metrics. II. *J. Differential Geom.* 61, No. 2, 173–193.
[53] Calabi, E. (2018). The space of Kähler metrics. In: Ji, Lizhen (ed.), *Complex geometry from Riemann to Kähler–Einstein and Calabi–Yau.* Somerville, MA: International Press; Beijing: Higher Education Press. Adv. Lect. Math. (ALM) 38, 209–210. (Republication of [8].)
[54] Calabi, E. (2018). On Kähler manifolds with vanishing canonical class. In: Ji, Lizhen (ed.) *Complex geometry from Riemann to Kähler–Einstein and Calabi–Yau.* Somerville, MA: International Press; Beijing: Higher Education Press. Adv. Lect. Math. (ALM) 38, 211–220. (Republication of [11].)

An Essay on Eugenio Calabi

Shing-Tung Yau[1]

Simon Donalson asked me to write an essay in honor of Eugenio Calabi. I am happy to do that.

Eugenio Calabi has been my mentor and is my good friend. I have known him since 1970 when I was a graduate student in Berkeley. I was told that a great geometer had just arrived on campus, and that he was full of ideas. So I was eager to meet him. The mathematics department was in Campbell Hall, and I met Calabi in the coffee room there. He smoked a pipe and wore thick glasses. He was an extremely friendly professor. When I went up to him to say hello, he did not seem to notice that I was merely a graduate student. Instead, he talked with me in the most personal manner.

I was working on the fundamental group of manifolds with negative curvature and was starting to look at works on flat manifolds. At the time, I was reading Wolf's book on spaces with constant curvature, and I had come across a lot of discussions on flat manifolds where a construction by Calabi was mentioned. Calabi studied the structure of manifolds with non-negative Ricci curvature by reducing it to those with trivial first Betti number. His work in this area was highly influential. In the beginning, it was based on the Bochner method. (Note: Bochner was Calabi's thesis advisor.) When that idea is coupled with the splitting principle of Cheeger–Gromoll, it reduces to the consideration of the fundamental group instead of the first Betti number. Such a construction was later readily applied to Calabi–Yau manifolds and called the structure theorem for Calabi–Yau manifolds.

When I was still a student, studying this work of Calabi led me to write a paper on flat manifolds although I forgot what I did, since it was really just a little exercise I paid scant attention to. But at the same time, I found out from him that he had classified all flat four-dimensional manifolds in a

[1] Shing-Tung Yau
Department of Mathematics, Harvard University, Cambridge, Massachusetts, USA
e-mail: yau@math.harvard.edu

paper that he has still not published! There was a question as to whether some manifolds in the list might be isomorphic to one another.

Calabi was then working on minimal surfaces – a beautiful subject that I gradually took up. On a piece of paper he wrote down for me many examples of minimal surfaces in closed form. Over the last forty years, I've often been curious as to how many examples in geometry were known to him at the time. He clearly knew some form of twistor construction way back then – long before most others considered such a thing.

I did not know anything about the Calabi conjecture when I first met him. I found out about the conjecture while browsing in the math library in Campbell Hall, where I stumbled upon a talk he'd given at the International Congress of Mathematicians.

As the years went by, I found out how much Calabi and I had in common in our ways of thinking. Much if not all of this might have been due to *his* influence on me. In addition to the deep subject of Kähler geometry, Calabi was interested in affine geometry, he was interested in maximal space-like hypersurfaces,... Although we did not see each other often, I have nevertheless modeled much of my taste after his. A good example is a paper I wrote on non-compact manifolds with non-negative Ricci curvature when I was at Stanford in 1974. After I wrote the paper using ideas of the Buseman function, I was surprised to learn that Calabi also worked out the same theorem at the same time. I remembered that it took me a full day to convince Gromov about the validity of this theorem when he visited Berkeley around that period. I needed to explain the geometry behind the Buseman function. A few years later S. Y. Cheng, Peter Li and I explored and extended a similar idea to give an estimate of the injectivity radius in terms of the volume of a geodesic ball, which became an important lemma for the work of Hamilton on Ricci flows.

When I thought I had found a counter-example to the Calabi conjecture, he was the one who said that my construction had some merit, though he later pointed out some questions in my argument. His confidence in the conjecture also gave me confidence to work on this tough problem at a time where comparatively little was known about it.

When I finally proved the Calabi conjecture in 1976, the first thing I did was to fly to the University of Pennsylvania to explain the proof in detail to Calabi and the geometers there. It was a rather pleasant trip as they were all pleased with my proof, although it took an entire day to present the details. I wanted to play safe and so arranged a meeting with Calabi and Louis Nirenberg to go through the proof again. It turned out that the only time that we could manage to get together was on Christmas of that year. Eugenio came all the way from Philadelphia to meet in Louis' office. At the end of that meeting, I was relieved to learn that Eugenio was really pleased to know that the problem was finally settled. I am truly grateful for his support.

He was fond of telling me his truly original ideas in many subjects, and I was always eager to listen. During the Special Year on Geometry at the

Institute for Advanced Study in Princeton, we had many occasions to exchange ideas. He invited me for dinner many times, and the conversations I had with him and his wife Giuliana were always warm and inspiring. He told me about the ideas in his thesis on isometric embeddings of Kähler manifolds into Hilbert spaces, the interpretation of the proof of Matsushima's theorem on the vanishing of the first Betti number of locally symmetric spaces of higher rank – a vanishing theorem that depends on a quadratic manipulation of the curvature tensor. This idea was later used to reprove the super-rigidity theorems of Margulis using harmonic maps – a result due independently to Mok–Siu–Yeung and Jost–Yau.

During that special year in Princeton, Calabi developed several important concepts on extremal Kähler metrics, and his lectures were always inspiring. Back then, I was already thinking about canonical metrics on holomorphic vector bundles in relation to the concept of stability. At the same time, Peter Li and I were working on the concept of balanced position of algebraic manifolds in complex projective spaces in relation to estimates for the first eigenvalue of the Laplacian. With Calabi I discussed similar notions in real projective and affine geometries.

I urged him to write a book on affine geometry. I think he took up the idea, but, so far as I know, he has not yet finished the job.

I have read many times his papers on hypersurfaces in seven-dimensional Euclidean space (which must be looked at as G2-manifolds) and his work related to Eckmann's construction of non-Kähler complex manifolds. They are all fascinating, and, at each reading, I find new things to admire in them. Like most of Calabi's ideas, it takes many years for us to fully digest them. I may say that I was always flattered when Calabi praised me. I remember one instance in particular. I was explaining to a student of mine about my conjecture that the first eigenvalue of an embedded minimal surface in the three-sphere is equal to two. While my student was puzzled by my conjecture, Calabi said that it showed true geometric intuition. Since Calabi is really the most original geometer I know of, such a statement from him made me feel truly proud of myself.

As a personal friend, he has always been extremely cordial to my wife and myself. He visited us in Los Angeles, just after we'd gotten married. We always enjoyed having meals with him. He invited me to visit Italy several times. But until 1989, I was stateless, holding no passport. The Italian consulate was especially difficult to deal with. Despite Calabi's great efforts, I was not able to get a visa to go to Italy until much later when I became an American citizen. But it was well worth the wait.

When my first son was born, my wife and son came to Princeton to join me for one semester. We lived in Philadelphia. Both Eugenio and Giuliana made a great effort to make sure that our life was comfortable. Eugenio helped to build the bed for my baby. Time has passed by quickly: That baby is now over 30 years old and Eugenio over 90. I wish a happy, healthy, and great future to both of them.

Part I
Commentaries on Calabi's Life and Work

Reflections on the Early Work of Eugenio Calabi

Blaine Lawson[1]

I first met Gene[2] Calabi at Stanford in 1967 while I was still a graduate student. That encounter had special significance in my mathematical life. Most importantly it began a long friendship with a man whose mathematical originality and character I have always deeply admired. I can't remember a conversation with Gene in which I did not learn something fascinating. He has always been open and free with his ideas and wonderfully supportive of young mathematicians.

Much is made in this volume, and in the mathematical community in general, about Gene's role in the subjects of Calabi–Yau manifolds and extremal metrics. While his ideas and theorems about special metrics in complex geometry have certainly been revolutionary, his earlier work was also of transforming originality. For personal reasons I am particularly fond of that early work, and I would like to present it here and discuss its impact over the years.

I want to express my thanks and indebtedness to Gopal Prasad for the detailed comments below concerning the seminal impact of the work of Calabi and Vesentini on the subject of rigidity of lattices in Lie groups.

Note: Naturally I submitted a preliminary version of this article to Gene for his comments. Several of his responses are included at appropriate points in the text below. The reader should find them interesting.

---

[1] Blaine Lawson
Department of Mathematics, Stony Brook University, Stony Brook, New York, USA
e-mail: `blaine@math.stonybrook.edu`

[2] 'Gene' is the nickname given to Eugenio Calabi by the author.

1. Calabi's Thesis: Isometric Holomorphic Embeddings

Gene wrote his Ph.D. thesis at Princeton under the direction of Salomon Bochner. It was concerned with the question:

> *Which complex hermitian manifolds (X,h) can be holomorphically and isometrically embedded into $\mathbf{C}^N$ (or more generally into a complex space form)?*

Calabi's answers to this problem constitute a gorgeous contribution to mathematics. It pre-dated by only a few years the work of Nash [21, 22] on the analogous riemannian embedding problem. As you shall see, these two cases are quite different in nature.

Note to begin that if (X,h) can be isometrically and holomorphically embedded into $\mathbf{C}^N$, then the metric h must be Kähler and real analytic. This holds also if $\mathbf{C}^N$ is replaced by any separable complex Hilbert space. Let $\mathbf{E}$ denote the indefinite complex Hilbert space of sequences $(z_1, z_{-1}, z_2, z_{-2}, \ldots)$ such that $\sum_k |z_k|^2 < \infty$, provided with the indefinite hermitian form $\mathrm{d}s^2 = \sum_k \operatorname{sgn}(k)|z_k|^2$.

Calabi's Universal Embedding Theorem. *Every complex manifold with an analytic Kähler metric can be locally embedded isometrically and holomorphically into* $\mathbf{E}$.

For embeddings into $\mathbf{C}^N$ (with positive definite metric) the story is quite different. The metrics that can be embedded are *highly restricted* and *intrinsically characterized.* Furthermore, when embeddings exist, they are unique up to ambient isometries. All this is true with $\mathbf{C}^N$ replaced by the complex space form $\mathbf{M}^N(c)$ of constant holomorphic sectional curvature c (complex projective, euclidean or hyperbolic spaces). Note that there are totally geodesic embeddings $\mathbf{M}^N(c) \subset \mathbf{M}^{N+N'}(c)$ unique up to isometries.

Calabi's Local Rigidity Theorem. *Suppose $f : X \to \mathbf{M}^N(c)$ and $f' : X \to \mathbf{M}^{N+N'}(c)$ are holomorphic isometric embeddings. Then $f = F \circ f'$ for some isometry F of $\mathbf{M}^{N+N'}(c)$.*

This is of course in stark contrast to the riemannian case. Let $\gamma : \mathbf{R} \hookrightarrow \mathbf{R}^2$ be a curve parameterized by arc-length. Then $(x, s) \mapsto (x, \gamma(s))$ is an isometric embedding of euclidean space $\mathbf{R}^n \hookrightarrow \mathbf{R}^{n+1}$.

Calabi's Intrinsic Characterization of Existence. *Let (X,h) be a connected Kähler manifold with a real analytic metric. Then there exists a holomorphic isometric embedding of a neighborhood of a point $p \in X$ into $\mathbf{M}^N(c)$ if and only if the metric satisfies an intrinsic, explicitly computable condition at p. Moreover, if this condition holds at one point, then it holds at all points of X.*

This intrinsic condition on the metric can be expressed in terms of Calabi's *diastatic function*, which we shall now examine. Recall that in holomorphic coordinates $z = (z_1, ..., z_n)$ a Kahler metric $\mathrm{d}s^2 = \sum_{i,j} h_{i\bar{j}} \mathrm{d}z_i \mathrm{d}\bar{z}_j$ can be written locally in terms of a real-valued potential function φ, that is,

$$h_{i\bar{j}} = \frac{\partial^2 \varphi}{\partial z_i \partial \bar{z}_j}.$$

This potential is not unique, but any other such potential $\widetilde{\varphi}$ is of the form

$$\widetilde{\varphi}(z,\bar{z}) \;=\; \varphi(z,\bar{z}) + f(z) + \overline{f(z)} \qquad \text{where } f \text{ is holomorphic.} \tag{1.1}$$

Recall now that a real analytic function $\varphi(z,\bar{z})$ can be expressed as a convergent power series in z and $\bar{z}$ and therefore extends to a holomorphic function of $2n$ variables $\varphi(z,w)$. The expression in (1.1) becomes

$$\widetilde{\varphi}(z,w) \;=\; \varphi(z,w) + f(z) + \overline{f}(w) \qquad \text{where } \overline{f}(w) = \overline{f(\bar{w})}. \tag{1.2}$$

From all of this one can deduce the following.

Calabi's Diastatic Function D. *Let $U \subset X$ be an open set with local holomorphic coordinates $z : U \xrightarrow{\cong} V \subset \mathbf{C}^n$ (V simply-connected). Then the* **diastatic function**

$$D : U \times U \longrightarrow \mathbf{R},$$

defined by

$$D(p,q) \;\equiv\; \varphi(z(p),\bar{z}(p)) + \varphi(z(q),\bar{z}(q)) - \varphi(z(p),\bar{z}(q)) - \varphi(z(q),\bar{z}(p)),$$

is independent of the choice of potential and independent of the choice of local coordinates. Hence it is **intrinsically defined** *in terms of the Kähler metric on U. It is real-valued, $D(p,q) = D(q,p)$, and $D(p,p) = 0$. Furthermore, if $d(p,q)$ is the intrinsic (riemannian) distance function on X, then $D(p,q) = d(p,q)^2 + O(d(p,q)^4)$.*

This function can be placed in a global context. Given any complex manifold X, there is a complex conjugate manifold $\overline{X}$ obtained by taking the atlas $\{(U_\alpha, z_\alpha)\}_\alpha$ and replacing it with the atlas $\{(U_\alpha, \bar{z}_\alpha)\}_\alpha$ whose transition functions are also holomorphic. The assertions above show that for real analytic Kähler metrics, the diastatic function D is well-defined in a neighborhood of the diagonal in $X \times \overline{X}$.

The Fundamental Property of D. *Let Y be a Kähler manifold with real analytic metric and diastatic function D_Y. Suppose $f : X \to Y$ is a holomorphic isometric embedding. Then D_Y pulls back to be the diastatic function of X, i.e., $D_X(p,q) = D_Y(f(p), f(q))$.*

Thus if $f : X \to \mathbf{C}^N$ is a holomorphic isometric immersion, then

$$D(p,q) \;=\; |f(p) - f(q)|^2.$$

Therefore, for fixed q,

$$D_q(p) \equiv D(p,q) = \sum_{k=1}^{N} |f_k(p)|^2 \quad \text{where } f_1,\ldots,f_N \text{ are holomorphic functions.}$$

Taking power series expansions in local coordinates z we have

$$D_q = \sum_{\alpha,\beta} C_{\alpha\beta} z^\alpha \bar{z}^\beta \qquad \text{and} \qquad f_k = \sum_{\alpha} a_{k,\alpha} z^\alpha \qquad \text{for each } k.$$

Enumerating the multi-indices we get the following identity of $\infty \times \infty$-matrices

$$C = a_1 \cdot \bar{a}_1^t + \cdots + a_N \cdot \bar{a}_N^t,$$

where a_k is considered as an infinite column vector. Thus

$$C \text{ is a positive semidefinite matrix of rank } \leq N. \tag{1.3}$$

This is precisely the necessary and sufficient condition for the euclidean case in the Intrinsic Characterization Theorem above.

For $f : X \to \mathbf{M}^N(c)$ with $c \neq 0$, the function D_q takes the form $\frac{1}{c}\log(1 + c\sum_k |f_k|^2)$ in appropriate local coordinates, so one can exponentiate and apply the above.

These general results have the following very nice application.

Theorem 1.1. *Let $X \subset \mathbf{M}^N(c)$ be a (local) n-dimensional complex submanifold whose induced metric has constant holomorphic sectional curvature $\bar{c}$.*

(*i*) *If $c \leq 0$, then $\bar{c} = c$ and X is totally geodesic*

(*ii*) *If $c > 0$, then $\bar{c} = \frac{1}{k}c$ for some integer k with $\binom{n+1}{k} \leq N+1$, and (after normalizing to $c = 1$), X is congruent to a piece of the "round" Veronese embedding*

$$\Phi_k : \mathbf{P}^n \hookrightarrow \mathbf{P}^N$$

given in homogeneous coordinates $[Z] = [Z_0,\ldots,Z_n]$ by

$$[Z] \mapsto \left[Z_0^k,\ldots,\sqrt{\binom{n}{\alpha}}Z^\alpha,\ldots,Z_n^k,0,\ldots,0\right] \equiv [W],$$

so that $|W|^2 = |Z|^{2k}$.

2. The Special Case of Holomorphic Curves

Suppose now that $\dim_{\mathbf{C}}(X) = 1$, that is, we restrict attention to holomorphic isometric immersions

$$f : \Sigma \to \mathbf{M}^n(c)$$

of a metric Riemann surface Σ. This is the most transparent and beautiful case. It is also the basic one. The general results as stated above follow directly from the 1-dimensional versions.

In many ways it is intuitively correct to think of $f : \Sigma \to \mathbf{M}^n(c)$ as a complex curve – generalizing real curves in, say, euclidean space. Note, however, that by the Rigidity Theorem, *all the extrinsic invariants of this curve are intrinsically determined by the metric.* More specifically, for any such holomorphic curve, one can establish Frenet formulas in complete analogy with the real case. The normal bundle decomposes generically into a flag of osculating bundles with associated curvature functions $K_k > 0$, $k = 1, \ldots, n-1$. In the real case these curvature functions can be chosen freely. In the complex case, Calabi shows that each K_k can be computed explicitly in terms of the metric $\mathrm{d}s^2$ on Σ. Furthermore, points where the kth osculating bundles fail to be defined directly (some $K_j = 0$ for $j \leq k$) are isolated, and the bundle has a holomorphic extension across these points. These points determine a divisor on the Riemann surface Σ, and in the case of an algebraic curve $\Sigma \to \mathbf{P}^n$ one can deduce the classical Plücker formulas directly from the metric and these divisors.

To present these explicit calculations we work locally on the Riemann surface. Let

$$\mathrm{d}s^2 = 2F|\mathrm{d}z|^2$$

be a real analytic metric on the unit disk $\Delta = \{z : |z| < 1\}$. In [3, 4] (see also [8]) Calabi proves the following.

Theorem 2.1. *The metric* $\mathrm{d}s^2$ *is induced by a full homomorphic map*

$$f : \Delta \to \mathbf{M}^n(c)$$

if and only if the recursively defined functions

$$F_0 = 1, \qquad F_1 = F,$$

$$F_{k+1} = \frac{F_k^2}{F_{k-1}}\left(\frac{\mathrm{d}}{\mathrm{d}z}\frac{\mathrm{d}}{\mathrm{d}\bar{z}}\log F_k + (k+1)c\right) \tag{2.1}$$

satisfy the conditions:

(1) For $k \leq n$ the function F_k is ≥ 0 and vanishes only at isolated points.
(2) The succeeding function F_{k+1} is defined by (2.1) away from those points, but extends to a real analytic function on all of Δ.
(3) $F_{n+1} \equiv 0$.

(A "full" map is one whose image does not lie in any totally geodesic $\mathbf{M}^m(c) \subset \mathbf{M}^n(c)$ for $m < n$.)

If $(\Sigma, \mathrm{d}s^2)$ is a connected Riemann surface with an analytic metric, then the conditions in Theorem 2.1 hold in some local coordinate system on Σ if and only if they hold in every local coordinate system. In this case there exists a global holomorphic isometric immersion of the universal covering of Σ into $\mathbf{M}^n(c)$.

Associated to the F_k is a globally defined sequence of *curvature functions*

$$K_k \equiv \frac{F_{k+1}F_{k-1}}{FF_k} = \tfrac{1}{2}\{\Delta \log F_k + (k+1)c\}, \tag{2.2}$$

where $\Delta \equiv \frac{2}{F}\frac{\mathrm{d}}{\mathrm{d}z}\frac{\mathrm{d}}{\mathrm{d}\bar{z}}$. Note that each $K_k \geq 0$ and $K_1 = c - K$, where K is the Gauss curvature of the metric $\mathrm{d}s^2$.

To see the meaning of all this we first consider the euclidean case $c = 0$. Given a holomorphic map

$$f : \Delta \to \mathbf{C}^n$$

with induced metric $\mathrm{d}s^2 = 2F|\mathrm{d}z|^2$, we have

$$F = |\varphi|^2 = \sum_{j=1}^{n} |\varphi_j|^2,$$

where $\varphi(z) \equiv f'(z)$. Calculation shows that

$$\begin{aligned} F_1 &= |\varphi|^2 \\ F_2 &= |\varphi \wedge \varphi'|^2 \\ F_3 &= |\varphi \wedge \varphi' \wedge \varphi''|^2 \\ &\vdots \\ F_k &= |\varphi \wedge \varphi' \wedge \varphi'' \wedge \cdots \wedge \varphi^{(k)}|^2 \\ &\vdots \end{aligned}$$

from which the necessity of the conditions in Theorem 2.1 is clear, as is the relation of the functions F_k to the kth osculating curve $\Phi_k \equiv [\varphi \wedge \varphi' \wedge \cdots \wedge \varphi^{(k-1)}] : \Delta \to \mathbf{P}^{\binom{n}{k}-1}$ of f.

The functions $K_1, K_2, K_3, \ldots$ are exactly the curvature functions which appear in the "Frenet formulas" for this given curve. (See [13] for details.)

The function K_k can also be expressed as the ratio $\mathrm{d}\sigma_k^2/\mathrm{d}s^2$ where $\mathrm{d}\sigma_k^2$ is the metric induced from $\mathbf{P}^{\binom{n}{k}-1}$ by the map Φ_k.

The projective case is similar. Let

$$f : \Delta \to \mathbf{P}^n$$

(with holomorphic sectional curvature 1) be represented in homogeneous coordinates by a holomorphic map

$$\varphi = (\varphi_0, \ldots, \varphi_n) : \Delta \to \mathbf{C}^{n+1}.$$

Then the functions in Theorem 2.1 are seen to be

$$F_k \;=\; \frac{|\varphi\wedge\varphi'\wedge\cdots\varphi^{(k)}|^2}{|\varphi|^{2(k+1)}}.$$

Again there is a well-defined osculating curve

$$\Phi_k \;\equiv\; [\varphi\wedge\varphi'\wedge\cdots\wedge\varphi^{k)}] : \Sigma \;\to\; \mathbf{P}^{\binom{n+1}{k+1}-1}$$

and again K_k can also be expressed as the ratio $\mathrm{d}\sigma_k^2/\mathrm{d}s^2$ where $\mathrm{d}\sigma_k^2$ is the metric induced by the map Φ_k. When the Riemann surface Σ is compact, the projective degree ν_k of the Kth osculating curve Φ_k is given by

$$\nu_k \;=\; \frac{1}{2\pi}\int_\Sigma K_k\, dA.$$

For further details see [13].

Final Note. In the case of curves Calabi's Rigidity Theorem has an elementary proof. Suppose $f:\Delta\to\mathbf{C}^n$ and $g:\Delta\to\mathbf{C}^{n+m}$ are isometric holomorphic maps. We claim that $f = T\circ g$, where T is a holomorphic isometry of $\mathbf{C}^{n+m}$ (a unitary transformation followed by a translation). Setting $\varphi = f'$ and $\psi = g'$ we have $|\varphi|^2 \equiv |\psi|^2$ on Δ since the induced metrics agree. Assume the first coordinate function $\varphi_1(0)\neq 0$ and set $\Phi = \frac{1}{\varphi_1}\varphi$, $\Psi = \frac{1}{\varphi_1}\psi$ on a neighborhood of zero. Then $1+|\Phi_2|^2+|\Phi_3|^2+\cdots = |\Psi|^2$. Taking $\frac{\mathrm{d}}{\mathrm{d}z}\frac{\mathrm{d}}{\mathrm{d}\bar z}$ gives $|\Phi_2'|^2+|\Phi_3'|^2+\cdots = |\Psi'|^2$. By induction on n there exists a unitary transform U_0 with $\Phi' = U_0\Psi'$. Integrating and then multiplying through by φ_1 and checking constants of integration gives $\varphi = U\psi$ for a unitary matrix U.

Calabi's Comments on his Thesis: "The original idea came in the summer of 1949. I knew from recent lecture notes about Bieberbach's imbedding of the complex unit disc with the hyperbolic metric in the Hilbert space ℓ^2 (square-summable complex sequences), and was calculating, as an exercise, what happens when one extends real analytic curves in $\mathbf{R}^n$, with their Frenet–Serret formulas, to their corresponding extensions as holomorphic Riemann surfaces in $\mathbf{C}^n$. The thesis itself grew out of elaborating from the original drafts."

3. Minimal Surfaces in Euclidean Space

Calabi understood that his work on holomorphic curves had direct application to the theory of minimal surfaces. Every minimal surface in euclidean space can be realized locally by a conformal immersion

$$f:\Delta \;\to\; \mathbf{R}^n$$

whose coordinate functions are harmonic. In particular, we have

$$\frac{\mathrm{d}}{\mathrm{d}z}\frac{\mathrm{d}}{\mathrm{d}\bar{z}}f = 0 \tag{3.1}$$

and conformality implies that

$$\left|\frac{\mathrm{d}f}{\mathrm{d}z}\right|^2 = \left|\frac{\mathrm{d}f}{\mathrm{d}\bar{z}}\right|^2 = F \quad \text{and} \quad \frac{\mathrm{d}f}{\mathrm{d}z}\cdot\frac{\mathrm{d}f}{\mathrm{d}z} = 0 \tag{3.2}$$

where the induced metric is $\mathrm{d}s^2 = 2F|\mathrm{d}z|^2$.

Let $\widetilde{f} : \Delta \to \mathbf{R}^n$ be the harmonic conjugate mapping. Then $\left|\frac{\mathrm{d}f}{\mathrm{d}z}\right|^2 = \left|\frac{\mathrm{d}\widetilde{f}}{\mathrm{d}z}\right|^2$, and we obtain a **holomorphic** mapping

$$\mathbf{f} \equiv \frac{1}{\sqrt{2}}\left(f + i\widetilde{f}\right) : \Delta \to \mathbf{C}^n \tag{3.3}$$

giving the **same** induced metric.

These remarks apply globally to the universal covering of any minimal surface in $\mathbf{R}^n$.

Calabi's results above can now be applied to:

(1) Intrinsically characterize the metrics which can appear on minimal surfaces in euclidean space.
(2) Classify the minimal surfaces up to congruence which are intrinsically isometric to a given one.

Part (1) is clear. These metrics are exactly the ones which can be realized on holomorphic curves in $\mathbf{C}^n$. For Part (2) the key observation is that by Calabi's Rigidity Theorem (in the simply-connected case)

> *Each class of isometric non-congruent minimal surfaces in euclidean space contains exactly one holomorphic curve.*

Starting with this holomorphic curve, all minimal surfaces isometric to it are obtained by applying an appropriate complex linear transformation and then taking the real part. This leads to the following theorem. Let Σ be a simply connected Riemann surface.

Calabi's Classification Theorem. *The space of non-conguent minimal immersions $f : \Sigma \to \mathbf{R}^m$, which are isometric to a given (full) holomorphic immersion $\mathbf{f} : \Sigma \to \mathbf{C}^m$, is naturally described as the set of all complex symmetric $m \times m$ matrices P such that*

(a) $1_m - P\overline{P} \geq 0$, *and*
(b) $(\mathbf{f})^t P \mathbf{f}' \equiv 0$,

that is, P annihilates all tangent vectors to the curve $\mathbf{f}$. Furthermore, let n be the dimension of the smallest affine subspace of $\mathbf{R}^m$ containing the minimal surface corresponding to P. Then

$$n - m = \operatorname{rank}(1_m - P\overline{P})$$

so that $m \le n \le 2m$, *and* $n = 2m \iff 1_m - P\overline{P} > 0$.

Examples. Let $f_\theta = \sqrt{2}\mathrm{Re}(e^{i\theta}\mathbf{f})$, where $\mathbf{f}$ is given as in (3.3). This is the family of *associate minimal surfaces* for f.

The classical case $m = 3$ is interesting. Here the intrinsic characterization goes back to Ricci–Curbastro. A metric $\mathrm{d}s^2 = 2F|\mathrm{d}z|^2$ appears on a minimal surface in $\mathbf{R}^3$ if and only if its Gauss curvature K is ≤ 0 with isolated analytic zeros, and the metric

$$\mathrm{d}\hat{s}^2 \equiv \sqrt{-K}\mathrm{d}s^2 \mathrm{isflat}.$$

All the other minimal surfaces in euclidean space which are isometric to this given one in $\mathbf{R}^3$ are either associate surfaces or they are linearly full in $\mathbf{R}^6$ ([14]).

4. Some Personal Reminiscences

My first encounter with Gene Calabi was an event that transformed my graduate – and indeed my mathematical – career. I had been working for my thesis on the problem of fully understanding the theorem of Ricci–Curbastro mentioned above. Doing this led me to the discovery of Theorem 1.1 and some of its applications. Checking with some experts led me to believe the result was new, and I was joyfully writing up the results when Calabi came to give the colloquium at Stanford. My advisor mentioned to Gene what I had been doing and learned the bad news, which he passed on to me in the early afternoon.

Later that day I went to Gene's colloquium, which turned out to be one of the most astonishing and beautiful mathematical lectures I had heard. It pulled me directly into the study of minimal submanifolds of spheres, which became the subject of my Ph.D. thesis.

For me it was quite a day.

What follows is a description of the results that Gene presented in that colloquium.

5. Minimal Surfaces in the Euclidean Sphere

By a minimal surface in $S^n = \{x \in \mathbf{R}^{n=1} : |x| = 1\}$ I mean an immersion of a smooth surface

$$f : \Sigma \to S^n$$

with mean curvature $\equiv 0$. Such immersions are real analytic. We take the induced metric on Σ which gives it the structure of a Riemann surface with an analytic metric.

Consider the case where Σ is compact (without boundary). In 1967 almost nothing was known about such surfaces, except for a theorem of Fred Almgren [1], which states that when $n=3$ and Σ has genus zero, the surface must be totally geodesic.

Calabi's Theorem on Minimal 2-spheres in S^n. *Let*

$$f: S^2 \to S^n \tag{5.1}$$

be a minimal 2-sphere which does not lie in any totally geodesic (equatorial) S^{n-1}. Then

(*i*) $n=2m$ *for some integer* m,
(*ii*) $\mathrm{Area}(\Sigma)=2\pi N$, *where* N *is an integer with*

$$N \geq m(m+1).$$

(*iii*) *There exists a holomorphic map*

$$\Psi : S^2 \to \mathcal{I}_m$$

into the hermitian symmetric space $\mathcal{I}_m \equiv \mathrm{SO}(2m+1)/\mathrm{U}(m)$ *such that*

$$f = \pi \circ \Psi,$$

where

$$\pi : \mathcal{I}_m \to S^{2m} \tag{5.2}$$

is the projection $\mathrm{SO}(2m+1)/\mathrm{U}(m) \to \mathrm{SO}(2m+1)/\mathrm{SO}(2m)$. *Furthermore, this map* Ψ *is everywhere orthogonal to the fibres of* π *and the metric induced on* S^2 *by* f *coincides with the one induced by* Ψ.

This theorem asserts that minimal 2-spheres in the euclidean sphere are in a strong sense algebraic, they correspond explicitly to certain rational curves in the rational variety $\mathcal{I}_m$. Note that $\mathcal{I}_m$ can be realized as the space of complex m-planes in $\mathbf{C}^{2m+1}=\mathbf{R}^{2m+1}\otimes_{\mathbf{R}}\mathbf{C}$ which are totally isotropic with respect to the complex quadratic form on $\mathbf{C}^{2m+1}$ obtained by extending the metric on $\mathbf{R}^{2m+1}$.

Note that part (i) in Calabi's theorem generalizes Almgren's result.

The first non-trivial case of this result, where $n=4$, is already interesting. Here the map (5.2) is the twistor fibration

$$\pi : \mathbf{P}^3_{\mathbf{C}} \to S^4$$

whose horizontal complex 2-plane field is a homogeneous contact structure on $\mathbf{P}^3_{\mathbf{C}}$. Every holomorphic curve tangent to this structure pushes down to a minimal surface in S^4. By the theorem above every minimal 2-sphere in S^4 which is non-totally geodesic has area $2\pi N$ for $N \geq 6$. Furthermore, every twisted cubic in $\mathbf{P}^3_{\mathbf{C}}$ is isometric to a minimal 2-sphere in S^4 whose area is exactly 12π.

The construction of this lifting $\varPsi$ is gorgeous and quite explicit. Let's realize the map (5.1) in local coordinates as a conformal map

$$\psi : \varDelta \to S^n$$

whose induced metric is

$$\mathrm{d}s^2 = 2F|\mathrm{d}z|^2.$$

Then ψ is a minimal surface if and only if

$$\frac{\mathrm{d}}{\mathrm{d}z}\frac{\mathrm{d}}{\mathrm{d}\bar{z}}\psi = -F\psi, \tag{5.3}$$

where we are considering ψ as a map $\psi : \varDelta \to \mathbf{R}^{n+1}$ with $|\psi|^2 \equiv 1$. The conformality of ψ is equivalent to

$$\frac{\mathrm{d}\psi}{\mathrm{d}z} \cdot \frac{\mathrm{d}\psi}{\mathrm{d}z} = 0. \tag{5.4}$$

Note that by (5.3) and the fact that $\psi \cdot \frac{\mathrm{d}\psi}{\mathrm{d}z} = 0$ we have

$$\frac{\mathrm{d}}{\mathrm{d}\bar{z}}\left(\frac{\mathrm{d}\psi}{\mathrm{d}z} \cdot \frac{\mathrm{d}\psi}{\mathrm{d}z}\right) = 0.$$

Therefore

$$\left(\frac{\mathrm{d}\psi}{\mathrm{d}z} \cdot \frac{\mathrm{d}\psi}{\mathrm{d}z}\right) \mathrm{d}z^2$$

is a global holomorphic 2-form on the Riemann surface $\varSigma$, and therefore if $\varSigma = S^2$, it must be identically zero. Proceeding by induction shows that

$$\frac{\mathrm{d}^k\psi}{\mathrm{d}z^k} \cdot \frac{\mathrm{d}^\ell\psi}{\mathrm{d}z^\ell} = 0 \qquad \text{for all } k+\ell > 0. \tag{5.5}$$

Therefore, the complex k-plane corresponding to the simple vector

$$\tau_k \equiv \frac{\mathrm{d}\psi}{\mathrm{d}z} \wedge \frac{\mathrm{d}^2\psi}{\mathrm{d}z^2} \wedge \cdots \wedge \frac{\mathrm{d}^k\psi}{\mathrm{d}z^k}$$

is totally isotropic. Furthermore, using (5.3) again one sees that the $2k$-jet of ψ at every point is spanned by ψ and the vectors in τ_k and $\overline{\tau}_k$. It now follows that the image of ψ is linearly full in some $\mathbf{R}^{2m+1}$.

If we now set

$$F_k = |\tau_k|^2$$

one can show that $\frac{1}{F_m}\tau_m$ is holomorphic, and the mapping $\varPsi$ is then defined in Plücker coordinates by

$$\Psi = \frac{1}{F_m}\tau_m.$$

Incidentally, the functions F_k satisfy the recursion relations

$$F_0 = 1, \qquad F_1 = F,$$

$$F_{k+1} = \frac{F_k^2}{F_{k-1}}\left(\frac{\mathrm{d}}{\mathrm{d}z}\frac{\mathrm{d}}{\mathrm{d}\bar{z}}\log F_k + F\right) \tag{5.6}$$

for $1 \le k \le m$ (and $F_{m+1} \equiv 0$), very close to those given in (2.1) above.

This spectacular, highly original work engendered well over a decade of intense research on minimal surfaces and harmonic mappings by people such as Barbosa, Bryant, Burns, Chern, Eells, Guest, Lemaire, Salamon, Wood, the author, and many others (See the Bourbaki article [16] for references).

6. Rigidity of Arithmetic Groups

In 1960, Gene Calabi and Edoardo Vesentini wrote a paper which was foundational in the modern theory of arithmetic groups and rigidity. They considered smooth compact quotients

$$X = D/\Gamma$$

by a discrete group of holomorphic isometries, where D is a Cartan domain, that is, a bounded symmetric, homogeneous domain in $\mathbf{C}^n$. These domains are simply connected, and decompose into products of irreducible ones. Their main technical result is a vanishing theorem for the cohomology groups $H^q(X,\Theta)$, where Θ is the tangent sheaf. They proved that when D is irreducible

$$H^q(X,\Theta) = 0 \qquad \text{for } 0 \le q < \gamma(D) - 1,$$

where $\gamma(D)$ is an integer, independent of Γ, which satisfies

$$\sqrt{2n} < \gamma(D) \le n+1,$$

where $n = \dim_{\mathbf{C}}(D)$.

By the deformation theory of Frölicher–Nijenhuis [11] and Kodaira–Spencer [12], this proves

The Calabi–Vesentini Rigidity Theorem. *When D is irreducible with $\dim_{\mathbf{C}}(D) > 1$, the complex structure on D/Γ is rigid for any co-compact discrete subgroup of holomorphic isometries Γ acting freely and properly discontinuously on D. This rigidity continues to hold when D is reducible provided that no irreducible factor of D has dimension 1.*

The assertion is famously false when $\dim_{\mathbf{C}}(D) = 1$, i.e., when D is the unit disk in $\mathbf{C}$.

The proof of the main technical result proceeds by the Bochner method. It involves a masterful calculation of the curvature term appearing in the formulas of Bochner–Kodaira–Nakano type for harmonic tangent-bundle-valued forms.

Combined with results of Atle Selberg, this theorem led to the following.

The Algebraicity Theorem. *Let D be a Cartan domain with no irreducible factors of dimension 1. Let G be the group of all holomorphic (isometric) transformations of D, represented as a group of real matrices. Let $\Gamma \subset G$ be a discrete subgroup acting on D with smooth compact quotient. Then there exists an inner automorphism of G such that the components of all matrices representing the transform of Γ generate a simple algebraic extension of the rational number field.*

A little before the Calabi–Vesentini work, Selberg announced a related rigidity result for co-compact discrete subgroups of $\mathrm{SL}_n(\mathbf{R})$. A full length paper [25] appeared in 1960 with references to [10]. A summary of this and some subsequent work appeared in [26].

The work of Calabi–Vesentini greatly influenced Andrei Weil who in his 1962 paper [29] proved a best possible deformation rigidity result for co-compact lattices in real semi-simple Lie groups. The work of Selberg, Calabi–Vesentini and Weil marked the beginning of one of the major developments in modern mathematics, with early contributions by A. Borel, J. Tits, J.-P. Serre, Harish-Chandra, H. Garland, M. S. Raghunathan, among many others. (See [24] for an account in 1972.) This field, concerned with lattices in Lie groups, culminated in famous work of G. Mostow and then G. Margulis.

The notion of Mostow rigidity is stronger than the deformation rigidity discussed above. Deformation rigidity says that under appropriate assumptions on $\Gamma \subset G$, the only continuous deformations $\Gamma_t \subset G$, $|t| < 1$, come from families of inner automorphisms of G. Mostow's *strong rigidity* asserts that given $\Gamma \subset G$ and $\Gamma' \subset G'$ (again with appropriate assumptions), any abstract isomorphism $\Gamma \cong \Gamma'$ extends to an isomorphism $G \cong G'$. Mostow was more explicitly influenced by the work of Selberg than the analytic approach of Calabi–Vesentini. However, Mostow says on page 6 of his monograph [20] that "The phenomenon of strong rigidity for arbitrary lattices first turned up in 1965 in my search for a geometric explanation of deformation-rigidity". The earlier results of Selberg, Calabi–Vesentini, and Andrei Weil on deformation rigidity had a decisive influence on him and in his formulation of strong rigidity.

Calabi's Comments: "The rigidity theorem for discrete, uniform transformation groups of uniform, bounded domains came from a coincidence of contacts. In 1958–59 I was spending a year at the I.A.S., and listened to lectures by Kodaira, in which I finally understood for the first time (after about six years of hearing about them) what cohomology with coefficients in a linear space sheaf is about. At the same time, I remember talking with

A. Selberg about compact Riemannian manifolds with constant negative sectional curvature: to my surprise, he told me then that, *even in 3 dimensions*, he believed that they might be rigid, while I was actually dreaming of finding moduli. I got then the idea of calculating the cohomology with coefficients in the sheaf of germs of Killing vector fields. While writing it up and discussing it with Vesentini, we got the idea of applying the 'vanishing theorem' idea to the cohomology in the sheaf of holomorphic vector fields in the irreducible, locally symmetric Kähler manifolds. The case of the real hyperbolic manifolds then fell by the wayside, because, by about 1961, Dan Mostow had proved their global rigidity."

7. Complex Structures on Products of Spheres

Together with Beno Eckmann, Gene Calabi wrote a paper in 1953 which was transformational in complex geometry. At the time very little was known about complex manifolds outside the Kähler domain. Aside from the examples of H. Hopf, obtained by dividing $\mathbf{C}^n - \{0\}$ by powers of a homothety, most known compact complex manifolds were either projective or deformations of projective. Calabi and Eckmann showed that for all integers $p, q \geq 1$ there exist complex structures on the product of spheres:

$$M_{p,q} \equiv S^{2p+1} \times S^{2q+1}.$$

These were the first known examples of *simply-connected* compact complex manifolds which were **not algebraic**. (Kodaira's classification of surfaces came over a decade later.) They have remarkable properties. There is a holomorphic fibration

$$M_{p,q} = S^{2p+1} \times S^{2q+1} \xrightarrow{\pi} \mathbf{P}^p_{\mathbf{C}} \times \mathbf{P}^q_{\mathbf{C}},$$

given by the product $\pi = \pi_1 \times \pi_2$ of the standard Hopf fibrations, whose fibres are elliptic curves. These fibres are homologous to zero, and therefore $M_{p,q}$ cannot carry a Kähler metric. There are many complex subvarieties of $M_{p,q}$, but they are all of the form $\pi^{-1}(V)$, where V is a subvariety of $\mathbf{P}^p_{\mathbf{C}} \times \mathbf{P}^q_{\mathbf{C}}$. (Taking V to be smooth gives a wealth of new compact complex manifolds.) Similarly the meromorphic functions on $M_{p,q}$ are all pull-backs of meromorphic functions on $\mathbf{P}^p_{\mathbf{C}} \times \mathbf{P}^q_{\mathbf{C}}$.

The subdomain

$$\left(S^{2p+1} - \{x_0\}\right) \times \left(S^{2q+1} - \{y_0\}\right)$$

is particularly interesting. It is a complex manifold diffeomorphic to $\mathbf{R}^{2(p+q+1)}$ which contains an enormous family of compact complex analytic subvarieties. For example, take $\pi^{-1}(W \times W') = \pi_1^{-1}(W) \times \pi_2^{-1}(W')$, where W, W' are subvarieties of $\mathbf{P}^p_{\mathbf{C}}$ and $\mathbf{P}^q_{\mathbf{C}}$ respectively with $x_0 \notin W$ and $y_0 \notin W'$. Any entire holomorphic function must be constant, so $M_{p,q}$ admits no global coordinate chart.

The construction of the complex structure on $M_{p,q}$ is as follows. Note that $S^{2p+1} \subset \mathbf{C}^{p+1}$ carries a field of tangent complex p-planes $\mathcal{H}$ which is invariant under the family of diffeomorphisms given by scalar multiplication by e^{it}. Let H be the "Hopf" vector field which generates this flow. Let $\mathcal{H}'$ and H' be the corresponding objects on S^{2q+1}. At each point $(x,y) \in S^{2p+1} \times S^{2q+1}$ the subspace $\mathcal{H} \times \mathcal{H}'$ is already provided with an almost complex structure J. We extend it to the complement by defining $J(H) = H'$ and $J(H') = -H$. The invariance of J under the flows generated by H and H' is the key to proving its integrability, that is, the vanishing of the Nijenhuis tensor.

This construction of Calabi–Eckmann can be directly generalized to a large and interesting class of spaces, as pointed out by Sid Webster. Let $\Sigma^{2p+1} \subset \mathbf{C}^N$ be a compact submanifold which arises as the local boundary of a complex $(p+1)$-dimensional submanifold. Let $\mathcal{H}$ be the tangential complex p plane field to Σ^{2p+1}. Suppose there is an S^1-action on Σ^{2p+1} generated by a vector field H which is transversal to $\mathcal{H}$ and whose flow preserves $\mathcal{H}$ and its almost complex structure. Let Σ^{2q+1} be another such creature of dimension $2q+1$. Then the construction above gives an integrable complex structure on

$$\Sigma^{2p+1} \times \Sigma^{2q+1}.$$

Such manifolds Σ^{2n-1} can be easily constructed by taking

$$\Sigma^{2n-1} = S^{2n+1} \cap \{Z \in \mathbf{C}^{n+1} : P(Z) = 0\},$$

where P is a weighted homogeneous polynomial in $(Z_0, \ldots, Z_n)$, such as $Z_0^{a_0} + \cdots + Z_n^{a_n}$. These include the Brieskorn polynomials whose links Σ are exotic spheres.

Over the years there have been many further constructions of compact complex non-Kähler manifolds. Notable among them is the work of Taubes [27] in complex dimension 3, involving twistor geometry, and the work of Lopez de Medrano, Meersseman, and Verjovsky [17], [18], [19] which represents a highly non-trivial extension of the Calabi–Eckmann construction.

8. Generalizing the Hopf Maximum Principle

In a paper entitled *An extension of E. Hopf's maximum principle with an application to Riemannian geometry*, written in 1958, Calabi introduced the notion of a weak subsolution for a linear elliptic partial differential equation and established the Strong Maximum Principle for such functions. This theorem enabled him to prove a beautiful theorem in global riemannian geometry.

What is particularly remarkable about this paper is that it presages modern viscosity theory (cf. [2]) more than twenty years before its emergence as a major branch of analysis.

Calabi considered differential operators in $\mathbf{R}^n$ of the form

$$L[u] \;=\; \sum_{j,k} a_{jk}(x)\frac{\partial^2 u}{\partial x_j \partial x_k} + \sum_j b_j(x)\frac{\partial u}{\partial x_j} \qquad \text{with } a \;>\; 0,$$

and extended the notion $L[u] \geq 0$ to upper semi-continuous functions u as follows. We say that a C^2-function φ defined near x is a *lower (upper) test function for u at x* if $u - \varphi$ has minimum (resp. maximum) value zero at x. Calabi defined u to be a weak solution to $L[u] \geq c$ at x if for each $\epsilon > 0$ there is a lower test function φ for u at x with $L[\varphi](x) \geq c - \epsilon$.

In modern viscosity theory one requires that for *every upper* test function φ at x, one has $L[\varphi](x) \geq c$. (If upper test functions do not exist, the condition is considered to be satisfied.) This viscosity criterion follows easily from Calabi's and is slightly more general. It allows functions to assume the value $-\infty$ for example. Using this fundamental approach, the school of Crandall, Ishii, Lions, Evans, and others, developed a substantial theory for fully nonlinear equations with important wide-ranging applications. It is an example of the depth and originality of Gene Calabi that he saw this important idea and its usefulness twenty years "before its time".

The first part of Gene's paper proved that any upper semi-continuous function u which satisfies $L[u] \geq 0$ (in his sense) on a domain in $\mathbf{R}^n$, and has an interior maximum point, must be constant. He then showed that on a riemannian n-manifold of non-negative Ricci curvature, the geodesic distance ρ from a point satisfies the inequality $\Delta\rho \leq (n-1)\rho$ in the weak sense. From this he then deduced that on such a manifold, assumed complete, there are no global solutions to the differential inequality $\Delta u \geq f(u)$ where $f(t)$ is a positive increasing function satisfying

$$\int_0^\infty \left\{ \int_0^s f(t)\,\mathrm{d}t \right\}^{\frac{1}{2}} \mathrm{d}s \;<\; \infty.$$

Calabi's Comments: The paper on the maximum principle was really an auxiliary result. The leitmotif of my work at the time was to complete the proof of what was called the Calabi conjectures, looking at the function space of all Kähler metrics in a compact, complex manifold, in a fixed cohomology class, and formulating variational problems that might yield the most 'harmonious' metric in such a space. By the time I announced the problems at the A.M.S. meeting in Baltimore in December 1952 (and again in Amsterdam in 1954), I had learned how little was known about nonlinear elliptic p.d.e.s., and was trying to find *a priori* estimates, using relations motivated by the geometry. The real inspiration came from a paper, presented by Bob Osserman, at an A.M.S. meeting around 1957, on the existence of solutions of the equation $\Delta u = f(u)$ in Euclidean $\mathbf{R}^n$. That was the incident that triggered the result that I am most proud of, "Improper Affine Hyperspheres...", in the Mich. Math. J., 1958. In that paper I needed to interpret sub-harmonic inequalities for non-smooth functions, like Riemannian geodesic distance and Finsler norms. I wrote the paper [5] separately, simply because I thought that it might have broader applications. By the way, the operator and inequality $L[u] = \sum_{jk} a_{jk}\partial^2 u/\partial x_j \partial x_k + \sum_j b_j \partial u/\partial x_j \geq 0$ was first considered by Eberhard Hopf. I first studied it in the book by Bochner and Yano, "Curvature and Betti Numbers". For the rest, I used Osserman's argument literally, except that, while Osserman considered only the Laplace operator in a Euclidean metric, I realized that it applied also in the case of metrics with non-negative Ricci curvature."

Author's Comments: The interaction between Bob Osserman and Gene Calabi at the A.M.S. meeting referred to above, was, in itself, one of the more interesting mathematical exchanges. The two were scheduled to give short presentations in the same session. Gene, however, was delayed and arrived only after Osserman had already left. However, Bob's final result remained suspended on a higher blackboard. Looking at it, Gene recognized exactly what he needed for the problem he was working on.

References

[1] Almgren Jr., F. J. (1966). Some interior regularity theorems for minimal surfaces and an extension of Bernstein's theorem. *Ann. of Math.* (2) 84, 277–292.

[2] Crandall, M. G., Ishii, H., Lions, P. L. (1992). User's guide to viscosity solutions of second order partial differential equations. *Bull. Amer. Math. Soc. (N. S.)* 27, 1–67.

[3] Calabi, E. (1953). Isometric imbedding of complex manifolds. *Ann. of Math.* (2) 58, 1–23.

[4] Calabi, E. (1953). Metric Riemann surfaces. In: Ahlfors, L. V., Calabi, E., Morse, M., Sario, L., Spencer, D. C. (eds.) *Contributions to the Theory of Riemann Surfaces*, 77–85. Annals of Mathematics Studies, no. 30. Princeton University Press, Princeton, NJ.

[5] Calabi, E. (1958). An extension of E. Hopf's maximum principle with an application to Riemannian geometry. *Duke Math. J.* 25, 45–56.

[6] Calabi, E. (1958). Improper affine hyperspheres of convex type and a generalization of a theorem by K. Jörgens. *Michigan Math. J.* 5, 105–126.

[7] Calabi, E. (1967). Minimal immersions of surfaces in Euclidean spheres. *J. Differential Geometry* 1, 111–125.

[8] Calabi, E. (1968). Quelques applications de l'analyse complexe aux surfaces d'aire minima. In: Rossi, H. *Topics in Complex Manifolds*. Séminaire de Mathématiques Supérieures, No. 30 (Été, 1967). Les Presses de l'Université de Montréal, Montréal, Que.

[9] Calabi, E. & Eckmann, B. (1953). A class of compact, complex manifolds which are not algebraic. *Ann. of Math.* (2) 58, 494–500.

[10] Calabi, E. & Vesentini, E. (1960). On compact, locally symmetric Kähler manifolds. *Ann. of Math.* (2) 71, 472–507.

[11] Frölicher, A. & Nijenhuis, A. (1957). A theorem of stability of complex structures. *Proc. Nat. Acad. Sci., U.S.A.* 43, 239–241.

[12] Kodaira, K. & Spencer, D. C. (1958). On deformations of complex analytic structures, I and II. *Ann. of Math.* 67, 328–460.

[13] Lawson, H. B. (1970). The Riemannian geometry of holomorphic curves. In: *Holomorphic Mappings and Minimal Surfaces.* Carolina Conference Proceedings, 86–107. Chapel Hill, North Carolina.
[14] Lawson, H. B. (1971). Some intrinsic characterizations of minimal surfaces. *Journal d'Analyse Mathématique* 24, 151–161.
[15] Lawson, H. B. (1980). *Lectures on Minimal Submanifolds.* I.M.P.A. Press, Rio de Janeiro, 1973. Publish or Perish Press, Berkeley.
[16] Lawson, H. B. (1985). Surfaces minimales et la construction de Calabi–Penrose. In: *Séminaire Bourbaki* 1983–84, 197–212. Astérisque 121–122. Société Mathématique de France, Paris.
[17] López de Medrano, S. & Verjovsky, A. (1997). A new family of complex, compact, non-symplectic manifolds. *Bol. Soc. Mat. Bra.* 28, 2, 253–269.
[18] Meersseman, L. (2000). A new geometric construction of compact complex manifolds in any dimension. *Math. Ann.* 317, 79–115.
[19] Meersseman, L. & Verjovsky, A. (2004). Holomorphic principal bundles over projective toric varieties. *J. Reine Angew. Math.* 572, 57–96.
[20] Mostow, G. D. (1973). *Strong rigidity of locally symmetric spaces.* Annals of Mathematics Studies, No. 78. Princeton University Press, Princeton, N.J.; University of Tokyo Press, Tokyo.
[21] Nash, J. (1954). C^1 isometric imbeddings. *Ann. of Math.* (2) 60, 383–396.
[22] Nash, J. (1956). The imbedding problem for Riemannian manifolds. *Ann. of Math.* (2) 63, 20–63.
[23] Osserman, R. (1957). On the inequality $\Delta u \geq f(u)$. *Pacific J. Math.* 7, 1641–1647.
[24] Raghunathan, M. S. (1972). *Discrete Subgroups of Lie Groups.* Springer-Verlag, Berlin.
[25] Selberg, A. (1960). On discontinuous groups in higher-dimensional symmetric spaces. In *Contributions to Function Theory* (internat. Colloq. Function Theory, Bombay, 1960), 147–164. Tata Institute of Fundamental Research, Bombay.
[26] Selberg, A. (1970). Recent developments in the theory of discontinuous groups of motions of symmetric spaces. In: *Proceedings of the Fifteenth Scandinavian Congress* (Oslo, 1968). Lecture Notes in Mathematics, Vol. 118, 99–120. Springer, Berlin.
[27] Taubes, C. H. (1992). The existence of anti-self-dual conformal structures. *J. Differential Geom.* 36, No. 1, 163–253.
[28] Weil, A. (1960). Discrete subgroups of Lie groups I. *Annals of Math.* 72, 369–384.
[29] Weil, A. (1962). Discrete subgroups of Lie groups II. *Annals of Math.* 75, 578–602.

Encounter with a Geometer: Eugenio Calabi

Marcel Berger[1]

P. de Bartolomeis, F. Tricerri, E. Vesentini (Eds.),
Manifolds and Geometry, Cambridge University Press, 20–60 (1996)

To Eugenio Calabi, for his seventieth birthday

"It is not to be denied that the advantages of an analytical treatment over a geometrical treatment, its concision, simplicity, uniformity, and especially its generality, usually become more and more decisive as the investigation becomes more and more complicated. However, it is always very important to continue to cultivate the geometric method. ... In particular we must praise the work under consideration for its great clarity ..., and therefore recommend its study as nourishing intellectual substance, by which undoubtedly much can be contributed to the revival and conservation of the genuine geometric spirit, sometimes missing in the mathematics of these times." (Gauss as quoted on page 123 of (Dombrowski 1979), in what was a review of Monge's Descriptive Geometry book, just out at the time).

"It is in the nature of mathematics of modern times (in contrast to that of antiquity), that we possess a lever in the form of our symbolic language and nomenclature, whereby most complicated reasonings are reduced to a certain mechanism. In this way Science has gained infinitely in richness, but has lost equivalently ... in beauty and character. How often is this lever applied purely mechanically .... I demand that one should whenever using calculus, ... remain aware of the original stipulations." (Gauss as quoted in page 124 of (Dombrowski 1979)

"En conclusion, peut-on dire que la Géométrie a perdu son identité? Au contraire, je pense qu'en éclatant au-delà de ses frontières traditionnelles, elle a révélé ses pouvoirs cachés, sa souplesse et sa faculté d'adaptation extraordinaire, devenant ainsi l'un des outils les plus universels et les plus utiles dans tous les secteurs des mathématiques." Jean Dieudonné (Dieudonné 1981)

"Obviously everything in differential geometry can be translated into the language of analysis, just as everything in algebraic geometry can be expressed in the language of algebra. Sometimes mathematicians, following their personal inclination or perhaps misled by a false sense of rigor, have turned their

---

[1] * 14 April 1927; † 15 October 2016
Institut des Hautes Études Scientifiques, Université Paris-Saclay, Le Bois-Marie, Bures-sur-Yvette, France

minds wholly to the translation and lost sight of the original text. It cannot be denied that this has led occasionally to work of great value; nevertheless, further progress has invariably involved going back to geometric concepts. The same has happened in our times with topology. Whether one considers analytic geometry at the hands of Lagrange, tensor calculus at those of Ricci, or more modern examples, it is always clear that a purely formal treatment of geometric topics would invariably have killed the subject if it had not been rescued by true geometers, Monge in one instance, Levi-Civita and above all Elie Cartan in another.

The psychological aspects of true geometric intuition will perhaps never be cleared up. At one time it implied primarily the power of visualization in three-dimensional space. Now that higher-dimensional spaces have mostly driven out the more elementary problems, visualization can at best be partial or symbolic. Some degree of tactile imagination seems also to be involved. Whatever the truth of the matter, mathematics in our century would not have made such impressive progress without the geometric sense of Elie Cartan, Heinz Hopf, Chern, and a very few more. It seems safe to predict that such men will always be needed if mathematics is to go on as before." André Weil (Weil 1992)

As the title indicates, the present text is not a systematic report on Calabi's work but merely a choice of some of his papers. I left out of course what concerns the Calabi conjecture, Calabi–Yau manifolds, etc since this will be reported on by J.-P. Bourguignon in this volume. With great regret I am unable to report on a very important part of Calabi's work, namely that dedicated to affine differential geometry, more or less in connection with real Monge–Ampère type equations.

Eugenio Calabi is a geometer. What geometry is being hard to define today, I have quoted above three masters on the subject. Note that Weil is referring to Monge as well as to Elie Cartan.

Let me add that a number theorist is said to be the personal friend of every prime number. I can risk saying that Calabi is the personal friend of every geometric problem and of every geometrical concept. I think the choice I am making will show this. But I do not pretend to exhaust the splendours of a diamond by only looking at only a few of its facets. The following eight items could also be called "eight gems".

I apologize for giving these eight topics an unequal treatment. It only reflects only the fact that I am unfamiliar with some of them.

Finally I am glad to have the opportunity here to thank many colleagues for their help when I was writing this text. I am sorry not to mention their names explicitly, the reason being that I am afraid of forgetting some of them.

Contents

The diastasis story

The article (Calabi 1953) is basically Calabi's dissertation. To our knowledge this wonderful text remains quite unknown and almost unused, with an amusing but not surprising exception: it is used in the paper (Calabi and Vesentini 1960) to compute the curvature tensor of Kähler symmetric spaces (see also (Bochner 1947) and (Borel 1960)). The diastasis was rediscovered more or less recently by various authors, see below. The underlying aim of this work is to study the Riemannian geometry of real analytic Kähler manifolds.

The description of geometric properties of Kähler manifolds is in fact a very hard subject. We have decisive conditions on Betti numbers from Hodge and Lepage theory, but thereafter it seems that one had to wait until the late 70's, when Y. T. Siu was able to obtain restrictions for their fundamental groups. Very recently, Gromov obtained very severe restrictions for that group: (Gromov 1989). An interesting desire of Gromov is also to measure the defect of a general Riemannian metric from being Kähler, see (Gromov 1992).

Why is this less obvious than it appears? Of course Kähler manifolds can be characterized among Riemannian manifolds by various conditions, such as integrability of the underlying almost complex structure, the closedness of the symplectic canonical form, identities for the curvature tensor. But for a geometer this is quite unsatisfactory. In particular no global invariant comes directly out of these conditions. In what follows we will consider only real analytic Kähler metrics.

A metrically minded geometer will look at normal geodesic coordinates as introduced by Riemann and see what extra properties they enjoy in Kähler manifolds. Alas! Riemann's normal geodesic coordinates are never complex analytic unless the manifold is flat, that is locally euclidean or equivalently with zero sectional curvature. This was first noticed by Bochner in (Bochner 1947). In this paper Bochner seems to have been the first to tackle the problem. For example he shows the following surprising fact. Let p be a point in a Kähler manifold and look for complex analytic coordinates such that the metric element g^*_{ij} has an expansion in the $z_i, \bar{z}_j$ starting with δ_{ij} and only after terms of order > 1 in z and in $\bar{z}$. Then, such a complex analytic coordinate system is unique up to a unitary transformation of the z_i at p.

In this paper Bochner also shows the following. Denote by $F(n, b)$ the Fubini-Study space of complex dimension n and whose holomorphic curvature is the constant $4b$. For $b < 0$ it is a complex hyperbolic space (diffeomorphic to $\mathbf{R}^{2n}$), for $b = 0$ it is the complex vector space $\mathbf{C}^n$ and for $b > 0$ it is the complex projective space $\mathbf{C}P^n$ of diameter $\pi/\sqrt{b}$. Bochner then shows that an isometric complex analytic embedding of $F(n, b)$ in a space $F(n', b')$ can exist only if $b < b'$; and that, when $b < 0$, the space $F(n, b)$ cannot be embedded in an $F(n', 0)$ with finite n', no matter how large. This might be surprising because Bochner had already remarked that complex analytic imbeddings always make the sectional curvature smaller than or equal to that of the ambient one (today one knows that this comes from the deeper fact that, by Wirtingers' trick, complex submanifolds of Kähler manifolds are always minimal).

Bochner left open the question of finally deciding in which $F(n', b')$ spaces an $F(n, b)$ can admit an isometric complex analytic imbedding. Calabi solves the question completely in (Calabi 1953). He does this, gaining at the same time a better understanding of the nature of the Riemannian metric in a Kähler analytic manifold, by introducing a new invariant, the diastasis. It is, in general, defined only locally as follows. The Kähler condition for the metric g is equivalent to saying that locally there exists some analytic real valued function f such that $g = \partial\bar{\partial} f$. Such a function is defined uniquely up to addition of $h + \bar{h}$ with h holomorphic. Consequently, when one dissociates the complex coordinates and their complex conjugates, one can define:

$$D(p, q) = f(p, \bar{p}) + f(q, \bar{q}) - f(p, \bar{q}) - f(q, \bar{p})$$

which depends only on p and q. It is called the diastasis of g between p and q. It is symmetric in p and q. If $d(p, q)$ is the ordinary Riemannian distance then $D(p, q) = d^2(p, q) + O(d^4(p, q))$. The two coincide for the complex vector spaces, namely the spaces $F(n, 0)$. Finally the diastasis has a canonical analytic expansion, unique up to unitary transformation, coming from Bochner coordinates.

One drawback of the diastasis is that it is not defined everywhere, but only locally. In a canonical coordinate system centered at some $p \in F(n, b)$, one has in fact

$$D(p,q) = \frac{1}{b}\log(1 + \sum_i b|z_i|^2)$$

with

$$\sum_i b|z_i|^2 < -\frac{1}{b} \qquad \text{(when } b < 0\text{)}$$

So the diastasis is everywhere defined in the $F(n,b)$ when $b < 0$. The diastasis $D(p,q)$ blows up in the complex projective spaces $F(n,b)$ $(b > 0)$ exactly when $d(p,q) = \pi/2\sqrt{b}$, that is when q is in the cut-locus of p. Note moreover that in a complex projective space $D(p,q)$ can be computed only as a function of $d(p,q)$.

The basic property of the diastasis is that: **it is hereditary under complex analytic embeddings.** Otherwise stated: if M is a complex analytic submanifold of a Kähler manifold N, one has: $D_M(p,q) = D_N(p,q)$, this of course for the Kähler structure that M inherits from N.

Using the diastasis Calabi first obtains a criterion for a Kähler manifold to be locally imbeddable in some $F(n',b')$. When $b' = 0$ the condition is that the Mac-Laurin expansion of the diastasis D in terms of $z_j = z_j(q) - z_j(p)$ and $\bar{z_j}$ has an (infinite Hermitian) matrix of coefficients that is positive semidefinite and of rank $< n'$. More precisely, the function $D(p,q)$ can be written at any point p as a finite sum of formal series h_σ as follows

$$D(p,q) = \sum_\sigma |h_\sigma(q) - h_\sigma(p)|^2.$$

For $b \neq 0$ one has the same condition but for the expansion of the modified diastasis

$$D' = \frac{1}{b}(\exp(bD) - 1)$$

This transformation is not mysterious, it is just the analogue of stereographic projection. In every case such embeddings are rigid up to unitary transformations.

Putting all this together, Calabi can now completely determined when $F(n,b)$ can be isometrically and complex analytically embedded in $F(n',b')$. The possible embeddings are given by explicit polynomials, known today as the generalized Veronese embeddings (suitably normalized). Specialized to the real projective subspaces of the complex $F(n,b)$, one even obtains new isometric embeddings of the $\mathbf{R}P^d$ spaces into various $\mathbf{R}P^{d'}$. The necessary and sufficient conditions read as follows

$$b > 0, \quad b' = kb \quad \text{with } k \text{ integer and } n' \geq \binom{n+k}{k} - 1.$$

Note that complex analyticity is absolutely needed if one wants to conclude in such an isometric embedding, since any compact Riemannian manifold can always be isometrically embedded in any other Riemannian manifold whose dimension is large enough. (This is Nash's theorem, see for example (Gromov 1986), section 3.3.5.)

It seems to me that the notion of diastasis should make a comeback. As Calabi remarked, for complex algebraic varieties in $\mathbb{C}P^n$ the diastasis $D(p, \bullet)$ blows up exactly at the hyperplane section dual of the point p. So that one wonders for example if there are some relations between the loci $F(p)$ where $D(p, \bullet)$ blows up and the cut-loci of the Riemannian geometry of our Kähler manifold. This was done for hermitian symmetric spaces of compact type in (Tasaki 1985). But one cannot expect too much in general. Calabi wrote to us he can prove that, generically, the locus where the diatasis blow up can be as close as one wishes to any set given a priori in the manifold. A good reason for that: the diastasis exists only for real analytic Riemannian metrics, unlike the cut-locus. In other words: the diastasis is an object which depends on the totality of the Kähler manifold. For example, it would be interesting to compare the diastasis with the various types of Kobayashi metrics (when they exist).

References using the diastasis and extensions of it are (Romero 1987; Umehara 1988). I thank Dominique Hulin for providing them.

Constant sectional curvature spaces (space forms)

Since the end of the last century, the will to correctly define and study space forms has been a driving force in mathematics, not only in geometry but also in number theory, group theory, etc. By "space forms" we understand here the restricted family of Riemannian manifolds of constant (sectional) curvature. It is reasonable to look only for complete ones, then to restrict oneself to the case of compact or finite volume ones. Natural problems are legion. They stem out of the two most natural questions: how to classify these spaces and how to study their nature (e.g. their existence, their arithmeticity, their volumes, their topology, etc.). For lack of space (not of interest) below we will only consider **compact space forms**.

Things being well understood conceptually thanks to the Hopf–Rinow theorem, it boils down to studying quotients of the simply connected spaces of constant curvature (namely spheres, euclidean spaces and hyperbolic spaces) by suitable discrete groups. It is these groups which make the link with group theory and especially with number theory in the hyperbolic case. Now these simply connected spaces are in particular symmetric spaces (of rank one). So that classically under the name "space form" we could also have been considering discrete quotients of symmetric spaces, first of rank one but also of any other rank. Such locally symmetric spaces are then **the natural geometric spaces** generalizing (reasonably) euclidean geometry (locally). Note that this includes semi-simple Lie groups for themselves. One could call them **generalized space forms**. We will treat Calabi–Vesentini contribution to those spaces in the next section.

In many respects Calabi's pioneering paper (Calabi 1961) addressed the case of constant (sectional) curvature. It had a tremendous influence. We will try to do it justice at reasonable minimum length. Before going on, a word about the bibliography. To our knowledge only two books are available that treat space forms in detail. The first by J. Wolf (Wolf 1972) and the second by Vinberg in the second part of (Vinberg 1993). But one should add to them the important (Gromov and Pansu 1990) and (Pansu 1993-94): these two texts deal with space forms (in the strict sense as well as the generalized ones of the following section).

Some remarks are in order. Except for generalities, Wolf's book treats only the case of positive and zero curvature; the negative case is absent. The recent book (Vinberg 1993) is very complete and treats all three cases. Consequently, we will be sometimes very concise and give no reference. The reader will find almost everything he may need in these three basic references.

Space forms are often called *Clifford–Klein space forms*. Klein, in 1890, inspired by a number of examples due to Clifford, formulated the problem of describing all connected compact Riemannian manifolds of constant curvature. In 1881 Killing published a book in which he again posed the problem, adding the fact that such a space was always a quotient X/Γ of one of the simply connected Riemannian manifolds X of constant curvature by a discrete group Γ of motions of X acting freely. Because the primitive state of general topology in those times, one has to wait the 30's and the Hopf–Rinow theorem to get a very modern statement. The insistence of Heinz Hopf in clarifying the notion of space forms and that of Elie Cartan were very influential. Note also that Hopf–Rinow theorem was stated (in 1931) only for surfaces. The immediate generalization was done by Myers in 1935 only.

Motivations for studying space forms are numerous. Perhaps the strongest one is that they are the most natural geometric objects. More precisely, we look for generalizations of our geometric world, namely the euclidean plane and the euclidean three-dimensional space. Their basic property is free mobility. This can be formulated in three equivalent ways. First two triples of points having the same mutual respective distances are congruent, i.e. can be carried into one another by a global motion of the space. Second double isotropy, i.e. two pairs of directions making the same angle can again be carried into one another. Third: there is an universal formula (namely

$a^2 = b^2 + c^2 - 2bc\cos a$ in the euclidean case) giving the distance between points seen from a third point with a given angle and whose distance from each point is given. Spaces which enjoy one of these equivalent properties are necessarily of constant (sectional) curvature. So that, if we look for a smooth and compact metric space which enjoys such a property *only locally*, we get exactly a space form.

Now, in the late 50's, the state of affairs concerning space forms was roughly as follows. The positive and the zero cases were very advanced. They are presented in (Wolf 1972) in great detail. Note that an explicit classification is still an open problem today for both cases. But, to give an example, in the flat case (zero curvature) it was known that the structure of the fundamental group determines the metric structure up to an affine diffeomorphism (think of flat tori): this is the famous Bieberbach theorem. On the contrary, in the positive case there are examples of manifolds with the same fundamental group but different topology. But today there is still the conjecture that on a given such manifold there is only one metric with constant (positive) sectional curvature. Mostow's rigidity famous theorem which goes back to 1968 says the answer is also positive in the negative curvature case. In passing we note that, for the flat case, Calabi found an induction argument ((Calabi 1957)) for going to higher dimensions: it is explained at large in Part II of (Wolf 1972). An interesting historical remark is that Hilbert, in his famous list of problems, asked for the classification of space forms, but considered the zero case as the difficult one, see page 25 of (Hilbert 1974). The reason was that Hilbert considered difficult to prove (or disprove) that, in the flat case, only a finite number of possibilities could occur. Since an infinite number was known in the negative case, Hilbert did not pay too much attention to the problem.

We resume our history in the late 50's. From what we have said above the main problem is seen to lie in discussing the negative curvature case. Many things were known for a dimension equal to two. And for a good reason: this is the realm of compact Riemann surfaces, a subject connected with hyperbolic geometry, uniformization, Fuchsian groups, number theory, algebraic curves, etc.

First, examples were known for surfaces of any genus ($\geq$ 2). Recall that (see above) now "elementary" Riemannian geometry implies that such a space is a compact quotient of the form $\Gamma \mathrm{SL}(2;\mathbb{R})/SO(2) = \mathrm{Hyp}^2/\Gamma$ of the hyperbolic plane Hyp^2. Some were obtained algebraically using suitable discrete subgroups of the groups of isometries of the hyperbolic plane, namely $\mathrm{SL}(2;\mathbb{R})$. Among these examples, some are arithmetic. For the exact definition of arithmeticity, one can consult (Vinberg 1993) and (Borel 1981). On the other hand examples were obtained by polygonal tilings of the hyperbolic plane (the tile being strongly connected with the Dirichlet-Voronoi domain of the associated group G). For example Klein in (Fricke and Klein 1897) found many triangles of Hyp^2 with angles equal to π/p, π/q, π/r which, by successive reflections, give an arithmetic Riemann surface. The complete classification of these arithmetic triangles was given by Takeuchi in 1977: (Takeuchi 1977). There are 76 such triples (p, q, r).

Secondly, the space of all Riemannian metrics of constant curvature -1 on "the" orientable surface of genus g started to be mastered in the course of a long history reaching from Riemann to Teichmller in the 30's (and people are still working on it). In particular this so-called moduli space has a real dimension equal to $6\gamma - 6$.

And now how about higher dimensions? The exact history is not clear to the author, especially when space forms of any dimension were obtained for the first time. Many things are in (Vinberg 1993) but this text seems to be silent about the existence question ; the introduction of (Borel 1963) and (Vinberg 1983) should be added. And see also Weil's notes (Weil 1958). We just now mention briefly important things.

The first point is that such space forms could be gotten in essentially two distinct ways (at least before (Gromov and Piatetski-Shapiro 1988)): the algebraic one and the geometric one. The algebraic way consists in finding suitable discrete subgroups of the group of motions of Hyp^d (arithmetic or not). The geometric one extends to higher dimension Klein's idea of taking a suitable polygon in Hyp^2 and using successive reflections along its side. The algebraic one is well explained in the Chapter 6 of (Vinberg 1993) and works in any dimension; it uses quaternion algebras and quadratic forms. So that one can say, in some sense, that existence in any dimension and by methods of number theory was in the folklore, say before the 60's. Selberg's work (Selberg 1960) is basic in the field to many respect. One of his lemmas (not too hard but very useful, going back in fact to Minkowski but much more general), says that any group acting discretely in Hyp^d (for any d) with a compact quotient, has a finite index subgroup which acts freely, yielding a quotient which is a manifold.

The geometric way was initiated by Schlegel in (Schlegel 1883). He classified the regular polytopes which are regularly tiling the spaces of constant curvature of dimensions 3 and 4. It seems that he also realized that there are no such ones starting dimension 5. His work was generalized by Im Hof in (Im Hof 1990). But tiling does not imply in general that the quotient is a manifold. The first examples in dimension 3 were given by Löbell in 1931. In dimension 4 a first example was given by Davis only in 1985. It is worth noting that geometric reflection groups generated by hyperbolic compact polyhedra cannot exist in every dimension but only in low ones. The first statement was by Vinberg in (Vinberg 1984). Drawing the exact line between arithmetic and non-arithmetic space forms is still a basic question.

An intermediate constructive way, mixing algebra and geometry, is given in (Gromov and Piatetski-Shapiro 1988). In dimension 3, one should also mention Thurston's construction as explained in (Gromov 1981).

Here our story begins. During 58-59 Calabi and Vesentini visited the Princeton Institute for Avanced Study. The big event that year was the seminar given by Kodaira on the theory of deformation of complex structures. We will come back to a very direct relation to it in the next section. But for the time being Selberg happened to tell Calabi that he thought that the two-dimensional case for negative space forms was exceptional and that, starting $d \geq 3$, there were no moduli space, i.e. those space forms were always rigid. Calabi thought that Selberg was wrong and he embarked on working to prove it, i.e. to find explicit non-trivial deformations. In fact he succeeded in proving that there is never such non-trivial deformation of a space form when the dimension is ≥ 3. Selberg was in fact right. The whole story was never published, only the beginning of it. This is the content of (Calabi 1961) but the complete result was spread all over the world more or less informally. See for example the footnote in (Weil 1962). Weil got his information from Kodaira by mail. It consisted in notes taken by Kodaira from informal lectures given by Calabi. Let us now sketch Calabi's proof. The first part of the proof is lucidly exposed in (Bérard Bergery, Bourguignon et al. 1973). In 1967 Mostow's rigidity theorem of course dramatically confirmed and extended Calabi's result.

The first thing to prove is that the space of non-trivial deformations of structures of constant curvature on a given compact space form M is $H^1(M;\mathcal{K})$, the first cohomology group of M with values in the sheaf $\mathcal{K}$ of germs of Killing vector fields on M. Then, to compute this group, Calabi managed to find a fine exact resolution of the sheaf of germs of deformations of the initial Riemann structure. For simplification, we will write everything at the level of tensors on M. Calabi's beautiful exact sequence is:

$$0 \to VF(M) \to \mathrm{Sym}^2(M) \to \mathrm{CurvTens}(M) \to \ldots \to \ldots$$

where the first term is the set of vector fields on M (this is exactly the tangent space to the space of all diffeomorphisms of M), the second the set of symmetric bilinear forms on M (this is exactly the tangent space to the space of all Riemannian metrics on M), the third the set of tensors on M which have the same symmetry properties as a general curvature tensor. The first differential operator (which is of order 1) is the Lie derivative and then its image in $\mathrm{Sym}^2(M)$ is nothing but the trivial deformations obtained by the diffeomorphisms of M. Calabi's nice discovery is the second differential operator, which we can call Der_R, which is of order 2; it is precisely the translation at that level of taking the derivative of the curvature operator.

Exactness at that level is hard to prove (see (Bérard Bergery, Bourguignon et al. 1973) together with (Calabi 1961)). Thereafter, to compute the various $H^q(M;\mathcal{K})$ Calabi used a method "à la Bochner-Weitzenböck" proving then that the zero-order term (involving the curvature) was of the right sign to finally yield the non-existence of non-trivial deformations. For Bochner's technique, here as well as in the following section, we refer the reader to the expository papers: (Wu 1987; Bérard 1988; Bourguignon 1988). Then Calabi introduced adjoint operators, followed by elliptic operators in order to have a Hodge theory. The originality here is that one had to consider an iterated Laplacian at the $\mathrm{Sym}^2(M)$ level because one operator was of order 2 instead of 1 in the standard situations. This was announced in (Calabi 1961), but, as we have said, not published. A more general fact, based on the work of Koiso, can be found in (Besse 1987), 12.73, page 357, in the case of Einstein metrics.

Let us now describe the heritage of (Calabi 1961). I quote D. Sullivan: 'it is after Kodaira together with Calabi that mathematicians realized that the moduli of deformations of some type, say "A", on a given compact manifold M, are directly linked to $H^1(M;\text{sheaf of germs of "A"})$'. In particular, as soon as this H^1 vanishes, one is sure that there is no deformation at all. The Kodaira–Spencer theory is in the case of holomorphic vector field s, something infinite dimensional but abelian. Calabi's case is for Killing vector fields, something finite dimensional but non abelian. In fact Chern soon became very interested with Calabi's results and asked P. Griffiths for his PhD to understand the common sheme to both theories. The outcome came soon: (Griffiths 1964) and (Griffiths 1965). Another legacy of Calabi's space form paper is to be found in (Weil 1962) and in (Matsushima 1962). In the first paper Weil is thanking Calabi for using a Calabi's idea in proving the rigidity of discrete groups of Lie groups with compact quotients, see the footnotes of pages 578 and 583. This idea was definitively not in (Calabi 1961), see above for historical details. In the second paper Matsushima proved the vanishing of the first Betti number for a lot of generalized space forms. In the footnote on page 316, he said he used some of Calabi's ideas. More precisely, he was able to use Calabi's idea through Weil's manuscript of (Weil 1962) and he thanks Weil in the footnote on page 316 for sending him his manuscript before its publication.

A less direct, but interesting, outcome of Calabi's theory is the following. It concerns what is called "infinitesimal rigidity" by Gasqui and Goldschmidt but we prefer to call it here "periodic geodesic-rigidity". We are concerned with Riemannian manifolds all of whose geodesics are periodic. On the various projective spaces $\mathbb{K}P^n$ (real, complex, quaternionic, and Cayley), the only known metrics all of whose geodesics are periodic are the standard ones. Are there others? A negative answer will solve the, still outrageously open, question which is: on which Riemannian manifolds do the two most simple metric invariants, namely the diameter and the injectivity radius, coincide? They do this for the $\mathbf{K}P^n$ and the standard spheres. And the obvious guess is that there are no others. The guess is known to be right for $\mathbf{K} = \mathbf{R}$, see for example the appendix D of (Besse 1978). Since for the remaining cases $\mathbf{C}$, $\mathbf{H}$, $\mathbf{Ca}$ we are still in a fix today, we might look for infinitesimal deformations of the standard metric on $\mathbf{K}P^n$. It was proved for $\mathbf{K} = \mathbf{R}$ by R. Michel in 1972 and for every $\mathbf{K}$ by C. Tsukamoto in 1981 that there is no such non-trivial deformation. Michel's proof (for the real case) was quite involved. Bourguignon noticed that Calabi's result immediately gives a very simple and conceptual proof as follows.

As a the first step the exactness of the above sequence can be rephrased: any variation of the constant curvature metric (of positive or negative sign) has to be trivial as soon as it is trivial for the derivative of all the sectional curvatures. Now, consider $\mathbf{R}P^n$ and a deformation conserving the length of the periodic geodesics. On the various projective planes, one knows that such a deformation has to be trivial by using the conformal representation theorem for S^2 and Radon's transform. Finally, our deformation is trivial on any two-dimensional subspace of $\mathbf{R}P$: this implies the total triviality by Calabi's exactness. We refer the reader to section 5.H of (Besse 1978) for details.

How about the other fields $\mathbf{K}$? For this one would need an extension of Calabi's exact sequence to more general Riemannian manifolds. J.-P. Bourguignon has remarked that in some sense such an extension cannot exist in all generality. Here is why. We consider the case of Einstein metrics. The differential operator to be considered ithen is the linearization of the map

$$g \longrightarrow \mathrm{Ricci}_g - \frac{1}{n}\mathrm{scalar}_g .$$

Unfortunately, the existence of non trivial moduli on K3 surfaces with Ricci flat metrics as proved by Yau shows that one cannot get a Poincaré lemma at the level $\mathrm{Sym}^2 M$ in the complex. Indeed, on such a K3 surface the sheaf of germs of Killing vector fields is reduced to the zero sheaf.

But such an extension was made for symmetric spaces by Gasqui and Goldschmidt in a series of papers (see (Gasqui and Goldschmidt) for a very recent reference). This time there appears a new differential operator of order 3:

$$\ldots \to \mathrm{Sym}^2(M) \to E(M) \to \ldots$$

It mimics the derivative of the covariant derivative of the curvature tensor. Here the space $E(M)$ is a suitable vector bundle of tensors over M involving curvature type tensors together with their covariant derivatives.

Gasqui and Goldschmidt have many applications of their theory to the periodic-geodesic-rigidity of various symetric spaces. First they recover Tsukamoto rigidity "à la Bourguignon". But they also prove such a rigidity property for the complex quadric. Such a result is much stronger since the complex quadric has still a lot of periodic geodesics but certainly many fewer than the $\mathbf{K}P^n$.

They also extend the following theorem of Michel obtained by him for $\mathbf{R}P^n$, to various symmetric spaces. Consider a 1-form α on such a real projective space and its integral over projective lines g (which are the same as the periodic geodesics):

$$\int_\gamma \alpha(\gamma'(t)) = I(\alpha, \gamma).$$

If α is exact (i.e. the differential $\alpha = df$ for some global function f on $\mathbf{R}P^n$) then $I(\alpha; \gamma) = 0$ for every g. Michel proved the converse, a nice criterion of exactness. Gasqui and Goldschmidt extend this criterion for the various $\mathbf{K}P$ and the complex quadric. For references look at the recent (Gasqui and Goldschmidt).

For the state of affairs today for (real) space forms we refer the reader to the already quoted (Vinberg 1993) and to (Gromov and Pansu 1990) and the references therein. A lot of basic questions are now quite well understood: arithmeticity, generation by crystallographic groups, repartition of their volumes, topology (e.g. the first Betti number). For the geometer what is happening for volumes is fascinating. Fixing of course the value of the curvature to be -1, possible volumes (area) in dimension two are the positive integers, thanks to the Gauss-Bonnet formula. Starting with dimension four and above, the set of possible volumes is always a discrete subset of the real line, a result of H.C. Wang in 1972. The three-dimensional case stands out: the set of possible volumes does have many accumulation points and, this is due to Jörgensen and Thurston in 1977. More details and the proof will be found in (Gromov 1981).

Even for the simplest case, that of surfaces, arithmeticity is much more interesting than might appear at first glance. I cannot resist mentioning the fascinating text (Sarnak 1993). For the volumes in the arithmetic cases see (Borel and Prasad 1990).

A third contribution of Calabi to space forms is a joint article (Calabi and Markus 1962) with L. Markus in which they classify four-dimensional space forms carrying a flat Lorentzian metric. See page 291 of (Calabi and Markus 1962).

Generalized space forms

We now explain the content of the joint paper (Calabi and Vesentini 1960). The most natural extension of the notion of space forms is the following. Looking for geometries which are locally like our euclidean space, we can ask for more general ones, replacing the free mobility axiom by the weaker condition: the space is isotropic (i.e. any two pointed-directions can be carried into one another by a suitable motion) or, equivalently, the space is (locally) two-point transitive, i.e. there is always a motion carrying any pair of points into another one as soon as their respective distances are equal. It is classical that only the following spaces enjoy that property: euclidean spaces, standard spheres, the various $\mathbf{K}P^n$ (endowed with their symmetric space canonical metrics) and their negatively curved pendants, which we are going to denote by $\mathrm{Hyp}^n_{\mathbf{K}}$. They are the same as the symmetric space of rank equal to one. Proofs are usually extremly lengthy. Today there is a very short proof of the fact such spaces are necessarily symmetric: (Szabo 1991). It remains to find the right symmetric spaces. For the general case one looks at the classification of these spaces, today of reasonable length. But we note that, in the positively curved case, Karcher found a direct proof: (Karcher 1988). Moreover this reference gives the simplest known construction of the Cayley plane, a notoriously difficult object to define.

We will call smooth and compact quotients of these spaces $\mathbf{K}$-space forms. Again we will consider only compact ones. The other more generalized space forms are compact andsmooth quotients of various symmetric spaces; this include semi-simple Lie groups for themselves.

A small remark is in order. When $\mathbf{K}$ is not $\mathbf{R}$ then Synge's theorem in Riemannian geometry implies that the $\mathbf{K}P$ have at most only one quotient, by $\mathbf{Z}_2$, because they are positively curved and of even dimension. So that the classification problem is ended. This explains why we will consider below only the $\mathrm{Hyp}^n_{\mathbf{K}}$, i.e. the negative curvature case.

The study of these space forms is important. We have already mentioned various types of questions: existence, classification, structure (arithmeticity, rigidity, topology e.g. the first Betti number to start with -, etc.). The case $\mathbf{K} = \mathbf{C}$ and $n = 1$ coincides with that of Riemann surfaces: $\mathbf{K} = \mathbf{C}$ and $n = 2$. But the case $n = 2$ and $\mathbf{K} = \mathbf{C}$ appeared as soon as 1988 in Picard's work under the guise of questions of complex analysis, e.g. singularities of complex ODE, see (Mostow 1980). These space forms are still under intensive study. They connect many fields, including: discrete groups, ergodic theory, geometry, complex analysis, number theory, algebraic topology. And already in the present volume they appear in the texts of D. Burns and F. Labourie. It seems that there is no complete review of the state of affairs concerning these generalized space forms. However (Gromov and Pansu 1990) was complete at the time it was written and more recently the Bourbaki seminar lecture by Pansu is directly related with our subject: see (Pansu 1993-94). References can be found there.

In 1958-59 Kodaira was lecturing on deformations of complex structures in Princeton IAS. What was the state of affairs for generalized space forms in the late 50's? It was much worse even that for the real case. A baffling fact is that (with the sole exception - outside the real case - of $\mathbf{K} = \mathbf{C}$ and $n = 2$) no examples were known to exist. This did not stop people from studying these compact space forms. Because there was in the air a more than strong belief in their general existence. Existence which appears only with A. Borel's founding paper (Borel 1963). Faith was great in those times, as we shall see below, for example, and as we have seen above.

In particular Calabi and Vesentini worked together on those forms, and among other things they proved the rigidity of every generalized space form of a symmetric space when it is Kähler (with the obvious exception of $\mathbf{K} = \mathbf{C}$, $n = 1$). Following earlier techniques of Bochner as extended by, for instance, Nakano to bundles over a Kähler manifold, they computed the curvature term in the Bochner–Weitzenböck formula giving, via extended Hodge-Dolbeault theory, the harmonic objects in $H^1(M;\theta)$, the first cohomology group of the compact Kähler manifold with values in the sheaf of germs of holomorphic vector fields of M. Things are here reduced to pure algebra since M is locally symmetric. It was the first example of a vanishing result "à la Bochner" for manifolds of negative curvature. In the present case, the curvature tensor term is positive and dominates the Ricci curvature one (which is negative). So they proved that $H^1(M;\theta) = 0$ always. By the simplest result in Kodaira–Spencer theory (due to Fröhlicher–Nijenhuis), this implies the rigidity of the complex structure. Coupled with results of Selberg this in turn implies the rigidity of the geometric structure.

The article by Calabi and Vesentini was of course very influential. The first page of (Gromov and Pansu 1990) is a perfect introduction to this. The generalized rigidity result of Mostow in 1973 of course supersedes their result. We mention extremely briefly a few things about $\mathbf{K}$-space forms. All the results quoted are deep and involve many modern techniques. Among others: Bochner's type formulas, the T-Kazdhan property and harmonic maps (together with, adapted to this non-linear case, Bochner's type formulas). References can be found in Pansu's text.

First, for the arithmeticity question, the situation is this. For $\mathbf{R}$ there are arithmetic as well as non-arithmetic examples in any dimension (but this does not end the question). On the contrary for **H** and **Ca**, things are forced to be arithmetic, see (Gromov and Schoen 1992). In the $\mathbf{C}$ case there are non-arithmetic examples for n equal to 2 and 3 (see (Mostow 1988)) but things are open for higher dimensions.

The set of volumes, for a given type, is always discrete in $\mathbb{R}$, with the sole exception of $\mathbb{K} = \mathbb{R}$ and $d = 3$.

When $\mathbb{K} = \mathbb{R}$ or $\mathbb{C}$, one has some examples with a non-zero first Betti number (by Millson and Kazhdan respectively). For $\mathbb{K} = \mathbf{H}$ or $\mathbf{Ca}$, the first Betti number is always zero. As for arithmeticity, separating the cases where the first Betti number is zero or not is an open (and interesting) problem.

Embedding minimal spheres into spheres

In 1951, H. Hopf proved that a surface of the topological type of the sphere immersed in $\mathbb{R}^3$ with constant mean curvature is a round sphere. Note that this result is of interest in the equality case of the isoperimetric inequality, a notoriously difficult topic. He gave two proofs of this. In one of them the fundamental idea was this.

We can apply the conformal representation theorem and consider our surface as a 1-dimensional complex manifold. In isothermal (a classical alternative name for complex $u + iv$) coordinates (u, v) of the surface, let $L.du^2 + 2M du.dv + N.dv^2$ be the second fundamental form and consider the *complex* valued expression $\frac{L-N}{2} + iM$. The constancy of the mean curvature implies that in fact the quadratic differential $(\frac{L-N}{2} + iM)(du + i\,dv)^2$ is a holomorphic quadratic differential on our surface. On a topological sphere it has, classically, to vanish identically. Which implies that everywhere $L = N$ and $M = 0$. This means exactly that every point is an umbilic. Classically (even locally) only round spheres can afford it.

The paper (Calabi 1967) (to be completed by (Calabi 1968) and (Chern 1970)) is a dramatic extension of Hopf's idea to the case of minimal immersions of the 2-dimensional sphere into round spheres of higher dimensions S^{n-1} in $\mathbb{R}^n$. Some underlying ideas come also from the first section on the diastasis, as is plain in (Calabi 1967). A modern and very lucid exposition of Calabi's article (Calabi 1967) can be found in the lecture by M. L. Michelsohn in (Michelsohn 1987).

Let us now consider a minimal immersion Ψ of the topological two sphere S^2 into some euclidean round sphere $S^N \subset E^{N+1}$. The idea of Calabi is to use the complex structure of S^2 to compute successive derivatives of y and not to stop at the second one (as was done for the second fundamental form). With obvious notations, the above Hopf differential was the quantity Ψ given by the scalar product $\Psi = (\Psi_{zz}.\Psi_{zz})dz^4$. In all that follows one uses the standard complex extension to $\mathbb{C}^{N+1}$ of the scalar product denoted by $(\alpha, \beta) \to \alpha.\beta$.

First here the condition of Ψ to be minimal translates into the fact that Φ is a quadratic differential on S^2, hence identically zero. This says that the vector $\Psi_{zz} \in \mathbb{C}^{N+1}$ is isotropic. We now take further derivatives and get the successive $2i$ -forms:

$$(\frac{\partial^i \Psi}{\partial z^i}) \cdot (\frac{\partial^i \Psi}{\partial z^i}) dz^{2i} \qquad (i > 2)$$

Using minimality of the immersion, the fact that the target space is a round sphere, one proves by induction at the same time that these $2i$-forms are holomorphic (and then vanish identically) and that all extended scalar products

$$\frac{\partial^i \Psi}{\partial z^i} \cdot \overline{\frac{\partial^j \Psi}{\partial z^j}}$$

also vanish whenever $i + j \geq 2$.

Calabi then introduces the exterior products

$$\Xi_k = \frac{\partial \Psi}{\partial \bar{z}} \wedge \ldots \wedge \frac{\partial^k \Psi}{\partial \bar{z}^k}$$

and the largest dimension m for which Ξ_m is not identically zero. Geometrically, this means that Ψ is a it full immersion in S^{2m}, namely not contained in a strict subsphere of S^{2m}. Moreover, the vanishing of the above scalar products means exactly that Ξ_m defines a holomorphic map from S^2 into the complex projective space $\mathbf{P}$ defined by the mth-exterior power of $\mathbb{C}^{2m+1}$.

But beware! This map makes sense only outside the points of S^2 where Ξ_m is not zero. Now Calabi, with a highly non trivial analysis using local PDE properties, managed to prove that these singularities eventually disappear. This yields a completely well defined holomorphic map into $\mathbf{P}$. It implies that S^2 is now complex-immersed into $\mathbf{P}$ and consequently, by a classical result, an algebraic curve in that space. Conclusion: minimal immersions of S^2 into round spheres are "algebraic". This implies for example that the total area of our S^2 is an integral multiple of 4π. If, moreover, one asks the Riemannian metric induced on S^2 by the immersion to yield the round sphere, then the immersion is completely determined for any given m. It is in fact given by spherical harmonics.

In the second paper (Calabi 1968) Calabi did not fail to note explicitly that the situation is conceptually the following one. Denote by $I_m = \mathrm{SO}(2m+1)/U(m)$ the homogeneous space made of totally isotropic m-spaces of $\mathbf{P}$. Then, the map Ξ_m takes its values in I_m thanks to the vanishing of all derivatives established above. This map fibers as shown in the following diagram:

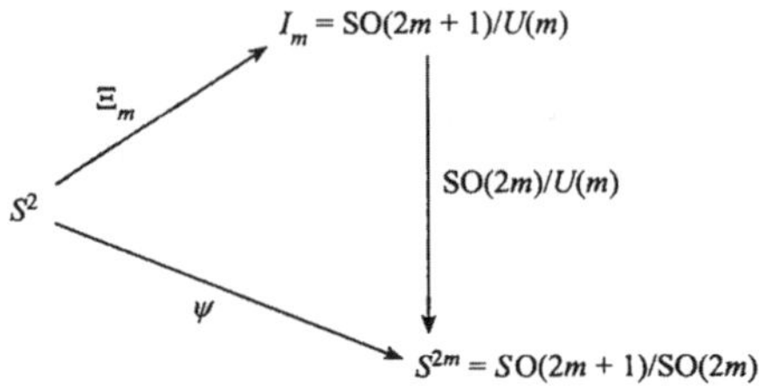

where the typical fiber is the Kählerian symmetric space $SO(2m)/U(m)$ of Lagrangian subspaces of $\mathbb{R}^{2m}$. Moreover, we have a Riemannian submersion, and the map Ξ_m is horizontal with respect to it. We first note that Calabi's result is opposite to Wirtinger's theory, which says that a complex and horizontal submanifold of I_m for this fibration is necessarily minimal.

Calabi's article was remarkably seminal. We first refer the reader to the four lectures following in the same book (Michelsohn 1987). And just sketch, few aspects of its posterity. In the 80's, physicists started to be extremely interested in minimal immersions of S^2 into complex projective spaces. Calabi's technique provides the answer, which is geometrically beautiful. Any full harmonic map from S^2 into $\mathbb{C}P^n$ can be written as the curve built up by the mth-normal lines to some holomorphic rational curve $(0 \leq m \leq n)$ (see Lawson's lecture in the volume (Michelsohn 1987)). Here we used in disguise the strong connection, in dimension two, between harmonic and minimal maps.

More important is the fact that the fibration $I_m \to S^{2m}$ is nothing but the twistor fibration of the sphere S^{2m}. For this again, see the four lectures by Gauduchon and Lawson in the volume (Michelsohn 1987).

Harmonic 1-forms

The paper (Calabi 1969) is quite spectacular. It addresses the problem of characterizing closed differential 1-forms α on a compact differentiable manifold M for which there exists a Riemannian metric g on M with respect to which α is harmonic. Even if 1-forms are always simpler to handle than those of higher degree (this because they are always decomposable or, say, dual to vectors), it required some courage to attack such a problem. Calabi completely solves the question in this article.

One of Calabi's motivations was in the air as early as 1955. Because of the Riemann–Roch (or Hopf–Stiefel) theorem in dimension 2, zeroes of harmonic 1-forms enjoy a combinatorial property involving only the genus of the surface. For example, on a 2-real dimensional torus, harmonic 1-forms never vanish (this by the way gives a simple proof of the conformal representation for tori). So the question was: do there exist topological-combinatorial restrictions in higher dimensions? A consequence of Calabi's results is that such a conjecture is completely excluded in higher dimensions.

To solve this problem Calabi found two conditions, one local and one global. Both of them are extremely simple to state. First, locally a closed 1-form α is the differential $\alpha = df$ of a function f (which is defined only up to a constant). When α does not vanish at $p \in M$ finding a Riemannian metric for which α is co-closed (hence harmonic) is trivial. Let us now assume that $\alpha(p) = 0$. We will say that p is a non-degenerate zero point of α if it is a non-degenerate critical point of f. This means that the second differential of f has maximal rank. Morse's famous lemma gives then special coordinates in which f is a sum of squares (with plus and minus signs). In these coordinates, df is co-closed after a suitable normalization with respect to an affinely flat metric. This means that the $(d-1)$-form $*\alpha$ obtained from α by euclidean duality is also closed: $d(*\alpha) = 0$. When the dimension is equal to 2 this starred form is nothing but the form obtained from α by a rotation of $p/2$.

This solves the local problem for forms all of whose zeroes are non degenerate, with the extra proviso that the indices of the zeroes are never equal to 0 or to the dimension of M. This necessary (and sufficient) condition comes simply from the minimum principle for harmonic functions. Calabi does not treat the general case where the zeros are more complicated. The reason is that the closed 1-forms, all of those zeroes are non degenerate, are dense in the space of all closed 1-forms in a reasonable sense.

The global condition is still simple to state. But the proof is considerably more difficult and we will not talk about it. Let α be a closed 1-form in M and p a point of M. The *upland of p with respect to α* is the set of points q for which there exists a non-trivial, differentiable, oriented path from p to q along which the form a is everywhere strictly positive except possibly at p itself. The *lowlands* are defined accordingly. The closed 1-form α is said to be *transitive* if, for every point in M, its upland and lowland are everywhere dense in M. This is the necessary and sufficient condition (added to the non-degeneracy of zeros) for such a 1-form to be harmonic in a suitable Riemannian metric on M.

Let us just show why some global condition is necessary. Around an axis of "revolution" we draw the picture below, and on this surface we consider the 1-form α which is the differential of the angle of rotation: $\alpha = df$ where

$f(x,y,z)$ is the argument modulo $2p$ of the complex number $x+\sqrt{-1}y$. Consider now the "middle" cycle γ in the picture, namely the middle component of the three components of $f^{-1}(0)$: it is such that $\int_\gamma *\alpha > 0$. But the cycle γ is homologous to zero since it divides the surface into two parts. Then if α were co-closed one would have $\int_\gamma *\alpha = 0$. The reader will enjoy discovering the upland and the lowland of a point $p \in \gamma$ with respect to α.

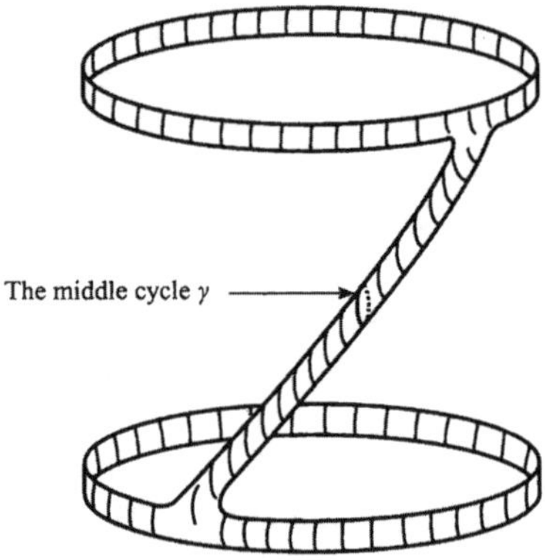

As said before it is known that harmonic 1-forms on any Riemannian torus T^2 never vanish. Calabi uses his sufficiency theorem to obtain 1-forms on tori T^d with $d \geq 3$ which admit effectively isolated zeroes in number equal to any even number.

Concerning the heritage of this result of Calabi, we note that it was only in 1981 that Eells and Lemaire used Calabi's paper to remark that there are smooth Riemannian metrics on T^3 such that any harmonic map into a flat torus always has singularities: (Eells and Lemaire 1988).

As for other possible follow ups, according to Calabi himself, there is a strange ressemblance between the transitivity condition above and D. Sullivan's duality condition, on a 1-dimensional foliation in a compact manifold, for this foliation to admit a Riemannian metric which makes it geodesic.

Periodic geodesics on surfaces

Why is it so important to study geodesics on surfaces? Let us quote Poincaré:

"Dans mes Méthodes nouvelles de la Mécanique céleste j'ai étudié les particularités des solutions du problème des trois corps et en particulier des solutions périodiques et asymptotiques. Il suffit de se reporter à ce que j'ai écrit à ce sujet pour comprendre l'extrême complexité du problème; à côté de la difficulté principale, de celle qui tient au fond même des choses, il y a une foule de difficultés secondaires qui viennent encore compliquer la tâche du chercheur. Il y aurait donc intérêt à étudier d'abord un problème où on rencontrerait cette difficulté principale, mais où on serait affranchi de toutes les difficultés secondaires. Ce problème est tout trouvé, c'est celui des lignes géodésiques d'une surface; c'est encore un problème de dynamique, de sorte que la difficulté principale subsiste; mais c'est le plus simple de tous les problèmes de dynamique; d'abord il n'y a que deux degrés de liberté, et puis si l'on prend une surface sans point singulier, on n'a rien de comparable avec la difficulté que l'on rencontre dans les problèmes de dynamique aux points où la vitesse est nulle; dans le problème des lignes géodésiques, en effet, la vitesse est constante et peut donc être regardée comme une des donnés de la question.

M. Hadamard l'a bien compris, et c'est ce qui l'a déterminé à étudier les lignes géodésiques des surfaces à courbure opposées; il a donné une solution complète du problème dans un mémoire du plus haut intérêt. Mais ce n'est pas aux géodésiques des surfaces à courbures opposées que les trajectoires du problème des trois corps sont comparables, c'est au contraire aux géodésiques des surfaces convexes." (Introduction of (Poincaré 1905))

In fact the study of periodic geodesics on spheres (2-dimensional ones) turned out to be extremely subtle. As well as many other very simple geometric questions on Riemannian metrics on S^2. Poincaré again considers periodic geodesics as basic. We take Poincaré's words from a Hadamard quotation in the paper Poincaré is precisely alluding to above, namely (Hadamard 1898):

"En second lieu, l'importance que ce géomètre a reconnue aux solutions périodiques, dans son Traité de Mécanique céleste, s'est manifestée également dans la question actuelle. Ici encore, elles se sont montrées la seule brèche par laquelle nous puissions essayer de pénétrer dans une place jusqu'ici réputée inabordable. D'une façon plus précise, elles ont joué pour nous le rôle d'une sorte de système de coordonnées auquel nous avons rapporté toutes les autres géodésiques."

The paper (Calabi and Cao 1992) brings many important contributions to the geometry of spheres (embedded ones or abstract spheres S^2 endowed with some Riemannian metric g). We briefly recall that the abstract case where the curvature is positive is the same as the case of an embedded convex surface bringing us back to the classical (but very hard) theorem of Weyl, Alexandrov, Nirenberg, Pogorelov. Let us first see why things are so subtle.

The first thing to do is to look for periodic geodesics on (S^2, g) of positive curvature in Poincaré's case ; but we can consider the general case as well. A trivial note: we always exclude of course the trivial periodic geodesics consisting of a single point. Poincaré gave three different ways to find at least one periodic geodesic but he realized honestly that he could not complete any of the proofs. One had to wait Birkhoff's fundamental contribution in 1917 to be able to always secure at least one periodic geodesic on any (S^2, g). Birkhoff's method yields a simple periodic geodesic, where 'simple' means a curve without self-intersection. In 1927, Lusternik and Schnirelman found a way to get three simple periodic geodesics, but it was only in 1978 that a complete proof for this claim was published by W. Ballmann: (Ballmann 1978). Of course everybody thought that there should be an infinite number of periodic geodesics, but this was only proved in 1992 with the joint efforts of Bangert and Franks. However, mathematicians as well as physicists (for energy reasons, etc.), are interested in knowing more. For example: what is the rate of growth of the lengths of the periodic geodesics (when infinite in number)? Very recently, N. Hingston succeeded in showing that they always grow at least as fast as prime numbers. More precisely this means that the number of periodic geodesics of length smaller than L grows at least as fast as $c\frac{L}{\log L}$ when L grows to infinity and for some positive constant c: see (Hingston 1993). It does not seem clear today whether or not the growth need or not to be always at least quadratic (as in flat tori for example).

The question is hard for spheres but, on the contrary, for compact negatively curved surfaces, one knows that the geodesic flow is ergodic. Hence ergodic theory shows that the length of periodic geodesics grows exponentially with L. The coefficient of the exponential is essentially equal to the entropy of the geodesic flow.

Let us come back to the method used by Birkhoff to get a periodic geodesic on an arbitrary (S^2, g). He invented a simultaneous shortening process applied to complete tapestries of the sphere obtained by a family of closed curves as shown in the figure below.

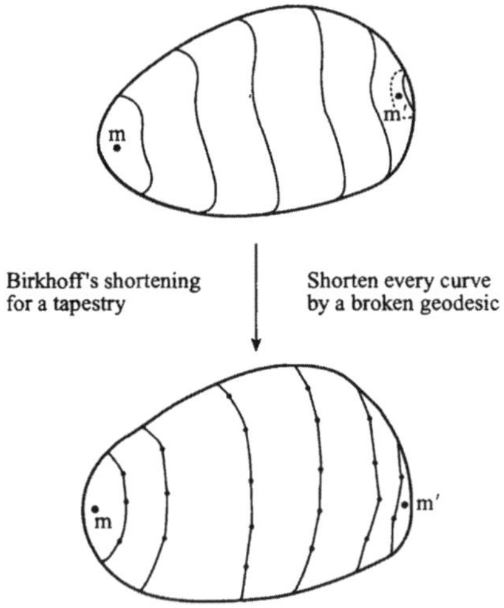

At the beginning, such a tapestry can be very bad anywhere in the sense that its curves are ridiculously "too long". Now shorten them and keep on doing so. Since a complete tapestry cannot *tear*, this implies that at the end some closed curve will be trapped in the middle and will have to be a periodic geodesic. Of course the proof has to be rigourous in convergence problems but it is today of reasonable difficulty. More precisely: any given tapestry has a curve of maximum length. One takes a "limit" tapestry such that this maximal length is a minimum among all tapestries. For this tapestry a curve of maximal length is a periodic geodesic.

It looks at first obvious to any reasonable person that a periodic geodesic obtained this way is simple, i.e. does not cross itself. But it becomes less obvious when one looks at 3 - (or more-) legged starfishes

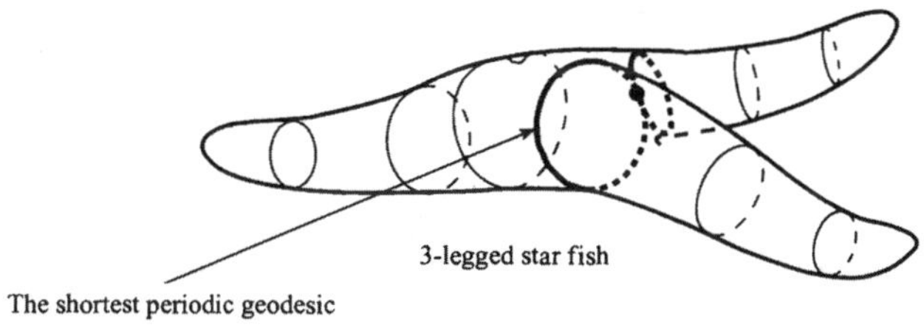

So that we now come to a very simple geometric questions: "is a periodic geodesic of shortest length always simple ?" Before looking at an answer to this question, we mention also other natural problems. We can introduce invariants, such as: the smallest length for periodic geodesics, the smallest one for simple periodic geodesics. Again we can try to get universal inequalities between those lengths and the other natural invariants which are the inner diameter $D = D(g)$ (this is not the ordinary diameter when the surface is embedded in $\mathbb{R}^3$) and the area $A = A(g)$. You can add the first non-zero eigenvalue of the Laplacian (the fundamental tone) $\lambda_1(g) = \lambda_1$ if you like analysis. Why not ask if there is an upper bound for the length of any simple periodic geodesic (in term of the diameter for example...)?

Before we get too enthusiastic, here are three facts. The first is due to Pitts who showed in 1981 that there are (S^2, g) whose shortest periodic geodesic is not simple. The shape is that of a 3-legged star fish, see pictures below. Note that these are not positively curved. The second is that the following conjecture is still open, namely, the ratio $A(g)/L^2(g)$ of the area to the square of the smallest length for periodic geodesics is always bounded by $\frac{1}{2\sqrt{3}}$ and that this bound cannot be achieved for a smooth (S^2, g). The conjectured limiting object is the double equilateral triangle. There is only a weaker inequality proved by C. Croke: (Croke 1988).

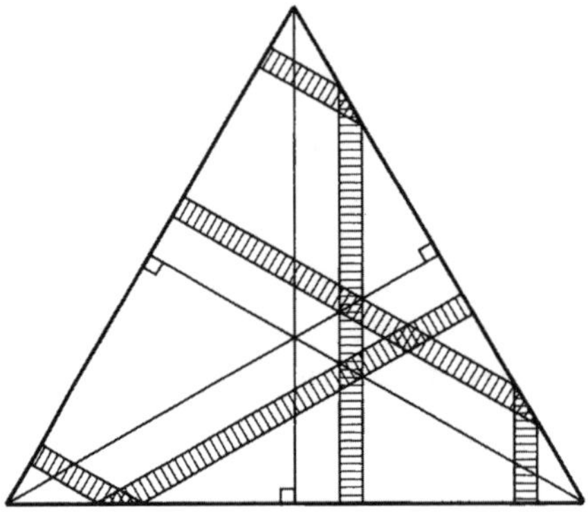

A doubled equilateral triangle

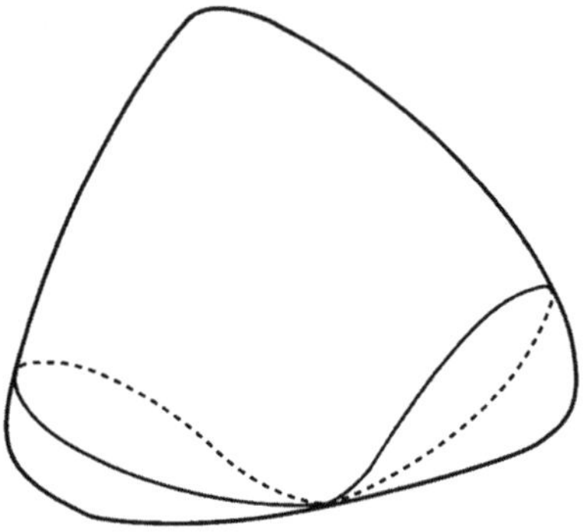

A very flat surface closed to a doubled equilateral triangle

Note that in this doubled triangle there are three continuous families of smallest periodic geodesics and that, except for the three doubled altitudes, none of them are simple.

The third fact is that the following conjecture of Alexandrov is still open: namely that for every (S^2, g) with non-negative curvature the minimum of the ratio $A(g)/D^2(g)$ is bounded from above by $\pi/2$. This value will never be attained in the smooth case, and the limiting object should be the double flat disk.

Looking at a surface very close to the double equilateral triangle one is not so sure about the conjecture claiming that smallest periodic geodesics are necessarily simple for positively curved surfaces. This is the major result of the Calabi–Cao paper: a periodic geodesic of smallest length in a smooth surface of non-negative curvature is always simple. The paper contains many other theorems, which treat problems of the type mentioned above. We will now only sketch ideas used in the proof of this main theorem.

For an (S^2, g) one introduces a new invariant $c(g)$ as follows. It is essentially the minimax associated to the set of all tapestries of S^2, as in Birkhoff, except that we now work not among the homotopy tapestries but in the integral 1-homology tapestries. We consider the set of 1-parameter families $\{\sigma(t)\}_{t\in[0,1]}$ made up by 1-$\mathbb{Z}$-cycles $\sigma(t)$ which start with a $\sigma(0)$ made up of a finite set of point curves and end in the same way. With the basic condition that this family represents a non-zero element in the fundamental group of the 0-pointed space of integral 1-cycles in S^2. The picture below is one example (3-legged star fish). But beware that in general a periodic geodesic can be non simple (Pitts' example mentioned above) or of length smaller than $c(g)$ as in the dumb-bell example below. The proof of this, for a surface of revolution, is elementary.

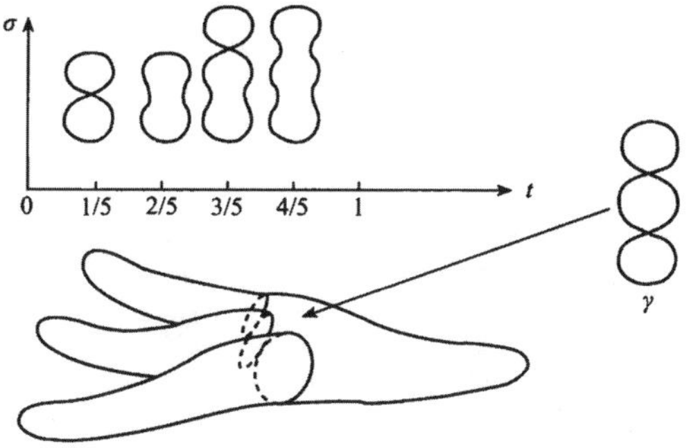

4-legged star fish

A result of Almgren–Pitts establishes the existence of $c(g)$ *and* its realization by a simple or figure-of-eight periodic geodesic. A double figure-of-eight-geodesic, as in a 4-legged starfish, never realize $c(g)$.

Now for surfaces with non-negative curvature, Calabi and Cao prove two things in one shot. A periodic geodesic of shortest length among periodic geodesics is first simple and secondly of length equal to $c(g)$.

The proof goes as follows. Start with a figure-of-eight periodic geodesic of length $c(g)$. If the curvature is identically zero along it, move it continuously up to the point where it will meet the positive curvature part of the surface. Then apply the second variation formula to it to get nearby curves of smaller length, call one of them g. The curve g is not a geodesic but can be used to build up a 3-legged starfish tapestry of the kind which enters into the definition of $c(g)$ and is of length everywhere smaller than before. This is the desired contradiction: a figure-of-eight cannot exist when the curvature is non-negative.

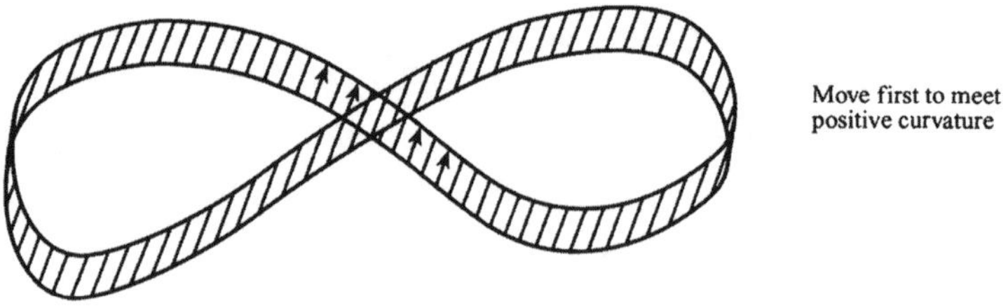

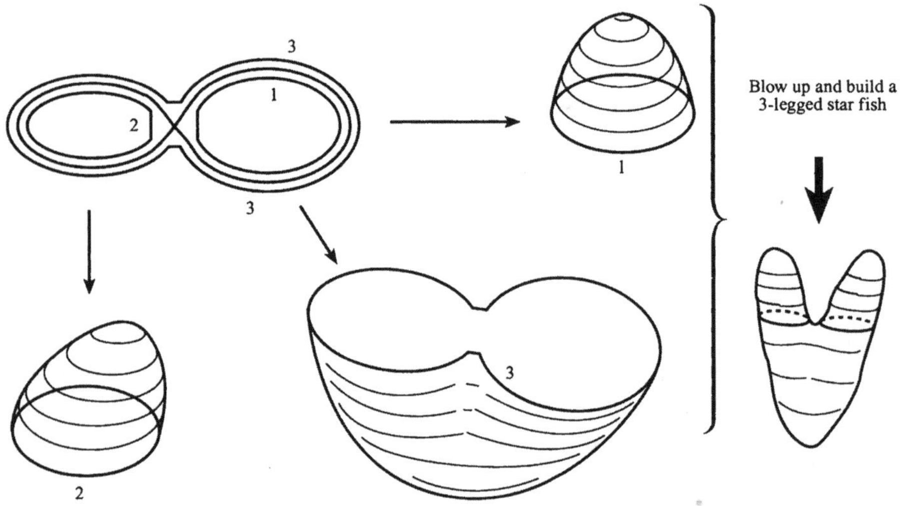

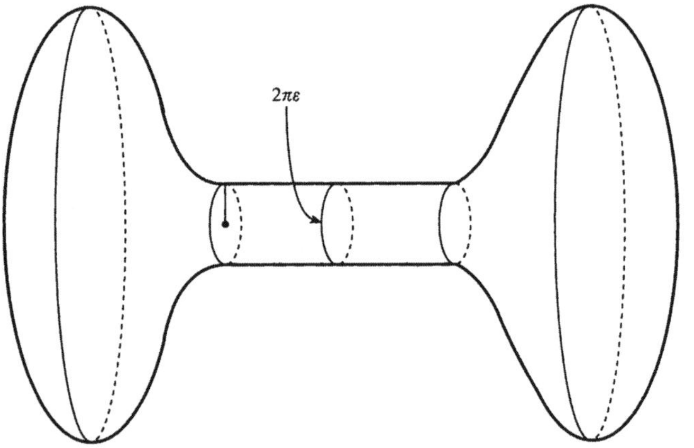

A dumb-bell of revolution proves that the Birkhoff minimax value can be (much) longer than the shortest periodic geodesic.

Extremal systolic surfaces of higher genus

We start with Loewner's theorem (1949, unpublished). Consider a 2-dimensional torus T^2. If it is embedded in $\mathbb{R}^3$ it inherits a Riemannian metric g. But we will also consider abstract Riemannian tori (T^2, g). In any case we have two natural numbers attached to them: the total area Area(g), and the systole Sys(g), which is by definition the infimum of the lengths of the closed non contractible curves traced on T^2. For these sections, here are some references: (Berger 1993), (Calabi 1993), (Gromov 1983), (Gromov 1992), (Gromov 1993).

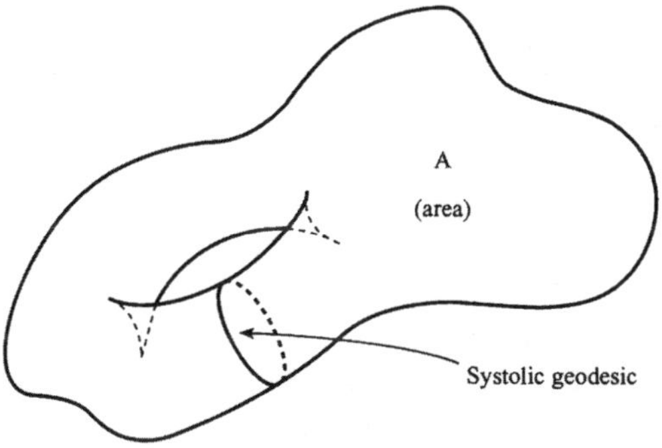

It seems reasonable that, when Sys(g) is not too small, the area is not too small either. We look, if you prefer, for a kind of isoperimetric inequality, where the systole represents the length of the boundary. To put it better: a "co-isoperimetric " inequality. Or: "an a priori inequality for the geometry of 2-dimensional tori".

The answer was given in an optimal form by Loewner. Whatever g is on T^2, one has $\text{Area}(g)/\text{Sys}^2(g) \geq \sqrt{3}/2$. The equality is achieved if and only if (up to similitudes of course) we are dealing with the so-called equilateral flat torus. This object is the quotient of the euclidean plane $\mathbb{R}^2$ by the group of translations generated by 1 and $\frac{1}{2}(1 + i\sqrt{3})$.

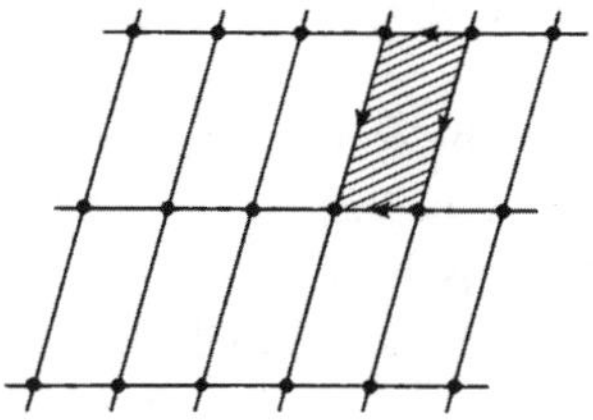

General torus (flat)

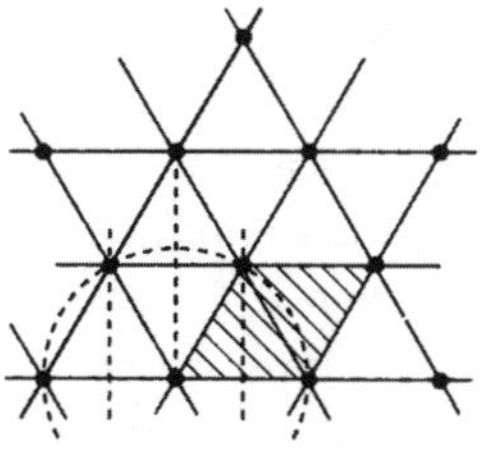

The flat equilateral torus

How about surfaces of higher genus, say, oriented of genus γ? This corresponds for the embedded case to surfaces with γ holes.

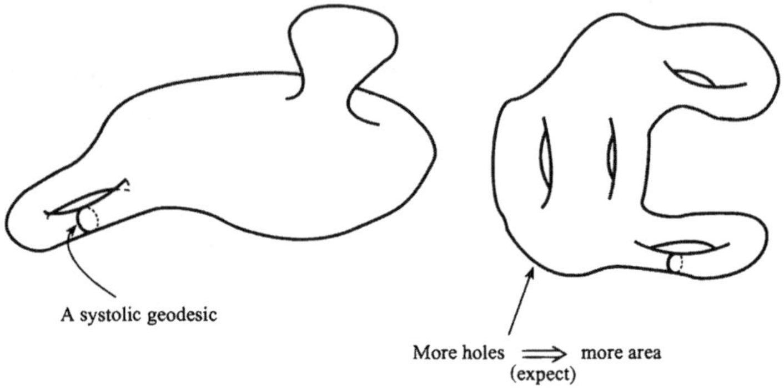

From the figure one can guess that the ratio $\mathrm{Area}(g)/\mathrm{Sys}^2(g)$ is this time bounded from below by some constant, which moreover should increase with the genus γ. But a statement of this sort was not made until M. Gromov's work in 1983. More precisely, he proved that there is a universal number c such that $\mathrm{Area}(g)/\mathrm{Sys}^2(g) \geq c\frac{\gamma}{\log^2 \gamma}$ for any surface of genus γ and any Riemannian metric on it. This estimate is optimal considering what enters into it. Only the constant c might be improved in the future.

We are now left with the question: a genus γ being given (typically asmall one: 0, 1, 2, 3, 4,...), what is the optimal lower bound $c(g)$ for $\mathrm{Area}(g)/\mathrm{Sys}^2(g)$? If $c(g)$ is known, what are the metrics for which it is achieved (if any)? Today, the question is solved completely only for the real projective space and the Klein bottle. For the real projective space $\mathbb{R}P^2$ the complete answer is due to Pu in (Pu 1952): the lowest ratio is attained for the standard metric alone (up to a scalar of course). The case of the Klein bottle is due to Bavard in (Bavard 1988) (see also (Blatter 1962)). We present the solution for the Klein bottle to show the subtlety of the problem (as soon as one leaves T^2 and $\mathbb{R}P^2$).

The best constant is that obtained for the following unique Riemannian Klein bottle. It is a singular Riemannian space, which is of constant curvature everywhere except on a singular line. Here is how to build it. Consider the part of the standard sphere between latitude 45° and latitude −45°. Take its quotient by the restriction of the antipodal map. One gets a Möbius strip. Glue two identical copies of this object along their boundary and you have done. The former common boundary will yield the line of singularity.

Note that singularities are needed if the curvature is (almost) everywhere positive, thanks to the Gauss–Bonnet theorem which requires distributional type negative curvature for the total curvature to be zero.

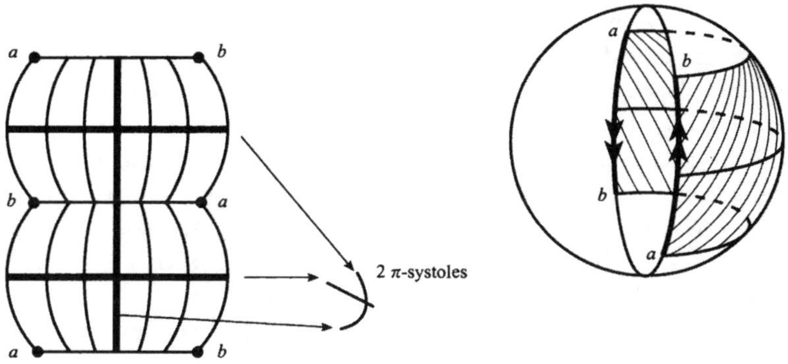

Exercise: find all π-systoles

Now consider, to start with, a surface of genus 2. If one is naive (not having thought enough about the Klein bottle example), one may think of looking for candidate surfaces among those of constant negative curvature. There are many reasons for that. The first is that the proof of Loewner's theorem rests essentially on conformal representation. The second is that the most natural metrics on surfaces of higher genus are those of constant negative curvature. Let us immediately kill that hope by considering the figure below:

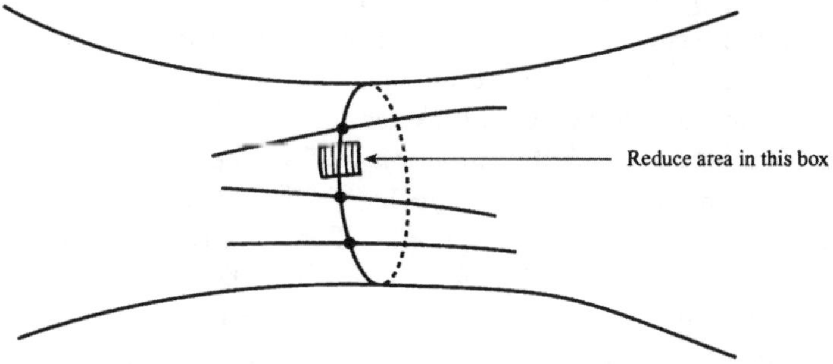

This proves that a surface of constant negative curvature can never realize the minimum of $\mathrm{Area}(g)/\mathrm{Sys}^2(g)$. In fact let c be a periodic geodesic whose length is equal to the systole. By the collar theorem and elementary considerations, the only other possible periodic geodesics realizing the systole and crossing c are finite in number. Pick up a point $p \in c$ not in such a systolic geodesic. You see immediately that you can strictly diminish the surface area by a small modification around p without changing the length of c. This was noticed by N. Kuiper as early as the 70's. It leads one to think, in fact, that through every point passes at least a one-parameter family of systolic geodesics (to be called a band in the sequel).

Since we are in a fix, we can still try to proceed by at least assuming the problem solved, namely that there exists a surface (M, g) of genus γ minimizing the ratio $\mathrm{Area}(g)/\mathrm{Sys}^2(g)$. An existence theorem can be found in (Gromov 1983), theorem 5.6.C'. The above remark is a good starting point for a systematic study of the geometry of such a minimizing object. But people who have tried it seem to have given up. The paper (Calabi 1993) is really the first to attack the problem. In fact the task is formidable, both conceptually and technically. It is reasonable to think that the results in it, still partial, will be followed up in other papers.

The paper brings various contributions. Calabi has good reason to think that, except perhaps in genus equal to 3, the extremal surfaces will have open pieces with curvature of the three possible signs: 1, 0, -1. And the geometry of such a piece is what has to be studied. It is extremely subtle. The argument we have presented above for throwing out constant negative curvature shows immediately that through any non-singular point of the surface there cannot be only one band of systolic periodic geodesics. If not, the above process shows how one can reduce the area without changing the systole.

The problem starts with two bands of systolic geodesics. Then Calabi shows that this can happen only if the two bands are orthogonal and the metric flat (at such points). This will be the general situation in the flat parts. The next case is that which can be seen in the equilateral torus: three bands of systolic geodesics (locally at every point). But then Calabi realizes that a (local) extremum for the area need not force the three bands to meet at angles of $\pi/3$ nor the metric to be flat.

The situation for three bands looks as follows. To every band of systolic geodesics is attached a potential function (defined up to a constant). So that we have three potential functions u, v, w. With them, one gets a surface in $\mathbb{R}^3$ under the map $p \rightarrow (u(p), v(p), w(p))$. The condition for the extremality of the area translates for this image in R^3 into a PDE of a type close to that of minimal surfaces. Alternatively, for two of the potentials (say u, v), the third w satisfies a PDE $w = F(u, v)$ which is quasi-linear of second order and of Monge–Ampère type. The whole story is subtle and we refer the reader to (Calabi 1993).

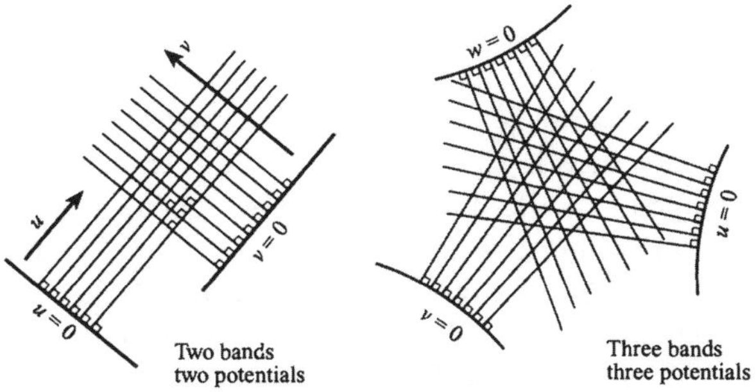

Moreover, it cannot be excluded that a fourth (or further) band of systolic geodesics crosses a 3-region. Then, one has to study whether it is possible to have two distinct PDE (or even more) of the above type with a common solution. Calabi put this question to R. Bryant in Oberwolfach in May 1993 and, for the answer, we refer the reader to Bryant's article in this volume.

In (Calabi 1993), one finds two examples of surfaces of genus 3 which are locally flat with only conical singularities (say: polyhedral manifolds). The pictures are given below. The first one is extremal, possibly a local minimum, for the systolic ratio; but it is not excluded that some smoothing will ameliorate the ratio. The second one cannot be deformed (locally, since in the large things might be different) to give a better ratio. In these pictures the shaded bands are the systolic ones.

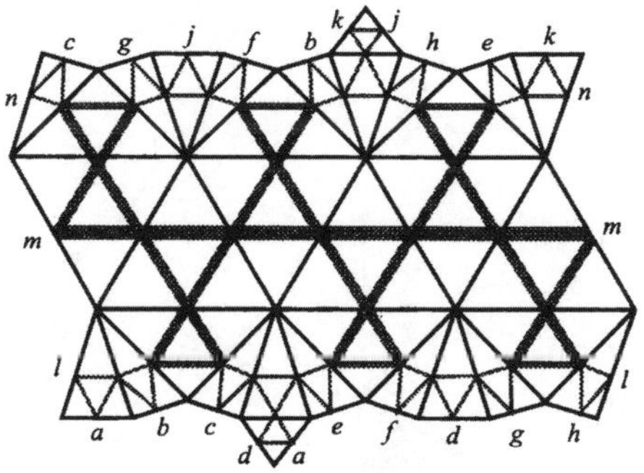

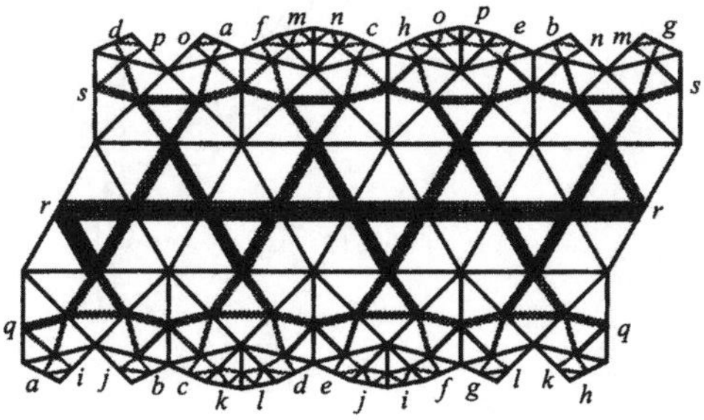

Finally, in (Calabi 1993) Calabi studies also the following problem. Besikovic's Lemma (see for example (Gromov 1983), section 7) says that, for any Riemannian metric on a cube (in particular a square), the total volume is greater than or equal to the product of the distance between the opposite faces. Here, the distance is the infimum of the distance between any two points of these opposite faces. The flat "square" metric achieves this bound. Calabi's idea is to consider the analogous problem for a planar hexagon, in particular to look for the minimal (or extremal at first) metric. The hope is that such hexagonal pieces will be the basic building blocks for minimal (extremal) systolic metrics on surfaces of genus ≥ 2.

The answer is not known. Calabi thinks that the minimal metric is made up of six pieces of zero curvature (where there are two systolic bands) surrounding a kind of circle. This circle will be a singular line and the inside will have three systolic bands. Such a hexagon will look, according to Calabi, like a fried egg.

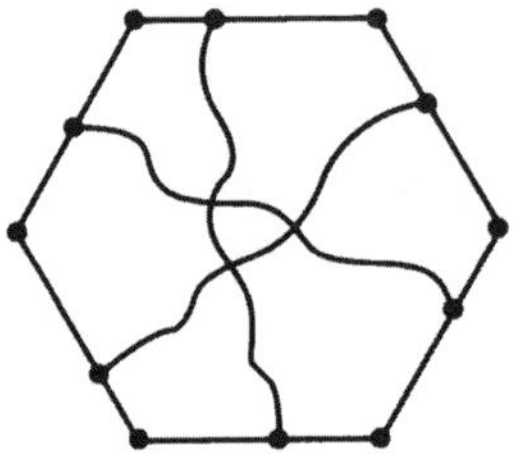

Calabi's hexagonal problem

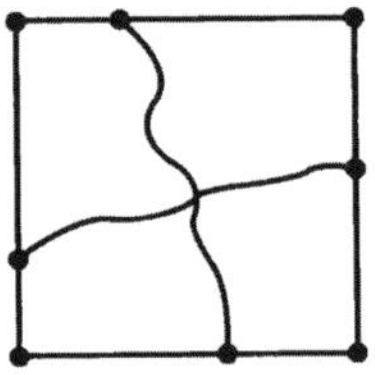

Besikovitch' case

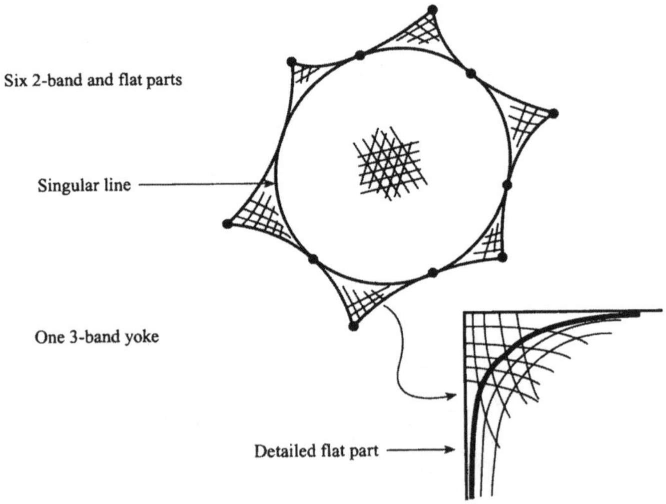

Convex particles between two fluids

Let me end this encounter with a very personal story which I will treat in a more familiar way. Two physicists came to me in 1991 with the following elementary geometric question. Let C be a convex compact body in the euclidean plane, and let $\alpha \in [0, p/2]$ be a given angle. Find a line D cutting C with this angle α at the two points of intersection. The physical motivation for the problem is simple. Consider a long prismatic particle (an example would be a log of wood floating in plane water) of convex section at a liquid/liquid interface in a gravity-free situation. The equilibrium positions (stable or not) are given by the solution of the above problem. Denote by C the orthogonal cross-section of the prismatic particle and by α the angle given by the formula $\cos\alpha = \frac{\gamma_{13}-\gamma_{23}}{\gamma_{12}}$ where the γ_{ij}'s are the interfacial tensions characterizing the system: the indices are 1 and 2 for the two liquids and 3 for the particle.

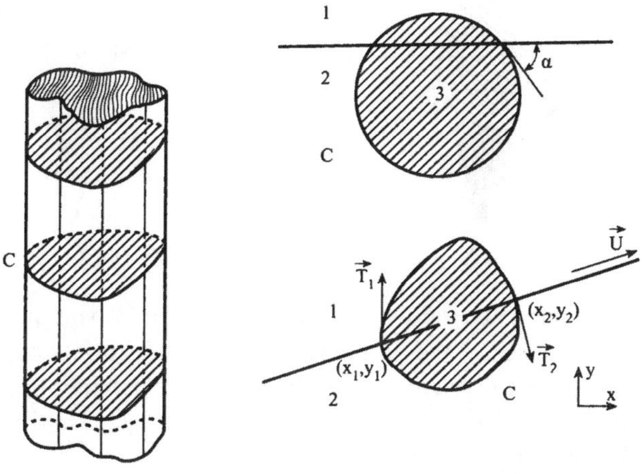

This looks like a trivial extremality problem. So I did not have too hard a time in proposing the following solution: consider the set of hatched regions as in the figure, where the curvilinear triangle is made up of two tangents to C (assume C is smooth) and with an angle at the vertex equal to $\pi - 2\alpha$. Calculus or elementary heuristic considerations (see figure) show that the critical values for the area of such regions, when it turns around C, occur when the two sides of the triangle are equal. This implies that the two base angles are equal and, by construction, necessarily then to the desired value α.

The physicists were smart. They called me a week later and said, "we know that for plane curves the curvature admits four critical values, not only two (this is the so-called four-vertex theorem, a not completely obvious fact). So we guess that there are also four solutions to our problem". The following week, I went to Oberwolfach and, during three days painfully tried to prove the required assertion between lectures. Eventually I gave up and turned the question over to Calabi who was at the meeting. This was after dinner. He immediately said he was certainly able to prove it, embarked in computations with the Euler equation for a convex body in the plane. I left him to get some sleep. Next morning, well before breakfast time he was in the dining-room and immediately came to me: "Marcel, I found four different solutions to your question overnight." For lack of time (or some other reason) I could not follow him so quickly. But before leaving he promised to send me a detailed written proof. This was done within a week. I passed it on to the physicists and what came out of it is (Raphaël, di Meglio et al. 1992).

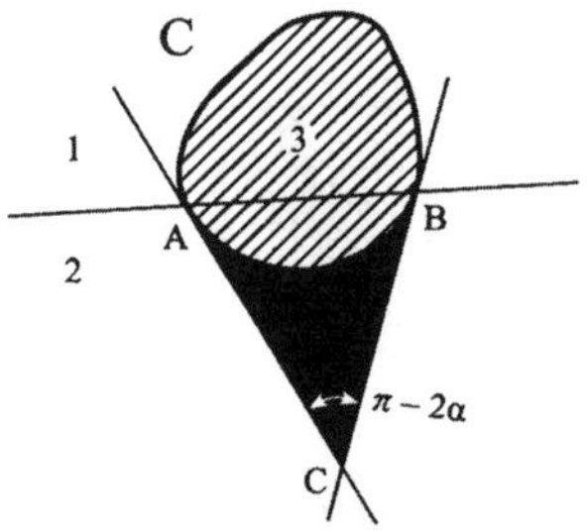

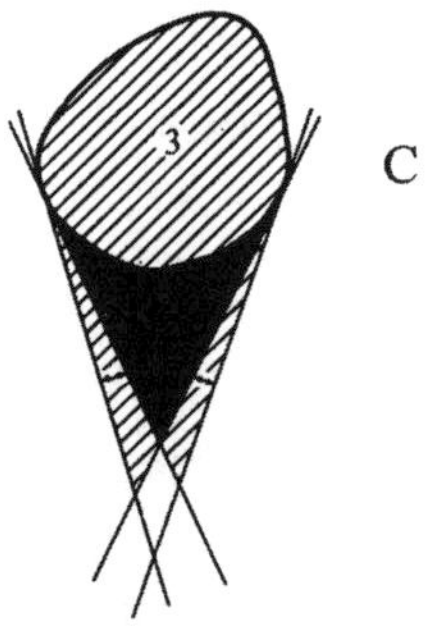

Here is the solution preferred by Calabi. I cannot resist the pleasure of giving it because of its elegance. We define the convex set C by its support function $u(\theta)$. This means that C is the curve whose tangent at the point of parameter θ is the line which makes with a fixed origin the angle θ and whose distance to the base point is equal to $u(\theta)$.

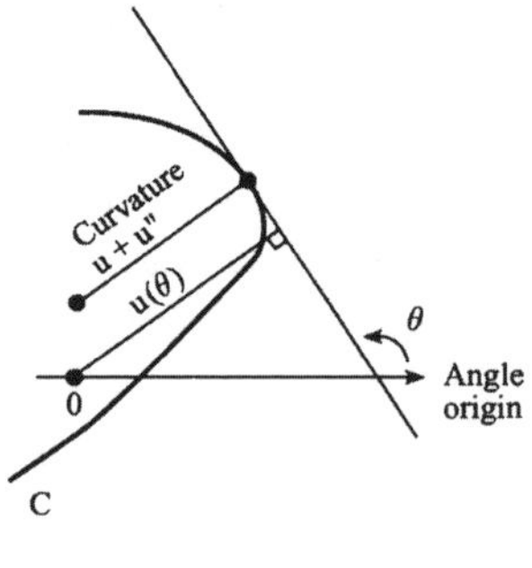

Euler equation for C

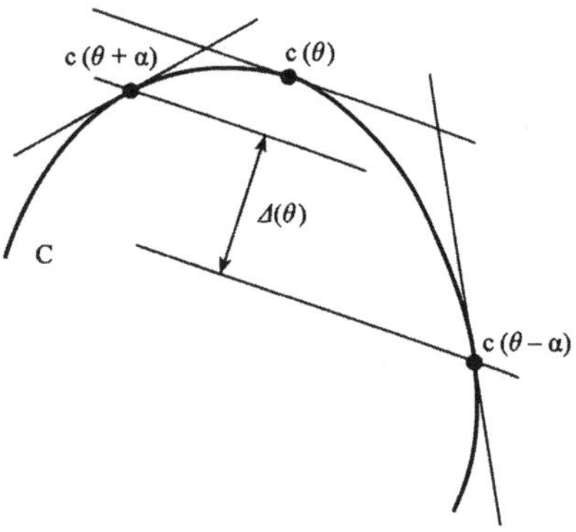

With every θ we associate the two points of C whose parameters are $\theta + a$ and $\theta - \alpha$, and we draw the two lines passing through these points and of angle θ with the origin. The picture shows that the parameter value θ will be the one sought if and only if these two lines coincide. So Calabi computed their common distance $\Delta(\theta)$ and found the two formulas:

$$\Delta(\theta) = -(\hat{u}'''(\theta) + \hat{u}'(\theta)) \int_{\alpha}^{\alpha} (\cos\phi - \cos\alpha) df$$

where $\hat{u}$ is the following natural weighted average of the function u:

$$\hat{u}(\theta) = \frac{\int_{\alpha}^{\alpha} (\cos\phi - \cos\alpha) u(\theta - \phi) df}{\int_{\alpha}^{\alpha} (\cos\phi - \cos\alpha) d\phi}.$$

Let us look at the correspondence $u \longrightarrow \hat{u}$ as a correspondence between convex plane curves given by their Euler equation. And recall that the curvature of the curve given by u is $u + u''$. The above formula shows that Δ vanishes if and only if the curvature of the curve given by the function $\hat{u}$ is critical. So the four solutions we are looking for are just those coming from the four-vertex theorem as applied to the transformed curve $\hat{u}$.

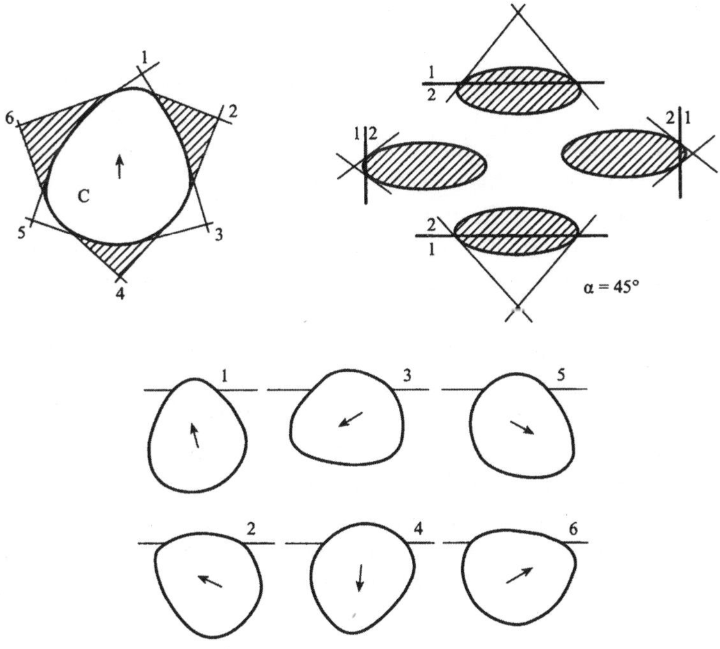

A convex cross-section C with six Young positions (α = 45°)

The article ends by showing that half of the solutions are stable. Let us add the question: Can we have six vertices very often? We refer the reader to (Osserman).

Now some geometer could have an inferiority complex when studying such "elementary" questions of geometry. Gromov passed me the following quotation from Constance Reid's book on Hilbert:

"... Lecturing on topology, Minkowski brought up the Four Color Theorem - a famous unsolved problem in that field of mathematics. He said: 'This theorem has not yet been proved, but that is because only mathematicians of third rank have occupied themselves with it', Minkowski announced to the class in a rare burst of arrogance. 'I believe I can prove it.'

He began to work out his demonstration on the spot. By the end of the hour he had not finished. The project was carried over to the next meeting of the class. Several weeks passed in this way. Finally, one rainy morning, Minkowski entered the lecture hall, followed by a crash of thunder; at the rostrum, he turned toward the class, a deeply serious expression on his gentle round face. 'Heaven is angered by my arrogance', he announced. 'My proof of the Four Color Theorem is also defective.'

He then took up the lecture on topology at the point where he had dropped it several weeks before."

Conclusions

For lack of space and time I have made a quite restrictive selection here. Looking at the bibliography of Calabi one can see that he attacked almost every geometrical problem that was general enough. Here are some more examples. Again let me mention how sorry I am not to have been able to report on his work on affine differential geometry and the Monge–Ampère equation. His interest for it was never ending: a Monge–Ampère type equation appears in (Calabi 1993). See also the articles by R. Bryant and by Y. T. Siu in this volume.

According to Dieudonné there is basically no mathematical theory which does not use some quadratic form. Calabi did not miss out in this respect: he has a paper on linear families of quadratic forms.

Calabi did not publish his solution of the differentiable pinching problem, which was solved in the late 60's simultaneously by D. Gromoll and by Y. Shikata: see Chapter 7 of (Cheeger and Ebin 1975).

As early as 1970 he set out to study the group of symplectomorphisms of a compact symplectic manifold. On the way to his results he invented what is now called the *Calabi invariant*.

In 1967 he studied manifolds with lower bounded Ricci curvature when they are non compact. The 1957 paper extending Hopf's maximum principle. You will already find in the 1967 paper already uses subharmonic functions and Busemann functions. These two ingredients are used in the same spirit in the Cheeger-Gromoll work (Cheeger and Gromoll 1971) on the splitting of manifolds of non-negative Ricci curvature. For another type of application of this work of Calabi see also page 475 of (Eells and Lemaire 1988).

The ability to built up examples and counter-examples is a characteristic of good mathematicians. V. Guillemin told me that Calabi always was for him an unfailing source of examples and counter-examples. The Calabi–Eckmann manifolds (1953), which are complex but not algebraic, was a pioneering work: see G. Tian's article in this volume. The same is true for the construction of examples of Riemannian manifolds with special holonomy, see J.-P. Bourguignon's lecture in this volume. And we have also seen the above in two other instances: explicit isometric imbeddings of space forms and that of minimal subspheres of spheres.

Another remark is in order. A good number of Calabi's papers either use complex numbers or are really living in the world of complex geometry. This is typical of many geometers since, say, Euler and Gauss. We have seen various examples above. Among them, the work on minimal embeddings of 2-dimensional spheres into spheres is especially impressive.

To finish by coming back to my introduction, let me say that Calabi also showed his personal friendship for the 6-dimensional sphere in two papers, one in 1958 and one jointly with H. Gluck in 1989.

Bibliography

Ballmann, W. (1978). *Doppelpunktfreie geschlossene Geodätische auf kompackten Flächen*. Math. Z. **161**: 41–46.

Bavard, C. (1988). *Inégalités isosystoliques conformes pour la bouteille de Klein*. Geometriae Dedicata: 146–166.

Bérard Bergery, L., J.-P. Bourguignon, et al. (1973). Déformations localement triviales des variétés Riemanniennes. Differential Geometry, Stanford, American Mathematical Society.

Bérard, P. (1988). *From vanishing theorems to estimating theorems; the Bochner technic revisited*. Bull. Amer. Math. Soc. **19**: 371–406.

Berger, M. (1993). Systoles et applications selon Gromov. Séminaire Bourbaki. Société Mathématique de France.

Besse, A. (1978). Manifolds all of whose geodesics are closed. Springer.

Besse, A. L. (1987). Einstein manifolds. Springer.

Blatter, C. (1962). *Zur Riemannschen Geometrie im Grossen auf dem Möbius band.* Compositio Math. **15**: 88–107.

Bochner, S. (1947). *Curvature in Hermitian metrics.* Bull. Amer. Math. Soc. **53**: 179–195.

Borel, A. (1960). *On the curvature tensor of Hermitian symmetric manifolds.* Annals of Math. **71**: 508–521.

Borel, A. (1963). *Compact Clifford–Klein forms of symmetric spaces.* Topology **2**: 111–122.

Borel, A. (1981). *Commensurability classes and volumes of hyperbolic 3-manifolds.* Ann. Sc. Norm. Super. Pisa IV, **8**: 1–33.

Borel, A. and G. Prasad (1990). *Finiteness theorems for discrete subgroups of bounded covolume in semi-simple groups.* Public. Math. I H E S **71**: (173–177):

Bourguignon, J.-P. (1988). The "magic" of Weitzenböck formulas. Variational methods, Paris, Birkhaser.

Calabi, E. (1953). *Isometric embedding of complex manifolds.* Annals of Math. **58**: 1–23.

Calabi, E. (1957). *Closed, locally euclidean, 4-dimensional manifolds.* Bull. of the A. M.S. **63**: 135.

Calabi, E. (1961). On compact Riemannian manifolds with constant curvature. Proc. Sympos. Pure Math., Amer. Math. Soc.

Calabi, E. (1967). *Minimal immersions of surfaces in Euclidean spheres.* J. Differential Geometry **1**: 111–125.

Calabi, E. (1968). Quelques aplications de l'analyse complexe aus surfaces d'aire minima. Topics in complex manifolds. Les presses de l'Université de Montréal.

Calabi, E. (1969). An intrinsic characterization of harmonic one-forms. Papers in Honor of K. Kodaira. Princeton University Press. 101–117.

Calabi, E. (1993). *Extremal isosystolic metrics for compact surfaces.* à paratre:

Calabi, E. and J. Cao (1992). *Simple closed geodesics on convex surfaces.* J. Differential Geometry **36**: 517–549.

Calabi, E. and L. Markus (1962). *Relativistic space forms.* Annals of Math. **75**: 63–76.

Calabi, E. and E. Vesentini (1960). *On compact locally symmetric Kählker manifolds.* Annals of Math. 1960: 472–507.

Cheeger, J. and D. Ebin (1975). Comparison theorems in Riemannian Geometry. North-Holland.

Cheeger, J. and D. Gromoll (1971). *The splitting theorem for manifolds of nonnegative Ricci curvature.* J. Differential Geometry **6**: 119–128.

Chern, S.-S. (1970). On the minimal immersions of the two-sphere in a space of constant curvature. Problems in Analysis: a symposium in honour of Salamon Bochner. Princeton University Press.

Croke, C. (1988). *Area and the length of shortest closed geodesics.* J. Differential Geometry **27**: 1–22.

Dieudonné, J. (1981). *Domination universelle de la Géométrie.* non publié:

Dombrowski, P. (1979). 150 years after Gauss. Société mathématique de France.

Eells, J. and L. Lemaire (1988). *Another report on harmonic maps.* Bull. London Math. Soc. **20**: 385–524.

Fricke, R. and F. Klein (1897). Vorlesungen ber die Theorie der automorphen Funktionen. Leipzig, Teubner.

Gasqui, J. and H. Goldschmidt (1994). *The infinitesimal rigidity of the complex quadrics of dimension four.* American J. of Math.

Griffiths, P. (1964). *Deformations of G-structures, Part A: General theory of deformations.* Math. Annalen 155: 292–315.

Griffiths, P. (1965). *Deformations of G-structures, Part B: deformations of geometric G-structures.* Math. Annalen 158: 326–351.

Gromov, M. (1981). Hyperbolic manifolds according to Thurston and Jorgensen. Lecture Notes in Math. Springer. 40–53.

Gromov, M. (1983). *Filling Riemannian manifolds.* J. of Diff. Geometry **18**: 1–147.

Gromov, M. (1986). Partial Differential Relations. Springer.

Gromov, M. (1989). *Sur le groupe fondamental d'une variété kählérienne.* C. R. Acad. Sc. Paris **308**: 67–70.

Gromov, M. (1992). Metric invariants of kähler manifolds. Differential geometry and topology, Alghero, World scientific.

Gromov, M. (1992). *Systoles and intersystolic inequalities.*:

Gromov, M. (1993). *Systoles ...?* à paraître:

Gromov, M. and P. Pansu (1990). Rigidity of lattices: an introduction. Geometric topology; recent developments, Montecatini, Springer.

Gromov, M. and I. Piatetski-Shapiro (1988). *Non-arithmetic groups in Lobachevskij spaces.* Public. Math. Inst. Hautes Etud. Sci. **66**: 93–103.

Gromov, M. and R. Schoen (1992). *Harmonic maps into singular spaces and p-adic superrigidity for lattices in groups of rank one.* Publications math. IHES **76**: 165–246.

Hadamard, J. (1898). *Les surfaces à courbure opposées et leurs lignes géodésiques.* J. Math. pures et appl. **4**: 27–73.

Hilbert, D. (1974). Mathematical developments arising from Hilbert problems. American Math. Soc.

Hingston, N. (1993). *On the growth of the number of closed geodesics on the two-sphere.* to appear:

Im Hof, H.-C. (1990). *Napier cycles and hyperbolic Coxeter groups.* Bull. Soc. Math. Belgique **42**: 523–545.

Karcher, H. (1988). A geometric classification of positively curved symmetric spaces and the isoparametric construction of the Cayley plane. On the geometry of differentiable manifolds. Soc. Math. France.

Matsushima, Y. (1962). *On the first Betti number of compact quotient spaces of higher-dimensional symmetric spaces.* Annals of Math. **75**: 312–330.

Michelsohn, M. L. (1987). *Surfaces minimales dans les sphères.* Astérisque Théorie des variétés minimales et applications: 154–155: 131–150.

Mostow, G. D. (1980). *A remarkable class of polyhedra in complex hyperbolic space.* Pacific J. of Math. **86**: 171–276.

Mostow, G. D. (1988). *On discontinuous action of monodromy groups on the complex n-ball.* J. Amer. Math. Soc. **1**: 555–586.

Osserman, R. (1985). *The four-or-more vertex theorem.* Amer. Math. Monthly **92**: 332–337.

Pansu, P. (1993-94). Sous-groupes discrets des groupes de Lie: rigidité, arithméticité. Séminaire Bourbaki, Paris, Société math. France.

Poincaré, H. (1905). *Sur les lignes géodésiques des surfaces convexes.* Trans. Amer. Math. Soc. **6**: 237–274.

Pu, P. (1952). *Some inequalities in certain nonorientable manifolds.* Pacific J. Math. **2**: 55–71.

Raphal, E., J.-M. di Meglio, et al. (1992). *Convex particles at interfaces.* J. Phys. I. France **2**: 571–579.

Romero, A. (1987). *An extension of Calabis's rigidity theorem to complex submanifolds of indefinite complex space forms.* Manuscripta Math. **59**: 261–276.

Sarnak, P. (1993). Arithmetic Quantum Chaos. Stanford.

Schlegel, V. (1883). *Theorie der homogen zusammengesetzen Raumgebilde.* Nova Acta Leop.-Carol. Dt. Akad. Natf. **44**: 343–459.

Selberg, A. (1960). On discontinuous groups in higher-dimensional symmetric spaces. Contributions in Function Theory. Bombay, Tata Institute of Fundamental Research. 147–164.

Szabo, Z. I. (1991). *A short topological proof for the symmetry of 2-point homogeneous spaces.* Inventiones math. **106**: 61–64.

Takeuchi, K. (1977). *Arithmetic triangle groups.* J. Math. Soc; Japan **29**: 91–106.

Tasaki, H. (1985). *The cut-locus and the diastasis of a hermitian symmetric space of compact type.* Osaka J. Math. **22**: 863–870.

Umehara, M. (1988). *Diastases and real analytic functions on complex manifolds.* J. Math. Soc. Japan **40**: 519–539.

Vinberg, E. B. (1983). Discrete reflection groups in Lobachevskij spaces. Intern. Congr. Math., Warsaw,

Vinberg, E. B. (1984). *The absence of crystallographic groups of reflections in Lobachevskii spaces of large dimension.* Trans. Mosc. Math. Soc. 1985: 75–112.

Vinberg, E. B. (1993). Geometry II. Springer.

Weil, A. (1958). Discontinuous groups of classical groups. University of Chicago.

Weil, A. (1962). *On discrete subgroups of Lie groups (II).* Annals of Math. **75**: 578–602.

Weil, A. (1992). S.S. Chern as Geometer and Friend. Chern - A Great Geometer of the Twentieth Century. Hong-Kong, International Press. 72–78.

Wolf, J. (1972). Spaces of constant curvature. Berkeley, J. Wolf.

Wu, H. (1987). The Bochner technique in differential geometry. London,

AMS classification: 32 C, 5O XX, 51 XX, 52 XX 80, 53 XX

Eugenio Calabi and Kähler Metrics

Jean-Pierre Bourguignon[1]

P. de Bartolomeis, Paolo, F. Tricerri, E. Vesentini (Eds.),
Manifolds and Geometry, Cambridge University Press, 61–85 (1996)

Dedicated to Eugenio Calabi on his seventieth birthday

Kählerian Geometry*, to which this talk is devoted, has grown into a subject in itself. Through major contributions, Eugenio Calabi has been one of the agents of this evolution. In his work, he set the right perspectives for this field at an early stage. As he told us, he studied Kähler manifolds, because *"elles sont vraiment simples"*.

This subject now lies at the crossroads between many branches of Mathematics :

- *Differential Geometry*, since the basic objects of the theory are differential geometric in nature ;
- *Complex Analysis*, since holomorphic maps and holomorphic bundles play an important role in Kählerian Geometry, and geometric tools from Kählerian Geometry prove useful to tackle purely complex analytic problems ;
- *Algebraic Geometry*, since, via Hodge Theory, Kählerian Geometry gives insight into the cohomological structure of algebraic manifolds ;
- *Global Analysis*, since it is thanks to the very substantial progress made in this area that major conjectures of the field have been settled ;
- and finally, but this may turn out to be very important for the future, *Theoretical Physics*, since today many challenges in Kählerian Geometry, such as understanding the so-called mirror symmetries, have their origin in physical models.

In this report, we would like to give an idea of this intricate interplay.

The material presented here, and the same holds true for the bibliography given at the end of the text, is very far from representing a survey

° Part of this work has been supported by the EU contract GADGET CHRX CT92 0050. This survey article has been completed while the author was enjoying the hospitality of the Mathematical Sciences Research Institute in Berkeley.

* On the subject matter of this talk, we are grateful to Eugenio Calabi for discussions We were very pleased and honoured to be given an opportunity to present a broad image of this field taking Eugenio Calabi's main contributions as reference points, and in this way to be able to pay a tribute to him. Thanks are also due to Paul Gauduchon and Gang Tian for useful comments on earlier drafts of this article.

[1] Jean-Pierre Bourguignon
Institut des Hautes Études Scientifiques, Université Paris-Saclay, Le Bois-Marie, Bures-sur-Yvette, France
e-mail: JPB@ihes.fr

of all currently active subareas in Kählerian Geometry. It only corresponds to articles and monographs relevant for the guided tour that we propose using Calabi's major results as milestones. We tried to keep to a minimum the overlap with Marcel Berger's lecture in this volume, which is devoted to other sides of Calabi's contributions to Geometry. Topics have been taken up again only when calling attention to them in the more restrictive context of Kählerian Geometry added a new perspective.

1. An Introduction to Kähler Metrics following Erich Kähler

The notion of a *Kähler metric* has been introduced in 1933 by Erich Kähler in [46], a visionary article. (For earlier attempts, see [72], and [73].) It may be worthwhile to review some of the features contained in it, since Kähler opens a number of paths which have later turned out to be crucial for the development of Kählerian Geometry.

Kähler considers a complex manifold M of complex dimension m endowed with a Hermitian metric g. In a (complex) coordinate system (z^α) around some point, one has $g = \sum_{\alpha,\beta=1}^{m} g_{\alpha\bar\beta}\, dz^\alpha\, d\bar z^\beta$ where $(g_{\alpha\bar\beta})$ is a Hermitian matrix. He then introduces the exterior 2-form ω (this is his notation) defined as

$$\omega = i \sum_{\alpha,\beta=1}^{m} g_{\alpha\bar\beta}\, dz^\alpha \wedge d\bar z^\beta \ .$$

(In his notation, $d(z^\alpha, \bar z^\beta)$ stands for the exterior product of the basis covectors, showing that then exterior calculus was not as natural as it is now.)

In real terms (and using more intrinsic notations), the Kähler form ω can be deduced from the Hermitian scalar product g and the almost complex structure J that holomorphic charts determine on each tangent space by $\omega(X,Y) = g(JX,Y)$ for X, $Y \in T_pM$ and $p \in M$.

Kähler's main motivation in introducing an exterior form seems to have been to derive new invariants using Elie Cartan's *"elegant Kalkül der symbolischen Differentialformen"*. The first invariant of course is $d\omega$, the (exterior) differential of ω (that Kähler denotes by ω'). Repeating his terms, the case $d\omega = 0$ presents itself as a remarkable exception. Then, he considers such a situation, and describes the long list of miracles occuring then :

• the metric can be expressed from a function u (now often called a *Kähler potential*) by the local formula $g = \sum_{\alpha,\beta=1}^{m} (\partial^2 u/\partial z^\alpha \partial \bar z^\beta)\, dz^\alpha\, d\bar z^\beta$; in more intrinsic terms, this says that $\omega = i\,\partial\bar\partial u$ where ∂ (resp. $\bar\partial$) denotes the holomorphic (resp. antiholomorphic) derivation ; Kähler gives a proof of the so-called local $i\partial\bar\partial$-lemma, to the effect that, in a fixed coordinate system, any closed form of type (p,p) lies in fact in the image of the operator $i\partial\bar\partial$, i.e., can be integrated twice ;

• the first example of a metric satisfying the closedness condition that Kähler gives is a negatively curved one since in that case M is automatically contractible if simply connected. and one can find a globally defined Kähler potential ; he takes $u = -k \log(1 - \sum_{\alpha,\beta=1}^{m} z^\alpha \bar{z}^\beta)$ for $k > 0$, and goes on noticing that on the hyperball the metric associated to this potential is invariant under hyper-Fuschian transformations which fix the boundary of the hyperball. He also considers another extreme case, that he calls the *hyper-Abelian* one, where the potential is defined in the polydisk by $u = -\sum_{i=1}^{m} k_i \log(1 - z^i \bar{z}^i)$ where the k_i are some positive constants. He then notices that one can mix these two metrics to study intermediate cases, and hopes that these metrics can be used for the algebraic study of automorphic forms ;

• he immediately moves on to considering the case where the Kähler metric defined by a potential u is an Einstein metric, and arrives at the non-homogeneous Monge-Ampère equation

$$\det\left(\frac{\partial^2 u}{\partial z^\alpha \partial \bar{z}^\beta}\right) = e^{-\lambda u} \tag{1}$$

where λ is the Einstein constant (the equation takes such a simple form because the reference metric is flat and the coordinates used holomorphic) ; somewhat later in the article, he notices that this equation can be derived using a Variational Principle ;

• he then computes the Riemann curvature tensor R^ω of the Kähler metric ω using its Christoffel symbols, that he obtains from the Euler-Lagrange formulation of the geodesic equation, taking advantage of the fact that the coefficients of type (2,0) and of type (0,2) of the metric vanish ; for the components of the curvature that do not vanish, he gets the formula

$$R_{\alpha\bar{\beta}\gamma\bar{\delta}} = -\frac{\partial^4 u}{\partial z^\alpha \partial \bar{z}^\beta \partial z^\gamma \partial \bar{z}^\delta} + \sum_{\sigma,\rho=1}^{m} g^{\sigma\bar{\rho}} \frac{\partial^3 u}{\partial \bar{z}^\rho \partial z^\alpha \partial z^\gamma} \frac{\partial^3 u}{\partial z^\sigma \partial \bar{z}^\beta \partial \bar{z}^\delta},$$

which has the advantage of exhibiting directly all the extra symmetries that the Riemann curvature tensor enjoys on a Kählerian manifold ;

• he then notices that the Ricci curvature Ric^ω can be expressed in a *"sehr elegant"* manner, namely

$$\mathrm{Ric}^\omega = -\sum_{\alpha,\beta=1}^{m} \frac{\partial^2 \log(D(u))}{\partial z^\alpha \partial \bar{z}^\beta}\, dz^\alpha\, d\bar{z}^\beta$$

where* $D(u) = \det(\partial^2 u / \partial z^\gamma \partial \bar{z}^\delta)$; moreover, the exterior 2-form ρ^ω associated to the Ricci curvature Ric^ω in the same way as the Kähler form ω

* This formula is consistent with the currently favored convention for which the round sphere has positive Ricci curvature ; notice that Kähler has the opposite sign convention, according to an earlier convention followed by Eisenhart and later by Bochner, leading to the paradoxical remark in [14].

is associated to the Hermitian metric g, i.e., defined, for X, $Y \in T_pM$, by $\rho(X,Y) = \mathrm{Ric}(JX,Y)$, is closed ;

• he then combines the Kähler form ω and the Ricci form ρ^ω to define $m+1$ different closed forms of type (m,m), namely $\Omega_r = (\rho^\omega)^r \wedge \omega^{m-r}$ for $0 \le r \le m$: one has $\Omega_0 = m!\, v_g$ where v_g is the volume form of the metric ; he shows that $\Omega_1 = \rho^\omega \wedge \omega^{m-1} = (m-1)!\,\mathrm{Scal}^\omega\, v_g$; he then gives a primitive for the volume form in terms of the potential u, taking advantage of the fact that he is working on a contractible space ; he also makes the point that for an arbitrary metric (i.e., a non necessarily Kähler one) the total scalar curvature functional cannot be expressed as a boundary integral (for that purpose, he mentions that its Euler-Lagrange equations are the Einstein field equations) ; he notices an analogy with the Gauß-Bonnet formula, and, analyzing how these forms are affected by a holomorphic change of coordinates, he then suggests looking for primitives of the forms Ω_r as a problem with possible far-reaching consequences in the theory of automorphic forms ; he succeeds for $r = 1$ and $m = 1$, and obtains the transgression formula for the Gaussian curvature on surfaces, transforming the surface integral of the curvature into a line integral of the geodesic curvature on the boundary of the domain of integration.

Two comments seem in order here.

First, although the theory of characteristic classes involving differential forms was yet to be developed (Shiing Shen Chern started it some six years later, cf. [21]), Kähler turns precisely his attention to the right characteristic forms in relation with Kählerian Geometry. In particular, he anticipates the transgression calculations that Chern did later. Of those, Chern says that they were instrumental in assuring him that he was on the right track. In modern terms, the Ricci form is nothing but the curvature form of the holomorphic determinant bundle $\Lambda^m T^*_{\mathbf{C}}M$ over M so that the cohomology class that it defines, that we denote by $[\rho^\omega]$, satisfies $[\rho^\omega] = 2\pi\, c_1(M)$, where $c_1(M)$ denotes the first Chern class of the holomorphic tangent bundle of the complex manifold M, often called the ***canonical class*** of M. This implies of course that $[\rho^\omega]$ depends only on the underlying holomorphic structure of M and not on the particular Kähler metric ω chosen.

Second, it is quite remarkable that all examples he mentions, and all applications he contemplates, always deal with negatively curved manifolds. He never envisions that the Fubini-Study metric on the complex projective space $\mathbf{C}P^m$ is indeed a Kähler metric.

In fact, it seems that Kähler's article did not get all the attention it deserved. In the early 40s, according to Shiing Shen Chern, it was well known among experts that the Fubini-Study metric is Kähler, and the connection with Algebraic Geometry is made explicit in Chapter V of the celebrated work of Hodge ([45]). It is quite remarkable that the article [15] by S. Bochner, who was Calabi's advisor, does not quote Kähler's article explicitly, although he freely uses the notion of a Kähler metric by refering to it as the *"Kähler rectriction"*.

2. Prescribing the Ricci Curvature for Kähler Metrics

a) The Calabi conjecture

We saw in section 1 that the Ricci form ρ^ω of a Kähler form ω is closed, and that its cohomology class is fixed by the holomorphic structure of M. Then comes the natural question.

CALABI'S CONJECTURE.– *On a Kählerian manifold M is any closed form of type (1,1) whose cohomology class lies in $2\pi c_1(M)$ the Ricci form of a Kähler metric ?*

In this form, the conjecture seems rather optimistic since it says that the cohomology constraint that we found on the Ricci curvature is the only one. The key point is to elaborate on the way the Ricci form ρ^ω depends on the Kähler form ω. Indeed, if $\tilde{\omega}$ is another Kähler metric, one has

$$\rho^{\tilde{\omega}} = \rho^\omega - i\partial\bar{\partial}\log\frac{\tilde{\omega}^m}{\omega^m} . \tag{2}$$

Fortunately, on a compact manifold it is possible to compare globally Kähler forms, since the $i\partial\bar{\partial}$-lemma has a global version, which we now state.

GLOBAL $i\partial\bar{\partial}$-LEMMA.– *If, on a compact manifold, ω and $\tilde{\omega}$ are Kähler forms, in particular closed forms of type $(1,1)$ of maximal rank, which are cohomologous, then there exists a globally defined function u so that $\tilde{\omega} = \omega + i\partial\bar{\partial}u$.*

Of course, not every cohomology class of type $(1,1)$ contains a form which is positive definite. Any such class is called a ***Kähler class***, or, in an older terminology, ***principal*** class). It turns out to be convenient ***to work in a fixed Kähler class.***

In [Cal 3], [Cal 4], [Cal 5]), Calabi proposes the setting in which the problem can be solved by reducing it to a (highly) non-linear partial differential equation. Indeed, if $\tilde{\rho}$ is a form of type (1,1) which is cohomologous to ρ^ω, then, by the global $i\partial\bar{\partial}$-lemma, there exists a function f so that $\tilde{\rho} = \rho^\omega + i\partial\bar{\partial}f$. Together with formula (2), this gives a geometric formulation of the Calabi Conjecture which amounts to *proving that any (positive) volume element v, such that $\int_M v = \int_M v_g$, can be realized as the volume element of a Kähler metric belonging to the class $[\omega]$ of ω* as Calabi notices in a footnote in [Cal 5].

This makes the conjecture look much nicer, since in particular it reduces it to solving a scalar equation. Indeed, given a function u, so that $\tilde{\omega} =$

$\omega + i\partial\bar{\partial}u$ is a Kähler form, then $\tilde{\omega}$ satisfies $[\tilde{\omega}] = [\omega]$. One then has to solve a global Monge-Ampère equation whose local expression is

$$\det\left(g_{\alpha\bar{\beta}} + \frac{\partial^2 u}{\partial z^\alpha \partial \bar{z}^\beta}\right) = \mathrm{e}^f \ . \tag{3}$$

In [Cal 4], Calabi proves uniqueness of the solution to this equation by a nice argument involving the intermediate positive forms $\tilde{\omega}^r \wedge \omega^{m-r}$ for $0 \leq r \leq m$. To prove the existence of solutions, he proposes to use the *continuity method*, i.e., to imbed Equ. (3) in a one-parameter family of equations (3_t) defined as follows

$$\frac{(\omega + i\partial\bar{\partial}u)^m}{\omega^m} = \mathrm{e}^{tf}\, \frac{\int_M \mathrm{e}^f\, \omega^m}{\int_M \mathrm{e}^{tf}\, \omega^m} \ . \tag{3_t}$$

The main object to consider now is the sub-interval I of $[0,1]$ made of real numbers t for which Equ. (3_s) has a solution for all $s \in [0,t]$. The interval I is non empty since (3_0) has 0 as trivial solution. Calabi proved that I is open by using the local inversion theorem in an appropriate Banach space. By doing so, he therefore proved that *"any closed form of type (1,1) which is cohomologous and sufficiently close to the Ricci form of a Kähler metric ω is itself the Ricci form of a Kähler metric belonging to the same Kähler class as ω"*. The difficult point is to prove that I is closed, hence showing that $I = [0,1]$, and that Equ. (3) has a solution. In order to make sense of the equation, one needs that the solution has a regularity better than C^2. For that, one needs a priori estimates on the size of the solutions in a space of functions whose topology is stronger that the C^2-topology. Calabi, exploiting ideas that he developed in [Cal 6] for the purpose of studying Affine Geometry (another of his favorite subjects in which the notion of curvature is third order), proved how from C^2-estimates one can derive C^3-estimates. The hardest estimate is the C^0 one. Going from a C^0-estimate to a C^2-estimate also requires non trivial technicalities.

In [4], Thierry Aubin gave a partial solution of the Calabi conjecture in the case of Kählerian manifolds with positive bisectional curvature.

In 1976, the full solution of the conjecture was found by S.T. Yau (cf. [102]). A somewhat simpler argument for the C^0-estimate can be found in [47]. (For an overview, on can consult [16] and, for a pedagogical and somewhat more leasurely presentation of the solution, [74] and chapters 2 and 11 of [13].)

b) Some interesting geometric consequences

Some consequences of the conjecture were drawn by Calabi himself (before it was even proved). Let us begin by one consequence in Analytic Geometry that one finds explained in [Cal 4].

PROPOSITION.– *Any compact complex manifold M with vanishing canonical class and irregularity q which admits a Kähler metric has a finite covering which is the product of a complex torus of dimension q and a simply connected compact manifold with vanishing canonical class.*

For that purpose, one uses the solution of the Calabi conjecture which produces a Ricci-flat Kähler metric on M since the form 0 is a particular form in the cohomology class of $c_1(M)$ assumed to vanish. With respect to this metric, holomorphic 1-forms are parallel. In the universal cover, the metric splits and the factor along these directions is flat. The other factor must be compact, hence simply connected.

PROPOSITION.– *Any compact complex manifold with vanishing first and second Chern classes admits a flat metric, hence has a torus as finite covering.*

The proof relies on expressing the Kählerian Chern number

$$(c_2 \cup [\omega]^{m-2})[M]$$

in terms of the curvature (cf. [1] and [38]). The actual formula coincides with that of the integrand of the Euler characteristic for 4-dimensional manifolds. (Indeed, for complex surfaces the second Chern class coincides with the Euler class.) When the metric is Einstein, this quadratic polynomial reduces to a sum of squares of irreducible components of the curvature. On a manifold with vanishing first Chern class such metrics exist as we explained before. They are even Ricci-flat.

This proposition still retains some mystery. It shows that Kählerian manifolds are very special among general complex manifolds. Indeed, one does not expect so strict limitations on a general complex manifold by specifying only two Chern classes (cf. Problem 19 in [39]). The study of special properties that Kählerian manifolds enjoy as Riemannian manifolds is presently a subject of great interest (cf. [37]). The fundamental group of a Kählerian manifold is also very special whereas, as shown by Clifford Taubes in [83], any finitely presented group is the fundamental group of a compact complex 3-dimensional manifold.

3. Kähler-Einstein Metrics

In the preceding paragraph, we already met some examples of Kähler-Einstein metrics, but because of their importance, they are worthy of further attention.

a) Tube domains

In [Cal 7], Calabi considers a special case of Equ. (1) on *tube domains*, i.e., domains in $\mathbf{C}^m = \mathbf{R}^{2m}$ which can be written as $\Omega \times \sqrt{-1}\,\mathbf{R}^m$ where Ω

is bounded in $\mathbf{R}^m$. In the case of homogeneous complex tube domains, the Bergman metrics are Einstein on each irreducible component of the product. The challenge is to construct non locally homogeneous examples.

Calabi looks for special solutions for which $u(z) = U(\Re e(z))$ for some real valued function U defined in Ω. This transforms the complex Monge-Ampère equation into a real one.

PROPOSITION (cf. [Cal 7]).– *On tube domains based on the ball, there exist complete radially symmetric Kähler-Einstein metrics which are not locally homogeneous.*

The proof reduces to solving Equ. (1) with this Ansatz for a radially symmetric function U. A careful study of the behaviour of U has to be made at the origin and on the boundary in order to ensure that the metric is complete. Checking the non constancy of eigenvalues of the curvature shows that the metric is not locally homogeneous.

b) The compact case : general discussion

Since we are interested in Kähler-Einstein metrics ω on M, i.e., metrics such that $\rho_\omega = \lambda\,\omega$, this means that the first Chern class $c_1(M)$ must, according to the sign of the Einstein constant λ, either vanish or have a Dolbeault representative which is negative definite or positive definite. (One then says that c_1 is *positive*, resp. *negative*.) This latter case of course puts strong restrictions on the holomorphic structure on M, and also fixes the Kähler class up to homotheties. It is traditional to normalize λ by taking it to be either 0, +1 or −1.

The situation turns out to be quite different in these three cases, at least as far as the difficulty of the problem is concerned. Therefore, we discuss the results and problems for these cases in different subsections. We present results obtained by solving the inhomogeneous Monge-Ampère equation. Other methods have also been applied, e.g., the Ricci-flow method by Cao (cf. [20]) or more geometric constructions by Norihito Koiso and Y. Sakane (cf. [53]). Recall that irreducible Hermitian symmetric metrics are automatically Kähler-Einstein.

c) The case $c_1(M) = 0$

Notice that, as we noticed earlier, the case $\lambda = 0$ is already covered by the solution of the Calabi conjecture per se since for each Kähler class ω on M, a Ricci-flat metric is obtained by taking the particular form 0 in the cohomology class $c_1(M) \in H^{1,1}(M)$ which is assumed to vanish and solving the associated Monge-Ampère equation.

COROLLARY.– *Any compact complex manifold with vanishing Chern class admits a unique Ricci-flat metric in each Kähler class.*

Typical examples of m-dimensional complex manifolds M that have the property $c_1(M) = 0$ are hypersurfaces of degree $m+2$ in $\mathbf{C}P^{m+1}$. Since those are simply connected, they give examples of Ricci-flat metrics which are non locally homogeneous since Killing vector fields would give rise to harmonic 1-forms, hence to holomorphic forms.

Among them, one finds the quartic surfaces, which are examples of algebraic K3 surfaces. Recall that by definition *a* K3 *surface* is a complex surface with vanishing canonical class and no holomorphic 1-forms. The deformation space of such complex structures is 20-dimensional, and the algebraic structures form in it a very intricate subset, which is the union of countably many subvarieties of complex dimension 19, and nevertheless everywhere dense. Among them, one finds the Kummer surfaces which are traditionally obtained from a complex torus by blowing up the 16 fixed points of the antipodal involution. (Note that, in [95], another approach to getting Ricci-flat metrics on a Kummer surface is given.) One of the challenges concerning these surfaces was to prove that they are all Kählerian. This was proved by Y.T. Siu in [76]. On all of them, therefore, one can find Ricci-flat Kähler metrics. The deformation theory of Einstein metrics on these surfaces is very interesting (cf. [75] for a systematic account, [50] and [94]). The Ricci-flat Kähler metrics on K3 surfaces have holonomy SU_2, since both a complex volume form and the Kähler form are parallel. Since $SU_2 = Sp_1$, they can also be viewed as *hyper-Kähler metrics*, i.e., metrics with holonomy $Sp_{m/2}$. We discuss these metrics in more detail in Sect. 4.

Compact three-dimensional manifolds with holonomy SU_3 and vanishing first Betti number are nowadays called *Calabi-Yau manifolds*. The necessity of naming them came from the widespread use of them by theoretical physicists as target manifolds for non-linear σ-models. These 6-dimensional spaces occur as possible compactifications by dimensional reduction from 10 to 4 dimensions in $N = 2$ supersymmetric Conformal Field Theories. The necessity of having at hand as many of these spaces as possible led to a systematic search of those. For example, a lot of numerical invariants of hypersurfaces in weighted projective spaces which are Calabi-Yau manifolds and of certain natural bundles on them have been computed in order to find cases where they are small enough. Physicists want these to be compatible with their present knowledge of elementary particles, in particular the fact that they seem to appear in only 3 families, hence bounding dimensions of some spaces of holomorphic sections.

This hunt led to the discovery of an unexpected "symmetry" between certain of these Calabi-Yau manifolds, the so-called *mirror symmetry*. This correspondence was given this name because it pairs together two different Calabi-Yau manifolds M and N whose Hodge diamonds (i.e., the display of dimensions of the cohomology spaces of type (p,q) in the (p,q)-plane) are symmetric with respect to the parallel to the q-axis around the value $\frac{3}{2}$ on the p-axis. This has the effect of putting into correspondence the space of infinitesimal deformations of the complex structure of M and the (complexification of the) space of infinitesimal deformations of the Kähler class of N, spaces which have a priori not much in common. Besides numerical evidence in many explicit examples of algebraic manifolds, what led to believe in such a symmetry is the fact that one can pass very simply from a Conformal Field Theory defined on M to one defined on N without encountering any drastic change in the physical aspects of the theory (cf. [2]). The form degenerates when a wall is reached in the Kähler cone of one, and one can then pass to the other manifold by penetrating in the deformation space of the other (after modifying the topology in resolving the singularity). Any time a mirror symmetry can be established between two Calabi-Yau manifolds, this leads to interesting formulas counting the number of rational curves of a given degree on the manifold since they are particular expectation values of correlation functions attached to the Conformal Field Theories. (For more details on these developments, one can consult articles of different natures in [104].)

d) The negative case

In Analytic Geometry, it is traditional to call (complex) manifolds with negative first Chern class *manifolds with ample canonical bundle*. The Monge-Ampère equation for the case $c_1(M) < 0$ was solved independently by T. Aubin and S.T. Yau.

THEOREM (cf. [5] and [102]).– *On a compact manifold M with negative first Chern class there exists a unique Kähler-Einstein metric, up to homotheties.*

Again in this case, the most natural way of solving it is by applying the continuity method. The C^0-estimate is easy since it follows directly from the maximum principle applied to the unknown Kähler potential u.

Uniqueness was obtained by Calabi using the same argument as the one for the Conjecture. Therefore, on a compact complex manifold with negative canonical class, the Kähler-Einstein metric is a very distinguished metric. In particular the group of holomorphic transformations, known to be finite, acts by isometries.

Among the examples of m-dimensional manifolds with ample canonical bundle, one finds hypersurfaces of degree at least $m+3$ on $\mathbf{C}P^{m+1}$ for $m > 1$. They are simply connected, hence give examples of non locally homogeneous Einstein metrics with $\lambda = -1$.

COROLLARY (cf. [101]).– *For any compact complex surface with ample canonical bundle M, one has* $c_1^2(M) \leq 3\, c_2(M)$. *Moreover, equality occurs if and only if M is covered by the ball.*

Again, this inequality relies on obvious inequalities between the integrands, taking advantage of the absence of any Ricci-trace free part of the curvature which was giving the only negative contribution.

COROLLARY (cf. [101]).– *Any complex surface which has the homotopy type of* $\mathbf{C}P^2$ *is biholomorphically equivalent to it.*

The equality in the previous Corollary allows one to finish the proof of this Corollary. The only case left open in [40] concerns complex structures on the candidate fake $\mathbf{C}P^2$ with ample canonical bundle. Equality in the inequality implies that the fake $\mathbf{C}P^2$ would be covered by the ball, hence not be simply connected, a contradiction.

e) The positive case : obstructions and existence

The case $c_1(M) > 0$ is more subtle. Many things come together to make the analysis harder. First, there is the possibility that a non trivial continuous group of holomorphic transformations exists. This may introduce some degeneracies. Second, when one tries to apply the continuity method, the openness of the interval I, which can be easily proved in all other cases, turns out to be more difficult. Indeed, one then needs a sharp estimate on the first eigenvalue of the Laplacian. The bad cases exactly correspond to the presence of holomorphic vector fields as shown by Lichnerowicz in [59] (cf. also [60]).

In fact two obstructions to solving the problem are now known. The first one is due to Y. Matsushima and in the version we give to A. Lichnerowicz.

THEOREM (cf. [63], [60], and [64]).– *On a compact complex manifold with positive first Chern class admitting a Kähler metric with constant scalar curvature, the Lie algebra of holomorphic vector fields is reductive.*

Typical manifolds whose Lie algebra of holomorphic vector fields is not reductive are projective spaces with a few points blown up in generic position.

The second obstruction is more complicated to state, and has a more refined structure. It is also related to the Lie algebra of holomorphic vector fields.

THEOREM (cf. [28] and [29]).– *Let X be any holomorphic vector field on a compact m-dimensional Kähler manifold (M,ω) and f the function defined by $\Delta_\omega f = (\int_M \omega^m)\,\mathrm{Scal}_\omega - \int_M \mathrm{Scal}_\omega\,\omega^m$ where Scal_ω denotes the scalar curvature and Δ_ω the Laplace-Beltrami operator of the metric ω.*

Then, $\int_M X.f_\omega\,\omega^m$ is independent of the Kähler metric ω in its Kähler class $[\omega]$.

This identity, first established by A. Futaki in [28], and developed fully in [30], gives an obstruction, naturally called the ***Futaki obstruction***, to the existence of a Kähler metric with constant scalar curvature in a given Kähler class. A fortiori the vanishing of the Futaki invariant is required for the existence of a Kähler-Einstein metric. There are examples of Kähler metrics which satisfy the Matsushima obstruction and not the Futaki obstruction, such as the total space of the projective bundle obtained from the direct sum of the hyperplane bundles over $\mathbf{C}P^2 \times \mathbf{C}P^2$ (cf. [29] and [30]).

The Futaki invariant has been generalized in different ways. It can be integrated from the Lie algebra of holomorphic vector fields to the group of holomorphic transformations (cf. [31]), or as is done in work on the so-called *K-energy* by T. Mabuchi (cf. [62], [8]). This notion led to the proof of the uniqueness of Kähler-Einstein metrics modulo connected group actions by S. Bando and T. Mabuchi (cf. [11]). It can also be constructed for more general functions of the metric than the scalar curvature as suggested by the construction in [17], and used in a vast generalization in [26].

An important generalization is due to W. Ding and G. Tian (cf. [25]). It goes as follows. First, embed M by powers of sections of the anticanonical bundle in a complex projective space $\mathbf{C}P^N$. (Recall that the anticanonical bundle is assumed to be ample.) By deforming the image of M by a family of holomorphic transformations going to infinity in the group of holomorphic transformations of $\mathbf{C}P^N$, one forms a possibly singular subvariety of $\mathbf{C}P^N$, say M_∞. If M_∞ is normal and if M admits a Kähler-Einstein metric, then the real part of the generalized Futaki invariant of M_∞ is non-negative. To establish this, a generalization of the Futaki invariant to normal varieties is needed. In this way, some Kähler orbifolds that are cubic hypersurfaces in $\mathbf{C}P^3$ can be shown to be obstructed (cf. [25]).

As we suggested earlier, the existence theory requires some more subtle estimates. Possible rephrasing of the analytic estimates needed are given in [6], [7], and [24].

Various low dimensional cases have first been worked out. As a result of accumulated work in the period 1986-1990, G. Tian clarified the situation for complex surfaces

THEOREM (cf. [85] and [90]).– *On a compact complex surface with positive first Chern class, there exists a Kähler-Einstein metric as soon as the Lie algebra of its group of holomorphic transformations is reductive.*

In higher dimensions, he also obtains a number of cases in which he can prove existence of Kähler-Einstein metrics, refining the existence theory for the Monge-Ampère equation, (For a survey of his work, see [86].) In particular he has a sharp estimate on an analytic invariant $\alpha(M)$ attached to the Kählerian manifold M.

PROPOSITION (cf. [25]).– *If* $\alpha(M) > n/(n+1)$, *then* M *admits a Kähler-Einstein metric.*

This allows him for example to obtain Kähler-Einstein metrics on the Fermat surface of equation $v^4 + w^4 + x^4 + y^4 + z^4 = 0$ in $\mathbf{C}P^5$.

During the same period, Y.T. Siu extends ideas of Moser on the prescribed curvature problem on the real projective 2-plane to the Kählerian context. He shows that the isometric action of a finite symmetry group allows him to get the required analytic estimate (cf. [78] and [79]). He deduces from this the existence of Kähler-Einstein metrics on the Fermat cubic in $\mathbf{C}P^3$. A.M. Nadel gives a criterion to check whether $\alpha(M) > n/(n+1)$. In fact he studies in what way the required estimate may fail. By localizing this on the manifold (cf. [66] and [67]), he introduces the notion of a *multiplier sheaf ideal.* This enables him to give fairly general criteria for solving the problem. Just to give a specific example of cases which can be covered by his method, let us mention that he can prove that Fermat hypersurfaces of degree larger or equal to $(m+1)/2$ in $\mathbf{C}P^m$ admit Einstein-Kähler metrics.

In fact the right perspective seems to be to tie the problem with the whole circle of ideas and constructions around the notion of *stable bundles* (for the first appearances of the notion in a differential geometric context, see [51], [61], and for introductions to this very active area of research from the differential point of view, one can consult [52] and [77]). What seems to be the right necessary and sufficient condition for the existence of a Kähler-Einstein metric on a complex manifold with positive Chern class is a certain *"stability of the underlying manifold"*, i.e., a certain holomorphic indecomposability of its tangent bundle. (For a first step in this direction, see [86].) For that, one has to view the manifold M as embedded projectively in the space of sections of a positive bundle, such that the anticanonical bundle, as Ding and Tian do for their generalization of the Futaki invariant.

f) Kähler-Einstein metrics as induced metrics

The analytic methods used so far to solve the (modified) Calabi conjecture do provide us with new examples of Kähler-Einstein metrics but none of them are explicit.

One can look for explicit Kähler-Einstein metrics as *induced metrics on complex submanifolds of standard spaces.* Unfortunately, it seems that no explicit non trivial solutions can be obtained in this way. For hypersurfaces, one gets only totally geodesic projective spaces or hyperquadrics as was shown by S.S. Chern (cf. [22]) and by B. Smyth (cf. [80]). In codimension 2, the same is true as proved in dimension 2 by Y. Matsuyama (cf. [65]) and in all dimensions by K. Tsukada (cf. [90]). The case of general codimensions is the content of the following theorem by M. Umehara.

THEOREM (cf. [97]).– *Any complex submanifold of $\mathbf{C}^N$ or of complex hyperbolic space $\mathbf{C}P^{*N}$ on which the induced Kähler metric is Einstein is totally geodesic.*

The connection with the work of Calabi comes from the fact that the argument used by Umehara relies on the use of the notion of *diastasis*, introduced in [Cal 2]. Indeed, this notion whose main advantage is to be inherited by embedding, not only allows to study embedding problems as Calabi showed in his early work [Cal 2]*, but also to discuss specific equations on the Kähler metric such as Equ. (1). This comes about because

* This work is described in detail in the article by Marcel Berger in this volume.

this equation has a peculiar function theoretic structure as discovered by M. Umehara in [98] and independently by D. Hulin ([44]). They can exclude the exponential function from an algebra of functions generated by norms of holomorphic coordinates, showing in this way a contradiction.

g) The non-compact case

This is a very active area of research. Already in [102], S.T. Yau considers a version of the Calabi Conjecture in a non compact setting. Indeed, he gives also a solution of the Calabi conjecture for complements of divisors with peculiar growth conditions on the Kähler metric. Many more contributions have been made in this direction, with in mind the search for a canonical metric on the complement of analytic sets of interest to algebraic geometers in order to be able to apply transcendental methods. To give a sample of recent important contributions, we just quote [9], [10], [48], [49], [88], [89], [91], [92], [103]. The aim is to obtain extensions of the uniformization theorem for Riemann surfaces to higher dimensions via these normalized metrics.

4. Extremal Kähler Metrics

a) The notion of an extremal metric

As we explained before, looking for Kähler-Einstein metrics makes sense only for very special complex manifolds since $c_1(M)$ must be either zero or definite. On a general compact Kähler manifold, there is therefore room for another category of "special" Kähler metrics. With the development of Global Analysis, one of the natural conditions to impose on a Kähler metric is that it extremizes a functional involving the curvature. The simplest one is of course the *total scalar curvature*, but in Kählerian Geometry this functional is constant in a fixed Kähler class. It is therefore natural to turn to the quadratic functionals of the curvature. This brings us in the realm of Yang-Mills theory, though we should notice that the space of parameters will be metrics (or even Kähler potentials) and not connections. This difference will be reflected in the kind of PDEs that will occur later when we discuss the Euler-Lagrange equations and the Second Variation Formula.

The space of curvature tensors of Kähler metrics at any point splits into three irreducible components under the action of the unitary group on the tangent space. Therefore, there are a priori three quadratic local functionals to be considered. It turns out that the three integrands Scal_ω^2, $||\mathrm{Ric}_\omega||^2$ and $||\mathrm{R}_\omega||^2$ are independent as algebraic entities, hence can be used as a basis of the space of U_m-invariant quadratic forms on the space of Kählerian curvature tensors. But the three functionals on the space of Kähler metrics to which they give rise are related *when the Kähler class is fixed* since

$$(c_2(M) \cup [\omega]^{m-2})[M] = \int_M \left(||R_\omega||^2 - 4\,||\mathrm{Ric}_\omega||^2 + \mathrm{Scal}_\omega^2\right)\,\omega^m$$

and

$$(c_1^2(M) \cup [\omega]^{m-2})[M] = \int_M (||\mathrm{Ric}_\omega||^2 - \mathrm{Scal}_\omega^2)\, \omega^m ,$$

and since the left hand sides are constant when ω varies in its Kähler class.

For simplicity, one works with the functional $\mathbf{S}(\omega) = \int_M \mathrm{Scal}_\omega^2\, \omega^m$. This is indeed the functional with which Calabi works in [Cal 11] and [Cal 12]. Critical points of $\mathbf{S}$ are called *extremal* Kähler metrics by Calabi.

PROPOSITION (cf. [Cal 11]).– *A Kähler metric ω is extremal if and only if the (1,0)-part of the gradient of its scalar curvature is a holomorphic vector field.*

A priori the variational formula for $\mathbf{S}$ gives a fourth-order equation in the curvature (because the dependence of the scalar curvature on the Kähler potential is second order) but, by repeated integrations by parts, one comes to this nice geometric statement. Kähler metrics with constant scalar curvature (and among them Kähler-Einstein metrics) are special cases of extremal metrics. Extremal metrics are reduced to those special metrics on complex manifolds with no holomorphic vector fields. This is in particular the case of manifolds with negative first Chern class. (Notice that for those the existence problem for Kähler-Einstein metrics is non obstructed.)

b) Using the second variational formula

As usual, when one considers a variational problem, it is very important to consider the second variational formula at a critical point, and Calabi establishes the one for $\mathbf{S}$ in [Cal 11]. From this formula, and further arguments he obtains the following theorem.

THEOREM (cf. [Cal 11] and [Cal 12]).– *Any extremal metric on M is a local minimum of $\mathbf{S}$, a global minimum if the scalar curvature is constant.*

Any orbit under the group of holomorphic transformations of M is a critical level of $\mathbf{S}$.

Moreover, any critical point always has the largest possible invariance group.

In [Cal 11], Calabi gives examples of extremal metrics on manifolds admitting no metrics with constant scalar curvature, proving the usefulness of the concept of extremal metrics. These metrics are constructed on total spaces of complex projective line bundles over compact symmetric $(m-1)$-dimensional complex spaces, typically, $\mathbf{C}P^{m-1}$. The metrics can be found by taking advantage of the large symmetry group which reduces the search of the metric to solving an ordinary differential equation. The main point is to have precise enough information in order to show that the solution explicitly expressible in terms of an affine variable does behave properly at infinity. Calabi also obtains information on the group of holomorphic transformations of a manifold admitting an extremal metric. If its Lie algebra is of positive dimension, then the group must contain a compact real Lie subgroup of positive dimension as well.

Kähler metrics with zero scalar curvature are also examples of extremal Kähler metrics. A systematic study of surfaces admitting such metrics is given in [57].

c) Some non existence results

These results do give the impression that the variational theory of the functional **S** will be a universal tool to obtain especially interesting Kähler metrics on compact complex manifolds. Unfortunately, one cannot always find extremal metrics on a compact complex manifold as was shown by Marc Levine in [58]. He indeed exhibits complex manifolds whose group of holomorphic transformations does not have any nontrivial compact Lie subgroup, contradicting the property mentioned above. Levine shows for example that the connected component of the identity of Hirzebruch surfaces (i.e., complex surfaces obtained as projective bundles of the rank 2 vector bundles $\mathcal{O}_{\mathbf{C}P^1}(-k) \oplus \mathcal{O}_{\mathbf{C}P^1}$ over $\mathbf{C}P^1$) is a group of matrices isomorphic to the Abelian group $\mathbf{C}^k$. He also gives an example of a surface obtained by appropriately blowing up points on $\mathbf{C}P^2$ so that the connected component of the group of holomorphic transformations is the Heisenberg group, hence another Lie group containing no compact Lie subgroup.

In [19] Burns and de Bartolomeis describe a class of non singular compact surfaces which are non trivially ruled over curves of genus 2 admitting no holomorphic vector fields. Hence, any extremal metric would have to have zero scalar curvature. The existence of such metrics on these manifolds is excluded by a theorem of Derdzinski (cf. [23]).

5. Hyper-Kähler Metrics

The name *hyper-Kähler* metrics to designate metrics with holonomy Sp_q is due to Eugene Calabi. This topic has had several very interesting developments in recent years, mixing in a very intricate way arguments from Algebraic Geometry, Differential Geometry and the theory of Nonlinear Partial Differential Equations and Integrable Systems. (To get an idea of this interplay, one can consult the survey articles [3] or [42], or the more systematic introduction given in Chapter 14 of [13].)

The first example of such metrics on compact manifolds arose with the solution of the Calabi conjecture on K3 surfaces. This case cannot be considered as completely decisive because of the Lie group isomorphism between SU_2 and Sp_1. In fact, shortly after this discovery, a proof of the non existence of compact manifolds with holonomy Sp_q for $q \geq 2$ circulated. This did affect the discussion by Calabi in [Cal 10] in which he concentrates on non compact examples.

It is interesting to discuss briefly the various geometrical aspects of hyper-Kähler metrics. Depending on which geometric object is taken as fundamental, one gets one or the other.

A first possible approach takes the metric as fundamental object. This point of view is in accordance with the name chosen by Calabi. It stresses the fact that the metric is Kähler for more than one complex structure, in fact for a whole 2-sphere of complex structures, like in $\mathbf{H}$, the field of quaternions. (Recall that elements of $\mathbf{H}$ are sometimes called *hypercomplex* numbers.) The 2-sphere of complex structures is obtained as the sphere of radius one in a 3-dimensional subspace of the space of almost complex structures, admitting orthogonal endomorphisms I, J and K as an orthonormal basis verifying $IJ = K$. The differential condition imposed on the fields I, J and K is that they are all parallel, enforcing the reduction of the holonomy group.

If the complex structure is taken as primary object, then one needs to introduce a Kähler metric ω and a holomorphic form of type (2,0) φ which is nowhere singular. Of course, this forces the real dimension to be a multiple of 4. Furthermore, φ is assumed to be parallel with respect to the Kähler metric. Because of this, these structures are sometimes called *complex symplectic*. The connection with the other definition requires some technicalities and is detailed in [Cal 10].

Calabi gives in [Cal 9] a construction of hyper-Kähler metrics on the totalspace M of a holomorphic bundle $\pi : M \longrightarrow B$ where B is a Kählerian manifold. This can be considered as an extension to the holomorphic context of the Kaluza-Klein Ansatz connecting Einstein metrics on a 5-dimensional extended space-time to Einstein metrics coupled with some other fields in a 4-dimensional space-time. In a Riemannian context, the data are a metric on the base of a G-bundle, a G-invariant metric on the fibre and a connection. These allow to define a metric on the total space. Here, it is tempting to try and get the metric via its Kähler potential keeping in mind the bundle structure. On a holomorphic vector bundle endowed with a Hermitian inner product t, if b is a local Kähler potential around some point in B, then the Kähler potential on the total space of the bundle is taken to be $u = b \circ \pi + t$. Such a formula gives a Kähler metric on M in a neighbourhood of the zero section, or globally if the curvature form of the bundle is non negative.

Calabi only considers bundles which are obtained from vector bundles by taking, in each fibre, a domain defined purely in terms of the distance induced by the Hermitian form t on the underlying vector bundle. Conditions which ensure that the metric is complete are discussed in detail in [Cal 9].

THEOREM (cf. [Cal 9]).– *The Kähler metric of potential* $b \circ \pi + \zeta \circ t$ *on the total space of the holomorphic cotangent bundle* $\pi : T^*\mathbf{C}P^m \longrightarrow \mathbf{C}P^m$ *of* *complex projective space with*

$$\zeta(r) = \sqrt{1+4r} - \log(1+\sqrt{1+4r})$$

is hyper-Kähler and complete.

In the case $m = 1$, the Einstein metric found by Calabi had already been introduced in General Relativity by Eguchi and Hanson (cf. [27]) because of its asymptotically Euclidean behaviour at infinity. (In this context, it is rather called a *gravitational instanton.*) As a consequence, it still bears their name. Generalizations to more than one center were given by Gibbons and Hawking in [35]. The Eguchi-Hanson metric has a companion, the Taub-NUT metric which is asymptotically locally flat (for a definition of this metric, see [13]). This metric has also a generalization with k centers. All these metrics are defined on the minimal resolution of the singularity resulting from the action of $\mathbf{Z}/k\mathbf{Z}$ on $\mathbf{C}^2$.

In [Cal 9], Calabi is pessimistic about getting compact manifolds admitting hyper-Kähler metrics. The first example of this type goes back to Fujiki, and consists of a desingularization of the symmetric product of two K3 surfaces, i.e., the quotient of the product by the involution exchanging the factors. In [12], Beauville gives wide generalizations of this example by considering higher symmetric products of K3 surfaces or of complex tori. The desingularization is much harder and requires to look at objects from a different point of view.

Somewhat later, the connection was made with classical constructions in Symplectic Geometry using momentum mappings (for substantial contributions, see [34], [32], [43], [70], [68], [69], [81], and [82]). This lead to many more examples of hyper-Kähler metrics. For example, using this construction, Kronheimer constructs (cf. [55] and [56]) asymptotically locally Euclidean metrics for all quotient singularities of $\mathbf{C}^2$ by a discrete subgroup of SU_2. (For more examples, one can consult [33].)

6. Open Problems

A number of oustanding problems remain open in Kählerian Geometry. We just propose a few here, chosen because of their fairly direct connection with Calabi's work.

PROBLEM 1.– *Give an explicit construction of Ricci-flat metrics on K3 surfaces. (For a description of such a metric on a singular K3 surface as an L^2-metric on a moduli space, cf. [18].)*

PROBLEM 2.– *Give a geometric criterion for the existence of Kähler-Einstein metrics on compact manifolds with positive first Chern class.*

PROBLEM 3.– *Give a necessary and sufficient condition for the existence of an extremal Kähler metric in a given Kähler class.*

PROBLEM 4.– *Establish the general framework under which the Mirror Symmetry between Calabi-Yau manifolds is valid.*

PROBLEM 5.– *In [100], Witten calls our attention to the moduli space of deformations of metrics on 3-dimensional Kähler manifolds which are solutions of a modification of the Kähler-Einstein equation. The leading term of this equation is indeed the Ricci curvature and the first perturbation term is the complex Hessian of the integrand of the Euler characteristic in 6 real dimensions. He suggests that this moduli space of "string theory vacuum states" may be the appropriate notion for which a generalization of Donaldson's 4-dimensional theory of moduli can be developed for 6-dimensional manifolds. Make steps in this direction.*

Bibliography

[Cal 1] CALABI, E., *Metric Riemann Surfaces,* in *Contributions to the Theory of Riemann Surfaces,* Ann. Math. Studies **30** (1953), 77–85.

[Cal 2] CALABI, E., *Isometric Imbedding of Complex Manifolds,* Ann. Math. **58** (1953), 1–23.

[Cal 3] CALABI, E., *The Variation of Kähler Metrics,* Bull. Amer. Math. Soc. **60** (1954), 167–168.

[Cal 4] CALABI, E., *On Kähler Manifolds with Vanishing Canonical Class,* in *A Symposium in Honor of S. Lefschetz,* Princeton Univ. Press, Princeton, 1955, 78–89.

[Cal 5] CALABI, E., *The Space of Kähler Metrics,* in Proc. International Congress of Mathematicians, Amsterdam, **II** (1954), 206–207.

[Cal 6] CALABI, E., *Improper Affine Hyperspheres and a Generalization of a Theorem of K. Jörgens,* Michigan Math. J. **5** (1958), 105–126.

[Cal 7] CALABI, E., *A Construction of Nonhomogeneous Einstein Metrics,* in *Differential Geometry,* Stanford 1973, Proc. Amer. Math. Soc. Symp. Pure Math. **XXVII** 2 (1975), 17–24.

[Cal 8] CALABI, E., *Construction de métriques de Kähler-Einstein,* in *Première classe de Chern et courbure de Ricci : Solution de la conjecture de Calabi,* Astérisque **58** (1978), 129–136.

[Cal 9] CALABI, E., *Métriques kählériennes et fibrés holomorphes,* Ann. Ec. Norm. Sup. Paris **12** (1979). 269–294.

[Cal 10] CALABI, E., *Isometric Families of Kähler Structures,* in *The Chern Symposium,* Berkeley 1979, Springer-Verlag, New York, 1980, 23–39.

[Cal 11] CALABI, E., *Extremal Kähler Metrics,* in *Seminar on Differential Geometry* (ed. by S.T. Yau), Ann. Math. Studies **102**, Princeton Univ. Press, Princeton, 1982, 259–290.

[Cal 12] CALABI, E., *Extremal Kähler Metrics, II,* in *Differential Geometry and Complex Analysis, Rauch Memorial Volume* (ed. by I. Chavel, H.M. Farkas), Springer-Verlag, New York, 1985, 95–114.

[1] APTE, M., *Sur certaines classes caractéristiques des variétés kählériennes compactes,* C.R. Acad. Sci. Paris **240** (1955), 149–151.

[2] ASPINWALL, P.S., GREENE, B.R., MORRISON, D.R., *Spacetime Topology Change : the Physics of Calabi-Yau Moduli Space,* Preprint, Inst. Adv. Study, Princeton, 1993.

[3] ATIYAH, M.F., *Hyperkähler Manifolds,* in *Complex Geometry and Analysis,* Proc. Pisa 1988 (ed.by V. Villani), Lect. Notes in Math. **1422**, Springer-Verlag, Berlin-Heidelberg-New York, 1990.

[4] AUBIN, T., *Métriques riemanniennes et courbure,* J. Differential Geom. **4** (1970), 363–424.

[5] AUBIN, T., *Equations du type de Monge-Ampère sur les variétés kählériennes compactes,* C.R. Acad. Sci. Paris **283** (1976), 119–121.

[6] AUBIN, T., *Réduction du cas positif de l'équation de Monge-Ampère sur les variétés kählériennes compactes à la démonstration d'une inégalité,* J. Functional Anal. **57** (1984), 143–153.

[7] AUBIN, T., *Métriques d'Einstein-Kähler et exponentielles des fonctions admissibles,* J. Functional Anal. **88** (1990), 385–394.

[8] BANDO, S., *The K-Energy Map, Almost Kähler-Einstein Metrics and an Inequality of the Miyaoka-Yau Type,* Tôhoku Math. J. **39** (1987), 231–235.

[9] BANDO, S., KOBAYASHI, R., *Ricci-Flat Kähler Metrics on Affine Algebraic Manifolds,* in *Geometry and Analysis on Manifolds,* Proc. Katata-Kyoto 1987, Lec. Notes in Math. **1339**, Springer-Verlag, Berlin-Heidelberg-New York, 1988.

[10] BANDO, S., KOBAYASHI, R., *Ricci-Flat Kähler Metrics on Affine Algebraic Manifolds, II,* Math. Ann. **287**, (1990), 175–180.

[11] BANDO, S., MABUCHI, T., *Uniqueness of Kähler-Einstein Metrics Modulo Connected Group Actions,* in *Algebraic Geometry,* Sendai 1985, Adv. Studies in Pure Math. **10**, Kinokuniya, Tokyo, 1987, 11–40.

[12] BEAUVILLE, A., *Variétés kählériennes dont la première classe de Chern est nulle,* J. Differential Geom. **18** (1983), 755–782.

[13] BESSE, A.L., *Einstein manifolds,* Ergeb. Math. **10**, Springer-Verlag, Berlin-Heidelberg, 1987.

[14] BOCHNER, S., *Vector Fields and Ricci Curvature,* Bull. Amer. Math. Soc. **52** (1946), 776–797.

[15] BOCHNER, S., *Curvature in Hermitian Metrics,* Bull. Amer. Math. Soc. **53** (1947), 179–195.

[16] BOURGUIGNON, J.-P., *Premières formes de Chern des variétés kählériennes compactes,* in *Séminaire Bourbaki 1977-78,* Exposé 507, Lect. Notes in Math. **710**, Springer-Verlag, Berlin-Heidelberg, 1978, 1–21.

[17] BOURGUIGNON, J.-P., *Invariants intégraux fonctionnels pour des équations aux dérivées partielles d'origine géométrique,* in *Differential Geometry,* Peñiscola 1985, Lect. Notes in Math. **1209**, (1986), 100-108, Springer, Berlin-Heidelberg-New York.

[18] BRAAM, P.J., MACIOCIA, A., TODOROV, A., *Instanton Moduli as a Novel Map from Tori to K3 Surfaces,* Inventiones Math. **108** (1992), 419–451.

[19] BURNS, D., DE BARTOLOMEIS, P., *Stability of Vector Bundles and Extremal Metrics,* Inventiones Math. **92** (1988), 403–407.

[20] CAO, H.D., *Deformation of Kähler Metrics to Kähler-Einstein Metrics on Compact Kähler Manifolds,* Inventiones Math. **81** (1985), 359–372.

[21] CHERN, S.S., *Characteristic Classes of Hermitian Manifolds,* Ann. Math. **47** (1946), 85–121.

[22] CHERN, S.S., *On Einstein Hypersurfaces in a Kählerian Manifold of Constant Sectional Curvature,* J. Differential Geom. **1** (1967), 21-31.

[23] DERDZINSKI, A., *Self-Dual Kähler Manifolds and Einstein Manifolds of Dimension Four,* Compositio Math. **49** (1983), 405–433.

[24] DING, W., *Remarks on the Existence Problem of Positive Kähler-Einstein Metrics,* Math. Ann. **282** (1988), 463–471.

[25] DING, W, TIAN, G., *Kähler-Einstein Metrics and the Generalized Futaki Invariant,* Inventiones Math. **110** (1992), 315–335.

[26] DONALDSON, S.K., *Anti-Self-Dual Yang-Mills Connections over Complex Algebraic Surfaces and Stable Vector Bundles,* Proc. London Math. Soc. **50** (1985), 1–26.

[27] EGUCHI, T., HANSON, A., *Asymptotically Flat Solutions to Euclidean Gravity,* Phys. Lett. **54 B** (1978), 249–251.

[28] FUTAKI, A., *On Compact Kähler Manifolds of Constant Scalar Curvature,* Proc. Japan Acad. Sci. **59** (1983), 401–402.

[29] FUTAKI, A., *An Obstruction to the Existence of Kähler-Einstein Metrics,* Inventiones Math. **73** (1983), 437–443.

[30] FUTAKI, A., *Kähler-Einstein Metrics and Integral Invariants,* Lect. Notes in Math. **1314**, Springer, Berlin-Heidelberg-New York, 1988.

[31] FUTAKI, A., *On a Character of the Automorphism Group of a Compact Complex Manifold,* Inventiones Math. **87** (1987), 655–660.

[32] GALICKI, K., *A Generalization of the Momentum Mapping Construction for Quaternionic Kähler Manifolds,* Commun. Math. Phys. **108** (1987), 117–138.

[33] GALICKI, K., *Multi-Centre Metrics with Negative Cosmological Constant,* Class. Quantum Grav. **8** (1991), 1529–1543.

[34] GALICKI, K., LAWSON, H.B., *Quaternionic Reduction and Quaternionic Orbifolds,* Math. Ann. **282** (1989), 1–21.

[35] GIBBONS, G.W., HAWKING, S., *Gravitational Multi-Instantons,* Phys. Lett. **78B** (1978), 430–434.

[36] GREENE, B.R., KIRKLIN, K.H., MIRON, P.J., *Topology and Geometry in Superstring-Inspired Phenomenology,* in *String Theory* (ed. by S.T. Yau), World Scientific, Singapore, 1988, 441–487.

[37] GROMOV, M, *Metric Properties of Kähler Manifolds,* Preprint, Inst. Hautes Et. Sci., 1993.

[38] GUGGENHEIMER, H., *Über vierdimensionale Einsteinräume,* Experientia **8** (1952), 420–421.

[39] HIRZEBRUCH, F., *Some Problems on Differentiable and Complex Manifolds,* Ann. Math. **60** (1954), 210–236.

[40] HIRZEBRUCH, F., KODAIRA, K., *On the Complex Projective Spaces,* J. Math. Pures Appl. **36** (1957), 201–216.

[41] HITCHIN, N.J., *Metrics on Moduli Spaces,* in *Proc. Lefschetz Centennial Conference,* Contemp. Math. **58** (1986).

[42] HITCHIN, N.J., *Hyper-Kähler Manifolds,* in *Séminaire Bourbaki,* Exposé 748, Astérisque **206** (1992), 137–166.

[43] HITCHIN, N.J., KARLHEDE, A., LINDSTRÖM, U., ROCEK, M., *Hyper-Kähler Metrics and Supersymmetry,* Commun. Math. Phys. **108** (1987), 535–589.

[44] HULIN, D., *Sous-variétés complexes d'Einstein de l'espace projectif,* Preprint, Ecole Polytechnique, 1994.

[45] HODGE, W., *Theory and Applications of Harmonic Integrals,* Cambridge Univ. Press, New York, 1941.

[46] KÄHLER, E., *Über eine bemerkenswerte Hermitesche Metrik,* Abh. Math. Sem. Hamburg Univ. **9** (1933), 173–186.

[47] KAZDAN, J.L., *Remark on the Preceding Paper by S.T. Yau,* Commun. Pure Appl. Math. **31** (1978), 412–413.

[48] KOBAYASHI, R., *Ricci-Flat Kähler Metrics on Affine Algebraic Manifolds and Degeneration of Kähler-Einstein K3 Surfaces,* in *Adv. Studies in Pure Math.* **18–II**, Kinokuniya and Academic Press, Tokyo and New York, 1990.

[49] KOBAYASHI, R., *On the Existence of Complete Ricci-Flat Kähler Metrics on Affine Algebraic Manifolds and its Application to a Characterization of Complex Projective Spaces,* Preprint, Nagoya Univ., 1993.

[50] KOBAYASHI, R., TODOROV, A., *Polarized Period Map for Generalized K3-Surfaces and the Moduli of Einstein Metrics,* Tôhoku Math. J. **39** (1987), 145–151.

[51] KOBAYASHI, S., *Curvature and Stability of Vector Bundles,* Proc. Japan Acad. Sci. **58** (1982), 158–162.

[52] KOBAYASHI, S., *Differential Geometry of Complex Vector Bundles,* Publ. Math. Soc. Japan **15**, Princeton Univ. Press, Princeton, and Iwanami Shoten, Tokyo, 1987.

[53] KOISO, N., SAKANE, Y., *Non-Homogeneous Kähler-Einstein metrics on Compact Complex Manifolds,* in *Curvature and Topology of Riemannian Manifolds,* Proc. 1985, Lect. Notes in Math. **1201** (1986), 166–179.

[54] KOISO, N., SAKANE, Y., *Non-Homogeneous Kähler-Einstein metrics on Compact Complex Manifolds,* Osaka J. Math. **25** (1988), 933–959.

[55] KRONHEIMER, P.B., *Instantons gravitationnels et singularités de Klein,* C.R. Acad. Sci. Paris **303** (1986), 53–55.

[56] KRONHEIMER, P.B., *The Construction of ALE Spaces as Hyper-Kähler Quotients,* J. Differential Geom. **29** (1986), 665–683.

[57] LEBRUN, C., SINGER, M, *Existence and Deformation Theory for Scalar Flat Kähler Metrics on Compact Complex Surfaces,* Inventiones Math. **112** (1993), 273–313.

[58] LEVINE, M., *A Remark on Extremal Kähler Metrics,,* J. Differential Geom. **21** (1985), 73–77.

[59] LICHNEROWICZ, A., *Sur les transformations analytiques des variétés kählériennes,* C.R. Acad. Sci. Paris **244** (1957), 3011–3014.

[60] LICHNEROWICZ, A., *Géométrie des groupes de transformations,* Dunod, Paris, 1958.

[61] LÜBKE, M., *Stability of Einstein-Kähler Vector Bundles,* Manuscripta Math. **42** (1983), 245–257.

[62] MABUCHI, T. *K-energy Maps Integrating Futaki Invariants,* Tôhoku Math. J. **38** (1986), 245–257.

[63] MATSUSHIMA, Y., *Sur les espaces homogènes kählériens d'un groupe réductif,* Nagoya Math. J. **11** (1957), 53–60.

[64] MATSUSHIMA, Y., *Sur la structure du groupe d'homéomorphismes d'une certaine variété kählérienne,* Nagoya Math. J. **11** (1957), 145–150.

[65] MATSUYAMA, Y., *On a 2-Dimensional Einstein-Kähler Submanifold of a Complex Space Form,* Proc. Amer. Math. Soc. **95** (1985), 595–603.

[66] NADEL, A.M., *Multiplier Ideal Sheaves and Existence of Kähler-Einstein Metrics of Positive Scalar Curvature,* Proc. Nat. Acad. Sci. U.S.A. **86** (1989), 7299–7300.

[67] NADEL, A.M., *Multiplier Ideal Sheaves and Kähler-Einstein Metrics of Positive Scalar Curvature,* Ann. Math. **132** (1990), 549–596.

[68] PEDERSEN, H., *Hamiltonian Constructions of Kähler-Einstein Metrics and Kähler Metrics of Constant Scalar Curvature,* Commun. Math. Phys. **136** (1991), 309–326 .

[69] PEDERSEN, H., *Hyperkähler Metrics and Monopoles,* in *The Mathematical Heritage of C.F. Gauss,* World Publ., River Edge, 1991, 573–584.

[70] PEDERSEN, H., POON, Y.S., *Hyperkähler Metrics and a Generalization of the Bogomolny Equations,* Commun. Math. Phys. **117** (1988), 569–580.

[71] SAKANE,Y., *Examples of Compact Einstein-Kähler Manifolds with Positive Ricci Tensor,* Osaka J. Math. **23** (1986), 585–616.

[72] SCHOUTEN, J.A., *Über unitäre Geometrie,* Nederl. Akad. Wetensch. Indagationes Math. **32** (1929), 457–465.

[73] SCHOUTEN, J.A., VAN DANTZIG, D., *Über unitäre Geometrie,* Math. Ann. **103** (1930), 319–346.

[74] SÉMINAIRE PALAISEAU, *Première classe de Chern et courbure de Ricci : Solution de la conjecture de Calabi,* Astérisque **58** (1978).

[75] SÉMINAIRE PALAISEAU, *Géométrie des surfaces K3,* Astérisque **126** (1985).

[76] SIU, Y.T., *Every K3-surface is Kähler,* Inventiones Math. **73** (1983), 139–150.

[77] SIU, Y.T., *Lectures on Hermite-Einstein Metrics for Stable Bundles and Kähler-Einstein Metrics,* Deutscher Mathematiker Ver. Seminar **8**, Birkhäuser, Basel, 1987.

[78] SIU, Y.T., *Kähler-Einstein Metrics for the Case of Positive First Chern Class,* in *Complex Analysis III* (ed. by C.A. Berenstein), Lect. Notes in Math. **1277**, Springer-Verlag, Berlin-Heidelberg-New York, 1987, 120–130.

[79] SIU, Y.T., *The Existence of Kähler-Einstein Metrics on Manifolds with Positive Anticanonical Line Bundle and a Suitable Finite Symmetry Group,* Ann. Math. **127** (1988), 585–627.

[80] SMYTH, B., *Differential Geomery of Complex Hypersurfaces,* Ann. Math. **85** (1967), 246–266.

[81] SWANN, A.F., *Aspects symplectiques de la géométrie quaternionique,* C.R. Acad. Sci. Paris **308** (1989), 225–228.

[82] SWANN, A.F., *Hyperkähler and Quaternionic Kähler Geometry,* Math. Ann. **289** (1991), 421–450.

[83] TAUBES, C.E., *The Existence of Anti-Self-Dual Conformal Structures,* J. Differential Geom. **36** (1992), 163–253.

[84] TIAN, G., *On Kähler-Einstein Metrics on Certain Kähler Manifolds with* $c_1(M) > 0$, Inventiones Math. **89** (1987), 225–246.

[85] TIAN, G., *On Calabi's Conjecture for Complex Surfaces with Positive First Chern Class,* Inventiones Math. **101** (1990), 101–172.

[86] TIAN, G., *Kähler-Einstein Metrics on Algebraic Manifolds,* Proc. Int. Congress Math. Kyoto 1990, Springer-Verlag, Tokyo (1991), 587–598.

[87] TIAN, G., *On Stability of the Tangent Bundles of Fano Varieties,* Intern. J. Math. **3** (1992), 401–413.

[88] TIAN, G., *Degeneration of Kähler-Einstein Manifolds, I,* Proc. Amer. Math. Soc. Symposia Pure Math. **54** 2 (1993), 595–609.

[89] TIAN, G., YAU, S.T., *Existence of Kähler-Einstein Metrics on Complete Kähler Manifolds and their Applications to Algebraic Geometry,* in *Mathematical Aspects of String Theory* (San Diego, California, 1986), Adv. Series Math. Phys. **1**, World Sci. Publishing, Singapore, 1987, 574–628.

[90] TIAN, G., YAU, S.T., *Kähler-Einstein Metrics on Complex Surfaces with* $c_1(M)$ *positive,* Commun. Math. Phys. **112** (1987), 175–203.

[91] TIAN, G., YAU, S.T., *Complete Kähler Manifolds with Zero Ricci Curvature, I,* J. Amer. Math. Soc. **3** (1990), 579–609.

[92] TIAN, G., YAU, S.T., *Complete Kähler Manifolds with Zero Ricci Curvature, II,* Inventiones Math. **106** (1991), 27–60.

[93] TODOROV, A., *Applications of the Kähler-Einstein-Calabi-Yau Metric to Moduli of K3 Surfaces,* Inventiones Math. **61** (1980), 251–265.

[94] TODOROV, A., *Spinors and Moduli of Einstein Metrics on Kähler Simply Connected Manifolds with a Canonical Class* $K \equiv 0$, in *Lect. on Supermanifolds, Geometrical Methods and Conformal Groups,* World Sci. Publ., Singapore, 1989, 38–43.

[95] TOPIWALA, P., *A New Proof of the Existence of Kähler-Einstein Metrics on K3, I,* Inventiones Math. **89** (1987), 425–448 ; *idem, II,* ibidem, 449–454.

[96] TSUKADA, K., *Einstein-Kähler Submanifolds with Codimension 2 in a Complex Space Form,* Math. Ann. **274** (1986), 503–516.

[97] UMEHARA, M., *Einstein-Kähler Submanifolds of a Complex Linear or Hyperbolic Space,* Tôhoku Math. J. **39** (1987), 385–389.

[98] UMEHARA, M., *Diastases and Real Analytic Functions on Complex Manifolds,* J. Math. Soc. Japan **40** (1988), 519–539.

[99] WEIL, A., *Introduction à l'étude des variétés kählériennes,* Publ. Inst. Math. Univ. Nancago **VI**, Act. Sci. Indust. 1267, Hermann, Paris, 1958.

[100] WITTEN, E., *Holomorphic Curves on Manifolds of* SU_3 *Holonomy,* in *String Theory* (ed. S.T. Yau), World Scientific, Singapore, 1988, 145–149.

[101] YAU, S.T., *On Calabi's Conjecture and some New Results in Algebraic Geometry,* Proc. Nat. Acad. Sci. U.S.A. **74** (1977), 1798–1799.

[102] YAU, S.T., *On the Ricci-Curvature of a Complex Kähler Manifold and the Complex Monge-Ampère Equation, I,* Commun. Pure Appl. Math. **31** (1978), 339–411.

[103] YAU, S.T., *Métriques de Kähler-Einstein sur les variétés ouvertes,* in *Première classe de Chern et courbure de Ricci : Solution de la conjecture de Calabi,* Astérisque **58** (1978), 163–167.

[104] YAU, S.T., *Essay on Mirror Manifolds,* Int. Press, Hong Kong, 1993.

Eugenio Calabi and the Curvature of Kähler Manifolds

Claude LeBrun[1]

In 1954, Eugenio Calabi gave two visionary lectures, laying out a program that would eventually revolutionize Kähler geometry. His goal was to endow each non-singular complex projective variety with a natural Riemannian metric, in a manner that would generalize the uniformization theorem for Riemann surfaces. It is unclear how well either lecture's audience really grasped the essence of Calabi's program. But, fortunately, Gene was asked to provide written accounts of both lectures, and the resulting articles [8, 9] not only sparked a profound and enduring dialogue between algebraic and Riemannian geometry, but also eventually helped transform theoretical physics beyond recognition.

The better known of these articles [9] appeared in a Princeton Festschrift honoring Solomon Lefschetz on the occasion of his retirement. What Calabi outlines there is an audacious scheme for proving the existence of Ricci-flat Riemannian metrics on certain compact manifolds. Before this, the only known examples of Ricci-flat metrics on compact manifolds were the "trivial" flat ones on n-dimensional tori T^n or on quotients T^n/Γ, where Γ acts freely; and a theorem [4] of Calabi's advisor Bochner was known to imply that *any* Ricci-flat metric on T^n (and thus on T^n/Γ) is in fact necessarily flat. But Calabi had now suddenly announced [9] that *non-trivial* compact Ricci-flat manifolds actually exist in abundance:

Theorem 1 (Calabi/Yau). *Let (M,J) be a compact complex manifold of Kähler type. Then M carries a Ricci-flat, J-compatible Kähler metric if and only if $c_1^{\mathbb{R}}(M,J) \in H^2(M,\mathbb{R})$ vanishes. When this happens, moreover, there is a unique Ricci-flat Kähler metric in every Kähler class on (M,J).*

While Calabi stated this as a theorem, rather than a conjecture, he also immediately indicated that his proof was not rigorous, and conceded late in

[1] Claude LeBrun
Department of Mathematics, Stony Brook University, Stony Brook, New York, USA
e-mail: claude@math.stonybrook.edu

the article that the gaps in his argument were substantial enough to cast serious doubt on the validity of the result. Nonetheless, he had correctly outlined what we now call the *continuity method.* A quarter-century later, Yau [38] was able to turn this into a rigorous proof of Theorem 1 by providing sophisticated analytic arguments which bridge the gaps in Calabi's reasoning.

Calabi's strategy actually involved proving a more general result, of which Theorem 1 is merely a corollary:

Theorem 2. *Let $\Omega \in H^2(M,\mathbb{R})$ be a Kähler class on a compact complex m-manifold (M^{2m},J), and let α be any volume form on M. Then there is a unique Kähler metric g on M such that*

(*i*) *g is compatible with the given complex structure J;*
(*ii*) *the Kähler form ω of g belongs to the deRham class Ω; and*
(*iii*) *the Riemannian volume form of g is a positive constant times α.*

Calabi's strategy for proving this depends on the fact that if g_0 is an arbitrary smooth Kähler metric on M, with Kähler form $\omega_0 = g_0(J\cdot,\cdot)$, any smooth Kähler form ω in the de Rham class $\Omega = [\omega_0]$ is determined by a single smooth function u, via

$$\omega = \omega_0 + i\partial\bar{\partial}u,$$

and ω conversely determines the potential u up to an additive constant. Of course, requiring that $g = \omega(\cdot, J\cdot)$ be positive-definite imposes an additional constraint on the function u, but the set of potentials satisfying this constraint is convex, and open in the C^2 topology; thus, the space of smooth, J-compatible Kähler metrics in the Kähler class Ω can be identified with a connected open set in $C^\infty(M)/\mathbb{R}$. On the other hand, the set of volume forms on M modulo overall constant rescaling can also be identified with $C^\infty(M)/\mathbb{R}$, via

$$f \mapsto e^f \omega_0^{\wedge m},$$

so that Theorem 2 amounts to the statement that

$$u \longmapsto \log \frac{(\omega_0 + i\partial\bar{\partial}u)^{\wedge m}}{\omega_0^{\wedge m}} \tag{1}$$

defines a diffeomorphism from the open convex set of "Kähler potentials" in $C^\infty(M)/\mathbb{R}$ onto all of $C^\infty(M)/\mathbb{R}$.

While Calabi did not rigorously prove that (1) is a diffeomorphism, he did show that this map is both *open* and *injective*; this reduces the problem to showing that (1) is also *closed*, a feat that was only accomplished much later by Yau [38]. Openness follows from the Banach-space inverse function theorem, by an argument that frequent imitation has largely robbed of its surprise; the linearization of the Monge–Ampère operator (1) is the Laplace–Beltrami operator, and this an isomorphism between $C^{k+2,\alpha}(M)/\mathbb{R}$ and $C^{k,\alpha}(M)/\mathbb{R}$, for any k and α. But Calabi's injectivity argument still seems miraculous, even today. Assume there is a Kähler form

$$\omega = \omega_0 + i\partial\bar{\partial}u$$

such that

$$\omega^{\wedge m} = \omega_0^{\wedge m}.$$

Calabi then rewrites this as

$$\Phi \wedge i\partial\bar{\partial}u = 0, \tag{2}$$

where

$$\Phi = \sum_{j=0}^{m-1} \omega^{\wedge j} \wedge \omega_0^{\wedge m-j-1}.$$

Multiplying (2) by u and integrating by parts then yields

$$0 = i\int_M \Phi \wedge \partial u \wedge \bar{\partial}u = \|\bar{\partial}u\|_h^2,$$

where the L^2 norm on the right is calculated with respect to a Hermitian metric h which is uniquely determined by Φ. It follows that u is constant, and that $\omega = \omega_0$. Since the base-point ω_0 used in this argument is entirely arbitrary, this shows that (1) is actually injective.

The implication [Theorem 2 $\Longrightarrow$ Theorem 1] is also due to Calabi. A crucial ingredient, going back to Kähler [29], is that the Ricci tensor of a Kähler metric is calculable in terms of the complex structure J and the volume form α alone; indeed, if we define the *Ricci form* ρ by

$$\rho = r(J\cdot,\cdot),$$

where r is the Ricci tensor, then ρ is locally given by

$$\rho = -i\partial\bar{\partial}\log\left(\frac{\alpha}{\phi\wedge\bar{\phi}}\right) \tag{3}$$

for any non-vanishing local holomorphic form ϕ of top degree. Moreover, the closed (1,1)-form ρ automatically represents $2\pi c_1^{\mathbb{R}}(M,J)$ in deRham cohomology. In particular, if $c_1^{\mathbb{R}}(M,J) = 0$, then any given Kähler metric satisfies $\rho = i\partial\bar{\partial}u$ for some real-valued function u. However, Theorem 1 then allows us to produce a new Kähler metric with volume form $\widehat{\alpha} = e^u\alpha$, and (3) then guarantees that this new Kähler metric is actually Ricci-flat.

Calabi's article for the Lefschetz volume also points out specific examples of manifolds that Theorem 1 endows with Ricci-flat metrics. In particular, $K3$ surfaces are discussed at some length, and Calabi shows that the Ricci-flat Kähler metrics they carry cannot be flat. Perhaps it was this that prompted him to write André Weil, who was carrying out pioneering work on $K3$ surfaces. In a letter to the author [15], Gene recounts:

> ...I had written to A. Weil, asking him if some of the consequences (existence of Ricci-flat Kähler metrics [on] varieties with trivial canonical class) were known. It was Weil who made me aware, for the first time, that the existence part of the statement was unproved. It was then that I started seeking out and buttonholing people like Lipman Bers and L. Nirenberg, hoping to get some guidance or hints on the subject. It was on such an occasion that Louis said, "I am telling you, and repeat after me: 'One can't prove existence theorems without *a prioiri* estimates'."

But, here again, Gene's geometric talents were to see him in good stead. On the advice of Hans Lewy, Calabi began to try to understand whether geometric ideas might imply *a priori* estimates for equations of Monge–Ampère type. Rather than focus on the complex case, however, he decided to instead study the simpler *real* Monge–Ampère equation, which, as he learned from Blaschke's work, naturally arises in the context of affine differential geometry. This eventually allowed him to prove a remarkable C^3 estimate for solutions of the real Monge–Ampère equation

$$\det\left(\frac{\partial^2 u}{\partial x^j \partial x^k}\right) = 1$$

which he then used to prove a beautiful affine uniqueness theorem for complete convex graphs. The miraculous interplay between geometry and analysis that this unveiled so impressed Gene that he now describes the resulting paper [10] as his best work. Whether or not we demur on this point is unimportant. Instead, one should notice that this *a priori* estimate prefigures the C^3 estimate for the *complex* Monge–Ampère equation, which plays a key role in proving Theorem 2 and related results.

Calabi's other landmark paper of 1954 is a one-page résumé [8] of his lecture at the International Congress of Mathematicians in Amsterdam. Here, and in a pair of related AMS abstracts [6, 7], he makes a string of audacious assertions, each of which seems far ahead of its time. The first of these is intimately tied to the results we have already discussed:

Theorem 3. *Let Ω be any Kähler class on a compact complex manifold (M, J), and let ϱ be any closed $(1,1)$-form which belongs to the deRham class $2\pi c_1^{\mathbb{R}}(M, J)$. Then there is a unique $\omega \in \Omega$ whose Ricci form is given by $\rho = \varrho$.*

Indeed, this is actually a corollary of Theorem 2. If $\omega_0 \in \Omega$ is a Kähler form with Ricci-form ρ_0 and volume form α_0, then

$$\varrho = \rho_0 + i\partial\bar{\partial}u$$

for some real-valued smooth function u. Then (3) tells us that ω has Ricci-form $\rho = \varrho$ iff its volume form is a constant times $e^{-u}\alpha_0$, and Theorem 2 guarantees that there is exactly one $\omega \in \Omega$ that has this property,

The next claim made in Calabi's *précis* is not labeled as a theorem, but nevertheless helps explain the article's title, *The Space of Kähler Metrics.* A Kähler class Ω on a compact complex manifold (M, J) can be thought of as an infinite-dimensional Fréchet manifold, with tangent space consisting of the real smooth exact $(1,1)$-forms ψ on M. On the other hand, each $\omega \in \Omega$ corresponds to a Kähler metric g on M, and so determines an L^2 inner product on this tangent space. This gives Ω an interesting Riemannian structure $\psi \mapsto \|\psi\|^2_{L^2,g}$, which we shall call the *Calabi metric.*

Proposition 4. *When equipped with the Calabi metric, the infinite-dimensional manifold Ω has constant positive sectional curvature. Moreover, any two points of Ω are joined by a unique geodesic segment.*

While Calabi only hints at the proof, enough clues are given to see that this, too, follows from Theorem 2. Indeed, if we map Ω injectively to the real pre-Hilbert space of smooth half-densities on M by

$$\omega \overset{F}{\longmapsto} |\omega^{\wedge m}|^{1/2},$$

its image then lies in the Hilbert sphere of radius $\sqrt{\Omega^m}$; indeed, $F(\Omega)$ is exactly the orthant of this sphere consisting of those smooth half-densites which are everywhere positive. However, the Lefschetz decomposition implies that the Hodge star operator acts on $(1,1)$-forms by

$$\star\psi = \tfrac{1}{(m-1)!}\,(\psi,\omega)\,\omega^{\wedge(m-1)} - \tfrac{1}{(m-2)!}\,\psi\wedge\omega^{\wedge(m-2)}\,. \tag{4}$$

Hence

$$\|\psi\|^2_{L^2,g} = \int_M (\psi,\omega)^2 d\mu - \tfrac{1}{(m-2)!}\int_M \psi^{\wedge 2}\wedge\omega^{\wedge(m-2)}\,, \tag{5}$$

where the second term on the right-hand side vanishes by Stokes' theorem if ψ is exact. However, if we interpret an exact $(1,1)$-form ψ as an element of $T_\omega\Omega$, we have

$$F_*\psi = \frac{1}{2}(\psi,\omega)\;|\omega^{\wedge m}|^{1/2}.$$

Thus

$$\|F_*\psi\|^2_{L^2} = \tfrac{m!}{4}\|\psi\|^2_{L^2,g},$$

showing that Calabi's metric on Ω is $(4/m!)$ times the pull-back of the L^2 metric on the Hilbert sphere. It follows that the geodesics of the Calabi metric are just the pre-images of great circles, so that Theorem 2 guarantees that any two elements of Ω are joined by a unique geodesic segment. Moreover, any 2-plane in $T_\omega\Omega$ is tangent to a totally geodesic surface arising as the pre-image of a great 2-sphere, allowing us to read off that Calabi's metric has constant sectional curvature $1/4V$, where $V = \Omega^m/m!$ is the total volume of any $\omega \in \Omega$.

Since Calabi had so thoroughly foreseen the theory of Ricci-flat Kähler metrics on compact complex manifolds, many modern readers will assume that his next step must have been to explore the theory of Kähler–Einstein metrics. However, while the concluding portion of his ICM abstract does indeed address this problem, it only does so implicitly and indirectly, as part of a much larger program. Let us digress for a moment to clarify this point.

Recall that a Riemannian metric is said to be *Einstein* if it has constant Ricci curvature. This is equivalent to requiring that the Ricci tensor be a constant multiple of the metric, so that

$$r = \lambda g \tag{6}$$

for some real number λ, called the Einstein constant. Einstein metrics which are also Kähler are called *Kähler–Einstein metrics*. In this case, (6) is equivalent to requiring that the Ricci form be a constant multiple of the Kähler form:

$$\rho = \lambda\omega.$$

One consequence is that the scalar curvature

$$s = g^{jk} r_{jk} = 2(\omega, \rho)$$

of a Kähler–Einstein metric is constant; indeed, one has $s = 2m\lambda$. But, conversely, (4) implies that the Ricci form ρ of a compact Kähler manifold is *harmonic* iff the scalar curvature is constant. Because any cohomology class has a *unique* harmonic representative, and because the Kähler form is automatically harmonic, it follows that a compact Kähler manifold (M, J, g) is Einstein iff

- its scalar curvature s is constant; and
- its first Chern class $c_1^{\mathbb{R}}(M, J)$ is a multiple of the Kähler class $[\omega]$.

When this happens, the Einstein constant of a Kähler–Einstein metric is determined by the requirement that $2\pi c_1^{\mathbb{R}}(M, J) = \lambda[\omega]$. In particular, when $\lambda \neq 0$, the Kodaira embedding theorem implies that (M, J) must be projective algebraic. Moreover, when $\lambda < 0$, high powers of the canonical line bundle $K = \Lambda^{m,0}$ have many holomorphic sections, so that the manifold is, in particular, of *general type*. On the other hand, when $\lambda > 0$, one instead concludes that high powers of the anti-canonical line bundle K^* give rise to projective embeddings of M; complex manifolds with this special property are rare, and are called *Fano varieties*.

Thus, while Calabi's ICM paper does not explicitly discuss Kähler–Einstein metrics, his last claim ostensibly offers a definitive picture of them:

Alleged Theorem. *Let (M, J) be a compact complex manifold which carries no non-trivial holomorphic vector fields. Then any Kähler class Ω on M is represented by a unique Kähler metric of constant scalar curvature.*

A slight generalization of this statement appears in [8] as Theorem 2, albeit with little indication of how Calabi believed it could be proved. In point of fact, Gene had simply overreached this time. Counter-examples [5, 30, 35, 37] would eventually show that an additional *stability* hypothesis is needed before such a statement could be expected to hold; the point is that moduli spaces of Riemannian structures on a compact manifold are necessarily *Hausdorff*, while moduli spaces of polarized complex structures on a compact manifold often fail to be Hausdorff in the absence of a stability hypothesis. Nonetheless, Calabi's assertion was remarkably prescient, and the exploration of this conjecture would become a dominant theme of subsequent research in Kähler geometry.

The depth of Calabi's ideas first became apparent in the Kähler–Einstein case. When $-c_1^{\mathbb{R}}(M,J)$ is a Kähler class, (M,J) carries no holomorphic vector fields, and Calabi's claim predicts that (M,J) should admit a $\lambda < 0$ Kähler–Einstein metric; and indeed, this was proved in the late 1970s by Aubin [2] and Yau [38], using an easier analog of the continuity argument used by Yau used to prove Theorem 1. On the the other hand, Matsushima [34] showed in 1957 that not every Fano manifold admits a $\lambda > 0$ Kähler–Einstein metric; but Matsushima's obstruction only comes into force when the Lie algebra of holomorphic vector fields is non-trivial, making Calabi's intuition appear[2] to be exactly on target.

Calabi's ICM paper also introduces an important variational problem in connection with Kähler metrics of constant scalar curvature. This attracted little attention at the time, but Gene fortunately consented, a quarter-century later, to carefully explain the whole story in an article [13] solicited by Yau for a volume based on his seminar at the IAS. The key idea is to look for critical points of the squared L^2-norm of the scalar curvature

$$\mathcal{C}(g) = \int_M s_g^2 d\mu_g ,$$

considered as a functional on the space of Kähler metrics g in a fixed Kähler class. This choice of functional would hardly have seemed obvious at the time, and the paradigm established by the Einstein–Hilbert functional might have made it seem more tempting to instead study the integral of s rather than of its square. However, since

$$s \, d\mu = 2(\rho, \omega) d\mu = 2\rho \wedge \star\omega = 2\rho \wedge \frac{\omega^{\wedge(m-1)}}{(m-1)!},$$

the integral of the scalar curvature is actually a topological invariant of (M, J, Ω), where $\Omega = [\omega]$:

$$\int_M s \, d\mu = \frac{4\pi \; c_1 \cdot \Omega^{m-1}}{(m-1)!}.$$

[2] Of course, the truth of the matter is rather more complicated; a Fano manifold actually admits a Kähler–Einstein metric iff it is K-stable [19, 20, 21, 22].

Since the total volume of any metric in Ω is $(\Omega^m/m!)$, the Cauchy–Schwarz inequality therefore immediately implies that

$$\int_M s_g^2 d\mu_g \geq \frac{(\int s \, d\mu)^2}{\int d\mu} = \frac{16m\pi^2}{(m-1)!} \frac{(c_1 \cdot \Omega^{m-1})^2}{\Omega^m} ,$$

with equality iff the scalar curvature s is constant; in particular, constant-scalar-curvature Kähler metrics are absolute minimizers of the Calabi functional

$$\mathcal{C} : \Omega \to \mathbb{R}.$$

On the other hand, Gene computes the first variation of $\mathcal{C}$ for a 1-parameter family

$$\omega_t = \omega + t \, i\partial\bar{\partial}u + O(t^2)$$

of Kähler metrics in $\mathcal{C}$, and finds that it is given by

$$\left.\frac{d}{dt}\mathcal{C}(\omega_t)\right|_{t=0} = -4 \int_M u(\nabla_\nu \nabla^{\bar{\mu}} \nabla_{\bar{\mu}} \nabla^\nu s) \, d\mu.$$

Thus, the Euler–Lagrange equations for $\mathcal{C}$ are equivalent to requiring that $(\nabla s)^{1,0}$ be a holomorphic vector field on (M, J). In particular, if (M, J) admits no non-trivial holomorphic vector fields, the only critical points of $\mathcal{C}$ are the Kähler metrics of constant scalar curvature; and these, by our previous discussion, are absolute minimizers of the Calabi functional $\mathcal{C}$.

The Calabi functional $\mathcal{C}$ is a *quadratic curvature functional*, and thus a distant cousin of the Yang–Mills functional of gauge theory. But what about *other* quadratic curvature functionals? For example, we might have chosen to study the L^2-norm of the Ricci curvature in lieu of the scalar curvature. However, as Gene explains in [13], this only makes a cosmetic difference. Indeed, (5) tells us that

$$\int_M |r|^2 d\mu = 2\|\rho\|_{L^2}^2 = \frac{1}{2}\int_M s^2 d\mu - \frac{8\pi^2 \, c_1^2 \cdot \Omega^{m-2}}{(m-2)!},$$

so the squared L^2-norm of the Ricci curvature only differs from half the Calabi functional $\mathcal{C}$ by a constant depending on the Kähler class. Indeed, the variational problem introduced by Calabi in his early paper [8] was formulated in terms of Ricci curvature rather than scalar curvature! Similarly, Gene points out that the squared L^2-norm of the Riemann tensor differs from a quarter of the Calabi functional by a topologically-determined constant[3] once the Kähler class Ω has been fixed:

[3] A formula of this kind seems to have first appeared in a paper by Apte [1]. Unfortunately, however, "Theorem 1" of Apte's paper is wildly false, due to a strange confusion of effective divisors with primitive cohomology classes.

$$\int_M |\mathcal{R}|^2 \mathrm{d}\mu = \frac{1}{4}\int_M s^2 \mathrm{d}\mu + \frac{8\pi^2(c_2 - c_1^2)\cdot \Omega^{m-2}}{(m-2)!} .$$

Thus, it doesn't really matter which flavor of curvature we choose; for Kähler metrics in a fixed Kähler class, any L^2-norm of curvature will lead to an equivalent variational problem.

While Gene had surely understood much of this story long before the appearance of his landmark 1982 article [13], other results unveiled by that paper obviously describe recent discoveries. We have already noted that a Kähler metric is a critical point of $\mathcal{C} : \Omega \to \mathbb{R}$ iff the gradient of its scalar curvature is the real part of a holomorphic vector field. Kähler metrics satisfying this condition are called *extremal Kähler metrics.* While Calabi's program originally focused on Kähler metrics of constant scalar curvature, his 1982 paper constructs examples of extremal Kähler metrics on compact complex manifolds which have *non-constant* scalar curvature. In particular, he explicitly constructs such a metric in every Kähler class on the one-point blow-up $\mathbb{CP}_2\#\overline{\mathbb{CP}}_2$ of the complex projective plane. This illustrates a facet of Calabi's mathematical work which we have yet to discuss: namely, his passionate interest in, and remarkable talent for constructing, illuminating explicit examples. In this case, Gene looks for solutions which are $\mathbf{U}(2)$-invariant, so that the Kähler potential only depends on the radial coordinate of $\mathbb{C}^2 - \{\vec{0}\} \approx S^3 \times \mathbb{R}^+$; this reduces the relevant PDE to an ODE, which he then proceeds to solve explicitly. The same method also, for example, yields explicit extremal Kähler metrics on all of the Hirzebruch surfaces.

Of course, explicit constructions like this do not simply appear out of thin air. In fact, Gene had previously explored a similar method in the context of Kähler–Einstein metrics, notably resulting [11, 12] in the explicit construction, for each n, of complete Ricci-flat Kähler metrics on both the canonical line bundle and the cotangent bundle of $\mathbb{CP}_n$. In the cotangent case, these metrics turn out to have holonomy $Sp(n)$, and so, in the now-standard terminology coined by Calabi himself, are *hyper-Kähler.* When $n = 1$, Calabi's two metrics coincide, not only with each other, but also with the celebrated Eguchi–Hanson metric, named in honor of the physicists who described it in an article [26] submitted for publication in October, 1978. Although Gene's paper [11] was submitted for publication two months later, the radically different nature of his approach and the much greater length of his article both indicate that his discovery of this prototypical "gravitational instanton" must have been independent of, and essentially contemporaneous with, the work of Eguchi and Hanson. Still, it is probably all for the best that this metric bears someone else's name. If someone were to call it the Calabi metric, the inevitable response would be," *Which* Calabi metric?"

Calabi's 1982 article, *Extremal Kähler Metrics*, had opened up a new field, and raised a host of fundamental questions. Three years later, this paper was followed [14] by an even more remarkable sequel, *Extremal Kähler Metrics II*, which answered some of these questions, and shed a great deal of light on others. The earlier paper had shown, by example, that an extremal Kähler metric might not have constant scalar curvature, and this sequel therefore focuses on issues related to properties of these *strictly* extremal Kähler metrics. Such metrics of course can only exist on compact complex manifolds for which the identity component of the biholomorphism group is non-trivial, so understanding the interplay between a strictly extremal Kähler metric g and the Lie algebra $\mathfrak{h}$ of holomorphic vector fields was obviously of fundamental importance. Here, key guidance was provided by earlier results regarding the simpler case when the scalar curvature is constant. For Kähler–Einstein metrics with $\lambda > 0$, Matsushima [34] had shown that $\mathfrak{h}$ is the complexification of the Lie algebra of Killing fields, so that the automorphism group is the complexification of a compact Lie group; in particular, this implies that a Fano manifold can only admit a Kähler–Einstein metric if $\mathfrak{h}$ is reductive. Shortly thereafter, Lichnerowicz [31] then proved that essentially the same result also holds for compact Kähler manifolds of constant scalar curvature. Calabi's explicit examples, however, had shown that $\mathfrak{h}$ certainly need not be reductive when the manifolds admits a *strictly* extremal Kähler metric. If a holomorphic vector field ξ has a zero somewhere on M, it can be written as $\xi = (\nabla f)^{1,0}$ for a function f satisfying the fourth-order elliptic equation

$$\nabla_\nu \nabla^{\bar{\mu}} \nabla_{\bar{\mu}} \nabla^\nu f = 0, \tag{7}$$

and every solution of this equation conversely gives rise to a holomorphic vector field. When the scalar curvature is constant, the fourth-order operator in (7) is *real*, so that complex conjugation of solutions gives $\mathfrak{h}$ a natural real form. This no longer holds in the strictly extremal case. However, by carefully examining equation (7), Calabi was nonetheless able to prove the following:

Theorem 5. *The identity component* $\mathrm{Iso}_0(M,g)$ *of the isometry group of any compact extremal Kähler manifold* (M,g,J) *is a maximal compact subgroup of the identity component* $\mathrm{Aut}_0(M,J)$ *of the biholomorphism group. In particular, the complex structure* J *determines* $\mathrm{Iso}_0(M,g)$ *up to conjugacy.*

For example, this implies that every extremal Kähler metric on $\mathbb{CP}_2\#\overline{\mathbb{CP}}_2$ is isometric to one which is invariant under the standard action of $\mathbf{U}(\mathbf{2})$, and so arises from the explicit construction Gene had given in [13].

Many fundamental results proved in [14] were made possible by the recent discovery of the *Futaki invariant* [27, 28]. If (M,g,J) is a compact Kähler manifold with automorphism algebra $\mathfrak{h}$, the latter is the linear map $\mathfrak{F}_\Omega : \mathfrak{h} \to \mathbb{C}$ defined by

$$\xi \longmapsto -2\int_M \xi \Delta^{-1}(s-\overline{s})\, \mathrm{d}\mu$$

where $\overline{s}$ is the (topologically determined) average of the scalar curvature, and the Green's operator Δ^{-1} is the inverse of the Laplace–Beltrami operator. Remarkably, $\mathfrak{F}_\Omega$ only depends on the Kähler class $\Omega = [\omega]$, rather than the specific Kähler metric $g = \omega(\cdot, J\cdot)$. As emphasized by Futaki, if Ω contains a metric of constant scalar curvature, then $\mathfrak{F}_\Omega$ must vanish. On the other hand, Gene observed that if Ω contains a strictly extremal Kähler metric, then $\mathfrak{F}_\Omega\left((\nabla s)^{1,0}\right) \neq 0$. This then immediately implies the following:

Proposition 6. *A Kähler class can never contain both a strictly extremal Kähler metric and a Kähler metric of constant scalar curvature.*

All this begins to suggest that some sort of uniqueness must hold for extremal Kähler metrics. At the same time, results regarding the constant-scalar-curvature case offered hope that extremal Kähler metrics might always be absolute minima of the Calabi functional $\mathcal{C} : \Omega \to \mathbb{R}$. In [14], Calabi actually proves local versions of both these statements:

Theorem 7. *In a fixed Kähler class Ω, each connected component of the space of extremal metrics is an orbit of the identity component* $\mathrm{Aut}_0(M, J)$ *of the biholomorphism group.*

Theorem 8. *At any extremal Kähler metric, the Hessian of the Calabi functional $\mathcal{C} : \Omega \to \mathbb{R}$ is positive semi-definite, and is strictly positive definite in directions transverse to the corresponding orbit of of* $\mathrm{Aut}_0(M, J)$. *In particular, a Kähler metric is extremal iff it is a local minimum of the Calabi functional.*

Calabi went on to conjecture that, modulo the action of Aut_0, extremal metrics in a fixed Kähler class are actually *globally* unique, and that they are in fact the *absolute* minimizers of the Calabi functional. Both of these remarkable conjectures were later proved to be true [3, 18, 23, 25, 33]. A crucial step in this direction was supplied by yet another paper [16] by Calabi, this time working in collaboration with his former student Xiuxiong Chen.

The Calabi–Chen paper is entitled *The Space of Kähler Metrics II*, thus making it simultaneously the sequel to Calabi's ICM summary [8] and to a similarly titled paper by Chen [17]. Recall that Calabi had explored a natural L^2-metric on the space Ω of metrics in a fixed Kähler class; as we saw in Proposition 4, this metric is positively curved and locally symmetric. However, there is another natural metric on Ω that turns out to be better adapted to the study of extremal Kähler metrics. Recall that the tangent space $T_\omega\Omega$ is the space of exact $(1,1)$-forms, and that every such form can be written as $i\partial\bar{\partial}u$ for a unique function u with integral zero. Taking L^2-inner products of such potentials u, rather than of the corresponding $(1,1)$-forms, gives Ω a Riemannian structure which is analogous to, but distinct from, the Calabi metric. We will call this the *Mabuchi metric*, as it was discussed in an article by Mabuchi [32] long before its rediscovery by Semmes [36] and by Donaldson [24]. While the Calabi and Mabuchi metrics are, heuristically, both locally symmetric, their geometric properties are profoundly different: the Calabi metric is positively curved, while heuristic calculation indicates that the Mabuchi metric instead has *non-positive* sectional curvature. An important technical difference is that the geodesics of the Calabi metric are easy to construct, while the construction of geodesics for the Mabuchi metric is a difficult analytic problem that is best reformulated as a homogeneous Monge–Ampère equation. However, this technical challenge is well worth surmounting, because various versions of the K-energy are convex on geodesics of the Mabuchi metric, allowing one prove the uniqueness of extremal Kähler metrics by showing that any two extremal Kähler metrics are connected by a (sufficiently regular) Mabuchi geodesic. Moreover, K-stability can be restated as a condition regarding the behavior of the K-energy on geodesic rays for the Mabuchi metric. The geometry of the Mabuchi metric is therefore a matter of fundamental importance for the theory of extremal Kähler metrics. The Calabi–Chen paper makes an important contribution here by showing that lengths of Mabuchi geodesics realize distances for a metric-space structure on Ω which is non-positively curved in the sense of Aleksandrov. This helps lay the groundwork for Chen's later proofs of global versions of Theorems 7 and 8.

Eugenio Calabi's work on extremal Kähler metrics has led to a remarkable mathematical synthesis that connects Riemannian geometry, algebraic geometry, global analysis, and theoretical physics, in a way that has provoked fundamental paradigm shifts in all of these disciplines. Indeed, looking back, Gene's visionary early work truly seems like a piece of twenty-first century mathematics that somehow landed in the middle of the twentieth century! In Riemannian geometry, his program has lent powerful impetus to the study of canonical metric problems, not only by giving us systematic constructions of Einstein metrics and solutions of other geometric variational problems, but also by inspiring others to risk audacious attempts at proving general existence results. In algebraic geometry, his ideas have not only led to powerful classification theorems, but have also motivated the invention of new notions of stability that promise to become the focus of intense research activity in the years ahead. In global analysis, his program has led to systematic work on Monge-Ampère equations which continues unabated even today. And in theoretical physics, the astonishing impact of his work on string theory has been reciprocated with a torrent of physically-motivated conjectures, providing bold new directions of research for all the mathematical disciplines under discussion.

Thank you, Gene. Your visionary contributions have made our mathematical world a far richer and more interesting place!

References

[1] Apte, M. (1955). Sur certaines classes caractéristiques des variétés kählériennes compactes. *C. R. Acad. Sci. Paris*, 240, 149–151.

[2] Aubin, T. (1976). Équations du type Monge-Ampère sur les variétés kählériennes compactes. *C. R. Acad. Sci. Paris*, Sér. A-B, 283, Aiii, A119–A121.

[3] Bando, S. & Mabuchi, T. (1987). Uniqueness of Einstein Kähler metrics modulo connected group actions. In *Algebraic Geometry*, Sendai, 1985. vol. 10 of Adv. Stud. Pure Math, 11–40. North-Holland, Amsterdam.

[4] Bochner, S. (1946). Vector fields and Ricci curvature. *Bull. Amer. Math. Soc.*, 52, 776–797.

[5] Burns, D. & De Bartolomeis, P. (1988). Stability of vector bundles and extremal metrics, *Invent. Math.*, 92, 403–407.

[6] Calabi, E. (1954). The variation of Kähler metrics. I. The structure of the space. *Bull. Amer. Math. Soc.*, 60, 167–168.

[7] Calabi, E. (1954). The variation of Kähler metrics. II. A minimum problem. *Bull. Amer. Math. Soc.*, 60, 168.

[8] Calabi, E. (1956). The space of Kähler metrics. In: *Proceedings of the International Congress of Mathematicians*, 1954: Amsterdam, September 2–9. Vol. II: Short lectures, 206–207. E. P. Noordhoff, Groningen.

[9] Calabi, E. (1957). On Kähler manifolds with vanishing canonical class. In: Fox, R. H. (ed.) *Algebraic Geometry and Topology. A Symposium in Honor of S. Lefschetz*, 78–89. Princeton University Press, Princeton, NJ.

[10] Calabi, E. (1958). Improper affine hyperspheres of convex type and a generalization of a theorem by K. Jörgens. *Mich. Math. J.*, 5, 105–126.

[11] Calabi, E. (1979). Métriques kählériennes et fibrés holomorphes. *Ann. Sci. École Norm. Sup.* (4), 12, 269–294.

[12] Calabi, E. (1980). Isometric families of Kähler structures. In: *The Chern Symposium 1979* (Proc. Internat. Sympos., Berkeley, Calif., 1979), 23–39. Springer, New York-Berlin.

[13] Calabi, E. (1982). Extremal Kähler metrics. In: *Seminar on Differential Geometry*, 259–290. Ann. of Math. Stud., 102. Princeton Univ. Press, Princeton, NJ.

[14] Calabi, E. (1985). Extremal Kähler metrics. II. In: *Differential Geometry and Complex Analysis*, Vol. dedic. H. E. Rauch, 95–114. Springer, Berlin.

[15] Calabi, E. (2014). *Letter to Claude LeBrun.* February 3.

[16] Calabi, E. & Chen, X. X. (2002). The space of Kähler metrics. II. *J. Differential Geom.*, 61, 173–193.

[17] Chen, X. X. (2000). The space of Kähler metrics. *J. Differential Geom.*, 56, 189–234.

[18] Chen, X. X. (2009). Space of Kähler metrics. III. On the lower bound of the Calabi energy and geodesic distance, *Invent. Math.*, 175, 453–503.

[19] Chen, X. X., Donaldson, S. K., Sun, S. Kähler–Einstein metrics and stability. e-print arXiv:1210.7494 [math.DG], to appear in *Int. Math. Res. Notices.*

[20] Chen, X. X., Donaldson, S. K., Sun, S. Kähler–Einstein metrics on Fano manifolds, I. approximation of metrics with cone singularities. e-print arXiv:1211.4566 [math.DG], to appear in *J. Amer. Math. Soc.*

[21] Chen, X. X., Donaldson, S. K., Sun, S. Kähler–Einstein metrics on Fano manifolds, II: limits with cone angle less than 2π. e-print arXiv:1212.4714 [math.DG], to appear in *J. Amer. Math. Soc.*

[22] Chen, X. X., Donaldson, S. K., Sun, S. Kähler–Einstein metrics on Fano manifolds, III: limits as cone angle approaches 2π and completion of the main proof. e-print arXiv:1302.0282 [math.DG], to appear in *J. Amer. Math. Soc.*

[23] Chen, X. X. & Tian, G. (2008). Geometry of Kähler metrics and foliations by holomorphic discs. *Publ. Math. Inst. Hautes Études Sci.*, 1–107.

[24] Donaldson, S. K. (1999). Symmetric spaces, Kähler geometry and Hamiltonian dynamics. In: *Northern California Symplectic Geometry Seminar.* Vol. 196 of Amer. Math. Soc. Transl. Ser. 2, 13–33. Amer. Math. Soc., Providence, RI.

[25] Donaldson, S. K. (2001). Scalar curvature and projective embeddings. I. *J. Differential Geom.*, 59, 479–522.

[26] Eguchi, T. & Hanson, A. J. (1979). Self-dual solutions to Euclidean gravity. *Ann. Physics*, 120, 82–106.

[27] Futaki, A. (1983). An obstruction to the existence of Einstein Kähler metrics. *Invent. Math.*, 73, 437–443.
[28] Futaki, A. (1983). On compact Kähler manifolds of constant scalar curvatures. *Proc. Japan Acad. Ser. A Math. Sci.*, 59, 401–402.
[29] Kähler, E. (1933). Über eine bemerkenswerte Hermitesche Metrik. *Abh. Math. Sem. Univ. Hamburg*, 9, 173–186.
[30] LeBrun, C. (1995). Polarized 4-manifolds, extremal Kähler metrics, and Seiberg-Witten theory. *Math. Res. Lett.*, 2, 653–662.
[31] Lichnerowicz, A. (1957). Sur les transformations analytiques des variétés kählériennes compactes. *C. R. Acad. Sci. Paris*, 244, 3011–3013.
[32] Mabuchi, T. (1987). Some symplectic geometry on compact Kähler manifolds. I. *Osaka J. Math.*, 24, 227–252.
[33] Mabuchi, T. (2004). Uniqueness of extremal Kähler metrics for an integral Kähler class. *Internat. J. Math.*, 15, 531–546.
[34] Matsushima, Y. (1957). Sur la structure du groupe d'homéomorphismes d'une certaine variété kählérienne. *Nagoya Math. J.*, 11, 145–150.
[35] Ross, J. (2006). Unstable products of smooth curves. *Invent. Math.*, 165, 153–162.
[36] Semmes, S. (1992). Complex Monge–Ampère and symplectic manifolds. *Amer. J. Math.*, 114, 495–550.
[37] Tian, G. (1997). Kähler–Einstein metrics with positive scalar curvature. *Invent. Math.*, 130, 1–37.
[38] Yau, S.-T. (1978). On the Ricci curvature of a compact Kähler manifold and the complex Monge–Ampère equation. I. *Commun. Pure Appl. Math.*, 31, 339–411.

Calabi's Work on Affine Differential Geometry and Results of Bernstein Type

Xiuxiong Chen[1] and Simon Donaldson[2]

We will begin with a quick review of some concepts from affine differential geometry. For our purposes this is the study of geometric properties of locally convex hypersurfaces $M \subset \mathbf{R}^{n+1}$ which are invariant under the group of unimodular affine transformations of $\mathbf{R}^{n+1}$. Locally, in suitable affine co-ordinates, M can be represented as a graph

$$x_0 = u(x_1, \dots x_n)$$

of a convex function u. The Hessian $u_{ij} = \frac{\partial^2 u}{\partial x_i \partial x_j}$ has an affine invariant meaning as a quadratic form on TM with values in the normal bundle $\nu = \mathbf{R}^{n+1}/TM$, so it clearly defines a conformal structure on M. The first observation is that we can normalise the conformal factor using the isomorphism $\Lambda^n TM = \nu^*$ furnished by the fixed volume form on $\mathbf{R}^{n+1}$. That is (following Calabi's exposition in [13]) there is a unique way to write the Hessian as $g \otimes \omega$ where g is an $\mathbf{R}$-valued quadratic form on TM, $\omega \in \nu$ and $\omega = \det(g)^{1/2}$ under the above isomorphism, where $\det(g)^{1/2}$ is the usual volume element defined by a quadratic form. This gives an affine-invariant Riemannian metric on M (the affine metric). In a graphical representation as above, with $x_1, \dots x_n$ regarded as co-ordinates on M, this metric is

$$g_{ij} = \det(u_{kl})^{-1/n+2} u_{ij}.$$

[1] Xiuxiong Chen
Department of Mathematics, Stony Brook University, Stony Brook, New York, USA
e-mail: xiu@math.sunysb.edu

[2] Simon Donaldson
Simons Center for Geometry and Physics, Stony Brook University, Stony Brook, New York, USA
e-mail: sdonaldson@scgp.stonybrook.edu

The discussion above involves the 2-jet of the manifold M, which is to say the first and second derivatives of a local representation u. The second observation is that the 3-jet fixes a preferred normal to M at each point. At a given point p of M take any graphical representation as above with p at the co-ordinate origin and with $\frac{\partial u}{\partial x_i}$ vanishing at 0, so that TM_p is the hyperplane $x_0 = 0$. This corresponds to some arbitrary choice of a normal subspace, complementary to $TM \subset \mathbf{R}^{n+1}$. Set

$$L_i = \frac{\partial \mathrm{logdet}(u_{kl})}{\partial x_i} = \sum_{a,b} u^{ab} u_{abi},$$

where u^{ab} denotes the inverse matrix of u_{ij}. Any other choice of normal subspace is determined by a vector $(c_1, \ldots, c_n)$, with new affine co-ordinates $x_0, \tilde{x}_1, \ldots, \tilde{x}_n$ given by

$$\tilde{x}_i = x_i - c_i x_0.$$

In such co-ordinates the same submanifold M is given by the graph of a function $x_0 = \tilde{u}(\tilde{x}_1, \ldots, \tilde{x}_n)$. Simple calculus shows that, in an obvious notation, and with all derivatives evaluated at the origin:

$$\tilde{u}_{ij} = u_{ij}$$

while

$$\tilde{u}_{ijk} = u_{ijk} + \sum_r c_r (u_{ri} u_{jk} + u_{rj} u_{ki}).$$

This implies that

$$\tilde{L}_i = L_i + (n+1) \sum_j u_{ij} c_j.$$

Thus there is a unique choice of c_i such that $\tilde{L}_i = 0$. This defines the *affine normal* direction: in other words it is the unique choice of normal such that in a graphical representation $\det(u_{ij})$ is constant to first order. (A more geometric definition is that the affine normal at p is the tangent to the curve formed by the mass-centres of the sections of M cut out by translates of the tangent space at p.) With this discussion in place, we may as well suppose that our original co-ordinate system was chosen so that the affine normal of M at 0 is the x_0 axis. The cubic term $A = u_{ijk}$ is then another fundamental affine invariant of M. In invariant language it lies in the kernel of the contraction map

$$S^3 T^* M \to T^* M$$

defined by the affine metric in M.

We move on discuss Calabi's paper [3]. A hypersurface $M^n \subset \mathbf{R}^{n+1}$ is called an *affine sphere* if there is a point q in $\mathbf{R}^{n+1}$ such that the normal line at any point p of M passes through q. In a limiting case (for a point q at infinity), M is an "improper affine sphere" if all normal directions are parallel. This case, of improper affine spheres, is the subject of Calabi's paper. Our account above shows immediately that the condition is locally given by the real Monge–Ampère equation

$$\det(u_{ij}) = \text{constant},$$

and we can normalise the constant to be 1. Thus the affine metric becomes simply $g_{ij} = u_{ij}$. Calabi proved:

If $M \subset \mathbf{R}^{n+1}$ is an improper affine sphere and either

- $n \leq 5$ and M is defined globally as a graph over $\mathbf{R}^n$ or,
- M is complete with respect to the affine metric,

then M is the graph of a quadratic function on $\mathbf{R}^n$.

In the case of the first item, this says that the only convex solutions of the Monge–Ampère equation $\det(u_{ij}) = 1$ on $\mathbf{R}^n$ are the obvious quadratic functions. When $n = 2$ this was an earlier result of Jörgens, proved using complex function theory. (The restriction to $n \leq 5$ in Calabi's result was later removed by Pogerolov [19].) Here we should recall the classical *Bernstein Theorem*, which states that the only minimal surfaces in $\mathbf{R}^3$ which are graphs over the whole of $\mathbf{R}^2$ are planes. The striking thing is that no *a priori* growth conditions are imposed. This is in contrast to the case of linear equations and reflects essentially nonlinear phenomena. Results of this nature have important applications to singularity formation, through "blow-up" arguments.

Calabi's proofs of these results introduced a number of seminal ideas. The Riemannian curvature tensor of an affine metric can be expressed as a quadratic expression in the cubic form A. When the manifold is an improper affine sphere the Ricci curvature is positive and the scalar curvature R is $|A|^2$. There is also a crucial formula for the Laplacian of R:

$$\Delta R = |\mathrm{Ricci}|^2 + |\mathrm{Riem}|^2 + |\nabla A|^2. \tag{1}$$

From this, Calabi obtains

$$\Delta|A| \geq \frac{n+1}{n(n-1)}|A|^{3/2}. \tag{2}$$

Now the argument is to consider a suitable comparison function $f(r)$ where r is the geodesic distance from a point. The positivity of the Ricci curvature yields a fundamental comparison inequality

$$\Delta r \geq \frac{n-1}{r} \tag{3}$$

(in the flat case we would have equality). But there is an important point that the distance function r is not smooth, so this inequality has to be understood in a suitable generalised sense, introduced by Calabi. (Likewise for the $|A|$ in (2).) A rotationally-invariant solution of the equation $\Delta f^{1/2} = \frac{n+1}{n(n-1)} f^{3/2}$ on $\mathbf{R}^n$ yields a comparison function $f(r)$ which blows up at finite distance and Calabi is able to deduce that for an entire solution $|A|$ must be identically zero. This argument used an extension of the maximum principle, established by Calabi in the paper [2], for differential inequalities in this generalised sense.

We will refer to the article of Lawson in this volume for a discussion of the significance of Calabi's generalised maximum principle and connections with the subsequent theory of "viscosity solutions" in PDEs. Likewise, the comparison inequality (3) for the distance function has played a fundamental role in many subsequent developments in Riemannian geometry. For example it is the foundation of the Cheeger–Gromoll splitting theorem for manifolds of nonnegative Ricci curvature containing a "line" [10] and is ubiquitous in the Cheeger–Colding theory of limits of manifolds satisfying Ricci curvature bounds [9].

We move on to Calabi's 1971 paper [5], which considered proper affine spheres, where the normal lines all pass through a single point which we can take to be the origin. The volume form of the affine metric and the isomorphism $\Lambda^n TM = \nu$ defines a normal vector N so at a point $X \in M$ we have $N = -HX$ for a scalar H which turns out to be constant. The affine sphere is called elliptic if $H > 0$ and hyperbolic if $H < 0$, the basic models being respectively an ellipsoid and a sheet of a 2-sheeted hyperboloid. Calabi proved that there are no other complete solutions in the elliptic case. (The completeness condition was later established by Cheng and Yau [14].) In the hyperbolic case he proves, by global arguments related to those outlined above, that the Ricci curvature is negative (but bounded below by H). He also stated an important conjecture: that complete solutions should be asymptotic to the boundary of a convex cone and conversely any such cone determines a solution, unique up to scaling. He constructed explicit examples such as the hypersurface defined by the equation $x_0 x_1 \ldots x_{n+1} = 1$ (with $x_i > 0$), which is asymptotic to the boundary of the convex cone $\{x_i > 0\}$. If the cone is written as the cone over a convex subset $K \subset \mathbf{R}^n$ then the problem can be set up, via the Legendre transform, as finding a function f on the interior of K solving the equation

$$\det f_{ij} = f^{-n-2},$$

and vanishing on the boundary. This conjecture of Calabi was proved by Cheng and Yau in [13]. (A recent survey on the subject of affine spheres appears in [18].)

We will now move slightly back in time to Calabi's paper [4], which gives a general discussion of Bernstein-type theorems. He observed that similar techniques to those which he had used for the Monge–Ampère equation applied to maximal spacelike submanifolds M^n in the Lorentz space $\mathbf{R}^{n,1}$. For example, the Ricci curvature of such a manifold is positive (whereas the Ricci curvature of a minimal submanifold in the Euclidean case is negative). Calabi proved a Bernstein theorem in any dimension n assuming the hypothesis of completeness in the induced metric. A proof without this assumption was given by Cheng and Yau in [12]. More generally, Jost and Xin extended the result to arbitrary codimension, i.e. they proved that the only "space-like" maximal submanifolds of $\mathbf{R}^{n,m}$ which are graphs over $\mathbf{R}^n$ are flat [15]. These results are in contrast to the Euclidean case where even for hypersurfaces the Bernstein property fails once $n \geq 8$ – the indefinite case is better behaved because of the favourable signs which arise, such as the sign of the Ricci curvature. Jost and Xin also point out that, through an observation of Hitchin, the Monge–Ampère case can be embedded in the maximal submanifolds discussion, since solutions correspond to Lagrangian maximal submanifolds in $\mathbf{R}^{n,n}$. This elucidates many of the affine differential geometry formulae. For example (1) can be viewed as the Eells–Sampson formula from harmonic map theory for the Gauss map of the submanifold.

Another question raised by Calabi in [4] is the complex analogue for the complex Monge–Ampère equation

$$\det\left(\frac{\partial^2\phi}{\partial\bar{z}_a\partial z_b}\right) = 1 \tag{4}$$

on $\mathbf{C}^n$. Solutions correspond to Kähler metrics with zero Ricci curvature. LeBrun pointed out [16] that the obvious Bernstein statement is false: the Taub–NUT metric gives a non-trivial solution on $\mathbf{C}^2$. There are important open questions in this area. For example, one might expect that, if the metric is complete and has Euclidean volume growth, then the only solutions are flat.[3] (This is known when $n = 2$.) Complex Monge–Ampère equations are of course fundamental in Kähler geometry generally. In the paper [4], Calabi mentions the then-recent thesis of Aubin, published as [1]. Aubin adapted Calabi's third order estimate (i.e. for the norm of the cubic form A) to the complex case. This estimate became one of the important steps in Yau's proof of the Calabi conjecture on compact Kähler manifolds.

We now discuss another series of papers by Calabi on "affine maximal" submanifolds, beginning in 1982 with [6]. The affine metric on a hypersurface $M \subset \mathbf{R}^{n+1}$ induces a volume form on M. In a graphical representation by a convex function u this is given by $\det(u_{ij})^{1/n+2}\mathrm{d}x_1 \ldots \mathrm{d}x_n$. Affine maximal submanifolds are, by definition, critical points for the functional given by the total volume, with respect to compactly supported perturbations. In a graphical representation the functional is

[3] Added in proof: Developments since the time of writing show that this expectation is not correct, see Li, Y. A new complete Calabi–Yau metric on $\mathbf{C}^3$. *Invent. math.* 217, 1–34 (2019).

$$\int \det(u_{ij})^{1/n+2} \mathrm{d}x_1 \dots \mathrm{d}x_n,$$

and the Euler–Lagrange equation can be written as

$$\Delta \det(u_{ij})^{-1/n+2} = 0, \tag{5}$$

where Δ is the Laplacian of the affine metric. In more invariant terms, the equation can be written $\Delta Y = 0$, where Y is the co-normal vector, an $\mathbf{R}^{n+1}$-valued function on M.

One of the main concerns of the paper [6] is to show that solutions *maximise* the volume (leading to the designation "affine maximal"). Various forms of this statement are proved: for infinitesimal variations and for general compactly supported variations over graphical regions.

The equation (5) makes it clear that improper affine spheres are affine maximal. The local differential geometric formulae, worked out in [6], are more complicated in the general case. The Ricci curvature of the affine metric is not necessarily positive (because there is a contribution from $\nabla^2 Y$) but the scalar curvature is positive and given by the same expression $R = |A|^2$. Calabi uses this to prove a Bernstein theorem in dimension $n = 2$ assuming that the affine metric is complete and $M \subset \mathbf{R}^3$ is closed. Restricting for simplicity to the case of graphs, the argument is that $\det(u_{ij})^{-1/n+2}$ is a positive harmonic function on M. But by a result of Blanc and Fiala the condition $R \geq 0$ implies that the conformal structure on M is "parabolic" – equivalent to $\mathbf{C}$ – so the only positive harmonic functions are constants. This reduces the question to the Monge–Ampère case and u must be quadratic.

In the pair of papers [7], [8] Calabi studied affine maximal surfaces using a complex representation, analogous to the Weierstrass representation of minimal surfaces. If we choose local isothermal co-ordinates on M, then the co-normal vector can be written as the imaginary part of a holomorphic map Z from a domain in $\mathbf{C}$ to $\mathbf{C}^3$. The surface can be recovered from this map by integration:

$$X(z) = -i(Z \times \overline{Z} + \int^z Z \times \mathrm{d}Z - \int^z \overline{Z} \times \mathrm{d}Z),$$

where $\times$ is the usual cross-product. Calabi finds the explicit conditions for a map Z to correspond to a surface and uses complex variable methods to obtain various Bernstein type theorems, extending his previous result and using different techniques.

Calabi conjectured in [6] that the assumption of metric completeness could be removed, i.e. that the only closed affine maximal surfaces in $\mathbf{R}^3$ are paraboloids. This conjecture was proved by Trudinger and Wang [20] and Jin and Li [17]. Trudinger and Wang also obtained results in higher dimension assuming "uniform convexity". They also gave counterexamples. When $n = 10$ and writing a point in $\mathbf{R}^{10}$ as (x', x_{10}), where $x' \in \mathbf{R}^9$, the function

$$u=\sqrt{|x'|^9+x_{10}^2}$$

is not smooth but solves the equation (5) in a certain generalised sense.

The affine maximal equation serves as a prototype of a fully nonlinear fourth order PDE. Regularity questions can be very difficult, as the example above indicates. More generally, one can consider the Euler–Lagrange equations associated to the integral of $F(\det(u_{ij}))$ for suitable functions F. In [21], Trudinger and Wang considered the case when $F(J)=J^{\alpha}$. Another interesting case is when $F(J)=-\log J$. In the latter case, solutions of the Euler–Lagrange equation correspond to "toric" Kähler metrics of zero scalar curvature and a small variant gives toric Kähler metrics which are "extremal", as defined by Calabi. The relation between these equations on $\mathbf{R}^n$ and the general extremal equation is much like that between the real and complex Monge–Ampère equations. Techniques from affine geometry have been applied to extremal metrics with very good effect recently by Chen, Li and Sheng [11]. This surprising interweaving of widely different strands, both going back to Calabi's work, is another testament to his vision.

References

[1] Aubin, T. (1970). Métriques riemanniennes et courbure. *Jour. Differential Geometry* 4, 383–428.

[2] Calabi, E. (1958). An extension of E. Hopf's maximum principle, with an application to Riemannian geometry. *Duke Math. J.* 25, 45–56.

[3] Calabi, E. (1958). Improper affine hyperspheres of convex type and a generalisation of a theorem of K. Jörgens. *Michigan Math. J.* 5, 105–126.

[4] Calabi, E. (1970). Examples of Bernstein problems for some nonlinear equations. In: *Global Analysis* (Proc. Sympos. Pure Math., Vol. XV, Berkeley, Calif., 1968), 223–230. Amer. Math. Soc., Providence, RI.

[5] Calabi, E. (1972). Complete affine hyperspheres. I. In: *Symposia Mathematica*, Vol. X (Convegno di Geometria Differenziale, INDAM, Rome, 1971/1972), 19–38. Academic Press, London.

[6] Calabi, E. (1982). Hypersurfaces with maximal affinely invariant area. *Amer. J. Math.* 104 No. 1, 91–126.

[7] Calabi, E. (1988). Convex affine maximal surfaces. Reprinted in *Affine Differentialgeometrie* (Oberwolfach 1986) Tech. Univ. Berlin, 199–223.

[8] Calabi, E. (1990). Affine differential geometry and holomorphic curves. In: *Complex Geometry and Analysis*, Proc. Int. Symp. in Honour of E. Vesentini (Pisa, Italy, 1988), 15–21. Lecture Notes in Math., 1422. Springer, Berlin.

[9] Cheeger, J. & Colding, T. (1996). Lower bounds on Ricci curvature and the almost-rigidity of warped products. *Annals of Math.* 144(1), 189–237.

[10] Cheeger, J. & Gromoll, D. (1971/2). The splitting theorem for manifolds of non-negative Ricci curvature. *J. Differential Geometry* 6, 119–128.

[11] Chen, B., Li, A.-M., Sheng, L. Extremal metrics on toric surfaces. arXiv 1008.2607

[12] Cheng, S. & Yau, S.-T. (1976). Maximal spacelike hypersurfaces in the Lorentz–Minkowski space. *Annals of Math.* 104(2), 407–419.

[13] Cheng, S. & Yau, S.-T. (1977). On the regularity of the Monge–Ampère equation $\det(\frac{\partial^2 u}{\partial x^i \partial x^j} = F(x,u)$. *Commun. Pure and Applied Math.* 30, 41–68.

[14] Cheng, S. & Yau, S.-T. (1986). Complete affine hypersurfaces, I: the completeness of the affine metric. *Commun. Pure Applied Math.* 39 No. 6, 839–866.
[15] Jost, J. & Xin, Y. L. (2001). Some aspects of the global geometry of entire spacelike submanifolds. *Results in Math.* 40, 233–245.
[16] LeBrun, C. (1991). Complete Ricci-flat Kähler metrics on metrics on $\mathbf{C}^n$ need not be flat. *Proc. Symp. Pure Math.* 52(2), 297–304.
[17] Li, A.-M. & Jia, F. (2001). The Calabi conjecture on affine maximal surfaces. *Results in Math.* 40, 265–272.
[18] Loftin, J. (2010). Survey of affine spheres. In: *Handbook of Geometric Analysis No. 2*. Adv. Lectures in Math. 13. International Press, 161–196.
[19] Pogerolov, V. (1972). On the improper convex affine hyperspheres. *Geometriae Dedicata* 1(1), 33–46.
[20] Trudinger, N. & Wang, X.-J. (2000). The Bernstein property for affine maximal hypersurfaces. *Inventiones Math.* 140 No. 2, 399–422.
[21] Trudinger, N. & Wang, X. (2002). Bernstein–Jörgens theorems for a fourth order partial differential equation, *J. Partial Differential Equations* 15, 78–88.

Part II
Collected Works

Convergence- and Sum-Factors for Series of Complex Numbers

Eugenio Calabi and Aryeh Dvoretzky

Trans. Amer. Math. Soc. 70, 177–194 (1951)

1. **Introduction.**

1.1. Let $\sum_{n=1}^{\infty} a_n$ be a series of real numbers tending to zero, $\zeta^{(1)} \geqq 0$ and $\zeta^{(2)} \leqq 0$. Then it is obvious that a sequence $\zeta_1, \cdots, \zeta_n, \cdots$ with each ζ_n equal to either $\zeta^{(1)}$ or $\zeta^{(2)}$ may be chosen so that $\sum_{n=1}^{\infty} \zeta_n a_n$ be convergent. If, furthermore, $\sum |a_n|$ is divergent and $\zeta^{(1)}\zeta^{(2)} \neq 0$, the sequence (ζ_n) may be chosen so that $\sum_{n=1}^{\infty} \zeta_n a_n$ converges to any preassigned sum.

On passing from series of real numbers to series of complex ones, the second statement above clearly ceases to be true while the first one, though remaining valid (cf. below), is far from obvious. It is the primary object of the present paper to extend and generalize the above and related statements to series of complex numbers.

1.2. The quotation of the principal results will be facilitated if the following definitions are introduced.

Definition 1. *A set* Z *of complex numbers is called a convergence factor set if, given any sequence* $(a_n)_1^\infty$ *of complex numbers satisfying*

$$\lim_{n=\infty} a_n = 0, \tag{1}$$

there exists a sequence $(\zeta_n)_1^\infty$ *with* $\zeta_n \in \mathrm{Z}$ $(n=1, 2, \cdots)$ *for which* $\sum \zeta_n a_n$ *is convergent.*

Definition 2. *A set* Z *of complex numbers is called a sum factor set if, given any sequence* $(a_n)_1^\infty$ *of complex numbers satisfying* (1) *and*

$$\sum_{n=1}^{\infty} |a_n| = \infty \tag{2}$$

and any complex number s, *there exists a sequence* $(\zeta_n)_1^\infty$, *with* $\zeta_n \in \mathrm{Z}$ $(n=1, 2, \cdots)$ *for which* $\sum_{n=1}^{\infty} \zeta_n a_n$ *converges to* s.

The following two theorems give a simple characterization of bounded convergence and sum factor sets.

If X is any set of complex numbers, we denote by $\overline{X}$ its closure and by $\mathcal{C}X$ its convex hull, that is, the (not necessarily closed) intersection of all convex sets containing X.

Theorem 1. *A bounded set* Z *is a convergence factor set if and only if* $0 \in \overline{\mathcal{C}\mathrm{Z}}$.

Theorem 2. *A bounded set* Z *is a sum factor set if and only if* 0 *is an interior*

Presented to the Society, December 28, 1948; received by the editors June 1, 1949.

DOI: 10.2307/1990531

point of $\mathcal{C}\mathbf{Z}$.

The only cases of convergence factor sets $\mathbf{Z}$ known before seem to be those whose closure $\bar{\mathbf{Z}}$ contains a pair of points (ζ and $-\zeta$) symmetric about the origin. Even this is (for $\zeta \neq 0$) not trivial and due to Ch. Hanani and one of the present authors who proved [1, 3](1) that if $\sum a_n$ is any series of complex numbers tending to zero it is possible to change the signs so that the resulting series $\sum \pm a_n$ is convergent, that is, that $\{1, -1\}$ (or $\{\zeta, -\zeta\}$) is a convergence factor set.

The origin can be an interior point of $\mathcal{C}\mathbf{Z}$ only if it is an interior point of a nondegenerate triangle or a nondegenerate quadrilateral with vertices in $\mathbf{Z}$. Thus if a set $\mathbf{Z}$ is a sum factor set it necessarily contains a subset of three or four points having the same property. The only previously published instances of sum factor sets seem to be due to H. Hornich who proved [4] that for $k \geqq 3$ the set of kth roots of unity is a sum factor set.

1.3. The proof of Theorem 2 is straightforward but that of Theorem 1 is more tricky. Dvoretzky and Hanani [3] make some use of the group property of sign changes that if the signs of some terms in $\sum a_n$ are changed, and then a similar operation is applied to the resulting series, the final series is again of the form $\sum \pm a_n$. No similar property obtains if $\{1, -2\}$, say, are used as multipliers instead of $\{1, -1\}$. In order to overcome this difficulty it was found convenient to obtain Theorem 1 as a special case of a more general result relating to the case when the factors ζ_n instead of being taken from a fixed set $\mathbf{Z}$ are taken from different sets $\mathbf{Z}_n$.

DEFINITION 3. *A sequence* $(\mathbf{Z}_n)_1^\infty$ *of sets of complex numbers is called a convergence factor sequence if, given any sequence of complex numbers* $(a_n)_1^\infty$ *satisfying* (1), *there exists a sequence* $(\zeta_n)_1^\infty$ *with* $\zeta_n \in \mathbf{Z}_n$ $(n = 1, 2, \cdots)$ *for which* $\sum \zeta_n a_n$ *is convergent.*

In §2 we prove

THEOREM 3. *A sequence* $(\mathbf{Z}_n)_1^\infty$ *of uniformly bounded sets of complex numbers is a convergence factor sequence if and only if* $\sum_{n=1}^\infty d_n < \infty$, *where*

$$d_n = \inf_{\zeta \in \mathcal{C}\mathbf{Z}_n} |\zeta| = \min_{\zeta \in \overline{\mathcal{C}\mathbf{Z}_n}} |\zeta|$$

is the distance of the origin from the convex hull of $\mathbf{Z}_n$.

If all $\mathbf{Z}_n$ are identical, so are all d_n and the above condition reduces to the requirement that the common value of all d_n be zero. Thus, Theorem 3 specializes to Theorem 1 in this particular case.

In §3 a similar extension of Theorem 2 is given. We introduce the following definition.

DEFINITION 4. *A sequence* $(\mathbf{Z}_n)_1^\infty$ *of sets of complex numbers is called a sum*

(1) Numbers in brackets refer to the bibliography at the end of the paper.

factor sequence if, given any sequence of complex numbers $(a_n)_1^\infty$ *satisfying* (1) *and* (2), *and a complex number* s, *there exists a sequence* $(\zeta_n)_1^\infty$ *with* $\zeta_n \in Z_n$ $(n=1, 2, \cdots)$ *for which* $\sum_{n=1}^\infty \zeta_n a_n$ *converges to* s.

We obtain

THEOREM 4. *A sequence* $(Z_n)_1^\infty$ *of uniformly bounded nonempty sets of complex numbers is a sum factor sequence if and only if there exist an integer* N *and a positive* δ *such that*

$$\{ |\zeta| < \delta \} \subset \complement Z_n \quad (n = N+1, N+2, \cdots).$$

Thus (Z_n) is a sum factor sequence if and only if 0 is an interior point of the intersection of all but a finite number of $\complement Z_n$. Theorem 2 is clearly a special case.

1.4. In §4 we show (Proposition 4) that a set S of complex numbers is the set of limit points of the partial sums of some series $\sum_{n=1}^\infty a_n$ satisfying (1) if and only if it is a continuum or becomes one on adjoining the point at infinity (cf. also Proposition 5). We then prove (Theorem 5) a result which shows in particular that if S satisfies the above requirements, if (Z_n) satisfies the conditions of Theorem 4, and if (a_n) satisfies (1) and (2), then $\zeta_n \in Z_n$ may be chosen so that S coincides with the set of limit points of the partial sums of $\sum_{n=1}^\infty \zeta_n a_n$.

Thus a uniformly bounded sequence of sum factors can secure not only convergence to a preassigned sum, but also the oscillation of the partial sums through any set which can at all be a derived set of partial sums of a series satisfying (1). For the special case when all Z_n coincide with the set of kth roots of unity ($k \geqq 3$ fixed) a partial result of the same kind was obtained by Hornich [5].

In §5 we deal with nonbounded convergence and sum factor sets and sequences. A complete and easy characterization of convergence factor sets and sequences is obtained (Theorems 6 and 7) but we did not obtain a similar complete solution of the corresponding problem for sum fa ctor sequences (sum factor sets are characterized by Theorem 8).

In §6 an application to series of functions is given and §7 contains a remark on the value of a constant encountered in §2.

1.5. In conclusion we note that many of the results of this paper extend also to vectors in any Euclidean space and more general linear spaces. In particular, if we confine ourselves to real (scalar) factors hardly any change in the proofs is required in order to establish the analogues of Theorem 1 and 3 for arbitrary finite-dimensional Banach spaces. (For infinitely-dimensional spaces, even for Hilbert spaces, even Theorem 1 with $Z = \{1, -1\}$ obviously fails.)

2. **Convergence factors.**

2.1. In studying convergence factors we find it convenient to refer to the following

LEMMA. *If a segment in the plane of length less than or equal to 2 is at a distance less than or equal to 1 from the origin, then either at least one of its points is at a distance less than or equal to 1 from the origin or both end points are at a distance less than $5^{1/2}$ from it.*

The verification of the lemma being immediate, we proceed to apply it and prove the following propositions.

PROPOSITION 1. *Let $\{\alpha_n, \beta_n, \gamma_n\}$ $(n=1, 2, \cdots)$ be a sequence of triplets of (not necessarily distinct) complex numbers satisfying*

$$\max(|\beta_n - \alpha_n|, |\gamma_n - \beta_n|, |\alpha_n - \gamma_n|) \leqq 2 \qquad (n = 1, 2, \cdots) \tag{3}$$

and

$$0 \in \mathcal{C}\{\alpha_n, \beta_n, \gamma_n\} \qquad (n = 1, 2, \cdots). \tag{4}$$

Then, given a positive integer N, there exists a sequence $(\zeta_n)_{n=1}^N$ with $\zeta_n \in \{\alpha_n, \beta_n, \gamma_n\}$ which satisfies

$$\left| \sum_{\nu=1}^{n} \zeta_\nu \right| < 5^{1/2} \qquad (n = 1, 2, \cdots, N). \tag{5}$$

In order to establish this proposition we prove the somewhat stronger

PROPOSITION 2. *Under the assumptions of Proposition 1 the following conclusion holds: Given a positive integer N, there exist an integer $u(N)$ and two sequences $(\zeta'_{n,N})_{n=1}^N$ and $(\zeta''_{n,N})_{n=1}^N$ satisfying*

$$1 \leqq u(N) \leqq N \tag{6}$$

$$\{\zeta'_{n,N}, \zeta''_{n,N}\} \subset \{\alpha_n, \beta_n, \gamma_n\} \qquad (n = 1, 2, \cdots, N), \tag{7}$$

$$\zeta'_{n,N} = \zeta''_{n,N} \qquad \textit{for } n \neq u(N), \tag{8}$$

$$\left| \sum_{\nu=1}^{n} \zeta'_{\nu,N} \right| < 5^{1/2} \quad \textit{and} \quad \left| \sum_{\nu=1}^{n} \zeta''_{\nu,N} \right| < 5^{1/2} \qquad (n = 1, 2, \cdots, N-1), \tag{9}$$

and

$$\min_{0 \leqq \mu \leqq 1} \left| \mu \sum_{\nu=1}^{N} \zeta'_{\nu,N} + (1-\mu) \sum_{\nu=1}^{N} \zeta''_{\nu,N} \right| \leqq 1. \tag{10}$$

Because of (7), (9), (10), and the lemma, Proposition 2 clearly implies Proposition 1.

2.2. **Proof of Proposition 2.** The proof proceeds by induction on N. $u(1)=1$ by (6). Since the triangle $\mathcal{C}\{\alpha_1, \beta_1, \gamma_1\}$ contains the origin, it meets the unit disc $|z| \leqq 1$ and thus, by (3), at least one of its sides meets the unit disc. Let $\zeta'_{1,1}$ and $\zeta''_{1,1}$ denote the end points of this side, then all the contentions of Proposition 2 with $N=1$ hold.

Assume now the truth of Proposition 2 for some fixed $N \geqq 1$. Put

$$s'_{n,N} = \sum_{\nu=1}^{n} \zeta'_{\nu,N}, \qquad s''_{n,N} = \sum_{\nu=1}^{n} \zeta''_{\nu,N} \quad (n = 1, 2, \cdots, N)$$

and consider separately the mutually exclusive and jointly exhaustive cases A_1, A_2, B_1, B_2, and B_3.

Case A₁. $|s'_{N,N}| \leqq 1$.

Put $u(N+1)=N+1$ and $\zeta'_{n,N+1}=\zeta''_{n,N+1}=\zeta'_{n,N}$ $(n=1, 2, \cdots, N)$. These values satisfy (6) and (8) with N replaced by $N+1$ and, by the induction assumption, also (9). Because of (4), the triangle(²) $s'_{N,N+1}+\mathcal{C}\{\alpha_{N+1}, \beta_{N+1}, \gamma_{N+1}\}$ contains the point $s'_{N,N+1}=s'_{N,N}$ which is assumed to lie within the unit disc. By (3) a side of this triangle meets the unit disc. Choose $\zeta'_{N+1,N+1}$ and $\zeta''_{N+1,N+1}$ from among α_{N+1}, β_{N+1}, and γ_{N+1} so that $s'_{N,N+1}+\zeta'_{N+1,N+1}$ and $s'_{N,N+1}+\zeta''_{N+1,N+1}$ be the end points of such a side. Then the sequences $(\zeta'_{n,N+1})$, $(\zeta''_{n,N+1})$ $(n=1, 2, \cdots, N+1)$ satisfy (7) and (10) with N replaced by $N+1$. Thus, in this case, Proposition 2 holds for $N+1$.

Case A₂. $|s'_{N,N}|>1$ but $|s''_{N,N}| \leqq 1$.

This case reduces to the preceding one on interchanging the superscripts $'$ and $''$.

Case B. $|s'_{N,N}|>1$ and $|s''_{N,N}|>1$.

It follows from (3), (7), (8), (10), and the lemma that

$$\max(|s'_{N,N}|, |s''_{N,N}|) < 5^{1/2}. \tag{11}$$

Consider the set of six (not necessarily distinct) points

$$\{s'_{N,N}, s''_{N,N}\} + \{\alpha_{N+1}, \beta_{N+1}, \gamma_{N+1}\}$$

and its convex hull

$$P_N = \mathcal{C}\{s'_{N,N}, s''_{N,N}\} + \mathcal{C}\{\alpha_{N+1}, \beta_{N+1}, \gamma_{N+1}\}.$$

P_N is the union of two triangles and three parallelograms, namely

$$\begin{aligned} P_n = {} & (s'_{N,N} + \mathcal{C}\{\alpha_{N+1}, \beta_{N+1}, \gamma_{N+1}\}) \cup (s''_{N,N} + \mathcal{C}\{\alpha_{N+1}, \beta_{N+1}, \gamma_{N+1}\}) \\ & \cup (\mathcal{C}\{s'_{N,N}, s''_{N,N}\} + \mathcal{C}\{\beta_{N+1}, \gamma_{N+1}\}) \\ & \cup (\mathcal{C}\{s'_{N,N}, s''_{N,N}\} + \mathcal{C}\{\gamma_{N+1}, \alpha_{N+1}\}) \\ & \cup (\mathcal{C}\{s'_{N,N}, s''_{N,N}\} + \mathcal{C}\{\alpha_{N+1}, \beta_{N+1}\}). \end{aligned} \tag{12}$$

By (4) the two triangles in (12) contain respectively $s'_{N,N}$ and $s''_{N,N}$; since P_N is convex, it contains the segment joining $s'_{N,N}$ and $s''_{N,N}$, and hence, by (10), a point s_N^* with $|s_N^*| \leqq 1$. Since s_N^* must belong to one of the triangles or parallelograms of (12), this triangle or parallelogram meets the unit disc. Since no quadrangle all of whose sides are of length less than or equal to 2 can contain the unit circle in its interior, it follows that at least one of the

(²) Here and elsewhere in the paper, if A and B are sets of complex numbers we denote by $A+B$ their group theoretic sum, that is, the set of all $a+b$ with $a \in A$, $b \in B$.

nine segments constituting the boundaries of the two triangles and three parallelograms of (12) meets the unit disc.

Let these nine segments be arranged in some order and let σ_N be the first of these which meets the unit disc; we now distinguish three cases:

Case B_1. σ_N is either $s'_{N,N}+\mathcal{C}\{\beta_{N+1}, \gamma_{N+1}\}$ or $s'_{N,N}+\mathcal{C}\{\gamma_{N+1}, \alpha_{N+1}\}$ or $s'_{N,N}+\mathcal{C}\{\alpha_{N+1}, \beta_{N+1}\}$.

In this case we set $u(N+1)=N+1$, $\zeta'_{n,N+1}=\zeta''_{n,N+1}=\zeta'_{n,N}$ $(n=1, 2, \cdots, N)$ and $\{\zeta'_{N+1,N+1}, \zeta''_{N+1,N+1}\}$ equal respectively to either $\{\beta_{N+1}, \gamma_{N+1}\}$ or $\{\gamma_{N+1}, \alpha_{N+1}\}$ or $\{\alpha_{N+1}, \beta_{N+1}\}$. It is easy to verify then from (11) and the definition of σ_N that (6) through (10) are satisfied when N is replaced by $N+1$.

Case B_2. σ_N is either $s''_{N,N}+\mathcal{C}\{\beta_{N+1}, \gamma_{N+1}\}$ or $s''_{N,N}+\mathcal{C}\{\gamma_{N+1}, \alpha_{N+1}\}$ or $s''_{N,N}+\mathcal{C}\{\alpha_{N+1}, \beta_{N+1}\}$.

Upon interchanging the superscripts $'$ and $''$ this reduces to case B_1.

Case B_3. σ_N is either $\mathcal{C}\{s'_{N,N}, s''_{N,N}\}+\alpha_{N+1}$ or $\mathcal{C}\{s'_{N,N}, s''_{N,N}\}+\beta_{N+1}$ or $\mathcal{C}\{s'_{N,N}, s''_{N,N}\}+\gamma_{N+1}$.

In this case we set $u(N+1)=u(N)$; $\zeta'_{n,N+1}=\zeta'_{n,N}$ and $\zeta''_{n,N+1}=\zeta''_{n,N}$ $(n=1, 2, \cdots, N)$; and $\zeta'_{N+1,N+1}=\zeta''_{N+1,N+1}$ equal respectively to α_{N+1} or β_{N+1} or γ_{N+1}.

Again, because of (11) and the definition of σ_N, (6) through (10) hold with $N+1$ replacing N.

Thus the proof of Proposition 2 is completed.

2.3. It is worth while to present the two following corollaries, which are of interest in themselves, although we do not need them to establish the main results.

COROLLARY 1. *Under the assumptions of Proposition* 1 *there exists a sequence* $(\zeta_n)_{n=1}^{\infty}$ *with* $\zeta_n \in \{\alpha_n, \beta_n, \gamma_n\}$ $(n=1, 2, \cdots)$ *such that*

$$\left|\sum_{\nu=1}^{n} \zeta_\nu\right| < 5^{1/2} \qquad (n = 1, 2, \cdots). \tag{13}$$

Proof. In the inductive construction used to establish Proposition 2 we had $u(N+1) \geqq u(N)$. Consequently either $u(N)\to\infty$ or $u(N)=u<\infty$ for all $N\geqq N_0$. In the first case $\zeta_n=\lim_{N\to\infty}\zeta'_{n,N}$ exists for all n. In the second case such a limit exists for $n\neq u$, while for $n=u$ and all sufficiently large N, $\zeta'_{u,N}$ and $\zeta''_{u,N}$ oscillate each over at most two of the numbers $\alpha_u, \beta_u, \gamma_u$; let ζ_u be one of these two numbers. In both cases the sequence $(\zeta_n)_{n=1}^{\infty}$ thus defined satisfies (13).

Specializing by assuming $\gamma_n=\beta_n$ for all n we have

COROLLARY 2. *Let* $\{\alpha_n, \beta_n\}$ *be a sequence of pairs of complex numbers such that every one of the segments* $\mathcal{C}\{\alpha_n, \beta_n\}$ *has length not exceeding* 2 *and contains the origin. Then there exists a sequence* $(\zeta_n)_{n=1}^{\infty}$ *with each* ζ_n *equal to either* α_n *or* β_n, *and such that*

$$\left|\sum_{\nu=1}^{n} \zeta_\nu\right| < 5^{1/2} \qquad (n = 1, 2, \cdots).$$

2.4. Using Proposition 1 we now proceed to prove

PROPOSITION 3. *Let $(\mathrm{Z}_n)_{n=1}^{\infty}$ be a sequence of uniformly bounded sets of complex numbers such that*

$$0 \in \mathcal{C}\mathrm{Z}_n \qquad (n = 1, 2, \cdots). \tag{14}$$

Then (Z_n) is a convergence factor sequence.

Proof. It is known [6, p. 155] that, because of (14), each Z_n contains three (not necessarily distinct) points $\alpha_n, \beta_n, \gamma_n$ such that $0 \in \mathcal{C}\{\alpha_n, \beta_n, \gamma_n\} \subset \mathcal{C}\mathrm{Z}_n$ $(n=1, 2, \cdots)$. Furthermore, since $\mathrm{Z}_n \subset \{|\zeta| \leqq M\}$ $(n=1, 2, \cdots)$ we have

$$\max (|\gamma_n - \beta_n|, |\beta_n - \alpha_n|, |\alpha_n - \gamma_n|) \leqq 2M \qquad (n = 1, 2, \cdots).$$

Let $(a_n)_{n=1}^{\infty}$ be a given sequence of numbers satisfying (1). We can determine an increasing sequence of integers $(N_k)_{k=1}^{\infty}$ such that $|a_n| \leqq 2^{-k}M^{-1}$ for $n > N_k$. Let

$$\alpha_n^{(k)} = 2^k \alpha_{N_k+n} a_{N_k+n}; \quad \beta_n^{(k)} = 2^k \beta_{N_k+n} a_{N_k+n}; \quad \gamma_n^{(k)} = 2^k \gamma_{N_k+n} a_{N_k+n}$$

$(n=1, 2, \cdots, N_{k+1}-N_k;\ k=1, 2, \cdots)$. Then

$$0 \in \mathcal{C}\{\alpha_n^{(k)}, \beta_n^{(k)}, \gamma_n^{(k)}\}$$

and

$$\max (|\gamma_n^{(k)} - \beta_n^{(k)}|, |\beta_n^{(k)} - \alpha_n^{(k)}|, |\alpha_n^{(k)} - \gamma_n^{(k)}|) \leqq 2.$$

By Proposition 1 there exist sequences $(\zeta_n^{(k)})_{n=1}^{N_{k+1}-N_k}$ with

$$\zeta_n^{(k)} \in \{\alpha_n^{(k)}, \beta_n^{(k)}, \gamma_n^{(k)}\}$$

such that

$$\left|\sum_{\nu=1}^{n} \zeta_\nu^{(k)}\right| < 5^{1/2} \qquad (n = 1, 2, \cdots, N_{k+1} - N_k;\ k = 1, 2, \cdots).$$

We now determine a sequence $(\zeta_n)_{n=1}^{\infty}$ with $\zeta_n \in \mathrm{Z}_n$ $(n=1, 2, \cdots)$ as follows: $\zeta_n = \alpha_n$ if $n \leqq N_1$ or if $a_n = 0$; while $\zeta_n = (1/2^k a_n)\, \zeta_{n-N_k}^{(k)}$ if $a_n \neq 0$ and $N_k < n \leqq N_{k+1}$ $(k=1, 2, \cdots)$.

Clearly $\zeta_n \in \{\alpha_n, \beta_n, \gamma_n\} \subset \mathrm{Z}_n$ $(n=1, 2, \cdots)$ and

$$\left|\sum_{n=N_k+1}^{N} \zeta_n a_n\right| < \frac{5^{1/2}}{2^k} \qquad (N = N_k+1, N_k + 2, \cdots, N_{k+1};\ k = 1, 2, \cdots).$$

Hence for any two integers p, q such that $q \geqq p > N_k$ we have

$$\left| \sum_{n=p}^{q} \zeta_n a_n \right| < 2 \frac{5^{1/2}}{2^k} + \sum_{j=1}^{\infty} \frac{5^{1/2}}{2^{k+j}} = \frac{3(5^{1/2})}{2^k}$$

which, by Cauchy's criterion, establishes the convergence of $\sum_{n=1}^{\infty} \zeta_n a_n$.

2.5. We can now prove Theorems 1 and 3. As shown in §1 it suffices to prove the latter.

Proof of Theorem 3.

Sufficiency. Let $z_n \in \mathcal{C}Z_n$ be such that $|z_n| < d_n + 2^{-n}$ $(n = 1, 2, \cdots)$ and put

$$H_n = -z_n + Z_n.$$

Then, if $\sum_{n=1}^{\infty} d_n$ converges, the sets H_n are uniformly bounded and $0 \in \mathcal{C}H_n$ $(n = 1, 2, \cdots)$. By Proposition 3, $(H_n)_{n=1}^{\infty}$ is a convergence factor sequence; thus, given any sequence $(a_n)_{n=1}^{\infty}$ satisfying (1) there exist numbers $\eta_n \in H_n$ $(n = 1, 2, \cdots)$ such that $\sum_{n=1}^{\infty} \eta_n a_n$ converges.

Upon setting $\zeta_n = \eta_n + z_n$ $(n = 1, 2, \cdots)$ we have $\zeta_n \in Z_n$ and, since $\sum_{n=1}^{\infty} |z_n| < \infty$, the series $\sum_{n=1}^{\infty} \zeta_n a_n$ converges as required.

Necessity. Assume that $\sum_{n=1}^{\infty} d_n = \infty$. We shall proceed to construct a sequence $(a_n)_{n=1}^{\infty}$ satisfying (1) such that every series $\sum_{n=1}^{\infty} \zeta_n a_n$ with $\zeta_n \in Z_n$ diverges.

Since $\sum d_n$ is divergent there exists a null-sequence $(u_n)_{n=1}^{\infty}$ of real non-negative numbers such that

$$\sum_{n=1}^{\infty} d_n u_n = \infty \tag{15}$$

and we may assume that $u_n = 0$ whenever $d_n = 0$.

Now let n run through all values for which $d_n \neq 0$ and denote by η_n the point of $\overline{\mathcal{C}Z_n}$ which is nearest to the origin, then

$$\operatorname{Re}\,(\zeta \exp\,(-i \arg \zeta_n)) \geqq |\eta_n| = d_n > 0 \tag{16}$$

for $\zeta \in \overline{\mathcal{C}Z_n}$, thus, a fortiori, for all $\zeta \in Z_n$.

The sequence $(a_n)_{n=1}^{\infty}$ defined by $a_n = u_n \exp\,(-i \arg \eta_n)$ whenever $d_n \neq 0$, and $a_n = 0$ otherwise, clearly satisfies (1) and, because of (15) and (16), we have, for any choice of $\zeta_n \in Z_n$ $(n = 1, 2, \cdots)$,

$$\sum_{n=1}^{\infty} \operatorname{Re}\,(\zeta_n a_n) \geqq \sum_{n=1}^{\infty} d_n u_n = \infty$$

which completes the proof.

3. **Sum Factors.** The principal results are Theorems 2 and 4, and, of course, only the latter need be proved.

Proof of Theorem 4.

3.1. *Sufficiency*. Let $(\mathrm{Z}_n)_{n=1}^{\infty}$ be a sequence of uniformly bounded sets, that is, there exists a finite, positive Δ for which

$$\mathrm{Z}_n \subset \{ |\zeta| < \Delta \} \qquad (n = 1, 2, \cdots), \tag{17}$$

and assume that there exists a positive number δ and a positive integer N_0 such that

$$\{ |\zeta| < \delta \} \subset \complement \mathrm{Z}_n \qquad (n = N_0 + 1, N_0 + 2, \cdots). \tag{18}$$

We must prove that given any sequence $(a_n)_{n=1}^{\infty}$ of complex numbers, satisfying (1) and (2), and any number s, there exists a sequence $(\zeta_n)_{n=1}^{\infty}$ with $\zeta_n \in \mathrm{Z}_n$ $(n=1, 2, \cdots)$ for which

$$\sum_{n=1}^{\infty} \zeta_n a_n = s \tag{19}$$

holds. This will be done by constructing a sequence (ζ_n) having the required properties.

Choose $\zeta_m \in \mathrm{Z}_m$ arbitrarily for $m=1, 2, \cdots, N_0$ and, having chosen $\zeta_m \in \mathrm{Z}_m$ for $m=1, 2, \cdots, n$ $(n \geqq N_0)$, put $s_n = \sum_{m=1}^{n} \zeta_m a_m$ and choose $\zeta_{n+1} \in \mathrm{Z}_{n+1}$ so that[3]

$$\operatorname{Re} (\zeta_{n+1} \exp (i[\arg a_{n+1} - \arg (s - s_n)]) > \delta/2. \tag{20}$$

This is clearly possible because of (18). We contend that the sequence thus constructed satisfies (19).

In fact, because of (1) and (2), given any $\epsilon > 0$ we can determine $N'(\epsilon)$ so large that

$$|a_n| < \delta\epsilon/(2\Delta^2) \qquad \text{for } n \geqq N'(\epsilon) \tag{21}$$

and then $N''(\epsilon) > N'(\epsilon)$ so large that

$$\sum_{n=N'(\epsilon)+1}^{N''(\epsilon)} |a_n| \geqq \frac{4}{\delta} |s - s_{N'(\epsilon)}|. \tag{22}$$

Our contention (19) will be established once we prove

$$|s_n - s| \leqq \epsilon \qquad \text{for } n \geqq N''(\epsilon). \tag{23}$$

Put $\rho_n = |s - s_n|$; then for $n \geqq N'(\epsilon)$ we have

$$\begin{aligned} \rho_{n+1}^2 &= |s - s_n - \zeta_{n+1} a_{n+1}|^2 \\ &= \rho_n^2 - 2\rho_n |a_{n+1}| \operatorname{Re} (\zeta_{n+1} \exp i[\arg a_{n+1} - \arg (s - s_n)]) \\ &\quad + |\zeta_{n+1}|^2 |a_{n+1}|^2. \end{aligned} \tag{24}$$

Applying (17) and (20), we obtain

(3) For the sake of brevity we make the convention $\arg 0 = 0$.

$$\rho_{n+1}^2 \leqq \rho_n^2 - \delta\rho_n |a_{n+1}| + \Delta^2 |a_{n+1}|^2. \tag{25}$$

If $\rho_n \leqq \epsilon$, then, using (21), we have

$$\rho_{n+1}^2 \leqq \left(\Delta |a_{n+1}| - \frac{\delta}{2\Delta}\rho_n\right)^2 + \left(1 - \frac{\delta^2}{4\Delta^2}\right)\rho_n^2$$
$$\leqq \left(\frac{\delta\epsilon}{2\Delta}\right)^2 + \left(1 - \frac{\delta^2}{4\Delta^2}\right)\epsilon^2 = \epsilon^2.$$

Thus if $\rho_m \leqq \epsilon$ for some $m \geqq N'(\epsilon)$ then $\rho_n \leqq \epsilon$ for all $n > m$.

If $\rho_n > \epsilon$ we have, from (21), (25), and $\delta \leqq \Delta$,

$$\rho_{n+1}^2 \leqq \left(\rho_n - \frac{\delta}{4}|a_{n+1}|\right)^2 - |a_{n+1}|\left(\frac{\delta\rho_n}{2} - \Delta^2 |a_{n+1}|\right) - \frac{\delta^2}{16}|a_{n+1}|^2$$
$$\leqq \left(\rho_n - \frac{\delta}{4}|a_{n+1}|\right)^2.$$

This, combined with the previous result, gives

$$\rho_{n+1} \leqq \max\left(\epsilon, \rho_n - \frac{\delta}{4}|a_{n+1}|\right) \qquad \text{for } n \geqq N'(\epsilon). \tag{26}$$

By (22) this last inequality implies (23), thus completing the sufficiency part of the proof.

3.2. *Necessity*. Let $(\mathrm{Z}_n)_{n=1}^{\infty}$ be a sequence of uniformly bounded sets. We assume that, no matter how small $\delta > 0$ is taken, there does not exist a value of N_0 for which (18) holds, and prove that $(\mathrm{Z}_n)_{n=1}^{\infty}$ is not a sum factor sequence.

In fact, by our assumption, there exists an increasing sequence $(n_k)_{k=1}^{\infty}$ of positive integers for which

$$\{|z| < 2^{-k}\} \not\subset \mathcal{C}\mathrm{Z}_{n_k} \qquad (k = 1, 2, \cdots).$$

Consequently there exist real numbers ϑ_k $(k = 1, 2, \cdots)$ such that

$$\mathrm{Re}\,(\zeta e^{i\vartheta_k}) < 2^{-k} \quad \text{for} \quad \zeta \in \mathrm{Z}_{n_k} \qquad (k = 1, 2, \cdots). \tag{27}$$

We now define a sequence $(a_n)_{n=1}^{\infty}$ of numbers as follows:

$$a_{n_k} = \frac{e^{i\vartheta_k}}{k} \ (k = 1, 2, \cdots) \text{ and } a_n = 0 \text{ otherwise.}$$

Clearly $(a_n)_{n=1}^{\infty}$ satisfies (1) and (2), while for any choice of $\zeta_n \in \mathrm{Z}_n$ we have, from (26),

$$\mathrm{Re}\left(\sum_{\nu=1}^{n} \zeta_\nu a_\nu\right) < \sum_{j=1}^{\infty} \frac{2^{-j}}{j} = \log 2 \qquad (n = 1, 2, \cdots).$$

Thus, for no choice of $\zeta_n \in Z_n$ can we have $\sum_{n=1}^{\infty} \zeta_n a_n = 1$, say, and thus (Z_n) is not a sum factor sequence.

4. Oscillation of series.

4.1. We first give some general results on the oscillation of series of complex numbers.

A complex number p is called a limit point of a series $\sum_{n=1}^{\infty} a_n$ if there exists an increasing sequence of integers $(n_k)_{k=1}^{\infty}$ such that $(\sum_{\nu=1}^{n_k} a_\nu)_{k=1}^{\infty}$ converges to p.

The set of limit points of a series is called the *derived set* of that series.

DEFINITION. *A set D is called an oscillation set if there exists a series $\sum_{n=1}^{\infty} a_n$ of complex numbers satisfying* (1), *whose derived set is D.*

Next we characterize oscillation sets by the following

PROPOSITION 4. *A set D is an oscillation set if and only if it is a continuum, or becomes one by adjoining the point at infinity.*

By a *continuum* we understand a nonempty, closed, connected set.

This proposition will follow directly from Proposition 5, which also serves to characterize oscillation sets, but gives more information on the behavior of unbounded subsequences of partial sums. Before stating Proposition 5, some preliminaries are necessary.

Besides the ordinary compactification of the open complex plane Π to a sphere by adjoining a point at infinity, we use another compactification of Π to a closed disc Π^*. This is done by adjoining to Π a circle at infinity whose points are (∞, θ) with θ real mod 2π. We set $|(\infty, \theta)| = \infty$ and $\arg(\infty, \theta) = \theta$ and define the topology in Π^* as follows: in the finite part of Π^* it coincides with that of Π, while the sets $1/\epsilon < |z| \leqq \infty$; $|\arg z - \theta| < \epsilon$ for all $\epsilon > 0$ constitute a fundamental system of neighborhoods of (∞, θ).

DEFINITION 4. *A set D^* in Π^* is called a Π^*-oscillation set, if there exists a series $\sum_{n=1}^{\infty} a_n$ of complex numbers satisfying* (1) *whose derived set* (*in the topology of* Π^*) *is D^*.*

The characterization of Π^*-oscillation sets is given by

PROPOSITION 5. *A set D^* in Π^* is a Π^*-oscillation set if and only if it is a continuum in Π^*.*

If D^* is the derived set in Π^* of a series, then the derived set D of the same series is $D^* \cap \Pi$; since[4] $\Pi^* - \Pi$ is connected, it follows that Proposition 5 implies Proposition 4.

---

[4] The minus sign is used in the set theoretic sense.

As remarked in 1.5 analogous results hold for vectors in Euclidean n-space. There is, however, one exception. Proposition 4 does not hold when $n=1$; Proposition 5 is still valid but since $\Pi^* - \Pi$, for $n=1$, consists of two points it is not connected and Proposition 4 does not follow. Thus the set $\{x < -1\} \cup \{x > 1\}$ which is an oscillation set for complex series is not an oscillation set for real series.

4.2. **Proof of Proposition 5.** The derived set L in Π^* of any series $\sum_{n=1}^{\infty} a_n$ is evidently closed and (since Π^* is compact) nonempty. Assume (1) and suppose that L is not connected. Then there exists a closed set K in Π^* which separates two points of L and which does not meet L. For every open set $K' \supset K$, since $a_n \to 0$, there exist infinitely many partial sums which assume values in K'. Since Π^*, and hence K, is compact, these partial sums must have a limit point in K, thus contradicting the assumption that L is not connected. Thus L is a continuum.

Conversely suppose a set D^* is a continuum in Π^*. We proceed to construct a series $\sum_{n=1}^{\infty} a_n$ satisfying (1) whose derived set in Π^* is D^*.

Let Θ_n $(n=1, 2, \cdots)$ be the set of real numbers ϑ for which there exist z (in Π^*) satisfying

$$z \in D^*, \qquad n^{1/2} \leqq |z| \leqq \infty, \qquad \arg z = \vartheta,$$

and put

$$D_n = (D^* \cap \{|z| < n^{1/2}\}) \cup \{|z| = n^{1/2}, \arg z \in \Theta_n\}.$$

Then D_n is a continuous image of D^* by the mapping ϕ_n defined through $\phi_n(z) = z$ for $|z| < n^{1/2}$; $\phi_n(z) = n^{1/2} \exp(i \arg z)$ for $|z| \geqq n^{1/2}$; hence D_n is a continuum.

Let z^* be any point of D^* and define a sequence $(z_n)_{n=0}^{\infty}$ by $z_0 = 0$; $z_n = \phi_n(z^*)$ $(n \geqq 1)$.

For each $n = 1, 2, \cdots$ construct a finite sequence of points $s_{n,1}, s_{n,2}, \cdots, s_{n,v_n}$ satisfying

$$s_{n,1} = z_{n-1}; \quad s_{n,v_n} = z_n; \quad |s_{n,j+1} - s_{n,j}| < 1/n^{1/2} \qquad (j = 1, 2, \cdots, v_n - 1),$$

$$s_{n,j} \in D_n + \{|z| < 1/n^{1/2}\} \qquad (j = 1, \cdots, v_n),$$

$$D_n \subset \bigcup_{j=1}^{v_n} \{|z - s_{n,j}| < 1/n^{1/2}\}.$$

All the above conditions can be satisfied because each D_n is a bounded continuum, $z_n \in D_n$, and the distance of z_{n-1} from D_n is not greater than $|z_n - z_{n-1}| \leqq n^{1/2} - (n-1)^{1/2} \leqq 1/n^{1/2}$.

Put $m_k = \sum_{j=1}^{k} v_j$, $s_0 = 0$ and $s_n = s_{k,r}$ for $m_k < n = m_k + r \leqq m_{k+1}$. Then the sequence $(a_n)_{n=1}^{\infty}$ with $a_n = s_{n+1} - s_n$ is a null sequence whose partial sums are s_n and it is easily verified that D^* is the derived set of the series $\sum_{n=1}^{\infty} a_n$. q.e.d.

4.3. Let a sum factor sequence $(Z_n)_{n=1}^{\infty}$ of uniformly bounded sets and a series $\sum a_n$ satisfying (1) be given and consider the derived sets (in Π or in Π^*) of the series $\sum_{n=1}^{\infty} \zeta_n a_n$ for all possible choices of $\zeta_n \in Z_n$. We consider the problem of characterizing all these derived sets. Since the sets Z_n are uniformly bounded we clearly have $\zeta_n a_n \to 0$ as $n \to \infty$ and, if $\sum_{n=1}^{\infty} |a_n| < \infty$, also $\sum_{n=1}^{\infty} |\zeta_n a_n| < \infty$. Thus if (2) is not satisfied the only derived sets of

$\sum_{n=1}^{\infty} \zeta_n a_n$ consist of single points, and the set of points each of which is the limit of the series forms a bounded subset of the plane, depending on $(a_n)_{n=1}^{\infty}$ and $(\mathrm{Z}_n)_{n=1}^{\infty}$, the structure of which may be quite complicated. On the other hand we have already seen (Theorem 4) that if $(a_n)_{n=1}^{\infty}$ satisfies (2) then all sets consisting of a single point are certainly the derived sets of some series $\sum \zeta_n a_n$. The following theorem shows moreover that if the assumptions of Theorem 4 are satisfied then the ζ_n may be chosen so that any prescribed oscillation set be the derived set of $\sum \zeta_n a_n$.

THEOREM 5. *Let $(\mathrm{Z}_n)_{n=1}^{\infty}$ be a sum factor sequence of uniformly bounded sets. Then, given any sequence $(a_n)_{n=1}^{\infty}$ of complex numbers satisfying* (1) *and* (2) *and any continuum D^* in Π^*, there exists a sequence $(\zeta_n)_{n=1}^{\infty}$ with $\zeta_n \in \mathrm{Z}_n$ $(n=1, 2, \cdots)$ such that the derived set (in Π^*) of $\sum_{n=1}^{\infty} \zeta_n a_n$ coincides with D^*.*

NOTE. Since Z_n are uniformly bounded, it follows that $\lim_{n=\infty} \zeta_n a_n = 0$ for any choice of $\zeta_n \in \mathrm{Z}_n$. Therefore, by Proposition 5, this theorem completely characterizes the derived sets of all series of the form $\sum_{n=1}^{\infty} \zeta_n a_n$.

Proof. By Proposition 5 we can choose a null sequence $(a_n')_{n=1}^{\infty}$ such that the derived set in Π^* of $\sum_{n=1}^{\infty} a_n'$ coincides with D^*. By Theorem 4 there exists an integer N_0 and two positive numbers δ, Δ such that

$$\{\,|\zeta| \leqq \delta\} \subset \mathcal{C}\mathrm{Z}_n \subset \{\,|\zeta| \leqq \Delta\} \qquad (n = N_0 + 1, N_0 + 2, \cdots). \tag{28}$$

Let N_0 be the first element of an increasing sequence of integers $(N_k)_{k=0}^{\infty}$ (which we shall define later), choose $\zeta_m \in \mathrm{Z}_m$ arbitrarily for $m=1, 2, \cdots, N_0$, and, having chosen $\zeta_m \in \mathrm{Z}_m$ for $m=1, 2, \cdots, n$ $(N_{k-1} \leqq n < N_k)$, put

$$s_n = \sum_{\nu=1}^{n} \zeta_\nu a_\nu; \qquad s_k' = \sum_{j=1}^{k} a_j' \quad \text{and choose} \quad \zeta_{n+1} \in \mathrm{Z}_{n+1}$$

so that

$$\operatorname{Re}\,(\zeta_{n+1} \exp\,(i[\arg a_{n+1} - \arg\,(s_k' - s_n)])) > \delta/2 \qquad (N_{k-1} \leqq n < N_k,\ k = 1, 2, \cdots). \tag{29}$$

To complete the proof it is necessary only to show that if $(N_k)_{k=1}^{\infty}$ are properly chosen, then $(\zeta_n)_{n=1}^{\infty}$ has the desired property. The choice of (N_k) proceeds as follows: Having chosen N_j for $j=0, 1, 2, \cdots, k-1$, we choose an integer $M_k > N_{k-1}$ such that

$$a_n < \frac{\delta\,|a_{k+1}'|}{2\Delta^2} \qquad \text{for } n \geqq M_k \tag{30}$$

and then choose $N_k > M_k$ so large that

$$\sum_{n=M_k+1}^{N_k} |a_n| \geqq \frac{4}{\delta}\,|s_k' - s_{M_k}|. \tag{31}$$

On replacing in (29), (30), (31) s_k' by s, $|a_{k+1}'|$ by ϵ, M_k by $N'(\epsilon)$, and N_k by $N''(\epsilon)$, these inequalities reduce to (20), (21), (22) respectively.

From (23) we have $|s_{N_k}-s_k'|\leqq a_{k+1}'$, and hence $|s_{N_k}-s_{k+1}'|\leqq 2|a_{k+1}'|$ $(k=1, 2, \cdots)$; while from (26) we get

$$|s_n - s_{k+1}'| \leqq \max(2|a_{k+1}'|, |a_{k+2}'|) \qquad (N_k \leqq n \leqq N_{k+1},\ k = 1, 2, \cdots).$$

This last inequality and $a_n' \to 0$ clearly imply that $\sum_n \zeta_n a_n$ and $\sum_k a_k'$ have the same derived sets, thus establishing the theorem.

5. **The unbounded case.**

5.1. The set $\mathbf{Z}$ in Theorems 1 and 2 was assumed to be bounded, similarly the factor sequences were assumed uniformly bounded in Theorems 3 and 4. We now proceed to study what happens if the boundedness restrictions are omitted. It turns out that no essentially new convergence factor sets or sequences are added by passing to the unbounded case. The same holds for sum factor sets but for sum factor sequences the situation is considerably more complicated and we do not obtain a characterization of such sequences.

5.2. For convergence factors we have the following two theorems.

THEOREM 6. *A set $\mathbf{Z}$ of complex numbers is a convergence factor set if and only if it contains a bounded subset which is a convergence factor set.*

That is (by Theorem 1), if and only if there exists a bounded set $\mathbf{Z}'\subset\mathbf{Z}$ such that $0\in\overline{\mathcal{C}\mathbf{Z}'}$.

THEOREM 7. *A sequence $(\mathbf{Z}_n)_{n=1}^{\infty}$ of sets of complex numbers is a convergence factor sequence if and only if there exists a sequence of uniformly bounded subsets $\mathbf{Z}_n'\subset\mathbf{Z}_n$ $(n=1, 2, \cdots)$ which is itself a convergence factor sequence.*

That is (by Theorem 3), if and only if there exists a sequence of subsets $(\mathbf{Z}_n')_{n=1}^{\infty}$ satisfying $\sum_{n=1}^{\infty} d_n' < \infty$ where d_n' denotes the distance of the origin from $\mathcal{C}\mathbf{Z}_n'$.

Proof. It is enough to establish Theorem 7 and, evidently, only the necessity has to be proved.

We therefore assume that $(\mathbf{Z}_n)_{n=1}^{\infty}$ does not satisfy the conditions of the theorem and show that it cannot be a convergence factor sequence.

Put

$$\text{(32)} \qquad \mathbf{Z}_n^{(k)} = \mathbf{Z}_n \cap \{|\zeta| \leqq k\} \qquad (k = 1, 2, \cdots; n = 1, 2, \cdots),$$

and denote by $d_n^{(k)}$ the distance of the origin from $\mathcal{C}\mathbf{Z}_n^{(k)}$. By assumption $\sum_{n=1}^{\infty} d_n^{(k)} = \infty$ for every k, therefore there exists an increasing sequence $(N_k)_{k=1}^{\infty}$ of integers (with $N_0=0$) satisfying

$$\sum_{n=N_k+1}^{N_{k+1}} d_n^{(k)} \geqq 1 \qquad (k = 1, 2, \cdots).$$

Now put for $k=1, 2, \cdots$,

$$\mathrm{H}_n = \mathrm{Z}_n^{(k)}, \qquad \delta_n = d_n^{(k)} \qquad (N_k < n \leqq N_{k+1}).$$

Clearly $\sum \delta_n$ is divergent, therefore there exists a null sequence $(u_n)_{n=1}^{\infty}$ of real non-negative numbers such that

$$\sum_{n=1}^{\infty} \delta_n u_n = \infty, \tag{33}$$

and we may assume that $u_n=0$ whenever $\delta_n=0$. Next define a sequence of complex numbers $(a_n)_{n=1}^{\infty}$ through:

Whenever $\delta_n=0$ then $a_n=0$ and when $\delta_n\neq 0$ and $N_k<n\leqq N_{k+1}$ then $a_n=\max(u_n, 1/k)\exp(-i \arg \eta_n)$ where η_n denotes the point of $\mathcal{C}\overline{\mathrm{H}}_n$ which is nearest to the origin.

The sequence thus constructed satisfies (1) and (2); we claim moreover that for no choice of $\zeta_n\in\mathrm{Z}_n$ will $\sum\zeta_n a_n$ be convergent. Indeed, if $\zeta_n\in\mathrm{Z}_n$ but $\zeta_n\notin\mathrm{H}_n$ then, by (32), $|\zeta_n|\geqq k$ and thus $|\zeta_n a_n|\geqq 1$ unless $a_n=0$. Therefore, $\sum\zeta_n a_n$ cannot converge unless $a_n\neq 0$ implies $\zeta_n\in\mathrm{H}_n$ for all n sufficiently large. But if $\zeta_n\in\mathrm{H}_n$ and $a_n\neq 0$ then we have, cf. (16), Re $(\zeta_n a_n)\geqq\delta_n u_n$ and thus $\sum\zeta_n a_n$ is divergent by (33). q.e.d.

5.3. Theorem 2 remains valid if the boundedness condition is dropped.

THEOREM 8. *A set* Z *(bounded or not) of complex numbers is a sum factor set if and only if* 0 *is an interior point of* $\mathcal{C}\mathrm{Z}$.

Indeed, if 0 is an interior point then there exists a subset Z′ of three or four points of Z such that 0 is an interior point of $\mathcal{C}\mathrm{Z}'$. Thus Z is a sum factor set by Theorem 2. (This shows incidentally that Theorem 8 can be formulated in the same manner as Theorems 6 and 7.)

On the other hand if 0 is not an interior point then there exists a real ϑ with Re $(\zeta e^{i\vartheta})\leqq 0$ for all $\zeta\in\mathrm{Z}$. Thus taking $a_n=e^{i\vartheta}/n$ we see that it is impossible to choose $\zeta_n\in\mathrm{Z}$ so that $\sum\zeta_n a_n$ will converge to a sum with a positive real part.

5.4. The generalization of Theorem 4 to sequences of unbounded or not uniformly bounded sets is not as simple. Any sequence (Z_n) is clearly a sum factor sequence if there exists a sequence of uniformly bounded subsets satisfying the conditions of Theorem 4, that is, which is itself a sum factor sequence. These sufficient conditions are, however, far from being necessary. Thus

$$\mathrm{Z}_n = \{\zeta = 0\} \cup \{|\zeta| \geqq n\} \qquad (n = 1, 2, \cdots)$$

is easily seen to be a sum factor sequence though it fails to satisfy the above conditions. Another example, with each Z_n individually bounded, is furnished by

$$Z_n = (\{\mathrm{Re}(\zeta) < 0\} \cap \{|\zeta| < n\}) \cup \{|\zeta| = n\}.$$

The complications of the problem are considerable even if we confine ourselves to real numbers. A sequence $(\Xi_n)_{n=1}^{\infty}$ of sets of real numbers is said to be a sum factor sequence for real series if whenever a real number s and a series of real terms $\sum a_n$ satisfying (1) and (2) are given, there exist $\xi_n \in \Xi_n$ $(n=1, 2, \cdots)$ such that $\sum_{n=1}^{\infty} \xi_n a_n = s$.

From Theorem 3 (or directly by a similar proof) we have:

A sequence of *uniformly bounded* non-empty sets (Ξ_n) is a sum factor sequence for real series if and only if there exist a positive δ and an integer N_0 such that

$$\{-\delta < \xi < \delta\} \subset \mathcal{C}\Xi_n \qquad (n > N_0). \tag{34}$$

If the boundedness of the sets is retained but not the uniformity, the preceding result has to be modified as follows:

A sequence of *bounded* sets (Ξ_n) is a sum factor sequence for real series if and only if there exist positive δ, Δ, and an integer N_0 satisfying (34),

$$\{|\xi| \leqq \delta\} \cap \mathcal{C}\Xi_n \text{ not void} \qquad (n > N_0), \tag{35}$$

and

$$\sum_{n=1}^{\infty} d_n' < \infty$$

where d_n' denotes the distance of $\xi=0$ from $\mathcal{C}(\Xi_n \cap \{|\xi| \leqq \Delta\})$.

The significance of (34) is clear, also that of the last condition—it asserts (cf. Theorem 7) that (Ξ_n) is a convergence factor sequence for real series, but (35) is a condition of a new type. If even the boundedness is dropped then the characterization of the above sum sequences is again different. Thus (35) is no longer necessary as is shown by

$$\Xi_n = \{\xi = 0\} \cup \{\xi < -n\} \cup \{\xi > n\}, \qquad (n = 1, 2, \cdots).$$

6. **An application.**

6.1. To illustrate the way in which the above results may be used we give the following theorem.

THEOREM 9. *Let $(f_n(x))_{n=1}^{\infty}$ be a sequence of nonincreasing positive functions defined for all real x and satisfying, moreover,*

$$\frac{f_{n+1}(y)}{f_{n+1}(x)} \leqq \frac{f_n(y)}{f_n(x)} \quad \textit{whenever} \quad -\infty < x < y < \infty \qquad (n = 1, 2, \cdots). \tag{36}$$

Let $(c_n)_{n=1}^{\infty}$ be a sequence of complex numbers and $(\mathrm{Z}_n)_{n=1}^{\infty}$ a sequence of sets satisfying the conditions of Theorem 7. Then it is possible to choose $\zeta_n \in \mathrm{Z}_n$ $(n=1, 2, \cdots)$ so that $\sum_{n=1}^{\infty} \zeta_n c_n f_n(x)$ be convergent for $x>\alpha$ where α is the *infimum of all x for which*

$$\lim_{n=\infty} c_n f_n(x) = 0.$$

Proof. Let $(b_n)_{n=1}^{\infty}$ be any sequence of complex numbers, then an immediate application of Abel's lemma to (36) gives

$$\sum_{\nu=p}^{q} b_\nu f_\nu(y) \leqq \frac{f_p(y)}{f_p(x)} \max_{p\leqq m\leqq q} \left| \sum_{\nu=p}^{m} b_\nu f_\nu(x) \right|. \tag{37}$$

Since the $f_n(x)$ are monotone nonincreasing functions this shows that whenever a series of the form $\sum b_n f_n(x)$ converges for $x=x_0$ it does so also for $x>x_0$.

We may assume $\alpha<+\infty$ since the contention of the theorem becomes vacuous otherwise. Let $(x_m)_{m=1}^{\infty}$ be a decreasing sequence of (finite) real numbers satisfying $x_m\to\alpha$ as $m\to\infty$. Clearly $(c_n f_n(x_m))_{n=1}^{\infty}$ are null sequences. By Theorem 7 we can find for every m numbers $\zeta_n^{(m)}\in Z_n$ $(n=1, 2, \cdots)$ so that the series $\sum_n \zeta_n^{(m)} c_n f_n(x_m)$ is convergent(5). There exists therefore a sequence $(N_m)_{m=1}^{\infty}$ of increasing integers such that

$$\left| \sum_{n=N_m}^{q} \zeta_n^{(m)} c_n f_n(x_m) \right| < 2^{-m} \qquad (m = 1, 2, \cdots) \tag{38}$$

for every $q\geqq N_m$. Because of (37) this inequality holds a fortiori also if x_m is replaced by any $x>x_m$.

We claim that the sequence $(\zeta_n)_{n=1}^{\infty}$ defined by

$$\zeta_n = \zeta_n^{(m)} \qquad \text{for } N_m \leqq n < N_{m+1} \ (m = 1, 2, \cdots),$$

and ζ_n arbitrary (say equal to $\zeta_n^{(1)}$) for $n<N_1$, satisfies the requirement of the theorem. Indeed, if $x>\alpha$ then $z>x_m$ for all $m>m_0$, and hence we have by (38) for $q>p$ and $N_m\leqq p<N_{m+1}$

$$\left| \sum_{n=p}^{q} \zeta_n c_n f_n(x) \right| < 2\cdot 2^{-m} + 2^{-m-1} + 2^{-m-2} + \cdots = 3\cdot 2^{-m}.$$

Thus $\sum \zeta_n c_n f_n(x)$ converges for $x>\alpha$ as required.

6.2. If $(\lambda_n)_{n=1}^{\infty}$ is an increasing sequence of real numbers, the functions $f_n(x)=e^{-\lambda_n x}$ satisfy the conditions of Theorem 9. Therefore if (Z_n) is any convergence factor sequence it is possible to choose $\zeta_n\in Z_n$ so that the abscissa of convergence of the Dirichlet series $\sum_{n=1}^{\infty} \zeta_n c_n e^{-\lambda_n z}$ coincide with the abscissa of the vanishing of the general term, namely with $\alpha=\limsup_{n=\infty} \log |c_n|/\lambda_n$. The special case $Z_n=\{-1, +1\}$ for all n is contained in [1]; it is used there to obtain the following result:

(5) If α is finite and $(c_n f_n(\alpha))_{n=1}^{\infty}$ is itself a null sequence, the whole diagonal procedure is superfluous.

Let γ denote the abscissa of absolute convergence of $\sum c_n e^{-\lambda_n z}$, $(0<\lambda_n \uparrow \infty)$. Then given any number A satisfying $\alpha \leqq A \leqq \gamma$ it is possible to choose the $\pm$ signs so that $\sum_{n=1}^{\infty} \pm c_n e^{-\lambda_n z}$ converges for $\mathrm{Re}(z)>A$ and represents a function having the line Re $(z)=A$ as a natural boundary.

The proof of this result for $\beta \leqq A<\gamma$ where β denotes the abscissa of convergence of $\sum c_n e^{-\lambda_n z}$ has been sketched in [2]. The extension to $\alpha \leqq A \leqq \gamma$ follows immediately from Theorem 9.

7. **Remark.** It is somewhat easier to prove Propositions 1 and 2 if the constant $5^{1/2}$ in (5) and (9) is replaced by a larger one. Though any constant would do for the applications we preferred to prove it with $5^{1/2}$ since this is quite likely the best possible one. However we cannot prove this. All we can show is that no constant smaller than $(2^2+2^{-4})^{1/2}=(4.0625)^{1/2}$ will do; for reasons of brevity the proof of this is omitted.

In the special case of Corollary 2 for which $\alpha_n+\beta_n=0$ $(n=1, 2, \cdots)$ it is known [3] that $5^{1/2}$ may be replaced by $3^{1/2}$ and Dr. Ch. Hanani and Dr. Th. Motzkin have shown that $3^{1/2}$ is the best possible constant in this case.

For vectors in Euclidean spaces of higher dimensions the exact value of the constant is not known even in the special case just mentioned.

Bibliography

1. A. Dvoretzky, *Studies on general Dirichlet series*, Jerusalem, 1941. (Hebrew summary and English abstract of a thesis.)

2. ———, *Sur les changements de signe des coefficients des séries de Dirichlet*, C. R. Acad. Sci. Paris vol. 221 (1945) pp. 687–689.

3. A. Dvoretzky and Ch. Hanani, *Sur les changements des signes des termes d'une série a termes complexes*, C. R. Acad. Sci. Paris vol. 225 (1947) pp. 516–518.

4. H. Hornich, *Über Reihen deren Glieder gegen null konvergieren*, Monatshefte für Mathematik und Physik vol. 45 (1937) pp. 432–434.

5. ———, *Zu einer geometrischen Theorie der Reihen*, Monatshefte für Mathematik und Physik vol. 46 (1938) pp. 266–276.

6. E. Steinitz, *Bedingt konvergente Reihen und konvexe Systeme* I, J. Reine Angew. Math. vol. 143 (1913) pp. 128–175.

Princeton University,
Princeton, N. J.
Hebrew University,
Jerusalem, Israel.

Completely Integrable Almost Complex Manifolds

Eugenio Calabi and Donald C. Spencer

Bull. Amer. Math. Soc. 57, 254–255 (1951)

An almost complex manifold is defined as a real, differentiable manifold with a differentiable transformation H on its covariant vector bundle satisfying $H^2 = -1$. Such a manifold is always even-dimensional. An almost complex manifold is called *completely integrable* if there exists in a neighborhood of each of its points an allowable coordinate system $(x^1, \cdots, x^{2k})$, satisfying (1) $Hd(x^\alpha + (-1)^{1/2}x^{k+\alpha}) = (-1)^{1/2}d(x^\alpha + (-1)^{1/2}x^{k+\alpha})$ $(1 \leqq \alpha \leqq k)$. Any two systems of complex-valued coordinates, $z^\alpha = x^\alpha + (-1)^{1/2}x^{k+\alpha}$ and $z'^\alpha = x'^\alpha + (-1)^{1/2}x'^{k+\alpha}$, associated with two systems satisfying (1) in a common domain, are complex analytically related, so that every completely integrable almost complex manifold corresponds to a complex manifold in a natural way. In an almost complex manifold one defines a tensor T^l_{ij} called the *pseudo-conformal curvature tensor*, from the tensor h^j_i representing the transformation H by $T^l_{ij} = (1/4)(\delta^{ab}_{ij} - \delta^{rs}_{ij}\, h^a_r h^b_s)\partial h^l_b/\partial x^a$, which leads to the main theorem. *An almost complex manifold is completely integrable if and only if its pseudo-conformal curvature vanishes identically.* The necessity is proved by a simple computation of exterior differential forms; the sufficiency is proved by constructing locally a Riemannian metric satisfying Kähler's condition, from which one obtains the complex analytic coordinates by a variational method. (Received March 7, 1951.)

Metric Riemann Surfaces[1]

Eugenio Calabi

Contributions to the Theory of Riemann Surfaces, 77–85,
Annals of Mathematics Studies, No. 30,
Princeton University Press (1953)

1. INTRODUCTION

The complex analytic structure of a Riemann surface is introduced in F. Klein's book [4] by considering the isothermal parameters in a differentiable, orientable surface in Euclidean 3-space. This method of course can be applied more generally to any abstract 2-dimensional, orientable Riemannian manifold. Admittedly this purely auxiliary metric structure on the Riemann surface is somewhat artificial, as evidenced by the fact that, beyond the proof of existence of local isothermal parameters, the function theory on the Riemann surface on one hand and the Riemannian geometry of the surface on the other seem to come essentially to a parting of the ways (it is significant that both theories should bear the same man's name). However, the rapid and fruitful development of the theory of Kähler manifolds in the last twenty years demonstrates how much can be obtained from the natural interconnection between the two structures.

The present note is a specialization to the 2-dimensional case, with additional results that are characteristic of it, of the author's doctoral thesis [3], which deals precisely with problems of isometric, complex analytic imbedding of complex manifolds. The purpose is to characterize all metrics that can be induced on a Riemann surface $\mathfrak{M}$ by a regular analytic map into the unitary N-dimensional space C^N $(N \leq \infty)$. If $x_1, \ldots, x_\sigma, \ldots$ $(1 \leq \sigma \leq N)$ are unitary coordinates in C^N such that the line element is $ds^2 = \Sigma_{\sigma=1}^{N} |d\, x_\sigma|^2$, and if z is a local uniformizer of $\mathfrak{M}$, this is equivalent to considering sets of holomorphic functions $x_\sigma = f_\sigma(z)$, whose first derivatives vanish nowhere simultaneously. In case $N = \infty$ we must also assume that the map of $\mathfrak{M}$ into C^N is continuous with respect to the Hilbert space topology of C^∞, for it has not been possible to prove it from complex analyticity alone.[2] The continuity of the map implies that the real-valued function $\Phi(z,\bar{z}) = \Sigma_{\sigma=1}^{N} |f(z)|^2$ is bounded in the neighborhood of any point; from this one proves that $\Phi(z,\bar{z})$ is real analytic using Lebesgue's theorem on dominated convergence and Weierstrass's theorem on uniform limits of holomorphic functions. Thus one sees that the first fundamental form induced on $\mathfrak{M}$, given by

$$ds^2 = \Sigma_{\sigma=1}^{N} |dx_\sigma|^2 = g(z,\bar{z})\, |dz|^2 \tag{1}$$

DOI: 10.1515/9781400828371-009

where

$$g(z,\bar{z}) = \Sigma_{\sigma=1}^{N} |f'_{\sigma}(z)|^2 = \frac{\partial^2 \Phi(z,\bar{z})}{\partial z\, \partial \bar{z}}$$

is a real analytic, positive density of weight 1. In this paper we not only characterize all metrics $g|dz|^2$ that can be thus obtained, but indicate also, in the course of the proofs, how one can obtain all possible generating functions $f_{\sigma}(z)$, if the necessary conditions are satisfied.

One can similarly characterize the metrics induced on $\mathfrak{M}$ by mapping it into the Fubini-Study spaces (cfr. [2], [3]); the latter spaces are especially interesting in the elliptic case, for they are the only ones of this type into which one can imbed a compact Riemann surface. However, this part will be omitted, as the statements and proofs do not present any new essential difficulties.

2. UNITARY GEOMETRY OF COMPLEX CURVES

We shall understand by a unitary coordinate transformation in C^N a substitution of the type $x'_{\sigma} = \Sigma_{\tau=1}^{N} a_{\sigma\tau} x_{\tau} + b_{\sigma}$, where $a_{\sigma\tau}$ is a unitary $N \times N$ matrix; by a unitary motion we mean a point transformation in C^N that carries a coordinate frame into one similarly related to the stationary frame. The fundamental form, invariant under the group of motions is

$$ds^2 = \Sigma_{\sigma=1}^{N} |dx_{\sigma}|^2.$$

If a region $\mathcal{R}$ of a Riemann surface $\mathfrak{M}$ admitting a local uniformizer z is mapped regularly and analytically in C^N by

(2) $$x_{\sigma} = f_{\sigma}(z) = \Sigma_{\mu=0} a_{\sigma,\mu} z^{\mu},$$

and p is any point in $\mathcal{R}$, one can replace $z(q)$ in (2) by $z(q) - z(p)$; if $\mathcal{R}$ does not lie in any proper linear surface of C^N, there is a unique unitary transformation of coordinates into the <u>canonical coordinates</u> of C^N with respect to p and the uniformizer z; they are characterized as follows: $x_{\sigma}(z(p)) = 0$, hence $a_{\sigma,0} = 0$ for all σ and if $a_{\sigma,\ell_{\sigma}}$ is the first non-vanishing coefficient in the power series of (2), then $a_{\sigma,\ell_{\sigma}}$ is real and positive and $\ell_{\sigma} \geq \ell_{\sigma-1} + 1 \geq \sigma$. Since the mapping is regular, we have $a_{1,1} = \sqrt{g(0,0)} > 0$, i.e., $\ell_1 = 1$. The integers ℓ_{σ} are independent of the local uniformizer and so are the constants $a_{\sigma,\ell_{\sigma}}(a_{1,1})^{-\ell_{\sigma}}$. The canonical coordinates with respect to a given point p are uniquely determined regardless of the local uniformizer up to a substitution from the toroidal group,

$$x'_{\sigma} = e^{i\theta_{\sigma}} x_{\sigma}.$$

The map of $\mathcal{R}$ into the right coset space of unitary transformations modulo the toroidal subgroup is analytic. The point p is called <u>ordinary</u> if each

$\ell_\sigma = \sigma$; otherwise it is called a point of hyper-osculation; the latter points form at most a countable set in $\mathcal{R}$. For a fixed uniformizer z in $\mathcal{R}$ and variable base point p ranging over the ordinary points alone, the canonical coordinates in C^N with respect to p and z depend on p continuously.

We now define formally the higher curvatures $\rho_\sigma (1 \leq \sigma < N)$ of $\mathcal{R}$ at any point p, in complete analogy with the case of real curves in Euclidean space, in terms of the coefficients of (2), where x_σ are the canonical coordinates with respect to p, by

$$\rho_\sigma(p) = K \frac{a_{\sigma+1,\ell_{\sigma+1}}}{a_{1,1} a_{\sigma,\ell_\sigma}},$$

where

$$K = \left(1 + \frac{1}{\ell_\sigma}\right) \prod_{\mu=1}^{\sigma-1} \left(1 + \frac{1}{\ell_\sigma - \ell_\mu}\right).$$

The point p is ordinary if and only if $\rho_\sigma(p) \neq 0$ $(\sigma = 1,2,\ldots,N-1)$.

The curvatures $\rho_\sigma(p)$ are continuous, real scalars, i.e., they are independent of the local uniformizers; their squares are indeed real analytic functions. The totality of the functions ρ_σ together with the metric $g(z,\bar{z})$ determine the map of $\mathcal{R}$ into C^N up to a unitary motion. The essential property that shows the contrast between real curve theory and complex curve theory, however, is that, while in the case of real curves the curvatures ρ_σ are arbitrary differentiable functions, here they are all dependent on the metric g, as we shall prove.

3. CONDITIONS FOR ISOMETRIC IMBEDDING

We have already indicated that the metric induced on $\mathfrak{M}$ by a regular analytic imbedding in C^N is necessarily real analytic. There are only two other conditions imposed besides the one of analyticity; the first, a purely local one, is developed in the following theorems; the second one is a global one that insures the single-valuedness of the mapping functions.

THEOREM 1. Let $\mathfrak{M}$ be a metric Riemann surface. For any point p_0 of $\mathfrak{M}$ there exists a neighborhood $\mathcal{R}$ of p_0 that can be isometrically and complex analytically imbedded in the Hilbert space C^∞ if and only if the metric is, first of all, real analytic and, secondly, if z is a local uniformizer on $\mathcal{R}$ at p, and (1) is the metric, then the real functions $G_\nu(z,\bar{z})$ $(\nu = 0,1,2,\ldots)$ are all non-negative in $\mathcal{R}$, where

$$G_0(z,\bar z) = 1; \quad G_1(z,\bar z) = g(z,\bar z)$$

(3)
$$G_{\nu+1}(z,\bar z) = \frac{\left(G_\nu(z,\bar z)\right)^2}{G_{\nu-1}(z,\bar z)} \quad \frac{\partial^2}{\partial z\, \partial \bar z} \log G_\nu(z,\bar z) \qquad (\nu \geq 1).$$

PROOF. We assume first that a neighborhood $\mathcal{R}$ of p_o defined by $|z| < \rho$ $(z(p_o) = 0)$, can be isometrically imbedded in C^{N^o} by (2). Since the metric $g(z,\bar z)$ is real analytic, we can write

$$g(z,\bar z) = \Sigma_{\mu,\nu=1}^{\infty} A_{\mu\nu}\, z^{\mu-1}\, \bar z^{\,\nu-1}\,,$$

where

(4)
$$A_{\mu\nu} = \mu\nu \Sigma_{\sigma=1}^{\infty} a_{\sigma,\mu}\, \bar a_{\sigma,\nu} = \frac{1}{(\mu-1)!(\nu-1)!} \; \frac{\partial^{\mu+\nu-2}\, g(z,\bar z)}{\partial z^{\mu-1} \partial \bar z^{\,\nu-1}}\;.$$

The second member of (4) clearly indicates that the infinite matrix $A_{\mu\nu}$ is hermitian and positive semi-definite; on denoting by $g_{\mu,\nu}$ the derivative

$$\frac{\partial^{\mu+\nu} g(z,\bar z)}{\partial z^\mu \partial \bar z^{\,\nu}}\,,$$

this implies at least that at the point p_o the minor determinants of rank μ

$$\begin{vmatrix} g(p_o) & \cdots & g_{0,\mu-1}(p_o) \\ g_{1,0}(p_o) & \cdots & g_{1,\mu-1}(p_o) \\ \cdots & \cdots & \cdots \\ g_{\mu-1,0}(p_o) & \cdots & g_{\mu-1,\mu-1}(p_o) \end{vmatrix} = \prod_{\nu=1}^{\mu-1} (\nu!)^2 \cdot \begin{vmatrix} A_{11} & \cdots & A_{1\mu} \\ \vdots & & \vdots \\ A_{\mu 1} & \cdots & A_{\mu\mu} \end{vmatrix}$$

must be non-negative for each value of μ. An easy computation shows that these minors are precisely the values of the functions G_μ of (3) at p_o. Since the same argument is valid at every point $p \in \mathcal{R}$ and since the functions $G_\mu(z,\bar z)$ depend on the uniformizer as real densities of weight $\mu(\mu+1)$, we have proved both the necessity and invariance of the condition.

Conversely suppose that each $G_\mu(z,\bar z)$ is analytic and non-negative. Then the set of points p where each G_μ is either positive or the function itself identically zero, being the complement of a set of first Baire category, is everywhere dense in $\mathcal{R}$ and the domains of convergence of the power series of each G_μ around each such point p cover $\mathcal{R}$. Let p_o be one such point, z a uniformizer vanishing at p_o, and

$$g(z,\bar z) = \Sigma_{\mu,\nu=1}^{\infty} A_{\mu\nu}\, z^{\mu-1}\, \bar z^{\,\nu-1};$$

if none of the functions G_μ vanish, then $A_{\mu\nu}$ is a positive definite matrix; if G_N is the last non-vanishing member of the sequence, then the matrix $A_{\mu\nu}$ is positive semi-definite of rank N and the minor $|A_{\mu\nu}|_{\mu,\nu\leq N}$ is positive. In either case the matrix can be written uniquely in the form

$$\Sigma_{\sigma=1}^{N} \mu\nu a_{\sigma,\mu}\bar{a}_{\sigma,\nu} ,$$

where $a_{\sigma,\mu} = 0$ for $\mu < \sigma$ and $a_{\sigma,\sigma}$ is real and positive. Then we have

$$g(z,\bar{z}) = \Sigma_{\mu,\nu=1}^{\infty} (\Sigma_{\sigma=1}^{N} a_{\sigma,\mu}\bar{a}_{\sigma,\nu}) z^{\mu-1} \bar{z}^{\nu-1} = \Sigma_{\sigma=1}^{N} \left| \Sigma_{\mu=\sigma}^{\infty} \mu a_{\sigma,\mu} z^{\mu-1} \right|^2 ,$$

where the change of order in the summation is permissible by the absolute convergence of power series. It is clear that, if we set

$$f_\sigma(z) = \Sigma_{\mu=\sigma}^{\infty} a_{\sigma,\mu} z^{\mu},$$

then

$$g(z,\bar{z}) = \Sigma_{\sigma=1}^{N} |f_\sigma'(z)|^2 ,$$

so that the functions $x_\sigma = f_\sigma(z)$ are analytic in the same domain as the domain of convergence of the power series of $g(z,\bar{z})$ and represent canonical coordinate of the points of that domain with respect to p_o under an isometric, complex analytic imbedding of the domain into C^N, thus completing the proof of Theorem 1.

We now prove that the isometric imbedding of Riemann surfaces in C^N is essentially unique.

THEOREM 2. Let $\mathcal{R}$ be any domain in a metric Riemann surface $\mathfrak{M}$ that can be isometrically and analytically mapped into C^N so that the image does not lie in any proper linear subspace of C^N. Then N is uniquely determined and the map is unique up to a unitary motion in C^N.

PROOF. If

$$g(z,\bar{z}) = \Sigma_{\mu,\nu=1}^{\infty} A_{\mu\nu} z^{\mu-1} \bar{z}^{\nu-1},$$

we let

$$\Phi(z,\bar{z}) = \Sigma_{\mu,\nu=1}^{\infty} \frac{1}{\mu\nu} A_{\mu\nu} z^{\mu} \bar{z}^{\nu} .$$

Then $\Phi(z,\bar{z})$ is a real valued, analytic scalar satisfying

$$\frac{\partial^2 \Phi(z,\bar z)}{\partial z \partial \bar z} = g(z,\bar z) .$$

This last equation determines Φ uniquely modulo harmonic functions. If p, q are any two points in the domain U of convergence of the power series of Φ, we let

$$D(p,q) = \Phi\left(z(p), \overline{z(p)}\right) + \Phi\left(z(q), \overline{z(q)}\right) - \Phi\left(z(p), \overline{z(q)}\right) - \Phi\left(z(q), \overline{z(p)}\right) .$$

This functional element generates a symmetric function in a subregion of $\mathfrak{M} \times \mathfrak{M}$, called the <u>diastasis</u> of $\mathfrak{M}$; it is uniquely determined by the metric and represents, for any fixed q, a normalization of the functions $\Phi\left(z(p), \overline{z(p)}\right)$, since

$$\frac{\partial^2 D(p,q)}{\partial z \partial \bar z}(p) = g\left(z(p), \overline{z(p)}\right) .$$

If $x_\sigma(p) = f_\sigma\left(z(p)\right)$ are functions representing an isometric map of $\mathcal{R}$ into C^N and $y_\tau(p) = h_\tau\left(z(p)\right)$ another set of functions mapping $\mathcal{R}$ similarly into $C^{N'}$, then

$$D(p,q) = \Sigma_{\sigma=1}^{N} \left| f_\sigma\left(z(q)\right) - f_\sigma\left(z(p)\right) \right|^2 = \Sigma_{\tau=1}^{N'} \left| h_\tau\left(z(q)\right) - h_\tau\left(z(p)\right) \right|^2 .$$

It follows that finite Euclidean distances between the images of p and q in C^N and $C^{N'}$ respectively are identical; the conclusion of the theorem then follows from well known facts about Euclidean geometry. The numbers N and N' are also proved equal by the fact that both are equal to the rank of the matrix $A_{\mu,\nu}$ at least when the origin is an ordinary point.

We call a metric Riemann surface <u>resolvable of rank N</u> at a point p if the metric is analytic at p and if some neighborhood of p can be isometrically mapped into C^N but not in $C^{N'}$ for any $N' < N$. We next prove that if a metric Riemann surface is resolvable of rank N at any one point and if the metric is analytic, then first of all the metric is resolvable of rank N everywhere and from that we prove that the universal covering space $\tilde{\mathfrak{M}}$ of $\mathfrak{M}$ can be globally mapped into C^N.

THEOREM 3. If a Riemann surface with an analytic metric is resolvable of rank N at any one point p, then it is resolvable of rank N everywhere.

PROOF. It is clear that the set $\mathcal{R}$ of points where the metric is resolvable of rank N is open; hence all we have to prove is that it is also closed.

Let p_0 be the limit point of $\mathcal{R}$ and z a uniformizer at p_0

such that $z(p_0) = 0$. Consider the power series expansion of the diastasis $D(p_0,p) = \Sigma_{\mu,\nu=1}^{\infty} B_{\mu\nu} \left(z(p)\right)^{\mu} \left(\overline{z(p)}\right)^{\nu}$, which we assume to converge for p in some domain U. Then we can expand the matrix $B_{\mu\nu}$ of coefficients in the form

$$B_{\mu\nu} = \Sigma_{\sigma=1}^{\infty} b_{\sigma,\mu} \overline{b}_{\sigma,\nu} - \Sigma_{\sigma=1}^{\infty} c_{\sigma,\mu} \overline{c}_{\sigma,\nu}$$

in some way, so that the functions

$$h_{\sigma}(z) = \Sigma_{\mu=1}^{\infty} b_{\sigma,\mu} z^{\mu}$$

and

$$h_{-\sigma}(z) = \Sigma_{\mu=1}^{\infty} c_{\sigma,\mu} z^{\mu}$$

are analytic in some subdomain V of U containing p_0 and

$$\Sigma'^{\infty}_{\sigma=-\infty} |h_{\sigma}(z)|^2$$

converges uniformly in V. The functions $x_{\pm\sigma} = h_{\pm\sigma}(z)$ map V continuously and isometrically into what might be called a generalized Minkowski space H of infinite complex dimension, with coordinates $x_{\pm\sigma}$ with the topology of Hilbert space for the purpose of defining continuity, and with the fundamental form $ds^2 = \Sigma'_{\sigma} (\operatorname{sgn}\sigma) |dx_{\sigma}|^2$ and corresponding diastasis

$$d(p,q) = \Sigma'_{\sigma} (\operatorname{sgn}\sigma) \left|x_{\sigma}(q) - x_{\sigma}(p)\right|^2$$

inducing metric and diastasis in V. In H we view $D(p,q)$ as an hermitian form on the vectors with components $x_{\sigma}(q) - x_{\sigma}(p)$ or as a real quadratic form on the real components of such vectors.

From the theory of real quadratic forms we use the following statements:

a) Given a real linear subspace H' of H spanned linearly by a set E, the form $D(p,q)$ restricted to H' is positive semi-definite of rank K if and only if for any finite set of points $p_0, p_1, \dots, p_r \in E$ the symmetric matrix whose components are $\Delta_{ij} = D(p_0,p_i) + D(p_0,p_j) - D(p_i,p_j)$ is positive semi-definite and of rank $\leq \min(r,K)$, where equality can be achieved for every value of r for a suitable choice of points.

b) If H' is a complex linear subspace of H in which the diastasis is positive semi-definite and of rank K, then H' is uniquely a direct sum of two complex linear, mutually orthogonal subspaces L and H^0 (if we consider H as a Hilbert space) and such that $D(p,q) = 0$ for $p,q \in H^0$, and $D(p,q)$ defines a positive definite hermitian form on L; then L is a $K/2$-dimensional unitary space and the orthogonal projection of H' onto L along H^0 preserves the diastasis.

We now suppose V mapped into H such that p_0 is at the origin. Let W be a subdomain of V, at all of whose points the metric is resolvable of rank N. Let H' be the real linear subspace of H spanned by W, and hence by all of V. Then H' is complex linear and for all finite sets of $r+1$ points in W the matrix considered in statement a) is positive semi-definite of rank $\leq \min(r, 2N)$. Hence, by statement b) H' can be projected complex analytically onto a linear N-dimensional complex unitary subspace L preserving the diastasis. So all of V can be isometrically and complex analytically mapped onto L, whose completion is C^N. Hence the theorem is proved.

It is easy now to apply the last two theorems, and the monodromy principle to verify that if the metric on $\mathfrak{M}$ is analytic everywhere and resolvable of rank N at just one point, then the whole universal covering space $\widetilde{\mathfrak{M}}$ of $\mathfrak{M}$ can be globally immersed isometrically in C^N. The fundamental group of $\mathfrak{M}$ has thus a concrete representation into the group of motions of C^N. It is also to be noted that the diastasis $D(p,q)$ is then definable and real analytic in all pairs of points in the whole covering surface $\widetilde{\mathfrak{M}}$; it is everywhere ≥ 0 and becomes zero only for those pairs of points $\widetilde{\mathfrak{M}}$ that have a common image in C^N under the map. If $\tilde{p}$ is any point of $\widetilde{\mathfrak{M}}$, if σ is an element of the fundamental group $\pi_1(\mathfrak{M})$ of $\mathfrak{M}$, and $\tilde{p}^{\sigma}$ the translation of $\tilde{p}$ by σ, then the imbedding, considered as a multiple-valued map of $\mathfrak{M}$ into C^N, is single-valued, if, and only if $D(\tilde{p}, \tilde{p}^{\sigma}) = 0$ for all $\tilde{p}$ in a fundamental domain of $\mathfrak{M}$ in $\widetilde{\mathfrak{M}}$ and all generators σ of $\pi_1(\mathfrak{M})$. In this case $D(p,q)$ is single valued in $\mathfrak{M} \times \mathfrak{M}$. The map is a topological imbedding if, in addition, $D(p,q) = 0$ implies $p = q$. We thus arrive at our main theorem.

THEOREM 4. A metric Riemann surface can be isometrically and complex analytically imbedded in C^N if, and only if the metric is real analytic and resolvable of rank at least N at one point, and if the diastasis is everywhere single-valued. The mapping is a topological imbedding, provided that the diastasis between distinct points is positive.

4. REMARKS

There are some relations between the densities G_μ of Theorem 1 and the curvatures $\rho_1, \ldots, \rho_{N-1}$ at each point of $\mathfrak{M}$. In fact if $x_\sigma = f_\sigma(z)$ represent an isometric map of $\mathfrak{M}$ into C^N, we can verify that

$$G_\mu(z,\bar{z}) = \sum_{1 \leq \sigma_1 < \sigma_2 < \ldots < \sigma_\mu \leq N} \begin{vmatrix} f'_{\sigma_1}(z) & f'_{\sigma_2}(z) & \ldots & f'_{\sigma_\mu}(z) \\ f''_{\sigma_1}(z) & f''_{\sigma_2}(z) & \ldots & f''_{\sigma_\mu}(z) \\ \vdots & \vdots & & \\ f^{(\mu)}_{\sigma_1}(z) & f^{(\mu)}_{\sigma_2}(z) & \ldots & f^{(\mu)}_{\sigma_\mu}(z) \end{vmatrix}^2 = \left(g(z,\bar{z})\right)^{\frac{\mu(\mu+1)}{2}} \prod_{\nu=1}^{\mu-1} \rho_\nu^{2(\mu-\nu)}$$

From this we have

$$\frac{G_{\mu+1}(z,\bar{z})}{G_\mu(z,\bar{z})} = g^{\mu+1}\rho_1^2\rho_2^2\cdots\rho_\mu^2; \qquad \rho_\mu{}^2 = \frac{G_{\mu+1}\,G_{\mu-1}}{g\,G_\mu^2}.$$

From (3) we obtain the recursive formula for the curvatures, using Δ to denote the operator

$$\frac{1}{g}\,\frac{\partial^2}{\partial_z\partial_{\bar{z}}},$$

$$\rho_\mu^2 = \Delta \log\left[g^{\frac{\mu(\mu+1)}{2}}\rho_1{}^{2(\mu-1)}\rho_2{}^{2(\mu-2)}\cdots\rho_{\mu-1}^2\right].$$

Thus we have proved that all the curvatures of $\mathfrak{M}$ are computable from the metric. It is perhaps curious to note that $-\rho_1{}^2$ is The Gaussian curvature of $\mathfrak{M}$.

The methods outlined above can be directly applied to write criteria for any real analytic density of weight w to be developable into a series of squares of absolute values of holomorphic densities of weight w (including, of course, the case of scalars) or even to solve other functional equations of similar nature. The direct applications of these methods furnish new proofs of some specific statements, such as the follows. The hyperbolic (Poincaré) plane can be imbedded in Hilbert space and has complete mobility in itself there (originally proved by Bieberbach, [1]), but cannot be imbedded in a finite dimensional unitary space (cfr. Bochner, [2]). The generalized Minkowski space H is the "universal enveloping space" into which every metric Riemann surface can be locally imbedded. However, the map here is not uniquely determined (i.e. up to a diastasis-preserving transformation of H). In particular all algebraic curves with the metric induced from an elliptic Fubini (complex projective) space can be imbedded in H except for a hyperplane section of the curve.

---

[1] This paper was prepared while the author held a fellowship from the Office of Naval Research, Project NR 045050.

[2] The author has been shown an elegant counterexample by B. Lepson, in which the interior of the unit circle $|z| < 1$ is mapped complex analytically in Hilbert space, so that the map is discontinuous at every point where z is real.

REFERENCES

[1] BIEBERBACH, L., Eine singularitätenfreie Fläche konstanter negativer Krümmung im Hilbertschen Raum, Comm. Math. Helv. 4 (1932), pp. 248-255.

[2] BOCHNER, S., Curvature in Hermitian Metric, Bull. Amer. Math. Soc., 53, (1947), pp. 179-195.

[3] CALABI, E., Isometric, Complex Analytic Imbedding of Kähler Manifolds, Princeton Thesis, 1950. (Ann. of Math., 54 (1953), to appear)

[4] KLEIN, F., Über Riemanns Theorie der algebraischen Funktionen und ihrer Integrale, Leipzig, 1882.

Complex Analytic Manifolds Without Countable Base

Eugenio Calabi[1] and Maxwell Rosenlicht[2]

Proc. Amer. Math. Soc. 4, 335–340 (1953)

1. It has been shown by Radó[3] that every Riemann surface satisfies Hausdorff's second countability axiom. In the same paper Radó gives a construction due to Prüfer of a real 2-dimensional locally Euclidean space whose open sets do not have a countable base. The purpose of this paper is to confirm a conjecture of Bochner that there exist non second-countable complex manifolds of complex dimension $n>1$. The authors have independently constructed examples of such manifolds by two different methods, the first using the local quadratic transformations of algebraic geometry, the second by verifying that a modification of Prüfer's example has the structure of a 2-dimensional real analytic manifold that can be extended into the complex domain by formally replacing real parameters by complex ones. We describe these manifolds, the relations among them, and some of their topological properties.

2. Let V be an irreducible nonsingular algebraic variety of dimension $n>1$ in a complex projective space of dimension N. Let $(y_0, y_1, \cdots, y_N)$ be a general point of V. Suppose that V contains the point $O=(1, 0, \cdots, 0)$; then the quadratic transform V_O of V with respect to O is the algebraic variety in projective space of dimension $(1/2)(N^2+3N-2)$ whose general point is

$$(y_0y_1, y_0y_2, \cdots, y_0y_N, (y_1)^2, y_1y_2, \cdots, y_1y_N, (y_2)^2, \cdots, y_2y_N, \cdots, (y_N)^2.$$

It is easily verified that V_O is nonsingular and that the given correspondence of general points induces a complex analytic birational map T_O of V_O onto V. If U_O is the set of points of V_O all of whose coordinates after the first N vanish, then T_O maps U_O onto $\{O\}$ and is biregular as a map from V_O-U_O onto $V-\{O\}$. The points of U_O correspond in a natural one-one way to the (complex) directions at O in V, so that U_O is an irreducible nonsingular subvariety of

Presented to the Society, September 5, 1952; received by the editors June 26, 1952.

[1] Under contract with the Office of Naval Research, Contract Nonr-467(00), Project No. NR 045 050.

[2] National Research Fellow.

[3] T. Radó, *Über den Begriff der Riemannschen Fläche*, Acta Univ. Szeged. (II) vol. 2 (1925).

DOI: 10.2307/2032126

V_O, analytically equivalent to a complex projective space of dimension $n-1$.

Using a suitable projective coordinate transformation in the ambient space of V we may similarly define the quadratic transform V_P of V with respect to any point P of V, the analytic birational map $T_P: V_P \to V$ and the subvariety $U_P = T_P^{-1}\{P\}$. If W is an irreducible nonsingular algebraic curve in V and $P \in W$, then $T_P^{-1}(W) = U_P \cup W_P$, where W_P is an algebraic curve biregularly equivalent to W under the map T_P and where U_P and W_P have a unique simple intersection at the point of U_P corresponding to the direction of the complex line through P tangent to W.

Now for fixed V and W consider the totality of manifolds $V_P - W_P$ where P ranges over some nondenumerable subset of W. Let M be the set obtained from this union by identifying points that correspond to the same point of $V - W$. If P, Q are distinct points of W, then the map $T_Q^{-1}T_P$ is biregular from $V_P - (U_P \cup W_P)$ onto $V_Q - (U_Q \cup W_Q)$, while no point of $U_P - (U_P \cap W_P)$ is identified with any point of $V_Q - W_Q$. We define a Hausdorff topology on M by taking as a base for the open sets of M the images in M of open sets in each $V_P - W_P$. Local complex analytic coordinates at any point in any $V_P - W_P$ will clearly induce local complex coordinate functions on M. Each of the spaces $V_P - W_P$ induces an open set on M and the totality of such open sets covers M. Since no denumerable subset of these covers M, our manifold does not satisfy the second countability axiom of Hausdorff.

We remark that any meromorphic function on V induces a meromorphic function on each V_P and hence on M. Conversely, a meromorphic function on M can be restricted to each $V_P - W_P$; so, if $n > 2$, by the continuity theorem for functions of several complex variables, we obtain that all meromorphic functions on M are derivable in this manner. It is also directly verifiable that a holomorphic covariant tensor on any domain of V, such as for instance an abelian differential, gives rise to a tensor of the same type on M which is holomorphic in a corresponding domain of M, but this is not true in general for contravariant tensors.

Note that a manifold such as M could have been obtained from any analytic V (not necessarily algebraic) in projective space, and any nonsingular W on V. It is clear that the complex analytic structure of V_P depends only on V, and not on its particular embedding in projective space. Thus V_P can be constructed from an arbitrary complex manifold V and an arbitrary $P \in V$, and V_P is a complex manifold, compact if V is compact (this seems to have been first noticed by H. Hopf). Similarly, for an arbitrary complex manifold V and an arbitrary closed complex submanifold W one can construct a manifold M.

3. In order to study some of the topological properties of M it is perhaps well to consider its relation to Prüfer's example, which we reproduce here in a much simplified way.

For each real t let E_t be a real Euclidean plane with coordinates (x_t, y_t). Consider the set S_+ of equivalence classes defined by the relation $(x_s, y_s) \sim (x_t, y_t)$ whenever

$$(1) \qquad y_s = y_t$$

$$(2) \qquad \begin{aligned} x_s y_s + s &= x_t y_t + t && \text{if } y_s = y_t > 0, \\ s = t, \qquad x_s &= x_t && \text{if } y_s = y_t \leqq 0. \end{aligned}$$

Let V_t be the saturation of the plane E_t in S_+ under the above equivalence relation. Then if we put on V_t the topology induced by the topology of the plane E_t, it is easily seen that for any two values s, t the spaces V_s and V_t induce the same topology in $V_s \cap V_t$. Thus $S_+ = \bigcup_t V_t$ with the given topology is indeed a 2-dimensional analytic manifold with coordinate domains V_t parametrized by the coordinates (x_t, y_t) of E_t. S_+ is the surface attributed to Prüfer, but our presentation of it points out the real analytic structure it can admit.

A slightly different example of an analytic surface, which we denote by S, can be made from the E_t's by defining a new equivalence relation $(x_s, y_s) \sim (x_t, y_t)$ whenever

$$(1) \qquad y_s = y_t,$$

$$(2) \qquad \begin{aligned} x_s y_s + s &= x_t y_t + t && \text{if } s \neq t, \\ x_s &= x_t && \text{if } s = t, \end{aligned}$$

and continuing otherwise exactly as before. Neither S_+ nor S satisfies the second countability axiom since the point set in each space corresponding to the origins of the various planes E_t is discrete and nondenumerable.

We now construct a complex manifold M with no countable base for its open sets merely by letting, purely formally, t, x_t, y_t take on all possible values in the complex number field and defining the equivalence relation and topology exactly as in the case of S.

We shall show that the manifold M thus described is identical with one that is obtained by the method of §2. In fact let V be the plane of two complex variables, considered as the affine part of a complex projective plane; use in V the affine coordinates (x, y) (nonhomogeneous coordinates in the projective plane) and let W be the complex line $y=0$. For any point P_t ($x=t$, $y=0$) of W we construct the quadratic transform V_{P_t} of V with respect to P_t and form the space $M_t=V_{P_t}-W_{P_t}$. Points of M_t correspond one-one to points of $V-W$ and line elements at P_t not in the direction of W. It follows that M_t is covered by the single coordinate system (x_t, y_t), where $x_t=(1/y)(x-t)$, $y_t=y$, and x_t, y_t range over all complex numbers. If $s\neq t$, we identify points (x_s, y_s) of M_s and (x_t, y_t) of M_t if and only if they come from the same point of V, that is, if and only if $y=y_s=y_t$, $x=x_sy_s+s=x_ty_t+t$. Thus the previous equivalence relation is established among the various M_t's.

4. In this section we study some topological properties of manifolds without countable base. We shall confine our discussion of the manifolds obtained by the method of §2 to the specific example M described in the previous section (the generalizations to other manifolds being trivial in most cases); at the same time the close formal relations among M, the Prüfer surface S_+, and our modification S of Prüfer's example warrant a comparative study of the three spaces.

First of all, S is a subspace of M consisting of all points covered by the coordinate systems (x_t, y_t) in M for t real, for which x_t, y_t are also real. S_+ is not separable, that is it has no countable everywhere dense subset: in fact any countable subset can meet at most a countable number of lower half-planes of the E_t's, and each of the remaining ones corresponds to an open subset of S_+. On the other hand, S and M are separable. In fact any countable dense set in any one coordinate domain is clearly everywhere dense in the whole manifold.

We now show that the spaces S_+, S, M are not metrizable, since they are not normal. In fact, consider in any one of them the set A of points $\{x_t=y_t=0,\ t \text{ real and rational}\}$ and the set B of points $\{x_t=y_t=0,\ t \text{ real and irrational}\}$. A and B are disjoint closed sets in each S_+, S, M. We show that the sets A, B are not contained in disjoint open sets of S_+ or S (and therefore, a fortiori, of M). For this purpose, let K_+ be the open upper half-plane of S_+ (or S) in one (and therefore in any) coordinate system, and let $K=K_+\cup A\cup B$. Consider the one-one map σ of K onto the closed real upper half-plane (ξ, η), $\eta\geqq 0$, defined by $\sigma(x_t, y_t)=(\xi, \eta)$, where $\xi=x_ty_t+t$, $\eta=y_t^2$. The topology induced in the closed upper half-plane by the map σ is as follows: it is the ordinary Euclidean topology for $\eta>0$, while the point $(t, 0)$ has as a fundamental system of neighborhoods the set $\sigma(U_{t,\epsilon})$, where $U_{t,\epsilon}$ is the subset of E_t consisting of the point $x_t=y_t=0$ and points (x_t, y_t) satisfying $x_t^2+y_t^2<\epsilon$, $y_t>0$. Thus $\sigma(U_{t,\epsilon})$ consists of the point $(t, 0)$ and points (ξ, η) satisfying

$$\frac{1}{\eta}(\xi - t)^2 + \eta < \epsilon, \qquad \eta > 0,$$

or

$$(\xi - t)^2 + \left(\eta - \frac{\epsilon}{2}\right)^2 < \left(\frac{\epsilon}{2}\right)^2.$$

This is precisely the topology of the space described by Alexandroff and Hopf[4] as an example of a regular space that is not normal, and in which the closed sets A, B have no disjoint neighborhoods.

Finally we prove that of the three manifolds S_+, S, M only the first is contractible, because the second has a noncountable fundamental group and the last is simply connected but has a noncountable second homotopy group. To prove the first statement, for each s, $0 \leqq s \leqq 1$, and each coordinate neighborhood E_t define

$$g_s(x_t, y_t) = \begin{cases} \left(x_t\left(\dfrac{1 - s + sy_t}{1 + sy_t}\right)^{1/2}, \ y_t\left(\dfrac{1 + sy_t}{1 - s + sy_t}\right)^{1/2}\right), & y_t > 0, \\ (x_t(1 - s)^{1/2}, \ y_t(1 - s)^{1/2}), & y_t \leqq 0. \end{cases}$$

Because g_s preserves the equivalence relation defining S_+, if we let s run from 0 to 1, g_s continuously deforms S_+ onto the subspace K of the preceding paragraph. Then K may be drawn continuously into the open upper half-plane, where it may be contracted linearly to a point.

To show that S is not simply connected we map S into the real plane R^2 with coordinates x, y by setting $\phi(x_t, y_t) = (x, y)$, where $x = x_t y_t + t$, $y = y_t$. There exists a unique simple closed curve γ in S whose image in R^2 is any given simple closed curve γ', provided that γ' is differentiable wherever it intersects the line W given by $y = 0$, and that its tangent at any such intersection is distinct from that line. In particular any circle γ' intersecting W at two distinct points $(s, 0)$, $(t, 0)$ comes from a unique curve $\gamma = \phi^{-1}(\gamma')$ in S. If γ could be shrunk continuously to a point in S, there would exist a continuous map $h: \Gamma \times I \to S$ (where I is the unit interval and Γ a circle) such that $h(\Gamma \times 0) = \gamma$ and $h(\Gamma \times 1)$ is a single point P. Then ϕh is a homotopy of γ' into the point $\phi(P)$. But $\phi h(\Gamma \times I)$ covers all points of the closed segment $[(s, 0), (t, 0)]$ in R^2, so $h(\Gamma \times I)$ must cover at least one point in each of the lines $y_{t'} = 0$ in each $E_{t'}$ for all $t' \in [s, t]$, that is, $h(\Gamma \times I)$ includes a nondenumerable discrete subset of S, which is impossible, since $\Gamma \times I$ is compact. Similarly, if γ'' is a circle in R^2 meeting W at

[4] P. Alexandroff and H. Hopf, *Topologie*, Berlin, 1935, p. 69.

two points distinct from $\gamma' \cap W$, then $\phi^{-1}(\gamma'')$ cannot be deformed into $\phi^{-1}(\gamma')$; this proves the nondenumerability of the fundamental group of S.

We prove similarly that the second homotopy group of M is nondenumerable. Let ϕ map M onto the complex affine plane C^2 with coordinates x, y by the formula $\phi(x_t, y_t) = (x, y)$, where $x = x_t y_t + t$, $y = y_t$. Consider in C^2 a 2-sphere γ' intersecting the plane W $(y=0)$ at two distinct points and such that the tangent (real) planes at these intersections are complex affine lines in C^2. Then there exists a unique topological sphere γ in M such that $\phi(\gamma) = \gamma'$. Any circle in W which separates the points $\gamma' \cap W$ has linking number ± 1 with γ'. Therefore any deformation of γ' to a point in C^2 must make γ' intersect any such circle. There is a continuous family of such circles which are concentric, so that the reasoning of the preceding paragraph shows that in any homotopy $h: \Gamma \times I \to M$ (Γ here a sphere) of γ onto a point, $h(\Gamma \times I)$ must contain a nondenumerable discrete set of points of M, which is impossible. Similarly if γ'' is a sphere in C^2 satisfying the same conditions as γ' but with $\gamma'' \cap W \neq \gamma' \cap W$, $\phi^{-1}(\gamma'')$ is not homotopic to $\phi^{-1}(\gamma')$, so the second homotopy group of M is nondenumerable.

Finally, M is simply connected. In fact any closed curve in M may be covered by a finite number of coordinate neighborhoods (x_t, y_t) in M. In each such neighborhood it can be deformed away from the line $y_t = 0$; therefore it can be brought into a single coordinate neighborhood, where its contractibility is obvious.

PRINCETON UNIVERSITY

A Class of Compact, Complex Manifolds Which Are Not Algebraic[1]

Eugenio Calabi[2] and Beno Eckmann

Ann. of Math. (2) 58, 494–500 (1953)

1. Introduction

The main purpose of this note is to exhibit the existence of complex analytic structures in the Cartesian product $\mathbf{S}^{2p+1} \times \mathbf{S}^{2q+1}$ of any two odd-dimensional spheres. Since the case $p = q = 0$ is trivially equivalent to a closed Riemann surface of genus 1 and periods 1, τ (hereinafter denoted by $\mathbf{T}^2_{1,\tau}$), we can exclude it a priori by assuming $p > 0$ from now on. The complex structures applied to the special case $\mathbf{S}^{2p+1} \times \mathbf{S}^1$ (for $q = 0$) are equivalent (see §3) to those discovered by H. Hopf [3], who obtained them starting from the universal covering space of $\mathbf{S}^{2p+1} \times \mathbf{S}^1$ with the obvious complex structure of $\mathbf{C}^p - \{0\}$ ($\mathbf{C}^p$ = p-dimensional vector space over the complex field); this method can not be easily generalized to $\mathbf{S}^{2p+1} \times \mathbf{S}^{2q+1}$ $(q > 0)$, because these manifolds are simply connected.

We shall denote by $\mathbf{M}_{p,q}$ any complex manifold homeomorphic to $\mathbf{S}^{2p+1} \times \mathbf{S}^{2q+1}$ of the type described in this article. Leaving the actual construction of such manifolds to the next section, we list in the following theorems some properties which are consequences of their existence and construction and which are interesting in connection with some problems and conjectures on complex analytic structure in higher dimensions.

Theorem I. $\mathbf{M}_{p,q}$ *does not admit any Kaehler metric, and therefore can not be imbedded in any complex projective space; hence it is not algebraic.*

In particular, for $q > 0$, $\mathbf{M}_{p,q}$ is the first known example of *a compact, simply connected complex manifold which is not algebraic*, since all previously known examples of compact, complex manifolds are either algebraic, or their universal covering spaces are open subsets of algebraic manifolds.

To prove Theorem I it is sufficient to recall that in a Kaehler manifold of complex dimension m all even dimensional Betti numbers R^{2k} $(0 \leqq k \leqq m)$ are positive and all odd dimensional ones are even valued [4], [2]. In the case of $\mathbf{M}_{p,q}$ (which is $p + q + 1$-dimensional), on the contrary, we note that $R^{2k} = 0$ $(1 \leqq k \leqq p + q)$ and, if $p \neq q$, $R^{2p+1} = R^{2q+1} = 1$, so that at least one, and generally both necessary restrictions on the Betti numbers of a manifold, in order that it may be Kaehlerian, are violated.

Theorem II. $\mathbf{M}_{p,q}$ *admits a complex analytic fibering, the fibres being 2-dimensional tori (algebraic curves of genus 1 of fixed, but arbitrary conformal type) and the base space the product* $\mathbf{P}_p \times \mathbf{P}_q$ *of a p-dimensional by a q-dimensional complex projective space.*

---

[1] Presented to the American Mathematical Society Sept. 4, 1951, Paper #533.

[2] Office of Naval Research Fellow under Contract N-onR-09800, Project #nR-045-001 during part of the time of preparation of this article.

DOI: 10.2307/1969750

This property will follow immediately from the construction of $\mathbf{M}_{p,q}$ in §2.

THEOREM III. *There exist in $\mathbf{M}_{p,q}$ compact, complex analytic submanifolds, whose carrying cycles are homologous to 0 in $\mathbf{M}_{p,q}$, namely the tori of the fibering of $\mathbf{M}_{p,q}$, according to Theorem II.*

We note that this last property proves again that $\mathbf{M}_{p,q}$ does not admit a Kaehler metric, since otherwise no compact, complex submanifold could be homologous to zero [2].

THEOREM IV. *Every compact, complex subvariety of $\mathbf{M}_{p,q}$ is the set of all points that are mapped onto an algebraic subvariety of $\mathbf{P}_p \times \mathbf{P}_q$ by the fibre projection described in Theorem II; it is therefore also fibered by tori.*

THEOREM V. *Every meromorphic function in $\mathbf{M}_{p,q}$ is constant along the torus-like fibres of $\mathbf{M}_{p,q}$ described in Theorem II.*

The two theorems stated above are proved in §4; the latter theorem is a corollary of the former.

THEOREM VI. *Let a and a' be points of $\mathbf{S}^{2p+1}$ and $\mathbf{S}^{2q+1}$ respectively and denote by $\mathbf{E}_{p,q}$ the open region in $\mathbf{M}_{p,q}$ corresponding to $(\mathbf{S}^{2p+1} - \{a\}) \times (\mathbf{S}^{2q+1} - \{a'\})$. Then $\mathbf{E}_{p,q}$ is an open $p + q + 1$-dimensional complex manifold homeomorphic to a cell and has the property that, for $q > 0$, it neither admits a Kaehler metric, nor can be covered by a single complex coordinate system, nor contains any non-constant, entire holomorphic function.*

This last theorem follows from Theorem III; the detailed proof is in §5. The implication of this theorem is to answer negatively a question raised by S. Bochner [1] regarding the uniformization of complex analytic cells.

2. The Complex Manifolds $M_{p,q;\tau}$

Let $\mathbf{E}_{p+1}$ be the $2p + 2$-dimensional Euclidean space described by the $p + 1$ complex affine coordinates $z_0, z_1, \cdots, z_p$, and consider its unit sphere $\mathbf{S}$ of dimension $2p + 1$, defined by the equation $\sum_{j=0}^{p} |z_j|^2 = 1$. The coordinates $z_0, \cdots, z_p$ of any point $z \in \mathbf{S}$ can be considered as the homogeneous coordinates of a point w of a p-(complex) dimensional complex projective space $\mathbf{P}$ and the map π, carrying z onto w defines a projection of $\mathbf{S}$ onto $\mathbf{P}$. For each $z \in \mathbf{S}$ the set $\pi^{-1}\pi(z)$ of all points equivalent to z under the mapping π consists of all points $\lambda z = (\lambda z_0, \cdots, \lambda z_p)$ with $|\lambda| = 1$; $\pi^{-1}\pi(z)$ is therefore a great circle of $\mathbf{S}$ and π is the well known fibre map of $\mathbf{S}^{2p+1}$ onto $\mathbf{P}$. Similarly let $\mathbf{E}'$ be the $q + 1$-dimensional complex affine space with coordinates $z_0', \cdots, z_q'$, $\mathbf{S}'$ its unit sphere (of dimension $2q + 1$), and π' the fibre map of $\mathbf{S}'$ onto $\mathbf{P}'$, the q-dimensional complex projective space whose homogeneous coordinates are represented by the affine coordinates of the points of $\mathbf{S}'$.

The map Π of the Cartesian product $\mathbf{S} \times \mathbf{S}'$ onto $\mathbf{P} \times \mathbf{P}'$, defined by $\Pi(z, z') = (\pi(z), \pi'(z'))$ is again a fibre map, the fibres $\Pi^{-1}\Pi(z, z')$ being the 2-dimensional tori described by $(\lambda z, \lambda' z')$ ($|\lambda| = |\lambda'| = 1$). We shall define on each of these tori a conformal structure so as to yield, together with the complex analytic structure of $\mathbf{P} \times \mathbf{P}'$, a complex structure on $\mathbf{S} \times \mathbf{S}'$.

Let $\mathbf{V}_{\alpha\beta}$ ($\alpha = 0, 1, \cdots, p$; $\beta = 0, 1, \cdots, q$) be any one of the open subsets of

$\mathbf{S} \times \mathbf{S}'$ defined by $z_\alpha z'_\beta \neq 0$. It is clear that the $(p + 1)(q + 1)$ domains $\mathbf{V}_{\alpha\beta}$ cover $\mathbf{S} \times \mathbf{S}'$. For each point (z, z') of $\mathbf{S} \times \mathbf{S}'$ and each $\mathbf{V}_{\alpha\beta}$ to which it belongs we set

$$(1) \qquad {}_\alpha w_j = \frac{z_j}{z_\alpha} \; (j = 0, \cdots, p); \qquad {}_\beta w'_k = \frac{z'_k}{z'_\beta} \; (k = 0, \cdots, q).$$

Then ${}_\alpha w_j$ $(j \neq \alpha)$ and ${}_\beta w'_k$ $(k \neq \beta)$ are local in homogeneous coordinates in the open subdomains $\Pi(\mathbf{V}_{\alpha\beta})$ covering $\mathbf{P} \times \mathbf{P}'$. If τ is an arbitrary, fixed, complex constant such that $\operatorname{Im}(\tau) \neq 0$, we set in addition

$$(2) \qquad t_{\alpha\beta} \equiv \frac{1}{2\pi i} (\log z_\alpha + \tau \log z'_\beta) \bmod (1, \tau) \qquad ((z, z') \epsilon \mathbf{V}_{\alpha\beta}),$$

where $t_{\alpha\beta}$ will remain defined only within its congruence class modulo the discrete lattice generated by 1 and τ. The $p + q + 1$ complex valued functions ${}_\alpha w_j$, ${}_\beta w'_k$, $t_{\alpha\beta}$ map $\mathbf{V}_{\alpha\beta}$ differentiably into the complex manifold $\mathbf{C}^{p+q} \times \mathbf{T}^2_{(1,\tau)}$ equal to the product of a $p + q$-dimensional complex affine space $\mathbf{C}^{p+q}$ with a complex torus $\mathbf{T}^2$ with periods $1, \tau$: we shall now show that this map is a homeomorphism of $\mathbf{V}_{\alpha\beta}$ onto $\mathbf{C}^{p+q} \times \mathbf{T}^2_{(1,\tau)}$. In fact, given any $p + q$ complex numbers ${}_\alpha w_j$, ${}_\beta w'_k$ $(j \neq \alpha, k \neq \beta)$ and any point of $\mathbf{T}^2$, represented by a complex number $t_{\alpha\beta} \bmod (1, \tau)$, we can find a unique set of complex numbers

$$z_j, z'_k \; (0 \leqq j \leqq p; 0 \leqq k \leqq q),$$

depending differentiably on the given quantities, satisfying (1), (2), and

$$(3) \qquad \textstyle\sum_{j=0}^{p} |z_j|^2 = \sum_{k=0}^{q} |z'_k|^2 = 1; \; z_\alpha z'_\beta \neq 0$$

as follows. We set ${}_\alpha w_\alpha = {}_\beta w'_\beta = 1$; from (1) and (3) we have

$$\sum_{j=0}^{p} |{}_\alpha w_j|^2 = \frac{1}{|z_\alpha|^2}; \qquad \sum_{k=}^{q} |{}_\beta w'_k|^2 = \frac{1}{|z'_\beta|^2};$$

thus the uniqueness and continuity of $|z_\alpha|$ and $|z'_\beta|$ are established. From (2) we have

$$\begin{aligned} \arg z_\alpha + \tau \arg z'_\beta &\equiv -i[(\log z_\alpha + \tau \log z'_\beta) - (\log |z_\alpha| + \tau \log |z'_\beta|) \\ &\equiv 2\pi t_{\alpha\beta} + i(\log |z_\alpha| + \tau \log |z'_\beta|) \bmod (2\pi, 2\pi\tau), \end{aligned}$$

and, since $\arg z_\alpha$, $\arg z'_\beta$ are real (mod 2π), and τ is not real, $\arg z_\alpha$ and $\arg z'_\beta$ are uniquely determined mod 2π and depend differentiably on $t_{\alpha\beta}$, $|z_\alpha|$, and $|z'_\beta|$.

Having thus determined the existence, uniqueness, and continuity of z_α and z'_β, we obtain the other coordinates z_j, z'_k immediately by setting $z_j = {}_\alpha w_j \, z_\alpha$, $z'_k = {}_\beta w'_k \, z'_\beta$, as required by (1), completing the proof of the homeomorphism between each $\mathbf{V}_{\alpha\beta}$ and $\mathbf{C}^{p+q} \times \mathbf{T}^2_{(1,\tau)}$.

We can now consider the complex valued functions ${}_\alpha w_j$ $(0 \leqq j \leqq p, j \neq \alpha)$, ${}_\beta w'_k$ $(0 \leqq k \leqq q, k \neq \beta)$, $t_{\alpha\beta}$ (mod $1, \tau$) as local coordinates in $\mathbf{V}_{\alpha\beta}$. We next prove that these coordinate systems are biregularly related in the intersection of their domains of definition. In fact we have in each intersection $\mathbf{V}_{\alpha\beta} \cap \mathbf{V}_{\gamma\delta}$ of any two such domains, after putting ${}_\alpha w_\alpha = {}_\gamma w_\gamma = {}_\beta w'_\beta = {}_\delta w'_\delta = 1$,

$$(4) \qquad {}_\gamma w_j = {}_\alpha w_j / {}_\alpha w_\gamma \ (0 \leqq j \leqq p); \qquad {}_\delta w'_k = {}_\beta w'_k / {}_\beta w'_\delta \qquad (0 \leqq k \leqq q)$$

$$(5) \qquad t_{\gamma\delta} \equiv t_{\alpha\beta} + \frac{1}{2\pi i} (\log {}_\alpha w_\gamma + \tau \log {}_\beta w'_\gamma) \ (\mathrm{mod}\ 1, \tau).$$

Since the coordinate transformation from any $\mathbf{V}_{\alpha\beta}$ to any $\mathbf{V}_{\gamma\delta}$ are clearly complex analytic, we have introduced in the manifold $\mathbf{S} \times \mathbf{S}'$ a complex analytic structure. We shall call the complex manifold thus obtained $\mathbf{M}_{p,q;\tau}$, or, if the value of τ is unimportant, $\mathbf{M}_{p,q}$.

Let (z, z') be a point in the coordinate domain $\mathbf{V}_{\alpha\beta}$; then the functions ${}_\alpha w_j$ and ${}_\beta w'_k$ $(j \neq \alpha; k \neq \beta)$ are the inhomogeneous coordinates of $\Pi(z, z')$ in $\Pi(\mathbf{V}_{\alpha\beta})$. Thus the fibre projection map Π can be rewritten in terms of the local complex coordinates at (z, z') as $\Pi({}_\alpha w_j, {}_\beta w'_k, t_{\alpha\beta}) = ({}_\alpha w_j, {}_\beta w'_k)$; this shows that the map Π of $\mathbf{M}_{p,q}$ onto $\mathbf{P} \times \mathbf{P}'$ is complex analytic. The fibre at each point $({}_\alpha w_j, {}_\beta w'_k, t_{\alpha\beta})$ is given by $\Pi^{-1}\Pi({}_\alpha w_j, {}_\beta w'_k, t_{\alpha\beta}) = ({}_\alpha w_j, {}_\beta w'_k, \mathbf{T}^2_{(1,\tau)})$, so that the fibres appear as complex curves without singularities in the complex manifold; moreover equation (5) shows that the operations of the structure group on the fibre under coordinate transformations are translations, i.e. complex analytic automorphisms, and that they depend complex analytically on the coordinates $({}_\alpha w_j, {}_\beta w'_k)$ of the base point (see Steenrod, [5]); thus $\mathbf{M}_{p,q}$ is a complex fibre bundle space, which proves Theorem II. Since the fibre is isomorphic to its own structural group of automorphisms, the bundle is actually a principal fibre bundle.

It is worth mentioning also that the manifolds $\mathbf{M}_{p,q;\tau}$ admit transitive Lie groups of complex analytic automorphisms, namely the direct product of the unitary groups acting on $\mathbf{S}$ and on $\mathbf{S}'$ respectively. The verification of this fact is immediate.

3. The Complex Manifolds $M_{p,0;\tau}$

In the special case $q = 0$ the sphere $\mathbf{S}'$ becomes a circle, whose points are described by a single coordinate system z'_0 ($z'_0 = e^{2\pi i\rho}$, ρ real mod 1). Having constructed the complex manifold $M_{p,0;\tau}$ with any fixed constant τ ($\mathrm{Im}\,(\tau) \neq 0$), we set $a = e^{2\pi i\tau}$ ($| a | \neq 1$); in this manifold we consider the multi-valued functions $u_j = z_j a^\rho$ $(j = 0, 1, \cdots, p)$ expressed in terms of ρ and the affine coordinates $(z_0, z_1, \cdots, z_p)$ of points of $\mathbf{S}$. The various values of each of these functions at each point of the manifold differ from each other by constant factors a^N (N integer). It is easily seen that at any point in any coordinate domain $\mathbf{V}_{\alpha 0}(\alpha = 0, \cdots, p)$ one has

$$(6) \qquad u_j = z_j a^\rho = {}_\alpha w_j\, z_\alpha e^{\tau \log z'_0} = {}_\alpha w_j e^{2\pi i t_{\alpha 0}};$$

since each u_j can be expressed as a holomorphic function, locally, of the complex coordinates in $\mathbf{M}_{p,0;\tau}$, u_j is itself a regular analytic function, and because of the multiplicative relation between the various values it takes at each point, it is a multiplicative entire function on the manifold, and hence an entire function on the universal covering space of $\mathbf{M}_{p,0}$. The set of functions u_j map $\mathbf{M}_{p,0;\tau}$ complex analytically into the complex manifold $\mathbf{H}_{p,\tau}$ described as follows, and originally discovered by H. Hopf, [3]. Let $\mathbf{E}^*_{p+1}$ be the complex manifold obtained by deleting the origin from the $p + 1$-dimensional complex affine space with coordinates $u_0, \cdots, u_p$ and let $\mathcal{G}$ be the discontinuous group of analytic automorphisms of $\mathbf{E}^*_{p+1}$ generated by the complex dilation γ: $(u_j) \to (au_j)$, for $a = e^{2\pi i\tau}$. A fundamental domain with respect to this group can be described as the spherical shell of points (u_j) satisfying $1 \leqq \sum_{j=0}^{p} |u_j|^2 < \max(|a|^2, |a|^{-2}) = e^{4\pi|\mathrm{Im}(\tau)|}$ with the obvious identification γ between the two components of the boundary. The manifold $\mathbf{H}_{p,\tau}$ is the complex manifold obtained by this identification. The mapping of $\mathbf{M}_{p,0;\tau}$ into $\mathbf{H}_{p,\tau}$ described by (6) is clearly an analytic isomorphism. In fact the image of each $\mathbf{V}_{\alpha 0}$ in $\mathbf{H}_{p,\tau}$ is the set of points (u_j) satisfying $u_\alpha \neq 0$, and for any such point the inverse map is represented by

$$(7) \qquad t_{\alpha 0} \equiv \frac{1}{2\pi i} \log u_\alpha \bmod (1, \tau); \qquad {}_\alpha w_j = u_j/u_\alpha,$$

which is everywhere complex analytic. This completes the proof of the analytic equivalence between $\mathbf{M}_{p,0;\tau}$ and $\mathbf{H}_{p,\tau}$.

4. Subvarieties of $M_{p,q}$

Let $\mathbf{W}$ be a compact, irreducible, complex analytic subvariety of $\mathbf{M}_{p,q}$ (defined by local equations with possible algebraic singularities) of complex dimension $m < p + q + 1$, other than a point. $\mathbf{W}$ is mapped under the fibre map Π onto a compact subvariety $\Pi(\mathbf{W}) = \mathbf{W}_0$ of $\mathbf{P} \times \mathbf{P}'$ of complex dimension $m_0 \leqq m$. A theorem due to W. L. Chow states that $\mathbf{W}_0$ is an irreducible algebraic subvariety of $\mathbf{P} \times \mathbf{P}'$ [6]; we shall prove that $m_0 < m$. In fact assume first that $m_0 = m$: then, since $\mathbf{W}_0$ is orientable, Π maps the cycle carried by $\mathbf{W}$ onto the one carried by $\mathbf{W}_0$ with positive degree. This is impossible, since on one hand Π induces a homomorphism of the $2m^{\text{th}}$ homology group of $\mathbf{M}_{p,q}$ (which is zero) into the corresponding homology group of $\mathbf{P} \times \mathbf{P}'$, and on the other hand the cycle carried by $\mathbf{W}_0$, which is an image, can not be homologous to zero, because of a theorem by S. Lefschetz [4]. Thus it is true that $m_0 < m$; for any point y of $\mathbf{W}$ the set $\Pi^{-1}\Pi(y) \cap \mathbf{W}$ is a compact, positive dimensional, complex analytic subvariety of the complex analytic fibre $\mathbf{T}_y = \Pi^{-1}\Pi(y)$ containing y, hence it is $\mathbf{T}_y$ itself. Thus we have shown that every positive dimensional, compact, analytic subvariety of $\mathbf{M}_{p,q}$ is fibered by tori, the base space being the projection of the variety into $\mathbf{P} \times \mathbf{P}'$, completing the proof of Theorem IV.

The proof of Theorem V follows as a corollary. In fact, if $f(y)$ is any meromorphic function of the points y of $\mathbf{M}_{p,q}$, then the set defined by $f(y) = $ constant is a $p + q$-dimensional subvariety of $\mathbf{M}_{p,q}$, and hence it is a union of fibres. This implies that every meromorphic function is constant along the fibres, so that its values are determined by $\Pi(y)$; in other words every meromorphic function

$f(y)$ on $\mathbf{M}_{p,q}$ satisfies the equation $f(y) = g(\Pi(y))$ for some meromorphic function g defined on $\mathbf{P} \times \mathbf{P}'$.

5. The Complex Analytic Cells $E_{p,q;\tau}$

Let (a, a') be an arbitrary point of $\mathbf{S} \times \mathbf{S}'$. Since the sets $\mathbf{S} - \{a\}$ and $\mathbf{S}' - \{a'\}$ are homeomorphic to cells, their Cartesian product $(\mathbf{S} - \{a\}) \times (\mathbf{S}' - \{a'\})$ is an open subset of $\mathbf{S} \times \mathbf{S}'$ homeomorphic to a Euclidean space of dimension $2p + 2q + 2$. If we superimpose on $\mathbf{S} \times \mathbf{S}'$ the complex structure of $\mathbf{M}_{p,q;\tau}$, then the cell $(\mathbf{S} - \{a\}) \times (\mathbf{S}' - \{a'\})$ becomes an open complex manifold, homeomorphic to a cell, which we denote by $\mathbf{E}_{p,q;\tau}$, or, if the value of τ is unimportant, by $\mathbf{E}_{p,q}$. We denote as in §2 by π, π' the fibre maps of $\mathbf{S}$ and $\mathbf{S}'$ onto $\mathbf{P}$ and $\mathbf{P}'$ respectively and by Π_0 the restriction to $\mathbf{E}_{p,q}$ of the fibre map Π of $\mathbf{M}_{p,q}$ onto $\mathbf{P} \times \mathbf{P}'$.

The inverse images $\Pi_0^{-1}(w, w')$ of points $(w, w') \in \mathbf{P} \times \mathbf{P}'$ into $\mathbf{E}_{p,q}$ are described completely by the following four cases; $\Pi_0^{-1}(w, w')$ is homeomorphic to a plane, if $(w, w') = \Pi(a, a')$; to a cylinder, if $w \neq \pi(a)$, but $w' = \pi'(a')$; to a cylinder also, if $w = \pi(a)$, but $w' \neq \pi'(a')$; to the whole torus of the original fibering of $\mathbf{M}_{p,q}$, if $w \neq \pi(a)$ and $w' \neq \pi'(a')$. The last two cases are possible, of course, if and only if $\mathbf{P}'$ has more than one point, i.e. if $q > 0$. We observe immediately that, if $q > 0$, the complex analytic cell $\mathbf{E}_{p,q;\tau}$ contains an everywhere dense set of tori $\mathbf{T}^2_{(1,\tau)}$ as compact, complex analytic submanifolds. It follows from this that $\mathbf{E}_{p,q}$, for $q > 0$, can not be mapped analytically and topologically into any complex affine space, because the affine coordinates, being entire holomorphic functions on $\mathbf{E}_{p,q}$ must be constants along its compact submanifolds; thus $\mathbf{E}_{p,q}$ *can not be covered by any single analytic coordinate system;* it follows also that $\mathbf{E}_{p,q}$ *can not have a Kaehler metric*, because no compact, complex submanifold of a Kaehler manifold (open or not) can be homologous to zero, as was shown by Hodge, [2], while on the other hand the tori $\mathbf{T}^2_{(1,\tau)}$ analytically imbedded in the cell $\mathbf{E}_{p,q}$ are continuously contractible to a point, and hence homologous to zero. Let f be an entire holomorphic function in $E_{p,q}(q > 0)$: f is obviously constant along the tori $\Pi_0^{-1}(w, w')$ of the fourth case mentioned above, and hence, by continuity, along each of the cylinders and the plane making up the remaining cases of $\Pi_0^{-1}(w, w')$; thus f induces a holomorphic function over $\Pi_0(\mathbf{E}_{p,q})$, i.e. $\mathbf{P} \times \mathbf{P}'$, and hence f is a constant; thus we have proved the last contention of Theorem VI, namely that *the only entire holomorphic functions on* $\mathbf{E}_{p,q}$ *are constants*. One can similarly prove that all meromorphic functions on $\mathbf{E}_{p,q}$ can be extended by continuity to meromorphic functions on $\mathbf{M}_{p,q}$, and hence they are characterized by Theorem V and by the comment at the end of §4. The three properties we have proved about the cells $\mathbf{E}_{p,q}(q > 0)$ of complex dimension $p + q + 1$ answer negatively, at least in the case of complex dimension $\geqq 3$, some of the questions raised by Bochner, [1], regarding the possibility of uniformization of complex manifolds homeomorphic to cells.

In the case we expressly excluded above, namely $q = 0$, the cell $\mathbf{E}_{p,0}$, being a simply connected subregion of the non-simply connected manifold $\mathbf{M}_{p,0;\tau}$ (isomorphic to $\mathbf{H}_{p,\tau}$), can be "rolled out" univalently into the universal covering space $\mathbf{E}^*_{p+1}$ (see §3) of $\mathbf{H}_{p,\tau}$. Hence $\mathbf{E}_{p,0;\tau}$ is analytically equivalent to a cellular region in the complex affine space $\mathbf{E}_{p+1}$, so that it is clear that none of the conclusions of Theorem VI can be applied to the cellular regions in the Hopf manifolds, which therefore appear to have several properties which make them exceptional among the other manifolds $\mathbf{M}_{p,q;\tau}$.

Louisiana State University
Swiss Federal School of Technology

References

[1] Bochner, S., Contributions to the Theory of Riemann Surfaces (by L. Ahlfors et al.), Princeton University Press, 1953.
[2] Hodge, W. V. D., Theory and Applications of the Harmonic Integrals, Cambridge University Press, 1941.
[3] Hopf, H., Zur Topologie der komplexen Mannigfaltigkeiten, Courant Anniversary Volume, New York, 1948, p. 168.
[4] Lefschetz, S., L'analysis situs et la Géométrie algébrique, Paris, Gauthier-Villars, 1950.
[5] Steenrod, N. E., The Topology of Fibre Bundles, Princeton University Press, 1952.
[6] Chow, W. L., *On Compact Complex Analytic Varieties*, Amer. J. Math., Vol. 71 (1949), pp. 893–914.

Isometric Imbedding of Complex Manifolds

Eugenio Calabi[1]

Ann. of Math. (2) 58, 1–23 (1953)

Chapter 1. Introduction

It is a classical (and generally unsolved) problem in Riemannian geometry whether a given Riemannian manifold can be isometrically imbedded in a Euclidean space of sufficiently high dimension, or, more generally, in a space of constant curvature. The purpose of this paper is to consider the imbedding problem in the special case where the Riemannian manifold and the enveloping space have a complex analytic structure and the imbedding is required to be complex analytic. This restriction of the problem is a topic of a paper by Bochner [1],[2] the only place, to the author's knowledge, where any results in that direction are to be found, namely, several partial conditions for the isometric imbedding of a Kaehler manifold in a finite dimensional unitary space.

The methods and results described in this article are refinements of those in the paper by Bochner mentioned above, and are a generalization of the author's [2] results on the isometric imbedding of metric Riemann surfaces. After some preliminaries (Chapter 2), one arrives at the necessary and sufficient conditions for the isometric imbedding, in the small as well as in the large, of a complex manifold in a unitary space of finite or infinite dimension (Chapter 3); by means of these results one deduces analogous conditions for the imbedding in the Hermitian manifolds of Fubini and Study (Chapter 4), following which some applications and special cases are described (Chapter 5).

Chapter 2. The Diastatic Function

Let $\mathbf{M}^k$ be a k-dimensional complex manifold and $\overline{\mathbf{M}}^k$ its conjugate manifold; by this we mean a complex manifold related to $\mathbf{M}^k$ by a homeomorphism mapping each point $\mathbf{p} \in \mathbf{M}^k$ onto a point $\bar{\mathbf{p}} \in \overline{\mathbf{M}}^k$, such that for each coordinate system $(z^1, z^2, \cdots, z^k)$ in a neighborhood $\mathbf{U}$ of $\mathbf{p}$ there exists a coordinate system $(z^{k+1}, z^{k+2}, \cdots, z^{2k})$ in the image $\bar{\mathbf{U}}$ of $\mathbf{U}$ under the conjugation, satisfying

$$z^{k+\alpha}(\bar{\mathbf{q}}) = \overline{z^{\alpha}(\mathbf{q})} \qquad (\alpha = 1, 2, \cdots, k). \tag{1}$$

Consider the product complex manifold $\mathbf{M}^k \times \overline{\mathbf{M}}^k$ and within it the diagonal set of points $(\mathbf{p}, \bar{\mathbf{p}})$, forming a real $2k$-dimensional real analytic submanifold, naturally homeomorphic to $\mathbf{M}^k$; this submanifold has a neighborhood that can be covered by a special class of coordinate systems $(z^1, z^2, \cdots, z^{2k})$ in $\mathbf{M}^k \times \overline{\mathbf{M}}^k$ as follows: for each complex coordinate system $(z^1, z^2, \cdots, z^k)$ defined in any do-

[1] This paper is a revision of the author's doctoral dissertation (Princeton, 1950), presented to the AMS on April 29, 1951, at the Chicago meeting, and prepared in its present form while the author held a Fellowship under Contract N onR 09800, Project No. NR 045 001, Office of Naval Research.

[2] Numbers in brackets refer to the bibliography at the end of this paper.

DOI: 10.2307/1969817

main $\mathbf{U}$ of $\mathbf{M}^k$, let $\bar{\mathbf{U}}$ be its conjugate domain in $\overline{\mathbf{M}}^k$ and $(z^{k+1}, \cdots, z^{2k})$ the coordinate system defined in $\bar{\mathbf{U}}$ by (1); then the sets $\mathbf{U} \times \bar{\mathbf{U}}$ cover a neighborhood of the submanifold of points $(\mathbf{p}, \bar{\mathbf{p}})$, and for any one such set, made up of points $(\mathbf{p},\bar{\mathbf{q}})(\mathbf{p},\mathbf{q} \in \mathbf{U})$, let $z^\alpha(\mathbf{p},\bar{\mathbf{q}}) = z^\alpha(\mathbf{p})$ and $z^{k+\alpha}(\mathbf{p},\bar{\mathbf{q}}) = z^{k+\alpha}(\bar{\mathbf{q}}) = (z^\alpha(\mathbf{q}))^-$ $(\alpha = 1, 2, \cdots, k)$; the points $(\mathbf{p}, \bar{\mathbf{p}})$ of the diagonal submanifold are then characterized by the equation $z^{k+\alpha} = \overline{z^\alpha}$.

From now on, whenever a real analytic functional element occurs in $\mathbf{M}^k$, we shall automatically lift its domain of definition into $\mathbf{M}^k \times \overline{\mathbf{M}}^k$, first by mapping each point $\mathbf{p}$ of the domain onto $(\mathbf{p}, \bar{\mathbf{p}})$, and then, after expressing the given functional element as a power series in $z^1, \cdots, z^k, \overline{z^1}, \cdots, \overline{z^k}$, by identifying each $z^\alpha(\mathbf{p})$ and $\overline{z^\beta(\mathbf{p})}$ respectively with $z^\alpha(\mathbf{p}, \bar{\mathbf{p}})$ and $z^{k+\beta}(\mathbf{p}, \bar{\mathbf{p}})$ $(\alpha, \beta = 1, 2, \cdots, k)$. Thus the real analytic functional element in $\mathbf{M}^k$ becomes a complex analytic functional element in $\mathbf{M}^k \times \overline{\mathbf{M}}^k$, and it is justifiable to consider its partial derivatives with respect to z^α or $\overline{z^\beta}$, or to substitute in the functional element values where $z^{k+\alpha}$ is not necessarily the complex conjugate of z^α and to call its value the value of the function at a point of a well defined manifold, namely $\mathbf{M}^k \times \overline{\mathbf{M}}^k$, and, if possible, to extend the functional element analytically in $\mathbf{M}^k \times \overline{\mathbf{M}}^k$. We shall henceforth adopt the notation α^* to denote the value $k + \alpha$ of an index, if the range of the index α is from 1 to k.

We now assume that $\mathbf{M}^k$ has a positive definite Hermite-Kaehler metric tensor (for the definition see *inter al.* Bochner, [1])[3]

$$ds^2 = g_{\alpha\beta^*}(z, \bar{z})dz^\alpha\, dz^{\beta^*} \qquad ((z, \bar{z}) = (z^1, \cdots, z^k, \bar{z}^1, \cdots, \bar{z}^k)), \tag{2}$$

whose components are real analytic functions of the coordinates. This metric is characterized by the fact (which could be used as an alternate definition) that in a neighborhood of each point there exists a real-valued analytic scalar $\Phi(z, \bar{z})$, such that

$$g_{\alpha\beta^*}(z, \bar{z}) = \frac{\partial^2 \Phi(z, \bar{z})}{\partial z^\alpha\, \partial \bar{z}^\beta} \qquad (\alpha, \beta = 1, 2, \cdots, k). \tag{3}$$

The function Φ is not uniquely determined by the Hermitian form (2), but the class of all functions that satisfy (3) is easily characterized by knowing one of them; in fact, if $\Phi'(z, \bar{z})$ also satisfies (3), then

$$\frac{\partial^2 [\Phi'(z, \bar{z}) - \Phi(z, \bar{z})]}{\partial z^\alpha\, \partial \bar{z}^\beta} = 0,$$

which amounts to

$$\Phi'(z, \bar{z}) - \Phi(z, \bar{z}) = \phi(z) + \overline{\phi(z)}, \tag{4}$$

where $\phi(z)$ is a complex analytic function of $z^1, \cdots, z^k$. Let $\Phi(z, \bar{z}) = \Phi(z^1, \cdots, z^{2k})$ be expressed as a power series in the $2k$ variables around a point of the diagonal

[3] For typographical convenience we take the square distance element to be one half of what it has been previously defined to be; some numerical constants appearing later will consequently be also different from their equivalent ones recorded in previous articles.

submanifold with coordinates z_0^i ($i = 1, 2, \cdots, 2k$), which converges in a domain $\mathbf{U} \times \bar{\mathbf{U}}$ defined by $|z^i - z_0^i| < \rho$ ($i = 1, 2, \cdots, 2k$) for some positive ρ. Let $\mathbf{p}$ and $\mathbf{q}$ be two arbitrary points of $\mathbf{U}$ with coordinates $(z(\mathbf{p}))$ and $(z(\mathbf{q}))$ respectively; we then define the functional element

$$(5)\quad D(\mathbf{p}, \mathbf{q}) = \Phi(z(\mathbf{p}), \overline{z(\mathbf{p})}) + \Phi(z(\mathbf{q}), \overline{z(\mathbf{q})}) - \Phi(z(\mathbf{p}), \overline{z(\mathbf{q})}) - \Phi(z(\mathbf{q}), \overline{z(\mathbf{p})}).$$

The real analytic function generated by the functional element of (5) is a new invariant of the Kaehler metric; because of its essential role in the following treatment we propose for the function $D(\mathbf{p}, \mathbf{q})$ the name *diastasis of* $\mathbf{p}$ *and* $\mathbf{q}$ or *diastatic function*, from the Greek διάσταсις (distance). The following propositions will describe some of its elementary properties.

PROPOSITION 1. *The functional element of* $D(\mathbf{p},\mathbf{q})$ *defined by* (5) *is uniquely determined by the Kaehler metric and independent of the local complex analytic coordinate system.*
In fact any function $\Phi(z, \bar{z})$ satisfying (3) is independent of local coordinates and, if two such functions defined over a common domain are substituted in (5), then it follows trivially from (4) that one obtains the same functional element $D(\mathbf{p}, \mathbf{q})$ from either of them.

PROPOSITION 2. *The diastasis is real valued in its domain of (real) analyticity.*
In fact, since $\Phi(z, \bar{z})$ is real valued, it follows that $\Phi(z(\mathbf{p}), \overline{z(\mathbf{q})}) = \overline{\Phi(z(\mathbf{q}), \overline{z(\mathbf{p})})}$; thus the conclusion follows from (5).

PROPOSITION 3. *The diastasis* $D(\mathbf{p}, \mathbf{q})$ *is symmetric in* $\mathbf{p}$ *and* $\mathbf{q}$ *and* $D(\mathbf{p}, \mathbf{p}) = 0$.

PROPOSITION 4. *For* $\mathbf{q}$ *fixed,* $\partial^2 D(\mathbf{p},\mathbf{q})/\partial(z^\alpha(\mathbf{p}))\partial(z^{\beta^*}(\bar{\mathbf{p}})) = g_{\alpha\beta^*}(z(\mathbf{p}), \overline{z(\mathbf{p})})$, *for* $\mathbf{p}$ *fixed,* $\partial^2 D(\mathbf{p},\mathbf{q})/\partial(z^\alpha(\mathbf{q}))\partial(z^{\beta^*}(\bar{\mathbf{q}})) = g_{\alpha\beta^*}(z(\mathbf{q}), \overline{z(\mathbf{q})})$, *and, in general,*

$$\partial^2 D(\mathbf{p},\mathbf{q})/\partial(z^\alpha(\mathbf{p}))\partial(z^{\beta^*}(\bar{\mathbf{q}})) = -g_{\alpha\beta^*}(z(\mathbf{p}), \overline{z(\mathbf{q})}), \quad \partial^2 D(\mathbf{p},\mathbf{q})/\partial(z^\alpha(\mathbf{p}))\partial(z^\beta(\mathbf{q})) = 0.$$

The verification of the last two propositions being immediate from the definitions, we arrive to the one which justifies the name "Diastasis" selected for the function $D(\mathbf{p},\mathbf{q})$.

PROPOSITION 5. *If* $\rho(\mathbf{p},\mathbf{q})$ *is the geodesic distance between* $\mathbf{p}$ *and* $\mathbf{q}$ *and* $D(\mathbf{p},\mathbf{q})$ *the functional element of the diastasis, defined in* (5), *then* $D(\mathbf{p},\mathbf{q}) = (\rho(\mathbf{p},\mathbf{q}))^2 + 0((\rho(\mathbf{p},\mathbf{q}))^4)$.

PROOF. We fix $\mathbf{p}$ and select a local coordinate system with origin at $\mathbf{p}$, which is canonical in the sense of Bochner (see [1], p. 181), that is to say, such that $D(\mathbf{p}, \mathbf{q}) = \Phi(z(\mathbf{q}), \overline{z(\mathbf{q})}) = \sum_{\alpha=1}^k |z^\alpha(\mathbf{q})|^2 + \psi_{2,2}(z(q), \overline{z(\mathbf{q})})$, where $\psi_{2,2}(z, \bar{z})$ is a power series in $(z, \bar{z})$ with no term of degree <2 in either the variables (z) or $(\bar{z})$. Since $g_{\alpha\beta^*} = \delta_{\alpha\beta}$ at the origin, and since the first partial derivatives of the metric vanish at the origin, it follows that $(\rho(\mathbf{p}, \mathbf{q})) = \sum_{\alpha=1}^k |z^\alpha(\mathbf{q})|^2 + 0((\sum_{\alpha=1}^k |z^\alpha(\mathbf{q})|^2)^2)$. The conclusion then follows from the last two approximations.

REMARK. For a flat (i.e. unitary, if finite dimensional, or Hilbert, if infinite dimensional) space the diastasis is exactly equal to the square of the geodesic distance. In fact, if (z) are unitary coordinates, then

$$(6)\qquad D(\mathbf{p}, \mathbf{q}) = \sum_{\alpha=1}^k |z^\alpha(\mathbf{q}) - z^\alpha(\mathbf{p})|^2 \qquad (k \leqq \infty).$$

Note. In this last proposition one must be careful to specify that $D(\mathbf{p}, \mathbf{q})$ represents the original functional element of the diastasis, and not values obtained from its maximal analytic continuation, since for $\mathbf{p}$ fixed it may well happen that the diastasis is not single valued in $\mathbf{q}$, thus making the statement untrue.

PROPOSITION 6. *Let $\mathbf{S}^N$ be an N-dimensional ($N < \infty$) Kaehler manifold with an analytic metric and $\mathbf{M}^k$ a complex analytic submanifold of S^N. Then the diastatic function $D_{\mathbf{S}}(\mathbf{p}, \mathbf{q})$ of $\mathbf{S}^N$, restricted to pairs of points of $\mathbf{M}^k$, is the same as the diastasis $D_{\mathbf{M}}(\mathbf{p}, \mathbf{q})$ obtained from the metric induced on $\mathbf{M}^k$.*

PROOF. It is a well known and easily verifiable fact that the metric induced on $\mathbf{M}^k$ under the conditions stated above is an analytic Kaehler metric. Let $(x^1, \cdots, x^N)$ and $(z^1, \cdots, z^k)$ be local holomorphic coordinate systems respectively in $\mathbf{S}^N$ and in $\mathbf{M}^k$, and $x^\sigma = f^\sigma(z)$ ($\sigma = 1, 2, \cdots, N$) be the local imbedding functions. Then, if $h_{\sigma\tau^*}(x, \bar{x}) = \partial^2\Phi_{\mathbf{S}}(x, x)/\partial x^\sigma \partial x^{\tau^*}$ is the Kaehler metric tensor of $\mathbf{S}^N$, it follows that

$$\frac{\partial^2\Phi_{\mathbf{S}}(f(z), \bar{f}(z))}{\partial z^\alpha \partial z^{\beta^*}} = h_{\sigma\tau^*}(f(z), \bar{f}(z)) \frac{\partial f^\sigma(z)}{\partial z^\alpha} \left(\frac{\partial f^\tau(z)}{\partial z^\beta}\right)^- = g_{\alpha\beta^*}(z, \bar{z}) = \frac{\partial^2\Phi_{\mathbf{M}}(z, \bar{z})}{\partial z^\alpha \partial z^{\beta^*}},$$

which is the induced Kaehler metric in $\mathbf{M}^k$; consequently we can take $\Phi_{\mathbf{M}}(z, \bar{z}) = \Phi_{\mathbf{S}}(f(z), \bar{f}(z))$, from which the conclusion follows by virtue of (5).

The last proposition shows the characteristic property of the diastasis, which puts it in contrast with the geodesic distance function, whose square it approximates in the small, for, while the former is inductive on analytic submanifolds, the latter is not, except in some very special cases. One can see from this property the importance of the diastasis in connection with isometric imbedding, especially in the cases where the ambient manifold has the property that the diastasis depends on geodesic distance alone, as is the case with the unitary and the Fubini-Study spaces, to which our discussion will be limited.

It is worth adding, in conclusion to this chapter, that the only known cases of Kaehler manifolds with analytic metric have the property that, for fixed $\mathbf{p}$ the analytic continuation of the diastasis $D(\mathbf{p},\mathbf{q})$ with respect to $\mathbf{q}$ over the covering manifold has as singularities at most a complex analytic hypersurface, which, if the original manifold is compact and simply connected, is homologous to the fundamental differential form of degree 2 associated with the metric. One could formulate several conjectures over the behavior of the diastasis in the large, which, in the author's opinion would furnish ample material for future study.

CHAPTER 3. IMBEDDING IN FLAT SPACES

We consider now the problem of the isometric, complex analytic imbedding of complex manifolds in the following types of flat spaces:

(a) The unitary spaces $\mathbf{C}^N$, or spaces with the coordinate system $(x) = (x^1, x^2, \cdots, x^N)$ and the metric $ds^2 = \sum_{\sigma=1}^N |dx^\sigma|^2$,

(b) The countable infinite dimensional unitary, or complex Hilbert, space $\mathbf{C}^\infty$ of points with coordinates $(x) = (x^1, x^2, \cdots, x^\sigma, \cdots)$ such that $\sum_{\sigma=1}^\infty |x^\sigma|^2 < \infty$, with the metric $ds^2 = \sum_{\sigma=1}^\infty |dx^\sigma|^2$,

(c) The *indefinite Hilbert space* $\mathbf{E}$, defined as the space of points whose coordinates are $(x^1, x^{-1}, x^2, x^{-2}, \cdots, x^{\pm\sigma}, \cdots)$ such that $\sum'_\sigma |x^\sigma|^2 < \infty$,[4] with the indefinite Hermitian form $ds^2 = \sum'_\sigma(\text{sgn } \sigma)\,|dx^\sigma|^2$, corresponding to the diastasis $D(\mathbf{p},\mathbf{q}) = \sum'_\sigma(\text{sgn } \sigma)\,|x^\sigma(\mathbf{q}) - x^\sigma(\mathbf{p})|^2$ (for the purpose of inducing metrics on complex analytic submanifolds), and with the topology induced by the ordinary Hilbert space metric $\rho(\mathbf{p},\mathbf{q}) = (\sum'_\sigma |x^\sigma(\mathbf{q}) - x^\sigma(\mathbf{p})|^2)^{\frac{1}{2}}$ to which we shall refer when using such terms as "continuous", "bounded", etc.

The indefinite Hilbert space is introduced here, because, as we shall prove shortly, it constitutes a "universal" Kaehler manifold, i.e. one in which every Kaehler manifold with analytic metric can be isometrically imbedded in the small. After having established the conditions for the imbedding in the positive definite unitary spaces, we shall prove the main result of this chapter, that if the local conditions are satisfied, then the imbedding functions can be extended as holomorphic functions on the given complex manifold, or a suitable covering of it; the proof of this last result is greatly simplified by making use of the local imbedding in the indefinite Hilbert space.

In the cases where the ambient space is infinite dimensional, as in case (b) and (c) we shall understand once and for all by a complex analytic imbedding of a manifold with parameters $(z) = (z^i, \cdots, z^k)$ a set of functions $x^\sigma = f^\sigma(z)$ satisfying not only

$$\sum_\sigma |f^\sigma(z)|^2 < \infty \tag{7}$$

at each point of the coordinate domain (z), but also the stronger local boundedness condition

$$\sum_\sigma |f^\sigma(z)|^2 < R < \infty \text{ for } |z^\alpha| < \rho \text{ for some } \rho > 0,\ R > 0. \tag{8}$$

It is an unsolved problem whether (8) is a consequence of (7) in the infinite dimensional cases; being unable to prove (8), we must include it in the assumptions, since otherwise the imbedding map could not even be continuous[5]. It is also necessary to establish the fact that the Kaehler metric of the infinite dimensional flat spaces induces, as it does in the finite dimensional case, an analytic Kaehler metric in the analytic submanifolds; this is done as a consequence of the following Lemma.

LEMMA 1. *If a sequence $(f^\sigma(z^1, \cdots, z^k))_{\sigma=1}^\infty$ of complex analytic functions defined over a common domain satisfies* (8), *then the real valued function* $F(z, \bar z) = \sum_{\sigma=1}^\infty |f^\sigma(z)|^2$ *is analytic for* $|z^\alpha| < \rho$ $(\alpha = 1, 2, \cdots, k)$.

PROOF. The conclusion will follow from the stronger statement, that the series of holomorphic functions $\sum_{\sigma=1}^\infty f^\sigma(z^1, \cdots, z^k) f^{\sigma *}(z^{k+1}, \cdots, z^{2k})$, where

$$f^{\sigma *}(\bar z^1, \cdots, \bar z^k) = (f^\sigma(z^1, \cdots, z^k))^-, \tag{9}$$

[4] By $\sum'_\sigma$ we mean the infinite sum for σ ranging over all non-zero integers.

[5] (Added in proof). An ingenious counterexample has been constructed by B. Lepson. It consists of an analytic mapping of the disc $|z| < 1$ into $\mathbf{C}^\infty$ given by a sequence of holomorphic functions $x^\sigma = f^\sigma(z)$ $(\sigma = 1,2, \cdots;\ |z| < 1)$ such that $\sum_{\sigma=1}^\infty |f^\sigma(z)|^2$ converges everywhere to a function which is discontinuous for all real values of z.

converges uniformly in the domain $|z^i| \leqq \rho'$ $(i = 1, 2, \cdots, 2k)$ for any ρ' $(0 < \rho' < \rho)$; by the theorem of Weierstrass the limit is then a holomorphic function $F'(z^1, z^2, \cdots, z^{2k})$, whence the conclusion follows, since

$$F'(z^1, \cdots, z^k, \bar{z}^1, \cdots, \bar{z}^k) = F(z, \bar{z}).$$

In fact it is sufficient to consider the multiple Cauchy integrals

$$F'_J(z^1, \cdots z^{2k}) = \frac{1}{(2\pi\sqrt{-1})^{2k}} \oint \frac{\sum_{\sigma=1}^{J} f^\sigma(\zeta^1, \cdots, \zeta^k) f^{\sigma*}(\zeta^{k+1}, \cdots, \zeta^{2k})}{\prod_{i=1}^{2k}(\zeta^i - z^i)}\, d\zeta^1 \cdots d\zeta^{2k}$$
$$(J = 1, 2, \cdots)$$

over the $2k$-dimensional torus $\{|\zeta^i| = \rho''\}$ $(\rho' < \rho'' < \rho)$. The integrand is obviously uniformly and absolutely convergent as J becomes infinite; thus, by the Lebesgue theorem on dominated convergence, the integral converges uniformly, completing the proof of the Lemma.

This last result enables us to formulate the first theorem on complex analytic imbedding.

THEOREM 1. *A complex manifold* $\mathbf{M}^k$ *with a metric can be (locally) imbedded isometrically and complex analytically in the indefinite Hilbert space* $\mathbf{E}$, *if and only if the metric is an analytic Kaehler metric.*

PROOF. We first assume that

$$x^\sigma = f^\sigma(z) = f^\sigma(z^1, \cdots, z^k) \qquad (\sigma = \pm 1, \pm 2, \cdots) \tag{10}$$

represent a set of local imbedding equations satisfying (8) in the domain

$$\{|z^\alpha| < \rho_\alpha\} \ (\alpha = 1, 2, \cdots, k).$$

If we denote by $\mathbf{p}_0$ the point corresponding to $z^\alpha = 0$, we can write the diastasis of $\mathbf{E}$ restricted to $\mathbf{M}^k$ in terms of (z) as follows:

$$D(\mathbf{p}_0, \mathbf{p}) = \sum_\sigma{}' (\operatorname{sgn} \sigma)\, |f^\sigma(z(\mathbf{p})) - f^\sigma(0)|^2. \tag{11}$$

By Lemma 1 this function is real analytic at $\mathbf{p}$, so that by Propositions 4 and 6 one can differentiate it with respect to $(z, \bar{z})$ and obtain the induced metric on $\mathbf{M}^k$, thus verifying immediately that it is an analytic Kaehler metric.

Conversely, assume that the metric in $\mathbf{M}^k$ is an analytic Kaehler metric, whose local power series expansion in terms of $(z, \bar{z})$ converges for $|z^\alpha| < \rho_\alpha$. We can then write the power series expansion of the diastasis

$$D(\mathbf{p}_0, \mathbf{p}) = \sum_{\substack{m_1, \cdots, m_k, n_1, \cdots, n_k \geqq 0 \\ \Sigma_\alpha m_\alpha \geqq 1; \Sigma_\alpha n_\alpha \geqq 1}} B_{m_1, \cdots, m_k, n_1, \cdots, n_k} \prod_{\alpha=1}^{k} [(z^\alpha(\mathbf{p}))^{m_\alpha} (\overline{z^\alpha(\mathbf{p})})^{n_\alpha}].$$

Since we shall be dealing repeatedly with power series expansions in k variables or in the $2k$ variables obtained from the first k by adjoining their conjugates, it will be convenient to introduce here some abbreviated notation that will still indicate the relevant properties of the terms of the summation. For each dimension k we select a fixed enumeration of all k-tuples of non-negative integers $(m_1, \cdots, m_k)$, and write them in sequence form $(m)_0, (m)_1, (m)_2, \cdots$, where $(m)_J$ stands for the $(J + 1)^{\text{st}}$ k-tuple $(m_{1,J}, m_{2,J}, \cdots, m_{k,J})$; we take $(m)_0$ to denote the k-tuple of all zeros, denote by $|(m)_J|$ the sum $\sum_{\alpha=1}^{k} m_{\alpha,J}$ and assume $|(m)_J|$ to be non-decreasing with increasing J; we then denote by $(z)^{(m)_J}$ the monomial in k variables $\prod_{\alpha=1}^{k}(z^\alpha)^{m_{\alpha,J}}$ (of total degree $|(m)_J|$). Thus the functional element of the diastasis becomes

$$D(\mathbf{p}_0,\mathbf{p}) = \sum_{J,K=1}^{\infty} B_{J,K}(z(\mathbf{p}))^{(m)_J}(\overline{z(\mathbf{q})})^{(m)_K} \qquad (|z^\alpha(\mathbf{p})| < \rho_\alpha); \tag{12}$$

the fact that the diastasis is real valued means that the infinite matrix $(B_{J,K})$ of coefficients is Hermitian. We shall now construct a sequence of functions $x^\sigma = f^\sigma(z)$ $(\sigma = \pm 1, \pm 2, \cdots)$, which satisfy (8) and (11) in the domain $|z^\alpha| < \rho'_\alpha$ for any k-tuple $(\rho'_1, \cdots, \rho'_k)$ such that $0 < \rho'_\alpha < \rho_\alpha$. Le $r_1, \cdots, r_k$ be any arbitrary positive numbers (to be fixed later), and

$$\begin{aligned} f^\sigma(z) &= \frac{1}{2}\left[B_{\sigma,\sigma}(r)^{(m)_\sigma} + \frac{1}{(r)^{(m)_\sigma}}\right](z)^{(m)_\sigma} + \sum_{J=\sigma+1}^{\infty} B_{J,\sigma}(r)^{(m)_\sigma}(z)^{(m)_J}, \\ f^{-\sigma}(z) &= \frac{1}{2}\left[B_{\sigma,\sigma}(r)^{(m)_\sigma} - \frac{1}{(r)^{(m)_\sigma}}\right](z)^{(m)_\sigma} + \sum_{J=\sigma+1}^{\infty} B_{J,\sigma}(r)^{(m)_\sigma}(z)^{(m)_J} \end{aligned} \tag{13}$$

$$(\sigma = 1, 2, \cdots).$$

Then one verifies immediately the identity

$$|f^\sigma(z)|^2 - |f^{-\sigma}(z)|^2 = \sum_{\min(J,K)=\sigma} B_{J,K}(z)^{(m)_J}(\bar{z})^{(m)_K},$$

and therefore

$$\begin{aligned} \sum_{\sigma=1}^{\infty}[|f^\sigma(z(\mathbf{p}))|^2 - |f^{-\sigma}(z(\mathbf{p}))|^2] &= \sum_{J,K=1}^{\infty} B_{J,K}(z(\mathbf{p}))^{(m)_J}(\overline{z(\mathbf{p})})^{(m)_K} \\ &= D(\mathbf{p}_0,\mathbf{p}). \end{aligned} \tag{14}$$

This last identity implies (11), namely that the functions $f^\sigma(z)$ are the desired imbedding functions, provided only that they satisfy the absolute square convergence condition (8); the domain in which this condition is satisfied depends on the choice of $(r_1, \cdots, r_k)$. Given the k-tuple (ρ') such that $0 < \rho'_\alpha < \rho_\alpha$, where (ρ) is the k-tuple appearing in (12), choose (ρ''), such that $\rho'_\alpha < \rho''_\alpha < \rho_\alpha$. From the domain of convergence of the series in (12), we know that there exists a positive constant H, such that $|B_{J,K}| \leqq H/(\rho'')^{(m)_J}(\rho'')^{(m)_K}$; hence, by applying this inequality to (13), we obtain in the domain $\{|z^\alpha| \leqq \rho'_\alpha\}$ $(\alpha = 1, 2, \cdots, k)$

$$\begin{aligned} |f^{\pm\sigma}(z)| &\leqq \frac{1}{2}\left[\left(\frac{r\rho'}{\rho''^2}\right)^{(m)_\sigma} H + \left(\frac{\rho'}{r}\right)^{(m)_\sigma}\right] + H\left(\frac{r}{\rho''}\right)^{(m)_\sigma}\left[\sum_{J=\sigma+1}^{\infty}\left(\frac{\rho'}{\rho''}\right)^{(m)_J}\right] \\ &< \frac{1}{2}\left(\frac{\rho'}{r}\right)^{(m)_\sigma} + H\left(\frac{r}{\rho''}\right)^{(m)_\sigma}\left[\sum_{J=0}^{\infty}\left(\frac{\rho'}{\rho''}\right)^{(m)_J}\right] \\ &= \frac{1}{2}\left(\frac{\rho'}{r}\right)^{(m)_\sigma} + \frac{H(r/\rho'')^{(m)_\sigma}}{\prod_{\alpha=1}^{k}(1 - (\rho'_\alpha/\rho''_\alpha))} \qquad (\sigma = 1, 2, \cdots). \end{aligned}$$

We now pick $r_\alpha = \sqrt{\rho'_\alpha\rho''_\alpha}$, so that

$$|f^{\pm\sigma}(z)| < \left(\sqrt{\frac{\rho'}{\rho''}}\right)^{(m)_\sigma}\left[\frac{1}{2} + \frac{H}{\prod_{\alpha=1}^{k}(1 - (\rho'_\alpha/\rho''_\alpha))}\right] = K(\sqrt{\rho'/\rho''})^{(m)_\sigma}$$

and, consequently,

$$\sum'_{\sigma} | f^{\sigma}(z) |^2 < 2K^2 \left[\sum_{\sigma=0}^{\infty} \left(\frac{\rho'}{\rho''} \right)^{(m)^{\sigma}} \right] = \frac{2K^2}{\prod_{\alpha=1}^{K} (1 - (\rho'_{\alpha}/\rho''_{\alpha}))} = R < \infty$$

which verifies that the functions $f^{\sigma}(z)$ satisfy (8) in the domain $\{| z^{*\alpha} | < \rho'_{\alpha}\}$; this, in conjunction with (14), implies (11) in the same domain, thus completing the proof of the Theorem.

We now proceed to consider the slightly more complicated case, where the ambient space is a positive definite unitary space; the next theorem establishes that any isometric, complex analytic imbedding in such a unitary space is essentially unique, *i.e.* determined up to within a motion.

THEOREM 2. (LOCAL RIGIDITY). *Let* $\mathbf{M}^k$ *be a Kaehler manifold, and assume that a given, arbitrarily small region* $\mathbf{G}$ *of* $\mathbf{M}^k$ *can be isometrically and complex analytically imbedded in a unitary space* $\mathbf{C}^N$ *of dimension* N ($N \leqq \infty$), *so that it does not lie in any proper complex analytic linear subspace of* $\mathbf{C}^N$. *Then* N *is uniquely determined by the Kaehler metric in* $\mathbf{G}$, *and the imbedding is determined up to within the group of motions in* $\mathbf{C}^N$ *(unitary transformations followed by translations).*

PROOF. We assume that the region $\mathbf{G}$ is described by the inequalities $| z^{\alpha} | < \rho_{\alpha}$ for a set of parameters (z) of $\mathbf{M}^k$; we also assume that

$$x^{\sigma} = f^{\sigma}(z) \qquad (\sigma = 1, 2, \cdots, N)$$

and

$$y^{\tau} = g^{\tau}(z) \qquad (\tau = 1, 2, \cdots, N') \; (N, N' \leqq \infty)$$

are two distinct sets of holomorphic functions, each representing an isometric, complex analytic imbedding of $\mathbf{G}$ respectively in $\mathbf{C}^N$ and in $\mathbf{C}^{N'}$. By assumption of the Theorem, there exists no non-trivial, inhomogeneous linear relation among the functions $f^{\sigma}(z)$ (respectively $g^{\tau}(z)$).

We first show that $\mathbf{G}$ does not lie in any real hyperplane of dimension $2N - 1$ in $\mathbf{C}^N$ (respectively $2N' - 1$ in $\mathbf{C}^{N'}$). In fact, if this were the case, one could write a real linear relation between the functions $f^{\sigma}(z)$ and their complex conjugates, such as $\sum_{\sigma=1}^{N} a_{\sigma} f^{\sigma}(z) = c + \sum_{\sigma=1}^{N} b_{\sigma} \bar{f}_{\sigma}(z)$, with the coefficients a_{σ} and b_{σ} not all equal to zero; this is clearly impossible, as the last equation shows, since a holomorphic function equal to an antiholomorphic function is a constant, implying a linear relation among the functions $f^{\sigma}(z)$ alone.

We next prove that $N = N'$. Theorem 1 implies that the metric in $\mathbf{G}$ must be real analytic and, by Proposition 6 we know that the diastasis of $\mathbf{C}^N$ and $\mathbf{C}^{N'}$, given by (6), restricted to $\mathbf{G}$, is the same; thus

$$(15) \quad D(\mathbf{p}, \mathbf{q}) = \sum_{\sigma=1}^{N} | f^{\sigma}(z(\mathbf{q})) - f^{\sigma}(z(\mathbf{p}))|^2 = \sum_{\tau=1}^{N} | g^{\tau}(z(\mathbf{q})) - g^{\tau}(z(\mathbf{p})) |^2.$$

Let $\mathbf{p}_0, \mathbf{p}_1, \cdots, \mathbf{p}_n$ be any $n + 1$ points of $\mathbf{G}$. Suppose that their images in $\mathbf{C}^N$ are dependent (in the sense of Euclidean geometry), *i.e.* the n vectors $\overrightarrow{\mathbf{p}_0\mathbf{p}_1}, \cdots, \overrightarrow{\mathbf{p}_0\mathbf{p}_n}$ satisfy $\sum_{\nu=1}^{n} a_{\nu} \overrightarrow{\mathbf{p}_0\mathbf{p}_{\nu}} = 0$ for real constants a_{ν} not all equal to zero.

Since the space is positive definite, to say that a vector is zero is equivalent to saying that its real scalar product with itself is zero; thus

$$(16)\qquad (\textstyle\sum_{\mu=1}^{n} a_\mu \overrightarrow{\mathbf{p}_0\mathbf{p}_\mu})\cdot(\sum_{\nu=1}^{n} a_\nu \overrightarrow{\mathbf{p}_0\mathbf{p}_\nu}) = \sum_{\mu,\nu=1}^{n} a_\mu a_\nu (\overrightarrow{\mathbf{p}_0\mathbf{p}_\mu}\cdot\overrightarrow{\mathbf{p}_0\mathbf{p}_\nu}) = 0,$$

where $\overrightarrow{\mathbf{p}_0\mathbf{p}_\mu}\cdot\overrightarrow{\mathbf{p}_0\mathbf{p}_\nu}$ denotes the real scalar product of the two vectors in $\mathbf{C}^N$. But the scalar product of two vectors is a metric invariant of $\mathbf{C}^N$, according to the relations $\overrightarrow{\mathbf{p}_0\mathbf{p}}\cdot\overrightarrow{\mathbf{p}_0\mathbf{p}} = D(\mathbf{p}_0,\mathbf{p})$ and $\overrightarrow{\mathbf{p}_0\mathbf{p}}\cdot\overrightarrow{\mathbf{p}_0\mathbf{q}} = \frac{1}{2}[D(\mathbf{p}_0,\mathbf{p}) + D(\mathbf{p}_0,\mathbf{q}) - D(\mathbf{p},\mathbf{q})]$. Hence (16) can be rewritten as

$$(17)\qquad \tfrac{1}{2}\textstyle\sum_{\mu,\nu=1}^{n} a_\mu a_\nu[D(\mathbf{p}_0,\mathbf{p}_\mu) + D(\mathbf{p}_0,\mathbf{p}_\nu) - D(\mathbf{p}_\mu,\mathbf{p}_\nu)] = 0.$$

This last equation expresses the fact that the $n + 1$ points $\mathbf{p}_0, \cdots, \mathbf{p}_n$ have linearly dependent images in $\mathbf{C}^N$; but then the same must be true for the same points imbedded in any other unitary space $\mathbf{C}^{N'}$ because of Proposition 6; therefore a maximal set of points of $\mathbf{G}$ which are linearly independent when imbedded in $\mathbf{C}^N$ can be constructed inductively by using the intrinsic properties of the metric alone, and can consist only of $2N + 1$ points. Thus we have proved that, if $\mathbf{G}$ can be imbedded in $\mathbf{C}^N$ so as not to lie in a linear subspace, then the dimension N is uniquely determined by the Kaehler metric of $\mathbf{G}$.

In order to prove that any two imbeddings of $\mathbf{G}$ in $\mathbf{C}^N$, under the conditions stated, are equivalent under a unitary motion, it is enough to observe that the correspondence between the two images of $\mathbf{G}$ in $\mathbf{C}^N$ preserves distance, because of Proposition 6 and the remark at the end of Proposition 5; hence, by the axiom of free mobility of real Euclidean spaces, there exists a real orthogonal transformation which transforms the equations $x^\sigma = f^\sigma(z)$ into the set $y^\tau = g^\tau(z)$ ($\sigma, \tau = 1, 2, \cdots, N$). This motion is unique, since the images of $\mathbf{G}$ span $\mathbf{C}^N$ linearly in the real sense. By a suitable translation we may carry the origin of the parameter system (z) into the origin of $\mathbf{C}^N$ for each of the two mappings; thus the transformation can be written as $x^\sigma = \sum_{\tau=1}^{N} a_\tau^\sigma y^\tau + \sum_{\tau=1}^{N} b_\tau^\sigma \overline{y^\tau}$; but this implies

$$(18)\qquad f^\sigma(z) - \textstyle\sum_{\tau=1}^{N} a_\tau^\sigma g^\tau(z) = \sum_{\tau=1}^{N} b_\tau^\sigma \overline{g^\tau(z)} \qquad (\sigma = 1, 2, \cdots, N).$$

Since both sides of this set of equations are constants (and indeed zero, since the left member vanishes for $z^\alpha = 0$), the correspondence between the mappings can be written as $x^\sigma = f^\sigma(z) = \sum_{\tau=1}^{N} a_\tau^\sigma g^\tau(z) = \sum_{\tau=1}^{N} a_\tau^\sigma y^\tau$, and, since this complex linear transformation preserves distance, it is unitary, thus completing the proof of Theorem 2.

The next step consists of characterizing the conditions for a complex manifold to satisfy the assumptions of Theorem 2, namely, for it to admit an isometric imbedding in a suitably high dimensional $\mathbf{C}^N$. We have already established that a necessary condition on the metric is that it be an analytic Kaehler metric; the additional condition can be best stated in terms of the power series expansion of the diastasis, as in (12).

THEOREM 3. *Let $\mathbf{M}^k$ be a Kaehler manifold with analytic metric, $\mathbf{p}_0$ a point of $\mathbf{M}^k$, (z) a complex coordinate system with origin at $\mathbf{p}_0$. Let $\mathbf{G}$ be the neighborhood of*

$\mathbf{p}_0$ *in which the power series expansion of the diastasis* $D(\mathbf{p}_0, \mathbf{p})$ *in terms of* $(z(\mathbf{p}), \overline{z(\mathbf{p})})$ *converges. Then, for any given positive integer* $N \leqq \infty$, *there exists a neighborhood of* $\mathbf{p}_0$ *that can be isometrically imbedded in* $\mathbf{C}^N$, *if and only if the coefficient matrix* $B_{J,K}$ *of the power series expansion* $D(\mathbf{p}_0,\mathbf{p}) = \sum_{J,K=1}^{\infty} B_{J,K}(z(\mathbf{p}))^{(m)J}(\overline{z(\mathbf{p})})^{(m)K}$ *is positive semidefinite and of rank at most* N. *Furthermore, if the condition is satisfied, at least all of* $\mathbf{G}$ *can be isometrically imbedded in* $\mathbf{C}^N$.

PROOF. We first assume that there exists a sequence of functions $x^\sigma = f^\sigma(z)$ $(\sigma = 1, 2, \cdots, N)$ which represents an imbedding of a neighborhood of $\mathbf{p}_0$ in $\mathbf{C}^N$; after a suitable translation in $\mathbf{C}^N$ we may assume $\mathbf{p}_0$ to be mapped onto the origin of $\mathbf{C}^N$. By Theorem 1 the induced diastasis

$$D(\mathbf{p}_0, \mathbf{p}) = \sum_{\sigma=1}^{N} | f^\sigma(z(\mathbf{p})) |^2 \tag{19}$$

is analytic, say, in the region $\{| z^\alpha | < \rho_\alpha\}$. If we expand each $f^\sigma(z)$ in power series $f^\sigma(z) = \sum_{J=1}^{\infty} b_J^\sigma(z)^{(m)J}$ and substitute in (19), we obtain

$$\begin{aligned} D(\mathbf{p}_0, \mathbf{p}) &= \sum_{\sigma=1}^{N} \Big| \sum_{J=1}^{\infty} b_J^\sigma(z(\mathbf{p}))^{(m)J} \Big|^2 = \sum_{J,K=1}^{\infty} \Big(\sum_{\sigma=1}^{N} b_J^\sigma \bar{b}_K^\sigma\Big)(z(\mathbf{p}))^{(m)J}(\overline{z(\mathbf{p})})^{(m)K} \\ &= \sum_{J,K=1}^{\infty} B_{J,K}(z(\mathbf{p}))^{(m)J}(\overline{z(\mathbf{p})})^{(m)K}, \end{aligned}$$

where

$$B_{J,K} = \sum_{\sigma=1}^{N} b_J^\sigma \bar{b}_K^\sigma . \tag{20}$$

This last equation can be interpreted as expressing the Hermitian matrix of coefficients in the power series expansion of the diastasis as the product of an $\infty \times N$ matrix with its transposed conjugate. Hence the matrix $B_{J,K}$ is necessarily positive semidefinite and of rank at most N.

Conversely, suppose that the matrix $B_{J,K}$ of coefficients is positive semidefinite, and of rank N. We shall then construct explicitly the sequence of isometric imbedding functions, first as formal power series, and then prove that these functions are actually complex analytic and satisfy (8) at least in the domain $\mathbf{G}$ of convergence of the power series expansion of the diastasis. In fact, from the theory of Hermitian matrices, we know that, if $B_{J,K}$ is positive semidefinite and of rank N, then one can construct for each successive $\sigma = 1, 2, \cdots, N$ a unique infinite vector b_J^σ satisfying (20), subject to the following additional conditions: (a) for each fixed σ the infinite numerical vector b_J^σ (with components indexed by J) is different from zero; (b) if L_σ is the index of the first non-zero component of the vector b_J^σ, then $L_\sigma < L_{\sigma+1}$ and $b_{L_\sigma}^\sigma$ is real and positive. We then define from the N infinite vectors b_J^σ the formal power series

$$f^\sigma(z) = \sum_{J=1}^{\infty} b_J^\sigma(z)^{(m)J} \qquad (\sigma = 1, 2, \cdots, N).$$

Let $| z^\alpha | < \rho_\alpha$ $(\alpha = 1, \cdots, k)$ define any maximal polycylindrical subdomain of $\mathbf{G}$; then, for any k-tuple of positive real constants $\rho'_\alpha < \rho_\alpha$ there exists a positive constant R, such that $| B_{J,K} | \leqq R^2/(\rho')^{(m)J}(\rho')^{(m)K}$; and, since

$$| b_J^\sigma |^2 \leqq \sum_{\sigma=1}^{N} | b_J^\sigma |^2 = | B_{J,J} | \leqq R^2/[(\rho')^{(m)J}]^2,$$

we have $|b_J^\sigma| \leq R/(\rho')^{(m)_J}$. Hence each of the N formal power series defining $f^\sigma(z)$ converges for $|z^\alpha| < \rho'_\alpha$ for every $\rho'_\alpha < \rho_\alpha$, and hence for $|z^\alpha| < \rho_\alpha$. Since the domain $\mathbf{G}$, being the convergence domain of a power series, is a union of its maximal polycylinders, the functions $f^\sigma(z)$ are holomorphic in $\mathbf{G}$, and the sum of the squares of their absolute values, being dominated in $\mathbf{G}$ by the real analytic function $D(\mathbf{p}_0,\mathbf{p})$, to which they converge formally, converges actually to the given diastasis in $\mathbf{G}$. Thus all of $\mathbf{G}$ can be isometrically imbedded in $\mathbf{C}^N$, completing the proof of the Theorem.

The theorem just established shows also that the property of the Hermitian matrix $B_{J,K}$ being positive semidefinite does not depend on the particular system of local parameters, since it is equivalent to the invariant property that a neighborhood of the origin can be isometrically imbedded in a unitary space. In the next two theorems we shall prove that, if the Kaehler metric is analytic in the large and if this property can be verified at one point $\mathbf{p}_0$, then there exists a suitable covering of the whole manifold $\mathbf{M}^k$ that can be globally imbedded in the unitary space.

DEFINITION. A Kaehler manifold $\mathbf{M}^k$ with analytic metric is called *resolvable of rank* N at a point $\mathbf{p}_0$ if, for a coordinate system having $\mathbf{p}_0$ as origin, the local power series expansion of the diastasis $D(\mathbf{p}_0, \mathbf{p}) = \sum_{J,K=1}^{\infty} B_{J,K}(z(\mathbf{p}))^{(m)_J}(\overline{z(\mathbf{p})})^{(m)_K}$ has the property that the coefficient matrix $B_{J,K}$ is positive semidefinite and of rank N.

THEOREM 4 (GLOBAL CHARACTER OF RESOLVABILITY). *If a Kaehler manifold* $\mathbf{M}^k$, *whose metric is analytic, is resolvable of rank* N *at one point* $\mathbf{p}_0$, *then it is resolvable of rank* N *at each of its points.*

PROOF. Since $\mathbf{M}^k$ is a connected topological space, it is sufficient to prove that the set of points at which it is resolvable of rank N is both open and closed. It is clearly open, since for every point $\mathbf{p}$ with this property there exists a neighborhood of $\mathbf{p}$ that can be isometrically imbedded in $\mathbf{C}^N$, but not in $\mathbf{C}^{N'}$ for any $N' < N$; but since every open subset of such a neighborhood spans $\mathbf{C}^N$ linearly, it follows that $\mathbf{M}^k$ is resolvable of rank N at each point of such a neighborhood.

On the other hand the set is closed. In fact, let $\mathbf{p}$ be a limit point of the set of points at which $\mathbf{M}^k$ is resolvable of rank N; we shall complete the proof of the Theorem by showing that $\mathbf{M}^k$ is resolvable of rank N at $\mathbf{p}$. Since the metric is analytic, by Theorem 1 there exists a coordinate neighborhood $\mathbf{V}$ around $\mathbf{p}$ that can be isometrically imbedded in the indefinite Hilbert space $\mathbf{E}$. By assumption the domain $\mathbf{V}$ contains a point $\mathbf{p}'$ and a connected neighborhood $\mathbf{V}'$ of $\mathbf{p}'$, that can be isometrically imbedded in and span linearly $\mathbf{C}^N$. Let (z) be a coordinate system defined over all of $\mathbf{V}$, with origin at $\mathbf{p}'$, and let

$$x^\sigma = f^\sigma(z), \qquad f^\sigma(0) = 0 \qquad (\sigma = 1, 2, \cdots, N), \tag{21}$$

$$y^\tau = g^\tau(z), \qquad g^\tau(0) = 0 \qquad (\tau = \pm 1, \pm 2, \cdots) \tag{22}$$

be two systems of equations representing isometric imbeddings respectively of $\mathbf{V}'$ in $\mathbf{C}^N$ and of $\mathbf{V}$ in $\mathbf{E}$, each system satisfying (8) in the appropriate domain.

Consider the complex linear subspace $\mathbf{E}'$ of $\mathbf{E}$ which is spanned by $\mathbf{V}'$ (and hence by $\mathbf{V}$) under (22): we assert that the diastasis of $\mathbf{E}$ restricted to $\mathbf{E}'$ positive semi-definite, meaning that for any point $\mathbf{q} \in \mathbf{E}'$ the diastasis $D(\mathbf{p}', \mathbf{q}) \geqq 0$. In fact, suppose that this were not true; then there would exist a finite collection of points $\mathbf{p}_1, \mathbf{p}_2, \cdots, \mathbf{p}_n$ of $\mathbf{V}'$ in $\mathbf{E}$, such that, on taking a suitable real linear combination of the vectors $\overrightarrow{\mathbf{p}'\mathbf{p}_\nu}$, $(\nu = 1, 2, \cdots, n)$, say $\sum_{\nu=1}^n a_\nu \overrightarrow{\mathbf{p}'\mathbf{p}_\nu} = \overrightarrow{\mathbf{p}'\mathbf{q}}$, we would have $D(\mathbf{p}', \mathbf{q}) < 0$. But this is impossible, since, if one takes the same linear combination of the n vectors defined by the same points (imbedded by (21)) in $\mathbf{C}^N$, we obtain a new vector $\overrightarrow{\mathbf{p}'\mathbf{q}'}$ in $\mathbf{C}^N$, for which $D(\mathbf{p}', \mathbf{q}') \geqq 0$; on the other hand (cfr. the proof of Theorem 2) $D(\mathbf{p}', \mathbf{q}) = D(\mathbf{p}', \mathbf{q}')$, since both terms can be expressed in terms of the diastasis between the points $\mathbf{p}', \mathbf{p}_1, \cdots, \mathbf{p}_n$, namely (cfr. (16) and (17))

$$\begin{aligned} D(\mathbf{p}', \mathbf{q}) &= \tfrac{1}{2} \textstyle\sum_{\mu,\nu=1}^n a_\mu a_\nu [D(\mathbf{p}', \mathbf{p}_\mu) + D(\mathbf{p}', \mathbf{p}_\nu) - D(\mathbf{p}_\mu, \mathbf{p}_\nu)] = D(\mathbf{p}', \mathbf{q}') \\ &= \tfrac{1}{2} \textstyle\sum_{\mu,\nu=1}^n a_\mu a_\nu \{\sum_{\sigma=1}^N [| f^\sigma(z(\mathbf{p}_\mu))|^2 + | f^\sigma(z(\mathbf{p}_\nu))|^2 - | f^\sigma(z(\mathbf{p}_\mu)) - f^\sigma(z(\mathbf{p}_\nu))|^2]\} \\ &= \textstyle\sum_{\sigma=1}^N | \sum_{\nu=1}^n a_\nu f^\sigma(z(\mathbf{p}_\nu)) |^2 \geqq 0. \end{aligned}$$

Let $\mathbf{E}_0$ be the null-space of $\mathbf{E}'$, i.e. the linear manifold of points of $\mathbf{E}'$ with coordinates (y) (referred to $\mathbf{E}$), such that $\sum_\tau'(\operatorname{sgn} \tau) | y^\tau|^2 = 0$. If we introduce the unitary metric in $\mathbf{E}'$ defined by Hermitian form $\sum_\tau' | y^\tau |^2$, let $\mathbf{E}_+$ be the orthogonal complement of $\mathbf{E}_0$ in $\mathbf{E}'$, and let $\mathbf{T}$ be the orthogonal projection of $\mathbf{E}'$ onto $\mathbf{E}_+$ along $\mathbf{E}_0$ (if necessary the above three linear spaces should be replaced by their respective completions with respect to the new norm). This operation has the effect of mapping $\mathbf{E}'$ linearly onto a proper subspace, in which the old diastasis is positive definite, and in such a way as to preserve the diastasis of all pairs of points. As a result of the projection $\mathbf{T}$, the closed linear subspace $\mathbf{E}'$ of $\mathbf{E}$, which is spanned linearly by the points of $\mathbf{V}'$ under the imbedding defined by (22) is replaced by the linear subspace $\mathbf{E}_+$, in which the diastasis of $\mathbf{E}$ is positive definite, and the imbedding equations are replaced by a new set, so that the new image of $\mathbf{V}'$ lies in and spans linearly $\mathbf{E}_+$. But the projection carries actually all of the domain $\mathbf{V}$ into $\mathbf{E}_+$, so that we have proved that $\mathbf{V}$ can be imbedded isometrically into a unitary space. Since $\mathbf{V}$ obviously spans the same linear space as $\mathbf{V}'$, and since by Theorem 2 such a space must be N-dimensional, it follows that the Kaehler metric of $\mathbf{M}^k$ is resolvable of rank N at each point of $\mathbf{V}$, including $\mathbf{p}$, thus completing the proof of the theorem.

By virtue of Theorem 4 we can speak of a Kaehler manifold as being resolvable of rank N without specification of point. We shall now use the previous results to show that, if the Kaehler metric of a manifold $\mathbf{M}^k$ is analytic (this being the only global condition), and if the manifold is resolvable of rank N (which need be verified at only one point), then any imbedding of a small domain of $\mathbf{M}^k$ in $\mathbf{C}^N$ can be extended indefinitely.

THEOREM 5. *A simply connected Kaehler manifold* $\mathbf{M}^k$ *can be globally imbedded in* $\mathbf{C}^N$, *if and only if the metric is analytic and* $\mathbf{M}^k$ *is resolvable of rank at most* N. *If* $\mathbf{M}^k$ *is resolvable of rank* $N' \leqq N$, *then the image of* $\mathbf{M}^k$ *in* $\mathbf{C}^N$ *spans linearly a subspace of* $\mathbf{C}^N$ *of dimension* N'.

Proof. The necessity of the conditions is obvious from Theorem 3. To prove that the conditions are sufficient, we assume that $\mathbf{M}^k$ is resolvable of rank N. By Theorem 3 we can cover $\mathbf{M}^k$ by coordinate neighborhoods, each of which can be imbedded isometrically in and span linearly $\mathbf{C}^N$. By Theorem 2, for each component $\mathbf{U}$ of the intersection of any two such neighborhoods there exists a unique motion in $\mathbf{C}^N$, which carries one image of $\mathbf{U}$ in $\mathbf{C}^N$, as part of the first coordinate neighborhood, into the image of $\mathbf{U}$ as part of the second. Let $\mathbf{p}_0$ be the origin of the coordinate system in one such neighborhood $\mathbf{V}_0$, and $x^\sigma = f^\sigma(z)$ a definite set of equations representing an isometric imbedding of $\mathbf{V}_0$ in $\mathbf{C}^N$; let any simple path from $\mathbf{p}_0$ to any point $\mathbf{p}$ of $\mathbf{M}^k$ be given. This path can be covered by a finite succession of overlapping, open segments $\pi_0, \pi_1, \cdots, \pi_n$, each of which lies in the domain of a set of imbedding functions. At the overlap of each $\pi_{\nu-1}$ with π_ν ($\nu = 1, \cdots, n$) there exists a unique unitary motion, which transforms the imbedding functions defined in the neighborhood of π_ν into those defined at $\pi_{\nu-1}$. It is clear that the unitary motion can be applied to the neighborhood of the whole segment π_ν and not just to its overlap with $\pi_{\nu-1}$, and still give an isometric imbedding. Thus any set of imbedding functions defined at $\pi_{\nu-1}$ can be extended to a neighborhood of π_ν, and hence, by induction, over the whole path from $\mathbf{p}_0$ to $\mathbf{p}$. Since $\mathbf{M}^k$ is arcwise connected, the imbedding can be continued to any point of $\mathbf{M}^k$, and since the manifold is, by assumption, simply connected, it follows from the monodromy principle that the extensions of the imbedding as indicated above do not depend on the path followed. If $\mathbf{M}^k$ is resolvable of rank N' and one imbeds it into $\mathbf{C}^N$ for $N > N'$, then any coordinate neighborhood imbedded in $\mathbf{C}^N$ spans only a N'-dimensional subspace of $\mathbf{C}^N$, and obviously no extension of the imbedding ever leads out of that linear subspace; thus the proof of Theorem 5 is complete.

Corollary. *If a simply connected Kaehler manifold* $\mathbf{M}^k$ *is resolvable of any finite or infinite rank, then the diastasis* $D(\mathbf{p},\mathbf{q})$ *can be analytically extended to all pairs of points* $\mathbf{p}, \mathbf{q} \in \mathbf{M}^k$ *and is everywhere non-negative.*

We are now in a position to state and prove the main theorem about isometric imbedding in unitary spaces, which summarizes and completes the results obtained sofar.

Theorem 6. *A complex manifold* $\mathbf{M}^k$ *with any metric admits a global isometric, complex analytic imbedding in* $\mathbf{C}^N$, *if and only if the following conditions are fulfilled:*

(a) *the metric is an analytic Kaehler metric;*

(b) *the Kaehler manifold is resolvable of rank at most* N *at any point;*

(c) *for each point* $\mathbf{p}$ *the analytic extension of the diastasis* $D(\mathbf{p}, \mathbf{q})$ *with respect to* $\mathbf{q}$ *over* $\mathbf{M}^k$ *is single valued.*

The imbedding is, in addition, topological (i.e. without self-intersections), if and only if

(d) $D(\mathbf{p}, \mathbf{q}) = 0$ only for $\mathbf{p} = \mathbf{q}$.

Proof. Conditions (a) and (b) are necessary and sufficient for the universal covering manifold $\tilde{\mathbf{M}}^k$ of $\mathbf{M}^k$ to admit a global, isometric immersion in $\mathbf{C}^N$, according to Theorem 5. The necessity of condition (c) is obvious; to prove its sufficiency, consider the analytic extension of the diastasis over the whole covering manifold $\tilde{\mathbf{M}}^k$: if the diastasis is single valued over $\mathbf{M}^k$, then, for any two points of $\tilde{\mathbf{M}}^k$ which cover the same point of $\mathbf{M}^k$, the diastasis between them must, in particular, be zero. But this means that the images of these two points under the immersion in $\mathbf{C}^N$ must coincide, since the diastasis equals the square of the rectilinear distance in $\mathbf{C}^N$. Hence, if the diastasis is single valued in $\mathbf{M}^k$, the Poincaré group of $\mathbf{M}^k$ acting on $\tilde{\mathbf{M}}^k$ leaves the image of $\tilde{\mathbf{M}}^k$ in $\mathbf{C}^N$ pointwise fixed, so that the immersion is a single valued map defined on $\mathbf{M}^k$. The necessity and sufficiency of condition (d) are both trivial, since the condition, as stated, is equivalent to the requirement that distinct points of $\mathbf{M}^k$ have positive distance between them after the immersion of $\mathbf{M}^k$ in $\mathbf{C}^N$. Thus Theorem 6 is proved.

The concluding part of this chapter is devoted to some considerations connected with local canonical coordinates (see Bochner, [1], page 181). These are defined in any Kaehler manifold in a neighborhood of any point $\mathbf{p}_0$ to be a system of coordinates $(z^1, z^2, \cdots, z^k)$ with origin at $\mathbf{p}_0$, so that the power series expansion of the diastasis $D(\mathbf{p}_0,\mathbf{p})$ becomes

$$(23)\quad D(\mathbf{p}_0,\mathbf{p}) = \sum_{\alpha=1}^{k} | z^\alpha(\mathbf{p}) |^2 + \sum_{|(m)_J|,|(m)_K|\geqq 2}^{\infty} B_{J,K}(z(\mathbf{p}))^{(m)_J}(\overline{z(\mathbf{p})})^{(m)_K},$$

where the range of the last summation implies that each term is of degree at least 2 in the variables $(z(\mathbf{p}))$ and likewise in $(\overline{z(\mathbf{p})})$. Their existence in Kaehler manifolds was established by Bochner (loc. cit.); our purpose is to determine their relation with the canonical coordinates of an enveloping Kaehler manifold in the case of isometric imbedding, in order to demonstrate some characteristic geometric properties that they possess.

PROPOSITION 7. *The canonical coordinates at any point $\mathbf{p}_0$ of a Kaehler manifold with analytic metric are unique up to a homogeneous unitary transformation.*

PROOF. We assume to begin with that we have two systems of canonical coordinates at $\mathbf{p}_0 \in \mathbf{M}^k$, denoted by (z) and (z') respectively. Since both coordinate systems have a common origin, it follows that their transformation equations must be of the form

$$(24)\qquad z'^\beta = a^\beta_\alpha z^\alpha + \sum_{J=k+1}^{\infty} b^\beta_J (z)^{(m)_J} \qquad (\beta = 1, 2, \cdots, k),$$

where the matrix $a^\beta_\alpha = (\partial z'^\beta/\partial z^\alpha)_{\mathbf{p}_0}$ is non-singular. Since (z') is a canonical coordinate system, we must have

$$(25)\quad \begin{aligned} &\sum_{\beta=1}^{k} | z'^\beta |^2 + \sum_{J,K=k+1}^{\infty} B'_{J,K}(z')^{(m)_J}(\bar{z}')^{(m)_K} \\ &\qquad = \sum_{\alpha=1}^{k} | z^\alpha |^2 + \sum_{J,K=k+1}^{\infty} B_{J,K}(z)^{(m)_J}(\bar{z})^{(m)_K} \end{aligned}$$

and hence, on substituting (24) in the left hand side of (25),

$$\begin{aligned}(26)\quad &\sum_{\alpha,\gamma=1}^{k}\Big(\sum_{\beta=1}^{k} a^\beta_\alpha \overline{a^\beta_\gamma}\Big) z^\alpha \overline{z^\gamma} + \sum_{J=k+1}^{\infty}\sum_{\gamma=1}^{k}\Big(\sum_{\beta=1}^{k} b^\beta_J \overline{a^\beta_\gamma}\Big)(z)^{(m)_J}\overline{z^\gamma} \\ &+ \sum_{\alpha=1}^{k}\sum_{K=k+1}^{\infty}\Big(\sum_{\beta=1}^{k} a^\beta_\alpha \overline{b^\beta_K}\Big) z^\alpha (\bar{z})^{(m)_K} + \sum_{J,K=k+1}^{\infty}\Big(\sum_{\beta=1}^{k} b^\beta_J \overline{b^\beta_K}\Big)(z)^{(m)_J}(\bar{z})^{(m)_K} \\ &\qquad + \sum_{J,K=k+1}^{\infty} C_{J,K}(z)^{(m)_J}(\bar{z})^{(m)_K} = \sum_{\alpha,\gamma=1}^{k} \delta_{\alpha,\gamma} z^\alpha \overline{z^\gamma} + \sum_{J,K=k+1}^{\infty} B_{J,K}(z)^{(m)_J}(\bar{z})^{(m)_K}.\end{aligned}$$

Equating coefficients on either side of (26), we have, in particular, $\sum_{\beta=1}^{k} b_J^\beta \overline{a_\gamma^\beta} = 0$ and, since the matrix a_α^β is non-singular, $b_J^\beta = 0$ $(J > k)$; furthermore, since $\sum_{\beta=1}^{k} a_\alpha^\beta \overline{a_\gamma^\beta} = \delta_{\alpha\gamma}$, the matrix a_α^β is unitary. Thus we obtain what we set out to prove, namely, $z'^\beta = \sum_{\beta=1}^{k} a_\alpha^\beta z^\alpha$, where a_α^β is a unitary matrix.

We now shall prove a general theorem on the relation between the canonical coordinates of a Kaehler manifold and those of a complex analytic submanifold.

THEOREM 7. *Let* $\mathbf{M}^k$ *be a finite dimensional analytic submanifold of an N-dimensional* $(N \leqq \infty)$ *Kaehler manifold* $\mathbf{S}^N$, *whose metric is analytic. Then the induced metric on* $\mathbf{M}^k$ *is an analytic Kaehler metric and, if* $(z^1, z^2, \cdots, z^k)$ *is a canonical system of coordinates in* $\mathbf{M}^k$ *with respect to a point* $\mathbf{p}_0$ *of* $\mathbf{M}^k$, *then there exists a canonical coordinate system* (x) *in* $\mathbf{S}^N$ *with respect to* $\mathbf{p}_0$, *such that the imbedding equations are*

$$x^\sigma = z^\sigma \qquad (\sigma = 1, 2, \cdots, k)$$

$$x^\sigma = f^\sigma(z) = \sum_{J=k+1}^{\infty} b_J^\sigma (z)^{(m)_J} \qquad (\sigma = k+1, \cdots, N),$$

where the range of indices in the summation implies that the imbedding functions $f^\sigma(z)$ *contain no term of degree less than* 2 *in* (z).

PROOF. Since $\mathbf{S}^N$ can be locally imbedded isometrically in the indefinite Hilbert space, as proved in Theorem 1, any analytic submanifold of $\mathbf{S}^N$, such as $\mathbf{M}^k$ is then immersed also analytically in the indefinite Hilbert space, so that the induced metric is an analytic Kaehler metric; thus it is possible to talk about diastasis and canonical coordinates in $\mathbf{M}^k$. If (z) is a canonical coordinate system in $\mathbf{M}^k$ with respect to a point $\mathbf{p}_0$, then it is clear that, by a suitable unitary transformation, one can choose a canonical coordinate system (x) in $\mathbf{S}^N$ with respect to $\mathbf{p}_0$, such that the local imbedding equations become

$$x^\sigma = f^\sigma(z) = z^\sigma + \sum_{J=k+1}^{\infty} b_J^\sigma (z)^{(m)_J} \qquad (\sigma = 1, 2, \cdots, k),$$

$$x^\sigma = f^\sigma(z) = \sum_{J=k+1}^{\infty} b_J^\sigma (z)^{(m)_J} \qquad (\sigma = k+1, \cdots, N).$$

Since both systems (x) and (z) are canonical in the respective manifolds, we have, for a variable point $\mathbf{p}$ of $\mathbf{M}^k$,

$$\begin{aligned} D(\mathbf{p}_0, \mathbf{p}) &= \sum_{\sigma=1}^{k} |f^\sigma(z(\mathbf{p}))|^2 + \sum_{J,K=k+1}^{\infty} B'_{J,K}(z(\mathbf{p}))^{(m)_J}(\overline{z(\mathbf{p})})^{(m)_K} \\ &= \sum_{\alpha=1}^{k} |z^\alpha(\mathbf{p})|^2 + \sum_{J,K=k+1}^{\infty} B_{J,K}(z(\mathbf{p}))^{(m)_J}(\overline{z(\mathbf{p})})^{(m)_K}. \end{aligned}$$

By substituting the power series for $f^\sigma(z)$ in the last set of equations, it follows from a power series argument analogous to that in the proof of Proposition 7 (see (25), (26)), that $b_J^\sigma = 0$ $(\sigma = 1, \cdots, k; J \geqq k+1)$, proving that (x) and (z) are coordinate systems which verify the assertion of the theorem.

COROLLARY. *If a Kaehler manifold* $\mathbf{M}^k$ *is resolvable of any rank, then the canonical coordinates with respect to any point of* $\mathbf{M}^k$ *can be analytically extended as holomorphic functions over the whole universal covering manifold of* $\mathbf{M}^k$. *If the diastasis is, in addition, single valued in* $\mathbf{M}^k$, *then the functions which form a canonical coordinate system at any point of* $\mathbf{M}^k$ *are entire functions on* $\mathbf{M}^k$.

In fact, under the hypothesis of the Corollary, $\mathbf{M}^k$, or at least its universal covering space, can be isometrically imbedded in some $\mathbf{C}^N$, whose canonical coordinates are precisely the unitary coordinates of $\mathbf{C}^N$.

Chapter 4. Imbedding in Fubini-Study spaces

We shall now consider the problem of the isometric and complex analytic imbedding of complex manifolds in a wider class of Kaehler manifolds than the flat ones. The only types of ambient spaces for which the results could be as complete as in the case of the flat spaces are those that have the characteristic property that the diastasis between any two points depends on their geodesic distance alone. It is easy to verify that such is the case with the Fubini-Study manifolds described below (see also [1], [3], and [4]); it is also true that these manifolds are the only ones having this property; however the proof of this fact would lead us too far from our main topic and will be therefore omitted. The statements and proofs will follow closely the pattern set in the previous chapter in most cases, while in others the proofs make direct use of the analogous statements already established in the case of imbedding in flat spaces.

We shall define the Fubini-Study spaces as follows. Consider the N-dimensional ($N \leqq \infty$) Kaehler manifolds defined as the metric completion of the maximal domain of analyticity of the Kaehler metric defined in the canonical coordinate system $(x^1, x^2, \cdots, x^N)$ by the following diastasis:

$$D(\mathbf{p}_0, \mathbf{p}) = \frac{1}{b}\log\left(1 + \sum_{\sigma=1}^{N} b \mid x^\sigma(\mathbf{p}) \mid^2\right) \qquad (b \text{ real}, \neq 0). \tag{27}$$

In case $b > 0$ the domain of analyticity of the metric is the complete numerical unitary space $\sum_{\sigma=1}^{N} \mid x^\sigma(\mathbf{p}) \mid^2 < \infty$, but the region covered by the coordinate domain is not metrically complete. An elegant method to show the structure of the metrical completion consists of introducing the homogeneous coordinates $(\xi^0, \xi^1, \xi^2, \cdots, \xi^N)$, where $x^\sigma = \xi^\sigma/\sqrt{b}\xi^0$, so that the diastasis (27) in terms of (ξ) becomes

$$D(\mathbf{p}_0, \mathbf{p}) = \frac{1}{b}\log\frac{\sum_{\sigma=0}^{N} \mid \xi^\sigma(\mathbf{p}) \mid^2}{\mid \xi^0(\mathbf{p}) \mid^2}, \tag{28}$$

whence, from Proposition 4 and equation (5), we obtain:

$$D(\mathbf{p}, \mathbf{q}) = \frac{1}{b}\log\frac{\left(\sum_{\sigma=0}^{N} \mid \xi^\sigma(\mathbf{p}) \mid^2\right)\left(\sum_{\sigma=0}^{N} \mid \xi^\sigma(\mathbf{q}) \mid^2\right)}{\left|\sum_{\sigma=0}^{N} \xi^\sigma(\mathbf{p})\xi^{\sigma*}(\bar{\mathbf{q}})\right|^2}. \tag{29}$$

From this last expression one sees that the metric obtained from this diastasis is analytic, with respect to suitable inhomogeneous coordinate systems, in the domain $\{0 < \sum_{\sigma=0}^{N} \mid \xi^\sigma \mid^2 < \infty\}$, that is to say, in the whole N-dimensional complex projective space, which is obviously metrically complete. The unitary group acting on the space of homogeneous coordinates (ξ) is transitive on the homogeneous equivalence classes, and leaves (29) invariant, so that the N-dimensional complex projective space with the Kaehler metric derived from the diastasis (29) admits a transitive group of isometries, which are collineations. We shall call the Kaehler manifolds thus obtained the N-dimensional Fubini spaces of the elliptic kind and curvature constant b (>0), and denote them by $\mathbf{F}(N, b)$.

In case $b < 0$ the domain of analyticity of $D(\mathbf{p}_0, \mathbf{p})$ in (27) is described by $\sum_{\sigma=1}^{N} | x^\sigma(\mathbf{p}) |^2 < -1/b$. Again we introduce homogeneous coordinates $(\xi^0, \cdots, \xi^N)$, where $x^\sigma = \xi^\sigma/\sqrt{-b}\xi^0$, whence we obtain:

$$D(\mathbf{p}_0, \mathbf{p}) = \frac{1}{b} \log \frac{\sum_{\sigma=0}^{N} \varepsilon_\sigma | \xi^\sigma(\mathbf{p}) |^2}{| \xi^0(\mathbf{p}) |^2} \qquad (\varepsilon_\sigma = 1; \varepsilon_\sigma = -1, \sigma > 0), \tag{30}$$

and

$$D(\mathbf{p}, \mathbf{q}) = \frac{1}{b} \log \frac{\left(\sum_{\sigma=0}^{N} \varepsilon_\sigma | \xi^\sigma(\mathbf{p}) |^2\right)\left(\sum_{\sigma=0}^{N} \varepsilon_\sigma | \xi^\sigma(\mathbf{q}) |^2\right)}{\left| \sum_{\sigma=0}^{N} \varepsilon_\sigma \xi^\sigma(\mathbf{p}) \xi^{\sigma*}(\bar{\mathbf{q}}) \right|^2}. \tag{31}$$

This last expression is invariant under the linear group acting on the homogeneous space of (ξ), which leaves the Hermitian form $\sum_{\sigma=0}^{N} \varepsilon_\sigma | \xi^\sigma |^2$ invariant, the so-called complex Lorentz group; this group is transitive in the spherical cell $\sum_{\sigma=0}^{N} \varepsilon_\sigma | \xi^\sigma |^2 > 0$ or $\sum_{\sigma=1}^{N} | x^\sigma |^2 < -1/b$. Therefore this cell with the Kaehler metric derived from (30) admits a transitive group of complex analytic isometries, and is therefore a complete Kaehler manifold; it is called the N-dimensional Fubini space of the hyperbolic kind and curvature constant b (<0), and is denoted by $\mathbf{F}(N, b)$.

For reasons of completeness we shall call Fubini spaces of the parabolic kind, i.e. with curvature constant $b = 0$ ($\mathbf{F}(N, 0)$) the unitary spaces $\mathbf{C}^N$. The justification of this is evident, if one lets b approach zero in equation (27).

It will be useful in the following discussion to introduce the generalization of the stereographic projection, the "flattening out" of the Gaussian sphere onto the complex plane to all other Fubini spaces. Given an arbitrary reference point $\mathbf{p}_0$ of $\mathbf{F}(N, b)$, let (x) be a canonical coordinate system with respect to $\mathbf{p}_0$; we then put a new Kaehler metric on the complex manifold covered by (x) by letting this domain become a unitary space $\mathbf{F}(N, 0)$ (or a subdomain thereof, if $b < 0$) in which (x) are unitary coordinates. The identity map of a neighborhood of $\mathbf{p}_0$ with the above described change of metric is what we understand as stereographic projection. It is equivalent to the transformation of the diastasis $D(\mathbf{p}, \mathbf{q})$ for $\mathbf{p}$ and $\mathbf{q}$ in a neighborhood of $\mathbf{p}_0$ to a new diastasis $D'(\mathbf{p}, \mathbf{q})$ obtained from the equation

$$D'(\mathbf{p}_0, \mathbf{p}) = \frac{1}{b} (e^{bD(\mathbf{p}_0,\mathbf{p})} - 1) = \sum_{\sigma=1}^{N} | x^\sigma(\mathbf{p}) |^2. \tag{32}$$

We shall now proceed to establish the conditions for a complex manifold $\mathbf{M}^k$ to admit an isometric imbedding in any given Fubini space.

THEOREM 8. *Let $\mathbf{M}^k$ be a complex manifold with a metric, $\mathbf{p}_0$ a point of $\mathbf{M}^k$, and (z) a local complex coordinate system with origin at $\mathbf{p}_0$. The necessary and sufficient conditions for the existence of some neighborhood of $\mathbf{p}_0$ admitting an isometric, complex analytic imbedding in any given Fubini space $\mathbf{F}(N, b)$ are, first, that the metric of $\mathbf{M}^k$ be a Kaehler metric, analytic at $\mathbf{p}_0$; second, that if one takes the stereographic projection of $\mathbf{M}^k$ with respect to $\mathbf{p}_0$, i.e., introduce the new diastasis $D'(\mathbf{p}_0, \mathbf{p}) = (e^{bD(\mathbf{p}_0,\mathbf{p})} - 1)/b$, then the neighborhood of $\mathbf{p}_0$ with the new Kaehler metric is resolvable of rank at most N.*

PROOF. It is clear from Theorem 7 that the first condition, that the metric be an analytic Kaehler metric, is necessary for the isometric imbedding. Thus, once this condition is satisfied, it is possible to introduce a diastasis $D(\mathbf{p}, \mathbf{q})$. The existence of an isometric imbedding of a neighborhood of $\mathbf{p}_0$ in $\mathbf{F}(N, b)$ means the existence of functions $x^\sigma = f^\sigma(z)$ $(\sigma = 1, 2, \cdots, N)$ defined in that neighborhood and satisfying (8), such that

$$D(\mathbf{p}_0, \mathbf{p}) = b^{-1} \log \left(1 + \textstyle\sum_{\sigma=1}^N b \,|\, f^\sigma(z(\mathbf{p}))\,|^2\right);$$

this is obviously equivalent to the equation $D'(\mathbf{p}_0, \mathbf{p}) = \sum_{\sigma=1}^N |\, f^\sigma(z(\mathbf{p}))\,|^2$, which in turn is equivalent, by Theorem 3, to saying that the neighborhood of $\mathbf{p}_0$, after the stereographic projection, is resolvable of rank at most N, which proves our assertion.

Having established the conditions for the local isometric imbedding, we shall show that such an imbedding, when possible, is determined uniquely up to within a transformation of $\mathbf{F}(N, b)$ belonging to its group of complex analytic isometries. We need only the concept of a linear subspace of a Fubini space: the latter is defined to be the complex analytic submanifold of $\mathbf{F}(N, b)$ defined by any set of linear relations between the homogeneous coordinates (ξ) of $\mathbf{F}(N, b)$ as used in equations (28) through (31). If $\mathbf{p}_0$ is any point of such a linear subspace of $\mathbf{F}(N, b)$ and (x) a system of canonical coordinates with respect to $\mathbf{p}_0$, then the linear subspace can also be defined by a set of homogeneous linear equations in (x); from this one deduces that a linear subspace of $\mathbf{F}(N, b)$ is also a Fubini space having the same curvature constant as the enveloping space.

THEOREM 9 (LOCAL RIGIDITY). *Let $\mathbf{M}^k$ be a Kaehler manifold and assume that there exists a small region $\mathbf{G}$ in $\mathbf{M}^k$ that admits an isometric, complex analytic imbedding in a Fubini space $\mathbf{F}(N, b)$, so that it does not lie in any proper linear subspace of $\mathbf{F}(N, b)$. Then N is uniquely determined by the curvature constant b and by the Kaehler metric in $\mathbf{G}$, and the imbedding is determined up to within the group of motions of $\mathbf{F}(N, b)$, i.e. the group of complex analytic isometries of $\mathbf{F}(N, b)$ with itself.*

PROOF. As in the case of Theorem 2, we assume that we can imbed $\mathbf{G}$ by two sets of functions, respectively in $\mathbf{F}(N, b)$ and $\mathbf{F}(N', b)$. If $\mathbf{p}$ is any point of $\mathbf{G}$, we can take canonical coordinates with respect to the respective images of $\mathbf{p}$ in either Fubini space and perform the stereographic projection. This reduces then the proof exactly to that of Theorem 2, so that the local rigidity property is maintained also in the Fubini spaces.

DEFINITION. A Kaehler manifold whose metric is real analytic is called *b-resolvable of rank N* at a point $\mathbf{p}_0$ if there exists a neighborhood of $\mathbf{p}_0$ that admits an isometric, complex analytic imbedding in a Fubini space $\mathbf{F}(N, b)$, but not in $\mathbf{F}(N', b)$ for any $N' < N$.

THEOREM 10 (GLOBAL CHARACTER OF b-RESOLVABILITY). *If a Kaehler manifold whose metric is everywhere analytic is b-resolvable of rank N at any one point, then it is b-resolvable of rank N at each of its points.*

PROOF. We follow the pattern of the proof of Theorem 4, by showing that the set of points of a Kaehler manifold $\mathbf{M}^k$ with analytic metric, at which $\mathbf{M}^k$ is b-resolvable of rank N is both open and closed. This set is clearly open, as one deduces from the definition of b-resolvability. On the other hand one can show that it is also closed, since it contains all of its limit points. In fact, let $\mathbf{p}$ be one such limit point, and let $\mathbf{V}$ be a coordinate neighborhood of $\mathbf{p}$ such that $D(\mathbf{q}, \mathbf{q}')$ is analytic and single valued for all $\mathbf{q}, \mathbf{q}' \in V$. Let $\mathbf{p}'$ be a point of $\mathbf{V}$, in whose neighborhood the manifold is b-resolvable of rank N, and take the stereographic projection of $\mathbf{V}$ with respect to $\mathbf{p}'$, giving a new diastasis $D'(\mathbf{p}', \mathbf{q}) = (e^{bD(\mathbf{p}', \mathbf{q})} - 1)/b$. The new Kaehler metric is obviously analytic on $\mathbf{V}$ and is 0-resolvable at $\mathbf{p}'$. Since the diastasis $D'(\mathbf{q}, \mathbf{q}')$ is single valued in $\mathbf{V}$, by Theorem 6 $\mathbf{V}$ with the new metric can be globally imbedded isometrically in and span linearly $\mathbf{C}^N$; furthermore, if $b < 0$, $D'(\mathbf{p}', \mathbf{q}) < -1/b$, so that the imbedding functions $f^\sigma(z)$ satisfy

$$\sum_{\sigma=1}^{N} | f^\sigma(z(\mathbf{q})) - f^\sigma(z(\mathbf{p}')) |^2 = D'(\mathbf{p}', \mathbf{q}) < -1/b.$$

Therefore in both the elliptic and the hyperbolic case the same functions $f^\sigma(z)$ map V, with the original Kaehler metric, isometrically in $\mathbf{F}(N, b)$, so that no open subset of $\mathbf{V}$ spans any proper linear subspace. This proves that $\mathbf{M}^k$ is b-resolvable of rank N at $\mathbf{p}$, and hence at all of its points. With this result established we shall henceforth refer to a Kaehler manifold as being b-resolvable of rank N without specifying at what point.

THEOREM 11. *A simply connected Kaehler manifold* $\mathbf{M}^k$ *with analytic metric can be globally imbedded isometrically in* $\mathbf{F}(N, b)$, *if and only if it is b-resolvable of rank at most N.*

PROOF. The condition is necessary already from local considerations, as we proved in Theorem 8. To prove that it is also sufficient, we assume that $\mathbf{M}^k$ is b-resolvable of rank N. By Theorem 8 every point of $\mathbf{M}^k$ has a neighborhood that can be isometrically imbedded in and span linearly $\mathbf{F}(N, b)$. If two such neighborhoods overlap, then each component $\mathbf{U}$ of their intersection is mapped into $\mathbf{F}(N, b)$ in two different ways, one for each of the neighborhoods to which $\mathbf{U}$ belongs. By Theorem 9 the two isometric images of $\mathbf{U}$ are equivalent under a uniquely determined motion in $\mathbf{F}(N, b)$. Thus, as in the proof of the analogous assertion for the parabolic case (Theorem 5), the isometric imbedding can be stepwise extended along any prescribed path, and this extension is monogenic, so that the monodromy principle can be applied, and, since $\mathbf{M}^k$ is assumed to be simply connected, the extension of a local imbedding into $\mathbf{F}(N, b)$ is single valued over the whole manifold. Theorem 11 is thus proved.

COROLLARY 1. *If a simply connected Kaehler manifold* $\mathbf{M}^k$ *is b-resolvable of rank* N *for any* $b > 0$ *and any integer* N, *then for any point* $\mathbf{p}$ *of the manifold the diastasis* $D(\mathbf{p}, \mathbf{q})$ *is real analytic, single valued, and non-negative in* $\mathbf{q}$ *over the whole manifold, except perhaps at a* $(k-1)$*-dimensional complex analytic subvariety (possibly reducible and with singularities) depending on* $\mathbf{p}$, *where the diastasis becomes infinite.*

PROOF. In each Fubini space of the elliptic kind the diastasis $D(\mathbf{p}, \mathbf{q})$ is single valued, non-negative, and real analytic in $\mathbf{q}$ for each fixed $\mathbf{p}$, except at a hyperplane depending on $\mathbf{p}$, the so-called antipolar hyperplane (see Cartan, [5]), where the diastasis becomes infinite. This fact is readily verifiable by inspecting the equation for the diastasis in terms of the homogeneous coordinates (36). Under the assumptions of this corollary the manifold $\mathbf{M}^k$ can be isometrically immersed in $\mathbf{F}(N, b)$, so that the diastasis $D(\mathbf{p}, \mathbf{q})$ in $\mathbf{M}^k$ is single valued, non-negative, and, for any fixed $\mathbf{p}$, real analytic in $\mathbf{q}$, except at the intersection of $\mathbf{M}^k$ in $\mathbf{F}(N, b)$ with the antipolar hyperplane of $\mathbf{p}$, where $D(\mathbf{p}, \mathbf{q})$ becomes infinite. This intersection is clearly either empty or a $(k-1)$-dimensional subvariety of $\mathbf{M}^k$, which proved our assertion.

COROLLARY 2. *If a simply connected Kaehler manifold* $\mathbf{M}^k$ *is b-resolvable of any rank and for* $b < 0$, *then the diastasis is non-negative, single valued, and real analytic in all pairs of points of* $\mathbf{M}^k$.

COROLLARY 3. *If a simply connected Kaehler manifold* $\mathbf{M}^k$ *is b-resolvable of rank* N *for any* b, N, *then the canonical coordinates* (z) *with respect to any point* $\mathbf{p} \in \mathbf{M}^k$ *can be extended as meromorphic functions over the whole manifold. If* $b > 0$, *then the polar variety is the set of points* $\mathbf{q}$ *for which* $D(\mathbf{p}, \mathbf{q})$ *becomes infinite, while if* $b < 0$, *the meromorphic extension is actually holomorphic in* $\mathbf{M}^k$ *and the functions satisfy* $\sum_{\alpha=1}^{k} | z^\alpha |^2 < -1/b$.

PROOF. After immersing $\mathbf{M}^k$ in $\mathbf{F}(N, b)$, we select a system of canonical coordinates (x) with respect to $\mathbf{p}$ in $\mathbf{F}(N, b)$ as in Theorem 7, that is, such that the imbedding equations are, for the first k coordinates, $x^\sigma = z^\sigma$ $(\sigma = 1, 2, \cdots, k)$. In case $b > 0$, the functions x^σ are inhomogeneous coordinates of the complex projective space, and hence they are meromorphic and have as polar variety the antipolar hyperplane of $\mathbf{p}$. In case $b < 0$, the functions x^σ are holomorphic in $\mathbf{F}(N, b)$ and satisfy $\sum_{\sigma=1}^{N} | x^\sigma |^2 < -1/b$. It is clear then that in either case these properties go over to the restrictions of the functions x^σ to $\mathbf{M}^k$.

We thus arrive at the general criterion for the isometric, complex analytic imbedding of complex manifolds into the Fubini spaces.

THEOREM 12. *A complex manifold* $\mathbf{M}^k$ *with any metric admits an isometric, global, complex analytic imbedding in a given Fubini space* $\mathbf{F}(N, b)$ *if and only if the following conditions are satisfied:*

(a) *the metric is an analytic Kaehler metric;*

(b) *the manifold is b-resolvable of rank at most* N *at any point;*

(c) *for each point* $\mathbf{p}$ *of* $\mathbf{M}^k$ *the maximal analytic extension of the diastasis* $D(\mathbf{p}, \mathbf{q})$ *in* $\mathbf{q}$ *is single valued.*

The imbedding is, in addition, topological, if and only if

(d) $D(\mathbf{p}, \mathbf{q}) = 0$ *only for* $\mathbf{p} = \mathbf{q}$.

PROOF. The proof is an almost verbatim repetition of the one for Theorem 6, which covers the case of Fubini spaces of the parabolic kind. It will be noted also that with the assumption (c) above, all the corollaries of Theorem 11 become valid also when $\mathbf{M}^k$ is not simply connected.

The case of b-resolvable *compact* complex manifolds with single valued diastasis is especially interesting, because of its relation to algebraic geometry. Since the manifold can be globally imbedded in $\mathbf{F}(N, b)$, it follows clearly that this can happen only if $b > 0$, since the elliptic spaces are the only ones that can contain compact analytic subvarieties. It has been proved by W. L. Chow, [6] that in this case the manifold is algebraic. In this connection it is worth noting that one can determine intrinsically the linear systems of hyperplane sections of the variety simply as the system of singular loci in the manifold of the diastasis $D(\mathbf{p}, \mathbf{q})$ in $\mathbf{q}$, when $\mathbf{p}$ ranges over a linear subspace of the enveloping Fubini space.

CHAPTER 5. AN APPLICATION

We shall consider here an example in which the theorems established in the previous chapters can be easily applied: to determine under what conditions a Fubini space $\mathbf{F}(N, b)$ can be isometrically imbedded into another Fubini space $\mathbf{F}(N', b')$. Some partial solutions to this problem have been published previously: Bochner, [1], demonstrated that any unitary space $\mathbf{C}^N$ can be imbedded in the infinite dimensional Fubini space of the elliptic kind $\mathbf{F}(\infty, 1)$; he also proved that in the general case it is necessary for the imbedding to exist, that at least $b \leqq b'$ and that $\mathbf{F}(N, b)$ $(b < 0)$ can not be imbedded in any $\mathbf{C}^{N'}$ for N' finite; Bieberbach, [7], on the other hand had already shown that the real hyperbolic plane, interpreted as $\mathbf{F}(1, -1)$, can be imbedded in Hilbert space $\mathbf{C}^\infty$. We shall now give the complete solution to the problem.

THEOREM 13. *A Fubini space* $\mathbf{F}(N, b)$ *can be globally imbedded isometrically into another Fubini space* $\mathbf{F}(N', b')$, *if and only if:*

(A) $b \leqq b'$ *and in addition*

(B) *either:* (1) $b \leqq 0$ *and*

(2) $N' = \infty$,

or: (1') $b' - kb$ *for some positive integer* k *and*

(2') $N' \geqq \binom{N+k}{k} - 1$ *for* N *finite*, $N' = \infty$ *for* $N = \infty$.

PROOF. Given the space $\mathbf{F}(N, b)$ with the diastasis expressed in terms of canonical coordinates (x) by (27), the proof consists of finding for what values of b' the diastasis is b'-resolvable and, if so, of what rank. In the following computations we suppose b and b' to be different from zero, since in the parabolic case the formulas need only be modified by an elementary limiting process. We thus apply Theorem 8 and, accordingly, construct the b'-stereographic projection of $\mathbf{F}(N, b)$, obtaining

$$(33)\quad D'(\mathbf{p}_0, \mathbf{p}) = \frac{1}{b'}\left(e^{b'D(\mathbf{p}_0,\mathbf{p})} - 1\right) = \frac{1}{b'}\left[\left(1 + \sum_{\sigma=1}^{N} b\,|\,x^\sigma(\mathbf{p})\,|^2\right)^{b'/b} - 1\right].$$

This last expression can then be expressed as a power series in (x) and $(\bar{x})$ as follows:[6]

$$(34)\quad D'(\mathbf{p}_0, \mathbf{p}) = \sum_{J,K=1}^{\infty} B_{J,K}(x(\mathbf{p}))^{(m)_J}(\overline{x(\mathbf{p})})^{(m)_K} = \sum_{n=1}^{\infty} \frac{1}{b'}\binom{b'/b}{n} b^n \left(\sum_{\sigma=1}^{N} |x^\sigma(\mathbf{p})|^2\right)^n$$

$$= \sum_{J=1}^{\infty} \frac{\prod_{n=1}^{|(m)_J|-1} (b' - nb)}{\prod_{\sigma=1}^{N} m_{\sigma,J}!} (x(\mathbf{p}))^{(m)_J}(\overline{x(\mathbf{p})})^{(m)_J}$$

$$= \sum_{J=1}^{\infty} C_J (x(\mathbf{p}))^{(m)_J}(\overline{x(\mathbf{p})})^{(m)_J},$$

where

$$(35)\qquad C_J = \frac{\prod_{n=1}^{|(m)_J|-1} (b' - nb)}{\prod_{\sigma=1}^{N} m_{\sigma,J}!}$$

We thus verify that the matrix $B_{J,K}$ of coefficients in the power series expansion of (34) is the diagonal real matrix $C_J\delta_{JK}$, so that it is very easy to determine both its signature and rank, to verify, by virtue of Theorem 12, that the conditions stated in this theorem are necessary and, since all Fubini spaces are simply connected, sufficient for $\mathbf{F}(N, b)$ to admit an isometric imbedding in $\mathbf{F}(N', b')$. If (y) are canonical coordinates in $\mathbf{F}(N', b')$ with respect to $\mathbf{p}_0$, and if the conditions for the isometric imbedding are satisfied, then a set of imbedding equations is given by

$$(36)\qquad y^J = \sqrt{C_J}(x)^{(m)_J} \qquad (J = 1, 2, \cdots, N').$$

It is to be noted that this last set of equations is valid also in the special cases, when either b or b' vanish, so that we have proved the theorem with a constructive proof of the sufficiency part.

It is known that the part of the Fubini spaces which is covered by the real values of either their canonical or their homogeneous coordinates constitute the manifolds of the non-Euclidean geometries. Since the set of equations (36) maps the real part of $\mathbf{F}(N, b)$ into the real part of $\mathbf{F}(N', b')$, it follows that these equations give examples of isometric, real analytic imbedding of some Riemannian manifolds with constant curvature into others, some of which, to the author's knowledge, had not been reported before. These include the imbedding of hyperbolic spaces of arbitrary dimension and constant negative curvature b into the infinite dimensional spaces of arbitrary constant curvature greater than b (the imbedding into hyperbolic spaces with curvature $<b$ and one higher dimension is given by the hyperbolic hyperspheres) and the isometric imbedding of an

[6] We agree that any product of an empty family of numbers is 1.

elliptic space with constant curvature $b > 0$ into a finite dimensional elliptic space with constant curvature kb ($k = 2, 3, \cdots$).

In conclusion we shall exhibit some examples of negative statements. It is easy to construct some open Kaehler manifolds that can not be imbedded in any Fubini space: for example the Gaussian plane with the line element $ds^2 = [4 \cos (z - \bar{z}) - 1] |dz|^2$; the imbedding in any Fubini space is impossible since the diastasis

$$D(z, z') = 4[\cos(z - \bar{z}) + \cos(z' - \bar{z}') - \cos(z - \bar{z}') - \cos(z' - \bar{z})] - | z - z' |^2$$

takes on negative values, *e.g.* for $z' = z + 2\pi$. There actually exists an even simpler, compact Kaehler manifold, that can not be imbedded in any Fubini space: it is sufficient to take the product $\mathbf{F}(N_1, b_1) \times \mathbf{F}(N_2, b_2)$ of two Fubini spaces of the elliptic kind, such that b_2/b_1 is irrational. By Theorem 13 there exists no Fubini space that can contain both of the factor spaces simultaneously as isometrically imbedded submanifolds.

PRINCETON UNIVERSITY and
LOUISIANA STATE UNIVERSITY

REFERENCES

[1] BOCHNER, S., *Curvature in Hermitian Metric*, Bull. Amer. Math. Soc., 53 (1947), pp. 179–195.

[2] CALABI, E., Metric Riemann Surfaces, Contributions to the Theory of Riemann Surfaces, Annals of Mathematics Studies, Princeton, 1953.

[3] FUBINI, G., Teoria dei gruppi discontinui, Pisa, Spoerri, 1908.

[4] STUDY, E., *Küreste Wege im komplexen Gebiet*, Math. Ann., 60 (1905), pp. 321–378.

[5] CARTAN, E., Leçons de géométrie projective complexe, Paris, Gauthier-Villars, 1931.

[6] CHOW, W. L., *On Compact, Complex Analytic Varieties*, Amer. J. Math., 71 (1949), pp. 893–914.

[7] BIEBERBACH, L., *Eine singularitätenfreie Fläche konstanter negativer Krümmung im Hilbertschen Raum*, Comm. Math. Helv., 4 (1932), pp. 248–255.

The Space of Kähler Metrics

Eugenio Calabi

Proc. Int Congress Math., held in Amsterdam,
Netherlands, 1954, 206–207, E. P. Norrdhoff (1954)

Let M^n be a closed, n-dimensional complex manifold. We assume that M^n admits at least one Kähler metric $g_{\alpha\beta^*}$; its associated closed exterior form $\omega = \sqrt{-1}\, g_{\alpha\beta^*}\, dz^\alpha \wedge dz^{\beta^*}$ determines a real cohomology class, called the principal class of the metric. Consider the space Ω of all infinitely differentiable Kähler metrics in M^n with the same principal class; the topology of Ω is defined by the L^2 topology of the tensorial components of metrics in Ω in compact subregions of co-ordinate domains. If $R_{\alpha\beta^*}$ is the Ricci tensor of any metric in Ω, then the Ricci form $\sqrt{-1}\, R_{\alpha\beta^*}\, dz^\alpha \wedge dz^{\beta^*}$ is closed and its cohomology class is $2\pi C^{(1)}$ ($C^{(r)} = r$th Chern class).

Theorem 1. Given in M^n any real, closed, infinitely differentiable exterior form Σ of type (1, 1) and cohomologous to $2\pi C^{(1)}$, there exists exactly one Kähler metric in Ω whose Ricci form equals Σ.

The proof proceeds by joining the Ricci form of one metric in Ω with Σ by a differentiably parametrized arc in the same linear space of forms (for example by linear interpolation) and constructing a corresponding parametrized path in Ω, then by proving that it is unique, and that its terminal point is independent of the path.

One can introduce in Ω a generalized Riemannian structure by defining the metric form $ds^2 = \left(\frac{\partial \omega(t)}{\partial t}, \frac{\partial \omega(t)}{\partial t}\right) dt^2$, where $\omega(t)$ represents a differentiable path in Ω. One shows that between any two points in Ω there exists exactly one geodesic, that the topology induced by the geodesic distance coincides with the chosen one, and that Ω with this metric is isometric with a geodesically convex subset of a Hilbert sphere.

If $g_{\alpha\beta^*}$ belongs to Ω and Σ is its Ricci form, then (Σ, Σ) is stationary with respect to first order variations in Ω, if and only if the contravariant components of the gradient of the scalar curvature generate a group of complex analytic transformations of M^n. If M^n has the property that each one-parameter complex transformation group of M^n (if any exist) leaves no point fixed, or if it is the Gaussian sphere, then this implies that the scalar curvature is constant, that Σ is harmonic, and that (Σ, Σ) is minimized.

Theorem 2. If M^n has the restrictive property described above, then there exists a metric in Ω which minimizes (Σ, Σ), unique but for analytic transformations of M^n, and characterized by the property that the scalar curvature is constant.

The Variation of Kähler Metrics I
The Structure of the Space

Eugenio Calabi

Bull. Amer. Math. Soc., 60, 167–168 (1954)

Let M^n be a compact complex manifold admitting at least one Kähler metric $g_{\alpha\beta^*}$; one considers the closed, exterior form $\omega_0=(-1)^{1/2}g_{\alpha\beta^*}dz^\alpha\wedge dz^{\beta^*}$ and its (the principal) cohomology class (ω). If $R_{\alpha\beta^*}$ is the Ricci tensor, then $\Sigma_0=(-1)^{1/2}R_{\alpha\beta^*}dz^\alpha\wedge dz^{\beta^*}$, the Ricci form, is closed and cohomologous to 2π times the first Chern class. *Theorem*: Given any closed exterior form Σ of type (1, 1), of class $\mathcal{C}^\infty$, cohomologous to 2π times the first Chern class, there exists a unique Kähler metric of class $\mathcal{C}^\infty$, whose principal cohomology class is (ω), and whose Ricci form equals Σ. With the aid of this theorem one considers the set of all Kähler metrics of class $\mathcal{C}^\infty$, whose principal class is (ω). It is an infinite-dimensional "open manifold," in which there is a natural Riemannian metric $dS^2=(\eta, \eta)$ where η is the first variation of any given ω and (ϕ, ψ) is the integrated inner product of two forms ϕ, ψ. This manifold has constant, positive curvature, unique geodesics of bounded length between any pair of points, and its completion consists of the positive semidefinite Kähler metrics defining the same principal class. (Received October 28, 1953.)

The Variation of Kähler Metrics II
A Minimum Problem

Eugenio Calabi

Bull. Amer. Math. Soc., 60, 168 (1954)

Let M^n be a compact, n-dimensional Kähler manifold, and (ω) its principal cohomology class (see previous abstract). For each principal Kähler form $\omega \in (\omega)$ let Σ be its Ricci form, (Σ, Σ) the norm of that form, $R = -2\Lambda\Sigma$ the scalar curvature, and H the operator projecting forms into their harmonic part. The following constants depend only on (ω), not on ω: $\int_{M^n}\omega^n = V$, $HR = 2nV^{-1}\int_{M^n}\omega^{n-1}\wedge\Sigma$, $(H\Sigma, H\Sigma) = V(HR)^2 - n(n-1)\int_{M^n}\omega^{n-2}\wedge\Sigma^2$. *Theorem* 1: Σ is harmonic with respect to ω if and only if R is a constant; if this is the case, and for fixed (ω), the metric is determined within any transformation due to the complex Lie group of analytic deformation of M^n, and (Σ, Σ) is minimized. The converse is implied for most complex manifolds by *Theorem* 2: If, for a certain $\omega \in (\omega)$, (Σ, Σ) is minimized, then the contravariant vector $g^{\alpha\beta^*}R_{\alpha\beta^*}$ is holomorphic; this implies that Σ is harmonic if M^n has no analytic deformations. The existence of an $\omega \in (\omega)$ minimizing (Σ, Σ) is, except in special cases, still unproved. (Received October 28, 1953.)

On Kähler Manifolds With Vanishing Canonical Class

Eugenio Calabi

Fox, R. H. (ed.) *Algebraic Geometry and Topology, A Symposium in Honor of S. Lefschetz*, 78–89, Princeton University Press, Princeton, NJ (1957)

THE purpose of this note is to show certain properties of compact Kähler manifolds whose canonical class vanishes; an application of the results obtained leads to a partial classification of all such manifolds, reducing it to the problem of classifying a more restricted class of compact Kähler manifolds, namely, those whose canonical class vanishes and whose first Betti number is zero. A complete classification of manifolds of the latter type is claimed by Severi [9] in the case of algebraic surfaces.

We shall first recall that the canonical class of an algebraic manifold M_n of dimension n is the homology class of the divisor of any non-zero meromorphic n-uple differential. This definition is not suitable for generalization to compact, complex manifolds, since it is not known whether one can always find a non-trivial meromorphic n-uple differential in all manifolds of this wider class. An alternate definition which allows such a generalization is due to Chern [3]. Let M_n be a compact, n-dimensional complex manifold, in which let g be any real, differentiable, positive-valued density of weight 1; g is expressible in any coordinate domain $\{U; (z)\}$ $((z) = (z^1, z^2, \ldots, z^n))$ by means of a positive, differentiable function $g_U(z, \bar{z})$, so that, if $\{V; (z')\}$ is another coordinate domain, then at every point of $U \cap V$

$$g_V(z', \bar{z}') = g_U(z, \bar{z}) \left| \frac{\partial(z)}{\partial(z')} \right|^2, \tag{1}$$

where $\partial(z)/\partial(z')$ is the Jacobian of the coordinated transformation. It follows from (1) that

$$\log g_V = \log g_U + 2\mathrm{Re} \log ((\partial(z)/\partial(z')),$$

and hence the tensor components defined in each coordinate domain $\{U; (z)\}$ by

$$R_{\alpha\beta^*}(z, \bar{z}) = \frac{\partial^2(\log g_U(z, \bar{z}))}{\partial z^\alpha \partial \bar{z}^\beta} \quad (\alpha, \beta = 1, 2, \ldots, n) \tag{2}$$

DOI: 10.1515/9781400879915-006

define the same hermitian tensor over all of M_n; consequently the exterior form†

$$\Sigma = -\tfrac{1}{2}dCd(\log g) = \sqrt{(-1)}\, R_{\alpha\beta^*} dz^\alpha \wedge d\bar{z}^\beta \tag{3}$$

is a closed, real-valued exterior differential form of type (1, 1)‡ defined over M_n. If a different density g' is chosen, then g'/g is a positive, differentiable scalar; hence

$$\Sigma' - \Sigma = -\tfrac{1}{2}dCd\left(\log \frac{g'}{g}\right) \tag{4}$$

is a closed form cohomologous to zero. Thus the cohomology class of Σ depends only on the complex analytic structure of M_n, and not on the particular density chosen. It is shown by Chern [3] that whenever a non-trivial meromorphic n-uple differential exists in M_n, then the dual of its divisor cycle is cohomologous to $(1/2\pi)\Sigma$. Thus for any compact, complex manifold one can define the canonical class (or first Chern class) to be the cohomology class of $(1/2\pi)\Sigma$.

From this point on we shall assume that the complex manifold M_n admits an infinitely differentiable, positive-definite Kähler metric (for a definition, see [2])

$$ds^2 = 2g_{\alpha\beta^*}(z, \bar{z})\, dz^\alpha d\bar{z}^\beta; \tag{5}$$

then one can take the density g to be the determinant of the matrix $(g_{\alpha\beta^*})$, i.e. the volume distribution with respect to a local coordinate system. One of the properties of a Kähler metric which we shall use here is that the tensor $R_{\alpha\beta^*}$ defined by (2) from the volume element g is equal to the Ricci tensor of the metric (cf. [1], p. 789); we shall therefore call the form Σ defined by (3) the Ricci form relative to the metric. The property which is characteristic of and was originally used to define a Kähler metric is that the exterior form ω associated with (5) and defined by

$$\omega = \sqrt{(-1)}\, g_{\alpha\beta^*} dz^\alpha \wedge d\bar{z}^\beta \tag{5a}$$

is, like the Ricci form, closed, differentiable, of type (1, 1), and real-valued (see [6]); we shall call ω the principal form of the metric. Since $\omega^n/n!$ is the volume element of M_n,§ its integral over the cycle carried by M_n is positive, so that, as a closed differential form of type (n, n), ω^n cannot be cohomologous to zero; therefore ω belongs to a non-trivial, real cohomology class (see [4]), which we shall call the principal class.

† The notation here is the same as in Kodaira ([8], pp. 95 et seq.).

‡ An exterior form of type (p, q) $(0 \leqq p,\ q \leqq n)$ is an exterior form of degree $p+q$, which is p-fold linear in $dz^1, \ldots, dz^n$ and q-fold linear in $d\bar{z}^1, \ldots\ d\bar{z}^n$.

§ An exponent written over an exterior form denotes its power with respect to exterior multiplication.

The topology of compact, Kähler manifolds from the standpoint of homology theory has been studied extensively by Hodge [4], who generalized to this class of manifolds the homology properties with respect to real coefficients that Lefschetz [7] had discovered for algebraic varieties. In particular we shall use the following facts. All holomorphic and all antiholomorphic differentials in M_n are harmonic; those of degree 1 (Picard differentials of the first kind and their conjugates) for a complete set of representatives of cohomology classes with complex coefficients for dimension 1; thus the first Betti number B_1 of M_n is twice the irregularity g_1, or number of linearly independent Picard differentials of the first kind in M_n. If $\alpha_1, \ldots, \alpha_{g_1}$ form a basis for these differentials, we call the Albanese variety A_M of M_n the complex torus of (real) dimension $2g_1$ whose periods are those of the integrals $\oint \alpha_i$ $(i=1, \ldots, g_1)$ over the 1-cycles of M_n; we have also a complex analytic mapping of M_n into A_M, which we shall call the Jacobi map, defined up to a translation in A_M, described as follows: if $p_0 \in M_n$ and $a_0 \in A_M$ are fixed points and if p ranges over M_n,

$$J(p) = a_0 + \left(\int_{p_0}^{p} \alpha_i \right)_{i=1}^{g_1} \quad \text{(modulo periods of } \oint \alpha_i).$$

The map J induces the natural isomorphism of the translation invariant simple differentials of A_M with the Picard differentials of M_n. As in algebraic geometry we shall call M_n *regular*, if $B_1 = 2g_1 = 0$.

We shall now postulate a general result about compact, Kähler manifolds. While the proof of this result is not complete, we feel justified in assuming the truth of the statement for several concurrent, intuitive reasons. A sketch of a heuristic argument for the proposition is given in the Appendix.

PROPOSITION 1. *Let M_n be a compact, complex manifold admitting an infinitely differentiable Kähler metric with principal form ω and Ricci form Σ. If Σ' is any closed, real-valued, infinitely differentiable form of type* (1, 1) *and cohomologous to Σ, then there exists a unique Kähler metric with principal form ω' cohomologous to ω and Ricci form equal to Σ'. This metric is always infinitely differentiable; it is real analytic, if Σ' is analytic.*

We shall now assume that the canonical class of M_n is zero, or, equivalently, that the Ricci form Σ is cohomologous to zero. We shall call M_n with this restriction, for short, a *special Kähler manifold*, because (see Chern [3]) this property is equivalent to the reducibility of the structural group of the tangent bundle from the unitary to the special (or unimodular) unitary group.

Examples of special Kähler manifolds of dimension n are given: (a) by the n-dimensional complex tori (group manifolds of compact, Abelian, complex Lie groups); (b) by the non-singular hypersurfaces of order $n+2$ in a complex projective $(n+1)$-space; the latter examples are, for $n \geqq 2$, also regular; for $n=1$ of course the only possible examples are algebraic curves of genus 1.

The following theorem describes the first characteristic property of special Kähler manifolds:

THEOREM 1. *If M_n is a special Kähler manifold of dimension n and with irregularity g_1, then $0 \leqq g_1 \leqq n$ (equality holding only for a complex torus); more precisely the Jacobi map J of M_n into its Albanese variety A_M:*

(i) *is a map onto;*

(ii) *it is the projection map of a complex analytic fibre bundle with base space A_M and bundle space M_n;*

(iii) *the structural group of the bundle is finite and Abelian;*

(iv) *the fibre is a connected, special Kähler manifold.*

PROOF. Since Σ is cohomologous to zero, we can set $\Sigma' = 0$; then, by applying Proposition 1, we can replace the given Kähler metric by another one with the same principal class, for which the Ricci tensor vanishes identically; this new metric is unique and real analytic. All of the statements that will follow will be referred to this metric.

An immediate application of the theorems of Bochner ([1], Theorems 1, 2, 5, 6) to the case of vanishing Ricci tensor leads to the following statement: a vector function in M_n cannot be harmonic, nor satisfy Killing's equation, nor have holomorphic or antiholomorphic contravariant components, unless it is a parallel vector field (of course if it is parallel, then it trivially satisfies all three other conditions simultaneously). In other words, a real vector function in M_n is harmonic (i.e. the real part of a Picard differential of the first kind), if and only if, its contravariant components are the sum of a holomorphic and an antiholomorphic vector, or if, and only if, it satisfies Killing's equation (i.e. the vector generates a one-parameter group of complex analytic, isometric transformations of M_n). It follows from this that the mapping J induces a one-to-one linear correspondence between the parallel vector fields of M_n and those of A_M; hence the group G' of translations of A_M is locally isomorphic under a map induced by J to the component G of the identity in the group of analytic (and isometric) transformations of M_n. This shows that J maps M_n onto A_M; in the special case where $g_1 = n$, the map is locally one-to-one and without singularities, because of the properties of the group of translations of A_M; therefore M_n is a compact covering of A_M under J, whence it is itself a complex torus, i.e. its own Albanese variety, so that J becomes the identity map.

The inverse image $J^{-1}(a)$ of each point $a \in A_M$ is a complex analytic subvariety of M_n. If U is a geodesically convex, sufficiently small neighborhood of a, the set $\tilde{U}$ of translations of A_M mapping a into U is a neighborhood of the identity in the translation group G', constituting a local group which under the local isomorphism induced by J carries each point of $J^{-1}(a)$ into an analytic cross-section of $J^{-1}(U)$; the image of each element of $\tilde{U}$ under the local isomorphism maps $J^{-1}(a)$ analytically and isometrically onto $J^{-1}(a')$ for some $a' \in U$. One verifies easily that each cross-section over U thus defined is analytically isometric with U and intersects orthogonally the subvariety $J^{-1}(a')$ for each $a' \in U$. Thus $J^{-1}(U)$ is analytically isometric with $U \times J^{-1}(a)$. Since each local cross-section as constructed above is uniquely determined by one of its points but for its extent, we have established the complex fibre bundle structure of M_n over A_M. Furthermore, each component of a fibre is a complex submanifold without singularities, because of the local metric product structure of the bundle space.

It is clear that J induces a homomorphism of G onto G', and that the kernel consists of the analytic isometries of M_n which are analytically deformable to the identity, and map each $J^{-1}(a)$ for each $a \in A_M$ into itself. But it is well known that the group of all isometries of a compact Riemannian space is a compact Lie group; thus G is a compact, connected Lie group, locally isomorphic to the compact, Abelian group G', so that G is itself Abelian. The kernel K of the homomorphism of G onto G' is thus a discrete, and hence finite, Abelian subgroup of G. It is easy to see that K is the structural group of the fibre bundle of M_n over A_M, so that the third contention of the theorem is proved.

We have already verified that each component of a fibre is a complex submanifold of M_n without singularities; from the local metric product structure we see that it is totally geodesic, so that its Ricci curvature tensor is the tensor induced from the Ricci tensor of the ambient space. It follows that the Ricci curvature of each component of a fibre vanishes identically, so that it is a special Kähler manifold. In order to prove that each fibre is connected, we identify all points of M_n lying on the same component of any fibre; we obtain thus a complex manifold A', over which M_n is fibered by connected submanifolds, and which is in turn fibered over A_M by fibres consisting of a finite number of points, so that the composition of the two fibre projections is J. The manifold A' thus obtained is a finite covering of A_M; therefore it is a complex torus, in which the group of periods of Picard integrals of the first kind is a subgroup of finite index of the corresponding one for A_M. This would imply, if the covering of A_M by A' were not trivial, that the group of periods of the integrals in M_n would be a proper subgroup of the group of periods of A_M, which would contradict the definition of the Albanese variety. Theorem 1 is thereby proved.

The next theorem is a new interpretation of the first, but it is perhaps more helpful in describing the structure of special Kähler manifolds.

THEOREM 2. *If M is a special Kähler manifold, then there exists a finite covering manifold $\tilde{M}$ of M, which is a direct product of its own Albanese variety $A_{\tilde{M}}$ with a special Kähler manifold Y, which is regular, and possibly a point. The relative covering group of $\tilde{M}$ over M is solvable.*

PROOF. We construct the principal bundle (cf. Steenrod, [10], pp. 35–9) associated with the fibre bundle $\{M, A_M, J\}$; since the structural group of the bundle is finite, the principal bundle is given by a finite, necessarily Abelian, covering A_1 of A_M. Denote the fibre of M by Y_0 and set $M_0 = M$ and $M_1 = Y_0 \times A_1$. Then the principal map P: $M_1 \to M_0$ defines a finite, Abelian covering of M_0 by M_1. The complex torus A_1 appearing as a direct factor of M_1 is not necessarily the Albanese variety of M_1, since M_1 may have some Picard differentials of the first kind other than those induced from M under the covering map P; it is obvious that the additional differentials appearing in M_1 are precisely those induced from its other direct factor, namely, Y_0. Thus the Albanese variety of M_1 is the direct product of A_1 with the Albanese variety of Y_0.

Since it is clear that M_1 is again a special Kähler manifold, the process may be repeated with M_0 replaced by its covering manifold M_1, thus obtaining a special Kähler manifold M_2 as a finite, Abelian covering of M_1, and continue inductively. At the k'^{th} step one obtains a special Kähler manifold M_k, which is a finite, Abelian covering of M_{k-1}, such that the Albanese variety of M_k is the direct product of the corresponding finite covering of the Albanese variety of M_{k-1} with the Albanese variety of the fiber Y_{k-1} of M_{k-1} under J. Since

$$g_1(M_k) - g_1(M_{k-1}) = \dim(Y_{k-1}) - \dim(Y_k) = g_1(Y_{k-1}),$$

either the process stops when the fibre Y_k is reduced to a point, in which case the covering manifold M_k is a complex torus, or one reaches a stage when the irregularity of Y_k is zero, in which case the next inductive step will yield as covering manifold $M_{k+1}=Y_k\times A_{k+1}$ with the properties we set out to prove. Repeated application of the process will then yield only the identity covering. Thus we have obtained a finite covering manifold of M, which is the direct product of a complex torus (its own Albanese variety) with a special Kähler manifold that is regular. The relative covering group is solvable, because it has a normal series consisting of finite, Abelian groups. This completes the proof of Theorem 2.

By means of the result just established one could completely classify all special Kähler manifolds in terms of those that are regular. We shall describe briefly how this can be done.

Let A be a complex torus and G a finite soluble group of complex analytic transformations of A such that no element of G except the identity leaves any point of A fixed (for a discussion of these crystallographic groups, cf. H. Weyl [11] and H. Hopf [5]). Let Y be a regular, special Kähler manifold (this may even be one point) and (G_0, ϕ) a (possibly trivial) representation of G as a group of analytic transformations of Y. We then define a group G' of transformations of $Y\times A$, naturally isomorphic to G as follows: if $x\in G$ and x' is the corresponding element of G', and if $y\in Y$ and $a\in A$, then

$$x'(y,a)=(\phi(x)(y), x(a)).$$

It is clear that no element of G' other than the identity leaves any point of $M_0\times A$ fixed; the identification of points in $y\times A$ equivalent under G' produces a Kähler manifold, which is obviously special. As a result of Theorem 2 we know that every special Kähler manifold can be obtained in this way, so that the whole class of special Kähler manifolds can be contructed from a classification of the regular ones plus the family of all compact Kähler manifolds which are metrically flat.

It might be interesting to know immediately, given a special Kähler manifold M_n with its Ricci flat metric, what the dimension is of the complex torus $A_{\mathfrak{M}}$ and the complementary dimension of the regular manifold Y such that $Y\times A_{\mathfrak{M}}$ is a finite covering of M_n. It is clear that an upper bound for the dimension of $A_{\mathfrak{M}}$ is given by maximum number m such that at each point of M_n there exist m linearly independent (over complex coefficients) parallel vector fields; however, this upper bound is not the best possible, at least for $n\geqq 3$, since one can construct a compact, regular, 3-dimensional Kähler manifold, with Riemann curvature identically zero, admitting a complex torus as a finite covering. Another unanswered question in this connection is whether a special Kähler manifold with no local parallel vector fields with respect to its Ricci flat metric is not only regular, but also simply connected.

One interesting application of the above results is the following one, suggested to the author by G. Washnitzer. Let M_n be a special Kähler manifold with an arbitrary (i.e not necessarily Ricci flat) Kähler metric. We consider on M_n the density g' defined by

$$g' = g \exp(2\Lambda G\Sigma), \tag{6}$$

where g is the volume density, Λ is algebraic operator on exterior p-forms ϕ defined by $\Lambda\phi = (-1)^p * (\omega \wedge * \phi)$ (i.e. the negative of the one so denoted by Kodaira [8]), G is the Green's operator as in de Rham [8], and Σ is the Ricci form (3). From the well-known identities in the exterior calculus in Kähler manifolds, and from the assumption that the closed form Σ is cohomologous to 0, we obtain the identity for the functional expression $g'_U(z,\bar{z})$ of g' in a complex coordinate system $\{U; (z)\}$

$$dCd(\log g'_U(z,\bar{z})) = 0.$$

Consequently the density g' is locally equal to the squared absolute value of a complex analytic, nowhere vanishing, n-uple differential Θ, uniquely determined up to constant factors of modulus unity, i.e. a generalization of a Prym differential. The analytical continuations of Θ over closed paths depend on the homology class of the latter only, and hence furnish a multiplicative representation of the first homology group of M_n. We remark that, if M_n is regular, then Θ is single-valued; the same is true if M_n is a complex torus, for then Θ becomes, with a constant factor, the exterior product of a base for the Picard differentials of the first kind.

A direct application of Theorem 2, in connection with the considerations just made, produces the following result.

Corollary to Theorem 2. *Let M_n be a special Kähler manifold and let the smallest finite covering manifold $\tilde{M}_n$ of M_n that is a product of a complex torus with a regular, special Kähler manifold be k-sheeted. Then there exists a divisor d of k such that the m-genera of M_n ($m = 1, 2, \ldots$) are 1 or 0 according as to whether d divides m or not.*

Proof. We recall that the m-genus of a complex manifold is the maximum number of linearly independent single-valued, everywhere holomorphic densities of weight m; for $m = 1$ it is also called the geometric genus. If M_n is itself the product of a regular, special Kähler manifold with a complex torus, then $k = d = 1$; in this case we have already verified that there is a single-valued, holomorphic, n-uple differential Θ with no zeros. Each mth tensor power of Θ $(m = 1, 2, \ldots)$ is a holomorphic density of weight m; it is unique up to constant factors, since its divisor vanishes; thus the m-genera of M_n are all 1.

In the general case there exists a k-sheeted covering $\tilde{M}_n$ of M_n, which is a product of a regular, special Kähler manifold with a complex torus. The real density g' defined by (6) on M_n induces the corresponding one of $\tilde{M}_n$. Hence the generalized Prym differential Θ on M_n has multiplicative periods corresponding to a representation of the relative covering group of M_n by $\tilde{M}_n$, which is finite and of order k; thus the multiplicative periods of Θ are the totality of dth roots of unity for some divisor d of k. For each positive integer m the density Θ^m is an m-uple, holomorphic Prym density on M_n, obviously unique up to a constant factor, since its divisor vanishes, and single-valued, if and only if m is a multiple of d. This concludes the proof.

APPENDIX. We shall give here a heuristic argument to make the statement of Proposition 1 seem plausible. The proof of the uniqueness of the Kähler metric with prescribed Ricci form (subject to the necessary restrictions stated) will be rigorous; the existence part will rely on the assumption of the existence of a solution of an integro-differential equation.

We remark first that, if M_n has two distinct metrics whose principal forms ω and ω' belong to the same (principal) class, then there exists a real-valued scalar Φ defined over all of M_n and unique up to an additive real constant, such that

$$\omega' - \omega = dCd\Phi; \tag{7}$$

in fact, it is enough to set $\Phi = \Lambda G(\omega' - \omega)$, where Λ and G are taken relative to either (or any other) Kähler metric, and verify that Φ is a solution. If Φ' is any other solution, then $dCd(\Phi - \Phi') = 0$, meaning that the difference between the two solutions of (7) is the sum of an everywhere holomorphic with an everywhere antiholomorphic scalar, i.e. a constant. Since the conjugate of a solution is clearly a solution, it follows that the imaginary part of any solution of (7) is a constant, so that its real part will do.

We now prove that, if the principal forms ω and ω' are distinct, but cohomologous, then their corresponding Ricci forms Σ and Σ' are distinct. In fact, suppose that $\Sigma = \Sigma'$; from equation (4) it follows that g'/g is the real part of an everywhere holomorphic scalar; therefore it is a constant, namely, the ratio of the volumes of M_n as calculated by each of the two metrics. But the volume elements are given respectively by the forms $\omega^n/n!$ and $\omega'^n/n!$, which are clearly cohomologous, so that their integrals over M_n are equal. Thus we have $\omega'^n = \omega^n$.

Using the solution Φ of equation (6), we obtain

$$(8) \qquad \begin{aligned} 0 = *(\omega'^n - \omega^n) &= *[dCd\Phi \wedge (\textstyle\sum_{k=0}^{n-1} \omega^k \wedge \omega'^{n-k-1})] \\ &= \textstyle\sum_{k=0}^{n-1} *(\omega^k \wedge \omega'^{n-k-1} \wedge dCd\Phi). \end{aligned}$$

Each term in the last summation on the above equation can be written in the form

$$H_{(k)}^{\alpha\beta^*} \frac{\partial^2 \Phi}{\partial z^\alpha \partial \bar{z}^\beta},$$

where the coefficient matrices $H_{(k)}^{\alpha\beta^*}$ are, for each value of k $(0 \leqq k \leqq n-1)$, Hermitian contravariant tensors; using the local orthogonal developments of the two metrics, one can easily show that each tensor $H_{(k)}^{\alpha\beta^*}$ is positive-definite. Thus the tensor $H^{\alpha\beta^*} = \sum_{k=0}^{n-1} H_{(k}^{\alpha\beta^*}$ is positive-definite, and the scalar Φ satisfies the equation

$$(9) \qquad H^{\alpha\beta^*} \partial^2 \Phi / \partial z^\alpha \partial \bar{z}^\beta = 0;$$

these two facts imply that, since M_n is compact, Φ is a constant, because no solution of (9) can achieve a maximum value in the interior of its domain of existence, unless it is a constant (cf. Bochner and Yano [2], pp. 26–31). From this it follows trivially that $\omega' - \omega = 0$.

We shall now derive the integro-differential equation whose solution yields the Kähler metric with the prescribed Ricci form Σ', subject to the conditions stated in Proposition 1. Assume that for each such form Σ' there existed a corresponding Kähler metric with principal form ω'; suppose further that for each continuous system $\Sigma(t)$ of Ricci forms depending differentiably on a real parameter t $(0 \leqq t \leqq 1)$ the corresponding principal forms $\omega(t)$ depended also differentiably on t. Then, by using equations (3) and (4) we could compute the relationship between the derivatives of $\Sigma(t)$ and $\omega(t)$ with respect to t. In fact

$$\begin{aligned} \frac{\partial \Sigma(t)}{\partial t} &= \lim_{h \to 0} -\frac{1}{2h} dCd\left(\log \frac{g(t+h)}{g(t)}\right) = -\tfrac{1}{2} dCd\Lambda_t\left(\frac{\partial \omega(t)}{\partial t}\right) \\ &= -\tfrac{1}{2}\Delta_t\left(\frac{\partial \omega(t)}{\partial t}\right), \end{aligned}$$

where the subscript t under the operators Λ and Δ indicates that they are the ones associated with the metric corresponding to $\omega(t)$. Since both of the derivatives with respect to t of $\omega(t)$ and $\Sigma(t)$ are closed forms cohomologous to zero, using the Green operator G_t relative to $\omega(t)$, we obtain the integro-differential equation

$$\frac{\partial\omega(t)}{\partial t} = -2G_t\left(\frac{\partial\Sigma(t)}{\partial t}\right). \tag{10}$$

It can be shown analytically that, subject to sufficient conditions of differentiability of the function $\partial\Sigma(t)/\partial t$, the integro-differential equation above has a solution $\omega(t)$ with initial value given by the principal form $\omega(0)$ of an arbitrary, infinitely differentiable Kähler metric. In particular, on letting the initial values correspond to the Kähler metric assumed in the statement of the Proposition, and choosing $\partial\Sigma(t)/\partial t$ to be the form $\Sigma' - \Sigma$, the solution $\omega(t)$ of equation (10) such that $\omega(0) = \omega$ formally gives the principal form of the Kähler metric whose Ricci form $\Sigma(t)$ satisfies

$$\Sigma(t) = \Sigma + t(\Sigma' - \Sigma); \tag{11}$$

for $t = 1$ the solution becomes the metric whose existence is asserted.

The first essential step in rigorizing this argument consists in showing that, as t varies, equation (10) preserves its elliptic character. In fact the corresponding equation for the volume element $\omega^n(t)/n!$ is given by

$$\begin{aligned}\frac{\partial(\omega^n(t))}{\partial t} &= n\omega^{n-1}(t)\frac{\partial\omega(t)}{\partial t} = \omega^n(t)\,\Lambda_t\left(\frac{\partial\omega(t)}{\partial t}\right)\\ &= -2\omega^n(t)\,\Lambda_t G_t(\Sigma' - \Sigma).\end{aligned} \tag{12}$$

Remembering again that the operator ΛG acting on closed forms of type $(1, 1)$ and cohomologous to zero gives the same scalar modulo additive constants independently of the metric, and that the volume of M_n under any of the metrics considered is constant, we have the explicit solution of equation (12):

$$\omega^n(t) = \omega^n(0)\exp[-2\Lambda G(\Sigma' - \Sigma)]/H\{\exp[-2\Lambda G(\Sigma' - \Sigma)]\}, \tag{13}$$

where the operators Λ, G and H (harmonic projection) are taken relative to the given metric $\omega = \omega(0)$. This fact indicates that, since the determinant $g(t)$ of the metric remains positive for all values of t, the solutions of equation (10) can be continued locally with respect to t.† The proof of existence of the solution $\omega(t)$ of (10) for

$$\partial\Sigma/\partial t = \Sigma' - \Sigma(0)$$

† There seems to be some question as to whether the interval in t for which one can solve for $\omega(t)$ is unbounded. This essential gap in the proof of Proposition 1 makes the results of this paper depend on the conjecture that a compact Kähler manifold admits a Kähler metric with any assigned, positive, differentiable volume element.

can then be completed by evaluating formally from it $\partial^2\omega(t)/\partial t^2$ and obtaining a uniform *a priori* estimate of the latter expression in terms of $\omega(t)$ and $\partial\omega(t)/\partial t$, so that one can then apply the Cauchy-Peano method of polygonal approximations to a solution.

LOUISIANA STATE UNIVERSITY

REFERENCES

[1] S. BOCHNER, *Vector fields and Ricci curvature*, Bull. Amer. Math. Soc., 52 (1946), pp. 776–797.

[2] —— and K. YANO, Curvature and Betti numbers, Annals of Mathematics Studies, no. 32, Princeton, 1953.

[3] SHIING-SHEN CHERN, *Characteristic classes of Hermitian manifolds*, Ann. of Math., 47 (1946), pp. 85–121.

[4] W. V. D. HODGE, Theory and applications of harmonic integrals, Cambridge University Press, 1941.

[5] HEINZ HOPF, *Zum Clifford-Kleinschen Raumproblem*, Math. Ann., 95 (1925), pp. 313–339.

[6] E. KÄHLER, *Über eine bemerkenswerte Hermitesche Metrik*, Abh. Math. Sem. Hamburg Univ., 9 (1933), pp. 173–186.

[7] S. LEFSCHETZ, L'analysis situs et la géométrie algébrique, Paris, Gauthier-Villars, 1950.

[8] G. DE RHAM and K. KODAIRA, Harmonic integrals, Institute for Advanced Study Seminar, 1950.

[9] F. SEVERI, *Sulle superficie algebriche con curva canonica d'ordine zero*, Atti Ist. Veneto, 68 (1908–9), pp. 31–32.

[10] N. E. STEENROD, The topology of fibre bundles, Princeton University Press, 1951.

[11] H. WEYL, Group theory, Lecture Notes, Princeton University, Spring, 1949.

Construction and Properties of Some 6-Dimensional Almost Complex Manifolds

Eugenio Calabi[1]

Trans. Amer. Math. Soc. 87, 407–438 (1958)

1. **Introduction and summary.** It is well known that the six-dimensional sphere S^6 admits the structure of an almost complex manifold [8; 23, pp. 209–217]. The standard way to exhibit this structure is to imbed S^6 as the unit sphere in the Euclidean 7-space R^7 and to consider the latter as the space of "purely imaginary" Cayley numbers, then using the multiplicative properties of the Cayley algebra applied to the tangent space at each point of S^6. Ehresmann and Libermann [9] and Eckmann and Frölicher [7] showed that this particular almost complex structure is not derivable from a complex analytic one.

In the present paper we show (§4) that the algebraic properties of the imaginary Cayley numbers induce an almost complex structure on any oriented, differentiable hypersurface $\mathfrak{X} \subset R^7$ just as in the case of the unit sphere. The necessary and sufficient conditions for the almost complex structure thus obtained in $\mathfrak{X}$ to be *integrable* (i.e. derivable from a complex analytic one) are then calculated, and one sees that no closed hypersurface can satisfy these conditions everywhere; there do exist, however, many open hypersurfaces beside the hyperplanes that satisfy the integrability conditions (Theorem 6). In §5 we construct some special examples of hypersurfaces satisfying the integrability condition, with the additional property of being covering spaces (preserving the complex analytic structure under the covering transformations) of compact, complex manifolds. This new method of constructing compact, complex manifolds provides the first known counterexamples of the two following conjectures: (a) Are the Chern characteristic classes of a compact, complex analytic manifold uniquely determined by the real differentiable structure and the orientation? (b) If a compact, complex manifold is differentiably homeomorphic to a Kähler manifold, does it admit a Kähler metric?

The integrability condition for an almost complex manifold is given by the vanishing of a certain torsion tensor, which can be expressed, if the almost complex structure is induced by an imbedding as a hypersurface in the Cayley space, in terms of the second fundamental form. It would be interesting to

Presented to the Society, April 14, 1956 under the title *Examples of 6-dimensional almost-complex and complex manifolds and their properties*; received by the editors October 6, 1956.

(1) This paper was completed while the author had a contract with the office of Naval Research.

DOI: 10.2307/1993108

investigate the integral properties of this torsion tensor for closed hypersurfaces of the Cayley space in order to find out if there are any functionals that are invariant under differentiable isotopy. The discovery of such a functional would be a first step toward finding some conditions under which a given, nonintegrable almost complex structure on a manifold is homotopic to an integrable one, an important problem, which is as tempting as it is elusive. To be more specific, it is still unknown whether the sphere S^6 or any product manifold $\mathcal{V}^2 \times S^4$ ($\mathcal{V}^2$ = any closed, orientable surface) admit any complex analytic structure. These manifolds admit an almost complex structure, since they can be imbedded in R^7; on the other hand we know that the almost complex structure of the product manifolds $S^1 \times S^5$ and $S^3 \times S^3$ induced by imbedding them in the Cayley space are homotopic to integrable ones, since the manifolds are known to admit complex analytic structures [16; 2], and since their second homology groups are trivial, whence (cf. [8]) all almost complex structures inducing the same orientation are mutually homotopic.

2. **The Cayley space.** The term Cayley space denoted by $\mathcal{Y}$ is to be understood as the 7-dimensional space of Cayley numbers whose real part is zero. The algebraic structure of the Cayley algebra as an alternative division algebra over the real number field can be conveniently described in terms of invariants of the Cayley space alone; these invariants are described by operations that show the analogy of the Cayley space with the Euclidean 3-space of elementary vector analysis, where the latter is regarded as the space of purely imaginary quaternions. The notation used is an obvious adaptation of the Gibbs notation of vector analysis. Thus the Cayley space $\mathcal{Y}$ is defined for our purpose axiomatically as follows: $\mathcal{Y}$ is a real vector space over the real number field R with the two following operations (we shall denote vectors in $\mathcal{Y}$ by capital letters and real scalars by small ones):

(a) a *scalar product* (also called "inner," or "dot" product) $A \cdot B$ mapping $\mathcal{Y} \times \mathcal{Y}$ into R

(b) a *vector product* (also called "cross" product) $A \times B$ mapping $\mathcal{Y} \times \mathcal{Y}$ into $\mathcal{Y}$.

These two binary operations are subject to the following six axioms:

(1) The scalar product $A \cdot B$ is bilinear, and symmetric.

(2) The scalar product $A \cdot A$ is equivalent to a positive definite quadratic form.

(3) The vector product $A \times B$ is bilinear and skew-symmetric.

(4) The two trilinear, real valued functions $(A \times B) \cdot C$ and $A \cdot (B \times C)$ are identical, and consequently alternating; their common value is denoted hereinafter by (ABC) or (A, B, C), and called the *triple scalar product*:

$$(ABC) = (A \times B) \cdot C = A \cdot (B \times C). \tag{2.1}$$

(5) The two trilinear, vector valued functions $(A \times B) \times C - (A \cdot C)B + (B \cdot C)A$ and $-A \times (B \times C) + (A \cdot C)B - (A \cdot B)C$ are identical to each other,

and consequently alternating; their common value is denoted hereinafter by $[ABC]$ or $[A, B, C]$, and called the *triple vector product*:

$$(2.2)\qquad \begin{aligned}[ABC] &= (A \times B) \times C - (A\cdot C)B + (B\cdot C)A \\ &= -A \times (B \times C) + (A\cdot C)B - (A\cdot B)C.\end{aligned}$$

(6) The triple vector product (2.2) is not identically zero.

The vector space $\mathcal{Y}$ with the system of axioms just given is then uniquely determined up to an isomorphism. The structure theorem, whose proof can be found in [11; 19; 26; 27] will be stated as follows, without proof.

THEOREM 1. *A real vector space $\mathcal{Y}$ with the operation satisfying the six axioms of a Cayley space is necessarily 7-dimensional; one can select a basis $(I_1, \cdots, I_7)$ for $\mathcal{Y}$ with the two following properties:*

(a) *is orthonormal with respect to the inner product.*

(b) *The table of multiplication of the cross products $I_j \times I_k$ is the following*

(2.3) $I_j \times I_k =$

$j \backslash k$	1	2	3	4	5	6	7
1	0	I_3	$-I_2$	I_5	$-I_4$	I_7	$-I_6$
2	$-I_3$	0	I_1	I_6	$-I_7$	$-I_4$	I_5
3	I_2	$-I_1$	0	$-I_7$	$-I_6$	I_5	I_4
4	$-I_5$	$-I_6$	I_7	0	I_1	I_2	$-I_3$
5	I_4	I_7	I_6	$-I_1$	0	$-I_3$	$-I_2$
6	$-I_7$	I_4	$-I_5$	$-I_2$	I_3	0	I_1
7	I_6	$-I_5$	$-I_4$	I_3	I_2	$-I_1$	0

For the purpose of reference in the sequel a frame (I_j) $(j=1, 2, \cdots, 7)$ satisfying properties (a) and (b) of Theorem 1 is called a *canonical* frame for $\mathcal{Y}$. It can be constructed as follows: I_1 and I_2 are arbitrary, mutually orthogonal unit vectors; $I_3=I_1\times I_2$; I_4 is an arbitrary unit vector orthogonal to I_1, I_2, I_3, and then

$$I_5 = I_1 \times I_4, \qquad I_6 = I_2 \times I_4, \qquad I_7 = -I_3 \times I_4.$$

The group of automorphisms of the Cayley space $\mathcal{Y}$ is isomorphic to the matrix group of transformations of one canonical frame into another; it is known to be the exceptional compact, simple Lie group $\mathbf{G}_2$ (using E. Cartan's terminology, [3]). Of this group we need only the following structural property, also proved in [11] and elsewhere.

THEOREM 2. *The subgroup $\boldsymbol{H}$ of $\mathbf{G}_2$ consisting of the automorphisms of $\mathcal{Y}$ that leave the vector I_1 invariant for a fixed, canonical frame (I_j) $(1\leqq j\leqq 7)$ is isomorphic to the group $SU(3)$ of unitary 3×3 matrices with unit determinant.*

To make the isomorphism explicit, let (I_j) and (I_j') be two canonical frames such that $I_1' = I_1$. Then any vector $U = \sum_{\nu=2}^{7} a_\nu I_\nu$ orthogonal to I_1 can be represented under a mapping ϕ by an ordered triple U^ϕ of complex numbers

$$U^\phi = (u_1, u_2, u_3), \text{ where } u_\alpha = a_{2\alpha} + ia_{2\alpha+1}. \tag{2.4}$$

The corresponding representation ϕ' of U relative to the frame (I_j') is then described necessarily by a complex 3×3 matrix as follows:

$$u_\beta' = \sum_{\alpha=1}^{3} \Theta_\beta^\alpha u_\alpha$$

which, as one can verify directly, is an arbitrary matrix in $SU(3)$.

3. **Almost complex manifolds.** Let $\mathfrak{X} = \mathfrak{X}_{(r)}^n$ be an n-dimensional, differentiable manifold of class $\mathfrak{C}^r$ $(r \geqq 1)$. We shall review at first some elementary facts about abstract, differentiable manifolds, while presenting most of the notations and conventions to be used in the sequel.

A *domain* in $\mathfrak{X}$ will mean always a connected, open set. We say that a (real or complex valued) function f in a domain U of $\mathfrak{X}$ is $\mathfrak{C}^s$-equivalent to zero at a point $p \in U$, written $f \equiv 0 \bmod \mathfrak{A}_{(s)}(p)$, if f is at least of class $\mathfrak{C}^s$ at p $(0 \leqq s \leqq r)$, and if for one (and hence any other) coordinate system $(x^1, \cdots, x^n)$ of class $\mathfrak{C}^r$ at p such that $x^j(p) = 0$ $(1 \leqq j \leqq n)$ we have

$$f(q) = o\left(\sum_{j=1}^{n} |x^j(q)|^s\right) \text{ as } q \to p, \qquad q \in U;$$

in other words $f \equiv 0 \bmod \mathfrak{A}_{(s)}(p)$, if, expressing f analytically in terms of $x^1, \cdots, x^n$, f vanishes at p together with all partial derivatives of order $\leqq s$.

A tangent vector function u of class $\mathfrak{C}^s$ $(0 \leqq s \leqq r-1)$ in a domain $U \subset \mathfrak{X}$ is defined as a mapping which assigns to any function f of class $\mathfrak{C}^{s'}$ $(1 \leqq s' \leqq r)$ in a subdomain V of U a function denoted by udf of class $\mathfrak{C}^{s''}$ $(s'' = \min(s, s'-1))$ in V, called the derivative of f along u, if the following properties are satisfied, in terms of constants a, b, and differentiable functions f, g in $V \subset U$ [6, pp. 76–79]:

(1) linearity: $ud(af+bg) = audf + budg$;

(2) u is a derivation: $ud(fg) = (udf)\cdot g + f\cdot(udg)$;

(3) continuity: if $f \equiv 0 \bmod \mathfrak{A}_{(1)}(p)$ $(p \in U)$, then $udf(p) = 0$.

The tangent vector functions of class $\mathfrak{C}^s$ in any coordinate domain V in $\mathfrak{X}$ (with coordinates $x^1, \cdots, x^n$, say) form a free module over the algebra of functions of class $\mathfrak{C}^s$ in V, spanned linearly by the n vector functions e_j $(1 \leqq j \leqq n)$ defined by

$$e_j df = \partial f/\partial x^j. \tag{3.1}$$

To any function f of class as least $\mathfrak{C}^1$ in a domain U of $\mathfrak{X}$ one can naturally assign its *differential* df, considered as a covariant vector, i.e. as a linear functional over the tangent vectors at each point of U with values defined by the mapping $df: u \to udf$. The covariant vectors in any coordinate domain V with coordinates $x^1, \cdots, x^n$ form an n-dimensional free module over the functions in V, dual to the module of tangent vectors, spanned by the differentials $dx^1, \cdots, dx^n$, the latter constituting the dual base to the base (e_i) of tangent vectors. Thus, if $udx^i = u^i$, then $u = \sum_{i=1}^n u^i e_i$ and, for any function f of class $\mathfrak{C}^1$

$$udf = \sum_{i=1}^{n} u^i \frac{\partial f}{\partial x^i}.$$

A linear operator(2) T of class $\mathfrak{C}^s$ $(0 \leqq s \leqq r-1)$ in a domain $U \subset \mathfrak{X}$ is an endomorphism of the space of tangent vectors at each point of U, such that, if u is a vector function of class $\mathfrak{C}^{s'}$ $(0 \leqq s' \leqq s)$ in any subdomain V of U, then its image under T, written as uT, is a vector of class $\mathfrak{C}^{s'}$ in V. We regard a linear operator T as synonymous with its adjoint, properly denoted by T^*, the latter defined to be the endomorphism of the module of differentials, $T^*: df \to T^*df$, such that

$$T^*df: u \to u(T^*df) = (uT)df.$$

This way of writing allows us to drop the asterisk on T^* without ambiguity: thus we may write $(uT)df = u(Tdf)$ more simply $uTdf$. Every linear operator T in this sense can be identified with the tensor of rank 2, covariant in the first argument, contravariant in the second, to which it corresponds in a natural way. In particular we denote by $\mathbf{1}$ the identity linear operator.

DEFINITION 3.1. *A space $\mathfrak{X}$ is called an almost complex manifold of class $\mathfrak{C}^r$, if it is a differentiable manifold (also denoted by $\mathfrak{X}$) of class $\mathfrak{C}^{r+1}$ with a linear operator J on the tangent vectors, defined and continuous of class $\mathfrak{C}^r$ on the whole manifold, satisfying the operator equation*

$$J^2 = -\mathbf{1}. \tag{3.2}$$

An almost complex manifold is even-dimensional. In fact, if we recall that the determinant of a linear transformation is a scalar point function, and if the dimension of an almost complex manifold $\mathfrak{X}$ is n, we have

$$(-1)^n = \det(-\mathbf{1}) = \det(J^2) = (\det(J))^2 \geqq 0,$$

whence $(-1)^n = 1$, or $n = 2m$.

By an elementary inductive process one can construct in a neighborhood of any point of an almost complex manifold $\mathfrak{X}$ a local frame (basis) for the module of tangent vectors $e_1, \cdots, e_{2m}$ of class $\mathfrak{C}^r$ (ordinarily unrelated to the frames defined in terms of the coordinates, (3.1)), such that

(2) Here and in the rest of the article all linear transformations of contravariant vectors will be considered as operations on the right; all numerical representations of contravariant vectors will be considered as one-row, many-column matrices.

$$e_j J = (-1)^{j+1} e_{j'} \qquad (j' = j - (-1)^j;\ 1 \leq j \leq 2m): \tag{3.3}$$

in other words the frame (e_j) is characterized by the property that its image under J is given respectively by $(e_2, -e_1, e_4, -e_3, \cdots, e_{2m}, -e_{2m-1})$: A local frame (e_j) for the tangent vectors which satisfies (3.3) is then called *adapted to* J; we shall denote by $\mathfrak{U}$ the aggregate consisting of any domain U in an almost complex manifold $\mathfrak{X}$ of class $\mathfrak{C}^r$ together with a frame (e_j) of tangent vectors of class $\mathfrak{C}^r$ in U adapted to J; $\mathfrak{U}$ is then called a local frame system.

For each local frame system $\mathfrak{U}$ we define a 1-to-1 linear representation $\psi = \psi_{\mathfrak{U}}$ of the module of tangent vectors in any subdomain $V \subset U$ (and by the same token of the vector space of tangent vectors at any point $p \in U$) onto the m-dimensional module of complex-valued functions in V (respectively, onto the space C^m of ordered m-uples of complex numbers) by defining for any tangent vector $u = \sum_{j=1}^{2m} a^j e_j$ in V (resp. at p) the m components of its image $u^\psi = (u^1, \cdots, u^m)$:

$$u^\alpha = a^{2\alpha-1} + i a^{2\alpha} \qquad (i = (-1)^{1/2};\ 1 \leq \alpha \leq m). \tag{3.4}$$

It is easy to verify that for any tangent vector u (at a point in or in a subdomain of a frame system adapted to J) we have

$$(uJ)^\psi = i(u^\psi) = (iu^1, \cdots, iu^m). \tag{3.5}$$

Any other local frame system $\mathfrak{U}' = \{U', (e'_j)\}$ adapted to J, and such that U' meets U, is related to $\mathfrak{U}$ in $U \cap U'$ by a linear transformation of the following type:

$$\begin{aligned} e'_{2\beta-1} &= \sum_{\alpha=1}^{m} (\overset{\alpha}{\eta}_\beta e_{2\alpha-1} + \overset{\alpha}{\theta}_\beta e_{2\alpha}), \\ e'_{2\beta} &= \sum_{\alpha=1}^{m} (-\overset{\alpha}{\theta}_\beta e_{2\alpha-1} + \overset{\alpha}{\eta}_\beta e_{2\alpha}) \qquad (1 \leq \beta \leq m), \end{aligned} \tag{3.6}$$

or by the easily understandable formal abbreviation:

$$e'_{2\beta-1} - ie'_{2\beta} = \sum_{\alpha=1}^{m} (\overset{\alpha}{\eta}_\beta + i\overset{\alpha}{\theta}_\beta)(e_{2\alpha-1} - ie_{2\alpha}) \qquad (1 \leq \beta \leq m), \tag{3.6'}$$

where the complex matrix $\Theta = (\eta^\alpha_\beta + i\theta^\alpha_\beta)$ is nonsingular and of class $\mathfrak{C}^r$ in $U \cap U'$.

If the representation of the tangent vector module onto the module of ordered, complex m-uples corresponding to $\mathfrak{U}'$ is denoted by $\psi' = \psi_{\mathfrak{U}'}$, then for any tangent vector $u = \sum_{j=1}^{2m} a^j e_j = \sum_{j=1}^{2m} a'^j e'_j$ in $U \cap U'$ we have the following relation, easily derived from (3.4) and (3.6), between the two representations $u^\psi = (u^1, \cdots, u^m)$ and $u^{\psi'} = (u'^1, \cdots, u'^m)$ of u:

$$u^\alpha = \sum_{\beta=1}^{m} u'^\beta (\overset{\alpha}{\eta}_\beta + i\overset{\alpha}{\theta}_\beta) \qquad (\alpha = 1, 2, \cdots, m) \tag{3.7}$$

Thus one concludes that the nonsingular transformation matrices connecting one representation ψ of a local frame system $\mathfrak{U}$ adapted to J with another are precisely the nonsingular, complex $m\times m$ matrices

$$\Theta = (\eta^{\alpha}_{\beta} + i\theta^{\alpha}_{\beta}); \tag{3.7'}$$

furthermore the determinant of the matrix function carrying $(\mathbf{e}_1, \cdots, \mathbf{e}_{2m})$ into $(\mathbf{e}_1', \cdots, \mathbf{e}_{2m}')$ is identically equal to $|\det(\Theta)|^2$, which is positive, proving that any two frames adapted to J are similarly oriented. Thus we have shown that every almost complex manifold is necessarily orientable.

Until now we have considered J as an operator on tangent vectors alone. It is useful to extend the domain of action of J not only to the covariant vectors, but to the whole algebra of tensor functions in $\mathfrak{X}$ as well; this is done under the following postulates:

(a) J is the identity on all scalars and scalar densities.

(b) J is a linear operator mapping the space of all tensors of any given type into itself.

(c) J is distributive with respect to tensor multiplication.

The existence and uniqueness of the extension of J from acting on contravariant and covariant vectors to arbitrary tensors is an elementary exercise: we shall only remark in this connection that J is commutative with tensor transposition (permutation of factors in tensor products), and as a consequence of (3.2) is anti-commutative with simple contraction (contraction of a single contravariant argument with a covariant one): thus, if $\mathbf{W}$ is a tensor, $P:\mathbf{W}\to P[\mathbf{W}]$ a simple contraction, and $J:\mathbf{W}\to J[\mathbf{W}]$ the action of the extended operator J, then $JP[\mathbf{W}]=-PJ[\mathbf{W}]$. In particular, if we denote by J_1 the tensor representing the action of J on vectors, we have the identity $J[J_1]=-J_1$.

The term "almost complex" owes its origin to the natural occurrence of an almost complex structure in every complex analytic manifold. In fact, if $\mathfrak{X}$ is a m-dimensional complex manifold, let $z^1, \cdots, z^m$ be a system of complex analytic (holomorphic) coordinates in a domain $V\subset\mathfrak{X}$ and let

$$x^{2\alpha-1} = \operatorname{Re}(z^{\alpha}); \qquad x^{2\alpha} = \operatorname{Im}(z^{\alpha}) \qquad (\alpha = 1, \cdots, m); \tag{3.8}$$

then $(x^1, x^2, \cdots, x^{2m})$ form a system of real, analytic coordinates in V. We can then define J in V by

$$Jdx^j = (-1)^j dx^{j'}, \ (j' = j - (-1)^j, \ j = 1, 2, \cdots, 2m) \tag{3.9}$$

and correspondingly for tangent vectors and tensors: then the coordinate frame of tangent vectors $(\mathbf{e}_j)$ defined by (3.1) is adapted to J in V; in fact

$$\mathbf{e}_j J dx^k \begin{cases} =1, & \text{if } j \text{ is odd and } k = j+1, \\ =-1, & \text{if } j \text{ is even and } k = j-1, \\ =0 & \text{in all other cases} \qquad (i, k = 1, 2, \cdots, 2m). \end{cases} \tag{3.10}$$

Any other real coordinate system $(x'^1, \cdots, x'^{2m})$ in a domain $V' \subset \mathfrak{X}$ meeting V defines similarly by (3.1) a local frame system (e'_j); the latter is adapted to J in $V \cap V'$ (since J is not defined so far outside of V), if and only if $(x') = (x'^1, \cdots, x'^{2m})$ in terms of (x) satisfy the Cauchy-Riemann equations

$$(3.11) \qquad \frac{\partial x'^{2\beta-1}}{\partial x^{2\alpha-1}} = \frac{\partial x'^{2\beta}}{\partial x^{2\alpha}}; \qquad \frac{\partial x'^{2\beta-1}}{\partial x^{2\alpha}} = -\frac{\partial x'^{2\beta}}{\partial x^{2\alpha-1}} \qquad (\alpha, \beta = 1, \cdots, m),$$

i.e. if the complex-valued coordinates $z'^\beta = x'^{2\beta-1} + ix'^{2\beta}$ are holomorphic in $V \cap V'$. An immediate consequence of this is that, if we define J in every domain of holomorphic coordinates by (3.8) and (3.9), the definition is consistent throughout $\mathfrak{X}$, so that $\mathfrak{X}$ becomes an almost complex manifold in a natural way. Equation (3.9) characterizes those real coordinate systems that are related by (3.8) to a holomorphic system, since (3.9) is equivalent to the Cauchy-Riemann equations (3.11) for a coordinate transformation. We can also combine (3.8) and (3.9) in writing the equations characterizing the holomorphic coordinates $(z^1, \cdots, z^m)$ directly:

$$Jdz^\alpha = idz^\alpha \qquad (\alpha = 1, \cdots, m);$$

more generally, a complex-valued function w in a domain in $\mathfrak{X}$ is holomorphic, if and only if it satisfies the equation

$$(3.12) \qquad Jdw = idw,$$

which describes the Cauchy-Riemann equations for the real and the imaginary parts of w in terms of arbitrary, real coordinate systems. We shall refer to (3.12) as the generalized Cauchy-Riemann equations.

Now suppose that $\mathfrak{X}$ is a $2m$-dimensional (in the real sense), almost complex manifold. If there exist in a neighborhood of each point $p \in \mathfrak{X}$ m solutions of (3.12), such that their differentials at p are linearly independent with respect to complex coefficients, then these m functions can be taken as complex-valued coordinates in a (perhaps smaller) neighborhood of p. It is readily verified that the functional transformation from one such coordinate system to another, expressed in terms of the real and imaginary parts as in (3.8) satisfies the Cauchy-Riemann equations in the usual sense (3.11): thus $\mathfrak{X}$ becomes a complex analytic manifold, whose structure is uniquely determined by the given almost complex structure. We can say then that the almost complex structure of $\mathfrak{X}$ is *completely integrable* to a (unique) complex analytic one. However it is known that the assumption that there exist in the neighborhood of each point m solutions of (3.12) with linearly independent gradients requires some restrictive properties on J in the form of integrability conditions, and it has been shown [7] that these conditions are not satisfied in certain cases. We shall derive these integrability conditions, duplicating the already published works of several authors [7; 1; 9; 27], and in the next chapters we shall show how to construct a large class of almost complex manifolds, for which the integrability condition is not satisfied in most cases, and is satisfied in a few special ones; in the latter cases however the holomorphic coordinates are not given *a priori*; their functional properties are indeed unknown to the author.

Let us assume first that the almost complex manifold $\mathfrak{X}$ is completely integrable to a complex analytic one. Let Φ be any (real- or complex-valued) function of class $\mathfrak{C}^2$ in a coordinate domain with holomorphic coordinates $(z^1, \cdots, z^m)$; let $(x^1, x^2, \cdots, x^{2m})$ be the corresponding real coordinate system defined by (3.8): we then define the Wirtinger differential operators $\partial/\partial z^\alpha$ and $\partial/\partial \bar{z}^\alpha$ formally by

$$\frac{\partial}{\partial z^\alpha} = \frac{1}{2}\left(\frac{\partial}{\partial x^{2\alpha-1}} - i\frac{\partial}{\partial x^{2\alpha}}\right),$$

$$\frac{\partial}{\partial \bar{z}^\alpha} = \frac{1}{2}\left(\frac{\partial}{\partial x^{2\alpha-1}} + i\frac{\partial}{\partial x^{2\alpha}}\right).$$

Then, using the summation convention for repeated indices (where Greek indices run from 1 to m and Latin indices from 1 to $n=2m$), we have the identity

$$d\Phi = \frac{\partial\Phi}{\partial z^\alpha} dz^\alpha + \frac{\partial\Phi}{\partial \bar{z}^\alpha} d\bar{z}^\alpha$$

and, from (3.12),

$$Jd\Phi = \frac{\partial\Phi}{\partial z^\alpha}(Jdz^\alpha) + \frac{\partial\Phi}{\partial \bar{z}^\alpha}(Jd\bar{z}^\alpha) = i\left(\frac{\partial\Phi}{\partial z^\alpha} dz^\alpha - \frac{\partial\Phi}{\partial \bar{z}^\alpha} d\bar{z}^\alpha\right).$$

Taking the exterior derivative of this last expression we have

$$dJd\Phi = -2i\frac{\partial^2\Phi}{\partial z^\alpha \partial \bar{z}^\beta} dz^\alpha \wedge d\bar{z}^\beta,$$

where we use the symbol $\wedge$ to denote exterior products; the bilinear exterior differential form $dJd\Phi$ has the property that it is equal to its transform by J: in fact

$$JdJd\Phi = -2i\frac{\partial^2\Phi}{\partial z^\alpha \partial \bar{z}^\beta}(Jdz^\alpha) \wedge (Jd\bar{z}^\beta) = -2i\frac{\partial^2\Phi}{\partial z^\alpha \partial \bar{z}^\beta} dz^\alpha \wedge d\bar{z}^\beta = dJd\Phi.$$

Thus we obtain the following identity in a completely integrable almost complex manifold $\mathfrak{X}$ (denoting by $\mathbf{1}$ the identity operator),

$$(3.13) \qquad (\mathbf{1} - J)dJd\Phi = 0$$

for every function Φ of class $\mathfrak{C}^2$.

On the other hand, if we do not assume that the almost complex manifold

$\mathfrak{X}$ is completely integrable, we can not deduce (3.13) as an identity. In fact suppose that Φ is a function of class $\mathfrak{C}^2$ in a coordinate domain with coordinates $(x^1, \cdots, x^{2m}) = (x)$: the action of J on the differential

$$d\Phi = (\partial\Phi/\partial x^j)dx^j$$

can be described in terms of the components $h_i^j(x)$ of the mixed tensor of rank 2 corresponding to J, as follows:

$$Jd\Phi = h_i^j(x)\frac{\partial\Phi}{\partial x^j}dx^i.$$

Then the left-hand term of (3.13) takes the form

$$\begin{aligned}(1 - J)dJd\Phi &= \left(\frac{\partial h_j^k}{\partial x^i} - \frac{\partial h_i^k}{\partial x^j}\right)\frac{\partial\Phi}{\partial x^k}\frac{dx^i \wedge dx^j}{2} - h_i^p h_j^q\left(\frac{\partial h_q^k}{\partial x^p} - \frac{\partial h_p^k}{\partial x^q}\right)\frac{\partial\Phi}{\partial x^k}\frac{dx^i \wedge dx^j}{2}\\ &= T_{ij}^k(x)\frac{\partial\Phi}{\partial x^k}\frac{dx^i \wedge dx^j}{2},\end{aligned}$$

where

$$T_{ij}^k(x) = (\delta_i^p\delta_j^q - h_i^p h_j^q)\left(\frac{\partial h_q^k}{\partial x^p} - \frac{\partial h_p^k}{\partial x^q}\right).$$

Thus we verify that the left-hand term of (3.13) in the general case is a first order differential operator on Φ, whose coefficients depend on the tensor components $h_i^j(x)$ and their first order partial derivatives; the invariance of (3.13) also shows what one can verify directly from (3.2), $h_i^j(x)h_j^k(x) = -\delta_i^k$, namely that $T_{ij}^k(x)$ are the components of a tensor, called the *holomorphic torsion tensor* of the almost complex manifold $\mathfrak{X}$. Since the left-hand side of (3.13) is a first order differential operator on Φ, we have the identity for all functions Φ of class $\mathfrak{C}^2$

$$(3.13') \qquad (1 - J)dJd\Phi = \frac{\partial\Phi}{\partial x^k}(1 - J)dJdx^k.$$

In otner words as have the following useful criterion:

The equation (3.13) *is identically satisfied for all functions* ϕ *of class* $\mathfrak{C}^2$, *if and only if it is satisfied for any set of functions that includes, for each point* $p \in \mathfrak{X}$, *a set of local coordinates at* p.

At the same time we have also the following result:

A necessary condition in order that an almost complex manifold $\mathfrak{X}$ *be completely integrable to a complex analytic one is that the tensor with components* $T_{ij}^k(x)$ *vanishes identically, i.e.*

$$(3.14) \qquad T^k_{ij}(x) = (\delta^p_i \delta^q_j - h^p_i h^q_j)\left(\frac{\partial h^k_q}{\partial x^p} - \frac{\partial h^k_p}{\partial x^q}\right) = 0.$$

The vanishing of the torsion tensor T^k_{ij} is also equivalent to the complete integrability of the formal Pfaffian system (3.12). Hence, if we assume that the tensor components $h^j_i(x)$ are real analytic with respect to some coordinate system in the neighborhood of each point, then (3.14) is also sufficient for the integrability of $\mathfrak{X}$ to a complex manifold. The sufficiency theorem on the integrability condition for Pfaffian systems with analytic coefficients has been attributed at times to Frobenius; however the only references in the necessary degree of generality, that are known to the author, are in Goursat's *Équations aux dérivées partielles du premier ordre* (1921), where it is attributed to Bouquet and in Carathéodory's *Variationsrechnung*, where it is attributed to A. Mayer, both original papers having appeared in 1872. In the case at hand the theorem that (3.14) is necessary, and in the analytic case sufficient, for the integrability of $\mathfrak{X}$ to a complex manifold has been stated by Eckmann and Frölicher [7], Calabi and Spencer [1], Ehresmann and Libermann [9], and Yano [25]. The sufficiency problem in the nonanalytic case is still unsolved; however we shall not need it here, since the only cases we shall encounter where (3.14) is satisfied will be in real analytic manifolds.(3)

We shall now consider a symmetric, bilinear, scalar-valued functional on the tangent vector functions, of class $\mathfrak{C}^r$, denoted by Q and operating as follows:

$$Q: \mathbf{u}, \mathbf{v} \to Q(\mathbf{u}; \mathbf{v}).$$

It is clear that the tensorial transform $J[Q]$ of Q by J is again a symmetric, bilinear, scalar-valued form of class $\mathfrak{C}^r$: in fact we have

$$J[Q](\mathbf{u}; \mathbf{v}) = Q(\mathbf{u}J; \mathbf{v}J)$$

from the definition of the extension of J to the whole tensor algebra. Furthermore $J[Q]$ is positive definite, if and only if Q is positive definite: in this case the bilinear form $Q' = Q + J[Q]$ is also positive definite and has the additional property of being invariant under J,

$$J[Q'] = Q',$$

since $J^2[Q] = Q$ identically.

DEFINITION 3.2. *A symmetric, bilinear form Q in an almost complex manifold $\mathfrak{X}$ is called Hermitian with respect to J, if it is invariant under J, $J[Q] = Q$, i.e. if*

---

(3) *Added in proof.* The proof of the sufficiency of the integrability condition (3.14) has now been established by A. Newlander and L. Nirenberg (Ann. of Math. vol. 65 (1957) pp. 391–404) without assuming analyticity. This remarkable result makes it unnecessary in the sequel to prove the analyticity of the almost complex manifolds that we shall encounter, where (3.14) is satisfied.

$$Q(uJ;\, vJ) = Q(u;\, v); \tag{3.15}$$

it is called ANTI-HERMITIAN, *if* $J[Q] = -Q$, *i.e. if*

$$Q(uJ;\, vJ) = -\, Q(u;\, v)$$

for all pairs of tangent vectors $\boldsymbol{u}$, $\boldsymbol{v}$ *in* $\mathfrak{X}$.

For any symmetric, bilinear form Q in $\mathfrak{X}$ one can construct a new bilinear form (not necessarily symmetric) $L(u;\, v)$ by setting

$$L(u;\, v) = Q(uJ;\, v). \tag{3.16}$$

One verifies that Q is Hermitian with respect to J, if and only if L is an alternating (i.e. exterior) form, and that Q is anti-Hermitian, if and only if L is symmetric. In fact the following identities hold:

$$\begin{aligned} L(u;v) \pm L(v;u) &= Q(uJ;v) \pm Q(vJ;u) = Q(uJ;v) \mp Q(uJ^2;vJ) \\ &= (Q \mp J[Q])(uJ;\, v), \end{aligned}$$

where $Q - J[Q]$ (respectively $Q + J[Q]$) vanish, if and only if Q is Hermitian (respectively anti-Hermitian). We shall remark in passing that, if Q is Hermitian, then the alternating form L is also invariant under J, while, if Q is anti-Hermitian, then so is the symmetric form L.

From now on we shall assume that the almost complex manifold $\mathfrak{X}$ admits a fixed Riemannian metric, i.e. a positive definite, symmetric bilinear form Q defined and of class $\mathfrak{C}^r$ on the whole manifold: for the existence of such a metric globally, when $\mathfrak{X}$ is paracompact, cf. Steenrod [23, p. 58]. Without loss of generality, merely by replacing Q, if necessary, by $(1/2)(Q+J[Q])$, one shows that the Riemannian metric introduced may be chosen to be Hermitian. An almost complex manifold in which a fixed, Hermitian (positive definite) metric has been incorporated into its axiomatic structure shall be called an *Hermitian almost complex manifold*: the scalar product of any two tangent vectors u, v with respect to the metric is then abbreviated by the notation $u \cdot v$, while the alternating form defined by (3.16) from the metric is denoted by ω: thus

$$uJ \cdot v = \omega(u;\, v) = -\, \omega\,(v;\, u).$$

If we write the form ω in terms of the differentials dx^j of the coordinates in any coordinate system $(x) = (x^1, \cdots, x^{2m})$, we have

$$\omega = H_{jk}(x)dx^j dx^k = \sum_{1 \leq j < k \leq 2m} H_{jk}(x) dx^j \wedge dx^k \quad (H_{jk} = -\, H_{kj}),$$

where the tensor components $H_{jk}(x)$ of ω form a skew-symmetric matrix whose determinant is nowhere zero; the existence of such an exterior differential form on a differentiable (paracompact) manifold can be shown to be equivalent to the existence of an almost-complex structure [23, p. 214].

It is possible to construct inductively in a suitable neighborhood of any point in an Hermitian almost complex manifold $\mathfrak{X}$ a local frame system (e_j) $(1\leqq j\leqq 2m)$ of class $\mathfrak{C}^r$, which is at the same time adapted to J (Equation (3.3)), and orthonormal with respect to the inner product,

$$e_j\cdot e_k = \delta_{jk} \qquad (j, k = 1, 2, \cdots, 2m); \tag{3.17}$$

any such frame system, i.e. satisfying both (3.3) and (3.17) is called a *unitary* frame with respect to J and the Hermitian metric, the term being justified by the fact that under the representations ψ of tangent vectors by complex m-uples relative to any such frame system given by (3.4) the matrices $\Theta=(\eta^\alpha_\beta+i\theta^\alpha_\beta)$ describing the transformation from one representation to another by Equations (3.7) and (3.7′) are unitary.

If $\mathfrak{U}=\{U; (e_j)\}$ is a unitary frame system and $u=\sum_{j=1}^{2m} a^j e_j$, $v=\sum_{j=1}^{2m} b^j e_j$ are any two vectors in U, then $\omega(u; v)$ is the usual symplectic product of the components of u and v:

$$\omega(u; v) = \sum_{\alpha=1}^{m} (a^{2\alpha-1}b^{2\alpha} - a^{2\alpha}b^{2\alpha-1}).$$

It is easy to verify in terms of local unitary frame systems that the exterior differential form ω^m of degree $2m$ (i.e. the m-fold exterior product of ω with itself) is related to the volume element dV defined by the Riemannian metric and the natural orientation defined by all frames adapted to J, by the identity in all positively oriented frames,

$$\omega^m = m!dV. \tag{3.18}$$

DEFINITION 3.3. *An Hermitian almost complex manifold $\mathfrak{X}$ is called a special Hermitian (almost complex) manifold, if it is possible to cover $\mathfrak{X}$ by a family of unitary frame systems, such that the unitary matrices $\Theta=(\eta^\alpha_\beta+i\theta^\alpha_\beta)$ describing the transformation from one frame system to another by* (3.6), (3.6′), (3.7′) *have unit determinant, i.e. belong to the special unitary group $SU(m)$.*

It has been shown by Chern [4] that the property of an almost complex manifold $\mathfrak{X}$ admitting the additional structure of a special Hermitian manifold imposes on $\mathfrak{X}$ some topological restrictions *relative to the almost complex structure* already assumed: if $\mathfrak{X}$ is compact, the condition (which is both necessary and sufficient) is that the 2-dimensional cohomology class c^1 known as the first Chern class be zero. For our purpose it is sufficient to compute c^1 in the following two cases:

(a) If $\mathfrak{X}$ is a compact, complex analytic manifold and g is any everywhere positive, real, scalar density of weight 1 and class $\mathfrak{C}^2$ (viz.: the volume element of a Riemannian metric), then the exterior differential form Σ of degree 2 defined in each complex analytic coordinate system $(z)=(z^1, \cdots, z^m)$ by

$$\Sigma = \frac{\partial^2 \log g(z, \bar{z})}{\partial z^\alpha \partial \bar{z}^\beta} dz^\alpha \wedge d\bar{z}^\beta = \frac{i}{2} dJd \log g(z, \bar{z}) \tag{3.19}$$

is closed and independent of the complex coordinate system used; hence Σ can be regarded as a closed, exterior form on all of $\mathfrak{X}$. Furthermore, if Σ' were the corresponding form derived from another density g', then g'/g would be a real, positive scalar defined globally in $\mathfrak{X}$, and

$$\Sigma' - \Sigma = \frac{i}{2} dJd \log (g'/g)$$

would be an exact (i.e. cohomologous to zero) exterior form: hence by de Rham's theorem (cf. [21]) Σ defines a cohomology class $[\Sigma]$ of dimension 2, which is independent of the density g used in (3.19). One of Chern's results is that

$$\mathbf{c}^1 = -\frac{1}{2\pi i}[\Sigma]. \tag{3.20}$$

(b) If $\mathfrak{X}$ is an algebraic manifold (over the complex field), or, more generally if $\mathfrak{X}$ is a compact, complex manifold which admits a meromorphic, scalar density ϕ of weight 1 (here the concept of a meromorphic density refers to a transformation rule that multiplies ϕ by a power of the complex analytic Jacobian of a transformation of holomorphic coordinates, and not by the real Jacobian of a coordinate transformation), then $\mathbf{c}^1$ is the dual of the $(2m-2)$-dimensional homology class of the divisor of ϕ (total multiplicity of manifolds of zeros minus that of poles of ϕ); the invariance of this definition comes from the fact that the difference between the divisors of two densities ϕ and ϕ' of equal weight is the divisor of a meromorphic function, ϕ/ϕ', which by Liouville's theorem is homologous to zero: the divisor of a meromorphic density ϕ is known as the *canonical divisor* of $\mathfrak{X}$.

DEFINITION 3.4. *A manifold $\mathfrak{X}$ is called Kählerian, if it is an Hermitian almost complex manifold, and if the covariant derivative of the tensor J (representing J) with respect to the Riemannian metric connection vanishes, or, equivalently, if the covariant derivative of ω vanishes.*

It is known (cf. [1; 17]) and can be verified by direct computation that a necessary and sufficient condition for an Hermitian almost complex manifold $\mathfrak{X}$ to be Kählerian is that, first of all the integrability condition (3.14) in terms of J be satisfied, and secondly that the exterior differential form ω be closed. The necessity of the two conditions is immediate, since, if $\mathfrak{X}$ is quasi-Kählerian, taking for any point $p \in \mathfrak{X}$ a system of geodesic coordinates at p, all the first partial derivatives of the tensor components h_j^k of J and H_{jk} of ω vanish at p. The sufficiency is proved by a similar argument.

DEFINITION 3.5. *An Hermitian almost complex manifold $\mathfrak{X}$ is called Hermitian symplectic, if the exterior form ω determined by J and the metric is closed.*

It has been shown (cf. [17] also for earlier references) that the existence of a symplectic structure on a compact manifold imposes topological restrictions beyond those required for an almost complex structure; however not all the topological properties that Hodge [14] proved for Kähler manifolds have been shown to hold for symplectic manifolds: it would seem indeed surprising if this were the case; nevertheless there is no example known to the author of a compact, symplectic manifold that is not known *a priori* to admit a Kählerian structure.

Given an Hermitian almost complex manifold $\mathfrak{X}$, we shall consider, beside the usual inner product $u \cdot v$ of two tangent vectors, also their Hermitian product

$$(u, v) = u \cdot v - i\omega(u; v), \tag{3.21}$$

which takes the form, relative to the representation ψ given by (3.4) relative to any local unitary frame system, letting $u^\psi = (u^1, \cdots, u^m)$, $v^\psi = (v^1, \cdots, v^m)$,

$$(u, v) = \sum_{\alpha=1}^{m} u^\alpha \bar{v}^\alpha, \tag{3.22}$$

which is the canonical form for the Hermitian product. The Hermitian product thus has the following properties:

$$\begin{aligned} &\text{(a)}\ (u, v) = (v, u)^-; \\ &\text{(b)}\ (uJ, v) = -(u, vJ) = i(u, v); \\ &\text{(c)}\ (u, v) + (v, u) = 2u \cdot v; \\ &\text{(d)}\ (u, v) - (v, u) = -2i\omega(u; v). \end{aligned} \tag{3.23}$$

The final topic in this general review of almost complex manifolds concerns the relationships of linear operators T on the space of tangent vectors at each point with the structural operator J and with the metric, if the manifold is also Hermitian.

DEFINITION 3.6. *A linear operator T operating pointwise on the space of tangent vectors in a Riemannian manifold $\mathfrak{X}$ is called symmetric with respect to the Riemann metric, if for every pair of tangent vectors u, v it satisfies*

$$uT \cdot v = u \cdot vT; \tag{3.24}$$

if $\mathfrak{X}$ is almost complex, then T is called complex-linear (respectively: complex-antilinear), if

$$TJ = JT \quad (\textit{respectively}: TJ = -JT); \tag{3.25}$$

if $\mathfrak{X}$ is Hermitian almost complex, then T is called an Hermitian operator, if it is both complex-linear with respect to J and symmetric, i.e. if

$$(uT, v) = (u, vT). \tag{3.26}$$

Given any symmetric transformation T in an Hermitian almost complex manifold $\mathfrak{X}$ there exists at each point $p \in \mathfrak{X}$ an orthonormal frame (e_j)

($j=1, 2, \cdots, 2m$), ordinarily unrelated to J, such that each $\mathbf{e}_j$ is a characteristic vector, i.e. satisfies

$$\mathbf{e}_j T = \lambda_j \mathbf{e}_j \qquad (1 \leqq j \leqq 2m), \tag{3.27}$$

and the corresponding proper value λ_j is real; conversely, if all the proper values of T at each point are real and if their corresponding characteristic vectors can be chosen mutually orthogonal, then T is symmetric. If T in addition is Hermitian, then in particular it commutes with J; consequently it is possible to choose at each point a unitary frame ($\mathbf{e}_j$) satisfying (3.27) with $\lambda_{2\alpha}=\lambda_{2\alpha-1}=\mu_\alpha$, i.e.

$$\mathbf{e}_{2\alpha-1} T = \mu_\alpha \mathbf{e}_{2\alpha-1}; \qquad \mathbf{e}_{2\alpha} T = \mu_\alpha \mathbf{e}_{2\alpha} \qquad (\alpha = 1, 2, \cdots, m). \tag{3.28}$$

If the proper values of a symmetric operator T are all positive, we say that T is positive definite; this property is equivalent to the inequality

$$uT\cdot u > 0 \tag{3.29}$$

for all tangent vectors $u \neq 0$.

If S is an orthogonal linear operator on the tangent vectors, i.e. if it satisfies the identity

$$uS\cdot vS = u\cdot v$$

for all vectors u and v, and if T is symmetric, then STS^{-1} is also symmetric, and similar to T, that is to say, it has the same proper values as T including multiplicity; in particular, if T is positive definite, then so is STS^{-1}; in this last case the operator

$$T + STS^{-1}$$

is also positive definite, because of (3.29). We shall consider the special case where $S=J$, the latter being clearly orthogonal (equivalent to saying that the metric is Hermitian); because of (3.2) we have $J^{-1}=-J$. For any symmetric T the operator $JTJ^{-1}=-JTJ$ is symmetric and similar to T; furthermore the operator T' defined by

$$T' = T + JTJ^{-1} = T - JTJ \tag{3.30}$$

is obviously Hermitian; if T is positive definite, then so is T'. Conversely, if $T'=0$, then T can not be positive definite; indeed the trace $\mathrm{Sp}(T)$ (sum of its proper values, or contraction) of T vanishes, since $\mathrm{Sp}(T)=\mathrm{Sp}(JTJ^{-1}) = -\mathrm{Sp}(JTJ)$ and hence

$$\mathrm{Sp}(T) = \frac{1}{2}\,\mathrm{Sp}\,(T - JTJ) = \frac{1}{2}\,\mathrm{Sp}\,(T') = 0. \tag{3.31}$$

4. **Hypersurfaces in the Cayley space.** Let $\mathfrak{X}$ be a 6-dimensional, orientable, differentiable manifold and let F be a regular, differentiable mapping of $\mathfrak{X}$ into the Cayley space $\mathfrak{Y}$, whose structure is discussed in §2. By a regular

mapping F we mean a differentiable mapping of class $\mathcal{C}^2$, such that the 6×7 matrix of functions representing the first partial derivatives in any parametric representation of F has rank 6 at every point of $\mathfrak{X}$: in other words F is locally a differentiable imbedding; no assumption will be made about the behavior of F in the large, so that we admit self-intersections and self-overlaps of $\mathfrak{X}$ under F, as well as the possibility of discrete sets of points in $\mathfrak{X}$ having images in $\mathfrak{Y}$ accumulating at points (that may or may not belong to $F(\mathfrak{X})$).

Our differential-geometric study of $\mathfrak{X}$ will be based on the invariants induced by F from the invariants defined in $\mathfrak{Y}$ by the group G_2 extended by the group of translations. The Cayley space $\mathfrak{Y}$ is homogeneous under this transformation group and, since $G_2 \subset SO(7)$ (the group of rotations in Euclidean 7-space), the invariants to be considered will include those of classical differential geometry.

In analogy with the notation used in the last chapter, in which df, the differential of a scalar function f in $\mathfrak{X}$, is linear, scalar-valued functional on the tangent vectors u at each point $p \in \mathfrak{X}$, we regard $X = F(p)$ as the image under F of a variable point p, and $dX(p)$ as the linear functional on tangent vectors u at p, whose value udX is the derivative of X along u at p: hence dX is a $\mathfrak{Y}$-valued linear functional on the vectors u, and udX can then be regarded as the image of u induced by F, i.e. as the vector in $\mathfrak{Y}$ tangent to $F(\mathfrak{X})$ at $F(p)$ corresponding to u for each $p \in \mathfrak{X}$ where u is defined. The condition that F is regular can be formulated as saying that for each $p \in \mathfrak{X}$ the linear mapping $dX(p)$ is a monomorphism (isomorphism into) of the tangent space of $\mathfrak{X}$ at p into $\mathfrak{Y}$. Since the vectors udX thus obtained are invariant under the translations of F in $\mathfrak{Y}$, we can apply to them all the operations that represent the invariants of G_2, such as the scalar and vector products already discussed in Chapters 1 and 2.

As in classical differential geometry the mapping F induces a quadratic form, the *first fundamental form*

$$ds^2 = dX \cdot dX, \tag{4.1}$$

which is a positive definite quadratic form on the tangent vectors: the corresponding Riemannian inner product between tangent vectors u and v is

$$u \cdot v = udX \cdot vdX. \tag{4.2}$$

Since the image under $dX(p)$ of the tangent vectors at each $p \in \mathfrak{X}$ forms a hyperplane in $\mathfrak{Y}$, one can choose either one of two opposite unit vectors orthogonal to the tangent space image. We recall that $\mathfrak{Y}$ has a natural orientation given by a canonical frame (I_j) $(j=1, \cdots, 7)$, since the group G_2 preserves orientation (Corollary 1, Theorem 2, §2); since $\mathfrak{X}$ is orientable, to either orientation we choose for $\mathfrak{X}$ we shall choose the unit normal vector $N = N(p)$ at $F(p)$ such that, if $(e_1, \cdots, e_6)$ is a positively oriented frame of tangent vectors at p, then the frame $(N(p), e_1dX, \cdots, e_6dX)$ at $F(p)$ is positively oriented relative to a canonical frame for $\mathfrak{Y}$. Opposite orientations of $\mathfrak{X}$ determine opposite choices of N. If F is of class $\mathfrak{C}^r$, then $N(p)$ is of class $\mathfrak{C}^{r-1}$, and if $Y = Y(p)$ is a $\mathfrak{Y}$-valued, translation invariant vector function of p of class $\mathfrak{C}^\rho$, $0 \leqq \rho < r$, such that $N \cdot Y$ vanishes identically, then there exists a uniquely determined vector function $u = u(p)$, of class $\mathfrak{C}^\rho$, such that $udX = Y$.

The second fundamental form is defined as the component along N of the second order differential D^2X: it is also written as a symmetric, bilinear form on the tangent vectors in $\mathfrak{X}$ whose value is a scalar of odd kind (by this we mean that if the orientation of $\mathfrak{X}$ is reversed, the value is replaced by its negative), denoted by $(N \cdot D^2X)(u; v)$. We compute it formally as follows: consider the identities

$$N \cdot udX = 0 \tag{4.3}$$

and

$$N \cdot N = 1; \tag{4.4}$$

differentiating absolutely (i.e. regarding u formally as a constant) these two identities along a second vector v we obtain

$$(N \cdot D^2X)(v; u) + vdN \cdot udX = 0 \tag{4.5}$$

and

$$2N \cdot vdN = 0. \tag{4.6}$$

The last of these identities implies that for each v there exists a tangent vector in $\mathfrak{X}$, which depends linearly on v, and hence a linear operator $K: v \to vK$, such that

$$vdN = -vKdX. \tag{4.7}$$

From (4.5) and the identity $(N \cdot D^2X)(v; u) = (N \cdot D^2X)(u; v)$ we obtain (cf. Definition 3.6)

$$uK \cdot v = -udN \cdot vdX = (N \cdot D^2X)(u; v) = (N \cdot D^2X)(v; u) = vK \cdot u, \tag{4.8}$$

which shows that K is the operator corresponding to the second fundamental form (Weingarten's equations) and that it is symmetric with respect to the Riemannian structure defined by the first fundamental form. The proper values $\kappa_1, \cdots, \kappa_6$ of K are called the principal curvatures of $\mathfrak{X}$ at each point p and the corresponding normalized characteristic vectors $e_1', \cdots, e_6'$ are the respective principal directions. Thus

$$e_j' K = \kappa_j e_j' \qquad (j = 1, \cdots, 6). \tag{4.9}$$

In the sequel we shall deal chiefly with the mean curvature of $\mathfrak{X}$, defined as usual by

$$H = \frac{1}{6} \sum_{j=1}^{6} \kappa_j = \frac{1}{6} \operatorname{Sp}(K).$$

We shall now recall two theorems on hypersurfaces in Euclidean space.

THEOREM A. *If a hypersurface $\mathfrak{X}$ in Euclidean space is of class $\mathfrak{C}^2$, and if its mean curvature H vanishes identically, then $\mathfrak{X}$ is a real analytic hypersurface. The condition $H=0$ means that $\mathfrak{X}$ is locally a solution of the Plateau problem.*

The proof of the first part of this statement is contained in [20], [18], and [15], which deal with the differentiability of solutions of nonlinear, second order, elliptic partial differential equations. A proof of the second half of the statement can be found, for instance, in Eisenhart's textbook [10, pp. 176–179].

THEOREM B. *If $\mathfrak{X}$ is a closed hypersurface of class $\mathfrak{C}^2$ in Euclidean space, then there exists a nonempty region $U \subset \mathfrak{X}$, where the second fundamental form is definite (positive or negative, depending on the orientation), i.e. where all the principal curvatures κ_j are $\neq 0$ and have equal sign.*

This proposition, known as Hadamard's lemma, is proved for instance in [5, p. 23], by taking an arbitrary point A in the Euclidean space and letting U be a sufficiently small neighborhood relative to $\mathfrak{X}$ of the point $p \in \mathfrak{X}$ whose Euclidean distance from A is a maximum.

Thus far we have confined our discussion of the regular mapping F of the orientable manifold $\mathfrak{X}$ into the Cayley space $\mathfrak{Y}$ exclusively to the invariants of the Euclidean structure of $\mathfrak{Y}$, i.e. making use only of the scalar product and orientation of $\mathfrak{Y}$, and not of the properties of the vector product. We shall now take the latter into consideration.

Let $p \in \mathfrak{X}$, let $X = F(p)$ be its image in $\mathfrak{Y}$, let $u = u(p)$ any tangent vector at p, and let $N = N(p)$ the normal vector to $F(\mathfrak{X})$ at $F(p)$: then the vector

$$N(p) \times udX(p)$$

is tangent to $F(\mathfrak{X})$ at $F(p)$, since $N \cdot (N \times udX) = (N, N, udX) = 0$; hence there exists a unique tangent vector v at p such that

$$vdX = N \times udX.$$

Since the mapping $u \to v$ is obviously linear, we can denote it, as in the beginning of §3, by the letter J: thus J is defined by

$$uJdX = N \times udX. \tag{4.10}$$

It follows then from (2.2), (4.3), and (4.4) that

$$\begin{aligned} uJ^2dX &= N \times (N \times udX) \\ &= -[N, N, udX] + (N \cdot udX)N - (N \cdot N)udX = -udX; \end{aligned} \tag{4.11}$$

thus J satisfies Equation (3.2) characterizing the structural operator of an almost complex manifold. Since the immersion F of $\mathfrak{X}$ into $\mathfrak{Y}$ is assumed to be of class $\mathfrak{C}^2$, it follows that J is of class $\mathfrak{C}^1$. From (2.1) and (4.11) we have the further identity

$$\begin{aligned} uJ\cdot vJ &= (uJdX)\cdot(vJdX) = (N\times udX)\cdot(N\times vdX) \\ &= \{(N\times udX)\times N\}\cdot vdX = \{-N\times(N\times udX)\}\cdot vdX \\ &= udX\cdot vdX = u\cdot v. \end{aligned} \tag{4.12}$$

The last three equations imply the following important result.

THEOREM 3. *A regular mapping F of an oriented, 6-dimensional manifold $\mathfrak{X}$, of class $\mathfrak{C}^p$ ($p\geqq 1$) into the Cayley space $\mathfrak{Y}$ induces in $\mathfrak{X}$ the structure of an Hermitian, almost complex manifold under the linear operator J defined by* (4.10) *and the induced Riemannian metric.*

The exterior form of degree 2 $\omega(u; v)$ defined as in (3.16) from J and the metric can be written as follows:

$$\omega(u;v) = uJ\cdot v = (N, udX, vdX). \tag{4.13}$$

We shall now follow the procedure outlined in §3 to find the necessary and sufficient conditions on F in order that the induced almost complex structure on $\mathfrak{X}$ be completely integrable, or respectively that the exterior form ω define a symplectic structure in $\mathfrak{X}$.

THEOREM 4. *A 6-dimensional, oriented manifold $\mathfrak{X}$ with the almost complex structure induced on it by a regular mapping F of class $\mathfrak{C}^2$ of $\mathfrak{X}$ into the Cayley space $\mathfrak{Y}$ is integrable to a complex analytic manifold, if and only if the linear operator K on its tangent space defined by the second fundamental form satisfies everywhere on $\mathfrak{X}$ the equation*

$$K - JKJ = 0. \tag{4.14}$$

Comparing Equation (4.14) with Definition 3.2, one can rephrase the necessary and sufficient condition that the almost complex structure induced on $\mathfrak{X}$ by F be completely integrable is that the second fundamental form be anti-Hermitian with respect to J, or equivalently that the linear operator K associated with the second fundamental form be complex-antilinear (cf. Definition 3.6).

Proof. From Equations (3.13) and (3.13′) one sees that the integrability condition in the case at hand is equivalent to the equation

$$(1 - J)dJdX = 0, \tag{4.15}$$

where the left-hand member is an exterior, bilinear, differential form at each point $p\in\mathfrak{X}$, with values in $\mathfrak{Y}$, tangent to $F(\mathfrak{X})$ at $X = F(p)$. We shall compute this left-hand member in terms of the basic invariants of $\mathfrak{Y}$.

The vector-valued differential JdX maps a tangent vector u into

$$uJdX = N \times udX;$$

hence its exterior derivative $dJdX$ maps any two tangent vectors u, v into

$$(u;v)dJdX = udN \times vdX - vdN \times udX = -uKdX \times vdX - udX \times vKdX.$$

It is convenient to split this last expression into the components respectively tangent to $F(\mathfrak{X})$ and parallel to N, according to the formula $A = N\times(A\times N) + (N\cdot A)N$ for $A \in \mathcal{Y}$: thus, applying the identities of Chapter 1 and Equation (4.3), we have

$$\begin{aligned}(u;v)dJdX &= N \times \{N \times (uKdX \times vdX) + N \times (udX \times vKdX)\} \\ &\quad - \{(N, uKdX, vdX) + (N, udX, vKdX)\}N \\ &= N \times \{-[N, uKdX, vdX] - [N, udX, vKdX]\} \\ &\quad - \{(N, uKdX, vdX) + (N, udX, vKdX)\}N.\end{aligned} \tag{4.16}$$

Similarly, denoting by K' the linear operator $JKJ^{-1} = -JKJ$, we evaluate the vectorial exterior form $JdJdX$ as follows:

$$\begin{aligned}(u;v)JdJdX &= -uJKdX \times vJdX - uJdX \times vJKdX \\ &= -(N \times uK'dX) \times (N \times vdX) - (N \times udX) \times (N \times vK'dX) \\ &= N \times \{[N, uK'dX, vdX] + [N, udX, vK'dX]\} \\ &\quad - \{(N, uK'dX, vdX) + (N, udX, vK'dX)\}N.\end{aligned} \tag{4.17}$$

The normal components of the latter expression satisfy the identities

$$\begin{aligned}(N,\ & uK'dX, vdX) \\ &= (N \times uJKJ^{-1}dX)\cdot vdX = uJKdX\cdot vdX = uJK\cdot v = uJ\cdot vK \\ &= uJdX\cdot vKdX = (N \times udX)\cdot vKdX = (N, udX, vKdX)\end{aligned}$$

and similarly

$$(N, udX, vK'dX) = (N, uKdX, vdX).$$

Thus, subtracting equation (4.17) from (4.16), one verifies that the normal components cancel each other, leaving only a tangential term, whose vanishing is equivalent to (4.15):

$$\begin{aligned}&(u;v)(1-J)dJdX \\ &\quad = -N \times \{[N, u(K+K')dX, vdX] + [N, udX, v(K+K')dX]\}.\end{aligned} \tag{4.18}$$

In order to prove the assertion we must show that the right hand member of (4.18) vanishes, if and only if $K+K'$ itself is zero. Since K is a symmetric operator (Equation (4.8) and Definition 3.6), it follows that $K+K' = K - JKJ = K + JKJ^{-1}$ is Hermitian (cf. Equation (3.30)): hence one can choose at each point $p\in\mathfrak{X}$ a unitary frame $e_1, \cdots, e_6$ of tangent vectors that are characteristic vectors of $K+K'$ corresponding to three real proper values μ_1, μ_2, μ_3, i.e. such that

$$(4.19)\qquad e_{2\alpha-1}(K+K')=\mu_\alpha e_{2\alpha-1};\quad e_{2\alpha}(K+K')=\mu_\alpha e_{2\alpha}\qquad(\alpha=1,2,3).$$

Consider now a canonical frame $I_1, \cdots, I_7$ in $\mathfrak{Y}$ at $F(p)$ such that $I_1=N$ and the remaining ones are tangent to $F(\mathfrak{X})$: $I_j=e'_{j-1}dX$ $(2\leqq j\leqq 7)$, where $(e'_1, \cdots, e'_6)$ is a unitary frame at p, as one readily verifies from Theorems 1 and 2. Then the frame (e_j) of characteristic vectors of $K+K'$ is related to the frame (e'_j) as in (3.6′) by a unitary 3×3 matrix $\Theta=(\eta^\alpha_\beta+i\theta^\alpha_\beta)$; since it is obviously possible to replace e_1 and e_2 without loss of generality by $e_1\cos\tau+e_2\sin\tau$ and $-e_1\sin\tau+e_2\cos\tau$ respectively, where τ is any real number, one can choose τ so that the matrix Θ becomes unimodular. By Theorem 2 the frame (e_j) thus adjusted has the property that the 7-frame $(N, e_1dX, \cdots, e_6dX)$ at $F(p)$ in $\mathfrak{Y}$ is canonical in the sense of Theorem 1; consequently from (2.3) we have the identities

$$[N, e_{2\alpha}dX, e_{2\beta}dX]=e_{2\gamma}dX\ (\alpha,\beta,\gamma=\text{any cyclic permutation of }1, 2, 3),$$

whence from (4.18) and (4.19) we have, by replacing u and v by $e_{2\alpha}$ and $e_{2\beta}$ respectively,

$$(4.20)\qquad \begin{aligned}(e_{2\alpha};e_{2\beta})(1-J)dJdX&=-(\mu_\alpha+\mu_\beta)N\times[N, e_{2\alpha}dX, e_{2\beta}dX]\\&=(\mu_\alpha+\mu_\beta)e_{2\gamma-1}dX,\end{aligned}$$

where α,β,γ are cyclic permutations of 1, 2, 3. Thus a necessary condition for the integrability condition (4.15) is that $\mu_\alpha+\mu_\beta=0$ for α,β ranging over distinct pairs of values 1, 2, 3; this clearly implies that each $\mu_\alpha=0$, and hence that $K+K'=0$, or equation (4.14); the latter is clearly sufficient for the vanishing of both members of (4.18) identically in u and v. Thus we have just proved that the integrability condition (3.13) for the almost complex structure induced on $\mathfrak{X}$ by F is Equation (4.14).

In order to prove that in our case the integrability condition (3.13) is sufficient, it is sufficient to notice that, if $K+K'=0$, it follows from equation (3.31) that the trace $\mathrm{Sp}(K)$ vanishes identically, and hence that $\mathfrak{X}$ is mapped (locally) into a minimal hypersurface. Since the mapping F can be defined by functions satisfying second order elliptic equations with analytic coefficients, as we recalled in Theorem A earlier in this chapter, we deduce that $F(\mathfrak{X})$ is locally an analytic hypersurface, and in this case the sufficiency of the integrability condition (3.13) has been already established. This completes the proof of Theorem 4.

Corollary. *If $\mathfrak{X}$ is a compact, orientable, 6-dimensional manifold, and F a regular mapping of $\mathfrak{X}$ into $\mathfrak{Y}$, then there exists a nonempty domain $U\subset\mathfrak{X}$ such that there are no solutions of the generalized Cauchy-Riemann equations (relative* *to the induced almost complex structure) in any subdomain of U except constants.*

Proof. It follows from Hadamard's lemma (quoted earlier in this chapter as Theorem B) that, if $\mathfrak{X}$ is compact, then there exists a point in a neighborhood of which the second fundamental form is either positive or negative definite, depending on the orientation. Assuming as we well may that it is positive definite, in that domain the operator $K+K'$ is a positive definite Hermitian operator, as we verified at the end of §3, and hence the proper values μ_1, μ_2, μ_3 of $K+K'$ are all positive in some region U. Let w be any holomorphic function in a subdomain V of U, let p be any point in V, and let (e_j) $(1 \leqq j \leqq 6)$ be a unitary frame at p consisting of characteristic vectors of $K+K'$, adjusted so that $(N, e_1dX, \cdots, e_6dX)$ is a canonical frame for $\mathfrak{Y}$ at $F(p)$. Then on the one hand, since w is holomorphic, we have $dw=-iJdw$, and hence

$$(1 - J)dJdw = 0;$$

on the other hand we have from Equation (4.20)

$$(e_{2\alpha}; e_{2\beta})(1 - J)dJdw = (\mu_\alpha + \mu_\beta)e_{2\gamma-1}dw,$$

and similarly

$$(e_{2\alpha-1}; e_{2\beta-1})(1 - J)dJdw = (\mu_\alpha + \mu_\beta)e_{2\gamma}dw,$$

where α, β, γ are any cyclic permutation of 1, 2, 3: since the μ_α's are all positive at p, it follows from the last three equations that dw vanishes at p, and hence everywhere in V. Thus we have proved that, if w is holomorphic in V, it is a constant.

A further analysis of the result just established reveals that, if F is an analytic mapping of a compact $\mathfrak{X}$ into $\mathfrak{Y}$, then there are no nonconstant holomorphic functions anywhere in $\mathfrak{X}$.

We pass now to the second problem, namely that of deriving the conditions on the mapping F in order that it induce on $\mathfrak{X}$ a symplectic structure in terms of the exterior form ω defined by (4.13).

THEOREM 5. *A 6-dimensional, oriented, differentiable manifold $\mathfrak{X}$ is a symplectic manifold with respect to the Hermitian, almost complex structure induced by a regular, differentiable mapping F of $\mathfrak{X}$ into $\mathfrak{Y}$, if and only if the linear operator K defined by its second fundamental form satisfies the two equations*

$$K + JKJ = 0; \qquad \mathrm{Sp}(K) = 6H = 0, \tag{4.21}$$

where H is the mean curvature of $F(\mathfrak{X})$.

Proof. From formula (4.13) expressing $\omega(u; v)$ as a triple scalar product of vectors in $\mathfrak{Y}$ we obtain the formula for the exterior derivative $d\omega$ as follows:

$$d\omega(u;\, v;\, w)$$

$$(4.22)\quad \begin{aligned} &= (udN,\, vdX,\, wdX) + (udX,\, vdN,\, wdX) + (udX,\, vdX,\, wdN) \\ &= -(uKdX,\, vdX,\, wdX) - (udX,\, vKdX,\, wdX) - (udX,\, vdX,\, wKdX). \end{aligned}$$

We must verify that (4.21) is the necessary and sufficient condition for the vanishing of the right hand member of (4.22). Let $(\mathbf{e}_j)$ $(1 \leqq j \leqq 6)$ be a canonical tangent frame at any point of $\mathfrak{X}$, i.e. a frame such that $(N, \mathbf{e}_1 dX, \cdots, \mathbf{e}_6 dX)$ is a canonical frame at the corresponding point for $\mathfrak{Y}$. In the sequel of this proof it is convenient to replace the frame $(\mathbf{e}_j)$ formally by the corresponding complex frame $(\mathbf{e}'_j)$ defined by

$$\mathbf{e}'_\alpha = \frac{1}{2}(\mathbf{e}_{2\alpha-1} - i\mathbf{e}_{2\alpha}); \qquad \mathbf{e}'_{3+\alpha} = (\mathbf{e}'_\alpha)^- = \frac{1}{2}(\mathbf{e}_{2\alpha-1} + i\mathbf{e}_{2\alpha}) \qquad (\alpha = 1, 2, 3)$$

or by the inverse transformation

$$\mathbf{e}_{2\alpha-1} = \mathbf{e}'_\alpha + \mathbf{e}'_{3+\alpha}; \qquad \mathbf{e}_{2\alpha} = i(\mathbf{e}'_\alpha - \mathbf{e}'_{3+\alpha}).$$

In terms of the new frame it is easy to express the formula for the inner product, the action of J, the triple scalar product as it appears in (4.22), and the property of any vector (or tensor) being real; we shall use Latin indices in the range from 1 to 6, Greek indices from 1 to 3, and starring an index to express addition or subtraction of 3 to its value.

A vector $\mathbf{u} = u^j \mathbf{e}'_j$ (using the summation convention) is real, if and only if $u^{j*} = \bar{u}^j$; similarly a tensor is real, if and only if any two components related to each other by starring all the indices are mutually complex conjugates. The action of J is described by $\mathbf{e}'_\alpha J = i\mathbf{e}'_\alpha$, $\mathbf{e}'_{\alpha^*} J = -i\mathbf{e}'_{\alpha^*}$; the scalar product is derived from the identity $2\mathbf{e}'_j \cdot \mathbf{e}'_k = \delta_{jk^*}$. Finally, if $\mathbf{u} = u^i\mathbf{e}'_i$, $\mathbf{v} = v^j\mathbf{e}'_j$, $\mathbf{w} = w^i\mathbf{e}'_i$ are any three vectors, then the triple scalar product (udX, vdX, wdX) is given by

$$(udX, vdX, wdX) = \frac{1}{2}\begin{vmatrix} u^1 & u^2 & u^3 \\ v^1 & v^2 & v^3 \\ w^1 & w^2 & w^3 \end{vmatrix} + \frac{1}{2}\begin{vmatrix} u^4 & u^5 & u^6 \\ v^4 & v^5 & v^6 \\ w^4 & w^5 & w^6 \end{vmatrix}.$$

Let the tensor components of K be expressed by the 6×6 matrix K_j^k: then the matrix satisfies the condition that K is real, $K_{j^*}^{k^*} = (K_j^k)^-$, and symmetric, $K_k^j = K_{j^*}^{k^*} = (K_j^k)^-$. Hence the scalar K_α^α satisfies the identity

$$K_\alpha^\alpha = K_{\alpha^*}^{\alpha^*} = \frac{1}{2} K_j^j = 3H,$$

where H is the mean curvature.

From the above identities one can determine completely $d\omega$ by evaluating $d\omega(\mathbf{e}'_i;\, \mathbf{e}'_j;\, \mathbf{e}'_k)$ for $i<j<k$:

$$d\omega(e_1'; e_2'; e_3') = -K_\alpha^\alpha = -3H;$$

$$d\omega(e_\alpha'; e_\beta'; e_{\delta^*}') = -\epsilon_{\alpha\beta\gamma}K_{\delta^*}^\gamma$$

$$(\epsilon_{\alpha\beta\gamma} = \text{sign of of permutation of } 1, 2, 3 \text{ to } \alpha, \beta, \gamma);$$

$$d\omega(e_\delta'; e_{\alpha^*}'; e_{\beta^*}') = -\epsilon_{\alpha\beta\gamma}K_\delta^{\gamma^*};$$

$$d\omega(e_4'; e_5'; e_6') = -K_{\alpha^*}^{\alpha^*} = -3H,$$

where the last two equations are the complex conjugates of the first two. If we compute the tensor components of JKJ, we notice that they are equal or opposite to those of K, according as to whether the indices (written as Greek letters with or without star) are of opposite type (one starred, the other not) or of the same type respectively. Thus the linear operator $K+JKJ$ has tensor components equal to $2K$, if they are of opposite type, and zero, if they are of the same type. From these considerations and from the last set of equations it follows that (4.21) is necessary and sufficient in order that $d\omega=0$, completing the proof of the theorem.

The same argument as the one used in the Corollary to Theorem 4 leads to the analogous result for symplectic structures as follows.

COROLLARY 1. *If $\mathfrak{X}$ is a compact, differentiable manifold with the Hermitian, almost complex structure induced by a regular mapping F of $\mathfrak{X}$ into $\mathfrak{Y}$, then $\mathfrak{X}$ can not be a symplectic manifold.*

Combining the results of Theorems 4 and 5 we obtain the criterion for an Hermitian, almost complex structure induced by F to be Kählerian.

COROLLARY 2. *An oriented, 6-dimensional manifold $\mathfrak{X}$ with the Hermitian, almost complex structure induced by a differentiable immersion F of $\mathfrak{X}$ into $\mathfrak{Y}$ is Kählerian, if and only if F maps $\mathfrak{X}$ into a hyperplane.*

Proof. From the remarks following Definition 3.4 it is clear that $\mathfrak{X}$ is Kählerian, if and only if it is simultaneously symplectic and integrable to a complex manifold, and hence if it satisfies simultaneously Equations (4.14) and (4.21). The two equations combined imply that K must vanish identically, and therefore that N is constant. Since $d(N\cdot X)=N\cdot dX=0$ for N constant, it follows that F satisfies the linear equation $X\cdot N=\text{const}$. The converse of the argument being trivial, the statement is proved.

Having verified through the last corollary that F induces a Kählerian structure on $\mathfrak{X}$ only in the trivial case where $F(\mathfrak{X})$ is a hyperplane, it is natural to ask in which other cases, if any, the Hermitian almost complex structure induced on $\mathfrak{X}$ by F is merely symplectic, or, alternatively, integrable to a complex analytic one: a complete solution of these two problems amounts to a complete characterization and functional representation of all hypersurfaces in $\mathfrak{Y}$ that satisfy respectively (4.21) or (4.14), and is probably very difficult. As a matter of fact, for the first problem, we neither know as yet of any example, nor can show the impossibility of a hypersurface in $\mathfrak{Y}$, other than a hyperplane, whose induced Hermitian structure is symplectic. On the other hand we shall show presently that there is a simple representation, albeit far from exhaustive as far as we can guess, of a large class of hypersurfaces in $\mathfrak{Y}$, whose almost complex structure is integrable to a complex analytic one: these manifolds are closely related to the minimal surfaces in Euclidean 3-space.

Consider a linear subspace $\mathfrak{Y}'$ of $\mathfrak{Y}$, closed under the operation of cross product $A\times B$, but in which the triple vector product $[ABC]$ vanishes identically. By Theorem 1 $\mathfrak{Y}'$ is completely characterized as the linear subspace of $\mathfrak{Y}$ that is spanned by the first three vectors $I_1^{(0)}$, $I_2^{(0)}$, $I_3^{(0)}$ in some fixed, canonical frame for $\mathfrak{Y}$, $(I_j^{(0)})$ $(j=1, 2, \cdots, 7)$; denote by $\mathfrak{Y}''$ the orthogonal complement of $\mathfrak{Y}'$ in $\mathfrak{Y}$, i.e. the oriented, linear subspace spanned by $(I_j^{(0)})$ $(j=4, 5, 6, 7)$, and identify $\mathfrak{Y}$, considered as a linear space, with the Cartesian product $\mathfrak{Y}'\times\mathfrak{Y}''$.

Let $\mathfrak{X}'$ be an oriented, differentiable (2-dimensional) surface and F' a regular, $\mathfrak{C}^2$ mapping of $\mathfrak{X}'$ into the oriented 3-space $\mathfrak{Y}'$. Then one can naturally define from F' a regular mapping F of the oriented, 6-dimensional manifold $\mathfrak{X}=\mathfrak{X}'\times\mathfrak{Y}''$ into $\mathfrak{Y}=\mathfrak{Y}'\times\mathfrak{Y}''$ as the product mapping of F' in $\mathfrak{X}'$ with the identity in $\mathfrak{Y}''$. We shall determine the conditions on F' in order that the almost complex structure induced on $\mathfrak{X}$ by the product mapping F be integrable to a complex analytic one.

THEOREM 6. *Let $\mathfrak{X}$ be the product manifold $\mathfrak{X}'\times\mathfrak{Y}''$ of an oriented surface $\mathfrak{X}'$ with the 4-dimensional linear subspace $\mathfrak{Y}''$ of $\mathfrak{Y}$ that is spanned by the vectors $I_4^{(0)}, \cdots, I_7^{(0)}$ of a fixed, canonical frame $(I_j^{(0)})$ $(1\leqq j\leqq 7)$ in $\mathfrak{Y}$. Let F' be any regular $\mathfrak{C}^2$ mapping of $\mathfrak{X}'$ into the linear subspace $\mathfrak{Y}'$ of $\mathfrak{Y}$ spanned by $I_1^{(0)}$, $I_2^{(0)}$, $I_3^{(0)}$. Then the product mapping F of $\mathfrak{X}$ into $\mathfrak{Y}$ defines an almost complex structure in $\mathfrak{X}$ that is integrable to a complex analytic one, if and only if F' maps $\mathfrak{X}'$ locally as a minimal surface in the 3-space $\mathfrak{Y}'$.*

Proof. Since the normal vector N at each point $p\in\mathfrak{X}$ lies obviously in a linear 3-space parallel to $\mathfrak{Y}'$, if one starts to build a variable, canonical frame (I_j) for $\mathfrak{Y}$ at $X=F(p)$ by setting $I_1=N$ and $I_2=\boldsymbol{e}_1dX$, where $\boldsymbol{e}_1$ is a tangent vector in $\mathfrak{X}$ parallel to $\mathfrak{X}'$, then $I_3=I_1\times I_2=\boldsymbol{e}_2dX$ lies again in a 3-space parallel to $\mathfrak{Y}'$, and hence $\boldsymbol{e}_2$ is also tangent to $\mathfrak{X}'$. One can then continue the construction of the local canonical frame $I_j=\boldsymbol{e}_{j-1}dX$ $(j=4, 5, 6, 7)$, where these last four vectors come from $\mathfrak{Y}''$. It is also clear that the only contribution to the second fundamental form $uK\cdot v=u\cdot vK$ can come only from the 2-dimensional linear subspace of the tangent space to $\mathfrak{X}$ coming from $\mathfrak{X}'$. Since $(\boldsymbol{e}_j)$ is a unitary frame with respect to the Hermitian, almost complex structure of $\mathfrak{X}$, we can write the second fundamental form in matrix form as follows

$$\mathbf{e}_j K = a_{j1}\mathbf{e}_1 + a_{j2}\mathbf{e}_2 \qquad (j = 1, 2),$$
$$\mathbf{e}_j K = 0 \qquad (j = 3, 4, 5, 6),$$

where

$$a_{12} = a_{21}.$$

Now, since $\mathbf{e}_j J = -(-1)^j \mathbf{e}_{j'}$, where $j' = j - (-1)^j$, we can write directly the formulas for $\mathbf{e}_j(K - JKJ)$ whose vanishing is equivalent to the integrability condition (4.14) of Theorem 4:

$$\mathbf{e}_j(K - JKJ) = (a_{11} + a_{22})\mathbf{e}_j \qquad (j = 1, 2),$$
$$\mathbf{e}_j(K - JKJ) = 0 \qquad (j = 3, 4, 5, 6).$$

Thus we have proved that the almost complex structure induced on $\mathfrak{X}$ by F is integrable, if and only if

$$a_{11} + a_{22} = 0.$$

But since the matrix (a_{jk}) $(j, k = 1, 2)$ clearly represents the second fundamental form for $\mathfrak{X}'$ under F' in terms of the tangent frame $\mathbf{e}_1$, $\mathbf{e}_2$, the trace $a_{11} + a_{22}$ is twice the mean curvature of $F'(\mathfrak{X}')$ in $\mathfrak{Y}'$. This concludes the proof of the theorem.

It is important to bear in mind that the almost complex structure induced on $\mathfrak{X}$ by F is not the product structure, except in the trivial case where $F'(\mathfrak{X}')$ lies in a plane. The reason for this is that, although for each point $p' \in \mathfrak{X}'$ the submanifold $\{p'\} \times \mathfrak{Y}''$ of $\mathfrak{X}$ is a complex analytic submanifold of $\mathfrak{X}$ (even when the integrability condition in the latter is not satisfied) isomorphic to the space of 2 complex variables, as p' varies over $\mathfrak{X}'$ the corresponding translation of $F(\{p'\} \times \mathfrak{Y}'')$ in $\mathfrak{Y}$ is not complex analytic, owing to the fact that the normal vector N (which alone determines the complex structure in the tangent space) is variable.

We shall prove now one last general property of the Hermitian, almost complex structure induced on any 6-dimensional manifold $\mathfrak{X}$ by a regular mapping F of $\mathfrak{X}$ into $\mathfrak{Y}$, namely that (cf. Definition 3.3) $\mathfrak{X}$ becomes a special Hermitian manifold in all cases.

THEOREM 7. *Given an oriented, differentiable, 6-dimensional manifold $\mathfrak{X}$ and a regular, differentiable mapping F of $\mathfrak{X}$ into the Cayley space $\mathfrak{Y}$, then F induces on $\mathfrak{X}$ the structure of a* special *Hermitian (almost complex) manifold.*

Proof. We shall consider as allowable local tangent frames in the neighborhood of any point in $\mathfrak{X}$ any sextuple $(\mathbf{e}_j)$ of tangent vector functions such that the septuple $(N, \mathbf{e}_1 dX, \cdots, \mathbf{e}_6 dX)$ at the image point X is a canonical frame for $\mathfrak{Y}$. Then the frame $(\mathbf{e}_j)$ is unitary with respect to the Hermitian, almost complex structure induced on $\mathfrak{X}$. The statement of Theorem 2 is that any two overlapping tangent frames (I_j) $(I_j = \mathbf{e}_{j-1} dX,\ 2 \leq j \leq 7)$ are related to one another by a linear transformation σ, which, since $I_1 = N$ is invariant, under the representation ϕ defined by (2.4) is representable in a unique way by a matrix $\phi(\sigma) \in SU(3)$. The corresponding frame transformation $\tilde{\sigma}$ of one system $(\mathbf{e}_j)$ to another can be described uniquely by a unitary, 3×3 matrix $\Theta = (\eta^\alpha_\beta + i\theta^\alpha_\beta)$ as in (3.6′) under the representation ψ of the tangent space in $\mathfrak{X}$ defined by (3.4). From the definitions of the representations ϕ and ψ of vectors respectively at a point $p \in \mathfrak{X}$ and in the orthogonal complement of $N(p)$ in $\mathfrak{Y}$ we obtain the identity for corresponding frames $(\mathbf{e}_j)$ and $(N, \mathbf{e}_j dX)$

$$u^\psi = (udX)^\phi,$$

and hence the transformations of allowable frames satisfy the relation

$$\psi(\tilde{\sigma}) = \Theta = \phi(\sigma) \in SU(3).$$

Hence the manifold $\mathfrak{X}$ with the covering by allowable, local tangent frames is a special Hermitian manifold, as we set out to prove. In other words the first Chern characteristic class $\mathbf{c}^1$ of every almost complex manifold obtained by our method vanishes.

5. **Examples of compact, complex analytic manifolds.** We shall apply now the results of §4 to construct some compact, complex analytic manifolds with 3 complex dimensions, that exhibit some interesting properties.

Let $\mathfrak{X}'$ be a closed, hyperelliptic Riemann surface of odd genus $g = 2h+1$ $(h \geqq 1)$. Without loss of generality $\mathfrak{X}'$ can be considered as the algebraic curve defined by the equation in the complex field

$$w^2 = f_k(z), \tag{5.1}$$

where $f_k(z)$ is an arbitrary polynomial of degree $k = 2g+2 = 4h+4$, with no multiple roots. The Abelian differentials of the first kind in $\mathfrak{X}'$ are then given by

$$\omega = \phi(z) w^{-1} dz,$$

where $\phi(z)$ is an arbitrary polynomial of degree at most $g-1 = 2h$: we shall single out one such polynomial $\phi(z)$ of degree equal to h and in terms of it take the following three differentials of the first kind,

$$\begin{aligned} \omega_1 &= \frac{1}{2}(\phi^2(z) - 1) w^{-1} dz, \\ \omega_2 &= \phi(z) w^{-1} dz, \\ \omega_3 &= \frac{1}{2i}(\phi^2(z) + 1) w^{-1} dz. \end{aligned} \tag{5.2}$$

The above construction was made in order to exhibit a closed Riemann surface $\mathfrak{X}'$ with three Abelian differentials of the first kind ω_j $(j = 1, 2, 3)$ with the following properties:

(a) linear independence;
(b) the three differentials satisfy the quadratic equation

$$\omega_1^2 + \omega_2^2 + \omega_3^2 = 0; \tag{5.3}$$

(c) there is no place (point) in $\mathfrak{X}'$ where the three differentials ω_j vanish simultaneously.

It is clear that the three differentials chosen satisfy properties (a) and (b); to verify property (c) we point out that it is obvious that ω_1 and ω_3 can not vanish simultaneously at any place where z is finite, so that there remains only to check the two places where z becomes infinite: at each of those places dz has a pole of order 2, w^{-1} has a zero of order $2h+2$, and $\phi^2(z)$ has a pole of order $2h$; hence neither ω_1 nor ω_3 vanishes at either of the two places where z is infinite, completing the argument.

Consider now the three real-valued functions of a place $p' \in \mathfrak{X}'$

$$x^j(p') = \operatorname{Re} \int_{p_0'}^{p'} \omega_j \qquad (j = 1, 2, 3) \tag{5.4}$$

as defining for any fixed place $p_0' \in \mathfrak{X}'$ a real analytic mapping of the maximal Abelian covering $\tilde{\mathfrak{X}}'$ of $\mathfrak{X}'$ into the Euclidean 3-space with Cartesian coordinates x^j. This mapping is regular, since the differentials ω_j satisfy property (c) (see above); by a classical theorem attributed to Monge, Lie, and Weierstrass (cf. [24, p. 184]) property (b) is equivalent to the statement that the functions $x^j(\tilde{p}')$ (where $\tilde{p}'$ is a point of $\tilde{\mathfrak{X}}'$ covering $p' \in \mathfrak{X}'$) map $\tilde{\mathfrak{X}}'$ locally into a minimal surface (globally the image is, in general, everywhere dense); finally, because of property (a), it follows that $\mathfrak{X}'$ is not mapped into a plane. This method of generating minimal surfaces was studied extensively by H. A. Schwarz [22] in connection with the Plateau problem. We remark in addition that the covering transformations of the image of $\tilde{\mathfrak{X}}'$ in the Euclidean 3-space over the original Riemann surface $\mathfrak{X}'$ are given by translations of the space, corresponding to the real parts of the periods of $\int \omega_j$.

Let $\tilde{\mathfrak{X}}$ be the product manifold $\tilde{\mathfrak{X}}' \times \mathfrak{Y}''$ of the maximal Abelian covering $\tilde{\mathfrak{X}}'$ of $\mathfrak{X}'$ with the real linear, 4-dimensional subspace $\mathfrak{Y}''$ of $\mathfrak{Y}$ that is spanned by the last four vectors of a fixed, canonical frame $(I_j^{(0)})$; let F' be the regular mapping of $\tilde{\mathfrak{X}}'$ into the 3-dimensional subspace $\mathfrak{Y}'$ of $\mathfrak{Y}$ that is spanned by the first three vectors, defined by

$$F'(\tilde{p}') = \sum_{j=1}^{3} x^j(\tilde{p}') I_j^{(0)}, \tag{5.5}$$

where $x^j(\tilde{p}')$ is defined from the corresponding, multiple valued function on $\mathfrak{X}'$ given by (5.4), and let F be the product mapping of F' in $\tilde{\mathfrak{X}}'$ with the identity mapping in $\mathfrak{Y}''$, as in Theorem 6. Since F' maps $\tilde{\mathfrak{X}}'$ locally into a minimal surface in $\mathfrak{Y}'$, the almost complex structure induced by F on $\tilde{\mathfrak{X}}$ is integrable to a complex analytic one.

Let T^4 denote a 4-dimensional torus obtained by factoring from $\mathcal{Y}''$ a free module generated by any four linearly independent vectors in $\mathcal{Y}''$, and let $\mathfrak{X}$ be the product manifold $\mathfrak{X}' \times T^4$. Then $\mathfrak{X}$ is a compact, oriented, 6-dimensional manifold, whose maximal Abelian covering is $\tilde{\mathfrak{X}} = \tilde{\mathfrak{X}}' \times \mathcal{Y}''$. To every covering transformation of $\tilde{\mathfrak{X}}$ with respect to $\mathfrak{X}$ there corresponds a transformation of $F(\tilde{\mathfrak{X}})$ onto itself that comes from a translation in $\mathcal{Y}$. Hence the covering transformations of $\tilde{\mathfrak{X}}$ preserve the complex analytic structure, as well as the Hermitian metric and the allowable tangent frames (cf. Theorem 7); thus we arrive at the result that F induces on the compact manifold $\mathfrak{X}$ the structure of a special Hermitian, complex analytic manifold. We shall now prove that $\mathfrak{X}$ does not admit a Kähler metric. In fact, if $\mathfrak{X}$ did admit a Kähler metric, then, since the first Betti number B_1 of $\mathfrak{X}$ is $2g+4$ (because $B_1(\mathfrak{X}') = 2g$ and $B_1(T^4) = 4$), by Hodge's theorem (cf. [14, p. 200]) there would exist $B_1/2 = g+2$ linearly independent, holomorphic, closed differential forms on $\mathfrak{X}$, which together with their complex conjugates would span linearly (with complex coefficients) the whole, B_1-dimensional space of harmonic 1-forms on $\mathfrak{X}$; of these $g+2$ holomorphic forms g at least are known, namely the injection in $\mathfrak{X}$ of the g Abelian differentials of the first kind in $\mathfrak{X}'$, say $\omega_1, \cdots, \omega_g$; since these differentials vanish identically when restricted to any toral submanifold $\{p'\} \times \mathcal{Y}''$ (for each $p' \in \mathfrak{X}'$ this is a complex analytic submanifold of $\mathfrak{X}$), if $\mathfrak{X}$ admitted a Kähler metric, there would be two more linearly independent, holomorphic, closed 1-forms, say ω', ω'', with nonzero integrals along the cyles lying in $\{p'\} \times \mathcal{Y}''$. From the multiplicative structure of the cohomology ring of $\mathfrak{X}$ it follows that the exterior 6-form Ψ defined by

$$\Psi = -i\omega_1 \wedge \bar{\omega}_1 \wedge \omega' \wedge \bar{\omega}' \wedge \omega'' \wedge \bar{\omega}'' \tag{5.6}$$

has a strictly positive integral over $\mathfrak{X}$, from which we deduce that the holomorphic 3-form Ω defined by

$$\Omega = \omega_1 \wedge \omega' \wedge \omega'' \tag{5.7}$$

is not identically zero. Since $\mathfrak{X}$ has the structure of a special Hermitian manifold, the first Chern class $\mathbf{c}^1$ is zero, and hence the divisor of Ω must be homologous to zero. But the divisor of Ω is a union of complex analytic submanifolds counted with positive multiplicity, since Ω is holomorphic, and it is known (cf. [17]) that a cycle of this type can not be homologous to zero in a Kähler manifold unless it is itself zero. Hence the holomorphic 3-form Ω has no zeros. On the other hand the Abelian differential ω_1 appearing as a factor of Ω in (5.7) certainly has some zeros (precisely of total multiplicity $2g-2$ in $\mathfrak{X}'$), so that the divisor of Ω in $\mathfrak{X}$ is nonempty, whence our assumption that $\mathfrak{X}$ admits a Kähler metric leads to a contradiction.

Consider now a comparison between the manifold $\mathfrak{X}$ and the product complex manifold $\mathfrak{X}'\times T^4$ of the Riemann surface $\mathfrak{X}'$ with any complex analytic manifold T^4 homeomorphic to a torus (e.g. by factoring out of the vector space of ordered pairs of complex numbers a free module generated by any four linearly independent translations). Then, since both $\mathfrak{X}'$ and T^4 admit a Kähler metric, so does the product manifold $\mathfrak{X}'\times T^4$. Furthermore, since the canonical divisor of $\mathfrak{X}'$ (see the discussion following Definition 3.3) is homologous to a point $p'\in\mathfrak{X}'$ counted with multiplicity $2g-2$, and since the canonical divisor of T^4 is zero, it follows that the canonical divisor of $\mathfrak{X}'\times T^4$ is homologous to $(2g-2)(\{p'\}\times T^4)$. The two properties just described that distinguish the two complex manifolds $\mathfrak{X}$ and $\mathfrak{X}'\times T^4$ from each other furnish the first known examples of compact, complex manifolds that are differentiably homeomorphic (preserving the natural orientation induced by the complex analytic structure), but whose Chern characteristic classes are distinct, and such that one, but not the other, manifold admits a Kähler metric. Examples of differentiably homeomorphic manifolds with almost complex structures inducing different first Chern classes were known previously, but, the examples given here being complex analytic, we can give a negative answer to a question raised by Hirzebruch [13, pp. 222–224 and 233], whether the Chern classes of a complex manifold depend on the differentiable structure and the orientation alone, and not on the particular complex analytic structure. The previous examples by Hirzebruch [12], of complex manifolds that are differentiably homeomorphic, but not complex analytically isomorphic had all the same Chern classes, and, since they could all be realized as algebraic manifolds, they all admitted Kähler metrics. We can thus state our final result:

If a compact, oriented, differentiable manifold admits more than one complex analytic structure compatible with the differentiable structure and the orientation, then the Chern classes are not necessarily uniquely determined in the manifold, nor is the property invariant, that a complex analytic structure in the manifold admits a Kähler metric.

In conclusion it seems likely that the following conjecture may be true, at least in the cases when the Chern classes are different from zero: *If $\mathfrak{X}$ is a compact, complex manifold and F a differentiable homeomorphism (in the real sense) of $\mathfrak{X}$ onto an algebraic manifold $\mathfrak{X}'$ such that the Chern classes and the orientation are preserved, then does $\mathfrak{X}$ admit a Kähler metric?*

References

1. E. Calabi and D. C. Spencer, *Completely integrable, almost complex manifolds*, Bull. Amer. Math. Soc. Abstract 57-3-251.

2. E. Calabi and B. Eckmann, *A class of compact, complex manifolds which are not algebraic*, Ann. of Math. vol. 58 (1953) pp. 494–500.

3. É. Cartan, *Groupes réels, finis et continus*, Ann. École Norm. vol. 31 (1914) pp. 263–355; also: *Oeuvres complètes*, Partie I, vol. 1, Paris, Gauthier-Villars, 1952, pp. 399–491.

4. S. S. Chern, *Characteristic classes in Hermitian manifolds*, Ann. of Math. vol. 47 (1946) pp. 85–121.

5. ———, *Topics in differential geometry*, mimeographed notes, Institute for Advanced Study, Princeton, 1951.

6. C. Chevalley, *Theory of Lie groups* I, Princeton University Press, 1946.

7. B. Eckmann and A. Frölicher, *Sur l'intégrabilité des structures presque-complexes*, C. R. Acad. Sci. Paris vol. 232 (1951) pp. 2284–2286.

8. C. Ehresmann, *Sur les variétés presque-complexes*, Proceedings of the International Congress of Mathematicians, 1950, vol. II, Amer. Math. Society, 1952, pp. 412–419.

9. C. Ehresmann and P. Libermann, *Sur les structures presque hermitiennes isotropes*, C. R. Acad. Sci. Paris vol. 232 (1951) pp. 1281–1283.

10. L. P. Eisenhart, *Riemannian geometry*, Princeton University Press, 1949.

11. H. Freudenthal, *Oktaven, Ausnahmengruppen und Oktavengeometrie*, Utrecht, 1951.

12. F. Hirzebruch, *Über eine Klasse von einfach-zusammenhängenden komplexen Mannigfaltigkeiten*, Math. Ann. vol. 124 (1951) pp. 77–86.

13. ———, *Problems on differentiable and complex manifolds*, Ann. of Math. vol. 60 (1954) pp. 213–236.

14. W. V. D. Hodge, *The theory and applications of harmonic integrals*, Cambridge University Press, 1941.

15. E. Hopf, *Über den funktionalen, insbesondere den analytischen Charakter der Lösungen elliptischer Differentialgleichungen zweiter Ordnung*, Math. Zeit. vol. 34 (1932) pp. 194–233.

16. H. Hopf, *Zur Topologie der komplexen Mannigfaltigkeiten*, Studies and Essays presented to R. Courant, New York, 1948, pp. 167–185.

17. A. Lichnérowicz, *Généralisations de la géométrie kählérienne globale*, Colloque de géométrie differentielle, Louvain, 1951, pp. 99–122.

18. C. B. Morrey, *Second order elliptic systems of differential equations, Contributions to the theory of partial differential equations*, Ann. of Math. Studies, no. 33, Princeton University Press, 1954, pp. 101–159.

19. R. Moufang, *Zur Struktur von Alternativkörpern*, Math. Ann. vol. 110 (1935) pp. 416–430.

20. L. Nirenberg, *On a generalization of quasi-conformal mappings and its application to elliptic partial differential equations, Contributions to the theory of partial differential equations*, Ann. of Math. Studies, no. 33, Princeton University Press, 1954, pp. 95–100.

21. G. de Rham, *Variétés différentiables*, Paris, Hermann, 1955.

22. H. A. Schwarz, *Gesammelte mathematische Abhandlungen*, vol. I, Berlin, Springer, 1890.

23. N. Steenrod, *The topology of fibre bundles*, Princeton University Press, 1951.

24. D. J. Struik, *Differential geometry*, Cambridge, Addison-Wesley, 1950.

25. K. Yano, *Some remarks on almost complex manifolds*, Proceedings of the International Congress of Mathematicians, 1954, vol. II, p. 268.

26. M. Zorn, *Theorie der alternativen Ringe*, Abh. Math. Sem. Hamb. Univ. vol. 8 (1931) pp. 123–147.

27. ———, *Alternativkörper und quadratische Systeme*, Abh. Math. Sem. Hamb. Univ. vol. 9 (1933) pp. 395–402.

University of Minnesota,
Minneapolis, Minn.

Improper Affine Hyperspheres of Convex Type and a Generalization of a Theorem by K. Jörgens

Eugenio Calabi

Michigan Math. J. 5, 105–126 (1958)

1. INTRODUCTION

The purpose of this article is to derive certain inequalities satisfied by the elliptic solutions of the nonlinear partial differential equation

$$\det\left(\frac{\partial^2 u}{\partial x^i \partial x^j}\right) = 1 \tag{1.1}$$

with n independent variables $(x) = (x^1, \cdots, x^n)$. An elliptic solution of (1.1) is a function $u = f(x) = f(x^1, \cdots, x^n)$ whose Hessian matrix (symmetric matrix of second derivatives) is definite at each point; that is to say, u is locally either a convex or a concave function (the latter can occur only if n is even) of (x): without loss of generality we shall consider only convex solutions.

The motivation for studying (1.1) is that it represents a special but significant case of the more general equation

$$\det\left(\frac{\partial^2 u}{\partial x^i \partial x^j}\right) = \phi(x)\,, \tag{1.1a}$$

where $\phi(x)$ is a given, positive-valued function; this last equation occurs, for instance, as a result of an elementary transformation of the one arising from the Minkowski problem on closed, convex hypersurfaces in Euclidean (n + 1)-space. The outstanding question about the regularity of weak solutions of the Minkowski problem is reduced to the question, if the kth partial derivatives of $\phi(x)$ in (1.1a) satisfy a Hölder condition, whether there exists a convex solution u of (1.1a), locally at least, and whether any such solution has all of its (k + 2)nd partial derivatives satisfying a Hölder condition.

It has been shown that (1.1a) has at most one convex solution u in a bounded domain D, if the boundary values of u are prescribed; however a necessary condition for the existence of such a solution with arbitrarily given, but smooth, boundary values is that the domain D be strictly convex. Aside from this, in order to establish the existence of a solution of the boundary value problem by the method of continuity, one requires some *a priori* estimates of the bounds for u, for its second partial derivatives, and for their Hölder constants, in terms of the domain D, the function $\phi(x)$, the boundary data, and their Hölder constants.

The main results of this article (wishfully the first of a series), viewed in the context of the more general problem, provide information on the *a priori* bounds, irrespective of the boundary values of u; the dependence of the bounds on the properties of $\phi(x)$ is set aside for the time being, by considering equation (1.1) instead of the more general one (1.1a). Stated briefly, the inequalities established in the present

---

Received October 6, 1957.

This article was written while the author was supported financially by a research grant from the National Science Foundation.

DOI: 10.1307/mmj/1028998055

paper provide an *a priori* majorant for the third derivatives of a convex solution of (1.1) in a domain D at an interior point $(x) \in D$ in terms of any uniform bound for the second derivatives in any neighborhood of (x), and in terms of the distance of (x) from the boundary of D: this reduces the problem of the existence of a solution u for the boundary value problem associated with (1.1) to estimating uniform bounds for the second derivatives in D. Better still, when the number n of independent variables does not exceed 5, Theorem 2, which is a generalization of the same statement for 2 variables by K. Jörgens [6] (related to Bernstein's theorem on minimal surfaces), enables us to dispense with uniform bounds for the second derivatives in a neighborhood of (x): the bounds for the third derivatives at (x) depend only on the bounds for the second derivatives at (x) and on the distance of (x) from the boundary of D.

Although most of the results in this paper can be formulated and proved in the language of partial differential equations, it would seem unfair to the informed reader to hide the fact that the heuristic reasoning behind them is decidedly geometrical, pertaining more specifically to affine differential geometry; on the other hand I am reluctant to make a display of concepts unfamiliar to many nonspecialists, when they are not essential in understanding the article. For this reason I adopt a compromise course, using only the well known formal techniques of the Ricci calculus (as in L. P. Eisenhart's *Riemannian Geometry* or in [3]), but interjecting an occasional paragraph or restatement of a result, interpreting the context in geometrical terms with special reference to [1].

It is clear that equation (1.1) is invariant under unimodular linear transformations of the independent variables $x^1, \cdots, x^n$; therefore it should not be too surprising if most of the analysis to follow should be based on the affine invariants of the graph of $u = f(x)$ $((x) = (x^1, \cdots, x^n))$ in the $(n+1)$-dimensional affine space E_{n+1}, with this graph considered as a differentiable hypersurface V_n, immersed in E_{n+1}. As long as we allow only linear transformations of the independent variables (x), the array of all partial derivatives of u of any given order k can be interpreted as the components of a covariant tensor of valence k, symmetric in all pairs of indices: we are chiefly interested in the tensors defined by the second and third derivatives,

$$g_{ij}(x) = \frac{\partial^2 u}{\partial x^i \partial x^j} \qquad (i, j, \cdots = 1, 2, \cdots, n) \tag{1.2}$$

and

$$A_{ijk}(x) = -\frac{1}{2}\frac{\partial^3 u}{\partial x^i \partial x^j \partial x^k}. \tag{1.3}$$

Since u is convex and satisfies (1.1), it follows that the symmetric tensor g_{ij} defined in (1.2) is positive definite; it therefore determines, in the domain of definition of u, a Riemannian metric

$$ds^2 = g_{ij}\, dx^i\, dx^j, \tag{1.4}$$

associated with the given function in a natural way, as well as the contravariant tensor g^{ij} defined by the inverse of the matrix of g_{ij}. We observe immediately from (1.1) that the determinant g of g_{ij} is identically equal to 1; therefore by differentiation we obtain the relation

$$g^{ij} A_{ijk} = 0 \qquad (k = 1, \cdots, n). \tag{1.5}$$

[Both tensors g_{ij} and A_{ijk} are affine invariants of the graph V_n of u in E_{n+1}. For the special case of a solution of (1.1), the metric (1.4) is known as the Schwarz-Pick metric of V_n, and the cubic differential form $\psi = A_{ijk}dx^i dx^j dx^k$ is known as the Darboux-Fubini-Pick form. Both the quadratic and the cubic differential forms are defined in the general case of a differentiable hypersurface in E_{n+1}, but only in the case of the graph of a solution of (1.1) does their expression take the simple form of (1.2) and (1.3). The graphs of the solutions of (1.1) characterize by means of certain affinely invariant, differential geometric properties the so-called *improper affine hyperspheres*. For a complete exposition of this subject, see [1], with whose notation we have tried to conform.]

If the values of the second derivatives g_{ij} of u are known at a point (x), then a natural norm for the third derivatives at (x) is given by the Riemannian absolute value of the tensor A_{ijk}, namely by

$$|\psi| = (A^{ijk}A_{ijk})^{1/2}, \tag{1.6}$$

where $A^{ijk} = g^{ia}g^{jb}g^{kc}A_{abc}$. Similarly the most natural yardstick for measuring in some sense the distance of (x) from the ideal boundary of the domain of existence of u is in terms of the geodesic distance defined by (1.4). Let D denote the interior of the domain of existence of a convex solution u of (1.1) (we remark here that, for the purpose of this paper, D need not be a schlicht domain, nor u single-valued). For any point $(x)_0 \in D$, we define $\mathcal{S}_{x_0}$ to be the set of all rectifiable, parametrized arcs C in D, images of the semiclosed interval $\{0 \leq t < 1\}$, such that the initial point $t = 0$ corresponds to $(x)_0$ and the arc itself is a relatively closed set in the topology of D: in other words $\mathcal{S}_{x_0}$ is the set of rectifiable arcs C in D joining $(x)_0$ with the ideal boundary of D.

DEFINITION 1. *Let* u *be a convex solution of* (1.1) *in an open domain* D. *For each point* $(x)_0 \in D$, *we define the geodesic distance* $\gamma(x)_0$ *of* $(x)_0$ *from the boundary of* D (*relative to* u) *as the quantity*

$$\gamma(x)_0 = \inf_{C \in \mathcal{S}_{x_0}} \left(\int_C (g_{ij}\, dx^i dx^j)^{1/2} \right), \tag{1.7}$$

where g_{ij} *is defined by* (1.2), *and the values of* $\gamma(x)_0$ *can range over the interval* $0 < \gamma(x)_0 \leq \infty$. *Similarly we define the* affine distance $\delta(x)_0$ *of* $(x)_0$ *from the boundary of* D (*always relative to* u) *as the quantity*

$$\delta(x)_0 = \inf_{C \in \mathcal{S}_{x_0}} \left(\int_C (g_{ij}((x)_0)\, dx^i dx^j)^{1/2} \right), \tag{1.8}$$

where the metric tensor is the tensor whose components are constant and equal to the values of g_{ij} *at the initial point.*

We remark here incidentally that, if one replaces the independent variables (x) through a unimodular linear transformation followed by a translation, so that $(x)_0$ becomes the origin O and $g_{ij}(O) = \delta_{ij}$ (Kronecker delta) in the new coordinate system, then the affine distance of $(x)_0$ from the boundary is the same as the Euclidean distance of the origin from the nearest singularity, with respect to the new coordinate system. Thus $\delta(x)_0 = \infty$ if and only if u is a single-valued function, defined over the whole numerical n-space. In addition, we point out that since u satisfies (1.1), an upper bound for the second partial derivatives of u determines a positive lower bound for the eigenvalues of g_{ij}; hence, if one wants to estimate a lower bound for $\delta(x)$ for any given $(x) \in D$, one needs only the Euclidean distance of (x) from the boundary and an upper bound for the second derivatives of u at (x) itself, while, if one wants to estimate a lower bound for $\gamma(x)$, one needs in addition a uniform upper bound for the second derivatives in some neighborhood of (x). Having thus introduced the principal notations and explanations, we can now state the two main theorems with their respective corollaries.

THEOREM 1. *Let* $u = f(x^1, \cdots, x^n)$ *be a convex solution of* (1.1) *in an open domain* D, *and suppose that* u *is of class* $\mathscr{C}^5$. *Consider the Riemannian structure in* D *defined by* (1.4) *and* (1.2), *and the corresponding norm* $|\psi|$ *for the third derivatives of* u, *defined by* (1.3) *and* (1.6), *at each point* $(x) \in D$. *Then the inequality*

$$|\psi|(x) \le \frac{c_n}{\gamma(x)} \qquad (c_n < n\sqrt{2}), \tag{1.9}$$

holds, where $\gamma(x)$ *is the geodesic distance of* (x) *from the boundary of* D.

COROLLARY. *Let* $u = f(x^1, \cdots, x^n)$ *be a convex solution of* (1.1) *in an open domain* D, *and suppose that* u *is of class* $\mathscr{C}^5$. *If the Riemannian structure defined in* D *by* (1.4) *and* (1.2) *is that of a complete Riemannian manifold, then* u *is a quadratic polynomial in* $x^1, \cdots, x^n$, *defined over the whole numerical space.*

The corollary above, restated in the language of affine differential geometry, asserts the following (apparently new) uniqueness theorem.

COROLLARY. (restatement of the previous one). *Let* V_n *be an* n*-dimensional, improper affine hypersphere, immersed and of class* $\mathscr{C}^5$ *in the affine* $(n+1)$*-space* E_{n+1}. *If the Schwarz-Pick metric induced in* V_n *is definite, and if* V_n *is complete with respect to that metric, then* V_n *is a convex paraboloid in* E_{n+1}.

The other main theorem of this paper has been stated [6, 9] only in the case of two independent variables; here we present its generalization to up to four variables.

THEOREM 2. *Let* $u = f(x^1, \cdots, x^n)$ $(n \le 5)$ *be a convex solution of* (1.1), *defined and of class* $\mathscr{C}^5$ *in an open domain* D *containing the open Euclidean ball with radius* r *and center at the origin* O. *Let* $g_{ij}(x) = \partial^2 u / \partial x^i \partial x^j$, *and suppose that* $g_{ij}(O) = \delta_{ij}$. *Then*

$$\sum_{i,j,k=1}^{n} \left(\frac{\partial^3 u}{\partial x^i \partial x^j \partial x^k}(O) \right)^2 \le \frac{4 M_n^2}{r^2}, \tag{1.10}$$

where M_n *is a universal constant.*

In other words, under the restriction $n \le 5$ and the assumptions in Theorem 1, the inequality

$$|\psi|(x) \le \frac{M_n}{\delta(x)} \tag{1.11}$$

is valid at each point $(x) \in D$, where $|\psi|(x)$ and $\delta(x)$ are defined by (1.3), (1.6), and by (1.8) respectively. (For upper bounds on the M_n, see (4.14) and (4.15) at the end of the paper.)

COROLLARY. *Let* $u = f(x^1, \cdots, x^n)$ $(n \le 5)$ *be a convex solution of* (1.1), *defined and of class* $\mathscr{C}^5$ *in the whole numerical* n*-space. Then* u *is a quadratic polynomial.*

It is natural to conjecture that the assertion of Theorem 2 should remain valid also for $n \geq 6$: this guess is almost forced by the fortuitous way in which the condition $n \leq 5$ occurs in the proof of the theorem. It also seems very likely that the assertions of both Theorems 1 and 2 remain valid without any assumptions on the existence of fifth derivatives of u; for the sake of simplicity we have not tried to relax the assumptions in that direction.

2. PRELIMINARY RESULTS

Here and for the rest of the paper, we shall consider $u = f(x^1, \cdots, x^n)$ to be a convex solution of (1.1), admitting continuous partial derivatives at least up to order 5, in an open domain D draped over a region in the numerical n-space R^n (*domaine étalé*). We introduce in D the Riemannian metric (1.4) defined by (1.2) and the tensor A_{ijk} defined by (1.3). Both definitions require the fixed coordinate system $(x) = (x^1, \cdots, x^n)$ or one related to it by a unimodular affine transformation, since in other coordinate systems equation (1.1) and the definitions of the two fundamental tensors have a more complicated expression. If we confine our computations to the coordinate systems affinely related to (x), we verify immediately the equation

$$g^{hk}A_{ijh} = A_{ij}{}^k = -\Gamma^k{}_{ij}, \tag{2.1}$$

where $\Gamma^k{}_{ij}$ denotes the Christoffel symbol of the second kind. Since $g = \det(g_{ij}) = 1$ identically, we verify the fact already mentioned in (1.5),

$$\frac{\partial \log \sqrt{g}}{\partial x^j} = \Gamma^i{}_{ij} = -A_{ij}{}^i = g^{ik}A_{ijk} = 0, \tag{2.2}$$

and, by differentiating the last two members covariantly,

$$g^{ik}A_{ijk,\ell} = A_{ij}{}^i{}_{,\ell} = 0, \tag{2.3}$$

where the index of covariant differentiation is written to the right, following the comma.

We compute now the covariant derivative $A_{ijk,\ell}$ of A_{ijk} explicitely:

$$A_{ijk,\ell} = -\frac{1}{2}\frac{\partial^4 u}{\partial x^i \partial x^j \partial x^k \partial x^\ell} + g^{hm}(A_{hjk}A_{mi\ell} + A_{ihk}A_{mj\ell} + A_{ijh}A_{mk\ell}),$$

and we verify that it is symmetric in all pairs of indices; for instance

$$A_{ijk,\ell} = A_{ij\ell,k}. \tag{2.4}$$

Equations (2.3) and (2.4) imply that the "divergence" of A_{ijk} vanishes:

$$g^{i\ell}A_{ijk,\ell} = 0. \tag{2.5}$$

Continuing these formal calculations, one verifies that the Riemann curvature tensor $R_{ijk\ell}$ has the following simple expression in terms of the A_{ijk}:

$$R_{ijk\ell} = g^{hm}(A_{hi\ell}A_{mjk} - A_{hik}A_{mj\ell} \tag{2.6}$$

Using (2.2), we shall give here similarly the formulas for the Ricci tensor

(2.7) $$R_{ik} = g^{j\ell} R_{ijk\ell} = g^{j\ell} g^{hm} A_{h\ell i} A_{mjk},$$

and for the scalar curvature

(2.8) $$R = g^{ik} R_{ik} = A^{ijk} A_{ijk} = |\psi|^2,$$

where $|\psi|$ is the symbol introduced in (1.6).

It is clear from the last equations that the scalar curvature is nonnegative, and that it vanishes if and only if $A_{ijk} = 0$, that is, if the full Riemann tensor $R_{ijk\ell}$ vanishes. It is also easy to see that the Ricci curvature is nonnegative, a fact of fundamental importance in the sequel: indeed, if ξ^i is any contravariant vector, we have

(2.9) $$R_{ik} \xi^i \xi^k = g^{hm} g^{\ell j} (A_{h\ell i} \xi^i)(A_{mjk} \xi^k) \geq 0,$$

equality holding only if $A_{ijk} \xi^k = 0$. From this fact we obtain also the inequality

(2.10) $$R_{ik} \xi^i \xi^k \leq R g_{ik} \xi^i \xi^k,$$

as one may verify by choosing an orthonormal basis for the vectors at any point.

[All the equations derived here, (2.1) through (2.8), are special cases applied to improper affine hyperspheres of standard relations between the Schwarz-Pick metric and the Darboux-Fubini-Pick cubic form ψ. These two differential forms determine any differentiable hypersurface in E_{n+1} uniquely, in analogy with the first and second fundamental forms in classical differential geometry, while the existence is subject to the integrability conditions, which in the case studied here specialize to (1.5), (2.4) and (2.6). In other words, all the results to follow can also be formulated in the language of Riemannian geometry, under the assumption of a positive definite Riemannian metric (1.4) together with a symmetric tensor A_{ijk} satisfying (1.5), (2.4) and (2.6).]

We shall prove now an algebraic inequality about the curvature tensors, valid in every Riemannian manifold with positive definite metric.

LEMMA 1. *In any* n*-dimensional Riemannian manifold with a positive definite metric tensor* g_{ij} *of class* C^2*, the following inequalities pertaining to the Riemann tensor* $R_{ijk\ell}$*, the Ricci tensor* R_{ik}*, and the scalar curvature* R *are valid at each point:*

(2.11) $$\frac{1}{n} R^2 \leq R_{ij} R^{ij},$$

(2.12) $$\frac{2}{n-1} R_{ij} R^{ij} \leq R_{ijk\ell} R^{ijk\ell}.$$

Proof. Let

$$R'_{ij} = R_{ij} - \frac{1}{n} R g_{ij};$$

then $g^{ij} R'_{ij} = 0$, so that the decomposition of the Ricci tensor given by

$$R_{ij} = R'_{ij} + \frac{1}{n} R g_{ij}$$

is orthogonal with respect to the metric tensor. We have then immediately

$$0 \leq R'_{ij} R'^{ij} = R_{ij} R^{ij} - \left(\frac{1}{n} R g_{ij}\right) \left(\frac{1}{n} R g^{ij}\right) = R_{ij} R^{ij} - \frac{1}{n} R^2, \tag{2.13}$$

and this proves (2.11). We remark that equality holds in (2.11), at any point, if and only if R'_{ij} vanishes at that point; it holds identically, if and only if either $n = 2$ or the metric is that of an Einstein space.

Before proving (2.12) we observe that for $n = 2$ both terms of that inequality are identical, so that we can limit our consideration to $n \geq 3$. Consider the conformal curvature tensor of H. Weyl,

$$C_{ijk\ell} = R_{ijk\ell} - \frac{1}{n-2} (g_{ik} R_{j\ell} + g_{j\ell} R_{ik} - g_{i\ell} R_{jk} - g_{jk} R_{i\ell})$$

$$+ \frac{R}{(n-1)(n-2)} (g_{ik} g_{j\ell} - g_{i\ell} g_{jk}),$$

and let $R'_{ijk\ell}$ denote the difference $R_{ijk\ell} - C_{ijk\ell}$. Since the contraction of any two of the four indices of $C_{ijk\ell}$ with the contravariant component of the metric tensor gives the zero tensor, the two tensors $R'_{ijk\ell}$ and $C_{ijk\ell}$ are mutually orthogonal. Computing the inner product of $R'_{ijk\ell}$ with itself, we obtain

$$R'_{ijk\ell} R'^{ijk\ell} = \frac{4}{n-2} R_{ij} R^{ij} - \frac{2}{(n-1)(n-2)} R^2,$$

so that from the orthogonal decomposition $R_{ijk\ell} = C_{ijk\ell} + R'_{ijk\ell}$ we obtain

$$R_{ijk\ell} R^{ijk\ell} \geq R'_{ijk\ell} R'^{ijk\ell} = \frac{4}{n-2} R_{ij} R^{ij} - \frac{2R^2}{(n-1)(n-2)}, \tag{2.14}$$

equality holding if and only if $C_{ijk\ell} = 0$. The latter condition is verified identically if and only if either $n = 3$, or the space is conformally flat. Combining (2.14) linearly with (2.13) so as to eliminate the term containing R^2, we obtain the desired inequality (2.12). We observe that equality holds in (2.12), at any given point, if and only if it holds in both (2.13) and (2.14). Therefore, in that case, the metric is one of constant curvature at that point,

$$R_{ijk\ell} = \frac{R}{(n-1)(n-2)} (g_{ik} g_{j\ell} - g_{i\ell} g_{jk}),$$

a condition which, if verified identically for $n \geq 3$, implies by Schur's theorem that the Riemannian manifold has constant scalar curvature R whose value determines the manifold uniquely within a local isometry. This completes the proof of the lemma.

Before we state the next result, we must generalize the concept of an elliptic, linear differential inequality (of the type, for example that defines subharmonic functions with respect to the Laplace-Beltrami operator), so as to make it applicable to continuous, nondifferentiable functions. Such a generalization was introduced in [2], where it was shown that every function which is subharmonic in the generalized sense satisfies the strong maximum principle in the sense of E. Hopf.

DEFINITION 2. *Let* L *denote a linear, elliptic, second-order differential operator with* n *independent variables, of the form*

$$L = a^{ij}(x)\frac{\partial^2}{\partial x^i \partial x^j} + b^i(x)\frac{\partial}{\partial x^i},$$

where the only assumptions on the coefficients $a^{ij}(x)$ *and* $b^i(x)$ *are that in some neighborhood of each* (x), $a^{ij}(x)$ *and* $b^i(x)$ *are bounded, and that the* $a^{ij}(x)$ *form a symmetric, uniformly positive definite matrix. If* $v = v(x)$ *is any upper semicontinuous, real-valued function of* (x) *in an open domain* U, *then for any given point* $(x)_0 \in U$ *and any real number* a, *we say that*

$$L[v]((x)_0) \geq a$$

if and only if, for each $\varepsilon > 0$, *there exists a neighborhood* $V_\varepsilon \subset U$ *of* $(x)_0$ *and a function* $v_\varepsilon(x)$ *of class* $\mathscr{C}^2$ *in* V_ε *which satisfy the following two conditions:*

a) $v(x) - v_\varepsilon(x)$ *achieves its minimum value over* V_ε *at* $(x)_0$;

b) *at* $(x)_0$, *the function* $v_\varepsilon(x)$ *satisfies the inequality*

$$L[v_\varepsilon]((x)_0) = a^{ij}((x)_0)\frac{\partial^2 v_\varepsilon}{\partial x^i \partial x^j}((x)_0) + b^i((x)_0)\frac{\partial x_\varepsilon}{\partial x^i}((x)_0) > a - \varepsilon .$$

Similarly, if $w = w(x)$ *is a real-valued function defined in a subset* $K \subset U$, *we say that* $L[v] \geq w$ *everywhere in* K, *if for each point* $(x) \in K$, *have*

$$L[v](x) \geq w(x)$$

in the sense just specified. If v(x) *is lower semicontinuous in* U, *and if* w(x) *is defined in* $K \subset U$, *we say that* $L[v] \leq w$ *everywhere in* K *if and only if* $L[-v] \geq -w$ *in* K *in the same sense as above.*

We can state now the first result.

PROPOSITION 1. *Let* u *be a convex solution of* (1.1), *of class* $\mathscr{C}^5$ *in an open domain; let* g_{ij} *be the metric tensor defined in that domain by the second partial derivatives of* u; *and let* Δ *denote the Laplace-Beltrami operator with respect to* g_{ij}. *Then the scalar curvature* R *is of class* $\mathscr{C}^2$ *and is nonnegative, and its square root satisfies (in the sense of Definition 2) the partial differential inequality*

$$\Delta \sqrt{R} \geq \frac{n+1}{n(n-1)} R^{3/2}. \tag{2.15}$$

Proof. It is clear from (2.8) that the scalar curvature R is nonnegative; since it can be expressed algebraically in terms of g^{ij} and A_{ijk}, in a way which involves partial derivatives of u up to the third order, it follows that R is of class $\mathscr{C}^2$, and hence $\sqrt{R}$ is of class $\mathscr{C}^2$ everywhere, except where $R = 0$; at the latter set of points $\sqrt{R}$ is merely continuous. At the points where $R = 0$, $\sqrt{R}$ achieves its minimum value, so that, if we apply Definition 2 (rather trivially) with v_ε replaced by 0 and L by Δ, the verification of (2.15) is immediate. Thus we can limit our consideration in the sequel to the points where $R > 0$, that is, where $\sqrt{R}$ has the required derivatives for formal calculation.

Consider now the Laplace-Beltrami operator Δ applied to the tensor A_{ijk} defined by (1.3), and the resulting tensor

$$(\Delta A)_{ijk} = g^{\ell m} A_{ijk,\ell m}.$$

Since the first covariant derivative $A_{ijk,\ell}$ of A_{ijk} is symmetric in each pair of indices by (2.4), we can rewrite the Beltramian $(\Delta A)_{ijk}$ and then apply successively the Ricci identity and equations (2.5) and (2.6), with the following result:

$$\begin{aligned}(\Delta A)_{ijk} &= g^{\ell m} A_{\ell jk,im} = g^{\ell m} A_{\ell jk,mi} + g^{\ell m}(A_{\ell jk,im} - A_{\ell jk,mi})\\ &= (g^{\ell m} A_{\ell jk,m})_{,i} + g^{\ell m}(A_{hjk} R^h{}_{\ell im} + A_{\ell hk} R^h{}_{jim} + A_{\ell jh} R^h{}_{kim})\\ &= A_{hjk} R^h{}_i + A^m{}_{hk}(A^h{}_{am} A^a{}_{ji} - A^h{}_{ai} A^a{}_{jm}) + A^m{}_{jh}(A^h{}_{am} A^a{}_{ki} - A^h{}_{ai} A^a{}_{km})\\ &= A^{h\ell m}(A_{h\ell i} A_{mjk} + A_{h\ell j} A_{mki} + A_{h\ell k} A_{mij}) - 2A^a{}_{ib} A^b{}_{jc} A^c{}_{ka}.\end{aligned}$$

This last expression is useful in computing ΔR in terms of A_{ijk} and its derivatives from (2.8); the outcome is the following:

$$\begin{aligned}\tfrac{1}{2}\Delta R &= \tfrac{1}{2} g^{\ell m} R_{,\ell m} = A^{ijk}(\Delta A)_{ijk} + A^{ijk,\ell} A_{ijk,\ell}\\ &= 3A^{ijk} A^{h\ell m} A_{h\ell i} A_{mjk} - 2A^{ijk} A^a{}_{ib} A^b{}_{jc} A^c{}_{ka} + A^{ijk,\ell} A_{ijk,\ell} \qquad (2.16)\\ &= R_{ij} R^{ij} + R_{ijk\ell} R^{ijk\ell} + A^{ijk,\ell} A_{ijk,\ell};\end{aligned}$$

the last step can be verified by direct substitution of the curvature tensors by equations (2.6) and (2.7). Applying Lemma 1, we obtain the inequality

$$\tfrac{1}{2}\Delta R \geq \frac{n+1}{n(n-1)} R^2 + A_{ijk,\ell} A^{ijk,\ell}. \qquad (2.17)$$

Consider now the first-order Beltrami parameter applied to R: from (2.8), we have

$$g^{ij} R_{,i} R_{,j} = 4A^{abc,i} A_{abc} A_{jk\ell,i} A^{jk\ell}. \qquad (2.18)$$

We now introduce the seven-indexed tensor

$$B_{abcijk\ell} = \tfrac{1}{2}(A_{abc,i} A_{jk\ell} - A_{abc} A_{jk\ell,i}),$$

and we point out that its inner product with itself can be expressed in terms of the quantities occurring in (2.17) and (2.18), as follows:

$$\begin{aligned}B_{abcijk\ell} B^{abcijk\ell} &= \tfrac{1}{2} A^{abc,i} A^{jk\ell}(A_{abc,i} A_{jk\ell} - A_{abc} A_{jk\ell,i})\\ &= \tfrac{1}{2} R A_{abc,i} A^{abc,i} - \tfrac{1}{8} g^{ij} R_{,i} R_{,j} \geq 0.\end{aligned}$$

Combining this last inequality with (2.16) and (2.17), we obtain the following identity and inequality, at all points where $R > 0$:

$$\Delta \sqrt{R} = \frac{\Delta R}{2\sqrt{R}} - \frac{g^{ij}R_{,i}R_{,j}}{4R^{3/2}} = \frac{R_{ij}R^{ij} + R_{ijk\ell}R^{ijk\ell}}{\sqrt{R}} + \frac{2B_{abcijk\ell}B^{abcijk\ell}}{R^{3/2}}$$

$$\geq \frac{n+1}{n(n-1)} R^{3/2};$$

this verifies our assertion (2.15).

We observe that, in order to have equality in (2.15) at a point where $R > 0$, it is necessary first of all that all sectional curvatures at that point be equal to each other (this is seen from the remarks following the proof of Lemma 1). In addition, it is necessary that the tensor $B_{abcijk\ell}$ vanish. The first condition is impossible in any neighborhood, for $n \geq 3$; for it would imply by Schur's theorem that R be a constant, contrary to the proposition just proved. The second condition would lead to a contradiction, if verified even at a single point where $A_{ijk} \neq 0$, since one can show by algebraic manipulation that the vanishing of $B_{abcijk\ell}$ at such a point would contradict (2.5). For this reason it seems that one could improve somewhat the numerical estimates that will be obtained in the sequel; however, at the present stage it does not seem useful to seek the best possible estimates.

We quote now as Proposition 2 a well-known theorem, whose proof was first given by H. Hopf and W. Rinow [5], and later simplified by G. de Rham [11]. The present statement of the theorem differs from the one in the references in that it applies to all Riemannian manifolds with positive definite metric, instead of only to complete ones; the proof, however, is identical, so that it can be omitted; the applications in the present article will occur mainly in noncomplete manifolds. The definition of $\gamma(p)$ in Proposition 2 agrees with our definition of the geodesic distance of (x) from the boundary (see Definition 1, (1.7)).

PROPOSITION 2 (Hopf, Rinow, de Rham). *Let* X *be a Riemannian manifold with positive definite metric of class* $\mathscr{C}^1$, *and denote by* d(p, q) *the Riemannian distance between any two points* p, q ∈ X *(that is, the lower bound of the Riemannian length of all arcs joining* p *and* q*). Let* X'*denote the complement of* X *in its completion with respect to the distance function* d(p, q), *and* $\gamma(p)$ *the distance (extended from* X *to its completion) of any point* p ∈ X *from* X' *(if* X *is already complete, we set* $\gamma(p) = \infty$*). Then the following statements are true.*

(i) *The closed metric ball* $\overline{\Sigma}(p, r)$ *for any* p ∈ X *and real number* $r > 0$, *defined as the set of points* q ∈ X *such that* $d(p, q) \leq r$, *is compact if and only if* $r < \gamma(p)$.

(ii) *For each* p ∈ X, *every geodesic arc with* p *as initial point can be extended to any Riemannian length less than* $\gamma(p)$; *however, if* $\gamma(p) < \infty$, *there exists at least one geodesic arc, originating at* p, *whose maximal extension in* X *is of length* $\gamma(p)$, *and with no end point in* X.

(iii) *For any two points* p, q ∈ X *such that* $d(p, q) \leq \max(\gamma(p), \gamma(q))$, *there exists at least one arc* C, *necessarily a geodesic, which joins* p *and* q *and whose length achieves the minimum value* d(p, q).

(iv) *For any two points* p, q ∈ X, *the inequality*

$$\gamma(q) - d(p, q) \leq \gamma(p) \leq \gamma(q) + d(p, q)$$

is valid; in particular $\gamma(q) = \infty$ *if and only if* $\gamma(p) = \infty$.

Our next proposition deals with a property which the Riemannian distance function $d(p, q)$ has for those pairs of points p and q that can be joined by a geodesic of length $d(p, q)$, as for instance under the conditions of statement (iii) in Proposition 2. In order to avoid confusion, we adopt a different name and notation for the restricted distance to which this property gives rise. The proposition itself, related to a lemma by S. B. Myers [8], is proved in detail in [2]. Here we shall elaborate only the formal aspects of the proof, since part of it is needed in the proof of Theorem 2.

DEFINITION 3. *Let* X *be a Riemannian manifold with a positive definite metric tensor of class* $\mathscr{C}^1$. *For any two points* $p, q \in X$ *we define the* geodesic distance $r(p, q)$ *to be the Riemannian distance* $d(p, q)$, *provided that* p *and* q *can be joined by at least one geodesic in* X *of length* $d(p, q)$; *if* p *and* q *can not be joined by an arc in* X *of length* $d(p, q)$, *then the geodesic distance shall remain undefined.*

PROPOSITION 3. *Let* X *be an* n*-dimensional Riemannian manifold with positive definite metric of class* $\mathscr{C}^3$ *and with nonnegative Ricci curvature; that is, let*

$$R_{ij}\xi^i\xi^j \geq 0 \quad \text{for all } \xi^i \ (i = 1, \cdots, n); \tag{2.19}$$

let p_0 *be an arbitrary point in* X, *fixed for the sequel; let*

$$r = r(p) = r(p_0, p) \qquad (p \in Y = Y_{p_0})$$

denote the geodesic distance between p_0 *and* p, *where* p *ranges over the domain* Y *(depending on* p_0*) for which that geodesic distance is defined; and let* $\phi(t)$ *be a twice differentiable function of a real variable* t *for* $t \geq 0$, *such that* $\phi'(t) \geq 0$ *for all* t *and* $\phi'(0) = 0$. *Then the function* $v = v(p) = \phi(r(p))$, *continuous for* $p \in Y$, *satisfies the inequality*

$$v(p) \leq \phi''(r(p)) + \frac{n-1}{r(p)}\phi'(r(p)) \tag{2.20}$$

in the sense of Definition 2, where Δ *is the Laplace-Beltrami operator; for* $p = p_0$ *the conclusion is valid, if the right-hand member of* (2.24) *is replaced by its limit as* p *approaches* p_0.

Proof. Under the differentiability assumptions on the metric, there exists an open, everywhere dense subdomain Y' of Y, in which $r(p)$ is twice differentiable. In fact, if $p_1 \in Y$, let C_1 be a geodesic of length $d(p_0, p_1)$, joining p_0 and p_1; then for each point p in C_1, other than the end points, the segment C of C_1 between p_0 and p is the only arc of length $d(p_0, p)$ joining its end points (because of the unique continuation property of the geodesics). Furthermore, since p can not be a conjugate point of p_0 along C, there is a connected neighborhood $U \subset Y$ of p, such that for each point $p' \in U$ there is a unique geodesic C' of length $d(p_0, p')$ joining p_0 and p', and this geodesic can be constructed by the continuity method, that is, by joining p and p' by a differentiable arc in U, and by obtaining the variational formula by Jacobi's equation for the geodesic joining p_0 with a moving point on the arc. Thus one shows that the function r is differentiable in U, and that it satisfies the equation

$$g^{ij}r_{,i}r_{,j} = 1. \tag{2.21}$$

Since $\phi(t)$ is twice differentiable, it follows immediately from (2.21) that

$$\Delta v = \Delta\phi(r) = \phi''(r) + \phi'(r)\Delta r \tag{2.22}$$

in the open, dense subset Y' of Y where r satisfies (2.21). Thus the proof can be completed by showing that $\Delta r \leq (n-1)/r$, since $\phi'(t) \geq 0$.

By formal computations one verifies that in Y' the geodesic hyperspheres defined by the equation r = (positive constant) are locally like twice differentiable hypersurfaces. From (2.21) the vector $\xi^i = g^{ij} r_{,j}$ is the (outward) unit normal vector to each such hypersphere, and the restriction of the covariant tensor $-r_{,ij}$ to any such hypersphere defines its second fundamental form relative to the orientation of the normal vector ξ^i. Consequently, if we denote by κ_ν ($\nu = 1, \cdots, n-1$) the $n-1$ principal normal curvatures of the geodesic hypersphere passing through each point $p \in Y'$ and by H the mean curvature

$$H = \frac{1}{n-1}\left(\sum_{\nu=1}^{n-1} \kappa_\nu\right),$$

we obtain the identity in Y'

$$\Delta r = -(n-1)\,H. \tag{2.23}$$

Let p be a point in Y', C the shortest geodesic joining p_0 to p, parametrized by the geodesic distance r from p_0. Denote by $\eta_\nu{}^i(p)$ ($\nu = 1, \cdots, n-1$) a choice of unit vectors at p, each of them tangent to the geodesic hypersphere in the (mutually orthogonal) directions corresponding to the principal curvatures κ_ν respectively. We define the $n-1$ vectors $\eta_\nu{}^i$ elsewhere in C by parallel translation from p along C, and we let $\kappa_\nu{}'$ denote the normal relative curvature of any geodesic hypersphere in the direction $\eta_\nu{}^i$. Then the Jacobi equation along C yields immediately the following Riccati equation:

$$\frac{d\kappa_\nu{}'}{dr} = (\kappa_\nu{}')^2 + R_{ijk\ell}\, \xi^i\, \eta_\nu{}^j\, \xi^k \eta_\nu{}^\ell \quad (\nu = 1, \cdots, n-1).$$

Taking the average over ν in the above equation and recalling the elementary identity $\Sigma_{\nu=1}^{n-1} \kappa_\nu{}' = (n-1)H$, we obtain the inequality

$$\frac{dH}{dr} = \frac{1}{n-1}\left(R_{ij}\, \xi^i \xi^j + \sum_{\nu=1}^{n-1} (\kappa'_\nu)^2\right) \geq H^2 + \frac{1}{n-1} R_{ij}\, \xi^i\, \xi^j. \tag{2.24}$$

From the assumption (2.19) that the Ricci curvature is nonnegative and from the fact that H is asymptotic to $-1/r$ as r approaches zero, we obtain the inequality $H \geq -1/r$, which, combined with (2.23) and (2.22), proves (2.20) everywhere in Y'.

The extension of the proof to include the rest of Y involves an application of Definition 2, and, since the details are not needed in the sequel, we merely refer for them to [2]. The statement concerning the relation (2.20) in the limit as p approaches p_0 is a trivial exercise. This concludes the proof of Proposition 3.

3. PROOF OF THEOREM 1

We can rephrase the statement of Theorem 1 to read that there exists a positive constant c_n, depending only on the number n of variables, such that, if at any point $(x)_0 \in D$ we have

$$|\psi|((x)_0) > a > 0 \tag{3.1}$$

for any positive constant a, then $\gamma((x)_0) \leq c_n/a$ (without loss of generality we shall henceforth take $(x)_0$ to be the origin O); indeed we shall deduce a contradiction by assuming that

$$\gamma(O) > \frac{c_n}{a}. \tag{3.2}$$

The estimate $c_n < n\sqrt{2}$ is obtained from the numerical solution of a certain ordinary differential equation and by majorizing the result. Its chief interest lies in the fact that it probably describes the order of magnitude of c_n as n becomes large.

From (2.8) and (2.15) (Proposition 1) we obtain the inequality

$$\Delta|\psi| \geq \frac{n+1}{n(n-1)}|\psi|^3. \tag{3.3}$$

This relation is the starting point for the whole proof: the rest of the argument follows from results by E. K. Haviland [4], and from a method due to R. Osserman [10], generalized by the author [2] to spaces with nonnegative Ricci curvature. The essential point consists of constructing an auxiliary function $v = v(x)$ in an open subdomain $\Sigma \subset D$ with the following four properties:

(a) *The domain* Σ *contains the origin* O *in its interior, and the closure of* Σ *relative to* D *is compact in* D.

(b) *The function* v *satisfies the inequality*

$$\Delta v \leq \frac{n+1}{n(n-1)} v^3 \tag{3.4}$$

everywhere in Σ, *in the sense of Definition* 2.

(c) *The function* v *satisfies*

$$v(O) = a, \tag{3.5}$$

where a *is the constant occurring in* (3.1) *and* (3.2); *at other points in* Σ,

$$v(x) > 0. \tag{3.6}$$

(d) *As* (x) *approaches the boundary of* Σ, v(x) *becomes uniformly infinite.*

We shall prove first that, if there exists a function v with the four properties listed above, then we have a contradiction with (3.1). From properties (a) and (d) and the fact that the function $|\psi| - v$ becomes $-\infty$ uniformly on the boundary of the relatively compact domain Σ, it follows that the function attains its maximum value M at some interior point (x_1) of Σ. Since, by (3.1) and (3.5), $|\psi| - v$ is positive at the origin, it follows *a fortiori* that $M > 0$. But from (3.3) and (3.4) we would have the inequality

$$\Delta(|\psi| - v) \geq \frac{n+1}{n(n-1)}(|\psi|^3 - v^3) \geq \frac{3(n+1)}{n(n-1)}(|\psi| - v)v^2$$

(in the sense of Definition 2) wherever $|\psi| \geq v$, since both $|\psi|$ and v are nonnegative. In particular, we would have, in a neighborhood of the maximum value point, the inequality

$$\Delta(|\psi| - v) > 0,$$

contrary to the maximum principle [2] applied to solutions, in the sense of Definition 2, of the above inequality. This contradiction would then prove the theorem.

It is natural, in seeking to construct a function $v(x)$ with the four required properties, to make it a function whose values depend only on the geodesic distance $r(O, (x))$ from the origin (see Definition 3). We abbreviate $r(O, (x))$ by $r(x)$, and we let $v(x) = \phi(r(x))$, where $\phi(t)$ is a function of a real variable $t \geq 0$ to be determined. To ensure that $v(x)$ satisfy (3.4), we apply Proposition 3, using the fact that by equation (2.9) the Ricci curvature is nonnegative. Accordingly, it is sufficient that $\phi(t)$ satisfy the ordinary differential equation

$$\phi''(t) + \frac{n-1}{t}\phi'(t) = \frac{n+1}{n(n-1)}(\phi(t))^3 \qquad (\phi'(0) = 0), \tag{3.7}$$

provided that at the same time we have

$$\phi'(t) \geq 0 \qquad \text{for } t > 0. \tag{3.8}$$

The condition (3.5) that $v(O) = a$ is translated into the initial condition

$$\phi(0) = a. \tag{3.9}$$

If we can prove that the solution of (3.7) with the initial condition (3.9) automatically satisfies (3.8), and that it diverges to $+\infty$ as t approaches a positive number t_1 (depending on a), then, if $\gamma(0) > t_1$, the domain of definition Σ of the function $v(x) = \phi(r(x))$ has a compact closure in D, by the Hopf-Rinow-de Rham theorem (Proposition 2), and v has all the properties needed for the proof of the theorem. Therefore we examine the properties of equation (3.7). This equation, in a more general form, has been considered by R. Osserman [10] in connection with the inequality $\Delta u \geq f(u)$ in Euclidean space. It is worth remarking that, for $n = 3$, equation (3.7) is a modified form of Emden's equation, obtained from the spherically symmetric, polytropic expansion of gases [7, pp. 559-561].

It is convenient to normalize the differential equation (3.7) and the condition (3.9) by replacing t and $\phi(t)$ respectively by s and $f(x) = y$, where

$$t = \frac{s}{a}\sqrt{\frac{n(n-1)}{n+1}}, \qquad y = f(x) = \frac{1}{a}\phi(t). \tag{3.10}$$

Then y satisfies the differential equation

$$y'' + \frac{n-1}{s}y' = y^3 \qquad \left(y' = \frac{dy}{ds},\ y'' = \frac{d^2y}{ds^2}\right), \tag{3.11}$$

with initial conditions

$$y = 1, \quad y' = 0 \qquad \text{for } s = 0. \tag{3.12}$$

We shall now discuss this differential equation.

LEMMA 2. *The differential equation* (3.11) *with initial conditions* (3.12) *admits a unique, analytic solution in the neighborhood of* $s = 0$. *This solution is an even function; it is monotone increasing for all* $s > 0$ *in the domain of regularity, and becomes infinite as* s *approaches a certain number* k_n *which satisfies the conditions*

$$(3.13)\quad \begin{cases} \sqrt{n} \le k_n \le \sqrt{2n} & (n \ge 4), \\ \sqrt{2n} \le k_n \le \sqrt{n}\int_0^{\pi/2}\left(1 - \frac{1}{2}\sin^2\theta\right)^{-1/2} d\theta \cong 1.8541\,\sqrt{n} & (n = 2, 3). \end{cases}$$

One can show immediately that there is a recursion formula for the expansion of the solution of (3.11), (3.12) as a formal power series in s; also, that all coefficients of even powers of s are positive, and that those of odd powers vanish. Comparing the expansion for y with the formal power series expansion for the solution of $\frac{n\,dz}{s\,ds} = z^3$ with the same initial conditions, one sees that the latter converges to the actual solution $z = \left(1 - \frac{s^2}{n}\right)^{-1/2}$, and that its coefficients dominate those of the former. Thus the formal power series expansion of y in terms of s actually converges to the unique, analytic solution of (3.11) with (3.12). Since all of the coefficients of the power series are nonnegative, the series converges for all positive values of s up to the first real singularity; therefore, for all positive values of s on the maximal real interval of regularity of y, the solution y as well as all of its derivatives are positive. Comparing now the power series expansion for y with that of the solution w of the equation $nw'' = w^3$ with the same initial conditions, we find that the coefficients of the expansion of y dominate the corresponding coefficients for w. On the other hand, w can be evaluated by quadratures, and it can be shown to be the elliptic function defined by inverting the elliptic integral

$$s = \sqrt{n}\int_0^{\operatorname{arcsec} w} \frac{d\theta}{\sqrt{1 - \frac{1}{2}\sin^2\theta}}.$$

Since w becomes infinite as s approaches the value corresponding to the complete elliptic integral in (3.13), y has a singularity as s approaches a value k_n dominated by that integral. Therefore y is regular in a bounded, open interval $|s| < k_n$, where

$$(3.14)\qquad \sqrt{n} \le k_n \le 1.8541 \cdots \sqrt{n}.$$

We can sharpen the estimate for k_n as follows: for $n = 4$, we can express y in closed form by

$$(3.15)\qquad y = 8/(8 - s^2) \qquad (n = 4),$$

so that $k_4 = \sqrt{8}$; this is the only known case where the solution appears to be elementary. In the remaining cases, since all derivatives of y (and in particular the third) are positive for $s > 0$, and since $y' = 0$ for $s = 0$, we use the inequality, obtained from the law of the mean,

$$(3.16)\qquad y'' - \frac{1}{s}y' > 0 \qquad (s > 0).$$

Multiplying the left member of this inequality by $(n - 4)/4$ and adding it to (3.11), one obtains the differential inequality

$$\frac{n}{4}y'' + \frac{3n}{4s}y' \begin{cases} > y^3 & (n > 4), \\ < y^3 & (n < 4) \end{cases} \tag{3.17}$$

for all $s > 0$. If we replace the inequality signs by equality, we obtain a differential equation, whose solution with the same initial values as in (3.12) is

$$h(s) = 2n/(2n - s^2). \tag{3.18}$$

One can now apply the principle of the maximum (respectively, minimum) to the function $y - h(s)$ from the inequality (3.17) for $n > 4$ (respectively, $n < 4$); according to this, $y - h(s)$ can not attain its maximum (minimum) value in the interior of the segment $0 < s < \min(\sqrt{2n}, k_n)$. On the other hand, by comparing the coefficients of the power series expansions of y and of $h(s)$, one sees that, in a neighborhood of $s = 0$, $y > h(s)$ for $n > 4$, and $y < h(s)$ for $s < 4$. Therefore the same inequalities hold throughout the whole interval in which both y and $h(s)$ are regular. This yields immediately the inequalities

$$k_n \leq \sqrt{2n} \qquad (n \geq 4),$$

$$k_n \geq \sqrt{2n} \qquad (n < 4),$$

which together with (3.14) imply (3.13).

It remains to be proved that y becomes infinite as s approaches k_n. To this end we subtract (3.16) from (3.11), with the result of the following inequality (always for $s > 0$)

$$\frac{n}{s}y' < y^3; \tag{3.19}$$

for if y did not become infinite, then, being monotone increasing, it would converge to a finite number; (3.19), y' would also be bounded and, being monotone, it would converge to a finite number. This would imply that for $s = k_n$ the limiting values of y and y' could be taken as initial values for a further regular continuation, which is impossible. This completes the proof of the lemma.

We return now to the function $\phi(t)$ defined in terms of $y = f(s)$ by (3.10). This function, which is the solution of (3.7) and (3.9), has now been shown to be monotone increasing for $t > 0$ (see (3.8)), and to diverge to infinity as t approaches c_n/a, where

$$c_n = k_n \sqrt{n(n-1)/(n+1)}.$$

From the estimates (3.13) on k_n it follows that $c_n \leq n\sqrt{2}$, as was claimed in the statement of Theorem 1 in the Introduction. Thus the auxiliary function $v(x) = \phi(r(x))$ becomes infinite uniformly on the relative boundary in D of the subdomain $\Sigma \subset D$ which consists of those points $(x) \in D$ whose geodesic distance $r(x)$ from the origin is less than c_n/a. By Proposition 2 (Hopf, Rinow, de Rham), Σ is relatively compact in D, provided that $\gamma(0) > c_n/a$; hence, if this last condition is satisfied, then v meets the four requirements (a), (b), (c), (d), including (3.4), (3.5), (3.6); these show that (3.3), (3.1), and (3.2) are incompatible. This completes the proof of Theorem 1. The corollary is clearly a special case.

4. PROOF OF THEOREM 2

We shall rephrase the statement of Theorem 2 in a way analogous to the reformation of Theorem 1 at the beginning of Section 3. Assuming, (as we do in the Introduction) that the function u is a convex solution of (1.1) in a domain D containing the origin O in its interior; that

$$\frac{\partial^2 u}{\partial x^i \partial x^j}(O) = \delta_{ij}, \tag{4.1}$$

and that $|\psi|(O) \geq a > 0$; we shall seek to construct a curve in D, from the origin to the boundary of D, whose *Euclidean* length in terms of the rectangular coordinates (x) can be bounded in terms of the positive constant a. It follows from Theorem 1 and Proposition 2 that under these assumptions there exists a geodesic C with respect to the Riemannian metric (1.2) (namely a shortest possible curve with O as initial point and no end point in D) whose Riemannian length γ is finite and satisfies

$$\gamma = \gamma(O) \leq \frac{c_n}{a} \qquad (c_n < n\sqrt{2}). \tag{4.2}$$

It is a remarkable and fortunate fact that, for $n \leq 5$, this same geodesic C with respect to the Riemannian metric has bounded Euclidean length, as we shall prove. As in the case of Theorem 1, the ultimate step depends on a lemma in ordinary differential equations. For the proof of this lemma, the author is indebted to N. Levinson.

Let the shortest geodesic C with initial point O and no end point in D be parametrized by its Riemannian length s from the origin; let its total length be γ, where γ satisfies (4.2); and denote by $\sigma = \sigma(s)$ the corresponding Euclidean length. Then the following inequality holds.

$$0 < \frac{d\sigma}{ds} = \sqrt{\delta_{ij}\frac{dx^i}{ds}\frac{dx^j}{ds}} < \sqrt{g^{ij}\delta_{ij}} = \rho(s) = \rho.$$

Differentiating the distortion parameter ρ with respect to s, we obtain

$$\frac{d}{ds}(g^{ij}\delta_{ij}) = 2\rho\frac{d\rho}{ds} = 2\delta_{ij}A^{ij}{}_k\frac{dx^k}{ds} \tag{4.3}$$

directly from (2.1). We regard this last expression as the Riemannian inner product between the tensors δ_{ij} and $A^{ij}{}_k dx^k/ds$, and prepare to apply the Cauchy-Schwarz inequality, in majorizing that expression by the product of the Riemannian absolute values. Actually a slightly better inequality is obtained by exploiting the fact that the δ_{ij} form a positive definite matrix, while the $A^{ij}{}_k dx^k/ds$ form a matrix with trace zero with respect to the metric g_{ij}. Let

$$a_{ij} = \delta_{ij} - \frac{\rho^2}{n}g_{ij}, \qquad a^{ij} = g^{ik}g^{j\ell}a_{k\ell};$$

then the decomposition $\delta_{ij} = a_{ij} + (\rho^2/n)g_{ij}$ is orthogonal with respect to the metric g_{ij} at each point, and the second term is orthogonal to $A^{ij}{}_k dx^k/ds$. Thus we have

$$\left|\delta_{ij}A^{ij}{}_k\frac{dx^k}{ds}\right| = \left|a_{ij}A^{ij}{}_k\frac{dx^k}{ds}\right| \leq (a_{ij}a^{ij})^{1/2}\left(A^{ij}{}_k\frac{dx^k}{ds}A_{ij\ell}\frac{dx^\ell}{ds}\right)^{1/2}.$$

The norm of the a_{ij} is majorized as follows:

$$a_{ij}a^{ij} = g^{ik}g^{j\ell}\delta_{ij}\delta_{k\ell} - \frac{\rho^4}{n} < \rho^4\left(1 - \frac{1}{n}\right), \tag{4.4}$$

while the squared absolute value of $A^{ij}{}_k\, dx^k/ds$ can be computed from (2.9):

$$g_{i\ell}\, g_{jm}\, A^{ij}{}_h \frac{dx^h}{ds} A^{\ell m}{}_k \frac{dx^k}{ds} = R_{hk}\frac{dx^h}{ds}\frac{dx^k}{ds}. \tag{4.5}$$

In what follows, we shall denote by $K(s)$ the function

$$K(s) = \sqrt{R_{hk}\frac{dx^h}{ds}\frac{dx^k}{ds}}. \tag{4.6}$$

Combining (4.3) through (4.6), and applying the Cauchy-Schwarz inequality, we obtain

$$\rho\left|\frac{d\rho}{ds}\right| \leq \left(\frac{n-1}{n}\right)^{1/2} \rho^2 K(s),$$

from which we deduce the integral inequality

$$\sigma(s) < \sqrt{n}\int_0^s \exp\left[\int_0^{s_1}\left(\frac{n-1}{n}\right)^{1/2} K(s_2)\, ds_2\right] ds_1 \qquad (0 < s < \gamma), \tag{4.7}$$

because of the initial conditions $\sigma(O) = 0$, $\rho(O) = \sqrt{n}$, the latter being a consequence of (4.1). Our purpose now is to prove, by means of the available information on $K(s)$, that the integral (4.7) converges as s approaches γ.

From the definition (4.6) of $K(s)$ and from (2.10), (2.8), and Theorem 1, and using the obvious identity $\gamma(x(s)) = \gamma(O) - s = \gamma - s$, we obtain the inequality

$$0 \leq K(s) \leq \frac{c_n}{\gamma - s}. \tag{4.8}$$

If one could prove the stronger inequality $K(s) \leq \frac{1-\varepsilon}{\gamma - s}\left(\frac{n}{n-1}\right)^{1/2}$ for some $\varepsilon > 0$, then the convergence of (4.7) as $s \to \gamma$ would be immediate. It seems doubtful, however, that this hypothesis be true, especially for high values of n (as a matter of fact, we did not succeed in proving it even for $n = 2$). Therefore some other property of $K(s)$, restraining its rate of growth, has to be used: the crucial inequality in this case is the Riccati inequality (2.24), where the "unknown" function $H(s)$ is the relative mean curvature of the geodesic hypersphere with center O that passes through the point $(x(s))$. Thus in the present notation we have the following majorant for $K(s)$:

$$0 \leq K(s) \leq \sqrt{(n-1)\left(\frac{dH(s)}{ds} - H^2(s)\right)}, \tag{4.9}$$

where $H(s)$ is a differentiable function of s in the open interval $0 < s < \gamma$, satisfying

$$H'(s) = \frac{dH}{ds} \geq H^2(s). \tag{4.10}$$

We shall now quote Levinson's lemma.

LEMMA 3. *Let* $H = H(s)$ *be a continuously differentiable, real-valued function of* s $(0 < s < \gamma < \infty)$, *satisfying* (4.10). *Then, for every* $\varepsilon > 0$ *and every constant* $c' < 2$,

$$\int_\varepsilon^\gamma \exp\left[c' \int_\varepsilon^s \sqrt{H'(s_1) - H^2(s_1)}\, ds_1\right] ds < \infty. \tag{4.11}$$

Proof. We perform the substitution

$$s = \gamma(1 - e^{-t}), \tag{4.12}$$

and introduce the following two functions of t:

$$w = w(t) = \gamma e^{-t} H(s) - \frac{1}{2}$$

and

$$z = z(t) = \gamma e^{-t} \sqrt{\frac{dH(s)}{ds} - H^2(s)}\,.$$

These substitutions lead to the following restatement of the lemma: if $z = z(t)$ is a continuous, nonnegative function of t for all $t > 0$, such that the Riccati equation

$$\frac{dw}{dt} = w^2 + z^2 - \frac{1}{4} \tag{4.13}$$

has a continuous solution $w = w(t)$ on the whole half-line $t > 0$, then for every $\varepsilon' > 0$ and every constant $c' < 2$ the following statement is true:

$$\gamma \int_{\varepsilon'}^\infty \exp\left[c' \int_{\varepsilon'}^t z(t_1)\, dt_1\right] e^{-t} dt < \infty.$$

In order to see this, we consider $w = w(t)$ as a solution of the differential inequality

$$\frac{dw}{dt} \geq w^2 - \frac{1}{4}$$

obtained from (4.13), and we compare it with the solutions of the corresponding equation (obtained by replacing the inequality by equality) that are regular on the whole positive half-line of t. We obtain immediately the bounds

$$-\frac{1}{2} \coth \frac{t}{2} \leq w(t) \leq \frac{1}{2}.$$

From these we obtain the following inequality, for every $t \geq \varepsilon'$:

$$\int_{\varepsilon'}^t z(t_1)\, dt_1 = \int_{\varepsilon'}^t \left(w'(t_1) - w^2(t_1) + \frac{1}{4}\right)^{1/2} dt_1 \leq \int_{\varepsilon'}^t \left(w'(t_1) + \frac{1}{4}\right)^{1/2} dt_1$$

$$\leq \sqrt{t - \varepsilon'} \left[\int_{\varepsilon'}^{t} \left(w'(t_1) + \frac{1}{4} \right) dt_1 \right]^{1/2} = \left[(t - \varepsilon') \left(w(t) - w(\varepsilon') + \frac{t - \varepsilon'}{4} \right) \right]^{1/2}$$

$$\leq \frac{1}{2} \left[(t - \varepsilon')^2 + 2(t - \varepsilon') \left(1 + \coth \frac{\varepsilon'}{2} \right) \right]^{1/2} < \frac{t - \varepsilon'}{2} + \frac{e^{\varepsilon'}}{e^{\varepsilon'} - 1} .$$

Consequently the left-hand member of (4.11) can be majorized by direct computation as follows: for any constant $c' < 2$,

$$\int_{\varepsilon}^{\gamma} \exp \left[c' \int_{\varepsilon}^{s} \sqrt{H'(s_1) - H^2(s_1)}\, ds_1 \right] ds$$

$$= \gamma \int_{\varepsilon'}^{\infty} \exp \left[c' \int_{\varepsilon'}^{t} z(t_1)\, dt_1 \right] e^{-t} dt < \frac{2(\gamma - \varepsilon)}{(2 - c')} \exp (c'\gamma/\varepsilon) ,$$

where $\varepsilon' = \log \left(\frac{\gamma}{\gamma - \varepsilon} \right)$, according to the reverse of the substitution (4.12). This completes the proof of the Lemma.

We remark that the condition $c' < 2$ under which the assertion of Lemma 3 is valid is the best possible one; in fact, if we take, for example, $\gamma = 1$ and $H(s) = (2 - 2s)^{-1}$, then the integral (4.11) diverges for every $c' \geq 2$.

We now conclude the proof of Theorem 2. The function $K(s)$ occurring in the integral in (4.7) has two majorants, given by (4.8) and (4.9) respectively: using the former in the interval from 0 to ε, for any ε $(0 < \varepsilon < \gamma)$, and the latter for $s > \varepsilon$, and setting $c' = (n - 1)/\sqrt{n}$, we can evaluate the limit of the integral in (4.7) as s approaches γ in terms of $\varepsilon' = \log \frac{\gamma}{\gamma - \varepsilon}$. If $c_n > 1$ and $c' < 2$, that is, under the condition $n \leq 5$, the result of the computation from (4.9) is

$$\lim_{s \to \gamma} \sigma(s) = \sigma(\gamma) < \frac{\gamma \sqrt{n}}{c_n - 1} \left(e^{(c_n - 1)\varepsilon'} - 1 \right) + \frac{2\gamma \sqrt{n}}{2 - c'} \exp \left[(c_n - 1)\varepsilon' + \frac{c' e^{\varepsilon'}}{e^{\varepsilon'} - 1} \right].$$

In order to simplify the last expression, we use the inequality

$$\frac{e^{\varepsilon'}}{e^{\varepsilon'} - 1} < \frac{1}{\varepsilon'} + 1$$

and set $\varepsilon' = \sqrt{c'/(c_n - 1)}$; we write $\sqrt{c'(c_n - 1)} = C_n$ and use the assertion of Theorem 1 that $\gamma \leq c_n/a$; this gives

$$\delta(0) \leq \sigma(\gamma) \leq \frac{M_n}{a} ,$$

where, for $n = 2, 3, 4, 5$,

$$M_n < \frac{c_n \sqrt{n}}{c_n - 1} \left(e^{C_n} - 1 \right) + \frac{2\, c_n \sqrt{n}}{2 - c'}\, e^{(2C_n + c')} .$$

Using the upper bound from Theorem 1, $c_n < n\sqrt{2}$, and the definitions of the remaining constants, we have calculated approximately the following numerical estimates for M_n:

(4.14) $$M_2 < 127; \quad M_3 < 2660; \quad M_4 < 40100; \quad M_5 < 653000.$$

This concludes the proof of Theorem 2 and, as a special case, of its Corollary.

The extreme wastefulness of our method of estimating the constants M_n, resulting in (4.14), is evidenced by comparing our upper bound for M_2, the only M_n whose existence has been reported so far in the literature, with the bound calculated by J. Nitsche [9], by means of complex variable techniques. By direct computation in terms of the local complex parameter

$$\sigma = x^1 + \frac{\partial u}{\partial x^1} + \sqrt{-1}\left(x^2 + \frac{\partial u}{\partial x^2}\right)$$

and of the holomorphic generating function

$$f(\sigma) = x^1 - \frac{\partial u}{\partial x^1} + \sqrt{-1}\left(\frac{\partial u}{\partial x^2} - x^2\right),$$

where σ and $f(\sigma)$ are Nitsche's notations, one obtains the identity

$$|\psi| = \frac{4|f''(\sigma)|}{(1 - |f'(\sigma)|^2)^{3/2}}.$$

From this and from Nitsche's arguments one obtains the following, considerable improvement on the upper bound for the constant M_n appearing in (1.10) and (1.11) for $n = 2$:

(4.15) $$M_2 \leq 4.$$

It is not surprising that the results obtained by complex variable techniques for $n = 2$ should yield much sharper estimates; on the other hand, the methods used in the present paper are applicable to any number of variables; the weakness of our numerical estimates for $n = 2$ is a strong indication of the room for possible improvement for $n \geq 3$, including the very likely possibility, mentioned in the Introduction, that the assertion of Theorem 2 may be valid also for $n \geq 6$.

REFERENCES

1. W. Blaschke, *Vorlesungen über Differentialgeometrie und geometrische Grundlagen von Einsteins Relativitätstheorie*, vol. II, Julius Springer, Berlin, 1923.

2. E. Calabi, *An extension of E. Hopf's maximum principle with an application to Riemannian geometry*, Duke Math. J. 25 (1958), 45-56.

3. E. Cartan, *Leçons sur la géométrie des espaces de Riemann*, 2nd edition, Gauthier-Villars, Paris, 1946.

4. E. K. Haviland, *A note on unrestricted solutions of the differential equation* $\Delta u = f(u)$, J. London Math. Soc. 26 (1951), 210-214.

5. H. Hopf and W. Rinow, *Über den Begriff der vollständigen differentialgeometrischen Fläche*, Comment. Math. Helv. 3 (1931), 209-225.

6. K. Jörgens, *Über die Lösungen der Differentialgleichung* $rt - s^2 = 1$, Math. Ann., 127 (1954), 130-134.

7. E. Kamke, *Differentialgleichungen, Lösungsmethoden und Lösungen*, 3rd edition, Chelsea, New York, 1948.

8. S. B. Myers, *Riemannian manifolds with positive mean curvature*, Duke Math. J., 8 (1941), 401-404.

9. J. C. C. Nitsche, *Elementary proof of Bernstein's theorem on minimal surfaces*, Ann. of Math. (2) 66 (1957), 543-544.

10. R. Osserman, *On the inequality* $\Delta u \geq f(u)$, Pacific J. Math. 7 (1957), 1641-1647.

11. G. de Rham, *Sur la réductibilité d'un espace de Riemann*, Comment. Math. Helv., 26 (1952), 328-344.

University of Minnesota

An Extension of E. Hopf's Maximum Principle with an Application to Riemannian Geometry

Eugenio Calabi

Duke Math. J. 25, 45–56 (1958)

1. **Introduction.** There are numerous instances in function theory, partial differential equations, and differential geometry where one needs to apply a maximum principle. One complete statement of this principle, applied to twice differentiable functions, is due to E. Hopf [2] and quoted below as Theorem 1. In the course of recent investigations, whose results now are being prepared for publication, it has been necessary to prove Hopf's theorem in the case of functions that are not necessarily differentiable, but for which one verifies a certain weak, elliptic differential inequality. In the next section we give a precise definition of the inequality in question, showing that, if applied to twice differentiable functions, it reduces to the elementary notion of that inequality, and that in the extended sense it provides a more general condition under which E. Hopf's theorem is satisfied.

In §3 we review some known facts about geodesics in a Riemannian manifold, which are then applied in §4. In this last section we prove that in a Riemannian manifold whose Ricci curvature is non-negative, if r denotes the (shortest) geodesic distance from a fixed point $\mathbf{p}_0$, and Δ is the Laplace-Beltrami operator, the following weak, elliptic differential inequality is valid,

$$\Delta r \quad (\text{weakly}) \leq \frac{n-1}{r}, \tag{1.1}$$

and some geometrical and potential-theoretic conclusions are derived from this result. A simplified proof of our generalization of Hopf's theorem, actually reduced to a modification of the original proof by Hopf, was suggested by J. B. Serrin, whose help in this connection I wish to acknowledge gratefully.

2. **The maximum principle.** In this article we denote by L a linear, uniformly elliptic, second order differential operator in n dimensions, containing only first and second partial derivative terms: in other words, if $u = f(x)$ $((x) = (x^1, \cdots, x^n))$ is a function of n variables of class $\mathfrak{C}^2$, then

$$L[u] = a^{ij}(x) \frac{\partial^2 u}{\partial x^i \, \partial x^j} + b^i(x) \frac{\partial u}{\partial x^i}, \tag{2.1}$$

where we make use of the summation convention. The only assumptions we make about the coefficients a^{ij} and b^i are that in a neighborhood of each point

Received May 1, 1957. This article was prepared while the author was partly supported by a research grant from the Office of Naval Research.

DOI: 10.1215/S0012-7094-58-02505-5

there exist three positive constants C_1, C_2, C_3, such that at every point (x) the following inequalities hold,

$$|\, b^i(x)\xi_i \,| \leq C_1\Big(\sum_{i=1}^{n} (\xi_i)^2\Big)^{\frac{1}{2}} \tag{2.2}$$

and

$$a^{ij}(x)\xi_i\xi_j \geq C_2\Big(\sum_{i=1}^{n} (\xi_i)^2\Big); \qquad |\, a^{ij}(x) \,| \leq C_3 \tag{2.3}$$

for any set of constants $\xi_1, \cdots, \xi_n$. In order to interpret (2.1) invariantly with respect to coordinate transformations, the latter must be of class $\mathfrak{C}^2$, the coefficients $a^{ij}(x)$ must be components of a contravariant tensor, and $b^i(x)$ satisfy the identity

$$b^i(x) = -a^{jk}(x)\Gamma^i_{jk}(x)$$

for some affine connection $\Gamma^i_{jk}(x)$ with bounded coefficients.

We shall presently give a definition of the inequalities $L[u] \geq v$ and $L[u] \leq v$ for classes of functions u that include all continuous functions.

Definition 1. *Given an elliptic operator L of the type* (2.1) *satisfying* (2.2) *and* (2.3), *an upper semicontinuous function $u = f(x)$ in an open domain U and a function $v = \phi(x)$ in U with no restriction whatever, we say that*

$$L[u] \text{ (weakly) } \geq v \tag{2.4}$$

in any subset $K \subset U$, if for each point $(x_0) \in K$ and any given (small) constant $\epsilon > 0$, there exists a neighborhood $V_{x_0,\epsilon} \subset U$ of (x_0) in U and a function $u_{x_0,\epsilon} = f_{x_0,\epsilon}(x)$ of class $\mathfrak{C}^2$ in V_{x_0}, such that the difference function $u - u_{x_0,\epsilon}$ achieves its minimum value at (x_0), i.e.

$$f(x) - f_{x_0,\epsilon}(x) \geq f(x_0) - f_{x_0,\epsilon}(x_0) \qquad ((x) \in V_{x_0,\epsilon}), \tag{2.5}$$

and $u_{x_0,\epsilon}$ satisfies the inequality

$$L[u_{x_0,\epsilon}] \geq v - \epsilon \tag{2.6}$$

at the point (x_0) in the strict sense. Similarly, if u is lower semicontinuous, we say that

$$L[u] \text{ (weakly) } \leq v \tag{2.7}$$

if $L[-u]$ (weakly) $\geq -v$ in the sense just defined.

The above definition finds its first justification in the fact that it is, as the deliberately ambiguous notation suggests, an extension of the elementary notion of the inequality (2.4), read literally from (2.1), if u is of class $\mathfrak{C}^2$.

Lemma 1. *Given in an open domain U a function $u = f(x)$ of class $\mathfrak{C}^2$ and an* *arbitrary function* $v = \phi(x)$, *then the inequality* (2.4), *is valid in the elementary sense, i.e.*

$$a^{ij}(x) \frac{\partial^2 f(x)}{\partial x^i \partial x^j} + b^i \frac{\partial f(x)}{\partial x^i} \geq \phi(x), \tag{2.8}$$

in any subset K *of* U, *if and only if it is weakly valid in the sense of Definition* 1.

Proof. If u satisfies (2.8), then it trivially satisfies the properties of Definition 1, for one need only take $V_{x_0,\epsilon}$ to be U and $u_{x_0,\epsilon} = u$. Conversely, if u satisfies the conditions of Definition 1, then for each $(x_0) \varepsilon K$ and $\epsilon > 0$, since $u - u_{x_0,\epsilon}$ has its minimum at (x_0) relative to $V_{x_0,\epsilon}$, we have

$$\left[a^{ij} \frac{\partial^2}{\partial x^i \partial x^j} (u - u_{x_0,\epsilon}) + b^i \frac{\partial}{\partial x^i} (u - u_{x_0,\epsilon})\right](x_0) \geq 0.$$

Consequently the elementary inequality

$$\left[a^{ij}(x) \frac{\partial^2 u}{\partial x^i \partial x^j} + b^i \frac{\partial u}{\partial x^i}\right](x_0) \geq \phi(x_0) - \epsilon$$

is valid at (x_0) for every $\epsilon > 0$. Since (x_0) is arbitrary in K, the conclusion follows immediately.

A more important justification of Definition 1 is the fact that it permits the application of Hopf's maximum principle to upper semicontinuous functions. We state now the extension of Hopf's principle.

THEOREM 1. *If* $u = f(x)$ *is an upper semicontinuous function in a connected, open region* U, *satisfying*

$$L[u] \text{ (weakly) } \geq 0 \tag{2.9}$$

everywhere in U, *and, if* u *attains a local maximum value at some point* $(x) \varepsilon U$, *then* u *is constant in* U.

Proof. The proof is essentially the same as the one given by E. Hopf [2] and later reported by Yano and Bochner [7; 26–29].

Suppose that $u = f(x)$ satisfies the hypotheses of Theorem 1 but not the conclusion. Then one can find a closed, n-dimensional Euclidean ball $E_0 \subset U$ of radius R_1 and center (x_1), i.e. a set represented by the inequality

$$\sum_{i=1}^{n} (x^i - x_1^i)^2 \leq (R_1)^2,$$

such that u achieves its maximum value M relative to a neighborhood $U_1 \subset U$ of E_1 at a point (x_2), on the boundary of E_1 and at no other point of E_1.

Without loss of generality we can consider the center (x_1) of E_1 as the origin, $x_1^i = 0$; in what follows we denote by r the Euclidean distance of any point $(x) \varepsilon U_1$ from the origin (x_1), i.e.

$$r^2 = \sum_{i=1}^{n} (x^i)^2.$$

Let E_2 be a closed Euclidean ball with center at (x_2), contained in U_1, and of radius $R_2 \leq \frac{1}{2}R_1$. Then the minimum distance of E_2 from (x_1) is $R_1 - R_2 \geq \frac{1}{2}R_1$; the boundary S_2 of E_2 is decomposed into the union of two spherical caps S' and S'', where $S' = S_2 \cap E_1$ and S'' is the complement. Thus for every point $(x) \in S'$ we have

$$f(x) \leq M - \delta \qquad ((x) \in S') \tag{2.10}$$

for some positive constant δ, while in S'' we have merely $f(x) \leq M$.

Consider the function w defined in E_2 by

$$w = e^{-\alpha r^2} - e^{-\alpha (R_1)^2} \tag{2.11}$$

where α is a positive constant to be restricted later; we compute $L[w]$ and apply (2.2) and (2.3) to estimate the result

$$\begin{aligned} L[w] &= \left[4\alpha^2\left(\sum_{i,j=1}^{n} a^{ij} x^i x^j\right) - 2\alpha\left(\sum_{i=1}^{n} (b^i x^i + a^{ii})\right)\right]e^{-\alpha r^2} \\ &\geq (4C_2\alpha^2 r^2 - 2C_1\alpha r - 2n\alpha C_3)e^{-\alpha r^2} \\ &\geq 2\alpha r\left(C_2\alpha R_1 - C_1 - \frac{2nC_3}{R_1}\right)e^{-\alpha r^2}, \end{aligned} \tag{2.12}$$

since $r \geq \frac{1}{2}R_1$ in E_2. We now assign to α a fixed value

$$\alpha = \frac{R_1(C_1 + \epsilon_0) + 2nC_3}{C_2 R_1^2}, \ \epsilon_0 > 0, \tag{2.13}$$

so that

$$L[w] \geq \epsilon_1 > 0 \tag{2.14}$$

everywhere in E_2 for some positive constant ϵ_1. Furthermore, w satisfies the inequalities

$$\begin{cases} 0 \leq w < 1 & \text{in } E_2 \cap E_1 \\ w < 0 & \text{outside of } E_1 . \end{cases} \tag{2.15}$$

Denote by u' the function

$$u' = u + \tfrac{1}{2}\delta w; \tag{2.16}$$

from (2.10) and (2.15) it follows that $u' < M$ everywhere on the boundary S_2 of E_2, while $u' = M$ at the center (x_2) of E_2; since u' is upper semicontinuous, it must achieve its maximum value $M' \geq M$ at some interior point (x_0) of E_2. At the same time we have from (2.14)

$$L[u'] \text{ (weakly) } \geq \tfrac{1}{2}\delta\, L[w] \geq \tfrac{1}{2}\delta\, \epsilon_1 > 0.$$

Let $\epsilon = \frac{1}{4}\delta\, \epsilon_1$, so that

$$L[u'] \text{ (weakly) } \geq 2\epsilon > 0 \text{ at the point } (x_0). \tag{2.17}$$

An immediate application of Definition 1 to interpret (2.17) using the same number ϵ means that there exists a function $u'_{x_0,\epsilon}$ of class $\mathcal{C}^2$ in a neighborhood V of (x_0), such that

$$L[u'_{x_0,\epsilon}](x_0) \geq \epsilon > 0, \tag{2.18}$$

and, for all points $(x) \in V$, $u'(x) - u'_{x_0,\epsilon}(x) \geq u'(x_0) - u'_{x_0,\epsilon}(x_0)$. Since both u' and $u'_{x_0,\epsilon} - u'$ achieve a local maximum value at (x_0), it follows that (x_0) is a relative maximum point for $u'_{x_0,\epsilon}$, contradicting (2.18); this contradiction invalidates our initial supposition, thus proving Theorem 1.

3. Metric geometry of Riemannian manifolds. This section is mainly expository; the topics treated are chosen almost exclusively for application in the final section.

Consider an arbitrary n-dimensional Riemannian manifold $\mathbf{X}$ with (positive definite) metric of class $\mathcal{C}^3$. Such a space is naturally a metric space under the distance function $d(\mathbf{p}, \mathbf{q})$ for $\mathbf{p}, \mathbf{q} \in \mathbf{X}$ defined as the greatest lower bound of the Riemannian arc length of all rectifiable arcs joining $\mathbf{p}$ with $\mathbf{q}$. The space $\mathbf{X}$ may or may not be complete under this metric; in all cases it possesses a completion $\hat{\mathbf{X}}$, not necessarily a manifold, obtained by adjoining to $\mathbf{X}$ as "virtual points" the equivalence classes of divergent Cauchy sequences. The complete space $\hat{\mathbf{X}}$ is again a metric space. If $\mathbf{X}$ is not complete, we denote for each point $\mathbf{p} \in \mathbf{X}$ by $\gamma(\mathbf{p})$ the distance of $\mathbf{p}$ from the complement of $\mathbf{X}$ in $\hat{\mathbf{X}}$, i.e.

$$\gamma(\mathbf{p}) = d(\mathbf{p}, \hat{\mathbf{X}} - \mathbf{X}); \tag{3.1}$$

if $\mathbf{X}$ is complete, we define $\gamma(\mathbf{p})$ to be the formal symbol $+\infty$, defining the latter to be the upper bound of the positive real numbers, with the usual arithmetical properties.

For each point $\mathbf{p} \in \mathbf{X}$ and each positive real number r we denote by $\Sigma(\mathbf{p}, r)$ the open metric ball (solid sphere) in $\mathbf{X}$ with center p and radius r,

$$\Sigma(\mathbf{p}, r) = \{\mathbf{q} \mid \mathbf{q} \in \mathbf{X};\, d(\mathbf{p}, \mathbf{q}) < r\}, \tag{3.2}$$

and by $\bar{\Sigma}(\mathbf{p}, r)$ the closure of $\Sigma(\mathbf{p}, r)$ in $\mathbf{X}$. The relationship between the distance function $d(\mathbf{p}, \mathbf{q})$ with geodesics in $\mathbf{X}$ is best formulated in terms of the following theorem, a slight generalization of the one known as the H. Hopf-W. Rinow theorem.

THEOREM 2. *If* $\mathbf{X}$ *is a Riemannian manifold with metric of class* $\mathcal{C}^2$, *then for each point* $\mathbf{p} \in \mathbf{X}$ *the following statements are true:*

i) *The closed metric ball* $\bar{\Sigma}(\mathbf{p}, r)$ *is compact, if and only if* $r < \gamma(\mathbf{p})$.

ii) *Every geodesic segment originating at* $\mathbf{p}$ *can be continued to any length* $< \gamma(\mathbf{p})$; *however, there exists a geodesic segment of length* $\gamma(\mathbf{p})$, *originating at* $\mathbf{p}$, *which has no end-point in* $\mathbf{X}$.

iii) *Whenever* $d(\mathbf{p}, \mathbf{q}) \leq \gamma(\mathbf{p})$ *for* $\mathbf{p} \in \mathbf{X}$, $\mathbf{q} \in \hat{\mathbf{X}}$, *the greatest lower bound of the arc length of all rectifiable curves joining* $\mathbf{p}$ *and* $\mathbf{q}$ *can be achieved by at least one curve, necessarily a geodesic segment.*

iv) *If* $\mathbf{p}, \mathbf{q} \in \mathbf{X}$, *then*

$$\gamma(\mathbf{q}) - d(\mathbf{p}, \mathbf{q}) \leq \gamma(\mathbf{p}) \leq \gamma(\mathbf{q}) + d(\mathbf{p}, \mathbf{q}). \tag{3.3}$$

This theorem was proved originally by H. Hopf and W. Rinow [3] in case $\mathbf{X}$ is a surface and is complete, i.e. $\gamma(\mathbf{p}) = \infty$; the proof for higher dimension in the complete case is given by G. de Rham [5]; we refer the reader to the latter proof, since it is equally valid for our formulation, except for some trivial modification in terminology.

Consider now a fixed point $\mathbf{p}_0 \in \mathbf{X}$ and denote by $\mathbf{Y}$ (more properly by $\mathbf{Y}(\mathbf{p}_0)$) the subset of $\mathbf{X}$ consisting of those points $\mathbf{p} \in \mathbf{X}$ that can be joined to $\mathbf{p}_0$ by a curve (necessarily a geodesic) of length $d(\mathbf{p}_0, \mathbf{p})$; it is clear that the set $\mathbf{Y}$ is open and contains $\bar{\Sigma}(\mathbf{p}_0, \gamma(\mathbf{p}_0))$. We partition $\mathbf{Y}$ into two complementary subsets $\mathbf{Y}_1$ and $\mathbf{Y}_2$ in the following way; $\mathbf{Y}_1$ consists of those points $\mathbf{p} \in \mathbf{Y}$ whose shortest connection with $\mathbf{p}_0$, namely a geodesic $l(\mathbf{p}_0, \mathbf{p})$ of length $d(\mathbf{p}_0, \mathbf{p})$, is unique and has no conjugate points (since it is a shortest connection, the only possible conjugate points are the end points $\mathbf{p}_0$ and $\mathbf{p}$); $\mathbf{Y}_2$ then consists of those points $\mathbf{p} \in \mathbf{Y}$ that can be joined with $\mathbf{p}_0$ either by more than one geodesic of length $d(\mathbf{p}_0, \mathbf{p})$ or by at least one such geodesic, on which $\mathbf{p}_0$ and $\mathbf{p}$ are conjugate points.

For each geodesic originating at $\mathbf{p}_0$ and continued in $\mathbf{Y}$ there is at most one point of it that lies in $\mathbf{Y}_2$ before the geodesic ceases to be a shortest connection; this is a trivial consequence of the triangle inequality for distances and of the unique continuation property of geodesics. It is also easy to see that $\mathbf{Y}_1$ is an open subset of $\mathbf{Y}$; therefore $\mathbf{Y}_1$ is an open, everywhere dense set in $\mathbf{Y}$ and its complement $\mathbf{Y}_2$ has (n-dimensional) measure zero.

We denote by $r(\mathbf{p})$ (more properly $r_{\mathbf{p}_0}(\mathbf{p})$) the restriction of $d(\mathbf{p}_0, \mathbf{p})$, as a function of $\mathbf{p}$, to $\mathbf{p} \in \mathbf{Y}$, and call it the *geodesic distance* function from $\mathbf{p}_0$. We want to emphasize the distinction between the geodesic distance $r_{\mathbf{p}_0}(\mathbf{p})$ and the Riemannian distance function $d(\mathbf{p}_0, \mathbf{p})$ since the two may differ in their domain of definition, if $\mathbf{X}$ is not complete. From the considerations just made it follows that, if the metric is of class $\mathfrak{C}^3$, as we assumed at the outset, then *the function* $r(\mathbf{p})$ *is of class* $\mathfrak{C}^2$ *everywhere in* $\mathbf{Y}_1 - \{\mathbf{p}_0\}$, *while in* $\mathbf{Y}$ *it merely satisfies the Lipschitz condition*

$$| r(\mathbf{p}_1) - r(\mathbf{p}_2) | \leq d(\mathbf{p}_1, \mathbf{p}_2). \tag{3.4}$$

4. **Geodesic spheres in Riemannian geometry.** We apply here the concepts and notations of the last section to derive some differential geometric properties of the geodesic distance function

$$r(\mathbf{p}) = d(\mathbf{p}_0, \mathbf{p}) \qquad (\mathbf{p} \in \mathbf{Y}(\mathbf{p}_0)) \tag{4.1}$$

from a fixed point $\mathbf{p}_0 \in \mathbf{X}$. It has been pointed out that $r(\mathbf{p})$ is of class $\mathfrak{C}^2$ in the open, everywhere dense sub-region $\mathbf{Y}_1 - \{\mathbf{p}_0\}$ of $\mathbf{Y}$, and it is clear that $\mathbf{Y}_1$ is a neighborhood of $\mathbf{p}_0$, and that it is geodesically starlike with respect to $\mathbf{p}_0$.

Consider for any positive constant r_1 the subset $S(\mathbf{p}_0, r_1)$ of $\mathbf{Y}$, also denoted for the sake of simplicity by S_{r_1}, defined by

$$S_{r_1} = S(\mathbf{p}_0, r_1) = \{\mathbf{p} \mid \mathbf{p} \in \mathbf{Y}; r(\mathbf{p}) = r_1\}. \tag{4.2}$$

S_{r_1} is the boundary of $\Sigma(\mathbf{p}_0, r_1) \cap \mathbf{Y}$ relative to $\mathbf{Y}$, and is called the *geodesic sphere* of radius r_1 with center $\mathbf{p}_0$; the part $S_{r_1,1} = S_{r_1} \cap \mathbf{Y}_1$ of S_{r_1} is called its *regular part.* It is (only locally, as it is not necessarily connected) a hypersurface of class $\mathfrak{C}^2$, since the function r is of class $\mathfrak{C}^2$ in $\mathbf{Y}_1 - \{\mathbf{p}_0\}$ with non-zero gradient.

The unit tangent vectors at $\mathbf{p}_0$ in $\mathbf{X}$ form a unit Euclidean $(n-1)$-dimensional sphere, which we parametrize by a fixed family of local coordinates, such as $(y^1, \cdots, y^{n-1})$; then, if l_1 is any geodesic segment contained in $\mathbf{Y}_1$ with origin at $\mathbf{p}_0$ and initial direction represented by $(y_1) = (y_1^1, \cdots, y_1^{n-1})$, and if $\mathbf{p}_1$ is the point on l_1 at an arc distance r_1 from $\mathbf{p}_0$ along l_1, then $r(\mathbf{p}_0, \mathbf{p}_1) = r_1$, and the quantities $(y_1^1, \cdots, y_1^{n-1}, r_1)$ can be regarded as local coordinates of $\mathbf{p}_1$; more precisely, there exists a positive ϵ such that the set $C(\mathbf{p}_1, \epsilon)$ of points $\mathbf{p}$ similarly determined by quantities $(y^1, \cdots, y^{n-1}, r)$ satisfying

$$\sum_{\alpha=1}^{n-1} (y^\alpha - y_1^\alpha)^2 < \epsilon^2; \quad 0 < r < r_1 + \epsilon \tag{4.3}$$

is contained in $\mathbf{Y}_1$; in this set the quantities $(y^1, \cdots, y^{n-1}, r)$ form a system of local coordinates of class $\mathfrak{C}^2$. We call the region $C(\mathbf{p}_1, \epsilon)$ described by (4.3) an *open, conical neighborhood* of $\mathbf{p}_1$ with vertex at $\mathbf{p}_0$; without loss of generality we might choose the local coordinate system (x) in $C(\mathbf{p}_1, \epsilon)$ to be precisely $(y^1, \cdots, y^{n-1}, r)$; the part of each geodesic sphere S_r contained in $C(\mathbf{p}_1, \epsilon)$ then can be defined by the equation

$$x^n = r \qquad (0 < r < r_1 + \epsilon) \tag{4.4}$$

and the local parametric equations of S_r then can be taken as

$$x^\alpha = y^\alpha (\alpha = 1, 2, \cdots, n-1), \qquad x^n = r. \tag{4.5}$$

We adopt the convention of using Greek indices for the range from 1 to $n - 1$ and denote by $a_{\alpha\beta}(y, r)dy^\alpha dy^\beta$ and $b_{\alpha\beta}(y, r)dy^\alpha dy^\beta$, respectively the first and the second fundamental forms in $S_r \cap C(\mathbf{p}_1, \epsilon)$. Then a direct computation yields the following identities (see [6; 73, Exercise 8.8] and [1; 158, Exercise 2])

$$r_{,i} = \mathbf{N}_i,$$

where $\mathbf{N}_i$ are the covariant components of the unit normal vector to S_r, and

$$r_{,ij} = -b_{\alpha\beta}a^{\alpha\rho}a^{\beta\sigma}\frac{\partial x^k}{\partial y^\rho}\frac{\partial x^l}{\partial y^\sigma}g_{ik}g_{jl};$$

if we denote by $\partial/\partial r$ the operation of differentiation along the geodesic from $\mathbf{p}_0$, i.e. along the normal vector $\mathbf{N}^i$, or in the invariant notation,

$$\frac{\partial}{\partial r} = \mathbf{N}^i \frac{\partial}{\partial x^i},$$

we obtain

$$\frac{\partial}{\partial r} b_\alpha^\beta = b_\alpha^\gamma b_\gamma^\beta + a^{\beta\gamma} R_{ijkh} \frac{\partial x^i}{\partial y^\gamma} \mathbf{N}^j \frac{\partial x^k}{\partial y^\alpha} \mathbf{N}^h,$$

where $b_\alpha^\beta = b_{\alpha\gamma} a^{\gamma\beta}$; from this we have the identities

$$\Delta r = -b_\alpha^\alpha = -\sum_{\nu=1}^{n-1} \kappa_\nu = -(n-1)H, \tag{4.6}$$

where κ_ν represent the principal normal curvatures and H the mean curvature of S_r, and

$$\begin{aligned} \frac{\partial H}{\partial r} &= \frac{1}{n-1} (b_\alpha^\gamma b_\gamma^\alpha + R_{jih}^i \mathbf{N}^j \mathbf{N}^h) \\ &= \frac{1}{n-1} \left[\left(\sum_{\nu=1}^{n-1} (\kappa_\nu)^2 \right) + R_{jh} \mathbf{N}^j \mathbf{N}^h \right], \end{aligned} \tag{4.7}$$

where $R_{jh} = R_{jih}^i$ is the Ricci curvature tensor of $\mathbf{X}$.

We remark that as $r \to 0$ the geodesic spheres S_r are asymptotic to Euclidean spheres of radius r with the normal vector $\mathbf{N}^i$ pointing outward; consequently we have

$$\lim_{r \to 0} H = -\infty; \tag{4.8}$$

we also remark the equation (4.7) leads to the differential inequality

$$\frac{\partial H}{\partial r} \geq H^2 + \frac{1}{n-1} R_{jh} \mathbf{N}^j \mathbf{N}^h. \tag{4.9}$$

We state now the following theorem:

THEOREM 3. *Let* $\mathbf{X}$ *be an n-dimensional Riemannian manifold with metric of class* $\mathfrak{C}^3$, $\mathbf{p}_0$ *a point of* $\mathbf{X}$, *and* $\mathbf{Y} = \mathbf{Y}(\mathbf{p}_0)$ *the domain of definition of the geodesic distance function* $r(\mathbf{p}) = d(\mathbf{p}_0, \mathbf{p})$ *from* $\mathbf{p}_0$. *Then, if the Ricci curvature tensor* R_{ij} *is positive semidefinite and* Δ *denotes the Laplace-Beltrami operator, the inequality*

$$\Delta r \text{ (weakly)} \leq \frac{n-1}{r} \tag{4.10}$$

is valid everywhere in $\mathbf{Y}$. *In particular, if* $\mathbf{X}$ *is complete, then* (4.10) *holds everywhere in* $\mathbf{X}$.

Proof. We split $\mathbf{Y}$ into the two complementary subsets $\mathbf{Y}_1$ (open, everywhere dense), and $\mathbf{Y}_2$ (of measure zero) as before, and prove that (4.10) is valid in $\mathbf{Y}_1$ in the elementary sense. In fact, from the relation (4.8) and (4.9), using the positive character of R_{ij}, we have immediately

$$H \geq \frac{-1}{r}, \tag{4.11}$$

which can be coupled with (4.6) to give the desired result in $\mathbf{Y}_1$. Now let $\mathbf{p} \in \mathbf{Y}_2$. First of all r is continuous at $\mathbf{p}$; then, choosing one geodesic l_1 of shortest length $r(\mathbf{p})$ joining $\mathbf{p}_0$ with $\mathbf{p}$, let $\mathbf{p}_\epsilon$ be a point on l_1 at a distance $\epsilon > 0$, arbitrarily small, from $\mathbf{p}_0$. Then, replacing $\mathbf{p}_0$ by $\mathbf{p}_\epsilon$, we see that $\mathbf{p} \in \mathbf{Y}_1(\mathbf{p}_\epsilon)$, since the part of l_1 between $\mathbf{p}_\epsilon$ and $\mathbf{p}$ is the unique geodesic of length $r(\mathbf{p}) - \epsilon$ joining $\mathbf{p}_\epsilon$ with $\mathbf{p}$, and the end points are not conjugate. Let $V_{\mathbf{p},\epsilon}$ be a neighborhood of $\mathbf{p}$ contained in $\mathbf{Y}_1(\mathbf{p}_\epsilon) \cap \mathbf{Y}(\mathbf{p}_0)$; then, on letting $r_{\mathbf{p},\epsilon}(\mathbf{q}) = d(\mathbf{p}_\epsilon, \mathbf{q})$ for $\mathbf{q} \in V_{\mathbf{p},\epsilon}$ we have

$$\Delta r_{\mathbf{p},\epsilon} \leq \frac{n-1}{r_{\mathbf{p},\epsilon}}.$$

From the triangle inequality we have the relations

$$r - r_{\mathbf{p},\epsilon} \leq \epsilon$$

equality holding at $\mathbf{p}$; thus we have everywhere in $V_{\mathbf{p},\epsilon}$

$$\Delta r_{\mathbf{p},\epsilon} \leq \frac{n-1}{r-\epsilon},$$

and $r - r_{\mathbf{p},\epsilon}$ achieves its maximum value ϵ at $\mathbf{p}$. Since ϵ is arbitrary, the inequality (4.10) is then proved completely in $\mathbf{Y}$ in the sense of Definition 1.

Corollary. *If $\phi(t)$ is a twice differentiable, monotone increasing (respectively decreasing) function of t defined for $0 < a \leq t \leq b$, then under the same assumptions of Theorem 3 the function $v(\mathbf{p})$ defined by*

$$v(\mathbf{p}) = \phi(r(\mathbf{p}))$$

for

$$\mathbf{p} \in \mathbf{Y}(\mathbf{p}_0) \cap (\Sigma(\mathbf{p}_0, b) - \bar{\Sigma}(\mathbf{p}_0, a))$$

satisfies

$$\Delta v \text{ (weakly)} \leq \phi''(r) + \frac{n-1}{r}\phi'(r) \tag{4.12}$$

(*respectively*: Δv (*weakly*) $\geq \phi''(r) + \frac{n-1}{r}\phi'(r)$).

Proof. In $\mathbf{Y}_1 \cap (\Sigma(\mathbf{p}_0, b) - \bar{\Sigma}(\mathbf{p}_0, a))$ we obtain by formal calculation, (assuming without loss of generality that $\phi'(t) \geq 0$)

$$\Delta v = \phi''(r)\left(g^{ij}\frac{\partial r}{\partial x^i}\frac{\partial r}{\partial x^j}\right) + \phi'(r)\Delta r$$

$$\leq \phi''(r) + \frac{n-1}{r}\phi'(r),$$

while, if $\mathbf{p} \varepsilon \mathbf{Y}_2 \cap (\Sigma(\mathbf{p}_0, b) - \bar{\Sigma}(\mathbf{p}_0, a))$, the same argument used in the proof of Theorem 3 leads immediately to the desired inequality in the weak sense.

As an illustration of the corollary just proved suppose that $\mathbf{X}$ is complete and with non-negative Ricci curvature. If we form the radial function of r that equals the Green's function in Euclidean space

$$\phi(r)\begin{cases} = \dfrac{1}{2\pi}\log r & (n = 2) \\ = -\frac{1}{4}\pi^{-n/2}\Gamma\left(\dfrac{n-2}{2}\right)\cdot r^{2-n} & (n > 2), \end{cases}$$

we obtain in $\mathbf{X}$ an approximation of the local fundamental singularity of the Laplace-Beltrami equation, namely

$$g(\mathbf{p}_0, \mathbf{p}) = \phi(d(\mathbf{p}_0, \mathbf{p})),$$

which is everywhere superharmonic in each variable, i.e. satisfies

$$\Delta_{\mathbf{p}} g(\mathbf{p}_0, \mathbf{p}) \quad \text{(weakly)} \leq 0 \tag{4.13}$$

for all points $\mathbf{p}_0, \mathbf{p} \varepsilon \mathbf{X}$.

A further application of Theorem 3 is the following generalization of a recent theorem by R. Osserman [4] to complete manifolds with non-negative Ricci curvature.

THEOREM 4. *If $\mathbf{X}$ is a complete Riemannian manifold with metric of class $\mathfrak{C}^3$ and non-negative Ricci curvature, if $f(t)$ is a positive, monotone increasing, differentiable function, defined for all real t, satisfying*

$$\int_0^\infty \left(\int_0^s f(t)\,dt\right)^{-\frac{1}{2}} ds < \infty, \tag{4.14}$$

then there exists no global solution in $\mathbf{X}$ of the inequality

$$\Delta u \ (\textit{weakly}) \geq f(u). \tag{4.15}$$

Proof. We follow essentially the same proof as the one given by Osserman [4] in the case of a Euclidean space. Consider the singular ordinary differential equation

$$\frac{d^2\phi(t)}{dt^2} + \frac{n-1}{t}\frac{d\phi}{dt} = f(\phi(t)) \tag{4.16}$$

subject to the initial conditions

$$\phi'(0) = 0; \qquad \phi(0) = a, \tag{4.17}$$

where a is for the moment arbitrary. This equation has a unique, differentiable solution $\phi(t)$ in the neighborhood of $t = 0$; under the assumption (4.14) we shall first prove that there exists for each a a positive number R_a such that $\phi(t) \to \infty$ as $t \to R_a$.

The solution $\phi(t)$ of (4.16) and (4.17) is monotone increasing and strictly convex for $t > 0$. In fact, multiplying both sides of (4.16) by t^{n-1}, we have

$$\frac{d}{dt}(t^{n-1}\phi'(t)) = t^{n-1}f(\phi(t)) > 0,$$

whence by the mean value theorem we have $\phi'(t) > 0$. Next, if we multiply both sides of (4.16) by t and differentiate, we obtain

$$t\frac{d^3\phi}{dt^3} + n\frac{d^2\phi}{dt^2} = f(\phi(t)) + tf'(\phi(t))\phi'(t),$$

or

$$\frac{d}{dt}\left(t^n\frac{d^2\phi(t)}{dt^2}\right) = t^{n-1}f(\phi(t)) + t^nf'(\phi(t))\phi'(t) > 0,$$

whence we have again by the mean value theorem

$$\phi''(t) \geq 0 \quad \text{for} \quad t > 0.$$

Thus, for $t > 0$, $\phi'(t)$ is positive and monotone increasing. Multiplying now both sides of (4.16) by $\phi'(t)$ and integrating with respect to t, we have

$$\int_{\phi(t_1)}^{\phi(t)} f(s)\,ds = (n-1)\int_{t_1}^{t}\frac{(\phi'(s))^2\,ds}{s} + \tfrac{1}{2}(\phi'(t))^2 - \tfrac{1}{2}(\phi'(t_1))^2$$

$$\geq \tfrac{1}{2}(\varphi'(t))^2 + (\varphi'(t_1))^2\left((n-1)\log\frac{t}{t_1} - \frac{1}{2}\right)$$

for $0 < t_1 < t$. In particular, if $t \geq 2t_1$ (assuming $n \geq 2$), we have

$$\frac{\varphi'(t)}{\left(2\int_{\varphi(t_1)}^{\varphi(t)} f(s)\,ds\right)^{\frac{1}{2}}} \geq 1 \qquad (0 < 2t_1 \leq t);$$

integrating the two expressions in this last inequality with respect to t, we obtain

$$t - 2t_1 \leq \int_{\varphi(2t_1)}^{\varphi(t)}\left(2\int_{\varphi(t_1)}^{y} f(s)\,ds\right)^{-\frac{1}{2}} dy \qquad (0 < 2t_1 \leq t).$$

From the assumption (4.14) it follows that the right-hand member of this last inequality is bounded for all values of $\varphi(t)$ (keeping t_1 fixed); this establishes an upper bound for the domain of regularity of $\varphi(t)$. Thus there exists a constant R_a such that as $t \to R_a$ either $\varphi(t)$ or $\varphi'(t)$ diverges to infinity; but from the convexity of $\varphi(t)$ and the equation (4.16) one easily deduces that $\varphi(t)$ becomes infinite if $\varphi'(t)$ does: this proves our assertion.

We now assume that there exists in $\mathbf{X}$ a function u satisfying (4.15). Let $\mathbf{p}_0$ be any point in $\mathbf{X}$ and let a be any number such that

$$a < u(\mathbf{p}_0); \tag{4.18}$$

let $r(\mathbf{p})$ be the geodesic distance from $\mathbf{p}_0$, $r(\mathbf{p}) = d(\mathbf{p}_0, \mathbf{p})$, and consider the solution $\phi(t)$ of (4.16) and (4.17) with initial value a. By the argument just established there exists a finite number $R_a > 0$ such that $\phi(t) \to \infty$ as $t \to R_a$; let $v(\mathbf{p}) = \phi(r(\mathbf{p}))$ for $\mathbf{p} \in \Sigma(\mathbf{p}_0, R_a)$. Since $\mathbf{X}$ is complete, it follows that $\Sigma(\mathbf{p}_0, R_a)$ is compact, and $v(\mathbf{p})$ becomes infinite as $\mathbf{p}$ approaches the boundary. By the corollary of Theorem 3 we have,

$$\Delta v \quad (\text{weakly}) \leq f(v) \tag{4.19}$$

at all points of $\Sigma(\mathbf{p}_0, R_a)$; the point $\mathbf{p}_0$ is included, since $\phi'(0) = 0$, so that v is differentiable at $\mathbf{p}_0$. Consider the function $w = u - v$ in $\Sigma(\mathbf{p}_0, R_a)$. Since $w \to -\infty$ at the boundary, it must achieve its maximum value at an interior point $\mathbf{p}_1$; from (4.18) we have $w(\mathbf{p}_0) > 0$, so that the maximum value $w(\mathbf{p}_1)$ is positive. But from (4.15) and (4.19) we have

$$\Delta w \quad (\text{weakly}) \geq f(u) - f(v)$$

so that in a neighborhood of the point $\mathbf{p}_1$, because of the monotonicity of f we have

$$\Delta w \geq 0,$$

which is impossible by Theorem 1. This contradiction shows the impossibility of solving (4.15) globally in $\mathbf{X}$ under the stated assumptions, as we set out to prove.

References

1. L. P. Eisenhart, *Riemannian Geometry*, Princeton, 1949.
2. E. Hopf, *Elementare Bemerkungen über die Lösungen partieller Differentialgleichungen zweiter Ordnung vom elliptischen Typus*, Sitzungsberichte, Preussische Akademie der Wissenschaften, vol. 19(1927), pp. 147–152.
3. H. Hopf and W. Rinow, *Über den Begriff der vollständigen differentialgeometrischen Fläche*, Commentarii Mathematici Helvetici, vol. 3(1931), pp. 209–225.
4. R. Osserman, *On a differential inequality and conformal mapping of surfaces*, Technical Report no. 3, December 16, 1956, Stanford University. To appear in the Pacific Journal of Mathematics.
5. G. de Rham, *Sur la réductibilité d'un espace de Riemann*, Commentarii Mathematici Helvetici, vol. 26(1952), pp. 328–344.
6. J. A. Schouten and D. J. Struik, *Einführung in die neueren Methoden der Differentialgeometrie*, vol. II, Groningen, Noordhoff, 1938.
7. K. Yano and S. Bochner, *Curvature and Betti numbers*, Annals of Mathematics Studies, no. 32, Princeton, 1953.

University of Minnesota

Errata: An Extension of E. Hopf's Maximum Principle with an Application to Riemannian Geometry

Eugenio Calabi

Duke Math. J. 26, 706 (1959)

The proof of Theorem 4 is incorrect, because the inequality on page 55, line 6 from the bottom, is reversed. To correct the proof, read the last nine lines of page 55 as follows:

$$\int_{\phi(t_1)}^{\phi(t)} f(s)\,ds = (n-1)\int_{t_1}^{t} \frac{(\phi'(s))^2\,ds}{s} + \tfrac{1}{2}(\phi'(t))^2 - \tfrac{1}{2}(\phi'(t_1))^2$$
$$\leq \left(\tfrac{1}{2} + (n-1)\log\frac{t}{t_1}\right)(\phi'(t))^2$$

for $0 < t_1 \leq t$; this can be written in the form

$$\frac{\phi'(t)}{\left(\int_{\phi(t_1)}^{\phi(t)} f(s)\,ds\right)^{\frac{1}{2}}} \geq \left(\tfrac{1}{2} + (n-1)\log\frac{t}{t_1}\right)^{-\frac{1}{2}}.$$

Integrating the last two expressions with respect to t, we obtain

$$\int_{t_1}^{t_2} \left(\tfrac{1}{2} + (n-1)\log\frac{t}{t_1}\right)^{-\frac{1}{2}} dt \leq \int_{\phi(t_1)}^{\phi(t_2)} \left(\int_{\phi(t_1)}^{y} f(s)\,ds\right)^{-\frac{1}{2}} dy.$$

From the assumption (4.14) it follows that the right-hand member of this last inequality is bounded for all values of $\phi(t_2)$ (for t_1 fixed); since the left-hand integral diverges as $t \to \infty$, the inequality establishes an upper bound for the values of t_2 such that the solution $\phi(t)$ is regular up to t_2. Thus there exists $\cdots$ (*continue with the text*).

DOI: 10.1215/S0012-7094-59-02672-9

Sur les variétés complexes compactes localement symétriques

Eugenio Calabi[1] and Edoardo Vesentini

Bull. Soc. Math. France 87, 311–317 (1959)

Cet exposé donne un résumé préliminaire d'un travail en collaboration qui paraîtra prochainement. Son but est d'étudier les déformations des variétés complexes compactes $X = D/\Delta$ qu'on peut obtenir comme quotients d'un domaine borné symétrique D de l'espace $\mathbf{C}^n$ de n variables complexes par un groupe discontinu Δ d'automorphismes. Nous prouverons que, si D est irréductible, de dimension $n > 1$, et s'il appartient à une des quatre grandes classes de la classification de Élie Cartan (*voir*, par exemple, [1], p. 179-180), alors la structure complexe de X est localement stable. Cette conclusion subsiste même si D est un produit $D = D_1 \times D_2 \times \ldots \times D_l$ de l domaines bornés symétriques irréductibles D_i, pourvu que chaque D_i, appartienne à une des quatre grandes classes de la classification de É. Cartan et que : $\dim_{\mathbf{C}} D_i > 1$ pour $i = 1, \ldots, l$.

Pour les détails et les démonstrations nous renvoyons à notre travail commun.

1. Généralités sur les déformations. — Toutes les variétés considérées ici seront supposées paracompactes. Soit M une variété différentiable (de classe C^∞) connexe, et soit $\mathcal{V} \xrightarrow{\varpi} M$ un espace fibré C^∞-différentiable (localement trivial) sur la base M, tel que chaque fibre $V_t = \varpi^{-1}(t)$ $(t \in M)$, soit une variété analytique complexe, de dimension complexe n, connexe, dont la structure complexe soit compatible avec la structure différentiable induite par celle de $\mathcal{V}$.

(1) The research of this author was supported by the United States Air Force, through the Air Force Office of Scientific Researcb of the Air Research and Development Command under Contract n° 49 (638), p. 253.

DOI: 10.24033/bsmf.1527

On dit que la structure complexe de V_t *dépend différentiablement* de $t \in M$ et que l'espace fibré $\mathcal{V} \xrightarrow{\overline{w}} M$ est une famille différentiable de structures complexes si, pour tout point $x \in \mathcal{V}$, il existe un voisinage $\mathcal{U}$ et un homéomorphisme (de classe C^∞) de $\mathcal{U}$ dans $\mathbf{C}^n \times \overline{w}(\mathcal{U})$, de façon telle que la restriction de cet homéomorphisme à $\mathcal{U} \cap V_t$ $(t \in M)$ soit une transformation birégulière (au sens analytique complexe) de $\mathcal{U} \cap V_t$ dans $\mathbf{C}^n \times t$.

La famille $\mathcal{V} \xrightarrow{\overline{w}} M$ est dite *triviale* s'il existe une représentation différentiable (de classe C^∞) : $\mathcal{V} \to V_0 = \overline{w}^{-1}(o) = (o \in M)$ qui transforme birégulièrement chaque fibre $V_t = \overline{w}^{-1}(t) = (t \in M)$, sur la fibre V_0.

Soit $\mathbf{\Theta}$ le faisceau des germes des champs de vecteurs tangents holomorphes sur une variété complexe, compacte X.

Un théorème fondamental de Frölicher et Nijenhuis ([2] et [8], théorème 6.3, p. 365), affirme que, si $H^1(X, \mathbf{\Theta}) = o$, la structure complexe de X est ocalement stable; c'est-à-dire, pour chaque famille différentiable $\mathcal{V} \xrightarrow{\overline{w}} M$ à fibres compactes, contenant X comme fibre $X = V_0 = \overline{w}^{-1}(o) = (o \in M)$, il existe un voisinage U de o dans M, tel que la famille $\overline{w}^{-1}(U)$ soit triviale.

2. Opérateurs différentiels sur une variété complexe. — Soit E une structure fibrée analytique complexe à fibre vectorielle sur une variété complexe compacte X, et soient $\mathbf{A}^{p,q}(E)$ et $A^{p,q}(E)$ respectivement le faisceau des germes des (p, q)-formes différentielles extérieures, (à valeurs complexes et de classe C^∞) à coefficients dans E, et l'espace vectoriel complexe des sections globales de $\mathbf{A}^{p,q}(E)$.

$\mathbf{A}(E) = \Sigma \mathbf{A}^{p,q}(E)$ est un faisceau bigradué sur X. L'opérateur $\bar{\partial}$ de différentiation extérieure, par rapport aux conjuguées des coordonnées locales complexes, définit un opérateur cobord pour le complexe $\Gamma(\mathbf{A}(E)) = \Sigma A^{p,q}(E)$ des sections globales de $\mathbf{A}(E)$. Il est bien connu, d'après Dolbeault (*voir*, par exemple, [4], p. 116), que la composante de type (p, q), $H^{p,q}(X, \mathbf{E})$, de la $\bar{\partial}$-cohomologie de $\Gamma(\mathbf{A}(E))$ est isomorphe au groupe de cohomologie : $H^q(X, \mathbf{E} \otimes \mathbf{T}^{(p)})$, où $\mathbf{E}$ et $\mathbf{T}^{(p)}$ dénotent respectivement le faisceau des germes des sections holomorphes de E et le faisceau des germes des sections holomorphes de la structure $T^{(p)}$ fibrée des p-formes holomorphes sur X.

Soit E^* la structure fibrée duale de E. Lorsqu'on se donne une métrique hermitienne (de classe C^∞) sur les fibres de E, dépendant différentiablement des fibres, on peut définir un isomorphisme de $E \otimes T^{(p)} \otimes \overline{T^{(q)}}$ sur $E^* \otimes T^{(q)} \otimes \overline{T^{(p)}}$, et par conséquent un isomorphisme de $A^{p,q}(E)$ sur $A^{q,p}(E^*)$, qui, à chaque forme φ de type (p, q), à coefficients dans E, associe une forme φ^* de type (q, p), à coefficients dans E^*. Lorsqu'on choisit une métrique hermitienne (de classe C^∞) dans la structure tangente à X, on peut définir l'isomorphisme bien connu :

$$\bigstar : \quad E \otimes T^{(p)} \otimes \overline{T^{(q)}} \to E \otimes T^{(n-q)} \otimes \overline{T^{(n-p)}}$$

et par conséquent un isomorphisme de $A^{p,q}(E)$ sur $A^{n-q,n-p}(E)$, qui, à chaque (p, q)-forme φ à coefficients dans E, associe une forme $\bigstar\varphi$ de type $(n\text{-}q, n\text{-}p)$, à coefficients dans la même structure E.

On définit par

$$(\varphi, \psi) = \int_X \varphi \wedge \bigstar \psi^* \qquad (\bigstar \psi^* = \bigstar(\psi^*) = (\bigstar \psi)^*).$$

Le produit intérieur de deux formes $\varphi, \psi \in A^{p,q}(E)$.

Soient ∂, $\mathfrak{d}$ et $\bar{\mathfrak{d}}$ les opérateurs définis respectivement par les formules

$$(1) \qquad \begin{cases} (\partial\varphi)^* = \bar\partial(\varphi^*), & \mathfrak{d}\varphi = -\bigstar\partial\bigstar\varphi, \\ \bar{\mathfrak{d}}\varphi = -\bigstar\bar\partial\bigstar\varphi & (\varphi \in A^{p,q}(E)). \end{cases}$$

On vérifie que pour chaque $\varphi \in A^{p,q}(E)$, $\psi \in A^{p,q+1}(E)$, $\tau \in A^{p+1,q}(E)$ on a

$$(\bar\partial\varphi, \psi) = (\varphi, \mathfrak{d}\psi), \qquad (\partial\varphi, \tau) = (\varphi, \bar{\mathfrak{d}}\tau).$$

Soit $\{U_i\}$, un recouvrement ouvert fini suffisamment fin de X, et soit $z_i^\alpha = (\alpha = 1, \ldots, n = \dim_{\mathbf{C}} X)$, un système de coordonnées locales complexes dans U_i. Si h_i dénote la représentation locale dans U_i de la métrique hermitienne choisie sur les fibres de E, la forme de type $(1, 0)$.

$$\theta_i = h_i^{-1}\, d' h_i$$

(où d' dénote l'opérateur de différentiation extérieure par rapport aux coordonnées locales complexes z_i^α) définit une forme globale θ à valeurs dans l'algèbre de Lie du groupe de structure de E qui est associé à une connexion dans la structure fibrée principale associée à E. La forme de courbure de cette connexion est donnée par $\bar\partial\,\theta$ ([9]).

Soit :

$$\chi = \sqrt{-1}\,\bar\partial\theta.$$

On démontre que la différenciation covariante D par rapport à la connexion θ est donnée par :

$$D = \partial + \bar\partial$$

Les deux opérateurs :

$$\square' = \bar\partial\mathfrak{d} + \mathfrak{d}\bar\partial, \qquad \square'' = \partial\bar{\mathfrak{d}} + \bar{\mathfrak{d}}\partial$$

sont strictement elliptiques. Puisque X est compacte, les espaces vectoriels complexes $\mathbf{H}'^{p,q}(X, E)$ et $\mathbf{H}''^{p,q}(X, E)$ des formes $\varphi \in A^{p,q}(E)$ telles que :

$$\square'\varphi = 0 \qquad \text{ou, respectivement,} \qquad \square''\varphi = 0$$

ont une dimension finie. On obtient des (1) que :

$$\mathbf{H}''^{p,q}(X, E) \simeq \mathbf{H}'^{q,p}(X, E^*).$$

Kodaira a démontré que :

$$\mathbf{H}'^{p,q}(X, E) \simeq H^q(X, \mathbf{E} \otimes \mathbf{T}^{(p)}).$$

Lemme 1. — *Si la métrique choisie dans (la structure tangente à) X est kählérienne, pour chaque forme $\varphi \in A^{p,q}(E)$, on a*

$$(\varphi, \square'\varphi) - (\varphi, \square''\varphi) = (e(\chi)\Lambda\varphi - \Lambda e(\chi)\varphi, \varphi),$$

où $e(\chi)$ dénote le produit extérieur de χ considérée comme opérateur sur les formes vectorielles extérieures.

On en déduit le théorème suivant :

Théorème 2. — *Si la métrique choisie dans X est kählérienne, pour chaque $\varphi \in \mathbf{H}'^{p,q}(X, E)$, on a :*

$$(\Lambda e(\chi)\varphi - e(\chi)\Lambda\varphi, \varphi) = (\varphi, \square''\varphi) = (\partial\varphi, \partial\varphi) + (\bar{\mathfrak{d}}\varphi, \bar{\mathfrak{d}}\varphi) \geq 0.$$

3. Variétés complexes localement symétriques. — Les domaines bornés symétriques D de l'espace $\mathbf{C}^n$ de n variables complexes ont été classifiés complètement par É. Cartan. Selon la numérotation de C. L. Siegel ([**10**] et [**1**], p. 178-179) on a les types suivants :

Type $\mathrm{I}_{r,s}$. — D est l'ensemble des matrices Z complexes de type $r \times s$ telles que

$$I_s - {}^tZ.\bar{Z} > 0, \tag{2}$$

où I_s est la matrice unité de $GL(s, \mathbf{C})$. On a : $n = \dim_{\mathbf{C}} D = rs$.

Type II_r. — D est l'ensemble des $r \times r$ matrices complexes *antisymétriques* Z satisfaisant la (2) avec $s = r$. On a $n = \dim_{\mathbf{C}} D = r(r-1)/2$.

Type III_r. — D est l'ensemble des $r \times r$ matrices complexes *symétriques* Z à partie imaginaire définie positive. On a $n = \dim_{\mathbf{C}} D = r(r+1)/2$.

Type IV_n. — D est l'ensemble des $n \times 1$ matrices Z complexes telles que :

$$2({}^tZ.\bar{Z}) < 1 + |{}^tZ.Z|^2 < 2.$$

Pour $n = 2$, D est le produit de deux demi-plans de Poincaré.

Outre ces quatre grandes classes de domaines bornés symétriques, il y a deux types exceptionnels, de dimensions complexes respectives $n = 16$, et $n = 27$, dont les définitions sont plus compliquées que les précédentes et que nous n'indiquerons pas ici.

Les types $I_{1,1}$, II_2, III_1 et IV_1 sont équivalents au demi-plan de Poincaré. On a aussi les équivalences suivantes : $I_{1,3}$ et II_3; $I_{2,2}$ et IV_4; III_2 et IV_3; IV_6 et II_4; $I_{r,s}$ et $I_{s,r}$.

D'après un résultat de H. CARTAN, le groupe $\mathfrak{A}(D)$ des automorphismes d'un domaine borné D, muni de la topologie de la convergence compacte, est un groupe de Lie. Les groupes discontinus d'automorphismes de D sont exactement les sous-groupes discrets de $\mathfrak{A}(D)$.

Soit D un domaine borné symétrique *irréductible* appartenant à un des types I-IV et soit Δ un sous-groupe discontinu d'automorphismes de D, tel que :

1° le quotient $X = D/\Delta$ soit compact;

2° l'identité soit le seul élément de Δ ayant des points unis dans D.

D'après un résultat de Kodaira [7], $X = D/\Delta$ est une variété projective non singulière

A l'aide du théorème 2 (ou d'une inégalité de S. NAKANO [9]) on peut démontrer la proposition suivante, qui est le résultat central de cet exposé :

THÉORÈME 3. — *Soit* Θ *le faisceau des germes des champs de vecteurs holomorphes tangents à* $X = D/\Delta$. *On a :*

$$H^q(X, \Theta) = 0,$$

lorsque D *est irréductible et*

1° $q < r + s - 1$ *pour chaque domaine* D *de type* $I_{r,s}$;
2° $q < 2r - 3$ » » D » II_r $(r \geq 2)$;
3° $q < r$ » » D » III_r;
4° $q < n - 1$ » » D » IV_n $(n \geq 3)$.

En particulier sous les hypothèses précédentes, si $n > 1$,

$$H^1(X, \Theta) = 0. \tag{3}$$

Il suit de là, par le théorème de Frölicher-Nijenhuis, le

THÉORÈME 4. — *Si* $n > 1$, *et si* D *est irréductible, la structure complexe de* $X = D/\Delta$ *est localement stable.*

En vertu du théorème 3 et à l'aide d'une formule de type Künneth (*voir*, par exemple, [3], p. 244), on peut démontrer que :

$$H^1(X, \Theta) = 0 \qquad [\text{et } H^0(X, \Theta) = 0]$$

même si D est un produit $D = D_1 \times D_2 \times \ldots \times D_l$ de domaines bornés symétriques *irréductibles* $D_i (i = 1, \ldots, l)$ de type I-IV et de dimensions complexes > 1. Il découle que *le théorème 4 subsiste même pour des domaines bornés symétriques réductibles satisfaisants aux conditions ci-dessus.*

4. Un cas particulier. — Le théorème de Riemann-Roch ([4], p. 154) appliqué à la structure fibrée des vecteurs complexes tangents à $X = D/\Delta$ permet de calculer la somme alternée des dimensions des groupes de cohomologie $H^q(X, \boldsymbol{\Theta})$ non compris dans l'énoncé du théorème 3, mais naturellement il ne peut pas donner en général aucun renseignement sur les dimensions de tous ces groupes.

Il y a seulement un cas où le théorème 3 et le théorème de Riemann-Roch permettent d'achever le calcul des dimensions des groupes $H^q(X, \boldsymbol{\Theta})$. C'est le cas d'un domaine borné symétrique irréductible D de type $\mathrm{I}_{1,s}$; c'est-à-dire le cas où D est la boule ouverte de rayon un de $\mathbf{C}^n$ ($n = \dim_{\mathbf{C}} D = s$). Dans ce cas, en vertu du théorème 3, le seul groupe de cohomologie $H^q(X, \boldsymbol{\Theta})$ qui n'est pas nécessairement zéro est le groupe $H^n(X, \boldsymbol{\Theta})$. Par un résultat de Igusa ([6], théorème 8, p. 676 et [5]) on obtient, à partir du théorème de Riemann-Roch que la dimension de $H^n(X, \boldsymbol{\Theta})$ est complètement déterminée par la mesure invariante $v(X)$ de X dans son revêtement universel D. En effet, la caractéristique

$$\chi(X, \boldsymbol{\Theta}) = \sum_{q=0}^{n} (-1)^q \dim H^q(X, \boldsymbol{\Theta})$$

est égale à

$$\chi(X, \boldsymbol{\Theta}) = (-1)^n \frac{n(n+2)n!}{\pi^n} v(X)$$

ou même, en vertu d'une formule de Igusa ([6], p. 676), à

$$\chi(X, \boldsymbol{\Theta}) = \frac{n(n+2)}{n+1} E(X),$$

où $E(X)$ dénote la caractéristique d'Euler-Poincaré de X.

Par conséquent, en vertu du théorème 3,

$$(4) \qquad \dim H^n(X, \boldsymbol{\Theta}) = \frac{n(n+2)n!}{n} v(X) = (-1)^n \frac{n(n+2)}{n+1} E(X).$$

Puisqu'on a l'isomorphisme canonique ([8], (11.9), p. 386)

$$H^q(X, \boldsymbol{\Theta}) \simeq H^{n-q}(X, \mathbf{T}^{(1)} \otimes \mathbf{K},$$

où K dénote la structure linéaire canonique sur X, on obtient de (4) que la dimension de $H^0(X, \mathbf{T}^{(1)} \otimes \mathbf{K})$, c'est-à-dire le « nombre » des formes linéaires holomorphes à coefficients dans K, est donné par :

$$\dim H^0(X, \mathbf{T}^{(1)} \otimes \mathbf{K}) = \frac{n(n+2)n!}{\pi^n} v(X) = (-1)^n \frac{n(n+2)}{n+1} E(X).$$

BIBLIOGRAPHIE.

[1] Borel (A.). — *Les fonctions automorphes de plusieurs variables complexes Bull. Soc. math. France*, t. 80, 1952, p. 167-182).

[2] Frölicher (A.) and Nijenhuis (A.). — *A theorem of stability of complex structures* (*Proc. nat. Acad. Sc. U. S. A.*, *t.* 43, 1957, p. 239-241.)

[3] Godement (R.). — *Topologie algébrique et théorie des faisceaux.* — Paris, Hermann, 1958, (*Act. scient. et ind.*, 1252).

[4] Hirzebruch (F.) — *Neue topologische Methoden in der algebraischen Geometrie.* — Berlin, Springer-Verlag, 1956 (*Ergebnisse der Mathematik*, neue Folge, 9).

[5] Hirzebruch (F.). — *Automorphe Formen und der Satz von Riemann-Roch* (*Symposium on algebraic topology*, 1956, Mexico, p. 129-144). — Mexico, Universidad nacional autonoma, 1958.

[6] Igusa (Jun-Ichi). — *On the structure of a special class of Kähler varieties* (*Amer*, *J. Math.*, t, 76, 1954, p. 669-678).

[7] Kodaira (K.). — *On Kähler varieties of restricted type* (*Annals Math.*, t. 60, 1954, p. 28 48).

[8] Kodaira (K.) and Spencer (D. C.) — *On deformations of complex analytic structures* (*Annals Math.*, t. 67, 1958, p. 328-466).

[9] Nakano (S.). — *On complex analytic vector bundles* (*J. Math. Soc. Japan*, t. 7, 1955, p. 1-12).

[10] Siegel (C. L.). — *Analytic functions of several complex variables.* — Princeton, Institute for advanced Study, 1949 (Notes miméogr.).

[11] Yano (K.) and Bochner (S.). — *Curvature and Betti numbers.* — Princeton, Princeton University Press, 1953 (*Annals Math. Studies*, 32).

On Compact, Locally Symmetric Kähler Manifolds

Eugenio Calabi and Edoardo Vesentini*

Ann. of Math. (2) 71, 472–507 (1960)

Contents

Introduction and summary of results

The purpose of this paper is to study certain aspects of the structure of the compact, complex manifolds X that can be uniformized by bounded, symmetric, homogeneous subdomains of the space $\mathbf{C}^n$ of n complex variables (Cartan domains). We are chiefly concerned in gathering information on the cohomology groups of these manifolds with coefficients in the sheaf Θ of germs of holomorphic tangent vector fields, particularly because of the important role that the first cohomology group $H^1(X, \Theta)$ plays in the theory of deformation of complex analytic structures [15, 23].

* The first named author was supported during a portion of preparation of this paper by the U. S. Air Force Office of Scientific Research. Both authors were supported during the final phases of the preparation of the paper by the National Science Foundation.

DOI: 10.2307/1969939

Chapter 2 includes a compendium of all the general differential geometric methods applied in the sequel. After partially reviewing the relationship of the group $H^1(X, \Theta)$ to the deformation of complex analytic structure in a compact manifold X, we apply the Spencer-Kodaira representation [23] of the cohomology groups $H^q(X, \Theta)$ by certain global tensor fields in X, satisfying a strongly elliptic system of partial differential equations (harmonic fields). We derive then a number of inequalities, including one due to S. Nakano [24], which furnish criteria for the nonexistence of nontrivial harmonic tensor fields, to show that certain cohomology groups $H^q(X, \Theta)$ are zero.

Nakano's inequality, as well as the other similar ones, are all special cases of S. Bochner's general criterion [5], according to which a tensor of a specified type cannot satisfy a given "harmonic" equation globally on a compact manifold, unless it is identically zero, if the curvature tensor satisfies suitable inequalities at every point. Bochner's criterion finds its most natural application in the case of compact, locally homogeneous manifolds, because for these spaces the uniform estimates of the curvature that are needed may be calculated from the structure of the space at just one of its points.

In this paper we consider the case of compact Kähler manifolds that are uniformizable by Cartan domains, because for these spaces there are especially simple methods to calculate the curvature tensor. One method is described in detail in Chapter 3, § 2; the curvature is evaluated there by means of the classification theorem (reviewed in Chapter 3, § 1) and by the knowledge of an invariant metric (Bergmann) for the irreducible Cartan domains (this calculation has been omitted in the case of the two exceptional domains). Another method described by A. Borel [10] in an article following the present one, in this issue of *Annals of Mathematics*, calculates the necessary invariants of the curvature directly from the structure theory of the simple Lie groups. After these sections come the proofs of the main results. At the end of Chapter 3 we prove the first three of the theorems stated below.

THEOREM 1. *For each irreducible Cartan domain D there exists an integer $\gamma(D)$ satisfying $\sqrt{2n} < \gamma(D) \leqq n+1$ ($n = \dim_{\mathbf{C}} D$), depending only on the curvature of the invariant metric of D, with the following property. For any compact locally symmetric, complex manifold X, uniformizable by* (i. e., *whose universal covering space is*) *D, the cohomology groups $H^q(X, \Theta)$ vanish for all q satisfying $0 \leqq q < \gamma(D)-1$. In particular, the only case where $\gamma(D) = 2$ is when* $\dim_{\mathbf{C}} D = 1$, *that is to say, when D is equivalent to the unit disc.*

The values of $\gamma(D)$ for each irreducible Cartan domain D can be found in Table 1, Chapter 3, § 3. Noting that, as a result of Theorem 1, if an irreducible Cartan domain D has at least two complex dimensions, then $H^1(X, \Theta) = 0$ for every compact manifold X uniformized by D, we obtain the following consequence, using a criterion of Frölicher and Nijenhuis [15, 23].

THEOREM 2. *The complex analytic structure of any compact, complex manifold X, that is uniformized by an irreducible Cartan domain D of at least two complex dimensions, is locally rigid, even if the complex structure of the universal covering space is allowed to vary. In other words, the number of moduli of a complex manifold X with the stated properties is zero.*

The generalization of Theorems 1 and 2 to the case where the uniformizing domain D is a reducible Cartan domain is immediate: the corresponding statements are given in Chapter 3, § 3, as corollaries to the theorems stated above.

The local Riemannian geometry of the Cartan domains is closely related, in a formal sense, to that of the compact, symmetric Kähler manifolds. In fact there is a one-to-one correspondence between the equivalence classes (under complex analytic isomorphism) of the former and of the latter, such that the corresponding models for the two types (one open, one compact) of spaces have the same dimensionality and holonomy group, and their curvature tensors differ only in sign from one another. Using techniques similar to those used in proving Theorem 1, we find a new proof of the following result, which is a particular case of a theorem by R. Bott [11].

THEOREM 3 (Bott). *Let Y be an irreducible, compact, symmetric Kähler manifold. Then the cohomology groups $H^q(Y, \Theta)$ vanish for all $q \geqq 1$, while, for $q = 0$, $\dim_{\mathbf{C}} H^0(Y, \Theta) = \dim_{\mathbf{C}} G_{\mathbf{C}}(Y) = \dim_{\mathbf{R}} G(Y)$, where $G_{\mathbf{C}}(Y)$ and $G(Y)$ denote, respectively, the complex Lie group of all holomorphic transformations, and the compact Lie group of all isometric transformations of Y onto itself.*

The generalization of Theorem 3 to reducible manifolds is immediate, but omitted from the present article.

In Chapter 4 we apply Theorem 3 to supplement Theorem 1 by showing that, for the compact, complex manifolds X uniformized by irreducible Cartan domains D, the cohomology groups $H^q(X, \Theta)$, which vanish for $q < \gamma(D) - 1$, cannot vanish for all q with $\gamma(D) - 1 \leqq q \leqq n = \dim_{\mathbf{C}} X$. In fact, it follows from the Riemann-Roch theorem and from Hirzebruch's proportionality principle [18, 19, 20] that the following relation is valid.

THEOREM 4. *Let X be an n-dimensional, compact, complex manifold, whose universal covering space is a Cartan domain D. Then the Θ-characteristic number*

$$\chi(X, \Theta) = \sum_{q=0}^{n}(-1)^q \dim_{\mathbf{C}} H^q(X, \Theta) ,$$

the arithmetic genus (Todd genus)

$$\chi(X, \Omega^0) = \sum_{q=0}^{n}(-1)^q \dim_{\mathbf{C}} H^q(X, \Omega^0) = \sum_{q=0}^{n}(-1)^q \dim_{\mathbf{C}} H^{0,q}(X, \mathbf{C})$$

(where Ω^0 is the sheaf of germs of holomorphic functions), and the Euler-Poincaré characteristic

$$\chi(X) = \chi(X, \mathbf{R}) = \sum_{q=0}^{2n}(-1)^q \dim_{\mathbf{R}} H^q(X, \mathbf{R})$$

are mutually related by the following proportionality formula

$$\chi(X, \Theta) = a_1(D)\chi(X, \Omega^0) = a_2(D)\chi(X) = R^n(X)a_3(D)v(X) , \tag{1.1}$$

where $R(X)$ is the value of the (constant) scalar curvature of X, and $v(X)$ is the volume of X. Moreover, $a_1(D)$ is a positive integer, $a_2(D)$ is a positive rational number, $a_3(D)$ a positive real number and $R(X) < 0$. The same relation, in addition, is valid, with the same proportionality constants $a_j(D)$, if one replaces X with the compact, symmetric space Y corresponding to D, with positive scalar curvature $R(Y)$.

The above theorem, together with Theorems 1 and 3, enable us to calculate the dimensions of $H^q(X, \Theta)$ for all values of q, only when $\gamma(D) = n+1$ ($n = \dim_{\mathbf{C}} D$), that is to say, when $H^n(X, \Theta)$ is the only non-vanishing group. From the calculations in Chapter 3, one sees that this is the case only when D is isomorphic to the unit ball in $\mathbf{C}^n$, $D = \{z \mid z = (z^1, \cdots, z^n) \in \mathbf{C}^n, \sum_{\alpha=1}^{n} |z^\alpha|^2 < 1\}$. In one other series of cases, however, namely when D is of type IV (see Chapter 3, § 1), we have $\gamma(D) = n$; in this case we have only a positive, lower bound for $\dim_{\mathbf{C}} H^n(X, \Theta)$. The dimensionality of $H^n(X, \Theta)$ has a certain interest because, by the Serre duality theorem [18, p. 119], this group is isomorphic to the group of global holomorphic sections of the sheaf $\Omega^1 \otimes \Omega^n$, where Ω^q represents the sheaf of germs of holomorphic q-forms. The precise statement is given as a Corollary to Theorem 4, in Chapter 4, § 3.

The rigidity of the complex analytic structure for compact complex manifolds X uniformizable by certain types of Cartan domains D (Theorem 2 and Corollary) can be coupled with some unpublished results by A. Selberg to obtain the following proposition.

THEOREM. *Let D be a Cartan domain, reducible or not, whose irreducible factors have at least two complex dimensions; let G be the group of all holomorphic (isometric) transformations of D, represented as an algebraic group of real matrices, and let Γ be a discrete subgroup of G acting on D with compact quotient space. Then there is an inner automorphism of G, such that the components of all the matrices representing the transform of Γ generate a simple algebraic extension of the rational number field.*

The details of the arguments leading to the above result are to appear in a paper by A. Selberg in the forthcoming proceedings of the Bombay conference, 1960, on Number Theory.

In concluding this summary, the authors express their thanks to A. Borel, who checked the results in Chapter 3 on the curvature of the irreducible Cartan domains (other than the two exceptional ones), and supplied his own results [10] for the exceptional domains.

CHAPTER 2. DIFFERENTIAL GEOMETRIC PRELIMINARIES

1. Differentiable families of complex analytic manifolds

We begin by recalling several notions used in the theory of deformations of complex analytic structures, as treated by Kodaira and Spencer [23]. We shall always assume all manifolds to be paracompact and differentiable; the term "differentiable" shall be understood to mean of class C^∞.

Let M be a connected manifold and $\mathcal{V}$ a (differentiable) fibre bundle over M with projection π: $\mathcal{V} \to M$, with structure group equal to the group of all differentiable transformations.

DEFINITION 1. A differentiable fibre bundle $\mathcal{V} \xrightarrow{\pi} M$ is called a *differentiable family of n-dimensional complex manifolds* (or of complex analytic structures) if the following two conditions are satisfied:

(a) each fibre $V_t = \pi^{-1}(t) \subset \mathcal{V}$ $(t \in M)$ is a complex analytic manifold with n complex dimensions, and whose complex analytic structure is compatible with the differentiable structure induced on V_t from the differentiable structure of $\mathcal{V}$;

(b) for each point $\mathfrak{p} \in \mathcal{V}$ there exists a neighborhood $\mathcal{U}$ of $\mathfrak{p}$ and a differentiable isomorphism h of $\mathcal{U}$ onto an open subdomain of $\mathbf{C}^n \times \pi(\mathcal{U})$, whose restriction h_t to $\mathcal{U} \cap V_t$ for each $t \in \pi(\mathcal{U})$ is a biregular (i. e., biholomorphic) map of $\mathcal{U} \cap V_t$ into $\mathbf{C}^n \times \{t\}$. With reference to some given "base" point $o \in M$, any one of the complex manifolds $V_t = \pi^{-1}(t)$ is said to be a *deformation* of the complex manifold $V_o = \pi^{-1}(o)$.

DEFINITION 2. A differentiable family $\mathcal{V} \xrightarrow{\pi} M$ of n-dimensional complex manifolds is called *trivial*, if for some point $o \in M$ there exists a differentiable map $\mathcal{V} \to V_0 = \pi^{-1}(o)$ which maps each fibre $V_t = \pi^{-1}(t)$ $(t \in M)$ biholomorphically onto V_0. The family $\mathcal{V} \xrightarrow{\pi} M$ is called *locally trivial* at $o \in M$, if there exists a neighborhood U of o in M, such that the family $\pi^{-1}(U) \xrightarrow{\pi} U$ is trivial.

Let X be a compact, complex manifold, and let Θ denote the sheaf of germs of holomorphic tangent vector fields in X. Then the cohomology groups of X with coefficients in Θ, $H^q(X, \Theta)$ $(q=0, 1, \cdots, n;\ n = \dim_{\mathbf{C}} X)$ are all finite dimensional vector spaces over the complex number field $\mathbf{C}$. In particular, $H^0(X, \Theta)$ corresponds to the space of holomorphic tangent vector fields, globally defined in X, and can hence be identified with the additive structure of the complex Lie algebra of the group of all biholomorphic maps of X onto itself. Thus $\dim_{\mathbf{C}} H^0(X, \Theta)$ gives the dimension number of the Lie group of all complex analytic transformations [2] of X. The interpretation of $H^1(X, \Theta)$ is tied in with the deformations of the complex structure of X [15, 23]; all we need of it here, however, is its original application in the following form.

PROPOSITION 1. (Frölicher, Nijenhuis [15, 23]). *If X is a compact, complex manifold and $H^1(X, \Theta) = 0$, then the complex structure of X is locally rigid; that is to say, every family of complex manifolds $\mathcal{V} \xrightarrow{\pi} M$ with a fibre $V_0 = \pi^{-1}(o)$ $(o \in M)$ analytically isomorphic to X and with compact fibres is locally trivial at o.*

While there is some interpretation of the role of $H^2(X, \Theta)$ as a group of obstruction classes [23], the full geometric meaning of that group and of the higher dimensional ones is still unknown, except for the $H^n(X, \Theta)$, for which the Serre duality theorem [18] furnishes an interpretation that we shall use in Chapter 4.

2. Forms with coefficients in a complex vector bundle

Let X be an n-dimensional *compact* complex manifold, and let E be a complex analytic vector bundle over X, with fibre $\mathbf{C}^m$ and structure group $\mathbf{GL}(m, \mathbf{C})$. E can be presented in terms of a sufficiently fine, open covering $\{U_j\}$ of X with a system of transition functions w_{jk}, where each w_{jk} is a holomorphic map

$$w_{jk}\colon U_j \cap U_k \to \mathbf{GL}(m, \mathbf{C})$$

satisfying the condition

$$w_{ik} = w_{ij} w_{jk} \quad \text{in} \quad U_i \cap U_j \cap U_k \,.$$

We shall denote by $\mathbf{E}$ the sheaf of germs of holomorphic cross sections of E.

Let $\mathbf{A} = \sum_{p,q} \mathbf{A}^{p,q}$ denote the bigraded sheaf of germs of differentiable exterior forms of X with complex coefficients, where each $\mathbf{A}^{p,q}$ corresponds to the pure forms of type (p, q), and let $\mathbf{\Omega} = \sum_p \mathbf{\Omega}^p$ denote the graded sheaf of germs of holomorphic exterior forms, where $\mathbf{\Omega}^p$ corresponds to the forms of degree p. If $\bar\partial$ stands as usual for the exterior differential operator with respect to the conjugates of the local complex coordinates, then $\mathbf{\Omega}^p$ can be identified with the subsheaf of $\mathbf{A}^{p,0}$ consisting of the forms of type $(p, 0)$ that are annihilated by $\bar\partial$. We set the following abbreviated notations:

$$\mathbf{A}^{p,q}(E) = \mathbf{E} \otimes \mathbf{A}^{p,q}; \quad \mathbf{A}(E) = \mathbf{E} \otimes \mathbf{A} = \sum_{p,q} \mathbf{A}^{p,q}(E);$$
$$\mathbf{\Omega}^p(E) = \mathbf{E} \otimes \mathbf{\Omega}^p \ ; \quad \mathbf{\Omega}(E) = \mathbf{E} \otimes \mathbf{\Omega} = \sum_p \mathbf{\Omega}^p(E) \, ,$$

where the tensor products are over the sheaf of rings $\mathbf{\Omega}^0$.

We shall denote by $A^{p,q}(E)$ the complex vector space of global, differentiable vector forms of type (p, q), with coefficients in E, also called E-vectorial (p, q)-forms, i.e., $A^{p,q}(E) = H^0(X, \mathbf{A}^{p,q}(E))$, and set $A(E) = \sum_{p,q=0}^{n} A^{p,q}(E)$. In terms of the presentation of E by the covering $\{U_j\}$ and by the system $\{w_{jk}\}$ of transition functions, an E-vectorial (p, q)-form φ is defined by a system $\{\varphi_j\}$, also denoted by φ, of column vectors $\varphi_j = {}^t(\varphi_j^1, \cdots, \varphi_j^m)$, whose components are differentiable (p, q)-forms in U_j, such that in $U_j \cap U_k$ the vectors φ_j and φ_k are related by the linear transformation

$$\varphi_j = w_{jk}\varphi_k \, . \tag{2.1}$$

It follows from the fact that each w_{jk} is holomorphic and from (2.1) that $\bar\partial$ is a coboundary operator of degree (0,1) on the bigraded complex $A(E)$. Let $Z^{p,q}(E)$ denote the subspace of $A^{p,q}(E)$ consisting of the global E-vectorial forms that are annihilated by $\bar\partial$; then the cohomology groups of type (p, q) defined by $\bar\partial$ on the complex $A(E)$ are

$$H^{p,q}(X,E) = H_{\bar\partial}^{p,q}(X, E) = Z^{p,q}(E)/\bar\partial A^{p,q-1}(E) \, .$$

According to the theorem of Dolbeault [18, p. 116] we have the following isomorphism

$$H_{\bar\partial}^{p,q}(X, E) \simeq H^q(X, \mathbf{\Omega}^p(E)) \, ,$$

where $H^q(X, \mathbf{\Omega}^p(E))$ denotes the q-dimensional cohomology group of X with coefficients in the sheaf $\mathbf{\Omega}^p(E)$.

3. The dual of a complex analytic vector bundle

We introduce now the dual bundle E^* of E; this bundle can be presented by means of the same covering $\{U_j\}$ of X, used in defining E, but replacing the transition functions w_{jk} by the new systems $\{w^*_{jk}\}$ of holomorphic transition functions, where

$$w^*_{jk} = {}^t w_{jk}^{-1} \colon U_j \cap U_k \to \mathbf{GL}(m, \mathbf{C}) . \tag{2.2}$$

Then there is a pairing of the bigraded sheaves $\mathbf{A}(E)$ and $\mathbf{A}(E^*)$ into $\mathbf{A}$ defined as follows: if $\varphi_{\mathfrak{p}}$ and $\psi_{\mathfrak{p}}$ are germs respectively of an E-vectorial (p, q)-form and of an E^*-vectorial (p', q')-form at the same point $\mathfrak{p} \in X$, then the form ${}^t\varphi_{\mathfrak{p}} \wedge \psi_{\mathfrak{p}}$ is a germ of an ordinary $(p + p', q + q')$-form at $\mathfrak{p}$ with complex coefficients. This definition is consistent, since one sees immediately from (2.1) and (2.2) that the value of ${}^t\varphi_{\mathfrak{p}} \wedge \psi_{\mathfrak{p}}$ is independent of the local bundle coordinate system in which it is computed. The resulting product satisfies the relation

$$\bar\partial({}^t\varphi_{\mathfrak{p}} \wedge \psi_{\mathfrak{p}}) = {}^t(\bar\partial\varphi_{\mathfrak{p}}) \wedge \psi_{\mathfrak{p}} + (-1)^{p+q}\, {}^t\varphi_{\mathfrak{p}} \wedge \bar\partial\psi_{\mathfrak{p}}; \tag{2.3}$$

from this identity it follows that, if ψ is any E^*-vectorial *holomorphic* $(p, 0)$-form globally defined on X, then for any form $\varphi \in A^{n-p,n-1}(E)$, the identity

$$\int_X {}^t\bar\partial\varphi \wedge \psi = 0$$

is satisfied.

4. Hermitian metric in the fibres of a complex vector bundle

Suppose now that there is a positive definite Hermitian metric in each fibre of E, and that its coefficients are differentiable in X. Such a metric can be described in terms of $\{U_j\}$, $\{w_{jk}\}$ by a system $\{a_j\}$ of positive definite, Hermitian, $m \times m$ matrix functions a_j, defined and differentiable in each U_j, and such that

$$a_j = {}^t\overline{w}_{kj} a_k w_{kj} \text{ in } U_j \cap U_k .$$

Such a metric defines a conjugate linear isomorphism of each fibre of E onto the corresponding fibre in E^*; we denote this isomorphism by $\#$; this isomorphism has a natural extension to the bigraded complex $A(E)$, mapping $A^{p,q}(E)$ isomorphically onto $A^{q,p}(E^*)$, as follows: if $\varphi \in A^{p,q}(E)$ and is expressed by the system $\{\varphi_j\}$ of column vectors φ_j in U_j, then $\varphi\# = \{\varphi_j\#\}$, where

$$\varphi_j\# = \bar a_j \overline{\varphi_j} .$$

With this device one can define on $A(E)$ a new differential operator ∂ of degree (1, 0)

$$\partial\colon A^{p,q}(E) \to A^{p+1,q}(E)$$

by the formula

$$(\partial\varphi)\# = \bar{\partial}(\varphi\#) \,. \tag{2.4}$$

In the case of ordinary forms with complex coefficients, the conjugate linear map $\varphi \to \varphi\#$ is the ordinary conjugation and ∂ becomes the exterior differentiation with respect to the local holomorphic coordinates (and not their conjugates); however, in order to avoid confusion, we denote ∂ in this case by the symbol d', already used by other authors. The system of (1, 0) forms l_j in U_j given by

$$l_j = a_j^{-1} d' a_j \,,$$

with values in the Lie algebra of the group $\mathrm{U}(m)$, defines a connection l in the principal bundle associated to E: this is the *metrical connection* derived from the metric $\{a_j\}$. It is easy to see that if $\varphi = \{\varphi_j\} \in A(E)$, then $\partial\varphi$ is also defined by $(\partial\varphi)_j$, where

$$(\partial\varphi)_j = d'\varphi_j + l_j \wedge \varphi_j \,. \tag{2.5}$$

The curvature form of the connection l is expressed by the local system of (1,1)-forms $s = \{s_j\}$ given by

$$s_j = \bar{\partial} l_j \,.$$

One easily derives from (2.5) that the differential operator

$$\partial\bar{\partial} + \bar{\partial}\partial\colon A^{p,q}(E) \to A^{p+1,q+1}(E)$$

has the expression for $\varphi \in A^{p,q}(E)$

$$(\partial\bar{\partial} + \bar{\partial}\partial)\varphi = \{s_j \wedge \varphi_j\} = e(s)\varphi \,, \tag{2.6}$$

where $e(s)$ denotes exterior multiplication of the matrix s of local (1,1)-forms with the column vector φ (cf., also [24, 26]).

5. Hermitian metric in the base space

We next assume that X is equipped with a positive definite Hermitian metric on its tangent bundle. As in the theory of harmonic integrals, this metric establishes a complex linear isomorphism between the ordinary (complex valued) (p, q)-forms and the $(n-q, n-p)$-forms, denoted by $\varphi \to *\varphi$; the extension of this isomorphism to one of $A^{p,q}(E)$ onto $A^{n-q,n-p}(E)$ is immediate. We retain the same conventions as to signs as used mainly by de Rham, Kodaira and several other authors: thus the operation $\varphi \to *\varphi$ satisfies the relation $**\varphi = (-1)^{p+q}\varphi$ for $\varphi \in A^{p,q}(E)$.

If we use now both Hermitian metrics, the one along the fibres of E and the one on the tangent bundle, both maps $\varphi \to \varphi\#$ and $\varphi \to *\varphi$ are defined for $\varphi \in A(E)$, and are commutatively composable: thus the map $\varphi \to (*\varphi)\# = *(\varphi\#)$ can be written as $\varphi \to *\varphi\#$; it has been used by Kodaira and Spencer [26] and by Hirzebruch [18, p. 117] with a different notation.

This operation defines a conjugate linear isomorphism of $A^{p,q}(E)$ onto $A^{n-p,n-q}(E^*)$. This double "duality" operation allows us to define a complex valued, Hermitian inner product of two forms $\varphi, \psi \in A^{p,q}(E)$ by

$$(\varphi, \psi) = \int_X {}^t\varphi \wedge *\psi\#; \tag{2.7}$$

the sign conventions are such that, if $\varphi \neq 0$, then (φ, φ) is strictly positive. It is also easy to see that $(\varphi, \psi) = (*\varphi, *\psi) = (\psi\#, \varphi\#) = (*\psi\#, *\varphi\#)$.

We now introduce two more differential operators, denoted by $\mathfrak{d}$ and $\bar{\mathfrak{d}}$ of degrees $(0, -1)$ and $(-1, 0)$ respectively on the complex $A(E)$ by

$$\mathfrak{d} = - *\partial*, \quad \bar{\mathfrak{d}} = - *\bar{\partial}* , \tag{2.8}$$

where one can verify without difficulty that

$$\bar{\mathfrak{d}}(\varphi\#) = (\mathfrak{d}\varphi)\# .$$

The two new operators $\mathfrak{d}$, $\bar{\mathfrak{d}}$ are adjoint to $\bar{\partial}$, ∂ respectively, with respect to the inner product (2.7); in fact we have, for any choice of $\varphi \in A^{p,q}(E)$, $\psi \in A^{p,q+1}(E)$, $\chi \in A^{p+1,q}(E)$, the identities

$$(\bar{\partial}\varphi, \psi) = (\varphi, \mathfrak{d}\psi); \quad (\partial\varphi, \chi) = (\varphi, \bar{\mathfrak{d}}\chi) . \tag{2.9}$$

Similarly we have the dual formula to (2.6) given by

$$(\mathfrak{d}\bar{\mathfrak{d}} + \bar{\mathfrak{d}}\mathfrak{d})\varphi = -i(s)\varphi \tag{2.10}$$

where $i(s)\varphi = (-1)^{p+q} * e(s)(*\varphi)\,(\varphi \in A^{p,q}(E))$ is the adjoint of $e(s)$. The differential operators ∂, $\mathfrak{d}$, $\bar{\mathfrak{d}}$ as well as $\bar{\partial}$ are coboundary operators on the bigraded complex $A(E)$; from (2.4) and (2.8) we obtain the corresponding interpretations of their cohomology groups of type (p, q):

$$\begin{aligned} &H^{p,q}_{\partial}(X, E) \simeq H^{q,p}_{\bar{\partial}}(X, E^*) \simeq H^p(X, \mathbf{\Omega}^q(E^*)) \\ &H^{p,q}_{\mathfrak{d}}(X, E) \simeq H^{n-p,n-q}_{\bar{\partial}}(X, E^*) \simeq H^{n-q}(X, \mathbf{\Omega}^{n-p}(E^*)) \\ &H^{p,q}_{\bar{\mathfrak{d}}}(X, E) \simeq H^{n-q,n-p}_{\bar{\partial}}(X, E) \simeq H^{n-p}(X, \mathbf{\Omega}^{n-q}(E)) . \end{aligned} \tag{2.11}$$

We consider now the two self-adjoint, second order operators $\square' = \bar{\partial}\mathfrak{d} + \mathfrak{d}\bar{\partial}$ and $\square'' = \partial\bar{\mathfrak{d}} + \bar{\mathfrak{d}}\partial$, and denote by $\mathbf{H}'^{p,q}(X, E)$ and $\mathbf{H}''^{p,q}(X, E)$ the subspaces of $A^{p,q}(E)$ consisting of the forms $\varphi \in A^{p,q}(E)$ satisfying respectively $\square'\varphi = 0$ and $\square''\varphi = 0$. Since both $\square'$ and $\square''$ are strongly elliptic operators, both "harmonic" spaces $\mathbf{H}'^{p,q}(X, E)$ and $\mathbf{H}''^{p,q}(X, E)$ are finite dimensional. In view of (2.9), these harmonic spaces can be also defined respectively by

$$\bar{\partial}\varphi = \mathfrak{d}\varphi = 0 \tag{2.12'}$$

and by

$$\partial\varphi = \bar{\mathfrak{d}}\varphi = 0\ . \tag{2.12''}$$

We also obtain the Hodge theorems

$$H^{p,q}_{\bar{\partial}}(X, E) \simeq \mathbf{H}'^{p,q}(X, E) \simeq H^{p,q}_{\mathfrak{d}}(X, E)$$

and

$$H^{p,q}_{\partial}(X, E) \simeq \mathbf{H}''^{p,q}(X, E) \simeq H^{p,q}_{\bar{\mathfrak{d}}}(X, E)\ ,$$

which imply, from (2.11), the Serre duality theorem [18, p. 119]

$$H^q(X, \Omega^p(E)) \simeq H^{n-q}(X, \Omega^{n-p}(E^*))\ ,$$

and, most important of all, the finite dimensionality of each of the cohomology groups considered.

6. Hermitian, complex analytic vector bundles over a Kähler manifold

We shall deduce now more identities and relations from the additional assumption, valid from here on, that *the Hermitian metric on the tangent bundle of X is Kählerian.* The fundamental form $\omega = \sqrt{-1}\, g_{\alpha\bar{\beta}} dz^\alpha \wedge d\bar{z}^\beta$ of type (1, 1) defines the two operators L and Λ on $A^{p,q}(E)$ of degrees respectively (1,1) and $(-1, -1)$ by $L\varphi = \omega \wedge \varphi$, $\Lambda\varphi = (-1)^{p+q} {*}L{*}\varphi$ ($\varphi \in A^{p,q}(E)$); the Kählerian assumption that ω is d-closed makes L and Λ acquire importance in the theory of harmonic forms with coefficients in E, as we shall verify. The following identities are listed here for easy reference: $L(\varphi\#) = (L\varphi)\#$; $\Lambda(\varphi\#) = (\Lambda\varphi)\#$; $L({*}\varphi) = {*}\Lambda\varphi$; $(\Lambda L - L\Lambda)\varphi = (n - p - q)\varphi$ ($\varphi \in A^{p,q}(E)$).

The Kähler condition implies (and, in the case of the trivial bundle $E = \mathbf{C}$, is actually equivalent to) the following relation

$$\Lambda\bar{\partial} - \bar{\partial}\Lambda = -\sqrt{-1}\,\bar{\mathfrak{d}}\ . \tag{2.13}$$

Applying this last formula to $\varphi\#$, ${*}\varphi$ and ${*}\varphi\#$, we obtain the respective dual identities

$$\Lambda\partial - \partial\Lambda = \sqrt{-1}\,\mathfrak{d};\quad L\bar{\mathfrak{d}} - \bar{\mathfrak{d}}L = \sqrt{-1}\,\bar{\partial};\quad L\mathfrak{d} - \mathfrak{d}L = -\sqrt{-1}\,\partial. \tag{2.14}$$

We consider next the "interchange" operators $\partial\bar{\mathfrak{d}} - \mathfrak{d}\bar{\partial}$ and $\bar{\mathfrak{d}}\partial - \bar{\partial}\mathfrak{d}$; applying (2.13) and (2.14) in different ways, we obtain

$$\partial\bar{\mathfrak{d}} - \mathfrak{d}\bar{\partial} = \sqrt{-1}\,(\Lambda\partial\bar{\partial} - \partial\bar{\partial}\Lambda) = \sqrt{-1}(L\mathfrak{d}\bar{\mathfrak{d}} - \mathfrak{d}\bar{\mathfrak{d}}L)\,, \tag{2.15}$$

and

$$\bar{\mathfrak{d}}\partial - \bar{\partial}\mathfrak{d} = \sqrt{-1}(\Lambda\bar{\partial}\partial - \bar{\partial}\partial\Lambda) = \sqrt{-1}(L\bar{\mathfrak{d}}\mathfrak{d} - \bar{\mathfrak{d}}\mathfrak{d}L)\,. \tag{2.16}$$

Adding together the respective members of (2.15) and (2.16) and using (2.6) and (2.10), one obtains the relation

$$\square'' - \square' = \sqrt{-1}(\Lambda e(s) - e(s)\Lambda) = \sqrt{-1}(i(s)L - Li(s))\,. \tag{2.17}$$

We prove now the following

LEMMA 1 (Nakano [24]).

(a) *If* $\varphi \in \mathbf{H}'^{p,q}(X, E)$, *then*

$$\sqrt{-1}(\Lambda e(s)\varphi, \varphi) \geqq 0\,, \tag{2.18}$$

equality holding if and only if $\partial\varphi = 0$, *and*

$$\sqrt{-1}(e(s)\Lambda\varphi, \varphi) \leqq 0\,, \tag{2.19}$$

equality holding if and only if $\bar{\mathfrak{d}}\varphi = 0$;

(b) *if* $\varphi \in \mathbf{H}''^{p,q}(X, E)$, *then*

$$\sqrt{-1}\,(\Lambda e(s)\varphi, \varphi) \leqq 0\,, \tag{2.18'}$$

equality holding if and only if $\bar{\partial}\varphi = 0$, *and*

$$\sqrt{-1}\,(e(s)\Lambda\varphi, \varphi) \geqq 0\,, \tag{2.19'}$$

equality holding if and only if $\mathfrak{d}\varphi = 0$.

PROOF. If $\varphi \in \mathbf{H}'^{p,q}(X, E)$ then $\bar{\partial}\varphi = \mathfrak{d}\varphi = 0$; we have then, from (2.13) and (2.6)

$$\begin{aligned} 0 \leqq (\partial\varphi, \partial\varphi) &= (\bar{\mathfrak{d}}\partial\varphi, \varphi) = \sqrt{-1}((\Lambda\bar{\partial} - \bar{\partial}\Lambda)\partial\varphi, \varphi) \\ &= \sqrt{-1}(\Lambda(\bar{\partial}\partial + \partial\bar{\partial})\varphi, \varphi) - \sqrt{-1}(\Lambda\partial\bar{\partial}\varphi, \varphi) - \sqrt{-1}(\Lambda\partial\varphi, \mathfrak{d}\varphi \\ &= \sqrt{-1}(\Lambda e(s)\varphi, \varphi)\,, \end{aligned}$$

which proves (2.18). The remaining inequalities are proved in the same way by replacing φ respectively by $*\varphi\#$, $\varphi\#$ and $*\varphi$.

As a corollary of Lemma 1, we get

LEMMA 2. *If* $\varphi \in \mathbf{H}'^{p,q}(X, E)$, *then*

$$\sqrt{-1}((\Lambda e(s) - e(s)\Lambda)\varphi, \varphi) \geqq 0\,, \tag{2.20}$$

while, if $\varphi \in \mathbf{H}''^{p,q}(X, E)$, *then*

$$\sqrt{-1}((\Lambda e(s) - e(s)\Lambda)\varphi, \varphi) \leqq 0\,, \tag{2.20'}$$

equality holding in either case if and only if

$$\varphi \in \mathbf{H}'^{p,q}(X, E) \cap \mathbf{H}''^{p,q}(X, E) .$$

Lemmas 1 and 2 become equivalent, if either p or q takes the extreme values 0 or n. In practice, the computations involved in applying Lemma 2 are generally shorter than those for Lemma 1.

7. The tangent vector bundle in a Kähler manifold

We turn now to the case where X is, as in the preceding paragraph, a compact, complex manifold, with a Kähler metric on its tangent bundle Θ, but take, as a special case of vector bundle E over X, the tangent bundle Θ itself, with the same Kähler metric. In this paragraph we shall make use of indicial notation for the corresponding Θ-vectorial (p, q)-forms, using greek indices $\alpha, \beta, \cdots$ in the range 1 to $n = \dim_{\mathbf{C}} X$, related to the holomorphic coordinates, and barred indices $\bar\alpha = \alpha + n, \cdots$ in the range $n + 1$ to $2n$, referring to the conjugate holomorphic coordinates $z^{\alpha+n} = z^{\bar\alpha} = \overline{z^\alpha}$.

The local components of the metric being $g_{\alpha\bar\beta}$ and $g^{\alpha\bar\beta}$ as usual, the coefficients of the connection are the Christoffel symbols $\left\{ {\sigma \atop \alpha\gamma} \right\}$ and $\left\{ {\bar\sigma \atop \bar\alpha\bar\gamma} \right\}$, and the curvature tensor is defined by the form

$$R^\sigma{}_{\alpha\bar\beta\gamma} d\bar z^\beta = \bar\partial \left\{ {\sigma \atop \alpha\gamma} \right\} ,$$

while the curvature form s of the connection is expressed locally by a $n \times n$ matrix (s^ρ_α), with

$$s^\rho_\alpha = R^\rho{}_{\alpha\beta\bar\gamma} dz^\beta \wedge dz^{\bar\gamma} \cdot$$

We shall use also the Ricci tensor

$$R_{\alpha\bar\beta} = -R^\sigma{}_{\alpha\bar\beta\sigma} = R^\sigma{}_{\alpha\sigma\bar\beta}$$

and the scalar curvature $R = 2g^{\alpha\bar\beta} R_{\alpha\bar\beta}$.

If $\varphi \in A^{p,q}(\Theta)$, then the components of φ are given by the tensor $\varphi^\alpha{}_{\gamma_1\cdots\gamma_p\bar\beta_1\cdots\bar\beta_q}$ antisymmetric in the indices γ's and $\bar\beta$'s, and $\varphi\#$, say, by

$$\varphi\#_{\bar\alpha\cdot\beta_1\cdots\beta_q\bar\gamma_1\cdots\bar\gamma_p} = (-1)^{pq} g_{\omega\bar\tau} (\varphi^\tau_{\gamma_1\cdots\gamma_p\bar\beta_1\cdots\bar\beta_q})^- .$$

Following (2.17) we compute the difference $\Box'' - \Box'$ as follows

$$((\Box'' - \Box')\varphi)^\alpha_{\gamma_1\cdots\gamma_p\bar\beta_1\cdots\bar\beta_q} = \sqrt{-1}((\Lambda e(s) - e(s)\Lambda)\varphi)^\alpha_{\gamma_1\cdots\gamma_p\bar\beta_1\cdots\bar\beta_q}$$

$$(2.21) \quad = (-1)^{pq} g^{\bar\eta\alpha} \{ R_\eta{}^\rho \varphi\#_{\rho\cdot\beta_1\cdots\beta_q\bar\gamma_1\cdots\bar\gamma_p} + \textstyle\sum_{\mu=1}^p R_\eta{}^{\rho\bar\tau}{}_{\bar\gamma_\mu} \varphi\#_{\rho\cdot\beta_1\cdots\beta_q\bar\gamma_1\cdots(\bar\tau)_\mu\cdots\bar\gamma_p}$$

$$+ \textstyle\sum_{\nu=1}^q R_\eta{}^\rho{}_{\beta_\nu}{}^\sigma \varphi\#_{\rho\cdot\beta_1\cdots(\sigma)_\nu\cdots\beta_q\bar\gamma_1\cdots\bar\gamma_p} \}^- ,$$

where the symbols $(\sigma)_\nu$ and $(\bar\tau)_\mu$ mean respectively σ in place of β_ν and $\bar\tau$ in place of $\bar\gamma_\mu$. In order to apply Lemma 2, we must investigate the algebraic properties of the operator (2.21); in the right hand member it is divided into three terms, all depending on the Riemann curvature tensor in different ways. The first term depends on the action of the Ricci tensor $R_{\alpha}{}^{\rho}$ on $\varphi\sharp$; the second and third term depend on the actions of the Riemann tensor on $\varphi\sharp$ in two different ways, each being self-adjoint with respect to the Kähler metric. We are thus led to consider the proper values of the following linear operators at each point of X:

(a) In connection with the first term of (2.21), of the transformation

$$\zeta_{\alpha} \rightarrow R_{\alpha}{}^{\rho}\zeta_{\rho};$$

(b) in connection with the third term of (2.21), of the transformation

$$\xi_{\alpha\beta} \rightarrow R^{\rho}{}_{\alpha\beta}{}^{\sigma}\xi_{\rho\sigma} = - R^{\rho}{}_{\alpha}{}^{\sigma}{}_{\beta}\xi_{\rho\sigma} \qquad (\xi_{\alpha\beta} = \xi_{\beta\alpha}); \tag{2.22}$$

(c) in connection with the second term of (2.21), of the transformation

$$\eta_{\alpha\bar{\gamma}} \rightarrow R^{\rho}{}_{\alpha}{}^{\bar{\tau}}{}_{\bar{\gamma}}\eta_{\rho\bar{\tau}} = - R^{\rho}{}_{\alpha\bar{\gamma}}{}^{\bar{\tau}}\eta_{\rho\bar{\tau}} \ . \tag{2.23}$$

Since the three operations are self-adjoint, the proper values are all real-valued point invariants of the Kähler metric. We observe that the operator (2.22) acts on symmetric tensors alone, since it follows from the Bianchi identity that the symmetric tensor space is invariant while the antisymmetric space is annihilated. We observe, incidentally, that, if no restriction is placed on the metric, the operator (2.22) can be a completely arbitrary self-adjoint one at any one point, determining uniquely there the curvature tensor of a local Kähler metric, while the operator (2.23), which also describes the curvature tensor completely, is not arbitrary, because subject to the restrictions imposed by the Bianchi identity.

We observe from (2.21) that, if $p = 0$, the linear operator (2.23) need not be considered; the inequality (2.20) of Lemma 2 becomes identical with the Nakano inequality (2.18), while (2.19) is trivial; this brings a notable simplification. For $p > 0$, on the other hand, not only do the present methods fail to prove that $\mathbf{H}'^{p,q}(X, \Theta)$ vanishes for the compact manifolds considered in Theorem 1, but there is a strong indication that these groups may not be trivial. In fact, if one assumes the conjecture that for each Cartan domain D there exist some compact manifolds X, uniformized by D, that admit non-trivial, global, holomorphic p-forms for each p ($0 < p \leqq n = \dim_{\mathbf{C}} X$), then for such X all the cohomology groups $H^{p,q}(X, \mathbf{C})$ are nontrivial ($0 \leqq p, q \leqq n$). For some $p \geqq 1$ let $\psi \in \mathbf{H}^{p-1,q}(X, \mathbf{C})$ be a non-trivial harmonic $(p-1, q)$-form with coefficients $\psi_{\gamma_1 \cdots \gamma_{p-1}\bar{\beta}_1 \cdots \bar{\beta}_q}$; then the Θ-vectorial (p, q) form φ defined by its components

$$\varphi^{\alpha}{}_{\gamma_1\cdots\gamma_p\bar\beta_1\cdots\bar\beta_q} = \sum_{\mu=1}^{p}(-1)^{\mu}\delta^{\alpha}{}_{\gamma_\mu}\psi_{\gamma_1\cdots\gamma_{\mu-1}\gamma_{\mu+1}\cdots\gamma_p\bar\beta_1\cdots\bar\beta_q}$$

is, for $1 \leqq p \leqq n$, $0 \leqq q \leqq n$, a non-trivial form satisfying $\bar\partial\varphi = \mathfrak{d}\varphi = 0$, and hence we have a natural isomorphism of $\mathbf{H}^{p-1,q}(X, \mathbf{C})$ into $\mathbf{H}'^{p,q}(X, \Theta)$. For this reason we shall apply Lemmas 1 and 2 to determine the vanishing of groups $\mathbf{H}'^{0,q}(X, \Theta) \simeq H^q(X, \boldsymbol{\Theta})$ alone.

The linear transformation represented by the curvature tensor in (2.22) acts on the complex analytic fibre bundle over X of symmetric bilinear functionals on Θ; the fibre of this bundle being a complex vector space of $N = \frac{1}{2}n(n+1)$ dimensions, the linear operator has a set of N proper values, $\lambda_1, \cdots, \lambda_N$, all real, and their sum, the trace of the linear operator, is $\frac{1}{2}R$, where R is the scalar curvature. We shall now derive a formula relating the proper values of (2.22) with the corresponding action of the curvature tensor on a Θ-vectorial $(0, q)$-form.

LEMMA 3. *Let X be an arbitrary n-dimensional Kähler manifold, and $\mathfrak{p}_0$ an arbitrary point of X. Denote by $\lambda_1, \cdots, \lambda_N$ ($N = \frac{1}{2}n(n+1)$) the proper values of the linear transformation at $\mathfrak{p}_0$*

$$\xi_{\alpha\beta} \to R^{\rho}{}_{\alpha\beta}{}^{\sigma}\xi_{\rho\sigma} \qquad (\xi_{\alpha\beta} = \xi_{\beta\alpha}) \tag{2.22}$$

on the space of symmetric tensors $\xi_{\alpha\beta}$, and suppose that these proper values are arranged in increasing order $\lambda_1 \leqq \lambda_2 \leqq \cdots \leqq \lambda_N$. Let ψ be an arbitrary Θ-vectorial $(0, q)$-form at $\mathfrak{p}_0$, and denote by $|\psi|^2$ and $Q(\psi)$ respectively the values of the two following Hermitian forms on ψ,

$$|\psi|^2 = *({}^t\psi\# \wedge *\psi) = \frac{1}{q!}\psi\#_{\alpha\cdot\beta_1\cdots\beta_q}\psi^{\alpha\cdot\beta_1\cdots\beta_q}$$

and

$$Q(\psi) = \frac{1}{q!}\sum_{\mu=1}^{q}R^{\rho}{}_{\alpha\beta_\mu}{}^{\sigma}\psi\#_{\rho\cdot\beta_1\cdots(\sigma)_\mu\cdots\beta_q}\psi^{\alpha\cdot\beta_1\cdots\beta_q} \qquad (q > 0)\,, \tag{2.24}$$

$$Q(\psi) = 0 \qquad (q = 0)\,.$$

Then

$$Q(\psi)\begin{cases} \geqq \frac{1}{2}(q+1)\lambda_1|\psi|^2 & \text{for } \lambda_1 \leqq 0,\ q > 0 \\ \geqq 0 & \text{for } \lambda_1 > 0 \text{ or } q = 0\,. \end{cases} \tag{2.25}$$

PROOF. If $q = 0$, the statement is trivial, so we consider only the case $q > 0$. Let $Q'(\psi)$ be the auxiliary Hermitian form defined by (2.24) with $R^{\rho}{}_{\alpha\beta}{}^{\sigma}$ replaced by $R'^{\rho}{}_{\alpha\beta}{}^{\sigma} = \frac{1}{2}\lambda_1(\delta^{\rho}_{\alpha}\delta^{\sigma}_{\beta} + \delta^{\sigma}_{\alpha}\delta^{\rho}_{\beta})$; the latter tensor corresponds in (2.22) to the transformation $\xi_{\alpha\beta} \to \lambda_1\xi_{\alpha\beta}$. Since λ_1 is the smallest proper value of (2.22), one sees easily that

$$(2.26) \quad Q(\psi) \geqq Q'(\psi) = \frac{1}{2q!}\lambda_1 \sum_{\mu=1}^{q}(\psi\#_{\alpha\cdot\beta_1\cdots\beta_q} + \psi\#_{\beta_\mu\cdot\beta_1\cdots(\alpha)_\mu\cdots\beta_q})\psi^{\alpha\cdot\beta_1\cdots\beta_q} .$$

We observe that the decomposition $\psi = \psi' + \psi''$ defined by

$$\psi'\#_{\alpha\cdot\beta_1\cdots\beta_q} = \frac{1}{q+1}(\psi\#_{\alpha\cdot\beta_1\cdots\beta_q} - \sum_{\mu=1}^{q}\psi\#_{\beta_\mu\cdot\beta_1\cdots(\alpha)_\mu\cdots\beta_q}) ,$$

$$\psi''\#_{\alpha\cdot\beta_1\cdots\beta_q} = \frac{1}{q+1}\sum_{\mu=1}^{q}(\psi\#_{\alpha\cdot\beta_1\cdots\beta_q} + \psi\#_{\beta_\mu\cdot\beta_1\cdots(\alpha)_\mu\cdots\beta_q})$$

is an orthogonal one (relative to the form $|\psi|^2$), since $\psi'\#$ is alternating in all $q+1$ indices, while $\psi''\#$ is orthogonal to all alternating $(q+1)$-forms. Thus we can develop (2.26) as follows:

$$Q(\psi) \geqq Q'(\psi) = \frac{1}{2}\lambda_1(q+1) * ({}^t\psi''\# \wedge *\psi) = \frac{1}{2}\lambda_1(q+1)\,|\psi''|^2 ,$$

and

$$0 \leqq |\psi''|^2 \leqq |\psi''|^2 + |\psi'|^2 = |\psi|^2 .$$

The last two inequalities imply (2.25).

8. A criterion for the vanishing of certain cohomology groups

We shall later apply Lemma 3 to the compact, locally irreducible and symmetric Kähler manifolds X. Such manifolds are in particular Einstein spaces; that is to say, the Ricci tensor $R_{\alpha\bar\beta}$ satisfies

$$R_{\alpha\bar\beta} = \frac{1}{2n} R g_{\alpha\bar\beta} \qquad (n = \dim_{\mathbf{C}} X) ,$$

where R is a constant, equal to twice the trace of the linear operator (2.22). For Kähler-Einstein spaces we can prove the following "vanishing theorem".

LEMMA 4. *Let X be a compact Kähler-Einstein, n-dimensional manifold. Denote by $\lambda_1, \cdots, \lambda_N$ $(N = \frac{1}{2}n(n+1))$ the proper values at each point of X of the linear transformation* (2.22) *defined by the curvature tensor, arranged so that $\lambda_1 \leqq \lambda_2 \leqq \cdots \leqq \lambda_N$, and let $\lambda = \inf_X \lambda_1$. Then, if R denotes the constant scalar curvature,*

$$(2.27) \qquad R \geqq n(n+1)\lambda ,$$

and the following conclusions are valid.

(a) *If $\lambda \geqq 0$ and $R > 0$, then $H^q(X, \Theta)$ for all $q > 0$.*

(b) *If $\lambda < 0 < R$ and $R + n\lambda > 0$, then $H^q(X, \Theta) = 0$ for all $q > -n\lambda/(R + n\lambda)$.*

(c) *If $R < 0$ then $0 < R/\lambda \leqq n(n+1)$, and $H^q(X, \Theta) = 0$ for all q such that $n(q+1) < R/\lambda$.*

PROOF. We represent $H^q(X, \Theta)$ by $\mathbf{H}'^{0,q}(X, \Theta)$. Let φ be any Θ-vectorial $(0, q)$-form in $\mathbf{H}'^{0,q}(X, \Theta)$: since $\Box'\varphi = 0$, we have $\Box''(\varphi\sharp)=0$; writing this componentwise, we have

$$\begin{aligned} 0 &= (\Box''\varphi\sharp)_{\alpha\cdot\beta_1\cdots\beta_q} \\ &= -g^{\sigma\bar\tau}\nabla_{\bar\tau}\nabla_\sigma\varphi\sharp_{\alpha\cdot\beta_1\cdots\beta_q} + \sum_{\mu=1}^{q}(\delta^\rho_\alpha R_{\beta_\mu}{}^\sigma + R^\rho{}_{\alpha\beta_\mu}{}^\sigma)\varphi\sharp_{\rho\cdot\beta_1\cdots(\sigma)_\mu\cdots\beta_q} \\ &= -g^{\sigma\bar\tau}\nabla_{\bar\tau}\nabla_\sigma\varphi\sharp_{\alpha\cdot\beta_1\cdots\beta_q} + \frac{qR}{2n}\varphi\sharp_{\alpha\cdot\beta_1\cdots\beta_q} + \sum_{\mu=1}^{q}R^\rho{}_{\alpha\rho_\mu}{}^\sigma\varphi\sharp_{\rho\cdot\beta_1\cdots(\sigma)_\mu\cdots\beta_q}\,, \end{aligned}$$

where ∇_σ and $\nabla_{\bar\tau}$ denote successive covariant differentiations, read from right to left. The integral identify $(\varphi\sharp, \Box''(\varphi\sharp)) = 0$ can thus be broken up into a vanishing sum of three terms as follows:

$$(\varphi\sharp, -\nabla^\sigma\nabla_\sigma\varphi\sharp) + \frac{qR}{2n}(\varphi\sharp, \varphi\sharp) + (Q(\varphi), 1) = 0\,,$$

where $Q(\varphi)$ is defined by (2.24). Of these three terms, the first one is nonnegative since by Green's theorem it is equal to $(\nabla_\sigma\varphi\sharp, \nabla_\sigma\varphi\sharp)$ (the meaning of this unconventional notation being clear). Applying now Lemma 3 to the remaining two terms, we have

$$0 \geqq \frac{qR}{2n}(\varphi\sharp, \varphi\sharp) + (Q(\varphi), 1) \geqq \left[\frac{qR}{2n} + \frac{q+1}{2}\cdot\min(\lambda, 0)\right](\varphi, \varphi)$$

for $q > 0$.

Thus in case (a) we have a clear contradiction, unless φ vanishes identically; in case (b) we have a contradiction for $\varphi \neq 0$, if $qR+n(q+1)\lambda > 0$, that is to say, if $q > -n\lambda/(R+n\lambda)$. In case (c) we must use either Lemma 1 or 2 (equivalently); more precisely we use (2.18), which can be calculated from (2.21): the integrand of $\sqrt{-1}(\Lambda e(s)\varphi, \varphi)$ is

$$\frac{1}{q!}\varphi^{\alpha\cdot\beta_1\cdots\beta_q}(R_\alpha{}^\rho\varphi\sharp_{\rho\cdot\beta_1\cdots\beta_q} - \sum_{\mu=1}^{q}R^\rho{}_{\alpha\beta_\mu}{}^\sigma\varphi\sharp_{\rho\cdot\beta_1\cdots(\sigma)_\mu\cdots\beta_q}) = \frac{R}{2n}|\varphi|^2 - Q(\varphi)\,.$$

From Lemma 3 we obtain at each point of X

$$\frac{R}{2n}|\varphi|^2 - Q(\varphi) \leqq \Big(\frac{R}{2n} - \frac{1}{2}(q+1)\lambda\Big)|\varphi|^2\,.$$

If $R/2n - (q+1)\lambda/2 < 0$, i.e., if $q+1 < R/n\lambda$ (since R, and hence also λ, are assumed to be negative) then for $\varphi\neq 0$ we have $R|\varphi|^2/2n - Q(\varphi) \leqq 0$ at each point and not identically zero. Integrating this over X we obtain a contradiction to inequality (2.18) in Lemma 1. This completes the proof of Lemma 4.

Chapter 3. Irreducible, Symmetric Kähler manifolds

1. Classification

Let G be a connected, simple Lie group, whose center consists of the identity alone, and let H be the connected component of the identity in the set of elements of G that are invariant under an involution τ of G. In order that the resulting irreducible symmetric space G/H be Kählerian, it is necessary and sufficient that:

(a) H be compact,

(b) the center of H be a circle group, whose element of period 2 has τ as its adjoint representation on G.

From the structure of simple groups, the irreducible Kählerian symmetric spaces fall into the two following types:

A. (non-compact spaces). G is a non-compact, simple group and H is a maximal, connected, compact subgroup of G; in this case G/H is complex analytically equivalent to an irreducible, bounded symmetric subdomain $D \subset \mathbf{C}^n$ (Cartan domain), and G is also the group of all complex analytic homeomorphisms of D onto itself.

B. (compact case). G is a compact group, and H is of maximal rank. The group of all complex analytic transformations of the compact, complex manifold $Y = G/H$ is the complex extension $G_{\mathbf{C}}$ of G, of which G is the maximal compact subgroup.

The irreducible, compact, symmetric Kähler manifolds Y have been classified by É. Cartan [13]; these are all rational algebraic varieties. Using the notations of C. L. Siegel [25] (cf., also [8]), Cartan's classification is the following.

Type B-$\mathrm{I}_{m,m'}$: $Y = \mathbf{U}(m + m')/\mathbf{U}(m) \times \mathbf{U}(m')$ (factoring out the center of $\mathbf{U}(m + m')$ to obtain effective action). Y is holomorphically equivalent to the complex Grassmann manifold $G_{m,m'}$, consisting of the $(m - 1)$-dimensional, complex projective linear subspaces $P_{m-1}(\mathbf{C})$ of the complex projective $(m + m' - 1)$-space $P_{m+m'-1}(\mathbf{C})$; its dimensionality is

$$n = \dim_{\mathbf{C}} Y = mm' .$$

Type B-II_m: $Y = \mathbf{SO}(2m)/\mathbf{U}(m)$. Y is holomorphically equivalent to the complex analytic submanifold of $G_{m,m}$, consisting of all the $(m - 1)$-dimensional, complex projective linear subspaces $P_{m-1}(\mathbf{C})$ that lie in a non-singular quadric hypersurface $Q_{2m-2}(\mathbf{C})$ in $P_{2m-1}(\mathbf{C})$; its dimensionality is

$$n = \dim_{\mathbf{C}} Y = \frac{1}{2} m(m - 1) .$$

Type B-III_m: $Y = \mathbf{Sp}(m)/\mathbf{U}(m)$. Y is holomorphically equivalent to the complex subvariety of $G_{m,m}$, consisting of all the $(m-1)$-dimensional, complex projective linear subspaces $P_{m-1}(\mathbf{C}) \subset P_{2m-1}(\mathbf{C})$, all of whose projective lines are contained in a general (i.e., nonsingular) linear complex. In terms of homogeneous coordinates in $P_{2m-1}(\mathbf{C})$, Y is represented by the m-dimensional, complex vector subspaces of the vector space $\mathbf{C}^{2m}$, that are totally isotropic with respect to a nonsingular, alternating bilinear form on $\mathbf{C}^{2m}$; the dimensionality of Y is

$$n = \dim_{\mathbf{C}} Y = \frac{1}{2}m(m+1) .$$

Type B-IV$_m$: $Y = \mathbf{SO}(m+2)/\mathbf{SO}(m) \times \mathbf{SO}(2)$. Y is holomorphically equivalent to the nonsingular, m-dimensional, complex quadric hypersurface $Q_m \subset P_{m+1}(\mathbf{C})$, and real-analytically isomorphic to the real Grassmann manifold $G_{2,m}(\mathbf{R})$ of oriented, real projective lines in a real projective $(m+1)$-space. The dimensionality of Y is

$$n = \dim_{\mathbf{C}} Y = m .$$

In addition to these four "classical" types there are the following two exceptional types:

Type B- V: $Y = \mathbf{E}_6/\mathbf{Spin}\ (10) \times \mathbf{SO}(2)$; $\dim_{\mathbf{C}} Y = 16$;

Type B-VI: $Y = \mathbf{E}_7/\mathbf{E}_6 \times \mathbf{SO}(2)$; $\dim_{\mathbf{C}} Y = 27$.

We consider now the classification of the non-compact, irreducible, symmetric Kähler manifolds: any irreducible, bounded, symmetric domain $D \subset \mathbf{C}^n$ is complex analytically isomorphic to a homogeneous space G/H as described above under Case A; conversely, any symmetric homogeneous space of type A is isomorphic to an irreducible, bounded domain D. Furthermore, any metric on G/H that is invariant under the group G is equal (within a positive factor) to the Bergmann metric of D.

The different types of spaces of type A are uniquely obtainable in a one-to-one way from the corresponding ones of type B, as follows.

(1) One constructs the complex Lie group $G_{\mathbf{C}}$ generated by the compact group G.

(2) One splits the Lie algebra $\mathfrak{g}$ of G orthogonally with respect to the Killing-Cartan form into $\mathfrak{h} \oplus \mathfrak{m}$, where $\mathfrak{h}$ is the Lie algebra of H and $\mathfrak{m}$ is its orthogonal complement. In terms of the involution τ of G, leaving H pointwise invariant, $\mathfrak{h}$ and $\mathfrak{m}$ are the subspaces of $\mathfrak{g}$ that, under the involution of $\mathfrak{g}$ induced by τ are, $\mathfrak{h}$ pointwise fixed, and $\mathfrak{m}$ corresponding to the proper value -1. Since τ induces an automorphism of the Lie algebra, we have immediately

$$[\mathfrak{h}, \mathfrak{h}] \subset \mathfrak{h}, \qquad [\mathfrak{h}, \mathfrak{m}] \subset \mathfrak{m}, \qquad [\mathfrak{m}, \mathfrak{m}] \subset \mathfrak{h} .$$

(3) Imbedding $\mathfrak{g}$ naturally into the complex Lie algebra $\mathfrak{g}_{\mathbf{C}}$ of $G_{\mathbf{C}}$, the real vector subspace $\mathfrak{h} \oplus \sqrt{-1}\,\mathfrak{m}$ of $\mathfrak{g}_{\mathbf{C}}$ is a real Lie subalgebra of $\mathfrak{g} \otimes \mathbf{C}$; it generates a Lie group G' (not compact), in which H is the maximal compact subgroup, and the extension of τ to $G_{\mathbf{C}}$ is an involution of G', for which H is the maximal, connected subset of elements fixed under τ. Then G'/H is the corresponding space of type A.

The classification of the irreducible, symmetric Kähler manifolds D of type A from those of type B is found in [14]: we give it here, using the notations of [1, pp. 112–114] for the groups.

Type A–$\mathrm{I}_{m,m'}$: $D = \mathbf{U}^m(m+m')/\mathbf{U}(m) \times \mathbf{U}(m')$, $\dim_{\mathbf{C}} D = mm'$. D is equivalent to the subdomain of $\mathbf{C}^{mm'}$ obtained by identifying $\mathbf{C}^{mm'}$ with the space of complex $m \times m'$ matrices Z and taking the subset of such matrices such that the $m' \times m'$ hermitian matrix $\mathbf{1}^{(m')} - {}^t\overline{Z}Z$ is positive definite, i.e.,

$$(3.1)_{m'} \qquad \mathbf{1}^{(m')} - {}^t\overline{Z}Z > 0\ .$$

Type A–II_m: $D = \mathbf{SO}^*(2m)/\mathbf{U}(m)$, $\dim_{\mathbf{C}} D = \frac{1}{2}m(m-1)$. D is equivalent to the subdomain of $\mathbf{C}^{m(m-1)/2}$, obtained by identifying the latter with the space of complex, skew-symmetric $m \times m$ matrices Z, and taking the subset of such matrices that satisfy $(3.1)_m$.

Type A–III_m: $D = \mathbf{Sp}(m, \mathbf{R})/\mathbf{U}(m)$, $\dim_{\mathbf{C}} D = \frac{1}{2}m(m+1)$. D is the subdomain of $\mathbf{C}^{m(m+1)/2}$ obtained by identifying the latter with the space of complex, symmetric $m \times m$ matrices Z, and taking the subset of such matrices that satisfy $(3.1)_m$.

Type A–IV_m: $D = \mathbf{SO}^m(m+2, \mathbf{R})/\mathbf{SO}(m) \times \mathbf{SO}(2)$, $\dim_{\mathbf{C}} D = m$. D is the subdomain of $\mathbf{C}^m$, considered as space of $m \times 1$ matrices Z, consisting of the elements Z satisfying

$$(3.2) \qquad 1 + |{}^tZZ|^2 - 2\,{}^t\overline{Z}Z > 0, \quad {}^t\overline{Z}Z < 1\ .$$

REMARK. The first inequality above implies ${}^t\overline{Z}Z \neq 1$; thus the second inequality merely selects one of the two components of the set defined by the first one, namely the bounded one.

In addition to the above four classical types, we have the two following exceptional types:

Type A–V : $D = \mathbf{E}_6^3/\mathbf{Spin}(10) \times \mathbf{SO}(2)$, $\dim_{\mathbf{C}} D = 16$;

Type A–VI: $D = \mathbf{E}_7^3/\mathbf{E}_6 \times \mathbf{SO}(2)$, $\dim_{\mathbf{C}} D = 27$.

From the above classification one must exclude, both for the compact and non-compact forms, II_1 (zero-dimensional) and IV_2 (reducible). The following isomorphisms further reduce the list of distinct ones: $\mathrm{I}_{m,m'} \simeq \mathrm{I}_{m',m}$; $\mathrm{I}_{1,1} \simeq \mathrm{II}_2 \simeq \mathrm{III}_1 \simeq \mathrm{IV}_1$; $\mathrm{II}_3 \simeq \mathrm{I}_{1,3}$; $\mathrm{IV}_3 \simeq \mathrm{III}_2$; $\mathrm{IV}_4 \simeq \mathrm{I}_{2,2}$; $\mathrm{IV}_6 \simeq \mathrm{II}_4$ (these isomorphisms are valid for both the compact and non-compact forms).

By completing an argument of É. Cartan [13], A. Borel has shown [7], (see also [8]) that every symmetric space D of type A can be imbedded, in an essentially unique way, as an open subdomain of the corresponding compact manifold Y (of type B), so that each holomorphic transformation of D can be extended to a holomorphic (but not generally isometric) transformation of Y.

2. The curvature of irreducible, symmetric Kähler manifolds

We shall calculate now the curvature tensor for the non-exceptional, irreducible, symmetric Kähler manifolds, first for those of type A, then for those of type B; we want to determine, specifically, the proper values $\lambda_1, \cdots, \lambda_N$ ($N = \frac{1}{2}n(n+1)$) of the linear operator (2.22), defined by the curvature tensor for each of these spaces. The method is the following. The Kähler metric on D (respectively Y) is derived from a certain distance-like function on $D \times D$ (or $Y \times Y$), invariant under the action of G. This function, related to the Bergmann kernel function, can be essentially guessed by inspection from the structure of the symmetric space; it has been introduced in [12] and is called the *diastasis* $\mathrm{D}(\mathfrak{p}, \mathfrak{q})$. The relation of this function to the metric is that at each point $\mathfrak{q}$ (independently of $\mathfrak{p}$) one has $\partial^2\mathrm{D}(\mathfrak{p}, \mathfrak{q})/\partial z^\alpha(\mathfrak{q})\partial \overline{z^\beta}(\mathfrak{q}) = g_{\alpha\bar\beta}(\mathfrak{q})$. For any point $\mathfrak{p}$ one distinguishes certain coordinate systems $z = (z^1, \cdots, z^n)$ with origin at $\mathfrak{p}$, called *canonical at* $\mathfrak{p}$ [3, 12]: they are characterized by the fact that the diastasis $\mathrm{D}(0, z)$ of the origin and z has the power series expansion

$$\mathrm{D}(0, z) = \sum_{\alpha=1}^{n} | z^\alpha |^2 + \frac{1}{4} A_{\alpha\rho\bar\beta\bar\tau} z^\alpha z^\rho \overline{z^\beta} \overline{z^\tau} + \psi(z, \bar z) ,$$

where ψ contains only terms of order at least 5 in $z, \bar z$, and at least 2 in z, and $\bar z$ separately. It follows from this that one can calculate the components of the curvature tensor immediately from

$$R_{\alpha\bar\beta\rho\bar\tau}(0) = -R^{\beta}{}_{\alpha\rho}{}^{\tau}(0) = A_{\alpha\rho\bar\beta\bar\tau} .$$

Next, we observe that the representation of the isotropy subgroup H on the space of covariant, symmetric, complex tensor $\xi_{\alpha\beta}$ at the origin may split into a certain number of irreducible representations. Since H is also the holonomy group of the manifold, it follows from the invariance of the curvature under the holonomy group (because the space is symmetric) that each irreducible, invariant subspace of that tensor space belongs to a common proper value under the action of the linear operator (2.22). Finally, by choosing a suitable representative element of each irreducible invariant subspace of tensors $\xi_{\alpha\beta}$, one calculates the proper values explicitly. Now we come to specific cases.

Type A–$\mathrm{I}_{m,m'}$. Denote by $Z=(z^{ir})$ a rectangular, $m\times m'$ complex matrix; we use indices $i, j, k, \cdots$ in the range 1 to m; $r, s, t, \cdots$ in the range 1 to m'. The domain D is covered by (z^{ir}) as holomorphic coordinates in a neighborhood of the origin; hence a tensor algebra adapted to this system of coordinates should have all indices appearing as pairs, like ir, js, $\cdots$, with the usual conventions about barred pairs of indices. The diastasis $\mathrm{D}(Z_1, Z_2)$ may be given by

$$\begin{aligned}&\mathrm{D}(Z_1, Z_2)\\ &=\log\det[(1^{(m')}-{}^t\overline{Z}_1Z_2)(1^{(m')}-{}^t\overline{Z}_2Z_1)(1^{(m')}-{}^t\overline{Z}_1Z_1)^{-1}(1^{(m')}-{}^t\overline{Z}_2Z_2)^{-1}]\ ;\end{aligned}\tag{3.4}$$

in particular, for $Z_1=0$, $Z_2=Z=(z^{ir})$, we have

$$\begin{aligned}\mathrm{D}(0, Z)&=\log\det(1^{(m')}-{}^t\overline{Z}Z)^{-1}=\textstyle\sum_{\nu=1}^{\infty}\mathrm{tr}\left(\frac{1}{\nu}({}^t\overline{Z}Z)^{\nu}\right)\\ &=\textstyle\sum_{i,r}|z^{ir}|^2+\frac{1}{2}\sum_{i,j,r,s}\bar{z}^{ir}z^{is}\bar{z}^{js}z^{jr}+\text{(higher order terms)}.\end{aligned}$$

From the above expansion one sees that (z^{ir}) are in fact canonical coordinates at the origin; thus we see that the curvature tensor at the origin (given by differentiating four times the diastasis) is

$$R_{ir\,\overline{js}\,kt\,\overline{lu}}(0)=\delta_{ij}\delta_{kl}\delta_{ru}\delta_{st}+\delta_{il}\delta_{jk}\delta_{rs}\delta_{tu}=-R^{js}{}_{ir\,kt}{}^{lu}(0)\ .\tag{3.5}$$

The space of symmetric, covariant tensors must now be written in the form $\xi_{ir\,js}$, where

$$\xi_{ir\,js}=\xi_{js\,ir}\ .\tag{3.6}$$

The isotropy group H acts on the tensors of this type as follows

$$\xi_{ir\,js}\longrightarrow\theta_i{}^k\theta_j{}^l\theta'_r{}^t\theta'_s{}^u\xi_{kt\,lu}\ ,$$

where $(\theta_i{}^k)\in\mathrm{U}(m)$, $(\theta'_r{}^t)\in\mathrm{U}(m')$; thus one sees that the tensors satisfying (3.6) have an invariant decomposition into two irreducible spaces, characterized one by

$$\xi_{ir\,js}=\xi_{jr\,is}=\xi_{is\,jr}=\xi_{js\,ir}\ ,\tag{3.7a}$$

and the other by

$$\xi_{ir\,js}=-\xi_{jr\,is}=-\xi_{is\,jr}=\xi_{js\,ir}\ .\tag{3.7b}$$

The space satisfying (3.7a) has $mm'(m+1)(m'+1)/4$ complex dimensions; the action of the curvature on this space can be calculated as follows:

$$\xi_{ir\,js} \longrightarrow R^{kt}{}_{ir\,js}{}^{lu}\xi_{kt\,lu} = -(\delta^k_i\delta^l_j\delta^u_r\delta^t_s + \delta^l_i\delta^k_j\delta^t_r\delta^u_s)\xi_{kt\,lu}$$
$$= -\xi_{is\,jr} - \xi_{jr\,is} = -2\xi_{ir\,js}\ ;$$

similary the space satisfying (3.7b) has $mm'(m-1)(m'-1)/4$ complex dimensions, and the action of the curvature is given by

$$\xi_{ir\,js} \longrightarrow R^{kt}{}_{ir\,js}{}^{lu}\xi_{kt\,lu} = -\xi_{is\,jr} - \xi_{jr\,is} = 2\xi_{ir\,js}\ .$$

In conclusion, the proper values $\lambda_1, \cdots, \lambda_N$ ($N = mm'(mm'+1)/2$) on the spaces of type A-$\mathrm{I}_{m,m'}$ are

-2 with multiplicity $mm'(m+1)(m'+1)/4$, and

$+2$ with multiplicity $mm'(m-1)(m'-1)/4$.

Type A–II_m. Denote by $Z = (z^{ij})$ a skew-symmetric, $m \times m$ complex matrix; the components z^{ij} for $i < j$ constitute a system of coordinates; a tensor system adapted to these coordinates will contain as indices pairs of letters ij, kl, $\cdots$, each skew-symmetric in each pair of letters making up a composite index. The diastasis $\mathrm{D}(Z_1, Z_2)$ in the domain of type A–II_m may be taken as

$$\begin{aligned}&\mathrm{D}(Z_1, Z_2)\\ &= \frac{1}{2}\log\det\left[(1^{(m)} - {}^t\overline{Z}_1Z_2)(1^{(m)} - {}^t\overline{Z}_2Z_1)(1^{(m)} - {}^t\overline{Z}_1Z_1)^{-1}(1^{(m)} - {}^t\overline{Z}_2Z_2)^{-1}\right]\ ;\end{aligned}\tag{3.8}$$

in particular, for $Z_1 = 0$, $Z_2 = Z = (z^{ij})$,

$$\mathrm{D}(0, Z) = \frac{1}{2}\log\det(1^{(m)} - {}^t\overline{Z}Z)^{-1} = \sum_{\nu=1}^{\infty}\frac{1}{2\nu}\operatorname{tr}\left(({}^t\overline{Z}Z)^\nu\right)$$
$$= \sum_{i<j}|z^{ij}|^2 + \frac{1}{4}\sum_{i,j,k,l}\bar{z}^{ik}z^{il}\bar{z}^{jl}z^{jk} + (\text{higher order terms})\ .$$

From this, one verifies that (z^{ij}) for $i < j$ are canonical coordinates at the origin. In the tensor formalism adapted to this system of coordinates, we shall adopt the convention, whenever a contraction between two composite indices occurs, to add only those terms where the pairs of letters are in increasing order: thus $\eta^{ij}\zeta_{ij}$ means $\sum_{i<j}\eta^{ij}\zeta_{ij} = \frac{1}{2}\sum_{i,j}\eta^{ij}\zeta_{ij}$. Consequently the components of the curvature tensor at the origin are given by differentiating $\mathrm{D}(0, Z)$ four times, as follows: from the formula $\partial z^{ij}/\partial z^{kl} = \delta^{ij}_{kl} = \delta^i_k\delta^j_l - \delta^i_l\delta^j_k$, we obtain

$$\begin{aligned}R^{i_1i_2}{}_{j_1j_2k_1k_2}{}^{l_1l_2}(0) &= -\left.\frac{\partial^4\mathrm{D}(0, Z)}{\partial\bar{z}^{i_1i_2}\partial z^{j_1j_2}\partial z^{k_1k_2}\partial\bar{z}^{l_1l_2}}\right|_{Z=0}\\ &= -\delta^{i_1i_2}_{j_1k_2}\delta^{l_1l_2}_{k_1j_2} - \delta^{l_1l_2}_{j_1k_2}\delta^{i_1i_2}_{k_1j_2} + \delta^{i_1i_2}_{j_1k_1}\delta^{l_1l_2}_{k_2j_2} + \delta^{l_1l_2}_{j_1k_1}\delta^{i_1i_2}_{k_2j_2}\ .\end{aligned}\tag{3.9}$$

The space of symmetric covariant tensors $\xi_{\alpha\beta}$ at the origin is written now in the form ξ_{ijkl}, satisfying

(3.10) $$\xi_{ijkl} = -\xi_{jikl} \ ; \qquad \xi_{ijkl} = -\xi_{ijlk}; \qquad \xi_{ijkl} = \xi_{klij} \ ;$$

the action of the isotropy group $H = \mathbf{U}(m)$ on this space is given by $\xi_{ijkl} \to \theta_i{}^p\theta_j{}^q\theta_k{}^r\theta_l{}^s\xi_{pqrs}$ $((\theta_i{}^p) \in \mathbf{U}(m))$. Thus the tensor space in question splits into two tensor subspaces, each corresponding to an irreducible representation of $\mathbf{U}(m)$: these two corresponding subspaces are described by the additional symmetry conditions, respectively,

(3.11a) $$\xi_{ijkl} + \xi_{iklj} + \xi_{iljk} = 0$$

and

(3.11b) $$\xi_{ijkl} = \xi_{iklj}$$

(implying that ξ_{ijkl} is alternating in each pair of the four subscripts).

The tensors ξ_{ijkl} satisfying (3.10) alone, under the action analogous to (2.22) of the curvature tensor (3.9) are transformed as follows:

$$\begin{aligned}\xi_{ijkl} \to R^{pq}{}_{ijkl}{}^{rs}(0)\xi_{pqrs} &= -\xi_{ilkj} - \xi_{kjil} + \xi_{iklj} + \xi_{iljk} \\ &= 2(\xi_{iljk} + \xi_{iklj}) \ .\end{aligned}$$

Thus, if ξ_{ijkl} satisfies (3.11a) it is mapped into $-2\xi_{ijkl}$, while if it satisfies (3.11b), its image is $4\xi_{ijkl}$. Since the space of tensors satisfying (3.10) and (3.11a) has the same formal symmetries as (in a real vector space) the Riemann curvature tensor in m dimensions, that tensor space has $m^2(m^2-1)/12$ dimensions. The space of tensors satisfying (3.10) and (3.11b) is equi-dimensional to the space of exterior 4-forms in m dimensions; hence it has $\binom{m}{4}$ dimensions. Summarizing, the proper values λ_a of (2.22) in a symmetric space of type A-II$_m$ are

-2 with multiplicity $m^2(m^2-1)/12$, and

$+4$ with multiplicity $\binom{m}{4}$.

Type A-III$_m$. Denote by $Z = (z^{ij})$ a symmetric, $m \times m$ complex matrix; the functions z^{ij} for $i \leqq j$ form a system of coordinates; a tensorial index in this space, referred to such a coordinate system is composed of two letters, each ranging from 1 to m, it being always understood that it is to be symmetric. The contraction of two composite indices will be understood to be one-half the usual tensorial contraction: thus $\eta^{ij}\zeta_{ij}$ means $(1/2)\sum_{i,j}\eta^{ij}\zeta_{ij} = \sum_{i<j}\eta^{ij}\zeta_{ij} + (1/2)\sum_{i=1}^{n}\eta^{ii}\zeta_{ii}$. Thus $\partial z^{ij}/\partial z^{kl} = \delta^i_k\delta^j_l + \delta^i_l\delta^j_k = e^{ij}_{kl}$ is the identity operator on the tangent space. The diastasis $\mathrm{D}(Z_1, Z_2)$ in a domain of type A-III$_m$ may be taken as the same function as in (3.8),

replacing skew symmetric by symmetric matrices; in particular, for $Z_1=0$, $Z_2 = Z = (z^{ij})$ we have

$$D(0, Z) = \frac{1}{2}\log\det(1^{(m)} - {}^t\bar{Z}Z)^{-1} = \sum_{\nu=1}^{\infty} \frac{1}{2\nu}\operatorname{tr}(({}^t\bar{Z}Z)^{\nu})$$
$$= \frac{1}{2}\sum_{i,j} z^{ij}\bar{z}^{ij} + \frac{1}{4}\sum_{i,j,k,l} \bar{z}^{ik}z^{il}\bar{z}^{jl}z^{jk} + \text{(higher order terms)}.$$

From this we verify that z^{ij} for $i < j$ and $2^{-1/2}z^{ij}$ for $i = j$ constitute a canonical coordinate system. However, it is convenient to retain z^{ij} $(i \leq j)$ as such, for in this coordinate system the metric at the origin 0 is given by

$$g_{ij\overline{kl}}(0) = \delta_{ik}\delta_{jl} + \delta_{il}\delta_{jk} ,$$

so that pulling composite indices down (or up) and simultaneously changing their parity (from barred to not and *vice versa*) at the origin, in this coordinate system does not affect the value of the components at the origin.

The components of the curvature tensor at the origin can be calculated as before, with the result

$$\text{(3.12)}\qquad \begin{aligned} R^{i_1i_2}{}_{j_1j_2k_1k_2}{}^{l_1l_2}(0) &= -\left.\frac{\partial^4 D(0, Z)}{\partial\bar{z}^{i_1i_2}\partial\bar{z}^{l_1l_2}\partial z^{j_1j_2}\partial z^{k_1k_2}}\right|_{Z=0} \\ &= -e^{i_1i_2}_{j_1k_2}e^{l_1l_2}_{k_1j_2} - e^{i_1i_2}_{k_1j_2}e^{l_1l_2}_{j_1k_2} - e^{i_1i_2}_{j_1k_1}e^{l_1l_2}_{k_2j_2} - e^{i_1i_2}_{k_2j_2}e^{l_1l_2}_{j_1k_1} . \end{aligned}$$

The space of symmetric, covariant tensors $\xi_{\alpha\beta}$ at the origin can be written in the form ξ_{ijkl}, satisfying

$$\text{(3.13)}\qquad \xi_{ijkl} = \xi_{jikl}; \qquad \xi_{ijkl} = \xi_{ijlk}; \qquad \xi_{ijkl} = \xi_{klij} ,$$

and the representation of the isotropy group $H = \mathbf{U}(m)$ on this tensor space is given by

$$\xi_{ijkl} \to \theta_i{}^p\theta_j{}^q\theta_k{}^r\theta_l{}^s\xi_{pqrs} \qquad ((\theta_i{}^p) \in \mathbf{U}(m)) .$$

Thus the tensor space is split by this representation into two invariant subspaces, each corresponding to an irreducible representation of $\mathbf{U}(m)$: the systems of linear equations defining these two subspaces are, in addition to (3.13), respectively,

$$\text{(3.14a)}\qquad \xi_{ijkl} = \xi_{iklj}$$

(i.e., ξ_{ijkl} symmetric in each pair out of the four subscripts) and

$$\text{(3.14b)}\qquad \xi_{ijkl} + \xi_{iklj} + \xi_{iljk} = 0 .$$

The tensors ξ_{ijkl} satisfying (3.13) alone, under the action of the curvature tensor (3.12) as in (2.22), are transformed as follows:

$$\xi_{ijkl} \longrightarrow R^{pq}{}_{ijkl}{}^{rs}(0)\ \xi_{pqrs} = -\xi_{ilkj} - \xi_{kjil} - \xi_{iklj} - \xi_{ljik} = -2(\xi_{ikjl} + \xi_{iljk})\ ;$$

one sees from this that, if ξ_{ijkl} satisfies (3.14a), it is mapped into $-4\xi_{ijkl}$, while if it satisfies (3.14b), it is mapped into $2\xi_{ijkl}$. The dimensionality of these two irreducible invariant tensor spaces are respectively $\binom{m+3}{4}$ and $m^2(m^2-1)/12$, so that the proper values λ_a of the linear transformation (2.22) in a space of type A–III$_m$ are

-4 with multiplicity $\binom{m+3}{2}$, and

$+2$ with multiplicity $m^2(m^2-1)/12$.

Type A–IV$_m$. Here we use $Z = {}^t(z^1, \cdots, z^m)$ as coordinates for the domain defined by (3.2); we shall use, in addition, the function $w = \sum_{\alpha=1}^m (z^\alpha)^2 = {}^tZZ$ to abbreviate some of the formulas. For this domain the diastasis $\mathrm{D}(Z_1, Z_2)$ is given by

$$\begin{aligned}\mathrm{D}(Z_1, Z_2) &= \log\,[(1 + \overline{w}_1 w_2 - 2{}^t\overline{Z}_1 Z_2)(1 + \overline{w}_2 w_1 - 2{}^t\overline{Z}_2 Z_1)] \\ &\quad -\log\,[(1 + |\,w_1\,|^2 - 2{}^t\overline{Z}_1 Z_1)(1 + |\,w_2\,|^2 - 2{}^t\overline{Z}_2 Z_2)]\ .\end{aligned} \tag{3.15}$$

In particular, for $Z_1 = 0$, $Z_2 = Z$, we have

$$\begin{aligned}\mathrm{D}(0, Z) &= -\log\,(1 - 2{}^t\overline{Z}Z + |\,w\,|^2) \\ &= 2\sum_{\alpha=1}^m |\,z^\alpha\,|^2 - \sum_{\alpha,\beta} (z^\alpha)^2(\overline{z}^\beta)^2 + 2\sum_{\alpha,\beta} |\,z^\alpha\,|^2\,|\,z^\beta\,|^2 \\ &\qquad + \text{(higher order terms).}\end{aligned}$$

From this one sees that a canonical coordinate system at the origin should be given by $(z^\alpha\sqrt{2}\,)$; however, it is not necessary to effect the clumsy coordinate transformation, if we point out that at the origin the components of the metric are $g_{\alpha\bar\beta}(0) = 2\delta_{\alpha\beta}$; thus we have, by the usual routine,

$$\begin{aligned}R^\rho{}_{\alpha\beta}{}^\sigma(0) &= -\frac{1}{4} R_{\alpha\bar\rho\beta\bar\sigma}(0) = -\frac{1}{4}\, \frac{\partial^4 \mathrm{D}(0, Z)}{\partial z^\alpha \partial \overline{z}^\rho \partial z^\beta \partial \overline{z}^\sigma}\bigg|_{Z=0} \\ &= \delta_{\alpha\beta}\delta^{\rho\sigma} - \delta^\rho_\alpha \delta^\sigma_\beta - \delta^\sigma_\alpha \delta^\rho_\beta\ .\end{aligned} \tag{3.16}$$

The space of symmetric, covariant tensors at the origin gives the following representation of the isotropy group $H = \mathbf{SO}(m) \times \mathbf{U}(1)$

$$\xi_{\alpha\beta} \longrightarrow e^{2it}\theta_\alpha{}^\rho \theta_\beta{}^\sigma \xi_{\rho\sigma} \qquad (t \in \mathbf{R},\ (\theta_\alpha{}^\rho) \in \mathbf{SO}(m))\ .$$

This representation is reducible, the two invariant subspaces being defined by the symmetric tensors $\xi_{\alpha\beta}$ that satisfy respectively

$$\sum_{\alpha=1}^m \xi_{\alpha\alpha} = 0 \tag{3.17a}$$

and

(3.17b) $$\xi_{\alpha\beta} = x\delta_{\alpha\beta} \qquad (x \in \mathbf{C}) .$$

The linear operator (2.22) on the tensors $\xi_{\alpha\beta}$ at the origin, with the curvature tensor given by (3.16) yields

$$\xi_{\alpha\beta} \to R^{\rho}{}_{\alpha\beta}{}^{\sigma}\xi_{\rho\sigma} = -2\xi_{\alpha\beta} + \sum_{\rho=1}^{m} \xi_{\rho\rho}\delta_{\alpha\beta} ,$$

thus, if $\xi_{\alpha\beta}$ satisfies (3.17a), its image is $-2\xi_{\alpha\beta}$, while if it satisfies (3.17b), the image is $(m-2)\xi_{\alpha\beta}$. Taking into account the dimensionality of these two respective spaces, the proper values λ_a of (2.22) are the following:

-2 with multiplicity $(m-1)(m+2)/2$, and

$m-2$ with multiplicity 1.

The corresponding calculations for the compact, irreducible, symmetric Kähler manifolds are quite similar to the ones just completed. In fact, we can use as local coordinates the components of matrices for types B–I, –II, –III and –IV exactly as in the noncompact forms, and give the corresponding formulas for the diastasis: for the manifolds of types B–I, –II, –III,

(3.18) $$\mathrm{D}(Z_1, Z_2) = c \log\det[(1 + {}^t\overline{Z}_1 Z_1)(1 + {}^t\overline{Z}_2 Z_2)(1 + {}^t\overline{Z}_1 Z_2)^{-1}(1 + {}^t\overline{Z}_2 Z_1)^{-1}]$$

with $c = 1, 1/2$, and $1/2$ and $1 = 1^{(m')}$, $1 = 1^{(m)}$, $1 = 1^{(m)}$ for the three types respectively and, for spaces of type B-IV, the diastasis is

(3.19) $$\begin{aligned}\mathrm{D}(Z_1, Z_2) = {} & \log[(1 + 2{}^t\overline{Z}_1 Z_1 + |{}^t Z_1 Z_1|^2)(1 + 2{}^t\overline{Z}_2 Z_2 + |{}^t Z_2 Z_2|^2)] \\ & -\log[(1 + 2{}^t\overline{Z}_1 Z_2 + {}^t\overline{Z}_1\overline{Z}_1{}^t Z_2 Z_2)(1 + {}^t\overline{Z}_2 Z_1 + {}^t\overline{Z}_2\overline{Z}_2{}^t Z_1 Z_1)] .\end{aligned}$$

By the same arguments one sees that the curvature tensor at the origin in the canonical coordinates chosen, for the spaces of type B is the exact opposite of that for the corresponding space of type A. Thus the proper values of (2.22), written in the reverse order, are opposite in sign to, and with the same multiplicities as, the corresponding ones for the spaces of type A.

The curvatures for the two exceptional spaces (both the noncompact and compact forms of each) have not been computed by this method; however the authors are indebted to A. Borel for the results [10] of his calculation. The proper values λ_a for these spaces are as follows:

Type A–V: -1 with multiplicity 126, and
$+3$ with multiplicity 10.

Type A–VI: -1 with multiplicity 351, and
$+4$ with multiplicity 27.

For the spaces of type B-V and B–VI the proper values are just reversed in sign from the above.

It is remarkable to see that, except for the Fubini-Study spaces (Types A–$I_{1,m}$ and B–$I_{1,m}$), where all proper values coincide, in all other cases there are exactly two distinct, proper values, always opposite in sign.

3. Proof and Corollaries of Theorems 1 and 2

The proof of Theorem 1 is now immediate from Lemma 4, using the invariants of the curvature tensor calculated in the preceding paragraph. Since $R < 0$, it follows that only case (c) of Lemma 4 applies, and the statement of Theorem 1 in Chapter 1 needs only the calculation of $\gamma(D) = R/n\lambda_1$ (according to Lemma 4), using the fact that $R = 2\sum_{a=1}^{N} \lambda_a$ ($N = n(n+1)/2$, $n = \dim_{\mathbf{C}} D$). This calculation can be carried out, as in Table 1.

Table 1

Type of D	$\dim_{\mathbf{C}} D$	λ_1	multiplicity of λ_1	λ_N	multiplicity of λ_N
A–$I_{m\,m'}$ $(1 \leq m \leq m')$	mm'	-2	$\binom{m+1}{2}\binom{m'+1}{2}$	$+2$	$\binom{m}{2}\binom{m'}{2}$
A–II_m $(m \geq 2)$	$\frac{m(m-1)}{2}$	-2	$\frac{m^2(m^2-1)}{12}$	$+4$	$\binom{m}{4}$
A–III_m $(m \geq 1)$	$\frac{m(m+1)}{2}$	-4	$\binom{m+3}{4}$	$+2$	$\frac{m^2(m^2-1)}{12}$
A–IV_m $(m \geq 3)$	m	-2	$\frac{(m-1)(m+2)}{2}$	$m-2$	1
A–V	16	-1	126	$+3$	10
A–VI	27	-1	351	$+4$	27

Type of D	R	$\frac{R}{n\lambda_1} = \gamma(D)$
A–$I_{m,m'}$ $(1 \leq m \leq m')$	$-2mm'(m+m')$	$m+m'$
A–II_m $(m \geq 2)$	$-2m(m-1)^2$	$2(m-1)$
A–III_m $(m \geq 1)$	$-2m(m+1)^2$	$m+1$
A–IV_m $(m \geq 3)$	$-2m^2$	m
A–V	-192	12
A–VI	-486	18

Concerning the value of $\gamma(D)$ we make the following observations. The constant $\gamma(D)$ has the same values as the constants $\lambda(G/U)$ calculated by A. Borel and F. Hirzebruch [9] to express the degree of the canonical divisor in the corresponding compact form of a symmetric Kähler space.

The strongest implication of the statement of Theorem 1 occurs in the case where D is equivalent to the metric unit ball (type A–$I_{1,m}$); in this case and in this alone $\gamma(D) = n + 1$ ($n = \dim_C D$), so that all groups $H^q(X, \Theta)$ except for $q = n$ vanish for any compact manifold X uniformized by D. Next in interest from this standpoint are the domains D of type A–IV, for which $\gamma(D) = n = \dim_C D$; for all compact manifolds X uniformizable by such a domain D, all $H^q(X, \Theta)$ vanish except possibly for $q = n$ or $n - 1$. In all other cases the number of groups whose vanishing is asserted by Theorem 1 is comparatively much lower, relative to the dimensions of D. However, it is verified that, in all cases of irreducible Cartan domains D with n complex dimensions, the constant $\gamma(D)$ satisfies $\sqrt{2n} < \gamma(D) \leqq n + 1$. In particular, if $n \geqq 2$, then $\gamma(D) \geqq 3$, so that each compact, complex manifold, covered by D with at least two complex dimensions has at least $H^1(X, \Theta) = 0$. It follows from Proposition 1 that the complex analytic structure of each such manifold X is locally rigid, in sharp contrast with the one-dimensional case, consisting of algebraic curves of genus $g \geqq 2$, for which there are $3g - 3$ moduli of deformation of complex analytic structure. This completes the proof of both Theorems 1 and 2.

Suppose now that a compact complex manifold X is uniformizable by a *reducible* Cartan domain D, whose decomposition into irreducible factors is $D = D_1 \times \cdots \times D_l$. In order to apply Lemma 4 to this case, we must find a metric that is an Einstein-Kähler metric. This can be accomplished easily by using the Bergmann metric, for which $R_{\alpha\bar\beta} = -g_{\alpha\bar\beta}$. Then this metric is itself a product metric, splitting in each irreducible factor D_μ into a Bergmann metric; the latter is proportional by a constant, positive factor to the metric used in § 2 to calculate the curvature. The action of the curvature tensor described by (2.22) also splits as follows: let $\Theta^* = \sum_{\mu=1}^{l} \Theta^*_\mu$ be the direct sum decomposition of the covariant complex tangent bundle of D into the subspaces corresponding to $D_1, \cdots, D_l$ respectively, and denote by $\Theta^*_\mu \cdot \Theta^*_\nu$ the *symmetrized* tensor product of Θ^*_μ with Θ^*_ν. Then a covariant, symmetric tensor $\xi_{\alpha\beta}$ at a point of X can be lifted (in many ways) into a similar tensor at a point of D, i.e., to a point in the bundle $\Theta^* \cdot \Theta^*$. The latter bundle is a direct sum given by $\Theta^* \cdot \Theta^* = \sum_{1 \leqq \mu \leqq \nu \leqq l} \Theta^*_\mu \cdot \Theta^*_\nu$. Now the linear transformation (2.22) leaves each space $\Theta^*_\mu \cdot \Theta^*_\nu$ invariant, and annihilates all the ones where $\mu \neq \nu$; when $\mu = \nu$ it acts as in D_μ itself, so that it has a spectrum (up to a positive factor) equivalent to the one shown in Table 1. If λ_1 is the lowest proper value of the transformation in $\Theta^* \cdot \Theta^*$, it is negative, and equals that for some D_{μ_0}. Furthermore, since D is an Einstein space, so is each D_μ: in particular, if $n = \dim_{\mathbf{C}} D$, $n_\mu = \dim_{\mathbf{C}} D_\mu$ and R, R_μ are the scalar curvatures of D, D_μ respectively, we have $R/n = R_\mu/n_\mu = R_{\mu_0}/n_{\mu_0}$: thus $\gamma(D) = R/n\lambda_1 = R_{\mu_0}/n_{\mu_0}\lambda_1 = \gamma(D_{\mu_0}) = \min_\mu \gamma(D_\mu)$. Thus we have proved the following result:

COROLLARY TO THEOREM 1. *Let X be a compact, complex manifold whose universal covering space is holomorphically equivalent to a reducible Cartan domain D, and let $D = D_1 \times \cdots \times D_l$ be the decomposition of D into its irreducible factors. Then the cohomology groups $H^q(X, \Theta)$ with coefficients in Θ vanish for all $q < \min_\mu \gamma(D_\mu) - 1$.*

REMARK. The above result is similar to a special case of Künneth formula [16], but is actually somewhat stronger, since X need not be a product manifold. At best, there is a finite covering of X that has a *local* product structure.

Applying again Proposition 1 (Frölicher-Nijenhuis), we obtain from the above corollary the following one.

COROLLARY TO THEOREM 2. *Let X be a compact, complex manifold, whose universal covering space is holomorphically equivalent to a reducible Cartan domain D, and let $D = D_1 \times \cdots \times D_l$ be the direct product decomposition of D into irreducible Cartan domains. Then, if $\dim_{\mathbf{C}} D_\nu \geqq 2$ for $1 \leqq \nu \leqq l$, the complex analytic structure of X is rigid, even with no restriction on the complex analytic structure of its covering space. In other words, X admits no moduli of complex structure.*

4. Proof of Theorem 3

Having calculated the invariants λ_a of the curvature for the compact, symmetric Kähler manifolds Y (of type B), we can apply Lemma 4 to prove Bott's theorem [11] by our own method. Since the scalar curvature R is positive in all types of Y, we shall need to consider only cases (a) and (b) of Lemma 4, according as to whether $\lambda_1 \geqq 0$ or $\lambda_1 < 0$. The calculations are shown in Table 2, which is arranged as follows: we calculate λ_1; if this is positive or if it is zero, since $R > 0$, case (a) of Lemma 4 applies; if it is negative and $R/n + \lambda_1 > 0$, then case (b) of Lemma 4 applies, and $H^q(Y, \Theta) = 0$ for all $q > \mu(Y)$, where $\mu(Y) = -n\lambda_1(R + n\lambda_1)^{-1}$, so that this last constant is evaluated. It will be noted that, whenever applicable, $\mu(Y) < 1$, this being a consequence of the fact (apparently fortuitous) that $R + 2n\lambda_1 > 0$ for all compact, irreducible symmetric Kähler manifolds Y. Thus we see from Table 2 that the only non-vanishing cohomology group $H^q(Y, \Theta)$ is for $q = 0$. The zero-dimensional group is the group of global cross-sections of Θ over Y, i.e., the additive group of

Table 2

Type of Space Y	$n = \dim_{\mathbf{C}} Y$	λ_1	multiplicity of λ_1	λ_N (if $> \lambda_1$)	multiplicity of λ_N
B–$\mathrm{I}_{1,m}$ $(m \geqq 1)$	m	$+2$	$\frac{m(m+1)}{2}$	$(=\lambda_1)$	
B–$\mathrm{I}_{m,m'}$ $(2 \leqq m \leqq m')$	mm'	-2	$\binom{m}{2}\binom{m'}{2}$	$+2$	$\binom{m+1}{2}\binom{m'+1}{2}$
B–II_m $(m \geqq 4)$	$\frac{m(m-1)}{2}$	-4	$\binom{m}{4}$	$+2$	$\frac{m^2(m^2-1)}{12}$
B–III_m $(m \geqq 2)$	$\frac{m(m+1)}{2}$	-2	$\frac{m^2(m^2-1)}{12}$	$+4$	$\binom{m+3}{4}$
B–IV_m $(m \geqq 3)$	m	$2-m$	1	$+2$	$\frac{(m-1)(m+2)}{2}$
B–V	16	-3	10	$+1$	126
B–VI	27	-4	27	$+1$	351

Type of Space Y	R	$\frac{R}{n} + \lambda_1$	$\frac{-n\lambda_1}{R + n\lambda_1}$	$H^q(Y, \Theta) = 0$ for:
B–$\mathrm{I}_{1,m}$ $(m \geqq 1)$	(>0)			$q \geqq 1$
B–$\mathrm{I}_{m,m'}$ $(2 \leqq m \leqq m'$	$2mm'(m+m')$	$2(m+m'-1)$	$\frac{1}{m+m'-1}$ (<1)	$q \geqq 1$
B–II_m $(m \geqq 4)$	$2m(m-1)^2$	$4(m-2)$	$\frac{1}{m-2}$ (<1)	$q \geqq 1$
B–III_m $(m \geqq 2)$	$2m(m+1)^2$	$4m+2$	$\frac{1}{2m+1}$ (<1)	$q \geqq 1$
B–IV_m $(m \geqq 3)$	$2m^2$	$m+2$	$\frac{m-2}{m+2}$ (<1)	$q \geqq 1$
B–V	192	9	$\frac{1}{3}$	$q \geqq 1$
B–VI	486	14	$\frac{2}{7}$	$q \geqq 1$

the complex Lie algebra of the group $G_{\mathbf{C}}$ of all holomorphic transformations of Y onto itself. Since the group $G_{\mathbf{C}}$ is the complex Lie group generated by the compact group G of all isometries of Y, we have $\dim_{\mathbf{C}} G_{\mathbf{C}} = \dim_{\mathbf{R}} G$. Thus all cohomology groups $H^q(Y, \Theta)$ are known and Theorem 3 is thereby proved.

CHAPTER 4. CHARACTERISTIC OF THE SHEAF Θ

1. The Riemann-Roch theorem and the proportionality principle

Given any compact, complex manifold X, we shall denote by $c_i(X) \in H^{2i}(X, \mathbf{Z})$ the ith Chern class of X. Corresponding to any positive partition (μ) of $n = \dim_{\mathbf{C}} X$, $(\mu) = (\mu_1, \cdots, \mu_n)$ $(\mu_i \geqq 0, 1 \leqq i \leqq n,$

$\sum_i \mu_i = n$), the Chern number $c_{(\mu)}[X]$ is defined by the formula

$$c_{(\mu)}[X] = c_{\mu_1}(X)\cdot c_{\mu_2}(X)\cdot \cdots \cdot c_{\mu_n}(X)[X] ,$$

where the right hand member denotes the value of the cup product $c_{\mu_1}(X)\cdot \cdots \cdot c_{\mu_n}(X) \in H^{2n}(X, \mathbf{Z})$ on the fundamental cycle defined by X with its natural orientation. In particular the Euler-Poincaré characteristic $\chi(X) = \chi(X, \mathbf{Z})$ of X is equal to the Chern number $c_n(X)[X]$.

Assume X to be a nonsingular, projective algebraic manifold. In view of Hirzebruch's formulation of the Riemann-Roch theorem [18, Hauptsatz 20.3.2., p. 148; Hauptsatz 21.1.1, p. 154], the characteristic $\chi(X, \mathbf{\Theta}) = \sum_{q=0}^{n}(-1)^q \dim_{\mathbf{C}} H^q(X, \mathbf{\Theta})$ ($n = \dim_{\mathbf{C}} X$) as well as the arithmetic genus $\chi(X, \mathbf{\Omega}^0) = \sum_{q=0}^{n}(-1)^q \dim_{\mathbf{C}} H^q(X, \mathbf{\Omega}^0)$ are expressible as rational linear combinations of the Chern numbers of X. For example, the combinations occurring for low dimensionalities are given in the following table.

Table 3

$n = \dim_{\mathbf{C}} X$	$\chi(X, \Omega^\circ)$	$\chi(X, \Theta)$
1	$\frac{1}{2} c_1[X]$	$\frac{3}{2} c_1[X] = 3 - 3g$ (g=Riemannian genus of X)
2	$\frac{1}{12}(c_1^2 + c_2)[X]$	$\frac{1}{6}(7c_1^2 - 5c_2)[X] = 2c_1^2[X] - 10\chi(X, \Omega^\circ)$
3	$\frac{1}{24} c_1c_2[X]$	$\frac{1}{24}(12c_1^3 - 19c_1c_2 + 12c_3)[X] = \frac{1}{2} c_1^3[X] - 19\chi(X, \Omega^\circ) + \frac{1}{2}\chi(X)$
4	$\frac{1}{720}(-c_4 + c_3c_1 + 3c_2^2 + 4c_2c_1^2 - c_1^4)[X]$	$\frac{1}{180}(-31c_4 + 76c_3c_1 + 3c_2^2 - 71c_2c_1^2 + 29c_1^4)[X]$

Following Hirzebruch [19, 20], two n-dimensional, compact, complex manifolds X and Y are called *proportional*, if the Chern numbers $c_{(\mu)}[X]$ and $c_{(\mu)}[Y]$ corresponding to all the positive partitions of n are proportional, i.e., if there exists a rational number a such that, for any positive partition (μ) of n, $c_{(\mu)}[X] = ac_{(\mu)}[Y]$. We write this relation briefly

$$X = aY . \tag{4.1}$$

Let $\mathbf{\Theta}'$ denote the sheaf of germs of holomorphic tangent vector fields in Y. It follows immediately from the definition of proportionality and from the Riemann-Roch theorem that, if X and Y are nonsingular, projective algebraic varieties satisfying (4.1), then the Euler-Poincaré characteristics $\chi(X)$ and $\chi(Y)$, the arithmetic genera $\chi(X, \mathbf{\Omega}^\circ)$ and $\chi(Y, \mathbf{\Omega}^\circ)$

and the characteristics $\chi(X, \Theta)$ and $\chi(Y, \Theta')$ are related by

$$\chi(X) = a\chi(Y)$$
$$\chi(X, \Omega^0) = a\chi(Y, \Omega^0) \tag{4.2}$$
$$\chi(X, \Theta) = a\chi(Y, \Theta') .$$

2. Proof of Theorem 4

Let D be an irreducible, bounded, symmetric domain, let Y be the corresponding compact, symmetric manifold (see Chapter 3), and let $X = \Gamma\backslash D$ be a compact, complex manifold obtained as the quotient of D by a discrete group Γ of holomorphic transformations without fixed points. The metric induced in $X = \Gamma\backslash D$ by the Bergmann metric of D is a Hodge metric; thus [22] X is holomorphically equivalent to a nonsingular, projective algebraic manifold. Using the fact that the curvature tensors in X and Y differ, so to speak, only in sign, Hirzebruch [19] has shown that X and Y are proportional, i.e., $X = aY$; furthermore the constant of proportionality a is positive or negative, according to whether $n = \dim_{\mathbf{C}} X$ is, respectively, even or odd.

Since Y is a rational algebraic manifold[1], its arithmetic genus $\chi(Y, \Omega^0)$ is 1; hence one sees from (4.2) that the constant a of proportionality occurring in (4.1) is an integer

$$X = \chi(X, \Omega^0) \cdot Y = (-1)^n \,|\, \chi(X, \Omega^0) \,|\, \cdot Y \qquad (n = \dim_{\mathbf{C}} X) \tag{4.3}$$

and hence, also the Euler-Poincaré and the Θ-characteristics satisfy respectively

$$\chi(X) = (-1)^n \,|\, \chi(X, \Omega^0) \,|\, \chi(Y) ,$$
$$\chi(X, \Theta) = (-1)^n \,|\, \chi(X, \Omega^0) \,|\, \chi(Y, \Theta') . \tag{4.4}$$

The proof of Theorem 4 can be completed by calculating the constants $a_j(D)$ ($j = 1, 2, 3$) occurring in (1.1); the data to calculate the value of $a_3(D)$ will be found in [19]; the others depend on the fact that, for the compact, irreducible symmetric Kähler spaces Y, by Theorem 2 we have

$$\chi(Y, \Theta') = \dim_{\mathbf{C}} H^0(Y, \Theta') = \dim_{\mathbf{R}} G ,$$

where G is the group of isomorphisms of Y. The Euler-Poincaré characteristic $\chi(Y)$ is well known: thus, if D is the Cartan domain corresponding to Y,

$$a_1(D) = \dim_{\mathbf{R}} G$$
$$a_2(D) = a_1(D)/\chi(Y) \tag{4.5}$$

[1] This follows from a general result of Goto [17] or also from the classification of compact, symmetric Kähler spaces.

For the sake of completeness, we list the values of $\dim_{\mathbf{R}} G$ and of $\chi(Y)$ in the following table.

Table 4

Type of Y:	$\mathrm{I}_{m,m'}$	II_m	III_m	IV_m	V	VI
$\dim_{\mathbf{R}} G$	$(m+m')^2-1$	$m(2m-1)$	$m(2m+1)$	$\frac{1}{2}(m+1)(m+2)$	78	136
$\lambda(Y)$	$\binom{m+m'}{m}$	2^{m-1}	2^m	$m+2$ (m even) $m+1$ (m odd)	27	56

The above table together with (4.5) completes the proof of Theorem 4.

3. An application of Theorem 4

Returning to the remarks in Chapter 3, § 3 following the proof of Theorem 1, we notice that, if D is of type $\mathrm{I}_{1,n}$, the only non-trivial cohomology group $H^q(X, \Theta)$ occurs for $q = n = \dim_{\mathbf{C}} X$. Hence, using Theorem 4,

$$\dim_{\mathbf{C}} H^n(X, \Theta) = (-1)^n \chi(X, \Theta) = |\chi(X, \Theta)| = n(n+2)|\chi(X, \Omega^0)|$$

$$= \frac{n(n+2)}{n+1}|\chi(X)| = \frac{n(n+2)n!}{(4\pi n(n+1))^n}|R|^n v(X) ,$$

where $v(X)$ is the volume of X (cf. also [21]).

Similarly, if D is of type IV_n, then the only non-trivial groups $H^q(X, \Theta)$ occur for $q = n-1$ or n ($n = \dim_{\mathbf{C}} X$); hence we obtain only a lower bound in this case for $\dim_{\mathbf{C}} H^n(X, \Theta)$:

$$\dim_{\mathbf{C}} H^n(X, \Theta) \geqq \dim_{\mathbf{C}} H^n(X, \Theta) - \dim_{\mathbf{C}} H^{n-1}(X, \Theta) = |\chi(X, \Theta)|$$

$$= \binom{n+2}{2}|\chi(X, \Omega^0)| = \frac{(n+1)(n+2)}{4+4[n/2]}|\chi(X)| = \frac{(n+2)!}{4(4\pi n^2)^n}|R|^n v(X) .$$

The chief interest of these results lies in the fact that, by the Serre duality theorem we have, in particular,

$$H^n(X, \Theta) \simeq H^0(X, \Omega^1 \otimes \Omega^n) \qquad (n = \dim_{\mathbf{C}} X) ,$$

where the group on the right hand side represents the space of holomorphic covariant vector densities, defined globally on X. Thus we have proved the following existence statement.

COROLLARY TO THEOREM 4. *Let X be a compact, n-dimensional complex manifold, whose universal covering space is analytically isomorphic to an irreducible Cartan domain D of type* A–$\mathrm{I}_{1,n}$ *(respectively,* A–IV_n*). Then the space of holomorphic covariant vector densities globally defined on X, $H^0(X, \Omega^1 \otimes \Omega^n)$, has at least* $n(n+2)$ *(respectively*

$\frac{1}{2}(n+1)(n+2))$ *complex dimensions. More precisely,*

$$\dim_{\mathbf{C}} H^0(X, \Omega^1 \otimes \Omega^n) = kn(n+2), \qquad \text{if } D \text{ is of type A–I}_{1,n};$$

$$\dim_{\mathbf{C}} H^0(X, \Omega^1 \otimes \Omega^n) \geqq \frac{1}{2} k(n+1)(n+2), \qquad \text{if } D \text{ is of type A–IV}_n,$$

where k is a positive integer, equal to $(-1)^n \chi(X, \Omega^0)$.

UNIVERSITY OF MINNESOTA
UNIVERSITY OF PISA

BIBLIOGRAPHY

1. M. BERGER, *Les espaces symétriques non compacts*, Ann. Ec. Norm. Sup. 74 (1957), 85–177.
2. S. BOCHNER and D. MONTGOMERY, *Groups of differentiable and real or complex analytic transformations*, Ann. of Math., 46 (1945), 685–694.
3. S. BOCHNER, *Curvature in Hermitian metric*, Bull. Amer. Math. Soc., 53 (1947), 179–195.
4. ———, *Tensor fields with finite bases*, Ann. of Math.. 53 (1951), 400–411.
5. ———, *A new viewpoint in differential geometry*, Can. J. Math., 3 (1951), 460–470.
6. A. BOREL et J. DE SIEBENTHAL, *Les sous-groupes de rang maximum des groupes de Lie clos.*, Comm. Math. Helv., 23 (1949), 200–221.
7. A. BOREL, *Les espaces hermitiens symétriques*, Séminaire Bourbaki, May, 1952.
8. ———, *Les fonctions automorphes de plusieurs variables complexes*, Bull. Soc. Math. France, 80 (1952), 167–182.
9. ——— and F. HIRZEBRUCH, *Characteristic classes and homogeneous spaces* I, Am. J. Math., 80 (1958), 458–538.
10. ———, *On the curvature tensor of Hermitian symmetric manifolds*, Ann. of Math., 71 (1960), 508-521.
11. R. BOTT, *Homogeneous vector bundles*, Ann. of Math., 66 (1957), 203–248.
12. E. CALABI, *Isometric imbedding of complex manifolds*, Ann. of Math., 58 (1953), 1–23.
13. É. CARTAN, *Sur une classe remarquable d'espaces de Riemann*, Bull. Soc. Math. France, 54 (1926), 214–264; 55 (1927), 114–134; also, *Œuvres Complètes*, Partie I, vol. 2, Paris, Gauthier-Villars, 1952, 587–659.
14. ———, *Sur les domaines bornés homogènes de l'espace de n variables complexes*, Abh. Math. Sem. Hamb. Univ., 11 (1935), 116–162; also, *Œuvres Complètes*, Partie I, vol. 2, Paris, Gauthier-Villars, 1952, 1256–1305.
15. A. FRÖLICHER and A. NIJENHUIS, *A theorem of stability of complex structures*, Proc. Nat. Acad. Sci., U. S. A., 43 (1957), 239–241.
16. R. GODEMENT, Topologie algébrique et théorie des faisceaux, Paris, Hermann et Cie., 1958.
17. M. GOTO, *On algebraic homogeneous spaces*, Am. J. of Math., 76 (1954), 811–818.
18. F. HIRZEBRUCH, Neue topologische Methoden in der algebraischen Geometrie, Berlin, Springer, 1956.
19. ———, Automorphe Formen und der Satz von Riemann-Roch, Symposium internacional de Topologia Algebraica, Mexico, 1958, 129–144.
20. F. HIRZEBRUCH, Characteristic numbers of homogeneous domains, Seminars on Analytic Functions, Vol. 2, Princeton, 1957, 92–104.
21. J.-I. IGUSA, *On the structure of a special class of Kähler varieties*, Am. J. Math.,

76 (1954), 669–678.

22. K. KODAIRA, *On Kähler varieties of restricted type*, Ann. of Math., 60 (1954), 28–48.
23. ——— and D. C. SPENCER, *On deformations of complex analytic structures*, I and II, Ann. of Math., 67 (1958), 328–460.
24. S. NAKANO, *On complex analytic vector bundles*, J. Math. Soc. Japan, 7 (1955), 1–12.
25. C. L. SIEGEL, Analytic functions of several complex variables, Mimeographed notes, Institute for Advanced Study, Princeton, 1949.
26. D. C. SPENCER, Differentiable manifolds, sheaves and connections, Mimeographed notes, Princeton University, Princeton, 1955.
27. K. YANO and S. BOCHNER, Curvature and Betti numbers, Ann. of Math. Studies, No. 32, Princeton University Press, 1953.

On Compact, Riemannian Manifolds with Constant Curvature I

Eugenio Calabi[1]

C. B. Allendoerfer (Ed.) *Differential Geometry*,
Proc. Sympos. Pure Math., Vol. III, 155–180,
American Mathematical Society, Providence, RI (1961)

1. Introduction. The purpose of this paper is to calculate the cohomology groups $H^q(X, \Theta)$ of a compact, n-dimensional, Riemannian manifold X with constant curvature (Clifford-Klein space), with coefficients in the sheaf Θ of germs of Killing vector fields. In principle, one could calculate by similar techniques the corresponding cohomology groups for any compact, locally homogeneous Riemannian manifold; the special case under consideration, that of manifolds with constant curvature, is the simplest one to handle by the methods described herein, and the computations have been carried to a satisfactory degree of completion for this case alone.

The reason for our being interested in any of the groups $H^q(X, \Theta)$ of degree q lies solely in the geometrical meaning of these groups for $q = 0$ and 1. The groups of degree 0 represent the linear space of all global vector fields in X satisfying Killing's equation; such vector fields form a Lie algebra under the Poisson "bracket" operation, and as such they are the Lie algebra of the Lie group of all isometric transformations of X (in the more general case of locally homogeneous manifolds, one should specify "admissible" transformations, since not all isometries are necessarily compatible with the local homogeneous structure). If X has a fundamental group Γ and universal covering space $\tilde{X}$ then, since X is compact and hence complete, $\tilde{X}$ is the ordinary, simply connected space of n-dimensional, non-Euclidean geometry (of elliptic, parabolic, or hyperbolic kind, depending on the sign of the constant curvature); thus $\tilde{X}$ is a homogeneous space of Lie groups G/H, where G is either $SO(n + 1)$, or the group of all proper Euclidean motions in Euclidean n-space, or the connected, open subgroup $SO^+(n, 1)$ of the Lorentz group in $n + 1$ dimensions, and H is the common, closed subgroup $SO(n)$ of each of the groups; Γ is then a discrete subgroup of G, acting by left translations and without fixed points on $G/H = \tilde{X}$, and X is the resulting orbit space, in other words, the double coset space $\Gamma\backslash G/H$. The group of all (admissible) isometries of X is then naturally isomorphic to the quotient group $N_G(\Gamma)/\Gamma$ of the normalizer of Γ in G by its subgroup Γ, and hence $H^0(X, \Theta)$ can be identified with the additive structure of the Lie algebra of $N_G(\Gamma)/\Gamma$, which is the same as the Lie algebra of $N_G(\Gamma)$.

The group $H^1(X, \Theta)$ of degree 1 represents the space of infinitesimal deformations (moduli) of the global metric structure of X that preserve the local

[1] Supported by a research contract from the Air Force Office of Scientific Research during the earlier part of the preparation of this article.

DOI: 10.1090/pspum/003

homogeneous structure of constant curvature; this will be explained in full detail in §3. The groups $H^q(X, \Theta)$ of higher degree q so far lack a satisfactory geometrical interpretation; however, since it seems plausible that this interpretation may be found in the near future, and for the sake of completeness, these groups will be calculated here just the same, and by the same general method.

2. Differential geometric discussion. It is well to emphasize from the beginning the contrast between two mathematically equivalent, but conceptually distinct ways to define a Riemannian manifold with constant curvature or, more generally, a locally homogeneous space X. Consider first the associated, simply connected, homogeneous space $\tilde{X}$; on the one hand $\tilde{X}$ can be represented as a left coset space G/H of Lie groups (specifically, of those described in the first section), while on the other hand it can be characterized, as it is in most textbooks on Riemannian geometry, as a complete, simply connected, n-dimensional Riemannian manifold with metric tensor g_{ij}, whose Riemann curvature tensor R_{ijkl} satisfies the equation

$$R_{ijkl} + K(g_{il}g_{jk} - g_{ik}g_{jl}) = 0, \tag{1}$$

where the scalar K, the curvature, is a constant with value positive (elliptic case), zero (parabolic case), or negative (hyperbolic case). It is a well known theorem that, given the dimensionality n and the real number K, the manifold $\tilde{X}$ is determined by (1) uniquely up to an isometry, and that it is the homogeneous space of one of the three types described in the introduction, depending on the sign of K. The existence of a manifold satisfying (1) is usually demonstrated by exhibiting the metric as a set of explicit functions in a special coordinate system: for instance the following,

$$\begin{aligned} g_{ij}(x^1, \cdots, x^n) &= S^{-1}\,\delta_{ij} - KS^{-2}x^ix^j, \\ g^{ij}(x^1, \cdots, x^n) &= S(\delta^{ij} + Kx^ix^j), \end{aligned} \tag{2}$$

where $S = 1 + K\sum_{i=1}^n (x^i)^2$, and the range of the coordinates is limited, for $K < 0$, to the region where $S > 0$.

A locally homogeneous manifold of the type associated with the spaces of non-Euclidean geometry, also called a Clifford-Klein space, can be defined on the one hand as a (not necessarily complete or simply connected) Riemannian manifold whose metric satisfies (1); on the other hand it can be defined as a differentiable coordinate manifold with coordinate functions $(x)_\alpha = (x_\alpha^1, \cdots, x_\alpha^n)$ ranging, say, over domains defined by $\sum_{i=1}^n (x_\alpha^i)^2 < (r_\alpha)^2$, such that the coordinate transformations in the intersection of any two coordinate domains are equal to the restriction to suitable regions of the transformations between two coordinate systems for which the metric tensor takes the form (2): it is well known that in the case under consideration these coordinate transformations are defined by rational functions of a special type, and have the property of being a pseudogroup under composition. Thus a Riemannian manifold X with constant curvature can be characterized in the former sense as a manifold with a fundamental tensor satisfying a structural equation, and in the latter sense as a manifold with structure defined by a *pseudogroup* of coordinate transformations in the sense [**3**] of C. Ehresmann.

It will be more convenient at times in the present article to replace the "classical" tensor notation by the more fashionable Cartan notation, in which the metric is described locally by the selection of a local field of pointwise orthonormal frames of 1-forms ω_i ($i = 1, 2, \cdots, n$); the structural equation (1) in this notation is derived in terms of the affine connection matrix of 1-forms (ω_{ij}) and the curvature matrix of 2-forms (Ω_{ij}), the latter defined by the equations

$$(3) \qquad d\omega_i = \omega_{ij} \wedge \omega_j; \qquad \omega_{ij} + \omega_{ji} = 0; \qquad \Omega_{ij} = d\omega_{ij} - \omega_{ik} \wedge \omega_{kj},$$

as follows: the equation corresponding to (1) becomes

$$(4) \qquad \Omega_{ij} + K\omega_i \wedge \omega_j = 0;$$

similarly we select a "standard" field of frames ω_i in a coordinate system $(x^1, \cdots, x^n)$ to characterize the metric (2), for instance by the following equations

$$\omega_i = S^{-1/2}\, dx^i - \tfrac{1}{2}S^{-1}(\sqrt{S} + 1)^{-1}x^i\, dS.$$

3. Deformation of structure. The theory of deformations of manifolds with a pseudogroup structure, initiated by K. Kodaira [**6**] in a fairly general case, can be applied with certain modifications to the case of locally homogeneous, Riemannian manifolds, since these manifolds can be interpreted, as it has been done in the last section with Clifford-Klein spaces, as manifolds with a pseudogroup structure; the main difference between the pseudogroups considered in Kodaira's theory and those arising from locally homogeneous manifolds is that in the latter case the pseudogroups are, so to speak, finite dimensional and hence locally compact, while in the former case they are, in general, infinite dimensional.

The principal tool used in Kodaira's deformation theory is the sheaf Θ of germs of infinitesimal transformations in the pseudogroup; in our case the sheaf Θ is simply that of germs of vector fields $\theta = \theta^i\, \partial/\partial x^i$ with local covariant components $\theta_i = g_{ij}\theta^j$ that satisfy Killing's equation

$$(5) \qquad \theta_{i,j} + \theta_{j,i} = 0,$$

where the indices following the comma denote covariant differentiation with respect to the fundamental metric tensor g_{ij}. Whenever we shall be using the synthetic notation instead of the tensorial one, we shall express (5) equivalently by the equation

$$(6) \qquad [\theta, ds^2] = 0,$$

where the bracket, generalizing the Poisson "bracket" product, denotes the Lie derivative of any suitable entity on the right compartment inside the bracket with respect to the vector field (or germ of a vector field) on the left.

We consider now the cohomology structure of a Clifford-Klein space X with coefficients in the sheaf Θ of germs of Killing vectors, beginning with the simplest case, namely the one where X is the complete, simply connected space of non-Euclidean geometry.

Proposition 1. *Let X be a complete, simply connected, n-dimensional Riemannian manifold with constant curvature K. Then the sheaf Θ of germs of germs of Killing vectors in X is trivial; the cohomology groups $H^q(X, \Theta)$ are trivial for all degrees q $(0 < q < n)$, while the group of degree 0 is isomorphic to an $n(n+1)/2$-dimensional vector space over the real numbers.*

Proof. It is well known that the stalk of the sheaf Θ is an $n(n+1)/2$-dimensional real vector space. In fact, every germ of a Killing vector field at any given point of X has a unique extension to a global, Killing field on all of X, so that the stalk represents faithfully the Lie algebra of the group of all isometries of X, which is $n(n+1)/2$-dimensional. Thus the sheaf Θ is isomorphic to a product sheaf of X with a real vector space. It follows immediately that the cohomology groups $H^q(X, \Theta)$ are equivalent to the tensor product $H^q(X, R) \otimes_{\mathbf{R}} R^{n(n+1)/2}$. Since X is either a sphere (if $K > 0$) or an open cell (if $K \leqq 0$), we see that $H^0(X, \Theta)$ is $n(n+1)/2$-dimensional over the real numbers, $H^q(X, \Theta) = 0$ for $0 < q < n$ or for $q = n$ and $K \leqq 0$, and $H^n(X, \Theta)$ is $n(n+1)/2$-dimensional for $K > 0$. This completes the proof.

From now on we shall confine our attention to Clifford-Klein spaces X that are compact; this assumption implies, in particular, that every manifold X is complete, and hence is a double coset space $\Gamma\backslash G/H$, according to the remarks in the Introduction. We have already mentioned the geometrical meaning of the group $H^q(X, \Theta)$ for $q = 0$; in this connection, we can quote now a result, which is a special case of a theorem by S. Bochner [**1**], as follows.

Proposition 2 (Bochner). *Let X be a compact Clifford-Klein space with curvature constant and equal to K. Then, if $K < 0$, the cohomology group $H^0(X, \Theta)$ of degree 0 vanishes, while if $K = 0$, it represents the linear space of autoparallel vector fields, globally defined in X.*

Proof. It follows from (1) that the Ricci tensor R_{ik} in X, defined by $R_{ik} = g^{jl} R_{ijkl}$ from the Riemann tensor, takes the form

$$R_{ik} = (n-1)K g_{ik}. \tag{7}$$

Thus, if $K < 0$, the manifold X has negative definite Ricci curvature, and, from Bochner's theorem, admits no globally defined Killing vector fields other than the zero one. If $K = 0$, then again, from Bochner's theorem, we know that the only Killing vector fields on X are the autoparallel vector fields, concluding thus the proof of the statement.

We shall show now how the first cohomology group $H^1(X, \Theta)$ is related to deformations of Clifford-Klein structure.

DEFINITION 1. *A differentiable family of n-dimensional, compact Clifford-Klein spaces is a domain-map-range compositum* $\mathcal{V} \to^{\pi} M$, *where* $\mathcal{V}$ *and* M *are differentiable manifolds with* $m + n$ *and* m *dimensions respectively,* π *is a differentiable map, with the following properties:*

(i) *the mapping* π *is surjective (meaning that the jacobian matrix of* π *has everywhere rank equal to* m*);*

(ii) *the inverse image* $\pi^{-1}(t)$ *in* $\mathcal{V}$ *of each point* $t \in M$ *is a compact submanifold of* $\mathcal{V}$, *denoted by* X_t;

(iii) *there is defined on* $X_t \subset \mathcal{V}$ *for each* $t \in M$ *a Riemannian metric which, if regarded as a quadratic form on a bundle of linear subspaces of the tangent bundle of* $\mathcal{V}$, *is a differentiable function on* $\mathcal{V}$, *and as a Riemannian metric in each* X_t *has constant curvature in the sense of* (1), *or equivalently,* (4).

If 0 *is a given base point in* M, *then the manifold* X_t *for* t *in a neighborhood of* 0 *is called a deformation of* X_0.

In connection with the above definition we recall a fundamental theorem due to C. Ehresmann [2], stating that, if $\pi: \mathcal{V} \to M$ is any surjective, differentiable map of a manifold $\mathcal{V}$ onto a manifold M, and if for each $t \in M$ the inverse image $\pi^{-1}(t)$ is a (connected) compact submanifold of $\mathcal{V}$, then the family of differentiable manifolds $\mathcal{V} \to^{\pi} M$ defines a fibre bundle with total space $\mathcal{V}$, base space M, and structure group equal to the group of all differentiable transformations of a typical fibre. In other words, under the conditions stated, the family of differentiable manifolds, disregarding any further structure that these manifolds may possess, is locally equivalent to a product.

DEFINITION 2. *A differentiable family* $\mathcal{V} \to^{\pi} M$ *of compact, Clifford-Klein spaces is called trivial, if there exists a differentiable mapping* $P : \mathcal{V} \to X_0 = \pi^{-1}(0)$ *for some point* $0 \in M$, *whose restriction to* X_t *for each* $t \in M$ *is an isometric map with respect to the corresponding Riemannian structures. The family is called locally trivial at a point* $0 \in M$, *if the restriction of the family to* $\pi^{-1}(U)$ *for some neighborhood* U *of* 0 *in* M *is a trivial family over* U.

Since our investigation will be only local with regards to M, we shall use a single coordinate system $(t_1, \cdots, t_m)$ to describe points of M. Given a differentiable family $\mathcal{V} \to^{\pi} M$ of compact, Clifford-Klein spaces, (or of manifolds with any other pseudogroup structure), it is natural to parametrize the total space $\mathcal{V}$ by an atlas of local coordinate systems $(x^1, \cdots, x^n, t_1, \cdots, t_m) = (x; t)$, such that the local coordinate representation of π, also denoted by π without fear of ambiguity, becomes $\pi(x; t) = t = (t_1, \cdots, t_m)$. If $(x; t)$ and $(x'; t)$ are two overlapping coordinate systems in $\mathcal{V}$, then the coordinate transformation functions, in general, are of the form $x^i = \varphi^i(x'; t)$, $t_\alpha = t_\alpha$ $(i = 1, \cdots, n;$ $\alpha = 1, \cdots, m)$. In view of the structure of the total space $\mathcal{V}$, we shall consider alternatively either one of the two types of coordinate atlases:

(i) An *Eulerian atlas*, motivated by the property of $\mathcal{V}$ being differentiably equivalent to a product bundle $X_0 \times M$, where $X_0 = \pi^{-1}(0)$ for some $0 \in$ M; for such an atlas the functions $x^i = \varphi^i(x'; t)$ describing an arbitrary coordinate transformation in the atlas are independent of t; $\mathcal{V}$ is thus presented as a product manifold in terms of the coordinates, and the fundamental form in $\mathfrak{V}$ is expressed by

$$ds^2 = g_{ij}(x;\, t)\, dx^i\, dx^j \tag{8}$$

in terms of the local coordinates; the matrix function g_{ij} is differentiable in $(x;\, t)$, and for each fixed t it satisfies (1) with respect to the coordinates $x = (x^1, \cdots, x^n)$.

(ii) A *Lagrangian atlas*, suggested by the fact that for each $t \in \boldsymbol{M}$ the manifold $\boldsymbol{X}_t = \pi^{-1}(t)$ admits a transitive pseudogroup of (isometric) local transformation; such an atlas has the property that, for each coordinate system $(x;\, t)$ the range with respect to x for any given $t \in \boldsymbol{M}$ is independent of t, and the fundamental form (8) is defined by functions g_{ij} independent of t and even taking the same numerical value for the same values of x in the various coordinate systems. For instance, the coordinate range and value of g_{ij} in the system $(x;\, t)$ could be given by (2) with x ranging over the open ball $\sum_{i=1}^{n} (x^i)^2 < r^2$. For such an atlas, in general, the coordinate transition functions $\varphi^i(x';\, t)$ are dependent on t, but for each t they describe local transformations in x that leave the form defined by the tensor with components (2) invariant.

Referring to the two definitions of a manifold with constant curvature presented in §2, one sees easily that the Eulerian atlas for $\mathfrak{V}$ corresponds more closely to the first definition (using equation (1)), while the Lagrangian atlas follows more closely the idea of the second definition (pseudogroup structure). It is clear how the foregoing definitions 1 and 2 should be modified, if one prefers to use the notation of moving frames rather than the tensorial one.

Let $\mathfrak{V} \rightarrow^{\pi} \boldsymbol{M}$ be a differentiable family of n-dimensional, compact, Clifford-Klein spaces; let $\boldsymbol{M}$ be covered by a single coordinate system including the point $0 = (0, \cdots, 0)$. We wish to investigate the infinitesimal properties of the family about the point 0, by analysing the relationships between the tangent bundle of $\mathfrak{V}$ restricted to $\boldsymbol{X}_0 = \pi^{-1}(0)$ and the tangent space T_0 of $\boldsymbol{M}$ at the point 0: we shall do so first in terms of a Lagrangian atlas and then in terms of an Eulerian one, and obtain two corresponding, equivalent criteria for the family $\mathfrak{V} \rightarrow^{\pi} \boldsymbol{M}$ to be locally trivial at 0.

Let a Lagrangian atlas on the manifold $\mathfrak{V}$ be defined by a family $(\mathfrak{U}_\iota;\, (x_\iota;\, t))$ indexed by $\iota \in J$, where $\mathfrak{U}_\iota \subset \mathfrak{V}$ is the coordinate domain and $(x_\iota;\, t) = (x_\iota^1, \cdots, x_\iota^n, t_1, \cdots, t_m)$ are the corresponding coordinate functions and, whenever $\mathfrak{U}_\iota \cap \mathfrak{U}_\kappa \neq \emptyset$, let the corresponding coordinate transformation be expressed by $x_\iota^i = \varphi_{\iota\kappa}^i(x_\kappa;\, t)$. We suppose that in each coordinate system and for each t the components g_{ij} of the metric are the same functions appearing in (2); then the same equations can be interpreted as describing isometries of suitable subdomains of the standard model $\tilde{X}$ of the simply connected, complete space of constant curvature onto one another, since the metric has the same form in both coordinate systems. This local isometry, of course, depends on the m parameters of t in each connected component of an overlap of Lagrangian coordinates, so that, for any tangent vector $\partial/\partial t$ at any point of $\boldsymbol{M}$, and in particular at 0, one can express the "velocity" of the relative, rigid flow of x_ι with respect to x_κ for any pair of coordinate systems such that $\mathfrak{U}_\iota \cap \mathfrak{U}_\kappa \neq \emptyset$. This velocity is expressed by a local vector field $\theta_{\iota\kappa} = \theta^i_{\iota\kappa}\partial/\partial x^i_\iota$, where

$$\theta_{\iota\kappa} = \frac{\partial\varphi^i_{\iota\kappa}(x_\kappa; t)}{\partial t}\cdot\frac{\partial}{\partial x^i_\iota}. \tag{9}$$

It is immediately clear that each vector field $\theta_{\iota\kappa}$ is defined in $\mathfrak{U}_\iota \cap \mathfrak{U}_\kappa$, and that it satisfies there Killing's equation, since it represents an infinitesimal isometric transformation. From the chain rule of successive coordinate transformations it is easily seen that, in $\mathfrak{U}_\iota \cap \mathfrak{U}_\kappa \cap \mathfrak{U}_\lambda$, one verifies the relation

$$\theta_{\iota\lambda} = \theta_{\iota\kappa} + \theta_{\kappa\lambda}; \tag{10}$$

in other words, for each Lagrangian atlas for $\mathfrak{V}$ and for each tangent vector $\partial/\partial t$ at any point $t \in \boldsymbol{M}$ there corresponds a 1-dimensional cocycle in the presheaf with coefficients in Θ associated with the covering of $\boldsymbol{X}_t$ associated with the atlas restricted to $\boldsymbol{X}_t$. It is not hard to see that, under refinements of the covering, the corresponding 1-dimensional cocycles define a well determined limiting, 1-dimensional cohomology class with respect to the sheaf Θ over the the manifold $\boldsymbol{X}_t$; we denote this cohomology class by $\rho_t(\partial/\partial t)$, and one can easily verify that ρ_t is a linear mapping of the tangent space to $\boldsymbol{M}$ at t into the module $H^1(\boldsymbol{X}_t, \Theta)$.

Let us go over the same situation as before, using this time an Eulerian atlas for the manifold $\mathfrak{V}$. The geometrical setup is the same as if one had an n-dimensional, compact, differentiable manifold $\boldsymbol{X} = \boldsymbol{X}_0$, covered by a fixed atlas of coordinate systems, and a *family* of Riemannian metrics depending differentiably on m parameters $t = (t_1, \cdots, t_m) \in \boldsymbol{M}$, such that for each value of t the equation of constant curvature (1) or (4) is identically satisfied. Corresponding to any tangent vector $\partial/\partial t$ at any point $t \in \boldsymbol{M}$, it makes sense to consider the directional derivative $(\partial g_{ij}(x, t)/\partial t)\, dx^i\, dx^j$ of the fundamental form; accordingly, we shall introduce here a symmetric, bilinear differential form $a_{ij}(x, t)\, dx^i\, dx^j$, where

$$a_{ij}(x, t) = \frac{1}{2}\frac{\partial g_{ij}(x, t)}{\partial t}. \tag{11}$$

A straightforward calculation furnishes then the corresponding infinitesimal variation of the Riemann curvature tensor with the following result:

$$\frac{\partial}{\partial t}R_{ijkl} = a_{il,jk} + a_{jk,il} - a_{ik,jl} - a_{jl,ik} + a_{hi}R^h{}_{jkl} - a_{hj}R^h{}_{ikl},$$

where two (or more) indices following a comma denote successive covariant differentiation (contravariant, if written as superscripts) with respect to the metric. Since equation (1) is satisfied identically in t, we obtain its variational formula as an equation satisfied by the tensor a_{ij}:

$$\begin{aligned} a_{il,jk} + a_{jk,il} - a_{ik,jl} - a_{jl,ik} & \\ + K(g_{il}a_{jk} + g_{jk}a_{il} - g_{ik}a_{jl} - g_{jl}a_{ik}) &= 0. \end{aligned} \tag{12}$$

The family of tensors $a_{ij}(x, t)$ is determined by the choice of Eulerian atlas on $\mathcal{V}$, not by the family $\mathcal{V} \to^{\pi} M$ of Clifford-Klein spaces alone. In fact, if another Eulerian atlas contains coordinate systems (written without indices for brevity) such as $(y; t)$, the two atlases together in general do not constitute an Eulerian atlas: on the contrary, the transition functions from a coordinate system $(x; t)$ of the first atlas to an overlapping one $(y; t)$ of the second (a change of coordinates in M, as one can easily see, would have no effect) is expressed, say, by functions of the form $x^i = \psi^i(y; t)$, that are actually dependent on t. This defines a vector field $\theta = (\partial\psi^i(y; t)/\partial t)(\partial/\partial x^i)$ associated with the tangent vector $\partial/\partial t$ to M in the overlap of the domains of $(x; t)$ and $(y; t)$; since the two atlases considered are both Eulerian, the vector field θ thus obtained at each point of X_t for any $t \in M$ is independent of the choice of pair of coordinates there from the two respective atlases, so that for each t and as a linear function on the tangent vectors $\partial/\partial t$, we have a vector field θ, globally defined on X_t by the transition from one atlas to the other. If we denote by $b_{ij}\, dy^i\, dy^j = a'_{ij}\, dx^i\, dx^j$ the differential form defined in X_t by the second atlas, and set $\theta = \theta^i\, \partial/\partial x^i$, $\theta_i = g_{ij}\theta^j$, we obtain the relation

$$a'_{ij}\, dx^i\, dx^j - a_{ij}\, dx^i\, dx^j = \tfrac{1}{2}[\theta, ds^2], \tag{13}$$

or, expressing this componentwise,

$$a'_{ij} = a_{ij} + \tfrac{1}{2}(\theta_{i,j} + \theta_{j,i}). \tag{13'}$$

We shall show now that, at least locally in X_t, it is possible to find an Eulerian atlas by modifying a given one, so that in a neighborhood of any preassigned point in X_t the form $a'_{ij}\, dx^i\, dx^j$ vanishes identically. For this it is sufficient to point out the rather obvious fact that any single coordinate system from a *Lagrangian* atlas on $\mathcal{V}$ can be imbedded in an Eulerian atlas; but then in the domain of the coordinate system chosen the components of the metric are independent of t, and hence the tensor a'_{ij} vanishes identically there. This fact can be used to complete the proof of the following, otherwise completely elementary, proposition.

PROPOSITION 3. *Let X be a Clifford-Klein manifold with constant curvature equal to K. Then, given any symmetric, covariant tensor a_{ij} in X, there exists in a neighborhood of any given point a covariant vector field with components θ_i satisfying*

$$a_{ij} = \tfrac{1}{2}(\theta_{i,j} + \theta_{j,i}), \tag{14}$$

if and only if a_{ij} satisfies the integrability condition expressed by (12).

The main conclusion of the foregoing discussion is that, given a differentiable family $\mathcal{V} \to^{\pi} M$, then for any point $t \in M$ there is a well defined linear map σ_t of the tangent space to M at t, which assigns to each tangent vector $\partial/\partial t$ the element represented by $a_{ij}\, dx^i\, dx^j$ of the quotient space of the space of globally defined, symmetric, bilinear forms satisfying (12) by the linear subspace consisting of those forms that are derived from a global vector field θ by means of (14).

The main result of this section can now be stated and proved.

THEOREM 1. *Let $\mathcal{V} \to^{\pi} \boldsymbol{M}$ be an m-dimensional, differentiable family of n-dimensional, compact, Clifford-Klein spaces, and let 0 be a point in $\boldsymbol{M}$; then the following three statements are equivalent.*

(i) *The family $\mathcal{V} \to^{\pi} \boldsymbol{M}$ is locally trivial at 0.*

(ii) *The linear map ρ_t of the tangent space to $\boldsymbol{M}$ at t for each t in a neighborhood of 0 into $H^1(\boldsymbol{X}_t, \Theta)$ $(\boldsymbol{X}_t = \pi^{-1}(t))$ is trivial.*

(iii) *There exists an Eulerian atlas on $\mathcal{V}$ such that for each t in a neighborhood of 0 and for each tangent vector $\partial/\partial t$ to $\boldsymbol{M}$ at t the variational form $a_{ij}\, dx^i\, dx^j$ obtained from it on $\boldsymbol{X}_t$ is derivable from a vector field θ in $\boldsymbol{X}_t$ by the formula $a_{ij}\, dx^i\, dx^j = [\theta, ds^2]/2$.*

PROOF. It is obvious that (i) implies both (ii) and (iii): in fact, if the family is locally trivial at 0, then, for a sufficiently small neighborhood U of 0 in $\boldsymbol{M}$, the family restricted to $\pi^{-1}(U)$ is equivalent to the Cartesian product of $\boldsymbol{X}_0 = \pi^{-1}(0)$ with U; an atlas of product coordinates for this subfamily is then simultaneously Lagrangian and Eulerian, whence both maps ρ_t and σ_t are zero for each tangent vector $\partial/\partial t$ at each point $t \in U$.

Now suppose that the statement (ii) is true. We assume without loss of generality that there is a coordinate system $(t_1, \cdots, t_m)$ in a neighborhood U of 0 in $\boldsymbol{M}$ with origin at 0 and with range $\sum_{\alpha=1}^{m} (t_\alpha)^2 < c$ for some positive constant c. Consider the deformation retraction of U to 0 defined by $t_\alpha \to (1 - s)t_\alpha$ for a real parameter s, with $0 \leqq s \leqq 1$; the velocity of this deformation is given by $(\partial t_\alpha/\partial s)\, \partial/\partial t_\alpha = -t_\alpha\, \partial/\partial t_\alpha$ for every value of s. This velocity field induces, in terms of a Lagrangian atlas for $\mathcal{V}$ for each pair of overlapping coordinate domains $\mathfrak{U}_\iota$, $\mathfrak{U}_\kappa$, a field $\theta_{\iota\kappa}(x, t)$ of Killing vectors, tangent to $\boldsymbol{X}_t$ everywhere in $\boldsymbol{X}_t \cap \mathfrak{U}_\iota \cap \mathfrak{U}_\kappa$ satisfying (10) in $\boldsymbol{X}_t \cap \mathfrak{U}_\iota \cap \mathfrak{U}_\kappa \cap \mathfrak{U}_\lambda$. The assumed truth of (ii) implies that, as soon as the atlas on $\mathcal{V}$ is sufficiently refined, there exists a family $\theta_\iota(x; t)$ of Killing vector fields, each of them defined in $\boldsymbol{X}_t \cap \mathfrak{U}_\iota$ and depending differentiably on $(x; t)$ in $\mathfrak{U}_\iota$, such that

$$\theta_{\iota\kappa} = \theta_\kappa - \theta_\iota \quad \text{in} \quad \mathfrak{U}_\iota \cap \mathfrak{U}_\kappa. \tag{15}$$

We shall construct now a one-parameter family of Lagrangian atlases on $\mathcal{V}$, parametrized by s $(0 \leqq s \leqq 1)$, which for $s = 0$ coincides with the initial atlas, and for $s = 1$ gives the product atlas for the trivial family $\boldsymbol{X}_0 \times U$, preserving the projection π: $\pi^{-1}(U) \to U$. Denote by $\mathfrak{U}_\iota$, $(x_\iota; t)$ the coordinate domains and functions of the given atlas; we suppose that the range of each coordinate system is described by $\sum_{i=1}^{n} (x_\iota^i)^2 < (r_\iota)^2$ for some constants $r_\iota > 0$ and that in each system the components g_{ij} of the metric tensor are expressed by (2). The family of new Lagrangian atlases for any value s of the parameter is described by domains $\mathcal{W}_\iota$ and functions $(y_\iota; t)$ with the same range and expression for g_{ij} as the corresponding $(x_\iota; t)$ and is related by the equations $x_\iota^i = \psi_\iota^i(y_\iota; t; s) =$

$\psi^i_{\iota;t,s}(y_\iota)$ representing non-Euclidean motions in terms of the coordinates restricted to $\mathfrak{U}_\iota \cap \mathfrak{W}_\iota$, characterized by the conditions:

(a) for $s = 0$, $\mathfrak{W}_\iota = \mathfrak{U}_\iota$ and $\psi^i_\iota(y_\iota; t; 0) = y^i_\iota$;

(b) for $0 \leqq s \leqq 1$, the velocity of rigid flow of $(\mathfrak{W}_\iota, y_\iota)$ relative to $(\mathfrak{U}_\iota, x_\iota)$ is given by the Killing vector field defined in the overlap by

$$\frac{\partial\psi^i_\iota(y_\iota; t; s)}{\partial s}\frac{\partial}{\partial x^i_\iota} = \theta_\iota(x_\iota; (1-s)t), \tag{16}$$

where $(1 - s)t$ is the abbreviation for the point in $t' \varepsilon \boldsymbol{M}$ with coordinates $t'_\alpha = (1 - s)t_\alpha$. The fact that the right hand term of (16) is a Killing vector field in $\boldsymbol{X}_t$ follows from the property of the coordinates x_ι being intrinsically undistinguishable between $\boldsymbol{X}_t$ and $\boldsymbol{X}_{(1-s)t}$, since the atlas is Lagrangian, and since the right hand term is a Killing vector field in $\boldsymbol{X}_{(1-s)t}$. The transition functions between the new coordinates $(y_\iota; t)$ for a given value of s can be expressed as follows (omitting for simplicity the index i for the individual coordinates):

$$y_\iota = \chi_{\iota\kappa}(y_\kappa; t; s) = (\psi_{\iota;t,s})^{-1} \circ \varphi_{\iota\kappa;t} \circ \psi_{\kappa;t,s}(y_\kappa), \tag{17}$$

where the term on the right represents the composition of the transitions, applied successively as read from right to left, of y_κ to x_κ, to x_ι, and finally to y_ι.

We calculate now the rate of change with respect to s of the transition functions (17) (Lagrangian derivatives), as follows

$$\begin{aligned}\frac{\partial\chi^i_{\iota\kappa}(y_\kappa; t; s)}{\partial s}\frac{\partial}{\partial y^i_\iota} &= -\frac{\partial\psi^i_\iota(y_\iota; t; s)}{\partial s}\frac{\partial}{\partial y^i_\iota} + \frac{\partial\psi^j_\kappa(y_\kappa; t; s)}{\partial s}\frac{\partial}{\partial y^j_\kappa}\\ &= -\theta_\iota(x_\iota; (1-s)t) + \theta_\kappa(x_\kappa; (1-s)t)\\ &= \theta_{\iota\kappa}(x; (1-s)t),\end{aligned} \tag{18}$$

as a consequence of (16) and (15). From the theorem on uniqueness of the solutions of an initial value problem in ordinary differential equations and from the definition of $\theta_{\iota\kappa}$ from the retraction of $U \subset \boldsymbol{M}$, we conclude that

$$\chi^i_{\iota\kappa}(y_\kappa; t; s) = \varphi^i_{\iota\kappa}(y_\kappa; (1-s)t).$$

Thus the local family $\pi^{-1}(U) \to^\pi U$ with the modified atlas corresponding to a value s of the parameter is identical with the restricted family $\pi^{-1}((1 - s)U) \to^\pi (1 - s)U$ with the restriction of the original Lagrangian atlas, save for the change in scale for the parameters t for U. In the limit, as s approaches 1, the final atlas shows that the family of Clifford-Klein spaces over U is indeed trivial.

Let us assume now that (iii) is valid. As before, we immediately restrict the family from the set $\mathfrak{V} = \pi^{-1}(\boldsymbol{M})$ to $\pi^{-1}(U)$ and introduce the same deformation retraction of U to 0 as before, with the parameter s. The velocity field of this retraction being $-t_\alpha(\partial/\partial t_\alpha)$, the corresponding variational formula for the metric $g_{ij}(x; t)$ in any coordinate system of the given Eulerian atlas is then given by

$$a_{ij}(x; t) = -\tfrac{1}{2} t_\alpha \frac{\partial g_{ij}(x; t)}{\partial t_\alpha}.$$

The assumed validity of (iii) implies that there exists a family $\theta(x; t) = \theta^i(x; t)\partial/\partial x^i$, depending differentiably on $(x; t)$ everywhere in $\pi^{-1}(U)$, of tangent vector fields in each X_t for $t \in U$, such that, on denoting by $\theta_i(x; t)$ the corresponding covariant components $g_{ij}(x; t)\theta^j(x; t)$, equation (14) is satisfied everywhere in each X_t; noting that, since, for $t = 0$, $a_{ij}(x; 0) = 0$, it is not hard to verify that $\theta(x; t)$ can be so chosen that $\theta(x; 0) = 0$.

One constructs now, in a way analogous to the one above, a one-parameter family of Eulerian atlases for $\pi^{-1}(U)$, parametrized by s $(0 \leqq s \leqq 1)$, so that for $s = 0$ the atlas coincides with the given one, and otherwise, letting $(y; t)$ denote a typical coordinate system for any given value of s and $(x; t)$ the corresponding one for $s = 0$, the mutual transition functions $x^i = \psi^i(y; t; s)$ are characterized by the differential equation

$$\frac{\partial \psi^i(y; t; s)}{\partial s} \frac{\partial}{\partial x^i} = \theta^i(x'; (1 - s)t) \frac{\partial}{\partial x'^i}, \tag{19}$$

where x' refers to the corresponding coordinate system in $X_{(1-s)t}$. Repeating the same calculations as before, one sees that the transition functions between the different coordinate systems $(y; t)$ are independent of s, and hence identical with the corresponding ones for the systems $(x; t)$; on the other hand, one verifies immediately that the components of the metric in X_t relative to the coordinate system $(y; t)$ which we may denote by $g'_{ij}(y; t; s)$ are given precisely by

$$g'_{ij}(y; t; s) = g_{ij}(x; (1 - s)t) \quad \text{for} \quad y^i = x^i.$$

By letting $s = 1$, one sees readily that the components of the metric in X_t become identical with the corresponding ones for X_0. This concludes the proof of the theorem.

A trivial consequence of Theorem 1 is that, if a family $\mathcal{V} \xrightarrow{\pi} M$ of compact, Clifford-Klein spaces has the property that each X_t for all t in a neighborhood of $0 \in M$ satisfies $H^1(X_t, \Theta) = 0$, then the family is locally trivial at 0. Actually one reaches the same conclusion under the weaker assumption that $H^1(X_t, \Theta) = 0$ only for $t = 0$. This will be proved in a broader context in the second part of this article.

4. A fine resolution of the sheaf Θ. Let X be a compact, n-dimensional, Clifford-Klein space and let Θ be the sheaf of germs of Killing vector fields in X. A fine resolution of the sheaf Θ is a finite sequence of sheaves of modules, connected by successive homomorphisms, of the form

$$0 \to \Theta \xrightarrow{i} \Phi^0 \xrightarrow{D_0} \Phi^1 \xrightarrow{D_1} \Phi^2 \xrightarrow{D_2} \cdots \xrightarrow{D_{l-1}} \Phi^l \xrightarrow{D_l} 0, \tag{20}$$

provided that the sequence is exact and that each of the "resolvent" sheaves $\Phi^0, \Phi^1, \Phi^2, \cdots, \Phi^l$ is a fine sheaf, i.e., one that admits arbitrarily fine, differentiable partitions of unity as operators. The number l, improperly called the length of the resolution, is usually not greater than n, the dimensionality of the base space X; every sheaf on an n-dimensional (metrizable, second-countable) space admits a resolution of length at most n. For the more detailed definitions and general properties of sheaves, see for instance [5].

For any sheaf Φ over X, we shall denote by $L(\Phi)$ the space of all cross sections of Φ over X: if Φ is a sheaf of modules, then $L(\Phi)$ is a module. In the case of Θ and the resolvent sheaves occurring in (20), the operators i and D_q have natural extensions to operators, denoted by the same symbols, acting on the respective spaces of cross sections; with reference to (20), we denote by $Z_D(\Phi^q)$ the kernel of the homomorphism D_q of $L(\Phi^q)$ into $L(\Phi^{q+1})$. From the exactness of (20) it follows that $Z_D(\Phi^q) \supset D_{q-1}L(\Phi^{q-1})$ (for the sake of uniformity, we have identified Φ^{-1}, Φ^{l+1}, and D_{-1} respectively with Θ, 0, and i). Denoting by $\mathcal{S}$ the fine resolution (20) of Θ and by $L(\mathcal{S})$ the sequence of corresponding cross sections, the latter is a cochain complex under the operator $D = \sum_{\alpha=-1}^{l} D_\alpha$: we consider now the corresponding cohomology groups

$$H_D^q(L(\mathcal{S})) = Z_D(\Phi^q)/D_{q-1}L(\Phi^{q-1}), \qquad (0 \leqq q \leqq l).$$

A fundamental theorem in the theory of sheaves, the analogue of de Rham's theorem, asserts the isomorphism $H^q(X, \Theta) \approx H_D^q(L(\mathcal{S}))$. For this reason we shall construct explicitly a fine resolution of the sheaf Θ of germs of Killing vectors on a compact, n-dimensional Clifford-Klein space X; this resolution has length exactly n. Before we can do this, however, a brief discussion is needed on the algebraic properties of the bigraded module of double exterior forms in an arbitrary differentiable manifold.

Let $\mathfrak{A}^q$ denote the sheaf of germs of differentiable q-forms in an arbitrary, n-dimensional manifold, and let $\mathfrak{A}$ denote the direct sum $\sum_{q=0}^{n} \mathfrak{A}^q$ (a sheaf of Grassmann algebras). Consider the tensor product over the sheaf of rings $\mathfrak{A}^0$, $\mathfrak{A} \otimes \mathfrak{A} = \sum_{p,q=0}^{n} \mathfrak{A}^{p,q}$, where $\mathfrak{A}^{p,q} = \mathfrak{A}^p \otimes \mathfrak{A}^q$. $\mathfrak{A} \otimes \mathfrak{A}$ is a sheaf of bigraded modules, the summands $\mathfrak{A}^{p,q}$ representing the bihomogeneous components of bidegree (p, q). We introduce two endomorphisms T_R and T_L of $\mathfrak{A} \otimes \mathfrak{A}$, called respectively the right and the left transfer operators, bihomogeneous of bidegree $(-1, +1)$ and $(+1, -1)$ respectively, defined as follows: let $\varphi \in \mathfrak{A}^{p,q}$ be expressed in terms of a local coordinate system $x = (x^1, \cdots, x^n)$ by

$$(21) \qquad \varphi = \frac{1}{p!q!} \varphi_{i_1 \cdots i_p j_1 \cdots j_q} dx^{i_1} \wedge dx^{i_2} \wedge \cdots \wedge dx^{i_p} \otimes dx^{j_1} \wedge \cdots \wedge dx^{j_q},$$

where the coefficients $\varphi_{i_1 \cdots i_p j_1 \cdots j_q}$ are the components of a germ of a tensor, alternating with respect to the i-indices and the j-indices; then $T_R\varphi$ and $T_L\varphi$ are defined, respectively, by

$$T_R\varphi = \frac{1}{(p-1)!q!}\varphi_{i_1\cdots i_p j_1\cdots j_q}\, dx^{i_1}\wedge\cdots\wedge dx^{i_{p-1}}\otimes dx^{i_p}\wedge dx^{j_1}\wedge\cdots\wedge dx^{j_q},$$
(22)
$$T_L\varphi = \frac{1}{p!(q-1)!}\varphi_{i_1\cdots i_p j_1\cdots j_q}\, dx^{i_1}\wedge\cdots\wedge dx^{i_p}\wedge dx^{j_1}\otimes dx^{j_2}\wedge\cdots\wedge dx^{j_q}.$$

The following identity can be verified without difficulty: if $\varphi \in \mathfrak{A}^{p,q}$, then

(23) $$(T_LT_R - T_RT_L)\varphi = (p-q)\varphi;$$

thus, using the bracket to denote the commutator of operators, and denoting by $P_{p,q}$ the natural projection of $\mathfrak{A}\otimes\mathfrak{A}$ onto the component of bidegree (p, q), we can express (23) equivalently by

(23′) $$[T_L, T_R] = \sum_{p,q=0}^{n}(p-q)P_{p,q}.$$

The following relations between T_R and T_L are closely related to Lepage's relations between the operators L and Λ in symplectic manifolds [7, pp. 21–28]. By applying (23) to $T_R^s\varphi$ for $0 \leqq s < r$, followed by T_R^{r-s-1} and adding with respect to s, one obtains

(24) $$[T_L, T_R^r] = \sum_{s=0}^{r-1} T_R^{r-s-1}[T_L, T_R]T_R^s = \sum_{p,q=0}^{n} r(p-q-r+1)T_R^{r-1}P_{p,q}.$$

Similarly, one sees

(24′) $$[T_L^r, T_R] = \sum_{p,q=0}^{n} r(p-q+r-1)T_L^{r-1}P_{p,q}.$$

Definition 3. *An element $\varphi \in \mathfrak{A}^{p,q}$ is called right primitive, if $T_R\varphi = 0$; it is called left primitive, if $T_L\varphi = 0$.*

By applying (24) to a left primitive element $\varphi \in \mathfrak{A}^{p,q}$ and subjecting the result to T_L^{s-1}, we obtain, for $s \leqq r \leqq p - q$,

(25) $$T_L^sT_R^r\varphi = \frac{r!(p-q-r+s)!}{(r-s)!(p-q-r)!}T_R^{r-s}\varphi,$$

and zero, if either of the quantities whose factorial appears in the denominator is negative. This implies the following result as an immediate consequence.

Proposition 4. *If $\varphi \in \mathfrak{A}^{p,q}$ is left primitive, then $T_R^r\varphi = 0$ for all $r \geqq \max(p-q-r, 0)$.*

Lemma 1. *If $\lambda \in \mathfrak{A}^{p,q}$ with $p < q$ and $T_L^r\lambda = 0$ for any $r \leqq q - p$, then $\lambda = 0$.*

Proof. If $T_L^r\lambda = 0$ for $0 < r \leqq q - p$, then from (24′)

$$T_L^rT_R\,\lambda = [T_L^r, T_R]\lambda = -r(q-p-r+1)T_L^{r-1}\,\lambda,$$

so that

$$T_L^{r+1}(T_R\,\lambda) = 0.$$

Continuing inductively, one sees that, if

$$T_L^{r+s}T_R^s\,\lambda = 0,$$

then, applying (24′) to $T_R^s\lambda$, we have

$$T_L^{r+s}T_R^{s+1}\,\lambda = [T_L^{r+s}, T_R]T_R^s\,\lambda = -(r+s)(q-p-r+s+1)T_L^{r+s-1}T_R^s\,\lambda$$

$$= (-1)^{s+1}\frac{(r+s)!\,(q-p-r+s+1)!}{(r-1)!\,(q-p-r)!}T_L^{r-1}\,\lambda.$$

Since the above process can be continued up to $s = p$, since $T_R^{p+1}\lambda = 0$, we see from the above relation that $T_L^{r-1}\lambda = 0$. Thus from the hypothesis one deduces successively

$$T_R^{r-1}\,\lambda = T_R^{r-2}\,\lambda = \cdots = T_R\,\lambda = \lambda = 0, \qquad \text{proving the lemma.}$$

We can state and prove now the main results we need on double forms.

THEOREM 2. *Given an element $\varphi \in \mathfrak{A}^{p,q}$, it can be expressed in one and only one way in the form*

$$\varphi = \sum_{s=0}^{\min(p,q)} T_R^{q-s}\,\psi_s, \tag{26}$$

where each $\psi_s \in \mathfrak{A}^{p+q-s,s}$ is a left primitive element. In particular, if $p < q$, there exists an element $\varphi' \in \mathfrak{A}^{p+1,q-1}$, for instance

$$\varphi' = \sum_{s=0}^{p} T_R^{q-s-1}\,\psi_s,$$

satisfying $T_R\varphi' = \varphi$.

PROOF. We construct first an expansion (26) by defining, inductively on s, the left primitive elements ψ_s. It is clear that $T_L^q\varphi \in \mathfrak{A}^{p+q,0}$ is left primitive: let

$$\psi_0 = \frac{p!}{q!\,(p+q)!}T_L^q\varphi.$$

Then applying (25) to $\psi_0 \in \mathfrak{A}^{p+q,0}$ with $r = s = q$, we have

$$T_L^qT_R^q\psi_0 = T_L^q\varphi, \quad \text{so that}$$

$$T_L^{q-1}(\varphi - T_R^q\psi_0) \quad \text{is left primitive.}$$

Assume now that the left primitive elements $\psi_s \in \mathfrak{A}^{p+q-s,s}$ have been chosen for $0 \leqq s < r$, for $r \geqq 1$, so that

$$T_L^{q-r+1}\Big(\varphi - \sum_{s=0}^{r-1} T_R^{q-s}\psi_s\Big) = 0. \tag{27}$$

Then, if $r \leqq \min(p, q)$, we can set

$$\psi_r = \frac{(p-r)!}{(q-r)!\,(p+q-2r)!} T_L^{q-r}\left(\varphi - \sum_{s=0}^{r-1} T_R^{q-s}\psi_s\right) \varepsilon\, \mathfrak{A}^{p+q-r,r},$$

from the previous equation ψ_r is clearly left primitive, and thus, applying (25) to ψ_r after replacing r and s by $q - r$, we have

$$T_L^{q-r}T_R^{q-r}\psi_r = T_L^{q-r}\left(\varphi - \sum_{s=0}^{r-1} T_R^{q-s}\psi_s\right),$$

so that

$$T_L^{q-r-1}\left(\varphi - \sum_{s=0}^{r} T_R^{q-s}\psi_s\right)$$

is left primitive. The process stops when $r = \min(p, q) + 1$; if $q \leqq p$, (27) means that

$$\varphi - \sum_{s=0}^{q} T_R^{q-s}\psi_s = 0$$

completing the induction; if $q > p$, then the induction is complete when $r = p$. In fact, if we set

$$\lambda = \varphi - \sum_{s=0}^{p} T_R^{q-s}\psi_s \,\varepsilon\, \mathfrak{A}^{p,q},$$

we see that λ satisfies $T_L^{q-p}\lambda = 0$. It follows from Lemma 1 that $\lambda = 0$.

To prove the uniqueness of the expansion (26), suppose that

$$\sum_{s=0}^{r} T_R^{q-s}\,\psi_s = 0$$

for some sequence of left primitive elements $\psi_s \;\varepsilon\; \mathfrak{A}^{p+q-s,s}$ and for some $r (0 \leqq r \leqq \min(p,q))$; one can show then that each one of the ψ_s vanishes. In fact, suppose that this were not the case, and let m be the smallest value of s such that $\psi_s \neq 0$: then, using (28)

$$0 = T_L^{q-m}\left(\sum_{s=m}^{r} T_R^{q-s}\psi_s\right) = T_L^{q-m}T_R^{q-m}\psi_m = \frac{(q-m)!\,(p+q-2m)!}{(p-m)!}\psi_m.$$

This shows that $\psi_m = 0$, contrary to what was supposed, thus concluding the proof of Theorem 2.

Let τ be the involution of $\mathfrak{A} \otimes \mathfrak{A}$ mapping $\mathfrak{A}^{p,q}$ onto $\mathfrak{A}^{q,p}$ as follows: if φ is expanded as in (21), then

$$\tau\varphi = \frac{(-1)^{pq}}{p!\,q!}\varphi_{i_1\cdots i_p j_1\cdots j_q}\, dx^{j_1} \wedge \cdots \wedge dx^{j_q} \otimes dx^{i_1} \wedge \cdots \wedge dx^{i_p}.$$

Then one can verify immediately that $\tau\, T_R = T_L\, \tau$: consequently all the statements made in this discussion remain valid, if one interchanges "left", T_L, and p respectively with "right", T_R, and q.

We now can construct the fine resolution of the sheaf Θ for Clifford-Klein spaces. We shall start by defining the resolvent sheaves Φ^q $(0 \leqq q \leqq n)$; these sheaves are in each case those of the germs of differentiable cross sections in the bundle of tensors of the following respective types.

For $q = 0$: Tangent vectors;
For $q = 1$: Symmetric bilinear forms on the tangent space; in other words, right primitive elements in $\mathfrak{A}^{1,1}$;

For $2 \leqq q \leqq n$: Right primitive elements in $\mathfrak{A}^{2,q}$.

For instance, the sheaf Φ^2 can also be described as that of germs of differentiable, covariant tensor fields with the same symmetry properties (double skew-symmetry and Bianchi relation) as a general Riemann curvature tensor.

The next step consists of defining the homomorphisms D_q: $\Phi^q \to \Phi^{q+1}$. At this point the notation of moving orthonormal frames and of exterior differential calculus is the most satisfactory; accordingly, if (ω_i) $(1 \leqq i \leqq n)$ constitutes a local frame of orthonormal 1-forms in a Clifford-Klein space X, we write $ds^2 = \omega_i \otimes \omega_i$ and recall the local structural equations (3) and (4).

The injection map $i : \Theta \to \Phi^0$ is described by the inclusion map of the germs of Killing vectors among the germs of differentiable vector fields.

Let $\theta \in \Phi^0$: then $D_0\theta = \frac{1}{2}[\theta, ds^2]$, or

$$D_0\theta = \tfrac{1}{2}\{[\theta, \omega_i] \otimes \omega_i + \omega_i \otimes [\theta, \omega_i]\} \in \Phi^1. \tag{28}$$

It is convenient to represent elements $\eta \in \Phi^1$ in the form $\eta = \omega_i \otimes \eta_i$, where (η_i) is an n-tuple of 1-forms; the condition that η is right primitive can be stated equivalently by

$$\omega_i \wedge \eta_i = 0. \tag{29}$$

The operation $D_1\eta$ is defined in two steps as follows: η determines uniquely a skew-symmetric matrix of germs of 1-forms (ψ_{ij}) by the equation

$$d\eta_i - \eta_j \wedge \omega_{ji} = \omega_j \wedge \psi_{ji} \qquad (\psi_{ij} = -\psi_{ji}); \tag{30}$$

then $D_1\eta = \varphi = \frac{1}{2}\omega_i \wedge \omega_j \otimes \varphi_{ij}$, where

$$\varphi_{ij} = d\psi_{ij} - \omega_{ik} \wedge \psi_{kj} - \psi_{ik} \wedge \omega_{kj} + K(\omega_i \wedge \eta_j + \eta_i \wedge \omega_j). \tag{31}$$

If η is written in the tensor notation as $a_{ij}\, dx^i \otimes dx^j$ (and indeed η has the same geometrical interpretation as (11)), then the formulas (30) and (31) for $D_1\eta$ are motivated by (12), after the change in notation. It is clear that $\varphi = D_1\eta$, as defined above, is an element of $\mathfrak{A}^{2,2}$; to show that it is in Φ^2, we must prove that it is right primitive, or equivalently, since $\varphi_{ij} = -\varphi_{ji}$, that

$$\omega_j \wedge \varphi_{ij} = 0. \tag{32}$$

In fact, we see immediately that, by taking the exterior derivative of both terms of (29), the resulting equation is

$$0 = d(\omega_i \wedge \eta_i) = d\omega_i \wedge \eta_i - \omega_i \wedge d\eta_i = \omega_i \wedge (-d\eta_i + \omega_{ij} \wedge \eta_j)$$
$$= -\omega_i \wedge \omega_j \wedge \psi_{ji};$$

thus we have the relation

$$\omega_i \wedge \omega_j \wedge \psi_{ij} = 0. \tag{33}$$

Next, by taking the exterior derivative of the difference between the two members of (30), the outcome is (after applying (3) and (4)):

$$0 = d(-\omega_j \wedge \psi_{ij} + \omega_{ij} \wedge \eta_j)$$
$$= \omega_j \wedge d\psi_{ij} - d\omega_j \wedge \psi_{ij} - \omega_{ij} \wedge d\eta_j + d\omega_{ij} \wedge \eta_j$$
$$= \omega_j \wedge (d\psi_{ij} - \omega_{jk} \wedge \psi_{ik}) - \omega_{ij} \wedge (\omega_{jk} \wedge \eta_k + \omega_k \wedge \psi_{kj}) + d\omega_{ij} \wedge \eta_j$$
$$= \omega_j \wedge (d\psi_{ij} - \psi_{ik} \wedge \omega_{kj} - \omega_{ik} \wedge \psi_{kj}) + (d\omega_{ij} - \omega_{ik} \wedge \omega_{kj}) \wedge \eta_j$$
$$= \omega_j \wedge (d\psi_{ij} - \psi_{ik} \wedge \omega_{kj} - \omega_{ik} \wedge \psi_{kj}) - K\omega_i \wedge \omega_j \wedge \eta_j = \omega_j \wedge \varphi_{ij},$$

which proves (32). Equation (33) is not needed here, but is essential as an integrability condition later, in the proof of exactness of the sequence.

We now shall define the differential operator $D_q : \Phi^q \to \Phi^{q+1}$ for $2 \leqq q \leqq n$, as in the following. We represent any element $\varphi \in \Phi^q$ in the form $\varphi = 2^{-1}\omega_i \wedge \omega_j \otimes \varphi_{ij}$, where (φ_{ij}) is a skew-symmetric matrix of q-forms satisfying (32): then $D_q\varphi = 2^{-1}\omega_i \wedge \omega_j \otimes \varphi'_{ij}$, with

$$\varphi'_{ij} = d\varphi_{ij} - \omega_{ik} \wedge \varphi_{kj} + (-1)^q \varphi_{ik} \wedge \omega_{kj}. \tag{34}$$

It is clear that $D_q\varphi$, as it has been defined, is in $\mathfrak{A}^{2,q+1}$; to show that it belongs to the sheaf Φ^{q+1} is equivalent to proving that φ'_{ij} satisfies (32), since (φ'_{ij}) is clearly skew-symmetric. In fact,

$$\omega_j \wedge \varphi'_{ij} = \omega_j \wedge (d\varphi_{ij} - \omega_{ik} \wedge \varphi_{kj} + (-1)^q \varphi_{ik} \wedge \omega_{kj})$$
$$= -d(\omega_j \wedge \varphi_{ij}) + \omega_{ik} \wedge (\omega_j \wedge \varphi_{kj}) = 0,$$

proving the assertion.

The next thing to prove is that the sequence of sheaves Φ^q and homomorphisms D_q constitutes a cochain complex, i.e., that $D_q\, D_{q-1} = 0$ for $0 \leqq q \leqq n$. For $q = 0$, an element $\theta \in \Phi^0$ belongs to the subsheaf $\Theta \subset \Phi^0$, if and only if it is annihilated by the operator (28) from the definition of Θ (Killing's equation): this proves not only that $D_0 i = 0$, but that the sequence (20) is exact at the level 0.

The statement that, if $\eta = \omega_i \otimes \eta_i = \alpha_{ij}\, dx^i \otimes dx^j \in \Phi^1$, then there exists an element $\theta \in \Phi^0$ satisfying $D_0\theta = \eta$, if and only if $D_1\eta = 0$, is precisely a repetition of Proposition 3, since the integrability condition $D_1\eta = 0$ is equivalent to (12). This proves the exactness of the sequence (20) at the level $q = 1$.

In order to prove that $D_q D_{q-1} = 0$ for $q \geqq 3$, we observe that (34) is the formula for the absolute differential of φ_{ij}, if the latter is regarded as a tensorial q-form. The formula for the second absolute differential of any element $\varphi = 2^{-1}\omega_i \wedge \omega_j \otimes \varphi_{ij} \in \mathfrak{A}^{2,q}$ is given by

$$\begin{aligned} d^2\varphi &= \tfrac{1}{2}\omega_i \wedge \omega_j \otimes (-\Omega_{ik} \wedge \varphi_{kj} + \varphi_{ik} \wedge \Omega_{kj}) \\ &= \tfrac{1}{2}\omega_i \wedge \omega_j \otimes K(\omega_i \wedge \omega_k \wedge \varphi_{kj} - \varphi_{ik} \wedge \omega_k \wedge \omega_j): \end{aligned}$$

this clearly vanishes, if φ is right primitive; the hypothesis that X be a Clifford-Klein space is essential here.

The proof that $D_2 D_1 = 0$ is similar to, but slightly longer than, the one for the previous case; for this reason we shall introduce the notation d' for the absolute differential of a tensorial form, referred to its components relative to the local frame (ω_i). Thus, if

$$\mu = \omega_{i_1} \otimes \omega_{i_2} \otimes \cdots \otimes \omega_{i_p} \otimes \mu_{i_1 \cdots i_p}$$

is a tensorial q-form (i.e., the coefficients $\mu_{i_1 \cdots i_q}$ are q-forms), its absolute differential is the tensorial $(q+1)$-form

$$d\mu = \omega_{i_1} \otimes \cdots \otimes \omega_{i_p} \otimes d'\mu_{i_1 \cdots i_p},$$

where

$$d'\mu_{i_1 \cdots i_p} = d\mu_{i_1 \cdots i_p} - \sum_{\alpha=1}^{p} \omega_{i_\alpha k} \wedge \mu_{i_1 \cdots i_{\alpha-1} k i_{\alpha+1} \cdots i_p}.$$

With this notation, for instance, the first part of (3) becomes $d'\omega_i = 0$, and (30), (31), and (34) become respectively

$$\begin{aligned} d'\eta_i &= \omega_j \wedge \psi_{ji}, \\ \varphi_{ij} &= d'\psi_{ij} + K(\omega_i \wedge \eta_j + \eta_i \wedge \omega_j), \quad \text{and} \\ \varphi'_{ij} &= d'\varphi_{ij}. \end{aligned} \tag{35}$$

Let $\eta = \omega_i \otimes \eta_i$; then $D_1\eta = 2^{-1}\omega_i \wedge \omega_j \otimes \varphi_{ij}$, and $D_2 D_1 \eta = 2^{-1}\omega_i \wedge \omega_j \otimes \varphi'_{ij}$, where one verifies from (35) that

$$\begin{aligned} \varphi'_{ij} = d'\varphi_{ij} &= d'^2\psi_{ij} + K\, d'(\omega_i \wedge \eta_j + \eta_i \wedge \omega_j) \\ &= -\Omega_{ik} \wedge \psi_{kj} - \Omega_{jk} \wedge \psi_{ik} + K(-\omega_i \wedge d'\eta_j + d'\eta_i \wedge \omega_j) \\ &= K(\omega_i \wedge \omega_k \wedge \psi_{kj} + \omega_j \wedge \omega_k \wedge \psi_{ik} - \omega_i \wedge \omega_k \wedge \psi_{kj} + \omega_k \wedge \psi_{ki} \wedge \omega_j) \\ &= 0, \end{aligned} \tag{36}$$

proving that, in a Clifford-Klein space, $D_2 D_1 = 0$.

Having proved that the sequence (20) is a cochain complex, and that it is acyclic in the degrees 0 and 1, we must prove that it is exact at the levels $q \geqq 2$. This might be done by a general theorem on the sufficiency of integrability conditions for systems of first order partial differential equations; however we shall follow a less sophisticated method, requiring Poincaré's lemma on exterior forms as the only analytical tool. Our proof distinguishes the simpler case, where $K = 0$, from the one where $K \neq 0$; the latter is proved in part by using the proof of the former.

If $K = 0$, we consider a local, rectangular Cartesian coordinate system $(x^1, \cdots, x^n)$ and set $\omega_i = dx^i$ as the local frame of 1-forms. For this choice of frame the connection symbols ω_{ij} vanish identically, and thus the differential operator d', defined in (35), coincides with the exterior derivative component by component.

We shall prove now the exactness. Let $\varphi = 2^{-1}\, dx^i \wedge dx^j \otimes \varphi_{ij} \in \Phi^q$ for any $q \geqq 2$, and assume that $D_q\varphi = 0$. i.e. that each component φ_{ij} is closed. Then there exists a skew matrix (ψ'_{ij}) of germs of $(q-1)$-forms satisfying

$$d\psi'_{ij} = \varphi_{ij} \qquad \text{(Poincaré's lemma)}. \tag{37}$$

Since $dx^j \wedge \varphi_{ij} = 0$, the n q-forms $dx^j \wedge \psi'_{ij}$ are closed, so that there exists an n-tuple of $(q-1)$-forms χ_i satisfying

$$d\chi_i = dx^j \wedge \psi'_{ij}. \tag{38}$$

If $q \geqq 3$, $\chi = dx^i \otimes \chi_i$ belongs to $\mathfrak{A}^{1,q-1}$, and $q - 1 > 1$; from Theorem 2 one knows that there exists an element $\lambda = 2^{-1}\, dx^i \wedge dx^j \otimes \lambda_{ij} \in \mathfrak{A}^{2,q-2}$ such that $T_R\, \lambda = \chi$.

Here and in all analogous cases, we suppose without essential restriction that $\lambda_{ij} = -\lambda_{ji}$, and set

$$\psi_{ij} = \psi'_{ij} + d\lambda_{ij}.$$

Then the element $\psi = 2^{-1}\, dx^i \wedge dx^j \otimes \psi_{ij}$ belongs to Φ^{q-1}, since it belongs to $\mathfrak{A}^{2,q-1}$ and

$$dx^j \wedge \psi_{ij} = dx^j \wedge \psi'_{ij} - d(dx^j \wedge \lambda_{ij}) = d\chi_i - d\chi_i = 0.$$

Furthermore, $d\psi_{ij} = \varphi_{ij}$, whence $D_{q-1}\psi = \varphi$, which proves exactness at all levels q, $3 \leqq q \leqq n$.

For $q = 2$, the proof is the same up to (38); since $d(dx^i \wedge d\chi_i) = 0$, there exists an element $\mu = \mu_{ij}\, dx^i \wedge dx^j \in \mathfrak{A}^{0,2}$ satisfying

$$d\mu = dx^i \wedge d\chi_i.$$

Replacing ψ'_{ij} by $\psi_{ij} = \psi'_{ij} - d\mu_{ij}$, the new form $\psi = 2^{-1}\, dx^i \wedge dx^j \otimes \psi_{ij} \in \mathfrak{A}^{2,1}$ satisfies

$$d\psi_{ij} = \varphi_{ij}$$

as well as (33). In fact,

$$2T_R^2\psi = dx^i \wedge dx^j \wedge \psi_{ij} = dx^i \wedge dx^j \wedge (\psi'_{ij} - d\mu_{ij}) = dx^i \wedge d\chi_i - d\mu = 0.$$

Since $d(dx^j \wedge \psi_{ij}) = -dx^j \wedge \varphi_{ij} = 0$, there exists an n-tuple of germs of 1-forms η'_i satisfying

$$d\eta'_i = -dx^j \wedge \psi_{ij}.$$

Since ψ_{ij} satisfies (33), we have

$$d(dx^i \wedge \eta_i') = dx^i \wedge dx^j \wedge \psi_{ij} = 0,$$

there exists a germ of a 1-form $\xi = \xi_i\, dx^i$ such that

$$d\xi = dx^i \wedge \eta_i';$$

if we set $\eta_i = \eta_i' + d\xi_i$, the new element $\eta = dx^i \otimes \eta_i$ satisfies the equations

$$d\eta_i = d\eta_i' = -dx^j \wedge \psi_{ij}, \qquad d\psi_{ij} = \varphi_{ij}$$

and

$$dx^i \wedge \eta_i = dx^i \wedge \eta_i' + dx^i \wedge d\xi_i = dx^i \wedge \eta_i' - d\xi = 0.$$

Thus η is an element in Φ^1 satisfying $D_1\eta = \varphi$. This completes the proof of the exactness of the sequence (20) for $K = 0$.

Before proceeding to the case $K \neq 0$, we observe that the foregoing proof works in exactly the same way, if we consider a manifold with a locally flat, indefinite fundamental form, and a system of coordinates $(x^1, \cdots, x^n)$ such that $ds^2 = a_{ij}dx^i dx^j$, with (a_{ij}) a constant, symmetric, nonsingular matrix. The sheaf Θ consists of the germs of vectors $\theta = \theta^i\, \partial/\partial x^i$ satisfying

$$D_0\theta = \frac{1}{2}\left(\frac{\partial}{\partial x^i}(a_{jk}\theta^k) + \frac{\partial}{\partial x^j}(a_{ik}\theta^k)\right) = 0,$$

so that the resolvent sheaves Φ^q $(0 \leqq q \leqq n)$ have the same definition and differential operators as before, with $\theta_i = a_{ij}\theta^j$.

We consider now the case $K \neq 0$. Let $\boldsymbol{R}^{n+1}$ denote the Euclidean vector space with Cartesian coordinates $x = (x^0, x^1, \cdots, x^n)$ and the inner product formula

$$x \cdot y = (\operatorname{sgn} K)x^0y^0 + \sum_{i=1}^{n} x^i y^i; \tag{38}$$

we introduce in $\boldsymbol{R}^{n+1}$ the Riemannian structure (relativistic, if $K < 0$) defined by

$$ds^2 = dx \cdot dx = (\operatorname{sgn} K)(dx^0)^2 + \sum_{i=1}^{n} (dx^i)^2. \tag{39}$$

The n-dimensional simply connected space $\tilde{X}$ with constant curvature K can then be identified with the hypersurface in $\boldsymbol{R}^{n+1}$ defined by the equation

$$x \cdot x = K^{-1} \quad (\text{and} \quad x^0 > 0, \quad \text{if} \quad K < 0)$$

with the Riemannian structure induced by (39); a coordinate system $(\xi^1, \cdots, \xi^n)$ in $\tilde{X}$, formerly denoted by $(x^1, \cdots, x^n)$, with respect to which the induced Riemannian metric takes the form (2) is given by

$$\xi^i = \frac{x^i}{x_0\,|\sqrt{K}|}.$$

The connected group G of all the non-Euclidean motions of $\tilde{X}$ onto itself is isomorphic, by restriction from $\boldsymbol{R}^{n+1}$ to $\tilde{X}$, to the component of the identity in the orthogonal group $O(n + 1; x\cdot x)$ in $\boldsymbol{R}^{n+1}$ that leaves the quadratic form $x\cdot x$ invariant. Thus G can be identified with a subgroup of the group G' of all proper Euclidean (for $K > 0$) or Lorentzian (for $K < 0$) motions in $\boldsymbol{R}^{n+1}$, namely the subgroup that keeps the origin fixed.

Similarly, the sheaf $\Theta = \Theta(\tilde{X})$ of germs of Killing vectors in $\tilde{X}$ is related in the following way to the sheaf $\Theta(\boldsymbol{R}^{n+1})$. Let $\boldsymbol{E}$ denote the open region in $\boldsymbol{R}^{n+1}$ where $Kx\cdot x > 0$ and, if $K < 0$, $x^0 > 0$; let k designate the imbedding of $\tilde{X}$ into $\boldsymbol{E}$ as defined above and $j : \boldsymbol{E} \to \tilde{X}$ the surjective map defined by

$$j(x) = (Kx\cdot x)^{-1/2}x.$$

Clearly jk is the identity map of $\tilde{X}$; furthermore every rigid motion (finite or infinitesimal) of $\tilde{X}$ can be extended, by homogeneity, to a unique rigid motion of $\boldsymbol{E}$; hence one obtains a natural injection j^* of the sheaf $\Theta = \Theta(\tilde{X})$ into the sheaf $\Theta(\boldsymbol{E})$ of germs of Killing vector fields in $\boldsymbol{E}$; let Θ' denote the image $j^*(\Theta(\tilde{X}))$ in $\Theta(\boldsymbol{E})$. With reference to the sequence (20) of sheaves, one can construct uniquely some natural injections j^* (by homogeneity) of the sheaves $\Phi^q(\tilde{X})$ into $\Phi^q(\boldsymbol{E})$ $(0 \le q \le n)$, with images denoted respectively by Φ'^q, such that the restriction $k^*(\Phi'^q)$ of Φ'^q to $\tilde{X}$ coincides with $\Phi^q(\tilde{X})$, and the diagram

$$\begin{array}{ccccccccccccccc}
0 \to \Theta(E) & \xrightarrow{i} & \Phi^0(E) & \xrightarrow{D_0} & \Phi^1(E) & \xrightarrow{D_1} & \Phi^2(E) & \xrightarrow{D_2} & \cdots & \xrightarrow{D_{n-1}} & \Phi^n(E) & \xrightarrow{D_n} & \Phi^{n+1}(E) \to 0 \\
i\uparrow & & i\uparrow & & i\uparrow & & i\uparrow & & & & i\uparrow & & \uparrow \\
0 \to \Theta' & \xrightarrow{i} & \Phi'^0 & \xrightarrow{D_0} & \Phi'^1 & \xrightarrow{D_1} & \Phi'^2 & \xrightarrow{D_2} & \cdots & \xrightarrow{D_{n-1}} & \Phi'^n & \xrightarrow{D_n} & 0 \\
j^*\uparrow & & j^*\uparrow & & j^*\uparrow & & j^*\uparrow & & & & j^*\uparrow & & \\
0 \to \Theta(\tilde{X}) & \xrightarrow{i} & \Phi^0(\tilde{X}) & \xrightarrow{D_0} & \Phi^1(\tilde{X}) & \xrightarrow{D_1} & \Phi^2(\tilde{X}) & \xrightarrow{D_2} & \cdots & \xrightarrow{D_{n-1}} & \Phi^n(\tilde{X}) & \xrightarrow{D_n} & 0
\end{array}$$

is commutative. In this diagram the maps denoted by i stand for trivial inclusion maps; the top and bottom rows denote respectively the fine resolution of $\Theta(\boldsymbol{E})$ (already proved) and the corresponding sequence for $\tilde{X}$, which will be shown to be a fine resolution, as soon as we have proved that it is acyclic at the levels $q \ge 2$; this is done simply by describing the sequence in the middle row of the diagram explicitly, and by proving then that it is exact, since the stalks at points of $\tilde{X}$ for the two bottom rows are naturally isomorphic.

Let $\mathbf{r}$ denote the vector field $x^i\partial/\partial x^i$ in $\boldsymbol{R}^{n+1}$ or in $\boldsymbol{E}$. A tensor field $\mathfrak{J}$ of any type in $\boldsymbol{E}$ is called homogeneous of degree λ for any real number λ, if and only if $[\mathbf{r}, \mathfrak{J}] = \lambda\mathfrak{J}$, where the bracket denotes, as usual, the Lie derivative of $\mathfrak{J}$ with respect to $\mathbf{r}$. If $\theta = \theta^i\partial/\partial x^i$ is any vector field, or germ of a vector field in $\boldsymbol{R}^{n+1}$, let $\mathbf{r}\cdot\theta$ denote the inner product $(\operatorname{sgn} K)\, x^0\theta^0 + \sum_{i=1}^{n} x^i\theta^i$ corresponding to the Riemannian structure (39). Finally, if $\mu = (1/p!)\mu_{i_1\cdots i_p}\, dx^{i_1} \wedge \cdots \wedge dx^{i_p}$ is any (germ of a) p-form in $\boldsymbol{R}^{n+1}$, we denote (cf. [**4**]) by $\mathbf{r} \barwedge \mu$ the $(p - 1)$-form

$$\mathbf{r}\barwedge\mu = \frac{1}{(p-1)!}\, x^{i_1}\mu_{i_1 i_2\cdots i_p}\, dx^{i_2} \wedge \cdots \wedge dx^{i_p},$$

and recall the identity

$$[\mathbf{r}, \mu] = \mathbf{r}\barwedge\, d\mu + d(\mathbf{r}\barwedge\mu).$$

An easy computation yields the following explicit description of the sheaves Θ' and Φ'^{q} $(0 \leq q \leq n)$ appearing in the middle row of the diagram. The sheaf Θ' is the sheaf of germs of vector fields θ in E that satisfy

(a₀) $[\theta, ds^2] = 0$ (Killing's equation),

(b₀) $\mathbf{r}\cdot\theta = 0$ (i.e., θ is tangent to each hypersphere $x\cdot x =$ constant),

(c₀) $[\mathbf{r}, \theta] = 0$ (i.e., θ is homogeneous of degree 0)

(note that (c_0) is a consequence of (a_0) and (b_0), but is included for reasons of convenience). Φ'^{0} is the sheaf of germs of vector fields θ in E that satisfy only (b_0) and (c_0). Φ'^{1} is the sheaf of germs of double forms $\eta = dx^i \otimes \eta_i \in \mathfrak{A}^{1,1}(E)$ satisfying $dx^i \wedge \eta_i = 0$ (i.e., $\eta \in \Phi^1(E)$) and, in addition,

(a₁) $\mathbf{r}\barwedge\eta_i = 0, \quad (i = 0, 1, \cdots, n),$

(b₁) $x^i\eta_i = 0,$

(c₁) $[\mathbf{r}, \eta_i] = \eta_i.$

For $2 \leq q \leq n+1$, Φ'^{q} is the sheaf of elements $\varphi = 2^{-1}\, dx^i \wedge dx^j \otimes \varphi_{ij} \in \mathfrak{A}^{2,q}(E)$ satisfying $T_R\,\varphi = 0$ (or equivalently $dx^j \wedge \varphi_{ij} = 0$, i.e., $\varphi \in \Phi^q(E)$) and, in addition,

(a$_q$) $\mathbf{r}\barwedge\varphi_{ij} = 0,$

(b$_q$) $x^j\varphi_{ij} = 0,$

(c$_q$) $[\mathbf{r}, \varphi_{ij}] = 0.$

REMARKS. For $q = 1$, conditions (a₁) and (b₁) are equivalent, since, if λ and μ are exterior forms of degree p, q respectively,

$$\mathbf{r} \barwedge (\lambda \wedge \mu) = (\mathbf{r} \barwedge \lambda) \wedge \mu + (-1)^p\lambda \wedge (\mathbf{r} \barwedge \mu),$$

whence

$$0 = \mathbf{r} \barwedge (dx^i \wedge \eta_i) = (\mathbf{r} \barwedge dx^i) \wedge \eta_i - dx^i \wedge (\mathbf{r} \barwedge \eta_i) = x^i\eta_i - (\mathbf{r} \barwedge \eta_i)\, dx^i.$$

Similarly, for $q \geqq 2$, (a$_q$) implies (b$_q$), since

$$0 = \mathbf{r} \barwedge (dx^j \wedge \varphi_{ij}) = (\mathbf{r} \barwedge dx^j)\varphi_{ij} - dx^j \wedge (\mathbf{r} \barwedge \varphi_{ij}) = x^j\varphi_{ij} - dx^j \wedge (\mathbf{r} \barwedge \varphi_{ij}).$$

We show, to start, that $D_q(\Phi'^{q}) \subset \Phi'^{q+1}$ for $0 \leqq q \leqq n$. For $q = 0$, if $\theta \in \Phi'^{0}$, and if $D_0\theta = dx^i \otimes \eta_i$, then the 1-form $x^i\eta_i$ is characterized by its value at any tangent vector ξ at any point; we have the easily verified identity

$$\xi \barwedge (x^i\eta_i) = [\xi, \mathbf{r}\cdot\theta] + \xi\cdot[\mathbf{r}, \theta] = 0,$$

which proves that $D_0\theta$ satisfies (a₁) and, equivalently, (b₁). To prove (c₁), we use the identity

$$[\mathbf{r}, D_0\theta] = [\mathbf{r}, [\theta, ds^2]] = [[\mathbf{r}, \theta], ds^2] + [\theta, [\mathbf{r}, ds^2]]$$

but $[\mathbf{r}, ds^2] = 2ds^2$, since the form ds^2 is homogeneous of degree 2, so that

$$[\mathbf{r}, D_0\theta] = 2\, D_0\theta;$$

hence $D_0\theta$ is homogeneous of degree 2, so that the forms η_i are homogeneous of degree 1; this proves (c_1), so that $D_0\Phi'^0 \subset \Phi'^1$. Next, let $\eta = dx^i \otimes \eta_i \in \Phi'^1$ and consider $D_1\eta$; since $dx^i \wedge \eta_i = 0$, we have, successively,

$$0 = d(x^i\eta_i) = x^i\, d\eta_i = -x^i\, dx^j \wedge \psi_{ij} = dx^i \wedge (x^j\psi_{ij}),$$

$$(40) \qquad 0 = d(\mathbf{r} \barwedge \eta_i) = [\mathbf{r}, \eta_i] - \mathbf{r} \barwedge d\eta_i = \eta_i + x^j\psi_{ij} - dx^j \wedge (\mathbf{r} \barwedge \psi_{ij}),$$

$$d\eta_i = d[\mathbf{r}, \mu_i] = [\mathbf{r}, d\eta_i].$$

The second of the above equations shows that η_i is determined algebraically by ψ_{ij} through the equation

$$(41) \qquad \eta_i = (\mathbf{r} \barwedge \psi_{ij})\, dx^j - x^j\psi_{ij}.$$

Since $dx^i\wedge \eta_i = 0$ and by using the first equation (40), we have

$$(\mathbf{r} \barwedge \psi_{ij})\, dx^i \wedge dx^j = 0,$$

but since the coefficients of this 2-form are already skew-symmetric, we obtain the two following relations, using (41):

$$(42) \qquad \mathbf{r} \barwedge \psi_{ij} = 0, \qquad \eta_i = -x^j\psi_{ij}.$$

The third equation (40) expresses the fact that $dx^j\wedge \psi_{ij}$ is homogeneous of degree 1, whence the forms ψ_{ij} are homogeneous of degree zero, or

$$(43) \qquad [\mathbf{r}, \psi_{ij}] = 0.$$

Since $D_1\eta = 2^{-1}\, dx^i\wedge dx^j \otimes \varphi_{ij}$ for $\varphi_{ij} = d\psi_{ij}$, we see immediately the following:

$$\mathbf{r} \barwedge \varphi_{ij} = \mathbf{r} \barwedge d\psi_{ij} = [\mathbf{r}, \psi_{ij}] - d(\mathbf{r} \barwedge \psi_{ij}) = 0,$$

$$x^j\varphi_{ij} = x^j\, d\psi_{ij} = d(x^j\psi_{ij}) - dx^j \wedge \psi_{ij} = -d\eta_i + d\eta_i = 0,$$

$$[\mathbf{r}, \varphi_{ij}] = [\mathbf{r}, d\psi_{ij}] = d[\mathbf{r}, \psi_{ij}] = 0.$$

This verifies that φ_{ij} satisfies properties (a_2), (b_2) and (c_2).

Finally, we prove that, if $\varphi = 2^{-1}\, dx^i\wedge dx^j \otimes \varphi_{ij} \in \Phi'^q$ for any $q \geqq 2$, then $d\varphi_{ij}$ satisfies properties (a_{q+1}), (b_{q+1}), (c_{q+1}); in fact

$$\mathbf{r} \barwedge d\varphi_{ij} = [\mathbf{r}, \varphi_{ij}] - d(\mathbf{r} \barwedge \varphi_{ij}) = 0,$$

$$x^j\, d\varphi_{ij} = d(x^j\varphi_{ij}) - dx^j \wedge \varphi_{ij} = 0,$$

$$[\mathbf{r}, d\varphi_{ij}] = d[\mathbf{r}, \varphi_{ij}] = 0.$$

We conclude this part of the proof by the remark that, although $\Phi^{n+1}(E) \neq 0$, the sheaf $\Phi'^{(n+1)} = 0$, since the mapping of $(n+1)$-forms φ_{ij} into n-forms $\mathbf{r} \barwedge \varphi_{ij}$ is injective.

Since the sheaves Φ'^q are naturally isomorphic to $\Phi^q(\tilde{X})$ under the restriction mapping, we have to prove that the sequence

$$\xrightarrow{D_{q-1}} \Phi'^q \xrightarrow{D_q}$$

is exact at all levels $q \geqq 2$. Let $\varphi = 2^{-1}\, dx^i \wedge dx^j \otimes \varphi_{ij} \,\varepsilon\, \Phi'^q$ for any $q \geqq 2$, and suppose that $d\varphi_{ij} = 0$. Let $r = (\mathbf{r}\cdot\mathbf{r})^{1/2}$ denote the radial distance function from the origin, and let

$$\bar{\varphi}_{ij} = \frac{dr}{r} \wedge \varphi_{ij};$$

the new forms $\bar{\varphi}_{ij}$ are of degree $q + 1$, are closed, and satisfy

(a) $$\mathbf{r} \barwedge \bar{\varphi}_{ij} = \varphi_{ij},$$

(b) $$x^i \bar{\varphi}_{ij} = 0,$$

(c) $$[\mathbf{r}, \bar{\varphi}_{ij}] = 0,$$

so that $\bar{\varphi} = 2^{-1}\, dx^i \wedge dx^j \otimes \bar{\varphi}_{ij} \,\varepsilon\, \Phi^{q+1}(E)$ and $D_{q+1}\bar{\varphi} = 0$. From the proof of exactness in the flat case, we can find $\bar{\psi} = 2^{-1}\, dx^i \wedge dx^j \otimes \bar{\psi}_{ij} \,\varepsilon\, \Phi^q(E)$ satisfying $D_q\bar{\psi} = \bar{\varphi}$; it is easy to see that one can choose $\bar{\psi}_{ij}$ so that they are homogeneous of degree 0, i.e.,

$$[\mathbf{r}, \bar{\psi}_{ij}] = 0.$$

If $q = 2$, we set

$$\psi_{ij} = -\mathbf{r} \barwedge \bar{\psi}_{ij},$$

and

$$\eta_i = -x^j \psi_{ij}.$$

The form $\eta = dx^i \otimes \eta_i$ now belongs to Φ'^1 and $D_1\eta = \varphi$; in fact

$$\begin{aligned}
\mathbf{r} \barwedge \eta_i &= x^j\mathbf{r} \barwedge (\mathbf{r} \barwedge \bar{\psi}_{ij}) = 0,\\
x^i\eta_i &= -x^i x^j \psi_{ij} = 0,\\
dx^i \wedge \eta_i &= -dx^i \wedge x^j\psi_{ij} = x^i\, dx^j \wedge \psi_{ij} = -x^i\, dx^j \wedge (\mathbf{r} \barwedge \bar{\psi}_{ij})\\
&= x^i\mathbf{r} \barwedge (dx^j \wedge \bar{\psi}_{ij}) - x^i(\mathbf{r} \barwedge dx^j)\bar{\psi}_{ij} = -x^i x^j \bar{\psi}_{ij} = 0;
\end{aligned}$$

furthermore

$$\begin{aligned}
d\eta_i &= -dx^j \wedge \psi_{ij} - x^j\, d\psi_{ij}\\
&= -dx^j \wedge \psi_{ij} + x^j\, d(\mathbf{r} \barwedge \bar{\psi}_{ij})\\
&= -dx^j \wedge \psi_{ij} + x^j([\mathbf{r}, \bar{\psi}_{ij}] - \mathbf{r} \barwedge d\bar{\psi}_{ij})\\
&= -dx^j \wedge \psi_{ij} - \mathbf{r} \barwedge (x^j\bar{\varphi}_{ij}) = -dx^j \wedge \psi_{ij} - x^j\varphi_{ij}\\
&= -dx^j \wedge \psi_{ij},
\end{aligned}$$

and

$$d\psi_{ij} = -d(\mathfrak{r} \barwedge \bar{\psi}_{ij}) = -[\mathfrak{r}, \bar{\psi}_{ij}) + \mathfrak{r} \barwedge d\bar{\psi}_{ij}$$
$$= \mathfrak{r} \barwedge \bar{\varphi}_{ij} = \varphi_{ij},$$

which proves our assertion for $q = 2$.

If $q \geqq 3$, we proceed as before to construct $\bar{\psi} \in \Phi^q(E) \subset \mathfrak{A}^{2,q}$ as before; since $q > 2$, there exists, by Theorem 2, an element

$$\chi = 6^{-1}\, dx^i \wedge dx^j \wedge dx^k \otimes \chi_{ijk} \in \mathfrak{A}^{3,q-1} \quad \text{such that} \quad T_R\chi = \bar{\psi};$$

(χ_{ijk}) is supposed to be alternating in its indices; it is not hard to see that, since each $\bar{\psi}_{ij} = dx^k \wedge \chi_{ijk}$ is homogeneous of degree 0, χ_{ijk} can be chosen so as to be homogeneous of degree -1, i.e.,

$$[\mathfrak{r}, \chi_{ijk}] = -\chi_{ijk}.$$

Now set

$$\psi_{ij} = x^k(\mathfrak{r} \barwedge d\chi_{ijk}), \quad \psi = 2^{-1}\, dx^i \wedge dx^j \otimes \psi_{ij};$$

we claim that $\psi \in \Phi'^{q-1}$ and $D_{q-1}\psi = \varphi$. First of all, $\psi \in \mathfrak{A}^{2,q-1}$; then

$$dx^j \wedge \psi_{ij} = dx^j \wedge x^k(\mathfrak{r} \barwedge d\chi_{ijk}) = -x^j\, dx^k \wedge (\mathfrak{r} \barwedge d\chi_{ijk})$$
$$= -x^j\{\mathfrak{r} \barwedge (dx^k \wedge d\chi_{ijk}) - x^k\, d\chi_{ijk}\} = x^j(\mathfrak{r} \barwedge d\bar{\psi}_{ij}) + x^jx^k\, d\chi_{ijk}$$
$$= x^j\varphi_{ij} = 0,$$

so that $\psi \in \Phi^{q-1}(E)$. Next:

$$d\psi_{ij} = dx^k \wedge (\mathfrak{r} \barwedge d\chi_{ijk}) + x^k\, d(\mathfrak{r} \barwedge d\chi_{ijk})$$
$$= -\mathfrak{r} \barwedge (dx^k \wedge d\chi_{ijk}) + (\mathfrak{r} \barwedge dx^k)\, d\chi_{ijk} + x^k\, d[\mathfrak{r}, \chi_{ijk}]$$
$$= \mathfrak{r} \barwedge d(dx^k \wedge \chi_{ijk}) + x^k\, d\chi_{ijk} - x^k\, d\chi_{ijk}$$
$$= \mathfrak{r} \barwedge d\bar{\psi}_{ij} = \varphi_{ij},$$

proving that $D_{q-1}\psi = \varphi$. Finally, to show that $\psi \in \Phi'^{q-1}$, we calculate the following:

$$\mathfrak{r} \barwedge \psi_{ij} = x^k\mathfrak{r} \barwedge (\mathfrak{r} \barwedge d\chi_{ijk}) = 0,$$
$$x^j\psi_{ij} = x^jx^k(\mathfrak{r} \barwedge d\chi_{ijk}) = 0,$$
$$[\mathfrak{r}, \psi_{ij}] = d(\mathfrak{r} \barwedge \psi_{ij}) + \mathfrak{r} \barwedge d\psi_{ij}$$
$$= \mathfrak{r} \barwedge \varphi_{ij} = 0.$$

This completes the proof of the exactness of the sequence (Φ'^q) and, through the natural isomorphism, establishes the proof of the exactness of the resolution (20) of the sheaf Θ in all cases.

In a subsequent article this exact sequence and the consequent de Rham theorem will be supplemented by an analogue of Hodge's theorem, which will represent the cohomology groups $H^q(X, \Theta)$ for a compact, Clifford-Klein space faithfully by tensor fields of the type Φ^q, satisfying globally certain elliptic systems of equations.

References

1. S. Bochner, *Vector fields and Ricci curvature*, Bull Amer. Math. Soc. vol. 52 (1946) pp. 776–797.
2. C. Ehresmann, *Sur les espaces fibrés différentiables*, C. R. Acad. Sci. Paris vol. 224 (1947) pp. 1611–1612.
3. ———, *Structures locales*, Ann. Mat. Pura Appl. vol. 36 (1954) pp. 133–142.
4. A. Frölicher and A. Nijenhuis, *Theory of vector-valued differential forms*. I, Proc. Kon. Ned. Akad. Wet. Amsterdam vol. 59 (1956) pp. 338–359.
5. R. Godement, *Topologie algébrique et théorie des faisceaux*, Paris, Hermann, 1958.
6. K. Kodaira, *On deformations of some pseudogroup structures*, Ann. of Math. vol. 71 (1960) pp. 224–302.
7. A. Weil, *Introduction à la théorie des variétés kählériennes*, Paris, Hermann, 1958.

University of Minnesota,
Minneapolis, Minnesota

Relativistic Space Forms

Eugenio Calabi and Lawrence Markus

Ann. of Math. (2) 75, 63–76 (1962)

1. The perfect cosmological principle and spaces of constant curvature

In the physics of relativistic cosmology, the space-time continuum is taken to be a four dimensional differentiable manifold which bears a fundamental quadratic form with Lorentz signature. An interesting hypothesis is the perfect cosmological principle [2] which states that the basic geometry of the world is the same in the vicinity of any two events in space-time. In this paper we shall describe and classify such cosmological models.

A Lorentz manifold M^n is a differentiable n-manifold, $n \geqq 2$, (that is, a connected separable metric space with differentiable, C^∞, local coordinates) with a Lorentz fundamental tensor g_{ij} (that is, a non-singular, symmetric, covariant, differentiable tensor field having a signature of $-++\cdots+$). The Christoffel-Levi Civita linear connection is the unique torsionless linear connection for which the parallel displacement defines isometries (maps preserving inner products of tangent vectors) between the Minkowski tangent spaces of M^n. The Lorentz manifold M^n is called complete if each geodesic of M^n can be extended in M^n for all real values of the affine path parameter. A Lorentz manifold M^n is called isochronous (time sense conserved), isochirous (reversing time whenever orientation is reversed), orientable, or proper (isochronous and orientable, and hence isochirous), in case the holonomy group belongs to the corresponding subgroup of the Lorentz group.

A Lorentz manifold M^n has constant curvature K in case for each point $P \in M^n$ there exist local coordinates centered at P in which

$$(1)\qquad ds^2 = \frac{-(dx^1)^2 + (dx^2)^2 + (dx^3)^2 + \cdots + (dx^n)^2}{\left[1 + \frac{K}{4}(-x^{1^2} + x^{2^2} + \cdots + x^{n^2})\right]^2}$$

This is the case if and only if the curvature tensor satisfies

$$(2)\qquad R_{ijkl} = K(g_{ik}g_{jl} - g_{il}g_{jk}) .$$

If $K \neq 0$, we replace the fundamental form ds^2 by the equivalent one

* This research was partially supported by NSF contract G-10090 and OOR contract DA-11-022-ORD-3369.

DOI: 10.2307/1970419

$|K|\,ds^2$ to form a Lorentz manifold of constant curvature $K/|K| = \pm 1$. The modified Lorentz manifold is complete if and only if the original Lorentz manifold is complete. Thus in the study of Lorentz manifolds of constant curvature, one need only consider the cases $K = 0$, $K = +1$, $K = -1$.

A Lorentz manifold is said to satisfy the local perfect cosmological principle just in case each isometry between tangent spaces T_{P_1} and T_{P_2} of points P_1 and P_2 of M^n can be realized as the differential of a differentiable isometry between neighborhoods of P_1 and P_2. Thus the local perfect cosmological principle obtains for a Lorentz manifold M^n if and only if M^n is a Lorentz manifold of constant curvature, cf., [4].

DEFINITION. A relativistic space form is a complete Lorentz manifold of constant curvature.

The cosmological models considered in this paper are 4-dimensional complete Lorentz manifolds of constant curvature K. In such worlds the Einstein field equations for empty space are satisfied provided the cosmological constant

$$\lambda = \frac{R}{4} = -3K .$$

The global perfect cosmological principle postulates that each isometry between tangent spaces T_{P_1} and T_{P_2} of points P_1 and P_2 of a Lorentz manifold M^n can be realized by a global differentiable isometry of M^n onto itself.

The Minkowski space F^n is the real number space R^n (a differentiable manifold in the usual way) with the flat Lorentz tensor

$$ds^2 = -(dx^1)^2 + (dx^2)^2 + \cdots + (dx^n)^2 , \qquad n \geqq 2 .$$

Thus F^n is a Lorentz manifold of constant curvature $K=0$. The geodesics of F^n are all non-singular affine maps of R^1 into F^n, and each pair of points can be joined by a geodesic. The isometries (necessarily differentiable isometries) of F^n onto itself are all affine Lorentz transformations. Clearly F^n is a proper, complete, Lorentz manifold which satisfies the global perfect cosmological principle.

The De Sitter space D^n is the quadric

$$-(x^1)^2 + (x^2)^2 + (x^3)^2 + \cdots + (x^{n+1})^2 = +1 , \qquad n \geqq 2 ,$$

with the Lorentz tensor inherited from the ambient Minkowski space F^{n+1}. The isometries of D^n are precisely the restrictions of the isometries of F^{n+1} leaving the origin fixed. Also the isometries of D^n act transitively on D^n, and thus it is easy to see that D^n is a proper, complete, Lorentz manifold which satisfies the global perfect cosmological principle, with $K = +1$. Each geodesic on D^n is one component of the intersection of D^n with a two-dimensional plane in F^{n+1} through the origin. If the 2-plane inherits a positive definite metric from F^{n+1}, the corresponding geodesic on D^n is a space-like ellipse of length 2π. Each time-like geodesic is a branch of a hyperbola, and each light-like geodesic is a straight line in F^{n+1}, the case being determined by whether the 2-plane inherits an indefinite or a singular metric, respectively. Each pair of diametrically opposite points can be joined by infinitely many geodesics, all of them space-like. If P_1 and P_2 are distinct and not diametrically opposite points of D^n, they can be joined by a unique geodesic except in the case where the points P_1 and P_2 lie on different components of the intersection of D^n with the 2-plane in F^{n+1} passing through the origin, P_1, and P_2. The De Sitter space D^n is diffeomorphic with the product of an (usual) $(n-1)$-sphere and a line, $S^{n-1} \times R^1$. Thus D^n is simply-connected for $n \geqq 3$.

The anti-De Sitter space A^n is the quadric

$$-(x^1)^2 - (x^2)^2 + (x^3)^2 + \cdots + (x^{n+1})^2 = -1 , \qquad n \geqq 2 ,$$

with the Lorentz tensor inherited from the ambient flat space R^{n+1} bearing the ultra-hyperbolic quadratic form

$$ds^2 = -(dx^1)^2 - (dx^2)^2 + (dx^3)^2 + \cdots + (dx^{n+1})^2 .$$

Again the isometries of A^n are restrictions of the isometries of the flat space, which leave the origin fixed, and these isometries act transitively on A^n. Thus A^n is a proper, complete, Lorentz manifold satisfying the global perfect cosmological principle, with $K = -1$. Again each geodesic on A^n is a component of the intersection of A^n with a 2-plane through the origin in the ambient R^{n+1}. Every 2-plane in R^{n+1} through the origin O, meeting A^n, inherits a metric which is either negative definite, indefinite, or negative semi-definite of rank 1, since it contains a negative definite line. According to the three cases in their respective order, the corresponding geodesics are time-like (ellipses), space-like (branch of hyperbola), or light-like (straight lines). Not every pair of points of A^n can be joined by a geodesic.

For $n = 2$, A^2 and D^2 are similar when space and time are interchanged by a change of sign of ds^2, and we shall not consider D^2 exhaustively. We obtain theorems concerning the geometry of D^n, $n \geqq 2$, but our classification of relativistic spherical space forms is complete only for $n \geqq 3$ since we analyse only those manifolds which are covered by D^n.

Now A^n is diffeomorphic with $S^1 \times R^{n-1}$, and so it is not simply-connected. We call the universal (metric) covering space $\widetilde{A}^n$ of A^n the universal anti-De Sitter space. Clearly $\tilde{A}^n$ is diffeomorphic with R^n and is a proper, complete, Lorentz manifold satisfying the global perfect cosmological principle with $K = -1$. Points of $\tilde{A}^n$ cannot always be joined by a geodesic.

It is known [4, 10] that a simply-connected complete Lorentz manifold U^n, $n \geqq 3$, of constant curvature $K = 0$, $+1$, or -1 is isometric with F^n, D^n, or $\tilde{A}^n$, respectively. Now let π_1 be a properly discontinuous group of isometries of U^n; that is,

(1) each point $P \in U^n$ has a neighborhood N_P such that gN_P, $g \in \pi_1$, intersects N_P if and only if g is the identity element of π_1, also

(2) if P_1 and P_2 are points of U^n which are not congruent under π_1, then there exist corresponding neighborhoods N_{P_1} and N_{P_2} such that $gN_{P_1} \cap N_{P_2}$ is empty for every isometry $g \in \pi_1$.

Thus the quotient space U^n/π_1 is a Hausdorff space which is a complete Lorentz manifold of constant curvature. Clearly every complete Lorentz manifold, with $n \geqq 3$, and of constant curvature $K = 0$, $+1$, or -1 is isometric with some one of these U^n/π_1.

In an earlier work [1] one of the of the authors studied relativistic space forms with $K = 0$. In this paper we shall study relativistic space forms with $K = +1$, that is, relativistic spherical space forms.

2. Classification of relativistic spherical space forms. Isochronous and proper Lorentz manifolds

A relativistic spherical space form is a complete Lorentz manifold M^n, for $n \geqq 3$, with constant curvature $K = +1$. We shall effect a classification of these manifolds, up to Lorentz isometry, in terms of the classification of riemannian spherical space forms of $n-1$ dimensions. Thus for $n = 4$, as in relativistic cosmology, the classification rests on the knowledge of the riemannian spherical space forms of dimension 3 which are quite well understood [8].

The De Sitter space D^n, $n \geqq 3$, is a relativistic spherical space form, and moreover every relativistic spherical space form is D^n/π_1, where π_1 is a properly discontinuous group of isometries of D^n.

DEFINITION. Consider a riemannian spherical space form with fundamental group $\hat{\pi}_1 \subset \mathrm{O}(n)$, $n \geqq 3$. Consider the subgroup π_1 of the Lorentz group $\mathrm{L}(n+1)$ consisting of linear isometries of the Minkowski space F^{n+1} which leave the time axis fixed and which act on the euclidean n space $x^1 = 0$ as $\hat{\pi}_1$. Then π_1 is a properly discontinuous group of isometries of $D^n \subset F^{n+1}$. We call D^n/π_1 an elementary relativistic spherical space form.

This construction defines a one-to-one transformation from the set of all isometry classes of $(n-1)$-dimensional riemannian spherical space forms into the set of all isometry classes of isochronous relativistic spherical space forms of n dimensions, $n \geqq 3$. Our principal result (Theorem 2) is that this transformation furnishes a complete classification of the isochronous relativistic spherical space forms.

THEOREM 1. *Let $M^n = D^n/\pi_1$, $n \geqq 2$, where π_1 is a properly discontinuous group of isometries of D^n. Then π_1 is finite.*

PROOF. Suppose π_1 is infinite. Take a point $P \in D^n$ on the equatorial sphere S^{n-1} where $x^1 = 0$.

We first show that there exists a neighborhood N of P in D^n such that gN neets S^{n-1} for only a finite number of $g \in \pi_1$. For if there were no such neighborhood, then there would exist an infinite sequence $\{P_m\}$ of distinct points of D^n with $\lim_{m\to\infty} P_m = P$, and an infinite sequence of distinct elements $g_1, g_2, \cdots, g_m, \cdots$ in π_1 such that $g_m(P_m) \in S^{n-1}$ for $m = 1, 2, 3, \cdots$. Since S^{n-1} is compact, we can replace $\{P_m\}$ and $\{g_m\}$ by sub-sequences (without changing notation) so that

$$\lim_{m\to\infty} g_m(P_m) = Q \in S^{n-1} \subset D^n .$$

Furthermore Q is congruent to P under π_1, for otherwise the quotient manifold $D^n/\pi_1 = M^n$ would fail to satisfy the Hausdoff separation axiom. Thus there exists a $\hat{g} \in \pi_1$ such that $\hat{g}(Q) = P$.

Consider the points $\hat{g}g_1(P_1), \hat{g}g_2(P_2), \cdots, \hat{g}g_m(P_m), \cdots$, which approach P. We can assume that $\hat{g}g_m$ is never the identity of π_1. But this contradicts the hypothesis that π_1 is properly discontinuous. Thus there exists a neighborhood N of P in D^n whose orbit under π_1 meets the equatorial sphere S^{n-1} for only a finite number of elements of π_1.

Since the equator S^{n-1} is compact, there exists a neighborhood $\mathfrak{N}$ of S^{n-1} in D^n whose orbit meets S^{n-1} for only a finite number of elements of π_1. However, this is impossible if π_1 is infinite, since every isometry of π_1 maps the equatorial S^{n-1} to an image which meets the equator. Therefore π_1 is finite.

COROLLARY. *A relativistic spherical space form is not compact and has a finite fundamental group π_1.*

PROOF. A relativistic spherical space form M^n, $n \geqq 3$, is of the form D^n/π_1, where π_1 is a properly discontinuous group of isometries of the simply-connected De Sitter space D^n. Since D^n is not compact, and π_1 is finite, M^n is not compact.

LEMMA 1. *Let g be a linear isometry of the Minkowski space F^{n+1}, $n \geqq 2$, satisfying $g^2 = I$. If g has no space-like eigenvector with eigen-*

value $+1$, *then, after a suitable choice of a proper orthogonal basis in* F^{n+1}, *either*

$$g = \operatorname{diag}\{1, -1, -1, \cdots, -1\}$$

or

$$g = \operatorname{diag}\{-1, -1, -1, \cdots, -1\} .$$

PROOF. Since $g^2 = I$, the vector space F^{n+1} is the direct sum $T \oplus S$ where $g = I$ on T and $g = -I$ on S. Since g is an isometry, the subspaces T and S are mutually orthogonal. Since T contains no space-like vector, it is at most one-dimensional. If T is one-dimensional, it can not be light-like, since it is disjunct from its orthogonal complement; hence it must be time-like, and its orthogonal complement S is then space-like. If T is zero-dimensional, then $g = -I$ and so

$$g = \operatorname{diag}\{-1, -1, -1, \cdots, -1\}$$

in every basis for F^{n+1}. If T is one-dimensional, a proper basis e_1 for T together with a proper orthonormal basis $\{e_2, e_3, \cdots, e_{n+1}\}$ for S constitute a proper basis for F^{n+1}, for which g is represented by

$$\operatorname{diag}\{1, -1, -1, \cdots, -1\} .$$

LEMMA 2. *Let* π_1 *be a properly discontinuous group of isometries of* $D^n \subset F^{n+1}$, $n \geqq 2$. *Then each transformation* $g \in \pi_1$, *with* $g \neq I$ *and* $g \neq -I$, *has exactly one* (*real, independent*) *time-like eigenvector and no light-like eigenvectors.*

PROOF. Now g defines a topological automorphism of the double cone consisting of time-like and light-like rays through the origin of F^{n+1} and so, by the Brouwer fixed point theorem, g has a time-like or light-like eigenvector.

Suppose g has two real, independent eigenvectors which are time-like or light-like. Since π_1 is finite, the eigenvalues of g are ± 1. In any case the 2-plane spanned by these two eigenvectors is invariant under g, and contains a space-like vector. Furthermore g^2 is the identity in this 2-plane; then g^2 has a space-like fixed vector and, since π_1 is properly discontinuous, $g^2 = I$ on F^{n+1}. Then g satisfies the assumptions of Lemma 1 and, since $g \neq -I$, it has exactly one time-like eigenvector.

Now suppose g has just one real independent eigenvector v, among the time-like and light-like vectors of F^{n+1}. First suppose v is light-like. If g also had a real space-like eigenvector, then $g^2 = I$, and Lemma 1 contradicts the existence of v. Thus g has exactly one real, independent eigenvector in F^{n+1}.

If a non-singular linear transformation on an even dimensional real vector space has a unique eigenvector, the eigenvector cannot be simple. Thus if $n+1$ is even, v is not a simple eigenvector of g. But if $n+1$ is odd, consider g acting on the orthogonal complement of v, that is, the invariant n-plane tangent to the light cone of F^{n+1} along v. In either case g must generate an infinite cyclic group, which is impossible since π_1 is finite. Thus v cannot be light-like and so v must be time-like. In any case the lemma is proved.

By Lemmas 1 and 2, we see that if $g \in \pi_1$ and $g \neq I$, $g \neq -I$ and if g has a space-like eigenvector, then

$$g = \operatorname{diag}\{1, -1, -1, \cdots, -1\}$$

in proper Lorentz orthonormal coordinates in F^{n+1}.

THEOREM 2. *A complete, isochronous, Lorentz manifold M^n, $n \geqq 3$, of constant curvature $K = +1$ is an elementary relativistic spherical space form.*

PROOF. Certainly $M^n = D^n/\pi_1$ where π_1 is a properly discontinuous group of isometries of the De Sitter space $D^n \subset F^{n+1}$; and moreover each transformation $g \in \pi_1$ is isochronous (preserves time sense) on F^{n+1}. The statement of the theorem amount to saying that there exists a time-like vector which is fixed under all the transformations of π_1.

By Lemma 2 each element $g \neq I$ of π_1 has exactly one time-like eigenvector and no light-like eigenvectors. Suppose g_1 and g_2 (both $\neq I$) in π_1 have linearly independent, time-like fixed vectors v_1 and v_2, respectively.

Consider the connected component of the quadric

$$-(x^1)^2 + (x^2)^2 + \cdots + (x^{n+1})^2 = -1$$

defined by $x^1 > 0$ in F^{n+1}, which, under the induced metric, is the Lobachevsky space L^n with positive definite Riemann metric and constant curvature -1. The isochronous group of F^{n+1} can be identified with the group of all isometries of L^n onto itself. Thus g_1 and g_2 define isometries of L^n with distinct fixed points corresponding to v_1 and v_2, respectively.

Consider the subgroup $G \subset \pi_1$ generated by g_1 and g_2 acting as isometries of L^n. Each element, not I, in G is an isometry of L^n having just one fixed point. Since G is of finite order by Theorem 1, consider an element $g^* \neq I$ in G with fixed point v^* in L^n for which the non-euclidean distance of v^* from v_1 is the maximum among all the elements of G. The Lobachevsky $(n-1)$-plane through v^*, orthogonal to the geodesic line from v_1 to v^*, separates L^n in two components and no elements of G can have fixed points in the component not containing v_1.

But now consider v^* as the origin in L^n and note the orbit V of v_1 under powers of g^*. The orbit V of v_1 lies on a riemannian $(n-1)$-sphere S of constant positive curvature, and g^* acts on S as an orthogonal transformation without fixed points. Also each point of V is the fixed point for an element of G. It is easy to see that the orbit V of the point v_1, under the finite cyclic group generated by g^*, cannot lie entirely in the closed hemisphere of S centered at v. Thus some point of V must lie in the open hemisphere centered at the antipodal point to v_1 in S. This contradicts the choice of g^*, and so there exists a single unit time-like vector in the future cone of F^{n+1} which is fixed under all transformations of π_1.

Choose the fixed time-like vector of π_1 as the x^1-axis of F^{n+1} by a proper Lorentz change of basis. Then π_1 is isomorphic with a group $\hat{\pi}_1$ of orthogonal transformations acting on the euclidean n-plane $x^1 = 0$. Clearly $\hat{\pi}_1$ is a properly discontinuous group of orthogonal transformations on the equatorial sphere S^{n-1}, the intersection of D^n and $x^1 = 0$. Thus the relativistic spherical space form M^n is elementary and corresponds to the riemannian spherical space form $S^{n-1}/\hat{\pi}_1$. This completes the proof of Theorem 2.

If an isochronous Lorentz manifold is orientable or isochirous, it is proper. This remark leads to the following corollary.

COROLLARY. *A proper relativistic spherical space form is an elementary relativistic spherical space form corresponding to an orientable riemannian space form.*

Thus the only proper relativistic spherical space form M^n, for n odd, is the De Sitter space D^n.

3. Non-isochronous relativistic spherical space forms

A Lorentz manifold that is isochronous, isochirous or orientable but is not proper has a uniquely determined 2-sheeted covering by a proper Lorentz manifold. A Lorentz manifold that is not isochronous or isochirous or orientable has three uniquely determined 2-sheeted coverings that are respectively isochronous, isochirous, or orientable; the minimal covering that is a proper Lorentz manifold is then a 4-sheeted one.

In the preceding section we have classified completely the n-dimensional isochronous relativistic spherical space forms in terms of the $(n-1)$-dimensional riemannian space forms of curvature $+1$; each relativistic form is proper or not according to whether the corresponding riemannian space form is orientable or not.

Consider now the case of a non-isochronous relativistic spherical space form M^n; it has a 2-fold covering by an isochronous space M' which, by Theorem 2, is an elementary relativistic space form. A simple example is the elliptic space [7], or projective model, $\bar{D}^n = D^n/Z_2$, where Z_2 is the group consisting of the isometries I and $-I$, whose 2-sheeted isochronous covering is D^n itself. The subgroup π_1' of the fundamental group π_1 of M^n corresponding to M' is of index 2 and is either the identity (if $M' = D^n$) or has a single, fixed, time-like axis (Theorem 2). In the former case π_1 consists of the identity and another map g of period 2, which, by Lemma 1, must be $-I$; thus if $M' = D^n$, M^n must be $\bar{D}^n = D^n/Z_2$. If π_1' is not the trivial group, there must exist a time-reversing element $g \neq -I$ in π_1; by Lemma 1, $g^2 \neq I$, and so both g and g^2 have the same, unique, time-like axis, according to Theorem 2. Therefore all elements $g \neq \pm I$ in π_1 have a unique, common, time-like eigenvector v_1.

The equatorial sphere S^{n-1}, that is, the intersection of D^n with the orthogonal complement of v_1, is invariant under π_1; the restriction $\hat{\pi}_1$ of π_1 to S^{n-1} is thus a properly discontinuous group of orthogonal transformations of S^{n-1} with a subgroup $\hat{\pi}_1'$ of index 2 corresponding to π_1'. We shall denote also by $\hat{\pi}_1$ the isochronous group acting on F^{n+1} obtained by extending $\hat{\pi}_1$ from S^{n-1} to F^{n+1}, keeping the vector v_1 fixed. Let τ denote the reflection of F^{n+1} along the hyperplane spanned by S^{n-1}; it reverses v_1 and maps D^n onto itself with S^{n-1} as fixed points. Note that τ commutes with each element of π_1 and $\hat{\pi}_1$. For each $g \in \pi_1$ the corresponding element $\hat{g} \in \hat{\pi}_1$ is defined by $\hat{g} = g$ if $g \in \pi_1'$; $\hat{g} = g\tau = \tau g$ otherwise. Since $\tau \notin \pi_1$, this mapping is an isomorphism. Thus, given the properly discontinuous group $\hat{\pi}_1$ acting on S^{n-1} together with a subgroup $\hat{\pi}_1'$ of index 2, the group π_1 can be reproduced as the set of elements $g = \hat{g}$ if $\hat{g} \in \hat{\pi}_1'$; $g = \hat{g}\tau = \tau\hat{g}$ otherwise.

THEOREM 3. *The non-isochronous relativistic spherical space forms with n-dimensions are in one-to-one correspondence with the two-sheeted coverings of riemannian $(n-1)$-spherical space forms. If n is odd, the only non-isochronous relativistic spherical space form is the projective model $\bar{D}^n = D^n/Z_2$, which is orientable but not isochirous. If n is even, such relativistic spherical space forms are isochirous but not orientable.*

PROOF. If n is odd, then S^{n-1} is an even-dimensional sphere; the only fixed point free rotation of S^{n-1} is the identity; hence the only fixed point free group of motions are the identity and the group $\{\pm I\}$; the space form thus defined is the $(n-1)$-dimensional real projective space, which is non-orientable. The corresponding non-isochronous relativistic space form obtained by the construction described above is the space $\bar{D}^n$ which, for n odd, is orientable but not isochirous.

If n is even, then the only fixed point free rigid motions of S^{n-1} are (proper) rotations: hence either of the two extensions (one isochronous, the other not) of such a motion to D^n is isochirous.

REMARK. It is clear that, for n even, such non-isochronous relativistic spherical space forms are in one-to-one correspondence with the pairs $(\hat{\pi}_1, \hat{\pi}_1')$ of groups, where $\hat{\pi}_1$ is a finite group of (proper) rotations of S^{n-1} without fixed points, and $\hat{\pi}_1'$ is a subgroup of index 2 in $\hat{\pi}_1$.

COROLLARY. *Every relativistic space form (isochronous or not) of even dimensions is isochirous.*

We have proved that in a complete 4-dimensional relativistic world satisfying the local perfect cosmological principle (with curvature $K = +1$) there exists a global chirality. Any physical experiment distinguishing a chirality could be interpreted as a non-conservation of parity. However the physical interpretation of this geometrical theorem is debatable, since from local structures and general philosophical hypotheses we obtain information about the world as a whole. Such theories may not lie within the domain of physics, cf., [5, 9].

4. Geometry of relativistic spherical space forms

Every relativistic spherical space form M^n is diffeomorphic with a fiber bundle having a riemannian $(n-1)$-dimensional spherical space form as base and R^1 as fiber. Each isochronous relativistic space form is a product bundle, since these are all elementary relativistic spherical space forms. For n odd the projective model $\bar{D}^n$ is orientable and hence it cannot be the product of a real projective space P^{n-1} and R^1. Similary for n even the non-isochronous relativistic spherical space forms are non-orientable and thus they are not product bundles.

Since $\bar{D}^n$ has P^{n-1} as a deformation retract, $\bar{D}^n$ is not homeomorphic with D^n. But for n even, every isochronous relativistic spherical space form is orientable and every non-isochronous relativistic spherical space form is non-orientable. Hence no isochronous relativistic spherical space form is homeomorphic with a non-isochronous space form.

Consider the geodesics on an isochronous relativistic spherical space form M^n. The space-like geodesics are all periodic closed curves whereas the time-like and light-like geodesics are homeomorphs of R^1. This is easily seen by examining the geodesics on D^n and recalling that $M^n = D^n/\pi_1$ where π_1 consists of rotations about the x^1-axis of F^{n+1}. Since each non-isochronous relativistic spherical space form has a 2-fold isochronous covering, we note that every space-like geodesic, but no time-like or light-like geodesic is periodic on a non-isochronous relativistic spherical space form.

EXAMPLE. Consider D^2 as the quadric surface $-t^2 + x^2 + y^2 = 1$ in the Minkowski space F^3. Consider the isochronous Lorentz manifold $M' = D^2/\pi'_1$ where π'_1 is generated by $g: t \to t,\ x \to -y,\ y \to x$. Then the space-like geodesic of D^2 in the 2-plane $y + 2t = 0$ projects to a periodic space-like geodesic on M' which is self-intersecting (since the points $t = 0$, $x = -1$, $y = 0$ and $t = 0$, $x = 1$, $y = 0$ of D^2 become identified in M').

Now consider the non-isochronous Lorentz manifold $M^2 = D^2/\pi_1$ where π_1 is generated by $g\tau$, with g as above and $\tau: t \to -t,\ x \to x,\ y \to y$. The light-like geodesic in D^2 lying in the 2-plane $x - t = 0$, $y = 1$ projects to a light-like geodesic in M^2 which is self-intersecting (since the points $t = -1$, $x = -1$, $y = 1$ and $t = 1$, $x = 1$, $y = 1$ of D^2 become identified in M^2). Time-like geodesics in D^2 near the above light-like geodesic also project to self-intersecting geodesics in M^2.

Now consider the De Sitter space $D^n \subset F^{n+1}$ for $n \geqq 2$. Points P and Q in D^n can be considered as vectors in F^{n+1}. Take $P \neq \pm Q$ in D^n; then P and Q are joinable in D^n by a

(1) unique time-like geodesic if and only if $P \cdot Q > 1$;

(2) unique light-like geodesic if and only if $P \cdot Q = 1$;

(3) two arcs of a unique closed space-like geodesic if and only if $-1 < P \cdot Q < 1$.

Note that P and Q are not joinable by a geodesic in D^n if and only if

$$P \cdot Q \leqq -1\,, \qquad P \neq -Q\,.$$

DEFINITION. A Lorentz manifold M^n is geodesically convex in case each pair of distinct points $P, Q \in M^n$ can be joined by a geodesic in M^n. A Lorentz manifold M^n, in which there exists a point P which is joinable to each point of M^n by a geodesic, is called geodesically star-like with respect to P.

Clearly a geodesically convex Lorentz manifold M^n is geodesically star-like with respect to each of its points. Note that D^n, $n \geqq 2$, is not geodesically star-like from any point; this follows from the homogeneity, since $(0, 0, \cdots, 0, 1)$ and $(\sqrt{3}, 0, \cdots, 0, -2)$ are not joinable by a geodesic in D^n.

THEOREM 4. *Every relativistic spherical space form $M^n \neq D^n$ is geodesically star-like from some point in M^n. Furthermore M^n is geodesically convex if and only if M^n is non-isochronous.*

PROOF. First let $M^n = D^n/\pi_1$ be an elementary relativistic space form and let $P \in M^n$ be the projection of the point $\hat{P}_1 = (0, 0, \cdots, 0, 1)$ which lies on the equatorial sphere of D^n. Take $Q \neq P$ in M^n and consider any point $\hat{Q}_1 \in D^n$ projecting to Q. The orbit $\pi_1\hat{Q}_1 = \{\hat{Q}_1, \hat{Q}_2, \cdots, \hat{Q}_l\}$ has a barycenter in F^{n+1} lying on the time-axis $x^2 = x^3 = \cdots = x^{n+1} = 0$. The orbit $\pi_1\hat{P}_1$ has a barycenter at the origin of F^{n+1}. Then, as vectors in F^{n+1},

$$(\hat{P}_1 + \hat{P}_2 + \cdots + \hat{P}_l)\cdot(\hat{Q}_1 + \hat{Q}_2 + \cdots + \hat{Q}_l) = 0 .$$

Thus there exists at least one pair $\hat{P}_i$, $\hat{Q}_j$ with $\hat{P}_i\cdot\hat{Q}_j \geqq 0$. Then, in D^n, join $\hat{P}_i$ to $\hat{Q}_j$ by a geodesic which projects to the required geodesic in M^n joining P to Q. Thus M^n is geodesically star-like from P.

If M^n is non-isochronous let $P \in M^n$ be the projection of a point P' of the 2-fold isochronous covering M' such that M' is geodesically star-like from P'. Thus every relativistic spherical space form, excepting D^n, is geodesically star-like from some point.

Now let $M^n = D^n/\pi_1$ be a non-isochronous relativistic spherical space form corresponding to the pair $(\hat{\pi}_1, \hat{\pi}_1')$ where $\hat{\pi}_1$ is a finite group of proper rotations of the equatorial sphere S^{n-1} of D^n and $\hat{\pi}_1'$ is a subgroup of index 2 in $\hat{\pi}_1$. Since $\bar{D}^n$ is geodesically convex (either $\hat{P}\cdot\hat{Q} \geqq 0$ or $\hat{P}\cdot(-\hat{Q}) \geqq 0$ in F^{n+1}) we can take n even and suppose $M^n \neq \bar{D}^n$. Let $g = \hat{g}\tau \in \pi_1$, where $\hat{g} \neq I$ is in $\hat{\pi}_1'$ and τ is time-axis reversal in F^{n+1}. Then each point of D^n has an orbit under the subgroup of π_1 generated by g having a barycenter at the origin of F^{n+1}. Thus any two distinct points P, Q in M^n are the projections of points $\{\hat{P}_1, \cdots, \hat{P}_l\}$ and $\{\hat{Q}_1, \cdots, \hat{Q}_l\}$ of D^n which are invariant sets under g. But

$$(\hat{P}_1 + \cdots + \hat{P}_l)\cdot(\hat{Q}_1 + \cdots + \hat{Q}_l) = 0$$

in F^{n+1}, and hence $\hat{P}_i\cdot\hat{Q}_j \geqq 0$ for some pair of points which project to P and Q of M^n. Since $\hat{P}_i$ and $\hat{Q}_j$ can be joined by a geodesic in D^n, P and Q can be joined by a geodesic in M^n. Hence the non-isochronous relativistic spherical space form M^n is geodesically convex.

Finally let $M^n = D^n/\pi_1$ be an elementary relativistic spherical space form with $\pi_1 \neq I$. Now π_1, restricted to the invariant equatorial sphere S^{n-1}, is a finite group $\hat{\pi}_1$ of isometries acting without fixed points. Pick points $P_0 = (0, \bar{x}^2, \cdots, \bar{x}^{n+1})$ and $Q_0 = (0, \bar{y}^2, \cdots, \bar{y}^{n+1})$ in the sphere $S^{n-1} \subset D^n \subset F^{n+1}$ such that

$$g_1(P_0)\cdot g_2(Q_0) \leqq \lambda < 1$$

for all g_1 and g_2 in π_1. Now consider the points in D^n

$$P_t = (\sinh t, \bar{x}^2 \cosh t, \cdots, \bar{x}^{n+1} \cosh t)$$

and

$$Q_t = (\sinh t, \bar{y}^2 \cosh t, \cdots, \bar{y}^{n+1} \cosh t) .$$

Then

$$g_1(P_t)\cdot g_2(Q_t) \leqq -\sinh^2 t + \lambda \cosh^2 t = 1 - (1-\lambda)\cosh^2 t ,$$

and so

$$g_1(P_t)\cdot g_2(Q_t) < -1$$

for all $g_1, g_2 \in \pi_1$, whenever t is sufficiently large. In this case no point of the orbit of P_t under π_1 can be joined by a geodesic of D^n to a point of the orbit of Q_t. Thus M^n is not geodesically convex. This completes the proof of the theorem.

For the De Sitter space D^n, $n \geqq 2$, we can choose a compact, totally geodesic, space-like $(n-1)$ manifold and this will always be a sphere S^{n-1} which can serve as the equatorial sphere for appropriate proper Lorentz coordinates in F^{n+1}. The time-function t which accompanies the choice of equatorial S^{n-1} can be characterized as the unique differentiable real-valued function on D^n which is zero on S^{n-1}, positive on the component of $D^n - S^{n-1}$ designated as the future by the selection of a time sense in D^n, and with gradient vector which is everywhere non-vanishing and time-like and has the magnitude

$$|\operatorname{grad} t| = (1+t^2)^{1/2} .$$

The integral curves for grad t must be time-like geodesics orthogonal to S^{n-1}. If coordinates $(x^2, \cdots, x^{n+1})$ are chosen on S^{n-1} by any isometric imbedding of S^{n-1} as the unit sphere in the euclidean n-space, then $(t, x^2, \cdots, x^{n+1})$ define global coordinates on D^n. Any two such sets of global coordinates on D^n are related by a Lorentz transformation.

For an isochronous relativistic spherical space form M^n, other than D^n, the choice of a global time-function is uniquely determined by the intrinsic geometry of M^n, as we shall see below, cf., [9].

Let $M^n = D^n/\pi_1$ with $\pi_1 \neq I$ be an isochronous relativistic spherical space form. Then there exists a unique $(n-1)$-dimensional space-like submanifold B^{n-1} (without self-intersections) which is compact and which is totally geodesic in M^n. Here B^{n-1} is the riemannian space form which is the image of the equatorial sphere S^{n-1} of the De Sitter space D^n under the identifications of π_1. Now define the real differentiable function t on M^n so that $t=0$ on B^{n-1} and t is positive on the component of $M^n - B^{n-1}$ designated as the future. Furthermore require t to have a non-vanishing time-like gradient vector with magnitude satisfying

$$|\operatorname{grad} t| = (1+t^2)^{1/2} ,$$

and the integral curves of grad t are time-like geodesics orthogonal to B^{n-1}. Then the global time-function t satisfying these conditions is uniquely determined on M^n.

Among all relativistic spherical space forms we can characterize D^n and $\bar{D}^n$ as those which satisfy the global perfect cosmological principle (the group of isometries is transitive on the Lorentz orthonormal frames).

THEOREM 5. *The only relativistic spherical forms satisfying the global perfect cosmological principle are D^n and $\bar{D}^n$.*

PROOF. If $M^n = D^n/\pi_1$ satisfies the global perfect cosmological principle, then each transformation $g \in \pi_1$ must commute with entire linear Lorentz group on F^{n+1}. But the center of the Lorentz group consists of just I and $-I$. Thus either $\pi_1 = I$ so $M^n = D^n$, or $\pi_1 = \{\pm I\}$ so $M^n = \bar{D}^n$.

UNIVERSITY OF MINNESOTA

BIBLIOGRAPHY

1. L. AUSLANDER and L. MARKUS, Flat Lorentz manifolds, *Memoir* 30, Amer. Math. Soc., 1958.
2. H. BONDI, Cosmology, Cambridge, 1952.
3. E. CALABI and L. MARKUS, Relativistic space forms, Int. Math. Congress. Edinburgh, 1958.
4. L. P. EISENHART, Riemannian Geometry, Princeton, 1949.
5. E. HILL, Current Issues in the Philosophy of Science, New York, 1961.
6. C. MISNER and J. WHEELER, *Physics as geometry*, Ann. of Physics, 2 (1957), 524-603.
7. E. SCHRÖDINGER, Expanding Universes, Cambridge, 1956.
8. W. THRELFALL and H. SEIFERT, *Topologische Untersuchung der Diskontinuitätsbereiche endlicher Bewegungsgruppen des dreidimensionalen sphärischen Raumes*, Math. Ann., 104 (1931), 1-70, ibid. 107 (1933), 543-586.
9. H. WEYL, Space, Time, Matter, New York, 1950.
10. J. WOLF, *Sur la classification des variétés riemanniennes homogènes à courbure constante*, C. R. Acad. Sci. Paris, 250 (1960), 3443-3445.
11. ———, *Homogeneous manifolds of constant curvature*, to appear.

Linear Systems of Real Quadratic Forms

Eugenio Calabi[1]

Proc. Amer. Math. Soc. 15, 844–846 (1964)

1. It is natural to consider invariantive properties of k-dimensional linear systems of real quadratic forms over $\boldsymbol{R}^n$ (or quadratic forms over $\boldsymbol{R}^n$ with values in $\boldsymbol{R}^k$, $2 \leqq k \leqq \frac{1}{2}n(n+1)$), because of the occurrence of such objects at least in the following contexts: in the second fundamental form of classical differential geometry for n-dimensional manifolds differentiably imbedded in an $(n+k)$-space, and in the study of certain matrix differential equations. For this reason there may be some interest in the following result concerning pencils (i.e., 2-dimensional systems) of quadratic forms.

2. THEOREM 1.[2] *Let $P(\mathbf{x})$, $Q(\mathbf{x})$ be two real quadratic forms over $\boldsymbol{R}^n$ ($\mathbf{x} \in \boldsymbol{R}^n$), with $3 \leqq n < \infty$. If the only $\mathbf{x} \in \boldsymbol{R}^n$ satisfying $P(\mathbf{x}) = Q(\mathbf{x}) = 0$ is $\mathbf{x} = 0$, then there exists a linear combination of P and Q with real coefficients, that is a positive definite quadratic form over $\boldsymbol{R}^n$.*

PROOF. From the hypothesis, the mapping assigning to each $\mathbf{x} \in \boldsymbol{R}^n$, other than 0, the pair $(P(\mathbf{x}), Q(\mathbf{x})) \in \boldsymbol{R}^2 - \{(0, 0)\}$, being homogeneous (of weight 2), defines by passage to quotients a continuous mapping f: $\boldsymbol{P}^{n-1}(\boldsymbol{R}) \to \boldsymbol{P}^1(\boldsymbol{R})$, where $\boldsymbol{P}^k(\boldsymbol{R})$ denotes the k-dimensional real projective space.

Since $n \geqq 3$, the fundamental group $\pi_1(\boldsymbol{P}^{n-1}(\boldsymbol{R}))$ is the cyclic group $\boldsymbol{Z}_2$ with two elements, while $\pi_1(\boldsymbol{P}^1(\boldsymbol{R})) \cong \boldsymbol{Z}$, the infinite cyclic group. Hence the induced homomorphism f_*: $\pi_1(\boldsymbol{P}^{n-1}(\boldsymbol{R})) \to \pi_1(\boldsymbol{P}^1(\boldsymbol{R}))$ is trivial (a catamorphism, should one say?). This fact implies that one can factor the map f into the commutative diagram

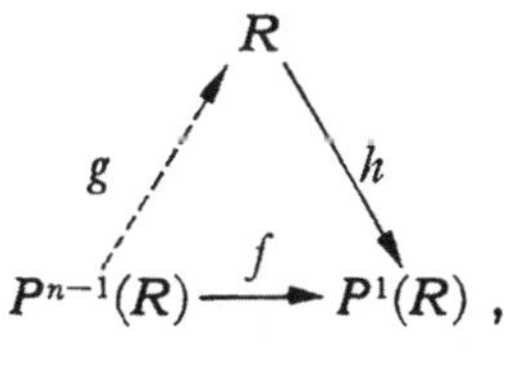

where h is the universal covering map of the real line $\boldsymbol{R}$ onto $\boldsymbol{P}^1(\boldsymbol{R})$. Clearly, since $n < \infty$, the image $g(\boldsymbol{P}^{n-1}(\boldsymbol{R})) \subset \boldsymbol{R}$ is a bounded, closed segment I: denote by p the restriction of the covering map h to I, so that $f = pg$.

Received by the editors April 4, 1963.

[1] J. S. Guggenheim Fellow, 1962–1963. Partially supported by the National Science Foundation.

[2] Cf. Remark 2 at the end of this paper.

DOI: 10.2307/2043553

Concerning the map $p: I \to P^1(R)$, we consider the two possibilities: either $p(I) = P^1(R)$ or $p(I) \subsetneq P^1(R)$. In the former case, there exists at least one point $\langle(a, b)\rangle \in P^1(R)$ (notation: $\langle(a, b)\rangle$ signifies the homogeneous equivalence class in $P^1(R)$ defined by $(a, b) \in R^2 - \{(0, 0)\}$), such that $p^{-1}(\langle(a, b)\rangle)$ contains at least two distinct points $t, t' \in I$: such being the case, the two subsets $g^{-1}(t)$, $g^{-1}(t')$ of $P^{n-1}(R)$ are nonempty, disjunct, relatively open and closed subsets of the real, $(n-2)$-dimensional, projective quadric in $P^{n-1}(R)$ defined by the homogeneous equation $bP(\mathbf{x}) - aQ(\mathbf{x}) = 0$. But it is well known that real projective quadric hypersurfaces, other than zero-dimensional ones, are either empty or connected point sets: hence this case cannot occur. We have shown thus that there exists a nonempty, open arc J in the circle $P^1(R)$, that is the complement of the image $p(I) = f(P^{n-1}(R))$. If (a, b) denotes the homogeneous coordinates of any point in J, then, for every nonzero vector $\mathbf{x} \in R^n$, the quadratic form $bP(\mathbf{x}) - aQ(\mathbf{x})$ takes a value different from zero; the quadratic form $bP - aQ$ is therefore either positive or negative definite (in which case $aQ - bP$ is positive). This completes the proof of the theorem.

3. A generalization of Theorem 1 to infinite-dimensional spaces leads only to a weaker conclusion. In fact, the two following quadratic forms in the Hilbert space of real sequences $(x_i)_{i \geq 1} = \mathbf{x}$ with summable squares, $P(\mathbf{x}) = \sum_{k=2}^{\infty} (1/k) x_k^2$ and $Q(\mathbf{x}) = x_1^2 - \sum_{i=2}^{\infty} x_i^2$ clearly satisfies the hypotheses of Theorem 1 (except $n < \infty$), but obviously no linear combination of P and Q is positive definite. This counterexample can be easily modified to work in any other type of infinite-dimensional real vector space (with or without norm). However, the following statement remains valid.

THEOREM 2. *Let $P(\mathbf{x})$ and $Q(\mathbf{x})$ be two quadratic forms in an infinite-dimensional, real vector space $\mathbf{V}$, such that $P(\mathbf{x}) = Q(\mathbf{x}) = 0$ implies $\mathbf{x} = 0$. Then there exists a linear combination of P and Q that is positive semi-definite in $\mathbf{V}$; if $\mathbf{W}$ is the subspace of $\mathbf{V}$ consisting of the vectors where this linear combination vanishes, then the restrictions of P and Q to $\mathbf{W}$ are both constant multiples, not both zero, of a common, positive definite quadratic form on $\mathbf{W}$.*

PROOF. One follows all the steps of the proof of Theorem 1; the only exception is that the interval $I = g(P_V(R)) \subset R$ is, in general, not compact, since the projective space $P_V(R)$ associated with an infinite-dimensional vector space V is not compact. On the other hand, the same argument as in Theorem 1 shows that the restriction p of the covering map $h: R \to P^1(R)$ to the interval I is a one-to-one map.

The linear combination $\pm(bP-aQ)$ with the desired property may be chosen for any (a, b) equal to homogeneous coordinates of any point $\langle(a, b)\rangle$ in the (nonempty) complement in $\boldsymbol{P}^1(\boldsymbol{R})$ of the open, proper subarc $p(\mathrm{Int}(I))$, where $\mathrm{Int}(I)$ denotes the interior of the interval I. This concludes the proof of the existence of a positive semidefinite linear combination of P and Q. The remaining conclusions follow trivially.

4. The statement of Theorems 1 and 2 is clearly false in the case of two-dimensional vector spaces, as shown by the example of the two binary quadratic forms $P(x, y)=x^2-y^2$, $Q(x, y)=2xy$. Similarly, corresponding statements fail in the case of k-dimensional linear systems of quadratic forms on $\boldsymbol{R}^n$ for $n\geqq 3$ and $3\leqq k<\frac{1}{2}n(n+1)$, even though the converse statements remain, in all cases, trivially valid. In fact, the linear system of quadratic forms with trace zero (referred to a suitable basis) form a maximal system, in the finite-dimensional case, containing no positive semidefinite forms; the system is precisely $(\frac{1}{2}n(n+1)-1)$-dimensional. From such a system, nevertheless, one can select, for $n\geqq 3$, three quadratic forms with no common zeros outside of the origin, for instance; $P_1(\mathbf{x})=x_1^2-x_2^2$, $P_2(\mathbf{x})=2x_1x_2$, $P_3(\mathbf{x})=x_1^2+x_2^2-(2/n-2)\sum_{i=3}^n x_i^2$.

REMARKS. 1. The whole content of Theorem 1 is an elementary property of the real number field. By the Tarski principle, the same theorem remains valid, if the real number field $\boldsymbol{R}$ is replaced by any other real-closed field. This furnishes one more example, in which topological methods are used to prove a statement of a purely algebraic nature. I do not know at present a purely algebraic proof.

2. Dr. Olga Taussky Todd has pointed out that, in essence, Theorem 1 is already known; it is equivalent to a theorem in Graeub's textbook [*Linear algebra*, Springer, 1958, pp. 158–163], stating that under the hypotheses of Theorem 1, the quadratic forms P and Q can be simultaneously diagonalized by a suitable choice of basis. The conclusion of Theorem 1 follows then from a trivial application of a result on linear inequalities [Stiemke, Math. Ann. **76** (1915), 340–342; also Annals of Mathematics Studies No. 24, *Contributions to the theory of games*, Princeton Univ. Press, Princeton, N. J.]; on the other hand, the simultaneous diagonalization of P and Q, given that they have a positive definite linear combination, is classical.

UNIVERSITY OF MINNESOTA

Quasi-Surjective Mappings and a Generalization of Morse Theory

Eugenio Calabi

Proc. U.S.-Japan Seminar in Differential Geometry, held in Kyoto, Japan, 1965, 13–16, Nippon Hyoronsha, Tokyo (1966)

Let X denote a connected, n-dimensional differential manifold, Y a p-dimensional one, and denote by $\tau(X)$, $\tau(Y)$ their respective tangent bundles. A differentiable mapping $f:X\to Y$ gives rise to the *jacobian bundle* over X associated with f, $J_f=\mathrm{Hom}\,(\tau(X), f^*\tau(Y))$, whose fibre over each $x\in X$ is the set of all linear maps $J_f(x)$: $\tau_x(X)\longrightarrow\tau_{f(x)}(Y)$ of the tangent space of X at x into that of Y at $f(x)$; the structure group of J_f is $GL(n, R)\times GL(p, R)$. The differential df of f is naturally a cross section in the bundle J_f called the *canonical cross section*. A homotopy of differentiable mappings f does not alter the structure of the jacobian bundle; it merely modifies the canonical cross section in a way that is restricted only by a local integrability and a global monodromy condition.

The jacobian bundle J_f can be naturally decomposed by the orbits under the action of the structure group on the fibres into a disjunct union of bundles.

$$J_f=\cup_{k\geq 0}J_{f,k}\,,$$

where for each $k\geqq 0$ and each $x\in X$ the fibre $J_{f,k}(x)$ consists of all linear maps of $\tau_x(X)$ into $\tau_{f(x)}(Y)$ that are of rank k; thus, for each non-negative integer $k\leqq\min(n, p)$, $J_{f,k}$ is a bundle of submanifolds of J_f of codimension $(n-k)(p-k)$, and empty for other values of k.

PROPOSITION 1 (H. Whitney, [3]). *Every differentiable mapping f: $X\longrightarrow Y$ can be approximated (uniformly with respect to any continuously pointwise defined norm) by a homotopically equivalent map g whose differential dg in $J_f=J_g$ is transversal to $J_{f,k}$ for each k.*

A mapping g of the type just described above is called 2-regular. Examples of such maps are i) *immersions*, or maps g such that $n\leqq p$ and $dg(X)\subset J_{g,n}$; ii) *surjections*, or maps g such that $p\leqq n$

and $dg(X)\subset J_{g,p}$; iii) *Morse functions*, for $p=1$, or proper maps of X into the non-negative real, closed half line $\boldsymbol{R}^+$, such that the only critical points are nondegenerate.

If $f: X \longrightarrow Y$ is a 2-regular map, then consider the decomposition $X=\bigcup_{k\geq 0} X_{f,k}$, where each $X_{f,k}$ is the set of all $x \in X$ such that $df(x)\in J_{f,k}$; then $X_{f,k}$ is either empty or a differentiably imbedded (not necessarily connected) submanifold of X of codimension $(n-k)(p-k)$; whence the nonempty ones can occur only for k in the range

$$\min(n,p)-\frac{2n}{|n-p|+\sqrt{(n-p)^2+4n}}\leq k \leq \min(n,p).$$

DEFINITION. *Let f: $X \longrightarrow Y$ be differentiable of class at least C^3 and suppose that* $\dim X=n$, $\dim Y=p\leq n+1$. *The mapping f is called quasisurjective, if and only if*

a) *f is 2-regular*

b) *$df(X)\subset J_{f,p}\cup J_{f,p-1}$ everywhere,*

c) *the critical set of f, $c_f=\{x \mid x\in X, \text{rank } df(x)<p\}$, which is, by the previous condition, a $(p-1)$-dimensional submanifold of X, has the property that the restriction of f to c_f is an immersion onto a hypersurface f in Y with possible self-intersection.*

REMARKS. For $Y=\boldsymbol{R}^+$, conditions b) and c) are vacuous, while a) characterizes Morse functions, provided that f is proper. For $p\geq 2$ quasisurjective maps are, in general, not dense in the space of differentiable maps, but form an open subset in an appropriate functional topology. For $p=n+1$ a quasisurjective map is merely an immersion.

If f: $X \longrightarrow Y$ is quasisurjective, then for any point $x_0\in c_f$ one can find systems of coordinates in X and Y, $(x^1,\cdots,x^n)$, $(y^1,\cdots,y^p)$ with origin at x_0 and $f(x_0)$ respectively, in terms of which the map f takes the form of the following p-tuple of functions

1) $y^\alpha=x^\alpha \qquad (1\leq\alpha\leq p-1)$,

2) $y^p=\frac{1}{2}(\sum_{j=p}^{n}\pm(x^j)^2)$.

The critical set is thus locally described by $x^p=\cdots=x^n=0$, its image by $y^p=0$. The quadratic form expressing y^p in terms of

$x^p, \cdots, x^n$ represents the *hessian* of f at the component of x_0 in c_f; the index of this quadratic form is independent of the choice of local coordinates as long as dy^p induces a fixed orientation on the normal line bundle of the local image of c_f in Y. Thus, with the possible exception of the components of c_f where the hessian has signature zero (equal number of positive and negative terms), the line bundle induced by the normal bundle of its local image in Y, is orientable. A quasisurjective map f is an interior map at all points except at those components of c_f where the hessian is positive or negative definite.

The generalization of Morse theory considered here applies to quasisurjective maps $f: X \longrightarrow \boldsymbol{R}^p$, $\boldsymbol{R}^p$ denoting Euclidean p-space, provided that f be a proper map and the complement of $f(X)$ in $\boldsymbol{R}^p$ have at least one unbounded component. In particular it applies to any quasisurjective $f: X \longrightarrow \boldsymbol{R}^p$ where X is compact. The main device is to consider real valued functions $g: \boldsymbol{R}^p \longrightarrow \boldsymbol{R}$ with the following properties:

1) g is a proper function without critical points on $\boldsymbol{R}^p$,
2) the restriction of g to $f(X)$ is nonnegative,
3) the restriction of $g \cdot f$ to the critical manifold c_f of f is a Morse function, i.e., a proper mapping of c_f into $\boldsymbol{R}$ with nonnegative values and all critical points nondegenerate.

A function g with the three properties described above induces a function $g \cdot f$ on X which is itself a Morse function, whose critical points coincide with those of the restriction of $g \cdot f$ to the critical set. The index of each critical point $x \in X$ with respect to $g \cdot f$ is related in a very simple way to the index of x considered as a critical point of c_f under the restriction of $g \cdot f$, and to the index of the hessian form of f at x. This leads, first of all to the following set of inequalities between the critical points.

THEOREM 1. *Let X be a compact n-dimensional manifold, $f: X \longrightarrow \boldsymbol{R}^p$ a quasisurjective map and $g: \boldsymbol{R}^p \longrightarrow \boldsymbol{R}$ a function without critical points, such that $g \cdot f$ is a proper map of X into $\boldsymbol{R}^+$ and its restriction, $g \cdot f | c_f$ to the critical set c_f of f is a Morse function. Let V_λ $(0 \leq \lambda \leq n-p+1)$ be the open and closed submanifold of c_f consisting of all components of c_f where the signature of the hessian is $\pm\lambda$ (or the index is $(n-p+1\pm\lambda)/2$) and let $m_{\lambda,\alpha}$ denote the number* *of critical points with index* α $(0\leqq\alpha\leqq p-1)$ *of the restriction of* $g\cdot f$ *to* V_λ. *Then the j-dimensional Betti numbers* $b_j(X)$ $(0\leqq j\leqq n)$ (*with respect to any coefficient field*) *satisfy the two inequalities*

$$\sum_{j=0}^{n} b_j(X) \leqq \sum_{\lambda=0}^{n-p+1} \sum_{\alpha=0}^{p-1} m_{\lambda,\alpha};$$

$$b_j(X) \leqq \sum_{\alpha=0}^{j} m_{n-p+1-2\alpha,\, j-\alpha} \qquad \left(0\leqq j\leqq \left[\frac{n-p+1}{2}\right]\right).$$

A more interesting implication is the following result specializing, for $p=1$, to Reeb's characterization of a combinatorial sphere.

THEOREM 2. *Let X be an n-dimensional, compact manifold and suppose that there exists a map* $f\colon X \longrightarrow \boldsymbol{R}^p$ *that is quasisurjective, such that the critical set* c_f *of f is* (*differentiably*) *homeomorphic to a* $(p-1)$*-sphere and g is a differentiable imbedding of* c_f *into* $\boldsymbol{R}^p$. *Then X is combinatorially equivalent to an n-sphere and the differential structure on X is derived from the Milnor group* $\Gamma_{n,p-1}$. *In particular, if the hypotheses are verified for* $p\geqq n-1$, *then X is differentiably equivalent to the usual sphere.*

The above theorem is used in another paper [2] in obtaining the corresponding restrictions in the differential structure of a Riemannian, complete, simply connected manifold X, whose sectional curvatures K are bounded by certain positive, estimated constants $\alpha_k\leqq K\leqq 1$ $(1/4=\alpha_0<\alpha_1\cdots<\alpha_k<\cdots<1, \lim_{k\to\infty}\alpha_k=1)$. For such manifolds, in fact, if $\alpha_k\leqq K\leqq 1$ for some k, one can construct a map $f\colon X\longrightarrow \boldsymbol{R}^{k+1}$ satisfying the hypotheses of Theorem 2.

References

[1] Calabi, E.: Quasisurjective Mappings and a generalization of Morse Theory, Topology (in preparation).

[2] ———: Differential structures on Riemannian manifolds with positive curvature, Bull. Amer. Math. Soc., (in preparation).

[3] Whitney, H.: Singularities of mappings of euclidean spaces, Symposium internacional de topologìa algebràica, Universidad Nacional Antònoma de Mexico, Mexico 1958, pp. 285-301.

Minimal Immersions of Surfaces in Euclidean Spheres

Eugenio Calabi

J. Differential Geometry 1, 111–125, (1967)

1. Introduction

The study of isolated singularities for minimal 3-dimensional varieties immersed in Euclidean n-space $\boldsymbol{E}^n$ requires as a first step a characterization of the tangent cone; the latter is the join of the origin 0 in $\boldsymbol{R}^n$ with a compact surface (the directrix) immersed in the unit Euclidean $(n-1)$-sphere S^{n-1} as a relatively minimal surface. Since comparatively little is known concerning such immersions, I propose to devote this as the first of a series of articles on the subject.

One may consider, for a start, a restricted type of singularity of a minimal 3-variety in $\boldsymbol{E}^n$, namely when this variety is topologically a manifold; in this case the directrix surface of the tangent cone is a 2-sphere immersed in S^{n-1} in a locally minimal way. This article is primarily devoted to minimal immersions of 2-spheres in Euclidean $(n-1)$-spheres. By this we mean immersions for which the total area is stationary with respect to variation, and minimal with respect to variation affecting sufficiently small portions of the surface at a time. Naturally, some of the conclusions developed here (through Lemma 5.3) apply to the minimal immersion of surfaces of positive genus as well; results pertaining to these will be collected elsewhere. In the case of minimal immersions of S^2 into the Euclidean sphere rS^{n-1} of radius r, the main result (Theorem 5.5) is that, if the image under such an immersion does not lie in any equatorial hyperplane section of rS^{n-1} then n is an odd integer and the area of the immersed S^2 is an integral multiple of $2\pi r^2$, at least equal to $\left(\frac{n^2-1}{8}\right)(4\pi r^2)$. There follow some discussion and examples to indicate why the above estimate is optimal.

2. Riemannian and Riemann surfaces

We denote by Σ an oriented surface, which, for the purposes of this article, may be assumed to be compact and either real analytic or differentiable. A differentiable Riemannian metric ds^2 on Σ together with the given orientation defines a covering of Σ by open domains with local (complex) isothermal parameters such as $w = u + iv$ $(i = \sqrt{-1})$ as well as its complex conjugate $\bar{w} = u - iv$. These parameters are defined up to a local holomorphic and holomorphically invertible transformation, and characterized by the following conditions.

Communicated June 6, 1967. This research was supported in part by NSF GP 4503.

DOI: 10.4310/jdg/1214427884

A complex valued function w in a domain $U \subset \Sigma$ is an isothermal parameter, if and only if

a) The map $w : U \to C$ is a topological imbedding.

b) The metric ds^2 can be expressed in U by the Hermitian differential form

$$ds^2 = 2F(w,\bar{w})(du^2 + dv^2) = 2F(w,\bar{w})|dw|^2 \tag{2.1}$$

where[1] $F(w, \bar{w})$ is a real analytic density, everywhere positive valued.

c) The real valued, exterior 2-form

$$\omega = 2F(w,\bar{w})\, du \wedge dv = iF(w,\bar{w})\, dw \wedge d\bar{w} \tag{2.2}$$

is positive with respect to the orientation of Σ. Thus Σ with its orientation and Riemannian metric is equivalent to a Riemann surface with a smoothly defined area element (2.2).

The tensor algebra bundle on Σ generated by the tangent bundle with the assigned Riemannian structure and orientation contains a complete set of irreducible representations of the structure group SO(2). In order to minimize redundancies, it is convenient to reduce this algebra as follows. First of all we consider the tensor algebra over the complex rather than the real field. Now let p, q be two rational numbers, unrestricted as to sign (more generally we could be bizarre and let p and q be complex numbers), such that $p - q$ is an integer; denote then by $E^{p,q}$ the complex line bundle whose elements are equivalence classes in the set of quadruples $(U, w, \boldsymbol{p}, v)$ where

a) U is an open domain in Σ and $\boldsymbol{p}$ a point in U;

b) w is a local isothermal parameter defined in U, and v a complex number;

c) $(U, w, \boldsymbol{p}, v)$ is equivalent to $(U', w', \boldsymbol{p}', v')$ if and only if:

(i) $$\boldsymbol{p} = \boldsymbol{p}' \in U \cap U',$$

(ii) $$\begin{aligned} v' &= v\left|\frac{\partial w}{\partial w'}(\boldsymbol{p})\right|^{2q}\left(\frac{\partial w}{\partial w'}(\boldsymbol{p})\right)^{p-q} \\ &= v\left|\frac{\partial w}{\partial w'}(\boldsymbol{p})\right|^{2p}\left(\overline{\frac{\partial w}{\partial w'}(\boldsymbol{p})}\right)^{q-p}, \end{aligned}$$

or, if p and q are integers,

[1] We use the convention that a real- or complex-valued function such as F in a domain $U \subset C$ is written as $F(w, \bar{w})$ for $w \in U$, when it is not assumed to be holomorphic. The same convention holds in the case of vector-valued functions.

$$v' = v\left(\frac{\partial w}{\partial w'}(p)\right)^p \left(\overline{\frac{\partial w}{\partial w'}(p)}\right)^q .$$

Thus, for each $(p, q), E^{p,q}$ is a real analytic, complex line bundle over Σ; $E^{p,0}$ is a holomorphic line bundle, while $E^{0,q}$ is an antiholomorphic line bundle. The tensor product is a pairing of $E^{p,q}$ and $E^{p',q'}$ into $E^{p+p',q+q'}$, so that $E^{0,0}$ is the trivial bundle and the dual of $E^{p,q}$ can be naturally identified with $E^{-p,-q}$. Finally the weak direct sum of all $E^{p,q}$ is a bundle of commutative algebras under addition and tensor multiplication, denoted by E. This algebra is bigraded by the values of (p, q); $E^{-1,0} \oplus E^{0,-1}$ is the complex tangent bundle $\tau_C(\Sigma) = \tau(\Sigma) \otimes_{\boldsymbol{R}} \boldsymbol{C}$, where $\tau(\Sigma)$ is the real tangent bundle. The bundle E admits a $(\boldsymbol{C} : \boldsymbol{R})$-semilinear involution (conjugation) mapping $E^{p,q}$ onto $E^{q,p}$ in the obvious way, the fixed elements of $E^{p,q} \oplus E^{q,p}$ (if $p \neq q$) and the elements of $E^{p,p}$ whose fibre component v is real valued are fixed under this conjugation and are hence called the *real* elements of E. For instance the Riemannian metric (2.1) is a real analytic cross section F, positive valued everywhere, in the bundle $E^{1,1}$. It is clear that the commutative tensor algebra thus defined is, for all computational purposes, equivalent to the more cumbersome, classical tensor algebra, in the sense that its irreducible spaces under the action of the structural group SO(2) × $\boldsymbol{R}^* = \boldsymbol{C}^*$ ($\boldsymbol{R}^*$ = multiplicative group of real numbers) contain a complete set of irreducible representations of the group.

3. Connexions

The metric (2.1) on Σ enables us to define a Levi-Civita connexion, i.e. a first order differential operator ∇ on differentiable cross sections of E. More precisely, the connexion is described axiomatically in terms of a splitting $\nabla = \nabla' + \nabla''$ of the absolute derivative, where ∇' and ∇'' are of bidegree respectively (1, 0) and (0, 1) on the bigraded algebra $E = \sum_{p,q} E^{p,q}$. These operators are characterized by the following four axioms:

1) The operators ∇' and ∇'' are C-linear derivations; in other words:

(3.1) $\nabla(cf) = c\nabla f$, where $c \in C$, f a differentiable cross section in $E^{p,q}$ (this implies, here and below, the same relation in terms of either ∇' or ∇'');

(3.2) $\nabla(f + g) = \nabla f + \nabla g$, where f, g are cross sections in $E^{p,q}$;

(3.3) $\nabla(f \otimes g) = f \otimes \nabla g + (\nabla f) \otimes g$, where f and g are cross sections, respectively in $E^{p,q}$ and $E^{p',q'}$.

2) If f is a cross section in $E^{p,0}$, locally described by the complex valued, differentiable functions $v = f(w, \bar{w})$ of local isothermal parameters w, then

$$\nabla'' f = \bar{\partial} f = \frac{\partial f(w, \bar{w})}{\partial \bar{w}}. \tag{3.4}$$

3) If $\bar{f}$ is the conjugate of f, then

$$\nabla' f = (\overline{\nabla'' \bar{f}}). \tag{3.5}$$

4) Finally, for the special cross section F in $E^{(1,1)}$ defining the metric (2.1), we have the identity

$$\nabla F \equiv 0. \tag{3.6}$$

It is easy to see that the six conditions (3.1) through (3.6) define ∇', ∇'' and hence $\nabla = \nabla' + \nabla''$ uniquely, as follows: if f is a differentiable cross section in $E^{p,q}$, locally defined by a differentiable function $v = f(w, \overline{w})$ in terms of a local isothermal parameter w, then

$$\nabla' f = \left(\frac{\partial f(w,\overline{w})}{\partial \overline{w}} - p \frac{\partial \log F(w,\overline{w})}{\partial w} \cdot f(w,\overline{w}) \right) \tag{3.7}$$

and

$$\nabla'' f = \left(\frac{\partial f}{\partial \overline{w}} - q \frac{\partial \log F(w,\overline{w})}{\partial \overline{w}} \cdot f(w,\overline{w}) \right). \tag{3.8}$$

The commutator of ∇' and ∇'' corresponds to the Ricci identity, which is expressed as follows:

If $f(w, \overline{w})$ represents locally a differentiable cross section in $E^{p,q}$, then

(3.9)

$$\begin{aligned} [\nabla', \nabla''] f &= (\nabla'\nabla'' - \nabla''\nabla') f = (p-q) f(w,\overline{w}) \frac{\partial^2 \log F(w,\overline{w})}{\partial w \partial \overline{w}} \\ &= (q-p) K F(w,\overline{w}) \otimes f(w,\overline{w}), \end{aligned}$$

where $K = K(w,\overline{w}) = -\frac{1}{F} \frac{\partial^2 \log F(w,\overline{w})}{\partial w \partial \overline{w}}$ is the Gaussian curvature of the metric (2.1).

4. Isometric immersions

Consider the n-dimensional Euclidean space $\boldsymbol{E}^n$. We denote by $\boldsymbol{R}^n$ the Euclidean vector space associated with $\boldsymbol{E}^n$; thus $\boldsymbol{R}^n$ can be regarded as the group of translations of $\boldsymbol{E}^n$, or else as the fibre in the tangent bundle of $\boldsymbol{E}^n$. In conjunction with $\boldsymbol{R}^n$ we consider the complex extension, $\boldsymbol{C}^n = \boldsymbol{R}^n \otimes_{\boldsymbol{R}} \boldsymbol{C}$ of $\boldsymbol{R}^n$, regarded as the fibre of the "complexified" tangent bundle of $\boldsymbol{E}^n$. We extend the orthogonal (inner product) structure from $\boldsymbol{R}^n$ both to an orthogonal and a unitary structure in $\boldsymbol{C}^n$ in the following, natural way: if $z' = (z'_1, \cdots\cdots, z'_n)$ and $z'' = (z''_1, \cdots, z''_n)$ are two vectors in $\boldsymbol{C}^n$, then the "dot" product is the symmetric bilinear form $z' \cdot z'' = \sum_{\alpha=1}^{n} z'_\alpha z''_\alpha$; moreover we denote by $\bar{z}'$ the complex conjugate of z'

$$\bar{z}' = (\bar{z}'_1, \cdots, \bar{z}'_n);$$

we also need the absolute value (or norm) of a vector $z = (z_1, \cdots, z_n)$ to be the square root of the positive definite hermitian form $z' \cdot \bar{z}''$, so that

$$|z| = \sqrt{z \cdot \bar{z}} = (\sum_{\alpha=1}^{n} |z_\alpha|^2)^{\frac{1}{2}}.$$

Thus the dot product, conjugation and norm on $\boldsymbol{C}^n$ are the invariants defining the real orthogonal (euclidean) group $O(n)$ acting on $\boldsymbol{C}^n$.

Consider now the Grassman algebra $\Lambda(\boldsymbol{C}^n)$ generated by $\boldsymbol{C}^n$ over C. If $\boldsymbol{Z} = \boldsymbol{z}_1 \wedge \cdots \wedge \boldsymbol{z}_p$ and $\boldsymbol{W} = \boldsymbol{w}_1 \wedge \cdots \boldsymbol{w}_p$ are two p-vectors in $\Lambda(\boldsymbol{C}^n)$ we can extend the $O(n)$-invariants of $\boldsymbol{C}^n$ to $\Lambda(\boldsymbol{C}^n)$ in the obvious way, by defining the dot product as the Gramian determinant

$$\boldsymbol{Z} \cdot \boldsymbol{W} = \det_{1 \le \alpha,\ \beta \le p} (\mathbf{Z}_\alpha \cdot \boldsymbol{w}_\beta),$$

the conjugation map by setting

$$\bar{\boldsymbol{Z}} = \bar{\boldsymbol{z}}_1 \wedge \cdots \wedge \bar{\boldsymbol{z}}_p,$$

and the positive valued norm $|\boldsymbol{Z}|$ of $\boldsymbol{Z}$ by means of the Hermitian product

$$|\boldsymbol{Z}|^2 = \boldsymbol{Z} \cdot \bar{\boldsymbol{Z}}.$$

With these notations, we let $X : \Sigma \to \boldsymbol{E}^n$ be a differentiable map of a surface Σ, as specified in §2, into the Euclidean n-space $\boldsymbol{E}^n$. We describe Σ locally in terms of a local isothermal parameter w with respect to the Riemannian structure induced on Σ by X; then X can be represented locally by an $\boldsymbol{E}^n$-valued differentiable function[2] $X(w, \bar{w})$ of w. The map is an isometric immersion, if and only if at each point

$$F(w, \bar{w}) = \left| \frac{\partial X}{\partial w} \right|^2 = \frac{\partial X}{\partial w} \cdot \frac{\partial X}{\partial \bar{w}}, \tag{4.1}$$

$$\frac{\partial X}{\partial w} \cdot \frac{\partial X}{\partial w} = 0 \qquad \left(\text{implying } \frac{\partial X}{\partial \bar{w}} \cdot \frac{\partial X}{\partial \bar{w}} = 0 \right), \tag{4.2}$$

where $F(w, \bar{w})$ is the coefficient of the Riemannian metric (2.1). We can now form the successive derivatives of X with respect to w or $\bar{w}$ either, from the local, (or analytical) standpoint

[2]See footnote 1.

$$\partial^p \bar{\partial}^q X = \frac{\partial^{p+q} X}{\partial w^p \partial \bar{w}^q} \qquad (p,\ q = 0, 1, 2, \cdots) \tag{4.3}$$

or in the Levi-Civita (or geometrical) sense such as an ordered sequence of (noncommuting) successive derivations containing, say, p times ∇' and q times ∇'' in some given order; in this case we have a partial operator of order $p+q$, whose principal part is $\partial^p \bar{\partial}^q X$. In the case of isometric immersion, several identities hold, involving dot products of derivatives of X. For instance, considering the absolute derivatives of both members in either (4.1) or (4.2), we obtain immediately[3]

$$\nabla' X \cdot \nabla' \nabla'' X = \nabla' X \cdot \nabla'^2 X = \nabla'' X \cdot \nabla'^2 X = 0, \tag{4.4}$$

showing that all second order covariant derivatives of X are orthogonal to all first order (and thereby define the second fundamental form).

5. The main theorems

We now make the basic assumptions of the article, namely that the isometric immersion $X : \Sigma \to \boldsymbol{R}^n$ maps Σ into a Euclidean $(n-1)$-dimensional sphere rS^{n-1} of radius r and that the resulting surface in the $(n-1)$-sphere with the induced Riemannian structure is locally a solution of the Plateau problem. For this purpose we identify $\boldsymbol{E}^n$ with $\boldsymbol{R}^n$ by the choice of an origin 0; the sphere rS^{n-1} of radius r and center 0 is the set of real vectors $\boldsymbol{x} \in \boldsymbol{R}^n$ satisfying $x \cdot x = r^2$.

Assumptions. We let Σ be an oriented surface and $X : \Sigma \to \boldsymbol{R}^n$ an analytic immersion of Σ satisfying the following five conditions

1) The image lies in the sphere rS^{n-1} of radius r, *i.e.*

$$X \cdot X = r^2 \qquad \text{at every point of } \Sigma. \tag{5.1}$$

2) We carry all calculations in terms of the local isothermal coordinate systems w defined in Σ by the given orientation and the Riemannian metric induced by X.

3) The immersion of Σ in rS^{n-1} is locally minimal; this means, in terms of X as an immersion in $\boldsymbol{R}^n$, that the curvature vector $\dfrac{1}{F(w,\overline{w})} \dfrac{\partial^2 X}{\partial w \partial \overline{w}}$ $F^{-1} \otimes \nabla' \nabla'' X$ is everywhere orthogonal to rS^{n-1}, *i.e.* proportional to the position vector X of the image.

4) The surface Σ is homeomorphic to a 2-sphere.

[3]The operations of dot product and Grassmann multiplication are understood to be naturally extended from C^n to its tensor product with the bundle E of bigraded tensor algebras over Σ.

5) Without loss of generality, the immersion X is linearly full, *i,e.* the image $X(\Sigma)$ is not contained in any Euclidean hyperplane in $\boldsymbol{R}^n$.

Remarks. The analyticity of X can be deduced as a consequence of 3), provided only that the map be of class C^3. The last two main assumptions will be used only later, and then the essential role that these assumptions play will be emphasized.

We draw immediately a couple of elementary conclusions from the first three of the five basic assumptions.

Lemma 5.1. *The mapping X satisfies the equation*

$$\partial\bar{\partial}X = \nabla'\nabla''X = -r^{-2}FX. \tag{5.2}$$

Proof. From Assumption 3 we have $\partial\bar{\partial}X = \lambda X$ for some real valued function λ; on the other hand we have also, from (5.1),

$$\begin{aligned} 0 &= \frac{1}{2}\partial\bar{\partial}(r^2) = \frac{1}{2}\partial\bar{\partial}(X\cdot X) = \partial\bar{\partial}X\cdot X + \bar{\partial}X\cdot\partial X \\ &= \lambda X\cdot X + F = r^2\lambda + F, \end{aligned}$$

whence $\lambda = -r^{-2}F$, as asserted.

Lemma 5.2. *The complex vector subspace of $\boldsymbol{C}^n$ spanned by the k'th order jet of X, at any point $\mathbf{p}\in\Sigma$ and for any positive integer k, is spanned by the $2k+1$ vectors X, ∇'^pX, ∇''^pX or equivalently X, ∂^pX and $\bar{\partial}^pX(1\le p\le k)$ evaluated at $\mathbf{p}$.*

Proof. The linear span over $\boldsymbol{C}$ of the k'th order jet of X can be generated equivalently by X and the $\frac{1}{2}k(k+3)$ partial derivatives of X of order $\le k$, or by X and the $2^{k+1}-2$ different Levi-Civita derivatives of X of order $\le k$; at the same time, in view of (5.2) any Levi-Civita derivative of X of order k in which each of ∇' and ∇'' appears at least once (or equivalently, any partial derivative $\dfrac{\partial^{p+q}X}{\partial w^p\partial\bar{w}^q}$ with $\min(p,q)\ge 1$ and $p+q=k$) can be expressed as a linear combination of X and derivatives of order $< k-1$; from this the conclusion follows immediately.

Until now we have not yet used assumptions 4) or 5), namely that Σ is homeomorphic to a sphere, or that X is linearly full. In the next lemma, however, Assumption 4 plays an essential role.

Lemma 5.3. *Under the basic assumptions* 1), 2), 3) *and* 4), *the complex subspace of $\boldsymbol{C}^n$ spanned by all the derivatives $\nabla'^pX(p\ge 1)$ at any point of Σ is totally isotropic with respect to the dot product, and is orthogonal to X.*

Proof. We set, formally, $\nabla'^oX = \nabla''^oX = X$; we must prove that, for any two nonnegative integers p, q with $p+q\ge 1$,

$$\nabla'^p X \cdot \nabla'^q X = 0.$$

This identity, for $1 \leq p+q \leq 3$, is true in the case of arbitrary immersions X of any surface Σ into a sphere, with no further assumptions; this follows easily from (4.2), (4.4) and (5.1). We shall prove it now for higher values of $p+q$, by induction on $p+q$. Suppose that $\nabla'^p X \cdot \nabla'^q X \equiv 0$ for all p, q with $1 \leq p+q \leq k-1$, for some integer $k \geq 2$; to prove it for $p+q=k$, we distinguish the case where k is even from that where k is odd. If k is odd, say $k = 2m+1 (m \geq 1)$, we have

$$\nabla'^m X \cdot \nabla'^{m+1} X = \frac{1}{2}\nabla'(\nabla'^m X \cdot \nabla'^m X) = 0$$

and, for any p such that $1 \leq p \leq m$

$$\nabla'^{m-p} X \cdot \nabla'^{m+p+1} X = (-1)^p \nabla' \left(\frac{1}{2}\nabla'^m X \cdot \nabla'^m X + \sum_{\mu=1}^{p} (-1)^\mu \nabla'^{m-\mu} X \cdot \nabla'^{m+\mu} X\right) = 0,$$

proving the induction step for k odd ; we remark that, in this case, we do not need the assumption 4) that Σ is homeomorphic to S^2.

If k is even, say $k = 2m$ $(m \geq 1)$, let $\Lambda_m = \nabla'^m X \cdot \nabla'^m X$. Then for $p+q = 2m$ and, without loss of generality, $p < q$,

$$\begin{aligned}\nabla'^p X \cdot \nabla'^q X &= (-1)^{m-p}\Lambda_m + (-1)^{m-p}\nabla'(\Sigma_{\mu=1}^{m-p}(-1)^\mu \nabla'^{m-\mu} X \cdot \nabla'^{m+\mu-1} X) \\ &= (-1)^{m-p}\Lambda_m.\end{aligned}$$

This reduces the remainder of the proof to showing that $\Lambda_m \equiv 0$.

First, we calculate $\nabla''\Lambda_m$. From the definition of Λ_m,

$$\nabla''\Lambda_m = 2\nabla''\nabla'^m X \cdot \nabla'^m X.$$

Next, we evaluate $\nabla''\nabla'^m X$; using the Ricci-identity (3.9) $m-1$ times, we get

(5.3)
$$\begin{aligned}\nabla''\nabla'^m X &= \textstyle\sum_{\mu=1}^{m-1} \nabla'^{\mu-1}(\nabla''\nabla' - \nabla'\nabla'')\nabla'^{m-\mu} X + \nabla'^{m-1}(\nabla''\nabla' X) \\ &= -r^{-2} F \nabla'^{m-1} X + \textstyle\sum_{\mu=1}^{m-1}(m-\mu) F \nabla'^{\mu-1}(K\nabla'^{m-\mu} X) \\ &= F\left(-r^{-2} + \binom{m}{2} K\right)\nabla'^{m-1} X + (\text{a linear combination of} \\ &\qquad \nabla' X, \cdots, \nabla'^{m-2} X),\end{aligned}$$

where K is the Gaussian curvature. From this, and using the induction assumption, we deduce immediately that

$$\nabla'' \Lambda_m \equiv 0.$$

This proves that, in the absence of assumption 4), if X satisfies the induction assumption $\nabla'^p X \cdot \nabla'^q X = 0$ for all $p,\ q < 2m$, then $\nabla'^m X \cdot \nabla'^m X = \Lambda_m$ is a holomorphic cross section in $E^{2m,0}$, *i.e.*, a holomorphic differential of weight $2m$. Since Σ is homeomorphic to S^2, *i.e.* a compact Riemann surface of genus zero, the only holomorphic cross section in $E^{p,0}$ for any $p > 0$ is the trivial one, so that in this case $\Lambda_m = 0$. This completes the proof of the lemma.

Remark. The above lemma and its proof are inspired by H. Hopf's analogous argument [1] in the case of closed surfaces of genus 0 immersed in $\boldsymbol{R}^3$ with constant mean curvature.

In what follows, it is useful to introduce the following differential geometric forms attached to the map X, analogous to the generalized Wronskians for differentiable curves in $\boldsymbol{E}^n$.

We denote by $T'_k X$, for any positive integer k, the following[4] cross section in $E^{k_2,0} \otimes_C \Lambda^k(\boldsymbol{C}^n)$, where $k_2 = \frac{1}{2}k(k+1)$,

$$T'_k X = \nabla' X \wedge \nabla'^2 X \wedge \cdots \wedge \nabla'^k X = \partial X \wedge \partial^2 X \wedge \cdots \wedge \partial^k X. \tag{5.4}$$

Clearly, T'_k vanishes at any point $\boldsymbol{p} \in \Sigma$, if and only if $\nabla' X, \nabla'^2 X, \cdots, \nabla'^k X$ at $\boldsymbol{p}$ are linearly dependent over the tensor algebra E. Since, by Lemma 5.1 the subspace of $\boldsymbol{C}^n$ spanned by the k vectors $\nabla' X, \nabla'^2 X, \cdots, \nabla'^k X$ at any point of Σ is totally isotropic with respect to the dot product, its dimensionality can not exceed $\frac{n}{2}$: thus $T'_k X \equiv 0$ as soon as $k > \frac{n}{2}$.

Lemma 5.4. *Under the basic assumptions, emphasizing here Assumption 5), the dimensionality n of the ambient space $\boldsymbol{R}^n$ is odd; more precisely, $n = 2m+1$, where m is the highest value for which $T'_m X$ is not identically zero.*

Proof. Let m denote the highest integer for which $T'_m X = \partial X \wedge \cdots \wedge \partial^m X = \nabla' X \wedge \cdots \wedge \nabla'^m X$ is not identically zero. We shall prove first that $n \geq 2m+1$. In fact, consider the exterior absolute value F_m of the $(2m+1)$-vector

$$X \wedge \nabla' X \wedge \nabla'^2 X \wedge \cdots \wedge \nabla'^m X \wedge \nabla'' X \wedge \cdots \wedge \nabla''^m X = X \wedge T'_m X \wedge T''_m X,$$

where $T''_m = \overline{T'_m}$. The square of F_m can be calculated from the Gramian determinant, which involves dot products such as $X \cdot X,\ X \cdot \nabla'^p X,\ X \cdot \nabla''^q X,\ \nabla'^p X \cdot \nabla'^q X,\ \nabla'^p X \cdot \nabla''^q X,\ \nabla''^p X \cdot \nabla''^q X$, etc., $(1 \leq p,\ q \leq m)$. Because of Lemma 5.3, this Gramian matrix splits into three blocks along the diagonal, namely $X \cdot X \oplus (\nabla'^p X \cdot \nabla''^q X) \oplus (\nabla''^p X \cdot \nabla'^q X)$. Thus we see that

[4] See footnote 3.

$$\begin{aligned} F_m &= |X| \det_{1\le p,\, q\le m} (\nabla'^{p}X \cdot \nabla''^{q}X) \\ &= r(\nabla'X \wedge \nabla'^{2}X \wedge \cdots \wedge \nabla'^{m}X)\cdot(\nabla''X \wedge \cdots \wedge \nabla''^{m}X) \\ &= r|T'_m X|^2. \end{aligned} \tag{5.5}$$

This shows that, wherever $T'_m X$ is not zero, F_m is strictly positive; the latter is the norm of a $(2m+1)$-vector in $\boldsymbol{C}^n$, showing that $n \geq 2m+1$.

We show now that, conversely, $n \leqq 2m+1$. Since the immersion X of Σ is real analytic and its image is not contained in any hyperplane, at each point of Σ the linear span of the jet of X of sufficiently high order is n-dimensional. By the definition of m, $T'_{m+1}X$ is identically zero, while $T'_m X$ is not; hence at any point $\boldsymbol{p} \in \Sigma$ where $T'_m X$ is not zero, $\nabla'^{m+1}X$ lies in complex m-dimensional subspace with basis $\nabla'X, \cdots, \nabla'^{m}X$ evaluated at $\boldsymbol{p}$. By considering inductively the vanishing of $\nabla'^{k}(T'_{m+1}X)(k=1,2,\cdots)$ it follows that each $\nabla'^{m+k}X$ at $\boldsymbol{p}$ lines in the same subspace. Therefore, for any nonnegative k, the $2m+2k+1$ vectors X, $\nabla'^{p}X$, $\nabla''^{2}X$ $(1 \leq p,\ q \leqq m+k)$ are a linear combination of the $2m+1$ vectors X, $\nabla'X, \cdots, \nabla'^{m}X$, $\nabla''X, \cdots, \nabla''^{m}X$. But the $2m+2k+1$ vectors span, by Lemma 5.2, the whole $(m+k)$'th order jet of X at $\boldsymbol{p}$. Therefore this jet has, for each k, a linear span of at most $2m+1$ dimensions. This shows that $n \leq 2m+1$, completing the proof.

We can state and prove now the main theorem of this article.

Theorem 5.1. *Let $X\colon \Sigma \to \boldsymbol{E}^n$ be an immersion of a surface Σ into $\boldsymbol{E}^n$, whose image is a locally minimal surface in a sphere rS^{n-1} of radius r, and is not contained in any hyperplane of $\boldsymbol{E}^n$. Then the following conclusions hold;*

i) *The area $A = A(X)$ of the image surface is an integer multiple of $2\pi r^2$.*

ii) *The dimension number n is odd, say $n = 2m+1$* and

$$A \geq 4\pi r^2 \binom{m\ +\ 1}{2} = 4\pi r^2 \left(\frac{n^2-1}{8}\right). \tag{5.6}$$

Proof. The proof of the two conclusions depend on certain calculations on the objects $T'_k X (1 \leq k \leq m)$ defined in (5.4) and on the related quantities

$$F_k = |X|\ |T'_k X|^2 \qquad (1 \leq k \leq m),$$

introduced in (5.5) for the special value $k = m$. Clearly $T'_k X$ is a real analytic cross section over Σ in $E^{\frac{1}{2}k(k+1),0} \otimes_C \boldsymbol{C}^{\binom{n}{k}}$, while $\mathbf{F}_k$ is a nonnegative cross section in E^{k_2,k_2}, where $k_2 = \binom{k+1}{2} = \frac{1}{2}k(k+1)$. It is convenient to replace F_k by its norm Φ_k under the Riemannian metric (2.1), since Φ_k is a scalar function. Thus we are led to consider the sequence of nonnegative real valued scalars $\Phi_k(k=1,2,\cdots)$, where

$$\Phi_k = F^{-\frac{1}{2}k(k+1)} F_k. \tag{5.7}$$

For the sake of conformity we also set $T_0'X = T_0''X$ identically equal to 1 and $F_0 = \Phi_0$ equal to the constant r. The following differential recursion formulas now hold and are easy to verify: For each $k \geq 1$

$$\nabla' T_k' X = T_{k-1}' X \wedge \nabla'^{k+1} X \, , \tag{5.8}$$

$$\begin{aligned} \nabla' T_k'' X &= \nabla' \nabla'' X \wedge \nabla''^2 X \wedge \cdots \wedge \nabla''^k X \\ &= -r^{-2} F X \wedge \nabla''^2 X \wedge \cdots \wedge \nabla''^k X \, , \end{aligned} \tag{5.9}$$

and similarly for the complex conjugates.

From the above, we can calculate formally the absolute differentials of each Φ_k (using k_2 as an abbreviation for $\binom{k+1}{2} = \frac{1}{2}k(k+1)$),

$$\begin{aligned} \nabla' \Phi_k &= r F^{-k_2} \nabla' (T_k' X \cdot T_k'' X) \\ &= r F^{-k_2} (\nabla' T_k' X \cdot T_k'' X + T_k' X \cdot \nabla' T_k'' X). \end{aligned}$$

Because of (5.8), (5.9), and since X, by Lemma 5.3, is orthogonal to each $\nabla'^p X$, $T_k' X \cdot \nabla' T_k'' X = 0$ identically; thus

$$\partial \Phi_k = \nabla' \Phi_k = r F^{-k_2} (T_{k-1}' X \wedge \nabla'^{k+1} X) \cdot T_k'' X \tag{5.10}$$

and similarly $\bar\partial \Phi_k = \nabla'' \Phi_k = r F^{-k_2} T_k' X \cdot (T_{k-1}'' X \wedge \nabla''^{k+1} X)$. We calculate now $\bar\partial \partial \Phi_k = \nabla'' \nabla' \Phi_k = \nabla' \nabla'' \Phi_k$ as follows:

$$\begin{aligned} \bar\partial \partial \Phi_k = &\; r F^{-k_2} \{ \nabla'' (T_{k-1}' X \wedge \nabla'^{k+1} X) \cdot T_k'' X + |\nabla' T_k' X|^2 \} \\ = &\; r F^{-k_2} (T_{k-1}' X \; \wedge \nabla'' \nabla'^{k+1} X) \cdot T_k'' X + r F^{-k_2} |T_{k-1}' X \wedge \nabla'^{k+1} X|^2. \end{aligned}$$

We apply the Ricci identity (3.9) in evaluating the term with $\nabla'' \nabla'^{k+1} X$, as it was done in (5.3) for the case $\nabla'' \nabla'^m X$; we get then

$$T_{k-1}' X \wedge \nabla'' \nabla'^{k+1} X = F(-\, r^{-2} + k_2 K) T_k' X \, .$$

Thus we have

$$\begin{aligned} \bar\partial \partial \Phi_k &= r F^{1-k_2} (-\, r^{-2} + k_2 K) F_k + r F^{-k_2} |\nabla' T_k' X|^2 \\ &= F(-\, r^{-2} + k_2 K) \Phi_k + r F^{-k_2} |T_{k-1}' X \wedge \nabla'^{k+1} X|^2. \end{aligned} \tag{5.11}$$

We now recall the algebraic identity (Lagrange)

$$\begin{aligned} &|T_{k-1}' X \wedge \nabla'^k X|^2 |T_{k-1}' X \wedge \nabla'^{k+1} X|^2 \\ &\qquad - (T_{k-1}' X \wedge \nabla'^{k+1} X) \cdot (T_{k-1}'' X \wedge \nabla''^k X) \\ &\qquad\qquad \otimes (T_{k-1}' X \wedge \nabla'^k X) \cdot (T_{k-1}'' X \wedge \nabla''^{k+1} X) \\ &\qquad = |T_{k-1}' X|^2 |T_{k-1}' X \wedge \nabla'^k X \wedge \nabla'^{k+1} X|^2 = F_{k-1} F_{k+1} \end{aligned}$$

and apply it in conjunction with (5.10) and (5.11); one obtains the following equation

$$\begin{aligned}\Phi_k \partial\bar\partial \Phi_k - \partial\Phi_k \bar\partial\Phi_k &= F(-r^{-2} + k_2 K)\Phi_k^2 + F^{-2k_2} F_{k-1} F_{k+1} \\ &= F(-r^{-2} + k_2 K)\Phi_k^2 + F\Phi_{k-1}\Phi_{k+1}\end{aligned}$$

or, if $\Phi_k \neq 0$, we obtain the recursion formula

(5.12)

$$\partial\bar\partial(\log \Phi_k) = F\left(\frac{\Phi_{k-1}\Phi_{k+1}}{\Phi_k^2} - r^{-2} + k_2 K\right) \qquad (\Phi_0 = \Phi_1 = r)\ .$$

Since each of the functions Φ_k is real analytic and nonnegative, we see that either Φ_k is identically zero, or else, by considering any locally defined, real valued, analytic function ϕ whose Laplacian satisfies

$$\Delta\phi = \frac{2}{F}\partial\bar\partial\phi = 2(k_2 K - r^{-2})\ ,$$

we see that $\log \Phi_k - \phi$ is subharmonic; in particular, the zeros of Φ_k can be at most isolated and, analytically, of (finite) even order. Let $2j_p^{(k)}$ denote the real analytic order of the zero of Φ_k at any $p \in \Sigma$ and, if Φ_k is not identically zero, set $N_k = \Sigma_{p\,\varepsilon\,\Sigma}\, j_p^{(k)}$. We can represent N_k analytically as a sum of the residues of the logarithmic singularities of $\log \Phi_k$ by the potential theoretic formula

$$-\ 2\pi N_k = \frac{1}{2}\lim_{\varepsilon\to 0}\int_{\Sigma_\varepsilon} (\Delta \log \Phi_k)\omega\ ,$$

where $\omega = iF\ dw \wedge d\bar w$ is the element of area, and Σ_ε denotes the complement in Σ of an ε-neighborhood of all points where Φ_k becomes zero.

Thus we have, in view of (5.12),

$$\lim_{\varepsilon\to 0}\int_{\Sigma_\varepsilon}(r^{-2} - k_2 K)\omega - 2\pi N_k = \lim_{\varepsilon\to 0}\int_{\Sigma_\varepsilon}\frac{\Phi_{k-1}\Phi_{k+1}}{\Phi_k^2}\omega \geq 0\ .$$

According to the Gauss-Bonnet formula, $\int_\Sigma K\omega = 4\pi$, so that, evaluating the above integrals and taking the limits as $\varepsilon \to 0$, we obtain the following estimate for the area $A = \int_\Sigma \omega$ of the images $X(\Sigma)$:

$$\begin{aligned}A &= (k_2 + \frac{1}{2}N_k)4\pi r^2 + \lim_{\varepsilon\to 0}\int_{\Sigma_\varepsilon}\frac{\Phi_{k-1}\Phi_{k+1}}{\Phi_k^2}\omega \\ &\geq (k_2 + \frac{1}{2}N_k)4\pi r^2\ ,\end{aligned}$$

equality holding if and only if Φ_k is not identically zero, while Φ_{k+1} vanishes everywhere; this happens, as we showed, precisely for $k = m = \frac{n-1}{2}$, according to Lemma 5.4. Thus we have proved the formula for the area A of Σ induced by the immersion X,

$$A = \left(\frac{m(m+1)+N_m}{2}\right)4\pi r^2 = \frac{n^2-1}{2}\pi r^2 + 2\pi N_m r^2, \tag{5.13}$$

where $2N_m$ is the total multiplicity of zeros of Φ_m. This completes the proof of the theorem.

Corollary. *Under the same assumptions of the theorem there exists a*
nonempty open set of Σ where the Gaussian curvature K satisfies

$$K \leq \frac{2}{m(m+1)r^2}. \tag{5.14}$$

Proof. Since the area A of the image of Σ is at least $4\pi m_2 r^2 = 2\pi m(m+1)r^2$ and, by the Gauss-Bonnet formula,

$$\int_\Sigma K\omega = 4\pi ,$$

the conclusion follows immediately.

The only explicit examples known of minimal immersions of 2-sphere Σ into rS^{n-1} are those where the Gaussian curvature of the induced metric on Σ is constant.

Theorem 5.2. *Let Σ be a 2-sphere with a Riemannian metric with constant curvature K, and let $X : \Sigma \to rs^{n-1} \subset \boldsymbol{E}^n (n = 2m+1 \geq 3)$ be an isometric, minimal immersion of Σ, such that the image is not contained in any hyperplane of $\boldsymbol{E}^n$. Then*

i) *The value of K is uniquely determined at the value,*

$$K = \frac{2}{m(m+1)r^2} ; \tag{5.15}$$

ii) *The immersion X is uniquely determined up to a rigid rotation of rS^{n-1}, and the n components of the vector X are a suitably normalized basis for the spherical harmonics of order m on Σ.*

Proof. If K is constant and $X : \Sigma \to rS^{n-1}$ an isometric, minimal immersion, the functions $\Phi_k (k = 0, 1, \cdots)$ can be calculated explicitly in a simple way: from the initial data $\Phi_0 = \Phi_1 = r$ and the recursion formula (5.12), we see that each Φ_k is constant, and hence

$$\Phi_{k+1} = \frac{\Phi_k^2}{\Phi_{k-1}}\left(r^{-2} - \frac{k(k+1)}{2}K\right);$$

since each Φ_k is then a positive constant for $k \leq m = \dfrac{n-1}{2}$ and $\Phi_{m+1} \equiv 0$, we see immediately that the determination (5.15) of the value of K is both necessary and sufficient. Therefore there exists an isothermal parameter w defined on all of Σ except for one point, such that

$$F(w, \bar{w}) = m(m+1)r^2(1 + w\bar{w})^{-2} .$$

The immersion function $X : \Sigma \to rS^{n-1}$ satisfies the equation $\partial\bar{\partial}X = -r^{-2}FX$, or equivalently, in terms of the Laplace operator $\Delta = 2F^{-1}\partial\bar{\partial}$,

$$\Delta X = -2r^{-2}X$$

thus each component of X is an eigenfunction of Δ (spherical harmonic) corresponding to the eigenvalue $-2r^{-2}$. It is known that on the 2-sphere with constant curvature K as in (5.15) the eigenvalues of Δ are

$$\lambda_k = -\frac{k(k+1)}{K} = -\frac{2k(k+1)}{m(m+1)r^2} \qquad (k = 0, 1, 2, \cdots)$$

and the eigenspace corresponding to each λ_k is $(2k+1)$-dimensional and generated as follows: map Σ isometrically onto the Euclidean sphere $r_0S^2 \subset \boldsymbol{R}^3$ with $r_0 = K^{-\frac{1}{2}} = \dfrac{m(m+1)}{2}r$ and consider the $2k+1$ linearly independent, homogeneous polynomials of degree k in the 3 Cartesian coordinates of $\boldsymbol{R}^3$, that satisfy Laplace's equation. Their restriction to r_0S^2 are the spherical harmonics of order k. Letting now $k = m$, we pick an orthogonal basis for all of them, and verify that the sum of their squares evaluated at each point of r_0S^2 is constant; hence they can be normalized so that the sum of their squares is r^2; these are then the $2m+1$ components of the imbedding function X. This completes the proof of the theorem.

It is natural to ask whether the example given above represents the only type of minimal immersion[5] of Σ into S^{n-1}. This, however, has not yet been settled, except in the trivial case $m = 1$ $(n = 3)$. Even a simplifying assumption that the area A of the image attains its lowest possible value $2m(m+1)\cdot\pi r^2$, which is equivalent to saying $\Phi_m > 0$ everywhere, and hence also each $\Phi_k > 0$ for $k \leq m$, and $\Phi_{m+1} \equiv 0$ does not seem to help. To show the difficulty, more explicitly, we consider the apparently easiest, nontrivial case, $m = 2$ (corresponding to $n = 5$). Since $\Phi_0 = \Phi_1 = r$, we apply (5.12) to calculate each Φ_k by the induction formula

[5] *Added in proof.* Since the completion of this paper, this question has been answered negatively; the details are to appear in a sequel.

$$\Phi_{k+1} = \frac{\Phi_k^2}{\Phi_{k-1}}\left(\frac{1}{F}\partial\bar{\partial} \log \Phi_k + r^{-2} - \frac{1}{2}k(k+1)K\right).$$

Thus $\Phi_2 = r(r^{-2} - K)$ (implying, incidentally, the condition $K \leq r^{-2}$ which, for other reasons as well is necessary), and

$$\Phi_3 = r(r^{-2} - K)^2(\frac{1}{2}\Delta \log (r^{-2} - K) + r^{-2} - 3K).$$

Thus a metric on S^2 is compatible with a linearly full, minimal immersion in rS^4, only if $K \leq r^{-2}$ everywhere, but with equality not holding identically, and

(5.16) $$\Delta \log(r^{-2} - K) = 6K - 2r^{-2} \text{ wherever } K < r^{-2};$$

the strict inequality $K < r^{-2}$ would then be satisfied everywhere, if and only if the total area of the immersed surface is $12\pi r^2$. For $m = 2$, the problem is equivalent to the question whether there are any Riemannian metrics that satisfy (5.16) other than the special case with constant curvature $K = 1/(3r^2)$.

Reference

[1] H. Hopf, *Über Flächen mit einer Relation zwischen den Hauptkrümmungen.* Math. Nachr. **4** (1950–51) 232–249.

UNIVERSITY OF PENNSYLVANIA.

On Ricci Curvature and Geodesics

Eugenio Calabi

Duke Math. J. 34, 667–676 (1967)

1. Introduction. The relationship in a Riemannian manifold between the numerical invariants of curvature at each point and the global metric properties of the manifold has been for a long time a subject for study by many people. In 1941 S. B. Myers [4], extending an earlier result by J. L. Synge, proved that, if the Ricci (mean) curvature of a manifold is uniformly bounded from below by a positive constant c^2, then every geodesic arc of length $\geq \pi/c$ contains conjugate points; in particular, if the manifold is complete, its diameter is bounded by π/c. A qualitative generalization of this result was published by W. Ambrose [1] in 1957, his statement reading as follows: Let $\mathbf{M}$ be a complete, n-dimensional Riemannian manifold ($n \geq 2$); suppose that there exists a point $\mathbf{p}_0 \in \mathbf{M}$ with the property that, for every infinite geodesic ray Γ issued from $\mathbf{p}_0$ and parametrized by its arc length s, the following improper integral diverges,

$$\liminf_{a\to\infty} \int_0^a K(s)\, ds = +\infty,$$

where $K(s)$ is the Ricci curvature at the tangent vector of Γ at the point corresponding to arc length s; then each Γ contains (infinitely many) conjugate points of $\mathbf{p}_0$, and consequently $\mathbf{M}$ is compact.

The present article contains a different sort of extension of Myers' theorem; the following statement arises as a consequence of the main result, which is stated as Theorem 2 in the next paragraph.

Theorem 1. *Let* $\mathbf{M}$ *be a complete, n-dimensional Riemannian manifold with nonnegative Ricci curvature everywhere. If, for some point* $\mathbf{p}_0 \in \mathbf{M}$, *every geodesic ray* Γ *issuing from* $\mathbf{p}_0$ *has the property that*

$$\limsup_{a\to\infty} \left\{\int_0^a \sqrt{K(s)}\, ds - \tfrac{1}{2}\log a\right\} = \infty, \tag{1.1}$$

then M is compact; $K(s)$ *denotes the Ricci mean curvature of* $\mathbf{M}$ *at the tangent vector of* Γ *over the point at arc distance s from* $\mathbf{p}_0$.

The above theorem and Ambrose's earlier one contain the same conclusion under two apparently similar assumptions; neither of these two sets of assumptions contains the other, so that these two theorems can be used jointly as independently sufficient conditions for compactness. The integral (1.1) over geodesic arcs without conjugate points, however, is of independent interest; a sharp estimate of such an integral has been used implicitly before [3, proof of

Received October 31, 1966. This paper was prepared while the author was supported in part by a grant from the National Science Foundation under Contract GP-4503.

DOI: 10.1215/S0012-7094-67-03469-2

Theorem 2] and can be applied again in at least two other instances known to the author. It is useful therefore to present here separately the results pertaining to estimates of that integral. Other geometric properties of Riemannian manifolds with nonnegative Ricci curvature can be found in [2], [5].

2. Geodesics without conjugate points. Throughout this paper **M** will denote a Riemannian manifold; completeness is not assumed. We use, in general the tensor notation of Eisenhart, with occasional abbreviations by means of more synthetic notations.

Let R^i_{jkl} denote the components of the Riemann curvature tensor; the Ricci tensor has components $R_{ij} = R_{ji}$ which we define with opposite sign from Eisenhart, by means of the contraction

$$R_{ij} = R^h_{ihj} = -R^h_{ijh} \, .$$

If $\xi = \xi^i(\partial/\partial x^i)$ is a nonzero tangent vector at a point of **M**, we define the Ricci, or mean, curvature of **M** at ξ, to be

$$Q(\xi) = \frac{1}{n-1} (R_{ij}\xi^i\xi^j)(g_{kl}\xi^k\xi^l)^{-1}$$

If $x^i = \varphi^i(s)$ describes locally, in terms of coordinates (x^i), a differentiable path Π in **M** parametrized by it arc length s, $a \leq s \leq b$, we let

$$K(s) = Q\left(\frac{dx^i}{ds}\frac{\partial}{\partial x^i}\right) = \frac{1}{n-1} R_{ij}\frac{dx^i}{ds}\frac{dx^j}{ds} \tag{2.1}$$

denote the Ricci curvature of **M** at the tangent vector of Π corresponding to the value s of the parameter. For the present purposes, the inclusion of an indication of the dependence of $K(s)$ on Π seems superfluous. We state now the main result of this paper.

THEOREM 2. *Let* **M** *be an n-dimensional Riemannian manifold with nonnegative Ricci curvature everywhere, and let* Γ *be an open geodesic arc in* **M**, *parametrized by its arc length s, the latter ranging over the open interval* (A, B) *where* $-\infty \leq A < B \leq \infty$. *Suppose that* Γ *contains no pair of conjugate points; then for every compact subinterval* $[a, b]$ *of* (A, B) *with* $A < a < b < B$

$$\int_a^b \sqrt{K(s)}\, ds \tag{2.2}$$

$$\leq \left\{\left(1 + \tfrac{1}{2}\log\frac{(B-a)(b-A)}{(B-b)(a-A)}\right)^2 - 1\right\}^{\frac{1}{2}} < 1 + \tfrac{1}{2}\log\frac{(B-a)(b-A)}{(B-b)(a-A)}.$$

The cross ratio whose logarithm occurs in (2.2) *is to be read as its limiting value, if either A or B or both are infinite.*

We recall that a closed segment $[\mathbf{p}_0, \mathbf{p}_1]$ of a geodesic arc Γ is said to contain no pair of conjugate points, if and only if the local variation of that segment of Γ in the space of arcs joining $\mathbf{p}_0$ and $\mathbf{p}_1$ has strictly positive second variation. Equivalently, this means that the exponential map $\exp_{\mathbf{p}_0}$ with center $\mathbf{p}_0$, and defined in a neighborhood of the segment $[\mathbf{0}, \mathbf{q}_1] \subset \mathbf{R}^n$ that maps onto the segment $[\mathbf{p}_0, \mathbf{p}_1] \subset \Gamma$, is nonsingular everywhere on that segment.

Before proving the two theorems stated above, a few preliminary concepts will be presented.

Definition 1. *Let $\Pi : (A, B) \to \mathbf{M}$ be a twice differentiable path, i.e. an immersion of an open segment (A, B) $(-\infty \leq A < B \leq \infty)$ into an n-dimensional Riemannian manifold $\mathbf{M}$. A tubular neighborhood $\mathfrak{A}_\Pi$ of Π is a composite object $(\mathbf{M}', \Pi', \pi)$, where $\mathbf{M}'$ is an n-dimensional Riemannian manifold, Π' : $(A, B) \to \mathbf{M}'$ a differentiable path in $\mathbf{M}'$, and $\pi : \mathbf{M}' \to \mathbf{M}$ a locally isometric map of $\mathbf{M}'$ into $\mathbf{M}$, satisfying the following conditions.*

1) *The immersion Π' is a differentiable imbedding, and is proper, i.e., for every compact $\mathbf{F} \subset \mathbf{M}$, $\Pi^{-1}(\mathbf{F})$ is a compact subset of (A, B);*

2) *For each $t \in (A, B)$, $\Pi(t) = \pi \circ \Pi'(t)$;*

3) *For each point $\mathbf{p}' \in \mathbf{M}' - \Pi'(A, B)$, there exists exactly one geodesic segment joining $\mathbf{p}'$ with the trace of Π', intersecting it at right angles; it also minimizes the geodesic distance between $\mathbf{p}'$ and the nearest point in $\Pi'(A, B)$. When no misunderstanding is anticipated, we may use the symbol $\mathbf{M}'$ as a synecdoche denoting the tubular neighborhood $(\mathbf{M}', \Pi', \pi)$.*

The proof of existence of tubular neighborhoods of a path is an easy exercise. The advantage of this concept is that statements such as that of Theorem 2 about an open differentiable path Π (e.g. a geodesic) are equivalent to the corresponding statement about the lift Π' of Π to a tubular neighborhood. Working then with Π' instead of Π, one avoids the topological difficulties that may arise from the fact that Π need not be either an imbedding or a proper map. An immediate application of this advantage is seen in the following statement.

Proposition 1. *Let Γ be an open geodesic segment in a Riemannian manifold $\mathbf{M}$, and let Γ be parametrized by its arc length s $(-\infty \leq A < s < B \leq \infty)$. Then a necessary and sufficient condition for Γ to contain no conjugate points is the following. There exists a tubular neighborhood $(\mathbf{M}', \Gamma', \pi)$ of Γ with a function $\Phi : \mathbf{M}' \to \mathbf{R}$ that is differentiable everywhere in $\mathbf{M}'$ and satisfying*

1) $\Phi \circ \Gamma'(s) = s$; $(A < s < B)$

2) *The gradient of Φ has absolute value* 1, *i.e.*

$$|\nabla\Phi|^2 = g^{ij}(\partial\Phi/\partial x^i)(\partial\Phi/\partial x^j) = 1 \text{ everywhere in } \mathbf{M}'. \tag{2.3}$$

Proof. We prove the necessity first. Suppose that Γ contains no pair of conjugate points and assume without loss of generality that $0 \in (A, B)$. Let $(\mathbf{M}'', \Gamma', \pi)$ be a tubular neighborhood of Γ. For any $a \in (A, B)$ consider the exponential map $\exp_a : \mathbf{R}^n \to \mathbf{M}''$ with center $\Gamma'(a)$; this map, by assumption, is defined and nonsingular in a neighborhood $\mathbf{U}_a$ of the open segment $\Gamma_a'' \subset \mathbf{R}^n$

that maps onto $\Gamma'(A, B)$ under this exponential map. We use a Cartesian coordinate system $(y_1, \cdots, y_n)$ in $\mathbf{R}^n$ so that, $\exp(y_1, 0, \cdots, 0) = \Gamma'(a + y_1)$ for $B - a < y_1 < A - a$; let $\mathbf{V}_a \subset \mathbf{U}_a$ be a neighborhood of Γ_a'' that is mapped diffeomorphically under $\exp_a$ onto a tubular neighborhood $\mathbf{M}_a' \subset \mathbf{M}''$ of $\Gamma'(A, B)$ and define, for $a \neq 0$, a function $\Phi_a : \mathbf{M}_a' \to \mathbf{R}$ by means of the equation

$$\Phi_a(\mathbf{q}') = a - (\operatorname{sgn} a)\cdot\left\{\left(\sum_{i=1}^{n} (y_i)^2\right)^{\frac{1}{2}} (\exp_a^{-1}(\mathbf{q}'))\right\}.$$

This map is differentiable everywhere except at $\Gamma'(a)$, and otherwise satisfies (2.3) everywhere. In particular there is a hypersurface $\Sigma_a \subset \mathbf{M}_a'$ corresponding to the zero set of Φ_a, intersecting $\Gamma'(A, B)$ at $\Gamma'(0)$ at right angles; the vector field $\nabla\Phi_a$, restricted to Σ_a defines the unit normal vector, oriented so that at $\Gamma'(0)$ it points in the direction of $\nabla\ \Gamma'(0)$. The second fundamental form of Σ_a, in terms of this orientation of the normal vector, is given by

$$\Theta_a = -\nabla\nabla\Phi_a .$$

We observe that, at the point $\Gamma'(0)$ $\Phi_a = 0$ and $\nabla\Phi_a = d\Gamma'_{\Gamma'(0)}(d/ds)$ thus Φ_a and $\nabla\Phi_a$ are both independent of a; furthermore, for any fixed $\mathbf{q}' \in \mathbf{M}_a'$, we have also $\mathbf{q}' \in \mathbf{M}_b'$ for b sufficiently close to a and, because of the triangle inequality,

$$\frac{\partial\Phi_a(\mathbf{q}')}{\partial a} > 0.$$

Thus, if $a > 0$ and a increases toward B, the quadratic form Θ_a is decreasing at $\Gamma'(0)$, while if $a < 0$ and a decreases toward A, Θ_a is monotone increasing. On the other hand, again because of the fact that the second variation of the arc length of Γ is positive between any two of its points, for any a, b satisfying $A < a < 0 < b < B$, we have

$$\Theta_a(\Gamma'(0)) < \Theta_b(\Gamma'(0)).$$

Hence there exists a quadratic form Θ on the tangent subspace of $\mathbf{M}''$ at $\Gamma'(0)$ orthogonal to Γ' such that, for all numbers a, b satisfying $A < a < 0 < b < B$,

$$\Theta_a(\Gamma'(0)) < \Theta < \Theta_b(\Gamma'(0)). \tag{2.3}$$

Now let Σ be any differentiable, oriented hypersurface imbedded in $\mathbf{M}''$, intersecting $\Gamma'(A, B)$ orthogonally and only in $\Gamma'(0)$, and with second fundamental form Θ at $\Gamma'(0)$. Consider the product $\mathbf{R} \times \Sigma$ and identify $\mathbf{R}$ with $\mathbf{R} \times \{\Gamma'(0)\} \subset \mathbf{R} \times \Sigma$, Σ with $\{0\} \times \Sigma$, and the "wedge" $\mathbf{R} \vee \Sigma$ with the union $\mathbf{R} \times \{\Gamma'(0)\} \cup \{0\} \times \Sigma$. Then there exists a differentiable map ψ_1 of a neighborhood $\mathbf{W}_1$ of $(A, B) \vee \Sigma \subset \mathbf{R} \times \Sigma$ into $\mathbf{M}''$ such that

i) ψ_1 is an extension of the imbedding of Σ in $\mathbf{M}''$

ii) for each $\mathbf{q} \in \Sigma$, $(\mathbf{R} \times \{\mathbf{q}\}) \cap \mathbf{W} = (A_\mathbf{q}, B_\mathbf{q}) \times \{\mathbf{q}\}$ and the map $\Gamma_\mathbf{q}' : (A_\mathbf{q}, B_\mathbf{q}) \to \mathbf{M}''$ defined by $\Gamma_q'(s) = \psi_1(s, \mathbf{q})$ is the geodesic arc, parametrized by its arc length s (positively oriented) and orthogonal to Σ at $\mathbf{q}$. In particular $\Gamma'_{(\Gamma'(0))} = \Gamma'$.

Because of (2.4), the map ψ_1 is regular everywhere in $\Gamma'(A, B)$; therefore there is a smaller neighborhood $\mathbf{W} \subset \mathbf{W}_1$ of (A, B) in $\mathbf{R} \times \Sigma$ such that the restriction ψ of ψ_1 to $\mathbf{W}$ is a differentiable imbedding of $\mathbf{W}$ onto an open tubular neighborhood $\mathbf{M}' \subset \mathbf{M}''$ of $\Gamma'(A, B)$; for $\mathbf{q} \in \psi(\mathbf{W})$, let $\Phi(\mathbf{q}) \in \mathbf{R}$ be defined to be the number t such that

$$\psi^{-1}(\mathbf{q}) = (t, \mathbf{q}'), \qquad \mathbf{q}' \in \Sigma;$$

then this function Φ, defined in the tubular neighborhood $(\mathbf{M}', \Gamma', \pi) \subset (\mathbf{M}'', \Gamma', \pi)$ of $\Gamma(A, B)$, is a function with the properties 1) and 2) required in the statement of the proposition.

Conversely, assume that a function Φ with these two properties exists in a tubular neighborhood $(\mathbf{M}', \Gamma', \pi)$ of Γ. Then, for any piecewise differentiable path Π in $\mathbf{M}'$ joining $\Gamma'(a)$ and $\Gamma'(b)$ with $A < a < b < B$, we have, because of (2.3),

$$\int_\Pi ds \geq \int_\Pi |d\Phi| \geq \Phi \circ \Gamma'(b) - \Phi \circ \Gamma'(a) = b - a.$$

Equality between the first two members can hold, if and only if Π is an integral curve of the gradient field of Φ; on the other hand, if the trace of Π does not coincide with $\Gamma'[a, b]$, then there exists a subinterval of Π of length at least δ for some $\delta > 0$, such that the angle α between the trace of Π and the normal vector of the hypersurfaces Φ = constant satisfies $|\sin \alpha| \geq \epsilon$ for some $\epsilon > 0$. From this it easily follows that

$$\int_\Pi ds \geq b - a + \delta(1 - |\cos \alpha|) \geq b - a + \tfrac{1}{2}\delta\epsilon^2.$$

This fact shows that the second variation of the length of any bounded, closed segment of Γ is positive definite. Thus the proof of the proposition is complete.

Definition 2. *Given a Riemannian manifold* $\mathbf{M}$ *and a geodesic arc* $\Gamma : (A, B) \to \mathbf{M}$ *parametrized by its arc length s, a geodesic field extension of* Γ *consists of a tubular neighborhood* $(\mathbf{M}', \Gamma', \pi)$ *of* Γ *and a smooth tangent vector field* θ *in* $\mathbf{M}'$ *satisfying*:

1) *For all* $s \in (A, B)$, $\theta(\Gamma'(s)) = d\Gamma'_s(d/ds)$;
2) *At each point* $\mathbf{q}' \in \mathbf{M}'$ $|\theta(\mathbf{q}')| = 1$;
3) *The vector field is irrotational; that is to say, if locally* $\theta = \theta^i(\partial/\partial x^i)$, *if the metric is* $ds^2 = g_{ij}\, dx_\lambda\, dx^i$ *and if* $\theta_i = g_{ij}\theta_\lambda$, *then*

$$\frac{\partial \theta_i}{\partial x^j} = \frac{\partial \theta_j}{\partial x^i}.$$

Since the domain $\mathbf{M}'$ is, by Definition 1, contractible and hence simply connected, a geodesic field extension is, globally in $\mathbf{M}'$, the gradient field of a function Φ that satisfies the conditions of Proposition 1 in order that Γ be without conjugate points.

3. Proof of the theorems. We begin by proving Theorem 2. Let $(\mathbf{M}', \Gamma', \pi)$ be a tubular neighborhood of Γ admitting a function $\Phi : \mathbf{M}' \to \mathbf{R}$ whose gradient $\nabla\Phi$ is a geodesic vector field extension of Γ'. Without loss of generality, we replace $\mathbf{M}$ and Γ by $\mathbf{M}'$ and Γ' and then drop the "prime" superscripts.

Let Σ_a for any $a \in (A, B)$ denote the hypersurface in $\mathbf{M}$ defined by the equation $\Phi = a$, The vector field $\theta = \nabla\Phi$ is, at each $\mathbf{q} \in \mathbf{M}$, the unit normal vector of the hypersurface $\Sigma_{(\Phi\mathbf{q})}$.

The first fundamental form induced on each Σ_a corresponds to the restriction of the metric to the tangent space of Σ_a ; the second fundamental form Θ is represented by

$$\Theta = -\nabla\nabla\Phi = -\Phi_{,ij}\, dx^i\, dx^j$$

in terms of local coordinates; we associate to it the Weingarten operator on the tangent space of each Σ_a with tensor components

$$B_i^j = -\Phi_{,ik}g^{kj};$$

the eigenvalues of (B_i^j) are the principal curvatures of Σ_a . Consider now the variation of the Weingarten operator along an integral curve of grad Φ, i.e. along the orthogonal trajectories of Σ_a as a varies. This is nothing else but the Lie derivative of the tensor (B_i^j) with respect to $\theta = \nabla\Phi$. Using the Ricci identities, we obtain

$$[\theta, B]_i^j = \theta^k B_{i,k}^j - \theta^j_{,k} B_i^k + \theta^k_{,i} B_k^j$$
$$= g^{jh}g^{kl}(-\Phi_{,l}\Phi_{ihk} + \Phi_{,hk}\Phi_{,il} - \Phi_{,li}\Phi_{,kh}) = -g^{jh}g^{kl}\Phi_{,l}(\Phi_{,ikh} + \Phi_{,m}R^m_{ihk}).$$

Furthermore, by taking first and second covariant derivatives on both terms of (2.3), we have successively

$$g^{kl}\Phi_{,l}\Phi_{,ki} = 0; \qquad g^{kl}\Phi_{,lh}\Phi_{,ki} = -g^{kl}\Phi_{,l}\Phi_{,kih} .$$

Substituting this in the previous equation, we get

$$[\theta, B]_i^j = g^{jh}g^{kl}(\Phi_{,lh}\Phi_{,ki} + R_{imhk}\theta^m\Phi_{,l})$$
$$= B_i^k B_k^j + R^j_{hik}\theta^h\theta^k.$$

This last equation is the Riccati matrix equation associated with the classical form of the Jacobi equation of Γ. Let us denote by $\kappa_1 , \cdots , \kappa_{n-1}$ the principal curvatures of each Σ_a , and by $H_{(1)}$, $H_{(2)}$ the first and second mean curvatures, i.e.

$$H = H_1 = \frac{1}{n-1}\sum_{\nu=1}^{n-1}\kappa_\nu = \frac{1}{n-1}B_i^i , \quad H_{(2)} = \frac{1}{n-1}\sum_{\nu=1}^{n-1}(\kappa_\nu)^2 = \frac{1}{n-1}B_i^k B_k^i .$$

By means of a contraction of the Riccati equation above, we obtain the equation

$$\theta^i\frac{\partial H}{\partial x^i} = H_{(2)} + \frac{1}{n-1}R_{hk}\theta^h\theta^k.$$

Because of the Cauchy–Schwarz inequality $H_{(2)} \geq H^2$, we conclude that the mean curvature H satisfies the differential inequation

$$\theta^i \frac{\partial H}{\partial x^i} \geq H^2 + Q(\theta) \qquad (Q(\theta) = \text{Ricci curvature at } \theta).$$

Restricting the above inequality to Γ, we set $H_1(s) = H(\Gamma(s))$ for $A < s < B$; we have thus

$$\frac{dH_1}{ds} \geq (H_1)^2 + K(s), \tag{3.1}$$

where $K(s)$ is defined in (2.1).

By comparison with the inequation (3.1) we deduce immediately the following statement, that we formalize as a lemma.

LEMMA. *Let $\Gamma : (A, B) \to \mathbf{M}$ be a geodesic without conjugate points in a Riemannian manifold $\mathbf{M}$, parametrized by its arc length s, $-\infty \leq A < s < B \leq \infty$, and denote by $K(s)$ the Ricci mean curvature of $\mathbf{M}$ at the tangent vector of Γ over $\Gamma(s)$. Then there exists a differentiable function $h(s)$, globally defined over the open interval (A, B), that satisfies the Riccati equation*

$$\frac{dh}{ds} = h^2 + K(s). \tag{3.2}$$

From the above lemma one obtains immediately Myers' theorem [4] under the assumption $K(s) \geq c > 0$; it is also possible, but not so easy, to deduce Ambrose's result. In proving our Theorem 2, we shall limit ourselves to the assumption that there is a function $h(s)$ in the open interval (A, B) satisfying the inequality

$$\frac{dh}{ds} \geq h^2: \tag{3.3}$$

the conclusion consists of an estimate of an upper bound for the integral

$$\int_a^b \sqrt{h'(s) - h^2(s)}\, ds \tag{3.4}$$

for any compact subinterval $[a, b]$ of (A, B), in terms of A, a, b, and B. We consider the four possible cases that can arise as A and B can be, independently, finite or infinite.

Case 1. $A = -\infty$ and $B = \infty$. It is clear that there can be no function $h(s)$ on the whole line satisfying (3.3) other than $h(s) \equiv 0$. Therefore in this case $K(s)$ must be identically zero.

Case 2. $A > -\infty$ and $B = \infty$. We make the following substitutions. Let $s = A + e^t$ and define $w(t)$ by means of

$$h(s) = e^{-t}(w(t) - \tfrac{1}{2}).$$

Then the interval (A, B) becomes, in terms of t, the whole real line; inequality (3.3) becomes

$$\frac{dw}{dt} \geq w^2 - \tfrac{1}{4} \qquad (-\infty < t < \infty) \tag{3.5}$$

and the integral (3.4) becomes

$$\int_a^b \sqrt{h'(s) - h^2(s)}\, ds = \int_{a'}^{b'} \sqrt{w'(t) - w^2(t) + \tfrac{1}{4}}\, dt \tag{3.6}$$

where the bounded interval $[a', b']$ corresponds to $[a, b]$ in terms of the substitution of t for s. It follows from (3.5) that $w(t)$ is uniformly bounded for all t by

$$-\tfrac{1}{2} \leq w(t) \leq \tfrac{1}{2}.$$

Applying the Schwarz inequality to (3.6), we have

$$\begin{aligned}\left(\int_{a'}^{b'} \sqrt{w'(t) - w^2(t) + \tfrac{1}{4}}\, dt\right)^2 &\leq (b' - a') \int_{a'}^{b'} (w'(t) - w^2(t) + \tfrac{1}{4})\, dt \\ &\leq (b' - a')(w(b') - w(a') + \tfrac{1}{4}(b' - a')) = (\tfrac{1}{2}(b' - a') + 1)^2 - 1.\end{aligned}$$

Thus we obtain, by rewriting in terms of s and $h(s)$,

$$\int_a^b \sqrt{h'(s) - h^2(s)}\, ds \leq \sqrt{(\tfrac{1}{2}(b' - a') + 1)^2 - 1} = \left\{\left(1 + \tfrac{1}{2} \log \frac{b - A}{a - A}\right)^2 - 1\right\}^{\frac{1}{2}}$$

Case 3. $A = -\infty$ and $B < \infty$. We replace s, A, B and $h(s)$ respectively by $\bar{s} = -s$, $\bar{A} = -B$, $\bar{B} = \infty$ and $h(s)$ by $\bar{h}(\bar{s}) = -h(-s)$, reducing this to Case 2.

Case 4. $-\infty < A < B < \infty$. We again make a substitution of the finite interval by the whole real line as follows: let $L = B - A$,

$$s = \frac{A + B}{2} + \frac{L}{2} \tanh t, \quad \text{and define } w(t) \text{ by means of}$$

$$h(s) = \frac{2}{L}\,(w(t) + \tanh t) \cosh^2 t.$$

Then inequality (3.3) is equivalent to

$$\frac{dw(t)}{dt} \geq w^2(t) - 1 \qquad (-\infty < t < \infty) \tag{3.7}$$

and (3.4) becomes, after appropriate substitution of the limits $[a', b']$ for $[a, b]$

$$\int_a^b \sqrt{h'(s) - h^2(s)}\, ds = \int_{a'}^{b'} \sqrt{w'(t) - w^2(t) + 1}\, dt. \tag{3.8}$$

It follows from (3.7) that, in this case, $w(t)$ is uniformly bounded by $-1 \leq w(t) \leq 1$. Applying then the Schwarz inequality to (3.8) we get

$$\begin{aligned}\left(\int_{a'}^{b'} \sqrt{w'(t) - w^2(t) + 1}\, dt\right)^2 &\leq (b' - a') \int_{a'}^{b'} (w'(t) + 1 - w^2(t))\, dt \\ &\leq (b' - a')(w(b') - w(a') + b' - a') \\ &\leq (b' - a')(1 + 1 + b' - a') \\ &= (1 + b' - a')^2 - 1.\end{aligned}$$

Thus we have

$$\begin{aligned}\int_{a'}^{b'} \sqrt{h'(s) - h^2(s)}\, ds &\leq ((1 + b' - a')^2 - 1)^{\frac{1}{2}} \\ &= \left\{\left(1 + \operatorname{arc\,tanh} \frac{2b - A - B}{L} - \operatorname{arc\,tanh} \frac{2a - A - B}{L}\right)^2 - 1\right\}^{\frac{1}{2}} \\ &= \left\{\left(1 + \tfrac{1}{2} \log \frac{(B - a)(b - A)}{(B - b)(a - A)}\right)^2 - 1\right\}^{\frac{1}{2}}.\end{aligned}$$

This concludes the proof of Theorem 2.

We now show that Theorem 1 is an easy consequence of Theorem 2. Suppose that **M** is complete, has positive semidefinite Ricci curvature everywhere, and is not compact. Then for every point $\mathbf{p}_0 \in \mathbf{M}$ there exists a geodesic ray issuing from $\mathbf{p}_0$, that achieves the shortest distance between any two of its points, and extends to infinity: in particular this ray has no conjugate points. Let it be parametrized by s, where $0 \leq s < \infty$. By virtue of Theorem 2, for each pair of real numbers ϵ, a with $0 < \epsilon < a < \infty$,

$$\int_\epsilon^a \sqrt{K(s)}\, ds < 1 + \tfrac{1}{2} \log \frac{a}{\epsilon},$$

which clearly contradicts (1.1).

We shall construct an example that shows the extent to which the estimate for the upper bound of (2.2) is sharp. Consider the standard unit euclidean $(n - 1)$-sphere S^{n-1} with the usual Riemannian metric form denoted by $d\sigma^2$, and let $M = (0, L) \times S^{n-1}$ be described as the set of pairs (t, ω) where $0 < t < L \leq \infty$ and $\omega \in S^{n-1}$. On M we define the Riemannian metric $ds^2 = dt^2 + t(1 - t/L)\, d\sigma^2$ (for $L = \infty$, we replace this last expression by its limit value); this metric has everywhere strictly positive curvature. The coordinate function t, globally defined on M, is a function whose gradient has length equal to unity everywhere; consequently the orthogonal trajectories of the hypersurfaces $\{t = \text{constant}\}$ are all geodesics without conjugate points, and the value increments of the parameter t along these geodesics correspond to the arc length of each segment. It is also easy to verify that along any one of these geodesics, using t as its parameter, we have

$$K(t) = \frac{L^2}{4t^2(L - t)^2},$$

so that, for $0 < a < b < \infty$,

$$\int_a^b \sqrt{K(t)}\, dt = \tfrac{1}{2} \log \frac{b(L - a)}{a(L - b)},$$

which comes within one unit of—and is asymptotically equal to—the estimate (2.2).

Other applications of Theorem 2 are contemplated in more specialized situations as mentioned in the introduction. In these the value $\frac{1}{2}$ of the coefficient preceding the logarithm plays a crucial role.

References

1. W. Ambrose, *A theorem of Myers*, this Journal, vol. 24(1957), pp. 345–348.
2. E. Calabi, *An extension of E. Hopf's maximum principle with an application to Riemannian geometry*, this Journal, vol. 25(1958), pp. 45–56 (Erratum, vol. 26(1959), p. 707.)
3. E. Calabi, *Improper affine hyperspheres of convex type and a generalization of a theorem by K. Jörgens*, Mich. Math. J., vol. 5(1958), pp. 105–126.
4. S. B. Myers, *Riemannian manifolds with positive mean curvature*, this Journal, vol. 8(1941), pp. 401–404.
5. K. Yano and S. Bochner, *Curvature and Betti Numbers*, Ann. of Math. Studies No. 32, Princeton, 1953.

University of Pennsylvania

On Differentiable Actions of Compact Lie Groups on Compact Manifolds

Eugenio Calabi*

Proc. Conf. on Transformation Groups,
(New Orleans, La., 1967), 210–213, Springer, New York (1968)

The present note is devoted to a new proof of a theorem by R. S. PALAIS and T. E. STEWART [1, 2] on the title subject and restated below, after the preliminary definitions.

Definition 1. Let G be a Lie group and X a differential manifold. A differentiable action F of G in X is a differentiable map $F: G \times X \to X$ satisfying the following relations:

1. $F(g_1, F(g_2, x)) = F(g_1 g_2, x)$,
$$(g_1, g_2 \in G,\ x \in X);$$
2. $F(1, x) = x$ $(1 = \text{identity of } G,\ x \in X)$.

Definition 2. Two actions F_1, F_2 of a Lie group G on a differential manifold X are called *(differentiably) equivalent*, if there exists a (differentiable) automorphism $\sigma: X \to X$ such that for all $g \in G$, and $x \in X$
$$F_2(g, \sigma(x)) = \sigma(F_1(g, x)).$$

Definition 3. Let G be a Lie group and M, X two differential manifolds. A differentiable family of actions of G on X parametrized by M is a differentiable map $\Phi: M \times G \times X \to X$ such that, for each $t \in M$ the restriction $\Phi_t: G \times X \to X$ defined by $\Phi_t(g, x) = \Phi(t, g, x)$ is a differentiable action of G.

Definition 4. Let $\Phi: M \times G \times X \to X$ be a differentiable family of actions of G on X parametrized by M and let t_0 be a point in M. This family is called *locally trivial at* t_0, if there exists an open neighborhood U of t_0 in M and a differentiable map $\Psi: U \times X \to X$ such that

1. Ψ is a *differentiable flow* on X parametrized by U with base point t_0, i.e.

a) For each $t \in U$ the map $\Psi_t: X \to X$ defined by $\Psi_t(x) = \Psi(t, x)$ is a differentiable automorphism of X, and

* Supported in part under NSF Contract GP 4509.

DOI: 10.1007/978-3-642-46141-5_14

b) for $t = t_0$, Ψ_t is the identity map on X;
2. Ψ *trivializes* Φ, i.e. for each $t \in U$ and each $g \in G$

$$\Phi(t, g, \Psi(t, x)) = \Psi(t, \Phi(t_0, g, x)).$$

A differentiable family Φ of group actions of G on X is called *locally trivial*, if and only if it is so at each point t of the parameter space M.

The statement of the theorem of PALAIS-STEWART is the following

Theorem 1. *A differentiable family Φ of actions of a compact Lie group G on a compact manifold X without boundary, parametrized by M, is locally trivial at each $t \in M$.*

Proof. Let γ_0 be a Riemannian metric on X. For each $t \in M$ and each $g \in G$ let $\Phi_{t,g}$ be the automorphism of X defined by $\Phi_{t,g}(x) = \Phi(t, g, x)$ and $\Phi^*_{t,g}(\gamma_0)$ the transformation of the metric γ_0 induced contragrediently by $\Phi_{t,g}$ and denote by $d\mu$ the normalized Haar measure on G. Then for each $t \in M$ we obtain a Riemannian metric

$$\gamma_t = \int_G \Phi^*_{t,g}(\gamma_0) \, d\mu$$

satisfying the two following properties: a) γ_t is invariant under the action Φ_t of G on X; b) γ_t depends differentiably on t.

We now apply some results from the potential theory in the manifold X with any Riemannian metric γ_t:

1. The Laplace-Beltrami operator $\Delta_t = -\nabla_t \cdot \nabla$ with respect to γ_t, acting on scalar functions is self-adjoint and has a non-negative, discrete spectrum. This fact, which is equivalent to the compactness of the Green's integral operator G_t on X (with respect to γ_t), means that, for any positive real number A, the invariant space of Δ_t corresponding to the segment $[0, A]$ in the spectrum of Δ_t is finite dimensional. Furthermore, the eigenfunctions corresponding to the point spectrum, which is contained in $[0, \infty]$, include a complete orthonormal basis for the space of functions f under any of the Hilbert norms $\|f\|_{k,t}$ with

$$(\|f\|_{k,t})^2 = (f, f)_{\gamma_t} + (\Delta_t^k f, f)_{\gamma_t}, \qquad k = 1, 2, \ldots,$$

where Δ_t^k denotes Δ_t iterated k times and the volume element in the integrals is with respect to the metric γ_t. From this and from the Sobolev inequalities, one can deduce the second result we need; it can be stated as follows.

2. Given any $t \in M$, there exists a sufficiently large, positive real number A (depending on t), such that an orthogonal basis for the eigenfunctions corresponding to the eigenvalues of Δ_t in the interval $[0, A]$ (say $u_1, u_2, \ldots, u_N$) defines a differentiable imbedding of X in $\mathbb{R}^N$.

Finally the continuity with respect to t of the Green's operator G_t corresponding to Δ_t implies the following (see [3]).

3. The spectrum of Δ_t depends continuously on t. More precisely, if a positive A is not in the spectrum of Δ_{t_0} for any given $t_0 \in M$, there exists a neighborhood U_0 of t_0 in M such that;

a) A is not in the spectrum of Δ_t for any $t \in U_0$;

b) for each differentiable arc Γ imbedded in U_0 and passing through t_0, there exists a family of differentiable functions $v_j: \Gamma \times X, \to \mathbb{R}$ $(j=1, \ldots, N)$ with the property that for each t the N functions $v_{j,t}: X \to \mathbb{R}$, $v_{j,t}(x) = v_j(t,x)$, are a basis for the eigenfunctions of Δ_t corresponding to all the eigenvalues in $[0, A]$.

Now let t_0 be any given point in M, and let U be a neighborhood of t_0 in M with the following properties:

a) U is in a local coordinate system with coordinates $\tau = (\tau_1, \ldots, \tau_m)$ with $\tau_\mu(t_0) = 0$ and range $\sum_{\mu=1}^{m} \tau_\mu^2 \leqslant \delta$ for some $\delta > 0$.

b) There exists an $A > 0$ which is not in the spectrum of Δ_t for any $t \in U$ and one can choose N continuous functions $v_j: U \times X \to \mathbb{R}$ $(1 \leqslant j \leqslant N)$ whose restriction to $\Gamma_{t_1} \times X$, where Γ_{t_1} is any diametral arc in U defined by $\tau(t) = \lambda \tau(t_1), (-1 \leqslant \lambda \leqslant 1)$, is differentiable, and such that their restrictions $v_{j,t}: X \to \mathbb{R}$ are an orthonormal basis for the eigenfunctions of Δ_t corresponding to all eigenvalues in $[0, A]$ under the Hilbert norm defined by γ_t.

c) For each $t \in U$ the N functions $v_{j,t}$ $(1 \leqslant j \leqslant N)$ define a family of differentiable imbeddings of X in $\mathbb{R}^N$ parametrized by t.

It is clear that, for each $t \in U$, the imbedding $(v_{j,t})$ of X into $\mathbb{R}^N$ is invariant (under the orthogonal group) under the action Φ_t of G on X, since the metric γ_t is invariant, and hence each eigenspace of Δ_t furnishes an orthogonal representation of G. It is well known that the set of equivalence classes of these representations is finite and that each is stable under deformations. Therefore there exists a map $\gamma: U \to O(N)$ such that the family $(v'_{j,t})_{t \in U}$ of imbeddings $(v'_{j,t}) = \lambda(t) \cdot (v_{j,t})$ has each of the images $v'_{j,t}(X)$ in $\mathbb{R}^N$ invariant under a fixed representation $\chi: G \to O(N)$ of G, the latter being equivariant with the action Φ_t of G on X. Consider now the (radial) contraction h of U to $\tau^{-1}(0) = t_0$, defined by $\tau_\mu(h(\sigma, t)) = \sigma \tau_\mu(t)$, $0 \leqslant \sigma \leqslant 1$, and for each $t \in U$, the restriction of the family $(v'_{j,t})$ to the orbit of t in U under h; this is a one-parameter family of imbeddings of X in $\mathbb{R}^N$. It is evident that the normal variation if this family with respect to the parameter σ is equivariant with the variation of the action Φ_t of G on X, since it is invariant under $\chi(G) \subset O(N)$. Integration of this normal variation of $v'_{j,t}$ with respect to σ finally establishes, for each $t \in U$ the family $(\Psi_t)_{t \in U}$, $\Psi_t: X \to X$, of diffeomorphisms of X onto itself that establish the equivalence of the action Φ_t of G with Φ_{t_0}. This completes the proof of the theorem.

The above theorem has an immediate generalization to compact manifolds X with boundary (e.g. consider the "double" $\tilde{X}$ of X and repeat the proof using the extension of actions of G by the reflection of $\tilde{X}$ across the boundary of X). The stability of differentiable actions of G on X fails, if G is not compact [1, 2]. On the other hand if G is compact and X is not compact, it seems likely that the conclusion still holds; in order to prove it, one should try to consider the family of orbit spaces under the actions of G and prove that such a family, considered as a family of singular real varieties, is locally trivial. If such be the case, the conjectured conclusion on X would follow by considering an open covering of the orbit space by open subdomains with compact closure and boundary smooth in the generic orbits and transversal to the singular ones, in such a way that the counter-images in X of these compact subsets are submanifolds with smooth boundary.

References

1. Palais, R. S., and T. E. Stewart: Deformations of compact, differentiable transformation groups, Amer. Journal of Math., **82**, 935—937 (1960).
2. — Equivalence of nearby differentiable actions of a compact group, A. M. S. Bull., **67**, 362—364 (1961).
3. Browder, F. E.: Families of linear operators depending upon a parameter, Amer. J. of Math., **87**, 752—758 (1965).

An Intrinsic Characterization of Harmonic One-Forms

Eugenio Calabi*

Global Analysis (Papers in Honor of K. Kodaira), 101–117,
Univ. Tokyo Press, Tokyo (1969)

I. Introduction

The author recalls hearing the question since about 1955, do the zeros of a harmonic 1-form in a compact riemannian manifold possess certain combinatorial properties that depend only on the differential topological structure of the manifold, and not on the particular riemannian metric. In the case of surfaces the Riemann-Roch theorem (or alternately the Hopf-Stiefel theorem) on the indices of the zeros of any harmonic 1-form on a closed surface provided a positive answer to the question, rendering an analogous conjecture at least plausible for perhaps some higher dimensional manifolds. The fact that such a conjecture fails completely is the main consequence of the results of this paper. The technique is to consider, in a given, compact, n-dimensional manifold $\mathbf{X}$, a given 1-form α, necessarily closed, and satisfying some reasonable technical condition on the points where it vanishes (only non-degenerate zeros admitted). Then the condition for the existence of a riemannian metric $\mathbf{g}$ with respect to which α is also co-closed are considered and settled completely, in terms of local properties (in § 2) and a new semi-global one (transitivity, §§ 4 and 5). Then in the sixth and concluding section it is shown how a harmonic 1-form can be perturbed by a closed 1-form in a small neighborhood, yielding a new closed 1-form with two additional zeros, such that under a different choice of metric, the resulting 1-form is again co-closed.

The proofs have undergone several modifications and corrections; the author is indebted to P. Deligne for suggesting a substantial simplification in the proof of Theorem 1 (precisely, Lemma 2, § 5).

Throughout this paper the term "differential manifold" denotes a connected manifold without boundary (unless otherwise

* Research partly supported by NSF under Grant GP 6895 and by the Institut des Hautes Etudes Scientifiques, 91 Bures-sur-Yvette, France

specified) of class C^r with $r \geqq 2$ (tensor fields are then of class C^{r-1}). The usual difficulty in dealing with exterior calculus in non-orientable manifolds is bypassed by using de Rham's idea of "forms of odd kind". A linear differential form (1-form, or pfaffian form) is called integrable, if it is locally a scalar multiple of a closed form by a factor which is nowhere zero. The adjoint of an exterior j-form φ in a riemannian n-manifold with metric $\mathbf{g}$, is an $(n-j)$-form of odd kind, which is denoted by $*_{\mathbf{g}}\varphi$, since the operation of adjunction depends on the particular riemannian metric, and the latter is subject to modification in the present context.

II. Local theory

We shall derive first a set of local conditions on a closed 1-form α in a differential manifold $\mathbf{X}$ for the existence of a local riemannian metric $\mathbf{g}$ with respect to which α is co-closed. Given a closed 1-form α on $\mathbf{X}$, for any point $\mathbf{p} \in \mathbf{X}$ and any simply connected neighborhood U of $\mathbf{p}$ we shall denote by $\int_{\mathbf{p}} \alpha|_U$ the uniquely defined scalar function f on U satisfying the equations

$$df = \alpha \qquad \text{in } U\,,$$
$$f(\mathbf{p}) = 0\,.$$

An equivalent formulation of the problem is that of characterizing locally the differentiable functions f that admit a local riemannian metric $\mathbf{g}$, with respect to which they satisfy the Laplace-Beltrami equation; in terms of local coordinates $\mathbf{x} = (x^1, \cdots, x^n)$, set $\mathbf{g} = g_{ij}(\mathbf{x})dx^i dx^j$, $(g_{ij}) = \mathfrak{g}$,

$$(1) \quad \Delta_{\mathbf{g}} f = -\sum_{i,j=1}^{n} (\det \mathfrak{g})^{-\frac{1}{2}} \frac{\partial}{\partial x^i}\Big((\det \mathfrak{g})^{\frac{1}{2}} g^{ij} \frac{\partial f}{\partial x^j}\Big) = 0\,;$$
$$(g^{ij}) = (g_{ij})^{-1} = \mathfrak{g}^{-1}\,.$$

In terms of the matrix valued function $G^{ij} = (\det \mathfrak{g})^{\frac{1}{2}}(g_{ij})^{-1}$, equation (1) is a first order, under-determined, linear homogeneous differential equation, whose coefficients are determined by the differential $df = \alpha$ and its first order jet (i.e., the second order jet of f, disregarding additive constants).

PROPOSITION 1. *Let α be a closed 1-form, not identically zero, in a differential manifold $\mathbf{X}$. In order for $\mathbf{X}$ to admit (locally or globally) a differentiable riemannian metric $\mathfrak{g}$ with respect to* *which α is co-closed, then α must satisfy at least the following conditions*

(i) *The zeros of α are all of finite order;*

(ii) *For any simply connected neighborhood U of any point $\mathbf{p} \in \mathbf{X}$ the integral $\int_{\mathbf{p}} \alpha|_U$ does not achieve a relative maximum or minimum value at any interior point of U.*

PROOF. Under the assumptions stated, the conditions on α are well-known properties of solutions of the Laplace-Beltami equation. Property (i) follows from the Carleman-Aronszajn estimates [1] leading to the unique continuation theorem for solutions of a strongly elliptic differential equation, while property (ii) restates the strong maximum principle of E. Hopf [2].

The conditions of Proposition 1 are not sufficient for the existence of a riemmannian metric $\mathbf{g}$ (example: on the manifold $\mathbf{R}$, the real number line, the differential x^2dx satisfies the two conditions everywhere, but cannot be co-closed with respect to any riemannian metric in any neighborhood of the point $x = 0$). Consider a local integral $f = \int_{\mathbf{p}} \alpha|_U$ of a closed 1-form: the zeros of α are synonymous with the critical points of f in the sense of Morse's theory; with this in mind we transfer to α the concepts attached in that theory to f.

Definition 1. Let α be a 1-form in an n-dimensional differential manifold $\mathbf{X}$. A zero point $\mathbf{p}$ of α in $\mathbf{X}$ is called *non-degenerate*, if the matrix of first order partial derivatives of α with respect to a local coordinate system $(x^1, \cdots, x^n)$ is non-singular at $\mathbf{p}$. If in addition α is closed, this matrix of partial derivatives is symmetric, and the non-degenerate quadratic form it defines at $\mathbf{p}$ (called the hessian form) on the tangent space is assigned an *index* $j = j_\alpha(\mathbf{p})$, defined to be the maximal dimension of a linear subspace of the tangent space at $\mathbf{p}$, in which the hessian form is negative definite.

We recall some of the classical, elementary properties of non-degenerate critical points of functions, as applied to α. The set of 1-forms (respectively, closed 1-forms or exact 1-forms) α all of whose zeros are non-degenerate is open and dense in Whitney's C^1-topology; non-degenerate zeros of 1-forms are always isolated; the index j of a non-degenerate zero a of closed 1-form $(0 \leqq j \leqq n)$ is invariant under transformation of the local coordinates in terms of which it is defined.

We shall specialize our attention now to 1-forms α on $\mathbf{X}$ all of whose zeros are non-degenerate: condition (i) of Proposition 1 is now trivially satisfied, since in that case each zero of α is of order exactly 1; condition (ii) means that the index j of each zero of α is restricted to the range $1 \leqq j \leqq n-1$, i.e., excluding the extreme values 0 and n. It it worth nothing here that a closed 1-form α on a *compact* manifold $\mathbf{X}$ satisfying condition (ii) can never be exact. The next proposition states that the two conditions of Proposition 1, applied to forms all of whose zeros are non-degenerate, are also sufficient for the existence of *local* metrics $\mathbf{g}$ with respect to which α is co-closed, i.e., for the existence of local solutions $\mathbf{g}$ of equation (1).

PROPOSITION 2. *Let* $\mathbf{X}$ *be an n-dimensional differential manifold and* α *a closed* 1-*form on* $\mathbf{X}$, *all of whose zeros are non-degenerate. Then there exists in some neighborhood* U *of any given point* $\mathbf{p} \in \mathbf{X}$ *a riemannian metric* $\mathbf{g}_{\mathbf{p}}$ *with respect to which* α *is co-closed, if and only if the index* j *of each zero of* α *is in the range* $1 \leqq j \leqq n-1$.

PROOF. The necessity of the condition on the index is obvious after Proposition 1; so we shall just prove the sufficiency.

Let $\mathbf{p} \in \mathbf{X}$. Suppose first that $\alpha(\mathbf{p}) \neq 0$; then there exists a simply connected neighborhood U_0 of $\mathbf{p}$ where $\alpha \neq 0$ everywhere and hence $f = \int_{\mathbf{p}} \alpha\,|_{U_0}$ has no critical points. By the implicit function theorem there exists a smaller neighborhood U of $\mathbf{p}$ with a local coordinate system $(x^1, \cdots, x^n)$ on U such that $f\,|_U = x^1$.

Put on U the local euclidean metric form

$$ds^2 = \sum_{i=1}^{n} (dx^i)^2\,; \tag{2}$$

then x^1 satisfies the Laplace equation, so that $\alpha\,|_U = dx^1$ is co-closed with respect to the metric (2).

Now suppose that $\alpha(\mathbf{p}) = 0$, and let j be the index of $\mathbf{p}$ as a zero point of α. Then from a lemma of Morse there exists a simply connected neighborhood U of $\mathbf{p}$ with a local coordinate system $(x^1, \cdots, x^n)$ with $x^i(\mathbf{p}) = 0$ such that

$$f = \int_{\mathbf{p}} \alpha\,|_U = -\frac{n-j}{2}\sum_{i=1}^{j} (x^i)^2 + \frac{j}{2}\sum_{i=j+1}^{n} (x^i)^2\,. \tag{3}$$

Again we introduce the euclidean metric (2) with respect to this coordinate system and observe that the function f of the coordinates satisfies Laplace's equation, so that $\alpha|_U = df$ is co-closed with respect to (2). We write $\alpha|_U$ in terms of these coordinates for future use as follows:

$$\begin{aligned} \alpha = df &= (n-j)\sum_{i=1}^{j} x^i dx^i - j\sum_{i=j+1}^{n} x^i dx^i \\ &= \Big(j\sum_{i=1}^{n} - n\sum_{i=1}^{j}\Big)x^i dx^i . \end{aligned} \tag{4}$$

This completes the proof of Proposition 2.

It is inevitable to guess at this point whether or not the local conditions of Proposition 2 are sufficient for the existence of a global riemannian structure on **X** with respect to which α is co-closed, particularly since the relevant equation can be linearized locally by using the matrix $G^{ij} = (\det(g_{ij}))^{\frac{1}{2}}(g^{ij})$, $(g^{ij}) = (g_{ij})^{-1}$, as the "unknown" in equation (1). It is clear that when the dimensionality n of **X** is 1, the problem has a trivial answer, since the conditions imply that α can have no zeros anywhere. In the higher dimensional cases the analogous conjecture becomes harder to prove, mainly because it is, in general, false, as one sees from the following example, in which **X** is a closed surface of genus 2.

III. A counter-example

Consider in euclidean 3-space $\mathbf{R}^3$ the two unit circles defined respectively by $\{x^2+y^2=1;\ z=1\}$, $\{x^2+y^2=1;\ z=-1\}$ in cartesian coordinates (x, y, z); join them by the arc of a circular helix, defined in terms of a parameter t by $\{x=\cos t,\ y=\sin t,\ z=t;\ -1\leq t\leq 1\}$. Visualize a tubular neighborhood of the resulting graph, with the boundary smoothed out at the two "Y-joints". The surface **X** obtained from the boundary is closed, of genus 2, and does not meet the z-axis $\{x=y=0\}$. We map the surface **X** onto the circle $\mathbf{R}/2\pi\mathbf{Z}$ by the map f, where f is defined in $\mathbf{R}^3$ outside the z-axis by $f(x, y, z) = \arg(x + y\sqrt{-1}) \pmod{2\pi}$ and let $\alpha = df$ induce on **X** the closed 1-form to be considered. Barring all unnecessary bumps on the surface, α has precisely two zeros, both non-degenerate, of index 1, one in each of the crotches of the Y-joints, and so satisfies the conditions of Proposition 2. On the other hand there can be no riemannian metric **g** on **X** with respect to which α is co-closed, for the reason given below. Suppose that **X** had a riemannian structure with respect to which α were co-closed, i.e., that the 1-form of odd kind $*\alpha$ were closed. Consider any integral curve of the exact differential equation

$\alpha = 0$, i.e., any component of $f^{-1}(\theta)$, $\theta \in \mathbf{R}/2\pi\mathbf{Z}$ these curves have a natural orientation on their normal bundle (positive side = transversal directions along which $\alpha > 0$), so that each component is a 1-cycle of odd kind; the restriction of $*\alpha$ to each component of such a 1-cycle of odd kind is strictly positive at each point, except at the zero points of α (where the curves have a singularity).

Let γ be the "middle" one of the three components of $f^{-1}(0)$, i.e., the 1-cycle of odd kind linked about the helix near the point (1, 0, 0); by the argument just given, $\int_\gamma *\alpha > 0$. On the other hand γ is homologous to 0, since it cuts the surface $\mathbf{X}$ into two components, so that by Stoke's theorem, if $*\alpha$ is to be closed $\int_\gamma *\alpha = 0$. This contradiction precludes the possibility that $*\alpha$ may be closed, no matter what metric is used in defining it. It is easy to produce similar examples also with surfaces of higher genus or manifolds of more dimensions.

The above example motivates the need for some global invariants attached to closed 1-forms α on a manifold, other than their de Rham cohomology class (periods of integrals).

IV. Transitive 1-forms

Definition 2. Let $\mathbf{X}$ be a differential manifold and α an integrable 1-form on $\mathbf{X}$. For any point $\mathbf{p} \in \mathbf{X}$, we define the *upland* (respectively, the *lowland*) in $\mathbf{X}$ relative to $\mathbf{p}$ with respect to α, denoted by $C_\alpha^+(\mathbf{p})$ (respectively, $C_\alpha^-(\mathbf{p})$) to be the set of points $\mathbf{q}$ for which there exists a non-trivial, differentiable, oriented path from $\mathbf{p}$ to $\mathbf{q}$, along which the form α is everywhere strictly positive (respectively, negative), except possibly at $\mathbf{p}$ itself. If, for every point $\mathbf{p} \in \mathbf{X}$ the set $C_\alpha^+(\mathbf{p})$ (respectively, $C_\alpha^-(\mathbf{p})$) is everywhere dense in $\mathbf{X}$, then α is called *transitive* on $\mathbf{X}$.

An easy result on the concepts just defined is the following.

PROPOSITION 3. *Given any differential manifold* $\mathbf{X}$ *with an integrable* 1-*form* α, *for each point* $\mathbf{p} \in \mathbf{X}$ *the upland and the lowland relative to* $\mathbf{p}$ *with respect to* α *are open subdomains of* $\mathbf{X}$; *their boundary sets consist of points where* α *vanishes and, otherwise, of non-singular, integral hypersurfaces of the total differential equation* $\alpha = 0$. *Furthermore, if all the zeros of* α *are non-degenerate, then* α *is transitive, if and only if there exists one point* $\mathbf{p} \in \mathbf{X}$ *such that both* $C_\alpha^+(\mathbf{p})$ *and* $C_\alpha^-(\mathbf{p})$ *are every-*

where dense in **X**, *in which case they coincide with the set of points where* $\alpha \neq 0$.

PROOF. It is clear that $C_\alpha^+(\mathbf{p})$ and $C_\alpha^-(\mathbf{p})$ can contain only points $\mathbf{q}$ such that $\alpha(\mathbf{q}) \neq 0$. In view of the proof of Proposition 2, for any such point $\mathbf{q}$ there exists a local coordinate system $\{U; x^1, \cdots, x^n\}$ in a neighborhood U of $\mathbf{q}$ such that $x^i(\mathbf{q}) = 0$ $(1 \leq i \leq n)$ and $\alpha|_U = e^f dx^1$, i.e., $x^1 = \int_{\mathbf{p}} e^{-f}\alpha|_U$. Without loss of generality we may assume that the range in $\mathbf{R}^n$ of such a coordinate system is an open, convex domain. It is then obvious that $C_\alpha^+(\mathbf{p}) \cap U$ (respectively, $C_\alpha^-(\mathbf{p}) \cap U$) is either empty or all of U or there exists a real number c such that the intersection is defined by the inequation $x^1 > c$ (respectively, $x^1 < c$). This shows that the uplands and lowlands are open domains and that all boundary points $\mathbf{q}$ such that $\alpha(\mathbf{q}) \neq 0$ are, in terms of the same local coordinate descriptions, defined locally by the equation $x^1 = c$, which represents locally the integral hypersurface of the differential equation $\alpha = 0$ near $\mathbf{q}$. There remains to show that, if for some $\mathbf{p} \in \mathbf{X}$ the sets $C_\alpha^+(\mathbf{p})$ and $C_\alpha^-(\mathbf{p})$ are both everywhere dense, and if all the zeros of α are non-degenerate, then α is transitive and both the upland and lowland relative to $\mathbf{p}$ coincide with the set $\{\mathbf{q} \mid \alpha(\mathbf{q}) \neq 0\}$.

Suppose that both $C_\alpha^+(\mathbf{p})$ and $C_\alpha^-(\mathbf{p})$ are everywhere dense; then for every $\mathbf{q}$ such that $\alpha(\mathbf{q}) \neq 0$, the whole coordinate neighborhood $\{U; x^1, \cdots, x^n\}$ of $\mathbf{q}$ with convex range and such that $\alpha|_U = e^f dx^1$ is contained in both $C_\alpha^+(\mathbf{p})$ and $C_\alpha^-(\mathbf{p})$, so that both domains coincide with the set where $\alpha \neq 0$. To show that α is transitive we have to consider an arbitrary point $\mathbf{q}$, and prove that $C_\alpha^+(\mathbf{q})$ and $C_\alpha^-(\mathbf{q})$ are again everywhere dense. First we see that, using the property that all zeros of α are non-degenerate, there exist oriented paths both from $\mathbf{p}$ to $\mathbf{q}$ and from $\mathbf{q}$ to $\mathbf{p}$, such that α along the path is everywhere strictly positive, except possibly at the end points, and that

(a) if α is non-zero at either $\mathbf{p}$ or $\mathbf{q}$, then the positivity condition on α can be retained at that end point as well;

(b) any path from $\mathbf{q}$ to $\mathbf{p}$ along which α is everywhere positive (respectively, negative) followed by another similar path from $\mathbf{p}$ to some given point $\mathbf{q}'$ can be smoothed out near $\mathbf{p}$ to give a positive (respectively, negative) path from $\mathbf{q}$ to $\mathbf{q}'$;

(c) if $\alpha(\mathbf{p}) \neq 0$, the smoothing process can be achieved without pulling the path away from $\mathbf{p}$ (a fact that is relevant to the corollary that follows).

These facts show that $C_\alpha^+(\mathbf{q}) \supset C_\alpha^+(\mathbf{p})$ and $C_\alpha^-(\mathbf{q}) \supset C_\alpha^-(\mathbf{p})$, completing the proof of the proposition.

COROLLARY. *Let* $\mathbf{X}$ *be a differential manifold and* α *transitive* 1*-form on* $\mathbf{X}$. *Then for each point* $\mathbf{p} \in \mathbf{X}$ *such that* $\alpha(\mathbf{p}) \neq 0$, *there passes a simple, oriented, closed, differentiable path* γ *along which* α *is strictly positive everywhere.*

PROOF. Since $C_\alpha^+(\mathbf{p})$ and $C_\alpha^-(\mathbf{p})$ are both open and everywhere dense, so is their intersection; let $\mathbf{q}$ be a point in the intersection; then, since α vanishes neither at $\mathbf{p}$ nor at $\mathbf{q}$, there exists oriented paths γ_1 from $\mathbf{p}$ to $\mathbf{q}$ and γ_2 from $\mathbf{q}$ to $\mathbf{p}$, along which α is strictly positive everywhere, including both end points. By the smoothing process mentioned at the end of the proof of Proposition 3, the combined closed path can be smoothed at both $\mathbf{p}$ and $\mathbf{q}$ to a differentiable path passing, in particular, still through $\mathbf{p}$. Finally, to obtain a simple path from this, in case it is not, let $\mathbf{q}'$ be the first self-intersection of the oriented, closed path after starting at $\mathbf{p}$, corresponding to the value t_1 of a global parameter, and let $t_2 > t_1$ be the last value of the parameter before the path becomes closed that corresponds again to $\mathbf{p}$. Skipping the segment from t_1 to t_2 we obtain again a closed path with a possible discontinuity of the tangent at $\mathbf{q}'$; this path can be once more smoothed out at $\mathbf{q}'$ locally without introducing new double points. The elimination of all double points requires at most finitely many repetitions of this process, yielding at the end a closed, oriented, simple path through $\mathbf{p}$ along which $\alpha > 0$. This completes the proof of the Corollary.

V. The main theorem

The results of the preceding sections lead up to the formulation of the main theorem.

THEOREM 1. *Let* $\mathbf{X}$ *be a compact, connected, n-dimensional, differential manifold and let* α *be a closed* 1*-form on* $\mathbf{X}$ *all of whose zeros are non-degenerate. Then a necessary and sufficient condition in order that* $\mathbf{X}$ *admit a riemannian metric* $\mathbf{g}$ *with respect to which* α *is co-closed is that* α *be transitive.*

Remark. The condition of transitivity on α implies immediately the local conditions of Proposition 2; in fact if $\mathbf{p}$ is a zero of α of index 0 (respectively n) then the set $C_\alpha^-(\mathbf{p})$ (respectively $C_\alpha^+(\mathbf{p})$) is clearly empty.

PROOF. We prove first that, if α is a harmonic 1-form, then it is transitive. Suppose, otherwise, that for a certain point $\mathbf{p} \in \mathbf{X}$ the set $C_\alpha^+(\mathbf{p})$ is not everywhere dense. Since the zeros of α are isolated, the boundary of $C_\alpha^+(\mathbf{p})$ consists, except for an at most finite set of points, of integral manifolds of the differential equation $\alpha = 0$. This set, with the transversal orientation assigning the positive "side" to the interior $C_\alpha^+(\mathbf{p})$, defines an $(n-1)$-cycle of odd kind γ, that is homologically trivial. Since we suppose that $*\alpha$ is closed, $\int_\gamma *\alpha = 0$; on the other hand the restriction of $*\alpha$ to the part of γ where $\alpha \neq 0$ is everywhere strictly positive and otherwise zero. Hence the boundary of $C_\alpha^+(\mathbf{p})$ is contained in the zero-set of α, which is at most a finite set. This means that $C_\alpha^+(\mathbf{p})$ is either everywhere dense (contrary to what we supposed to begin with) or empty; this last possibility is ruled out, since this can not be for any $\mathbf{p}$, since α has only non-degenerate zeros and under this condition it would mean that $\mathbf{p}$ is a zero of index n, contradicting E.Hopf's maximum principle.

We now assume that α is transitive and try to construct a riemannian metric $\mathbf{g}$ with respect to which α is co-closed. This can be accomplished after a few lemmas that we shall now formulate.

LEMMA 1. *Let* $\mathbf{X}$ *be a paracompact, n-dimensional manifold; let* α *be a* 1-*form on* $\mathbf{X}$, *all of whose zeros are non-degenerate, and let* ψ *be an* $(n-1)$-*form on* X *of odd kind satisfying the following conditions.*

(a) *At each point* $\mathbf{p} \in \mathbf{X}$ *where* $\alpha(\mathbf{p}) \neq 0$ *the n-form of odd kind* $\alpha \wedge \psi$ *is strictly positive;*

(b) *For each point* $\mathbf{p} \in \mathbf{X}$ *where* α *vanishes there exists in a neighborhood* $U_\mathbf{p}$ *of* $\mathbf{p}$ *a riemannian metric* $\mathbf{g}_\mathbf{p}$, *with respect to which* $\psi|_{U_\mathbf{p}} = *_{(\mathbf{g}_\mathbf{p})}\alpha|_{U_\mathbf{p}}$, *the neighborhoods* $U_\mathbf{p}$ *being pairwise disjunct.*

Then there exists on $\mathbf{X}$ *a riemannian metric* $\mathbf{g}$, *with respect to which* $*\alpha = \psi$. *Furthermore one can construct* $\mathbf{g}$ *so as to coincide with* $\mathbf{g}_\mathbf{p}$ *in any neighborhood* $V_\mathbf{p}$ *of a point* $\mathbf{p}$ *where* α *vanishes, where* $\bar{V}_\mathbf{p} \subset U_\mathbf{p}$.

Remark. One could replace condition (b) by a weaker one

involving the first order jets of α and ψ respectively at each point where α vanishes (ψ being also there, of necessity, zero); the metric $\mathbf{g}$ constructed in this case has a prescribed zero order jet. The details of this, however, are extraneous to the main result of this paper, and are omitted for the sake of brevity.

PROOF. Let A^j and A^j_- denote the vector bundles over $\mathbf{X}$ of j-forms and of j-forms of odd kind respectively, and consider the differentiable vector bundle Hom (A^1, A^{n-1}_-). A vector sub-bundle of this bundle, denoted by $\mathrm{Hom}_S\,(A^1, A^{n-1}_-)$ will be called that of symmetric homomorphisms, defined as follows. An element $\tau_{\mathbf{p}} \in \mathrm{Hom}\,(A^1, A^{n-1}_-)$ over $\mathbf{p} \in \mathbf{X}$ is called symmetric, if and only if for any two 1-forms $\beta_{\mathbf{p}}$, $\gamma_{\mathbf{p}}$ at $\mathbf{p}$ we have

$$\beta_{\mathbf{p}} \wedge \tau_{\mathbf{p}}(\gamma_{\mathbf{p}}) = \gamma_{\mathbf{p}} \wedge \tau_{\mathbf{p}}(\beta_{\mathbf{p}}) \,.$$

Furthermore a symmetric homomorphism $\tau_{\mathbf{p}}$ at $\mathbf{p} \in \mathbf{X}$ is called positive definite (respectivery semi-definite), if for every $\beta_{\mathbf{p}} \neq 0$

$$\beta_{\mathbf{p}} \wedge \tau_{\mathbf{p}}(\beta_{\mathbf{p}}) > 0 \qquad \text{(resp. } \geqq 0\text{)} \,,$$

where the inequality in the bundle of n-forms of odd kind has the obvious meaning. The bundle of positive definite, symmetric homomorphisms of A^1 into A^{n-1}_- is a differentiable, open sub-bundle of $\mathrm{Hom}_S\,(A^1, A^{n-1})$ whose fibres are convex: we denote this bundle by T. Since the fibres are contractible, given any differentiable cross-section τ of T over an open set $U \subset \mathbf{X}$, for any closed set $K \subset U$ there exists a global cross section of T over $\mathbf{X}$ that coincides with τ in K. A typical cross section of T is defined in terms of any riemannian structure $\mathbf{g}$ in $\mathbf{X}$ by the homomorphism $*_{(\mathbf{g})}$; indeed the map $\{\mathbf{g} \longrightarrow *_{\mathbf{g}}\}$ is a bundle homomorphism of the bundle of riemannian metrics into the bundle T. The interesting thing is that, for $n \neq 2$ this homomorphism is bijective. In fact, let $\omega^1_{\mathbf{p}}, \cdots, \omega^n_{\mathbf{p}}$ be a basis for A^1 at any point $\mathbf{p} \in \mathbf{X}$. In terms of this basis the components of a riemannian form at $\mathbf{p}_n$ are a symmetric, positive definite matrix $(g_{ij}) = \mathfrak{g}$, where if $\beta = \sum_{i=1}^n b_i \omega^i$ is a 1-form, we have

$$|\beta|^2 = \textstyle\sum_{ij} g^{ij} b_i b_j \qquad ((g^{ij}) = (g_{ij})^{-1}) \,,$$

Consequently in terms of the riemannian form $\mathbf{g}$, we have

$$(5) \quad *\beta = (\det(\mathfrak{g}))^{1/2} \textstyle\sum_{i,k=1}^n (-1)^{k-1} g^{ik} b_i |\omega^1 \wedge \cdots \vee^{(k)} \cdots \wedge \omega^n| \operatorname{sgn} \omega^k .$$

On the other hand, the most general homomorphism τ of A^1 into A^{n-1}_- is represented by a matrix (G^{ij}) in the formula

$$(6) \quad \tau(\beta) = \textstyle\sum_{i,k=1}^n (-1)^{k-1} G^{ik} b_i |\omega^1 \wedge \cdots \vee^{(k)} \cdots \wedge \omega^n| \operatorname{sgn} \omega^k$$

where τ is symmetric (respectively, positive definite or positive semi-definite) according to the applicability of the same properties to the matrix $(G^{ik}) = \mathfrak{G}$. From formulas (5) and (6) we see that, given a riemannian metric $\mathbf{g}$ represented by a matrix $\mathfrak{g}$, the matrix $\mathfrak{G}$ representing $*_{\mathbf{g}}$ is given by

$$\mathfrak{G} = (\det(\mathfrak{g}))^{1/2}\,\mathfrak{g}^{-1},$$

whence

$$\det(\mathfrak{G}) = (\det(\mathfrak{g}))^{n-2/2}.$$

Thus one sees that, for $n = 2$, $\det(\mathfrak{G}) = 1$, showing that the homomorphism is not surjective on T, while for $n \neq 2$ the map $\{\mathbf{g} \to \tau = *_{\mathbf{g}}\}$ is bijective, its inverse being given by

$$\mathfrak{g} = (\det((\mathfrak{G}))^{1/n-2}\,\mathfrak{G}^{-1}.$$

Let (W_ν) be a locally finite, open cover of X with the following properties: for each ν either α vanishes in precisely one point $p \in W_\nu$ and $\bar{V}_{\mathrm{p}} \subset W_\nu \subset U_{\mathrm{p}}$, or else α vanishes nowhere in W_ν; W_ν is contractible and $\bar{W}_\nu$ is disjunct from the closed neighborhoods $\bar{V}_{\mathrm{p}}$ of every zero point p of α prescribed in the statement; let (φ_ν) be a differentiable partition of unity subordinate to the locally finite cover (W_ν). If $W_\nu \supset V_{\mathrm{p}}$ for some zero point of α, we restrict the given riemannian metric structure $\mathbf{g}_{\mathrm{p}}$ from U_ν and define a positive semi-definite, differentiable cross section τ_ν over X in $\mathrm{Hom}_S(A^1, A_-^{n-1})$ by

$$\tau_\nu = \begin{cases} = \varphi_\nu *_{(\mathbf{g}_{\mathrm{p}})} & \text{in } W_\nu \\ = 0 & \text{in } X - W_\nu . \end{cases}$$

If W_ν is disjunct from all the neighborhoods V_{p}, since W_ν is contractible, there exists a differentiable frame $(\omega_\nu^1, \cdots, \omega_\nu^n)$ for A^1 over W_ν satisfying the following.

$$\omega_\nu^1 = \alpha,$$
$$\omega_\nu^i \wedge \psi = 0 \qquad (i = 2, \cdots, n),$$

and a "dual" basis $(\eta_\nu^1, \cdots, \eta_\nu^n)$, uniquely defined by

$$\eta_\nu^1 = \psi,$$
$$\omega_\nu^i \wedge \eta_\nu^k = \begin{cases} \alpha \wedge \psi > 0 & (i = k), \\ 0 & (i \neq k). \end{cases}$$

We then have again a positive semi-definite cross section τ_ν in

$\mathrm{Hom}_S(A^1, A_-^{n-1})$ defined by

$$\tau_\nu(\omega_\nu^i) = \begin{cases} = \varphi_\nu \eta_\nu^i & \text{in } W_\nu \\ = 0 & \text{outside}\,. \end{cases}$$

Finally the sum

$$\tau = \textstyle\sum_\nu \tau_\nu$$

is a differentiable, differentiable cross section over $\mathbf{X}$ in the bundle T of positive definite homomorphisms in $H_S(A^1, A_-^{n-1})$. By the previous argument, if $n \neq 2$, τ determines a riemannian structure $\mathbf{g}$ on $\mathbf{X}$ such that $*_{\mathbf{p}}\alpha = \psi$ and $\mathbf{g} = \mathbf{g}_{\mathbf{p}}$ in each neighborhood $V_{\mathbf{p}}$ of any zero point $\mathbf{p}$ of α.

In the remaining case $n = 2$ the proof is somewhat simpler. We cover $\mathbf{X}$ by the two open sets U_0, U_1, where

$$U_0 = \bigcup_{\alpha(\mathbf{p})=0} U_p\,, \qquad U_1 = X - \bigcup_{\alpha(\mathbf{p})=0} \bar{V}_p$$

and let $\{\varphi_0, \varphi_1\}$ be a differentiable partition of unity subordinate to $\{U_0, U_1\}$. In U_0 we have the riemannian structure $\mathbf{g}_0$ obtained from the given metrics $\mathbf{g}_{\mathbf{p}}$ each defined in the pairwise disjunct $U_{\mathbf{p}}$. In U_1 we define the riemannian metric g_1 defined by

$$\mathbf{g}_1 = \alpha \otimes \alpha + \psi \otimes \psi\,,$$

where $\mathbf{g}_1$ is interpreted as a cross section in the tensor product bundle $A^1 \otimes A^1 = A_-^1 \otimes A_-^1$. The riemannian metric

$$\mathbf{g} = \varphi_0 \mathbf{g}_0 = \varphi_1 g_1$$

can be readily verified to satisfy all the required properties stated by the lemma.

We return now to the proof of the main theorem. The preceding lemma reduces the problem, given α with the stated assumptions, to find a closed $(n-1)$-form of odd kind ψ in $\mathbf{X}$ such that, wherever $\alpha \neq 0$, $\alpha \wedge \psi > 0$ and at each zero point $\mathbf{p}$ of α there exists a local riemannian metric $\mathbf{g}_{\mathbf{p}}$ with respect to which $\psi = *_{(\mathbf{g}_{\mathbf{p}})}\alpha$. The latter property is satisfied as a consequence of Proposition 2, since we have already noted that there can be no zeros of α of index 0 or n. For each zero point $\mathbf{p}$ of α we choose a fixed coordinate neighborhood $\{U_{\mathbf{p}}; x^1, \cdots, x^n\}$ about $\mathbf{p}$ with $x^i(\mathbf{p}) = 0$, with respect to which α is locally represented by (4) (cf. Proposition 2); we assume furthermore that the neighborhoods $U_{\mathbf{p}}$ of the different zero points are pairwise disjunct. Let $\mathbf{p}_\nu(\nu = 1, 2, \cdots, N)$ be an enumeration of the zero points of α and let δ_ν be positive numbers such that the range of the coordinate domain $U_{\mathfrak{p}_\nu}$ includes the ball defined by

$$\sum_{i=1}^{n} (x^i)^2 \leqq \delta_\nu^2$$

and let V_t, for any real t in the interval $0 < t \leqq 1$, denote the union of the disjunct neighborhoods of the different $\mathfrak{p}_\nu$ defined in the respective coordinate systems by

$$\sum_{i=1}^{n} (x^i)^2 < t^2\delta_\nu^2 \, . \tag{7}$$

We next prove a lemma that is a sharper version of the corollary of Proposition 3, § 4.

LEMMA 2 (Deligne). *Let* **X** *be an n-dimensional differential manifold, α a closed 1-form that is transitive. For any point* $\mathfrak{p} \in \mathbf{X}$ *such that* $\alpha(\mathfrak{p}) \neq 0$ *there exists a closed* $(n-1)$*-form of odd kind* ψ *with compact support, such that* $\alpha \wedge \psi \geqq 0$ *everywhere in* **X** *and* $(\alpha \wedge \psi)(\mathfrak{p}) > 0$.

PROOF. Let $\mathfrak{p} \in \mathbf{X}$ be a point where α does not vanish. According to the corollary of Proposition 3 there exists a simple closed, differentiable, oriented curve in **X** passing through $\mathfrak{p}$, along which α is strictly positive everywhere. Let γ be such a curve: we construct a tubular neighborhood W of γ, such that the local integral manifolds defined by the equation $\alpha = 0$ (that are transversal to γ at each point) are everywhere transversal to the boundary of W and intersect W in an open $(n-1)$-ball meeting γ in exactly one point. Then W is a differentiable fibre bundle over γ, the fibres being the integral manifolds of α. By virtue of a theorem of Mazur the structure group of this bundle is reducible to the identity or the cyclic group of order 2 generated by a reflection along a hypersurface, according as to whether or not W is orientable. In other words the open submanifold W of **X** with the 1-form α in it is either isomorphic to the product $S^1 \times E^{n-1}$, or to the n-dimensional Möbius band, with fibre E^{n-1} and base S^1, where $S^1 = \mathbf{R}/c\mathbf{Z}$ for some positive constant $c = \int_\gamma \alpha$ and E^{n-1} is the interior of the unit euclidean ball in $\mathbf{R}^{n-1}$. Let π denote the projection of W onto S^1, with fibres along the integral manifolds of the equation $\alpha = 0$, let S^1 be parametrized by $x^1 \bmod c\mathbf{Z}$ so that $\pi^* dx^1 = \alpha$ and E^{n-1} be locally parametrized by $(x^2, \cdots, x^n) \in \mathbf{R}^{n-1}$ satisfying $\sum_{i=2}^{n} (x^i)^2 < 1$ and the group action, if required, represented by the involution $(x^2, x^3, \cdots, x^n) \to (-x^2, x^3, \cdots, x^n)$, and suppose that γ corresponds to $\{x^2 = x^3 = \cdots = x^n = 0\}$ and in γ the point $\mathbf{p}$ corresponds to $x^1 = 0 \bmod c\mathbf{Z}$.

Let $h(t)$ be an even, non-negative valued, differentiable function on $\mathbf{R}$ which is identically equal to 1 for $|t| \leqq \frac{1}{3}$ and identically zero for $|t| \geqq \frac{2}{3}$. Then the $(n-1)$-form $\psi_{\mathbf{p}}$ of odd kind can be represented by the equations

$$\begin{cases} \psi_p = h(\sum_{i=2}^n (x^i)^2)\,|dx^2 \wedge \cdots \wedge dx^n|\,(\operatorname{sgn} dx^1) & \text{inside } W; \\ \psi_{\mathbf{p}} = 0 & \text{in } X - W\,. \end{cases}$$

Clearly $\psi_{\mathbf{p}}$ is closed and, since

$$\alpha \wedge \psi_{\mathbf{p}} = h\left(\textstyle\sum_{i=2}^n (x^i)^2\right)|dx^1 \wedge \cdots \wedge dx^n| \geqq 0$$

everywhere in W and vanishes outside, it satisfies the properties required and completes thereby the proof of the lemma.

It is to be noted that the existence of such a $\psi_{\mathbf{p}}$ for every $\mathbf{p}$ with $\alpha(\mathbf{p}) \neq 0$ is equivalent to α being transitive, once one knows that the zeros of α are reasonable, i.e., for instance all non-degenerate. It would make matters easier in the sequel if we could produce an analogous $\psi_{\mathbf{p}}$, *mutatis mutandis* when $\alpha(\mathbf{p}) = 0$, if $\mathbf{p}$ is a non-degenerate zero of a transitive α; however this is not the case up to now.

We complete the proof of the theorem as follows. Consider the open neighborhood V_1 defined by (7), for $t = 1$, of the set of all zeros of α. To each $\mathbf{p} \in \mathbf{X} - V_1$ we assign a closed $(n-1)$-form of odd kind $\psi_{\mathbf{p}}$ with the properties asserted in Lemma 2; such a form satisfies $\alpha \wedge \psi_{\mathbf{p}} > 0$ in a neighborhood of $\mathbf{p}$ but $\psi_{\mathbf{p}} = 0$ identically in a neighborhood V_t of all the zeros of α for some positive $t \leqq 1$, depending on $\mathbf{p}$. Let $U_{\mathbf{p}}$ denote the open set of points where $\psi_{\mathbf{p}} \neq 0$, whence also by the construction $\alpha \wedge \psi_{\mathbf{p}} > 0$. Since $\mathbf{X} - V_1$ is compact, there exists a finite set of points $\mathbf{q}_1, \cdots, \mathbf{q}_M \in \mathbf{X} - V_1$ such that $\bigcup_{\mu=1}^M U_{\mathbf{p}_\mu} \supset \mathbf{X} - V_1$; in this case there is a positive $\delta_0 < 1$ such that $\bigcup_{\mu=1}^M U_{q_\mu} \subset \mathbf{X} - V_{\delta_0}$. Thus the form $\psi' = \sum_{\mu=1}^M \psi_{\mathbf{q}_\mu}$ has the following properties:

(i) ψ' is closed and vanishes identically in V_{δ_0},

(ii) $\alpha \wedge \psi' > 0$ at all points where $\psi' \neq 0$,

and is bounded away from zero in $\mathbf{X} - V_{\delta'}$ for some δ' in the interval $\delta_0 < \delta' < 1$.

Now in the neighborhood of each zero point $\mathbf{p}_\nu$ of α $(\nu = 1, \cdots, N)$ we carry out the following construction. Let $V_{\delta,\nu}$ be the component of V_δ containing $\mathbf{p}_\nu$ and refer to the same local coordinate $x^1, \cdots, x^n$ in $V_{1,\nu}$ used in (7), satisfying the relations (3) and (4) with respect to α, where j is the index of $\mathbf{p}_\nu$. Let η_ν be a non-negative, differentiable function on $\mathbf{X}$ which is identically 1 in $V_{\delta',\nu}$ and vanishing identically outside $V_{1,\nu}$; consider the local euclidean metric $\mathbf{g}_\nu$ defined by (2) in $V_{1,\nu}$ and with respect to it we form the $(n-2)$-form of odd kind

$$\begin{cases} \chi_\nu = \eta_\nu \sum_{j+1\leqq i\leqq n}^{1\leqq h\leqq j,} (-1)^{h+i} x^h x^i \\ \qquad \times \, dx^1 \wedge \cdots \vee^h \cdots \wedge dx^j \wedge dx^{j+1} \wedge \cdots \vee^i \cdots \wedge dx^n \\ \qquad\qquad \times \operatorname{sgn}(dx^h \wedge dx^i) & \text{in } V_{1,\nu}\,, \\ \chi_\nu = 0 & \text{outside } V_{1,\nu}\,. \end{cases}$$

A direct computation shows that in $V_{\delta',\nu}$ (i.e., in a set where $\eta = 1$ identically, $d\chi_\nu = *_{(\mathbf{g}_\nu)}\alpha$. Hence we can let $\psi_\nu = d\chi_\nu$ globally, since ψ_ν vanishes identically outside $V_{1,\nu}$; let

$$\psi'' = \sum_{\nu=1}^{N} \psi_\nu \,.$$

Then the $(n-1)$-form ψ'' of odd kind has the following properties:

(i) ψ'' is exact;

(ii) $\alpha \wedge \psi'' \geqq 0$ in $V_{\delta'} = \bigcup_{\nu=1}^{N} V_{\delta',\nu}$;

(iii) ψ'' vanishes outside V_1;

(iv) there exists a local riemannian structure $\mathbf{g}''$ (the union of the local euclidean metrics $\mathbf{g}_\nu$) in $V_{\delta'}$ with respect to which $\psi'' = *_{\mathbf{g}''}\alpha$.

Combining the above properties of ψ'' with those previously established for ψ', we see that for a sufficiently large, positive constant K, the form

$$\psi = K\psi' + \psi''\,,$$

which coincides with ψ'' in V_{δ_0}, satisfies $\alpha \wedge \psi > 0$ everywhere else. In other words ψ satisfies the conditions stated in Lemma 1 for the existence of a metric $\mathbf{g}$ on $\mathbf{X}$ for which $\psi = *_{\mathbf{g}}\alpha$. Since ψ is closed, α is co-closed with respect to this metric. Theorem 1 is thereby proved.

It was mentioned earlier that, given a closed 1-form α all of whose zeros are non-degenerate, in order that α be transitive it is necessary (but not sufficient) that each zero of α have index different from 0 and n. One can show by a method of continuity that transitivity, in a compact manifold $\mathbf{X}$, can be deduced from the additional condition that each zero of α have index different from 1 as well (or alternately different from $n-1$). In particular, every closed 1-form with no zeros whatever is transitive.

VI. An example

We show that one can construct some rather pathological-looking, closed 1-forms that are transitive. In particular, there exist in a 3-dimensional (or higher) torus a riemannian metric with respect to which some harmonic 1-forms can have isolated zeros. We shall construct in euclidean 3-space a smooth function $f(x, y, z)$ with the following properties:

(i) for $x^2 + y^2 + z^2 > M^2$, M sufficiently large, $f(x, y, z) = x$ identically;

(ii) $f(x, y, z)$ is an even function of both y and z;

(iii), $f(x, y, z)$ has exactly two critical points, $(x, y, z) = (\pm c, 0, 0)$, both non-degenerate and of index respectively 1 and 2.

An example of such a function is the following: let $h(t)$ be an even function of one real variable, of class C^∞, identically equal to 1 for $|t| \leq 1$, identically zero, for $|t| \geq M$, non-negative and with derivative less than $\frac{2}{M}$ in absolute value throughout. Then, for M sufficiently large, letting $r = (x^2 + y^2 + z^2)^{1/2}$,

$$f(x, y, z) = \frac{x(x^2 + y^2 + z^2 - 1) + (y^2 - z^2)h(r)}{h(r) + (x^2 + y^2 + z^2 - 1)(1 - h(r))}$$

the resulting function has the desired properties, with the constant $c = 3^{-1/2}$. Let $\mathbf{X}$ be an n-dimensional compact riemannian manifold ($n \geq 3$) and α a non-trivial harmonic 1-form on $\mathbf{X}$. Choose a point $\mathbf{p} \in \mathbf{X}$ such that $\alpha(\mathbf{p}) \neq 0$ and let $\{U, x^1, \cdots, x^n\}$ be a local coordinate system in a neighborhood U of $\mathbf{p}$ with $x^i(\mathbf{p}) = 0$, such that $\alpha|_U = dx^1$, and choose two positive integers j, k such that $j + k = n - 1$; we denote by x' the subset of coordinates $(x^2, \cdots, x^{j+1})$ and by x'' the coordinates $(x^{j+2}, \cdots, x^n)$ and by y, z respectively the sums $\left(\sum_{i=2}^{j+1} (x^i)^2\right)^{\frac{1}{2}}$ and $\left(\sum_{i=j+2}^{n} (x^i)^2\right)^{\frac{1}{2}}$. If the local coordinate system includes in its range the ball

$$\sum_{i=1}^{n} (x^i)^2 = (x^1)^2 + y^2 + z^2 < A\,,$$

we can perturb α by replacing it inside U by

$$\frac{A}{M} df\left(\frac{Mx^1}{A}, \frac{My}{A}, \frac{Mz}{A}\right),$$

which coincides with α near the boundary of U, and hence can be continued outside by the given α. The new differential has the same zeros, of course, as α outside U, while inside U, where α had no zeros, the new closed 1-form has precisely two new zeros, both non-degenerate, corresponding to the two critical points of $f(x, y, z)$; their indices are respectively k and $k+1$. A careful check of the transversal trajectories of the new form inside U with respect to the original metric as compared with those of α preserves the transitivity property. Therefore it is possible to increase the number of zeros of a closed form by two at a time, and still preserve the existence of a riemannian metric with respect to which the form is harmonic. In particular any n-torus ($n \geqq 3$) admits riemannian metric with respect to which harmonic 1-forms may have any given, even number of isolated zeros.

UNIVERSITY OF PENNSYLVANIA AND I.H.E.S., Bures-sur-Yvette

REFERENCES

[1] N. ARONSZAJN, *A unique continuation theorem for solutions of elliptic partial differential equations or any qualities of second order*, J. Math. Pures Appl. (9) **36** (1957), 235 249.

[2] E. HOPF, *Elementare Bemerkungen über die Lösungen partieller Differentialgleichungen zweiter Ordnung vom elliptischen Typus*, Sitzungsber. der Preuss. Akad. der Wissenschaften, **19**, (1927), 147–152.

(Received October 15, 1968)

On the Group of Automorphisms of a Symplectic Manifold

Eugenio Calabi[1]

Problems in Analysis,
Lectures at the Sympos. in honor of Salomon Bochner,
held at Princeton University, Princeton, NJ, USA, 1969, 1–26,
Princeton Univ. Press, Princeton, NJ (1970)

1. Introduction

Let $\mathbf{X}$ be a connected, differential manifold of $2n$ dimensions. A *symplectic structure* on $\mathbf{X}$ is the geometrical structure induced by a differentiable exterior 2-form ω defined on $\mathbf{X}$, satisfying the following conditions:

(i) The form ω is closed: $d\omega = 0$;
(ii) It is everywhere of maximal rank; this means that the $2n$-form ω^n (nth exterior power of ω) is everywhere different from zero, or equivalently, the skew-symmetric $(2n) \times (2n)$ matrix of coefficients of ω, in terms of a basis for the cotangent space, is everywhere nonsingular.

A classical theorem, ordinarily attributed to Darboux, states that a $2n$-dimensional symplectic manifold (i.e., a manifold with a symplectic structure) can be covered by a local, differentiable coordinate system $\{U; (x)\}$ where $(x) = (x^1, \ldots, x^{2n})\colon U \to \mathbf{R}^{2n}$, in terms of which the local representation of the structural form ω becomes

$$(1.1)\qquad \begin{aligned} \omega|_U &= dx^1 \wedge dx^2 + dx^3 \wedge dx^4 + \cdots + dx^{2n-1} \wedge dx^{2n} \\ &= \sum_{j=1}^{n} dx^{2j-1} \wedge dx^{2j}; \end{aligned}$$

such a system of coordinates is called a *canonical* system.

The purpose of this study is to describe the group $\mathbf{G}$ of automorphisms of a symplectic manifold, i.e., the group of all differentiable automorphisms of $\mathbf{X}$ which leave the structural 2-form ω invariant, and the invariant subgroups of $\mathbf{G}$. The group $\mathbf{G}$ can also be characterized as mapping canonical coordinate systems into canonical systems.

[1] The research reported here was supported in part by the National Science Foundation.

DOI: 10.2307/j.ctt130hk8g.4

Two normal subgroups of $\mathbf{G}$ are distinguished immediately as follows:

DEFINITION 1.1. Let $(\mathbf{X}, \omega)$ be a $2n$-dimensional symplectic manifold and let $\mathbf{G}$ be the group of all symplectic transformations of $\mathbf{X}$. If $\mathbf{X}$ is not compact, we denote by $\mathbf{G}_0$ the subgroup of $\mathbf{G}$ consisting of all symplectic transformations of $\mathbf{X}$ that have compact support; that is to say, a symplectic transformation $\mathbf{g} \in \mathbf{G}$ belongs to $\mathbf{G}_0$ if and only if $\mathbf{g}$ equals the identity outside a compact region of $\mathbf{X}$.

DEFINITION 1.2. Let $(\mathbf{X}, \omega)$ be a $2n$-dimensional symplectic manifold and let $\mathbf{G}$ be the group of all symplectic transformations of $\mathbf{X}$. We denote by $\mathbf{G}_{0,0}$ the subgroup of $\mathbf{G}$ called the *minigroup* generated by the so-called locally supported transformations, defined as follows: a transformation $\mathbf{g} \in \mathbf{G}$ is called locally supported if there exists a canonical coordinate system $\{U; (x)\}$ defined in a contractible domain U with compact closure, such that the support of $\mathbf{g}$ lies in U.

The minigroup $\mathbf{G}_{0,0}$ and its corresponding Lie algebras are introduced here merely for expository convenience. In Section 3 it will be shown that the commutator subgroup of the arc-component of the identity in $\mathbf{G}_0$ coincides either with $\mathbf{G}_{0,0}$ or with a normal subgroup of codimension 1 in $\mathbf{G}_{0,0}$ (see Theorem 3.7, Section 3).

An elementary example of a locally supported transformation is the following: let $\{U; (x)\}$ be a canonical coordinate system in $\mathbf{X}$; let its range $V = (x)(U) \subset \mathbf{R}^{2n}$ contain the ball $\left\{(t) \middle| \sum_{i=1}^{2n} (t^i)^2 < A\right\}$ for some $A > 0$; choose a real-valued differentiable function $\vartheta(r)$ of a real variable $r \geqq 0$ with support contained in a closed segment $[0, A']$ with $A' < A$. Then it is easily verifiable that the transformation $\mathbf{f}$ in $\mathbf{R}^{2n}$ (with the 2-form $\omega = \sum_{j=1}^{n} dt^{2j-1} \wedge dt^{2j}$),

$$\begin{aligned} (t) \to (t') &= \mathbf{f}(t) \text{ with} \\ t'^{2j-1} &= t^{2j-1} \cos \vartheta(r) - t^{2j} \sin \vartheta(r), \\ t'^{2j} &= t^{2j-1} \sin \vartheta(r) + t^{2j} \cos \vartheta(r), \\ &\left(r = \sum_{i=1}^{2n} (t^i)^2;\ 1 \leqq j \leqq n\right), \end{aligned}$$

is a symplectic transformation which equals the identity for $r \geqq A'$. Therefore its restriction to V defines via the coordinate map (x) a symplectic transformation in U that can be trivially extended by the identity map in $\mathbf{X} - U$ to a locally supported symplectic transformation in $\mathbf{X}$.

We shall state here the main results of this study in a preliminary form; more precise and stronger versions of these are repeated as theorems in the later sections.

STATEMENT 1. The groups $\mathbf{G}$, $\mathbf{G}_0$, and $\mathbf{G}_{0,0}$ are infinite dimensional Lie groups in terms of the Whitney C^∞ topology (in the case of $\mathbf{G}$) and the compactly supported C^∞ topology (in the case of $\mathbf{G}_0$ and $\mathbf{G}_{0,0}$).

STATEMENT 2. The minigroup $\mathbf{G}_{0,0}$ is a closed, normal subgroup of $\mathbf{G}_0$; the quotient group $\mathbf{G}_0/\mathbf{G}_{0,0}$ is locally isomorphic to the de Rham cohomology group $H_0^1(\mathbf{X}, \mathbf{R})$, the first cohomology group of $\mathbf{X}$ with real coefficients and compact support.

STATEMENT 3. If $\mathbf{X}$ is not compact, the completion $\mathbf{G}_{1,0}$ of $\mathbf{G}_{0,0}$ in the compact-open topology of presheaves (i.e., the group obtained by adjoining to $\mathbf{G}_{0,0}$ the infinite products of sequences of $\mathbf{g}_\nu \in \mathbf{G}_{0,0}$, where for each compact $K \subset \mathbf{X}$ only finitely many of the $\mathbf{g}_\nu$ differ from the identity in K) is a closed, normal subgroup of $\mathbf{G}$; the quotient group $\mathbf{G}/\mathbf{G}_{1,0}$ is locally isomorphic to the de Rham group $H^1(\mathbf{X}, \mathbf{R})$, i.e., to the first cohomology group with closed support.

STATEMENT 4. The group $\mathbf{G}_{1,0}$ has no connected, closed, normal subgroups other than the identity; in particular its commutator subgroup is an open subgroup. The same is true of $\mathbf{G}_{0,0}$, of course, if $\mathbf{X}$ is compact (in which case $\mathbf{G}_{0,0} = \mathbf{G}_{1,0}$). On the other hand, if $\mathbf{X}$ is not compact, the commutator subgroup $\mathbf{G}'_{0,0}$ of $\mathbf{G}_{0,0}$ is normal in $\mathbf{G}_0$, has codimension equal to 1, relative to $\mathbf{G}_{0,0}$, and has no connected, closed, normal subgroups other than the identity.

The next two sections deal with the Lie group structure of the groups $\mathbf{G}$ and $\mathbf{G}_0$, emphasizing the relationship with the corresponding Lie algebras; the main tools used here are due to J. Moser [1]. In Section 3 we prove the four main statements just given at the Lie algebra level, and in Section 4 we expand the results at the group level, trying as far as we have succeeded to obtain results on the global structure of these groups and their closed, normal subgroups.

2. Infinite dimensional Lie groups of differentiable transformations

We shall summarize in this section some of the known facts about infinite dimensional Lie groups or pseudogroups of differentiable transformations, especially with regard to their relationships with the corresponding Lie algebras of tangent vector fields.

We denote by $\mathbf{G}$, $\mathbf{H}$, and so forth, groups of differentiable transformations of a manifold; the corresponding pseudogroups of local transformations are denoted by G, H, and so forth; the associated Lie algebras of globally defined vector fields will be denoted by German capitals $\mathfrak{G}$, $\mathfrak{H}$, and so forth; the corresponding presheaves of local vector fields will be denoted by small German letters, such as $\mathfrak{g}$, $\mathfrak{h}$, etc.

An infinite dimensional Lie group **G** of global, differentiable transformations in a manifold **X** (or alternately a pseudogroup G of local transformations) is an infinite dimensional, differential manifold (naturally we do not exclude from this notion finite dimensional manifolds), in the following sense: for any finite dimensional, differential manifold **M** a map $\varphi: \mathbf{M} \to \mathbf{G}$ is defined to be differentiable, if and only if the corresponding evaluation map $\tilde{\varphi}: \mathbf{M} \times \mathbf{X} \to \mathbf{X}$, where $\tilde{\varphi}(t, \mathbf{x}) = (\varphi(t))(\mathbf{x})$, is differentiable (in the case of a pseudogroup, one requires also that the domain of definition of $\tilde{\varphi}$ be open in $\mathbf{M} \times \mathbf{X}$).

For our present purposes, it seems irrelevant to fix any topology on **G**; we regard it rather as a *compactoid*, that is to say, we allow ourselves to consider the equivalence class of topologies that are compatible with the category of differentiable maps $\varphi: \mathbf{M} \to \mathbf{G}$ just defined. The Lie algebra $\mathfrak{G}$ associated to **G** is then the set of tangent vector fields (respectively presheaves of local vector fields) each defined as the equivalence class of differentiable paths in **G** originating at the identity with the obvious equivalence relation. The question that ordinarily arises here is how to recapture the arc component of the identity in **G** from the sheaf of germs $\mathfrak{g}$ of Lie algebras determined by $\mathfrak{G}$.

Any local cross section ϑ in $\mathfrak{g}$ (that is to say, each local vector field belonging to $\mathfrak{g}$) defines a one-parameter subpseudogroup of differentiable transformations by integrating the vector field to the corresponding autonomous flow. The composition of such one-parameter local flows defines a pseudogroup $\Gamma_{\mathbf{G}}$ in the pseudogroup G associated to **G**. If **G** is finite dimensional, it is known classically that $\Gamma_{\mathbf{G}}$ defines an open subpseudogroup of G. In the infinite dimensional case the same conclusion holds, if **G** acts real analytically, at least by considering first local cross sections of $\mathfrak{g}$ and collating the resulting local transformations. In the C^{∞} case for infinite dimensional Lie groups, several authors have shown examples to the effect that one-parameter subpseudogroups are not dense in a neighborhood of the identity, and the details of some of these examples give a strong indication that even the composition of the elements of such one-parameter systems may not fill out any neighborhood of the identity in **G**. Some authors have suggested a method based on affine connections; this method permits one to attach to each local vector field in $\mathfrak{G}$ a differentiable *path* in **G** originating at the identity but not constituting, in general, a one-parameter pseudogroup, so that the union of such paths fill out a neighborhood of the identity in **G**. This method is satisfactory in the case of differentiably acting pseudogroups definable in terms of a first-order, integrable differential system in the coordinate transformations but would require higher order connections for pseudogroups of a more complicated nature.

The correspondence between presheaves of Lie algebras of local vector fields and pseudogroups of local differentiable maps can be established in a natural way only as a local one-to-one correspondence between differentiable paths. Thus the fundamental theorem on existence, uniqueness, and continuity with respect to initial data for ordinary differential equations establishes a one-to-one, locally biregular correspondence between differentiable, one-parameter families of local vector fields in $\mathfrak{g}$ (i.e., paths in $\mathfrak{g}$) and local flows in $\mathbf{X}$ belonging to G (i.e., paths in G): any topological structure in the stalks of the sheaf of germs of one-parameter families of vector fields in $\mathbf{X}$ belonging to $\mathfrak{g}$ (provided that its definition includes minimal regularity conditions) yields a well defined topology on the stalks of the corresponding sheaf of germs of G-flows. The corresponding topology on the sets of germs of elements of G is then obtained by passage to the quotient, assigning to each path in G (originating at the germ of the identity) the germ of the terminal element $\in G$ of the path. Thus one obtains from the sheaf of germs of Lie algebras $\mathfrak{g}$ a sheaf G of germs of diffeomorphisms. The group $\mathbf{G}$ of global cross sections in G can be obtained without any difficulty if the manifold $\mathbf{X}$ is compact.

In the case where $\mathbf{X}$ is not compact, the corresponding sheaves $\mathfrak{g}$ of germs of Lie algebras of vector fields and G of germs of pseudogroups of transformations lead to the corresponding global Lie algebras and groups in many ways, of which two are the most important: the ones with unrestricted support and the ones with compact support. They are obtained from the topologies of the corresponding stalks by "globalizing" them, in the former case either by the compact-open extension of a uniform topology on the stalks (compact-open topology) or by a Whitney topology and in the latter case by a uniform topology over uniformly compactly supported cross sections. It is worthwhile noting that the global groups obtained from the pathwise-connected sheaf of groups is not necessarily connected, as we know well in the case of the celebrated group of all differentiable, orientation-preserving automorphisms of spheres.

3. The Lie algebra of symplectic vector fields

We apply the concepts reviewed in the previous section to the case of the Lie algebra associated with the group $\mathbf{G}$ or $\mathbf{G}_0$ of automorphisms of a symplectic manifold $(\mathbf{X}, \omega)$.

If φ is any exterior p-form and ξ is a vector field in a manifold, we denote by $\xi \llcorner \varphi$ the interior product of ξ with φ; this is the $(p-1)$-form (identically zero if $p = 0$) whose value at a $(p-1)$-vector $\eta_1 \wedge \cdots \wedge \eta_{p-1}$ is given by

$$(\xi \llcorner \varphi)(\eta_1 \wedge \cdots \wedge \eta_{p-1}) = \varphi(\xi \wedge \eta_1 \wedge \cdots \wedge \eta_{p-1}). \tag{3.1}$$

The Lie derivative of φ with respect to ξ is then obtained from the formula

$$[\xi, \varphi] = d(\xi \llcorner \varphi) + \xi \llcorner d\varphi.$$

It is well known that any differentiable path in the sheaf of germs of diffeomorphisms, originating at the germ of the identity at any $\mathbf{x} \in \mathbf{X}$, leaves the form φ invariant along the orbit of $\mathbf{x}$, if and only if the path in the sheaf of germs of vector fields given by the differential of the given path yields a one-parameter family of germs of vector fields ξ satisfying $[\xi, \varphi] = 0$. Thus the Lie algebra of the symplectic group $\mathbf{G}$ or $\mathbf{G}_0$ is described locally by the vector fields ξ satisfying, since ω is closed,

$$d(\xi \llcorner \omega) = 0. \tag{3.2}$$

We now introduce the Lie algebras corresponding to the groups $\mathbf{G}$, $\mathbf{G}_0$, and $\mathbf{G}_{0,0}$ previously given in Definitions 1.1 and 1.2.

PROPOSITION 3.1. *The Lie algebras* $\mathfrak{G}$, $\mathfrak{G}_0$, *and* $\mathfrak{G}_{0,0}$ *corresponding, respectively, to the groups* $\mathbf{G}$, $\mathbf{G}_0$, *and* $\mathbf{G}_{0,0}$ *are given by the global vector fields* ξ *on* $\mathbf{X}$ *satisfying* (3.2) *and, in addition,*

(i) *satisfying no further conditions in the case of* $\mathfrak{G}$;
(ii) *having compact support in the case of* $\mathfrak{G}_0$;
(iii) *generated, in the case of* $\mathfrak{G}_{0,0}$, *by vector fields* ξ_ν, *where each* ξ_ν *has compact support contained in a contractible domain* U_ν *admitting a canonical coordinate system* $(x^1, \ldots, x^{2n})$, *that is to say, such that there exists a function* u_ν *with compact support in* U_ν *satisfying*

$$\xi_\nu \llcorner \omega = du_\nu,$$

or equivalently

$$\xi_\nu = \sum_{j=1}^{n} \left(\frac{\partial u_\nu}{\partial x^{2j}} \frac{\partial}{\partial x^{2j-1}} - \frac{\partial u_\nu}{\partial x^{2j-1}} \frac{\partial}{\partial x^{2j}} \right). \tag{3.3}$$

The proof of this proposition is clear.

Since the form ω is everywhere of maximal rank, the bundle map from the tangent to the cotangent vector bundle defined by assigning to ξ the 1-form $\xi \llcorner \omega$ is bijective. Thus, for instance, the Lie algebra $\mathfrak{G}$ is isomorphic, as a vector space over the real numbers, to the set of all closed Pfaffian forms on $\mathbf{X}$. This isomorphism induces a Lie algebra structure on the set of all Pfaffian forms called the "Poisson bracket"; more precisely, for any 1-form α we denote by $\alpha^\#$ the uniquely defined tangent vector ξ such that $\xi \llcorner \omega = \alpha$ (at the bundle, sheaf, local, or global level); then the Poisson bracket of two Pfaffian forms α and β is defined to be

$$\{\alpha, \beta\} = [\alpha^\#, \beta^\#] \llcorner \omega.$$

The Lie algebra of all C^∞ 1-forms (globally defined on $\mathbf{X}$) with the Poisson bracket is isomorphic under the map $\{\alpha \to \alpha^\#\}$ to the Lie algebra of all tangent vector fields; the vector subspace consisting of all closed 1-forms is a Lie subalgebra. It follows from (3.2) that this subalgebra is isomorphic to the algebra $\mathfrak{G}$ of all vector fields ξ such that $[\xi, \omega] = 0$. The subalgebras $\mathfrak{G}_0$ and $\mathfrak{G}_{0,0}$ of $\mathfrak{G}$ are similarly characterized. We shall now define some vector subspaces of these Lie algebras that will, in fact, turn out to be ideals.

DEFINITION 3.2. We denote by $\mathfrak{G}'$ the vector subspace of $\mathfrak{G}$ consisting of all germs of vector fields $\xi \in \mathfrak{G}$ such that $\xi \llcorner \omega$ is exact; that is to say, $\mathfrak{G}'$ consists of all the vector fields $(du)^\#$ where u is an arbitrary differentiable function on $\mathbf{X}$. Similarly we denote by $\mathfrak{G}_0'$ the vector subspace of $\mathfrak{G}_0$ consisting of all the vector fields $(du)^\#$ where u is an arbitrary function on $\mathbf{X}$ with compact support. Finally, we denote by $\mathfrak{G}_0''$ the vector subspace of $\mathfrak{G}_0'$ consisting of the vector fields $\xi = (du)^\#$, where u is a function with compact support on $\mathbf{X}$ satisfying, in addition,

$$\int_{\mathbf{X}} u\omega^n = 0. \tag{3.4}$$

REMARKS. In the case of the unrestricted Lie algebra $\mathfrak{G}$, one cannot define the algebra that corresponds to $\mathfrak{G}_0''$ in the case of $\mathfrak{G}_0$; clearly, if $\mathbf{X}$ is a compact manifold, then $\mathfrak{G}_0 = \mathfrak{G}$ and, in defining $\mathfrak{G}_0' = \mathfrak{G}'$ the function u is determined by the vector field $\xi = (du)^\#$ only up to an additive constant; therefore one can always choose u so as to satisfy (3.4); thus we have

$$\mathfrak{G}_0' = \mathfrak{G}_0'' = \mathfrak{G}'.$$

On the other hand, if $\mathbf{X}$ is not compact, the function u generating an element $\xi \in \mathfrak{G}_0'$ is uniquely determined by ξ, since it is required by definition to have compact support; therefore the condition (3.4) becomes meaningfully restrictive, and indeed

$$\dim_{\mathbf{R}} (\mathfrak{G}_0'/\mathfrak{G}_0'') = 1.$$

Similarly, using the theorems of de Rham, for all symplectic manifolds

$$\dim_{\mathbf{R}} (\mathfrak{G}/\mathfrak{G}') = \dim_{\mathbf{R}} H^1(X, \mathbf{R}) = b_1(\mathbf{X}),$$

where $b_1(\mathbf{X})$ denotes the first Betti number of $\mathbf{X}$ with respect to real coefficients (and homology with compact support) while in the case of noncompact $\mathbf{X}$

$$\dim_{\mathbf{R}} (\mathfrak{G}_0/\mathfrak{G}_0') = b_1^0(\mathbf{X}),$$

where b_1^0 denotes the first Betti number for homology with closed support.

We now show that the vector subspaces $\mathfrak{G}'$, $\mathfrak{G}_0'$, and $\mathfrak{G}_0''$ are indeed Lie algebra ideals of $\mathfrak{G}$ and $\mathfrak{G}_0$, respectively.

PROPOSITION 3.3 (R. S. Palais). *The commutator ideal of* $[\mathfrak{G}, \mathfrak{G}] = \mathfrak{G}^2$ *is contained in* $\mathfrak{G}'$, *while the commutators* $\mathfrak{G}_0^2 = [\mathfrak{G}_0, \mathfrak{G}_0]$ *and* $\mathfrak{G}_0^3 = [\mathfrak{G}_0, \mathfrak{G}_0^2]$ *are contained, respectively, in* $\mathfrak{G}_0'$ *and* $\mathfrak{G}_0''$.

PROOF. Let $\xi = \alpha^\#$, $\eta = \beta^\#$ be arbitrary elements of $\mathfrak{G}$, that is to say, let α and β be closed 1-forms. Then $\{\alpha, \beta\} = [\xi, \eta] \mathbin{\llcorner} \omega$ is not only closed but indeed exact, for, with no assumption on ξ, η, and ω, we have the following identities:

$$
\begin{aligned}
[\xi, \eta] \mathbin{\llcorner} \omega &= [\xi, \eta \mathbin{\llcorner} \omega] - \eta \mathbin{\llcorner} [\xi, \omega] \\
&= d(\xi \mathbin{\llcorner} (\eta \mathbin{\llcorner} \omega)) + \xi \mathbin{\llcorner} d(\eta \mathbin{\llcorner} \omega) - \eta \mathbin{\llcorner} [\xi, \omega] \\
&= d(\xi \mathbin{\llcorner} (\eta \mathbin{\llcorner} \omega)) + \xi \mathbin{\llcorner} [\eta, \omega] - \eta \mathbin{\llcorner} [\xi, \omega] - \xi \mathbin{\llcorner} (\eta \mathbin{\llcorner} d\omega) \\
&= dw + \xi \mathbin{\llcorner} [\eta, \omega] - \eta \mathbin{\llcorner} [\xi, \omega] - \xi \mathbin{\llcorner} (\eta \mathbin{\llcorner} d\omega),
\end{aligned}
\tag{3.5}
$$

where

$$w = \xi \mathbin{\llcorner} (\eta \mathbin{\llcorner} \omega);$$

thus, under the assumptions $d\omega = 0$, $[\xi, \omega] = [\eta, \omega] = 0$, we have

$$[\xi, \eta] \mathbin{\llcorner} \omega = dw.$$

This shows that $[\mathfrak{G}, \mathfrak{G}] \subset \mathfrak{G}'$ and, obviously, also $[\mathfrak{G}_0, \mathfrak{G}_0] \subset \mathfrak{G}_0'$, if ξ and η (and hence w) have compact support.

In order to show that $\mathfrak{G}^3 \subset \mathfrak{G}_0''$, we assume that $\xi = \alpha^\#$, $\eta = \beta^\#$ belong to $\mathfrak{G}_0$; if w is defined by (3.6), we have an elementary identity.

$$
\begin{aligned}
w\omega^n = (\xi \mathbin{\llcorner} (\eta \mathbin{\llcorner} \omega))\omega^n &= -n(\xi \mathbin{\llcorner} \omega) \wedge (\eta \mathbin{\llcorner} \omega) \wedge \omega^{n-1} \\
&= -n\alpha \wedge \beta \wedge \omega^{n-1}.
\end{aligned}
\tag{3.6}
$$

It follows that, if either of the two closed forms with compact support α or β is exact (say $\alpha = du$, where u has compact support),

$$\int_{\mathbf{X}} w\omega^n = -n \int_{\mathbf{X}} d(u\beta \wedge \omega^{n-1}) = 0.$$

This completes the proof of this proposition.

The next results show that the continuation of the commutator sequences yields no new ideals. For this purpose we recall the definition of the Lie algebra $\mathfrak{G}_{0,0}$ of the minigroup (Definition 1.2 and Proposition 3.1). Clearly, since $\mathfrak{G}_{0,0} \subset \mathfrak{G}_0'$, we have $[\mathfrak{G}_{0,0}, \mathfrak{G}_{0,0}] \subset \mathfrak{G}_{0,0}' \subset \mathfrak{G}_0''$, where $\mathfrak{G}_{0,0}'$ is defined to be $\mathfrak{G}_{0,0} \cap \mathfrak{G}_0'$; in other words $\mathfrak{G}_{0,0}'$ is generated by the vector fields $(du)^\#$, where u is a function which has compact support in a contractible domain U admitting a canonical coordinate system and satisfies (3.4).

LEMMA 3.4. *The Lie algebras* $\mathfrak{G}_{0,0}$ *and* $\mathfrak{G}'_{0,0}$ *coincide, respectively, with* $\mathfrak{G}'_0$ *and* $\mathfrak{G}''_0$. *The algebra* $\mathfrak{G}'$ *in the case of a noncompact manifold is generated by infinite sums with locally finite supports of elements of* $\mathfrak{G}''_0 = \mathfrak{G}'_{0,0}$.

PROOF. Let $\xi \in \mathfrak{G}'_0$. This means that $\xi = (du)^\#$, where u is a differentiable function on $\mathbf{X}$ with compact support. Let $\{U_\nu\}_{\nu \in J}$ be a locally finite, open cover of $\mathbf{X}$ by contractible open sets U_ν, each of which admits a canonical coordinate system $(U_\nu, {}_\nu(x))$ and let (φ_ν) be a differentiable partition of unity, where each φ_ν has compact support contained in U_ν; set $u_\nu = \varphi_\nu u$; then $(du_\nu)^\# \in \mathfrak{G}_{0,0}$ and all but a finite number of them vanish, since u has compact support. This shows that $(du)^\# \in \mathfrak{G}'_0$ can be expressed as a sum of elements in $\mathfrak{G}_{0,0}$. Now suppose that $\xi = (du)^\# \in \mathfrak{G}''_0$: since in this case $\mathbf{X}$ is assumed not to be compact and is connected, condition (3.4) is equivalent to the existence of a $(2n - 1)$-form ψ on $\mathbf{X}$ with compact support, such that $d\psi = u\omega^n$. Let $\psi_\nu = \varphi_\nu \psi$ be a decomposition of ψ by the partition of unity (φ_ν), and define u_ν by the condition

$$u_\nu \omega^n = d\psi_\nu = d(\varphi_\nu \psi).$$

Then clearly $(du_\nu)^\# \in \mathfrak{G}'_{0,0}$, showing that $\mathfrak{G}''$ is generated additively by $\mathfrak{G}'_{0,0}$.

The final assertion about $\mathfrak{G}'$ now has to be proved only in the case where $\mathbf{X}$ is not compact. In this case an element $\xi = (du)^\#$ is defined by a function u on $\mathbf{X}$ whose support is unrestricted. Since $\mathbf{X}$ is connected and not compact, the $2n$-dimensional real cohomology group of $\mathbf{X}$ with unrestricted support is trivial; therefore the $2n$-form $u\omega^n$ can be represented by $d\psi$, where ψ is a $(2n - 1)$-form. Decomposing ψ by means of the partition of unity φ_ν as in the case of $\mathfrak{G}''_0$, we show that ξ can be expressed as an infinite sum of elements $\xi_\nu \in \mathfrak{G}'_{0,0}$ with compact, locally finite supports; this completes the proof of the assertion.

LEMMA 3.5. *The Lie algebra* $\mathfrak{G}'_{0,0}$ *coincides with its own commutator.*

PROOF. Let $\xi = (du)^\#$ be a vector field in $\mathfrak{G}_{0,0}$ defined from a function u with compact support $K \subset U$ where U is a contractible domain in which canonical coordinates $(x) = (x^1, \ldots, x^{2n})$ can be defined. Expressing u in terms of these coordinates, the vector field ξ can be described by (3.3). In terms of these same coordinates, we have

$$\omega^n = n!\, dx^1 \wedge \cdots \wedge dx^{2n}.$$

Since $\xi \in \mathfrak{G}_{0,0}$, the function u satisfies (3.4); in other words we have

$$\int_{(x)(U)} u(x)\, dx' \cdots dx^{2n} = 0. \tag{3.7}$$

We shall show that there exists a set of $4k$ functions (v_i, w^i) $(i = 1, 2, \ldots, 2n)$ each with compact support contained in U and satisfying, like the function u, equation (3.7) and satisfying

$$du = \sum_{i=1}^{2n} \{dv_i, dw^i\}$$

or equivalently

$$\xi = (du)^{\#} = \sum_{i=1}^{2n} [(dv_i)^{\#}, (dw^i)^{\#}].$$

This means that u is determined from the other functions (using 3.3 and 3.5) by the formula

$$u(x) = \sum_{i=1}^{2n} \sum_{j=1}^{n} \left(\frac{\partial v_i}{\partial x^{2j}} \frac{\partial w^i}{\partial x^{2j-1}} - \frac{\partial v_i}{\partial x^{2j-1}} \frac{\partial w^i}{\partial x^{2j}} \right). \tag{3.8}$$

Since the function u satisfies (3.4), there are, by virtue of Gauss' theorem, $2n$ functions $z^1, \ldots, z^{2n}$ in $\mathbf{X}$ each with compact support in U such that

$$u(x) = \sum_{i=1}^{2n} \frac{\partial z^i(x)}{\partial x^i}. \tag{3.9}$$

As a matter of fact, the functions z^i can be chosen so that their support is contained in an arbitrarily small, connected, open neighborhood of the compact support K of the function u; in particular we let each of the functions z^i have support contained in a compact K_1 with $K \subset K_1 \subset U$.

Now consider the functions $\bar{v}_i = -(-1)^i z^{(i-(-1)^i)}$ and $\bar{w}^i = x^i$; disregarding the bars, they satisfy (3.8). In fact we see that, for each i,

$$\sum_{j=1}^{n} \left(\frac{\partial \bar{v}_i}{\partial x^{2j}} \frac{\partial \bar{w}^i}{\partial x^{2j-1}} - \frac{\partial \bar{v}_i}{\partial x^{2j-1}} \frac{\partial \bar{w}^i}{\partial x^{2j}} \right) = \frac{\partial z^{(i-(-1)^i)}}{\partial x^{(i-(-1)^i)}}.$$

Consequently, summing both members of this equation, we have a solution of (3.8); however, the functions $\bar{v}_i$ do not, in general, satisfy (3.4), nor do $\bar{w}_i = x^i$ even have compact support. This can be remedied in a second stage, as we shall now proceed to do.

Choose a non-negative valued, differentiable function $h(x)$ which is identically equal to 1 in K_1 and whose support is a compact domain $K_2 \subset U$. We can then replace the functions $\bar{w}^i = x^i$ by $h(x) \cdot x^i$; the new choice of $\bar{w}^i$ will not affect equation (3.8) and the resulting functions $h \cdot x^i$ have compact support, so that $(d(h \cdot x^i))^{\#} \in \mathfrak{G}_{0,0}$. Next, pick two more non-negative, nonzero, differentiable functions h' and h'' with compact supports K_3 and K_4 mutually disjunct and contained in $U - K_2$. Since

$\int_U h'\omega^n$ and $\int_U h''\omega^n$ are both positive, there exist real constants c_i' and c_i'' ($i = 1, 2, \ldots, 2n$) such that

$$\int_U (\bar{v}_i - c_i'h')\omega^n = \int_U (h \cdot \bar{w}^i - c_i''h'')\omega^n = 0 \qquad (1 \leqq i \leqq 2n).$$

Thus we can set

$$v_i = \bar{v}_i - c_i h'; \quad w^i = h(x)x^i - c_i''h'' \qquad (1 \leqq i \leqq 2n).$$

The replacement of the $4n$ functions $\bar{v}_i$ and $\bar{w}^i$ by v^i and w^i, respectively, does not affect the computation of the Poisson brackets $\{dv_i, dw^i\}$, so that (3.8) is still satisfied, while each of the functions v_i and w^i satisfies, as does the function u, equation (3.4). Hence

$$(dv_i)^\# \text{ and } (dw_0^i)^\# \subset \mathfrak{G}'_{0,0}$$

and

$$(du)^\# = \sum_{i=1}^{2n} [(dv_i)^\#, (dw^i)^\#].$$

This completes the proof of Lemma 3.5.

Before stating the first main result on the Lie algebras $\mathfrak{G}$ and $\mathfrak{G}_0$, we need one more notion, that of the *symplecting pairing*.

DEFINITION 3.6. In a symplectic manifold $(\mathbf{X}, \omega)$ we define the *symplectic pairing* to be the alternating, bilinear map of $H_0^1(\mathbf{X}, \mathbf{R})$ (first cohomology group with compact support and real coefficients) into the real numbers, denoted by $(\mathbf{u}, \mathbf{v})$ as follows: let α, β be closed 1-forms representing by means of de Rham's theorem the cohomology classes $\mathbf{u}$, $\mathbf{v}$, respectively. Then the symplectic pairing is given by

$$(\mathbf{u}, \mathbf{v}) = \int_{\mathbf{X}} \alpha \wedge \beta \wedge \omega^{n-1}. \tag{3.10}$$

It is well known that the symplectic pairing is nonsingular in the case of the symplectic structure subordinate to a compact Kähler manifold, while it is identically zero in the case of the natural symplectic structure of the cotangent bundle of any differential manifold or, more generally, in any (necessarily noncompact) symplectic manifold with $n \geqq 2$, where ω^{n-1} is cohomologous to zero.

THEOREM 3.7. *Let $(\mathbf{X}, \omega)$ be a symplectic manifold, let $\mathfrak{G}$ be the Lie algebra of all differentiable, symplectic, global vector fields, and let $\mathfrak{G}_0$ be the Lie subalgebra of $\mathfrak{G}$ consisting of vector fields with compact support. Then the commutator algebra $\mathfrak{G}^2 = [\mathfrak{G}, \mathfrak{G}]$ coincides with $\mathfrak{G}'$ and $\mathfrak{G}/\mathfrak{G}^2$ is isomorphic to $H^1(\mathbf{X}, \mathbf{R})$; furthermore, if $\mathbf{X}$ is not compact, the commutator* *algebra* $\mathfrak{G}_0^2 = [\mathfrak{G}_0, \mathfrak{G}_0]$ *coincides with either* $\mathfrak{G}_0'$ *or* $\mathfrak{G}_0''$ *depending on whether the symplectic pairing of* $H_0^1(\mathbf{X}, \mathbf{R})$ *is, respectively, nontrivial or identically zero; the third derived algebra* $\mathfrak{G}_0^3 = [\mathfrak{G}_0, \mathfrak{G}^2]$ *coincides in all cases with* $\mathfrak{G}_0''$. *The Lie algebras* $\mathfrak{G}'$ *and* $\mathfrak{G}_0''$ *are each equal to their own commutator algebras.*

PROOF. From Proposition 3.3 we have the inclusion relations

$$[\mathfrak{G}, \mathfrak{G}] \subset \mathfrak{G}', \qquad [\mathfrak{G}_0, \mathfrak{G}_0] \subset \mathfrak{G}_0', \qquad [\mathfrak{G}_0, \mathfrak{G}_0'] \subset \mathfrak{G}_0''.$$

In Lemma 3.5 it was established that $[\mathfrak{G}_0'', \mathfrak{G}_0''] \supset \mathfrak{G}_0''$. Since $\dim_{\mathbf{R}}(\mathfrak{G}_0'/\mathfrak{G}_0'') = 1$, this means that $\mathfrak{G}_0^2$ coincides with either $\mathfrak{G}_0'$ or $\mathfrak{G}_0''$. It follows from equation (3.6) in the proof of Proposition 3.3 that $\mathfrak{G}_0^2 = \mathfrak{G}_0'$ if and only if the symplectic pairing is nontrivial, and otherwise $\mathfrak{G}_0^2 = \mathfrak{G}_0''$ (this will be the case, in particular, if $H_0^1(\mathbf{X}, \mathbf{R}) = 0$ or if ω^{n-1} is cohomologous to zero). The corresponding conclusions in the case of $\mathfrak{G}$ can be obtained by showing that $\mathfrak{G}'$ is equal to its own commutator subalgebra. This can be done by the following argument. There exists an open cover $\{U_\nu\}_{\nu\in J}$ of $\mathbf{X}$ by contractible canonical coordinate domains with the following additional property: there exists a partition of the indexing system J into a finite collection of subsets, $J = \bigcup_{\mu=1}^{N} J_\mu$ such that, for each μ, the open sets $(U_\nu)_{\nu\in J_\mu}$ are pairwise disjunct. Let (φ_ν) be a partition of unity subordinate to $\{U_\nu\}_{\nu\in J}$ and let $e_\mu = \sum_{\nu\in J_\mu} \varphi_\nu$; then $(e_\mu)_{\mu=1}^N$ is a finite partition of unity; any vector field $\xi \in \mathfrak{G}'$ can be expressed as $(du)^\#$ for some function u such that (after adding a suitable constant to u, if $\mathbf{X}$ is compact) $u\omega^n = d\psi$ for some $(2n-1)$-form ψ, and let $\psi_\mu = e_\mu \cdot \psi$. Let u_μ be defined by the equation $u_\mu\omega^n = d\psi_\mu$ and consider the vector fields $\xi_\mu = (du_\mu)^\#$. By using the arguments of Lemma 3.5 in each U_ν for $\nu \in J_\mu$ we see that, since the domains $\{U_\nu\}_{\nu\in J_\mu}$ are pairwise disjunct, ξ_μ $(\mu = 1, 2, \ldots, N)$ can be expressed as a sum of commutators of $2n$-pairs of vector fields with support in $\bigcup_{\nu\in J_\mu} U_\nu$; the sum of these N sums of commutators reproduces the given $\xi \in \mathfrak{G}'$ represented as a finite sum of commutators of vector fields belonging to $\mathfrak{G}'$. This completes the proof of the asserted commutator relations. This concludes the proof of the theorem.

An immediate application of Theorem 3.7, to be used in the next section, is that we can obtain, using de Rham's theorem, a complete set of Abelian representations of the Lie algebras $\mathfrak{G}$ and $\mathfrak{G}_0$ as follows. For any differentiable, compact, integral 1-cycle γ and any $\xi \in \mathfrak{G}$, we define

$$J(\xi)(\gamma) = \int_\gamma \xi \,\llcorner\, \omega. \tag{3.11}$$

Similarly, if $\xi \in \mathfrak{G}_0$ then, for any integral, locally finite, differentiable, integral 1-cycle γ we define $J(\gamma)(\xi)$ by the same formula. It is clear that the value of the integral depends only on the homology class of γ: it establishes the homomorphism of $\mathfrak{G}$ (respectively $\mathfrak{G}_0$) into hom $({}_0H_1(X, \mathbf{R}), \mathbf{R}) = H^1(X, \mathbf{R})$ [respectively into hom $(H_1(X, \mathbf{R}), \mathbf{R}) = H_0^1(X, \mathbf{R})$].

We have seen that these two homomorphisms are surjective and are Lie algebra homomorphisms, if we regard the cohomology groups as Abelian Lie algebras. We can also construct geometrically a Lie algebra representation J' of $\mathfrak{G}_0'$ with kernel $\mathfrak{G}_0''$ onto the additive group $\mathbf{R}$ as follows. If $\mathbf{X}$ is not compact, for each point $\mathbf{x} \in \mathbf{X}$ let $\gamma_{\mathbf{x}}$ be an infinite, locally finite, differentiable 1-chain whose boundary $\partial\gamma_{\mathbf{x}}$ equals $\mathbf{x}$. If $\xi \in \mathfrak{G}_0'$, define the function $\psi(\xi): \mathbf{X} \to \mathbf{R}$ by the equation

$$\psi(\xi)(\mathbf{x}) = \int_{\gamma_{\mathbf{x}}} \xi \llcorner \omega. \tag{3.12}$$

Since $\gamma_{\mathbf{x}}$ is defined modulo the group of 1-cycles and $\xi \in \mathfrak{G}_0'$, the function $\psi(\xi)$ is independent of the choice of the chain; in fact it coincides with the function u with compact support such that $(du)^{\#} = \xi$. The representation of $\mathfrak{G}_0'$ into $\mathbf{R}$ with kernel $\mathfrak{G}_0''$ then is obviously given by the integral

$$J'(\xi) = \int_{\mathbf{X}} u\omega^n = \int_{\mathbf{X}} \psi(\xi)\omega^n.$$

We have at the moment no satisfactory, faithful geometrical representation of the non-Abelian nilpotent Lie algebra $\mathfrak{G}_0/\mathfrak{G}_0''$ in the cases where $[\mathfrak{G}_0, \mathfrak{G}_0] = \mathfrak{G}_0'$, i.e., where the symplectic pairing is not identically zero. The next result concerns the structure of Lie algebra ideals of $\mathfrak{G}$ and $\mathfrak{G}_0$ that give rise to normal subgroups. It is clear that any normal subgroup $\mathbf{H}$ of any (finite or infinite dimensional) Lie group $\mathbf{G}$ of differentiable transformations has as its Lie algebra an ideal in the Lie algebra of $\mathbf{G}$. The converse, however, is false, unless the germs of the actions of $\mathbf{G}$ at each point are uniquely determined by their local power series expansion; this would happen, for instance, when $\mathbf{G}$ is finite dimensional or if its action is real analytic. In the case at hand, however, when $\mathbf{G}$ is the group of all automorphisms of a symplectic manifold $(\mathbf{X}, \omega)$, we know that there are nontrivial elements of $\mathbf{G}$ that act trivially in a nonempty open set, and nontrivially in another. In particular let U be an open subdomain of $\mathbf{X}$ with $U \neq \varnothing$, $\overline{U} \neq \mathbf{X}$; then the Lie subalgebra $\mathfrak{G}_U$ of $\mathfrak{G}$ consisting of all vector fields $\xi \in \mathfrak{G}$ that vanish identically in U is locally transitive in $X - \overline{U}$; clearly it is an ideal in $\mathfrak{G}$; however, the group $\mathbf{G}_U$ of global transformations of $\mathbf{X}$ defined by $\mathfrak{G}_U$ is not a normal subgroup of $\mathbf{X}$, since U is not invariant under $\mathbf{G}$. This fact justifies the restrictive condition imposed below.

Definition 3.8. Let $\mathbf{H}$ be a Lie group of differentiable transformations of a manifold $\mathbf{X}$ and $\mathfrak{H}$ its Lie algebra of vector fields. A Lie algebra ideal

$\mathfrak{H}' \subset \mathfrak{H}$ is called stable, if it is invariant under the adjoint representation of $\mathbf{H}$ in $\mathfrak{H}$.

It is easy to verify that the local isomorphism classes of normal Lie subgroups of $\mathbf{H}$ are in one-to-one correspondence with the stable ideals of the Lie algebra of $\mathbf{H}$.

THEOREM 3.9. *Let* $(\mathbf{X}, \omega)$ *be a* $2n$*-dimensional symplectic manifold, and let* $\mathbf{G}$ *be the group of all global, symplectic automorphisms of* $(\mathbf{X}, \omega)$*; let* $\mathbf{G}_0$ *be the subgroup of* $\mathbf{G}$ *consisting of the transformations with compact support and* $\mathbf{G}_0''$ *its third derived group. Then every nontrivial ideal of the Lie algebra* $\mathfrak{G}$ *of* $\mathbf{G}$ *(or* $\mathfrak{G}_0$ *of* $\mathbf{G}_0$*) that is stable under the adjoint action of* $\mathbf{G}_0$ *(indeed even of* $\mathbf{G}_0''$ *alone) contains the Lie algebra* $\mathfrak{G}_0''$ *of* $\mathbf{G}_0''$.

The proof of this theorem requires a lemma which is the analogue of the lemma of Palais–Cerf in the case of symplectic transformations. It is formulated here in a form somewhat stronger than strictly necessary because it is of independent interest, and the proof given here is independent of that of the Palais–Cerf Lemma for the sake of completeness.

LEMMA 3.10. *The identity component* $\mathbf{G}_0''$ *of the third derived group of* $\mathbf{G}_0$ *is strongly transitive on* $\mathbf{X}$ *in the following sense. Let* $\bar{\mathbf{g}}$ *be a germ of a symplectic transformation at a point* $\mathbf{x}_0 \in \mathbf{X}$ *with target* $\bar{\mathbf{g}}(\mathbf{x}_0) = \mathbf{x}_1$*, let* γ *be a simple, differentiable path from* $\mathbf{x}_0$ *to* $\mathbf{x}_1$ *parametrized by* t $(0 \leqq t \leqq 1$, $\gamma(0) = \mathbf{x}_0$, $\gamma(1) = \mathbf{x}_1)$ *and let* $|\gamma|$ *be the corresponding point set. Then for any neighborhood* U *of* $|\gamma|$ *there exists a global transformation* $\mathbf{g} \in \mathbf{G}_0''$ *that represents the given germ* $\bar{\mathbf{g}}$ *at* $\mathbf{x}_0$ *and equals the identity outside of* U*. Indeed* $\mathbf{g}$ *can be achieved by a* $\mathbf{G}_0''$*-flow that is trivial outside* U.

PROOF. The proof consists of constructing a sequence of three successive differentiable flows in $\mathbf{X}$ (trivial outside of the given neighborhood U of $|\gamma|$); the composition of successive differentiable flows into a differentiable flow is elementary. The first stage will bring the point $\mathbf{x}_0$ to $\mathbf{x}_1$ along the path γ; the second one, leaving $\mathbf{x}_1$ fixed, will rotate the tangent space $\mathbf{X}_{\mathbf{x}_1}$ of $\mathbf{X}$ at $\mathbf{x}_1$ so that the composition of this flow with the first stage will match the first order jet of the given germ $\bar{\mathbf{g}}$ of a symplectic map; the third stage will be a symplectic flow, trivial as the previous ones outside U which leaves $\mathbf{x}_1$ and each tangent vector at $\mathbf{x}_1$ fixed and, after composing it with the two previous stages, matches the germ $\bar{\mathbf{g}}$ itself.

Stage 1. The flow carrying $\mathbf{x}_0$ to $\mathbf{x}_1$ can be actually imbedded in a one-parameter group, by constructing a symplectic vector field $\xi \in \mathfrak{G}_0''$ on $\mathbf{X}$ with compact support in U whose restriction to $|\gamma|$ coincides with the tangent vectors $d\gamma(t)$ along the path. Since ξ has compact support, it generates a one-parameter subgroup of $\mathbf{G}_0''$ acting globally on $\mathbf{X}$ and trivially outside U, whose action on $\mathbf{x}_0$ is an extension to all of $\mathbf{R}$ of the parameter domain of the path γ. The tangent vector $d\gamma(t)$ at each point $\gamma(t)$ of $|\gamma|$ yields, by contraction with ω, a 1-form $d\gamma(t) \llcorner \omega = \alpha(t)$ which is orthogonal to $d\gamma(t)$. Let u be a function on $\mathbf{X}$ with the following properties:

(i) At each point $\gamma(t)$ of $|\gamma|$, $du(\gamma(t)) = \alpha(t)$.
(ii) The function u has compact support contained in U.
(iii) The integral $\int_{\mathbf{X}} u\omega^n$ vanishes.

[Note that property (i) implies that u is constant along $|\gamma|$.] It is clear that such a function u exists and that the vector field $\xi = (du)^{\#}$ has all the required properties to achieve the flow carrying $\mathbf{x}_0$ to $\mathbf{x}_1$. This shows that $\mathbf{G}_0''$ is pointwise transitive on $\mathbf{X}$.

Stage 2. We show now that $\mathbf{G}_0''$ is transitive on the bundle of symplectic tangential frames. We are given a germ $\bar{\mathbf{g}}_1$ of a symplectic transformation at a point $\mathbf{x}_1$ with $\bar{\mathbf{g}}_1(\mathbf{x}_1) = \mathbf{x}_1$ and we want to exhibit a $\mathbf{G}_0''$-flow $(\mathbf{g}(t))_{0 \leqq t \leqq 1}$ in $\mathbf{X}$ leaving $\mathbf{x}_1$ and everything outside a neighborhood of U of $\mathbf{x}_1$ pointwise fixed, such that $\mathbf{g}(0)$ is the identity and $d\mathbf{g}(1)(\mathbf{x}_1) = d\bar{\mathbf{g}}_1(\mathbf{x}_1)$. The fibre map $d\bar{\mathbf{g}}(\mathbf{x}_1)$ of the tangent space $\mathbf{X}_{\mathbf{x}_1}$ at $\mathbf{x}_1$ belongs to the $2n$-dimensional, real symplectic group $Sp(\mathbf{X}_{\mathbf{x}_1}, \omega(\mathbf{x}_1))$ which is known to be arcwise connected. Let $S(t)$ $(0 \leqq t \leqq 1)$ be a differentiable path in this group from the identity to $d\bar{\mathbf{g}}_1(\mathbf{x}_1)$ and let $s'(t) = \dfrac{dS(t)}{dt} \circ S^{-1}(t)$ its logarithmic derivative, so that $s'(t)$ belongs to the Lie algebra $\mathfrak{sp}\ (\mathbf{X}_{\mathbf{x}_1}, \omega(\mathbf{x}_1))$. This means that the bilinear form $\omega(\mathbf{x}_1) \circ s'(t)$ on $\mathbf{X}_{\mathbf{x}_1}$ defined by $(\omega(\mathbf{x}_1) \circ s'(t))(v, v') = \omega(\mathbf{x}_1)(s'(t)(v), v')$ is symmetric for each t $(0 \leqq t \leqq 1)$. We construct now a differentiable function u: $[0, 1] \times \mathbf{X} \to \mathbf{R}$, i.e., a differentiable one-parameter of functions $(u_t)_{0 \leqq t \leqq 1}$ on $\mathbf{X}$ $(u_t(x) = u(t, x))$ with the following properties for each value of the parameter t:

(i) The function u_t has compact support, contained in U.
(ii) The differential du_t of u_t vanishes at $\mathbf{x}_1$.
(iii) The Hessian $\mathbf{H}_{\mathbf{x}_1}[u_t]$ of u_t at $\mathbf{x}_1$ equals the symmetric bilinear form $\omega(\mathbf{x}_1) \circ s'(t)$.
(iv) The integral $\int_{\mathbf{X}} u_t\omega^n$ vanishes.

The construction of the family of functions (u_t) with the four properties just described is a simple matter. We now consider the one-parameter family of vector fields $\xi_t = (du_t)^{\#}$ $(0 \leqq t \leqq 1)$ and look at the differential equation.

$$(3.13) \qquad \frac{\partial \mathbf{g}(t, \mathbf{x})}{\partial t} = \xi(t)(\mathbf{g}(t, \mathbf{x})); \qquad \mathbf{g}(0, \mathbf{x}) = \mathbf{x}.$$

Property (i) of u_t and the definition of $\xi(t)$ imply that $(\mathbf{g}_t)_{0 \leqq t \leqq 1}$ $(\mathbf{g}_t(\mathbf{x}) = \mathbf{g}(t, \mathbf{x}))$ has a global solution on $[0, 1] \times \mathbf{X}$, describing a $\mathbf{G}'_0$-flow in $\mathbf{X}$ that is trivial outside U; property (ii) implies that $\mathbf{x}_1$ is fixed throughout the flow. Property (iii) means that, at the fixed point $\mathbf{x}_1$ of this flow, the one-parameter family $(d\mathbf{g}_t(\mathbf{x}_1))$ of actions of $(\mathbf{g}_t)$ on the tangent space $\mathbf{X}_{\mathbf{x}_1}$ at $\mathbf{x}_1$ satisfies the same differential equation with the same initial condition as the path $S(t)$ in the linear symplectic group $Sp(\mathbf{X}_{\mathbf{x}_1}, \omega(\mathbf{x}_1))$; therefore $d\mathbf{g}_1(\mathbf{x}_1) = S(1) = d\bar{\mathbf{g}}(\mathbf{x}_1)$; the second stage of the process is then completed merely by remarking that property (iv) of u_t implies that the flow is actually in the commutator subgroup $\mathbf{G}''_0$ of $\mathbf{G}'_0$.

Stage 3. In order to complete the proof of the lemma, we now suppose that the given germ $\bar{\mathbf{g}}$ of a symplectic transformation leaves $\mathbf{x}_1$ fixed and acts as the identity on the tangent bundle $\mathbf{X}_{\mathbf{x}_1}$ of $\mathbf{x}_1$ and try to extend it to a global, symplectic $\mathbf{G}''_0$-flow on $\mathbf{X}$, trivial outside a given neighborhood U of $\mathbf{x}_1$.

Let $\{U', (x)\}$ be a local, canonical coordinate system in a neighborhood U' of $\mathbf{x}_1$, $U' \subset U$, with $x^i(\mathbf{x}_1) = 0$ $(1 \leqq i \leqq 2n)$ and including in its range the closed ball $\sum_{i=1}^{2n} (x^i)^2 \leqq a_0^2$ for some $a_0 > 0$; let U_r, for any positive $r \leqq a_0$, denote the subdomain of U' defined by $\sum_{i=0}^{2n} (x^i)^2 < r^2$ and let the germ $\bar{\mathbf{g}}$ be represented by a local symplectic map $x^i \to y^i = \bar{g}^i(x^1, \ldots, x^{2n})$ $(1 \leqq i \leqq 2n)$ in a neighborhood U_{a_1} of $\mathbf{x}_1$ $(0 < a_1 \leqq a_0)$. We have, thus, by assumption:

$$\sum_{j=1}^{n} dy^{2j-1} \wedge dy^{2j} = \sum_{j=1}^{n} dx^{2j-1} \wedge dx^{2j} \tag{3.14}$$

and

$$\bar{g}^i(x) = x^i + O(|x|^2), \quad \text{where } |x|^2 = \sum_{i=1}^{2n} (x^i)^2. \tag{3.15}$$

Let $h(s)$ be a monotone decreasing C^∞ function of $s \geqq 0$ which equals identically 1 for $0 \leqq s \leqq b$, is strictly positive for $b < s < 2b$, and vanishes for all $s \geqq 2b$, for some positive $b < \frac{1}{2}a_1$, sufficiently small for later purposes. Consider the map of $\mathbf{f}$: $[0, 1] \times \mathbf{X} \to \mathbf{X}$ defined as follows:

$$\mathbf{f}(t, \mathbf{x}) = \mathbf{x} \quad \text{if either } t = 0 \text{ or } \mathbf{x} \notin U_{2b}; \tag{3.16}$$

otherwise, in terms of the local coordinates $(x^1, \ldots, x^{2n})$, i.e., setting $f_t^i(x^1(\mathbf{x}), \ldots, x^{2n}(\mathbf{x})) = x^i(\mathbf{f}(t, \mathbf{x}))$,

(3.17)

$$f_t^i(x) = (t \cdot h(|x|)^{-1} \bar{g}^i(t \cdot h(|x|)x^1, \ldots, t \cdot h(|x|)x^{2n}), \qquad (i = 1, \ldots, 2n).$$

One verifies first that, because of (3.15), equations (3.16) and (3.17) match each other in a C^∞ way; furthermore, if the constant b is sufficiently small, the map $\mathbf{f}_t\colon \mathbf{X} \to \mathbf{X}$ ($\mathbf{f}_t(\mathbf{x}) = \mathbf{f}(t, \mathbf{x})$) is a diffeomorphism for each t (with $\mathbf{f}_0$ equal to the identity), while f_1 coincides with $\bar{g}$ in U_b and each $\mathbf{f}_t$ is symplectic in U_b. Thus the diffeotopy $(\mathbf{f}_t)_{0 \leqq t \leqq 1}$ is symplectic for each t except in the interior of $U_{2b} - U_b$ for $t > 0$. Furthermore, one verifies directly that in U_b we have $\dfrac{\partial f^i}{\partial t} = O(|x|^2)$ uniformly in t, so that the flow acts as the identity on the tangent space at $\mathbf{x}_1$. In order to replace $(\mathbf{f}_t)$ by a diffeotopy that is symplectic everywhere in $\mathbf{X}$ and for all t, we shall construct another diffeotopy $(\mathbf{g}'_t)_{0 \leqq t \leqq 1}$ acting as the identity for $t = 0$, and, for $t > 0$, outside $\bar{U}_{b'} - \bar{U}_{b/2}$ for some b' ($2b < b' < a_1$) so that the composite diffeotopy $(g_t)_{0 \leqq t \leqq 1}$ with $g_t = f_t \circ (g'_t)^{-1}$ satisfies $\omega \circ \mathbf{g}_t = \omega$ for each t everywhere in $\mathbf{X}$. Let $\omega_t = \omega \circ \mathbf{f}_t$; then $\omega_t = \omega_0 = \omega$ except in the spherical shell $U_{2b} - \bar{U}_b$. Let b_1 be arbitrary constant with $2b < b_1 < a_1$; then $\omega_t - \omega$ is closed and cohomologous to 0 with compact support in the open submanifold $U_{b_1} - \bar{U}_{b/2}$. One can therefore find a differentiable, one-parameter family of 1-forms α_t with compact support in $U_{b_1} - \bar{U}_{b/2}$ satisfying for each t ($0 \leqq t \leqq 1$)

$$d\alpha_t = \frac{\partial \omega_t}{\partial t} \quad \text{and} \quad \alpha_0 = 0.$$

Consider the corresponding one-parameter family of vector fields $(\eta_t)_{0 \leqq t \leqq 1}$ uniquely determined by the equation $\eta_t \llcorner \omega_t = \alpha_t$ and set up the differential equation $\dfrac{\partial \mathbf{g}'_t}{\partial t} = \eta_t \circ \mathbf{g}'_t$ with $\mathbf{g}'_0 =$ identity.

An easy calculation shows that, if we set $\omega'_t = \omega \circ \mathbf{g}'_t$, we have

$$\frac{\partial \omega'_t}{\partial t} = [\eta_t, \omega'_t] = \frac{\partial \omega_t}{\partial t}, \qquad \omega'_0 = \omega_0 = \omega.$$

Thus we have defined another flow $(\mathbf{g}'_t)_{0 \leqq t \leqq 1}$ which has the same effect on ω as $(\mathbf{f}_t)$ in $U_{b_1} - U_{b/2}$, i.e., $\omega_t = \omega \circ \mathbf{f}_t = \omega \circ \mathbf{g}'_t$ and leaves the complement pointwise fixed. Consequently, if we set

$$\mathbf{g}_t = \mathbf{f}_t \circ (\mathbf{g}'_t)^{-1} \qquad (0 \leqq t \leqq 1),$$

we obtain a flow that coincides with $\mathbf{f}_t$ in $U_{b/2} \cup (\mathbf{X} - U_{b_1})$ and leaves ω invariant. Since this flow belongs clearly to the minigroup $\mathbf{G}_{0,0}$ which, by Lemma 1 of Theorem 3.4 coincides with $\mathbf{G}'_0$, there exists a differentiable one-parameter family of real valued, C^∞ functions $(v_t)_{0 \leqq t \leqq 1}$ with support in U_{b_1} such that $(\mathbf{g}_t)$ is the solution of the differential equation

$$\frac{\partial \mathbf{g}_t}{\partial t} = (dv_t)^\# \circ \mathbf{g}_t.$$

In order to replace this flow by a $\mathbf{G}_0''$-flow, we need only to replace (v_t) by a modified family $(v_t + w_t)$, where w_t is a family of differentiable functions with support in $U_a - \overline{U}_{b_1}$ such that

$$\int (v_t + w_t)\omega^n = 0 \qquad (0 \leqq t \leqq 1).$$

This concludes the proof of the lemma.

Proof of Theorem 3.9. Suppose that $\mathfrak{I}$ is a nontrivial stable Lie algebra ideal of $\mathfrak{G}_0''$ other than 0 and let ξ_0'' be a nonzero vector field in $\mathfrak{I}$. Let U_0 be a neighborhood of a point $\mathbf{x}_0 \in \mathbf{X}$, in which ξ is everywhere $\neq 0$. Then there exists, in a smaller neighborhood U_1 of $\mathbf{x}_0$, a canonical coordinate system $(x^1, \ldots, x^{2n})$ such that $x^i(x_0) = 0$ and $\xi|_{U_1} = \dfrac{\partial}{\partial x^1}$. The existence of such a coordinate system can be verified readily from any of the known proofs of Darboux's theorem on the existence of canonical coordinates. In other words $\xi|_{U_1} = (dx^2)^\#$. Now suppose that η is any given vector field in $\mathfrak{G}_0''$. The theorem is proved if we can show that η is an element of the $\mathbf{G}_0''$-stable ideal of vector fields generated by ξ.

The condition $\eta \in \mathfrak{G}_0''$ means that the uniquely determined function v with compact support such that $(dv)^\# = \eta$ satisfies $\int v\omega^n = 0$; therefore there exists a $(2n-1)$-form ψ with compact support K in X such that $d\psi = v\omega^n$. We choose, using Lemma 3.10, a finite, open cover $\{V_\nu, {}_\nu(x)\}_{\nu \in J}$ of a connected, open neighborhood of K by canonical coordinate domains $({}_\nu(x) = ({}_\nu x^1, \ldots, {}_\nu x^{2n}))$, each with the same range $\sum_{i=1}^{2n} ({}_\nu x^i)^2 < a^2$, with a sufficiently small, so as to satisfy the following additional property: for each $\nu \in J$ and each index i $(1 \leqq i \leqq 2n)$ the local symplectic map $\mathbf{f}_{\nu,i}\colon V_\nu \to U_1$ defined by

$$x^1 \circ \mathbf{f}_{\nu,i} = {}_\nu x^i$$
$$x^2 \circ \mathbf{f}_{\nu,i} = -(-1)^i {}_\nu x^{(i-(-1)^i)}$$

and

$$x^{2j+1} \circ \mathbf{f}_{\nu,i} = {}_\nu x^{(i+2j)}$$
$$x^{2j+2} \circ \mathbf{f}_{\nu,i} = -(-1)^i {}_\nu x^{(i+2j-(-1)^i)}, \qquad (2 \leqq j \leqq n-1),$$

(where the indices $(i + 2j)$, $(i + 2j - (-1)^i)$ are to be interpreted as integers modulo $(2n)$) can be extended to a global $\mathbf{G}_0''$-automorphism of $\mathbf{X}$, which we denote again by $\mathbf{f}_{\nu,i}$. Let $(\varphi_\nu)_{\nu \in J}$ be a partition of unity in K subordinate to the covering (V_ν) and set up the corresponding partitions $\psi = \sum_\nu \psi_\nu$, $v = \sum_\nu v_\nu$, $\eta = \sum_\nu \eta_\nu$ defined by setting, respectively, $\psi_\nu = \varphi_\nu \psi$, $v_\nu \omega^n = d\psi_\nu$, $\eta_\nu = (dv_\nu)^\#$. Then each $\eta_\nu \in \mathfrak{G}_0''$. We can prove the theorem by showing that each $\eta_\nu \in \mathfrak{I}$.

We first look at the corresponding $(2n-1)$-form ψ_ν. In terms of the local coordinates ${}_\nu(x)$ we have

$$\psi_\nu = \sum_{i=1}^{2n} (-1)^{i+1}\psi_\nu^i({}_\nu(x))\, d({}_\nu x^1) \wedge \cdots \wedge d({}_\nu x^{i-1}) \wedge d({}_\nu x^{i+1}) \wedge \cdots \wedge d({}_\nu x^{2n}).$$

Each of the $2n$ components $\psi_\nu^i({}_\nu(x))$ of ψ_ν (extended by the constant zero outside of V_ν) gives rise to a vector field $\zeta_\nu^i = (d\psi_\nu^i({}_\nu(x)))^\# \in \mathfrak{G}_0'$; for each $\nu \in J$ the relationship between the $2n$ vector fields ζ_ν^i $(1 \leqq i \leqq 2n)$ and η_ν can be described by the equation

$$\eta_\nu = \sum_{i=1}^{2n} \left[\frac{\partial}{\partial({}_\nu x^i)}, \zeta_\nu^i\right].$$

We now apply to each term the mapping $\mathbf{f}_{\nu,i}$, which sends $\dfrac{\partial}{\partial({}_\nu x^i)}$ into the restriction of local vector field $\dfrac{\partial}{\partial x^i}$ to some subdomain of U_1, i.e., into a restriction of the given vector field ξ, and ζ_ν^i into a vector field $d\mathbf{f}_{\nu,i}(\zeta_{\nu,i}) = \zeta_{\nu,i}^i$ with compact support in $\mathbf{f}_{\nu,i}(V_\nu) \subset U_1$. In other words we have the representation of

$$\eta = \sum_{\nu \in J} \sum_{i=1}^{2n} [d(f_{v,i}^{-1})(\xi), \zeta_v^i],$$

where each $\zeta_\nu^i \in \mathfrak{G}_0'$. We can modify each ζ_ν^i by adding to it a suitably chosen vector field $\eta_{\nu,i}' \in \mathfrak{G}_0'$ with support disjunct[2] from the compact support of $d(f_{\nu,i}^{-1}) \circ \xi$, so that the sum $\zeta_\nu^i + \eta_{\nu,i}'$ belongs to $\mathfrak{G}_0''$; this modification of ζ_ν^i does not affect the Lie product. This shows that the given vector field $\eta \in \mathfrak{G}_0''$ belongs to the $\mathbf{G}_0''$-stable ideal generated by ξ, thereby completing the proof of the theorem.

An analogous result on the commutator algebra $\mathfrak{G}'$ without restriction at infinity would be desirable. This can be achieved in an elementary way by making suitable assumptions of completeness in an appropriate topology, using the results just established, as follows.

COROLLARY. *Let* $(\mathbf{X}, \omega)$ *be a noncompact symplectic manifold,* $\mathfrak{G}$ *the Lie algebra of all differentiable automorphisms of* $(\mathbf{X}, \omega)$ *and* $\mathfrak{G}'$ *its commutator algebra. Then any nonzero ideal in* $\mathfrak{G}'$ *that is both: (i) invariant under the adjoint action of* $\mathbf{G}'$, *(ii) either closed under infinite sums with locally finite support or complete under a compact-open extension of a function-space topology on the stalks of the sheaf of germs, must be all of* $\mathfrak{G}'$.

[2] The case where $\mathbf{X}$ is compact and the support of ξ is all of $\mathbf{X}$ does not present any essential difficulty, since ξ could be replaced without loss of generality by its Lie product (assumed $\neq 0$) with a vector field in $\mathfrak{G}_0''$ with smaller support.

The proof of this corollary can be omitted.

Summarizing the conclusions of this section, we see that the finite-dimensional representations of **G** and $\mathbf{G}_0$ are very limited. They were described completely at the end of the proof of Theorem 3.7.

4. Representation of the automorphism groups

In this section we shall show how and to what extent the finite-dimensional representations of the Lie algebras $\mathfrak{G}$, $\mathfrak{G}_0$, and $\mathfrak{G}_0'$ can be "lifted" to representations of the corresponding Lie groups **G**, $\mathbf{G}_0$, and $\mathbf{G}_0'$, respectively. Since these representations are either Abelian or (in the case of $\mathfrak{G}_0/\mathfrak{G}_0''$ when **X** is not compact and the symplectic pairing in $H_0^1(\mathbf{X}, \mathbf{R})$ is nontrivial) nilpotent, it is clear that they give rise naturally to representations of the respective universal covering groups (analytic groups, in the terminology of Chevalley) $\tilde{\mathbf{G}}$, $\tilde{\mathbf{G}}_0$, and $\tilde{\mathbf{G}}_0'$ into quotient groups of the corresponding Lie groups $H^1(\mathbf{X}, \mathbf{R})$, $H_0^1(\mathbf{X}, \mathbf{R}) \dotplus \mathbf{R}$ (this is the semidirect product defined by means of the symplectic pairing) and **R**. However, since we have at present little or no knowledge about either the group of components (zero-dimensional homotopy group) of **G**, $\mathbf{G}_0$, or $\mathbf{G}_0'$, such information about their universal covering is not especially interesting except as an intermediate step in attacking the representation problem of the groups themselves.

In the remarks after the proof of Theorem 3.7 we exhibited the geometric realization of some Abelian representations of the Lie algebras $\mathfrak{G}$, $\mathfrak{G}_0$, and $\mathfrak{G}_0'$, using de Rham's theorem. We shall apply these geometric realizations to obtain the corresponding Abelian representations at the first of the analytic groups.

DEFINITION 4.1. Given a Lie group **L** (finite- or infinite-dimensional) we denote by $\Omega(\mathbf{L})$ the right path-semigroup of **L**; this is the semigroup whose elements are equivalence classes under right translations of piecewise differentiable, oriented paths in **L**, composed by composition of paths, after a right translation to match points of transition from each path to the next. We denote by $\Omega_0(\mathbf{L})$ the normal sub-semigroup of **L** represented by closed paths that are nomotopically trivial.

It is clear that the quotient semigroup $\Omega(\mathbf{L})/\Omega_0(\mathbf{L})$ is naturally isomorphic to the universal covering group $\tilde{\mathbf{L}}$ of the identity component of **L**.

LEMMA 4.2. *Given a (finite- or infinite-dimensional) Lie group* **L** *of differentiable transformations of a manifold* **X** *and a differentiable representation* Φ *of* $\Omega(\mathbf{L})$ *into a Lie group* **H**, *this representation is trivial on* $\Omega_0(\mathbf{L})$ *(and thereby induces a representation of* $\tilde{\mathbf{L}}$*), if and only if the associated* *linear map $d\Phi$ of the Lie algebra $\mathfrak{L}$ of* **L** *into that of* **H** *is a Lie algebra homomorphism.*

PROOF. It is understood that a representation $\Phi\colon \Omega(\mathbf{L}) \to \mathbf{H}$ is differentiable, if and only if any (piecewise) differentiable, finite-dimensional parametric family of paths in **L** has as its image in **H** under Φ a corresponding (piecewise) differentiable family of elements in **H**. Given a differentiable representation Φ, then to any (piecewise) differentiable path $\sigma\colon [0, a] \to \mathbf{L}$ in **L** we associate the one-parameter family $\sigma_{|[0,t]}\colon [0, t] \to \mathbf{L}$ $(0 \leqq t \leqq a)$ of paths. The image of this family is a (piecewise) differentiable path in **H**, originating at the identity, and whose initial tangent

$$\dot{\Phi}_0 = \left.\frac{d\Phi(\sigma_{|[0,t]})}{dt}\right|_{t=0}$$

depends only on the initial tangent

$$\dot{\sigma}_0 = \left.\frac{d\sigma_{|[0,t]}}{dt}\right|_{t=0}$$

of the given path σ. The map of Lie algebras thus obtained is clearly linear. The conclusion of the proof is now reduced to a repetition of the standard proof of the first fundamental theorem of Lie groups, i.e., to the existence and uniqueness theorems for initial value problems in ordinary differential equations.

We shall now construct some specific representations, when **L** is either the group $\mathbf{G}$, $\mathbf{G}_0$, or $\mathbf{G}_0'$ of automorphisms of a symplectic manifold $(\mathbf{X}, \omega)$, of $\Omega(\mathbf{L})$, respectively, into the following Abelian Lie groups, respectively, $H^1(\mathbf{X}, \mathbf{R})$, $H_0^1(\mathbf{X} \cdot \mathbf{R})$, and $\mathbf{R}$ (written additively), satisfying the conditions of the lemma.

We consider, in their respective order, the groups: (i) $\mathbf{G}$, (ii) $\mathbf{G}_0$, and (iii) $\mathbf{G}_0'$ (in case **X** is not compact) of automorphisms of a symplectic manifold $(\mathbf{X}, \omega)$ and let $\sigma = \{\sigma(t)\}$ describe a differentiable path in each of these groups, parametrized by t $(0 \leqq t \leqq 1)$ with $\sigma(0) = \text{identity}$. In each case we consider also, respectively: (i) any finite singular (piecewise differentiable) 1-cycle γ with integer coefficients in **X**; (ii) any locally finite, integral, singular 1-cycle γ in **X**; (iii) for any given point **x** any choice of a singular, locally finite, infinite 1-chain $\gamma_{\mathbf{x}}$ in **X** whose boundary $\partial\gamma_{\mathbf{x}}$ is the 0-cycle **x**. In each case we introduce the finite 2-chains $\sigma(\gamma)$ [respectively $\sigma(\gamma_{\mathbf{x}})$ in case (iii)] obtained by the homotopy operator on **X** represented by σ and calculate the functionals $\overline{\Phi}(\sigma, \gamma)$ in cases (i) and (ii) and $\overline{\Phi}(\sigma, \gamma_{\mathbf{x}})$ in case (iii), defined respectively by

$$(4.1) \qquad \overline{\Phi}(\sigma, \gamma) = \int_{\sigma(\gamma)} \omega, \qquad \overline{\Phi}(\sigma, \gamma_{\mathbf{x}}) = \int_{\sigma(\gamma_{\mathbf{x}})} \omega.$$

PROPOSITION 4.3. *Let* $(\mathbf{X}, \omega)$ *be a symplectic manifold and consider, respectively: (i) the automorphism group* $\mathbf{G}$ *of* $(\mathbf{X}, \omega)$, *(ii) (if* $\mathbf{X}$ *is not compact) the compactly supported automorphism group* $\mathbf{G}_0$, *and (iii) (if* $\mathbf{X}$ *is not compact) the subgroup* $\mathbf{G}_0'$ *of* $\mathbf{G}_0$. *Let* $\overline{\Phi}(\sigma, \gamma)$ *and* $\overline{\Phi}(\sigma, \gamma_{\mathbf{x}})$ *be defined, in their respective cases, by* (4.1). *Then the following statements are valid. In cases (i) and (ii) the functional* $\overline{\Phi}(\sigma, \gamma)$ *depends only on the homotopy class of the path* σ *(relative to its end points) and on the real homology class of* γ *[with compact support in case (i), locally compact support in case (ii)]. In case (iii) the functional* $\overline{\Phi}(\sigma, \gamma_{\mathbf{x}})$ *depends only on* $\mathbf{x}$, *independent of the choice of* $\gamma_{\mathbf{x}}$*; the resulting function* $\Phi(\sigma, \mathbf{x})\colon \Omega(\mathbf{G}_0') \times \mathbf{X} \to \mathbf{R}$ *has, for each fixed* $\sigma \in \Omega(\mathbf{G}_0')$, *uniformly compact support in* $\mathbf{X}$ *and its integral*

$$\overline{\Psi}(\sigma) = \int_{\mathbf{X}} \Phi(\sigma, \mathbf{x})\omega^n \tag{4.2}$$

depends only on the homotopy class of the path $\sigma \in \Omega(\mathbf{G}_0')$ *relative to its end points; in other words* $\overline{\Phi}$ *and* $\overline{\Psi}$ *define, in case (i), a homomorphism* χ *of the universal covering group* $\tilde{\mathbf{G}}$ *of* $\mathbf{G}$ *onto* $H^1(\mathbf{X}, \mathbf{R})$*; in case (ii), an epimorphism* $\chi\colon \tilde{\mathbf{G}}_0 \to H_0^1(\mathbf{X}, \mathbf{R})$*; in case (iii), a homomorphism* $\chi\colon \tilde{\mathbf{G}}_0' \to \mathbf{R}$ *that is an epimorphism if and only if* $\mathbf{X}$ *is not compact.*

PROOF. It is clear that in case (i) the 2-chain $\sigma(\gamma)$ is finite; in case (ii) the 1-cycle γ may be, and in case (iii) the 1-chain $\gamma_{\mathbf{x}}$ has to be infinite, but since the chain homotopy operator represented by σ has compact support, the resulting 2-chains $\sigma(\gamma)$ and $\sigma(\gamma_{\mathbf{x}})$ are null at infinity, so that the functionals $\overline{\Phi}(\sigma, \gamma)$ and $\overline{\Phi}(\sigma, \gamma_{\mathbf{x}})$ are well defined.

We shall consider temporarily only case (ii). The integral $\int_{\sigma(\gamma)} \omega$ can be evaluated by reducing it to two successive, 1-dimensional integrals as follows. Denote by $\sigma_{|[0,s]}$ the path in $\mathbf{G}_0$ restricted to the interval $[0, s]$ of the parameter t of σ, and by σ^t the element $\sigma(t) \in G_0$ at the value t; let $\dot{\sigma}$ be the vector field $\dfrac{d(\sigma^t)}{dt}$ in $\mathbf{X}$ and by $\xi(t)$ the vector field $\dot{\sigma}^t \circ (\sigma^t)^{-1}$ (i.e. such that the path σ is the solution of the ordinary differential equation $\dfrac{d\sigma^t}{dt} = \xi(t) \circ \sigma^t$ with initial value $\sigma^0 =$ identity). Then both $\dot{\sigma}^t$ and $\xi(t)$ belong to the Lie algebra $\mathfrak{G}_0$. We calculate the derivative of $\overline{\Phi}(\sigma_{|[0,t]}, \gamma)$ with respect to t as follows:

$$\frac{d}{dt}\overline{\Phi}(\sigma_{|[0,t]}, \gamma) = \int_{\sigma^t\gamma} \dot{\sigma}^t \mathbin{\llcorner} \omega = \int_{\gamma} \xi(t) \mathbin{\llcorner} (\omega \circ (\sigma^t)^{-1}) = \int_{\gamma} \xi(t) \mathbin{\llcorner} \omega \tag{4.3}$$

where the last equality stems from the invariance of ω under the automorphism $(\sigma^t)^{-1}$. The integrand has compact support, so that the integral is well defined. $\bar{\Phi}(\sigma, \gamma)$ can thus be calculated by successive integrations as follows:

$$\bar{\Phi}(\sigma,\gamma) = \int_0^1 \left(\int_{\sigma^t\gamma} \dot{\sigma}^t \llcorner \omega\right) dt = \int_0^1 \left(\int_\gamma \xi(t) \llcorner \omega\right) dt. \tag{4.4}$$

The analogues of this formula in cases (i) and (iii) are of the same type and require the same proof. Since, in the three respective cases and for each t, the vector field $\xi(t)$ belongs to the Lie algebra $\mathfrak{G}$, $\mathfrak{G}_0$, and $\mathfrak{G}'_0$, respectively, the 1-form $\xi(t) \llcorner \omega$ is, respectively, closed, closed with compact support, and the differential of a function $u(t, x)$ with compact support in $[0, 1] \times \mathbf{X}$. Therefore the integral $\int_\gamma \xi(t) \llcorner \omega$ [resp. $\int_{\gamma_\mathbf{x}} \xi(t) \llcorner \omega$ in case (iii)] depends only on σ and, in case (i), on the compact homology class of the 1-cycle γ; in case (ii), on the (unbounded) homology class of γ; and, in case (iii), on the boundary $\mathbf{x}$ of $\gamma_\mathbf{x}$; in this case it merely reproduces the function $\{u(t, x)\}$ with compact supports whose differential with respect to x is $\xi(t) \llcorner \omega$. We denote by $\Phi'(\sigma, [\gamma])$ the resulting function resulting from Φ by passage to the quotient, where $[\gamma]$ denotes the homology class of γ [compact in case (i), noncompact in case (ii)], and by u_t the corresponding function of σ and x in case (iii), where $u_t(x) = u(t, x)$.

The maps $\Phi\colon \Omega(\mathbf{G}) \to H^1(\mathbf{X}, \mathbf{R})$ in case (i) and $\Phi\colon \Omega(\mathbf{G}_0) \to H_0^1(\mathbf{X}, \mathbf{R})$ in case (ii) defined by $\Phi(\sigma)([\gamma]) = \Phi'(\sigma, [\gamma])$ [where γ is the corresponding dual homology class] are clearly homomorphisms of the semigroup $\Omega(\mathbf{G})$ [resp. $\Omega(\mathbf{G}_0)$] into the corresponding cohomology groups. They are differentiable since they can be computed by integration over finite parametric families of paths in $\Omega(\mathbf{G})$ [or $\Omega(\mathbf{G}_0)$]. The differential of Φ at the identity reduces simply to the de Rham representation of the symplectic vector field $\xi(0) \in \mathfrak{G}$ (resp. $\mathfrak{G}_0$) by an element in $H^1(\mathbf{X}, \mathbf{R})$ [resp. $H_0^1(\mathbf{X}, \mathbf{R})$] as described in equation (3.7), after the proof of Theorem 3.4. This representation is a Lie algebra homomorphism, as a consequence of Proposition 3.3. Hence by Lemma 4.2 it induces a representation of the analytic (simply connected) group $\tilde{\mathbf{G}}$ into $H^1(\mathbf{X}, \mathbf{R})$ [resp. $\tilde{\mathbf{G}}_0$ into $H_0^1(\mathbf{X}, \mathbf{R})$], which proves the assertion in cases (i) and (ii). We remark that in these two cases the proof of the invariance of the map Φ on $\Omega(\mathbf{G})$ [or $\Omega(\mathbf{G}_0)$] under homotopy of paths could be obtained directly, without using Lemma 4.2.

In case (iii) we have a map $\bar{\chi}\colon \Omega(\mathbf{G}'_0) \to \bar{\mathbf{R}}$ defined by

$$\bar{\chi}(\sigma) = \int_\mathbf{X} \left(\int_0^1 u_t(\mathbf{x})\, dt\right) \omega^n = \int_\mathbf{X} \left\{\int_0^1 \left[\int_{\gamma_\mathbf{x}} \xi(t) \llcorner \omega\right] dt\right\} \omega^n.$$

Again one verifies in a straightforward way that this map is additive on $\Omega(\mathbf{G}'_0)$ and differentiable, and, since the associated map of $\mathfrak{G}'_0$ (the differential of $\bar{\chi}$ with respect to t at $t = 0$) is a Lie algebra homomorphism, the map $\bar{\chi}$ is invariant under homotopy of paths in $\Omega(\mathbf{G}'_0)$, and therefore defines by passage to quotient a map

$$\chi: \tilde{\mathbf{G}}'_0 \to \mathbf{R},$$

where $\tilde{\mathbf{G}}'_0 = \Omega(\mathbf{G}'_0)/\Omega_0(\mathbf{G}'_0)$ is the universal covering group $\tilde{\mathbf{G}}'_0$ of the identity component in $\mathbf{G}'_0$. It follows trivially from deRham's theorem that the representations of $\tilde{\mathbf{G}}$, $\tilde{\mathbf{G}}_0$, and $\tilde{\mathbf{G}}'_0$ into $H^1(\mathbf{X}, \mathbf{R})$, $H_0^1(\mathbf{X}, \mathbf{R})$, and $\mathbf{R}$, respectively, are surjective (except the third case, if applied to a compact manifold). This completes the proof of Proposition 4.3.

It would be interesting to investigate how the representation $\chi: \tilde{\mathbf{G}}'_0 \to \mathbf{R}$ must be modified to obtain a corresponding one of $\mathbf{G}'_0$, i.e., what open subgroups of $\mathbf{G}'_0$ have representations χ_1 into quotient groups of $\mathbf{R}$ that are locally equivalent to χ. This, however, has not been possible except in the case $n = 1$ of open surfaces $\mathbf{X}$ with an oriented area element ω; in this case one knows from the theory of Teichmüller–Ahlfors that the group $\mathbf{G}_0$ is homotopically equivalent to the group of automorphisms of the fundamental group of $\mathbf{X}$. In particular, the identity component of $\mathbf{G}_0$ is simply connected, so that the representation χ_1 of this group onto $\mathbf{R}$ exists and coincides with χ. It has been pointed out that, in the special case of the Cartesian plane $\mathbf{R}^2$ with $\omega = dx \wedge dy$, the representation χ of the group $\mathbf{G}_0 = \mathbf{G}'_0$ of measure-preserving diffeomorphisms with compact support onto $\mathbf{R}$ had been found by Levi Civita. (I have been unable, however, to trace the reference.) This representation, intuitively, measures for each such diffeomorphism $\mathbf{g}$, how much $\mathbf{g}$ "twists" the plane in the mean, in the clockwise sense.

We shall now give the partial modification of Proposition 4.3 that defines Abelian representations of the groups $\mathbf{G}$ and $\mathbf{G}_0$.

Definition 4.4. Let $(\mathbf{X}, \omega)$ be a symplectic manifold. We denote by $\Lambda = \Lambda_\omega$ the additive subgroup of $\mathbf{R}$ made up of the numbers $\omega[w]$ where w runs over all *integral*, compact, 2-dimensional homology classes, $w \in {}_0H_2(\mathbf{X}, \mathbf{Z})$. If Λ is discrete, we say that $(\mathbf{X}, \omega)$ is a *Hodgean* manifold.

Theorem 4.5. *Let $(\mathbf{X}, \omega)$ be a Hodgean manifold; let ${}_0\mathbf{G}$ denote the open subgroup of the automorphism group $\mathbf{G}$ of $(\mathbf{X}, \omega)$ whose elements act as the identity on ${}_0H_1(\mathbf{X}, \mathbf{Z})$; if $\mathbf{X}$ is not compact, let ${}_0\mathbf{G}_0$ denote the open subgroup of G_0 whose elements act as the identity on $H_1(\mathbf{X}, \mathbf{Z})$ (noncompact homology). Then the representation $\chi: \tilde{\mathbf{G}} \to H^1(\mathbf{X}, \mathbf{R})$ [resp. $\tilde{\mathbf{G}}_0 \to H_0^1(\mathbf{X}, \mathbf{R})$] of Proposition* 4.3 *is locally equivalent to a representation* ${}_0\chi: \mathbf{G} \to$ hom $({}_0H_1(\mathbf{X}, \mathbf{Z}), \mathbf{R}/\Lambda)$ [*resp.* ${}_0\chi: \mathbf{G}_0 \to$ hom $(H_1(\mathbf{X}, \mathbf{Z}), \mathbf{R}/\Lambda)$], *where Λ is the additive subgroup of $\mathbf{R}$ consisting of the integral periods of ω.*

PROOF. The two parallel statements (which are distinct if $\mathbf{X}$ is not compact) concerning $\mathbf{G}$ and $\mathbf{G}_0$ are entirely analogous in content and proof, so that we shall confine our statements mainly to the first case.

Let $\mathbf{g} \in {}_0\mathbf{G}$ and let γ be any compact, 1-dimensional, singular differentiable cycle in $\mathbf{X}$ with integer coefficients, its homology class being denoted by $[\gamma] \in {}_0H_1(\mathbf{X}, \mathbf{Z})$. Then $\mathbf{g}(\gamma) - \gamma$ is a cycle which, by the definition of ${}_0\mathbf{G}$, bounds an integral 2-chain τ, which is compact (since either γ, in the first case, or $\mathbf{g}$, in the second case, has compact support). τ is unique modulo the group ${}_0Z_2(\mathbf{X}, \mathbf{Z})$ of compact, integral 2-cycles, so that the integral $\int_\tau \omega$ is uniquely determined by $\mathbf{g}$, ω and γ, modulo the discrete group $\Lambda \subset \mathbf{R}$. In addition, if γ is homologous to zero, i.e., $\gamma = \partial\tau_0$, then we can take $\tau = \mathbf{g}(\tau_0) - \tau_0$ and, since ω is invariant under g, $\int_\tau \omega = \int_{(g(\tau_0)-\tau_0)} \omega = 0$, so that $\int_\tau \omega$ depends only on the homology class of γ. We thus obtain a map ${}_0\chi$ by passage to quotients, ${}_0\chi \colon {}_0\mathbf{G} \to \hom({}_0H_1(\mathbf{X}, \mathbf{Z}), \mathbf{R}/\Lambda)$, defined by

$$(4.5) \qquad {}_0\chi(\mathbf{g})[\gamma] = \int_\tau \omega; \qquad \partial\tau = \mathbf{g}(\gamma) - \gamma.$$

Now suppose that $\mathbf{g}$ is (differentiably) *homotopic* to the identity under a homotopy $h\colon I \times \mathbf{X} \to \mathbf{X}$. This is a considerably weaker assumption than saying that $\mathbf{g}$ lies in the component of the identity in $\mathbf{G}$ (in the case of $\mathbf{G}_0$ we assume the homotopy to be with compact support). Let $h_t\colon \mathbf{X} \to \mathbf{X}$ be the differentiable map defined by $h_t(x) = h(t, x)$, $h_0 =$ identity, $h_1 = \mathbf{g}$; then $h_t(\gamma) - \gamma$ is homotopic to the identity and the resulting chain homotopy h_* defines a 2-chain $\tau = h_*\gamma$ for which we can define

$${}_1\chi(\mathbf{g})([\gamma]) = \int_{h_*\gamma} \omega \in \mathbf{R}.$$

This map ${}_1\chi$ has values in the additive group $\mathbf{R}$ and is a lift of the homomorphism χ restricted to the open subgroup ${}_1\mathbf{G} \subset \mathbf{G}$ of automorphisms that are homotopic to the identity ((identity component of $\mathbf{G}$) $\subset {}_1\mathbf{G} \subset {}_0\mathbf{G} \subset \mathbf{G}$). In this case it is easy to see that by restricting χ to the identity component of $\mathbf{G}$ we obtain the representation characterized in Proposition 4.3.

This completes the proof of the theorem.

An analogous result in the case of open subgroups of $\mathbf{G}_0'$ has not been found, since the argument used in order to obtain the real-valued representation of its covering group $\mathbf{G}_0'$ requires in an essential way Lemma 4.2, i.e., that the homotopy h from the identity to $\mathbf{g}$ be within the group $\mathbf{G}_0'$ itself.

5. Conclusions and problems

The most interesting conclusions of this paper in terms of further developments are probably related to Lemma 3.10, the analogue of the Palais–Cerf Lemma. It would be interesting to describe a set of sufficient conditions for a pseudogroup Γ of differentiable transformations to have the Palais–Cerf property, i.e., that each germ of an element of Γ can be extended to a global transformation of $\mathbf{X}$ by an element of the group $\mathbf{G}$ of global cross-sections of Γ over $\mathbf{X}$ (or, better, an element in the identity component of $\mathbf{G}$). The corresponding question of the extension of a germ to global transformations belonging to a derived group $\mathbf{G}^{(k)}$ of $\mathbf{G}$ appears to be much more complicated (see, for example, the case where Γ is the pseudogroup preserving a positive volume element ω_n in an n-manifold). It is conjectured that the existence of a syzygean resolution of the sheaf Θ_Γ of germs of Γ-vector fields,

$$0 \leftarrow \Theta_\Gamma \leftarrow \Phi^{-1} \leftarrow \Phi^{-2} \leftarrow \cdots,$$

with each Φ^{-j} a fine sheaf, may be of significance here. In particular the length of such a resolution may describe the dimension of the primary obstruction to the extension of a cross-section in Θ_Γ over a compact $K \subset \mathbf{X}$ to all of $\mathbf{X}$. In the case of the symplectic group, Θ_Γ is isomorphic by Proposition 3.1 to the sheaf of germs of closed 1-forms, so that the syzygean resolution can be described by

$$0 \leftarrow \Theta_\Gamma \leftarrow \mathscr{A}^0 \leftarrow (\mathbf{R}) \leftarrow 0$$

($\mathscr{A}^p$ = sheaf of germs of p-forms), which has length 1. In the case where Γ consists of the germs of diffeomorphisms preserving a nowhere-vanishing n-form ω_n, the corresponding resolution is

$$0 \leftarrow \Theta_\Gamma \leftarrow \mathscr{A}^{n-1} \overset{d}{\leftarrow} \mathscr{A}^{n-2} \leftarrow \cdots \leftarrow \mathscr{A}^0 \leftarrow \mathbf{R} \leftarrow 0,$$

which has length $n - 1$. The primary obstructions to the extension of cross-sections in Θ_Γ from K to $\mathbf{X}$ are represented in these two special cases by the relative cohomology groups $H^2(\mathbf{X}, K; \mathbf{R})$ and $H^n(\mathbf{X}, K; \mathbf{R})$ ($n = \dim \mathbf{X}$), respectively.

UNIVERSITY OF PENNSYLVANIA
AND
INSTITUT DES HAUTES ETUDES SCIENTIFIQUES,
BURES-SUR-YVETTE

REFERENCE

1. MOSER, JÜRGEN, "On the volume elements on a manifold," *Trans. Amer. Math. Soc.*, *120*, No. 2 (1965), pp. 286–294.

On the Smoothness of Isometries

Eugenio Calabi and Philip Hartman

Duke Math. J. 37, 741–750 (1970)

1. Main result. The object of this note is to prove the following:

THEOREM (*). *Let $x, y \in R^n$. Consider the unit balls $|x| < 1$, $|y| < 1$ to be Riemann manifolds M, N with the respective positive definite elements of arclength*

$$ds^2 = g_{ij}(x)\,dx^i\,dx^j, \tag{1.1}$$

$$ds^2 = h_{ij}(y)\,dy^i\,dy^j, \tag{1.2}$$

where $g_{ij} = g_{ji}$, $h_{ij} = h_{ji}$ are of class C^k. Let $y = f(x)$ be an isometric map of an open neighborhood of $x = 0 \in M$ onto a neighborhood of $y = 0 \in N$.

(i) *If $k = 0$, then $f(x)$ need not be of class C^1.*
(ii) *If $k = 0$ and $g_{ij}(x)$, $h_{ij}(y)$ have (uniform) degrees of continuity satisfying a Dini condition* (e.g., *if $g_{ij}(x)$ and $h_{ij}(y)$ are uniformly Hölder continuous*), *then $f(x)$ is of class C^1.*
(iii) *If $1 \leq k \leq \infty$, then $f(x)$ is of class C^{k+1}.*

Part (iii) is a strengthened form of a result of Myers and Steenrod [7] which states that if $k = 1$, then $y = f(x)$ is of class C^1. Their proof is not correct for it employs normal coordinates, but if $(g_{ij}(x))$ is only assumed to be of class C^1, then normal coordinates need not exist; cf. [1] or [5]. A proof of Part (iii) for $k = \infty$ is given in [6; 169–172]; this proof for $2 \leq k \leq \infty$ only yields the conclusion that $y = f(x)$ is of class C^{k-1}. For if $(g_{ij}(x))$ is of class C^k, $k > 1$, then the corresponding exponential maps are of class C^{k-1}, and need not be of class C^k; [2]. The arguments of Palais [8] (cf. [6; 169–172]) do not overcome the objections raised here.

Part (iii) implies the positive definite case of a theorem of [3] which states that if (g_{ij}), (h_{ij}) are symmetric, non-singular (possibly indefinite), and of class C^k, $1 \leq k \leq \infty$; and if $y = f(x)$ is of class C^1 and satisfies

$$h_{pm}(y)(\partial f^p/\partial x^i)(\partial f^m/\partial x^j) = g_{ij}(x),$$

then $f(x)$ is of class C^{k+1}. In view of this result, it would suffice to show that $f(x)$ is of class C^1 in the proof of Part (iii). But the proof below will not use [3].

In Part (i), (1.1) can be of class C^∞ (or reduce to the Euclidean metric $ds^2 = |dx|^2$), while the geodesics of (1.2) are not even differentiable functions of arclength. In particular, the example below proving Part (i) shows that

Received December 2, 1968. Research partially supported by NSF Grant No. GP 6895 and by Air Force Office of Scientific Research, Office of Aerospace Research, United States Air Force, under AFOSR Contract No. F44620-67-C-0098.

DOI:10.1215/S0012-7094-70-03789-0

COROLLARY. *The geodesics of a continuous (positive definite) Riemann element of arclength need not have a tangent at every point.*

The proof of Part (ii) depends on showing that when $(g_{ij}(x))$ has a (uniform) degree of continuity satisfying a Dini condition, then the geodesics of (1.1) are of class C^1 with respect to arclength and, in fact, that the tangents to a geodesic have a uniform degree of continuity, independent of the geodesic.

The proof of (ii) will be given in §4 below and will depend on preliminary results of §§2 and 3. Part (iii) will then be proved in §5 using (ii). (It should be mentioned however that, since the differential equation for geodesics is available in (iii), the required preliminary results are trivial in this case.) Finally, (i) will be proved in §6.

2. Preliminaries. Let $q(x)$ be a function defined on some x-domain. A continuous, non-decreasing function $\omega(\delta)$ defined for $\delta \geq 0$, satisfying $\omega(0) = 0$ is called a (uniform) degree of continuity for $q(x)$ if

$$|q(x_0) - q(x_1)| \leq \omega(|x_0 - x_1|). \tag{2.1}$$

A continuous non-negative function $\omega(\delta)$, $\delta \geq 0$, is said to satisfy a Dini condition if

$$\int_{+0} \delta^{-1}\omega(\delta)\, d\delta < \infty. \tag{2.2}$$

LEMMA 2.1. *Let $q(s)$, $0 \leq s \leq L$, satisfy*

$$|q(s + 2h) - 2q(s + h) + q(s)| \leq |h|\, \omega(|h|) \tag{2.3}$$

(for all s, h for which (2.3) is meaningful), where $\omega(\delta) \geq 0$ is a continuous, non-decreasing function satisfying a Dini condition. Then $q(s)$ has a continuous derivative $q'(s)$ satisfying

$$|q(s + h) - q(s) - hq'(s)| \leq |h|\, \omega_0(|h|), \tag{2.4}$$

$$|q'(s + h) - q'(s)| \leq 2\omega_0(|h|), \tag{2.5}$$

where

$$\omega_0(\delta) = \int_0^\delta t^{-1}\omega(t)\, dt. \tag{2.6}$$

The particular case of this lemma where $\omega(\delta) = C\delta^\lambda$, $0 < \lambda \leq 1$, is given in [4; 62]. The proof of the general case is similar and, in view of its brevity, will be given here for completeness.

Proof. Consider the series

$$S = \sum_{m=0}^{\infty} 2^m\{q(s + 2^{-m}h) - 2q(s + 2^{-m-1}h) + q(s)\}. \tag{2.7}$$

The inequality (2.3) implies that the m-th term is majorized by

$$2^m(2^{-m-1}h)\omega(2^{-m-1}\,|h|) = (|h|/2)\omega(2^{-m-1}\,|h|) \le |h| \int_\alpha^\beta \delta^{-1}\omega(\delta)\,d\delta,$$

where $\alpha = 2^{-m-1}\,|h|$, $\beta = 2^{-m}\,|h|$. Hence the series is absolutely convergent, uniformly convergent with respect to h (and s), and

$$|S| \le |h|\,\omega_0(|h|). \tag{2.8}$$

The m-th partial sum is $q(s + h) - 2^{m+1}q(s + 2^{-m-1}h) + (2^{m+1} - 1)q(s)$, that is, $q(s + h) - q(s) - h[q(s + 2^{-m-1}h) - q(s)]/2^{-m-1}h$. The uniform convergence of S implies that $\lim h[q(s + 2^{-m-1}h) - q(s)]/2^{-m-1}h$, as $m \to \infty$, exists uniformly in h (and s). Hence $q'(s)$ exists and $S = q(s + h) - q(s) - hq'(s)$, so that (2.8) implies (2.4). Replacing s, h by $s + h$, $-h$, respectively, in (2.4) gives

$$|q(s) - q(s + h) + hq'(s + h)| \le |h|\,\omega_0(|h|).$$

This, together with (2.4), yields (2.5) and completes the proof.

3. Continuous metrics. Let $G(x) = (g_{ij}(x))_{i,j=1,\cdots,n}$ and write (1.1) as

$$ds^2 = G(x)\,dx \cdot dx, \tag{3.1}$$

where a dot denotes scalar multiplication in R^n. If $\xi \in R^n$, let $|\xi|$ be the Euclidean norm. If ξ is a tangent vector to M at x, its Riemann length will be written as $||\xi|| = (G(x)\xi\cdot\xi)^{\frac{1}{2}} = |G^{\frac{1}{2}}(x)\xi|$, where $G^{\frac{1}{2}}(x)$ denotes the symmetric, positive definite square root of the matrix $G(x)$.

Theorem (*) and the assertions below are "local". We can therefore suppose that there is a constant $C = C_G > 1$ with the property that

$$|\xi|^2/C^2 \le G(x)\xi\cdot\xi = ||\xi||^2 \le C^2\,|\xi|^2 \quad \text{for } |x| < 1 \quad \text{and all } \xi, \tag{3.2}$$

for otherwise the ball $|x| < 1$ is replaced by a smaller ball.

All geodesics below are parametrized by arclength s and $x' = dx/ds$. We shall repeatedly use the following known fundamental assertions.

Proposition 3.1 (Hilbert). *Let $G(x)$ be continuous for $|x| < 1$ and satisfy (3.2), and let $0 < r < 1$. Then there exists a $\delta = \delta(r) > 0$ and an $r_1 = r_1(r, \delta)$, $0 < r < r_1 < 1$, with the properties that if $|x_0| \le r$ and $|x_0 - x_1| \le \delta$, then there exists a geodesic $x = x(s)$, $0 \le s \le L$, of minimum length joining $x_0 = x(0)$ and $x_1 = x(L)$, and $|x(s)| \le r_1$ for $0 \le s \le L$. Furthermore, $r_1 \to r$ as $\delta(r) \to 0$.*

The uniform Archimedean property of minimizing geodesics and its consequences, given in [5; 144–148], have the following refinement.

Lemma 3.1. *Let $G(x)$ be a continuous, positive definite symmetric matrix for $|x| < 1$ satisfying (3.2), where $C = C_G > 1$. Let $\omega(\delta)$, $\delta \ge 0$, be a continuous,* *non-decreasing function satisfying*

$$|G^{\frac{1}{2}}(x_0)\xi - G^{\frac{1}{2}}(x_1)\xi| \leq \omega(|x_0 - x_1|)|\xi| \tag{3.3}$$

for $|x_0|$, $|x_1| < 1$ *and all* ξ. *Then there exists a constant* C_0 , *depending only on* ω *and* C *in* (3.2), *with the property that if* $x = x(s)$, $0 \leq s \leq L$, *is a minimizing geodesic, then the inequality*

$$|x(s + 2h) - 2x(s + h) + x(s)| \leq C_0 |h| \omega(4C |h|) \tag{3.4}$$

holds for $0 \leq s, s + h, s + 2h \leq L$.

It is clear that if (3.2) is satisfied and if

$$|g_{ij}(x_0) - g_{ij}(x_1)| \leq \omega(|x_0 - x_1|) \quad \text{for } i, j = 1, \cdots, n$$

and for $|x_0|$, $|x_1| < 1$, then there exists a constant C' such that (3.3) holds if ω is replaced by $C'\omega$.

Proof. (a) Consider the chord $x_t = x(0) + t[x(h) - x(0)]$, $0 \leq t \leq 1$, joining $x(0)$ and $x(h)$ and let $0 \leq v \leq h$. The first step in the proof will be an estimate for the ratio of h to $|G^{\frac{1}{2}}(x(v))[x(h) - x(0)]|$, the length of the chord with respect to the flat metric

$$ds^2 = G(x(v))\, dx \cdot dx \quad \text{for fixed } v, \qquad 0 \leq v \leq h; \tag{3.6}$$

namely, the estimate

$$|1 - |G^{\frac{1}{2}}(x(v))[x(h) - x(0)]|/h| \leq C\omega(2Ch). \tag{3.7}$$

In order to verify (3.7), note that (3.2) and

$$||x'(s)||^2 = G(x(s))x'(s)\cdot x'(s) = 1 \tag{3.8}$$

almost everywhere imply that

$$|x'(s)| \leq C \quad \text{almost everywhere.} \tag{3.9}$$

Hence the absolute continuity of $x(s)$ gives

$$|x(s) - x(0)| \leq Cs \quad \text{for} \quad 0 \leq s \leq L.$$

With respect to (1.1), the chord has a Riemann length not less than h,

$$h \leq \int_0^1 |G^{\frac{1}{2}}(x_t)[x(h) - x(0)]|\, dt. \tag{3.10}$$

For $0 \leq v \leq h$ and $0 \leq t \leq 1$,

$$|x_t - x(v)| \leq |x(0) - x(v)| + |x_t - x(0)| \leq 2Ch.$$

Thus, by (3.3) and (3.10),

$$h \leq |G^{\frac{1}{2}}(x(v))[x(h) - x(0)]| + \omega(2Ch)Ch. \tag{3.11}$$

On the other hand,

$$|G^{\frac{1}{2}}(x(v))[x(h) - x(0)]| = \left| \int_0^h G^{\frac{1}{2}}(x(v))x'(s)\, ds \right| \leq \int_0^h |G^{\frac{1}{2}}(x(v))x'(s)|\, ds.$$

Thus, by (3.3), (3.8) and (3.9),

$$(3.12) \qquad |G^{\frac{1}{2}}(x(v))[x(h) - x(0)]| \leq h + \omega(Ch)Ch.$$

The desired inequality (3.7) follows from (3.11) and (3.12).

(b) Let $0 < 2h \leq L$ and let

$$(3.13) \qquad d_{-1} = |G^{\frac{1}{2}}(x(h))[x(h) - x(0)]|, \qquad d_1 = |G^{\frac{1}{2}}(x(h))[x(2h) - x(h)]|$$

be the lengths of the chords C^{-1} joining $x(0)$, $x(h)$ and C^1 joining $x(h)$, $x(2h)$, respectively, with respect to the flat metric

$$(3.14) \qquad ds^2 = G(x(h))dx \cdot dx.$$

Define the vectors

$$(3.15) \qquad \xi_{-1} = G^{\frac{1}{2}}(x(h))[x(h) - x(0)]/d_{-1}\,, \quad \xi_1 = G^{\frac{1}{2}}(x(h))[x(2h) - x(h)]/d_1\,,$$

so that $|\xi_{\pm 1}| = 1$ and

$$x(h) - x(0) = d_{-1}G^{-\frac{1}{2}}(x(h))\xi_{-1}\,, \qquad x(2h) - x(h) = d_1 G^{-\frac{1}{2}}(x(h))\xi_1\,.$$

Let d_0 be the length of the chord C^0 joining $x(0)$, $x(2h)$ in the metric (3.14),

$$d_0 = |G^{\frac{1}{2}}(x(h))[x(2h) - x(0)]|\,.$$

Consider the perpendicular, in the metric (3.14), from $x(h)$ to C^0 and let α_{-1}, α_1 denote the angles, in this metric, between C^{-1}, C^1 and C^0, respectively. Let e_{-1}, e_1 be the lengths, with respect to (3.14), of the two segments of C^0, so that $e_{-1}, + e_1 = d_0$ and

$$\sin \alpha_{-1} + \sin \alpha_1 = (e_{-1}/d_{-1}) + (e_1/d_1).$$

By (3.7),

$$(3.16) \qquad |h/d_{\pm 1} - 1| \leq 2C\omega(2Ch) \quad \text{if} \quad 2C\omega(2Ch) \leq 1$$

and, correspondingly,

$$(3.17) \qquad |d_0/h - 2| \leq 2C\omega(4Ch).$$

Thus, if $2C\omega(2Ch) \leq 1$,

$$|\sin \alpha_{-1} + \sin \alpha_1 - 2| \leq 8C\omega(4Ch);$$

so that

$$0 \leq 1 - \sin \alpha_{\pm 1} \leq 8C\omega(4Ch).$$

Hence, there exists a constant C', independent of h, such that

$$|\alpha_{\pm 1} - \pi/2| \leq C'\omega(4Ch) \quad \text{if} \quad 16C\omega(4Ch) \leq 1.$$

The Euclidean angle between the unit vectors ξ_{-1}, ξ_1 is $\alpha_{-1} + \alpha_1$. Hence the constant C' can be chosen so that

$$|\xi_{-1} - \xi_1| \le C'\omega(4Ch) \quad \text{if} \quad 16C\omega(4Ch) \le 1.$$

By (3.15) and (3.7),

$$|G^{\frac{1}{2}}(x(h))[x(h) - x(0)] - h\xi_{-1}| \le Ch\omega(2Ch),$$

$$|G^{\frac{1}{2}}(x(h))[x(2h) - x(h)] - h\xi_1| \le Ch\omega(2Ch).$$

Consequently, if $16C\omega(4Ch) \le 1$, then

$$|G^{\frac{1}{2}}(x(h))[x(2h) - 2x(h) + x(0)]| \le (C' + 2C)h\omega(4Ch),$$

so that, by (3.2)

$$|x(2h) - 2x(h) + x(0)| \le C_0 h\omega(4Ch),$$

where $C_0 = C(C' + 2C)$. This proves (3.4) for $s = 0$ and hence for $0 \le s \le L$.

Lemmas 2.1 and 3.1 have the following consequence.

THEOREM 3.1. *If ω in Lemma 3.1 satisfies the Dini condition* (2.2), *then (minimizing) geodesics $x = x(s)$, $0 \le s \le L$, are of class C^1 and*

$$|x'(s + h) - x'(s)| \le 2\omega_1(|h|), \tag{3.18}$$

$$|x(L) - x(0) - Lx'(0)| \le L\omega_1(L), \tag{3.19}$$

where

$$\omega_1(\delta) = C_0 \int_0^{4C\delta} t^{-1}\omega(t)\, dt. \tag{3.20}$$

Remark. The proof of Theorem (*) will depend on Theorem 3.1 and the following inequality

$$L \le C|x(L) - x(0)|, \tag{3.21}$$

which is a consequence of (3.2) and (3.10).

4. Proof of Theorem (*), Part (ii). Let $G(x) = (g_{ij}(x))$ as above and, correspondingly, $H(y) = (h_{ij}(y))$. We shall assume (3.2), with $C = C_G$, and an analogous inequality for $H(y)\xi\cdot\xi$, with $C = C_H$. Let $\omega(\delta)$, $\delta \ge 0$, be a continuous, non-decreasing function satisfying a Dini condition. We shall assume (3.3) and the analogous inequality

$$|H^{\frac{1}{2}}(y_0)\xi - H^{\frac{1}{2}}(y_1)\xi| \le \omega(|y_0 - y_1|)|\xi|.$$

We can also suppose that $C = C_G = C_H$.

Without loss of generality, we can suppose that $f(x)$ is defined on $|x| < 1$. Since f is an isometry, (3.2) and its analogue imply that $f(x)$ is uniformly Lipschitz continuous. By a theorem of Rademacher [9], $f(x)$ has a differential almost everywhere,

$$df(x_0) = F(x_0)dx, \tag{4.1}$$

for almost all x_0. In (4.1), $F(x_0) = (\partial f^i(x_0)/\partial x^j)$ is the Jacobian matrix of $y = f(x)$ at $x = x_0$. It is clear that

$$F(x) : T_x(M) \to T_y(N), \quad \text{where} \quad y = f(x),$$

is a unitary operator; that is,

$$F^*(x)H(y)F(x) = G(x). \tag{4.2}$$

This relation, (3.2) and its analogue imply

$$|F(x_0)\xi| \le C^2|\xi| \quad \text{for all } \xi. \tag{4.3}$$

Let f have a differential at the point $x = x_0$. Let $x = x(s)$, $0 \le s \le L$, be a minimizing geodesic in M joining $x_0 = x(0)$ and $x_1 = x(L)$. Then $y(s) = f(x(s))$, $0 \le s \le L$, is a minimizing geodesic in N joining $y_0 = f(x_0)$ and $y_1 = f(x_1)$. In view of (4.1), $y'(0)$ exists and is $y'(0) = F(x_0)x'(0)$. The analogue of (3.19) gives

$$|f(x_1) - f(x_0) - LF(x_0)x'(0)| \le L\omega_1(L).$$

By (3.19) and (4.3),

$$|f(x_1) - f(x_0) - F(x_0)[x_1 - x_0]\,| \le (1 + C^2)L\omega_1(L).$$

Consequently, (3.21) implies that

$$|f(x_1) - f(x_0) - F(x_0)[x_1 - x_0]| \le C_1|x_1 - x_0|\omega_1(C|x_1 - x_0|), \tag{4.4}$$

where $C_1 = (1 + C^2)C$, provided that the differential (4.1) exists at $x = x_0$.

There is no condition on x_1 in (4.4). Thus x_1 can be replaced by any other point, say x_2,

$$|f(x_2) - f(x_0) - F(x_0)[x_2 - x_0]| \le C_1|x_2 - x_0|\omega_1(C|x_2 - x_0|).$$

Adding this relation to (4.4) gives

$$|f(x_2) - f(x_1) - F(x_0)[x_2 - x_1]| \le C_1\{|x_1 - x_0|\omega_1(C|x_1 - x_0|) + |x_2 - x_0|\omega_1(C|x_2 - x_0|)\}.$$

Assuming that $df(x_1) = F(x_1)dx$ exists, the analogue of (4.4) is

$$|f(x_2) - f(x_1) - F(x_1)[x_2 - x_1]| \le C_1|x_2 - x_1|\omega_1(C|x_2 - x_1|).$$

Consequently,

$$|[F(x_0) - F(x_1)](x_2 - x_1)| \le \tfrac{1}{2}C_1 \sum_{i=0}^{2}\sum_{j=0}^{2} |x_i - x_j|\,\omega_1(C\,|x_i - x_j|). \tag{4.5}$$

Let $|\xi| = 1$, $x_2 = x_1 + t\xi$, and $0 < t = |x_1 - x_2| = |x_1 - x_0|$, so that $|x_2 - x_0| \leq 2t = 2|x_0 - x_1|$. Dividing (4.5) by $t = |x_2 - x_1|$ gives, for all ξ,

$$|[F(x_0) - F(x_1)]\xi| \leq 6C_1\omega_1(2C|x_0 - x_1|)|\xi| . \tag{4.6}$$

It follows that $F(x)$ is uniformly continuous on its domain of definition [with a (uniform) degree of continuity $6C_1\omega_1(2C\delta)$]. Thus $F(x)$ can be extended to a continuous function on $|x| < 1$. Hence $f(x)$ is of class C^1. This proves (ii) in Theorem (*).

5. Proof of Theorem (*), Part (iii). Part (iii) follows from Part (ii) and [3], but we can give a simpler derivation of (iii).

Since $G(x)$ and $H(y)$ are of class C^1, they can be supposed to be uniformly Lipschitz continuous on $|x| < 1$, that is, (3.3) and its analogue hold if $\omega(\delta) = K\delta$ for some constant K. Hence, by (3.20), $\omega_1(\delta) = 4KC_0C\delta$ and, by (4.6),

$$|[F(x_0) - F(x_1)]\xi| \leq (6C_1)(4KC_0C)(2C|x_0 - x_1|)|\xi| . \tag{5.1}$$

It follows that the partial derivatives of f are uniformly Lipschitz continuous.

The relation (4.2), that is,

$$g_{ij}(x) = h_{pm}(y)(\partial f^p/\partial x^i)(\partial f^m/\partial x^j), \quad \text{where} \quad y = f(x),$$

can be differentiated almost everywhere by the chain rule to give the transformation rule

$$\Gamma^p_{ij}(\partial f^m/\partial x^p) - \Delta^m_{pq}(\partial f^p/\partial x^i)(\partial f^q/\partial x^j) = \partial^2 f^m/\partial x^i\, \partial x^j \tag{5.2}$$

for the Christoffel symbols $\Gamma^p_{ij}(x)$, $\Delta^m_{ij}(y)$ of the second kind for (1.1), (1.2), respectively. Since Γ^p_{ij}, Δ^m_{ij} are continuous, it follows that $\partial^2 f^m/\partial x^i\, \partial x^j$, which are defined almost everywhere, can be extended to continuous functions satisfying (5.2) on $|x| < 1$. Hence $f(x)$ is of class C^2. This proves Part (iii) if $k = 1$.

Finally, if $G(x)$, $H(y)$ are of class C^k, $1 < k \leq \infty$, then Γ^p_{ij}, Δ^m_{ij} are of class C^{k-1} and an induction on k, employing (5.2), shows that $f(x)$ is of class C^{k+1}. This completes the proof of Part (i).

6. Proof of Theorem (*), Part (i). In order to give an example proving (i), let $n = 2$ and let (1.1) be the Euclidean metric

$$ds^2 = du^2 + dv^2, \tag{6.1}$$

where $(u, v) = (x^1, x^2)$. Define a map $(\alpha, \beta) = f(u, v)$ of the $(u, v) -$ plane into the $(\alpha, \beta) -$ plane, in terms of polar coordinates (r, θ) in the $(u, v) -$ plane and (ρ, ϕ) in the $(\alpha, \beta) -$ plane, by

$$f(0, 0) = (0, 0), \tag{6.2}$$

$$\rho = r, \qquad \phi = \theta - g(r), \tag{6.3}$$

where $g(r) > 0$ is a continuously differentiable, decreasing function for $0 < r < \infty$

satisfying

$$g(r) \to \infty \quad \text{as} \quad r \to +0, \tag{6.4}$$

$$rg'(r) \to 0 \quad \text{as} \quad r \to +0, \tag{6.5}$$

where $g' = dg/dr$. (For example, let $g(r) = \log \log 1/r$ for $0 < r < \frac{1}{2}$.)

The inverse of (6.3) is

$$r = \rho, \qquad \theta = \phi + g(\rho), \tag{6.6}$$

so that (6.1) becomes

$$ds^2 = (1 + \rho^2 g'^2)d\rho^2 + 2\rho^2 g' \, d\rho \, d\phi + \rho^2 \, d\phi^2$$

or, since

$$d\rho = (\alpha d\alpha + \beta d\beta)/\rho \quad \text{and} \quad d\phi = (-\beta d\alpha + \alpha d\beta)/\rho^2,$$

we have

$$ds^2 = h_{11}(\alpha, \beta) \, d\alpha^2 + 2h_{12}(\alpha, \rho) \, d\alpha \, d\beta + h_{22}(\alpha, \beta) \, d\beta^2, \tag{6.7}$$

where

$$\begin{aligned} h_{11} &= 1 + \alpha^2 g'^2 - 2\alpha\beta g'/\rho, \\ h_{12} = h_{21} &= \alpha\beta g'^2 + (\alpha^2 - \beta^2)g'/\rho, \\ h_{22} &= 1 + \beta^2 g'^2 + 2\alpha\beta g'/\rho, \end{aligned} \tag{6.8}$$

for $\rho > 0$. If the $h_{ij}(0, 0)$ are defined by $h_{11}(0, 0) = h_{22}(0, 0) = 1$ and $h_{12}(0, 0) = h_{21}(0, 0) = 0$, then (6.5) shows that $H(\alpha, \beta) = (h_{ij}(\alpha, \beta))$ is continuous for all (α, β) and is positive definite for small $\alpha^2 + \beta^2$.

If we restrict (u, v) to a small disk, $u^2 + v^2 < \epsilon^2$, then (6.2) and (6.3) (that is, (6.2) and $\alpha = u \cos g(r) + v \sin g(r)$, $\beta = v \cos g(r) - u \sin g(r)$, $r = (u^2 + v^2)^{\frac{1}{2}} \geq 0$) define an isometry f of (6.1) onto (6.7). But $f(u, v)$ is not of class C^1; in fact, the geodesic $\theta = \text{const.}$, $0 \leq r < \epsilon$, for (6.1) is mapped onto the spiral (6.3). Note that this spiral is therefore a geodesic for (6.7), but has no tangent at the origin.

References

1. P. Hartman, *On the local uniqueness of geodesics*, Amer. J. Math., vol. 72(1950), pp. 723–730.
2. P. Hartman, *On geodesic coordinates*, Amer. J. Math., vol. 73(1951), pp. 949–954.
3. P. Hartman, *On isometries and on a theorem of Liouville*, Math. Z., vol. 69(1958), pp. 202–210.
4. P. Hartman, *Hölder continuity and non-linear elliptic partial differential equations*, Duke Math. J., vol. 25(1958), pp. 57–66.
5. P. Hartman and A. Wintner, *On the problem of geodesics in the small*, Amer. J. Math., vol. 73(1951), pp. 132–148.
6. S. Kobayashi and K. Nomizu, *Foundations of Differential Geometry I*, New York, 1963.
7. S. B. Myers and N. Steenrod, *The group of isometries of a Riemann manifold*, Ann. of Math., vol. 40(1939), pp. 400–416.

8. R. S. Palais, *On the differentiability of isometries*, Proc. Amer. Math. Soc., vol. 8(1957), pp. 805–807.
9. H. Rademacher, *Ueber partielle und total Differenzierbarkeit von Funktionen mehrer Variablen, und über die Transformation der Doppelintegrale*, Math. Ann., vol. 97(1918), pp. 340–359.

University oy Pennsylvania
and
Institut des Hautes Études Scientifiques
The Johns Hopkins University

Examples of Bernstein Problems for Some Nonlinear Equations

Eugenio Calabi[1]

Global Analysis, Proc. Sympos. Pure Math.,
Vol. XV, held in Berkeley, Calif., USA, 1968, 223–230,
Amer. Math. Soc., Providence, RI (1970)

1. **Introduction.** We consider the following examples of nonlinear partial differential equations. In each of these examples either the equation is absolutely elliptic, or we shall confine our attention exclusively to elliptic solutions; each of these equations arises in the context of some differential geometric problem.

(a) The minimal hypersurface equation in $\mathbf{R}^{n+1}$, for $x_{n+1} = u(x_1, \ldots, x_n)$, $u: U \to \mathbf{R}$, U open in $\mathbf{R}^n$ to satisfy

$$(1 + |\nabla u|^2) \sum_{j=1}^{n} \frac{\partial^2 u}{\partial x_j^2} - \sum_{i,j=1}^{n} \frac{\partial u}{\partial x_i} \frac{\partial u}{\partial x_j} \frac{\partial^2 u}{\partial x_i \partial x_j} = 0;$$

this equation is absolutely elliptic; trivial solutions are given when u is a linear polynomial in $x_1, \ldots, x_n$.

(b) The maximal space-like hypersurface equation in the Lorentz-Minkowski $(n + 1)$-space with coordinates $(x_1, \ldots, x_n, t)$ and fundamental form $dx^2 = \sum_{j=1}^n (dx_j)^2 - dt^2$; the equation is for $t = u(x_1, \ldots, x_n)$ to satisfy

$$(1 - |\nabla u|^2) \sum_{j=1}^{n} \frac{\partial^2 u}{(\partial x_j)^2} + \sum_{i,j=1}^{n} \frac{\partial u}{\partial x_i} \frac{\partial u}{\partial x_j} \frac{\partial^2 u}{\partial x_i \partial x_j} = 0,$$

a solution u is elliptic if and only if the fundamental form induced on its graph by the enveloping space

$$ds^2 = \sum_{j=1}^{n} (dx_j)^2 - \left(\sum_{j=1}^{n} \frac{\partial u}{\partial x_j} dx_j\right)^2 = g_{ij} dx^i dx^j, \tag{1}$$

is positive definite; equivalently, it means that $|\nabla u|^2 < 1$, and the graph of u is an extremal, among functions for which $1 - |\nabla u|^2 > 0$ under interior variation, for the volume integral.

$$V = \int w \, dx_1 dx_2 \ldots dx_n, \qquad w = (1 - |\nabla u|^2)^{1/2} > 0. \tag{2}$$

Again, trivial solutions of this equation are linear polynomials.

(c) The unimodular Hessian equation for $u: U \to \mathbf{R}$, U open in $\mathbf{R}^n$:

$$\det(\partial^2 u / \partial x_i \partial x_j) = 1;$$

---

[1] Research supported by NSF Grant GP 6895.

a solution u is elliptic, if and only if the function $u(x)$ (possibly after replacing u by $-u$) is locally convex, thereby defining the Riemannian metric

$$ds^2 = g_{ij}dx^i dx^j, \qquad \text{where } g_{ij} = \partial^2 u/\partial x_i \partial x_j. \tag{3}$$

Trivial solutions, as in the previous cases, are characterized by the fact that the components g_{ij} of the metric are constant, i.e. u is a quadratic polynomial.

(d) The unimodular Levi invariant equation for $u: U \to \mathbf{R}$, U open in $\mathbf{C}^n$;

$$\det(\partial^2 u/\partial z_\alpha \partial \bar{z}_\beta) = 1; \tag{4}$$

a solution u is elliptic, if and only if the function $u(z, \bar{z})$ (possibly after replacing u by $-u$) is plurisubharmonic, or equivalently the Levi form

$$ds^2 = \sum_{\alpha,\beta=1}^{n} \frac{\partial^2 u}{dz_\alpha d\bar{z}_\beta} dz^\alpha d\bar{z}^\beta = g_{\alpha\bar{\beta}} dz^\alpha d\bar{z}^\beta \tag{5}$$

is positive definite, thereby defining a Kähler metric. This metric, as a result of (4), has its Ricci tensor identically zero. A solution of (4) is called trivial, if, in addition, the full curvature tensor of the metric (5) vanishes, i.e. if the components $g_{\alpha\bar{\beta}}$ become constant after a suitable local holomorphic transformation of coordinates with jacobian $= 1$ (such a transformation leaves equation (4) invariant).

The four types of equations described above, coupled with the respective stronger equations characterizing the trivial solutions, are typical examples of systems for which a *Bernstein problem* can be meaningfully formulated. In each case one can introduce in the domain of a solution a metric structure in several ways (geometrically related to the data), including the euclidean metric on the coordinate space; a Bernstein problem consists of the following question: when the domain of a solution is sufficiently "big" in terms of a convenient metric (e.g. complete with respect to such a metric), is then the solution automatically a trivial one. A more refined statement of a Bernstein problem is the following: one chooses a convex, nonnegative valued functional on the derivatives of a function at each point, in such a way that, when applied to any solution of one of the differential equations under consideration, its vanishing characterizes the trivial solutions; one can then find an "a priori" bound for this functional at any interior point of the domain of a solution, in terms of its suitably defined distance from a (possibly ideal) boundary. Clearly a positive answer to the latter question implies a positive answer to the former one.

The aim of the paper is to review some of the important known methods and results on Bernstein problems in each case.

2. **The two-dimensional case.** We begin with some elementary facts about the four equations considered, in the case of two independent variables.

For $n = 2$ the techniques of complex analysis on a surface lead to several possible metrics, each of which is useful in obtaining Bernstein's classical result. As matter of fact there is a natural one-to-one correspondence, for $n = 2$, between all solutions of problem (a) with those of (b) and, if one allows also a type of simple branch-point singularity for solutions of either type of equation, a correspondence between them and solutions of problem (c) as follows: Let $u'(x, y)$ be a solution of (a) in a simply connected domain $U \subset \mathbf{R}^2$; denote by p', q', r', s', t' the partial derivatives $\partial u'/\partial x$, $\partial u'/\partial y$, $\partial^2 u'/\partial x^2$, $\partial^2 u'/\partial x\partial y$ and $\partial^2 u'/\partial y^2$ respectively and let $w' = (1 + p'^2 + q'^2)^{1/2}$. Then, setting $p'' = q'/w'$ and $q'' = -p'/w'$, we verify that equation (a) is equivalent to the integrability of the differential $du'' = p''dx + q''dy$, where u'' is a solution of (b), which is elliptic, since $1 - p''^2 - q''^2 = w'^{-2} > 0$. Similarly, setting

$$\bar{r} = \frac{1 + p'^2}{w'} = \frac{1 - p''^2}{w''} \qquad \text{(where } w'' = w'^{-1} = (1 - p''^2 - q''^2)^{1/2}\text{)}$$

$$\bar{s} = \frac{p'q'}{w'} = \frac{-p''q''}{w''},$$

$$\bar{t} = \frac{1 + q'^2}{w'} = \frac{1 - q''^2}{w''},$$

equation (a) for u' (or equivalently, (b) for u'', with the ellipticity condition) is equivalent to the integrability of the differential system

$$\bar{r} = \frac{\partial^2 \bar{u}}{\partial x^2}, \bar{s} = \frac{\partial^2 \bar{u}}{\partial x \partial y}, \bar{t} = \frac{\partial^2 \bar{u}}{\partial y^2},$$

for a function $\bar{u}$ on U (unique, modulo linear polynomials) which is a convex solution of (c); moreover every such $\bar{u}$ determines u' and u'' uniquely modulo additive constants and a change of sign, except for possible branching at the (necessarily isolated) points where $\bar{r} = \bar{t} = 1, \bar{s} = 0$.

The classical Bernstein theorem can be established by using the Riemannian metrics induced by any of the corresponding solutions of (a), (b) or (c) as well as the Euclidean metric in the (x, y)-plane.

3. **The general case.** For $n \geq 3$ there exists no known relation between solutions of (a), (b) and (c). For each value of n, each solution u of (c) in a domain $U \subset \mathbf{R}^n$ gives rise to a solution $\bar{u}$ of (d) in the tubular domain $U_T = U + i\mathbf{R}^n \subset \mathbf{C}^n$ by setting $\bar{u}(z_1, \ldots, z_n) = u(\operatorname{Re} z_1, \ldots, \operatorname{Re} z_n)$; the function $\bar{u}$ is a trivial solution of (d), if and only if u is a trivial solution of (c); however only a small class of solutions of (d) is representable in this way. This is essentially all that is known so far about growth properties of solutions of (d) up to now.

The Bernstein problem for the minimal hypersurface equation (a) has been settled, for $3 \leq n \leq 7$ successively by De Giorgi, Almgren and Simons,[2] with no information for $n \geq 8$.[3] The Bernstein problem for (b) with $n = 3, 4$ and for (c)

[2] See Lecture Notes, Symposium on Global Analysis, University of California, Berkeley, 1968.

[3] ADDED IN PROOF. De Giorgi and Bombieri have recently announced an example of a nontrivial solution of problem (a) defined over the whole numerical space $\mathbf{R}^n$ for $n \geq 8$, disproving the Bernstein conjecture as usually formulated in these cases.

with $3 \leq n \leq 5$ can be solved by using essentially the same method for both problems. A detailed exposition on problem (c) is given in [2]. The method has, in the first stage, some analogy with Simons' proof of the Bernstein theorem in the case of equation (a).

In the next section we shall review an earlier solution [2] of the Bernstein problem for equation (c), and demonstrate simultaneously how the same method essentially can be applied in the case of equation (b).

4. **Outline of solution of problems** (b) **and** (c). We consider an elliptic solution $u(x_1, \ldots, x_n)$ of equation (b) or (c) in a domain $U \subset \mathbf{R}^n$ and introduce the natural Riemannian metric associated with the differential geometric structure of the graph of u, namely (1) for equation (b) and (3) for equation (c). In the case of (b), the second fundamental form

$$B_{ij}dx^i dx^j = w^{-1}\sum_{i,j}\frac{\partial^2 u}{\partial x^i \partial x^j}dx^i dx^j,$$

where

$$w = (1 - \sum_i (\frac{\partial u}{\partial x_i})^2)^{1/2},$$

is the invariant whose vanishing in an open set characterizes a trivial solution; in the case of equation (c) the analogous role is played by the cubic form

$$\sum_{i,j,k} A_{ijk}dx^i dx^j dx^k, \qquad -2A_{ijk} = \frac{\partial^3 u}{\partial x_i \partial x_j \partial x_k}.$$

The above tensors satisfy, respectively, three sets of equations namely (in tabular form)

	equation (b)	equation (c)
(i) the basic equation	$g^{ij}B_{ij} = 0$	$g^{ij}A_{ijk} = 0$
(ii) A "Codazzi" equation	$\nabla_k B_{ij} = \nabla_j B_{ik}$	$\nabla_l A_{ijk} = \nabla_k A_{ijl}$
(iii) A "Gauss" equation	$R_{ijkl} = B_{il}B_{jk} - B_{ik}B_{jl}$	$R_{ijkl} = g^{hm}(A_{hil}A_{mjk} - A_{hik}A_{mjl})$

where ∇_k denotes the components of covariant differentiation with respect to the Riemannian metric and R_{ijkl} the components of the Riemann curvature tensor The important conclusions that follow from the above equations are the following

(i) Let R = scalar curvature; we have, in the case of (b) and (c) respectively $R = B^{ij}B_{ij}$ and $R = A^{ijk}A_{ijk}$; either way $R \geq 0$. The function $\psi = \sqrt{R}$ vanishes in an open set, if and only if the solution is a trivial one, and always satisfies the differential inequality in terms of the Beltrami operator $\Delta = \nabla^i \nabla_i$

(6) $$\Delta\psi \geq c\psi^3,$$

where c is a positive constant depending only on n.

(ii) The Ricci curvature is nonnegative: this fact alone implies two lemmas (for proofs, see respectively **[1]**, **[3]**).

LEMMA 1. **[1]** *Let X be a Riemannian manifold with nonnegative Ricci curvature, and let $r(x)$ denote the geodesic distance of x from a fixed base point x_0, restricted to those x for which this distance is achieved by at least one geodesic. If $\phi(t)$ is a differentiable, monotone increasing function of a real $t \geq 0$ with $\phi'(0) = 0$, then for $v(x) = \phi(r(x))$, we have*

(7) $$\Delta v \leq \phi''(r(x)) + \frac{n-1}{r(x)}\phi'(r(x)).$$

LEMMA 2. **[3]** *Let X be a Riemannian manifold with nonnegative Ricci curvature and let Γ be an open geodesic segment in U not containing any conjugate points; let it be parametrized by its arc length s in the range $-\infty \leq A < s < B \leq \infty$. Then, for any compact subinterval $[a, b]$ of (A, B) and letting*

$$Q(s) = \left[(n-1)^{-1} R_{ij}\frac{dx^i}{ds}\frac{dx^j}{ds}\right]^{1/2}, \qquad (R_{ij} = \text{Ricci tensor}),$$

we have the inequality

(8) $$\int_a^b Q(s)ds < 1 + \tfrac{1}{2}\log\frac{(B-a)(b-A)}{(B-b)(a-A)}.$$

We establish a first Bernstein theorem for all n as follows:

THEOREM 1. *Let u be a solution of* (b) *or* (c) *in a domain $U \subset \mathbf{R}^n$; let $x_0 \in U$ be an arbitrary point. There is a universal constant c' depending only on n, such that, if $\psi(x_0) > a > 0$, there exists a geodesic segment Γ in U, originating at x_0, without endpoint in U and of length less than c'/a. In particular, if U is complete with respect to the Riemannian structure, the solution of the equation is trivial and $U = \mathbf{R}^n$.*

The proof follows from an argument by R. Osserman **[4]**. In view of (7), suppose $\psi(x_0) > a > 0$: solve the differential equation with initial condition at $t = 0$, using the positive constant c appearing in (6)

$$\phi''(t) + \frac{n-1}{t}\phi'(t) = c\phi^3(t); \qquad \phi(0) = a, \quad \phi'(0) = 0;$$

the solution is monotone increasing for $t > 0$ and blows up to $+\infty$ as $t \to c'/a$

for some constant depending only (as c does) on n. Hence the function $v = \phi(r(x))$ satisfies the inequality resulting from Lemma 1

$$\Delta v \leq cv^3, \tag{10}$$

the differential inequality opposite to the one satisfied by ψ in (6); in addition, we have

$$0 < v(x_0) = a < \psi(x_0).$$

Looking at the function $\psi - v$, the latter becomes $-\infty$ on the boundary (relative to U) of the geodesic ball $\{r(x) \leq c'/a\}$.

Suppose that every path issuing from x_0 and of length c'/a has an endpoint in U: then the geodesic ball $\{r(x) \leq c'/a\}$ is compact and hence $\psi - v$ attains its absolute maximum value somewhere in the interior. This maximum value is certainly positive, since at x_0 we have already $\psi(x_0) - v(x_0) = \psi(x_0) - a > 0$. But this is impossible because, looking at (6) and (10),

$$\Delta(\psi - v) \geq c(\psi^3 - v^3) > 0,$$

implying by the maximum principle that $\psi - v$ cannot attain any interior, positive valued relative maximum. Hence U must have a singularity within the geodesic distance $< c'/a$ of the point x_0. This proves Theorem 1.

The singularity of a nontrivial solution of equation (b) or (c) established by Theorem 1 is, *a priori*, a virtual one, from the analytic (or "nonparametric") standpoint, since in actual fact it could be conceivably at infinity in terms of the coordinate parameters $x_1, \ldots, x_n$. This possibility, however, is ruled out by the next theorem, which establishes a comparison between the Riemannian metric induced on U by a solution of the equation and (for any fixed $x_0 \in U$) the euclidean metric with constant coefficients

$$d\sigma^2 = \sum_{i,j=1}^{n} g_{ij}(x_0)dx^i dx^j. \tag{11}$$

Both metrics coincide at x_0 and they are identical elsewhere, if and only if the solution of (b) or (c) is a trivial one. For each of the two metrics we define, $\gamma(x)$ and $\delta(x)$ to be the shortest distance respectively relative to the Riemannian metric ds^2 defined by (1) or (3) (the one used up to now) and the Euclidean metric (11) based on any fixed $x_0 \in U$. Since the distortion function $\psi = \sqrt{R}$ determines a bound on the distortion rate of $d\sigma$ relative to ds, it is easy to deduce from Theorem 1 an upper bound for $\gamma(x_0)$ in terms of $\delta(x_0)$; we must, however, establish an upper bound for $\delta(x_0)$ in terms of $\gamma(x_0)$. More precisely, given $x_0 \in U$, suppose that $\gamma(x_0)$ is known and finite. Then there exists a geodesic path Γ of length $\gamma(x_0)$ originating in x_0 and without endpoint in U; let it be parametrized by its Riemannian arc length s $(0 \leq s < \gamma(x_0))$ and denote by $\sigma = \sigma(s)$ the euclidean arc length of the segment from x_0 to $x(s)$ defined by (11). We obtain the following differential inequalities bounding the distortion of σ relative to s (in order to get best results from the structure equations, we treat equations (b) and (c) separately). In the case of (b) we have

$$\sigma(0) = 0, \qquad \sigma'(0) = 1, \qquad 1 \leq \sigma'(s) \leq \rho(s),$$

where

$$\rho(s) = \Big(1 - \sum_{j=1}^{n} \frac{\partial u}{\partial x_j}\Big|_{x_0} \frac{\partial u}{\partial x_j}\Big|_{x(s)}\Big) \cdot w^{-1}(x_0) w^{-1}(x(s)) \geq 1,$$

$$\text{equality holding only if } \frac{\partial u}{\partial x_j}\Big|_{x(s)} = \frac{\partial u}{\partial x_j}\Big|_{x_0},$$

and

$$|d\rho(s)/ds| \leq [(n-1)(\rho^2(s) - 1)]^{1/2} Q(s),$$

where $w(x(s))$ is defined in (2) and $Q(s)$ in Lemma 2. In the case of equation (c) we have directly $\sigma(0) = 0$, $\sigma'(0) = 1$, $d\sigma/ds > 0$ and

$$\left|\frac{d^2\sigma}{ds^2}\right| \leq \frac{n-1}{n^{1/2}} \frac{d\sigma}{ds} Q(s).$$

Integrating both of the differential inequalities one gets, respectively: for solutions of equation (b)

$$\rho(s) \leq \cosh \int_0^s (n-1)^{1/2} Q(s_1) ds_1,$$

while in the case of (c)

$$d\sigma/ds \leq \exp\left\{\frac{n-1}{n^{1/2}} \int_0^s Q(s_1) ds_1\right\}.$$

We estimate $\int_0^s Q(s_1) ds_1$ for $0 < s < \gamma(x_0)$ as follows: for $0 \leq s_1 \leq \alpha$ for some fixed $\alpha < \gamma(x_0)$ we can use Theorem 1, whereby $Q(s_1) < c'/\gamma(x(s_1)) = c'(\gamma(x_0) - s_1)^{-1}$ for some positive constant c' depending only on n. For the interval $\alpha \leq s_1 \leq s$ we use Lemma 2, which gives an essentially sharp estimate of upper bounds for the integral of $Q(s_1)$ over compact subintervals of Γ. Integrating the estimated upper bound for $\sigma'(s)$ we see that $\sigma(s)$ is bounded as $s \to \gamma(x_0)$ for $n < 5$ in the case of equation (b) while in the case of (c) we get boundedness for $(n-1)/n^{1/2} < 2$, or in other words $n \leq 5$. This completes the proof of the Bernstein theorem for equation (b) for $n \leq 4$ and (c) for $n \leq 5$.

It is conjectured that a conceivable example disproving an analogous Bernstein statement for some higher value of n could be constructed by the method of axially symmetric functions

$$u(x_1, \ldots, x_n) = \bar{u}\left(x_1, \left(\sum_{j=2}^{n} (x_j)^2\right)^{1/2}\right),$$

where $\bar{u}(x, y)$ is a function defined in $\mathbf{R}^2$, satisfies the symmetrized form of (b) or (c) and such that the positive x_1-axis is the geodesic arc Γ of length $\gamma(0) < \infty$ in terms of the induced Riemannian metric.

In the case of problem (d) the question now open is to estimate the rate of growth of a norm of the curvature tensor of the Kähler metric (5), in such a way that, if the curvature is not zero at some point x_0, then a singularity of the solution of equation (4) must be encountered within a distance from x_0 that one can majorize *a priori*. It is possible that the recent results by Thierry Aubin (Thèse, Collège de France, 1969) may furnish the necessary technique.

BIBLIOGRAPHY

1. E. Calabi, *A generalization of E. Hopf's maximum principle with an application to Riemannian geometry,* Duke Math. J. **25** (1958), 45–56; Erratum, *ibid.* **26** (1959), 707.

2. ———, *Improper affine hyperspheres of convex type etc.,* Michigan Math. J. **5** (1958), 105–126.

3. ———, *On Ricci curvature and geodesics,* Duke Math. J. **34** (1967), 667–676.

4. R. Osserman, *On the inequality* $\Delta u \geq f(u)$, Pacific J. Math. **7** (1957), 1641–1647.

UNIVERSITY OF PENNSYLVANIA

Über singuläre symplektische Strukturen

Eugenio Calabi

In Tagungsbericht 23/1969,
Differentialgeometrie im Großen 13.7. bis 19.7.1969,
Mathematisches Forschungsinstitut Oberwolfach, p. 4.

Man betrachtet das Problem der Existenz einer symplektischen Struktur auf einer 2n-dimensionalen kompakten Mannigfaltigkeit X, wo vorausgesetzt ist, daß die zwei bekannten notwendigen Bedingungen schon erfüllt sind, nämlich daß

1) es eine fast komplexe Struktur in X gibt,

2) es eine 2-dimensionale Cohomologieklasse α in X gibt mit $\alpha^n \neq 0$.

Es sei ω eine geschlossene 2-Form, die die Klasse α darstellt, also $\int_X \omega^n > 0$ (in Bezug auf eine geeignete Orientierung). Dann ist der Ort der Nullstellen (=S) der 2n-Form ω^n ein Träger der Hindernisse für die Existenz einer symplektischen Struktur. Wenn man zunächst Transversalitätsbedingungen für S voraussetzt, dann gibt es das Problem, die Keime dieser Singularitäten von ω zu charakterisieren. Durch die von MARTINET und dem Verfasser gewonnenen Resultate ist die Klassifizierung dieser Singularitätenkeime im Fall n=2 vollständig gelöst.

Complete Affine Hyperspheres I(*) (**)

Eugenio Calabi

Symposia Mathematica, Vol. X
Convegno di Geometria Differenziale
INDAM, Rome, 1971/1972, 19–38,
Academic Press, London (1972)

1. Introduction.

Affine hyperspheres are a class of hypersurfaces in an affine $(n+1)$-space, considered from the viewpoint of invariants under the group G of unimodular affine transformations. For $n = 2$ they were first treated by G. Tzitzèica [7, 8] who called them S-surfaces and later by Blaschke [2], who renamed them affine spheres, redefining them in terms of affine invariants.

A hypersurface in the unimodular affine $(n+1)$-space E is regarded as a differentiable immersion $\boldsymbol{X}$ of an n-dimensional manifold M in E. The immersion $\boldsymbol{X}$ defines on M a quadratic form G on its tangent bundle with values in the line bundle of n-forms, defined in terms of local coordinates $(u^1, \ldots, u^n)$ by the equation

$$G = \left[\frac{\partial^2 \boldsymbol{X}}{\partial u^\alpha \partial u^\beta}, \frac{\partial \boldsymbol{X}}{\partial u^1}, \ldots, \frac{\partial \boldsymbol{X}}{\partial u^n}\right] (du^\alpha du^\beta) \otimes du^1 \wedge \ldots \wedge du^n, \tag{1.1}$$

where $[\boldsymbol{A}, \boldsymbol{B}, \ldots, \boldsymbol{C}]$ denotes the volume of the $(n+1)$-parallelotope spanned by $\boldsymbol{A}, \boldsymbol{B}, \ldots, \boldsymbol{C}$. If this quadratic form is assumed to be nonsingular, it defines a pseudoriemannian structure tensor g with corresponding volume element ω, uniquely defined by the equation

$$G = g \otimes \omega, \tag{1.2}$$

whence the discriminant $|G|$ of G is equal to the $(n+2)^{nd}$ tensor power of ω, up to choice of sign depending on the orientation of M. Throughout this paper we consider only the cases where g is positive definite:

(*) I risultati conseguiti in questo lavoro sono stati esposti nella conferenza tenuta il 24 maggio 1971.

(**) Research partly supported by the National Science Foundation, Contract No. GP-29258.

this means that $\mathbf{X}(M)$ is locally strongly convex, and oriented so that G is positive valued. With this Riemannian structure and the corresponding affine connection, one defines the affine normal vector $\boldsymbol{N}$ at each point of M by the equation

$$\boldsymbol{N}=\frac{1}{n}\Delta\mathbf{X}\,, \tag{1.3}$$

where $\Delta = g^{\alpha\beta}\nabla_\alpha\nabla_\beta$ denotes the Laplace-Beltrami operator with respect to g. Synthetic geometric interpretations for $\boldsymbol{N}$, ω and g are given by Blaschke [1, pp. 115, 125-27, 128 Problem 2] for surfaces in affine 3-space, with an immediately obvious generalization to higher dimensions.

DEFINITION 1.1: *An immersed, locally strongly convex hypersurface* $\mathbf{X}(M)\subset E$ *is called an affine hypersphere, if the affine normal lines through each point of the image either all intersect at one point, called the* center *of* $\mathbf{X}(M)$, *or else are all mutually parallel* (*center at infinity*). *An affine hypersphere is called of elliptic, parabolic* (*or improper*) *or hyperbolic type, according to whether the center is, respectively, on the concave side of* $\mathbf{X}(M)$, *at infinity or on the convex side.*

The prototypes of affine hyperspheres of the three types are given by the locally strongly convex quadrics, namely the ellipsoids, convex paraboloids and 2-component hyperboloids, after which the types of hyperspheres are named. However there are many others beside these, except in the case of plane curves ($n = 1$), where the nonsingular conic sections are the only « affine circles ».

From the general structure equations for the invariants of locally strongly convex hypersurfaces we have the following characterization of affine hyperspheres.

a) A locally strongly convex hypersurface $\mathbf{X}(M)$ is an affine hypersphere with center $\mathbf{0}$, *i.e.* of elliptic or hyperbolic type (« proper affine sphere » in Blaschke's terminology), if and only if the affine normal vector $\boldsymbol{N}$, defined by (1.3), satisfies the equation

$$\boldsymbol{N}=-H(\mathbf{X}-\mathbf{0})\,, \tag{1.4}$$

where H, called the affine mean curvature, is a nonzero constant. Its sign, positive or negative, determines whether the resulting hypersphere is, respectively, of elliptic or of hyperbolic type.

b) A locally strongly convex hypersurface is a hypersphere of parabolic type, or improper hypersphere, if and only if its affine normal vector $\boldsymbol{N}$ is constant; the affine mean curvature, in this case, is identically zero.

Affine hyperspheres of the three types have also a nonparametric characterization; this representation has the advantage of reducing the classification of affine hyperspheres to that of the solutions of a certain partial differential equation. The letter has its most convenient formulation in terms of the Legendre transformation, as defined below.

Let $(x_1, \ldots, x_{n+1})$ be an affine coordinate system with normalized volume element for the space E, and let a strongly locally convex hypersurface $\mathbf{X}: M \to E$ be represented in a neighborhood of a point by a strongly locally convex function

$$x_{n+1} = f(x_1, \ldots, x_n) = f(x)$$

for $(x_1, \ldots, x_n)$ ranging in a domain $D \subset \mathbb{R}^n$. Let $\hat{D} \subset \mathbb{R}^n$ be the image of D under the locally invertible differentiable map

$$(x_1, \ldots, x_n) \xrightarrow{\nabla f} \left(\frac{\partial f}{\partial x_1}, \ldots, \frac{\partial f}{\partial x_n}\right).$$

The Legendre transform of f is defined to be the function $u(\xi_1, \ldots, \xi_n)$: $\hat{D} \to \mathbb{R}$ that is uniquely determined by the equation

$$u\left(\frac{\partial f}{\partial x_1}, \ldots, \frac{\partial f}{\partial x_n}\right) = -f(x_1, \ldots, x_n) + \sum_{\alpha=1}^{n} x_\alpha \frac{\partial f(x)}{\partial x_\alpha}. \tag{1.5}$$

PROPOSITION 1.2: *Let $\mathbf{X}: M \to E$ be an immersed hypersurface, represented locally by a locally strongly convex function $x_{n+1} = f(x_1, \ldots, x_n)$. Then we have the following characterization of $\mathbf{X}$ as affine hypersphere.*

a) *The hypersurface $\mathbf{X}: M \to E$ is a proper affine hypersphere with center at the origin and mean curvature H, if and only if the Legendre transform function $u(\xi) = u(\xi_1, \ldots, \xi_n)$ satisfies the equation*

$$\det\left(\frac{\partial^2 u(\xi)}{\partial \xi_\alpha \partial \xi_\beta}\right) = (Hu(\xi))^{-n-2}. \tag{1.6}$$

b) *The hypersurface $\mathbf{X}: M \to E$ is affine hypersphere of parabolic type (improper affine hypersphere), with the affine normal vector everywhere equal to $(0, 0, \ldots, 0, 1)$, if and only if the function $x_{n+1} = f(x)$ satisfies the equation*

$$\det\left(\frac{\partial^2 f(x)}{\partial x_\alpha \partial x_\beta}\right) = 1; \tag{1.7}$$

this condition is equivalent also, in terms of the Legendre transform $u(\xi)$ of f, to the equation

$$\det\left(\frac{\partial^2 u(\xi)}{\partial\xi_\alpha\,\partial\xi_\beta}\right)=1\,. \tag{1.7'}$$

Part a) of this proposition is proved in the next chapter; part b) is stated and proved, at least for $n=2$, in Blaschke [1, pp. 172, 216] (its generalization to higher dimensions is trivial). The discussion of hyperspheres of parabolic type was made in two earlier articles [4, 5]; therefore we shall be primarily interested only in the proper types.

The problems of interest in this paper are the representation, properties and (to some extent) classification of the affine hyperspheres that are complete with respect to the affine invariant metric (1.1), (1.2). In this case one easily deduces from the structure equations that the complete hyperspheres of elliptic type are necessarily closed (and, indeed, must be ellipsoids). The analogous result for the parabolic case was also proved earlier [4], that they are paraboloids. However it is possible to construct many other types of complete affine hyperspheres of hyperbolic type for $n\geqslant 2$; of these, the simplest to describe is given by any of the 2^n components of the affine algebraic hypersurface

$$x_1\,x_2\,\ldots\,x_{n+1}=c\,,\qquad c=\text{constant}\neq 0. \tag{1.8}$$

This last example is discussed in Chapters 3 and 4. In Chapter 3 it is shown how affine hyperspheres can be composed together to obtain higher dimensional ones; the example defined by (1.8) results from the composition of trivial « 0-dimensional hyperspheres »; in Chapter 4 another method is introduced to construct affine hyperspheres that are homogeneous under the action of certain Lie groups. The final chapter is mainly devoted to the theorem that, in a complete affine hypersphere of hyperbolic type, the Ricci curvature must be everywhere nonpositive. This theorem will be used in a subsequent paper to construct more general types of hyperspheres, working toward the verification of the following conjecture.

CONJECTURE: *Every complete, n-dimensional affine hypersphere with mean curvature $H<0$ is asymptotic to the boundary of a convex cone with vertex at the center; conversely, every pointed, nondegenerate convex cone K determines an affine hypersphere of hyperbolic type, asymptotic to the boundary of K, and uniquely determined by the value $H<0$ of its mean curvature.*

2. The structure equations.

Let $X: M \to E$ be a locally strongly convex hypersurface in the $(n+1)$-space E. For each local coordinate system $(u_1, \ldots, u_n)$ on M we have a field of frames at each point of its immersed image, made up of the $n+1$ vectors $(X_{|1}, \ldots, X_{|n}, N)$, where $X_{|\alpha} = \partial X / \partial u_\alpha$ and N is the affine normal vector. The alternating scalar product of the frame $[X_{|1}, \ldots, X_{|n}, N]$ gives the density w of the affine volume element with respect to the coordinate system, while the product $[X_{|1}, \ldots, X_{|n}, X_{|\alpha\beta}]$, $X_{|\alpha\beta} = \partial^2 X / \partial u_\alpha \partial u_\beta$, gives the coefficients of the quadratic form (1.1). It follows that the vector-valued matrix $X_{|\alpha\beta}$ of second derivatives, modulo the vectors in the tangent hyperplane are equivalent to $g_{\alpha\beta} N$, where $g_{\alpha\beta}$ are the coefficients of the affine invariant metric (1.2). Introducing the Christoffel symbols $\Gamma^\gamma_{\alpha\beta}$ of the affine connection of this metric, we obtain the structure equation

$$X_{|\alpha\beta} = \Gamma^\gamma_{\alpha\beta} X_{|\gamma} + A^\gamma_{\beta\alpha} X_{|\gamma} + g_{\alpha\beta} N . \tag{2.1}$$

The second structure equation has to do with the differential of the affine normal vector N

$$N_{|\alpha} = -B^\beta_\alpha X_{|\beta} ,$$

which turns out to be in the tangent hyperplane because of integrability conditions.

We shall henceforth assume that $X: M \to E$ is an affine hypersphere of elliptic or hyperbolic type, with mean curvature $H \neq 0$ and center at the origin in E. This assumption simplifies the second structure equation, replacing it by the condition

$$N = -HX , \tag{2.2}$$

whence the tensor (B^β_α) is equal to H times the identity operator on the tangent space.

The structure equations (2.1) and (2.2) depend on the intrinsic data on M consisting of the Riemannian structure $(g_{\alpha\beta})$ and the tensor $(A^\beta_{\alpha\beta})$, which are related by the following integrability conditions.

a) The trilinear form $A_{\alpha\beta\gamma} du^\alpha \otimes du^\beta \otimes du^\gamma$ where $A_{\alpha\beta\gamma} = g_{\gamma\lambda} A^\lambda_{\alpha\beta}$ is symmetric in all pairs of indices; it is called the fundamental cubic form.

b) The cubic form $A_{\alpha\beta\gamma} du^\alpha du^\beta du^\gamma$ is apolar with respect to the metric, meaning that its contracted linear form, whose coefficients are $g^{\alpha\beta}A_{\alpha\beta\gamma}$ vanishes identically;

(2.4) $$g^{\alpha\beta}A_{\alpha\beta\gamma} = A^{\alpha}_{\alpha\gamma} \equiv 0 .$$

c) The coefficient tensor $A_{\alpha\beta\gamma}$ of the cubic form satisfies a « Codazzi » condition

(2.5) $$\nabla_\lambda A_{\alpha\beta\gamma} = \nabla_\gamma A_{\alpha\beta\lambda} ,$$

meaning that its covariant derivative defines a symmetric quadrilinear form.

d) The curvature tensor is uniquely determined by the cubic form tensor under the equation

(2.6) $$R_{\alpha\beta\gamma\delta} = g^{\lambda\mu}(A_{\lambda\alpha\delta}A_{\mu\beta\gamma} - A_{\lambda\alpha\gamma}A_{\mu\beta\delta}) + H(g_{\alpha\gamma}g_{\beta\delta} - g_{\alpha\delta}g_{\beta\gamma}) .$$

The detailed proofs of these structure equations, without the assumption that X be an affine hypersphere, can be found in [1, pp. 167-172]; the latter assumption simplifies somewhat the equations. The cubic form $A_{\alpha\beta\gamma}du^\alpha du^\beta du^\gamma$ vanishes identically and exclusively in the case of the non-singular quadric hypersurfaces, so that it gauges the extent to which the third order jet of $X(E)$ at each point deviates from that of a quadric hypersurface.

From the integrability conditions (2.4) and (2.6) we obtain the following relation involving the Ricci tensor $R_{\alpha\gamma} = g^{\beta\delta}R_{\alpha\beta\gamma\delta}$ and the cubic form:

(2.7) $$R_{\alpha\beta} = A^{\lambda}_{\alpha\mu}A^{\mu}_{\beta\lambda} + (n-1)Hg_{\alpha\beta} .$$

This relation leads immediately to the following:

PROPOSITION 2.1: *In affine hypersphere with mean curvature H, the Ricci curvature in any tangential direction, represented in terms of a tangent vector $\xi = \xi^\alpha(\partial/\partial u_\alpha) \neq 0$ by the expression*

(2.8) $$Q(\xi) = \frac{1}{n-1}(R_{\alpha\beta}\xi^\alpha\xi^\beta)(g_{\alpha\beta}\xi^\alpha\xi^\beta)^{-1} ,$$

has the constant H as an absolute lower bound.

The proof of the above proposition is immediate from (2.7). In view of this relation, we find that a convenient definition of a pointwise norm ψ of the cubic form is the following one.

Definition 2.2: *At each x of affine hypersphere $X\colon M \to E$ we define the norm of the cubic form at x to be the following expression*

$$\psi(x) = \sup_{\xi\in T_x} \left\{ \left(\frac{A^{\alpha}_{\beta\lambda} A^{\beta}_{\alpha\mu} \xi^{\lambda} \xi^{\mu}}{n-1} \right)^{\frac{1}{2}} \middle| \, g_{\alpha\beta}\xi^{\alpha}\xi^{\beta} = 1 \right\}. \tag{2.9}$$

Another important consequence of the structure equations of an affine hypersphere is a simple formula for the Laplacian of the cubic form, that plays an important role in the next section.

Proposition 2.3: *The Laplacian of the cubic form in an affine hypersphere with mean curvature H, defined to be the cubic form*

$$(\Delta A)_{\alpha\beta\gamma} du^{\alpha} du^{\beta} du^{\gamma} = (\nabla^{\lambda}\nabla_{\lambda} A_{\alpha\beta\gamma}) du^{\alpha} du^{\beta} du^{\gamma},$$

has as its coefficient tensor the following expression

$$(\Delta A)_{\alpha\beta\gamma} = A^{\lambda}_{\alpha\mu} A^{\mu}_{\lambda\varrho} A^{\varrho}_{\beta\gamma} + A^{\lambda}_{\beta\mu} A^{\mu}_{\lambda\varrho} A^{\varrho}_{\gamma\alpha} + A^{\lambda}_{\gamma\mu} A^{\mu}_{\lambda\varrho} A^{\varrho}_{\alpha\beta} - 2 A^{\lambda}_{\alpha\mu} A^{\mu}_{\beta\varrho} A^{\varrho}_{\gamma\lambda} + (n+1) H A_{\alpha\beta\gamma}. \tag{2.10}$$

The proof of the above formula is a straightforward computation from the integrability conditions (2.4), (2.5), (2.6) and the standard Ricci identities on the second order covariant derivatives.

We conclude this section with a proof of part *a*) of Proposition 1.2. If we use $\xi_1, \ldots, \xi_n$ as local coordinates for M and introduce the Legendre transform function $u(\xi_1, \ldots, \xi_n)$, the latter is strongly convex, its partial derivatives are the original affine coordinate functions x_i ($1 \leqslant i \leqslant n$, $x_i = \partial u/\partial \xi_i$); therefore the mapping function X is given by the column vector

$$X = \begin{pmatrix} u_{|1} \\ \vdots \\ u_{|n} \\ -u + \sum_{i=1}^{n} \xi_i u_{|i} \end{pmatrix} \qquad u_{|i} = \partial u/\partial \xi_i. \tag{2.11}$$

One then calculates the first and second partial derivatives of X to represent the affine invariant metric

$$g_{ij} = \varrho u_{|ij},$$

where $u_{|ij} = \partial^2 u/\partial\xi_i \partial\xi_j$ and $\varrho = (\det(u_{|ij}))^{1/n+2}$. Continuing with the calculation of the Christoffel symbols, one obtains the following formula for the affine normal vector

$$(2.12)\qquad N = \begin{pmatrix} \varrho^{-2}\varrho_{|1} \\ \vdots \\ \varrho^{-2}\varrho_{|n} \\ \varrho^{-1} + \varrho^{-2}\sum_{i=1}^{n} \xi_i \varrho_{|i} \end{pmatrix}.$$

Substituting the last two numbered formulas in the equation (1.4) defining an affine hypersphere, we see that a necessary and sufficient condition in order that $N = -HX$ is that

$$\varrho = \frac{1}{Hu},$$

which is equivalent to (1.6), completing the proof of part *a*) of Proposition 1.2.

3. A composition formula.

In this chapter we show a construction whereby any two given affine hyperspheres of hyperbolic type can be composed to yield a third one, whose dimensions exceed by one unit the sum of the dimensions of the two original ones.

Let $\boldsymbol{Y}\colon M' \to E'$ and $\boldsymbol{Z}\colon M'' \to E''$ be affine hyperspheres of p and q dimensions respectively (in affine $(p+1)$- and $(q+1)$-space), of hyperbolic type and with affine mean curvatures respectively H' and H''; we shall endow E' and E'' with the structure of unimodular vector spaces rather than affine spaces by fixing in each an origin, namely the centers of $\boldsymbol{Y}(M')$ and $\boldsymbol{Z}(M'')$.

We shall now construct a $(p+q+1)$-dimensional affine hypersphere of hyperbolic type in $E' \oplus E''$ as follows: the abstract manifold M is the Cartesian product $M' \times \mathbb{R} \times M''$ with the orientation induced by the indicated ordering of the factors; to define the mapping $\boldsymbol{X}\colon M \to E' \oplus E''$, we let t denote the canonical parameter of $\mathbb{R}$, choose two positive constants c' and c'' (arbitrarily for the moment) and set

$$(3.1)\qquad \boldsymbol{X} = \begin{pmatrix} c'\boldsymbol{Y}\exp\dfrac{-t}{p+1} \\ c''\boldsymbol{Z}\exp\dfrac{t}{q+1} \end{pmatrix}.$$

Direct calculation of the first and second derivatives of $\boldsymbol{X}$ leads to the following formula for the affine invariant metric

$$(3.2)\qquad g = c\left\{(p+1)(-H')g' \oplus \frac{p+q+2}{(p+1)(q+1)}\,dt^2 \oplus (q+1)(-H'')g''\right\},$$

where g' and g'' are the affine invariant metrics defined by $\boldsymbol{Y}$ and $\boldsymbol{Z}$ respectively, and c is a positive constant, defined by the following, hideous equation,

$$(3.3)\qquad c^{p+q+3} = \frac{c'^{2p+2}c''^{2q+2}}{(p+q+2)(p+1)^{p+1}(q+1)^{q+1}(-H')^{p+2}(-H'')^{q+2}}\,.$$

Thus the metric induced by $\boldsymbol{X}$ on M is the cartesian product, up to constant factors, of the metrics on $\boldsymbol{Y}$, $\mathbb{R}$ and $\boldsymbol{Z}$. The affine normal vector $\boldsymbol{N}$ induced by $\boldsymbol{X}$ can now be readily calculated from (3.1) and (3.2), yielding the following formula:

$$(p+q+1)\boldsymbol{N} = \Delta\boldsymbol{X} =$$

$$= \frac{1}{c}\left(\begin{array}{l}\dfrac{c'\exp[(-t)/(p+1)]}{(p+1)(-H')}\Delta\boldsymbol{Y} + \dfrac{(q+1)c'\exp[(-t)/(p+1)]}{(p+1)(p+q+2)}\boldsymbol{Y}\\[2ex] \dfrac{c''\exp[t/(q+1)]}{(q+1)(-H'')}\Delta\boldsymbol{Z} + \dfrac{(p+1)c''\exp[t/(q+1)]}{(q+1)(p+q+2)}\boldsymbol{Z}\end{array}\right) =$$

$$= \frac{(p+q+1)}{c(p+q+2)}\boldsymbol{X}\,.$$

This shows that the resulting immersion $\boldsymbol{X}$ is an affine hypersphere of hyperbolic type, with center at the origin and mean curvature

$$H = \frac{-1}{(p+q+2)c}$$

with the positive constant c defined by (3.3). It is now easy to normalize the constants c' and c'' to assign to H a reasonable value; for instance, if one takes $c' = c'' = 1$, the affine mean curvature H of $\boldsymbol{X}$ is given by the following,

$$(3.4)\qquad (p+q+2)^{p+q+2}(-H)^{p+q+3} =$$

$$= (p+1)^{p+1}(-H')^{p+2}(q+1)^{q+1}(-H'')^{q+2}.$$

One verifies that the construction just described defines a composition in the set of isomorphism classes of affine hyperspheres of hyperbolic type: this composition is naturally associative and commutative, in view of (3.1) and (3.4). The construction remains valid, even if one takes as one or more of the factor « spheres » Y, Z, ... a zero-dimensional one; in this case the « affine hypersphere » Y is merely a point $Y \in \mathbb{R} - \{0\}$ and H' an arbitrary negative constant.

The composition of $n+1$ copies of this trivial 0-dimensional hypersphere yields precisely the example (1.8) of an n-dimensional complete, affine hypersphere of hyperbolic type: the affine metric in $\mathbb{R}^n$ and the cubic form, being translation invariant, is a covariant constant, and is different from zero for $n \geqslant 2$, since the hypersurface is not a quadric except for $n = 1$. On the other hand each component of an n-dimensional 2-component hyperboloid, say the one defined by the equation

$$x_{n+1} = \Big(\sum_{i=1}^{n} x_1^2\Big)^{\frac{1}{2}}$$

is a complete affine hypersphere whose affine invariant metric, for $n \geqslant 2$, has constant, negative curvature. This shows that the quadrics for $n \geqslant 2$ are irreducible with respect to the composition just described while the (1-dimensional) hyperbola component $x_1 x_2 =$ constant is representable as the composition of two 0-dimensional affine hyperspheres. More generally, one can easily verify that the composition of p-dimensional, locally convex quadrics ($p \geqslant 2$) and 0-dimensional hyperspheres yields as many affinely inequivalent, n-dimensional, complete, affine hyperspheres of hyperbolic type as there are partitions of $n+1$ into nonnegative integers $\neq 2$.

4. Homogeneous affine hyperspheres.

In this section we describe another method of constructing affine hyperspheres of hyperbolic type producing additional isomorphism classes of complete affine hyperspheres.

Let K be an open, convex cone in $\mathbb{R}^{n+1}$; we say that K is nondegenerate, if it is not empty and contains no (complete) affine line; this means that K is not a cartesian product of a lower dimensional cone with a euclidean space. There are certain cones K of this type that are *homogeneous*, meaning that there exists a closed subgroup of the full matrix group $GL(n+1, \mathbb{R})$ whose action on $\mathbb{R}^{n+1}$ has K as one of its orbits. For the existence and classification of such groups, see [6] [9]. Without loss of generality we assume that $\mathcal{G}$ is the largest subgroup of $GL(n+1, \mathbb{R})$ that maps K onto itself. Since $\mathcal{G}$ evidently contains in its center the group $\mathfrak{Z}$ of positive scalar multipliers, we can consider the quotient group $\mathcal{G}/\mathfrak{Z} = P\mathcal{G}$ with its action on the open domain $\hat{K}$ in the projective n-space $\mathbb{R}P^n$, defined as the quotient space of K under the decomposition by rays through the origin.

Definition 4.1: *A closed subgroup of $GL(n+1, \mathbb{R})$ that has an open, nondegenerate cone $K \subset \mathbb{R}^{n+1}$ as one of its orbits is said to operate primitively if there is no equivalence relation other than the two trivial ones in the projective image $\hat{K} \subset \mathbb{R}P^n$ of K, that is preserved under the action of $P\mathcal{G}$.*

An equivalent definition of a primitive action of $\mathcal{G}$ on a cone is that subgroup of $P\mathcal{G}$ that maps a point $r_0 \in \hat{K}$ into itself is a maximal closed, proper subgroup of $P\mathcal{G}$.

If a group $\mathcal{G}$ operates transitively on a proper convex cone $K \subset \mathbb{R}^{n+1}$, then we denote by $S\mathcal{G}$ the unimodular part of $\mathcal{G}$, that is to say normal subgroup $\mathcal{G} \cap SL(n+1, \mathbb{R})$ of $\mathcal{G}$. Clearly $\mathcal{G}$ is a direct product of $S\mathcal{G}$ with $\exp(\mathbb{R})$, the multiplicative positive scalar group, and $S\mathcal{G}$ is isomorphic as an abstract Lie group to $P\mathcal{G}$.

Theorem 4.2: *Let $\mathcal{G}$ be a closed subgroup of $GL(n+1, \mathbb{R})$ that operates transitively and primitively on a nondegenerate open cone $K \subset \mathbb{R}^{n+1}$, this cone being invariant under $\mathcal{G}$. Then the orbit of any point $x_0 \in K$ under the action of the unimodular subgroup $S\mathcal{G} = SL(n+1, \mathbb{R}) \cap \mathcal{G}$ of $\mathcal{G}$ is an affine hypersphere in $\mathbb{R}^{n+1}$, homogeneous under the action of $S\mathcal{G}$.*

Proof: We show first that $S\mathcal{G}$ can not operate transitively on K, so that orbit of a point is necessarily a hypersurface, since $S\mathcal{G}$ has codimension 1 in $\mathcal{G}$. In fact, for any $x_0 \in K$ consider the hyperplanes through x_0 that are transversal to every ray in the closure of K and hence bound, together with the boundary of K a convex body (truncated cone). Among these hyperplanes there is a unique one, call it Π_{x_0}, for fixed $x_0 \in K$, that minimizes the volume of this truncated cone; x_0 lies in the centroid of the intersection $\Pi_{x_0} \cap \bar{K}$ therefore it is obvious that for every point x_1 in the orbit of x_0 under the action of $S\mathcal{G}$ the corresponding hyperplane Π_{x_1} through x_1 bounds inside K another truncated cone with volume equal to the one bounded by Π_{x_0}; therefore x_1 can not lie in the closure of the truncated cone bounded by Π_{x_0} unless $x_1 = x_0$; this shows that the orbit of x_0 under $S\mathcal{G}$ lies entirely outside the truncated cone and Π_{x_0} is the hyperplane tangent to the orbit at x_0, and this orbit is a strictly convex hypersurface.

Now to prove the assertion of the theorem, consider the family of hyperplanes parallel to Π_{x_0} and separated by Π_{x_0} from the origin. Each of them intersects the closure of the cone K in a bounded region, and the union of the centroids of these intersections is the part of the ray $\{tx_0 | t \in \mathbb{R}^+\}$ for which $t > 1$. Next, consider the orbit Σ cf x_0 under the action of $S\mathcal{G}$ and perform a similar operation: take the locus of the centroids of all the interections of the convex hull of Σ with hyperplanes parallel to Π_{x_0} that intersect Σ in a closed, $(n-1)$-dimensional, convex hypersurface. The result will be a path originating at x_0, called the barycentric path from x_0. This curve is asymptotic to the ray through x_0 as one extends it to infinity; its initial tangential direction (cf. [1, pp. 114-116]) is the affine normal line. On the other hand the barycentric path through x_0 spans a 1-parameter family of rays that is obviously invariant under the subgroup $\mathcal{K}_{x_0}$ of $S\mathcal{G}$ that leaves x_0 fixed. Since $S\mathcal{G}$ acts primitively on the set of rays, the barycentric path through x_0 must necessarily coincide with the ray through x_0. This shows that the affine normal vector of Σ through x_0 is parallel to the vector from the origin to x_0 and hence Σ is an affine hypersphere, as asserted.

We list some examples of nondegenerate cones which are orbits of subgroups of $GL(n+1, \mathbb{R})$ with primitive action.

a) The quadric cones $(x_{n+1} = (x_1^2 + \ldots + x_n^2)^{\frac{1}{2}})$ with the (isochronous subgroup of the) Lorentz group on them; the hyperspheres then are merely the component inside K of the 2-sheeted hyperboloid asymptotic to the boundary.

b) The positive quadrant $\{x_i > 0,\ 1 \leqslant i \leqslant n+1\}$, with $\mathcal{G}$ equal to the group of permutations of the coordinates extended by the positive diagonal group (in this case the identity component, the diagonal group, acts transitively but not primitively on K). The affine hypersphere is again the set $\left\{\prod_{i=1}^{n} x_i = \text{const},\ (x_i > 0)\right\}$.

c) The groups $GL(m, \mathbb{R})$, $GL(m, \mathbb{C})$ and $GL(m, \mathbb{H})$ have representations respectively in $\mathbb{R}^{m(m+1)/2}$, $\mathbb{R}^{m^2}$ and $\mathbb{R}^{m(2m-1)}$ where the latter are considered as the space of real quadratic, complex-hermitian and quaternion-hermitian forms on $\mathbb{R}^m$, $\mathbb{C}^m$ and $\mathbb{H}^m$ respectively; in each case the positive definite forms constitute orbits under the action of $GL(m, *)$ and this action is primitive.

d) The cartesian product of k equal copies $(k \geqslant 2)$ of a convex cone K which is an orbit under a closed subgroup $\mathcal{G}$ of the full linear group, the subgroup acting primitively on the cone; the cone K^k is an orbit under the action of the Kth cartesian power $\mathcal{G}^k$ of $\mathcal{G}$; the group $\mathcal{G}^k$ does not act primitively on K^k, but its extension by the symmetric group of all permutations of the factors acts primitively. The resulting affine hypersphere is the same as the composition of k copies of the hypersphere in K homogeneous under $\mathcal{G}$.

The fact that the affine hyperspheres can be composed to yield hyperspheres inside a homogeneous cone even when the resulting group does not act primitively, as is the case when not all the factors are homogeneous, indicates that the assumption of primitivity in Theorem 4.2 can be weakened. Indeed it has been shown recently by S. Gigena that any cone K that is an orbit under a closed subgroup $\mathcal{G} \subset GL(n+1, \mathbb{R})$ determines an hypersphere by the same construction, with no further assumption on the action of $\mathcal{G}$ on K.

5. A growth property of the cubic form.

In this chapter we wish to generalize to the affine hyperspheres of hyperbolic type some of the results estalished in earlier papers for parabolic type. The main result is the following

THEOREM 5.1: *Let $X\colon M \to E$ be an n-dimensional hypersphere of hyperbolic type. If it is complete with respect to the affine invariant metric, then its Ricci curvature is non-positive in every direction, at every point. The only complete affine hyperspheres of elliptic or parabolic types are quadrics.*

The proof of this theorem follows from two lemmas.

LEMMA 5.2: *Let $X\colon M \to E$ be an n-dimensional affine hypersphere with mean curvature H (positive, negative or zero) and let the function ψ denote the norm of the cubic form at each point as defined in (2.9). Then ψ satisfies the differential inequality*

$$\Delta\psi \geqslant (n+1)(\psi^3 + H\psi)\,. \tag{5.1}$$

PROOF: Since ψ as defined is continuous but not differentiable, the above inequality must be proved in the weak sense as defined in [3]. Let $x_0 \in M$ and let $\xi(x_0)$ be a unit tangent vector at x_0 that maximizes the quadratic form

$$A_{\alpha\mu} A^{\mu}_{\beta\lambda} \xi^{\alpha} \xi^{\beta}$$

occurring in (2.9). Extend $\xi(x_0)$ to a vector field η in a geodesically convex neighborhood U of x_0 by translating $\xi(x_0)$ by Levi-Civita parallelism along the geodesics from x_0 and consider the differentiable function

$$\bar{\psi} = \left(\frac{A^{\lambda}_{\alpha\mu} A^{\mu}_{\beta\lambda} \eta^{\alpha} \eta^{\beta}}{n-1}\right)^{\frac{1}{2}};$$

then, since η is clearly a unit vector field, we have $\bar{\psi} \leqslant \psi$ everywhere in U with equality holding at x_0. According to the definition of the inequality (5.1) in the weak sense, it is sufficient to prove it in the usual sense with the function ψ replaced by $\bar{\psi}$, at the point x_0 alone. In calculating $\Delta\bar{\psi}$ at x_0 we have the advantage of knowing that, since η is self-parallel along every geodesic from x_0, we have

$$\nabla\eta(x_0) = \Delta\eta(x_0) = 0 .$$

Consequently we have, if $\psi(x_0) = \bar{\psi}(x_0) \neq 0$,

$$\begin{aligned}
(n-1)\,\Delta\bar{\psi}(x_0) &= \frac{1}{2\bar{\psi}(x_0)}\,\Delta(A^{\lambda}_{\alpha\mu} A^{\mu}_{\beta\lambda})(x_0)\,\eta^{\alpha}(x_0)\,\eta^{\beta}(x_0) - \\
&\quad - \frac{1}{4(n-1)\,\bar{\psi}^3(x_0)}\,\nabla^{\gamma}(A^{\lambda}_{\alpha\mu} A^{\mu}_{\beta\lambda})\,\nabla_{\gamma}(A^{\lambda'}_{\alpha'\mu'} A^{\mu'}_{\beta'\lambda'})\,\eta^{\alpha}\eta^{\beta}\eta^{\alpha'}\eta^{\beta'}(x_0) = \\
&= \frac{1}{\bar{\psi}(x_0)}\,A^{\lambda}_{\alpha\mu}(\Delta A)^{\mu}_{\beta\lambda}\eta^{\alpha}\eta^{\beta} + \frac{1}{(n-1)\,\bar{\psi}^3(x_0)}\,\cdot \\
&\quad \cdot\left\{A^{\lambda}_{\alpha\mu} A^{\mu}_{\beta\lambda}(\nabla^{\gamma} A^{\lambda'}_{\alpha'\mu'})(\nabla_{\gamma} A^{\mu'}_{\beta'\lambda'}) - A^{\lambda}_{\alpha\mu}(\nabla^{\gamma} A^{\mu}_{\beta\lambda})\,A^{\lambda'}_{\alpha'\mu'}(\nabla_{\gamma} A^{\mu'}_{\beta'\lambda'})\right\}\eta^{\alpha}\eta^{\beta}\eta^{\alpha'}\eta^{\beta'} \geqslant \\
&\geqslant \frac{1}{\bar{\psi}(x_0)}\,A^{\lambda}_{\alpha\mu}(\Delta A^{\mu}_{\beta\lambda})\,\eta^{\alpha}\eta^{\beta}(x_0) .
\end{aligned}$$

The Laplace-Beltrami operator in the last expression is the one treated in Proposition 2.3; the result is given by (2.10). Thus we have

$$\begin{aligned}
(n-1)\,\bar{\psi}\Delta\bar{\psi}(x_0) &\geqslant A^{\lambda}_{\alpha\mu}\eta^{\alpha}\Big(A^{\mu}_{\lambda\gamma} A^{\gamma\lambda'}_{\mu'} A^{\mu'}_{\lambda'\beta} + A^{\mu}_{\beta\gamma} A^{\gamma\lambda'}_{\mu'} A^{\mu'}_{\lambda'\lambda} + A_{\lambda\beta\gamma} A^{\lambda\gamma'}_{\mu'} A^{\mu'\mu}_{\lambda'} - \\
&\quad - 2A^{\mu\lambda'}_{\mu'} A^{\mu'}_{\lambda\gamma} A^{\gamma}_{\lambda'\beta} + (n+1)\,H A^{\mu}_{\lambda\beta}\Big)\eta^{\beta} = \\
&= (\eta^{\alpha} A^{\lambda}_{\alpha\mu} A^{\mu}_{\lambda\gamma})(\eta^{\beta} A^{\lambda'}_{\beta\mu} A^{\mu'\gamma}_{\lambda'}) + (n^2-1)\,H\bar{\psi}^2 + \\
&\quad + \big(\eta^{\alpha}(A^{\lambda}_{\alpha\mu} A^{\mu'}_{\lambda'\lambda} - A^{\lambda}_{\alpha\lambda'} A^{\mu'}_{\mu\lambda})\big)\big(\eta^{\beta}(A^{\mu}_{\beta\gamma} A^{\gamma\lambda'}_{\mu'} - A^{\lambda'}_{\beta\gamma} A^{\gamma\mu}_{\mu'})\big) \geqslant \\
&\geqslant (n-1)^2\bar{\psi}^4 + (n^2-1)\,H\bar{\psi}^2 + 2(n-1)\,\bar{\psi}^4 = \\
&= (n-1)(n+1)(\bar{\psi}^4 + H\bar{\psi}^2) ;
\end{aligned}$$

the last inequality above is a consequence of an elementary quadratic inequality of norms in tensor spaces. On the other hand, if $\psi(x_0) = 0$ the inequality $(\Delta\psi)(x_0) \geq 0$ is trivial, so that (5.1) is completely proved as asserted.

The second lemma is a generalization of one previously proved [3] in the case of nonnegative Ricci curvature.

LEMMA 5.3: *Let M be a Riemannian manifold and x_0 a given, fixed point of M. We make the following two assumptions on M and x_0:* a) *the Ricci curvature has a fixed lower bound H (positive, negative or zero) everywhere, i.e.*

$$R_{\alpha\beta}\xi^\alpha\xi^\beta \geq (n-1)Hg_{\alpha\beta}\xi^\alpha\xi^\beta \tag{5.2}$$

for every tangent vector ξ; b) *for every point $x \in M$ there is a geodesic joining it to x_0 which is shortest among all paths from x_0 to x. Then, if $f(t)$ is any differentiable monotone nondecreasing function of a real variable $t \geq 0$ and $r: M \to \mathbb{R}$ denotes the Riemannian geodesic distance from x_0, we have*

$$\Delta f(r) \leq f''(r) + (n-1)\tau(H, r)f'(r)\,, \tag{5.3}$$

where

$$\tau(H, r) = \begin{cases} \sqrt{H}\cot\left(r\sqrt{H}\right) & (H > 0) \\ \dfrac{1}{r} & (H = 0) \\ \sqrt{-H}\coth\left(r\sqrt{-H}\right) & (H < 0) \end{cases} \tag{5.4}$$

PROOF: The inequality (5.3) comes from the inequality

$$\Delta f(r) \leq f''(r) + f'(r)\Delta r$$

in the weak sense, proved in [3], for f nondecreasing, and the inequality

$$\Delta r \leq (n-1)\tau(H, r) \tag{5.5}$$

which follows from an integral estimate of the Jacobi equation for the variation of geodesics; the method is entirely similar to that used in [3] for the case $H = 0$, so that the details need not be included here.

We now return to the proof of Theorem 5.1.

If $\mathbf{X}: M \to E$ is a complete, n-dimensional affine hypersphere with affine mean curvature H, we distinguish the three cases determined by the sign of H. In all cases the Ricci curvature has H as an absolute lower bound, as stated in Proposition 2.1.

If $H>0$, we can apply Myers' theorem, so that M, being complete and of finite geodesic diameter, is a closed manifold. On the other hand, if the cubic form is not identically zero, its norm ψ satisfies (5.1) and is therefore strictly subharmonic at points where it is positive; this makes it impossible for ψ to attain a positive maximum value anywhere in M, contradicting the compactness, unless ψ vanishes identically. Therefore the cubic form must vanish identically and the hypersphere is an ellipsoid.

If $H=0$ the proof that the hypersphere is a convex paraboloid was established in [4] and is therefore omitted here.

If $H<0$ we have seen that there are affine hyperspheres of hyperbolic type that are homogeneous under the action of a linear Lie group; therefore they must be complete with respect to the affinely invariant metric; since there exist several for $n\geqslant 2$ that are not quadrics, the cubic form is not necessarily zero.

Suppose that the Ricci curvature $Q(\xi)(x_0)$ were positive for some tangent vector ξ at some $x_0\in M$. Because of (2.7), (2.8) and (2.9), the norm $\psi(x)$ of the cubic form satisfies near x_0 the inequality

$$(\psi(x))^2>-H\,.$$

Let a be a constant satisfying

$$\sqrt{-H}<a<\psi(x_0)\,; \tag{5.6}$$

we solve the following ordinary differential equation for a function $f(t)$ $(t\geqslant 0)$,

$$f''(t)+(n-1)\sqrt{-H}\coth\left(t\sqrt{-H}\right)\cdot f'(t)=(n+1)\left(f^3(t)+Hf(t)\right), \tag{5.7}$$

with the initial conditions

$$f(0)=a\,,\qquad f'(0)=0\,. \tag{5.8}$$

We show first that the solution $f(t)$ of (5.7), (5.8) is a monotone increasing function of t for $t\geqslant 0$; in fact, multiplying both members of (5.7) by $\sinh^{n-1}(t\sqrt{-H})$, we get

$$\frac{d}{dt}\left(f'(t)\sinh^{n-1}\left(t\sqrt{-H}\right)\right)=(n+1)\sinh^{n-1}\left(t\sqrt{-H}\right)\left(f^3(t)+Hf(t)\right)$$

and, since the right hand member is positive for $f(t)\geqslant a$ because of (5.6), it follows that $f'(t)$ is strictly positive for $t>0$.

We now show that, when the constant a satisfies (5.6), the solution $f(t)$ of the differential equation becomes infinite as t approaches a positive constant δ that we shall estimate. For this purpose we shall choose the comparison function $v(t)$

$$v(t) = \frac{ab}{b+1-\cosh\left(t\sqrt{-H}\right)} \tag{5.9}$$

satisfying the same initial conditions (5.8) as $f(t)$, where b is a positive constant to be choses so that $v(t)$ satisfies the following inequation

$$v''(t) + (n-1)\sqrt{-H}\coth\left(t\sqrt{-H}\right)v'(t) < (n+1)\left(v^3(t) + Hv(t)\right) \tag{5.10}$$

in its interval of regularity, $|t| < (-H)^{-\frac{1}{2}}$ arc cosh $(b+1)$. Direct substitution of (5.9) in (5.10) yields, after clearing common positive factors, the following inequation

$$-H\left(3y^2 - (n+2)(b+1)y + (n+1)(b+1)^2 - 2\right) \leqslant (n+1)a^2b^2, \tag{5.11}$$

where $y = \cosh(t\sqrt{-H})$. Since the left hand member of (5.11) is a convex function of y, it is sufficient to satisfy the inequality at the two end points of the relevant interval $1 \leqslant y \leqslant b+1$; substituting, we get the following admissibility condition on b,

$$b \geqslant \max\left\{\frac{4(-H)}{(n+1)a^2+2H}, \quad \frac{n(-H)}{(n+1)(a^2+H)}\right\}$$

for instance, we may take the following, simple value for b, satisfying the above inequality

$$b = \frac{\max\{n, 4\}(-H)}{(n+1)(a^2+H)}.$$

Using (5.7) and (5.10) one sees that $f(t) - v(t)$ can not achieve a maximum value in the interior of their common interval of regularity. Clearly both $f(t)$ and $v(t)$ are even functions of t; hence $f(t)$ must become infinite at the ends of an interval $(-\delta, \delta)$, where

$$\delta < \frac{1}{\sqrt{-H}} \operatorname{arc\,cosh}\left(1 + \frac{\max\{n, 4\}(-H)}{(n+1)(a^2+H)}\right). \tag{5.12}$$

To complete the proof of Theorem 5.1, we shall actually prove the following, more general statement.

THEOREM 5.4: *Let* $\mathbf{X}: M \to E$ *be an n-dimensional affine hypersphere of hyperbolic type with affine mean curvature* $H < 0$; *assume that for some unit tangent vector* $\xi(\boldsymbol{x}_0)$ *at a point* $\boldsymbol{x}_0 \in M$ *the Ricci curvature at* $\xi(\boldsymbol{x}_0)$ *is strictly positive*

$$Q(\xi(x_0)) = \frac{1}{n-1} R_{i\alpha}(x_0)\xi^i \xi^j = a_1^2 > 0 .$$

Then there exists a geodesic ray originating at $\boldsymbol{x}_0$, *without end point in M and of length*

$$\gamma(x_0) \leqslant \frac{1}{\sqrt{-H}} \operatorname{arc\,cosh}\left(1 + \frac{\max\{n, 4\}(-H)}{(n+1)a_1^2}\right). \tag{5.13}$$

In fact, assume that (5.13) were false: then for any radius γ short of the right hand member of (5.13) the closed geodesic ball $B(\boldsymbol{x}_0, \gamma)$ in M with center $\boldsymbol{x}_0$ and radius γ, by virtue of the Hopf-Rinow theorem, is compact. Consider the norm $\psi(\boldsymbol{x}_0)$ of the cubic form at $\boldsymbol{x}_0$, as defined in (2.9): according to (2.7) we have

$$(\psi(\boldsymbol{x}_0))^2 \geqslant (a_1^2 - H) > -H .$$

For any positive constant a satisfying (5.6) and arbitrarily close to $\psi(\boldsymbol{x}_0)$, solve the equation (5.7) with initial values (5.8) and consider the function

$$z(\boldsymbol{x}) = f(r(\boldsymbol{x}))$$

where r denotes the geodesic distance from $\boldsymbol{x}_0$ to $\boldsymbol{x}$ in the open ball such that $r(x)$ is in the domain of regularity of the solution of the differential equation (5.7). This function, since $f'(t) > 0$ for $t > 0$, satisfies the conditions of Lemma 5.3, and therefore

$$\Delta z \leqslant f''(r) + (n-1)\sqrt{-H}\coth\left(r\sqrt{-H}\right) f'(r) \leqslant (n+1)(z^3 + Hz) , \tag{5.14}$$

which is the opposite of the differential inequality (5.1) satisfied by ψ, according to Lemma 5.2. Furthermore z becomes infinite on the boundary of a geodesic ball centered at $\boldsymbol{x}_0$ with radius δ, where

$$\delta < \frac{1}{\sqrt{-H}} \operatorname{arc\,cosh}\left(1 + \frac{\max\{n, 4\}(-H)}{(n+1)(a^2 + H)}\right)$$

and, since a may be arbitrarily close to $\psi(x_0)$, we have also

$$\delta \leqslant \frac{1}{\sqrt{-H}} \operatorname{arc\,cosh}\left(1+\frac{\max\{n,4\}(-H)}{(n+1)(\psi^2(x_0)+H)}\right) = \frac{1}{\sqrt{-H}} \operatorname{arc\,cosh}\left(1+\frac{\max\{n,4\}(-H)}{(n+1)a_1^2}\right).$$

Therefore the function $\psi(x)-z(x)$ tends to $-\infty$ uniformly on the boundary of a compact domain; consider a point x_1 where $\psi - z$ attains its maximum value; this value is positive, since $\psi(x_0) > a = z(x_0)$. On the other hand, because of (5.1) and (5.14), we have also

$$\Delta(\psi - z)(x_1) \geqslant (n+1)\big(\psi^3 - z^3 + H(\psi - z)\big)(x_1) > 0\,,$$

contradicting the maximum principle for the Laplacian [3]. This contradiction proves the statement of Theorem 5.4 and, by direct implication, also Theorem 5.1.

The example of an affine hypersphere given by (1.8) is mapped transitively onto itself under the diagonal subgroup of $SL(n+1,\mathbb{R})$ acting on the imbedding coordinates; hence the manifold is complete with respect to the invariant metric, and since the group is abelian and acts in a simply transitive way, this metric has curvature identically zero. This shows that there is not much room for improving the statement of Theorem 5.1. However it does not seem probable, under the same assumptions and for $n \geqslant 3$, that all sectional curvatures be nonpositive. It would be desirable to verify whether a more conservative conjecture, like the absence of conjugate points on every geodesic, is true. For the moment, the conclusion reached has strong implications only in the case $n = 2$. In this case one can also use the powerful tools of complex analysis: this will be the subject of a subsequent paper.

Testo pervenuto il 29 novembre 1971.
Bozze licenziate il 10 luglio 1972.

REFERENCES

[1] W. Blaschke, *Vorlesungen über Differentialgeometrie*, II, Berlin, J. Springer, 1923.
[2] W. Blaschke, Berichte der Leipziger Akad., 69 (1917), 167; 70 (1918), 27.
[3] E. Calabi, *An extension of E. Hopf's maximum principle with an application to Riemannian geometry*, Duke Math. Jour., 25 (1958), 45-56; 26 (1959), 707.

[4] E. Calabi, *Improper affine hyperspheres of convex type and a generalization of a theorem by K. Jörgens*, Mich. Math. Jour., 5 (1958), 105-126.

[5] E. Calabi, *Examples of Bernstein problems for some nonlinear equations*, A.M.S. Proc. Symp. Pure Math. XV, Global Analysis (1970), 223-230.

[6] J. Tits, *Sur certaines classes d'espaces homogènes de Groupes de Lie*, Mém. Acad. Roy. Belg., 29, fasc. 3 (1955).

[7] G. Tzitzèica, *Sur certaines surfaces réglées*, C. R. Acad. Paris, 145 (1907), 132-133; 146 (1908), 165-166.

[8] G. Tzitzèica, *Sur une nouvelle classe de surfaces*, Rend. Circ. Mat. Palermo, 25 (1908), 180-187; 28 (1909), 210-216.

[9] E. B. Vinberg, *Homogeneous Cones*, Dokl. Akad. Nauk, Ser. Mat., 133, no. 1 (1960), 9-12 (A.M.S. Transl., 1, no. 4 (1960), 787-790).

A Construction of Nonhomogeneous Einstein Metrics

Eugenio Calabi*

Differential Geometry,
Proc. Sympos. Pure Math., Vol. XXVII, held at
Stanford University, Stanford, Calif., USA, 1973, Part 2, 17–24,
Amer. Math. Soc., Providence, RI (1975)

A continuing interest in the problem of the existence, in a compact complex manifold admitting some Kähler metric, of a new Kähler metric with prescribed volume element [4], [5], has prompted the author to present this method of constructing in a simple way a class of Einstein metrics, and especially Ricci flat metrics in certain types of open manifolds.

The manifolds under consideration are complex tubular domains $M = \frac{1}{2}D \oplus i\boldsymbol{R}^n \subset \boldsymbol{C}^n$, where D denotes any connected, open subset of $\boldsymbol{R}^n$. The resulting complex manifolds are domains of holomorphy, if and only if D is convex [3, pp. 90–92]; the metrics constructed in M are Kähler metrics, invariant under the group of translations along $i\boldsymbol{R}^n$. Among the metrics constructed in M it is possible to impose additional restrictions, such as Einstein's equation $R_{i\bar{j}} = Kg_{i\bar{j}}$ ($K =$ constant), or, in particular, Ricci flat metrics, $R_{i\bar{j}} = 0$. In the Ricci flat case it turns out that M with the metric of the type considered can never be complete, except in the trivial case where $D = \boldsymbol{R}^n$ and $M = \boldsymbol{C}^n$ with a flat, translation invariant metric.

Let $f : D \to \boldsymbol{R}$ be a strongly convex, differentiable, real valued function on D, i.e. a function whose Hessian is everywhere positive definite, and denote by $F(z, \bar{z})$ the strongly plurisubharmonic function on the tube $M = \frac{1}{2}D \oplus i\boldsymbol{R}^n$ defined by

$$F(z, \bar{z}) = f(z_1 + \bar{z}_1, z_2 + \bar{z}_2, \cdots, z_n + \bar{z}_n). \tag{1}$$

Consider in M the Kähler metric defined by the Levi invariant of F, i.e.

AMS (MOS) subject classifications (1970). Primary 53C25, 35J60, 32A07, 32F05.
*Supported in part by NSF grant 29258.

Zbl: 0309.53043

(2) $$ds^2 = 2g_{\alpha\bar\beta}(z, \bar z)\, dz^\alpha d\bar z^\beta,$$

where

(3) $$g_{\alpha\bar\beta}(z, z) = \frac{\partial^2 F(z, \bar z)}{\partial z_\alpha \partial \bar z_\beta} = \frac{\partial^2 f(x)}{\partial x_\alpha \partial x_\beta}\bigg|_{x=z+\bar z};$$

its volume element is $dv = i^n \det(g_{\alpha\bar\beta})\, dz_1 \wedge d\bar z_1 \wedge \cdots \wedge dz_n \wedge d\bar z_n$. If the convex function $f(x)$ satisfies, for instance, the differential equation

(4) $$\det\left(\frac{\partial^2 f}{\partial x_i\, \partial x_j}\right) = 1,$$

already considered elsewhere **[6]**, **[8]**, then

$$g(z, \bar z) = \det(g_{\alpha\bar\beta}(z, \bar z)) = 1,$$

and consequently the Ricci tensor **[2]**

$$R_{\alpha\beta} = -\frac{\partial^2}{\partial z_\alpha\, \partial \bar z_\beta}(\log g(z, \bar z))$$

vanishes identically. On the other hand, the full Riemann curvature tensor can be computed from the following formula, letting $(K^{\alpha\beta}(z, \bar z))$ denote the transposed inverse matrix function of $(g_{\alpha\bar\beta}(z, \bar z))$ (cf. **[2]**, deduce from Lemma 5):

$$R^\alpha{}_{\beta\gamma\bar\delta} = -\frac{\partial}{\partial \bar z_\delta}\left(K^{\alpha\bar\mu}(z, \bar z)\frac{\partial^3 F(z, \bar z)}{\partial z_\beta \partial z_\gamma \partial \bar z_\mu}\right).$$

It is easy to verify that $R^\alpha{}_{\beta\gamma\bar\delta}$ vanishes identically, in view of (2), if and only if all third order derivatives of $f(x)$ vanish identically, i.e. if $f(x)$ is a quadratic polynomial. Thus the tubular domain M with the Kähler metric (3) is a Ricci flat manifold that is not locally flat, except in the case where $f(x)$ is one of the trivial solutions of (4).

It has already been reported that the Ricci flat Kähler metrics obtained in M by this method described above can never be complete **[6]** nor can they be defined in all of $\boldsymbol{C}^n$ **[8]** except for the trivial solutions. However we can give a simple example of a nontrivial solution $f(x)$ of (4) in $\boldsymbol{R}^n - \{0\}$ yielding a metric in $M = \boldsymbol{C}^n - i\boldsymbol{R}^n$ $(n \geqq 2)$, namely

$$f(x) = \int_0^{|x|} (c + r^n)^{1/n}\, dr, \quad |x| = \left(\sum_{\alpha=1}^n (x_\alpha)^2\right)^{1/2}, \quad c > 0.$$

Other solutions can be obtained by direct sums of nontrivial and at most one trivial solution in lower dimensional spaces. For $n = 2$ there are several known representations of solutions of (4) (cf. **[1**, p. 216**]**).

One can similarly construct, more generally, Kähler metrics in a tubular domain $M = \frac{1}{2}D \oplus i\boldsymbol{R}^n$ satisfying an Einstein equation

$$R_{\alpha\bar\beta} = K g_{\alpha\bar\beta} \qquad (K = \text{constant})$$

by letting $f(x)$ be a solution of the equation

$$\det\left(\frac{\partial^2 f(x)}{\partial x_i \partial x_j}\right) = e^{-Kf(x)}, \tag{5}$$

in the place of (4). In this case it is clear from Myers' theorem [7] that, for $K > 0$, no solution of (5) can yield a complete metric in M. The case $K < 0$, however, seems to be quite different. For example, if D is a convex domain containing no complete straight line and $M = \frac{1}{2}D \oplus i\boldsymbol{R}^n$, in addition is a homogeneous complex manifold, then M is a Cartesian product of irreducible, homogeneous, Bergman manifolds. For each of these irreducible factors the Bergman metric satisfies Einstein's equation with $K < 0$; consequently the product of a suitable renormalization of these metrics satisfies Einstein's equation in M; thus M admits an Einstein metric, invariant under a transitive group of holomorphic transformations, so that M is complete with respect to this metric. This fact and others lead to the following conjecture.

CONJECTURE. Let D be an open convex domain in $\boldsymbol{R}^n$ containing no complete straight line. Then there exists in D, for any $K < 0$, a unique, convex function $f : D \to \boldsymbol{R}$ satisfying (5), approaching $+\infty$ uniformly in each bounded portion of the boundary, and bounded from below by $-c \log |x|$ for some $c > 0$ as $x \to \infty$ if D is unbounded.

Assuming, for the moment, the conjecture to be true for one value of K, say K_1, yielding a solution $f_1(x)$ of (5) with the asserted properties, one has a solution $f(x)$ for any other negative value of K by setting

$$f(x) = \frac{1}{K}(K_1 f_1(x) - n \log(K_1/K)).$$

The uniqueness of f with the properties required can be proved by means of the strong maximum principle [11] for the difference of two solutions of (5) with $K < 0$; similarly one should prove that the boundary condition on f is sufficient to show that the resulting Kähler metric on M makes the tubular domain complete.

We shall now construct an explicit example of a solution of equation (5) in a particular bounded, convex domain D, yielding on $M = \frac{1}{2}D + i\boldsymbol{R}^n$ an Einstein-Kähler metric with $K = -1$, with the properties that M is complete with respect to this metric and the metric is not locally homogeneous. To our best recollection, this is the first known example of a complete Einstein metric that is not locally homogeneous (incomplete examples were given explicitly in $\boldsymbol{R}^+ \times \boldsymbol{R}^3$ by T. J. Willmore [9], [10]).

The convex domain D is an open ball in $\boldsymbol{R}^n$ of unspecified radius $r_1 > 0$ and center at the origin. We seek a solution of (5) in D, that is a convex function diverging uniformly to $+\infty$ at the boundary. The uniqueness of the solution $f(x)$ and the symmetry of D under the orthogonal group $O(n)$ imply that $f(x)$ can be expressed as a radial function, $f(x) = y(r)$, $r = (\sum_{\alpha=1}^n x_\alpha^2)^{1/2}$, where y satisfies the ordinary differential equation (taking $K = -1$, without loss of generality)

$$(y'/r)^{n-1} y'' = e^y. \tag{6}$$

Regularity at the origin requires the following initial conditions

$$y'(0) = 0, \qquad y''(0) = e^{y(0)/n}, \tag{7}$$

whence one obtains by recursion a solution expressed as a formal power series with respect to r^2. In order to prove that the functions $y(r)$ that are solutions of (6) and (7) actually generate an Einstein metric with all the asserted properties, one has to study the solution with some detail.

Replacing $y(r)$ by $y_\lambda(r) = y(\lambda r) + 2n \log \lambda$ for any $\lambda > 0$ yields a one-parameter family of solutions from one solution, showing the effect of different initial values $y(0)$ on the solution globally. It is also reasonable, in looking for comparison functions, to seek them with the same type of logarithmic homogeneity property.

Definition. In view of equations (6), (7) we define a function $z(r)$ to be a supersolution (respectively, a subsolution) in an interval $-a < r < a$ for any $a > 0$ if $z'(0) = 0$ and, for all r in the interval, $z''(r) \geqq 0$ and $e^{-z(r)}(z'(r)/r)^{n-1} z''(r) \leqq 1$ (respectively $\geqq 1$). A solution of (6), (7) is the same thing as a simultaneous sub- and supersolution.

The following two elementary functions will be used as comparison functions in any open interval $(-a, a)$

$$z_l(r) = z_l(a, r) = -(n+1)\log(1 - r^2/a^2) + n \log((2n+2)/a^2)$$

and

$$z_u(r) = z_u(a, r) = (n+1)\log \sec \pi r/2a + n \log (n+1)\pi^2/4a^2.$$

One verifies that z_l and z_u are respectively a subsolution and a supersolution. In fact both z_l and z_u are differentiable and convex in the interval $(-a, a)$, and

$$e^{-z_l}(z_l'/r)^{n-1} z_l'' = 1 + r^2/a^2 \geqq 1,$$
$$e^{-z_u}(z_u'/r)^{n-1} z_u'' = ((2a/\pi r)[\sin(\pi r/2a)]^{n-1}) \leqq 1.$$

We apply now the maximum principle [11, pp. 6–7] as follows: if $g_1(r)$ and $g_2(r)$ are respectively a subsolution and a supersolution of (6), (7) in a common open interval, then $g_1(r) - g_2(r)$ cannot achieve a nonnegative relative maximum value at any interior point, unless they coincide in the whole interval (and are therefore a solution).

Let $y(r)$ be the solution of (6), (7) with initial value $y(0)$. We can show that $y(r)$ becomes $+\infty$ at the end points of an interval $(-A, A)$, where

$$(2(n+1))^{1/2} e^{-y(0)/2n} \leqq A \leqq \frac{\pi}{2}(n+1)^{1/2} e^{-y(0)/2n}.$$

In fact, if the upper bound were violated, choose a constant a strictly between the asserted upper bound and A and consider $y(r) - z_u(a, r)$: this function becomes $-\infty$ as $r \to a$, so that the difference would achieve an absolute maximum value at some interior point; but this maximum value would be positive, since $y(0) > z_u(a, 0)$, contradicting the maximum principle. The lower bound is proved similarly by considering $z_l(a, r) - y(r)$ for suitable a.

Now let a be the exact upper bound of the domain of regularity of $y(r)$. The same arguments can be used to show that, for $-a < r < a$, $z_l(a, r) \leqq y(r) \leqq z_u(a, r)$.

In order to evaluate the sharpness of the above estimate, one verifies that, in the interval $|r| < a$, the difference function $z_u(a, r) - z_l(a, r)$ is the bounded, positive, even, convex function

$$(n+1)\log\frac{a^2-r^2}{a^2\cos(\pi r/2a)} + n\log\frac{\pi^2}{8}$$

taking its minimum value $n\log(\pi^2/8)$ at $r = 0$ and approaching the upper bound $n\log(\pi^2/8) + (n+1)\log(4/\pi)$ as $r \to \pm a$.

We estimate next the asymptotic behavior of the derivatives of y by comparing them with the corresponding ones of z_u and z_l. Denote by $v(r)$, $w_u(r)$ and $w_l(r)$ respectively the functions $n^{-1}y'^n(r), n^{-1}z_u'^n(r), n^{-1}z_l'^n(r)$. Then one verifies directly that, in the interval $0 < r < a$,

$$\frac{v''(r)}{v'(r)} - \frac{n-1}{r} - (nv(r))^{1/n} = 0,$$

$$\text{(8)} \qquad \frac{w_u''(r)}{w_u'(r)} - \frac{n-1}{r} - (nw_u(r))^{1/n} = (n-1)\left(\frac{\pi}{2a}\cot\frac{\pi r}{2a} - \frac{1}{r}\right) \leqq 0,$$

$$\frac{w_l''(r)}{w_l'(r)} - \frac{n-1}{r} - (nw_l(r))^{1/n} = \frac{2r}{a^2+r^2} \geqq 0,$$

where the allowable functions are known to be strictly monotone increasing in the semiclosed interval $0 \leqq r < a$, vanishing of order n at $r = 0$; v is a solution of the equation derived from (6), w_u and w_l are respectively a super- and a subsolution. The conditions for the maximum principle are again valid for the differences $v - w_u$ and $w_l - v$, so that neither of these two difference functions can achieve local, positive valued maximum value at any interior point; the boundary conditions then imply that

$$w_l(r) \leqq v(r) \leqq w_u(r) \qquad (0 \leqq r < a)$$

and the equation (8) itself implies that $v(r)$ is strictly convex. As a corollary we have the following estimates for $y'(r)$:

$$z_l'(r) = \frac{2(n+1)r}{a^2-r^2} \leqq y'(r) \leqq z_u'(r) = \frac{(n+1)\pi}{2a}\tan\frac{\pi r}{2a},$$

where the gap between the upper and lower bounds is again uniformly bounded

$$0 \leqq z_u'(r) - z_l'(r) = (n+1)\left(\frac{\pi}{2a}\tan\frac{\pi r}{2a} - \frac{2r}{a^2-r^2}\right) \leqq n+1 \quad (0 \leqq r < a).$$

Having estimated both upper and lower bounds for $y(r)$ and $y'(r)$, we obtain immediate bounds for $y''(r)$ by reading (6), namely $y'' = e^y(r/y')^{n-1}$ whence

$$\exp\{z_l(r/z_u')^{n-1}\} \leqq y'' \leqq \exp\{z_u(r/z_l')^{n-1}\}$$

In particular, we see that

$$y''(r) \geqq \exp\{z_l(r)(z_u'(r)/r)^{-n+1}\} = \frac{2^{2n-1}(n+1)\,a^{n+1}\,r^{n-1}\cot^{n-1}(\pi r/2a)}{\pi^{n-1}(a^2-r^2)^{n+1}},$$

the lower bound being asymptotic to $c^2/(a^2-r^2)^2$ for some positive constant c as $r \to a$. An important consequence of this fact is that, as $r \to a$, the following improper integral diverges logarithmically,

$$\int_0^r (y''(\rho))^{1/2}\, d\rho \geqq \frac{c}{a} \operatorname{arc\,tanh} \frac{r}{a}. \tag{9}$$

The solution $y(r)$ of (6), (7), which becomes infinite as $r \to a$ gives rise to the radial function $f(x)$ in the n-dimensional ball D of radius a in R^n, by setting

$$f(x) = y(r), \qquad r = \left(\sum_{\alpha=1}^{n} (x_\alpha)^2\right)^{1/2};$$

the function $f(x)$ is smooth in the interior, diverges uniformly to $+\infty$ at the boundary, is convex in the interior and satisfies the partial differential equation (5) for $K = -1$. Consequently the Kähler metric defined in the tube domain $M = \frac{1}{2}D + iR^n$ by (3) is an Einstein metric.

One verifies now without difficulty that M is complete with respect to this metric. In fact if we map M into the strip $[0, a] \times [0, \infty]$ by mapping the point $\frac{1}{2}(x+it) \in M$ $(x \in D, t \in R^n)$ to the point $(r, \rho) = (|x|, |t|)$ we obtain the following lower bound for the metric (3)

$$ds^2 = \frac{\partial^2 f(x)}{\partial x_\alpha \partial x_\beta}(dx^\alpha dx^\beta + dt^\alpha dt^\beta) \geqq y''(|x|)(dr^2 + d\rho^2) \geqq y''(r)\, dr^2.$$

Now suppose that we have a path Γ in M of finite length A, originating at the origin and parametrized by its arc length s $(0 \leqq s < A)$. Because of (9) we see that for each s the point $\frac{1}{2}(x_s + it_s)$ must satisfy

$$|x_s| \leqq a \tanh(as/c) < a \tanh(aA/c) < a$$

for the same constant c appearing in (9) so that x_s remains in a compact subdomain $D' \subset D$ defined by $|x| \leqq a' < a$. Let $\delta^2 = \inf_{0 \leqq r \leqq a'} y''(r)$; then one sees that the oscillation of $|t|$ along Γ is bounded by A/δ from above, so that the range of Γ is in a compact subdomain of M, proving the completeness.

Finally we show that, for $n \geqq 2$, the Einstein-Kähler metric just constructed in the tube $M = \frac{1}{2}D + iR^n$ where D is a euclidean n-ball is not locally homogeneous. For this purpose we carried on a long, tedious calculation on the Kählerian invariants of the metric.

We recall that the manifold M is the open submanifold of C^n defined by

$$M = \left\{z = (z_1, \cdots, z_n) \in C^n \,\middle|\, \sum_{\alpha=1}^{n} (z_\alpha + \bar{z}_\alpha)^2 < a^2\right\};$$

we have in M the plurisubharmonic potential defined by

$$F(z,\bar z) = f(x) = y(r) \qquad \left(x = (x_1,\cdots,x_n);\ x_\alpha = z_\alpha + \bar z_\alpha;\ r^2 = \sum_\alpha x_\alpha^2\right),$$

where $y(r)$ is the solution of (6), (7) which is regular in the interior and unbounded at the end points of the interval $(-a, a)$.

The metric given by (3) is then reduced to the expression

$$g_{\alpha\bar\beta}(z,\bar z) = \frac{y'(r)}{r}\delta_{\alpha\beta} + \left(y'' - \frac{y'}{r}\frac{x_\alpha x_\beta}{r^2}\right)$$

when $r > 0$, and to its limit value $y''(0)\,\delta_{\alpha\beta}$ at $r = 0$. The components $R^\lambda{}_{\alpha\bar\beta}{}^{\bar\mu}(z,\bar z)$ can be calculated formally for $r > 0$, using the formula

$$R^\lambda{}_{\alpha\bar\beta}{}^{\bar\mu} = -g^{\mu\bar\delta}(z,\bar z)\frac{\partial}{\partial \bar z_\delta}\left(g^{\lambda\bar\gamma}\frac{\partial^2 g_{\alpha\bar\gamma}}{\partial z_\beta}\right)$$

with the result

$$\begin{aligned}
R^\lambda{}_{\alpha\bar\beta}{}^{\bar\mu} = {} & \left(\frac{1}{ry'} - \frac{y''}{y'^2}\right)(\delta^\lambda{}_\alpha\delta^\mu{}_\beta + \delta^\lambda{}_\beta\delta^\mu{}_\alpha) + \left(-\frac{1}{ry'} + \frac{1}{r^2y''}\right)\delta^{\lambda\mu}\delta_{\alpha\beta} \\
& + \left(-\frac{1}{r^4y''} - \frac{1}{r^3y'} - \frac{y'''}{r^2y'y''} + \frac{2y''}{r^2y'^2}\right) \\
& \cdot(\delta^\lambda{}_\alpha x_\beta x_\mu + \delta^\lambda{}_\beta x_\alpha x_\mu + \delta^\mu{}_\alpha x_\beta x_\lambda + \delta^\mu{}_\beta x_\alpha x_\lambda + \delta^{\lambda\mu}x_\alpha x_\beta) \\
& + \left(\frac{1}{r^4y''} + \frac{1}{r^3y'} - \frac{2y'}{r^5y''^2} - \frac{y'y'''}{r^4y''^3}\right)\delta_{\alpha\beta}x_\lambda x_\mu \\
& + \left(-\frac{y''''}{r^4y''^2} + \frac{y'''^2}{r^4y''^3} + \frac{5y'''}{r^4y'y''} - \frac{8y''}{r^4y'^2}\right. \\
& \qquad \left. + \frac{2y'}{r^7y''^2} + \frac{y'y'''}{r^6y''^3} + \frac{3}{r^5y'} + \frac{3}{r^6y''}\right)x_\lambda x_\mu x_\alpha x_\beta .
\end{aligned}$$

We interpret the curvature tensor at each point as a linear, selfadjoint endomorphism of the bundle $S^2(T^{(1,0)})$ of symmetric products of the holomorphic tangent bundle $T^{1,0}$. Owing to the rotational symmetry of M, it is clear that the following 2-dimensional (over C) subspace of this bundle is invariant under the rotational group of isometries and hence under the curvature action: the space spanned by

$$\sum_{\alpha=1}^n \frac{\partial}{\partial z_\alpha}\otimes\frac{\partial}{\partial z_\alpha} \quad\text{and}\quad \sum_{\alpha,\beta=1}^n x_\alpha x_\beta\frac{\partial}{\partial z_\alpha}\otimes\frac{\partial}{\partial z_\beta}.$$

If M with this metric were locally homogeneous, then all eigenvalues of the linear action of the curvature tensor would be constant throughout M, and in particular the symmetric functions of the eigenvalues restricted to any invariant subspace, such as the one just exhibited. One can calculate now that the trace of the corresponding 2×2 matrix is

$$-\frac{y''''}{y''^2} + \frac{y'''^2}{y''^3} - \frac{2y''}{y'^2} + \frac{n-1}{r^2y''} - \frac{n-3}{ry'},$$

and one readily verifies that the system of two equations consisting of equation (6) and the above expression equated to a constant is strictly overdetermined and inconsistent for $n \geqq 2$.

The example just exhibited and the estimates on its asymptotic behavior near the boundary will be probably of some value in establishing existence proofs of Kähler-Einstein metrics in other tube domains $M = \frac{1}{2}D + iR^n$, where D is any smoothly bounded, strongly convex domain in R^n.

References

1. W. Blaschke, *Differentialgeometrie*. II. *Affine Differentialgeometrie*, Julius Springer, Berlin, 1923.
2. S. Bochner, *Vector fields and Ricci curvature*, Bull. Amer. Math. Soc. **52** (1946), 776–797. MR **8**, 230.
3. S. Bochner and W. T. Martin, *Several complex variables*, Princeton Math. Ser., vol. 10, Princeton Univ. Press, Princeton, N.J., 1948. MR **10**, 366.
4. E. Calabi, *On Kähler manifolds with vanishing canonical class*, Algebraic Geometry and Topology (A Sympos. in Honor of S. Lefschetz), Princeton Univ. Press, Princeton, N.J., 1957, pp. 78–98. MR **19**, 62.
5. ———, *The space of Kähler metrics*, Proc. Internat. Congress Math., Amsterdam, 1954, pp. 206–207.
6. ———, *Improper affine hyperspheres of convex type and a generalization of a theorem by K. Jörgens*, Michigan Math. J. **5** (1958), 105–126. MR **21** #5219.
7. S. B. Myers, *Riemannian manifolds with positive mean curvature*, Duke Math. J. **8** (1941), 401–404. MR **3**, 18.
8. A.V. Pogorelov, *On the improper convex affine hyperspheres*, Geometriae Dedicata **1** (1972), 33–46.
9. T. J. Willmore, *An introduction to differential geometry*, Clarendon Press, Oxford, 1959, p. 238. MR **28** #2482.
10. ———, *On compact Riemannian manifolds with zero Ricci curvature*, Proc. Edinburgh Math. Soc. (2) **10** (1956), 131–133. MR **17**, 783.
11. M.H. Protter and H.F. Weinberger, *Maximum principle in differential equations*, Prentice-Hall, Englewood Cliffs, N. J., 1967.

University of Pennsylvania

On the Sequential and Random Selection of Subspaces Over a Finite Field

Eugenio Calabi* and Herbert S. Wilf*

J. Combinatorial Theory Ser. A 22, no. 1, 107–109 (1977)

Let V_n be an n-dimensional vector space over $GF(q)$, and let $0 \leqslant k \leqslant n$. We consider the question of generating sets of basis vectors for k-dimensional subspaces V_k of V_n in two ways: (a) the sequential generation of a set of basis vectors for *every* such subspace and (b) the random selection of such a subspace V_k in such a way that all subspaces have equal a priori probabilities of being chosen. Such questions are of interest in coding theory.

The answer to the first question has been known for some time and comes from the row-echelon, or "Schubert," form of the $k \times n$ matrix B whose rows are the basis vectors: We fix a k-subset $(a_1, \ldots, a_k) = S$ of $\{1, 2, \ldots, n\}$. In the columns of B which correspond to S we enter a $k \times k$ identity matrix. In row i of B we enter 0 in all columns $j > a_i$, for $i = 1, \ldots, k$. Finally the

$$N = \sum_{i=1}^{k} a_i - \binom{k}{2} = N(S)$$

remaining entries of B may be assigned independently to elements of the field. Hence for the given subset S we obtain q^N sets of basis vectors, each describing a different subspace V_k. By varying S we obtain all of the desired V_k, and of course

$$\begin{bmatrix} n \\ k \end{bmatrix}_q = \frac{(q^n - 1)(q^{n-1} - 1) \cdots (q^{n-k+1} - 1)}{(q^k - 1)(q^{k-1} - 1) \cdots (q - 1)} = \sum_S q^{N(S)} \qquad (1)$$

is the total number of such V_k.

Before discussing our algorithm for uniformly random selection, we mention, briefly, two others which may come immediately to mind.

* Supported in part by the National Science Foundation.

DOI: 10.1016/0097-3165(77)90069-3

First, we might imagine fixing a $k \times k$ identity matrix in the first k columns of a $k \times n$ matrix, and choosing the remaining $k(n-k)$ elements independently from the field. This proposal results in nonuniform distribution, and in an essential way. For example, with $k = 2$, $n = 3$ over $GF(2)$, we would choose from the four basis matrices

$$\begin{pmatrix}100\\010\end{pmatrix} \quad \begin{pmatrix}100\\011\end{pmatrix} \quad \begin{pmatrix}101\\010\end{pmatrix} \quad \begin{pmatrix}101\\011\end{pmatrix}$$

with equal probability. Of these four vector spaces, the last has all nonzero vectors of weight (number of 1's) at least 2, while the other three spaces have minimum weight 1. Actually, however, the probability that a random $V_2 \subseteq V_3$ over $GF(2)$ will have minimum weight 2 is 1/7.

A second proposal might be to select the rows of the matrix independently, and check each one for independence from its predecessors. Here the distribution is uniform, but the independence check of the jth vector will require Cjn operations, for a total of Cnk^2 units of labor altogether. Our algorithm operates in time Cnk which is minimal for the problem since there are nk items of output: the matrix elements. Indeed, as will be seen, at each stage of the algorithm we either fill in a whole row and column with zeros and a single 1, or else we fill in a whole column with independently randomly chosen elements from the field. We, therefore, can output the basis matrix essentially just as fast as we can write it down!

We begin with the well known identity

$$\begin{bmatrix} n \\ k \end{bmatrix}_q = \begin{bmatrix} n-1 \\ k-1 \end{bmatrix}_q + q^k \begin{bmatrix} n-1 \\ k \end{bmatrix}_q \tag{2}$$

and we think of the two terms on the right as counting complementary subsets of the totality of basis matrices B in echelon form: the first term counts those $k \times n$ matrices $B = (b_{ij})_{i=1,k;j=1,n}$ for which $b_{11} = 1$, $b_{i1} = 0$ $(1 < i \leqslant k)$, $b_{1j} = 0$ $(1 < j \leqslant n)$ and for which necessarily the remaining $(k-1) \times (n-1)$ array is the basis matrix of a $V_{k-1} \subseteq V_{n-1}$ (i.e., has rank $k-1$). The second term counts the other basis matrices B, i.e., those whose first column is an arbitrary vector of V_k, and for which the remaining $k \times (n-1)$ array is the basis matrix of a $V_k \subseteq V_{n-1}$ (i.e., has rank k).

The fraction of all subspaces which are of the first kind is

$$p = p_{k,n} = \begin{bmatrix} n-1 \\ k-1 \end{bmatrix}_q \Big/ \begin{bmatrix} n \\ k \end{bmatrix}_q = (q^k - 1)/(q^n - 1), \tag{3}$$

and so if n, k, q are given, the full algorithm is as follows:

(A) $r \leftarrow 1$; $s \leftarrow 1$.

(B) Choose a random number ξ. If $\xi < p_{k+1-r,n+1-s}$, go to (C); Otherwise, choose independently at random from the field the elements b_{is} $(i = r, k)$, set $s \leftarrow s + 1$ and go to (D).

(C) Set $b_{rj} \leftarrow 0$ $(j = s + 1, n)$; $b_{is} \leftarrow 0$ $(i = r + 1, k)$; $b_{rs} \leftarrow 1$; $r \leftarrow r + 1$; $s \leftarrow s + 1$.

(D) If $r \leqslant k$, go to (B); exit. ∎

References

1. W. V. D. Hodge and D. Pedoe, "Methods of Algebraic Geometry," Vol. II, pp. 321–326, Cambridge, 1947.
2. J. W. Milnor and J. Stasheff, Characteristic classes, *in* "Annals of Math., Studies #76," pp. 72–81, Princeton Univ. Press.

Métriques kählériennes et fibrés holomorphes(*)

Eugenio Calabi(**)

Ann. Sci. École Norm. Sup. (4) 12, n° 2, 269–294 (1979)

1. Introduction

Nous rappelons le fait suivant, qui est bien connu. Lorsqu'on se donne un fibré différentiable quelconque, l'espace de base et la fibre étant munis de métriques riemanniennes et le groupe structural G opérant isométriquement sur les fibres, alors la donnée d'une connexion sur le fibré compatible avec l'action du groupe structural détermine dans l'espace total du fibré une métrique riemannienne, uniquement caractérisée par les propriétés suivantes :

1° la métrique induite sur la fibre au-dessus de n'importe quel point de la base équivaut à la métrique donnée dans la fibre abstraite par n'importe quel isomorphisme structural entre les deux;

2° en chaque point du fibré, tout vecteur tangent qui est horizontal par rapport à la connexion est orthogonal à la fibre passant par ce point;

3° la projection du fibré sur l'espace de base est une surjection (submersion) riemannienne, compatible partout avec la métrique donnée dans l'espace de base.

D'autre part, si la structure riemannienne de l'espace de base comme celle des fibres sont des structures kählériennes, la même construction de structure métrique sur l'espace total du fibré, n'amène qu'à une structure presque hermitienne, qui en général n'est pas intégrable à une structure complexe et, même si elle l'est, la métrique définie n'est kählérienne que dans des cas qui sont très rares; par l'exemple, dans le cas des fibrés vectoriels holomorphes, une structure hermitienne ainsi construite n'est kählérienne que si la courbure de la connexion est identiquement nulle.

(*) Exposé d'une conférence tenue à Palaiseau au Centre de Mathématiques de l'École Polytechnique, le 5 juin 1978.

(**) Research supported in part by the National Science Foundation, Contract GP 75-07917 and by the Institut des Hautes Études scientifiques, Bures-sur-Yvette.

DOI: 10.24033/asens.1367

Il est toutefois possible, au moins dans certains cas, d'introduire dans l'espace total d'un fibré holomorphe une forme kählérienne déterminée par des données correspondantes dans l'espace de base et dans la fibre, si on sacrifie en échange la troisième des propriétés énoncées ci-dessus, en la remplaçant par une condition plus faible, mais moins facile à énoncer hors de ce contexte. De telles métriques ont été introduites sans doute ailleurs, mais aucune étude systématique n'a été portée jusqu'ici à l'attention de l'auteur. Une conséquence d'une analyse de telles métriques kählériennes dans certains fibrés analytiques complexes nous permet de construire de nombreux exemples explicites de métriques kählériennes, parfois complètes, jouissant de certaines propriétés remarquables, parmi lesquelles nous allons trouver :

1° des métriques kählériennes (section 4) à courbure de Ricci constante (condition d'Einstein) et en particulier à courbure de Ricci nulle;

2° des métriques kählériennes (section 5) dont le groupe d'holonomie est le groupe compact $Sp(n)$ $(n=1, 2, \ldots)$, où $2n$ est la dimension complexe de la variété fibrée; ce sont apparemment les premiers exemples connus de telles structures, qui ont été appelées « structures tangentielles quaternioniennes », mais pour lesquelles nous préférons l'appellation de « structures hyperkählériennes », pour des raisons que nous expliquerons au moment opportun.

2. Formes kählériennes dans les fibrés vectoriels

Soit M une variété complexe à n dimensions complexes, compacte ou non, munie d'une métrique kählérienne, que l'on peut représenter localement, soit par sa forme riemannienne définie positive

(2.1) $$ds^2 = 2\, g_{\alpha\bar\beta}(z, \bar z)\, dz^\alpha\, d\bar z^\beta,$$

soit par sa forme extérieure

$$F = -(2\pi i)^{-1} g_{\alpha\bar\beta}(z, \bar z)\, dz^\alpha \wedge d\bar z^\beta.$$

La condition de Kähler s'exprime, d'une manière équivalente, soit par le fait que la forme F est fermée, soit par l'existence locale d'une fonction potentielle de Kähler Φ, unique à une fonction pluriharmonique près, telle que

(2.2) $$g_{\alpha\bar\beta} = \frac{\partial^2 \Phi}{\partial z^\alpha . \partial \bar z^\beta} \qquad \text{ou bien} \qquad F = (2\pi i)^{-1}\, \bar\partial \partial \Phi.$$

D'une manière plus générale nous dirons qu'une forme extérieure ω dans une variété complexe quelconque est *de type kählérien* si c'est une forme de bidegré (1, 1), fermée et ne prenant que des valeurs réelles sur n'importe quelle paire de vecteurs tangents réels.

Soit $L \xrightarrow{\pi} M$ un fibré vectoriel holomorphe sur M de rang (dimension complexe de la fibre) m : si L est muni d'une structure hermitienne définie positive, cela signifie que le fibré L peut être considéré comme muni alternativement de deux groupes structuraux :

1° le groupe GL$(m, \mathbf{C})$, par rapport auquel on peut exprimer le fait que L est un fibré holomorphe;

2° le groupe $\mathrm{U}(m)\subset \mathrm{GL}(m, \mathbf{C})$, la réduction au groupe $\mathrm{U}(m)$ étant définie par le choix de structure hermitienne.

Soit U un ouvert de M servant à la fois comme domaine d'un système de coordonnées holomorphes (z^α) $(1\leqq\alpha\leqq n)$ de M et comme domaine d'existence d'un système de coordonnées holomorphes linéaires (ζ^λ) $(1\leqq\lambda\leqq m)$ sur chaque fibre dans $\pi^{-1}(\mathrm{U})$, de façon que $(z^1, \ldots, z^n; \zeta^1, \ldots, \zeta^m)$ constitue un système de coordonnées holomorphes dans l'ouvert $\pi^{-1}(\mathrm{U})\subset \mathrm{L}$. On peut représenter localement la structure hermitienne dans L par une matrice hermitienne $(a_{\lambda\bar\mu}(z, \bar z))$ de fonctions sur U telle que la forme hermitienne s'exprime par

$$t=a_{\lambda\bar\mu}(z, \bar z)\, \zeta^\lambda\bar\zeta^\mu \tag{2.3}$$

dont la valeur t, positive sauf en la section nulle de L, est une fonction invariante, définie globalement dans L. Dans la suite on désignera par $\mathscr{K}$ et $\mathscr{A}$ respectivement une structure kählérienne sur M, comme dans (2.1) ou (2.2), et une structure hermitienne sur $\mathrm{L} \xrightarrow{\pi} \mathrm{M}$, comme dans (2.3).

Une structure hermitienne $\mathscr{A}$ sur L équivaut à une réduction du groupe structural de L de GL$(m, \mathbf{C})$ à $\mathrm{U}(m)$, dans le sens qu'elle entraîne la formule de substitution remplaçant l'atlas des coordonnées *holomorphes* des fibres (ζ^λ) par des coordonnées correspondantes orthonormées, notées par $(w_1, \ldots, w_m)$ pour lesquelles la fonction norme t s'exprime par

$$t=\sum_{\lambda=1}^{m} |w_\lambda|^2.$$

Les transformations qui relient une carte de coordonnées orthonormées à une autre appartiennent donc au sous-groupe $\mathrm{U}(m)\subset \mathrm{GL}(m, \mathbf{C})$ pour la fibre $\pi^{-1}(x)$ au-dessus de n'importe quel point $x\in \mathrm{M}$, mais la dépendance de la transformation par rapport au point x ne sera plus holomorphe comme elle l'est dans le cas des coordonnées (ζ^λ). C'est pour cette raison principalement que dans la suite nous allons préférer des coordonnées holomorphes.

On introduit la connexion de Cartan dans le fibré, déterminée canoniquement par $\mathscr{A}$ comme dans [7] et définie par les formes locales associées à chaque carte locale holomorphe $(z, \zeta)=(z^1, \ldots, z^n; \zeta^1, \ldots, \zeta^m)$:

$$\Theta^\lambda_\nu=\mathrm{L}^\lambda{}_{\nu\alpha}\, dz^\alpha \qquad (1\leqq\lambda,\ \nu\leqq m),$$

où

$$\mathrm{L}^\lambda{}_{\nu\sigma}=a^{\lambda\bar\mu}\, \frac{\partial a_{\nu\bar\mu}}{\partial z^\alpha}, \qquad (a^{\lambda\bar\mu})={}^t(a_{\lambda\bar\mu})^{-1};$$

on en dérive la forme de courbure $\Omega^\lambda{}_\nu$, définie par les formules

$$\Omega^\lambda{}_\nu=\mathrm{S}^\lambda{}_{\nu\alpha\bar\beta}\, dz^\alpha \wedge d\bar z^{\beta}, \qquad \mathrm{S}^\lambda{}_{\nu\alpha\bar\beta} = \frac{\partial}{\partial \bar z^{\beta}}\, \mathrm{L}^\lambda{}_{\nu\alpha}.$$

On dit que le fibré $L \xrightarrow{\pi} M$ muni de la structure hermitienne $\mathscr{A}$ est positif (respectivement non négatif, non positif ou négatif) si *la forme bihermitienne de courbure* $a_{\lambda\bar{\mu}} S^{\lambda}{}_{\nu\alpha\bar{\beta}} \zeta^{\nu} \bar{\zeta}^{\mu} dz^{\alpha} d\bar{z}^{\beta}$ vérifie, en chaque point du fibré L hors de la section nulle, l'inégalité

$$(2.4) \qquad a_{\lambda\bar{\mu}} S^{\lambda}{}_{\nu\alpha\bar{\beta}} \zeta^{\nu} \bar{\zeta}^{\mu} Z^{\alpha} \bar{Z}^{\beta} > 0 \qquad ((\zeta) \neq 0, (Z) \neq 0)$$

(respectivement $\geqq 0$, $\leqq 0$ ou < 0).

Nous allons maintenant définir, en partant de M munie d'une structure kählérienne (2.1), (2.2), et d'un fibré vectoriel holomorphe $L \xrightarrow{\pi} M$ avec une structure hermitienne $\mathscr{A}$ donnée, une forme de type kählérien g_L sur le fibré tangent de l'espace total L. Cette forme est uniquement déterminée par les propriétés suivantes :

1° pour tout point $x \in M$ la restriction de la forme à la sous-variété $\pi^{-1} x$ coïncide avec la métrique hermitienne plate induite par le potentiel kählérien $t|_{\pi^{-1}x}$;

2° pour chaque point y du fibré L et pour tout vecteur tangent X à M au point πy, le relevé horizontal $\tilde{X}_y$ de X à y au sens défini par la connexion de Cartan de $\mathscr{A}$ est orthogonal à la fibre passant par y;

3° la restriction de g_L à l'espace tangent à la section nulle de L (identifiée canoniquement avec M) coïncide avec la métrique kählérienne donnée dans M.

Il n'est point difficile de vérifier que la seule forme de type kählérien dans L qui vérifie les axiomes ci-dessus est déterminée par le potentiel local

$$(2.5) \qquad \Psi = \Phi \circ \pi + t,$$

où Φ est une fonction locale de potentiel dans M pour la métrique kählérienne donnée et t est comme dans (2.3). On vérifie également le fait suivant : étant donnée la fibre $\pi^{-1}(x_0)$ au-dessus d'un point quelconque $x_0 \in M$, alors, pour tout chemin différentiable $x(s)$ dans M partant de x_0, la famille des relevés horizontaux de $x(s)$ dans L partant de n'importe quel point $\zeta_0 \in \pi^{-1}(x_0)$ définit pour chaque s une isométrie linéaire complexe de $\pi^{-1}(x_0)$ sur $\pi^{-1}(x(s))$. Ce fait entraîne que, E étant un ouvert connexe dans L où la forme $\partial\bar{\partial}\Psi$ de type kählérien est définie positive, pour tout $x \in \pi(E) \subset M$, l'intersection $\pi^{-1}(x) \cap E$ est totalement géodésique dans E.

Pour calculer explicitement la forme $\partial\bar{\partial}\psi$ partant de (2.5), il est utile parfois de remplacer le repère complexe naturellement associé aux coordonnées holomorphes locales $(z^1, \ldots, z^n; \zeta^1, \ldots, \zeta^m)$ par le repère adapté à la connexion de L, consistant en les champs de vecteurs

$$\left(\nabla_{z^1}, \ldots, \nabla_{z^n}; \frac{\partial}{\partial \zeta^1}, \ldots, \frac{\partial}{\partial \zeta^m}\right),$$

où

$$\nabla_{z^\alpha} = \frac{\partial}{\partial z^\alpha} - L^{\lambda}{}_{\nu\alpha} \zeta^{\nu} \frac{\partial}{\partial \zeta^{\lambda}} \qquad (1 \leqq \alpha \leqq n)$$

et, pour exprimer les formes différentielles, d'employer le repère dual $(dz^1, \ldots, dz^n;$ $\nabla\zeta^1, \ldots, \nabla\zeta^m)$, où

$$\nabla\zeta^\lambda = d\zeta^\lambda + L^\lambda{}_{\nu\alpha}\zeta^\nu dz^\alpha \qquad (1 \leqq \lambda \leqq m).$$

Ainsi on trouve, en partant de (2.4) :

$$\partial\Psi = \pi^* \partial\Phi + \frac{\partial a_{\lambda\bar{\mu}}}{\partial z^\alpha}\zeta^\lambda\zeta^{\bar{\mu}} dz^\alpha + a_{\lambda\bar{\mu}}\bar{\zeta}^\mu d\zeta^\lambda = \frac{\partial\Phi}{\partial z^\alpha} dz^\alpha + a_{\lambda\bar{\mu}}\bar{\zeta}^\mu \nabla\zeta^\lambda,$$

d'où on obtient, par une dérivation supplémentaire,

$$\partial\bar{\partial}\Psi = (g_{\alpha\bar{\beta}} + a_{\nu\bar{\mu}} S^\nu{}_{\lambda\alpha\bar{\beta}}\zeta^\lambda\bar{\zeta}^\mu)\, dz^\alpha d\bar{z}^\beta + a_{\lambda\bar{\mu}}\nabla\zeta^\lambda \nabla\zeta^{\bar{\mu}}. \tag{2.6}$$

Cette dernière formule montre que la forme déterminée dans L par les axiomes 1, 2, 3 dépend explicitement de la forme bihermitienne de courbure, telle qu'elle apparaît dans (2.4).

PROPOSITION 2.1. — *La forme* (2.6) *de type kählérien induite dans le fibré* L *par les données d'une métrique kählérienne dans* M *et d'une structure hermitienne* $\mathscr{A}$ *dans le fibré* $L \xrightarrow{\pi} M$ *est définie positive dans un voisinage de la section nulle de* L. *Elle est définie positive globalement dans* L, *si et seulement si la forme de courbure* (2.4) *de la connexion associée à* $\mathscr{A}$ *est partout non négative.*

Démonstration. — La première partie de l'énoncé est évidente d'après la formule (2.6). Si la forme de courbure est non négative au sens de (2.4), ceci entraîne immédiatement que la forme (2.6) est définie positive partout dans L. Réciproquement, si la forme (2.4) est négative à n'importe quel point $x \in M$ pour une valeur convenable de (ζ^λ) et (Z^α), alors on voit que, si on remplace le vecteur (ζ^λ) au besoin par un multiple scalaire à valeur absolue assez élevée, la forme (2.6) devient indéfinie au point ayant $(z^\alpha, \zeta^\lambda)$ pour coordonnées locales.

3. Fibrés holomorphes associés à des fibrés vectoriels

Nous allons maintenant généraliser la construction d'une forme de type kählérien à certains fibrés holomorphes sur une variété kählérienne donnée, où les fibres n'ont plus de structure d'espace vectoriel.

Nous n'allons considérer qu'une classe très restreinte de fibrés holomorphes, principalement parce qu'une construction aussi générale que possible entraînerait des complications que les applications envisagées ne justifient pas et deuxièmement parce que, d'après P. Blanchard [2], il y a des conditions qui sont nécessaires de toute façon pour l'existence d'une métrique kählérienne de quelque forme que ce soit dans tout fibré holomorphe. Les fibrés ne seront que des fibrés holomorphes associés à un fibré holomorphe vectoriel $L \xrightarrow{\pi} M$, considérant ce dernier de groupe structural U(m) grâce à une structure hermitienne donnée $\mathscr{A}$. Les fibres seront des ouverts de la fibre abstraite $\mathbf{C}^m$, invariants par l'action du groupe U(m), donc effectivement définis par des intervalles dans la demi-droite non négative $\mathbf{R}^+$, représentant des valeurs admissibles pour la fonction norme $t : L \to M$ définie par $\mathscr{A}$ (éventuellement la valeur $t=\infty$ aussi peut être admissible si l'on considère des fibrés d'espaces projectifs complexes). Si on garde sur de tels fibrés la connexion de Cartan induite par $\mathscr{A}$ sur L par restriction et si le fibré envisagé contient la section nulle de L, on n'a aucune difficulté à adapter les axiomes 1, 2, 3 à la situation actuelle, en remplaçant la métrique plane des fibres du premier axiome par une métrique quelconque, invariante par l'action du groupe structural $U(m)$; une telle métrique admet pour potentiel de Kähler une fonction, invariante elle aussi par $U(m)$, donc représentable par la composée $u \circ t$, où $u(x)$ est une fonction différentiable d'une variable réelle définie dans un intervalle de nombres non négatifs. La forme de type kählérien dans l'espace total E du fibré sera donc dérivée d'un potentiel, comme en (2.5) :

$$\Psi = \Phi \circ \pi + u \circ t, \tag{3.1}$$

où la fonction $u(x)$ d'une variable non négative doit satisfaire aux conditions qui entraînent que la forme hermitienne déduite de Ψ est définie positive dans E. La dérivation est analogue à celle qui donne la forme hermitienne (2.6) dans le cas des fibrés vectoriels

$$\partial\bar{\partial}\Psi = (g_{\alpha\bar{\beta}} + (u' \circ t)\, a_{\nu\bar{\mu}} S^{\nu}{}_{\lambda\alpha\bar{\beta}} \zeta^{\lambda}\bar{\zeta}^{\mu})\, dz^{\alpha}\,\overline{dz^{\beta}} + ((u' \circ t)\, a_{\lambda\bar{\mu}} + (u'' \circ t)\, a_{\lambda\bar{\nu}} a_{\kappa\bar{\mu}} \zeta^{\kappa}\bar{\zeta}^{\nu})\, \nabla\zeta^{\lambda}\,\overline{\nabla\zeta^{\mu}}. \tag{3.2}$$

Si l'intervalle des valeurs de t définissant le fibré $E \to M$ ne comprend pas la valeur $t=0$, on prendra (3.1) comme potentiel d'une structure de type kählérien par définition, bien qu'il ne soit pas aussi simple de justifier un tel choix par des axiomes le caractérisant uniquement.

PROPOSITION 3.1. — *La forme* (3.2) *restreinte aux directions verticales, c'est-à-dire ne tenant compte que des termes en $\nabla\zeta^{\lambda}$, est définie positive si et seulement si, dans le cas général, pour toute valeur $x \geqq 0$ dans l'intervalle de définition de $u(x)$, cette fonction satisfait aux conditions*

$$u'(x) > 0, \qquad u'(x) + x u''(x) > 0. \tag{3.3}$$

On a une exception à l'énoncé dans le cas des fibrés dont la fibre a une dimension complexe, la condition $u'(x) > 0$ cesse alors d'être nécessaire si la section nulle n'est pas comprise dans E.

Démonstration. — On vérifie la nécessité de la seconde inégalité (3.3) immédiatement pour tout $m \geqq 1$ en remplaçant chaque $\nabla\zeta^{\lambda}$ par la valeur ζ^{λ} elle-même et en tenant compte de l'identité $t = a_{\lambda\bar{\mu}}\zeta^{\lambda}\bar{\zeta}^{\mu}$. La première inégalité (3.3) est nécessaire pour $m \geqq 2$, car on peut donner à $\nabla\zeta^{\lambda}$ une valeur telle que $a_{\lambda\bar{\mu}}\nabla\zeta^{\lambda}\bar{\zeta}^{\mu} = 0$. Pour $m=1$ la première inégalité est évidemment nécessaire partout sur la section nulle de L et, puisque $xu'(x)$ est une fonction strictement croissante en raison de la seconde inégalité (3.3) $u'(x)$ doit être positive partout. On rencontre donc la seule exception à la nécessité des deux inégalités dans le cas $m=1$. Supposons que l'intervalle des valeurs admises de t soit ouvert à sa borne inférieure, que nous noterons x_0 ($x_0 \geqq 0$), alors les conditions nécessaires qui remplacent (3.3) se réduisent à la seule deuxième inégalité. Nous laissons au lecteur la démonstration de la suffisance des inégalités, car elle est immédiate.

On va vérifier maintenant qu'en supposant l'existence d'une certaine borne inférieure uniforme pour la courbure de la connexion (condition vide, en particulier, si M est compacte), il existe des fonctions u telles que la forme (3.2) soit définie positive partout dans la variété L tout entière.

DÉFINITION. — On dit que la forme de courbure $\Omega^\lambda{}_\nu = S^\lambda{}_{\nu\alpha\bar\beta}\, dz^\alpha\, \overline{dz}^\beta$ de la connexion de Cartan d'un fibré holomorphe $L \xrightarrow{\pi} M$ muni d'une structure hermitienne a en un point donné $p_0 \in M$, admet un nombre réel c pour borne inférieure si, quelles que soient les $m+n$ variables complexes $Z^1, \ldots, Z^n, \zeta^1, \ldots, \zeta^m$, le tenseur de courbure $(S^\lambda{}_{\nu\alpha\bar\beta})$ satisfait à l'inégalité

$$a_{\lambda\bar\mu} S^\lambda{}_{\nu\alpha\bar\beta}\, \zeta^\nu \bar\zeta^\mu Z^\alpha \bar Z^\beta \geqq c\, a_{\lambda\bar\mu} \zeta^\lambda \bar\zeta^\mu g_{\alpha\bar\beta} Z^\alpha \bar Z^\beta,$$

où les $g_{\alpha\bar\beta}$ sont les composantes de la métrique kählérienne donnée au point p_0. On dit que c est une borne inférieure uniforme pour la courbure dans M, si quel que soit $p \in M$, la constante c est une borne inférieure pour la courbure à p.

Nous n'aurons pas l'occasion d'utiliser ici la notion analogue de borne supérieure.

THÉORÈME 3.2. — *Soit* $L \xrightarrow{\pi} M$ *un fibré vectoriel holomorphe sur* M *muni d'une structure hermitienne* $\mathscr{A}$ *et soit* M *munie d'une métrique kählérienne* $\mathscr{K}$. *Si la courbure de la connexion de Cartan de* $\mathscr{A}$ *a une borne inférieure uniforme quelconque, alors il existe une métrique kählérienne complète* $\mathscr{C}_y$ *dans la fibre* $\mathbf{C}^m$, *invariante par l'action de* U(m), *telle que la forme de type kählérien, déterminée sur* L *par la métrique kählérienne* $\mathscr{K}$ *de* M, *par la connexion de Cartan associée à* $\mathscr{A}$ *et par la métrique* $\mathscr{C}_y$ *est définie positive partout sur* L. *En plus, si la structure kählérienne* $\mathscr{K}$ *de* M *est complète, il en est de même pour la structure kählérienne ainsi définie dans* L.

Démonstration. — Si la borne inférieure de la courbure est $\geqq 0$, alors la métrique du fibré linéaire, c'est-à-dire en prenant $u(x) = x$ est elle-même définie positive partout (prop. 2.1) dans L. Il n'y a aucune difficulté à vérifier que, la métrique donnée dans M étant complète, la métique (2.6) l'est aussi.

Supposons donc que la courbure ait comme borne inférieure uniforme une constante $-b$ strictement négative ($b>0$). On peut choisir une fonction $u(x)$ engendrant la forme (3.2) de façon que cette dernière soit partout définie positive et définisse sur L une structure métrique complète; par exemple on peut faire le choix suivant :

$$u(x) = \frac{2}{3b} \log(c+x) - \frac{1}{3} \log\log(c+x) \qquad \text{où} \quad c = e^b. \tag{3.4}$$

Pour montrer que la métrique définie par la forme hermitienne (3.2) avec $u(x)$ explicitée en (3.4) restreinte à chaque fibre est définie positive, il suffit d'après la proposition 3.1, de vérifier les inégalités (3.3). On a en effet, pour tout $x \geqq 0$,

$$u'(x) = \frac{1}{3(c+x)}\left(\frac{2}{b} - \frac{1}{\log(c+x)}\right) \geqq \frac{1}{3b(c+x)},$$

$$xu''(x)+u'(x)=\frac{1}{3}(c+x)^{-2}\left(\frac{2c}{b}-\frac{c}{\log(c+x)}+\frac{x}{(\log(c+x))^2}\right)$$

$$\geqq \frac{c}{3b(c+x)^2}+\frac{x}{3(c+x)^2(\log(c+x))^2}.$$

On voit ensuite que chaque fibre, munie de la métrique kählérienne $\partial\bar\partial(u\circ t)$, est complète, car l'intégrale $\int_0^\infty (xu''(x)+u'(x))^{1/2}\,dx$ est divergente.

Nous montrons maintenant que la forme hermitienne totale (3.2) est définie positive partout dans L en démontrant que sa partie « horizontale » majore partout la forme $(1/3)\,g_{\alpha\bar\beta}\,dz^\alpha\,d\bar z^\beta$. D'après la définition de la borne inférieure de la courbure, on voit en effet que

$$(g_{\alpha\bar\beta}+(u'\circ t)\,a_{\nu\bar\mu}\,S^\nu{}_{\lambda\alpha\bar\beta}\,\zeta^\lambda\bar\zeta^\mu)\,dz^\alpha\,d\bar z^\beta\geqq(1-bt\circ(u'\circ t))\,g_{\alpha\bar\beta}\,dz^\alpha\,d\bar z^\beta$$

et, d'après (3.4), pour tout $x\geqq 0$,

$$1-bxu'(x)=1-\frac{2x}{3(c+x)}+\frac{bx}{3(c+x)\log(c+x)}=\frac{1}{3}+\frac{1}{3(c+x)}\left(2c+\frac{bx}{\log(c+x)}\right)\geqq\frac{1}{3}.$$

Cette dernière inégalité démontre donc que la métrique kählérienne ainsi définie dans L majore la forme hermitienne suivante dans L, compatible avec la structure de fibré au sens réel, c'est-à-dire dans le sens qui a été précisé dans l'introduction,

$$\frac{1}{3}g_{\alpha\bar\beta}(z,\bar z)\,dz^\alpha\,d\bar z^\beta+((u'\circ t)\,a_{\lambda\bar\mu}+(u''\circ t)\,a_{\lambda\bar\nu}\,a_{\varkappa\bar\mu}\,\zeta^\varkappa\bar\zeta^\nu)\,\nabla\zeta^\lambda\,\nabla\bar\zeta^\mu.$$

Il est évident que cette dernière structure hermitienne dans L est complète si et seulement si sa restriction à chaque fibre ainsi qu'à la section nulle sont complètes elle-mêmes; le théorème est donc démontré.

Corollaire. — *Il existe une métrique kählérienne complète dans le complémentaire d'un seul point dans l'espace projectif complexe* $P^n(\mathbf{C})$.

Démonstration. — Le complément d'un point p_0 dans $P^n(\mathbf{C})$ est l'espace total d'un fibré linéaire holomorphe ayant $P^{n-1}(\mathbf{C})$ comme base. Le théorème que nous venons de démontrer assure l'existence d'une métrique kählérienne complète dans la variété $P^n(\mathbf{C})\setminus\{p_0\}$.

Ce dernier résultat nous mène à proposer la conjecture que toute variété complexe ouverte admettant une métrique kählérienne en admet une complète.

4. Variétés d'Einstein-Kähler

On va appliquer ici des idées que nous avons développées dans la section précédente : en partant d'une variété donnée et munie d'une métrique d'Einstein-Kähler, c'est-à-dire avec courbure de Ricci constante, nous allons en construire d'autres, qui seront des fibrés holomorphes associés à des fibrés en droites, de groupe structural U(1).

Soit M une variété complexe munie d'une métrique kählérienne, satisfaisant à l'équation d'Einstein :

$$R_{\alpha\bar{\beta}} = k_0\, g_{\alpha\bar{\beta}}, \qquad R_{\alpha\bar{\beta}}, \qquad R_{\alpha\bar{\beta}} = -\frac{\partial^2}{\partial z^\alpha\, \partial \bar{z}^\beta} \log \det(g_{\gamma\bar{\delta}}),$$

où k_0 est une constante quelconque. Cette équation est équivalente à la relation locale

$$\det(g_{\alpha\bar{\beta}}) = |\text{hol}|^2\, e^{-k_0 \Phi}, \tag{4.1}$$

où Φ représente un potentiel de Kähler et $|\text{hol}|^2$ la valeur absolue carrée d'une fonction holomorphe non spécifiée, mais qui peut se réduire à l'unité, soit (si $k_0 \neq 0$) en modifiant le choix de potentiel Φ, soit (si $k_0 = 0$) par une restriction dans le choix de coordonnées holomorphes locales à celles pour lesquelles la densité de l'élément de volume $\det(g_{\alpha\bar{\beta}})$ devient identiquement égale à 1 (l'annulation du tenseur de Ricci est précisément la condition d'existence de tels systèmes de coordonnées).

On va construire des métriques d'Einstein-Kähler de courbure de Ricci égale à une constante k, *a priori* arbitraire, qui est compatible avec une structure de fibré vectoriel holomorphe $L \xrightarrow{\pi} M$, où M est une variété d'Einstein-Kähler à courbure de Ricci égale à k_0. Nous nous limiterons ici au cas de fibrés linéaires, c'est-à-dire avec fibre de dimension complexe égale à 1. Le cas des fibrés où la dimension des fibres est plus grande ne présente que des complications typographiques, qu'on laissera résoudre au lecteur. Pour fixer les idées, soit $\dim_C L = n\, (n \geqq 2)$ et par conséquent $\dim_C M = n-1$. Une restriction bien plus sérieuse vient de ce qu'on va supposer L muni d'une structure hermitienne $\mathscr{A}$ à courbure égale à une constante l. Pour préciser les notations, la forme hermitienne (2.3) se réduit à

$$t = a(z, \bar{z})\zeta\bar{\zeta} = a|\zeta|^2 \tag{4.2}$$

et la forme de courbure explicitée dans (2.4) devient

$$a.S_{\alpha\bar{\beta}}|\zeta|^2 Z^\alpha \bar{Z}^\beta = t\, S_{\alpha\bar{\beta}} Z^\alpha \bar{Z}^\beta$$

avec

$$S_{\alpha\bar{\beta}} = \frac{\partial^2 \log(a(z, \bar{z}))}{\partial z^\alpha\, \partial \bar{z}^\beta}.$$

L'hypothèse de courbure constante et égale à l, $S_{\alpha\bar{\beta}} = l g_{\alpha\bar{\beta}}$, équivaut à la condition

$$a(z, \bar{z}) = |\text{hol}|^2 e^{l\Phi},$$

Φ étant un potentiel local de la métrique kählérienne donnée dans M. Par des choix convenables de coordonnées holomorphes ζ dans la fibre on réduit la dernière condition à

$$a(z, \bar{z}) = e^{l\Phi}.$$

L'existence de tels fibrés linéaires holomorphes est garantie par le théorème de Hodge-Kodaira, selon lequel une condition nécessaire et suffisante est que la classe de cohomologie de De Rham de la (1, 1)-forme l F soit une classe entière, où $F = -(2\pi\sqrt{-1})^{-1} g_{\alpha\bar\beta}\, dz^\alpha \wedge d\bar z^\beta$. Or on sait déjà que k_0 F est une classe entière, puisqu'elle représente la courbure du fibré linéaire canonique, la structure étant d'Einstein-Kähler; donc l'hypothèse de courbure constante de $\mathscr{A}$ entraîne qu'il faut que la métrique de M soit, à un facteur constant près, une métrique de Hodge (condition qui est satisfaite automatiquement si $k_0 \neq 0$); si la classe de De Rham de F n'est pas nulle, les valeurs admissibles de l seront limités aux éléments d'un sous-groupe discret de **R**; si au contraire la classe de F est nulle (par exemple, si M est une variété de Stein) alors l peut prendre une valeur réelle arbitraire.

PROPOSITION 4.1. — *Soit* M *une variété complexe de dimension* $n-1$, *munie d'une métrique de Hodge avec courbure de Ricci constante et égale à* k_0. *On se donne un fibré linéaire holomorphe* $L \xrightarrow{\pi} M$ *avec une structure hermitienne* $\mathscr{A}$ *dont la courbure, en relation avec la métrique de* M, *a une valeur constante et égale à* l. *Pour toute constante* $x_0 > 0$ *soit* H_{x_0} *l'hypersurface réelle* $\{q \in L \mid t(q) = x_0\} \subset L$ *et considérons une fonction* $u(x)$ *définie dans un intervalle ouvert* I *autour de* x_0. *Alors la forme de type kählérien* $\partial\bar\partial\psi$ *déterminée par* $u(x)$ *d'après* (3.2) *dans le voisinage* $t^{-1}(I) \subset L$ *de* H_{x_0} *est définie positive, si et seulement si* $u(x)$ *satisfait aux conditions*

$$1 + lxu'(x) > 0, \qquad u'(x) + xu''(x) > 0. \tag{4.3}$$

Si cette condition est satisfaite dans I *et si* $l \neq 0$, *alors la métrique kählérienne* $\partial\bar\partial\Psi$ *définie dans* $t^{-1}(I)$ *est à courbure de Ricci égale à une constante donnée* k, *si et seulement si* $u(x)$ *satisfait à l'équation différentielle*

$$(1 + lxu'(x))^{n-1}(xu''(x) + u'(x)) = cx^{l^{-1}(k_0 - k - l)} e^{-ku(x)} \qquad (l \neq 0), \tag{4.4}$$

où c *est une constante positive quelconque; si* $l = 0$, *on a forcément* $k = k_0$ *et* (4.4) *est remplacée par*

$$\begin{cases} u(x) = 2k^{-1} \log(1 + c_0^2 k x^c) + c_1 \log x + c_2 & (l = 0,\ k_0 = k \neq 0), \\ u(x) = c_0^2 x^c + c_1 \log x + c_2 & (k = k_0 = l = 0), \end{cases} \tag{4.4'}$$

où c, c_0, c_1, c_2 *sont des constantes réelles arbitraires, et* $c_0 \neq 0$.

Démonstration. — La forme de type kählérien dans le voisinage de l'hypersurface H_{x_0} est donnée par (3.2); avec les hypothèses présentes (fibre de dimension 1 et courbure du fibré linéaire de valeur égale à l) elle se réduit à

$$\partial\bar\partial\Psi = (1 + lxu'(x))\, g_{\alpha\bar\beta}\, dz^\alpha d\bar z^\beta + c_0 e^{l\Phi}(xu''(x) + u'(x)) \,|\nabla\zeta|^2, \tag{4.5}$$

où x représente les valeurs près de x_0 de la forme hermitienne t dans le fibré L. La densité de l'élément de volume correspondant à cette forme par rapport aux coordonnées holomorphes $(z^1, \ldots, z^{n-1}; \zeta)$, tenant compte de la condition réduite d'Einstein dans M, devient donc

$$g_L = c_0 (1 + lxu'(x))^{n-1} e^{-k_0\Phi} e^{l\Phi}(xu''(x) + u'(x)).$$

Ceci permet de vérifier que les conditions (4.3) et $c_0 > 0$ sont nécessaires et suffisantes afin que la forme (4.5) soit définie positive. La condition (4.1), c'est-à-dire que la métrique qu'on vient d'introduire dans L ait courbure de Ricci constante et égale à une quantité k arbitrairement donnée devient donc

$$(4.6) \qquad (1+lxu'(x))^{n-1}(xu''(x)+u'(x)) = |\mathrm{hol}|^2 \exp\{(k_0-k-l)\Phi - ku(x)\}.$$

On doit maintenant distinguer le cas général du fibré L non trivial ($l \neq 0$) du cas spécial $l=0$ où le fibré est localement plat. Si $l \neq 0$ on peut faire disparaître le potentiel local Φ de la métrique de M^n, en utilisant l'identité suivante, valable partout dans L sauf sur la section nulle du fibré,

$$|\mathrm{hol}|^2 e^{\Phi} = |\mathrm{hol}|^2 t^{l^{-1}};$$

l'équation (4.6) se transforme donc de la manière suivante :

$$x^{-l^{-1}}(k_0-k-l)(1+lxu'(x))^{n-1}(xu''(x) \to u'(x))\, e^{ku(x)} = |\mathrm{hol}|^2.$$

Nous allons montrer que le membre de droite de cette dernière équation se réduit à une constante. En effet, supposons qu'il en était autrement et soit $\varphi = \varphi(z, \zeta)$ une fonction holomorphe locale, non nulle et non constante, dont la valeur absolue carrée représente le membre de droite; alors, à tout point où $d\varphi \neq 0$, $|\varphi|^2$ est constante sur chaque hypersurface réelle d'une famille régulière à un paramètre; l'invariant de E. Levi de chacune de ces hypersurfaces est représenté, à un facteur non nul près, par une restriction de la forme hermitienne $\partial\bar{\partial} \log |\varphi|^2$, qui est identiquement nulle. D'autre part, tenant compte du membre de gauche de l'équation, les mêmes hypersurfaces sont déterminées par l'équation $t = \mathrm{Cte}$ et leur invariant de Levi, calculé par la restriction de $\partial\bar{\partial} \log t$, à chaque point, au sous-espace vectoriel complexe maximal de l'espace tangent de H_t, se réduit, d'après (2.4) et (4.2), à la forme de courbure $S_{\alpha\bar{\beta}}\, dz^\alpha\, dz^{\bar{\beta}} = lg_{\alpha\bar{\beta}}\, dz^\alpha\, d\bar{z}^\beta \neq 0$. Cette contradiction entraîne que $\varphi(z, \zeta)$ est effectivement une constante, dont nous désignerons la valeur absolue carrée par c. On en déduit immédiatement que $u(x)$ doit satisfaire à l'équation différentielle (4.4). La constante c est nécessairement positive, à cause de (4.3).

Si d'autre part on a $l=0$, la forme (4.2) est localement réductible à la somme directe de la métrique d'Einstein-Kähler de M à courbure de Ricci k_0 avec une métrique kählérienne à symétrie circulaire dans la fibre. Cette somme directe n'aura une courbure de Ricci constante et égale à k que si $k = k_0$ et si la métrique dans la fibre de potentiel $u(x)$ a une courbure de Gauss constante et égale elle aussi à k_0. On peut vérifier facilement que les seules fonctions $u(x)$ qui engendrent une telle métrique dans une couronne circulaire sont données par (4.4′). Ceci complète la démonstration.

On va formuler maintenant les conditions supplémentaires auxquelles la fonction $u(t)$ doit satisfaire pour qu'une métrique d'Einstein-Kähler déterminée par une solution $u(x)$ de (4.4) ou (4.4′) puisse se prolonger régulièrement sur la section nulle de $L \xrightarrow{\pi} M$.

Proposition 4.2. — *On se donne un fibré linéaire holomorphe* $L \xrightarrow{\pi} M$ *avec les mêmes hypothèses que dans la proposition précédente. Une métrique d'Einstein-Kähler dans* L *du type* (4.5) *à courbure de Ricci égale à* k*, engendrée par une fonction* $u(x)\,(0 \leq x < x_1)$ *se prolonge* *régulièrement sur la section nulle, si et seulement si* $k=k_0-l$ *et* $u'(0)>0$ *dans le cas général* ($l\neq 0$)*; dans le cas particulier* $l=0$ *la condition se traduit simplement par la restriction suivante sur les fonctions* $u(x)$ *données dans* (4.4′) *:*

$$u(x)=2k^{-1}\log(1+c_0^2kx)+c_2 \qquad (k=k_0\neq 0,\ c_0\neq 0,\ l=0)$$

ou

$$u(x)=c_0^2x+c_2 \qquad (k=k_0=l=0,\ c_0\neq 0).$$

Démonstration. — Le cas particulier $l=0$ étant élémentaire, nous nous bornerons au cas général $l\neq 0$. L'équation différentielle (4.4) admet le facteur intégrant

$$l^{-1}(k_0-k)-kxu'(x).$$

En multipliant les deux membres de (4.4) par ce facteur et en intégrant on obtient l'équation du premier ordre

$$(4.7)\quad \frac{k_0}{nl^2}[(1+lxu'(x))^n-1]-\frac{k}{(n+1)l^2}[1+lxu'(x))^{n+1}-1]=cx^{l^{-1}(k_0-k)}e^{-ku(x)}+c_2.$$

Le membre de gauche est un polynôme $P(xu'(x))$ en $xu'(x)$ à coefficients réels dépendant de k_0, k et l, sans terme constant; on a en effet

$$\begin{aligned}P(w)&=\frac{k_0}{nl^2}((1+lw)^n-1)-\frac{k}{(n+1)l^2}((1+lw)^{n+1}-1)\\&=\frac{k_0-k}{l}w+\sum_{j=2}^{n+1}\frac{n(k_0-k)-k_0(j-1)}{n(n+1)}\binom{n+1}{j}l^{j-2}w^j.\end{aligned}$$

En comparant $P(xu'(x))$ avec le membre de droite de (4.7) dans un voisinage de $x=0$ on voit *a priori* deux possibilités

(*a*) $$k_0=k \quad \text{et} \quad c_2=-ce^{-ku(0)}$$

ou bien

(*b*) $$k_0-k=l \quad \text{et} \quad c_2=0.$$

La première de ces possibilités mène à une contradiction, car en supposant qu'on a une solution $u(x)=u(0)+xu'(0)+O(x^2)$, avec $u'(0)>0$, de l'équation (4.7), on trouve alors le membre de gauche égal à $O(x^2)$, tandis que le membre de droite est égal à $-cku'(0)e^{-ku(0)}.x+O(x^2)$, ceci n'étant possible que si $k=k_0=0$; mais alors, le membre de droite étant une constante, ceci entraîne $u'(x)=c/x$, qui n'a pas de solutions admissibles pour $x=0$.

Il ne nous reste donc que la possibilité (*b*); pour démontrer qu'il existe effectivement dans ce cas une solution $u(x)$ qui, pour $0\leqq x<x_1$, engendre une forme définie positive dans le

domaine correspondant de L, on transcrit l'équation (4.7) de la manière suivante :

$$P(xu'(x))=cxe^{-ku(x)} \tag{4.8}$$

avec $k=k_0-l$ et P désignant le polynôme

$$P(w)=\frac{k_0}{nl^2}[(1+lw)^n-1]-\frac{k_0-l}{(n+1)l^2}[(1+lw)^{n+1}-1]=w+\sum_{j=2}^{n+2}\frac{nl-(j-1)k_0}{n(n+1)}\binom{n+1}{j}l^{j-2}\,w^j.$$

Par le théorème des fonctions implicites, il existe une fonction analytique réelle et une seule $F(v)$ dans un voisinage de $v=0$ telle que $F(P(w))=w$. En introduisant le paramètre auxiliaire

$$\xi=F(cxe^{-ku(x)})$$

dans un voisinage de $\xi=0$, l'équation (4.8), vue comme $xu'(x)=\xi$, se transforme en l'équation suivante :

$$P'(\xi)\,d\xi=P(\xi)(1-k\xi)\frac{dx}{x}$$

et, puisque $P'(w)=(1+lw)^{n-1}(1-kw)$, on résout l'équation (4.8) par les quadratures suivantes, en fonction du paramètre ξ,

$$x=x_0\exp\int_{\xi_0}^{\xi}\frac{(1+lw)^{n-1}}{P(w)}dw,\qquad u(x)=u_0+\int_{\xi_0}^{\xi}\frac{w(1+lw)^{n-1}}{P(w)}dw \tag{4.9}$$

qui donne la solution $u(x)$ de (4.8) telle que $u(x_0)=u_0$, pourvu que x_0 et u_0 soient des constantes réelles ($x_0>0$), telles que $cx_0\,e^{-ku_0}$ soit dans l'intervalle d'analycité de F et que l'on pose en plus

$$\xi_0=F(cx_0\,e^{-ku_0}) \tag{4.9'}$$

Ceci complète la démonstration.

Il vaut la peine de remarquer que le paramètre ξ en fonction duquel on a exprimé les solutions (4.9) de l'équation différentielle (4.8) a une interprétation géométrique; en effet, si on considère la métrique d'Einstein-Kähler dans le domaine maximal $E\subset L$ où cette métrique est régulière, alors on peut associer à chaque point $(z,\zeta)\in E$ le disque dans la fibre $\pi^{-1}\pi(z,\zeta)\cap E$ défini par $D(z,\zeta)=\{(z,\zeta')\,|\,t(z,\zeta')\leq t(z,\zeta)\}$; l'aire du disque $D(z,\zeta)$ par rapport à cette métrique est égale à $2\pi\,(t.u'\circ t)(z,\zeta)=2\pi\xi(t(z,\zeta))$.

On va signaler maintenant trois cas particuliers de l'équation différentielle (4.8), dans lesquels les solutions locales dans un voisinage de $x=0$ ont une forme particulièrement élémentaire en dehors du cas $l=0$, où le fibré est localement plat, qui n'est pas intéressant :

1° si $k_0=0$, c'est-à-dire si M est à courbure de Ricci nulle, et $k=-l$, alors

$P(w)=(n+1)^{-1}l^{-1}[(1+lw)^{n+1}-1]$ et on a

$$(4.10)\quad x=x_0 \exp \int_{\xi_0}^{\zeta} \frac{(n+1)l(1+lw)^{n-1}}{(1+lw)^{n+1}-1}dw = x_0 \exp \sum_{j=0}^{n}\left(\omega^{-j}\log\frac{1-\omega^j+l\xi}{1-\omega^j+l\xi_0}\right) \qquad (\omega=e^{2\pi i/(n+1)})$$

et

$$(4.10')\quad u(x)=u_0+\int_{\xi_0}^{\xi}\frac{(n+1)lw(1+lw)^{n-1}}{(1+lw)^{n+1}-1}dw = u_0-\frac{1}{l}\log\frac{x\,P(\xi_0)}{x_0\,P(\xi)}=\frac{1}{l}\log\frac{P(\xi)}{cx};$$

2° si $l\neq 0$, $k_0=-nl$ et $k=-(n+1)l$, alors $P(w)=w(1+lw)^n$; alors les intégrales (4.9) s'expriment sous forme élémentaire, d'où on déduit

$$(4.11)\qquad x=x_0\frac{\xi(1+l\xi_0)}{\xi_0(1+l\xi)},\qquad u(x)=u_0+l^{-1}\log\left(\frac{1+l\xi}{1+l\xi_0}\right),$$

la métrique induite sur chaque fibre $\pi^{-1}(z)\cap E$ est une métrique à courbure constante et égale à $-2l$;

3° si $l=k_0$ et $k=0$, c'est le cas le plus intéressant, donnant lieu à une métrique à courbure de Ricci nulle dans un sous-domaine maximal de régularité $E\subset L$. Dans ce cas $P(w)=(nl)^{-1}[(1+lw)^n-1]$; en substituant ce polynôme dans les intégrales (4.9) on obtient :

$$(4.12)\quad \begin{cases} x=x_0\dfrac{P(\xi)}{P(\xi_0)}; \\ u(x)=u_0+n(\xi-\xi_0)-\dfrac{1}{l}\displaystyle\sum_{j=1}^{n-1}(1-\omega^j)\log\left(\frac{1-\omega^j+l\xi}{1-\omega^j+l\xi_0}\right) \qquad (\omega=e^{2\pi i/n}). \end{cases}$$

On va considérer maintenant la question globale de savoir si le domaine maximal $E\subset L$, où la forme $\partial\bar\partial(\Phi+u\circ t)$ est définie positive et y définit une métrique d'Einstein-Kähler, est complet par rapport à cette métrique.

THÉORÈME 4.3. — *Soit* M *une variété kählérienne à $n-1$ dimensions, à courbure de Ricci constante et égale à k_0 et soit* $L\to M$ *un fibré linéaire holomorphe muni d'une structure hermitienne $\mathscr{A}$ à courbure de valeur constante l. Alors la métrique d'Einstein-Kähler, adaptée à ces données dans* L, *dans un voisinage de la section nulle et définie positive dans un domaine maximal* $E\subset L$, *définit une structure d'Einstein-Kähler complète dans* E *si et seulement si :* 1° M *est complète par rapport à la métrique donnée;* 2° *la constante k de* L *est égale à k_0-l et* 3° *$l\geqq 0$ et $k\leqq 0$.*

Démonstration. — Nous démontrerons d'abord la nécessité des trois conditions énoncées ci-dessus. La nécessité que M soit complète par rapport à la métrique kählérienne donnée est évidente, puisque la section nulle de L est un ensemble fermé dans L isométrique à M. La valeur k de la courbure de Ricci constante dans L est forcément égale à k_0-l d'après la proposition 4.2. La courbure de Ricci k dans L ne peut pas être positive car toute variété complète à courbure de Ricci uniformément strictement positive est compacte (th. de Myers [9]).

Il nous reste encore à démontrer la nécessité de la condition $l \geqq 0$. En raisonnant par l'absurde, supposons $l<0$ et reportons-nous à la solution paramétrique (4.9) de l'équation différentielle (4.8). La solution $u(x)$ qui en résulte satisfait à la relation $xu'(x)=\xi$ avec le paramètre ξ et reste valable pour $0<\xi<\alpha_1$, où α_1 représente la racine positive la plus petite du polynôme $P(w)$ s'il y en a une, ou pour $0<\xi<\infty$ si $P(w)$ n'a pas de racines positives. D'autre part, en regardant l'expression (4.5) de la forme de type kählérien qui en résulte, on voit qu'elle n'est définie positive que lorsque ξ est dans l'intervalle $]0, \min\{-l^{-1}, \alpha_1\}[$ correspondant à la condition $1+lxu'(x)>0$; la condition supplémentaire $xu''(x)+u'(x)>0$ est satisfaite automatiquement, puisque

$$xu''(x)+u'(x)=\frac{d}{dx}(xu'(x))=\left(\frac{dx}{d\xi}\right)^{-1}=x^{-1}\,P(\xi)(1+l\xi)^{-n+1}.$$

Pour étudier le comportement asymptotique de la métrique lorsque $\xi \to \min\{-l^{-1}, \alpha_1\}$, on doit distinguer deux cas selon que $\alpha_1<-l^{-1}$ ou $\alpha_1 \geqq l^{-1}$; on observe que le polynôme $P(w)$ vérifie l'identité algébrique

$$P(w)=\frac{k_0}{nl^2}[(1+lw)^n-1]-\frac{(k_0-l)}{(n+1)l^2}[1+lw)^{n+1}-1]=w(1+lw)^n-\frac{k_0+nl}{n(n+1)}w^2\sum_{j=1}^{n} j(1+lw)^{j-1};$$

donc $P(w)$ a une racine dans l'intervalle $]0, -l^{-1}[$ pour $l<0$ si et seulement si $k_0>-nl$. Considérons maintenant le chemin défini dans une fibre quelconque par $z=\mathrm{Cte}$; $\zeta=\tau\,(\tau>0)$. On donne ci-dessous l'expression de l'élément différentiel de sa longueur

$$\left(\frac{ds}{d\tau}\right)^2=2(xu''(x)+u'(x)).a(z,\bar{z})=2\,a(z,\bar{z})\frac{d\xi}{dx},$$

où $x=t=a(z,\bar{z})\tau^2$, donc

$$\frac{d\tau}{d\xi}=\frac{1}{2\sqrt{a(z,\bar{z})x}}\,\frac{dx}{d\xi}.$$

Par conséquent

$$\left(\frac{ds}{d\xi}\right)^2=2\,a(z,\bar{z})\frac{d\xi}{dx}\,\frac{1}{4\,a(z,\bar{z})x}\left(\frac{dx}{d\xi}\right)^2=\frac{1}{2x}\,\frac{dx}{d\xi}=\frac{(1+l\xi)^{n-1}}{2\,P(\xi)};$$

la longueur de l'arc donc est donnée par l'intégrale définie

$$s=\int_0^{\xi}\left(\frac{(1+lw)^{n-1}}{2\,P(w)}\right)^{1/2}dw. \tag{4.14}$$

Cette dernière expression montre le résultat désiré par les raisonnements suivants : 1° si $k_0>-nl$, alors il y a une racine α_1 de $P(w)$ et une seule dans l'intervalle $]0, -l^{-1}[$ et, comme cette racine est simple, l'intégrale converge si $\xi \to \alpha_1$; 2° si $k_0=-nl$, alors $P(w)=w(1+lw)^n$ et l'intégrale (4.13) converge vers $\pi/\sqrt{-2l}$ lorsque $\xi \to -l^{-1}$; 3° si $k_0<nl$, alors l'intégrale

(4.13) converge trivialement pour $\xi \to -l^{-1}$. On a donc démontré que les conditions énoncées sont nécessaires pour qu'il résulte de la construction une métrique d'Einstein-Kähler complète.

Pour montrer la suffisance, supposons que M est complète par rapport à la métrique donnée, que $l \geqq 0$ et que $k = k_0 - l \leqq 0$. Alors le polynôme $P(\xi) = \int_0^{\xi} (1-kw)(1+lw)^{n-1}\,dw$ est strictement positif pour tout $\xi > 0$. La solution (4.9) de l'équation différentielle (4.8) est donc valable pour tout $\xi \geqq 0$; la variable x en fonction de ξ est bornée ou non selon que $k<0$ ou $k=0$, de telle sorte que pour $k<0$ la métrique d'Einstein-Kähler est définie dans une partie de L où la norme hermitienne t de L est bornée par une constante $\beta\,(0<\beta<\infty)$ dépendant de k_0 et l, telle que $\beta \to \infty$ lorsque $k \to 0^-$; si $k^0 = l \geqq 0$ et par conséquent $k=0$, alors la métrique est définie dans la variété L tout entière. Dans chacun de ces cas l'intégrale (4.13) diverge si $\xi \to \beta$. Par un raisonnement analogue à celui de la démonstration de la proposition 3.2, on montre que le domaine d'existence de la métrique d'Einstein-Kähler à courbure de Ricci égale à k dans le domaine $E = \{(z, \zeta) \in L \mid t(z, \zeta) < \beta\}$ est complet par rapport à cette métrique. Le théorème 4.3 est donc démontré.

Il vaut la peine de signaler deux cas particuliers, où les métriques d'Einstein-Kähler complètes prévues par le théorème 4.3 ont une forme particulièrement élémentaire. Ils correspondent aux exemples 2° et 3° qu'on a décrits après la proposition 4.2 :

1° si la variété à $n-1$ dimensions complexes M est munie d'une métrique d'Einstein-Kähler complète à courbure de Ricci $k_0 \neq 0$ et si L est le fibré linéaire sur M admettant une structure hermitienne à courbure de valeur constante égale à $l = -k_0/n$, alors l'espace total du fibré associé à L où la fibre est une surface simplement connexe et complète à courbure constante égale à $-l$ prend de ces données une métrique d'Einstein-Kähler à courbure de Ricci égale à $k = ((n+1)/n)\,k_0$; si $k_0 < 0$ ce fibré est complet, tandis que si $k_0 > 0$, la fibre est la droite projective $P_1(\mathbf{C})$; dans ce cas la métrique est dégénérée à la section infinie, mais en considérant cette dernière comme un seul point, on obtient un espace complexe avec un point singulier isolé, muni d'une structure métrique complète et d'Einstein-Kähler en dehors de la singularité. La singularité elle-même disparaît dans le seul cas où $M = P_{n-1}(\mathbf{C})$ muni d'une métrique de Fubini et l'espace fibré singulier qu'on obtient n'est autre que l'espace $P_n(\mathbf{C})$ avec la métrique de Fubini ayant la même courbure sectionnelle holomorphe constante k_0/n que M;

2° si M est une variété complexe compacte à n-1 dimensions, munie d'une métrique d'Einstein-Kähler à courbure de Ricci $k_0 > 0$ et si $L \xrightarrow{\pi} M$ est le fibré linéaire *canonique*, muni de la norme hermitienne canoniquement associé à l'élément de volume de M, à courbure de valeur constante k_0, alors la construction de la métrique d'Einstein-Kähler décrite dans la démonstration du théorème 4.3 introduit sur l'espace L tout entier une métrique kählérienne complète et à courbure de Ricci nulle. En général une telle métrique est caractérisée par la propriété que son groupe d'holonomie local est contenu dans le groupe $SU(n)\,(n = \dim_{\mathbf{C}} L)$; dans le cas des métriques construites par notre méthode on peut même montrer très simplement que le groupe d'holonomie global est contenu dans $SU(n)$ (l'égalité ayant lieu dans presque tous les cas) par le raisonnement suivant. Puisque L est le fibré canonique sur M, c'est-à-dire le fibré linéaire holomorphe des $(n-1)$-formes holomorphes dans M. Soit $(z^1, \ldots, z^{n-1})$ un système de coordonnées holomorphes locales dans M : on associe à ce système la coordonnée locale holomorphe et linéaire sur les fibres ζ (c'est-à-dire la section holomorphe locale dans le fibré dual) prenant la valeur 1 sur la $(n-1)$-forme $dz^1 \wedge \ldots \wedge dz^{n-1}$. Alors la n-forme $\theta = dz^1 \wedge \ldots \wedge dz^{n-1} \wedge d\zeta$ dans L est indépendante du choix de coordonnées locales $(z^1, \ldots, z^{n-1})$; il est encore facile de vérifier que l'élément de volume de la métrique à courbe de Ricci nulle est égal, à un facteur constant près, à $\theta \wedge \overline{\theta}$, de sorte que la forme θ est de dérivée covariante nulle; le transport parallèle dans L du co-repère $(dz^1, \ldots, dz^{n-1}, d\zeta)$ évalué en n'importe quel point de départ amène à un co-repère dont le produit extérieur coïncide avec θ; ceci montre que le groupe d'holonomie est effectivement contenu dans le groupe SU(n). La formule (4.9) qui détermine la fonction u qui engendre la métrique selon (3.1) et (3.2), dans le cas $k_0 = l > 0$, se prête à une simplification. Le polynôme $P(w)$ est égal à $1/nl\,[(1+lw)^n - 1]$, de sorte que, suivant (4.9), on a

$$x = x_0 \frac{(1+l\xi)^n - 1}{(1+l\xi_0)^n - 1},$$

$$u(x) = u_0 + \int_{\xi}^{\xi} \frac{nlw(1+lw)^{n-1}\, dw}{(1+lw)^n - 1} = u_0 + \frac{1}{l} \int_{1+l\xi_0}^{1+l\xi} \left(n - \frac{n(\eta^{n-1} - 1)}{\eta^n - 1} \right) d\eta.$$

Par un passage à la limite, on peut poser $x_0 = 0$, $u_0 = u(0)$, d'où on tire la formule suivante pour la fonction $u(x)$:

$$(4.14) \quad u(x) = u_0 + \frac{n}{l}(\sqrt[n]{1+cx} - 1) - \sum_{j=1}^{n-1} \left(\frac{1-\omega^j}{l} \right) \log \left(\frac{\sqrt[n]{1+cx} - \omega^j}{1-\omega^j} \right) \qquad (\omega = e^{2\pi i/n}).$$

Pour des valeurs de n petites on peut réduire cette formule encore à des expressions élémentaires. En particulier, pour $n=2$ on part de l'espace de base $M = P_1(\mathbf{C})$ muni d'une métrique à courbure constante; le fibré canonique sur $P_1(\mathbf{C})$ est biholomorphiquement équivalent au cône quadratique complexe $\{(x, y, z) \in \mathbf{C}^3 \mid z^2 = xy\}$ avec la singularité de l'origine résolue par un éclatement; dans ce cas la fonction $u(x)$ se réduit à

$$u(x) = u_0 + \frac{2}{l}(\sqrt{1+cx} - 1) - \frac{2}{l} \log \frac{\sqrt{1+cx} + 1}{2}.$$

La même métrique, exprimée par d'autres cartes locales, a été trouvée récemment par plusieurs autres auteurs [1], [4], [6] et [17] qui y voient une réalisation de modèles d'instantons. D'un point de vue plus général, S. T. Yau vient d'annoncer des résultats analogues sous la forme de théorèmes d'existence de métriques kählériennes complètes à courbure de Ricci nulle dans des variétés kählériennes compactes [13] à première classe de Chern réelle nulle, ainsi que dans certaines variétés algébriques ouvertes (résultat à paraître).

5. Variétés hyperkählériennes

Nous donnons d'abord quelques définitions afin de fixer la terminologie dans un sujet où souvent elle n'est ni sémantiquement justifiable ni consistante.

Définition 5.1 : On dit qu'une variété différentiable V à m dimensions réelles est munie d'une structure *presque quaternionnienne* par la donnée de deux champs continus d'automorphismes linéaires I et J du fibré tangent T(V) sur lui-même (notés comme agissant à droite) et vérifiant : les relations $I^2 = J^2 = -\mathrm{id}$, $I \circ J + J \circ I = 0$.

Si $\{V; I, J\}$ est une variété presque quaternionnienne, notons par K le champ d'automorphismes défini par $K = I \circ J = -J \circ I$; pour tout quaternion $q = q_0 + q_1 i + q_2 j + q_3 k \in \mathbf{H}$ on définit l'action de q (à droite) sur tout vecteur tangent X par

$$X q = q_0 X + q_1 XI + q_2 XJ + q_3 XK, \tag{5.1}$$

les relations algébriques entre les opérateurs de structure I, J, K impliquent que l'action de $\mathbf{H}$ sur T(V) définie par (5.1) détermine sur T(V) une structure de fibré en $\mathbf{H}$-modules à droite; en particulier la dimension réelle m de V est divisible par 4. Notons $\mathscr{J}$ l'ensemble de tous les quaternions q satisfaisant à l'équation $q^2 = -1$, c'est-à-dire $q \in \mathscr{J}$ si $q = q_1 i + q_2 j + q_3 k$ avec $q_1^2 + q_2^2 + q_3^2 = 1$. Pour tout $q \in \mathscr{J}$ l'action de q sur T(V) correspond donc à une structure presque complexe; par conséquent toute variété presque quaternionnienne V contient dans sa structure une famille de structures presque complexes, paramétrée par la 2-sphère $\mathscr{J} \subset \mathbf{H}$.

On peut introduire, grâce à une partition de l'unité, dans toute variété presque quaternionnienne $\{V; I, J\}$, une métrique hermitienne, c'est-à-dire une structure riemannienne que I et J conservent; ceci équivaut à une réduction du groupe structural de T(V) au groupe $\mathrm{Sp}(n) = O(4n) \cap GL(n; \mathbf{H})$ $(m = 4n)$. L'existence d'une telle réduction dans une variété V quelconque est équivalente à l'existence d'une structure presque quaternionnienne.

On dit qu'une structure presque quaternionnienne différentiable dans une variété V est *intégrable*, si quel que soit $q \in \mathscr{J}$, la structure presque complexe définie par q est intégrable au sens de Eckmann-Frölicher [3]; en utilisant le formalisme de Frölicher-Nijenhuis [5], [11], ceci équivaut à la condition que les tenseurs structurels I et J, considérés comme 1-formes à valeurs dans T(V) satisfont aux relations

$$[I, I] = [I, J] = [J, J] = 0.$$

On en déduit du théorème de Newlander-Nirenberg [10] que toute variété V ayant une structure presque quaternionnienne intégrable admet par conséquent une famille de structures complexes naturellement paramétrées par $\mathscr{J}$; pour cette raison on propose ici de désigner toute structure presque quaternionnienne intégrable comme structure *hypercomplexe* (l'usage antérieur de l'adjectif pour décrire les algèbres sur un corps étant tombé en désuétude); il faut remarquer qu'une telle structure n'a aucune relation apparente avec une structure locale définie par des cartes locales à valeurs dans $\mathbf{H}^n$.

Supposons maintenant qu'une variété presque quaternionnienne différentiable à $4n$ dimensions $\{V; I, J\}$ puisse être munie aussi d'une structure hermitienne telle que, en considérant la connexion de Christoffel de V par rapport à la structure riemannienne sous-jacente, les dérivées covariantes des tenseurs représentant I et J [et par conséquent l'action à droite de tout quaternion q sur T(V)] s'annulent identiquement. Alors la structure presque quaternionnienne est automatiquement intégrable et, pour n'importe quel point $q \in \mathcal{J}$, la métrique donnée est kählérienne par rapport à la structure complexe correspondant à la structure complexe résultant de l'intégrabilité de q; pour cette raison on propose de désigner la structure définie par I, J et la métrique comme une *structure hyperkählérienne*. Une autre caractérisation d'une telle structure parmi les structures riemanniennes est que le groupe d'holonomie locale définie par le transport de Levi-Civita soit contenu dans le groupe $\mathrm{Sp}(n) \subset \mathrm{O}(4n)$.

Les seuls exemples connus jusqu'à maintenant de structures hyperkählériennes localement irréductibles et non symétriques étaient les variétés kählériennes à deux dimensions complexes à courbure de Ricci nulle décrites à la fin de la section 4. Que ces exemples aient des structures hyperkählériennes (c'est-à-dire de structures riemanniennes avec groupe d'holonomie $\mathrm{Sp}(n) \subset \mathrm{O}(4n)$) vient du fait que pour $n=1$, $\mathrm{Sp}(1)=\mathrm{SU}(2)$, tandis que, pour $n \geqq 2$, $\mathrm{Sp}(n)$ est strictement plus petit que $\mathrm{SU}(2n) \subset \mathrm{O}(4n)$. La construction donnée ici à la fin du chapitre 4, d'une struxture kählérienne complète à courbure de Ricci nulle dans le fibré linéaire canonique (fibré cotangent holomorphe) sur la droite projective en est un autre exemple élémentaire non compact. Ce sont justement ces exemples, généralisés aux dimensions supérieures d'une manière différente de celle du chapitre 4, qui inspirent la construction de structures hyperkählériennes que nous allons effectuer.

Soit V une variété hyperkählérienne à $4n$ dimensions réelles et, pour commencer, fixons notre attention sur une des structures kählériennes qui lui sont propres, par exemple celle définie par la métrique avec l'opérateur I. On considère en chaque point $x \in \mathrm{V}$ la décomposition de l'espace tangent de V en x complexifié $\mathrm{T_V}(x) \otimes \mathbf{C} = \mathrm{T}'(x) \oplus \mathrm{T}''(x)$, où $\mathrm{T}'(x)$ est engendré par les opérateurs $(\partial/\partial z^j)(x)(1 \leqq j \leqq 2n)$ pour un système quelconque de coordonnées autour de x qui sont holomorphes par rapport à la structure I et $\mathrm{T}''(x)$ est le sous-espace complémentaire conjugué de $\mathrm{T}'(x)$ [rappelons que ceci veut dire

$$\frac{\partial}{\partial z^j}(x)\mathrm{I} = \sqrt{-1}\frac{\partial}{\partial z^j}(x) \quad \text{et} \quad \frac{\partial}{\partial \bar{z}^j}(x)\mathrm{I} = -\sqrt{-1}\frac{\partial}{\partial \bar{z}^j}(x)\Big] \quad ([8], [12]).$$

Alors tout opérateur linéaire réel, tel que J, sur l'espace $\mathrm{T_V}(x)$ qui anticommute à I, rapporté à la base complexe $(\partial/\partial z^j, \partial/\partial \bar{z}^j)$ se décompose en deux opérateurs, notés par J′ et J″, conjugués entre eux, tels que $\mathrm{J}' : \mathrm{T}'(x) \to \mathrm{T}''(x)$ et $\mathrm{J}'' : \mathrm{T}''(x) \to \mathrm{T}'(x)$; on a naturellement $\mathrm{J}'' \circ \mathrm{J}' = -\mathrm{id}|_{\mathrm{T}'(x)}$ et $\mathrm{J}' \circ \mathrm{J}'' = -\mathrm{id}|_{\mathrm{T}''(x)}$; on peut donc écrire :

$$\frac{\partial}{\partial z^j}\mathrm{J}' = \mathrm{H}_j^{\bar{k}}(z, \bar{z})\frac{\partial}{\partial \bar{z}^k} \qquad \text{et} \qquad \frac{\partial}{\partial \bar{z}^j}\mathrm{J}'' = \mathrm{H}_{\bar{j}}{}^{k}(z, \bar{z})\frac{\partial}{\partial z^k},$$

où $\overline{\mathrm{H}_j^{\bar{k}}} = \mathrm{H}_{\bar{j}}^{k}$ et $\mathrm{H}_j^{\bar{k}}\,\mathrm{H}_{\bar{k}}^{l} = -\delta_j^{\,l}$ et les dérivées covariantes de $(\mathrm{H}_j^{\bar{k}})$ et $(\mathrm{H}_{\bar{j}}^{k})$ par rapport à la métrique kählérienne s'annulent identiquement. Si on exprime la métrique kählérienne par

$$ds^2 = g_{j\bar{k}}(z, \bar{z})\, dz^j d\bar{z}^k + g_{\bar{j}k}(z, \bar{z})\, d\bar{z}^j dz^k$$

avec

$$g_{\bar{j}k} = g_{k\bar{j}} = \overline{g_{j\bar{k}}},$$

et si on introduit le champ contravariant hermitien :

$$g^{j\bar{k}} \frac{\partial}{\partial z^j} \otimes \frac{\partial}{\partial \bar{z}^k} \text{ adjoint de } ds^2 \,(g^{j\bar{k}} g_{l\bar{k}} = \delta^j_l),$$

on obtient les tenseurs antisymétriques à dérivée covariante nulle

$$(\mathrm{H}_{jk}(z, \bar{z})) \quad \text{et} \quad (\mathrm{H}^{jk}(z, \bar{z})),$$

où

$$\mathrm{H}_{jk}(z, \bar{z}) = g_{k\bar{l}} \mathrm{H}_j{}^{\bar{l}} \quad \text{et} \quad \mathrm{H}^{jk}(z, \bar{z}) = g^{j\bar{l}} \mathrm{H}_{\bar{l}}{}^{k}\,; \tag{5.1}$$

en particulier la (2,0)-forme différentielle

$$\mathrm{F_J} = \frac{1}{2} \mathrm{H}_{jk} \, dz^j \wedge dz^k \tag{5.2}$$

et le champ de 2-vecteurs adjoint

$$\mathrm{F_J^*} = \frac{1}{2} \mathrm{H}^{jk} \frac{\partial}{\partial z^j} \wedge \frac{\partial}{\partial z^k}, \qquad \mathrm{H}^{jk} = g^{j\bar{l}} g^{k\bar{l}} \overline{\mathrm{H}_{il}} \tag{5.3}$$

sont tous les deux holomorphes; en particulier toute variété hyperkählérienne est automatiquement une variété kählérienne par rapport à I et symplectique complexe par rapport à la forme $\mathrm{F_J}$. Une caractérisation réciproque d'une structure hyperkählérienne est donnée par la proposition suivante :

PROPOSITION 5.1. — *Une variété symplectique complexe et kählérienne* $\{\mathrm{V}; \mathrm{F}, ds^2\}$ *à* $2n$ *dimensions complexes est hyperkählérienne par rapport à la métrique donnée et aux opérateurs* I *de la structure complexe et* J *relié à* F *et* ds^2 *par les relations* (5.1) *et* (5.2) *si et seulement si le champ de 2-vecteurs*

$$\mathrm{F}^* = \frac{1}{2} \mathrm{H}^{jk} \frac{\partial}{\partial z^j} \wedge \frac{\partial}{\partial z^k}$$

associé à F *et à* ds^2 *par* (5.3) *est, premièrement, holomorphe et, deuxièmement, coïncide avec la forme adjointe de* F; *en d'autres termes le tenseur* (H^{jk}) *vérifie la relation*

$$\mathrm{H}_{jk} \mathrm{H}^{jl} = \delta^l_k. \tag{5.4}$$

La démonstration de cette proposition est immédiate.

On va utiliser la proposition ci-dessus pour construire des structures hyperkählériennes, d'abord localement et ensuite globalement, compatibles avec une structure de fibré vectoriel holomorphe sur une variété kählérienne convenable M à n dimensions complexes. Parmi les fibrés vectoriels holomorphes sur M le choix le plus naturel d'un fibré admettant une structure symplectique holomorphe est évidemment celui du fibré cotangent holomorphe $T^{*\prime}(M)$. A chaque système de coordonnées locales holomorphes $(z^1, \ldots, z^n)$ dans un domaine $U \subset M$ on va associer le système de coordonnées holomorphes $(z^1, \ldots, z^n; \zeta_1, \ldots, \zeta_n)$ dans $\pi^{-1}(U)$, où pour chaque point $f_x \in \pi^{-1}(x)$ $(x \in U)$:

$$z^\alpha(f_x) = z^\alpha(x), \qquad \zeta_\lambda(f_x) = \left\langle f_x, \frac{\partial}{\partial z^\lambda}(x) \right\rangle \qquad (\alpha, \lambda = 1, 2, \ldots, n).$$

A toute métrique kählérienne dans M avec potentiel local de Kähler égal à $\Phi(z, \bar{z})$ dans U on associe la forme hermitienne dans les fibres de $\pi^{-1}(U)$ définie par la métrique même; ceci s'exprime en fonction des coordonnées locales de $\pi^{-1}(U)$ par

$$t = g^{\lambda\bar{\mu}} \zeta_\lambda \bar{\zeta}_\mu \qquad \text{où} \quad g_{\mu\bar{\lambda}} = \frac{\partial^2 \Phi}{\partial z^\mu \partial \bar{z}^\lambda} \quad \text{et} \quad g^{\lambda\bar{\mu}} g_{\nu\bar{\mu}} = \delta^\lambda_\nu.$$

Suivant la méthode du chapitre 3, soit $u(x)$ une fonction différentiable d'une variable réelle x dans un intervalle positif relativement ouvert dans la demi-droite non négative $I =]a, b[\cap [0, \infty[$ et soit W_I la partie ouverte de $T^*(M)$ où la fonction t prend ses valeurs dans I; d'après (3.2) on considère dans W_I la forme de type kählérien dans W_I d'après (3.2) :

$$\text{(5.5)} \quad \partial\bar{\partial}\psi = \partial\bar{\partial}(\Phi \circ \pi + u \circ t) = (g_{\alpha\bar{\beta}} + (u' \circ t)\, R^{\lambda\bar{\mu}}{}_{\alpha\bar{\beta}}\, \zeta_\lambda \bar{\zeta}_\mu)\, dz^\alpha d\bar{z}^\beta + ((u' \circ t)\, g^{\lambda\bar{\mu}} + (u'' \circ t)\, g^{\varkappa\bar{\mu}} g^{\lambda\bar{\nu}} \zeta_\varkappa \bar{\zeta}_\nu)\, \nabla\zeta_\lambda \overline{\nabla\zeta_\mu}$$

où

$$\nabla\zeta_\lambda = d\zeta_\lambda - \Gamma^\nu_{\lambda\alpha} \zeta_\nu dz^\alpha, \qquad \nabla\bar{\zeta}_\mu = d\bar{\zeta}_\mu - \overline{\Gamma^\nu_{\mu\beta} \zeta_\nu dz^\beta}$$

et $\Gamma^\nu_{\lambda\alpha}$ sont les symboles de Christoffel de la métrique kählérienne de M :

$$\Gamma^\nu_{\lambda\alpha}(z, \bar{z}) = g^{\nu\bar{\mu}} \frac{\partial g_{\lambda\bar{\mu}}(z, \bar{z})}{\partial z^\alpha}$$

et $R^{\lambda\bar{\mu}}{}_{\alpha\bar{\beta}}$ désigne les composantes correspondantes de la courbure de Riemann :

$$R^{\lambda\bar{\mu}}{}_{\alpha\bar{\beta}} = -g^{\nu\bar{\mu}} \frac{\partial}{\partial \bar{z}^\beta} \Gamma^\lambda_{\nu\alpha}.$$

La variété complexe $T^*(M)$ admet la structure symplectique holomorphe canoniquement définie par la 2-forme holomorphe

$$\text{(5.6)} \qquad F = \sum_{\alpha=1}^{n} dz^\alpha \wedge d\zeta_\alpha = \sum_{\alpha=1}^{n} dz^\alpha \wedge \nabla\zeta_\alpha;$$

la structure symplectique réelle correspondant à la structure différentiable réelle sous-jacente, si on identifie canoniquement le fibré $T^*(M)$ avec $T^{*\prime}(M)$, est celle qui est définie par la forme $F + \bar{F}$.

Nous allons appliquer maintenant la proposition 5.1 pour trouver des conditions sous lesquelles la forme de type kählérien (5.5), supposée définie positive, et la structure symplectique que nous venons de décrire peuvent déterminer une structure hyperkählérienne. Dans ce but nous allons abréger la formule (5.5) en écrivant :

$$\partial\bar{\partial}\Psi = G_{\alpha\bar{\beta}}(z,\zeta;\bar{z},\bar{\zeta})\,dz^{\alpha}d\bar{z}^{\beta} + P^{\lambda\bar{\mu}}(z,\zeta,\bar{z},\bar{\zeta})\,\nabla\zeta_{\lambda}\nabla\bar{\zeta}_{\mu}, \tag{5.7}$$

où

$$G_{\alpha\bar{\beta}} = g_{\alpha\bar{\beta}}(z,\bar{z}) + (u'\circ t)\,R^{\lambda\bar{\mu}}{}_{\alpha\bar{\beta}}\zeta_{\lambda}\bar{\zeta}_{\mu} \tag{5.7'}$$

et

$$P^{\lambda\bar{\mu}} = (u'\circ t)\,g^{\lambda\bar{\mu}} + (u''\circ t)\,g^{\kappa\bar{\mu}}g^{\lambda\bar{\nu}}\zeta_{\kappa}\bar{\zeta}_{\nu}. \tag{5.7''}$$

Proposition 5.2. — *Si la forme* (5.7) *de type kählérien est définie positive dans un ouvert quelconque* $U\subset T^{*\prime}(M)$, *alors la structure kählérienne qui en résulte dans* U *et la structure symplectique holomorphe* (5.6) *déterminent par la proposition* 5.1 *une structure hyperkählérienne si et seulement si la forme* (5.7) *vérifie l'équation*

$$G_{\alpha\bar{\beta}}P^{\lambda\bar{\beta}} = \delta^{\lambda}_{\alpha} \tag{5.8}$$

identiquement dans U.

Démonstration. — Ayant défini la forme F par (5.6), le champ de 2-vecteurs F* associée à F par (5.3) est donnée par

$$F^* = G^{\alpha\bar{\beta}}P_{\lambda\bar{\beta}}\nabla_{z^{\alpha}}\wedge\frac{\partial}{\partial\zeta^{\lambda}} \qquad ((G^{\alpha\bar{\beta}}) = {}^t(G_{\alpha\bar{\beta}})^{-1},\ (P_{\lambda\bar{\mu}}) = {}^t(P^{\lambda\bar{\mu}})^{-1})$$

où

$$\nabla_{z^{\alpha}} = \frac{\partial}{\partial z^{\alpha}} + \Gamma^{\lambda}_{\mu\alpha}\zeta_{\lambda}\frac{\partial}{\partial\zeta_{\mu}} \qquad (1\leqq\alpha\leqq n) \quad \text{et} \quad \left(\nabla_{z^1},\ldots,\nabla_{z^n};\frac{\partial}{\partial\zeta_1},\ldots,\frac{\partial}{\partial\zeta_n}\right)$$

forment la base de $T'(T^{*\prime}(M))$ duale de la base $(dz^1,\ldots,dz^n;\ \nabla\zeta_1,\ldots,\nabla\zeta_n)$ de l'espace $T'^*(T^{*\prime}(M))$. On se rapporte maintenant à la seule condition (5.4) de la proposition 5.1 pour vérifier que, pour que la structure en question soit hyperkählérienne, il est au moins nécessaire que

$$G^{\alpha\bar{\beta}}P_{\lambda\bar{\beta}} = \delta^{\alpha}_{\lambda},$$

cette condition est équivalente à l'équation $G^{\alpha\bar{\beta}} = P^{\alpha\bar{\beta}}$, ou bien à $P_{\lambda\bar{\mu}} = G_{\lambda\bar{\mu}}$, donc à (5.8).

Réciproquement, si (5.8) est satisfaite, ceci implique que

$$F^* = \sum_{\alpha=1}^{n}\nabla_{z^{\alpha}}\wedge\frac{\partial}{\partial\zeta^{\alpha}} = \sum_{\alpha=1}^{n}\frac{\partial}{\partial z^{\alpha}}\wedge\frac{\partial}{\partial\zeta_{\alpha}}.$$

la dernière expresssion indiquant immédiatement que F^* est holomorphe. Ayant démontré que l'équation (5.8) entraîne que F^* est holomorphe et satisfait à (5.4) avec F, ceci montre la suffisance de (5.8) pour que la structure soit hyperkählérienne.

Il est évident que l'équation (5.8) entraîne une restriction considérable non seulement sur le choix de la fonction u déterminant la structure kählérienne dans $T^{*\prime}(M)$ mais aussi sur celui de la structure kählérienne de M même. Par exemple, on note que d'après (5.7″) la matrice $(P^{\lambda\bar{\mu}})$ est égale à un multiple scalaire de $(g^{\lambda\bar{\mu}})$ plus une matrice de rang au plus égal à 1; donc, si (5.8) est satisfaite, ceci entraîne que $(G_{\alpha\bar{\beta}})$ aussi est un multiple scalaire de $(g_{\alpha\bar{\beta}})$ plus une matrice de rang au plus égal à 1, ce qui implique une forte restriction sur le tenseur de courbure de M tel qu'il intervient dans (5.7′). Nous proposons à ce propos la conjecture que, pour $n \geqq 3$, si une métrique kählérienne localement irréductible dans M satisfait à la condition que, pour tout (ζ_λ), $(R^{\lambda\bar{\mu}}{}_{\alpha\bar{\beta}}\,\zeta_\lambda\,\bar{\zeta}_\mu)$ est égal à un multiple scalaire de $(g_{\alpha\bar{\beta}})$ plus une matrice hermitienne de rang $\leqq 1$, alors la métrique est à courbure sectionnelle holomorphe constante. On est donc amené à limiter notre construction de structures hyperkählériennes au cas où la variété de base M est un espace de Fubini-Study; naturellement on peut éliminer *a priori* le cas trivial de courbure nulle. On est donc dans la situation où le potentiel local de Kähler Φ dans M est donné, pour des coordonnées (z^α) convenables, par

$$\Phi = K^{-1}\log\left(1 + K\sum_{\alpha=1}^{n}|z^\alpha|^2\right),$$

où K est une constante $\neq 0$ et; si $K<0$, z est limité à la boule $\left\{z \,\middle|\, \sum_{\alpha=1}^{n}|z^\alpha|^2 < -K^{-1}\right\} \subset \mathbf{C}^n$.

Dans ce cas le tenseur de courbure $R^{\lambda\bar{\mu}}{}_{\alpha\bar{\beta}}$ prend la forme

$$R^{\lambda\bar{\mu}}{}_{\alpha\bar{\beta}} = K\,(\delta^\lambda_\alpha\,\delta^\mu_\beta + g^{\lambda\bar{\mu}}\,g_{\alpha\bar{\beta}})$$

et par conséquent

$$G_{\alpha\bar{\beta}} = (1 + K\,t\,(u'\circ t))\,g_{\alpha\bar{\beta}} + K\,(u'\circ t)\,\zeta_\alpha\,\bar{\zeta}_\beta.$$

L'équation (5.8) est encore apparemment surdéterminée, car elle équivaut à

$$\begin{aligned}\delta^\lambda_\alpha = G_{\alpha\bar{\beta}}\,P^{\lambda\bar{\beta}} &= [(1 + K\,t\,(u'\circ t))\,g_{\alpha\bar{\beta}} + K\,(u'\circ t)\,\zeta_\alpha\,\bar{\zeta}_\beta]\,[(u'\circ t)\,g^{\lambda\bar{\beta}} + (u''\circ t)\,g^{\varkappa\beta}\,g^{\lambda\nu}\,\zeta_\varkappa\,\bar{\zeta}_\nu]\\ &= [(u'\circ t) + K\,t\,(u'\circ t)^2]\,\delta^\lambda_\alpha + (g^{\lambda\bar{\nu}}\,\zeta_\alpha\,\bar{\zeta}_\nu)\,[(u''\circ t)\,(1 + K\,t\,(u'\circ t)) + K\,(u'\circ t)^2 + K\,t\,(u'\circ t)(u''\circ t)];\end{aligned}$$

elle se réduit donc aux deux équations différentielles que la fonction $u(x)$ doit vérifier simultanément

(5.9)
$$u'(x) + K\,x\,(u'(x))^2 = 1,$$
$$u''(x) + (1 + K\,xu'(x)) + K\,u'(x)\,(u'(x) + K\,xu''(x)) = 0.$$

Par une heureuse coïncidence, la deuxième équation est la dérivée de la première; on obtient alors les deux solutions de (5.9) pour $u'(x)$,

$$u'(x)=\frac{-1\pm\sqrt{1+4\mathrm{K}x}}{2\mathrm{K}x}=\frac{2}{1\pm\sqrt{1+4\mathrm{K}x}},$$

la seule solution qui nous intéresse doit satisfaire à $u'(x)>0$ pour $x>0$; on calcule donc la fonction $u(x)$ par une quadrature élémentaire.

$$u(x)=\mathrm{Cte}+\mathrm{K}^{-1}[\sqrt{1+4\mathrm{K}x}-\log(1+\sqrt{1+4\mathrm{K}x})].$$

Il nous reste encore à vérifier si la forme de type kählérien définie dans le fibré cotangent holomorphe d'un espace de Fubini par la fonction $u(x)$ ci-dessus est définie positive dans un domaine à déterminer et, si oui, si la structure métrique qu'elle y détermine est complète, mais ceci est un calcul bien élémentaire, dont nous ne donnerons que le résultat.

Théorème 5.3. — *Soit* M^n *l'espace de Fubini à n dimensions complexe à courbure sectionnelle holomorphe égale à une constante* $2\mathrm{K}\neq 0$. *On considère la forme de type kählérien dans le fibré vectoriel cotangent holomorphe* $\mathrm{T}^*(\mathrm{M})\xrightarrow{\pi}\mathrm{M}$, *définie par*

$$\partial\bar\partial(\Phi\circ\pi+u\circ t),$$

où Φ *est un potentiel local de la métrique kählérienne de* M, t *est la fonction norme (carré de la longueur) dans* $\mathrm{T}^{*\prime}(\mathrm{M})$ *définie par la métrique et*

$$u(t)=\mathrm{K}^{-1}(\sqrt{1+4\mathrm{K}t}-\log(1+\sqrt{1+4\mathrm{K}t})).$$

Alors, si $\mathrm{K}>0$, *la forme est définie positive partout sur* $\mathrm{T}^{*\prime}(\mathrm{M})$ *et la structure métrique qui en résulte est complète; si* $\mathrm{K}<0$, *la forme n'est définie positive que dans le domaine où* $t<-(4\mathrm{K})^{-1}$ *et la métrique n'est pas complète dans ce domaine. Dans les deux cas la structure associée à la structure kählérienne et à la 2-forme holomorphe canonique (de Hamilton) est une structure hyperkählérienne.*

On va conclure par quelques remarques supplémentaires. On vérifie facilement que le groupe d'holonomie des variétés décrites dans le théorème ci-dessus est effectivement égal au groupe $\mathrm{Sp}(n)$ tout entier. Il est bien possible qu'il en existe d'autres exemples, mais il semble très difficile d'en trouver parmi des variétés compactes.

Une autre question curieuse qui se pose consiste à trouver une description aussi explicite que possible de la famille de structures complexes paramétrée par la 2-sphère $\mathscr{J}$ qu'entraîne la structure hyperkählérienne qu'on vient de construire, par exemple dans $\mathrm{T}^{*\prime}(\mathrm{P}^n_{\mathbf{C}})$. En désignant la structure complexe originelle par I, le point $-\mathrm{I}\in\mathscr{J}$ représente la structure conjuguée (qui est isomorphe à la structure I par une involution antiholomorphe). Pour ces deux structures complexes (de fibré vectoriel holomorphe) la section nulle est une sous-variété compacte maximale qui, du point de vue de la géométrie riemannienne, est une sous-variété compacte, orientée, totalement géodésique et d'aire minimale dans sa classe d'homologie.

Comme cette sous-variété ne peut rester holomorphe pour aucune structure complexe de $\mathscr{J}$ hors de I et $-$I, les autres structures complexes n'admettent aucune sous-variété holomorphe compacte à n-dimensions. D'autre part en calculant la déformation des structures complexes du point de vue local on peut conclure qu'elles sont toutes équivalentes au produit $P^n_C \times P^n_C$ avec l'excision de l'hypersurface constituée des paires de points de P^n_C qui sont polaires l'un à l'autre par rapport à une quadrique non singulière.

Pour terminer, nous signalons que les exemples de variétés hyperkählériennes que nous venons de décrire, fournissent la première démonstration d'existence de structures riemanniennes dont le groupe d'holonomie (homogène) est égal à Sp (n). En tenant compte de la liste des groupes d'holonomie possibles [14] on vérifie que, avec les théorèmes d'existence de Yau [13], les résultats d'Alekseevskiĭ [15] et ceux de Brown-Gray [16], les seuls groupes d'holonomie pour lesquels le problème de l'existence non triviale est encore ouvert sont G_2 et Spin (7).

BIBLIOGRAPHIE

[1] V. A. Belinskiĭ, G. W. Gibbons, D. N. Page et C. N. Pope, *Asymptotically Euclidean Bianchi IX Metrics in Quantum Gravity* (*Phys. Lett.*, vol. 76 B, 1978, p. 433-435).

[2] A. Blanchard, *Sur les variétés analytiques complexes* (*Ann. scient. Éc. Norm. Sup.*, Paris, vol. 73, 1956, p. 157-202).

[3] B. Eckmann et A. Frölicher, *Sur l'intégrabilité des structures presque complexes* (*C.R. Acad. Sc.*, Paris, t. 237, 1951, p. 2284-2286).

[4] T. Eguchi et A. J. Hanson, *Asymptotically Flat Self-dual Solutions to Euclidean Gravity* (*Phys. Lett.*, vol. 74 B, 1978, p. 249-251).

[5] A. Frölicher et A. Nijenhuis, *Theory of Vector-Valued Differential Forms Part I* (*Proc. Kon. Nad. Akad. v. Wet.*, vol. 59 A, 1956, p. 338-359).

[6] G. W. Gibbons et C. N. Pope, *CP^2 as a Gravitational Instanton* (*Comm. Math. Phys*, vol. 61, 1978, p. 239-248).

[7] S. Kobayashi et K. Nomizu, *Foundations of Differential Geometry*, New York, John Wiley-Interscience, vol. II, 1969, p. 178-185.

[8] A. Lichnerowicz, *Théorie globale des connexions et des groupes d'holonomie*, Roma, Cremonese, 1955, p. 218-232 et p. 258-261.

[9] S. B. Myers, *Riemannian Manifolds with Positive Mean Curvature* (*Duke Math. J.*, vol. 8, 1941, p. 401-404).

[10] A. Newlander et L. Nirenberg, *Complex Analytic Coordinates in Almost Complex Manifolds* (*Ann. Math.*, vol. 65, 1957, p. 391-404).

[11] A. Nijenhuis, *Jacobi-type Identities for Bilinear Differential Concomitants of Certain Tensor Fields* (*Proc. Kon. Nad. Akad. v. Wet.*, vol. 58 A, 1955, p. 390-403).

[12] A. Weil, *Introduction à l'étude des variétés kählériennes*, Paris, Hermann, *Act. Sc. Ind.*, vol. 1267, 1958, p. 30-43 et p. 83-102).

[13] S. T. Yau, *On the Ricci Curvature of a Compact Kähler Manifold and the Complex Monge-Ampère Equation* (*Comm. Pure Appl. Math.*, vol. 31, 1978, p. 339-411).

[14] M. Berger, *Remarques sur les groupes d'holonomie des variétés riemanniennes*, (*C.R. Acad. Sc.*, Paris, t. 262, série A, 1966, p. 1316-1318).

[15] D. V. Alekseevskiĭ, *Classification of Quaternionic Spaces with Transitive Solvable Groups of Motions* (*Math. Izv. U.S.S.R.*, vol. 9, n° 2, 1975, p. 297-339).

[16] R. B. Brown et A. Gray, *Riemannian Manifolds with Holonomy Group Spin* (9), *Differential Geometry* (in honoi of Kentaro Yano), Tokyo, Kinokuniya, 1972, p. 41-59.
[17] N. Hitchin, *Polygons and Gravitons* [*Proc. Cambr. Phil. Soc.* (à paraître), 1979].

Isometric Families of Kähler Structures[1]

Eugenio Calabi*

The Chern Symposium 1979, Proc. Internat. Sympos.,
held in Berkeley, Calif., USA, 1979, 23–39,
Springer, New York-Berlin (1980)

1. Introduction

The present study is motivated by the classical problem, loosely stated, of determining and describing, for any given family of complex analytic structures on a fixed, underlying real manifold, which subfamily consists of algebraic manifolds. This problem, which is probably inaccessible in its full generality, even when correctly stated with its missing qualifications, is treated here in a very restricted case, where each of the complex structures of the families under consideration admits a Kähler metric, such that the complex manifolds of the same family are isometric (though not complex analytically, in general). Clearly, this situation is highly restrictive compared to the general problem; however the information that one can extract from the two typical classes of such isometric families that can occur provides some interesting results on the global structure of some moduli spaces of nonalgebraic varieties, such as the $K3$-surfaces, the complex tori, and a special type of $2n$-dimensional, rational affine variety.

2. The Holonomy Group

Let (M, g) be a connected, $2n$-dimensional Riemannian manifold, and for some fixed $x \in M$ let H_x denote the linear holonomy group, based at x, relative to the Levi–Civita connection of (M, g). Then $H_x \subset O(M_x)$, where M_x denotes the tangent space of M at x and $O(M_x)$ is the group of linear transformations of M_x, isomorphic to $O(2n)$, that leaves the quadratic form $g(x)$ invariant. It is well known (Lichnérowicz [12], Newlander and Nirenberg [13]) that (M, g) is the Riemannian structure underlying a Kähler structure $(M; g, J)$ if and only if there exists an imbedding of the linear group $U(n)$ in $O(M_x)$, its image denoted by $U(M_x)$, such that $H_x \subset U(M_x) \subset O(M_x)$. The imbedding is uniquely determined by the image $J(x) \in O(M_x)$ of the element $\sqrt{-1}\,1 \in U(n)$ (1 = identity); the element $J(x)$ is, *a priori*, an arbitrary element of $O(M_x)$ satisfying $(J(x))^2$

[1] Supported in part by NSF grant No. MCS 78-02285.
*Department of Mathematics, University of Pennsylvania, Philadephia PA 19104, USA, and Institute for Advanced Study, Princeton, NJ 08540, USA.

DOI: 10.1007/978-1-4613-8109-9_3

$= -1$, and the group $\mathrm{U}(M_x)$ is then the centralizer of $J(x)$ in $\mathrm{O}(M_x)$. The necessary and sufficient condition for (M, g) to be "Kählerizable" is that the centralizer $Z(H_x)$ of H_x in $\mathrm{O}(M_x)$ contain an element $J(x)$ with $J(x)^2 = -1$. In this case the parallel transport of $J(x)$ from x to any other point of M is independent of the path and thus generates the structure tensor J whose covariant derivative is zero, so that the almost complex structure defined by J is integrable.

If there is one tensor J with the properties just described, then also $-J$ satisfies the same conditions, corresponding to the conjugate complex structure. The interesting cases occur when the centralizer of H_x contains a continuum of elements $J(x) \in \mathrm{O}(M_x)$ satisfying $(J(x))^2 = -1$. For such manifolds the centralizer group of H_x in $\mathrm{O}(M_x)$ is a compact group with more than one dimension. Keeping in mind Marcel Berger's classification of the admissible holonomy groups of locally irreducible, non-symmetric Riemannian manifolds [4] and subsequent later contributions (Brown and Gray [5], Alekseevskiĭ [1]) to that topic, the only irreducible holonomy groups that can occur, with some isolated exceptions, are the compact groups $\mathrm{Sp}(n)$ acting in $4n$-space, endowing it with a structure of a quaternionic module. We are thus limited to examining only two typical cases of manifolds admitting Kählerian structures in infinitely many complex structures with mutually isometric Kähler metrics: the locally flat ones, and the $\mathrm{Sp}(n)$-manifolds. Among the former, we consider only the tori, because of the importance of the function theory in abelian varieties and other, nonalgebraic, tori; among the latter, we deal especially with the $K3$-surfaces, since they are the only compact manifolds of $4n$ real dimensions admitting $\mathrm{Sp}(n)$ as holonomy group, according to a recent result announced by Bogomolov [6]. In higher dimensions we consider only one isolated case in each dimensionality $4n$, the manifold that can be described briefly as the cotangent vector bundle over the complex projective n-space; we include it because it helps in describing an interesting "vanishing cycle" that occurs in the resolution of a certain type of singularity.

3. Complex Structures in $2n$-Dimensional Tori

We recall the classical theorem on the period relations for n-dimensional abelian varieties; we restate it here in the form used by A. Weil [15]. A $2n$-dimensional torus T (i.e. the affine space R^{2n} modulo the discrete translation group Z^{2n}) with a translation-invariant complex structure J, abbreviated the pair (T, J), is an algebraic manifold (indeed, after Kodaira, a projective algebraic manifold), if and only if there exists a 2-dimensional, integral cohomology class on T, uniquely representable by a translation invariant, exterior 2-form ω, with the following properties. If $(z_1, \ldots, z_n)$ is any complex system of complex-valued, translation-flat, local coordinates on T that is holomorphic relative to J, then the form ω is purely of bidegree (1, 1) with respect to $(z_1, \ldots, z_n; \bar{z}_1, \ldots, \bar{z}_n)$, i.e.

$$\omega = (2\pi i)^{-1} \sum_{\alpha,\beta=1}^{n} a_{\alpha\bar{\beta}}\, dz^{\alpha} \wedge dz^{\bar{\beta}}, \tag{3.1}$$

and the matrix $(a_{\alpha\bar{\beta}})$ of coefficients of $2\pi i\omega$, necessarily Hermitian, is positive definite.

If we disregard, for the moment, the complex structure J, the integral cohomology class ω is simply one whose nth cup power is not zero: such a class is called a *polarization* of T. While the set of all translation-invariant complex structures J on T is equivalent to the homogeneous space $\mathrm{GL}(2n, R)/\mathrm{GL}(n, C)$ the set of those J's for which the translation-invariant form ω representing a given polarization is of bidegree (1, 1), called complex structures compatible with ω, belong to $n+1$ disjunct, $n(n+1)$-dimensional (over R) components, according to the signature of the matrix $(a_{\alpha\bar{\beta}})$ occurring in (3.1); the component where $(a_{\alpha\bar{\beta}})$ is of index j $(0 \leqslant j \leqslant n)$ is equivalent to the complex symmetric space $\mathrm{Sp}(n, R)/\mathrm{U}(n; j)$, where $\mathrm{U}(n; j)$ denotes the subgroup of $\mathrm{GL}(n, C)$ that leaves invariant a Hermitian form of the type $-\sum_{\alpha=1}^{j} |z_\alpha|^2 + \sum_{\alpha=j+1}^{n} |z_\alpha|^2$.

While the generic structure in the components corresponding to the index j $(1 \leqslant j < n)$ is not algebraic,[2] there is some interest in the particular complex structures arising there, because such families of complex tori, or some holomorphic subfamilies of them, occur naturally as intermediate Jacobi varieties of certain appropriate families of higher-dimensional algebraic manifolds.

Every n-dimensional complex torus (T, J) admits a family of locally flat Kähler metrics g equivalent to the space $\mathrm{GL}(n, C)/\mathrm{U}(n)$ of all positive definite Hermitian forms. Disregarding again the complex structure, any such metric is merely one of the set of translation-invariant Riemannian metrics on T, the total set being equivalent to the symmetric space $\mathrm{GL}(2n, R)/\mathrm{O}(2n)$; however, for each such metric g, the set of complex structures J that are *compatible with* g, i.e. with respect to which the symmetric bilinear form derived from g is the real part of a Hermitian form, is equivalent to the compact, complex symmetric space $\mathrm{O}(2n)/\mathrm{U}(n)$.

We now look at the combined problem; we assume that we are given a translation-invariant Riemannian metric g on T and a polarization, represented by a unique real-valued, translation-invariant form

$$\omega = \sum_{1 \leqslant i < j \leqslant 2n} a_{ij}\, dx_i \wedge dx_j,$$

with a nonsingular coefficient matrix (a_{ij}), unrelated to g. We pose the question of describing the set of complex structures J that are compatible simultaneously with g and with ω, and how this set is partitioned according to the index of ω with respect to J.

Theorem 3.1. *Given a $2n$-dimensional torus T with an arbitrary translation-invariant Riemannian metric g and a polarization represented by a translation-invariant, nondegenerate, exterior 2-form ω, for each integer j $(0 \leqslant j \leqslant n)$ there are, in general, exactly $\binom{n}{j}$ complex structures J on T with respect to which g is Hermitian and ω is a form of bidegree purely* (1, 1) *and of index j. More precisely, there is always a unique J corresponding to a positive polarization* $(j = 0)$; *then the conjugate complex structure $-J$ is the unique one corresponding to the index*

[2] I owe this remark to P. Griffiths.

$j = n$; for the remaining, intermediate values of j, the set $\mathcal{J}(g, \omega, j)$ of complex structures J compatible with both g and ω, for which ω is of index j, depends in a certain way on a certain partition of n, $n = \sum_{\rho=1}^{r} \{m_\rho \mid 1 \leqslant m_1 \leqslant \cdots \leqslant m_r \leqslant n\}$, depending on the pair (g, ω), namely, $\mathcal{J}(g, \omega; j)$ is the disjunct union of the product of real Grassmannians as follows:

$$\mathcal{J}(g, \omega; j) = \bigcup_{j\,:\,0 < j_\rho < m_\rho} \left(\prod_{\rho=1}^{r} \mathrm{Gr}(m_\rho; j_\rho) \right) \quad \textit{with} \ \sum_{\rho=1}^{r} j_\rho = j, \tag{3.2}$$

where $\mathrm{Gr}(m; k)$ *denotes the Grassmann manifold describing the set of all k-dimensional vector subspaces of R^m.*

Proof. We identify each translation-invariant complex structure J on T with the one induced by J, and also denoted by J, on the tangent space V at any point of T; V in turn is naturally identified with the universal covering group of the translation group of T. Following the viewpoint described by Atiyah [2], each complex structure J on V is uniquely defined by any n-dimensional (over C) vector subspace W of $V_C = V \otimes_R C$ that is transveral to V (equivalently, transversal to the complex conjugate subspace $\overline{W}$). Then V_C naturally splits as a direct sum $W \oplus \overline{W}$, and the complex structure induced by W on V is the one defined by restricting to V the canonical isomorphism $V_C \to V_C / W$, so that the complex structure is induced by the resulting bijective map of V onto $\overline{W}$, and the almost complex structural operator J on V is the one whose C-linear extension to V_C has W and $\overline{W}$ as the eigenspaces corresponding to the eigenvalues $-\sqrt{-1}$ and $\sqrt{-1}$ respectively.

Consider the symmetric bilinear form obtained from g, as well as the alternating bilinear form ω, on V as bijective, R-linear maps of V onto the dual vector space V^*, and extend them by C-linearity to bijective operators of V_C onto $V_C^* = \mathrm{Hom}_C(V_C, C)$. Then J is compatible with g (respectively ω) if and only if $g(W) = \overline{W}^*$ (respectively, $\omega(W) = \overline{W}^*$, $\overline{W}^* = \mathrm{Hom}_C(\overline{W}, C) \subset V_C^*$); thus the complex structure defined by W is compatible with g (respectively ω) if only if W—and hence also $\overline{W}$—is totally isotropic with respect to G (respectively ω). Consider now the automorphism $P = g^{-1}\omega : V_C \to V_C$; first of all, if we express P in terms of a real basis (that is to say, belonging to V) that is orthonormal with respect to g, the matrix of P is real and skew-symmetric; therefore P is diagonalizable and its eigenvalues are purely imaginary, nonzero, and conjugate in pairs, say $(\pm\lambda'_\alpha\sqrt{-1}\,)$ $(1 \leqslant \alpha \leqslant n,\ 0 < \lambda'_1 \leqslant \lambda'_2 \leqslant \cdots \leqslant \lambda'_n)$. Let us say that, among the n positive constants λ'_α, there are r distinct ones, λ_ρ $(1 \leqslant \rho \leqslant r$, $0 < \lambda_1 < \lambda_2 < \cdots < \lambda_r)$, respectively with multiplicity m_ρ, so that $\lambda_\rho\sqrt{-1}$ is the eigenvalue corresponding to an m_ρ-dimensional eigenspace $E_\rho \subset V_C$, and $\overline{E}_\rho$ is the eigenspace for the eigenvalue $-\lambda_\rho\sqrt{-1}$. For the sake of simplicity let us write $E_{-\rho}$ for $\overline{E}_\rho$ and denote by $\lambda_{-\rho}$ the negative quantity $-\lambda_\rho$, so that $\lambda_\rho\sqrt{-1}$ is the eigenvalue for E_ρ for $\rho = \pm 1, \ldots, \pm r$.

If W is totally isotropic with respect to both g and ω, then $g^{-1}\omega(W) \subseteq g^{-1}(\overline{W}^*) \subseteq W$, so that W is an invariant space of $P = g^{-1}\omega$ and hence W is spanned by n out of $2n$ possible linearly independent eigenvectors of P. Let

$\dim_C(W \cap E_{-\rho}) = j_\rho$ $(1 \leqslant \rho \leqslant r)$, where $0 \leqslant j_\rho \leqslant m_\rho$; then $j_\rho = \dim_C(\overline{W} \cap E_\rho)$ as well. We can choose a basis $(e'_1, \ldots, e'_n)$ for W and set $e'_{-\alpha} = \bar{e}'_\alpha$ $(1 \leqslant \alpha \leqslant n)$ as a basis for $\overline{W}$ with the following properties:

(a) Each e'_α is an eigenvalue for $P = g^{-1}\omega$, belonging, say to E_{ρ_α} $(\rho_\alpha = \pm 1, \ldots \pm r)$.
(b) For each pair of indices α, β $(1 \leqslant \alpha, \beta \leqslant n)$, $g(e'_\alpha, \bar{e}'_\beta) = \delta_{\alpha\beta}$.

Then $\omega(e'_\alpha, \bar{e}'_\beta) = \lambda_{\rho_\alpha}\sqrt{-1}\, g(e'_\alpha, \bar{e}'_\beta)$ and $\omega(e'_\beta, \bar{e}'_\alpha) = \lambda_{\rho_\alpha}\sqrt{-1}\, g(e'_\beta, \bar{e}'_\alpha)$. Thus the Hermitian form $\sqrt{-1}\, g(\bar{z}, \zeta)$ for $z, \zeta \in W$ is diagonal with respect to the basis $(e'_1, \ldots, e'_n)$, and we verify that

$$\sqrt{-1}\, \omega(\bar{e}'_\alpha, e'_\beta) = \lambda_{\rho_\alpha} g(e'_\beta, \bar{e}'_\alpha) = \lambda_{\rho_\alpha}\delta_{\alpha\beta},$$

showing that there are as many negative numbers in the sequence $(\lambda_{\rho_1}, \ldots, \lambda_{\rho_n})$ as there are dimensions in $W \cap (\sum_{\rho=1}^{e} E_{-\rho})$, that is to say $\sum_{\rho=1}^{r} j_\rho$.

The conclusion now follows by straightforward verification. We remark now that, in general, the eigenvalues of P are all distinct, so that the set of complex structures J compatible with both G and ω, and for which ω has index j, is limited to the choice of subspaces W spanned by j out of the n eigenspaces $\overline{E}_1, \ldots, \overline{E}_n$, and by the conjugates of the remaining ones. On the other hand the uniqueness of the choice $W = \sum_{\rho=1}^{n} E_\rho$ that achieves the index $j = 0$ is obvious. □

Corollary. *The pseudo-Kählerian, complex symmetric spaces*

$$\mathrm{O}(2n)/\mathrm{U}(n; j) \qquad (0 \leqslant j \leqslant n)$$

admits a proper, branched (not holomorphic) covering over the Kählerian symmetric space $\mathrm{O}(2n)/\mathrm{U}(n)$ *with* $\binom{n}{j}$ *sheets, with branch sets equivalent to products of real Grassmannians,* $\prod_{\rho=1}^{r} \mathrm{Gr}(m_\rho; j_\rho)$ *with* $0 \leqslant j_\rho \leqslant m_\rho$, $\sum_{\rho=1}^{r} j_\rho = j$ *for each partition* $\sum_{\rho=1}^{r} m_\rho$ *of* n.

The statement of the above is simply a rewording of the previous theorem in a less precise form.

4. Hyper-Kähler Manifolds

The only known locally irreducible Kähler manifolds that admit a continuum of complex structures all admitting a common Kähler metric are the *hyper-Kähler manifolds* (also called by other authors Hamiltonian Kähler manifolds or Kähler manifolds with an intergrable quaternionic structure). These are manifolds that are $4n$-dimensional and whose holonomy group with respect to a given Riemannian connection is contained in (and in general equals) $\mathrm{Sp}(n)$, so that the tangent bundle has the structure of a quaternionic module, the quaternion scalar operations (for instance, represented as acting on the right) are preserved under Levi–Civita parallel transports.

Hyper-Kähler manifolds may be presented as Kähler manifolds (M_0, g)

admitting holomorphic 2-form φ_0, whose coefficient matrix $(P_{\alpha\beta})$ ($\varphi_0 = P_{\alpha\beta} dz^{\alpha} \wedge dz^{\beta}$ in local coordinates) is everywhere nonsingular (hence $\dim_C M_0$ is even) and with identically vanishing covariant derivative with respect to the given Kähler metric. Let $\dim_C M_0 = 2n$; then the nth exterior power of φ_0 is a holomorphic $2n$-form on M_0 whose squared absolute value is a positive multiple of the volume density μ of the Kähler metric. Hence the Ricci tensor of M_0, $R_{\alpha\bar{\beta}} = -\partial^2(\log \mu)/\partial z^{\alpha}\, \partial \bar{z}^{\beta}$, vanishes identically.

Let T' and $T'' = \overline{T}'$ denote the subspaces of the complexified tangent bundle $T(M) \otimes C$ that are locally spanned, respectively, by $(\partial/\partial z^1, \ldots, \partial/\partial z^n)$ and by $(\partial/\partial \bar{z}^1, \ldots, \partial/\partial \bar{z}^n)$, and denote by $(\ ,\ )$ the Hermitian inner produce of any two (real- or complex-valued) tangent vectors. Thus, if we denote by $u = u' + u''$ the decomposition of any complex tangent vector according to the decomposition $T(M) \otimes C = T' \oplus T''$, and by $\langle\ ,\ \rangle$ the Riemannian inner product in $T(M_0)$ extended by C-linearity to complex tangents, then for any two real or complex tangents u, v, we have

$$(u, v) = \langle u', v'' \rangle, \qquad \langle u, v \rangle = \langle u', v'' \rangle + \langle v', u'' \rangle.$$

Furthermore we have the C-linear operators K_1' and K_1'', mapping $T(M_0) \otimes C$ respectively onto T'' and T', defined as follows: for any tangent vector $u = u' + u''$, $K_1' u'' = K_1'' u' = 0$ and $K_1' u'$, $K_1'' u''$ are characterized, respectively, by the equations, for any $v = v' + v''$,

$$\begin{aligned} (K_1' u', v) &= \langle v', K_1' u' \rangle = \varphi_0(u', v'), \\ (K_1'' u'', v) &= \langle K_1'' u'', v'' \rangle = \bar{\varphi}_0(u'', v''). \end{aligned} \tag{4.1}$$

It follows that K_1' maps T' bijectively onto T'', K_1'' maps T'' bijectively onto T', K_1'' is the complex conjugate of K_1', and thus the map $K_1 = K_1' + K_1''$ is a real, bijective map of $T(M_0)$ onto itself; its matrix with respect to any (real) orthonormal basis with respect to the metric is antisymmetric, so that K_1 is diagonalizable over $T(M_0) \otimes C$, and its spectrum is on the purely imaginary axis and symmetric about 0. Furthermore K_1 has covariant derivative zero. In addition, $K_1^2 = K_1' K_1'' + K_1'' K_1'$ is symmetric with respect to the underlying Riemannian structure; it is negative definite, and has T' and T'' as invariant subspaces. Replacing the 2-form φ_0 by $\sqrt{-1}\, \varphi_0$, we obtain likewise a real, bijective map

$$K_2 : T(M_0) \rightarrow T(M_0),$$

$$K_2 = \sqrt{-1}\, (K_1' - K_1''),$$

with the same properties described for K_1 and such that $K_2^2 = K_1^2$.

We construct now from K_1 and K_2 two analogous operators, denoted respectively by J_1 and J_2,

$$J_\nu = \frac{1}{\pi} K_\nu \int_{-\infty}^{\infty} \left(t^2 - K_\nu^2\right)^{-1} dt, \qquad \nu = 1, 2. \tag{4.2}$$

From the spectral decomposition of K_ν, one deduces, for either value of ν, that the integral above converges, that J_ν commutes with K_ν, and that J_ν maps

T' and T'' each bijectively onto the other, so that

$$J_\nu = J'_\nu + J''_\nu, \qquad J''_\nu u' = J'_\nu u'' = 0.$$

Furthermore

$$J_\nu^2 = J'_\nu J''_\nu + J'_\nu J''_\nu = -1 \qquad (\nu = 1, 2),$$

$$J'_2 = \sqrt{-1}\, J'_1, \qquad J''_2 = -\sqrt{-1}\, J''_1,$$

and both J_1 and J_2 have vanishing covariant derivatives with respect to the metric, so that they define integrable complex analytic structures with respect to the metric, and the metric is Kählerian with respect to either of the two resulting new complex structures.

If we denote, in addition, by $J_0 = J'_0 + J''_0$ the almost complex structure subordinate to the given complex structure of M_0, where

$$J'_0 = \sqrt{-1}\,\Big|_{T'}, \qquad J''_0 = -\sqrt{-1}\,\Big|_{T''},$$

then J_0, J_1, J_2 represent three pairwise anticommutative, almost complex structures, with $J_1 J_0 = -J_0 J_1 = J_2$, so that the algebra generated by adjoining J_0, J_1, J_2 to the real scalars (as operators on $T(M_0)$) is isomorphic to the division algebra of quaternions. For any $\alpha = (\alpha_0, \alpha_1, \alpha_2) \in S^2 \subset R^3$ (i.e. such that $\alpha_0^2 + \alpha_1^2 + \alpha_2^2 = 1$), the almost complex structural tensor

$$J(\alpha) = \sum_{\nu=0}^{2} \alpha_\nu J_\nu$$

is integrable to a complex analytic structure on M_0, with the same given metric on M_0 for a Kähler metric, thus describing a family of isometric (with the common metric being Kählerian), complex analytic structures. We restate this fact formally as follows:

Theorem 4.1. *Any hyper-Kählerian manifold $(M_0;\ g_0, \varphi_0)$ has, in a cononical way, a family of complex analytic structures $J(\alpha)$ $(\alpha \in S^2)$ including the given Kähler structure of M_0 and its conjugate, each of them admitting the same metric g_0 as a Kähler metric; for antipodal points, $J(-\alpha) = -J(\alpha)$ is the conjugate complex structure of $J(\alpha)$. For any $\beta \in S^2$ that is at right angles to $\alpha \in S^2$, the 2-form $\psi_\beta(u, v)$ defined by the equation*

$$2\psi_\beta(u, v) = \langle J(\beta)u, v\rangle - \sqrt{-1}\,\langle J(\beta)u, J(\alpha)v\rangle$$

is holomorphic with respect to the complex structure defined by J_α.

We shall examine the S^2-parameterized family of isometric Kähler structures thus constructed in two typical cases, one compact and one not.

A recent result announced by Bogomolov [6] states that there are no compact, $2n$-complex dimensional, locally irreducible hyper-Kähler manifolds for $n \geqslant 2$. For $n = 1$, using Kodaira's classification theorem for compact, complex surfaces, the only nonflat, compact, complex manifolds with first Chern class equal to zero (a necessary condition in a Kähler manifold to admit a nowhere vanishing, holomorphic form of maximum degree) that can arise are the so-called $K3$-surfaces, a family of compact, complex manifolds homologically equivalent to the nonsingular quartic surfaces in the complex projective 3-space, or to those quartics with isolated, ordinary quadratic singularities (up to 16 of them are possible), if removed by a simple "blowing up". The latter family, disregarding the biregular equivalence under projective transformations, has complex structures depending on 19 complex parameters or *moduli*. On the other hand Kodaira and Spencer [9] calculated that the number of complex moduli for small deformation of complex structure of the $K3$-surfaces is 20, revealing that there is a complex analytic family of nonalgebraic $K3$-surfaces, in which the algebraic ones constitute analytic subsets of codimension at least one, much in the same way that the 2-complex-dimensional abelian varieties form analytic subsets of codimension at least one in the 4-(complex)-parameter family of complex analytic structures in a 4-torus.

Let M_0 be a $K3$-surface with any given Kähler metric g'. This surface admits a holomorphic 2-form φ_0 which is nowhere zero; we normalize φ_0 so that $\int_{M_0}\varphi_0 \wedge \bar{\varphi}_0$ equals the total volume of the M_0 under the given Kähler metric. A fundamental theorem due to S. T. Yau [16], to which T. Aubin [3] has made earlier contributions, asserts that there exists another Kähler metric in the same Kähler cohomology class as g', whose volume element coincides with $\varphi_0 \wedge \bar{\varphi}_0$ (such a metric is then unique); we call such a metric a Yau metric on the $K3$-surface. Thus there are as many Yau metrics in M_0 as there are real, 2-dimensional cohomology classes representable by principal Kählerian (positive definite) forms of bidegree $(1, 1)$. Each such metric in M_0 induces in it a structure of a hyper-Kähler manifold.

The second real cohomology group of M_0 is 22-dimensional (since M_0 is simply connected and has Euler characteristic 24); within the space of all real-valued, harmonic 2-forms on M_0, we distinguish the 3-dimensional subspace spanned by the principal Kähler form ω_0 $(=\sqrt{-1}\, g_{\alpha\bar{\beta}}\, dz^{\alpha} \wedge dz^{\bar{\beta}}$ in terms of local holomorphic coordinates), and by

$$\omega_1 = \varphi_0 + \bar{\varphi}_0, \qquad \omega_2 = \sqrt{-1}\,(\varphi_0 - \bar{\varphi}_0),$$

where φ_0 is the holomorphic 2-form in M_0 used earlier.

These three harmonic 2-forms have equal cup squares $\frac{1}{2}\omega_\nu \wedge \omega_\nu$ identical to the volume element of the Yau metric (by definition). The orthogonal complementary space of harmonic forms (19-dimensional over R) is made up of the $(1, 1)$-forms η that are *primitive* with respect to the Kähler class, i.e. satisfy $\eta \wedge \omega_0 = 0$ identically (and, trivially, also $\eta \wedge \omega_1 = \eta \wedge \omega_2 = 0$). The primitive $(1, 1)$-forms are also characterized as being skew-self-adjoint, i.e. ${}^*\eta = -\eta$; in other words $\eta \wedge \eta$ is a 4-form that is negative at each point where $\eta \neq 0$. Thus the signature of M_0 is $3 - 19 = -16$.

Consider now the 2-parameter family of complex structures in M_0, obtained by integration of the structure tensors $J(\alpha)$ $(\alpha = (\alpha_1, \alpha_2, \alpha_3) \in S^2)$ arising from the hyper-Kähler or Yau metric structure of M_0. As in the case of complex tori, in order to pick a reasonable analytic subfamily of algebraic structures in the moduli space of M_0, we shall introduce a *polarization*. This is defined by an integral, 2-dimensional cohomology class u in M_0, represented by a real-valued, closed, exterior 2-form Φ in M_0 (of mixed bidegree, in general), whose exterior product with itself has a strictly positive integral over M_0; such a polarization is called *weakly positive*. If, in addition, the polarization class u can be represented by a real-valued, closed 2-form Φ such that $\Phi \wedge \Phi$ is strictly positive at each point (relative to the natural orientation of M_0), we say that the polarization is *strongly positive*. Finally, if the polarization class u can be represented by a closed, real-valued 2-form Φ which, in terms of the given complex structure in M_0, is purely of bidegree (1, 1) and also satisfies everywhere the condition $\Phi \wedge \Phi > 0$, we say that u is a *strictly positive* polarization. Disregarding for the moment the assumption that u is assumed to be an integral cohomology class, each of the three sets consisting of, respectively, weakly positive, strongly positive, and strictly positive, real, 2-dimensional cohomology classes in M_0, is an open cone in $H^2(M_0, R)$, each contained in the previous one, and containing the nonzero elements in the 3-dimensional subspace spanned by the de Rham classes of ω_0, ω_1, and ω_2 (these three forms span a maximal vector subspace in the space of 2-forms Φ for which the quadratic form $\Phi \wedge \Phi$ is strictly positive definite everywhere). The boundaries of the last two of these three cones are unknown, as is also the answer to the question whether any two of the cones coincide.

Lemma 4.2. *Let Φ be any real-valued, harmonic 2-form in M_0 such that $\int_{M_0} \Phi \wedge \Phi > 0$. With reference to a given Ricci flat Kähler metric in M_0, let the Hodge decomposition of Φ be expressed by $\Phi = a\omega_0 + c\varphi_0 + \bar{c}\bar{\varphi}_0 + \eta$, where ω_0 is the principal Kähler* (1, 1)*-form, $a \in R$, φ_0 is a holomorphic 2-form in M_0, normalized so that $\int_{M_0} \varphi_0 \wedge \bar{\varphi}_0 = \frac{1}{2}\int_{M_0} \omega_0 \wedge \omega_0$, $c \in C$, and η is a primitive, harmonic* (1, 1)-*form, then the following conclusions hold*:

(a) *At least one of the coefficients a, c is different from zero.*

(b) *The corresponding Hodge decomposition of Φ with respect to each of the complex structures induced by each $J(\alpha)$ ($\alpha \in S^2$), takes the form*

$$\Phi = c_\alpha \varphi_{(\alpha)} + \bar{c}_\alpha \bar{\varphi}_{(\alpha)} + a_\alpha \omega_{(\alpha)} + \eta,$$

where η is the same as for the originally assumed complex structure.

(c) *There exists a unique $\alpha \in S^2$ for which $c_\alpha = 0$ and $a_\alpha > 0$; for the conjugate complex structure we have $a_\alpha < 0$.*

Proof. Since η is a primitive (1, 1)-form, this means that $\int_{M_0} \eta \wedge \eta < 0$ unless η is identically zero; since $\int_{M_0} \Phi \wedge \Phi > 0$, it follows that Φ cannot equal η. This proves assertion (a).

In order to prove assertion (b), we refer to the calculations leading to the statement of Theorem 4.1. Observe that the operators K_1 and K_2 already satisfy $K_\nu^2 = -1$ ($\nu = 1, 2$); hence from Equation (4.2) we have $J_\nu = K_\nu$ ($\nu = 1, 2$).

In terms of the complex structure defined in M_0 by $J(\alpha) = \sum_{\nu=0}^2 \alpha_\nu J_\nu$, $\sum_{\nu=0}^2 \alpha_\nu^2 = 1$, we observe that the principal Kähler form of the Yau metric is

$$\omega_0(\alpha) = \sum_{\nu=0}^{2} \alpha_\nu \omega_\nu,$$

where

$$\omega_1 = \varphi_0 + \bar{\varphi}_0, \qquad \omega_2 = \sqrt{-1}\,(\varphi_0 - \bar{\varphi}_0);$$

in fact we have, from the definition of J_ν $(\nu = 0, 1, 2)$,

$$\omega_0(\alpha)(u, v) = \langle J(\alpha)u, v\rangle = \sum_{\nu=0}^{2} \alpha_\nu \langle J_\nu u, v\rangle = \sum_{\nu=0}^{2} \alpha_\nu \omega_\nu(u, v).$$

Consider the 22-dimensional vector space $H^2(M_0, g)$ over the reals, consisting of all the real-valued, harmonic 2-forms Φ in M_0: this space is defined only in terms of the underlying Riemannian structure of M_0, independently of any of the superimposed, complex structures determined by any $J(\alpha)$. For each $\alpha \in S^2$, the Hodge decomposition of Φ is again of the form $a(\alpha)\omega_0(\alpha) + c(\alpha)\varphi_0(\alpha) = \overline{c(\alpha)\varphi_0(\alpha)} + \eta(\alpha)$, where $\omega_0(\alpha)$ is the principal (1, 1)-form (4.2) and $\varphi_0(\alpha)$ is a holomorphic (2, 0)-form normalized by $\int_{M_0} \varphi_0(\alpha) \wedge \overline{\varphi_0(\alpha)} = \frac{1}{2}\int_{M_0} \omega_0(\alpha) \wedge \omega_0(\alpha)$. It is easy to verify that, since the orientation of M_0 is fixed, for each $\Phi \in H^2(M_0, g)$, the dual form $*\Phi$ has the Hodge decomposition, in terms of $J(\alpha)$, $*\Phi = a(\alpha)\omega_0(\alpha) + c(\alpha)\varphi_0(\alpha) + \overline{c(\alpha)\varphi_0(\alpha)} - \eta(\alpha)$. Hence the anti-self dual part $\eta(\alpha)$,

$$\eta(\alpha) = \tfrac{1}{2}(\Phi - *\Phi),$$

is determined independently of the complex structure $J(\alpha)$, proving assertion (b).

For each $\alpha \in S^2$ the choice of the normalized, holomorphic 2-form $\varphi_0(\alpha)$ is unique up to a complex constant factor with unit absolute value. A real basis for the self-dual subspace of $H^2(M_0, g)$ can be made up of $\omega_0(\alpha)$, $\omega_1(\alpha) = \varphi_0(\alpha) + \overline{\varphi_0(\alpha)}$, $\omega_2(\alpha) = \sqrt{-1}\,(\varphi_0(\alpha) - \overline{\varphi_0(\alpha)})$; such a basis is orthornormal both globally on M_0 and pointwise, since the pointwise norm of any self-dual harmonic form in M_0 is constant (a consequence of the metric being a Yau metric). Consequently the set of all real bases $\omega_\nu(\alpha)$ $(\nu = 0, 1, 2)$ is related to the initial basis ω_ν by a linear transformation belonging to the group SO(3), with the first row consisting of the coordinates $(\alpha_0, \alpha_1, \alpha_2)$ of α.

Finally assertion (c) is proved as follows. At any one point where $\Phi \neq 0$ there are exactly two antipodal values of $\alpha \in S^2$ such that Φ at that point is purely of bidegree (1, 1) with respect to the complex structure $J(\alpha)$. At that point therefore, and for the chosen values of α, the (2, 0)- and (0, 2)-components of Φ vanish; but, since in a compact, complex manifold with vanishing first Chern class, a harmonic (2, 0)-form (or its conjugate) cannot vanish at a point unless such a form is identically zero, there are exactly two antipodal values of $\alpha \in S^2$, where Φ is purely of bidegree (1, 1). Since Φ cannot be a primitive form, there is in either case a part in the decomposition that is a nonzero multiple of the principal form, with positive coefficient for one of the values of α, and negative for the other. This completes the proof of the lemma. □

We now consider a strongly positive polarization u in the manifold M_0 and repeat the argument of the lemma just proved. We recall that, according to Kodaira's imbedding theorem, a compact Kähler surface is a projective algebraic variety if and only if it admits a strictly positive polarization (with integer coefficients) representable by a closed 2-form Φ of bidegree purely (1, 1) and of positive type at each point; the sense in which such a surface becomes a projective variety is that any sufficiently high, positive integer multiple of the polarization class becomes the divisor class of a projective hyperplane intersection with M_0, as a result of a holomorphic imbedding of the given surface into a complex projective space. We have thus proved the following theorem.

Theorem 4.3. *Let M_0 be any analytic $K3$-surface. Given any strictly positive polarization u in M_0 and any Yau metric (i.e. a Ricci flat Kähler metric) in M_0, there is a unique complex structure $J(\alpha)$ on M_0 with respect to which given Yau metric is a Kahler metric and the polarization class representable by a harmonic form.*

The question whether the conclusion of this theorem still holds if the polarization class u is assumed to be only strongly positive appears still uncertain. A recently announced result by A. N. Todorov [14] seems to imply it, by making use of a strong formulation of a Torelli theorem (Horikawa [8], Kulikov [11]) for $K3$-surfaces.

A second class of examples of hyper-Kähler structures is furnished by the holomorphic cotangent vector bundle $T^{*\prime}(CP^n)$ of the n-dimensional complex projective space CP^n. Let $(u_0, \ldots, u_n) \in C^{n+1}\backslash\{0\}$ denote the complex homogeneous coordinates in CP^n; let $z^\alpha = u_\alpha/u_0$ $(1 \leqslant \alpha \leqslant n)$ denote the inhomogeneous coordinate functions in the domain $U_0 = \{u_0 \neq 0\} \subset CP^n$, and ζ_α $(1 \leqslant \alpha \leqslant n)$ the fibre coordinates of $T^{*\prime}(CP^n)|_{U_0}$ at the point represented by the holomorphic form $\sum_{\alpha=1}^n \zeta_\alpha dz^\alpha$; and let the Fubini–Study metric in CP^n be represented by the Hermitian form in U_0

$$ds^2 = 2g_{\alpha\bar\beta}(z, \bar z)\, dz^\alpha\, d\bar z^\beta,$$

where

$$g_{\alpha\bar\beta} = \frac{1}{2}\,\frac{\partial^2}{\partial z^\alpha\,\partial \bar z^\beta}\log\left(1 + \sum_{\lambda=1}^n |z^\lambda|^2\right).$$

Then the transposed inverse matrix $(g^{\alpha\bar\beta})$ of $(g_{\alpha\bar\beta})$ is expressed by

$$g^{\alpha\bar\beta} = 2\left(1 + \sum_{\lambda=1}^n |z^\lambda|^2\right)(\delta^{\alpha\beta} + z^\alpha \bar z^\beta).$$

The manifold $M_0 = T^{*\prime}(CP^n)$ is covered by $n+1$ subdomains $W_a = T^{*\prime}(U_a)$ $(0 \leqslant a \leqslant n)$, where $U_a = \{u_a \neq 0\}$, each subdomain being equivalent to $T^{*\prime}(U_0)$ under permutations of the homogeneous coordinates $(u_0, \ldots, u_n)$. The norm function of M_0 is expressed in $T^{*\prime}(U_0)$ by the real-valued function

$$t = g^{\alpha\bar\beta}\zeta_\alpha\bar\zeta_{\bar\beta} = (1 + |z|^2)(|\zeta|^2 + |z \cdot \zeta|^2),$$

where

$$|z|^2 = \sum_{\alpha=1}^n |z^\alpha|^2, \qquad |\zeta|^2 = \sum_{\alpha=1}^n |\zeta_\alpha|^2, \qquad z \cdot \zeta = \sum_{\alpha=1}^n z^\alpha \zeta_\alpha.$$

Then the real-valued function $\Psi(z,\bar{z};\zeta,\bar{\zeta})$ in $T^{*\prime}(U_0)$, expressed in terms of the coordinates $(z^1,\ldots,z^n;\zeta_1,\ldots,\zeta_n)$ by

$$\Psi(z,\bar{z};\zeta,\bar{\zeta}) = \log(1+|z|^2) + \sqrt{1+4t} - \log(1+\sqrt{1+4t}\,), \tag{4.3}$$

$$t = (1+|z|^2)(|\zeta|^2 + |z\cdot\zeta|^2), \tag{4.4}$$

(cf. Calabi [7, §5]) is the Kähler potential for a metric that is hyper-Kählerian with respect to the given complex structure and the Hamiltonian form

$$\varphi_0 = \sum_{\alpha=1}^{n} dz^\alpha \wedge d\zeta_\alpha.$$

We shall describe now the complex structures that occur in this open variety, induced by the structural tensors $J(\alpha)$ for $\alpha = (\alpha_0,\alpha_1,\alpha_2) \in S^2$ other than the points $(\pm 1,0,0)$ corresponding to the originally given complex structure and its conjugate.

Let $\mathcal{V}$ be the $(2n+1)$-dimensional complex analytic set in $C^{(n+1)^2}$ consisting of the complex $(n+1)\times(n+1)$ matrices (x_{ij}) $(0 \leqslant i,j \leqslant n;\ n \geqslant 1)$ that are of rank one or zero,

$$x_{ik}x_{jl} - x_{il}x_{jk} = 0 \qquad (0 \leqslant i,j,k,l \leqslant n),$$

and the map $\mathrm{tr}: \mathcal{V} \to C$ defined by the trace; for each $s \in C$, $V_s = \mathrm{tr}^{-1}(s)$ is an irreducible, $2n$-dimensional complex space, nonsingular for $s \neq 0$ and with an isolated singularity at the origin for $s = 0$. The absence of singularities in V_s away from the origin stems from the fact that for each $s \neq 0$, V_s, as well as $V_0\backslash\{0\}$ for $s = 0$, is homogeneous under the adjoint action of the matrix group $\mathrm{GL}(n+1,C)$ on $C^{(n+1)^2}$. Since the variety $\mathcal{V}$ is a cone, the inhomogeneous hyperplane sections V_s $(s \neq 0)$ are projectively equivalent to the infinite part of $\mathcal{V}\backslash V_0$ in the projective completion of $C^{(n+1)^2}$. Thus the infinite part of $\mathcal{V}$ has an alternative analytic realization as the product of CP^n with the dual projective space, i.e. as the space of pairs (p,q^*) consisting of points p and hyperplanes q^* in CP^n, while the infinite part of V_0 is the subset consisting of incident pairs $\{(p,q^*)\mid p \in q^*\}$. Thus V_s, for each $s \neq 0$, is a nonsingular affine algebraic variety, holomorphically equivalent to the set of pairs (p,q^*) with $p \notin q^*$.

The singularity of $\mathcal{V}$ at the origin can be resolved in two distinct ways, mutually equivalent under duality; to each matrix $X = (x_{ij}) \in \mathcal{V}\backslash\{0\}$ we associate the row-augmented matrix $(u;X)$, obtained by adjoining to X an extra column whose entries $(u_0,\ldots,u_n)$ are homogeneous coordinates, proportional to any one of the nonzero columns of X, say $(u_0,\ldots,u_n) = \rho(x_{0b},\ldots,x_{nb})$ for some $\rho \neq 0$. Thus we obtain the subvariety of $CP^n \times (\mathcal{V}\backslash\{0\})$ that can be represented by $(n+2)\times(n+1)$ matrices of rank 1, where the first column $(u_0,\ldots,u_n)$ is never zero, and identified under proportional equivalence. This variety is obviously biregularly equivalent to $\mathcal{V}\backslash\{0\}$, but its closure in $CP^n \times \mathcal{V}$ includes $CP^n \times \{0\}$ in correspondence with the singularity of $\mathcal{V}$. The resulting space is therefore a complex analytic subvariety of $CP^n \times C^{(n+1)^2}$; we shall denote it by $\mathcal{V}_1$, and show presently that it is a desingularized model for $\mathcal{V}$ under the obvious holomorphic map $\pi: \mathcal{V}_1 \to \mathcal{V}$. An equivalent desingularization can be obtained by interchanging the roles of rows and columns in $\mathcal{V}$, while the composite of the two dilatations, replacing $\{0\}$ by $CP^n \times CP^n$, is equivalent to the result on $\mathcal{V}$ of the standard dilatation of $C^{(n+1)^2}$ with respect to the origin.

Lemma 4.4. *The variety $\mathcal{V}_1$ obtained from $\mathcal{V}$ by replacing the origin by an n-dimensional complex projective space is nonsingular; moreover the map $\pi_1 = \mathrm{tr} \circ \pi : \mathcal{V}_1 \to C$ is a surjection with no critical points. In particular the inverse image $\pi^{-1}(V_0) = \pi_1^{-1}(0)$ of V_0 is a desingularized model of V_0; moreover it is holomorphically equivalent to $T^{*\prime}(CP^n)$ and diffeomorphic to $V_s = \mathrm{tr}^{-1}(s)$ for each $s \neq 0$.*

Proof. Consider the open cover of $\mathcal{V}_1$ made up of the $n+1$ sets $\mathcal{U}_a$ $(0 \leqslant a \leqslant n)$, where each $\mathcal{U}_a$ is the subset such that the homogeneous coordinate u_a occurring in the augmented matrix $(u_i; x_{ij})$ is $\neq 0$. We show that each $\mathcal{U}_a$ can be mapped bijectively onto C^{2n+1} by the $2n+1$ functions $(z_i; \zeta_j; s)$ $(0 \leqslant i \leqslant n,\ 0 \leqslant j \leqslant n;\ i \neq a,\ j \neq a)$, where z_i and ζ_j are defined by

$$z_i = \frac{u_i}{u_a}, \qquad \zeta_j = x_{aj}.$$

In order to show that this is indeed a local coordinate system for $\mathcal{V}_1$, it is sufficient to show that all the extrinsic coordinates x_{ij} of the point corresponding to any assigned set of values $(z_i; \zeta_j; s)$ are uniquely determined and are holomorphic functions of $(z_i; \zeta_j; s)$ defined in C^{2n+1}; we omit the homogeneous coordinates (u_k), since their mutual ratios are described by the values of $z_i = u_i/u_a$ alone. In fact, since the augmented matrix $(u_i; x_{ij})$ is, by definition, of rank one, it follows that, for $j \neq a$,

$$x_{ij} = \frac{u_i}{u_a} x_{aj} = z_i \zeta_j,$$

$$x_{aj} = \zeta_j,$$

while for $j = a$,

$$x_{aa} = s - \sum_{i \neq a} x_{ii} = s - \sum_{i \neq a} z_i \zeta_i,$$

$$x_{ia} = z_i x_{aa} = z_i \Big(s - \sum_{j \neq a} z_j \zeta_j \Big) \qquad (i \neq a).$$

In addition this argument shows that, for each fixed value of $s \in C$ (whether equal to zero or not), the $2n$ functions $(z_i; \zeta_j)$ constitute a system of local coordinates in $\pi_1^{-1}(x) \cap \mathcal{U}_a$, so that the submanifolds $\pi_1^{-1}(s) = \pi^{-1}(V_s)$ (trivially isomorphic to V_s for $s \neq 0$) are diffeomorphic to $\pi_1^{-1}(0) = \pi^{-1}(V_0)$.

There remains to be shown that $\pi_1^{-1}(0)$ is isomorphic (holomorphically) to $T^{*\prime}(CP^n)$. Indeed, we shall prove not only this fact, but also that, for each $s \neq 0$, $\pi_1^{-1}(s)$, and hence V_s, is isomorphic to a holomorphic bundle of affine complex n-spaces over CP^n, obtained by a deformation of, but inequivalent to, $T^{*\prime}(CP^n)$. We define the holomorphic differential $\psi_{a,s}$ in $\mathcal{U}_a \cap \pi_1^{-1}(s)$ for each a $(0 \leqslant a \leqslant n)$ and each $s \in C$,

$$\psi_{a,s} = \sum_{i \neq a} \zeta_i \, dz_i.$$

In $\mathfrak{U}_a \cap \mathfrak{U}_b \cap \pi_1^{-1}(s)$ we calculate the discrepancy between $\psi_{a,s}$ and $\psi_{b,s}$ as follows: $\mathfrak{U}_a \cap \mathfrak{U}_b$ is characterized by the condition that $u_a u_b \neq 0$; under this condition we have

$$\begin{aligned}
\psi_{b,s} &= \sum_{i\neq b} x_{bi}\, d\left(\frac{u_i}{u_b}\right) = \sum_{i \neq b} x_{bi}\left(\frac{u_a}{u_b}\, d\left(\frac{u_i}{u_a}\right) + \frac{u_i}{u_a}\, d\left(\frac{u_a}{u_b}\right)\right) \\
&= \sum_{i\neq a,b} x_{ai}\, d\left(\frac{u_i}{u_a}\right) + \left(\sum_{i\neq b} x_{ii}\right) d\log\left(\frac{u_a}{u_b}\right) \\
&= \sum_{i\neq a,b} x_{ai}\, d\left(\frac{u_i}{u_a}\right) + (s - x_{bb})\, d\log\frac{u_a}{u_b} \\
&= \sum_{i\neq a,b} x_{ai}\, d\left(\frac{u_i}{u_a}\right) + x_{bb}\frac{u_a}{u_b}\, d\left(\frac{u_b}{u_a}\right) - s\, d\log\left(\frac{u_b}{u_a}\right) \\
&= \psi_{a,s} - s\, d\log\left(\frac{u_b}{u_a}\right).
\end{aligned}$$

Thus we have, in $\pi_1^{-1}(s) \cap \mathfrak{U}_a \cap \mathfrak{U}_b$,

$$\psi_{b,s} - \psi_{a,s} = -s\, d\log\left(\frac{u_b}{u_a}\right). \tag{4.5}$$

This shows that, for $s = 0$, the $n+1$ differential forms, defined each in one of the covering coordinate domains, blend into a global 1-form ψ_0. The latter defines a contact structure in $\pi^{-1}(V_0)$, since, locally $d\psi_0 = \sum_{i\neq a} d\zeta_i \wedge dz^i$, which is a Hamiltonian 2-form. It is easy to verify that the canonical map of $\pi^{-1}(V_0)$ onto $T^{*\prime}(CP^n)$ defined locally by the evaluation of ψ in terms of local holomorphic coordinates is a global isomorphism.

For $s \neq 0$, the manifolds $\pi_1^{-1}(s) = \pi^{-1}(V_s)$, canonically isomorphic to V_s, are obviously holomorphic bundles over CP^n under the map π_s assigning to the augmented matrix $(u_i; x_{ij})$ the point in CP^n with homogeneous coordinates (u_i); in terms of the local coordinates (z_i, ζ_j) for any $\mathfrak{U}_a \cap \pi_1^{-1}(s)$, the first n parameters (z_i) represent the base point, the remaining ones (ζ_j) the fibre coordinates ranging over C^n, and (4.5) determines an affine transformation from one system of fibre coordinates to another. This bundle is clearly a deformation of the one for $s = 0$, which was shown to be isomorphic to $T^{*\prime}(CP^n)$, but is inequivalent to it for $s \neq 0$, since V_s is an affine variety in $C^{(n+1)^2}$, so that it has no compact complex subvarieties of positive dimension; in particular V_s, as a bundle over CP^n, has no holomorphic cross section for $s \neq 0$, unlike the cotangent bundle: which has the zero section. This completes the proof of the lemma. □

We shall conclude by showing that the family $(V_s)_{s\in C}$, where $V_s = \pi_1^{-1}(s) \subset \mathfrak{V}_1$, describes *locally* a model of the deformation of complex structure of $T^{*\prime}(CP^n)$ defined by the S^2-parameterized family of integrable almost complex structures $J(\alpha)$ associated to the hyper-Kählerian structure on the cotangent bundle.

Theorem 4.5. *The two deformations of the holomorphic cotangent bundle $T^{*\prime}(CP^n)$ of the n-dimensional complex projective space defined, respectively, by the S^2-parameterized family of isometric complex structures connected with the hyper-Kählerian structure, and by the desingularization of the family defined by the complex $(n+1)\times(n+1)$-matrices of rank $\leqslant 1$, fibered by the trace mapping over C, are locally isomorphic.*

Proof. The proof consists, roughly speaking, of introducing in each manifold $V_s = \mathrm{tr}^{-1}(s)$ $(s \neq 0)$, consisting of the complex $(n+1)\times(n+1)$ matrices with rank 1 and trace s, a hyper-Kählerian structure invariant under a maximal compact group of automorphisms, and converging to the one quoted earlier from [7], as $s \to 0$. We shall only outline the proof, since the details are not essentially different from the ones in the case just quoted. The maximal compact group of automorphisms of V_s, as a bundle over CP^n, acts equivariantly with the elliptic group action on the base space: it can be described easily by the adjoint representation of $U(n+1)$ on the space $C^{(n+1)^2}$ of all $(n+1)\times(n+1)$ matrices. This group action leaves invariant not only the rank and the trace function, but also the Hermitian norm function, assigning to each matrix $X = (x_{ij})$ its Hermitian norm $|X|^2 = \mathrm{tr}({}^t\overline{X}X) = \sum_{i,j=0}^n |x_{ij}|^2$. Indeed, the orbits of the action of $U(n)$ in each V_s are simply the sets defined by the values that this Hermitian norm takes, ranging over all real values $\geqslant |s|^2$. Thus the space of orbits under the action of $U(n)$ in each V_s has one real dimension. In terms of the local coordinates (z^i, ζ_j) defined in each of the $n+1$ domains $\mathfrak{U}_a \cap \pi_1^{-1}(s)$ covering $\pi^{-1}(V_s)$ used in Lemma 4.4, the Hermitian norm can be expressed by the function

$$\begin{aligned} t &= \sum_{i,j=0}^{n} |x_{ij}|^2 = \sum_{i,j\neq a} |z^i\zeta_j|^2 + \sum_{j\neq a} |\zeta_j|^2 + |x_{aa}|^2\Big(1 + \sum_{i\neq a} |z_i|^2\Big) \\ &= (1+|z|^2)(|\zeta|^2 + |s - z\cdot\zeta|^2), \end{aligned} \tag{4.6}$$

where $|z|^2 = \sum_{i\neq a} |z^i|^2$, $|\zeta|^2 = \sum_{i\neq a} |\zeta_i|^2$, $z\cdot\zeta = \sum_{i\neq a} z^i\zeta_i$. An easy calculation reduces this expression to a form analogous to (4.4) in the case of $T^{*\prime}(CP^n)$. Consider the family, indexed by s, of analytic (nonholomorphic) cross section in each V_s considered as a complex affine space bundle over CP^n and converging as $s \to 0$ to the zero section in $T^{*\prime}(CP^n)$, defined in terms of the local coordinates by

$$\zeta_i = s(1+|z|^2)^{-1}\overline{z^i} \qquad (0 \leqslant i \leqslant n, i \neq a).$$

Also introduce the auxiliary coordinates in V_s replacing the holomorphic fibre coordinates ζ_i by η_i, where

$$\eta_i = \zeta_i - s(1+|z|^2)^{-1}\overline{z^i}. \tag{4.7}$$

Then the hermitian norm t, expressed in terms of (z^i, η_j), takes the form

$$t = |s|^2 + (1+|z|^2)(|\eta|^2 + |z\cdot\eta|^2). \tag{4.8}$$

We now consider the holomorphic bundle Kähler metric [7] in V_s derived from the local Kähler potential function Ψ_a defined in any of the $n+1$ open domains $V_{s,a} = \pi(\mathfrak{U}_a \cup \pi_1^{-1}(s))$ $(0 \leq a \leq n)$ covering V_s, with holomorphic coordinates (z^i, ζ_j), as follows:

$$\Psi_a = \log(1 + |z|^2) + u(t), \tag{4.9}$$

where t is the norm function (4.6), and $u(t)$ a suitable, smooth function of one real variable t $(t \geq |s|^2)$, and calculate the Kähler form $\partial\bar{\partial}\Psi_a$. The calculation is facilitated if one uses the local complex tangent frame in $V_{s,a}$ defined in terms of the local holomorphic coordinates as follows:

$$\nabla_{z^i} = \frac{\partial}{\partial z^i}\bigg|_{\eta=\text{const.}} + \left\{ {k \atop j \quad i} \right\} \eta_k \frac{\partial}{\partial \zeta_j}\,;\quad \frac{\partial}{\partial \zeta_j}\,;\quad \overline{\nabla_{z^i}}\,;\quad \frac{\partial}{\partial \bar{\zeta}_j}\,,$$

where all indices range from 0 to n, excluding the value a, the summation convention holds, and $\{ {k \atop j \; i} \}$ is the Christoffel symbol for the connection of the Fubini–Study metric in CP^n with respect to the inhomogeneous coordinates (z^i), and where (η_k) are the auxiliary functions defined by (4.7). The dual coframe is then made up of the differentials

$$dz^i,\quad D\eta_j,\quad \overline{dz^i},\quad \overline{D\eta_j}\,,$$

where

$$D\eta_j = d\eta_j - \eta_k \left\{ {k \atop j \quad i} \right\} dz.$$

The differential operators ∇_{z^i} and the differentials $D\zeta_j$ are conveniently used here, because they arise naturally from a connection in the holomorphic bundle $V_s \underset{2m}{\overset{\pi}{\to}} CP^n$. In terms of this basis, the Hamiltonian form φ_0 in V_s can be expressed equivalently locally in any of the following three ways:

$$\varphi_0 = d\zeta_i \wedge dz^i = D\zeta_i \wedge dz^i = \partial\eta_i \wedge dz^i.$$

With this formalism, one obtains an ordinary differential equation for the function $u(t)$ in (4.9) whose solutions generate a Kähler metric for which the covariant derivative of φ_0 is identically zero. The details are essentially the same as in the concluding pages of [7], and the solution is

$$u(t) = \sqrt{1 + 4(t - |s|^2)} - \log\left(1 + \sqrt{1 + 4(t - |s|^2)}\right).$$

The resulting Kähler metric is then a hyper-Kähler metric in V_s, converging to the one already described in the limiting case of $T^{*\prime}(CP^n)$, the desingularized model for V_0. Except for the few remaining routine computational details, the proof of Theorem 4.5 is thus complete. □

References

[1] D. V. Alekseevskiı, Classification of quaternionic spaces with transitive, solvable groups of motions. *Izv. Akad. Nauk SSSR* **39** (1975), *A.M.S. Transl.* **9**, 297–339 (1975).

[2] M. F. Atiyah, *Some Examples of Complex Manifolds*. Bonner Mathematische Schriften, No. 6, Bonn, 1958.

[3] T. Aubin, Equations du type Monge–Ampère sur les variétés kählériennes compactes. *Bull. Sci. Math.* **102**, 63–95 (1978).

[4] M. Berger, Remarques sur les groupes d'holonomie des variétés riemanniennes. *C. R. Acad. Sci. Paris Sér. A* **262**, 1316–1318 (1966).

[5] R. B. Brown and A. Gray, Riemannian manifolds with holonomy group *Spin*(9), in *Differential Geometry (Symposium in Honor of Kentaro Yano)*. Konokuniya, Tokyo, 1972, pp. 41–59.

[6] F. A. Bogomolov, Hamiltonian Kähler manifolds. (in Russian), *Dokl. Akad. Nauk SSSR* **243** (5), 1101–1104 (1978).

[7] E. Calabi, Métriques kählériennes et fibrés holomorphes. *Ann. Sci. Ec. Norm. Sup. Paris, 4me Sér.* **12**, 269–294 (1979).

[8] E. Horikawa, Surjectivity of the period map of *K*3 surfaces of degree 2. *Math. Ann.* **228**, 113–146 (1977).

[9] K. Kodaira and D. C. Spencer, On deformations of complex structures, II. *Ann. Math.* **67** (1958), esp. pp. 403–408.

[10] V. Kulikov, Epimorphicity of the period mapping for *K*3 surfaces. (in Russian) *Usp. Mat. Nauk* **32**, (4; 196), 257–258 (1977).

[11] V. Kulikov, Degeneration of *K*3 surfaces and Enriques surfaces. *Izv. Akad. Nauk SSSR* **41**, 1008–1042 (1977); *AMS Transl.* **11**, 957–989 (1977).

[12] A. Lichnérowicz, *Théorie Globale des Connexions et des Groupes d'Holonomie*. Rome, Cremonese, 1955, esp. pp. 250–251.

[13] A. Newlander and L. Nirenberg, Complex analytic coordinates in almost complex manifolds. *Ann. Math.* **65**, 391–404 (1957).

[14] A. N. Todorov, *Moduli of kählerian K-3 surfaces*. Mathematische Arbeitstagung, Bonn, 1979.

[15] A. Weil, Théorèmes fondamentaux de la théorie des fonctions thêta (d'après des mémoires de Poincaré et Frobenius), in *Sém. Bourbaki*, No. 16, Mai 1949; *Oeuvres scientifiques—Collected papers*. Springer, New York, Heidelberg, Berlin, 1979, I, pp. 414–421.

[16] S. T. Yau, On the Ricci curvature of compact Kähler manifolds and complex Monge–Ampère equations, I. *Comm. Pure Appl. Math.* **31**, 339–411 (1978).

Géométrie différentielle affine des hypersurfaces

Eugenio Calabi

Bourbaki Seminar, Vol. 1980/81, 189–204, Lecture Notes in Math., 901, Springer, Berlin-New York (1981)

I. La géométrie différentielle affine est un domaine de la géométrie différentielle classique, c'est-à-dire de la géométrie des sous-variétés d'un espace ambiant donné, ordinairement un espace numérique réel, où les structures fondamentales sont induites par les invariants du groupe unimodulaire affine $ASL(n,\mathbb{R})$ agissant sur l'espace ambiant $\mathbb{R}^n$, remplaçant ainsi le groupe euclidien $ASO(n)$ de la géométrie élémentaire par un groupe plus étendu. Le choix du groupe $ASL(n,\mathbb{R})$ d'ailleurs n'est pas absolument rigide, puisqu'on le remplace à l'occasion soit par le groupe affine total $AGL(n,\mathbb{R})$ soit par les sous-groupes respectifs laissant un point-origine 0 fixe (géométrie centro-affine). Dans cette géométrie la plupart des résultats connus concerne les hypersurfaces ; on va donner une sélection de ces résultats, qui paraissent intéressants pour leurs applications à la géométrie des corps convexes, au calcul des variations, aux équations non-linéaires aux dérivées partielles et aux systèmes différentiels.

La bibliographie ci-jointe donne une liste des exposés détaillés les plus importants de la théorie, dont nous ne donnerons ici qu'un précis limité aux notions requises pour chaque résultat présenté. Dans la suite l'espace ambiant sera supposé de $n+1$ dimensions, le groupe d'équivalence étant $ASL(n+1,\mathbb{R})$ sauf avis contraire ; les sous-variétés seront des hypersurfaces, donc des variétés à n dimensions plongées (immergées) d'une manière *admissible* dans $\mathbb{R}^{n+1}$. Une immersion est appelée *admissible* (pour la géométrie affine) si elle est différentiable de classe au moins C^5, si elle est localement injective et si, quelle que soit une structure euclidienne surimposée dans $\mathbb{R}^{n+1}$ (définie par une forme quadratique définie positive sur $\mathbb{R}^{n+1}$), la deuxième forme fondamentale de l'hypersurface en question est partout régulière (c'est-à-dire de rang maximal) ; on vérifie que cette dernière condition est invariante par rapport au groupe affine, puisque sa validité ne dépend point du choix de la structure euclidienne.

DOI: 10.1007/BFb0097198

II. Nous allons désigner l'espace ambiant par E et le groupe d'équivalence affine par $ASL(E)$. L'*espace vectoriel associé* à E , que nous désignerons par TE est l'espace vectoriel des translations de E sur lui-même ; il s'identifie naturellement à l'espace tangent à n'importe quel point de E et l'espace vectoriel dual de TE sera désigné par T^*E , étant naturellement identifié à l'espace cotangent de E à n'importe quel point. L'algèbre des invariants du groupe $ASL(E)$ est engendrée par le produit alterné : soient $X_0,X_1,\ldots,X_n$ $n+1$ vecteurs dans TE ; leur produit alterné est noté $[X_0,X_1,\ldots,X_n]$; de même pour $\varphi_0,\varphi_1,\ldots,\varphi_n \in T^*E$ leur produit alterné est également noté $[\varphi_0,\varphi_1,\ldots,\varphi_n]$; on a alors l'identité

$$[\varphi_0,\varphi_1,\ldots,\varphi_n][X_0,X_1,\ldots,X_n] = \det_{0\leq i,j\leq n}(\langle\varphi_i,X_j\rangle) ,$$

où $\langle\varphi_i,X_j\rangle$ désigne le produit scalaire naturel entre T^*E et TE . En plus, on désigne par $\{X_1,\ldots,X_n\}$ (respectivement $\{\varphi_1,\ldots,\varphi_n\}$) le vecteur dans T^*E (resp. TE) déterminé uniquement par la relation

$$\langle\{X_1,\ldots,X_n\},X_0\rangle = [X_0,X_1,\ldots,X_n] ,$$

resp.

$$\langle\varphi_0,\{\varphi_1,\ldots,\varphi_n\}\rangle = [\varphi_0,\varphi_1,\ldots,\varphi_n]$$

quels que soient respectivement X_0 et φ_0 .

Soit Σ une variété différentiable abstraite à n dimensions, $(u^1,\ldots,u^n)$ un système de paramètres locaux et $X : \Sigma \to E$ une immersion admissible de X dans E ; alors la notation $\{dX\ldots dX\}$ signifie l'expression globale de l'objet qui se réduit localement aux expressions suivantes

$$\{dX \wedge \ldots \wedge dX\} = \{\frac{\partial X}{\partial u^i}\,du^i \wedge \ldots \wedge \frac{\partial X}{\partial u^k}\,du^k\} = n!\{\frac{\partial X}{\partial u^1},\ldots,\frac{\partial X}{\partial u^n}\}\,du^1 \wedge \ldots \wedge du^n :$$

c'est donc une n-forme sur Σ à valeurs dans $T^*E\setminus\{0\}$, qui annule à chaque point $p \in \Sigma$ l'hyperplan tangent à son image $X(p) \in E$. On appelle la direction des vecteurs-images $\{dX \wedge \ldots \wedge dX\} \in T^*E\setminus\{0\}$ pour chaque $p \in \Sigma$ la direction *co-normale à* X *au point* p . On peut choisir un générateur Y^* de la direction co-normale d'une manière invariante à un changement de signe près, de la manière suivante : quel que soit un champ de vecteurs $\varphi_0 : \Sigma \longrightarrow T^*E\setminus\{0\}$ engendrant la direction co-normale, la troisième condition d'admissibilité de l'immersion X s'exprime alternativement par la condition que la n-forme

$[\varphi_0,d\varphi_0 \wedge \ldots \wedge d\varphi_0] = n!\left[\varphi_0,\frac{\partial\varphi_0}{\partial u^1},\ldots,\frac{\partial\varphi_0}{\partial u^n}\right]du^1 \wedge \ldots \wedge du^n$ est partout $\neq 0$; on définit alors Y^* uniquement par l'équation

$$\{dX \wedge \ldots \wedge dX\} = Y^*[Y^*,dY^* \wedge \ldots \wedge dY^*] \tag{1}$$

si n est impair ; si n est pair la définition reste unique à un changement de signe près par

$$\{dX \wedge \ldots \wedge dX\} = \pm Y^*[Y^*,dY^* \wedge \ldots \wedge dY^*] ; \tag{1'}$$

dans le cas où $X(\Sigma)$ est localement (strictement) convexe, on a un choix canonique

190

de l'orientation de Y^*, consistant avec (1), indépendamment de la parité de n, par la condition supplémentaire, à chaque $p_o \in \Sigma$,

(1")
$$\langle Y^*(p_o), X(p) - X(p_o)\rangle > 0$$

quel que soit $p \neq p_o$ dans un voisinage assez restreint de p_o. On appelle l'application $Y^* : \Sigma \longrightarrow T^*E$ *l'indicatrice co-normale de* X.

Le premier problème qui se pose à ce point d'une façon naturelle, c'est celui de l'inversion de l'indicatrice co-normale : soit $Y^* : \Sigma \longrightarrow T^*E$ une application différentiable donnée et telle que la n-forme

$$[Y^*, dY^* \wedge \ldots \wedge dY^*]$$

est partout $\neq 0$ (condition de transversalité radiale) : trouver une application admissible $X : \Sigma \to E$ telle que la signature de la deuxième forme fondamentale (au sens de la géométrie euclidienne par rapport à n'importe quelle forme quadratique) ait une valeur donnée et dont l'indicatrice co-normale coïncide avec Y^*.

THÉORÈME 1.— *Le problème de l'inversion d'indicatrice co-normale est équivalent, tant du point de vue local que du global, au problème de Minkowski ; en particulier, si* Σ *est fermée et nécessairement difféomorphe à la sphère* S^n, *la condition d'intégrabilité pour le problème global se réduit à la suivante*

(2)
$$\int_\Sigma Y^*[Y^*, dY^* \wedge \ldots \wedge dY^*] = 0 \ ;$$

l'interprétation géométrique de cette condition est que le corps étoilé $B \subset T^*E$ *dont la frontière est* $Y^*(\Sigma)$ *doit avoir l'origine comme centre de masse.*

Démonstration.— On choisit une forme quadratique définie positive quelconque dans TE, qui détermine un isomorphisme canonique entre TE et T^*E et définit dans E une géométrie euclidienne subordonnée : quelle que soit l'hypersurface admissible $X : \Sigma \to E$, sa courbure de Gauss K est partout $\neq 0$ et, en notant N le vecteur unitaire normal, le vecteur co-normal affine Y^* s'exprime en fonction des invariants de la géométrie différentielle élémentaire par l'équation

$$\langle Y^*, A\rangle = \pm |K|^{-1/n+2} N.A$$

quel que soit le vecteur $A \in TE$. On a donc

$$N = \pm (Y^*.Y^*)^{(1/2)} Y^* , \quad N \in S^n$$

et l'indicatrice co-normale est le graphe en coordonnées polaires de la fonction strictement positive $(Y^*.Y^*)^{1/2}$ dans un domaine donné dans S^n, dont la $(n+2)$-ième puissance donne la réciproque de $|K|$ en fonction de N, c'est-à-dire la donnée du problème de Minkowski. La condition d'intégrabilité (2) signifie que la n-forme $Y^*[Y^*, dY^* \wedge \ldots \wedge dY^*]$ est exacte, condition qui s'ensuit de (1), car pour n'importe quel point $X_o \in E$,

$$\{dX \wedge \ldots \wedge dX\} = d\{X - X_0, dX \wedge \ldots \wedge dX\} .$$

Le problème de Minkowski peut donc s'interpréter comme un problème de géométrie affine ; l'avantage de cette interprétation est que, pour tout problème explicite de Minkowski, exprimé par une fonction strictement positive donnée (locale ou globale) dans S^n , on sait *a priori* comment il faut transformer la donnée pour que les solutions correspondent à une transformation affine donnée des solutions du problème originaire : en effet, soit A une transformation affine de E , dA sa partie homogène, dA^* la co-adjointe de dA (dans le langage des matrices dA^* est la transposée de la réciproque de dA) et $\det(A)$ le déterminant de dA ; alors quelle que soit l'hypersurface admissible $X : \Sigma \to E$ avec indicatrice co-normale $Y^* : \Sigma \longrightarrow T^*E$, l'indicatrice co-normale de $A(X)$ est égale à

$$Y^*_A = |\det(A)|^{2n+2/n+2} dA^*(Y^*) .$$

Une application de ce résultat mène à un type de réduction de chaque problème de Minkowski dans l'espace E à $n+1$ dimensions (on supposera $n \geq 2$, le cas $n = 1$ étant élémentaire) à une famille de problèmes de Minkowski en dimension inférieure, de façon que chaque famille de problèmes ainsi réduits au cas inductivement plus élémentaire détermine à son tour le problème de départ dans E .

Dans ce but nous identifions E avec TE par un choix arbitraire d'origine 0; soient E' et E'' deux sous-espaces vectoriels supplémentaires de E ($= TE$) à $k+1$ (resp. $n-k$) dimensions ($1 \leq k \leq n-1$) , soit A l'automorphisme linéaire de E admettant E' et E'' comme sous-espaces invariants et dont la restriction à E' (resp. E'') est égale à la multiplication scalaire par $k-n$ (resp. k), introduisons une forme quadratique définie positive (donc, une structure euclidienne) dans E telle que E' et E'' soient orthogonaux l'un à l'autre et désignons par S^n la sphère euclidienne de rayon 1 et de centre 0 par rapport à cette structure. Alors chaque problème de Minkowski dans E , dans le cas généralisé, est déterminé par la donnée d'une mesure de Borel positive μ dans S^n , telle que 1° le support de μ n'est contenu dans l'intersection de S^n avec aucun hyperplan de E , 2° le centre de masse de S^n par rapport à μ est au centre 0 . Le problème de Minkowski est dit régulier, si la mesure μ , comparée avec la mesure de Lebesgue ordinaire, équivaut à une densité strictement positive partout sur S^n et différentiable (par exemple, d'ordre C^3), que nous noterons ρ_0 . Soit ρ_0 donc une telle densité admissible pour le problème de Minkowski, soit $X : \Sigma \to E$ une solution globale, considérons la famille à un paramètre de données de Minkowski ρ_t $(t \geq 0)$ dont les solutions respectives sont les hypersurfaces

$$X_t = \exp(tA).(X) .$$

On vérifie facilement que l'aire à n dimensions de la solution $X_t(\Sigma)$ (toujours par rapport à la structure euclidienne introduite) est une fonction décroissante de

192

t avec borne inférieure strictement positive, grâce au choix de valeurs propres de A ; de plus, en augmentant la valeur de t de plus en plus, la solution $X_t(\Sigma)$ s'allonge et converge comme ensemble compact vers le sous-espace E'' tout entier pour $t \to \infty$. Au même temps, la donnée de Minkowski ρ_t de X_t, considérée comme densité d'une mesure lisse, converge par rapport à la topologie des distributions vers une mesure ρ' dont la masse totale équivaut à la limite des aires de $X_t(\Sigma)$, mais qui est inadmissible comme donnée du problème de Minkowski, car son support coïncide avec l'intersection $E' \cap S^n$. Cependant, la mesure ρ' correspond à une densité structement positive et lisse dans la k-sphère $S' = E' \cap S^n$, que nous noterons également ρ' et le centre de masse de S' par rapport à ρ' est à l'origine de façon que ρ' représente une donnée admissible et régulière pour pour le problème de Minkowski dans l'espace $E' \subset E$ à $k+1$ dimensions.

Considérons pour un entier fixe k $(1 \leq k \leq n-1)$ la variété ouverte $W(E,k)$ de toutes les paires de sous-espaces vectoriels supplémentaires E',E'' de E à $k+1$ (resp. $n-k$) dimensions (alternativement, la grassmannienne $G(E,k)$, qui consiste en les paires (E',E'') orthogonales par rapport à une structure euclidienne fixée) et pour chaque paire (E',E'') la réduction d'un problème de Minkowski régulier dans E au problème correspondant dans E' comme ci-dessus. Les trois problèmes suivants se présentent à ce propos d'une manière naturelle.

1° En partant de la famille de données de Minkowski réduites, $\rho' : S^n \cap E' \to \mathbb{R}_+$, paramétrées par la famille $W(E,k)$ ou $G(E,k)$, comment peut-on récupérer la donnée du problème de Minkowski originaire dans E ?

2° En partant de la famille de solutions du problème de Minkowski - chaque problème réduit à E' comme ci-dessus - paramétrée par les paires (E',E'') parcourant $W(E,k)$ ou $G(E,k)$, peut-on récupérer la solution du problème originaire dans E sans faire appel aux méthodes connues pour résoudre les problèmes de Minkowski à $n+1$ dimensions ?

3° En partant d'un corps convexe $B \subset E$ et pour toute décomposition donnée en somme directe $E = E' \oplus E''$ avec $\dim_{\mathbb{R}} E' \geq 2$, trouver une construction géométrique du corps convexe $B' \subset E'$ dont la frontière est la solution du problème de Minkowski venant de celui correspondant à la frontière de B par la réduction définie par la paire (E',E'') ; une telle construction géométrique ne devant pas faire intervenir la solution d'un problème de Minkowski dans E'.

Le premier des trois problèmes énoncés ci-dessus est complètement résolu ; en effet, la famille de données réduites $\rho' : S^n \cap E' \to \mathbb{R}_+$, paramétrée même par la seule grassmannienne $G(E,k)$ équivaut à une transformation de Radon généralisée de la donnée $\rho : S^n \to \mathbb{R}_+$ du problème de départ. Pour l'inversion de ce problème de Radon, nous nous remettons à l'exposé par S. Helgason [5]. Le deuxième des problèmes ci-dessus est complètement inconnu ; il serait très intéressant d'en trouver une solution, car elle mènerait à une méthode alternative pour résoudre le problème de Minkowski régulier, par induction sur la dimensionalité $n+1$ de E, et engendrant une hypersurface différentiable d'emblée, c'est-à-dire ne passant pas par les solutions faibles comme dans la méthode connue, qui exige un grand travail pour établir la régularité des solutions partant de données régulières.

Quant au troisième problème, on n'a trouvé de solution que dans le cas particulier où $k = 1$. On rappelle la notion de *somme* introduite elle aussi par H. Minkowski, de deux parties B, C ordinairement supposées compactes et convexes, d'un espace vectoriel E : c'est l'ensemble $B+C$ constitué de tous les éléments $b+c \in E$ tels que $b \in B$, $c \in C$. De même, $B_1,\ldots B_r$ étant une famille finie de r ensembles (compacts et convexes) et $\alpha_1,\ldots,\alpha_r$ une famille correspondante de scalaires positifs, la combinaison $\sum_{\nu=1}^{r} \alpha_\nu B_\nu$ est l'ensemble des combinaisons linéaires $\sum_{\nu=1}^{r} \alpha_\nu b_\nu$ pour tout choix de vecteurs $b_\nu \in B_\nu$. Encore, soit $(B_y)_{y\in\Omega}$ une famille de parties compactes et convexes et non-vides $B_y \subset E$ paramétrées par un ensemble Ω mesuré par μ avec $\mu(\Omega) < \infty$, alors on définit la somme de Minkowski de la famille notée $\int_\Omega B_y d\mu(y)$, comme partie convexe de E constituée par les $x = \int_\Omega \psi(y)d\mu(y)$ pour toute application mesurable $\psi : \Omega \to E$, telle que, pour chaque $y \in \Omega$, $\psi(y) \in B_y$.

Or soit B un corps convexe dans l'espace vectoriel E à $n+1$ dimensions, soient E' et E'' deux sous-espaces vectoriels supplémentaires de E respectivement à $k+1$ et $n-k$ dimensions, $\pi : E \to E''$ la surjection linéaire de E sur E'' annulant E' et laissant chaque vecteur $x \in E''$ fixe. Alors, y étant un élément arbitraire de E'', l'ensemble convexe $(B+\{-y\}) \cap E' \subset E'$ est non-vide si et seulement si y appartient au corps convexe $\pi(B)$, image de B dans E'' par la projection affine π. On définit le corps convexe $\mathcal{S}_{E',E''}(B) \subset E'$, appelé la *tomographie* de B par rapport au couple (E',E'') par la somme de Minkowski suivante

$$\mathcal{S}_{E',E''}(B) = \int_{\pi(B)} (B+\{-y\}) \cap E' |dy| ,$$

$|dy|$ dénotant une mesure de Lebesgue pour E'' restreinte dans ce cas à $\pi(B)$. La solution du troisième problème énoncé auparavant, dans le cas particulier $k = 1$ est donnée par le théorème suivant, dont la démonstration ne présente aucune difficulté.

THÉORÈME 2.— *Soit* E *un espace vectoriel sur* $\mathbb{R}$ *à* $n+1$ *dimensions* $(n \geq 2)$ E' *et* E'' *deux sous-espaces vectoriels supplémentaires de* E, *respectivement à* 2 *et à* $n-1$ *dimensions. Soit* B *un corps convexe dans* E *dont la frontière* $V = \partial B$ *est une solution d'un problème de Minkowski avec donnée* $\rho : S^n \to \mathbb{R}_+$, *où* S^n *est la sphère unitaire de* E *par rapport à une métrique euclidienne pour laquelle* E' *et* E'' *sont orthogonaux,* $V' \subset E'$ *un ovale dans le plan* E' *qui est solution du problème de Minkowski réduit à* E' *par rapport à la paire* (E',E'') *et*

$B' \subset E'$ *la tomographie de* B *par rapport à* (E',E'') . *Alors la frontière de* B' *coïncide, à une homothétie près, avec* V' .

III. Nous reprenons l'exposé sur les hypersurfaces admissibles $X : \Sigma \to E$ en général. L'indicatrice co-normale $Y^* : \Sigma \longrightarrow T^*E$ est donc une immersion différentiable et radialement transverse de Σ dans l'espace vectoriel T^*E dual de TE , signifiant que la n-forme dans Σ

$$\frac{1}{n!}[Y^*,dY^* \wedge \ldots \wedge dY^*]$$

est partout $\neq 0$. Il existe alors une nouvelle application $Y : \Sigma \longrightarrow TE$ caractérisée uniquement par les équations

(3) $$\langle Y^*,Y\rangle = 1 \quad , \quad \langle dY^*,Y\rangle = 0 \; ;$$

en appliquant la première de ces deux équations à la définition (1), (1') de Y^* , on voit que, à chaque point de Σ ,

(4) $$\frac{1}{n!}[Y,dX \wedge \ldots \wedge dX] = \pm \frac{1}{n!}[Y^*,dY^* \wedge \ldots \wedge dY^*] \neq 0 \; ;$$

le vecteur Y est donc partout transverse à l'hyperplan tangent au point correspondant de $X(\Sigma)$: ce fait et l'invariance de la définition de Y par rapport au groupe de transformations affines $ASL(E)$ justifie l'appellation de $Y(p)$, pour tout $p \in \Sigma$, de *vecteur normal affine* de X à p ; on appelle la droite $\{t \in \mathbb{R} \longrightarrow X(p) + tY(p)\}$ la *droite normale affine*.

La valeur absolue de la n-forme (4) induite sur Σ est donc une densité de mesure sur Σ , strictement positive et lisse partout sur Σ , que l'on appelle l'*élément d'aire affine* induit par X sur Σ . Si on calcule la différentielle des deux côtés de la première des deux équations (3) et en tenant compte de la deuxième équation, on obtient la relation

(5) $$\langle Y^*,dY\rangle = 0$$

la différentielle de Y est donc restreinte à l'hyperplan que le vecteur co-normal annule, c'est-à-dire, à l'hyperplan, pour chaque $p \in \Sigma$, tangent à $X(\Sigma)$ au point p . Considérons maintenant la forme quadratique sur Σ représentée par l'expression

$$\langle Y^*,d^2X\rangle$$

ou, en nous rapportant à un système de paramètres locaux $(u^1,\ldots,u^n)$ pour Σ

$$\langle Y^*, \frac{\partial^2 X}{\partial u^i \partial u^j}\rangle du^i du^j \; .$$

Cette forme, évidemment invariante par rapport à l'action du groupe affine $ASL(E)$ sur X est conforme à la deuxième forme fondamentale induite par n'importe quelle structure euclidienne surimposée sur E ; elle est donc régulière, c'est-à-dire de rang maximum (c-à-d $= n$) partout sur Σ et, en dérivant les deux côtés de l'identité $\langle Y^*,dX\rangle = 0$, on vérifie que

(6) $$\langle Y^*,d^2X\rangle = - \langle dY^*,dX\rangle \; .$$

L'expression ci-dessus sert comme définition de la *première forme fondamentale* de la géométrie différentielle affine des hypersurfaces ; elle représente une structure pseudo-riemannienne sur Σ induite par X, analogue dans un sens imprécis à la première forme fondamentale de la géométrie euclidienne (ou riemannienne) induite sur les sous-variétés immergées. Le cas particulier des hypersurfaces admissibles localement convexes devient le plus intéressant de tous à partir de ce point. En effet la convention (1") qui fixe l'orientation de Y^* entraîne la positivité de la forme $\langle Y^*, d^2X\rangle$ et, puisque cette forme est régulière, elle est donc définie positive partout : donc, si $X(\Sigma)$ est localement convexe dans E alors elle est strictement localement convexe et la structure déterminée sur Σ par la première forme fondamentale (6) est une structure riemannienne. Il faut remarquer aussi que l'élément de volume dans Σ associé à la structure pseudo-riemannienne (6) (quelle qu'en soit la signature) coïncide avec l'élément d'aire affine défini antérieurement par les valeurs absolues des n-formes (4)

$$dV = \frac{1}{n!}\left|[Y^*, dY^* \wedge \ldots \wedge dY^*]\right| = \frac{1}{n!}\left|[Y, dX \wedge \ldots \wedge dX]\right| .$$

Les invariants qu'on vient d'introduire nous permettent de définir une classe importante d'hypersurfaces : les *hypersphères affines*.

DÉFINITION.— *Une hypersurface* $X : \Sigma \to E$, *admissible pour la géométrie différentielle affine s'appelle une hypersphère affine, si toutes ses droites normales affines passent par un point commun, dit le centre de l'hypersphère, ou bien si les normales affines sont parallèles deux-à-deux (c'est-à-dire, le centre se trouve à l'infini) ; dans ce dernier cas, on dit que l'hypersphère affine est impropre.*

Il y a plusieurs manières alternatives pour définir les hypersphères affines, tant du point de vue de la géométrie euclidienne que de celui de la géométrie centro-affine [2], [6], [7].

Les hypersphères affines propres avec centre au point 0 sont donc caractérisées par l'équation

$$Y + H(X - X_0) = 0 , \tag{7}$$

tandis que les impropres par l'équation

$$Y = cY_0 ,$$

où H et c sont des scalaires $\neq 0$ et Y_0 est un vecteur constant. Une conséquence immédiate de l'identité (5) est que H et c sont constants. Une conséquence géométrique élémentaire de ce fait donne les caractérisations suivantes des hypersphères affines.

PROPOSITION 2.— *Pour toute hypersurface admissible* $X : \Sigma \to E$ *les trois conditions suivantes sont équivalentes l'une à l'autre :*

a) $X(\Sigma)$ *est une hypersphère affine impropre ;*

b) *l'indicatrice co-normale* $Y^* : \Sigma \longrightarrow T^*E$ *de* X *a son image* $Y^*(\Sigma)$ *contenue*

196

dans un hyperplan ;

c) *la normale affine* Y *de* X *est constante.*

PROPOSITION 3.— *Une hypersurface admissible* $X : \Sigma \to E$ *est une hypersphère affine propre avec centre au point* 0 *, considéré comme origine de coordonnées affines, si et seulement si elle satisfait à la relation suivante avec son indicatrice co-normale* Y^* *: soit* H *la constante de* (7) *; alors*

$$[Y^*, dY^* \wedge \ldots \wedge dY^*] = -H[X-0, dX \wedge \ldots \wedge dX] .$$

En d'autres termes le volume de la clôture convexe de l'origine 0 *avec n'importe quelle région compacte assez petite est proportionnel au volume de la clôture convexe de la partie correspondante de son indicatrice co-normale dans* T^*E *avec l'origine de* T^*E .

La caractérisation des hypersphères affines du point de vue analytique est la suivante. Pour les hypersphères impropres on choisit un système de coordonnées affines $(x_0,\ldots,x_n)$ tel que les composantes du vecteur normal affine Y pour un point arbitraire $p_0 \in \Sigma$ sont $(1,0,\ldots,0)$. Alors Y est constant si et seulement si l'image $X(\Sigma)$ est localement le graphe d'une fonction de la forme

$$x_0 = u(x_1,\ldots,x_n) = u(x)$$

telle que

$$\det(\partial^2 u(x)/\partial x_i \partial x_j) = \pm 1 .$$

Quant aux hypersphères affines propres, on fixe le centre à l'origine 0 d'un système de coordonnées affines $(x_0,x_1,\ldots,x_n)$ pour E et on se limite, pour commencer, aux hypersphères admissibles $X : \Sigma \to E$ de passant pas par 0 et telles que pour tout point $p \in \Sigma$ l'hypersurface soit transverse au rayon vecteur de 0 à $X(p)$; toute hypersurface avec cette propriété peut se représenter localement par des paramètres $t_i = -\frac{x_i}{x_0}$ $(1 \le i \le n)$ après une transformation linéaire des (x_i) convenablement choisie et on pose

$$X = \begin{pmatrix} x_0 \\ x_1 \\ \cdot \\ \cdot \\ \cdot \\ x_n \end{pmatrix} = \begin{pmatrix} -1/u(t) \\ t_1/u(t) \\ \cdot \\ \cdot \\ \cdot \\ t_n/u(t) \end{pmatrix}$$

pour une fonction différentiable $u(t) = u(t_1,\ldots,t_n) \neq 0$. L'hypersurface ainsi tracée est admissible, si et seulement si $\det(\partial^2 u/\partial t_i \partial t_j) \neq 0$ partout ; elle est une hypersphère affine avec centre à l'origine, si et seulement si elle satisfait à l'équation aux dérivées partielles

$$(9) \qquad \det\left(\frac{\partial^2 u(t)}{\partial t_i \partial t_j}\right) = \pm \left|\frac{H}{u(t)}\right|^{n+2}$$

Les équations (8) et (9) sont des exemples d'équations du type dit de

Monge-Ampère. Les méthodes connues pour la construction de leurs solutions sont limitées à présent aux seuls cas où les solutions sont strictement elliptiques, c'est-à-dire au cas où les fonctions $u(x)$ et $u(t)$ sont strictement convexes ou concaves, sauf quelques résultats particuliers concernant les solutions hyperboliques en deux variables indépendantes. Nous faisons référence aux travaux de L. Nirenberg et C. Loewner, de A.V. Pogorelov et de S.T. Yau et S.Y. Cheng [4] sur l'existence et la régularité des solutions des problèmes des valeurs à la frontière pour les équations (8) et (9) [4]. Les théorèmes d'existence les plus intéressants concernent la classification des hypersphères affines localement convexes qui sont *complètes*, c'est-à-dire globalement convexes et plongées dans E. Le cas le plus simple est celui où $H > 0$: une simple application du principe du maximum montre que les ellipsoïdes épuisent les possibilités. Même le cas limite $H = 0$ (correspondant aux hypersphères affines impropres) mène à un résultat analogue : les seules hypersphères affines impropres, localement convexes et complètes sont les paraboloïdes convexes. Par contre, pour $H < 0$, on a le théorème suivant de Cheng et Yau [4] :

Toute hypersphère affine, convexe et complète avec $H < 0$ *est asymptotique à la frontière d'un cône convexe non-dégénéré avec sommet au centre de l'hypersphère ; réciproquement, pour tout cône convexe non-dégénéré et* avec frontière assez lisse *il existe une hypersphère affine, convexe et complète, asymptotique à la frontière du cône vers l'infini, avec centre au sommet du cône et cette hypersphère est unique, dès qu'on fixe la valeur de la constante* $H < 0$.

La solution de ce dernier problème est encore incomplète en ce qui concerne la régularité de l'hypersphère-solution si on enlève des hypothèses de régularité et de convexité de la frontière du cône donné.

IV. Une autre sorte de problème qui se présente de manière naturelle est celui des valeurs extrémales de l'aire affine d'une hypersurface admissible, où on fixe la frontière dans un sens raisonnable. L'aire affine d'une partie compacte K d'une hypersurface, représentée localement par le graphe d'une fonction de n variables réelles par rapport à un système de coordonnées affines de E,

$$x_0 = u(x) = u(x_1, x_2, \ldots, x_n)$$

s'exprime par l'intégrale

$$A = \int_{K_x} \left|\det\left(\frac{\partial^2 u}{\partial x_i \partial x_j}\right)\right|^{1/n+2} |dx_1 \wedge \ldots \wedge dx_n| ,$$

où K_x dénote la région compacte dans l'espace des paramètres $(x_1, \ldots, x_n)$ correspondant à la région K de l'hypersurface. Cette dernière est admissible si et seulement si la matrice hessienne $\left(\frac{\partial^2 u}{\partial x_i \partial x_j}\right)$ de la fonction $u(x)$ est partout régulière, c'est-à-dire avec déterminant $\neq 0$. L'équation d'Euler-Lagrange exprimant la stabilité infinitésimale de A pour toute déformation de $u(x)$ avec support compact

198

dans l'intérieur de K_X est

$$(10) \qquad \sum_{i,j=1}^{n} \frac{\partial}{\partial x_i}\left(G^{ij}\frac{\partial w}{\partial x_j}\right) = 0 ,$$

où (G^{ij}) dénote la matrice inverse de $\left(\frac{\partial^2 u}{\partial x_i \partial x_j}\right)$ et

$$w = \left|\det\left(\frac{\partial^2 u}{\partial x_i \partial x_j}\right)\right|^{1/n+2} .$$

Encore une fois, on vérifie que la condition pour qu'une fonction $u(x)$ satisfaisant à (10) soit une solution elliptique est que cette fonction soit strictement convexe ou concave, c'est-à-dire que la première forme fondamentale affine de l'hypersurface en question soit définie (positive, sans défaut de généralité). Les théorèmes suivants vont paraître ailleurs prochainement avec des démonstrations détaillées [3].

THÉORÈME 4.— *Soit une hypersurface* X *dans* E *qui est le graphe (globalement) d'une fonction* $x_0 = u(x)$ *strictement convexe et supposons que cette fonction satisfasse à l'équation d'Euler-Lagrange* (10). *Alors toute autre hypersurface localement convexe* X' *dans* E *qui coïncide avec* X *au dehors d'une partie relativement compacte a une aire affine strictement plus* petite *que* X .

Ce fait justifie l'appellation que l'on propose pour ces hypersurfaces, comme hypersurfaces d'aire affine maximale.

THÉORÈME 5.— *Soit* X *une surface localement convexe dans l'espace affine* E *à* 3 *dimensions et supposons qu'elle satisfasse à l'équation d'Euler-Lagrange* (10) *localement. Alors l'aire affine de toute partie compacte de* X *est strictement plus grande que celle de toute autre surface admissible obtenue par une déformation intérieure suffisamment petite de cette partie.*

On a enfin un résultat faiblement analogue au théorème de Bernstein pour les surfaces d'aire minima (dans le sens classique).

THÉORÈME 7.— *Soit* X *une surface admissible et convexe dans l'espace* E *à trois dimensions et supposons que* X *soit une surface d'aire affine maxima. Si* X *est complète aussi bien dans le sens géométrique, étant une surface globalement convexe et plongée proprement, que dans le sens métrique par rapport à la structure riemannienne induite par la première forme fondamentale de la géométrie affine, alors* $X(\Sigma)$ *est un paraboloïde convexe.*

V. Nous allons donner maintenant une formulation assez simple des équations de structure pour les hypersurfaces admissibles ; ce n'est que pour rendre le formalisme moins encombrant que nous nous limiterons au cas convexe, où la première forme fondamentale est définie positive. Soit $X : \Sigma \to E$ une hypersurface admissible localement strictement convexe ; quel que soit $p \in \Sigma$ on associe à p une famille de repères affines dans E de la manière suivante : prenant $X(p)$ comme origine, on peut choisir une base unimodulaire de TE $(e_0,e_1,\dots,e_n)$ en mettant $e_0 = Y(p)$, le vecteur normal affine à $X(p)$ et $(e_1,\dots,e_n)$ une base pour l'espace tangent à $X(\Sigma)$ au point p qui est orthonormée par rapport à la première forme fondamentale et $(e_0,e_1,\dots,e_n)$ proprement orienté pour TE . Soit $(e_0^*,\dots,e_n^*)$ la base duale de T*E : alors $e_0^* = Y^*$, le vecteur co-normal à $X(p)$ et, en considérant les autres vecteurs e_i^* comme 1-formes sur E , et en désignant par ω_i les "pullbacks" respectifs sur Σ , $\omega_i = X^*(e_i^*)$, on a l'équation

$$dX = \omega_i e_i \, .$$

On introduit maintenant les matrices anti-symétriques de 1-formes (ϑ_{ij}) représentant la connexion riemannienne induite par les co-repères (ω_i) sur Σ , définies par $d\omega_i = \vartheta_{ij} \wedge \omega_j$; alors, tenant compte des définitions de $Y^* = e_0^*$ et de $Y = e_0$, on a les équations exprimant les différentielles des repères,

$$\text{(11)} \qquad \begin{aligned} de_i &= (\vartheta_{ij} + P_{ij})e_j + \omega_i e_0 \\ de_0 &= - Q_i e_i \end{aligned}$$

où, de façon équivalente, les formules analogues pour le repère dual

$$\text{(11')} \qquad \begin{aligned} de_0^* &= - \omega_i e_i^* \\ de_i^* &= (\vartheta_{ij} - P_{ij})e_j^* + Q_i e_0^* \, . \end{aligned}$$

Ici les objets (P_{ij}) et (Q_i) sont des 1-formes avec valeurs respectivement dans les tenseurs symétriques et dans les vecteurs tangents de Σ et satisfont aux conditions d'intégrabilité suivantes

a) $P_{ij} \wedge \omega_j = 0$, autrement dit la forme trilinéaire $P_{ij} \otimes \omega_i \otimes \omega_j$ invariante est totalement symétrique, donc associée à une *forme cubique* sur Σ , appelée la deuxième forme fondamentale

b) $Q_i \wedge \omega_i = 0$, c'est-à-dire, la forme bilinéaire $Q_i \otimes \omega_i$ invariante sur Σ est symétrique, donc associée à une forme quadratique sur Σ , appelée la troisième forme fondamentale

c) On a l'identité $\sum_{i=1}^{n} P_{ii} = 0$, autrement dit, la forme cubique est apolaire par rapport à la première forme fondamentale

d) Si on dénote par Ω_{ij} la forme de courbure de la structure riemannienne déterminée par la première forme fondamentale,

$$\Omega_{ij} = d\vartheta_{ij} - \vartheta_{ik} \wedge \vartheta_{kj} \, ,$$

on a la relation suivante, analogue à l'équation de Gauss dans la géométrie classique

$$\text{(12)} \qquad \Omega_{ij} = - P_{ik} \wedge P_{kj} + \frac{1}{2}(\omega_i \wedge Q_j - \omega_j \wedge Q_i) \, ;$$

e) les deux types d'équations ci-dessous, analogues aux équations de Codazzi-Mainardi complètent les équations de structure

(13) $$dP_{ij} - \vartheta_{ik} \wedge P_{kj} - P_{ik} \wedge \vartheta_{kj} = -\frac{1}{2}(\omega_i \wedge Q_j + \omega_j \wedge Q_i)$$

(14) $$dQ_i - \vartheta_{ij} \wedge Q_j = Q_j \wedge P_{ij}$$

La forme cubique exprime pour chaque $p \in \Sigma$ la différence entre le jet d'ordre 3 de X à p et celui d'une hypersurface quadratique ayant contact avec $X(\Sigma)$ à p d'ordre maximal. La troisième forme fondamentale exprime la variation du vecteur normal affine ; elle jour donc un rôle tout à fait semblable à celui de la deuxième forme fondamentale dans la géométrie euclidienne des hypersurfaces ; en effet, la valeur moyenne de ses valeurs propres (par rapport à la première forme), appelée la *courbure moyenne affine* est précisément la fonction dont l'annulation équivaut à la propriété décrivant $X(\Sigma)$ comme hypersurface d'aire affine maxima.

Nous allons conclure l'exposé par une application des équations de structure pour caractériser intrinsèquement les surfaces dans l'espace à trois dimensions qui sont images non-singulières d'une immersion harmonique ; ce calcul a été fait indépendamment par R.L. Bryant.

Soit Σ une surface et $X : \Sigma \to E$ une immersion différentiable dans l'espace affine E à trois dimensions. On cherche des conditions génériques sur X pour l'existence d'une structure conforme (vue comme classe d'équivalence de structures riemanniennes) sur Σ telle que la composition de X avec toute fonction linéaire sur E est fonction harmonique sur Σ par rapport à la structure conforme introduite.

Il est évident que toute condition sur X qu'on peut dériver doit être invariante par les transformations affines de E et par conséquent devrait pouvoir s'exprimer par les invariants affines de X, pourvu que X soit admissible. D'un côté une surface est admissible si et seulement si, en se rapportant à une structure euclidienne auxiliaire sur E, la courbure de Gauss K de $X(\Sigma)$ est partout $\neq 0$; d'autre part, une condition nécessaire pour que X soit harmonique vient du principe du maximum, qui empêche la possibilité que K soit strictement positif, d'où on arrive à la restriction que K doit être strictement négatif partout et par conséquent la première forme fondamentale est indéfinie. Dans ce cas les systèmes de paramètres locaux dans Σ qui sont les plus utiles sont les systèmes assymptotiques, c'est-à-dire les systèmes (u,v) tels que, à chaque point où ils sont définis, les deux vecteurs $\partial^2X/\partial u^2$ et $\partial^2X/\partial v^2$ sont tous les deux tangents à $X(\Sigma)$. Si on introduit maintenant les champs de repères (non-unimodulaires) formés par $\partial X/\partial u = X_u$, $\partial X/\partial v = X_v$ et le vecteur normal affine Y on exprime les trois formes fondamentales de la manière suivante : la première, $g = 2F du dv$, où $F = F(u,v) > 0$; la deuxième, II, réduite aux seuls termes $A du^3 + D dv^3$, à cause de la condition d'apolarité par rapport à g ; la troisième, $\mathrm{III} = L du^2 + 2HF du dv + N dv^2$, où H dénote la courbure moyenne affine. Les équations de structure sont les suivantes :

$$\begin{aligned}
dX &= X_u du + X_v dv \\
dX_u &= \left(\frac{F_u}{F} X_u + \frac{A}{F} X_v\right) du + FYdv \\
dX_v &= \left(\frac{D}{F} X_u + \frac{F_v}{F} X_v\right) dv + FYdu \\
dY &= -\left(HX_u + \frac{L}{F} X_v\right) du - \left(\frac{N}{F} X_u + HX_v\right) dv ,
\end{aligned} \tag{15}$$

où les paramètres u , v écrits comme indices inférieurs dénotent des dérivées partielles respectives. Les conditions d'intégrabilité sont représentées par les cinq équations suivantes, qui correspondent à (12), (13) et (14) : désignons par K_a la courbure de Gauss-Riemann associée à la structure pseudo-riemannienne définie par $2Fdudv$, c'est-à-dire $K_a = -F^{-1}(\log F)_{uv}$; alors on a

$$K_a = \frac{AD}{F^3} + H \qquad \text{(équation de Gauss)}$$

$$A_v = -FL \ ; \ D_u = -FN \qquad \text{(équations de Codazzi de première espèce)}$$

$$L_v - FH_u = \frac{AN}{F} \ ; \ N_u - FH_v = \frac{DL}{F} \qquad \text{(équations de Codazzi de deuxième espèce)}$$

L'existence d'une structure conforme sur Σ par rapport à laquelle X est harmonique équivaut à l'existence, localement, de trois fonctions e , f , g de u et v , telles que

$$eg - f^2 = 1 \ , \quad e > 0 \ , \quad (gX_u - fX_v)_u + (-fX_u + eX_v)_v = 0 .$$

Tenant compte des équations de structure (15), la dernière équation se transcrit en

$$(g_u + \frac{F_u}{F} g - f_v + \frac{D}{F} e)X_u + (\frac{A}{F} g - f_u + e_v + \frac{F_v}{F} e)X_v - 2FfY = 0 ,$$

d'où $f = 0$ identiquement ; on réduit alors les trois inconnues à une seule $\varphi = \varphi(u,v)$ en posant $e = e^{\varphi}$, $f = 0$, $g = e^{-\varphi}$, satisfaisant au système de deux équations.

$$\begin{aligned}
\varphi_u &= \frac{F_u}{F} + \frac{D}{F} e^{2\varphi} \\
\varphi_v &= -\frac{F_v}{F} - \frac{A}{F} e^{-2\varphi}
\end{aligned} \tag{16}$$

Ce système nécessite une condition d'intégrabilité venant de

$$\left(\frac{F_u}{F} + \frac{D}{F} e^{2\varphi}\right)_v + \left(\frac{F_v}{F} + \frac{A}{F} e^{-2\varphi}\right)_u = 0 ,$$

c'est-à-dire

$$e^{2\varphi}\left((\frac{D}{F})_v + 2 \frac{D}{F} \varphi_v\right) + 2(\log F)_{uv} + e^{-2\varphi}\left((\frac{A}{F})_u - 2 \frac{A}{F} \varphi_u\right) = 0$$

ou bien

$$F(\frac{D}{F^3})_v e^{2\varphi} - 2K_a - 4 \frac{AD}{F^3} + F(\frac{A}{F^3})_u e^{-2\varphi} = 0 . \tag{17}$$

Cette dernière équation montre que, dans une situation générique et précisément où

$$(\frac{A}{F^3})_u (\frac{D}{F^3})_v \neq 0 \tag{17'}$$

il y a au plus *deux* fonctions ψ satisfaisant l'équation quadratique

$$F(\frac{D}{F^3})_v \psi^2 - 2\left(K_a + 2 \frac{AD}{F^3}\right) \psi + F(\frac{A}{F^3})_u = 0 ,$$

parmi lesquelles les solutions strictement positives sont les seules candidates pour une solution

202

$\varphi = \frac{1}{2}\log\psi$ de (17) ; cette dernière est plus faible que le système de deux équations (16) qui correspondent au problème en question. Pour vérifier si au moins une des deux solutions de (17) satisfait au système (16), on dérive (17) par rapport à u et v, avec les résultats respectifs suivants :

$$(18')\quad 2D\left(\frac{D}{F^3}\right)_v e^{4\varphi} + F^{-2}\left(F^3\left(\frac{D}{F^3}\right)_v\right)_u e^{2\varphi} - 2\left(K_a + \frac{2AD}{F^3}\right)_u - 2D\left(\frac{A}{F^3}\right)_u + F^2\left(F^{-1}\left(\frac{A}{F^3}\right)_u\right)_u e^{-2\varphi} = 0 ,$$

$$(18'')\quad F^2\left(F^{-1}\left(\frac{D}{F^3}\right)_v\right)_v e^{2\varphi} - 2\left(K_a + \frac{2AD}{F^3}\right)_v - 2A\left(\frac{D}{F^3}\right)_v + F^{-2}\left(F^3\left(\frac{A}{F^3}\right)_u\right)_v e^{-2\varphi} + 2A\left(\frac{A}{F^3}\right)_u e^{-4\varphi} = 0 .$$

Le résultat du calcul est qu'il existe une solution ψ correspondant à une structure conforme sur Σ pour laquelle X est une immersion harmonique, si et seulement si $e^{2\varphi}$ est une solution positive simultanée de l'équation quadratique (17) et des deux équations cubiques (18') et (18"). Dans le cas générique (17') au moins une des deux solutions $\psi = e^{2\varphi}$ de (17) satisfait aux équations (18') et (18") si et seulement si la matrice suivante à 7 lignes et à 5 colonnes est de rang au plus égal à 4 ;

$$(19)\quad \begin{pmatrix}
2D(\frac{D}{F^3})_v & F^{-2}\left(F^3(\frac{D}{F^3})_v\right)_u & -2\left(K_a+\frac{2AD}{F^3}\right)_u - 2D(\frac{A}{F^3})_u & F^2\left(F^{-1}(\frac{A}{F^3})_u\right)_u & 0 \\
0 & 2D(\frac{D}{F^3})_v & F^{-2}\left(F^3(\frac{D}{F^3})_v\right)_u & -2\left(K_a+\frac{2AD}{F^3}\right)_u - 2D(\frac{A}{F^3})_u & F^2\left(F^{-1}(\frac{A}{F^3})_u\right)_u \\
F(\frac{D}{F^3})_v & -2K_a - 4\frac{AD}{F^3} & F(\frac{A}{F^3})_u & 0 & 0 \\
0 & F(\frac{D}{F^3})_v & -2K_a - 4\frac{AD}{F^3} & F(\frac{A}{F^3})_u & 0 \\
0 & 0 & F(\frac{D}{F^3})_v & -2K_a - 4\frac{AD}{F^3} & F(\frac{A}{F^3})_u \\
F^2\left(F^{-1}(\frac{D}{F^3})_v\right)_v & -2\left(K_a+\frac{2AD}{F^3}\right)_v - 2A(\frac{D}{F^3})_v & F^{-2}\left(F^3(\frac{A}{F^3})_u\right)_v & 2A(\frac{A}{F^3})_u & 0 \\
0 & F^2\left(F^{-1}(\frac{D}{F^3})_v\right)_v & -2\left(K_a+\frac{2AD}{F^3}\right)_v - 2A(\frac{D}{F^3})_v & F^{-2}\left(F^3(\frac{A}{F^3})_u\right)_v & 2A(\frac{A}{F^3})_u
\end{pmatrix}$$

en plus, les deux solutions de l'équation quadratique (17) en $e^{2\varphi}$, supposées strictement positives, sont aussi des solutions de (18') et (18"), si et seulement si la matrice carrée à 4 colonnes est de rang 2 ,

$$(20)\quad \begin{pmatrix}
2D(\frac{D}{F^3})_v & F^{-2}\left(F^3(\frac{D}{F^3})_v\right)_u & -2\left(K_a+\frac{2AD}{F^3}\right)_u - 2D(\frac{A}{F^3})_u & F^2\, F^{-1}(\frac{A}{F^3})_{u\ u} \\
F(\frac{D}{F^3})_v & -2K_a - 4\frac{AD}{F^3} & F(\frac{A}{F^3})_u & 0 \\
0 & F(\frac{D}{F^3})_v & -2K_a - 4\frac{AD}{F^3} & F(\frac{A}{F^3})_u \\
F^2\left(F^{-1}(\frac{D}{F^3})_v\right)_u & -2\left(K_a+\frac{2AD}{F^3}\right)_v - 2A(\frac{D}{F})_v & F^{-2}\left(F^3(\frac{A}{F^3})_u\right)_v & 2A(\frac{A}{F^3})_u
\end{pmatrix}$$

En tenant compte de l'hypothèse générique (17'), la première condition portant sur la matrice (19) équivaut à l'annulation des deux seuls mineurs consistant des cinq premières et des cinq dernières lignes ; la condition que la matrice (20) a un rang égal à 2 équivaut au système de 4 équations indépendantes, correspondant à l'annulation des mineurs à trois colonnes qu'on obtient en effaçant la première ou la quatrième ligne et la première ou la quatrième colonne.

Il faut remarquer aussi que les deux matrices (19) et (20) dépendent des invariants affines déterminés par le jet de X d'ordre 5 . Ceci montre donc qu'une surface à courbure de Gauss strictement négative et satisfaisant l'inégalité (17') est une surface harmonique si elle satisfait à un système de deux équations d'ordre 5 , tandis que si elle satisfait au système de 4 équations d'ordre 5 définies par la matrice (20) étant de rang 2 , elle est harmonique par rapport à deux structures conformes. Ces cas particuliers où la condition (17') n'est pas satisfaite simplifient les calculs qui suivent, mais exigent un travail détaillé cas par cas. Il nous suffit de signaler que les quadriques, caractérisées par $A = D = 0$ ne sont des surfaces harmoniques que si elles sont paraboloïdes. Les surfaces qui sont harmoniques de deux manières distinctes, du point de vue des surfaces analytiques réelles ou complexes, ont été étudiées par plusieurs auteurs depuis Darboux ; elles sont importantes dans la théorie des fonctions thêta en trois variables.

BIBLIOGRAPHIE

[1] W. BLASCHKE *Vorlesungen über Differentialgeometrie ; III. Affine Differentialgeometrie*, Berlin, Springer, 1923.

[2] E. CALABI *Complete Affine Hyperspheres* I Symposia Mathematica X, Roma (1971), 19-38.

[3] E. CALABI *Hypersurfaces with maximal affinely invariant area*, Ann. Math. Jour., à paraître.

[4] S.Y. CHENG and S.T. YAU *On the Regularity of the Monge-Ampère Equation* $\det(\partial^2 u/\partial x_i \partial x_j) = F(x,u)$, Comm. P. Appl. Math., 30(1977), 41-68.

[5] S. HELGASON *The Radon Transform*, Basel, 1980.

[6] E. SALKOWSKI *Affine Differentialgeometrie*, Berlin, de Gruyter, 1934.

[7] P.A. SHIROKOV (Shirokow) and A.P. SHIROKOV *Affine Differentialgeometrie*, Leipzig, Teubner, 1962.

Linear Systems of Real Quadratic Forms II

Eugenio Calabi[1]

Proc. Amer. Math. Soc. 84, no. 3, 331–334 (1982)

ABSTRACT. Consider a mapping F: $\mathbf{R}^n \to \mathbf{R}^3$ ($n > 3$) defined by an ordered triple of real-valued quadratic forms; if some linear combination of these three quadratic forms is positive definite, then the image set under the map is a convex cone.

This note is a sequel to an earlier one [1] with the same title, dealing with maps of $\mathbf{R}^n$ into $\mathbf{R}^k$ defined by k quadratic forms in n real variables. One of the problems to which this note is devoted is that of describing properties of the image set under such a mapping, especially when $n > k$.

We begin with a few elementary remarks.

1. A set of k given quadratic forms is linearly dependent, if and only if the image set under the mapping that they define into $\mathbf{R}^k$ lies in a vector subspace, so that it is often convenient to assume without loss of generality that the k given quadratic forms are linearly independent.

2. If F: $\mathbf{R}^n \to \mathbf{R}^k$ is defined by k quadratic forms $u_\alpha = f_\alpha(x)$ ($x \in \mathbf{R}^n$; $\alpha = 1, 2, \ldots, k$), then the image set $F(\mathbf{R}^n) \subset \mathbf{R}^k$ is always a cone, that is to say a set of points (vectors) in $\mathbf{R}^k$ closed under multiplication by nonnegative scalars. This set lies in a closed half-space, defined, let us say, by a homogeneous linear inequality

$$\sum_{\alpha=1}^{k} c_\alpha u_\alpha \geqslant 0 \qquad ((c_1, \ldots, c_k) \neq (0, 0, \ldots, 0))$$

if and only if the quadratic form

$$f_c = \sum_{\alpha=1}^{k} c_\alpha f_\alpha \tag{1}$$

is positive semidefinite.

3. If, for some real constants $(c_1, \ldots, c_k)$ the quadratic form (1) is positive definite, and if $n < \infty$, then the mapping F: $\mathbf{R}^n \to \mathbf{R}^k$ defined by $u_\alpha = f_\alpha(x)$ $(1 \leqslant \alpha \leqslant k)$ is a proper map into the half-space $\{0\} \cup \{u = (u_\alpha) \in \mathbf{R}^k | \Sigma_{\alpha=1}^k c_\alpha u_\alpha > 0\}$.

Our main result is the following one.

THEOREM. *If F: $\mathbf{R}^n \to \mathbf{R}^k$ is defined by k quadratic forms $u_\alpha = f_\alpha(x)$ ($x \in \mathbf{R}^n$, $\alpha = 1, 2, \ldots, k$) and satisfies the following assumptions*:

(a) $n > k = 3$.

(b) *For a certain set of k constants $c_1, \ldots, c_k$ the quadratic form $f_c = \Sigma_{\alpha=1}^k c_\alpha f_\alpha$ on $\mathbf{R}^n$ is positive definite. Then the image cone $K = F(\mathbf{R}^n)$ in $\mathbf{R}^k$ is a convex cone.*

Received by the editors April 10, 1981.
1980 *Mathematics Subject Classification*. Primary 15A63; Secondary 52A15.
[1] Supported in part by NSF Grant #MCS 78-02285.

0002-9939/81/0000-1015/$02.00

DOI: 10.2307/2043553

Proof. It follows from Remarks 2 and 3 above that the image cone $K = F(\mathbf{R}^n)$ satisfies at least the following properties:

(a) K is contained in the closed half-space $\overline{H}_c$ defined by the linear inequality $\sum_{\alpha=1}^k c_\alpha u_\alpha \geqslant 0$.

(b) If $n < \infty$, then the map F is a proper map; indeed there is constant $\varepsilon > 0$ such that, for all $x \in \mathbf{R}^n$, $\sum_{\alpha=1}^k c_\alpha f_\alpha(x) \geqslant \varepsilon |x|^2$; therefore the cone K is contained in a closed, convex cone in $\mathbf{R}^k$ that is nondegenerate, in the sense that it contains no complete straight line.

Assume now, for the moment, that $n < \infty$ and suppose that the conclusion fails: since the image cone $K = F(\mathbf{R}^n)$ is closed and not convex, there must exist a ray $\Lambda = \{\lambda\eta\}_{\lambda \geqslant 0}$, $\eta = (\eta_1, \eta_2, \eta_3)$, which is outside of K but interior to its convex hull; therefore for each plane Π containing Λ there are points of K that lie on each side of Π. Let two such planes be defined by $\sum_{\alpha=1}^3 a_\alpha u_\alpha = 0$ and $\sum_{\alpha=1}^3 b_\alpha u_\alpha = 0$, that is to say, assume that $(a) = (a_1, a_2, a_3)$ and $(b) = (b_1, b_2, b_3)$ are linearly independent solutions of the equation with unknowns u_1, u_2, u_3,

$$\sum_{\alpha=1}^{3} \eta_\alpha u_\alpha = 0; \tag{2}$$

then (a) and (b) are a basis for all the solutions of (2) and therefore, for any pair of real numbers $(\lambda, \mu) \neq (0, 0)$ the quadratic form

$$g(x) = \sum_{\alpha=1}^{3} (\lambda a_\alpha + \mu b_\alpha) f_\alpha(x) \tag{3}$$

vanishes precisely at the points $x \in \mathbf{R}^n$ such that $F(x) = (f_1(x), f_2(x), f_3(x))$ lies in the plane

$$\left\{ (u) = (u_1, u_2, u_3) \in \mathbf{R}^3 \;\middle|\; \sum_{\alpha=1}^{3} (\lambda a_\alpha + \mu b_\alpha) u_\alpha = 0 \right\} \tag{4}$$

passing through the half-line Λ generated by (η_1, η_2, η_3).

On the one hand for each pair (λ, μ) the plane (4) admits points of K on each of its two sides; therefore the quadratic form $g(x)$ in (3) is indefinite for each (λ, μ). On the other hand the complete line generated by Λ meets the image cone K only at the origin, since $\Lambda \cap K = \{0\}$ and $K \setminus \{0\}$ is contained in an open half-space, while $-\Lambda$ is in the opposite half-space. Thus the two quadratic forms $\sum_{\alpha=1}^3 a_\alpha f_\alpha(x)$, and $\sum_{\alpha=1}^3 b_\alpha f_\alpha(x)$ have no common zeros except $x = 0$. This contradicts the main theorem of [**1**], see also [**2**] according to which, if two quadratic forms on $\mathbf{R}^n$, for $n \neq 2$, $n < \infty$ have only 0 as a common null-vector, then there is a linear combination of them that is positive definite. Therefore, if $n < \infty$, we have proved that the image cone must be convex.

In the case of an infinite dimensional linear space $\mathbf{R}^\infty$, with any topological structure, the same conclusion holds, since the image cone under F of any finite dimensional vector subspace is convex. The theorem is thereby proved.

The following two examples show the futility of attempts to prove certain naive generalizations of the theorem.

(A) Consider in $\mathbf{R}^n$ ($3 \leqslant n < \infty$) the three quadratic forms

$$f_1(x) = (x_1)^2, \quad f_2(x) = x_1 x_2, \quad f_3(x) = (x_2^2) - \sum_{j=3}^{n} (x_j)^2:$$

the quadratic form f_1 is positive semidefinite (but not definite); yet the image cone under the map $F = (f_1, f_2, f_3)\colon \mathbf{R}^n \to \mathbf{R}^3$ is the set

$$\{(u) | u_1 > 0, u_1 u_3 \leqslant (u_2)^2\} \cup \{(u) | u_1 = u_2 = 0\},$$

which is not convex.

(B) Consider in $\mathbf{R}^n$ ($3 \leqslant n < \infty$) the four quadratic forms

$$f_1(x) = (x_1)^2 + \sum_{j=3}^{n} (x_j)^2, \quad f_2(x) = (x_1)^2 - (x_2)^2,$$

$$f_3(x) = 2x_1 x_2, \quad f_4(x) = x_2^2 - \sum_{j=3}^{n} (x_j)^2:$$

the quadratic form $2f_1 + f_4$ is positive definite; yet the image cone $F(\mathbf{R}^n) \subset \mathbf{R}^4$ under the map $F = (f_1, f_2, f_3, f_4)$ is the 3-dimensional semi-algebraic set

$$\left\{(u) = (u_1, u_2, u_3, u_4) \mid u_1 + u_4 = \sqrt{(u_2)^2 + (u_3)^2}, u_1 \geqslant |u_2 + u_4|, u_1 \geqslant u_2 + |u_4|\right\}$$

which is not convex.

We conclude with an application. Let $\mathcal{H}$ be any linear function space, in which three quadratic forms are given. For instance, let $\mathcal{H}$ be the function space on an open Riemannian manifold consisting of all real-valued functions ϕ that are square-summable, along with their first and second derivatives; let $f_\alpha(\phi)$ ($\alpha = 1, 2, 3$) denote respectively the squared L_2-norms $\|\phi\|^2$, $\|\nabla\phi\|^2$, $\|\Delta\phi\|^2$ of ϕ, of its gradient, and of its Laplacian, with respect to the Riemannian metric, for any $\phi \in \mathcal{H}$. It is known that $f_2(\phi)$ has an upper bound in terms of $f_1(\phi)$ and $f_3(\phi)$, that f_1 has a positive lower bound in terms of $f_2(\phi)$ and $f_3(\phi)$, and that f_3 has a positive lower bound in terms of $f_1(\phi)$ and $f_2(\phi)$ (if $f_2(\phi)$ is sufficiently large); however it is not easy to determine in general what the best estimates are, nor what other constraints there are (if any) on the possible values of $(f_1(\phi), f_2(\phi), f_3(\phi)) \in \mathbf{R}^3$. The main theorem of this paper proves that the set of possible values of these triple functionals form a convex cone K in $\mathbf{R}^3$; from this one easily deduces that the set of values of (f_1, f_2, f_3) is not enlarged if we replace real-valued functions with complex-valued ones (or $\mathbf{R}^k$-valued ones for any euclidean $\mathbf{R}^k$). Finally, since the image cone K is convex, it is completely determined by its dual cone K^* consisting of all the triples $(v_1, v_2, v_3) \in \mathbf{R}^3$ such that, for all $\phi \in \mathcal{H}$, $\sum_{\alpha=1}^{3} v_\alpha f_\alpha(\phi) \geqslant 0$. This reduces the question to the apparently simpler one of determining the set of all triples (v_1, v_2, v_3) for which a certain selfadjoint completion of the differential operator

$$\phi \to v_3 \Delta^2 \phi - v_2 \Delta \phi + v_1 \phi$$

(determined by asymptotic conditions of ϕ at "infinity", or by boundary conditions if the manifold is a compact manifold with boundary) is positive semidefinite.

This last problem is far from trivial even in one of the simplest conceivable cases, that of functions ϕ on an open interval $[0, a] \subset \mathbf{R}$; let

$$f_1(\phi) = \frac{1}{a}\int_0^a |\phi(t)|^2\,dt, \quad f_2(\phi) = a\int_0^a |\phi'(t)|^2\,dt, \quad f_3(\phi) = a^3\int_0^a |\phi''(t)|^2\,dt$$

(note that the coefficients are chosen so that the functionals f_α are independent of a). In this case it has been determined that the image cone of $\mathcal{H}$ under (f_1, f_2, f_3) is bounded by the convex hull of an infinite sequence of analytic, convex, conical surfaces, each one obtained by solving a certain one-parameter family of fourth order, ordinary differential equations with appropriate (but unusual) boundary conditions.

References

1. E. Calabi, *Linear systems of real quadratic forms*, Proc. Amer. Math. Soc. **15** (1964), 844–846.
2. P. Finsler, *Über das Vorkommen definiter und semidefiniter Formen in Sharen quadratischer Formen*, Comment. Math. Helv. **9** (1937), 188–192.

Department of Mathematics, University of Pennsylvania, Philadelphia, Pennsylvania 19104

Extremal Kähler Metrics

Eugenio Calabi*

Seminar on Differential Geometry, 259–290,
Ann. of Math. Stud., 102, Princeton Univ. Press, Princeton, NJ (1982)

Introduction

The gauge field problem in its most general formulation seeks a connection (vector potential) in a given vector bundle with orthogonal group structure (the gauge group), which minimizes the L_2 norm (action), defined by integrating a natural pointwise squared absolute value of the curvature (gauge field). In recent years the problem has attracted the attention of geometers in a setting that goes beyond the scope that made it interesting to physicists. The most natural geometric problem that falls in this classification deals with the special case of the tangent bundle of a compact differentiable manifold, where one may choose an arbitrary Riemannian metric, which fixes a unique torsion-free connection (the Levi-Civita connection), subject to some reasonable constraints (for instance, fixing the total volume, or within a pre-assigned conformal class).

A special case of the Riemannian problem was proposed several years ago by the author [3, 4]+. It deals with the set of all Kähler metrics that can be introduced in a given, compact complex manifold M. A natural, additional restriction (involving at most a finite number of parameters)

*Supported in part by N.S.F., Contract No. MCF 78-02285.

+Arabic numerals in brackets refer to the bibliography of research papers; Roman numerals to textbooks and monographs.

that is appropriate to impose is to fix the cohomology class of the Kähler metrics. This amount to using an initially given Kähler metric, expressed by

$$ds^2 = 2g_{\alpha\overline{\beta}}(z,\overline{z})\,dz^\alpha dz^{\overline{\beta}}$$

in terms of a local system of holomorphic parameters $z = (z^1, \cdots, z^n)$ and their complex conjugates $\overline{z}$; a new metric tensor $g'_{\alpha\overline{\beta}}(z,\overline{z})$ is cohomologous to $(g_{\alpha\overline{\beta}})$, if and only if there exists a real valued scalar function u, globally defined on the manifold M, such that, at each point

$$g'_{\alpha\overline{\beta}} = g_{\alpha\overline{\beta}} + \frac{\partial^2 u}{\partial z^\alpha \partial z^{\overline{\beta}}}\,. \tag{1}$$

On the other hand, given $g_{\alpha\overline{\beta}}$, the function u in (1), which may be called a *metric distortion potential*, may be chosen arbitrarily, subject only to a condition of differentiability (for the present purposes, differentiability of class C^6 is sufficient), and with appropriate bounds on the second derivatives to ensure that the resulting Kählerian form $g'_{\alpha\overline{\beta}}dz^\alpha dz^{\overline{\beta}}$ be positive definite everywhere.

We present here some preliminary results on the variational problem of minimizing the functional on the function space of all Kähler metrics $(g_{\alpha\overline{\beta}})$, which assigns to each metric the L_2-norm of the corresponding curvature tensor (in terms of the metric itself), integrated over M with respect to the volume element of the metric. The first section is devoted to setting up the notations and conventions followed in the sequel, and to reducing the variational problem of minimizing the L_2-norm of the Riemann curvature to the corresponding, equivalent problems of minimizing the L_2-norm of either the Ricci curvature or of the scalar curvature; since the formal computations with the scalar curvature are more tractable than the corresponding ones for the Ricci or the Riemannian tensors, the second and third sections mention the problem only in terms of the scalar curvature. In the second section we derive the Euler equation for the variational problem considered and prove that the second variation is essentially positive definite. The third section describes some special types of compact, complex manifolds, for which the extremal Kähler metrics for our variational problem are explicitly constructible; what makes the examples interesting is the fact that, for these manifolds, the extremal Kähler metric does not (and cannot) have constant scalar curvature.

1. *Preliminary results*

All manifolds in this paper are understood to be connected. Let M be a compact, complex manifold with n (complex) dimensions, admitting at least one Kähler metric, expressed locally, in terms of a holomorphic system of parameters $z = (z^1, \cdots, z^n)$ and their conjugates $\overline{z} = (\overline{z^1}, \cdots, \overline{z^n})$, by

(1.1)
$$ds^2 = 2g_{\alpha\overline{\beta}}\, dz^{\alpha}\, dz^{\overline{\beta}}\,.$$

We use Greek indices, ranging in values from 1 to n, in relation with the local parameters $(z^1, \cdots, z^n)$, barred Greek indices referring to the conjugate parameters $\overline{z}$, and Latin indices ranging from 1 to 2n referring to both z and $\overline{z}$, where formally $\overline{z^{\alpha}}$ is identified with $z^{n+\alpha}$, and $n+\alpha$ is denoted by $\overline{\alpha}$. The Kähler condition on the metric is expressed by the existence of a locally defined, real valued, smooth function $\Phi = \Phi(z,\overline{z})$, independent of the local coordinates $(z,\overline{z})$ but not unique, such that

(1.2)
$$g_{\alpha\overline{\beta}}(z,\overline{z}) = \frac{\partial^2\Phi}{\partial z^{\alpha}\partial z^{\overline{\beta}}}\,.$$

The complex analytic structure of M has an intrinsic invariant, the function space of all Kähler metrics (we assume this space to be nonempty); to each such metric, the associated, real valued, exterior form of bidegree (1.1) in dz, $d\overline{z}$,

$$\omega = \sqrt{-1}\, g_{\alpha\overline{\beta}}\, dz^{\alpha} \wedge dz^{\overline{\beta}}$$

is closed, because of (1.2), and therefore determines a 2-dimensional deRham cohomology class $[\omega]$, called the *principal cohomology class* of the metric. The set of all deRham classes admissible as principal cohomology classes for some Kähler metric is an open, convex cone in the Hodge-Dolbeault cohomology group $H^{1,1}(M, R)$ (deRham classes representable by closed (1,1)-forms), of a shape that is unknown in general, but conjectured to be determined by a finite number of real analytic inequalities.

In this paper we are interested only in the space of Kähler metrics in a fixed principal class $[\omega]$, determined for instance by one given Kähler metric (1.1). Any other Kähler metric $(g'_{\alpha\overline{\beta}})$ in this class is then obtained from $(g_{\alpha\overline{\beta}})$ be means of a globally defined, suitable, real valued function u on M by means of the local equations

$$g'_{\alpha\overline{\beta}} = g_{\alpha\overline{\beta}} + \frac{\partial^2 u}{\partial z^\alpha \partial \overline{z^\beta}} .$$

For further details on the elementary properties of Kähler manifolds, we refer to any one of the available texts on the subject, such as [I] through [VIII].* The notations that we use in this paper represent a compromise between the different ones followed by the various authors to whom we refer. The point of departure of this paper lies in a preliminary abstract [4] concerning the existence and properties of a Kähler metric that is extremal with respect to certain functionals involving the Ricci tensor. In order to fix the notations, we recall that the local components of the metric connection (Γ^i_{jk}), of the Kahler metric (1,1), that are not identically zero are

$$\Gamma^\lambda_{\alpha\beta} = g^{\lambda\overline{\mu}} \frac{\partial g_{\alpha\overline{\mu}}}{\partial z^\beta} = g^{\lambda\overline{\mu}} \frac{\partial^3 \Phi}{\partial z^\alpha \partial z^\beta \partial \overline{z^\mu}} \qquad ((g^{\lambda\overline{\mu}}) = {}^t(g_{\lambda\overline{\mu}})^{-1})$$

---

*Roman numerals in brackets refer to the general textbook references; Arabic numerals, to the bibliography on related research papers.

and

$$\Gamma^{\bar\lambda}_{\bar\alpha\bar\beta} = (\overline{\Gamma^{\alpha}_{\alpha\beta}}) .$$

The curvature tensor $(R_{ijk\ell})$ is skew-hermitian in each of the pairs (i, j) and (k, ℓ) of indices; in other words $\Omega_{ij} = R_{ijk\ell} dz^k \wedge dz^\ell = R_{k\ell ij} dz^k \wedge dz^\ell$ is a tensorial form of bidegree $(1, 1)$, and the tensor $(R_{ijk\ell})$ is completely determined by the components

$$R_{\bar\mu\alpha\bar\beta\lambda} = -R_{\alpha\bar\mu\bar\beta\lambda} = g_{\gamma\bar\mu} \frac{\partial\Gamma^{\gamma}_{\alpha\lambda}}{\partial z^{\bar\beta}} ; \tag{1.3}$$

the Ricci tensor $(R_{ij}) = (g^{\ell m} R_{i\ell jm})$ is hermitian and satisfies a Kähler condition as follows

$$R_{\alpha\bar\beta} = g^{\lambda\bar\mu} R_{\alpha\bar\mu\bar\beta\lambda} = -\frac{\partial^2 \log\det(g_{\lambda\bar\mu})}{\partial z^\alpha \partial \overline{z^\beta}} , \tag{1.4}$$

and the scalar curvature (which we renormalize to be one half of the usual value found in Riemannian geometry texts) is expressed by the equation

$$R = \frac{1}{2} g^{ij} R_{ij} = g^{\alpha\bar\beta} R_{\alpha\bar\beta} .$$

To any Kähler metric (1.1), that we can symbolize by (g), we associate the following invariants:

a) the volume element

$$dV(g) = \det(g_{\alpha\bar\beta}) \bigwedge_{\lambda=1}^{n} (\sqrt{-1}\, dz^\lambda \wedge \overline{dz^\lambda}) = \omega^{[n]} ,$$

where $\omega^{[0]} = 1$, $\omega^{[k]} = \frac{1}{k}\, \omega \vee \omega^{[k-1]}$, denotes the divided exterior powers of the form ω, and

b) the following real numbers:

i) The total volume

$$V(g) = \int_M dV(g)$$

ii) the curvature integrals

$$\overline{R}(g) = \int_M R\,dV(g)\,, \qquad S(g) = \int_M R^2 dV(g)\,,$$

$$Q(g) = \frac{1}{2}\int_M R_{ij}R^{ij}dV(g) = \int_M g^{\alpha\bar{\mu}}g^{\lambda\bar{\beta}}R_{\alpha\bar{\beta}}R_{\lambda\bar{\mu}}dV(g)\,,$$

and

$$K(g) = \frac{1}{4}\int_M R^{ijk\ell}R_{ijk\ell}dV(g) = \int_M g^{\alpha\bar{\alpha}'}g^{\beta'\bar{\beta}}g^{\lambda\bar{\lambda}'}g^{\mu'\bar{\mu}}R_{\alpha\bar{\mu}\beta\bar{\lambda}}R_{\mu'\bar{\alpha}'\bar{\lambda}'\beta'}dV(g)\,.$$

Thus $V(g)$, $\overline{R}(g)$, $S(g)$, $Q(g)$ and $K(g)$ are continuous functionals on the space of all Kähler metrics on M, where this function space may be viewed with any topology T at least as strong as the uniform C^2-topology. We shall restrict our attention to metrics (g) in a fixed principal class $[\omega]$, with a topology T that is at least as strong as the C^4-topology, denoting it by $\mathfrak{g}(\omega, T)$ or more briefly by $\mathfrak{g}$.

PROPOSITION 1.1. *The following functionals are constant over* $\mathfrak{g}$: $V(g)$, $\overline{R}(g)$, $Q(g) - S(g)$, $K(g) - Q(g)$.

Proof. The volume $V(g)$ of M with respect to the metric g is given, as we have mentioned earlier, by

$$V(g) = \int_M \omega^{[n]}\,,$$

where $\omega^{[n]}$ denotes the n'th divided exterior power of ω (the n-fold exterior product of ω with itself divided by $n!$). Since ω is a closed 2-form in a fixed de Rham cohomology class, the independence of $V(g)$ from g for $g \in \mathfrak{g}$ is verified.

In order to prove the invariance of $\overline{R}(g)$, we recall that the Ricci form $\Sigma_1 = \sqrt{-1}\, R_{\alpha\overline{\beta}}\, dz^{\alpha} \wedge dz^{\overline{\beta}}$ is closed, because of (1.4); furthermore its deRham class is fixed, since for any two Kähler metrics (g) and (g'), the difference between the corresponding Ricci forms Σ_1 and Σ'_1 is

$$\Sigma'_1 - \Sigma_1 = -\sqrt{-1}\, \frac{\partial^2}{\partial z^{\alpha}\partial \overline{z}^{\beta}} \left(\log \frac{\det (g'_{\lambda\overline{\mu}})}{\det (g_{\lambda\overline{\mu}})}\right) dz^{\alpha} \wedge dz^{\overline{\beta}}$$

and $\log \dfrac{\det (g'_{\lambda\mu})}{\det (g_{\lambda\overline{\mu}})}$ is the local representation of a globally defined, real valued scalar (indeed, the deRham class of $\frac{1}{2\pi}\Sigma_1$ is the first Chern class of M).

Next, we recall the algebraic identity

$$\Sigma_1 \wedge \omega^{[n-1]} = (\Lambda\Sigma_1)\omega^{[n]},$$

where, for any p-form $\phi = \phi_{i_1,\cdots,i_p}\, dz^{i_1} \wedge \cdots \wedge dz^{i_p}$ with $p \geq 2$, $\Lambda\phi$ is the (p–2)-form

$$\Lambda\phi = -\sqrt{-1}\, g^{\alpha\overline{\beta}} \phi_{\alpha\overline{\beta} i_3,\cdots,i_p}\, dz^{i_3} \wedge \cdots \wedge dz^{i_p},$$

so that $\Lambda\Sigma_1 = R$. Therefore

$$\overline{R}(g) = \int_M \Lambda\Sigma_1 \omega^{[n]} = \int_M \Sigma_1 \wedge \omega^{[n-1]},$$

and the deRham class of $\Sigma_1 \wedge \omega^{[n-1]}$ is fixed. For $n = 1$ (closed Riemann surfaces) the proof of the proposition is now complete, since in that case the integrands in the definitions of $S(g)$, $Q(g)$ and $K(g)$ are identical.

We next prove the constancy of $S(g) - Q(g)$ as (g) varies in $\mathbf{g}$. This follows from the fact that the deRham class of $\Sigma_1^{[2]} \wedge \omega^{[n-2]}$ is fixed and from the two identities

$$(1.5)\qquad \Sigma_1^{[2]}\wedge\omega^{[n-2]} = \frac{1}{2}(\Lambda^2\Sigma_1^{[2]})\omega^{[n]}$$

and

$$(1.6)\qquad \Lambda^2\Sigma_1^{[2]} = (\Lambda\Sigma_1)^2 - g^{\alpha\bar{\mu}}g^{\lambda\bar{\beta}}R_{\alpha\bar{\beta}}R_{\lambda\bar{\mu}} = R^2 - R_{\alpha\bar{\beta}}R^{\alpha\bar{\beta}} .$$

Therefore

$$S(g) - Q(g) = 2\int_M \Sigma_1^{[2]}\wedge\omega^{[n-2]} ,$$

where the last expression is clearly constant as (g) varies in $\mathfrak{g}$.

Finally, in order to prove that $Q(g)-K(g)$ is constant, we introduce the closed, exterior (2,2)-form Σ_2, defined from the curvature form, such that $(2\pi)^{-2}\Sigma_2$ represents the second Chern class

$$\Sigma_2 = (R^{\lambda}{}_{\mu\alpha\bar{\beta}}R^{\mu}{}_{\lambda\gamma\bar{\delta}} - R^{\lambda}{}_{\lambda\alpha\bar{\beta}}R^{\mu}{}_{\mu\gamma\bar{\delta}})\,dz^{\alpha}\wedge d\bar{z}^{\beta}\wedge dz^{\gamma}\wedge d\bar{z}^{\delta}$$

and apply the algebraic identities similar to (1.5) and (1.6)

$$\Sigma_2\wedge\omega^{[n-2]} = \frac{1}{2}(\Lambda^2\Sigma_2)\omega^{[n]}$$

and

$$\Lambda^2\Sigma_2 = 2R_{\alpha\bar{\beta}}R^{\alpha\bar{\beta}} - R_{\bar{\mu}\alpha\lambda\bar{\beta}}R^{\alpha\bar{\mu}\beta\lambda} - R^2 .$$

Since the integral of $\Sigma_2\wedge\omega^{[n-2]}$ over M is obviously constant as (g) varies in $\mathfrak{g}$, it follows that $2Q(g)-K(g)-S(g)$ is constant, and since $S(g)-Q(g)$ was just proved to be constant, so is $Q(g)-K(g)$. This completes the proof of the proposition.

COROLLARY. *Let* V *and* $\bar{R}$ *denote respectively the constant values of* $V(g)$ *and* $\bar{R}(g) = \int_M R\,dV(g)$ *as* (g) *varies in the function space* $\mathfrak{g}$, *and let* S_0, Q_0 *and* K_0 *denote respectively the greatest lower bounds of the nonnegative functionals* $S(g)$, $Q(g)$ *and* $K(g)$. *Then we have the following estimates for* S_0, Q_0 *and* K_0:

a) $S_0 \geq \frac{\bar{R}^2}{V}$,

b) $Q_0 \geq S_0 - \left(1 - \frac{1}{n}\right) \frac{\bar{R}^2}{V}$,

c) $K_0 \geq \frac{2(2n+1)}{(n+1)(n+2)} S_0 - \frac{4(n-1)}{n(n+2)} \frac{\bar{R}^2}{V}$.

Proof. The first inequality is an immediate application of the Schwarz inequality. In fact, for any $g \in \mathfrak{g}$,

$$S(g)V = \int_M R^2 dV(g) \int_M dV(g) \geq \left(\int_M R dV(g)\right)^2 = \bar{R}^2 .$$

The inequality b) is seen by looking at the Hodge orthogonal decomposition of the closed (1,1)-form Σ_1,

$$\Sigma_1 = H\Sigma_1 + (\Sigma_1 - H\Sigma_1) ,$$

where $H\phi$ denotes the Hodge orthogonal projection of any given exterior form ϕ into the space of harmonic forms. Recalling that, in a compact Kähler manifold, H commutes with the contraction operator Λ and substituting $\Sigma_1 - H\Sigma_1$ for Σ_1 in (1.5) and (1.6), we have the following inner product identity, since $\Sigma_1 - H\Sigma_1$ is an exact (1,1)-form

$$(\Sigma_1 - H\Sigma_1 , \; \Sigma_1 - H\Sigma_1) = (\Lambda(\Sigma - H\Sigma_1), \; \Lambda(\Sigma_1 - H\Sigma_1)) =$$

$$= (R, R) - (HR, HR) \geq S_0 - \frac{\bar{R}^2}{V} .$$

At the same time, we note that, for the harmonic 2-form $H\Sigma_1$ we have a further orthogonal decomposition

$$H\Sigma_1 = \left(H\Sigma_1 - \frac{1}{n} \Lambda(H\Sigma_1)\omega\right) + \frac{1}{n} (\Lambda H\Sigma_1)\omega .$$

Therefore we have the inequality, since $\Lambda H\Sigma_1$ is equal to the constant $V^{-1}\bar{R}$,

$$Q(g) = (\Sigma_1 , \Sigma_1) \geq \left(\tfrac{1}{n} (\Lambda H\Sigma_1)\omega , \ \tfrac{1}{n}(\Lambda H\Sigma_1)\omega\right) + (\Sigma_1 - H\Sigma_1 , \Sigma_1 - H\Sigma_1)$$

$$\geq \frac{\overline{R}^2}{nV} + \left(S_0 - \frac{\overline{R}^2}{V}\right)$$

which is the estimate asserted.

We observe, incidentally, that the estimate a) is sharp and attainable, if and only if there exists a Kähler metric $(g) \in \mathfrak{g}$ with constant scalar curvature R. This condition is equivalent to the Ricci form Σ_1 being harmonic. In fact, if R is constant,

$$-*d*\Sigma_1 = -\sqrt{-1}\, g^{\lambda\overline{\mu}}(\nabla_{\overline{\mu}} R_{\lambda\overline{\beta}}\, dz^{\overline{\beta}} - \nabla_\lambda R_{\alpha\overline{\mu}}\, dz^{\alpha})$$

$$= \sqrt{-1}\left(\frac{\partial R}{\partial z^\alpha} dz^\alpha - \frac{\partial R}{\partial z^{\overline{\beta}}} dz^{\overline{\beta}}\right) = 0\ ;$$

conversely, if Σ_1 is harmonic, then so is $\Lambda\Sigma_1 = R$, and therefore R is constant. Inequality b) is sharp, only if Σ_1 is cohomologous to a real multiple of ω but has been proved attainable, under that hypothesis, [10], by S. T. Yau, only when Σ_1 is cohomologous to a nonpositive multiple of ω.

Finally, we prove the estimate c) as follows. We introduce the Kählerian complex projective curvature tensor [VII]

$$C_{\alpha\overline{\mu}\overline{\beta}\gamma} = R_{\alpha\overline{\mu}\overline{\beta}\gamma} - \frac{1}{n+2}(R_{\alpha\overline{\beta}}\, g_{\gamma\overline{\mu}} + R_{\alpha\overline{\mu}}\, g_{\gamma\overline{\beta}} + R_{\gamma\overline{\beta}}\, g_{\alpha\overline{\mu}} + R_{\gamma\overline{\mu}}\, g_{\alpha\overline{\beta}}) +$$

$$+ \frac{1}{(n+1)(n+2)} R(g_{\alpha\overline{\beta}}\, g_{\gamma\overline{\mu}} + g_{\gamma\overline{\mu}}\, g_{\gamma\overline{\beta}})$$

and the trace-free part of the Ricci curvature tensor

$$P_{\alpha\overline{\beta}} = R_{\alpha\overline{\beta}} - \tfrac{1}{n} R g_{\alpha\overline{\beta}}\,.$$

From these two tensors we obtain a pointwise orthogonal, 3-way decomposition of the full curvature tensor

$$R_{\alpha\bar{\mu}\beta\bar{\gamma}} = C_{\alpha\bar{\mu}\beta\bar{\gamma}} + \frac{1}{n+2}(P_{\alpha\bar{\mu}}g_{\gamma\bar{\beta}} + P_{\alpha\bar{\beta}}g_{\gamma\bar{\mu}} + P_{\gamma\bar{\beta}}g_{\alpha\bar{\mu}} + P_{\gamma\bar{\mu}}g_{\alpha\bar{\beta}})$$

$$+ \frac{R}{n(n+1)}(g_{\alpha\bar{\mu}}g_{\gamma\bar{\beta}} + g_{\alpha\bar{\beta}}g_{\gamma\bar{\mu}}) = I + II + III .$$

We take the pointwise squared absolute value of $(R_{\alpha\bar{\mu}\beta\bar{\gamma}})$ thus decomposed and, neglecting the first term, integrate it over M with respect to $dV(g)$, with the following inequality as a result

$$K(g) \geq \frac{4}{n+2}\int_M P_{\alpha\bar{\mu}}P^{\alpha\bar{\mu}}dV(g) + \frac{2}{n(n+1)}S(g) . \tag{1.7}$$

Since $P_{\alpha\bar{\mu}}P^{\alpha\bar{\mu}} = R_{\alpha\bar{\mu}}R^{\alpha\bar{\mu}} - \frac{1}{n}R^2$ and because of the estimate b) for Q_0, we reduce (1.7) to the following

$$K(g) \geq \frac{4}{n+2}\left(Q(g) - \frac{1}{n}S(g)\right) + \frac{2}{n(n+1)}S(g)$$

$$\geq \frac{4(n-1)}{n(n+2)}\left(S_0 - \frac{\bar{R}^2}{V}\right) + \frac{2}{n(n+1)}S_0 =$$

$$= \frac{2(2n+1)}{(n+1)(n+2)}S_0 - \frac{4(n-1)}{n(n+2)}\frac{\bar{R}^2}{V} ,$$

which is the estimate asserted in c). This last estimate is probably not sharp, because there are other inequalities satisfied by the curvature tensor in any Kähler manifold, which are not applied here. The corollary is now proved.

A consequence of Proposition 1.1 is that the variational problems of minimizing the three functionals $K(g)$, $Q(q)$ and $S(g)$ for $(g) \in \mathbf{g}$ are pairwise equivalent. For this reason we shall limit our consideration in the sequel to the formally simpler problem of minimizing $S(g)$. We will conclude this section by showing that the estimate $\frac{\bar{R}^2}{V}$ for a lower bound S_0 of $S(g)$, that is estimate a) of the corollary just proved, is not sharp in the case of some special compact Kähler manifolds. This fact is a sort of a converse of a theorem proved first, in a weaker form, by Y. Matsushima [8] and shortly after, in the form needed here, by A. Lichnérowicz [7; IV].

THEOREM 1.2. *Let* M *be a compact, n-dimensional complex manifold admitting Kahler metrics, and assume that* M *admits a nontrivial,* con*nected Lie group of holomorphic transformations. If the maximal* connected *group* G *of holomorphic transformations of* M *is not reductive, then the lower bound* $\frac{\bar{R}^2}{V}$ *cannot be achieved by the functional* S(g) *for any Kähler metric in any cohomology class.*

Proof. We recall the theorem of Bochner and Montgomery [1, 2] to the effect that the maximal connected group of holomorphic automorphisms of any compact, complex manifold is a complex Lie group, which we denote by G. The group G is reductive, if it is the smallest complex Lie group containing its maximal compact subgroup (as a real Lie subgroup). If the lower bound $\bar{R}^2/V$ for S(g) is achieved by some Kähler metric $(g) \in \mathfrak{g}$, this means that the scalar curvature R with respect to the metric (g) is constant. The theorem of Matsushima-Lichnérowicz [7], [IV, pp. 153-158] asserts that, if a compact Kähler manifold M has constant scalar curvature, then the group G of holomorphic transformations of M is reductive, contradicting our assumption. The information that this theorem does *not* provide, under the stated assumptions, is a positive lower bound for $S_0 - (\bar{R}^2/V)$, although very probably such a bound could be found in terms of the structure of the Lie algebra of G (as a Lie algebra of holomorphic tangent vector fields in M) and of the cohomology class $[\omega]$ of the Kähler metrics under consideration.

2. *Variational formulas*

Let M be a compact, n-dimensional complex manifold with a principal class $[\omega]$ determined by the principal (1,1)-form of a given Kähler metric (1.1), and let $\mathfrak{g}$ be the function space of all Kähler metrics belonging to the same class, with a sufficiently strong function space topology (for instance, the uniform C^4 topology). A one-parameter family of metrics in $\mathfrak{g}$, parametrized by a real variable $t\ (t_0<t<t_1,\ t_0<0<t_1)$, is representable, locally, as a one-parameter family of positive definite hermitian forms (g_t) with coefficient matrices

$$g_{\alpha\overline{\beta}}(t;z,\overline{z}) = g_{\alpha\overline{\beta}}(z,\overline{z}) + \frac{\partial^2 u(t;z,\overline{z})}{\partial z^\alpha \partial z^{\overline{\beta}}}$$

where $u(t;z,\overline{z})$ is a smooth, real-valued function, globally defined in $(t_0,t_1)\times M$. Without loss of generality we may set $g_{\alpha\overline{\beta}}(z,\overline{z}) = g_{\alpha\overline{\beta}}(0;z,\overline{z})$ and $u(0;z,\overline{z})=0$. The infinitesimal variational formulas represent the derivatives with respect to t, at $t=0$, of various geometric objects (connections, curvature, etc.) attached to (g_t), expressed as the result of some linear differential operators on the infinitesimal distortion potential $u(z,\overline{z}) = \left.\frac{\partial u(t;z,\overline{z})}{\partial t}\right|_{t=0}$.

Thus, denoting the infinitesimal variations by the usual prefix δ, we have

$$\delta g_{\alpha\overline{\beta}} = u_{,\alpha\overline{\beta}}$$

$$\delta(\det g_{\alpha\overline{\beta}}) = \det(g_{\alpha\overline{\beta}})\, g^{\lambda\overline{\mu}} u_{,\lambda\overline{\mu}}\ ;$$

therefore the variation of the Ricci curvature is given by

$$\delta R_{\alpha\overline{\beta}} = \frac{\partial^2}{\partial z^\alpha \partial z^{\overline{\beta}}}\,\Delta u\ ,$$

where Δ denotes the Laplace operator $-g^{\lambda\overline{\mu}}\nabla_\lambda \nabla_{\overline{\mu}}$, and

$$\begin{aligned}\delta R &= g^{\alpha\overline{\beta}}\delta R_{\alpha\overline{\beta}} - g^{\lambda\overline{\beta}} g^{\alpha\overline{\mu}}(\delta g_{\lambda\overline{\mu}}) R_{\alpha\overline{\beta}}\\ &= -\Delta^2 u - R_\mu{}^\lambda u_{,\lambda}{}^\mu = -u_{,\alpha}{}^\alpha{}_\beta{}^\beta - u_{,\lambda}{}^\mu R_\mu{}^\lambda\\ &= (-u_{,\alpha\beta}{}^\alpha + u_{,\lambda} R^\lambda{}_{\alpha\beta}{}^\alpha)^{,\beta} - u_{,\lambda}{}^\mu R_\mu{}^\lambda\\ &= -u_{,\alpha\beta}{}^{\alpha\beta} + g^{\lambda\overline{\mu}} u_{,\lambda} R_{,\overline{\mu}}\,.\end{aligned}$$

From the last expression we obtain the variational formula for $S(g) = \int_M R^2 dV(g)$

$$\delta S(g) = \int_M \delta(R^2 \det(g_{\alpha\bar{\beta}}))(-1)^{\frac{n}{2}} dz^1 \wedge d\bar{z}^1 \wedge \cdots \wedge dz^n \wedge d\bar{z}^n ,$$

where

$$\delta(R^2 \det(g_{\alpha\bar{\beta}})) = \det(g_{\alpha\bar{\beta}})(-R^2 \Delta u + 2R(-u_{,\alpha\beta}{}^{\alpha\beta} + g^{\lambda\bar{\mu}} u_{,\lambda} R_{,\bar{\mu}}))$$

$$= \det(g_{\alpha\bar{\beta}})(-2Ru_{,\alpha\beta}{}^{\alpha\beta} + g^{\lambda\bar{\mu}}(R^2 u_{,\lambda})_{,\bar{\mu}}) .$$

Since the integral of the second term on M vanishes, we have the formula, expressed in terms of functional inner products

(2.2) $$\delta S(g) = \delta(R, R) = -2(R, u_{,\alpha\beta}{}^{\alpha\beta}) - 2 \int_M Ru_{,\alpha\beta}{}^{\alpha\beta} dV(g) ;$$

we obtain thus the following result

THEOREM 2.1. *The Euler-Lagrange equation for a Kähler metric* $g \in \mathfrak{g}$ *for which* $S(g) = (R, R)$ *is stationary is the following condition on the scalar curvature*

(2.3) $$R_{,\bar{\alpha}\bar{\beta}} = 0 \text{ (or equivalently } R_{,\alpha\beta} = 0 \text{)}.$$

In other words, the Euler equation expresses the condition that the real vector field on M

$$X = g^{\lambda\mu}\left(R_{,\lambda} \frac{\partial}{\partial z^{\mu}} + R_{,\mu} \frac{\partial}{\partial z^{\lambda}}\right)$$

generates a holomorphic flow (possibly trivial, if R *is constant).*

Proof. We start with the variational formula (2.2) for $S(g)$. Equating it to zero for all real valued functions u, and applying Gauss' theorem four times, we obtain the equation

$$(R^{,\alpha\beta}{}_{\alpha\beta}, u) = 0$$

which, if satisfied for all functions u, means

$$R^{,\alpha\beta}{}_{\alpha\beta} = 0 . \tag{2.4}$$

Next we prove the following lemma

LEMMA. *If a (real or complex valued) function* ϕ *satisfies the equation* $\phi^{,\alpha\beta}{}_{\alpha\beta} = 0$ *in a compact Kähler manifold, then* $\phi_{,\overline{\alpha}\overline{\beta}} \equiv 0$.

Proof. If $\phi^{,\alpha\beta}{}_{\alpha\beta} \equiv 0$ identically, then

$$0 \le g^{\alpha\overline{\lambda}} g^{\beta\overline{\mu}} \phi_{,\overline{\lambda}\overline{\mu}} \overline{(\phi_{,\overline{\alpha}\overline{\beta}})} = (\phi^{,\alpha\beta}\overline{\phi}_{,\alpha} - \phi^{,\alpha\beta}{}_{\alpha}\overline{\phi})_{,\beta} + \phi^{,\alpha\beta}{}_{\alpha\beta}\overline{\phi}$$

$$= *d*(g_{\beta\overline{\mu}}(\phi^{,\alpha\beta}{}_{\alpha}\overline{\phi} - \phi^{,\alpha\beta}\overline{\phi}_{,\alpha})dz^{\overline{\mu}}) ;$$

were the last expression is written as a divergence; thus, because of Gauss' theorem, $\phi_{,\overline{\lambda}\overline{\mu}} = 0$, proving the lemma.

We apply the lemma to the scalar curvature R, obtaining from (2.4), the asserted equation

$$R_{,\overline{\alpha}\overline{\beta}} = 0 ,$$

which is equivalent to the statement that the gradient vector field of R, $\operatorname{grad} R = g^{\alpha\overline{\beta}}\left(\frac{\partial R}{\partial z^{\overline{\beta}}}\frac{\partial}{\partial z^{\alpha}} + \frac{\partial R}{\partial z^{\alpha}}\frac{\partial}{\partial z^{\overline{\beta}}}\right)$, generates a one-parameter group of holomorphic transformations of M. This completes the proof of the theorem.

In the next section we shall exhibit an example of a class of Kähler manifolds with metrics satisfying this Euler equation, but where the scalar curvature is not constant.

COROLLARY. *If a Kähler metric satisfies the Euler equation (2.3), then the nonnegative functionals* Q(g) *and* K(g) *are stationary, as well as* S(g), *under infinitesimal deformations of the Kähler metric preserving the principal class.*

This is an immediate consequence of Proposition 1.1.

There is strong intuitive reason to believe that every Kähler metric satisfying (2.3) achieves a minimum value for $S(g)$ under changes of the metric, not only in the "trivial" cases where R is constant. The main indication comes from a computation of the second variation

THEOREM 2.2. *Let* M *be an* n*-dimensional, compact Kähler manifold with a fixed class* $[\omega]$ *of Kähler metrics. Let* $(g) \in [\omega]$ *be any metric on* M *satisfying the Euler equation*

$$R_{,\bar{\alpha}\bar{\beta}} = 0$$

for a critical point of the functional $S(g) = (R, R)$. *Then the second variation of* $S(g)$ *with respect to any infinitesimal deformation with*

$$\delta g_{\alpha\bar{\beta}} = u_{,\alpha\bar{\beta}}$$

is positive semi-definite, and is zero, if and only if $\delta g_{\alpha\bar{\beta}}$ *is induced by a holomorphic flow.*

Proof. From the first variation of (R, R)

$$\delta(R, R) = -2(R_{,\bar{\alpha}\bar{\beta}}{}^{\bar{\alpha}\bar{\beta}}, u)$$

$$= -2 \int_M R_{,\bar{\alpha}\bar{\beta}}{}^{\bar{\alpha}\bar{\beta}}\, u\, dV(g) ,$$

one calculates the second variation $\delta^2(R, R)$. This calculation is tedious since it involves the variation of the connection in the fourth order covariant derivatives. However, the final terms are greatly simplified, owing to the fact that $R_{,\bar{\alpha}\bar{\beta}} = 0$. The final outcome is given by the following expression, after which one may apply Gauss's theorem

$$\delta^2(R,R) = 2(u_{,\lambda\mu}{}^{\lambda\mu\alpha\beta}{}_{\alpha\beta}, u)$$
$$= 2(u_{,\lambda\mu}{}^{\lambda\mu}, u_{,\overline{\lambda}\overline{\mu}}{}^{\overline{\lambda}\overline{\mu}})$$

to show that $\delta^2(R,R) > 0$ except if $u_{,\overline{\lambda}\overline{\mu}}{}^{\overline{\lambda}\overline{\mu}} = 0$. The latter equation is satisfied, according to the lemma of Theorem 2.1, if and only if

$$u_{,\overline{\lambda}\overline{\mu}} = 0$$

or, equivalently, since u is real valued, if the real vector field

$$U = g^{\lambda\overline{\mu}}\left(u_{,\overline{\mu}}\frac{\partial}{\partial z^\lambda} + u_{,\lambda}\frac{\partial}{\partial z^{\overline{\mu}}}\right)$$

generates a holomorphic flow. It is then immediate to prove that

$$\delta(g_{\alpha\overline{\beta}}) = \frac{1}{2}\mathfrak{L}_U(g_{\alpha\overline{\beta}}),$$

the last expression denoting the Lie derivative of the metric with respect to the gradient field U of u. This proves the theorem.

3. *Some examples*

The theorem of Lichnérowicz-Matsushima (quoted here as Theorem 1.2) states that any compact Kählerian manifold M whose maximal connected group G of automorphisms is not reductive does not admit any Kähler metric with constant scalar curvature. Therefore, if the variational problem of minimizing (R, R) over the space of all metrics with a given principal class has a solution, it must be one for which the vector field $X = g^{\alpha\overline{\beta}}\left(R_{,\overline{\beta}}\frac{\partial}{\partial z^\alpha} + R_{,\alpha}\frac{\partial}{\partial z^{\overline{\beta}}}\right)$ (generating a one-parameter group of holomorphic transformations) is nontrivial. It is interesting to note that, in the case of one complex dimension (compact Riemann surfaces), the variational problem of minimizing (R, R) (here, fixing the principal class is equivalent to fixing the total area of M) has R = constant as the only stationary metric. A lemma equivalent to this fact was proved, independently, also by J. Kazdan and F. Warner [6].

THEOREM 3.1. *The only solution of the Euler equation for metrics satisfying* $R_{,\bar{\alpha}\bar{\beta}} = 0$ *on a compact Riemann surface are metrics of constant scalar curvature* R.

Proof. In the case of genus ≥ 2, there are no holomorphic tangent vector fields, so that the condition $R_{,\bar{\alpha}\bar{\beta}} = 0$, equivalent to the vector field $g^{\alpha\bar{\beta}} R_{,\bar{\beta}} \frac{\partial}{\partial z^\alpha}$ being holomorphic, implies $R_{,\bar{\beta}} = 0$ and hence $R = \text{constant}$. In the case of genus one, the nontrivial holomorphic tangent vector fields are nowhere vanishing and hence $g^{\alpha\bar{\beta}} R_{,\bar{\beta}} \frac{\partial}{\partial z^\alpha}$, which vanishes at the critical points of R, cannot be holomorphic unless it vanishes identically.

Finally, in the case of the sphere (genus zero), if there existed a metric with $g^{\alpha\bar{\beta}} R_{,\bar{\beta}} \frac{\partial}{\partial z^\alpha}$ holomorphic and not identically zero, then this vector field should have at least two zeros (critical points of R), but then also at most two zeros (fixed points of a one-parameter group of conformal transformations). Furthermore the rotated vector field

$$JX = g^{\alpha\bar{\beta}} \left(\sqrt{-1} \frac{\partial R}{\partial \overline{z^\beta}} \frac{\partial}{\partial z^\alpha} - \sqrt{-1} \frac{\partial R}{\partial z^\alpha} \frac{\partial}{\partial \overline{z^\beta}} \right)$$

generates a compact group of isometries, or rotations of the sphere. A 2-sphere with a one-parameter group of isometries can be parameterized (as a Riemannian manifold) by two real parameters (u, v), where $0 \leq u \leq a$ for some constant $a > 0$, and $v \in \mathbf{R}/2\pi\mathbf{Z}$, with each of the boundary circles, represented respectively by $u = 0$ and $u = a$, collapsed to a corresponding point. The isometry group is represented by the translations of the parameter v and the metric is expressed in terms of these parameters by

$$ds^2 = du^2 + (\phi(u))^2 dv^2 ,$$

where $\phi(u)$ is a smooth, odd, periodic function of $u \in \mathbf{R}$ with period $2a$, satisfying the following conditions in the closed interval $[0, a]$:

a) $\phi(u) > 0$ for $0 < u < a$

b) $\phi(0) = \phi(a) = 0$, $\phi'(0) = -\phi'(a) = 1$.

The Gaussian (scalar) curvature R is a function depending only on u, namely $R = -\frac{\phi''(u)}{\phi(u)}$ for $0 < u < a$, with R bounded at the end points. One shows that, under these conditions, R must have more than two critical points on the sphere. In fact, the function

$$\psi(u) = (\phi'(u))^2 + R(\phi(u))^2$$

has, by assumption, the same value (unity) for $u = 0$ or a, and hence $\psi'(u)$ must vanish at least at one point $u_0 (0 < u_0 < a)$. But

$$\begin{aligned}\psi'(u_0) &= 2\phi'(u_0)(\phi''(u_0) + R(u_0)\phi(u_0)) + R'(u_0)(\phi(u_0))^2 \\ &= R'(u_0)(\phi(u_0))^2\end{aligned}$$

and $\phi(u_0) > 0$, so that $R'(u_0) = 0$, verifying that R, as a function on the sphere, must have critical points not only at each of the two "poles," $u = 0$, $u = a$, of the sphere, but also at each point of some circle $\{u = u_0,\ 0 \leq v < 2\pi\}$. Thus the gradient field of R cannot generate a one-parameter group of conformal transformations of M. This completes the proof of the theorem.

An analogous statement for extremal Kähler metrics in higher dimensions is easy to prove in the case of Kähler manifolds of a general type (complex manifolds M with no holomorphic tangent vector fields) or in the case of products of a manifold of this general type with a complex torus, since each one-parameter group of holomorphic transformations in this case has no fixed point. However, in the case of manifolds with transformation groups admitting fixed points, there are considerable difficulties in proving the existence or any local compactness property of the space of Kähler metrics with $R_{,\overline{\alpha}\overline{\beta}} = 0$. For instance, it is not known, in the case of compact, higher dimensional, *homogeneous*, simply connected, complex manifolds admitting Kähler metrics, whether the equation $R_{,\overline{\alpha}\beta} = 0$ implies that R = constant.

We shall conclude this paper with a construction of compact, complex manifolds M each admitting a Kähler metric g satisfying $R_{,\overline{\alpha}\beta} = 0$ with R not constant; the idea of this construction is to assume M to be sufficiently "symmetric" to reduce the equation to an ordinary differential equation.

The manifolds M, with n complex dimensions, are in all cases complex projective line bundles of a certain type over a compact, symmetric, (n-1)-dimensional Kähler manifold; for the sake of simplicity we shall limit the description to bundles over the (n-1)-dimensional, complex projective (n-1)-space $P^{n-1}_{\mathbf{C}}$. The latter is then represented as the quotient of the homogeneous coordinate space $\mathbf{C}^n \setminus \{0\}$ (with numerical coordinates $(x_1, \cdots, x_n)$) by the action of the group of nonzero scalar multiplications. The base space $P^{n-1}_{\mathbf{C}}$ is covered by n coordinate domains $U_\lambda (1 \leq \lambda \leq n)$; each U_λ is characterized by the condition $x_\lambda \neq 0$ and covered by the (nonhomogeneous) holomorphic coordinate system $({}_\lambda z^\alpha)$ $(1 \leq \alpha \leq n;\ \alpha \neq \lambda)$, ranging over $\mathbf{C}^{n-1}$, where ${}_\lambda z^\alpha = \frac{x_\alpha}{x_\lambda}$ (we use this equation also, with $\alpha = \lambda$ so that, formally, ${}_\lambda z^\lambda = 1$).

We describe a bundle $M \xrightarrow{\pi} P^{n-1}_{\mathbf{C}}$ by introducing a projective, holomorphic fibre coordinate $y_\lambda \in \mathbf{C} \cup \{\infty\}$ in $\pi^{-1}(U_\lambda)$ with the transition relation in $\pi^{-1}(U_\lambda \cap U_\mu)$

$$y_\mu = \left(\frac{x_\mu}{x_\lambda}\right)^k y_\lambda = ({}_\lambda z^\mu)^k y_\lambda, \tag{3.1}$$

for any fixed, integral exponent k; without loss of generality we set $k \geq 0$ (otherwise one may replace each y_λ by y_λ^{-1} and k by $-k$). The resulting manifold will be denoted by $M_k (k = 0, 1, 2, \cdots)$. The manifold M_0 is the trivial bundle $P^{n-1}_{\mathbf{C}} \times P^1_{\mathbf{C}}$ and will be excluded from our consideration; for $k \geq 1$ the bundle $M_k \xrightarrow{\pi} P^{n-1}_{\mathbf{C}}$ has two distinguished cross sections $s_0, s_\infty : P_C^{n-1} \to M_k$, whose images in M_k, denoted by S_0 and S_∞, are described respectively by the equations $y_\lambda = 0$ and $y_\lambda = \infty$ for each λ. The complement $M' = M_k \setminus (S_0 \cup S_\infty)$ of these two cross sections can be globally parametrized by the homogeneous coordinate space $C^n \setminus \{0\}$ under the k-to-one map that assigns to any point $(x_1, \cdots, x_n)$ with, let us say, $x_\lambda \neq 0$, the point in $M' \cap \pi^{-1}(U_\lambda)$ with coordinates

$$((_\lambda z^a);\, y_\lambda) = \left(\left(\frac{x_a}{x_\lambda}\right);\, (x_\lambda)^k\right) \quad (1 \leq a \leq n;\ a \neq \lambda)\ . \tag{3.2}$$

We shall describe now the maximal connected group G of holomorphic transformations of M_k. This group is a complex Lie group [1, 2]; since it is connected, its action preserves the fibration $\pi : M_k \to P_C^{n-1}$, and hence has a homomorphism $\bar{\pi}$ into the complex projective group PGL(n, C) acting on P_C^{n-1}. On the other hand G contains a compact subgroup G_K, represented by the action of the unitary group U(n) on the complex parameters $(x_1, \cdots, x_n)$ occurring in (3.2), where the action of U(n) on M_k is effective, modulo the central subgroup Z_k consisting of scalar multiplications by the k'th roots of unity. Hence G contains the complexification of G_K, which is isomorphic to $GL(n, C)/Z_k$, and, since $\bar{\pi}$ obviously maps the latter group onto PGL(n, C), the homomorphism $\bar{\pi} : G \to PGL(n, C)$ is surjective.

The kernel $B \subset G$ of $\bar{\pi}$ is the subgroup of G mapping each fibre $\pi^{-1}(z) \subset M_k$ $(z \in P_C^{n-1})$ onto itself; each element can be described by extending (by continuity) its action in any coordinate domain $\pi^{-1}(U_\lambda)$, in terms of the coordinates $((_\lambda z^a);\, y_\lambda)$ of (3.2), mapping the point $(_\lambda z^a), y_\lambda)$ into the point $(_\lambda z^a), y'_\lambda)$ with

$$(y'_\lambda)^{-1} = c(y_\lambda)^{-1} + P_k(_\lambda z^a)\ ,$$

where c is any complex constant $\neq 0$ and P_k any polynomial of degree $\leq k$ in the n-1 variables $(_\lambda z^a)$ $(a \neq \lambda)$. Thus B is a complex, solvable

Lie group with $\binom{n+k-1}{k}+1$ complex dimensions, G is $\left(n^2+\binom{n+k-1}{n}\right)$-dimensional, has trivial center and has the n^2-(real)-dimensional subgroup G_K as its maximal compact subgroup. Therefore G is not reductive.

We shall now construct a Kähler metric in M_k satisfying the Euler equation $R_{,\bar{\alpha}\bar{\beta}} = 0$, by a method similar to one developed in [5] for other purposes, and where the construction is described in greater detail. We begin by assuming that such a metric is, in addition, invariant under the action of a maximal compact subgroup G_K ($G_K \cong U(n)/Z_k$) of transformations of M_k. It is unknown (but plausible to speculate) whether this restriction is necessary for the solution of the Euler equation in general for M_k. The calculations are carried out most conveniently in terms of the coordinate covering $(x_1, \cdots, x_n)$ of $M_k \setminus (S_0 \cup S_\infty)$, since the action of G_K is the easiest to describe in terms of these coordinates. The orbits under the action of G_K are parametrized by the real parameter $t = \log \sum_{\alpha=1}^{n} |x_\alpha|^2$; any Kähler metric $g_{\alpha\bar{\beta}}(x, \bar{x})$ in $\mathbb{C}^n \setminus \{0\}$ invariant under the action of U(n) can be generated by a Kähler potential $\Phi(x, \bar{x})$ $g_{\alpha\bar{\beta}} = \dfrac{\partial^2\Phi}{\partial x_\alpha \partial \bar{x}_\beta}$ such that

$$\Phi(x, \bar{x}) = u(t) \quad \left(t = \log\left(\sum_{\alpha=1}^{n} |x_\alpha|^2\right)\right)$$

for some function u(t) which is differential for all real t. Conversely, any function u(t) generates a positive definite Kähler metric in $\mathbb{C}^n \setminus \{0\}$ (obviously invariant under U(n)), and hence in $M' = M_k \ (S_0 \cup S_\infty)$, if and only if it satisfies the differential inequations

$$u'(t) > 0, \quad u''(t) > 0. \tag{3.3}$$

In fact, for any smooth function u(t), for $t = \log \sum_{\alpha=1}^{n} |x_\alpha|^2$,

(3.4) $$g_{\alpha\bar{\beta}} = \frac{\partial}{\partial x_\alpha \partial \bar{x}_\beta} u\left(\log \sum_\lambda |x_\lambda|^2\right) = e^{-t}u'(t)\delta_{\alpha\beta} + e^{-2t}\bar{x}_\alpha x_\beta(u''(t) - u'(t))$$

and

(3.5) $$\det(g_{\alpha\bar{\beta}}) = e^{-nt}(u'(t))^{n-1}u''(t) ,$$

so that (3.3) is necessary and sufficient in order that $g_{\alpha\bar{\beta}}$ be positive definite. The property that any Kähler metric (3.4) in $C^n \setminus \{0\}$, pulled into $M' = M_k \setminus (S_0 \cup S_\infty)$, be extendable by continuity to a positive definite, smooth metric in all of M_k can be translated into the following asymptotic properties of $u(t)$ as $t \to \pm\infty$:

a) There exists a real constant $a > 0$ such that the function $\bar{u}_0(r)$, defined for all $r > 0$ by the equation

$$u_0(e^{kt}) = u(t) - at ,$$

is extendable by continuity to a smooth function at $r = 0$, satisfying the additional condition

(3.6) $$u_0'(0) > 0 .$$

b) There exists a real constant $b > 0$ such that the function $\overline{u_\infty}(\bar{r})$, defined for all positive $\bar{r}$ by the equation

$$\overline{u_\infty}(e^{-kt}) = u(t) - bt ,$$

is extendable by continuity to a smooth function at $\bar{r} = 0$, satisfying the additional condition

(3.7) $$u_\infty'(0) > 0 .$$

It is easy to verify the necessity and sufficiency of the above conditions by calculating the metric in a neighborhood of S_0 and S_∞ respectively.

We note that the positive constants a, b describe the cohomology class of the resulting Kähler metric. In fact, the second homology group of M_k with real coefficients is generated by the two 2-cycles represented by two complex projective lines lying one in each of the two cross sections S_0 and S_∞; the restriction of the Kähler metric to each of these cross sections is a Fubini-Study metric with scalar curvature respectively $\frac{n(n-1)}{a}$ and $\frac{n(n-1)}{b}$. Therefore the integral of $\omega = \sqrt{-1}\, g_{\alpha\overline{\beta}}\, dz^\alpha \wedge d\overline{z^\beta}$ on each of these two projective lines is respectively $2\pi a$ and $2\pi b$. Furthermore $b > a$, since $a = \lim_{t\to-\infty} u'(t)$, $b = \lim_{t\to\infty} u'(t)$, and $u''(t) > 0$ for all real t.

For any function $u(t)$ satisfying (3.3) and the asymptotic conditions (3.6), (3.7) in terms of preassigned constants a, b $(0<a<b)$, define the function (cf. (3.5)).

$$v(t) = -\log\det(g_{\alpha\overline{\beta}}(x,\overline{x})) = nt - (n-1)\log u'(t) - \log u''(t)\,; \tag{3.8}$$

from this one calculates the Ricci tensor

$$R_{\alpha\overline{\beta}}(x,\overline{x}) = \frac{\partial^2}{\partial x_\alpha \partial\overline{x}_\beta}\, v\left(\log \sum_{\lambda=1}^{n} |x_\lambda|^2\right) = e^{-t} v'(t)\delta_{\alpha\beta} + e^{-2t}\,\overline{x}_\alpha x_\beta (v''(t) - v'(t)).$$

Next, one computes the contravariant components of the metric from (3.4) and (3.5),

$$g^{\alpha\overline{\beta}}(x,\overline{x}) = e^t (u'(r))^{-1}\,\delta^{\alpha\beta} + x_\alpha \overline{x}_\beta ((u''(t))^{-1} - (u'(t))^{-1})\,;$$

these are used in computing the scalar curvature $R = g^{\alpha\overline{\beta}} R_{\alpha\overline{\beta}}$; we get $R = w(t)$ where

$$w(t) = (n-1)\frac{v'(t)}{u'(t)} + \frac{v''(t)}{u''(t)}\,. \tag{3.9}$$

Finally the equation $R_{,\overline{\alpha}\overline{\beta}} = 0$, equivalent to the condition that the components of the tangent vector field $g^{\alpha\overline{\beta}} \frac{\partial R}{\partial \overline{x}_\beta} \frac{\partial}{\partial x_\alpha}$ be holomorphic, is expressed in terms of $w(t) = R$ as follows: since

$$g^{\alpha\overline{\beta}} \frac{\partial R}{\partial \overline{x}_\beta} = g^{\alpha\overline{\beta}}(x,\overline{x})e^{-t} x_\beta w'(t) = x_\alpha \frac{w'(t)}{u''(t)}$$

and since x_α is holomorphic and $\frac{w'(t)}{u''(t)}$ real valued, the Euler equation $R_{,\overline{\alpha}\overline{\beta}} = 0$ is equivalent to the equation

$$\frac{w'(t)}{u''(t)} = \text{constant};$$

we set the constant (for typographical convenience) to be $(n+1)(n+2)c_1$; the differential equation

$$w' = (n+1)(n+2)c_1 u''$$

is immediately integrable to

$$R = w'(t) = (n+1)(n+2)c_1 u'(t) + n(n+1)c_2 . \tag{3.10}$$

Replacing $w(t)$ by its expression in terms of $u(t)$, $v(t)$ and their derivatives according to (3.9), we obtain a second order differential relation between u and v that admits $(u'(t))^{n-1}u''(t)$ as an integrating factor. The new integration yields

$$(u'(t))^{n-1}v'(t) = (n+2)c_1(u'(2))^{n+1} + (n+1)c_2(u'(t))^n + c_3 . \tag{3.11}$$

We replace $v'(r)$ by the derivative of the equivalent expression of v in terms of $u(t)$ according to (3.8) and multiply both sides by the integrating factor $u''(t)$, obtaining the equation

$$n(u'(t))^{n-1}u''(t) - (n-1)(u'(t))^{n-2}(u''(t))^2 - (u'(t))^{n-1}u''(t)u'''(t) =$$

$$= \{(n+2)c_1(u'(t))^{n+1} + (n+1)c_2(u'(t))^n + c_3\} u''(t) .$$

This equation is integrable directly to the first order differential equation with respect to $\psi(t) = u'(t)$

$$(u'(t))^n - (u'(t))^{n-1}u''(t) = c_1(u'(t))^{n+2} + c_2(u'(t))^{n+1} + c_3u'(t) + c_4 ,$$

or

$$\frac{d\psi(t)}{dt} = -c_1\psi^{n+2} - c_2\psi^{n+1} + \psi^n - c_3\psi - c_4 .$$

The separation of the variables t and ψ gives

$$dt = \frac{\psi^{n-1}\,d\psi}{-c_1\psi^{n+2} - c_2\psi^{n+1} + \psi^n - c_3\psi - c_4} .$$

The integration of both sides establishes an implicit functional relation and $\psi(t) = u'(t)$. The solution is a function $u(t)$ satisfying the inequalities (3.3), if and only if the constants c_1, c_2, c_3, c_4 are such that the rational function of ψ,

$$f(\psi) = \frac{\psi^{n-1}}{-c_1\psi^{n+2} - c_2\psi^{n+1} + \psi^n - c_3\psi - c_4} , \tag{3.12}$$

is strictly positive valued in an interval I in the half-line $\psi > 0$. The completeness of the range of values of the "independent" variable t in the whole real line requires that the interval I in which ψ varies be bounded by two poles of the rational function (3.12) of ψ. Finally the asymptotic conditions of $u(t)$ as $t \to -\infty$ and as $t \to +\infty$, expressed respectively by (3.6) and (3.7) in terms of the positive constants a, b $(0<a<b)$, can be translated into the following properties of (3.12):
i) the rational function (3.12) has simple poles at $\psi = a$ and at $\psi = b$ with residues respectively equal to $\frac{1}{k}$ and $-\frac{1}{k}$; ii) the rational function of ψ,

$$\frac{\psi^{n-1}(b-\psi)(\psi-a)}{-c_1\psi^{n+2} - c_2\psi^{n+1} + \psi^n - c_3\psi - c_4}$$

is analytic and strictly positive valued in the closed interval $[a, b]$. These conditions correspond, for any given constants a, b $(0<a<b)$ to a system of four linear equations in c_1, c_2, c_3, c_4 with a unique set of solutions, which we include for the record:

(3.13)*

$$c_1 = \frac{(k-1)b^{2n}+n(n-k)a^{n-1}b^{n+1}-2(n^2-1)a^nb^n+n(n+k)a^{n+1}b^{n-1}-(k+1)a^{2n}}{b^{2n+2}-(n+1)^2a^nb^{n+2}+2n(n+2)a^{n+1}b^{n+1}-(n+1)^2z^{n+2}b^n+a^{2n+2}},$$

$$c_2 = \frac{(-k+2)b^{2n+1}-(n+1)(n-k)a^{n-1}b^{n+2}+(n+2)(n-k-1)a^nb^{n+1}+(n+2)(n+k-1)a^{n+1}b^n-(n+1)(n+k)a^{n+2}b^{n-1}+(k+2)a^{2n+1}}{b^{2n+2}-(n+1)^2a^nb^{n+2}+2n(n+2)a^{n+1}b^{n+1}-(n+1)^2a^{n+2}b^n+a^{2n+2}},$$

$$c_3 = \frac{(n-k)a^{n-1}b^{2n+2}+(n+1)(k-2)a^nb^{2n+1}-(n+2)(k-1)a^{n+1}b^{2n}+(n+2)(k+1)a^{2n}b^{n+1}-(n+1)(k+2)a^{2n+1}b^n+(n+k)a^{2n+2}b^{n-1}}{b^{2n+2}-(n+1)^2a^nb^{n+2}+2n(n+2)a^{n+1}b^{n+1}-(n+1)^2a^{n+2}b^n+a^{2n+2}},$$

$$c_4 = \frac{-(n-k-1)a^nb^{2n+2}-n(k-2)a^{n+1}b^{2n+1}+(n+1)(k-1)a^{n+2}b^{2n}-(n+1)(k+1)a^{2n}b^{n+2}+n(k+2)a^{2n+1}b^{n+1}-(n+k-1)a^{2n+2}b^n}{b^{2n+2}-(n+1)^2a^nb^{n+2}+2n(n+2)a^{n+1}b^{n+1}-(n+1)^2a^{n+2}b^n+a^{2n+2}}.$$

* In each of the following four fractions the numerators are divisible by $(b-a)^3$ and the denominators by $(b-a)^4$; the simplified form, however, is clumsy to express.

The verification of all the inequalities required of the rational function (3.12) of ψ,

$$f(\psi) = \frac{\psi^{n-1}}{-c_1\psi^{n+2}-c_2\psi^{n+1}+\psi^n-c_3\psi-c_4},$$

is an exceedingly tedious task to carry out in its full generality. However, in the case of surfaces $(n=2)$, i.e. when the base space is a projective line, the rational function reduces to

$$\frac{\psi}{-c_1\psi^4-c_2\psi^3+\psi^2-c_3\psi-c_4} =$$

$$= \frac{b-a}{k(\psi-a)(b-\psi)} + \frac{(b-a)((k-1)b+(k+1)a)}{k((k-1)b+(k+1)a)(\psi^2+ab)+k(b^2+2kab-a^2)\psi},$$

which trivially satisfies the inequalities for all choices of integers $k \geq 1$ and $0 < a < b$. While the solution of the equation (3.10) to determine $u(r)$ in closed form is impossible (except perhaps for some special values of a, b, k, none of which have been found), nevertheless one can calculate some of the properties of the resulting manifolds. In particular we are interested in the total volume V of M_k under the metric g determined by the cohomology class represented by the pair (a,b) $(0<a<b)$, and in the integrals

$$\bar{R} = \int_{M_k} R\,dV(g), \quad S(g) = \int_{M_k} R^2\,dV(g).$$

Let σ_n denote the volume of the unit Euclidean ball in $2n$ (real) dimensions, $\sigma_n = \pi^n(n!)^{-1}$; then we refer to (3.5) in expressing the volume element induced by the metric in $C^n\setminus\{0\}$,

$$dV_0 = \det(g_{\lambda\bar{\mu}}(x,\bar{x})) \bigwedge_{\alpha=1}^{n} (\sqrt{-1}\,dx_\alpha \vee d\overline{x_\alpha}) = 2^n n\sigma_n(u'(t))^{n-1}u''(t)\,dt.$$

Recalling that the parametrization map of $C^n\setminus\{0\}$ onto $M_k\ (S_0 \cup S_\infty)$ is k-to-one, we obtain the volume formula for M_k,

$$(3.14)\qquad V = 2^n k^{-1}\sigma_n \int_{-\infty}^{\infty} n(u'(t))^{n-1}u''(t)\,dt = 2^n k^{-1}\sigma_n(b^n - a^n)\,.$$

The scalar curvature R is expressed conveniently by (3.9), so that

$$(3.15)\quad RdV = 2^n k^{-1} n\sigma_n((n-1)v'(t)(u'(t))^{n-2}d(u'(t)) + (u'(t))^{n-1}d(v'(t)) =$$
$$= 2^n k^{-1} n\sigma_n d\{(u'(t))^{n-1}v'(t)\}\,.$$

The integral of the above expression for $-\infty < t < \infty$ is obtainable from (3.11) and (3.13). Thus

$$(3.16)\quad \bar{R} = \int_{M_k} RdV = 2^n k^{-1} n\sigma_n\{(n+2)c_1(b^{n+1}-a^{n+1}) + (n+1)c_2(b^n - a^n)\} =$$
$$= 2^n k^{-1} n\sigma_n((n+k)b^{n-1} - (n-k)a^{n-1})\,.$$

Similarly we can calculate the stationary value of $S(g) = \int_{M_k} R^2 dV(g)$ from the metric just constructed as follows. We use the differential form of (3.14) to express the element of volume $dV(g)$ and (3.10) to express R; this yields

$$(3.17)\quad \int R^2 dV = 2^n \tfrac{n}{k}\sigma_n \int_{-\infty}^{\infty} ((n+1)(n+2)c_1 u'(t) + n(n+1)c_2)^2 (u'(t))^{n-1} d(u'(t))$$
$$= \frac{2^n n(n+1)}{k}\sigma_n\{(n+1)(n+2)c_1^2(b^{n+2}-a^{n+2}) + 2n(n+2)c_1c_2(b^{n+1}-a^{n+1}) + n(n+1)c_2^2(b^n - a^n$$

and substitute for the constants c_1, c_2 the values in terms of n, k, a, b given by (3.13).

It is very likely that this last value represents the absolute minimum value of $\int R^2 dV(g)$ for any metric Kähler metric (g) in M_k in the cohomology class represented by (a, b).

4. *Conclusion*

The problem of minimizing $\int_M R^2 dV(g)$ for all Kähler metrics of any fixed cohomology class $[\omega]$ in a given, compact complex manifold M can be viewed as a generalization of the uniformization problem from closed Riemann surfaces to higher dimensions. We believe that several interesting geometrical consequences may be derived, both from its solution (assuming that there is one), as well as from the conceivable methods of obtaining it. Concerning the existence problem of extremal Kähler metrics in our sense in general, the following observations can be made.

1. If the first Chern class of M is negative (i.e. if the canonical bundle is ample) Yau has proved [10] that the extremal metric exists, and it has necessarily constant (negative) scalar curvature, at least when the principal class is chosen to be proportional to the first Chern class.

2. More generally, for closed Kählerian manifolds admitting no holomorphic tangent vector fields other than zero, the parabolic equation with respect to a real parameter $t \geq 0$

$$\frac{\partial g_{\alpha\bar{\beta}}(t; z, \bar{z})}{\partial t} = \frac{\partial^2}{\partial z^\alpha \partial \bar{z}^\beta} R(g(t))$$

with $g_{\alpha\bar{\beta}}(0; z, \bar{z})$ equal to any given Kähler metric is one whose solutions decrease $\int R^2 dV(g)$ as a function of t. In estimating whether the solution exists and converges, as $t \to \infty$, to the one with R = constant, it is important to keep control of the lowest eigenvalue $\lambda(t)$ of the self-adjoint positive definite, fourth order elliptic operator $\square$ of Lichnérowicz, defined, for each Kähler manifold, by

$$\square u = \Delta^2 u - R^{\alpha\bar{\beta}} u_{,\alpha\bar{\beta}} = u_{,\alpha\beta}{}^{\alpha\beta} - g^{\alpha\bar{\beta}} u_{,\alpha} R_{,\bar{\beta}}$$

$$= u^{,\alpha\beta}{}_{\alpha\beta} - g^{\alpha\bar{\beta}} R_{,\alpha} u_{,\bar{\beta}} .$$

The crucial estimate which is still lacking is one showing that, for some suitable Sobolev norm $\| \ \|_g$ defined in terms of any Kähler metric (g), the space of all Kähler metrics (g) in a fixed cohomology class, and for which $\|R(g)\|_g$ is bounded by any given positive constant, is compact with respect to a similar norm on the metrics (g) themselves.

3. For compact Kähler manifolds M admitting holomorphic tangent vector fields $\neq 0$, the analogous problem seems much harder. In this case the space of all Kähler metrics (in a fixed cohomology class) that are equivalent to any given one under the maximal connected, complex Lie group G of all holomorphic automorphisms of M, is not compact; however it is locally compact in view of [1, 2]. In this case one needs to devise a transformation semigroup on the function space $\mathfrak{g}$ of all admissible Kähler metrics, decreasing the functional $S(g) = \int_M R^2(g)\, dV(g)$, and with relatively compact trajectories, and hopefully converging to metrics (g) for which S(g) is stationary.

BIBLIOGRAPHY AND REFERENCES

[I] Chern, S. S., *Complex Manifolds*, Universitade do Recife, Recife (Braz.), *Textos de Matematica*, N. 5, 1959.

[II] ———, *Complex Manifolds without Potential Theory*, Princeton, N. J., Van Nostrand, 1967.

[III] Flaherty, E. J., *Hermitian and Kählerian Geometry in Relativity*, Berlin-New York, Springer-Verlag, *Lecture Notes in Physics*, N. 46, (1976).

[IV] Lichnérowicz, A., *Géométrie des groupes de transformations*, Paris, Dunod, 1958.

[V] Weil, A., *Introduction à l'étude des variétés kählériennes*, Paris, Hermann, *Actualités Scientifiques et Industrielles*, N. 1267, 1971.

[VI] Wells, R. O., *Differential Analysis on Complex Manifolds*, Englewood Cliffs, N. J., Prentice Hall, 1973.

[VII] Yano, K. and Bochner, S., Curvature and Betti Numbers, Princeton, N. J., Princeton University Press, *Annals of Mathematics Studies*, No. 32, 1953.

[VIII] Séminaire Palaiseau, *Première classe de Chern et courbure de Ricci*, Soc. Math. de France, Astérisque, *58*, 1978.

[1] Bochner, S. and Montgomery, D., *Locally Compact Groups of Differentiable Transformations*, Ann. of Math., 47(1946), 639-653.

[2] ———, *Groups on Analytic Manifolds*, Ann. of Math., *48*(1947), 659-669.

[3] Calabi, E., *The Variation of Kähler Metrics*, II *A Minimum Problem*, Amer. Math. Soc., Bull., *60*(1954) Abstract N. 294, p. 168.

[4] ———, *The Space of Kähler Metrics*, Proc. Intern. Congress of Math., Amsterdam, 1954, vol. II, 206-207.

[5] ———, *Métriques kählériennes et fibrés holomorphes*, Ann. Sci. Ec. Norm. Sup. Paris, 4me série, *12*(1979), 269-294.

[6] Kazdan, J.L. and Warner, F.W., *Surfaces of Revolution with Monotonic Increasing Curvature ...*, Amer. Math. Soc., Proc., *32* (1972), 139-141.

[7] Lichnérowicz, A., *Sur les transformations analytiques des variéties kählériennes*, C.R. Acad. Sci. Paris, *244*(1957), 3011-3014.

[8] Matsushima, Y., *Sur la structure du groupe d'homéomorphismes analytiques d'une certaine variété kählérienne*, Nag. Math. J., *11* (1957), 145-150.

[9] Myers, S.B. and Steenrod, N.E., *The Group of Isometries of a Riemannian Manifold*, Ann. Math., *40*(1939), 400-416.

[10] Yau, S.T., *On the Ricci Curvature of a Compact Kähler Manifold and the Complex Monge-Ampère Equation*, Comm. P. & Appl. Math., *31* (1978), 339-411.

Hypersurfaces with Maximal Affinely Invariant Area

Eugenio Calabi*

Amer. J. Math. 104, No. 1, 91–126 (1982)

1. Introduction. Let $X:\Sigma \to E$ be a smooth immersion of an n-dimensional, connected, differential manifold Σ into an $(n+1)$-dimensional, real affine space ($n \geq 1$). We shall deal here with properties of the resulting hypersurface $X(\Sigma)$ (parametrized by Σ) that are invariant under the group $ASL(E)$ of all affine transformations of E that are unimodular, i.e. preserve volume and orientation. Since the subject of affine differential geometry is not as widely studied as it deserves to be, we shall give a preliminary exposition here of some of its basic invariants, needed in the next section; for a more detailed treatment, we refer to the items [2], [6], [7], [8] of the bibliography.

We limit our consideration here to those hypersurfaces in E that are locally strongly convex. We introduce, temporarily, a euclidean structure in E, defined by a positive definite quadratic form F on the tangent space $T(E)$ of E ($T(E)$ is, more exactly, the fibre in the tangent bundle of E and, at the same time, the group of global translations of E); thus we may calculate from the better known, classical invariants of $X(\Sigma)$ under the euclidean group $ASO(E, F)$ (group of translations and proper rotations) some of the more elementary invariants under the larger group $ASL(E)$. Since the hypersurface $X(\Sigma)$ is locally strongly convex, its euclidean unit normal N can be oriented so as to be on the concave side of $X(\Sigma)$ and the resulting second fundamental form $\mathrm{II} = \langle N, d^2X\rangle = -\langle dN, dX\rangle$ is positive definite, where $\langle\ ,\ \rangle$ denotes the symmetric bilinear pairing derived from the quadratic form F. Locally, $X(\Sigma)$ can be described in the so-called non-parametric form; for any point $p \in \Sigma$ there exists a neighborhood U of p and a cartesian coordinate system in E (rectangular with respect to F), $(x_0, x_1, \ldots, x_n)$, so that $X(U)$ is the graph of a function $x_0 = z(x_1, x_2, \ldots, x_n) = z(x)$, that is smooth and strongly convex in a corresponding domain $D \subset \mathbf{R}^n$, meaning that the Hessian matrix (z_{ij}),

---

Manuscript received June 17, 1980.
*Supported in part by N.S.F. Grant MCF 78-02285.

American Journal of Mathematics, Vol. 104, No. 1, pp. 091-126
0002-9327/82/1041-0091 $01.50

DOI: 10.2307/2374069

$$z_{ij} = \frac{\partial^2 z(x)}{\partial x_i \partial x_j} \quad (1 \le i, j \le n),$$

is positive definite everywhere in D.

We recall how one expresses certain euclidean invariants of $X(U)$ in terms of the non-parametric representation:

a) the element of euclidean area of $X(U)$ is given by the density

$$(1.1) \quad d\bar{A} = \mu(x)\, |dx^1 \wedge \cdots \wedge dx^n|, \qquad \mu(x) = \left(1 + \sum_{i=1}^{n} \left(\frac{\partial z(x)}{\partial x_i}\right)^2\right)^{1/2};$$

b) the unit normal N, the second fundamental form II and the Gaussian curvature K (product of the principal curvatures), are expressed respectively by

$$N = (\mu(x))^{-1} \begin{pmatrix} 1 \\ -z_1 \\ \cdot \\ \cdot \\ \cdot \\ -z_n \end{pmatrix}, \qquad \left(z_i = \frac{\partial z(x)}{\partial x_i}\right),$$

$$\mathrm{II} = (\mu(x))^{-1} \sum_{i,j=1}^{n} z_{ij}(x)\, dx^i dx^j,$$

$$K = (\mu(x))^{-n-2} \det(z_{ij}(x)).$$

From the four euclidean invariants just displayed, we construct some others, in which the function $\mu(x)$ no longer appears explicitly, as follows:

i) The one-dimensional vector subspace of $T(E)$ generated, for each $p \in U$, by $N(p)$ can be regarded, under the identification of N with the linear functional $\langle N, \ \rangle$ on $T(E)$, as the set of all linear functionals on $T(E)$ that annihilate the tangent space $dX(p)$ of X at p; this set has thus an affinely invariant meaning, if interpreted as a one-dimensional vector subspace of the dual vector space $T^*(E)$ of $T(E)$, where $T^*(E)$ in the affine geometry, is a completely distinct vector space from $T(E)$. We choose as a generator of this one-dimensional subspace, instead of $\langle N, \ \rangle$, the vector in $Y^* \in T^*(E)$, where

$$(1.2) \qquad Y^* = K^{\left(\frac{-1}{n+2}\right)} \langle N, \ \rangle.$$

ii) We define the following, positive definite quadratic form on Σ, conformally equivalent to II and denoted by $G = g_{ij}(u)du^i du^j$ (with the summation implicitly understood over repeated indices) in terms of any local system of parameters $(u) = (u^1, \ldots, u^n)$ in Σ, including the system $x = (x^1, \ldots, x^n)$ in case of a non-parametric representation, where

$$G = g_{ij}(u)du^i du^j = K^{\left(\frac{-1}{n+2}\right)} \text{II}; \tag{1.3}$$

iii) using G formally as a Riemannian structure, we introduce its corresponding element of area dA (not to be confused with the euclidean area element (1.1) $d\bar{A}$),

$$dA = (\det(g_{ij}(u)))^{1/2} |du^1 \wedge \cdots \wedge du^n| = K^{\frac{1}{n+2}} d\bar{A}. \tag{1.4}$$

Proposition 1.1. *The linear functional $Y^*(p)$ on $T(E)$ defined for each $p \in \Sigma$ by* (1.2), *the positive definite quadratic form G defined by* (1.3) *and the positive area density dA defined by* (1.4) *are invariant under the affine group $ASL(E)$ of unimodular affine transformations of E.*

Proof. It is clear from the definitions that Y^*, G and dA are invariant under the subgroup $ASO(E, F)$ of euclidean motions in E relative to the euclidean norm F; in order to show their invariance under the larger group $ASL(E)$, we express each of those geometric entities, locally, in terms of a non-parametric representation of X. We obtain, respectively,

$$Y^*(x) = (h(x))^{-1} \begin{pmatrix} 1 \\ -z_1(x) \\ \cdot \\ \cdot \\ \cdot \\ -z_n(x) \end{pmatrix}, \tag{1.5}$$

where

$$h(x) = (\det(z_{ij}(x)))^{\frac{1}{n+2}},$$

$$g_{ij}(x) = (h(x))^{-1} \frac{\partial^2 z(x)}{\partial x^i \partial x^j}, \tag{1.6}$$

(1.7) $$dA(x) = h(x)\,|dx^1 \wedge \cdots \wedge dx^n|.$$

Looking at the three expressions above, one easily verifies that the three geometric entities that they represent are also invariant, in addition, under the connected, one-parameter transformation group M of E mapping the point of E with coordinates $(x^0, x^1, \ldots, x^n)$ onto the one with coordinates $(t^{-n}x^0, tx^1, tx^2, \ldots, tx^n)$ for any $t > 0$. In addition, they are also invariant under the subgroup $\Pi \subset ASL(E)$, isomorphic to $ASL(\mathbf{R}^n)$, that leaves invariant the coordinate function x^0 and the direction of the x^0-axis. Since the direction of the x^0-axis and of the hyperplanes $\{x^0 = \text{constant}\}$ are, loosely speaking, arbitrary, the group under which Y^*, G and dA are invariant, includes the subgroup of $ASL(E)$ generated by $ASO(E)$, M, and Π. It is clear that this group is the whole group $ASL(E)$ (actually $ASO(E, F)$ and M alone suffice to generate $ASL(E)$). This completes the proof of the proposition.

We call the vector $Y^*(p) \in T^*(E)$ associated to each point $p \in \Sigma$ the *affine co-normal* of X at p. The positive definite quadratic form G defines in Σ a Riemannian structure, induced by X and invariant under transformations of X by $ASL(E)$, the *Berwald-Blaschke* metric of X or first (affine) fundamental form induced by X. The integral of its associated element of n-dimensional area dA over any compact region $D \subset \Sigma$ defines the so-called *affinely invariant* area of D.

The purpose of this paper is to study the variational problems associated with extremal values of $\int dA$ under interior deformations of X preserving the property of local strong convexity. Since the integrand dA depends on derivatives of X of up to second order, the allowable deformations of X, heuristically, seem to be the following. Let Σ be an n-dimensional, compact manifold with boundary $\partial\Sigma$. Given a locally strongly convex immersion $X_0 : \Sigma \to E$, the allowable deformations of X are locally strongly convex immersions X_1 with the following three properties:

1) the restriction of X_1 to $\partial\Sigma$ coincides with that of X_0;

2) for each $p \in \partial\Sigma$, the tangent hyperplane $dX_1(p)$ to $X_1(\Sigma)$ at p coincides with $dX_0(p)$;

3) there exists a smooth, one-parameter family $X_t (0 \leq t \leq 1)$ of locally strongly convex immersions of Σ in E, satisfying properties 1) and 2) for each t, and coinciding with X_0 and X_1 for $t = 0, 1$ respectively.

In the case ($n = 1$) of plane curves, the variational problem has the following complete solution. The data provided by an initially given, locally strongly convex curve $X_0:\Sigma \to E$ (Σ is a closed segment) are divided in two cases: either there exists at most one tangent line to $X(\Sigma)$ parallel to any given line or there exist two distinct points $p, q \in \Sigma$ (possibly boundary points) such that $dX(p)$ and $dX(q)$ are parallel. The former case is characterized by the property that $X_0(\Sigma)$ is a simple, convex arc, contained in the triangle ABC, where A and B are the images of the end points of Σ in E and AC, BC the corresponding tangent lines. In this case, any allowable deformation of X_0 yields a convex curve $X(\Sigma)$, similarly inscribed in the triangle ABC. The affinely invariant arc length $\int dA$ of $X(\Sigma)$, on the one hand, can be arbitrarily small, and, on the other hand, achieves its maximum value (equal to twice the cube root of the area of the triangle ABC) exclusively in the case of the arc of the parabola passing through the points A, B with tangents respectively AC and BC [2, pp. 37–40]. In the second case, the allowable curves can have affine arc length either arbitrarily large or arbitrarily small, and in no case stationary.

In the general case ($n \geq 2$), it is not hard to verify that one can deform the map X in the interior of Σ, so as to make the resulting affinely invariant area arbitrarily small. The variational problem of determining X so that its affinely invariant area is stationary under infinitesimal deformations leads to an Euler-Lagrange equation of fourth order, interpreted (in analogy with the euclidean case) as the vanishing of a real valued invariant of X called the *affine mean curvature* and denoted by H. In general, one should expect hypersurfaces with vanishing affine mean curvature to achieve maximum affinely invariant area under interior deformations. This fact is verified at least in the following important case, proved in Section 2.

THEOREM 1.2. *Let $X:\Sigma \to E$ be a locally strongly convex hypersurface immersed in E and satisfying the Euler-Lagrange equation for stationary affinely invariant area under interior, infinitesimal deformations. Assume, in addition, that X is locally a graph, meaning that there exists a straight line (viz. the x^0-axis of a suitably chosen affine coordinate system $(x^0, x^1, \ldots, x^n)$ in E) whose parallel passing through $X(p)$ for each $p \in \Sigma$ is transversal there to X. Then X has strictly larger affinely invariant area than any other allowable hypersurface in E obtained by an interior deformation of X.*

In view of the above result, we suggest naming the locally strongly convex hypersurfaces in E with vanishing affine mean curvature *affine maximal hypersurfaces*, this denomination being more descriptive than that of "Affinminimalflächen" used by Blaschke [2] and others (in the case $n = 2$). The question arises now whether such hypersurfaces $X(\Sigma)$ actually achieve maximum affinely invariant area under all interior deformations (or at least small ones), even if there is no straight line in E all of whose parallels are transversal to it $X(\Sigma)$. For $n = 1$ this question does not arise, since the only "affine geodesic plane curves," or curves with vanishing affine (mean) curvature are parabolas, while for $n \geq 2$ one can exhibit some (e.g. affine maximal hypersurfaces that are axially symmetric), by solving an appropriate, ordinary differential equation, that are not graphs in any direction. After developing the complete system of invariants for hypersurfaces (Section 3), we devote Section 4 to a calculation of the second variation of affinely invariant area of affine maximal hypersurfaces under small interior deformation and prove the following result.

THEOREM 1.3. *If $X:\Sigma \to E$ is any affine maximal surface in affine 3-space E ($n = 2$), then the second variation of the affinely invariant area under all interior deformations of X is negative definite.*

Until now it has not been possible to generalize the above statement to n-dimensional hypersurfaces for $n \geq 3$, without assuming that they are graphs; similarly, we have not succeeded in showing that, for $n = 2$, the Euler-Lagrange equation implies absolute maximality of affine area, as it does in the case of local graphs, although one could hardly conceive the possibility of a counterexample.

The most elementary solutions of the Euler-Lagrange equation for stationary affinely invariant area are the elliptic (i.e. convex) paraboloids, each of which is described, in suitable unimodular affine coordinates, as the graph of the function

$$x^0 = \frac{1}{2}\sum_{i=1}^{n} (x^i)^2; \tag{1.8}$$

a more general class is made up of the so-called (locally convex) *improper affine hyperspheres* [2, pp. 209, 216], which are also graphs, locally, of solutions of the second order partial differential equation

$$x^0 = z(x^1, \ldots, x^n) = z(x); \qquad \det\left(\frac{\partial^2 z}{\partial x^i \partial x^j}\right) = 1. \tag{1.9}$$

The latter class of affine maximal hypersurfaces has the strong Bernstein property, in the sense that the only solutions of (1.9), which either (a) are complete with respect to the Berwald-Blaschke metric (1.6), or (b) extend, as functions $z(x)$ satisfying (1.9), to the whole n-space $\mathbf{R}^n$, reduce to paraboloids ([4] for $n = 2$, [3], [5] for $n \geq 3$). In the general case of affine maximal hypersurfaces, one naturally is led to posing the Bernstein problem of determining suitable asymptotic boundary conditions (such as completeness) under which the only affine maximal hypersurfaces are improper affine hyperspheres, and hence paraboloids. A modest result in this direction is proved in Section 5, and stated here.

THEOREM 1.4. *Let $X:\Sigma \to E$ be an affine maximal surface ($n = 2$) in 3-space. If, in addition, X is both* geometrically complete (*i.e. a properly imbedded, globally convex surface without finite boundary points*) *and* metrically complete (*with respect to the Berwald-Blaschke metric*), *then it is a paraboloid.*

Again, it has not been possible to establish either an analogous result for $n \geq 3$ or, in the case of affine maximal surfaces, whether either type of completeness implies the other. Theorems of the type of the Bernstein result above are just one of the steps that can be used in solving the problem of existence of an affine maximal hypersurface with assigned boundary data. Another problem in this area that seems still beyond our grasp is that of finding adequate, and suitably general, conditions for the boundary data to be compatible with a solution of the variational problem of maximizing affinely invariant area of a locally strongly hypersurface satisfying the assigned boundary conditions.

2. Proof of Theorem 1.2. We consider, at first, an n-dimensional, arbitrary, locally strongly convex, immersed hypersurface $X:\Sigma \to E$, where the parameter manifold Σ may be compact or not, and have a smooth boundary $\partial\Sigma$, possibly empty.

Definition 2.1. An allowable interior deformation of X is a differentiable map $\mathcal{X}:I \times \Sigma \to E$, where I is an open interval $(-\epsilon', \epsilon)$ ($\epsilon, \epsilon' > 0$) with the following properties:

i) For each $t \in I$ the map $X_t:\Sigma \to E$ defined by $X_t(p) = \mathcal{X}(t, p)$ is a locally strongly convex immersion and for $t = 0$ X_0 coincides with X.

ii) There exists a compact subdomain $\Sigma' \subset \Sigma$ (closure of a connected, open subset of Σ) with smooth boundary $\partial\Sigma'$, where $\partial\Sigma'$ may contain, meet, or be disjunct from $\partial\Sigma$, such that, for each $p \in \Sigma\backslash\Sigma'$ and all $t \in I$, $\mathcal{X}(t, p) = X(p)$.

iii) For each $p \in \partial\Sigma'$ and for all $t \in I$, $\mathcal{X}(t, p) = X(p)$ and the tangent hyperplane $dX_t(p)$ coincides with $dX(p)$ (this condition is essential only for $p \in \partial\Sigma' \cap \partial\Sigma$, and is contained in ii) otherwise). A locally strongly convex immersion $X':\Sigma \to E$ is said to be interior-homotopic to X, if there exists an allowable interior deformation $\mathfrak{X}:I \times \Sigma \to E$ with $I = (-\epsilon', 1 + \epsilon)$ ($\epsilon, \epsilon' > 0$) such that $X_0 = X$ and $X_1 = X'$.

In the sequel, when we study variations of geometric invariants of X under interior deformation, we may replace Σ, without loss of generality, by the compact subdomain $\Sigma' \subset \Sigma$, or else assume from the beginning that Σ is compact and with smooth boundary. It is worthwhile to remark that two locally strongly convex immersions $X, X':\Sigma \to E$ that have same values and tangent hyperplanes on $\partial\Sigma$ (and outside of a compact set) need not be interior-homotopic, even when $n \geq 2$.

A locally strongly convex hypersurface $X:\Sigma \to E$ is said to be locally a graph in a given direction A, if, for each $p \in \Sigma$, the line parallel to A through $X(p)$ is transversal to the tangent hyperplane $dX(p)$ at p. Let $(x^0, \ldots, x^n)$ be an allowable (i.e. unimodular), affine coordinate system in E such that the x^0-axis is parallel to A. Then, if X is locally a graph in the direction A, the map $\pi:\Sigma \to \mathbf{R}^n$ defined by $\pi(p) = (x^1, \ldots, x^n)(X(p))$ (omitting the coordinate x^0) is an open immersion. By the implicit function theorem, for each point $p \in \Sigma$ there exists a connected neighborhood U of p in Σ, such that $\pi(U)$ is a local coordinate system for U. Denote by D_U the image $\pi(U) \subset \mathbf{R}^n$; then $X(U)$ is the graph of a single valued function

$$x^0 = z(x^1, \ldots, x^n) = z(x) \quad ((x) \in D_U)$$

which may be convex or concave; in the latter case, replacing the coordinate system $(x^0, x^1, \ldots, x^n)$ by $(-x^0, -x^1, x^2, \ldots, x^n)$, makes the modified function $z(x)$ graphically represented by $X(\pi^{-1}(D_U))$ strongly convex, and, since Σ is connected, the orientation of the direction A making the local functions $z(x)$ convex is unique. Globally, the function $z(x)$ need not be single valued in $\mathbf{R}^n$, but Σ can be regarded as a draped domain (domaine étalé) over $\mathbf{R}^n$, so that we will often treat $X(\Sigma)$ as the graph of a multiple-valued, locally convex function in the draped domain $D = \pi(\Sigma)$ over $\mathbf{R}^n$.

Let Σ be now a compact, n-dimensional manifold with smooth boundary, draped by an immersion map $\pi:\Sigma \to \mathbf{R}^n$ onto a domain (singular, positive n-chain) $D = \pi(\Sigma)$ with boundary $\partial D = \pi(\partial\Sigma)$, such that $X(\Sigma)$ is the graph of a strongly convex, multiple valued function over D,

$$x_0 = z(x).$$

We have shown in (1.4), (1.7) how the element of affinely invariant area of $X(\Sigma)$ is expressed in terms of the non-parametric representation $x^0 = z(x)$. Thus the total area of $X(\Sigma)$ is given by

$$A(X) = A = \int_D h(x)\,|dx|,$$

where

$$h(x) = \left(\det \frac{\partial^2 z}{\partial x^i \partial x^j}\right)^{\frac{1}{n+2}}, \quad \text{and} \quad |dx| = |dx^1 \wedge \cdots \wedge dx^n|.$$

If $\mathfrak{X}: I \times \Sigma \to E$ is an interior deformation of X, then, for each $t \in I$, the mapping $\pi_t: \Sigma \to \mathbf{R}^n$ assigning to each $p \in \Sigma$ the $(x^1, \ldots, x^n)$-coordinates of $X_t(x) = \mathfrak{X}(t, p)$ is an open immersion of Σ in the fixed draped domain $\pi(\Sigma) = \pi_0(\Sigma) = D$. Thus the deformation can be represented "non-parametrically" as a family of functions $z_t(x^1, \ldots, x^n)$ $(t \in I)$ in a common draped domain D, with values and gradients independent of t on the boundary $\partial D = \pi(\partial\Sigma)$. Conversely, any locally strongly convex immersion $X': \Sigma \to E$ is interior-homotopic to X, if and only if X' is locally a graph in the same direction as X and has a non-parametric representation as the graph of a locally convex function $x^0 = z'(x)$ over the same, compact, draped domain $D = \pi(\Sigma) \subset \mathbf{R}^n$ as X, and $z'(x)$ takes the same values and gradient as $z(x)$ at each boundary point $x \in \partial D$. In this case, the deformation from $z(x)$ to $z'(x)$ can be chosen in the simplest way by the linear interpolation

$$x^0 = z_t(x) = z(x) + t(z'(x) - z(x)).$$

This fact shows that the whole function space of allowable, locally strongly convex hypersurfaces that are interior-homotopic to X is explicitly contractible by linear interpolation to X. The variation space can be represented in the function space $\mathcal{V}$ of all real valued functions $v(x)$ of class C^k $(k \geq 2)$ in the draped domain D that vanish with their gradient on the boundary; for any such function v there is a maximal open interval $I \subset \mathbf{R}$ containing the point 0 and with end points depending continuously on v in the uniform C^2-topology such that, for every $t \in I$, the function

$$z_t(x) = z(x) + tv(x)$$

is locally strongly convex everywhere in D. In order to compare the area of the hypersurface X_t represented as the graph of z_t with that of X, we introduce the positive, definite matrix function $(K^{ij}(x))$ in D defined as the inverse matrix $(z_{ij}(x))$, $z_{ij} = \partial^2 z/\partial x^i \partial x^j$. Let $(\lambda_1(x), \ldots, \lambda_n(x))$, for each point $x \in D$, denote the eigenvalues of the matrix $(V^i_{\ j}(x))$, where

$$V^i_{\ j}(x) = K^{ik}(x) \frac{\partial^2 v}{\partial x^j \partial x^k};$$

for each t in the interval I of all real numbers such that $z_t = z + tv$ is locally strongly convex, the numbers

$$1 + t\lambda_\nu(x) \quad (\nu = 1, \ldots, n)$$

are the solutions of the algebraic equation in μ

$$\det\left(\frac{\partial^2 z_t(x)}{\partial x^i \partial x^j} - \mu \frac{\partial^2 z(x)}{\partial x^i \partial x^j}\right) = 0,$$

and hence have to be positive.

The element of area dA_t of X_t is expressed, in terms of $(x^1, \ldots, x^n)$, by

$$\text{(2.1)} \quad dA_t = \left(\det \frac{\partial^2 z_t}{\partial x^i \partial x^j}\right)^{\frac{1}{n+2}} |dx| = \prod_{\nu=1}^{n} (1 + t\lambda_\nu(x))^{\frac{1}{n+2}} dA;$$

we regard the product in the last expression, evaluated at each x, as the geometric mean of $n + 2$ positive numbers, two of which are identically unity, and the other n are the quantities $1 + t\lambda_\nu(x)$. A well-known inequality states that the geometric mean of any finite family of positive numbers is dominated by their arithmetic mean, and equal to it, if and only if all the numbers of the family coincide. Thus

(2.2)

$$dA_t(x) \leq \frac{1}{n+2}\left(2 + \sum_{\nu=1}^{n} (1 + t\lambda_\nu(x))\right) dA = dA + \frac{t\,dA}{n+2}\left(\sum_{\nu=1}^{n} \lambda_\nu(x)\right),$$

equality holding, if and only if $t = 0$ or $\lambda_1(x) = \cdots = \lambda_n(x) = 0$. The right-hand member of the above inequality can be interpreted as follows: differentiating the terms of (2.1) with respect to t at $t = 0$, we have

$$\frac{\partial dA_t(x)}{\partial x}\bigg|_{t=0} = \frac{dA}{n+2}\left(\sum_{\nu=1}^{n} \lambda_\nu(x)\right); \tag{2.3}$$

thus (2.2) reduces to

$$dA_t(x) \leq dA(x) + t\frac{\partial(dA_t)(x)}{\partial t}\bigg|_{t=0}.$$

Now assume that X satisfies the Euler-Lagrange equation for stationary affinely invariant area. Integrating the terms of the above inequality over D, we obtain the inequality

$$A_t \leq A_0 + t\cdot\frac{dA_t}{dt}\bigg|_{t=0};$$

but the Euler equation means precisely that $dA_t/dt\,|_{t=0}$ vanishes for all allowable variation functions $\nu(x)$, so that the affinely invariant area $A = A_0$ of $X(\Sigma)$ is indeed maximal among all locally strongly convex hypersurfaces X' that are interior-homotopic to X. The condition for equality between A_t and A_0 implies that all the elementary symmetric functions $t\lambda_\nu(x)$ vanish almost everywhere, meaning that $(\partial^2\nu/\partial x_i\partial x_j)$ is identically zero; but since the boundary values of ν are zero, this means that ν is zero, and hence the deformation of X is trivial. This completes the proof of Theorem 1.2.

The proof of this theorem can be used at the same time to derive the Euler equation for stationary A.

The infinitesimal variation of A is given by integrating both terms of (2.3) over D:

$$\frac{dA_t}{dt}\bigg|_{t=0} = \frac{1}{n+2}\int_D \sum_{\nu=1}^{n} \lambda_\nu(2)dA$$

$$= \frac{1}{n+2}\int_D \left(K^{ij}(x)\frac{\partial^2\nu(x)}{\partial x^i\partial x^j}\right)h(x)|dx|.$$

Since $v(x)$ is assumed to vanish with its first partial derivatives on the boundary ∂D of D and since D is compact, we can apply Stokes' theorem twice yielding

$$\left.\frac{\partial A_t}{\partial t}\right|_{t=0} = \frac{1}{n+2}\int_D v(x)\frac{\partial^2}{\partial x^i \partial x^j}(K^{ij}(x)h(x))\,|dx|.$$

Thus, the right-hand member of the above equation vanishes for all smooth v with compact support, if and only if $(\partial^2/\partial x^i \partial x^j)(K^{ij}(x)h(x))$ is identically zero. Using the definitions of $K^{ij}(x)$ and $h(x)$, we have the useful identities

$$\frac{\partial}{\partial x^i}h(x) = \frac{1}{n+2}h(x)K^{jk}\frac{\partial^3 z(x)}{\partial x^j \partial x^k \partial x^i} = -\frac{h(x)}{n+2}\frac{\partial^2 z}{\partial x^i \partial x^j}\frac{\partial K^{jk}(x)}{\partial x^k}.$$

Thus the Euler equation may be written in the form

$$(2.4)\quad 0 = \frac{\partial^2}{\partial x^i \partial x^j}(K^{ij}(x)h(x)) = -\left(\frac{n+1}{n+2}\right)\frac{\partial}{\partial x^i}\left(K^{ij}\frac{\partial h(x)}{\partial x^j}\right).$$

Finally we recall that the Berwald-Blaschke metric can be expressed as in (1.6) as the quadratic form

$$(2.5)\qquad ds^2 = (h(x))^{-1}\frac{\partial^2 z(x)}{\partial x^i \partial x^j}dx^i dx^j;$$

the Laplace-Beltrami operator Δ determined by this metric can be expressed in terms of the special local coordinates $(x^1, \ldots, x^n)$, in the form

$$\Delta f = \frac{1}{h(x)}\frac{\partial}{\partial x^i}\left(h^2(x)K^{ij}\frac{\partial f}{\partial x^j}\right);$$

thus equation (2.4) is equivalent to the condition

$$(2.6)\qquad \Delta\left(\frac{1}{h(x)}\right) = 0, \qquad h(x) = \left(\det\left(\frac{\partial^2 z(x)}{\partial x^i \partial x^j}\right)\right)^{\frac{1}{n+2}}.$$

In the next section we shall show, in particular, that the quantity

$$\frac{(h(x))^2}{n}\Delta\left(\frac{1}{h(x)}\right) \tag{2.7}$$

whose vanishing is equivalent to the Euler equation (2.4), can be introduced as a scalar invariant, because its value is independent of the coordinates; it expresses the so-called *affine mean curvature* at each point of Σ.

3. Review of the Structure Equations. Let $X:\Sigma \to E$ be smooth, locally strongly convex immersion of an n-dimensional manifold Σ in the affine $(n + 1)$-space E. Having defined some of the invariants of $X(\Sigma)$ under the unimodular affine group $ASL(E)$, we shall complete the task by introducing the remaining invariants, that describe the resulting hypersurface up to an equivalence under transformations in the group $ASL(E)$.

The affine co-normal Y^*, introduced in the last section, has the following affinely invariant definition. Let Z^* be any smoothly defined, nowhere vanishing, dual vector function on Σ, such that $\langle Z^*, dX\rangle = 0$ identically, meaning that, for each $p \in \Sigma$, $Z^*(p)$ generates the one-dimensional vector subspace of $T^*(E)$ that annihilates all tangent vectors of $X(\Sigma)$ at p. Then, since the hypersurface $X(\Sigma)$ is locally strongly convex everywhere, the n-form ω_{Z^*} in Σ defined as follows, in terms of any local system of parameters $(u^1, \ldots, u^n) = (u)$ in Σ,

$$\omega_{Z^*} = [Z^*, (dZ^*)^n] = n!\det\left(Z^*, \frac{\partial Z^*}{\partial u^1}, \ldots, \frac{\partial Z^*}{\partial u^n}\right) du^1 \wedge \cdots \wedge du^n,$$

is nowhere zero. Any other choice of Z^* would differ from Z^* by a nowhere vanishing scalar factor; replacing Z^* by φZ^*, where φ is a non-vanishing, smooth function in Σ, gives

$$\omega_{\varphi Z^*} = \varphi^{n+1}\omega_{Z^*}.$$

For any fixed vector $V \in T(E)$ we consider the n-form

$$[V, (dX)^n] = n!\det\left(V, \frac{\partial X}{\partial u^1}, \ldots, \frac{\partial X}{\partial u^n}\right) du^1 \wedge \cdots \wedge du^n;$$

since this n-form is a linear functional on V that vanishes at any point $p \in \Sigma$, if and only if V is tangent to $X(\Sigma)$ at p ($V \in dX(p)$), there is a non-zero function λ on Σ, uniquely determined by the choice of Z^* such that, for every $V \in T(E)$,

$$[V, (dX)^n] = \lambda\langle Z^*, V\rangle[Z^*, (dZ^*)^n];$$

there is an $ASL(E)$-invariant normalization of Z^*, namely $\pm|\lambda|^{(-1/n+2)}Z^*$, called the affine co-normal, denoted by Y^*, uniquely determined by the two following properties:

i) for every $V \in T(E)$,

$$[V, (dX)^n] = \pm\langle Y^*, V\rangle[Y^*, (dY^*)^n] \tag{3.1}$$

ii) for a suitable neighborhood U of any point $p \in \Sigma$ and for any point $q \neq p$, $q \in U$

$$\langle Y^*(p), X(q) - X(p)\rangle > 0 \tag{3.2}$$

(this relation makes sense, since $X(U)$ is a convex hypersurface, so that $X(U\backslash\{p\})$ lies on one side of the tangent hyperplane $dX(p)$ at p). It follows from the identity $\langle Y^*, dX\rangle = 0$ and from (3.2) that the following bilinear differential form is symmetric and positive definite everywhere

$$\langle Y^*, d^2X\rangle = -\langle dY^*, dX\rangle; \tag{3.3}$$

this form is the Berwald-Blaschke form, or first (quadratic) fundamental form of the affine geometry of $X(\Sigma)$. This form is adopted as a definition of a Riemannian structure induced on Σ by X; the corresponding element of area (or n-dimensional volume) is also expressible as

$$dA = \frac{1}{n!}|[Y^*, (dY^*)^n]|. \tag{3.4}$$

Since the n-form $[Y^*, (dY^*)^n] = \pm n!dA$ is nowhere zero, there exists a vector function $Y:\Sigma \to T(E)$, uniquely determined by the relation

$$\langle Y^*, Y\rangle = 1, \qquad \langle dY^*, Y\rangle = 0. \tag{3.5}$$

This vector function Y is called the *affine normal* of $X(\Sigma)$; it satisfies the relations

$$\langle Y^*, dY\rangle = 0, \qquad [Y, (dX)^n] = \pm n!dA; \tag{3.6}$$

the former relation shows that the affine normal indicatrix map, at each point $p \in \Sigma$, varies infinitesimally along a hyperplane parallel to the tangent $dX(p)$ at p; the second one shows that, for any local system of parameters $(u^1, \ldots, u^n)$ in a domain $U \subset \Sigma$, the $(n + 1)$-uple of vectors for each point $p \in U$

$$(Y(p), X_1(p), \ldots, X_n(p)), \qquad X_i = \frac{\partial X}{\partial u^i}$$

is linearly independent, and is a suitable choice for a local frame. If we express the Berwald-Blaschke metric locally in the familiar form

$$ds^2 = g_{ij}(u)du^i du^j$$

with $g = \det(g_{ij})$ and $dA = \sqrt{g(u)}\,|du^1 \wedge \cdots \wedge du^n|$, we have

$$\det(Y, X_1, \ldots, X_n) = \pm\sqrt{g}; \tag{3.7}$$

thus we can regard the $(n + 1)$-uple $(Y, X_1, \ldots, X_n)$ as a local field of moving frames. The structure of X is determined, after the method of moving affine frames, by expressing its derivatives as combinations of the vectors in the frame itself; in order to keep track of the tensorial invariance properties of the intervening coefficients, it is better to write these equations in terms of the covariant derivatives of the tensorial quantities with respect to the Levi Civita connection derived from the Berwald-Blaschke metric; thus we shall denote by $X_{i,j} = X_{,ij}$ the second covariant derivative of the position vector X,

$$X_{i,j} = \frac{\partial^2 X(u)}{\partial u^i \partial u^j} - \Gamma_{ij}{}^h X_{,h},$$

where $\Gamma_{ij}{}^h$ are the Christoffel symbols of the connection. The structure equations can be written as follows: from the definitions of the frame one can show that

$$X_{,ij} = A_{ij}^{\ h} X_{,h} + g_{ij} Y \tag{3.8}$$

$$Y_{,j} = -B_j^{\ h} X_h, \tag{3.9}$$

where $A_{ij}^{\ h}$ is a tensor field ($A_{ij}^{\ h}\, du^i \otimes du^j \otimes \partial/\partial u^h$, in the invariant form) $B_j^{\ h}$ another tensor field, and (g_{ij}) the components of the metric. From equations (3.3), (3.5) and (3.6) one sees that $(Y^*, Y^*_1, \ldots, Y^*_n)$ can be regarded as a sort of a dual frame field of $(Y, X_1, \ldots, X_n)$; its structure equation corresponding to (3.8), (3.9) are determined as follows:

$$Y^*_{,ij} = \frac{\partial^2 Y^*}{\partial u^i \partial u^j} - \frac{\partial Y^*}{\partial u^h} \Gamma_{ij}^{\ h} = -A_{ij}^{\ h} Y^*_{,h} - B_i^{\ h} g_{hj} Y^*. \tag{3.10}$$

The structure equations (3.8), (3.9), or alternatively (3.10), are thus determined by the three tensors (g_{ij}), $(A_{ij}^{\ h})$ and $(B_i^{\ h})$, where (g_{ij}) defines a Riemannian structure. The other two tensors are not arbitrary, in the same way that, in the euclidean geometry of immersed hypersurfaces, the second fundamental form is related to the first, through a set of integrability conditions. These conditions are all derived from the identities $Y_{,ij} = Y_{,ji}$, $X_{,ijk} - X_{,ikj} = R^h_{\ ijk} X_h$, or alternatively $Y^*_{,ijk} - Y^*_{,ikj} = R^h_{\ ijk} Y^*_{,h}$, where $(R^h_{\ ijk})$ is the Riemann-Christoffel curvature tensor derived from (g_{ij}). Taking the corresponding covariant derivatives on the right-hand members of (3.8) and (3.9), after some routine computations, one obtains the following relations:

a) The trilinear covariant tensor $A_{ijk} = g_{hk} A_{ij}^{\ h}$ is symmetric in each pair of indices, and satisfies the algebraic relation

$$g^{ij} A_{ijk} = 0; \qquad (\text{or } A_{ij}^{\ i} = 0) \tag{3.11}$$

the resulting cubic differential form $A_{ijk}\, du^i du^j du^k$ is called the *second fundamental form* of the immersed hypersurface $X(\Sigma)$; the algebraic relation (3.11) is referred to as the *apolarity* of the second fundamental form with respect to the first.

b) The bilinear covariant tensor $B_{ij} = g_{jk} B_i^{\ h}$ is symmetric; the resulting quadratic differential form

$$B_{ij} du^i du^j = \langle dY^*, dY \rangle, \qquad B_{ij} = \left\langle \frac{\partial Y^*}{\partial u^i}, \frac{\partial Y}{\partial u^j} \right\rangle,$$

is called the third fundamental form. In analogy with the second fundamental form of euclidean geometry, the eigenvalues of $(B_i{}^j)$ and the corresponding eigenvectors tangent to Σ are called respectively the principal (affine) curvatures $\kappa_\nu (1 \le \nu \le n)$ and principal directions; the latter are pairwise orthogonal (necessarily when the corresponding curvatures are distinct, by choice if they coincide) each principal curvature and direction is a solution of the equation $dY + \kappa dX = 0$. The arithmetic mean of the principal affine curvatures, denoted by H, is called the affine mean curvature of $X(\Sigma)$,

$$H = \frac{1}{n} \sum_{\nu=1}^{n} \kappa_\nu = \frac{1}{n} g^{ij} B_{ij} = \frac{1}{n} B_i{}^i. \tag{3.12}$$

c) The three fundamental forms are related by the following three systems of integrability conditions, with the following respective denominations, each reminiscent of similar relations between the two fundamental forms in the classical euclidean geometry of hypersurfaces:

i) the Codazzi-Mainardi equations of the first kind

$$A_{ijk,\ell} - A_{ij\ell,k} = \frac{1}{2}(g_{ik}B_{j\ell} - g_{i\ell}B_{jk} + g_{jk}B_{i\ell} - g_{j\ell}B_{ik}); \tag{3.13}$$

ii) the Codazzi-Mainardi equations of the second kind

$$B_{ij,k} - B_{ik,j} = A_{ij}{}^h B_{hk} - A_{ik}{}^h B_{hj}; \tag{3.14}$$

iii) the Gauss equations

$$R_{ijk\ell} = g^{hm}(A_{hjk}A_{mi\ell} - A_{hj\ell}A_{mik}) + \frac{1}{2}(g_{ik}B_{j\ell} - g_{i\ell}B_{jk} - g_{jk}B_{i\ell} + g_{j\ell}B_{ik}). \tag{3.15}$$

The above equations are frequently applied in a contracted form; thus the Codazzi equations of the first kind, (3.14), contracted by multiplying both sides by $g^{i\ell}$ and adding with respect to i, ℓ, and recalling the apolarity (3.11) of the cubic form, become

$$A_{jk}{}^h{}_{,h} = \frac{n}{2}(Hg_{jk} - B_{jk}). \tag{3.16}$$

The Codazzi equations of the second kind (3.14), multiplying both sides by g^{ik} and adding, reduce to

$$g^{ik}B_{ij,k} = nH_{,j} + A_{ij}{}^{h}B_{h}{}^{i}. \tag{3.17}$$

The Gauss equations (3.15) can be contracted in two successive stages, first with $g^{j\ell}$, then with $g^{ik}g^{j\ell}$, yielding respectively

$$R_{ik} = A_{i\ell}{}^{h}A_{hk}{}^{\ell} + \frac{1}{2}(nHg_{ik} + (n-2)B_{ik}), \tag{3.18}$$

$$R = A_{ij}{}^{k}A_{k}{}^{ij} + n(n-1)H. \tag{3.19}$$

4. Normal Variational Formulas. We refer again to Definition 2.1, describing an allowable, interior deformations of a locally strongly convex, immersed hypersurface $X:\Sigma \to E$. In order to formalize an intuitive notion of equivalent deformation, we need the concept of an interior isotopy of Σ (onto itself).

Definition 4.1. Let Σ be a smooth, n-dimensional, connected manifold, compact or not, with boundary $\partial\Sigma$ (possibly empty). An *interior isotopy* of Σ is a differentiable map $\mathcal{H}: I \times \Sigma \to \Sigma$, where I is an open interval $(-\epsilon', \epsilon) \subset \mathbf{R}$ with $\epsilon, \epsilon' > 0$ with the following properties:

i) For each $t \in I$ the map $H_t:\Sigma \to \Sigma$, defined by $H_t(p) = \mathcal{H}(t, p)$, is an autodiffeomorphism of Σ, and, for $t = 0$, H_0 is the identity map;

ii) The restriction of H_t to the boundary $\partial\Sigma$ and to the complement of fixed, compact subdomain $\Sigma' \subset \Sigma$ is equal to the identity for each $t \in I$.

Definition 4.2. Two allowable interior deformations $\mathcal{X}$, $\mathcal{X}':I \times \Sigma \to E$ of a given, locally strongly convex immersion of a hypersurface $X:\Sigma \to E$ are said to be equivalent, if there exists an interior isotopy $\mathcal{H}:I \times \Sigma \to \Sigma$ such that, for each $t \in I$,

$$X'_t = X_t \circ H_t,$$

or in other words,

$$\mathcal{X}'(t, p) = \mathcal{X}(t, \mathcal{H}(t, p)).$$

Definition 4.3. An allowable interior deformation $\mathfrak{X}: I \times \Sigma \to E$ of a given, locally strongly convex hypersurface immersion $X: \Sigma \to E$ is said to be *normal*, if and only if, for each (t, p), the tangent hyperplane $dX_t(p)$ to $X_t(\Sigma)$ at p is parallel to $dX_0(p) = dX(p)$, i.e., to the corresponding one for the given hypersurface. Similarly, two locally strongly convex hypersurface immersions X and $\bar{X}$ of Σ in E are said to be in normal correspondence, if:

i) X and $\bar{X}$ coincide in $\partial\Sigma$ and outside of a compact subdomain $\Sigma' \subset \Sigma$;

ii) for each $p \in \Sigma$, the tangent hyperplanes $dX(p)$ and $d\bar{X}(p)$ of X and $\bar{X}$ respectively, are mutually parallel, and, for some neighborhood U of p, the images $X(U)$ and $\bar{X}(U)$ lie on similarly placed sides of the respective, mutually parallel tangent hyperplanes corresponding to p.

PROPOSITION 4.1. *Every allowable deformation $\mathfrak{X}: I \times \Sigma \to E$ of a given, locally strongly convex hypersurface immersion $X: \Sigma \to E$ is equivalent to a unique normal deformation. Furthermore, any two locally strongly convex hypersurface immersions X, $\bar{X}: \Sigma \to E$ that are in normal correspondence can be mutually connected by a (normal) allowable interior deformation by means of the pointwise linear interpolation between X and $\bar{X}$, provided that X and $\bar{X}$ coincide in $\partial\Sigma$ and outside of a compact subset of Σ.*

Proof. Given any allowable deformation $\mathfrak{X} = (X_t)_{t \in I}$ of a locally strongly convex immersed hypersurface $X = X_0: \Sigma \to E$, consider, for each $t \in I$, the co-normal indicatrix map $Y^*_t: \Sigma \to T^*(E)$; each such map is, because of (3.1), a locally strongly starlike (or radially transversal) immersion. Choose in the vector space $T^*(E)$ a strongly convex norm denoted by $\| \;\|$ (perhaps, but not necessarily euclidean), denote by S the unit sphere with respect to $\| \;\|$, so that S is a smooth, convex hypersurface. The condition that each X_t is locally strongly convex implies that the renormalized co-normal map

$$N_t = \| Y^*_t \|^{-1} Y^*_t$$

is an open immersion of Σ in the differentiable sphere S, and hence a local diffeomorphism, and the collective map $\mathfrak{N}: I \times \Sigma \to S$ defined by setting $\mathfrak{N}(t, p) = N_t(p)$ is a smooth, surjective map. Since the derived linear bundle map dN_t from the tangent bundle $T(\Sigma)$ to $T(S)$ is bijective on each fibre, for each $(t, p) \in I \times \Sigma$ there is a unique tangent vector $L_t(p) \in T(\Sigma)$ such that

$$dN_t(L_t(p)) = d\mathfrak{N}\left(\frac{\partial}{\partial t}\right) = \frac{\partial N_t}{\partial t}, \tag{4.1}$$

and $L_t(p)$ depends smoothly on (t, p). Furthermore one verifies easily that $L_t(p) = 0$ for p in $\partial\Sigma$ or outside the same compact subdomain $\Sigma' \subset \Sigma$ that includes all points where the family $\mathfrak{X}$ is nontrivial. Consequently, there exists a unique interior isotopy $\mathfrak{H}: I \times \Sigma \to \Sigma$ such that, for $H_t(p) = \mathfrak{H}(t, p)$,

$$\frac{\partial H_t}{\partial t}(p) = dH_t(-L_t(p)), \qquad H_0(p) = p.$$

Construct now a new interior deformation $\bar{\mathfrak{X}}: I \times \Sigma \to E$, equivalent to $\mathfrak{X}$, by setting

$$\bar{\mathfrak{X}}(t, p) = \mathfrak{X}(t, H_t(p));$$

it is easy to verify now that the renormalized co-normal indicatrix $\bar{N}_t$ of each $\bar{X}_t$ $(\bar{X}_t(p) = \bar{\mathfrak{X}}(t, p))$ is given by

$$\bar{N}_t = N_t \circ H_t; \tag{4.2}$$

differentiating both sides of this last equation with respect to t one obtains

$$\frac{\partial \bar{N}_t}{\partial t} = \frac{\partial N_t}{\partial t} \circ dH_t + dN_t \circ \frac{\partial H}{\partial t} = dN_t \circ dH_t(L_t - L_t) = 0.$$

Therefore, $\bar{N}_t$ is independent of t; it follows that, for each $(t, p) \in I \times \Sigma$, the affine co-normal $\overline{Y^*}_t$ of $\bar{X}_t$ at p is a positive scalar multiple of $Y^*(p)$, implying that the tangent hyperplanes $d\bar{X}_t(p)$ and $dX(p)$ are parallel, since they have the same annihilator in $T^*(E)$. Hence $\bar{\mathfrak{X}}$ is a normal deformation.

The uniqueness of the normal deformation equivalent to a given deformation $\mathfrak{X}$ follows from the transitivity of equivalence; assume now that $\mathfrak{X}$, $\bar{\mathfrak{X}}$ are two equivalent, interior deformations of X with $\bar{X}_t = X_t \circ H_t$ for an interior isotopy $\mathfrak{H}$ with $\mathfrak{H}(t, p) = H_t(p)$. If both $\mathfrak{X}$ and $\bar{\mathfrak{X}}$ are normal, it follows that the respective families of normalized co-normal maps N_t and $\bar{N}_t$ coincide, since $N_t = \bar{N}_t = N_0$ for $t \in I$. Looking at (4.2) and recalling that N_0 is a local diffeomorphism, it follows from continuity considerations that H_t is the identity map for each t.

Finally, assume that X and $\bar{X}$ are two locally strongly convex immersions of Σ in E in normal correspondence. Define the map $\mathcal{X}: I \times \Sigma \to E$ by setting simply

$$\mathcal{X}(t, p) = (1 + t)X(p) + t\bar{X}(p) = X_t(p)$$

The main thing to prove here is that there exist positive constants ϵ', ϵ such that, for each $t \in I = (-\epsilon', 1 + \epsilon)$, X_t is an immersion and that it is locally strongly convex. This is an easy consequence of some well known properties of convex bodies. Given any $p \in \Sigma$, let U be a neighborhood of p such that $X(U)$ and $\bar{X}(U)$ are globally convex, i.e., each image is contained in the boundary of its convex hull, respectively $\mathcal{C}X(U)$ and $\mathcal{C}\bar{X}(U)$. From the assumption that X and $\bar{X}$ are in normal correspondence, it follows that, for each $q \in \Sigma$ (and especially $q \in U$), the affine co-normal $\bar{Y}^*(q)$ of $\bar{X}$ at q is a strictly positive scalar multiple of $Y^*(q)$, say

$$\bar{Y}^*(q) = \varphi(q)Y^*(q), \qquad \varphi(q) > 0 \quad (q \in \Sigma);$$

Consider the support functions $M_G: T^*(E) \to \mathbf{R}$ for any convex body $G \subset E$, and especially for $G = \mathcal{C}X(U)$ or $\mathcal{C}\bar{X}(U)$; thus, for any $Z^* \in T^*(E)$,

$$M_G(Z^*) = \sup_{V \in G}\langle Z^*, V - 0\rangle,$$

where 0 is a fixed choice of "origin" in E. Because of property (3.2) in the definition of the affine co-normal, and since $X(U)$ and $\bar{X}(U)$ are globally convex, for $G_0 = \mathcal{C}X(U)$ (respectively $G_1 = \mathcal{C}\bar{X}(U)$) and for any $q \in U$,

$$M_{G_0}(-Y^*(q)) = -\langle Y^*(q), X(q) - 0\rangle,$$

$$M_{G_1}(-\bar{Y}^*(q)) = -\langle \bar{Y}^*(q), X(q) - 0\rangle,$$

and therefore also

$$M_{G_1}(-Y^*(q)) = -\langle Y^*(q), \bar{X}(q) - 0\rangle.$$

For $0 \leq t \leq 1$ we construct the Minkowski sum

$$G_t = (1-t)G_0 + tG_1 = \{(1-t)V_0 + tV_1 \mid V_0 \in G_0,\ V_1 \in G_1\}.$$

It follows that the support function M_{G_t} of G_t is again determined by the equation

$$M_{G_t}(Z^*) = (1-t)M_{G_0}(Z^*) + tM_{G_1}(Z^*) \quad (Z^* \in T^*E)$$

and, in particular, for any $q \in U$

$$M_{G_t}(-Y^*(q)) = -\langle Y^*(q), (1-t)X(q) + t\bar{X}(q) - 0\rangle$$

$$= -\langle Y^*(q), X_t(q) - 0\rangle,$$

for $X_t(q) = (1-t)X(q) + t\bar{X}(q)$. The consequence is that, for $0 \leq t \leq 1$, the differentiable map $X_t : \Sigma \to E$ has the property that, for each $q \in U$, $X_t(q)$ is an extreme point of the convex body G_t; hence $X_t(U)$ is a convex hypersurface; the regularity and strong convexity now are easy consequences of a calculation involving the envelope of the differentiable family of supporting hyperplanes $\{X_t(q) + dX(q)\}$ that one can recover from the gradient mapping of the support function M_{G_t} restricted to the cone generated by $\{-Y^*(q) \mid q \in U\}$, where it is smooth and has a hessian form of maximal (for a linearly homogeneous function) rank n. The verification of the other required properties for the maps X_t to constitute (at least for $0 \leq t \leq 1$) an allowable deformation is a trivial matter, that we shall omit. The extendibility of the family to one in which $t \in (-\epsilon, 1 + \epsilon')$ for ϵ, $\epsilon' > 0$ follows from the fact that the properties of differentiable maps X_t to be locally strongly convex immersion is an open property in the C^2-topology of such maps. Proposition 4.1 is thereby proved.

A consequence of Proposition 4.1 is that we can restrict our consideration, as we will from now on except when we state otherwise, to normal deformations, because several formal relations are thereby greatly simplified.

Let $X : \Sigma \to E$ be a locally strongly convex, immersed hypersurface and $\mathfrak{X} : I \times \Sigma \to E$ an allowable, interior deformation of X. For each $(t, p) \in I \times \Sigma$ we express the displacement vector $\mathfrak{X}(t, p) - X(p)$ in terms of a local, coordinate frame $(Y, X_1, \ldots, X_n)$ at p $(X_i = \partial X / \partial u^i)$ for some local parameter system $(u) = u^1, \ldots, u^n)$ defined in a neighborhood of p in Σ, as follows

(4.3) $$\mathfrak{X}(t,p) = X(p) + \bar{v}(t,u)Y(p) + \bar{w}^i(t,u)X_{,i}(u),$$

where $\bar{v}$, $\bar{w}^i$ are one-parameter families (with parameter t) respectively of scalar functions and tangent vector field components in Σ, for $t \in I$.

Proposition 4.2. *A given deformation $\mathfrak{X}: I \times \Sigma \to E$ of a locally strongly convex $X: \Sigma \to E$ is normal, if and only for each $(t, p) \in I \in \Sigma$, the local deformation components $\bar{v}$, $\bar{w}^i$ of $\mathfrak{X}$ with respect to any local parameter system $(u^1, \ldots, u^n)$ as in (4.3) satisfy the relation*

(4.4) $$\bar{w}^i(t,u)g_{ij}(u) + \frac{\partial \bar{v}(t,u)}{\partial u^j} = 0,$$

where $g_{ij}(u)$ are the components of the Berwald-Blaschke metric.

Proof. For each $t \in I$, the locally strongly convex immersion X_t, where $X_t(p) = \mathfrak{X}(t,p)$, has the tangent hyperplane at p spanned by the n vectors $X_{t,i} = \partial X_t/\partial u^i$ $(1 \le i \le n)$; thus we have

$$X_{t,i} = \frac{\partial X_t}{\partial u^i} = \frac{\partial X}{\partial u^i} + \frac{\partial \bar{v}}{\partial u^i} Y + \bar{v} Y_{,i} + \bar{w}^h{}_{,i} X_{,h} + \bar{w}^j X_{,ji}.$$

We expand the derived vectors $Y_{,i}$ and $X_{,ji}$ according to the structure equations (3.8), (3.9); as a consequence we obtain the following,

(4.5) $$X_{t,i} = X_{,i} + \left(\frac{\partial \bar{v}}{\partial u^i} + \bar{w}^j g_{ji}\right) Y + (\bar{w}^h{}_{,i} + \bar{w}^\ell A_{\ell i}{}^h - \bar{v} B_i{}^h) X_{,h}.$$

From the above expression, it is clear that the tangent hyperplane to X_t at p is parallel to that of $X = X_0$, if and only if the normal components of each $X_{t,i}$ vanish identically, thus establishing (4.4) as the necessary and sufficient condition, as asserted.

It follows that any normal deformation of X can always be represented, uniquely, by the family of functions $\bar{v}: I \times \Sigma \to \mathbf{R}$, satisfying certain boundary conditions, that we shall specify below.

Proposition 4.3. *Each normal, allowable deformation $\mathfrak{X}$ of a locally strongly convex immersion $X: \Sigma \to E$ is uniquely representable by a one-parameter family of functions $\bar{v}: I \times \Sigma \to \mathbf{R}$ by means of the formula*

(4.6) $$\mathfrak{X}(t,p) = X(p) + \bar{v}(t,p)Y(p) - g^{ih}(u(p)) \frac{\partial \bar{v}}{\partial u^i} X_{,h}(u).$$

Conversely, any function $\bar{v}: I \times \Sigma \to \mathbf{R}$ *describes by* (4.5) *a normal, allowable, interior deformation of* X, *if and only if it satisfies the following conditions:*

i) I *is an open interval* $(-\epsilon', \epsilon) \subset \mathbf{R}$ $(\epsilon, \epsilon' > 0)$ *and* $\bar{v}$ *is a differentiable function in* $I \times \Sigma$;

ii) $\bar{v}(t, p)$ *vanishes identically if either* $t = 0$, *or* $p \in \partial\Sigma$, *or for* p *outside of some compact subdomain* $\Sigma' \subset \Sigma$;

iii) *the first derivatives of* $\bar{v}$ *vanish identically in* $\partial\Sigma$ *for each* $t \in I$;

iv) *for each* $(t, p) \in I \times \Sigma$ *and for any local coordinate system* $(u) = (u^1, \ldots, u^n)$ *near* p, *the function* $\bar{v}$ *satisfies the following condition (invariant under coordinate transformations).*

$$\det(\delta_i^{\,h} - C_i^{\,h}(t, u)) \neq 0, \tag{4.7}$$

where $(\delta_i^{\,h})$ *is the unit matrix and* $C_i^{\,h}$ *are components of a tensor defined by*

$$C_i^{\,h} = g^{hj}(\bar{v}_{,ji} + \bar{v}_{,\ell}A_{ji}^{\,\ell} + \bar{v}B_{ji}) = \bar{v}^{\cdot h}{}_{;i} + \bar{v}^{\cdot \ell}A_{\ell i}^{\,h} + \bar{v}B_i^{\,h} \tag{4.8}$$

Proof. We know from Proposition (4.1) that every allowable interior deformation of X is equivalent to a unique normal deformation, and the latter, decomposed locally according to local $(n + 1)$-frames, as in (4.3), must satisfy the condition $\bar{w}^i g_{ij} + \partial\bar{v}/\partial u^j = 0$ of Proposition 4.2. Thus the "tangential components" $(\bar{w}^i)$ of the deformation are uniquely determined by the "affine normal" component $\bar{v}$, by setting

$$\bar{w}^i(t, u) = -g^{ij}(u)\frac{\partial\bar{v}(t, u)}{\partial u^j}; \tag{4.9}$$

we abbreviate the right hand member of the above equation by means of the symbol for the contravariant derivative, $\bar{v}^{\cdot i}(t, u)$. Substitution of this tangent vector field for $\bar{w}^i$ in (4.3), gives (4.6), as asserted.

Conversely, let $\bar{v}(t, u)$ be any one-parameter family of scalar functions $v_t: \Sigma \to \mathbf{R}$ $(v_t(p) = \bar{v}(t, p))$ for $t \in I = (-\epsilon', \epsilon)$ satisfying the four conditions i), ii), iii), iv) stated above, and let $X_t(p) = \mathfrak{X}(t, p) = X(p) + \bar{v}(t, p)Y(p) - g^{ih}\bar{v}_{,i}X_{,h}$ as in (4.6). It is clear that the first three conditions are equivalent to the fact that $\mathfrak{X}$ represents a differentiable deformation of X, trivial, outside Σ' and leaving fixed each point and tangent hyperplane of $X(\partial\Sigma)$. There remains the task of showing that property iv) is equivalent to the condition that, for each $t \in I$, X_t is a locally strongly convex immersion. Substituting for $\bar{w}^i$ in (4.5) according to (4.9), we have

$$X_{t,i} = X_{,i} - (\bar{v}^{\cdot h}{}_i + \bar{v}^{\cdot \ell} A_{\ell i}{}^h + \bar{v} B_i{}^h) X_{,h} = X_{,i} - C_i{}^h X_{,h},$$

where $(C_i{}^h)$ is the tensor defined by (4.8). The above equation shows that the determinant inequality (4.7) is equivalent to the condition that X_t be an immersion for any given t.

We continue now in deriving the variational formulas for the affine invariants of X by the same procedure that is followed in Section 3. Since the tangent hyperplane $dX_t(p)$ of X_t at p is obviously parallel to $dX(p)$ for each p, one verifies that the affine co-normal Y^*_t of X_t is proportional to Y^* as follows,

$$Y^*_t(p) = (\det(\delta_i{}^h - C_i{}^h(t,p)))^{\frac{1}{n+2}}\, Y^*(p)$$

It is convenient at this point to introduce the following abbreviated notation

$$\bar{w}(t,p) = (\det(\delta_i{}^h - C_i{}^h(t,p)))^{\frac{-1}{n+2}}, \tag{4.10}$$

so that

$$Y^*_t(p) = \bar{w}^{-1} Y^*. \tag{4.11}$$

With this notation, it is easier to differentiate Y^*_t along Σ,

$$Y^*_{t,i}(p) = \bar{w}^{-2}(-\bar{w}_{,i} Y^* + \bar{w} Y^*_{,i}), \tag{4.12}$$

from which we can calculate the distorsion formula for the Berwald-Blaschke form

$$\begin{aligned} g_{t.ij} &= -\langle Y^*_{t,i}, X_{t,j} \rangle = -\bar{w}^{-1} \langle Y^*_{,i}, (\delta_j{}^h - C_j{}^h) X_{,h} \rangle \\ &= \bar{w}^{-1}(g_{ij} - C_{ij}(t,p)) = \langle Y^*_t, X_{t,ij} \rangle. \end{aligned} \tag{4.13}$$

Since $C_{ij}(t,p)$ is identically zero (and $\bar{w} = 1$) for $t = 0$, and $(g_{ij} - C_{ij})$ is nonsingular for $t \in I$, it follows by continuity that the Berwald-Blaschke metric tensor $(g_{t.ij})$ is positive definite for all $t \in I$, everywhere in Σ, and, because it equals $\langle Y^*_t, X_{t.ij}\rangle$, it shows that X_t is locally strongly convex. This completes the proof of Proposition 4.3.

The next proposition gives an affinely invariant, geometric interpretation for the Euler equation corresponding to stationary, affinely invariant area under all infinitesimal, allowable interior deformations. Without loss of generality, we may assume Σ to be a compact manifold with smooth boundary $\partial\Sigma$.

If $\mathfrak{X}: I \times \Sigma \to E$ is any allowable interior deformation of X, we denote by dA_t the affinely invariant area element induced by X_t for each $t \in I$, and by A_t the total affinely invariant area of $X_t(\Sigma)$,

$$A_t = \int_\Sigma dA_t.$$

The infinitesimal variation of A induced by $\mathfrak{X}$, denoted by δA is then defined to be the derivative of A_t with respect to t, evaluated at $t = 0$,

$$\delta A = \left.\frac{\partial A_t}{\partial t}\right|_{t=0}$$

PROPOSITION 4.4. *Let $X: \Sigma \to E$ be a locally strongly convex immersion of a compact, n-dimensional manifold Σ with smooth boundary $\partial\Sigma$ in the $(n + 1)$-dimensional affine space E. Then the affine area of Σ induced by X is stationary, i.e., $\delta A = 0$ for every allowable interior deformation $\mathfrak{X}$ of X, if and only if the affine mean curvature H of $X(\Sigma)$ is identically zero.*

Proof. Let $\mathfrak{X}: I \times \Sigma \to E$ be any allowable, interior deformation of X with $X_t(p) = \mathfrak{X}(t, p)$. Since for every deformation $\bar{\mathfrak{X}}: I \times \Sigma \to E$ $(\bar{X}_t(p) = \bar{\mathfrak{X}}(t, p))$ equivalent to $\mathfrak{X}$, the immersed hypersurface $\bar{X}_t(\Sigma)$ is obviously the same as $X_t(\Sigma)$ for each $t \in I$, we may assume, without loss of generality, that $\mathfrak{X}$ is a normal deformation, represented by a function $\bar{v}: I \times \Sigma \to \mathbf{R}$ satisfying the four conditions imposed by Proposition 4.3. We refer now to equations (4.10), (4.11), (4.13) in the proof of the same proposition. From these, one deduces that the area element dA_t induced by X_t on Σ is given by

$$dA_t = \bar{w}^{-n-1} dA. \tag{4.14}$$

The corresponding infinitesimal variation is then

$$\delta dA = \delta(\bar{w}^{-n-1})dA = -(n+1)(\delta\bar{w})dA,$$

since $\bar{w}(t, p) = 1$ for $t = 0$.

We calculate $\delta\bar{w}$ in terms of $\delta\bar{v}$ by differentiating both members of (4.10) with respect to t and substituting the expression (4.8) for $(C_i{}^h)$ in the result, recalling that $C_i{}^h(0, p) = 0$ identically

$$\delta\bar{w} = \frac{1}{n+2}\delta C_i^i = \frac{1}{n+2}((\delta\bar{v})^{\cdot i}{}_i + (\delta\bar{v})^{\cdot \ell}A_{\ell i}{}^i + (\delta\bar{v})B_i{}^i)$$

$$= \frac{1}{n+2}(\Delta(\delta\bar{v}) + nH(\delta\bar{v})).$$

The function $\delta\bar{v}:\Sigma \to \mathbf{R}$, $\delta\bar{v}(p) = \partial\bar{v}(t, p)/\partial t\,|_{t=0}$ represents the infinitesimal, normal deformation of X; all first order infinitesimal variational formulas of differential geometric invariants of X can be expressed in terms of linear differential operations on $\delta\bar{v}$. We shall replace the function $\delta\bar{v}$ by the (typographically) more convenient notation $(n+2)v$. Thus the function $v:\Sigma \to \mathbf{R}$ represents the infinitesimal, normal deformation of X, according to the formula

$$\delta X = (n+2)(vY - v^{\cdot h}X_{,h}), \tag{4.14}$$

from which we derive, as before,

$$\delta X_{,i} = (n+2)(-v^{\cdot h}{}_i - v^{\cdot \ell}A_{\ell i}{}^h - vB_i{}^h)X_{,h};$$

then, introducing the function $w = \delta\bar{w}$,

$$w = \delta\bar{w} = \frac{1}{n+2}(\Delta(\delta\bar{v}) + nH\bar{v}) = \Delta v + nHv, \tag{4.15}$$

we have, successively

$$\begin{aligned} \delta Y^* &= -wY^*, \\ \delta Y^*_{,i} &= -w_{,i}Y^* - wY^*_{,i}, \end{aligned} \tag{4.16}$$

whence

$$\delta g_{ij} = \langle -\delta Y^*_{,i}, X_{,j}\rangle - \langle Y^*_{,i}, \delta X_{,j}\rangle$$

$$= -wg_{ij} - (n+2)(v_{,ij} + v_{,k}A_{ij}{}^k + vB_{ij}),$$

and $\delta(dA) = (1/2)g^{ij}\delta g_{ij}dA = -(n+1)wdA$, from which we obtain, finally,

(4.17)
$$\delta A = \int_{\Sigma'} \delta dA = -(n+1)\int_\Sigma (\Delta v + nHv)dA$$

$$= -n(n+1)\int_\Sigma HvdA;$$

the last equality follows from the fact that, since v vanishes identically (as well as dv) on $\partial\Sigma$, $\int_\Sigma \Delta v dA = 0$. Since v, with this last restriction, is otherwise an arbitrary function in Σ, it follows from (4.17) that $\delta A = 0$ for every allowable interior deformation of X, if and only if the affine mean curvature H of $X(\Sigma)$ vanishes identically, thus proving the proposition.

In order to verify whether the Euler equation $H = 0$ implies that the affine area functional achieves a local maximum value, we must calculate the second variation $\delta^2 A = (d^2A_t/dt^2)|_{t=0}$ of the affinely invariant area for any affine maximal hypersurface. In view of (4.17), it follows that, if $H = 0$ identically,

(4.18)
$$\delta^2 A = -n(n+1)\int_\Sigma (\delta H)vdA;$$

this reduces our task to that calculating δH. If we recall the defining equations (3.5) of the affine normal Y, we see that the infinitesimal variation δY of Y is characterized by the equations

$$\langle \delta Y^*, Y\rangle + \langle Y^*, \delta Y\rangle = 0, \qquad \langle \delta Y^*_{,i}, Y\rangle + \langle Y^*_{,i}, \delta Y\rangle = 0;$$

applying (3.3) and (4.16), we have

(4.19)
$$\delta Y = wY - w^{,h}X_{,h}.$$

Differentiating this last equation, we have, on the one hand,

$$\delta Y_{,i} = (wY - w^{\cdot h}X_{,h})_{,i} = -(w^{\cdot h}{}_{i} + w^{\cdot \ell}A_{\ell i}{}^{h} + wB_i{}^{h})X_{,h};$$

on the other hand, from the structure equation (3.9), we have also

$$\delta Y_{,i} = -(\delta B_i{}^{h})X_h - B_i{}^{k}\delta X_{,k}$$

$$= \{-\delta B_i{}^{h} + (n+2)B_i{}^{k}(v^{\cdot h}{}_{k} + v^{\cdot \ell}A_{\ell k}{}^{h} + vB_k{}^{h})\}X_{,h}.$$

Comparing the coefficient matrices of $X_{,h}$ in the last two sets of equations, we obtain

$$\delta B_i{}^{h} = w^{\cdot h}{}_{i} + w^{\cdot \ell}A_{\ell i}{}^{h} + wB_i{}^{h} + (n+2)B_i{}^{k}(v^{\cdot h}{}_{k} + v^{\cdot \ell}A_{\ell k}{}^{h} + vB_k{}^{h}),$$

(4.20)

and by contraction of the indices i, h,

$$(4.21)\quad n\delta H = \Delta w + nHw + (n+2)B_i{}^{k}(v^{\cdot i}{}_{k} + A_{k\ell}{}^{i}v^{\cdot \ell} + vB_k{}^{i}).$$

Since $H = 0$ identically by assumption and $w = \Delta v + nHv$ by definition, the last formula simplifies somewhat as follows, using also the contracted form (3.17) of the Codazzi equations of the second kind,

$$(4.21)\qquad n\delta H = \Delta^2 v + (n+2)(B^{ij}v_{,ij} + B^{ij}{}_{,j}v_{,i} + B^{ij}B_{ij}v)$$

$$= \Delta^2 v + (n+2)((B^{ij}v_{,i})_{,j} + (B^{ij}B_{ij})v).$$

Returning to (4.18), we obtain the second variational formula for the affine area

(4.22)

$$\delta^2 A = -(n+1)\int_\Sigma (n\delta H)v\,dA$$

$$= -(n+1)\int_\Sigma v\{\Delta^2 v + (n+2)((B^{ij}v_{,i})_{,j} + B^{ij}B_{ij}v)\}dA$$

$$= -(n+1)\int_\Sigma \{(\Delta v)^2 - (n+2)B^{ij}v_{,i}v_{,j} + (n+2)B^{ij}B_{ij}v^2\}dA.$$

The last expression is deceptively suggestive that $\delta^2 A$ may be always negative definite; as we shall se presently, this conclusion is far from easy to prove, even in the only case where we have been able to do so, for $n = 2$. The formal statement to this effect is in the first section, labeled as Theorem 1.3.

Proof of Theorem 1.3. We shall reduce the formula (4.22) for $\delta^2 A$ (disregarding as far as possible the assumption $n = 2$) to a quadratic functional on the functions v with compact support in Σ, expressed as a combination of scalar products among three elements of the pre-Hilbert space $S_0{}^2T^*$ of symmetric, bilinear forms in Σ, with trace zero with respect to the Berwald-Blaschke metric; each of these bilinear forms corresponds to the result of a linear differential operation on v, of order respectively 2, 1 and 0, as follows:

$$(Lv)_{ij} = v_{.ij} - \frac{1}{n}(\Delta v)g_{ij}$$

$$(Av)_{ij} = A_{ij}{}^h v_{.h}$$

$$(Bv)_{ij} = B_{ij}v.$$

The fact that each of the resulting bilinear forms has trace zero with respect to (g^{ij}) is obvious from the first expression; for the second one it follows from the apolarity property (3.11) of the cubic form, and for the third one, it corresponds to the vanishing of the affine mean curvature. Thus, for instance, we shall denote by (Lv, Av) the Hilbert inner product

$$\int_\Sigma (Lv)_{ij}(Av)_{k\ell}g^{ik}g^{j\ell}dA.$$

In order to reduce $\delta^2 A$ to the form just outlined, we shall deal separately with the various terms in the integrals (4.22). The integral $\int_\Sigma (\Delta v)^2 dA$ can be reduced as follows: applying Stokes' theorem and the Ricci identities,

$$\int_\Sigma (\Delta v)^2 dA = \int_\Sigma v\Delta^2 v\,dA = \int_\Sigma vg^{ik}g^{j\ell}v_{.ikj\ell}dA$$

$$= \int_\Sigma v g^{ik} g^{j\ell} (v_{.ijk} + v_{,h} R^h{}_{ikj})_{,\ell} dA$$

$$= \int_\Sigma (g^{ik} g^{j\ell} v_{.ij} v_{.k\ell} - v(R^{h\ell} v_{.h})_{,\ell}) dA$$

$$= \int_\Sigma (g^{ik} g^{j\ell} v_{.ij} v_{.k\ell} + R^{ik} v_{,i} v_k) dA.$$

On the one hand we use the (pointwise) orthogonal decomposition of $v_{.ij}$

$$v_{.ij} = \left(v_{.ij} - \frac{1}{n} \Delta v g_{ij} \right) + \frac{1}{n} \Delta v g_{ij} = (Lv)_{ij} + \frac{1}{n} (\Delta v) g_{ij};$$

on the other hand we use the contracted form (3.18) of the Gauss equation, yielding

$$\int_\Sigma (\Delta v)^2 dA = (Lv, Lv) + \frac{1}{n} \int_\Sigma (\Delta v)^2 dA$$

$$+ \int_\Sigma \left(A_j^{\ hi} A_h^{\ jk} + \frac{n-2}{2} B^{ik} \right) v_{,i} v_{,k} dA$$

$$= \frac{n}{n-1} (Lv, Lv) + \frac{n}{n-1} (Av, Av)$$

$$+ \frac{n(n-2)}{2(n-1)} \int_\Sigma B^{ik} v_{,i} v_{,k} dA.$$

Thus $\delta^2 A$ can be written as follows

$$\delta^2 A = -(n+1) \left\{ \frac{n}{n-1} (Lv, Lv) + \frac{n}{n-1} (Av, Av) + (n+2)(Bv, Bv) \right\}$$

$$+ \frac{(n+1)(n^2+4n-4)}{2(n-1)} \int_\Sigma B^{ik} v_{,i} v_{,k}.$$

The integral in the last expression can be reduced as follows: applying

Stokes' theorem and the contracted form (3.17) of the Codazzi equation of the second kind, we have

$$-\int_\Sigma B^{ik}v_{,i}v_{,k}dA = \int_\Sigma (B^{ik}v_{,i})_{,k}\,v dA = \int_\Sigma (B^{ik}v_{,ik} + B^{ik}{}_{,k}v_i)v dA$$

$$= \int_\Sigma v(B^{ik}(Lv)_{ik} + B^{hk}A_{hk}{}^{i}v_{,i})dA = (Bv, Lv + Av).$$

This last identity enables to express $\delta^2 A$ entirely as a combination of inner products as follows

(4.23)

$$\delta^2 A = -\frac{n+1}{n-1}\{n(Lv, Lv) + n(Av, Av) + (n-1)(n+2)(Bv, Bv)\}$$

$$-\frac{(n+1)(n^2+4n-4)}{2(n-1)}(Bv, Lv + Av).$$

The above formula for $\delta^2 A$ expresses it as a negative semidefinite quadratic functional of v when $n = 2$, namely

$$\delta^2 A = -6(Lv + Bv, Lv + Bv) - 6(Av + Bv, Av + Bv).$$

Thus we have proved that $\delta^2 A$ is negative semidefinite for all infinitesimal deformations of X represented by functions $v:\Sigma \to \mathbf{R}$ with compact support, except perhaps for those that satisfy the following, overdetermined, hyperbolic system of equations

$$Lv = -Bv = Av,$$

or, more explicitly,

$$v_{,ij} - \frac{1}{2}\Delta v g_{ij} = A_{ij}{}^{h}v_{,h} = -B_{ij}v. \tag{4.24}$$

To show that there is no function in Σ, vanishing with its gradient on the boundary of Σ, (recalling that Σ is compact) satisfying the above system equations, let (P^{ij}) be any contravariant, non-degenerate symmetric tensor satisfying $g_{ij}P^{ij} = 0$; (P^{ij}) is the symbol of a second order hyperbolic operator on functions of two variables, whose characteristic directions, at any given point, may be an arbitrary pair of line elements that are mutually orthogonal with respect to (g_{ij}). For any orthogonal net of curves in Σ there exists a non-singular, symmetric tensor P^{ij} in Σ satisfying $P^{ij}g_{ij} = 0$, such that the characteristic net of the differential operator $v \to P^{ij}v_{,ij}$ is the given net; in particular, any given, simple curve, smoothly imbedded in Σ is nowhere characteristic with respect to the operator $P^{ij}v_{,ij}$ for a suitable choice of the tensor P^{ij}. If γ is such a curve (including, if necessary, parts of $\partial\Sigma$), along which v is identically zero, but touching the support of a function v satisfying (4.24) at an interior point p of γ, let P^{ij} be a non-singular, symmetric tensor in a neighborhood of γ, satisfying $P^{ij}g_{ij} = 0$, and for which γ is nowhere in a characteristic direction; then, contracting the three members of (4.24) with P^{ij}, we obtain the system

$$P^{ij}v_{,ij} = P^{ij}A_{ij}{}^{h}v_{,h} = -p^{ij}B_{ij}v,$$

with v vanishing with its gradient along a non-characteristic curve γ, touching the support of v at point p. These facts contradict the uniqueness, in particular, of the solution of the Cauchy problem, for instance, of $P^{ij}v_{,ij} + P^{ij}B_{ij}v = 0$ with zero Cauchy data on γ. This contradiction shows that (4.24) has only one solution, the trivial one, vanishing with its gradient, on $\partial\Sigma$. Thus for any v representing an allowable, infinitesimal normal deformation of X, $\delta^2 A$ is negative valued, thus completing the proof of Theorem 1.3.

We conclude this section with a proof of Theorem 1.4, a weak form of a statement of the type of Bernstein's theorem on complete minimal surfaces in euclidean 3-space.

Proof of Theorem 1.4. Let $X:\Sigma \to E$ be a locally strongly convex hypersurface, immersed in $(n + 1)$-space (again we wish to emphasize the special role that the case $n = 2$ plays on the proof at the crucial point). If $X(\Sigma)$ is geometrically complete and locally strongly convex, then it follows from Hadamard's geometrical lemma that $X(\Sigma)$ is either a closed, hypersurface bounding a convex body, or an open, globally convex hypersurface, properly imbedded (i.e., with no finite boundary point) in E, and therefore globally the graph of a single-valued, strongly convex function, say

(4.25) $$x^0 = z(x^1, \ldots, x^n) = z(x)$$

defined in a convex, open domain $D \subset \mathbf{R}^n$, for suitable affine coordinates $(x^0, x^1, \ldots, x^n)$. If, in addition, we assume that X is an affine maximal hypersurface, then the closed (compact) case is excluded, because then, for any fixed point $0 \in E$, the real valued function on Σ, $\varphi = \langle Y^*, X - 0\rangle$ must achieve a positive maximum value at some point $p_0 \in \Sigma$; for such a point the origin 0 lies then on the affine normal line $\{t \to X(p_0) + tY(p_0) \mid t \in \mathbf{R}\}$, since $\varphi_{,i}(p_0) = 0$, and $\varphi_{ij}\xi^i\xi^j \leq 0$. But since $\varphi_{.ij} = -B_{ij}\varphi + g_{ij}$, it shows that (B_{ij}) must be positive definite at p_0, contradicting the assumption that $H = 0$.

If, on the other hand, $X(\Sigma)$ is the graph of a strongly convex function (4.25), we may assume again (by a suitable unimodular affine change of coordinates $(x^0, \ldots, x^n)$), that

(4.26) $$\begin{cases} z(0, \ldots, 0) = z(0) = 0, \\ \dfrac{\partial z}{\partial x^i}(0) = 0, \qquad \dfrac{\partial^2 a}{\partial x^i \partial x^j}(0) = \delta_{ij} \quad (1 \leq i, j \leq n). \end{cases}$$

Consider the $n + 1$ components $(y_0, \ldots, y_n)$ of the co-normal vector Y^* with respect to the dual coordinate system in $T^*(E)$. Then each y_α is a real valued function on Σ, and since $X(\Sigma)$ is a graph, the function y_0 is strictly positive valued (cf. (1.5)). At the same time, it follows from the structure equation (3.10) and from the assumption $H = 0$, that each component y_α is a harmonic function, since

$$\Delta Y^* = -g^{ij}(A_{ij}{}^h Y^*_{,h} + B_{ij}Y^*) = 0.$$

If we assume now that Σ is metrically complete, i.e., complete in the Riemannian sense with respect to the Berwald-Blaschke metric, and using the contracted froms (3.18), (3.19) of the Gauss equation, if $n = 2$ and $H = 0$, it follows that $R \geq 0$ (for all n) and the Ricci tensor (R_{ij}) is positive semidefinite (for $n = 2$). It follows from the theorem of Blanc-Fiala [1] that, since Σ is complete and has nonnegative curvature with respect to the metric, it is of parabolic type and hence the positive harmonic function y_0 is constant. From equation (1.5) we know that

$$y_0 = (\det(z_{ij}(x)))^{\frac{-1}{n+2}} = (\det(z_{ij}(0)))^{\frac{-1}{n+2}} = 1.$$

Therefore the function $z(x_1, x_2)$ whose graph describes $X(\Sigma)$ is a solution of the Monge-Ampère equation

$$z_{11}z_{22} - (z_{12})^2 = 1.$$

The completeness of the surface allows us to apply the theorem of K. Jörgens [4] according to which $z(x)$ is a quadratic polynomial, and, because of the special choice of coordinates satisfying (4.26),

$$x^0 = \frac{1}{2}(x^1)^2 + \frac{1}{2}(x^2)^2.$$

This result completes the proof of Theorem 1.4. A refinement of this theorem enables us to estimate *a priori* the third derivatives

$$\frac{\partial^3 z}{\partial x_i \partial x_j \partial x_k}$$

of the function $z(x)$ representing any *incomplete* solution of the equation $H = 0$ (for example, occurring as the solution of a boundary value problem) in terms of bounds of the lower order derivatives. This would be but one step in solving the affine analogue of the Plateau problem.

UNIVERSITY OF PENNSYLVANIA
INSTITUTE FOR ADVANCED STUDY

REFERENCES

---

[1] Blanc, Ch. and F. Fiala, Le type d'une surface et sa courbure totale, *Comm. Math. Helv.* **14** (1941–42), 230–233.
[2] Blaschke, W., Vorlesungen über Differentialgeometrie; II-Affine Differentialgeometrie, Berlin, Springer, 1923.

[3] Calabi, E., Improper Affine Hyperspheres of Convex Type and a Generalization of a Theorem by K. Jörgens, *Mich. Math. J.* **5** (1958), 105–126.
[4] Jörgens, K., Über die Lösungen der Differentialgleichung $rt - s^2 = 1$, *Math. Ann.* **127** (1954), 180–184.
[5] Pogorelov, A. V., On the Improper, Convex, Affine Hyperspheres, *Geometriae Dedicata* **1** (1972), 33–46.
[6] Salkowski, E., Affine Differentialgeometrie, Berlin, de Gruyter, 1934.
[7] Shirokov (Shirokow), P. A. & Shirokov, A. P., Affine Differentialgeometrie, Leipzig, Teubner, 1962.
[8] Hsiung, C. C. and J. K. Shahin, Affine Differential Geometry of Closed Hypersurfaces, *Proc. London Math. Soc.*, 3rd ser., **17** (1967), 715–735.

Extremal Kähler Metrics II

Eugenio Calabi[1]

Differential Geometry and Complex Analysis,
Vol. dedic. H. E. Rauch, 95–114, Springer, Berlin (1985)

Abstract

Given a compact, complex manifold M with a Kähler metric, we fix the deRham cohomology class Ω of the Kähler metric, and consider the function space $\mathscr{G}_\Omega$ of all Kähler metrics in M in that class. To each $(g) \in \mathscr{G}_\Omega$ we assign the non-negative real number $\Phi(g) = \int_M R_g^2\, dV_g$ (R_g = scalar curvature, dV_g = volume element). Aiming to find a $(g) \in \mathscr{G}_\Omega$ that minimizes the function Φ, we study the geometric properties in M of any $(g) \in \mathscr{G}_\Omega$ that is a critical point of Φ, with the following results.

1) Any metric (g) that is a critical point of Φ is necessarily invariant under a maximal compact subgroup of the identity component $\mathfrak{H}_0(M)$ of the complex Lie group of all holomorphic automorphisms of M.

2) Any critical metric $(g) \in \mathscr{G}_\Omega$ of Φ achieves a local minimum value of Φ in $\mathscr{G}_\Omega$; the component of (g) in the critical set of Φ coincides with the orbit of (g) under the action of the group $\mathfrak{H}_0(M)$, it is diffeomorphic to an open euclidean ball, and the critical set is always non-degenerate in the sense of $\mathfrak{H}_0(M)$-equivariant Morse theory.

3) If there exists a $(g) \in \mathscr{G}_\Omega$ with constant scalar curvature R, then it achieves an absolute minimum value of Φ; furthermore every critical metric in $\mathscr{G}_\Omega$ has constant R, and achieves the same value of Φ.

4) Whenever the existence of a critical Kähler metric (g) can be guaranteed (i.e., always, according to a conjecture [2]), then Futaki's obstruction determines a necessary and sufficient condition for the existence of a $(g) \in \mathscr{G}_\Omega$ with constant scalar curvature.

1. Introduction

The previous article with the same title [2] introduces the following variational problem. We consider a compact, connected, complex n-dimensional manifold M

---

[1] Research supported by NSF Grant No. MCS 81-15107 and by the Institute for Advanced Study, Princeton, NJ.

[2] (Added in proof) This conjecture has just been disproved by a class of counterexamples due to M. Levine [6]

DOI: 10.1007/978-3-642-69828-6_8

without boundary, and assume that M admits a Kähler metric, locally expressible in the form

$$(g)\colon ds^2 = 2 g_{\alpha\bar\beta}\, dz^\alpha\, d\overline{z^\beta};$$

we fix the deRham cohomology class Ω of the real valued, closed exterior (1, 1)-form

$$\omega = \sqrt{-1}\; g_{\alpha\bar\beta}\, dz^\alpha \wedge d\overline{z^\beta}$$

associated to (g), and denote by $\mathscr{G}_\Omega$ the function space of all differentiable Kähler metrics (g) with $\omega \in \Omega$. In this function space we introduce the (non-negative) real-valued functional Φ, which assigns to each (g) the integral

$$\Phi(g) = \int_M R^2\, dV, \tag{1.1}$$

where $dV = dV_g = (\sqrt{-1})^n \det(g_{\lambda\bar\mu}) \bigwedge_{\alpha=1}^{n} (dz^\alpha \wedge d\overline{z^\alpha})$ denotes the volume element in M associated with (g) and

$$R = -\, g^{\alpha\bar\beta} \frac{\partial^2}{\partial z^\alpha\, \partial \overline{z^\beta}} \log\det(g_{\lambda\bar\mu})$$

the scalar curvature.

The variational problem introduced in [2] is that of minimizing the functional $\Phi(g)$ over all $(g) \in \mathscr{G}_\Omega$. A preliminary justification for considering this problem is the fact that, as (g) varies in $\mathscr{G}_\Omega$, both the total volume

$$V = V_g = \int_M dV$$

and the total scalar curvature

$$S_g = \int_M R\, dV$$

remain constant; thus, by virtue of the Schwartz inequality, the function $\Phi(g)$ has a non-negative lower bound S_g^2/V_g; the latter can be achieved, if and only if there exists a $g \in \mathscr{G}_\Omega$ with constant scalar curvature. It is known that this is not always the case, so that any Euler-Lagrange equation should include not only $R = \text{const}$ as a solution, but also a more general conditions, to include the cases where there is an obstruction to constant R.

The Euler-Lagrange equation, characterizing metrics $(g) \in \mathscr{G}_\Omega$ that are critical points of Φ, was described in [2] and its derivation is briefly outlined in the next section; it takes the following form: for any real or complex valued function φ on M, denote by $\uparrow\bar\partial\varphi$ the differentiable section in the holomorphic tangent bundle T'_M of M, locally represented by

$$\uparrow\bar\partial\varphi = \left(g^{\alpha\bar\mu} \frac{\partial\varphi}{\partial\overline{z^\mu}}\right) \frac{\partial}{\partial z^\alpha},$$

and denote by L the second order differential operator

$$L\varphi = \bar\partial \uparrow \bar\partial\varphi = \frac{\partial}{\partial\overline{z^\beta}} \left(g^{\alpha\bar\mu} \frac{\partial\varphi}{\partial\overline{z^\mu}}\right) \frac{\partial}{\partial z^\alpha} \otimes d\overline{z^\beta}. \tag{1.2}$$

Thus $L\varphi$ is a cross section in the vector bundle $T'_M \otimes A_M^{0,1}$, where $A_M^{p,q}$ is the bundle of complex valued exterior (p,q)-forms in M, and $L\varphi = 0$, if and only if $\uparrow \bar{\partial}\varphi$ is a holomorphic section in T'_M. It is convenient to introduce also the formal adjoint operator L^* of L with respect to the hermitian metric (g); thus L^* is a map from $\Gamma_M(T'_M \times A_M^{0,1})$ into $\Gamma(A_M^{0,0})$, where $\Gamma_M(E)$ denotes the $A_M^{0,0}$-module of differentiable sections over M in any vector bundle E. The Euler-Lagrange equation then takes the form

$$L^* LR = 0, \tag{1.3}$$

whose global solutions, in a compact manifold M without boundary, automatically satisfy the equation $LR = 0$.

The existence of an extremal metric in $\mathscr{G}_\Omega$, that is to say of a metric (g) for which $\Phi(g)$ assumes its minimum possible value in $\mathscr{G}_\Omega$, is not yet proved for all possible pairs (M, Ω); however, its existence in a fairly wide collection of special cases, and other heuristic reasons, lead us to believe in its general validity. The purpose of this paper is to describe general properties of *critical* Kähler metrics (g), that is to say metrics $(g) \in \mathscr{G}_\Omega$ that are critical points of the function Φ. Such a description serves several purposes; it may first of all suggest methods of attacking the existence problem of a minimizing metric by furnishing some useful estimates; it provides some very explicit information about the structure of the group $\mathfrak{H}(M)$ of all holomorphic transformation of M, and especially of its identity component $\mathfrak{H}_0(M)$; finally the results of the study of the local behavior of the function Φ at or near the critical points are likely to be extended to the total function space $\mathscr{G}_\Omega$. At a later stage it will also be interesting to investigate how the minimum value behaves as Ω itself varies in the open cone of Kähler classes in M, or perhaps even on the cone's boundary, corresponding to classes of degenerate Kähler metrics, or (by varying also the complex structure of M) on neighboring Kählerian structures.

We list below some of the important results of this paper.

Theorem 1. Let M be a closed, Kählerian manifold, Ω a given Kähler class on M, and $(g) \in \mathscr{G}_\Omega$ a Kähler metric that is a critical point of the function Φ. Then one can define in terms of (g) a unique semidirect sum splitting of the Lie algebra $\mathfrak{h}(M)$ of all holomorphic tangent vector fields of M as follows,

$$\mathfrak{h}(M) = \mathfrak{a}(M) \oplus \mathfrak{h}'(M), \tag{1.4}$$

where $\mathfrak{a}(M)$ is the Lie subalgebra of $\mathfrak{h}(M)$ consisting of the autoparallel, holomorphic vector fields of M, and $\mathfrak{h}'(M)$ is the Lie algebra ideal in $\mathfrak{h}(M)$ consisting of the image under $\uparrow \bar{\partial}$ of the kernel of L in $\Gamma_M(A^{(0,0)})$. The ideal $\mathfrak{h}'(M)$ contains the derived Lie algebra of $\mathfrak{h}(M)$.

The above theorem generalizes to metrics satisfying (1.3) a similar decomposition, first discovered by Y. Matsushima [5] in the case of Einstein-Kähler manifolds, and then by A. Lichnérowicz [4] in the case of Kähler manifolds with constant R.

Corollary. Let M be a closed Kähler manifold, $\mathfrak{A}_M$ its Albanese torus with is structural map $J: M \to \mathfrak{A}_M$, $\mathfrak{H}_0(M)$ the identity component of the group $\mathfrak{H}(M)$ of all holomorphic transformations of M, $J_*: \mathfrak{H}(M) \to \mathfrak{H}(\mathfrak{A}_M)$ the induced holomorphic homomorphism of the respective transformation groups. Then, if M admits a critical Kähler metric $(g) \in \mathscr{G}_\Omega$ for some given Kähler class Ω, the decomposition (1.4) of the Lie algebra $\mathfrak{h}(M)$ of $\mathfrak{H}_0(M)$ induces the following decomposition of $\mathfrak{H}_0(M)$: the Lie subgroup $\mathfrak{A}'_M \subset \mathfrak{H}_0(M)$ generated by $\mathfrak{a}(M)$ is a complex torus, in which the homomorphism J_* has a kernel of finite order, while the total kernel $\mathfrak{H}'(M)$ of J_* in $\mathfrak{H}_0(M)$ has at most finitely many components, of which the one containing the identity is the group generated by $\mathfrak{h}'(M)$.

The above corollary is likely to be valid without assuming the existence of an extremal Kähler metric, at least in the case of non-singular, projective algebraic varieties; in that case, however, the decomposition statement would be existential, rather than explicit.

The open subgroup $\mathfrak{H}_\Omega(M)$ of $\mathfrak{H}(M)$ consisting of all holomorphic transformations of M preserving the Kähler class Ω has $\mathfrak{H}_0(M)$ as its identity component; it has a natural action on the function space $\mathscr{G}_\Omega$ of Kähler metrics, and thus can be regarded as the "gauge group" action for the variational problem of minimizing Φ, since the value of Φ is invariant under the action of $\mathfrak{H}_\Omega(M)$. The isotropy subgroup of this action for any Kähler metric $(g) \in \mathscr{G}_\Omega$ is the group $\mathfrak{I}_H(M,(g))$ of the holomorphic isometries of $(M,(g))$, which, for generic (g) is the trivial group. Thus the orbit of each $(g) \in \mathscr{G}_\Omega$ under the action of $\mathfrak{H}_\Omega(M)$ has $2\dim_{\mathbf{C}} \mathfrak{H}_\Omega(M) - \dim_{\mathbf{R}} \mathfrak{I}_H(M,(g))$ dimensions. Since the maximum dimension of $\mathfrak{I}_H(M,(g))$ for all possible $(g) \in \mathscr{G}_\Omega$ is that of any maximal, compact subgroup of $\mathfrak{H}_\Omega(M)$, one obtains for any (M,Ω) (in fact, independently of Ω) a lower bound for the dimension of the orbit of any (g). A consequence of the next two theorems is, in fact, that any critical Kähler metric in $\mathscr{G}_\Omega$ achieves this lower bound: in other words, each critical Kähler metric includes in its holomorphic isometry group the identity component (at least) of a maximal compact subgroup of $\mathfrak{H}_\Omega(M)$.

Theorem 2. The hessian form of Φ in $\mathscr{G}_\Omega$ at any critical point (g), that is to say the second variation of Φ, is positive semidefinite with finite co-rank. More precisely, let $(g)_t$ be any smooth, regular family of Kähler metrics in $\mathscr{G}_\Omega$, depending on a real parameter t, such that, for $t = 0$, $(g)_0$ is a critical point of Φ. Then $\left.\dfrac{d^2\Phi[(g)_t]}{dt^2}\right|_{t=0} \geqq 0$, and is > 0 whenever the path $(g)_t$ at $t = 0$ is transversal to the orbit of $(g)_0$ under the action of the group $\mathfrak{H}_0(M)$. Consequently, the orbit of $(g)_0$ under the action of $\mathfrak{H}_0(M)$ coincides with the component of $(g)_0$ in the critical set of Φ, and with the component of $(g)_0$ in the level set of $(g)_0$ for the function Φ.

The final statement says, in other words, that every critical metric (g) achieves a local, non-degenerate minimum value of Φ relative to the action of the gauge group $\mathfrak{H}_\Omega(M)$.

Perhaps the most remarkable property of the critical Kähler metrics is that they automatically exhibit essentially the greatest degree of symmetry compatible with the complex structure of M. This is stated formally as follows.

Theorem 3. For any critical Kähler metric (g) in a compact, complex manifold M, the identity component of the holomorphic isometry group $\mathfrak{I}_{H,0}(M,g)$ coincides with a maximal compact, connected subgroup of the group of holomorphic transformations $\mathfrak{H}(M)$ of M.

Corollary. Each connected component in $\mathscr{G}_\Omega$ of the critical set of the functional Φ is an imbedded submanifold, diffeomorphic to a finite dimensional euclidean space, and homogeneous under the action of the identity component subgroup $\mathfrak{H}_0(M)$ of the group $\mathfrak{H}(M)$ of all holomorphic transformations of M.

The fourth and last section includes an application to our theory of a recent result by A. Futaki [3] and S. Bando [1], describing an obstruction to the existence of a Kähler metric with constant scalar curvature.

Theorem 4. In any function space $\mathscr{G}_\Omega$ of Kähler metrics in M in a fixed Kähler class Ω, the existence of a Kähler metric (g) with R = const and of another one, also critical for Φ but with R not constant, are mutually exclusive. In other words, if for each pair (M, Ω) there is a Kähler metric minimizing (g), then Futaki's obstruction to the existence of a $(g) \in \mathscr{G}_\Omega$ with R = const is not only a necessary condition but also a sufficient one.

An example is available, showing that Futaki's condition for the existence of a metric with R = const is satisfied for some Kähler classes, while for others it is not, so that the question, whether an extremal Kähler metric for the functional Φ satisfies R = const or not (assuming, of course, that such a metric exists), can not necessarily be determined solely by the complex structure of the manifold M.

In addition to the main problem of existence of an extremal Kähler metric in each $\mathscr{G}_\Omega$, there are several obvious questions arising in connection with the variational problem itself; we may mention some of the most conspicuous ones: whether the critical value of the functional Φ is unique in each case; whether it necessarily corresponds to the globally minimum value of Φ; and finally, whether the critical set is always a connected manifold. Each of these questions, as well as the existence question, would be settled affirmatively, if one could show that, under a suitable topology on the function space $\mathscr{G}_\Omega$, the function Φ could be treated as a Morse function invariant under $\mathfrak{H}_\Omega(M)$. It is more conceivable, however, that, by using other methods, one may find partial solutions to these problems and by doing so gain more insight in the rest.

2. Decomposition Theorems for Critical Kähler Manifolds

We will recall some notations from the introductory section and introduce some new ones, while reviewing some elementary, well known facts. Throughout this article M denotes a fixed, compact, connected, complex n-dimensional, nonsingular manifold without boundary; we assume that M admits a Kähler metric denoted by (g), while Ω denotes the deRham class of the associated (1.1)-form, also called the Kähler class of (g). We denote by $\mathscr{G}_\Omega$ the function space of all Kähler metrics (g) in the Kähler class Ω. For any other Kähler metric $(g') \in \mathscr{G}_\Omega$, represented by the local components $g'_{\alpha\bar\beta}$, there exists a real valued function u: $M \to \mathbf{R}$, unique up to an additive constant, called the Kähler distortion potential from (g) to (g'), such that

$$g'_{\alpha\bar\beta} = g_{\alpha\bar\beta} + \frac{\partial^2 u}{\partial z^\alpha\, \partial \overline{z^\beta}}.$$

Given any Kähler metric $(g) \in \mathscr{G}_\Omega$, it defines the following objects, in which the dependence on (g) may be indicated as a subscript when necessary, but usually omitted whenever convenient:

a) the Laplace-Beltrami operator $\Delta = \Delta_g$,

$$\Delta\varphi = g^{\alpha\bar\beta}\frac{\partial^2 \varphi}{\partial z^\alpha\, \partial \overline{z^\beta}} = \varphi_{,\alpha}{}^{\alpha} = \varphi^{,\alpha}{}_{\alpha}$$

(here we denote covariant or contravariant derivatives by lower or upper indices following a comma, read from left to right; the summation convention is assumed throughout);

b) the first-order complex derivation $\uparrow\bar\partial$, where

$$\uparrow\bar\partial\varphi = g^{\alpha\bar\mu}\frac{\partial\varphi}{\partial \overline{z^\mu}}\frac{\partial}{\partial z^\alpha} = \varphi^{,\alpha}\frac{\partial}{\partial z^\alpha};$$

c) the second order operator L,

$$L\varphi = \bar\partial\uparrow\bar\partial\varphi = \frac{\partial}{\partial \overline{z^\beta}}\left(g^{\alpha\bar\mu}\frac{\partial\varphi}{\partial \overline{z^\mu}}\right)\frac{\partial}{\partial z^\alpha}\otimes d\overline{z^\beta} = \varphi^{,\alpha\lambda}g_{\lambda\bar\beta}\frac{\partial}{\partial z^\alpha}\otimes d\overline{z^\beta};$$

thus L is an operator from the sheaf of sections in $A_M^{0,0}$ (complex valued functions) to that of sections in $T'_M \otimes A_M^{0,1}$ (holomorphic-tangent-vector-valued (0, 1)-forms);

d) The adjoint operator L^* of L, from germs of sections in $T'_M \otimes A_M^{0,1}$ to those of $A_M^{0,0}$:

$$\begin{aligned} L^*\left(\psi^\alpha{}_{\bar\beta}\frac{\partial}{\partial z^\alpha}d\overline{z^\beta}\right) &= g^{\gamma\bar\beta}\,\psi^\alpha{}_{\bar\beta,\gamma\alpha} \\ &= g^{\lambda\bar\mu}\frac{\partial}{\partial z^\lambda}\left[g^{\gamma\bar\beta}\frac{\partial}{\partial z^\gamma}(g_{\alpha\bar\mu}\psi^\alpha{}_{\bar\beta})\right]; \end{aligned}$$

e) The composite operator $D = L^* L$; this is a complex valued, self-adjoint operator on real or complex valued functions; it can be expressed in various ways, using the Ricci identities: for instance

$$\begin{aligned} D\varphi &= \varphi^{,\alpha\beta}{}_{\beta\alpha} = \varphi^{,\alpha\beta}{}_{\alpha\beta} = \varphi^{,\alpha}{}_{\alpha}{}^{\beta}{}_{\beta} + (\varphi^{,\alpha}R_\alpha{}^\beta)_{,\beta} \\ &= \Delta^2\varphi + \varphi^{,\alpha\bar\beta}R_{\alpha\bar\beta} + \varphi^{,\beta}R_{,\beta}, \end{aligned} \tag{2.1}$$

where $R_{\alpha\bar\beta} = -\dfrac{\partial^2}{\partial z^\alpha\, \partial \overline{z^\beta}}\log\det(g_{\lambda\bar\mu})$ represents the covariant components of the Ricci tensor, and $R_\alpha{}^\beta = g^{\beta\bar\mu}R_{\alpha\bar\mu}$ the mixed tensor components. Alternatively, when the dependence of D on (g) must be explicitly emphasized, we may write in the displayed form

$$D\varphi = g^{\alpha\bar{\nu}} \frac{\partial}{\partial z^\alpha} \left\{ g^{\beta\bar{\gamma}} \frac{\partial}{\partial z^\beta} \left[g_{\lambda\bar{\nu}} \frac{\partial}{\partial \bar{z}^{\gamma}} \left(g^{\lambda\bar{\mu}} \frac{\partial \varphi}{\partial \bar{z}^\mu} \right) \right] \right\}. \tag{2.2}$$

f) The conjugate complex operator $\bar{D}$ of D is also used; it is defined by $\bar{D}\varphi = \overline{(D\bar{\varphi})} = \varphi_{,\alpha\beta}{}^{\beta\alpha}$, and, like D, can be similarly expressed in several equivalent ways.

The hermitian product between two global cross sections over M in the same hermitian, complex vector bundle is denoted by round parentheses. Thus, for instance, for any two functions φ, ψ

$$(\varphi, \psi) = \int_M \varphi(z, \bar{z})\, \overline{\psi(z, \bar{z})}\, dV$$

and

$$\begin{aligned}(L\varphi, L\psi) &= \int_M \varphi^{,\alpha}{}_{\bar{\beta}}\, \overline{(\psi^{,\lambda}{}_{\bar{\mu}})}\, g_{\alpha\bar{\lambda}} g^{\mu\bar{\beta}}\, dV \\ &= \int_M \varphi^{,\alpha\beta} (\bar{\psi})_{,\alpha\beta}\, dV = (D\varphi, \psi) = (\varphi, D\psi).\end{aligned} \tag{2.3}$$

The last identity with $\psi = \varphi$ shows that D is a positive semidefinite, self-adjoint operator on functions in M, and $D\varphi = 0$ only if $L\varphi = 0$, i.e. if $\uparrow\bar{\partial}\varphi$ is a holomorphic vector field.

The relation (2.1) between the operator D and Δ^2 shows also that D, and hence also L, are strongly elliptic operators; using the analogous relation between $\bar{D}$ and Δ^2, and taking the difference between the two respective parts, one obtains following comparison between $\bar{D}$ and D:

$$(\bar{D} - D)\varphi = R^{,\alpha}\varphi_{,\alpha} - R_{,\alpha}\varphi^{,\alpha}; \tag{2.3'}$$

this relation implies immediately the following lemma due to A. Lichnérowicz [4].

Lemma 2.1. The operators D and $\bar{D}$ coincide, if and only if the scalar curvature R is constant. Moreover, in the special case where $\varphi = R$, $DR = \bar{D}R$ for any Kähler metric.

To any *real* valued function φ on M we associate the tangent vector field δ_φ in the function space manifold $\mathscr{G}_\Omega$, that assigns to each $g \in \mathscr{G}_\Omega$ the infinitesimal variation

$$\delta_\varphi(g) = 2 \frac{\partial}{\partial t}\bigg|_{t=0} \left(g_{\alpha\bar{\beta}} + t \frac{\partial^2 \varphi}{\partial z^\alpha \partial \bar{z}^\beta} \right) dz^\alpha\, d\bar{z}^\beta = 2 \frac{\partial^2 \varphi}{\partial z^\alpha \partial \bar{z}^\beta} dz^\alpha\, d\bar{z}^\beta.$$

It is clear that, for any two real valued functionals φ, ψ on M, the associated vector fields δ_φ and δ_ψ on $\mathscr{G}_\Omega$ commute; further calculations yield the following identities

$$\delta_\varphi(dV) = \Delta\varphi\, dV = \varphi^{,\alpha}{}_{\alpha}\, dV, \tag{2.4}$$

$$\delta_\varphi(R) = -\Delta^2\varphi - \varphi^{,\alpha\bar{\beta}} R_{\alpha\bar{\beta}} = -D\varphi + \varphi^{,\alpha} R_{,\alpha} = -\bar{D}\varphi + \varphi_{,\alpha} R^{,\alpha}, \tag{2.5}$$

and hence

$$\begin{aligned}\delta_\varphi(\Phi(g)) &= \int_M (2R\,\delta_\varphi(R) + R^2\,\Delta\varphi)\, dV = \int_M (-2R\, D\varphi + (R^2 \varphi^{,\alpha})_{,\alpha})\, dV \\ &= -2(D\varphi, R) = -2(\varphi, DR).\end{aligned} \tag{2.6}$$

This last identity shows that $DR = 0$ is precisely the Euler-Lagrange equation characterizing Kähler metrics (g) that are critical points for the function Φ on $\mathscr{G}_\Omega$. Equation (2.3) specialized to the case $\varphi = \psi = R$, shows that $DR = 0$ is equivalent to the equation $LR = 0$, or to the property that $\uparrow \bar\partial R$ is a holomorphic vector field. We can now proceed to prove the first decomposition theorem.

Proof of Theorem 1. We assume that a certain metric $(g) \in \mathscr{G}_\Omega$ on M has the property that its scalar curvature R satisfies $R^{,\alpha\beta} = 0$. Let $X = X^\alpha \dfrac{\partial}{\partial z^\alpha}$ be any holomorphic, tangent vector field on M, and denote by $\downarrow$ the bijective, $\mathbf{C}$-linear vector bundle homomorphism of T'_M onto $A_M^{0,1}$ determined by the metric, and by $\uparrow$ the inverse mapping. Consider the Hodge decomposition of the $(0,1)$-form $\downarrow X = X_{\bar\beta}\, d\overline{z^\beta}$, $X_{\bar\beta} = g_{\alpha\bar\beta} X^\alpha$, consisting of the two parts

$$X_{\bar\beta}\, d\overline{z^\beta} = (HX)_{\bar\beta}\, d\overline{z^\beta} + X'_{\bar\beta}\, d\overline{z^\beta},$$

where $(HX)_{\bar\beta}\, d\overline{z^\beta} = H \downarrow X$ is a harmonic $(0,1)$-form, i.e. a closed, conjugate-holomorphic 1-form, and $X'_{\bar\beta}\, d\overline{z^\beta}$ a $\bar\partial$-exact one: this means that there exists a complex valued function ψ on M, unique up to an additive constant, such that $X'_{\bar\beta} = \partial\psi/\partial\overline{z^\beta}$; applying $\uparrow$ to both sides, we have

$$X = \uparrow HX + \uparrow \bar\partial\psi. \tag{2.7}$$

Since X is annihilated by $\bar\partial$, we can express $D\psi = L^*\bar\partial \uparrow \bar\partial\psi$ as follows:

$$\begin{aligned} D\psi &= -L^*\bar\partial(\uparrow HX) = -g^{\alpha\bar\beta}(HX)_{\bar\beta}{}^{,\gamma}{}_{\gamma\alpha} \\ &= -g^{\alpha\bar\beta}(HX)_{\bar\beta}{}^{,\gamma}{}_{\alpha\gamma} = -g^{\alpha\bar\beta}((HX)_{\bar\beta,\alpha}{}^{\gamma}{}_{\gamma} + ((HX)_{\bar\tau} R^{\bar\tau}{}_{\bar\beta}{}^{\gamma}{}_{\alpha})_{,\gamma}) \end{aligned}$$

where $(HX)_{\bar\beta,\alpha} = 0$ identically, since (HX) is conjugate-holomorphic. Thus

$$D\psi = -((HX)_{\bar\tau} R^{\bar\tau\gamma})_{,\gamma} = -(HX)_{\bar\tau} R^{,\bar\tau}.$$

The last expression $(HX)_{\bar\tau} R^{,\bar\tau}$ is the inner product of a conjugate-holomorphic $(0,1)$-form and a conjugate-holomorphic section in the tangent bundle; therefore $(HX)_{\bar\tau} R^{,\bar\tau}$ is a complex constant; but $D\psi$ can be written as a divergence, so that $\int_M D\psi\, dV = 0$ identically; therefore $D\psi$, being constant, is identically zero. Thus $\uparrow \bar\partial\psi$ is a holomorphic vector field, and so is $\uparrow HX$; the latter statement means that $(HX)_{\bar\beta,\bar\tau} = 0$; however, we have also $(HX)_{\bar\beta,\gamma} = 0$, since the $(0,1)$-form HX is conjugate-holomorphic, so that the complex vector field $\uparrow HX$ is an autoparallel vector field. It is clear that the decomposition (2.7) is unique; therefore (1.4) represents a decomposition of modules into a direct sum. It is easy to verify now that the Lie "bracket" products of the parts (1.4) of the decomposition satisfy the following relations:

$$[\mathfrak{a}(M), \mathfrak{a}(M)] = \{0\}$$
$$[\mathfrak{h}(M), \mathfrak{h}(M)] \subset \mathfrak{h}'(M):$$

the latter identity is verified by considering any two holomorphic vector fields X, Y and their respective decompositions

$$X = \uparrow HX + \uparrow\bar\partial\psi, \qquad Y = \uparrow HX + \uparrow\bar\partial\eta$$

just established; one verifies then that the holomorphic vector field $[X, Y]$ can be expressed by

$$[X, Y] = \uparrow \bar{\partial}(X^{\beta} \eta_{,\beta} - Y^{\beta} \psi_{,\beta}).$$

This completes the proof of Theorem 1. This proof follows essentially the same idea as Lichnérowicz's in the special case $R = \text{const}$ [4].

Proof of the Corollary of Theorem 1. We recall that the Albanese torus of a Kähler manifold M has the following definition. Let $2q$ be the first Betti number of M with real coefficients, i.e., $q = \dim_{\mathbf{C}} H^{1,0}(M, \mathbf{C})$; let $(\gamma_1, \gamma_2, \ldots, \gamma_{2q})$ be a set of 1-cycles in M representing a system of generators of $H_1(M, \mathbf{Z})$ modulo the torsion subgroup, and $(\omega_1, \ldots, \omega_q)$ a basis for the holomorphic (necessarily closed) 1-forms in M. Then $\mathfrak{A}_M$, the Albanese torus of M, is the complex torus group manifold consisting of $\mathbf{C}^q$, modulo the discrete, additive subgroup Λ generated by the $2q$ vectors ζ_k $(1 \leqq k \leqq 2q)$, where $\zeta_k = (\int_{\gamma_k} \omega_1, \int_{\gamma_k} \omega_2, \ldots, \int_{\gamma_k} \omega_q)$. Choosing an arbitrary "base" point $p_0 \in M$, one defines the Albanese (or Jacobi) map $J \colon M \to \mathfrak{A}_M$ by setting $J(p) = \left(\int_{p_0}^{p} \omega_1, \ldots, \int_{p_0}^{p} \omega_q \right)$ (modulo Λ). This map is a holomorphic map of M into $\mathfrak{A}_M$, such that the induced homomorphism of $H_1(M, \mathbf{Z})$ into $H_1(\mathfrak{A}_M, \mathbf{Z})$ (the latter is naturally identified with Λ) is surjective, the kernel being the torsion group. Denote by $\mathfrak{H}'(M)$ the open subgroup of the full group $\mathfrak{H}(M)$ of holomorphic automorphisms of M, that has trivial action on the first homology group $H_1(M, \mathbf{Z})$. If $F \in \mathfrak{H}'(M)$, then $J[F(p)] = J(p) + J[F(p_0)]$; thus the map $J_* \colon \mathfrak{H}'(M) \to \mathfrak{A}_M$ defined by $J_*(F) = J[F(p_0)]$ is a holomorphic homomorphism of the first group into the second.

Let us assume now that M has a Kähler metric $(g) \in \mathscr{G}_{\Omega}$ that is a critical point of Φ, and consider the resulting semidirect sum splitting of the Lie algebra $\mathfrak{h}(M)$ of the holomorphic tangent vector fields into $\mathfrak{a}(M) \oplus \mathfrak{h}'(M)$, as in Theorem 1. Since the elements of $\mathfrak{h}'(M)$ may be characterized by the property that their inner products with each holomorphic 1-form in M are identically zero, the local group generated by $\mathfrak{h}'(M)$ is the local group of the kernel of J_* in the open subgroup $\mathfrak{H}'(M) \subset \mathfrak{H}(M)$ with trivial action of $H_1(M, \mathbf{Z})$; therefore the restriction J_{*0} of J_* to the identity component $\mathfrak{H}_0(M)$ in $\mathfrak{H}(M)$ has the property that its kernel $\mathfrak{H}_0'(M) \subset \mathfrak{H}_0(M)$ is a normal, closed subgroup, containing the commutator subgroup of $\mathfrak{H}_0(M)$, since $\mathfrak{h}'(M)$ contains the derived ideal $[\mathfrak{h}(M), \mathfrak{h}(M)]$ of $\mathfrak{h}(M)$.

As to the other term in the decomposition (1.4), the subalgebra $\mathfrak{a}(M) \subset \mathfrak{h}(M)$ consists of the autoparallel, holomorphic vector fields in M. Considering the underlying Riemannian structure of M, the action of the local group generated by $\mathfrak{a}(M)$ on M is that of the local group generated by all autoparallel, real vector fields. It follows from the deRham decomposition theorem that the connected group generated by the autoparallel vector fields of any closed Riemannian manifold is a closed, abelian subgroup of the group of isometries. Hence $\mathfrak{a}(M)$ generates a complex analytic, toral subgroup of $\mathfrak{H}_0(M)$, which we denote by $\mathfrak{A}_M'$. Since $\mathfrak{A}_M'$ is obviously transversal to $\mathfrak{H}_0'(M)$, it follows that the intersection group $\mathfrak{A}_M' \cap \mathfrak{H}_0'(M)$ is a finite abelian group, namely the kernel of the restriction to $\mathfrak{A}_M'$ of $J_* \colon \mathfrak{H}_0(M) \to \mathfrak{A}_M$. This completes the proof of the corollary.

3. The Second Variational Formula

We shall calculate the second variational formula for Φ at the critical points, as a preliminary to the proof of Theorem 2.

Proposition 3.1. Given any two real valued functions φ, ψ in a compact Kähler manifold M, whose metric (g) is a critical point of Φ, the second variation (or Hessian form) of Φ at (g) in the directions δ_φ, δ_ψ is

$$\delta_\psi \delta_\varphi \Phi(g) = 2(D\bar{D}\psi, \varphi). \tag{3.1}$$

Proof. Recall the remark following Lemma 2.1 to the effect that, for any two real valued functions φ, ψ on M, the associated tangent vector fields δ_φ, δ_ψ on the function space $\mathscr{G}_\Omega$ are mutually commutative. Therefore, we know *a priori* that, for any Kähler metric $(g) \in \mathscr{G}_\Omega$, critical or otherwise, the hessian form $\delta_\psi \delta_\varphi \Phi(g)$ is going to be a *real valued*, symmetric bilinear functional on φ, ψ.

The first derivative $\delta_\varphi \Phi(g)$ is given by (2.6): we write it in the form

$$\delta_\varphi \Phi(g) = -2 \int_M \varphi\, D_g R_g\, dV_g;$$

therefore the derivative of $\delta_\varphi \Phi$ in the direction δ_ψ can be written as a sum of three parts, as follows:

$$\delta_\psi \delta_\varphi \Phi(g) = -2 \int_M \varphi \{D(\delta_\psi R_g)\, dV + (\delta_\psi D_g)(R)\, dV + DR(\delta_\psi\, dV)\}. \tag{3.2}$$

We then substitute for $\delta_\psi R_g$ in the first part according to the last expression for it in (2.5),

$$\delta_\psi R = -\bar{D}\psi + \psi_{,\lambda} R^{,\lambda} = -\psi_{,\lambda\mu}{}^{\lambda\mu} + \psi_{,\lambda} R^{,\lambda}. \tag{3.3}$$

In order to calculate $\delta_\psi D_g$ in the second part, we express the operator D in the displayed form (2.2), showing the explicit dependence of D on (g)

$$D = g^{\alpha\bar{\nu}} \frac{\partial}{\partial z^\alpha} \left\{ g^{\beta\bar{\gamma}} \frac{\partial}{\partial z^\beta} \left[g_{\lambda\bar{\nu}} \frac{\partial}{\partial z^{\bar{\gamma}}} \left(g^{\lambda\bar{\mu}} \frac{\partial}{\partial z^{\bar{\mu}}} \right) \right] \right\}; \tag{3.4}$$

we recall the variational formula for $g_{\alpha\bar{\beta}}$ and $g^{\alpha\bar{\beta}}$, respectively $\delta_\psi(g_{\alpha\bar{\beta}}) = \psi_{,\alpha\bar{\beta}}$, $\delta_\psi(g^{\alpha\bar{\beta}}) = -\psi^{,\alpha\bar{\beta}}$; we then superpose the appropriate variational formula for each explicit occurence of (g) in (3.4) from right to left. Thus the second part of (3.2) can be written as a sum of four terms: using a test function u (independent of (g)), we obtain, using successively two common notations in tensor analysis,

$$\begin{aligned} \delta_\psi D u &= -\nabla_\alpha \nabla_\beta \nabla^\beta (\psi^{,\alpha\bar{\mu}} \nabla_{\bar{\mu}} u) + \nabla_\alpha \nabla_\beta (\psi^{,\alpha\bar{\mu}} \nabla^\beta \nabla_{\bar{\mu}} u) \\ &\quad - \nabla_\alpha (\psi^{,\beta\bar{\gamma}} \nabla_\beta \nabla_{\bar{\gamma}} \nabla^\alpha u) - \psi^{,\alpha\bar{\nu}} \nabla_\alpha \nabla_\beta \nabla^\beta \nabla_{\bar{\nu}} u \\ &= -(u^{,\lambda} \psi_{,\lambda}{}^{\alpha})^{,\beta}{}_{\beta\alpha} + (u^{,\lambda\alpha} \psi_{,\lambda}{}^{\beta})_{,\beta\alpha} \\ &\quad - (u^{,\alpha\lambda}{}_{\beta} \psi_{,\lambda}{}^{\beta})_{,\alpha} - u^{,\alpha\beta}{}_{\lambda\alpha} \psi_{,\beta}{}^{\lambda} \\ &= -(u^{,\lambda} \psi_{,\lambda}{}^{\alpha})^{,\beta}{}_{\beta\alpha} + (u^{,\lambda\alpha} \psi_{,\lambda}{}^{\beta}{}_{\beta})_{,\alpha} - u^{,\alpha\beta}{}_{\beta\lambda} \psi_{,\alpha}{}^{\lambda}. \end{aligned}$$

The third part of (3.2) includes the variational term for dV, namely $\delta_\psi(dV) = \psi_{,\lambda}{}^{\lambda}\, dV$, according to (2.4). We combine these substitutions in the three parts of (3.2), with the following outcome, replacing u by R in (3.5),

$$\begin{aligned}\tfrac{1}{2}\delta_\psi\delta_\varphi\Phi(g) &= \int_M \varphi\{D\bar{D}\psi - (\psi_{,\lambda}R^{\cdot\lambda})^{\cdot\alpha\beta}{}_{\beta\alpha} + (\psi_{,\lambda}{}^{\alpha}R^{\cdot\lambda})^{\cdot\beta}{}_{\beta\alpha} \\ &\quad - (\psi_{,\lambda}{}^{\beta}{}_{\beta}R^{\cdot\lambda\alpha})_{,\alpha} + R^{\cdot\alpha\beta}{}_{\beta\lambda}\psi_{,\alpha}{}^{\lambda} - R^{\cdot\alpha\beta}{}_{\beta\alpha}\psi_{,\lambda}{}^{\lambda}\}\,dV \\ &= (D\bar{D}\psi,\varphi) + \int_M \varphi\{(-\psi_{,\lambda}R^{\cdot\lambda\alpha})^{\cdot\beta}{}_{\beta\alpha} - (\psi_{,\lambda}{}^{\beta}{}_{\beta}R^{\cdot\lambda\alpha})_{,\alpha} \\ &\quad + R^{\cdot\alpha\beta}{}_{\beta\lambda}\psi_{,\alpha}{}^{\lambda} - R^{\cdot\alpha\beta}{}_{\beta\alpha}\psi_{,\lambda}{}^{\lambda}\}\,dV \\ &= \int_M R^{\cdot\alpha\beta}\{\varphi_{,\alpha\lambda}{}^{\lambda}\psi_{,\beta} + \varphi_{,\alpha}\psi_{,\beta}{}^{\lambda}{}_{\lambda} + (\varphi\psi_{,\alpha}{}^{\lambda})_{,\lambda\beta} - (\varphi\psi_{,\lambda}{}^{\lambda})_{,\alpha\beta}\}\,dV \\ &\quad + (D\bar{D}\psi,\varphi). \end{aligned} \tag{3.6}$$

This last formula for the second derivative of Φ at an arbitrary Kähler metric (g) will be used again in later investigations. For the moment, however, it suffices to show that, at a critical (g), i.e., if $R^{\cdot\alpha\beta}$ vanishes identically, the part in the integral sign is eliminated, and thus the assertion (3.1) is proved.

We shall prove now a simple consequence of Proposition 3.1 that is as peculiar as it is crucial in the sequel.

Corollary. If (g) is a critical Kähler metric in a compact manifold M, then the operators D and $\bar{D}$ commute with one another; their composition $D\bar{D}$, as a result, is a self-adjoint, positive semidefinite operator.

Proof. Recall the preliminary remark in the proof of Proposition 3.1 to the effect that, for any Kähler metric (g), $\delta_\psi\delta_\varphi\Phi(g)$ is a real valued, symmetric bilinear form on the real valued functions φ, ψ. Since, when (g) is a critical Kähler metric, $\delta_\psi\delta_\varphi\Phi$ takes the simple form (3.1), it follows that, in that case, for every pair of real valued functions φ, ψ, the hermitian product $(D\bar{D}\psi,\varphi)$ is real valued; thus $D\bar{D}\psi$ is a real valued functions for every real valued ψ; hence $D\bar{D} = \bar{D}D$ as a differential operator; this establishes the commutativity relation. Since D (and $\bar{D}$) are both self-adjoint, positive semidefinite operators, their commutativity implies that their composition $D\bar{D}$ is, likewise, self-adjoint and positive semidefinite. This proves the corollary. We can complete now the proof of Theorem 2.

Proof of Theorem 2. The assertion of the corollary of Proposition 3.1 just proved establishes the fact that the second variation of Φ at a critical point (g) of the function space $\mathscr{G}_\Omega$ in the direction δ_φ is expressible as

$$(\delta_\varphi)^2\,\Phi(g) = 2(\varphi, D\bar{D}\varphi) \geqq 0. \tag{3.7}$$

Now $D\bar{D}$ is a strongly elliptic operator of order 8 (it can be expressed as Δ^4 plus lower order terms), so that the co-rank of the hessian form, equal to the dimension of the space of functions annihilated by $D\bar{D}$, is finite. Since D and $\bar{D}$ are commuting, self-adjoint, strongly elliptic operators, they can be diagonalized by a simultaneous eigenvalue-eigenspace expansion, so that the annihilator space of $D\bar{D}$ is the sum of the annihilators of D and $\bar{D}$. The next part of the reasoning is somewhat trivial if $R = \text{const}$ (since this is equivalent to saying that D and $\bar{D}$ coincide). The annihilator space of D is the ($\mathbf{C}$-linear) vector space of *complex*-valued functions that are annihilated by $L = \bar{\partial}\uparrow\bar{\partial}$, while the annihilator of $\bar{D}$ is the space of conjugates of the annihilators of D (the intersection of the two spaces contains at least the constants and the function R). In fact, the space of real valued annihilators of $D\bar{D}$ is precisely the space of real parts of annihilators of D, and hence of L. This fact is to be used later, in the proof of Theorem 3.

Consider now any tangent vector δ_φ to the function space $\mathscr{G}_\Omega$ at a critical metric (g), for which the second variation (3.7) of Φ vanishes; this means that φ is in the finite dimensional space of real valued functions annihilated by $D\bar{D} = \bar{D}D$. Therefore there exists a complex valued functions $\psi \in \ker D$ such that $\varphi = \psi + \bar{\psi}$; moreover ψ is uniquely determined by φ modulo a purely imaginary valued function $\chi\sqrt{-1}$ (that is to say, χ is a real valued function) in $\ker D \cap \ker \bar{D}$. We next look at the two one-parameter, real Lie groups generated by the holomorphic vector fields $X = \uparrow\bar{\partial}\varphi$ (respectively, $Y = \uparrow\bar{\partial}(\chi\sqrt{-1})$); naturally, as transformation groups on the real manifold underlying M, they should be regarded as generated by the real tangent vector fields $X + \bar{X}$ (respectively $Y + \bar{Y}$). The effect of the infinitesimal generator on the metric (Lie derivative of the metric with respect to the vector field, or Killing operator on the vector field) is given by

$$L_{(X+\bar{X})}(g) = 2\,\frac{\partial^2(\psi + \bar{\psi})}{\partial z^\alpha\,\partial \overline{z^\beta}}\,dz^\alpha\,d\overline{z^\beta} = \delta_\varphi(g),$$

or respectively

$$L_{(Y+\bar{Y})}(g) = 2\sqrt{-1}\,\frac{\partial^2(\chi - \bar{\chi})}{\partial z^\alpha\,\partial \overline{z^\beta}}\,dz^\alpha\,d\overline{z^\beta} = 0.$$

Thus we see that *the only directions δ_φ at a critical metric (g) in which* (3.7) *vanishes are the directions tangent to the orbit of (g) under the action of $\mathfrak{H}_\Omega(M)$*. Conversely, for any one-parameter family of metrics $(g)_t$ parametrized by a real variable t, such that $(g)_0$ is a critical metric and the family transversal to $\mathfrak{H}_\Omega(M)\,((g)_0)$ at $(g)_0$, the second derivative of $\Phi((g)_t)$ with respect to t at $t = 0$ is strictly positive. This complete the proof of Theorem 2.

Proof of Theorem 3. We shall now prove that, for any critical Kähler metric (g), the identity component of the holomorphic isometry group $\mathfrak{I}_{H,0}(M,(g))$ coincides with a maximal compact subgroup $\mathfrak{H}_c(M,g)$ of the identity component $\mathfrak{H}_0(M,G)$ of the group of all holomorphic transformations of M.

Let us recall the decomposition of the complex Lie algebra $\mathfrak{h}(M)$ established by Theorem 1, determined by a critical Kähler metric (g),

$$\mathfrak{h}(M) = \mathfrak{a}(M) \oplus \mathfrak{h}'(M),$$

where $\mathfrak{a}(M)$ is the abelian Lie subalgebra consisting of all holomorphic, autoparallel vector fields on M, $\mathfrak{h}'(M) = \uparrow\bar{\partial}(\ker L)$, and $\mathfrak{h}'(M)$ is a Lie algebra ideal of $\mathfrak{h}(M)$ containing the derived ideal. The kernel of the operator L, which coincides with the set of all complex valued functions annihilated by D, is mapped onto $\mathfrak{h}'(M)$ by $\uparrow\bar{\partial}$, with the constants being the only elements going to the zero element. For this reason, for the remainder of this proof, whenever we mention "functions on M", real or complex valued, we shall implicitly understand the qualification "modulo constants"; in addition, we shall usually omit the indication of M in mentioning $\mathfrak{h}(M)$ and its vector subspaces; thus the decomposition asserted in

Theorem 1 will read as $\mathfrak{h} = \mathfrak{a} \oplus \mathfrak{h}'$. The remainder of the proof becomes substantially simpler in the special cases where R is constant, treated by A. Lichnérowicz [4]; nonetheless, it is that proof that motivates what follows.

Assume now that the Kähler metric is a critical point of the functional Φ, i.e., that the scalar curvature R satisfies $R \in \ker D$ (or equivalently, $R \in \ker \bar{D}$, $R \in \ker L$, or $R \in \ker \bar{L}$). We look at $\ker D$ as a **C**-module of complex functions, on which $\bar{D}$ operates as a self-adjoint operator with a finite, non-negative spectrum (since D and $\bar{D}$ commute, $\ker D$ is invariant under $\bar{D}$). Denote by $E_\lambda \subset \ker D$ the eigenspace of $\bar{D}$ in the kernel of D for each λ in the spectrum; thus for instance, constants and R belong to E_0; if R is constant, then D and $\bar{D}$ coincide, whence the whole $\ker D$ is E_0. The space $E_0 = \ker D \cap \ker \bar{D}$ is invariant under complex conjugation; therefore it can be decomposed as an **R**-module into real- and purely imaginary-valued functions; we denote these two subspaces respectively by $E_{0,r}$ and $E_{0,i}$ with $E_{0,i} = E_{0,r}\sqrt{-1}$.

Consider now the following direct sum decomposition of $\ker D$, $\ker D = E_{0,i} \oplus E_{0,r} \oplus \sum_{\lambda > 0} E_\lambda$, and examine the effect of the map $\uparrow \bar{\partial}$: $\ker E \to \mathfrak{h}' \subset \mathfrak{h}$ on each direct summand. We obtain the following direct sum decomposition of $\mathfrak{h}$:

$$\mathfrak{h} = \mathfrak{a} \oplus \mathfrak{k}' \oplus \mathfrak{m} \oplus \sum_{\lambda > 0} \mathfrak{h}_\lambda, \tag{3.9}$$

where $\mathfrak{a} = \mathfrak{a}(M)$ is as before, $\mathfrak{k}' = \uparrow \bar{\partial} E_{0,i}$ (and $E_{0,i} \cap \ker(\uparrow \bar{\partial}) = \mathbf{R}\sqrt{-1}$), and $\mathfrak{m} = \uparrow \bar{\partial} E_{0,r}$ (here, $\uparrow \bar{\partial}$ annihilates just the real constants); for each $\lambda \in \operatorname{spec}(\bar{D}|_{\ker D})$ with $\lambda > 0$ we set $\mathfrak{h}_\lambda = \uparrow \bar{\partial} E_\lambda$, a bijective mapping; for $\lambda = 0$ we set, exceptionally, $\mathfrak{h}'_0 = \uparrow \bar{\partial} E_0 = \mathfrak{k}' \oplus \mathfrak{m}$, $\mathfrak{h}_0 = \mathfrak{a} \oplus \mathfrak{h}'_0$ and $\mathfrak{k} = \mathfrak{a} \oplus \mathfrak{k}'$. This decomposition is justified by the following lemma, which describes the adjoint action of $X_0 = \uparrow \bar{\partial} R$ on $\mathfrak{h}$.

Lemma 3.2. Let (g) be a critical Kähler metric in M and introduce the decomposition (3.9) of the Lie algebra $\mathfrak{h}$ of holomorphic vector fields in M, and consider the special element $X_0 = \uparrow \bar{\partial} R \in \mathfrak{h}$. Then we have the following relations:

a) For each $\lambda \in \operatorname{spec}(\bar{D}|_{\ker D})$ including $\lambda = 0$, and for each $Y \in \mathfrak{h}_\lambda$,

$$[X_0, Y] = \lambda Y;$$

in other words, its operator ad X on $\mathfrak{h}$ and $\bar{D}$ on $\ker D$ have the same spectrum, and (disregarding constants in E_0 and $\mathfrak{a}$ in $\mathfrak{h}_0$) bijectively corresponding eigenspaces under the map $\uparrow \bar{\partial}$.

b) For each pair of numbers λ, μ in $\operatorname{spec}(\bar{D}|_{\ker D})$ $[\mathfrak{h}_\lambda, \mathfrak{h}_\mu] \subset \mathfrak{h}_{\lambda+\mu}$, with the usual convention that $\mathfrak{h}_{\lambda+\mu} = \{0\}$, if $\lambda + \mu$ is not in the spectrum.

c) The subspaces $\mathfrak{a}$, $\mathfrak{k}$ and $\mathfrak{m}$ of $\mathfrak{h}_0$ satisfy the further relations

(i) $\mathfrak{a}$ is in the center of $\mathfrak{h}_0$

(ii) $[\mathfrak{k}, \mathfrak{k}] \subset \mathfrak{k}$, $[\mathfrak{k}, \mathfrak{m}] \subset \mathfrak{m}$, $[\mathfrak{m}, \mathfrak{m}] \subset \mathfrak{k}$.

Proof of Lemma 3.2. Let w be a complex valued function belonging to E_λ; then [cf. (2.3)]

$$\lambda w = \bar{D} w = (\bar{D} - D) w = R^{,\beta} w_{,\beta} - R_{,\beta} w^{,\beta}.$$

Writing, as usual, $\uparrow \bar{\partial} w = w^{,\alpha} \dfrac{\partial}{\partial z^\alpha}$ and recalling that $w^{,\beta\alpha} = R^{,\beta\alpha} = 0$ (equivalently $Lw = LR = 0$),

$$\begin{aligned}\lambda w^{,\alpha} &= (R^{,\beta} w_{,\beta} - R_{,\beta} w^{,\beta})^{,\alpha} = R^{,\beta} w^{,\alpha}{}_{\beta} - w^{,\beta} R^{,\alpha}{}_{\beta}\\ &= [X_0, \uparrow \bar{\partial} w]^\alpha;\end{aligned}$$

this proves conclusion a) for any $Y \in \mathfrak{h}_\lambda$ except for the case $\lambda = 0$, $Y \in \mathfrak{a}$. This case, however, was already included in the proof of Theorem 1.

Assertion b) follows from a standard argument used in a different context, in the structure theory of Lie algebras. Let $Y_1 \in \mathfrak{h}_\lambda$, $Y_2 \in \mathfrak{h}_\mu$ and $X_0 = \uparrow \bar{\partial} R$; then using part a) just established,

$$[X_0, [Y_1, Y_2]] = [[X_0, Y_1], Y_2] + [Y_1, [X_0, Y_2]] = \lambda [Y_1, Y_2] + \mu [Y_1, Y_2],$$

proving that $[Y_1, Y_2] \in \mathfrak{h}_{\lambda+\mu}$.

Assertion c) is proved also as part of Theorem 1 insofar as the assertion on the abelian algebra $\mathfrak{a}$ is involved. For the other relations, they are proved similarly to each other. For instance, let $Y_1, Y_2 \in \mathfrak{m}$; that amounts to assuming $Y_\nu = \uparrow \bar{\partial} w_\nu$ $(\nu = 1, 2)$ with w_ν real valued and in $\ker D \cap \ker \bar{D}$. Then

$$\begin{aligned}[Y_1, Y_2]^\alpha &= w_1{}^{,\beta} w_2{}^{,\alpha}{}_{\beta} = w_2{}^{,\beta} w_1{}^{,\alpha}{}_{\beta} = (w_1{}^{,\beta} w_{2,\beta} - w_2{}^{,\beta} w_{1,\beta})^{,\alpha}\\ &= 2[\mathrm{Im}(g^{\beta\bar{\gamma}} w_{1,\bar{\gamma}} w_{2,\beta}) \sqrt{-1}]^{,\alpha};\end{aligned}$$

this shows that $[Y_1, Y_2]$ is in the image under $\uparrow \bar{\partial}$ of a purely imaginary function, which is necessarily also in $\ker D \cap \ker \bar{D}$. This completes the proof of the lemma.

We shall complete now the proof of Theorem 3. In the proof of Theorems 1 and 2 it was established respectively that the real Lie subalgebras $\mathfrak{a}$ and $\mathfrak{k}'$ of $\mathfrak{h}$ consists of Killing vector fields with respect to a critical (g); it is easy to see that they generate indeed the Lie algebra

$$\mathfrak{k} = \mathfrak{a} \oplus \mathfrak{k}';$$

of the group $\mathfrak{I}_H(M, (g))$ of holomorphic isometries of $(M, (g))$; denote by $\mathfrak{I}_{H,0}(M, (g))$, or more briefly $\mathfrak{I}_{H,0}$, the component of the identity in $\mathfrak{I}_H(M, (g))$, and let us suppose that $\mathfrak{I}_{H,0}$ is not a maximal compact, connected subgroup of $\mathfrak{H}_\Omega(M)$; thus there exists a compact, connected subgroup $\tilde{\mathfrak{K}} \subset \mathfrak{H}_0(M)$ that properly contains $\mathfrak{I}_{H,0}$; let Y be an element of the Lie algebra $\tilde{\mathfrak{k}}$ of $\tilde{\mathfrak{K}}$ that is not in $\mathfrak{k}$; let

$$Y = H + \sum_{\lambda \geqq 0} Y_\lambda \quad \text{and} \quad Y_0 = Y_0' + Y_0''$$

be the decomposition of Y with $H \in \mathfrak{a}$, $Y_\lambda \in \mathfrak{h}_\lambda$ for each $\lambda \in \mathrm{spec}(\mathrm{ad}\, X_0)$, $Y_0' \in \mathfrak{k}'$, $Y_0'' \in \mathfrak{m}$. Since $Z_0 = \uparrow \bar{\partial}(R \sqrt{-1}) \in \mathfrak{k}'$, we can consider the adjoint action of the one-parameter group of isometries generated by Z_0 on Y. We then have

$$\mathrm{ad}\exp(t Z_0)(Y) = H + Y_0' + Y_0'' + \sum_{\lambda > 0} e^{\lambda t \sqrt{-1}}\, Y_\lambda \in \tilde{\mathfrak{k}}$$

therefore, by taking appropriate linear combinations of the resulting elements for sufficiently many values of t, we obtain the result that

$$H + Y_0' + Y_0'' \in \tilde{\mathfrak{k}}$$

as well as

$$Y_\lambda \in \tilde{\mathfrak{k}} \qquad \text{for each } \lambda > 0.$$

If, for any $\lambda > 0$ and Y_λ is $\neq 0$, this implies that $\bar{D} \neq D$ and, since R is not constant, $Z_0 = \uparrow \bar{\partial}(R\sqrt{-1}) \neq 0$, whence by Lemma 3.2, Z_0 and Y_λ generate a solvable, non-abelian Lie subalgebra of $\tilde{\mathfrak{k}}$, which is impossible, since $\tilde{\mathfrak{k}}$ generates, as assumed, a compact group. Hence $\sum_{\lambda > 0} Y_\lambda = 0$, and

$$Y = H + Y_0' + Y_0'' \in \tilde{\mathfrak{k}}, \quad \text{with} \quad H + Y' \in \mathfrak{a} + \mathfrak{k}' = \mathfrak{k};$$

therefore, since $\tilde{\mathfrak{k}} \supset \mathfrak{k}$, $\tilde{\mathfrak{k}} \neq \mathfrak{k}$ and $Y \notin \mathfrak{k}$, the $\mathfrak{m}$-component Y_0'' of Y must be a non-trivial element of $\tilde{\mathfrak{k}}$; let $Y_0'' = \uparrow \bar{\partial} u$ for some real valued, non-constant function $u \in \ker D \cap \ker \bar{D}$. Then the one-(real)parameter Lie group $\{\exp t Y_0'' \mid t \in \mathbf{R}\}$ generated by Y_0' has the property that, along any of its non-trivial orbits parametrized by t, the function u depends on t in a strictly increasing way; therefore the semigroup cannot be contained in any compact transformation group. This contradiction shows that there exists no compact, connected Lie group of holomorphic transformations of M that properly contains the connected isometry group of (g). This completes the proof of Theorem 3.

The corollary of Theorem 3 follows immediately now from Mal'čev's theorem. Having established that each connected component of the critical set of Φ in $\mathscr{G}_\Omega$ is the orbit of one of its points [a critical Kähler metric (g)] under the action of $\mathfrak{H}_0(M)$, and that a maximal compact, connected subgroup of $\mathfrak{H}_0(M)$ leaves (g) fixed, Mal'čev's theorem states that any maximal compact subgroup $\mathfrak{K}$ of $\mathfrak{H}_0(M)$ is homotopically equivalent to $\mathfrak{H}_0(M)$, and hence automatically connected; the left coset space is a homogeneous space of $\mathfrak{H}_0(M)$, diffeomorphic both to a finite dimensional euclidean space and to the orbit of (g) under $\mathfrak{H}_0(M)$. The problem whether the critical set of Φ in $\mathscr{G}_\Omega$ can have more than one component, for the time being at least, remains unsolved.

4. Consequences of Futaki's Theorem

Earlier this year A. Futaki [3] announced a new obstruction for a compact, complex manifold M to admit an Einstein-Kähler metric in the absence of the primary obstruction, i.e., assuming *a priori* that there exists in M a Kähler metric in the deRham class equal to the first Chern class. His idea consists of introducing a new invariant of Kählerian structures in M, the form of a $\mathbf{C}$-linear functional f_g on the Lie algebra $\mathfrak{h}(M)$ of holomorphic tangent vector fields in M, and proving the surprising fact that this functional is independent of the Kähler metric (g), as long as (g) varies in a fixed Kähler class Ω; he proves also that this functional f_g vanishes on the derived Lie algebra of $\mathfrak{h}(M)$, and therefore constitutes a $\mathbf{C}$-character, or a Lie algebra homomorphism of $\mathfrak{h}(M)$ into $\mathbf{C}$. It is not difficult to extend the definition of f_g to Kähler metrics (g) in any cohomology class, and to prove a similar invariance property; this has been announced informally by this author, as well as by S. Bando [1] and by A. Futaki (manuscript submitted, Proc. Jap. Acad.); the vanishing of f_g for any (g) in any given cohomology class Ω

is necessary for the existence of another Kähler metric in the same class Ω with constant R.

We shall reproduce below a somewhat modified proof of Futaki's theorem, for the sake of completeness; to this end we shall first recall a few facts about the Hodge theory of Kähler manifolds. In order to be consistent with conventions followed in the rest of this paper, we define the Hodge operator Δ for exterior (p,q)-forms to be

$$\Delta = -\bar\partial\bar\partial^* - \bar\partial^*\bar\partial = -\partial\partial^* - \partial^*\partial = -\tfrac{1}{2}(d\delta + \delta d),$$

where ∂, $\bar\partial$ are the Dolbeault differential operators of bidegree $(1,0)$ and $(0,1)$ respectively and ∂^*, $\bar\partial^*$ their respective adjoints. Denote by H the orthogonal projection operator mapping $\Gamma_M A^{p,q}$ [the space of smooth (p,q)-forms in M] onto the space of harmonic (p,q)-forms; the latter are automatically annihilated by all four operators $\partial, \bar\partial, \partial^*, \bar\partial^*$. We need to use also the Hodge-Green integral operator G on (p,q)-forms, characterized by the identities

a) $G\Delta = \Delta G = (\text{identity}) - H$

b) $HG = GH = 0.$

Furthermore, G is self-adjoint and (together with H, Δ) commutes with each of the four differential operators ∂, $\bar\partial$, ∂^*, $\bar\partial^*$.

For each Kähler metric $(g) \in \mathscr{G}_\Omega$ consider the corresponding scalar curvature $R = R_g$ and introduce the real valued scalar function $F = F_g$ defined by

$$F = G(R)$$

and observe that, since the $(1,1)$-form $\partial\bar\partial F = \dfrac{\partial^2 F}{\partial z^\alpha\, \partial \overline{z^\beta}}\, dz^\alpha \wedge d\overline{z^\beta}$ is exact, and its contraction is $\Delta F = R - H(R)$, it follows that the closed $(1,1)$-form

$$\left(R_{\alpha\bar\beta} - \frac{\partial^2 F}{\partial z^\alpha\, \partial \overline{z^\beta}}\right) dz^\alpha \wedge d\overline{z^\beta}$$

is harmonic, i.e., is the harmonic form with respect to (g) representing the first Chern class multiplied by $(-2\pi\sqrt{-1})$. In other words, the Hodge decomposition of $R_{\alpha\bar\beta}\, dz^\alpha \wedge d\overline{z^\beta}$ consists of two parts

$$R_{\alpha\bar\beta}\, dz^\alpha \wedge d\overline{z^\beta} = (HR)_{\alpha\bar\beta}\, dz^\alpha \wedge d\overline{z^\beta} + F_{,\alpha\bar\beta}\, dz^\alpha \wedge d\overline{z^\beta}, \tag{4.1}$$

the first part being harmonic, and the second one exact.

We now define Futaki's functional $f = f_g\colon \mathfrak{h}(M) \to \mathbf{C}$ in terms of any $X = X^\alpha \dfrac{\partial}{\partial z^\alpha} \in \mathfrak{h}(M)$ by the formula

$$f_g(X) = -\int_M X(F)\, dV = -\int_M X^\alpha F_{,\alpha}\, dV. \tag{4.2}$$

Proposition 4.1 (Futaki). The linear mapping $f_g\colon \mathfrak{h}_M \to \mathbf{C}$ is invariant as (g) varies in the function space $\mathscr{G}_\Omega$ defined by a fixed Kähler class Ω. Furthermore it is invariant under adjoint action on $\mathfrak{h}(M)$ of the group $\mathfrak{H}_\Omega(M)$, consisting of all holomorphic transformations preserving the Kähler class Ω.

Proof. The invariance of f_g is proved by showing that, for any smooth function $\varphi: M \to \mathbf{R}$, the tangent vector δ_φ that it determines on $\mathscr{G}_\Omega$ has the property that

$$\delta_\varphi(f_g) = 0;$$

for this purpose we must first evaluate $\delta_\varphi(F)$.

Lemma. If $F = F_g = G(R)$ and δ_φ is any tangent vector to the function space $\mathscr{G}_\Omega$ at (g), then, up to an additive constant (immaterial for our purposes),

$$\delta_\varphi(F) = -\Delta\varphi + G(\partial^*\eta), \tag{4.3}$$

where η is the $(1,0)$-form defined by

$$\eta = \eta_\alpha\, dz^\alpha = \varphi_{,\beta}(HR)_\alpha{}^\beta\, dz^\alpha = g^{\beta\bar\mu}\,\varphi_{,\beta}(HR)_{\alpha\bar\mu}\, dz^\alpha, \tag{4.4}$$

and $(HR)_{\alpha\bar\beta}$ is the harmonic tensor appearing in (4.1).

Proof of the Lemma. From the relation defining F (up to an additive constant)

$$\Delta F = R - H(R)$$

we obtain by applying δ_φ to both sides, on the one hand

$$\delta_\varphi(\Delta F) = \Delta(\delta_\varphi F) + (\delta_\varphi\Delta)F = \Delta(\delta_\varphi F) - \varphi_{,\beta}{}^\gamma\, F_{,\gamma}{}^\beta,$$

and on the other, from (2.5),

$$\delta_\varphi(R - H(R)) = \delta_\varphi R = -\Delta^2\varphi - \varphi_{,\beta}{}^\gamma\, R_\gamma{}^\beta.$$

Equating the right-hand members of the last two equations we have, using again (4.1),

$$\begin{aligned}\delta_\varphi F &= -\Delta\varphi - G[\varphi_{,\beta}{}^\gamma(R_\gamma{}^\beta - F_{,\gamma}{}^\beta)] + \text{const}\\ &= -\Delta\varphi - G[\varphi_{,\beta}{}^\gamma(HR)_\gamma{}^\beta] \quad (\text{modulo const}).\end{aligned}$$

Furthermore, since the $(1,1)$-form $(HR)_{\alpha\bar\beta}\, dz^\alpha \wedge d\bar z^\beta$ is harmonic,

$$(HR)_\gamma{}^{\beta,\gamma} = 0,$$

$$\delta_\varphi F = -\Delta F - G((\varphi_{,\beta}(HR)_\gamma{}^\beta)^{,\gamma}) = -\Delta\varphi + G(\partial^*\eta) \quad (\text{mod. const}).$$

which proves the lemma.

We return to proving Proposition 4.1: we need only to calculate

$$\begin{aligned}\delta_\varphi f_g(X) &= -\delta_\varphi \int_M X^\alpha F_{,\alpha}\, dV = \int_M X^\alpha[(-\delta_\varphi F)_{,\alpha} - \varphi_{,\beta}{}^\beta\, F_{,\alpha}]\, dV\\ &= \int_M X^\alpha\{\varphi_{,\beta}{}^\beta{}_\alpha - [G(\partial^*\eta)]_{,\alpha} - \varphi_{,\beta}{}^\beta\, F_{,\alpha}\}\, dV.\end{aligned}$$

We then apply standard identities in developing $\varphi_{,\beta}{}^\beta{}_\alpha$ as follows:

$$\begin{aligned}\varphi_{,\beta}{}^\beta{}_\alpha &= \varphi_{,\beta\alpha}{}^\beta - \varphi_{,\beta}R^\beta{}_\alpha = \varphi_{,\beta\alpha}{}^\beta - \varphi_{,\beta}[F_{,\alpha}{}^\beta + (HR)_\alpha{}^\beta]\\ &= \varphi_{,\beta\alpha}{}^\beta - \varphi_{,\beta}F_{,\alpha}{}^\beta - \eta_\alpha,\end{aligned}$$

so that

$$\delta_\varphi f_g(X) = \int_M X^\alpha(\varphi_{,\beta\alpha} - \varphi_{,\beta}F_{,\alpha})^{,\beta}\, dV - \int_M X^\alpha[\eta_\alpha + (G\partial^*\eta)_{,\alpha}]\, dV, \tag{4.5}$$

where $\eta = \eta_\alpha\, dz^\alpha$ is defined by (4.4). The first integral in (4.5) vanishes because, by Gauss's theorem,

$$\int_M X^\alpha(\varphi_{,\beta\alpha} - \varphi_{,\beta} F_{,\alpha})^\beta\, dV = \int X^{\alpha,\beta}(-\varphi_{,\beta\alpha} + \varphi_{,\beta} F_{,\alpha})\, dV$$

and $X^{\alpha,\beta} = g^{\beta\bar\mu} \dfrac{\partial}{\partial \overline{z^\mu}} X^\alpha = 0$, since X is holomorphic. In order to see that the second integral in (4.5) vanishes as well, we introduce the (1,0)-form $\downarrow \bar X = g_{\alpha\bar\beta} \overline{X^\beta}\, dz^\alpha$ and observe that $\partial \downarrow \bar X = 0$, since X is holomorphic; the second integral in (4.5) then can be written as the following hermitian product (integrated over M) as follows

$$\begin{aligned}\int (\eta_\alpha + (G\partial^*\eta)_{,\alpha}) X^\alpha\, dV &= (\eta + \partial\partial^* G\eta, \downarrow \bar X)\\ &= (\eta - \Delta G\eta + \partial^*\partial G\eta, \downarrow \bar X) = (H\eta, \downarrow \bar X) + (\partial G\eta, \partial \downarrow \bar X).\end{aligned}$$

Now $H(\eta) = 0$, since $\eta = \bar\partial^*(\varphi (HR)_{\alpha\bar\beta}\, dz^\alpha \wedge d\overline{z^\beta})$; the vanishing of $\partial \downarrow \bar X$ completes the proof that $\delta_\varphi f_g = 0$ identically. Therefore, one is justified in denoting f_g by f_Ω.

The last statement about the invariance of f_Ω under the action of $\mathfrak{H}_\Omega(M)$ follows by the same argument used by Futaki: if $T\colon M \to M$ is a holomorphic transformation, then $f_g(X) = f_{(T^*g)}(T^*X)$; if, in addition, $T \in \mathfrak{H}_\Omega(M)$, i.e., if both (g) and $(T^*g) \in \mathscr{G}_\Omega$, then, by the invariance just proved, we have $f_{(T^*g)}(T^*X) = f_{(g)}(T^*g)$; therefore

$$f_\Omega(X) = f_\Omega(T^*X).$$

This completes the proof of Futaki's theorem. In passing from $T \in \mathfrak{H}_\Omega(M)$ to infinitesimal group elements, represented by a second, holomorphic vector field Y, the "pullback" action of T^* is replaced by the Lie derivation, so that

$$f_\Omega([Y, X]) = 0 \qquad \text{identically.}$$

We propose to refer to the functional $f_\Omega\colon \mathfrak{h}(M) \to \mathbf{C}$ as the *Futaki character* of (M, Ω).

It is now clear that, if there exists in $\mathscr{G}_\Omega$ a Kähler metric (g) with $R = \text{const}$, then $f_g(X)$ must be identically zero for each $X \in \mathfrak{h}(M)$ and each $(g) \in \mathscr{G}_\Omega$. On the other hand, if one could prove that there exists an extremal Kähler metric (or just a critical one) in each $\mathscr{G}_\Omega$, then this would imply that the vanishing of f_Ω is also sufficient for the existence of a (g) with constant R. This follows from Theorem 4, which we shall now prove.

Proof of Theorem 4. Given a Kähler class Ω in M, assume that we can find a critical Kähler metric $(g) \in \mathscr{G}_\Omega$ for the functional Φ. For such a metric the vector field $X_0 = \uparrow \bar\partial R$ is holomorphic, as we have seen, so that one can evaluate the Futaki character at X_0. We have

$$\begin{aligned}f(X_0) &= -\int_M X_0(F)\, dV = -\int_M R^{,\alpha} F_{,\alpha}\, dV\\ &= \int_M R\Delta F\, dV = \int_M R(R - HR)\, dV = \int (R - HR)^2\, dV \geqq 0.\end{aligned}$$

This shows that $f_M(X_0)$ is zero or strictly positive, according to whether $R = H(R)$ identically (i.e., $R = \text{const}$) or not. It follows from Futaki's theorem that there cannot be two critical Kähler metrics in $\mathscr{G}_\Omega$, one with constant R and the other not; if a critical Kähler metric is known to exist, then it will satisfy $R = \text{const}$ or not, according to whether the Futaki character is zero or not. This proves the theorem.

The question now arises, how does the Futaki character depend on the cohomology class Ω of the metric (g). It is obvious that some form of dependence must exist, for if we merely multiply a Kähler metric (g) (and hence its cohomology class Ω) by a constant $c > 0$, then on the one hand $F = G_g(R_g)$ is unaffected (modulo additive constants); on the other hand dV is homogeneous of weight n ($n = \dim_{\mathbf{C}} M$) with respect to (g), so that f, under such a change of metric, is replaced by $c^n f$.

A more interesting example on the effect of varying Ω on the Futaki character is the following: let $P_{\mathbf{C}}^2$ denote the complex projective plane and let M be the rational surface obtained by blowing up three non-collinear points of $P_{\mathbf{C}}^2$; since any two ordered, non-collinear triples in $P_{\mathbf{C}}^2$ are equivalent under the complex projective group, we may assume that they correspond to a basis for the vector space on which the homogeneous coordinates of $P_{\mathbf{C}}^2$ are linear functionals. The identity component $\mathfrak{H}_0(M)$ in the group $\mathfrak{H}(M)$ of holomorphic transformations of M maps injectively into a subgroup of the projective group, described by the action of complex diagonal (3×3)-matrix group on the homogeneous coordinates. The total automorphism group $\mathfrak{H}(M)$ of M is then generated by extending $\mathfrak{H}_0(M)$ by means of the permutation group acting on the three homogeneous coordinates selected, as well as by the holomorphic involution of M induced by the birational involution $\{(x, y, z) \to (yz, zx, xy)\}$ of $P_{\mathbf{C}}^2$. Now any Kähler class Ω on M which takes the same value on each of the three exceptional cycles arising from blowing up the three points $(1,0,0)$, $(0,1,0)$ and $(0,0,1)$ must have Futaki character identically zero, because there is no linear functional, other than zero, on the Lie algebra $\mathfrak{h}(M)$ of $\mathfrak{H}_0(M)$, that is invariant under cyclic permutation of the homogeneous coordinates.

On the other hand, consider a family of Kähler classes on M that have fixed, positive values on one non-exceptional cycle and on one of the three exceptional ones mentioned above, let us say on the one obtained from the point $(1,0,0) \in P_{\mathbf{C}}^2$, while the values at the other two exceptional cycles approach zero. In some sense the Kähler class Ω then approaches a degenerate class, pulled back from a Kähler class Ω' on the surface M' obtained by recollapsing the cycles in M obtained from $(0,1,0)$ and $(0,0,1)$ (or by blowing up just $(1,0,0)$ in $P_{\mathbf{C}}^2$). One can prove by a continuity argument that f_Ω in M approaches the character induced by pulling back the Futaki character $f_{\Omega'}$ from M' to M. In M' the existence of a critical Kähler metric (g') with $R_{g'}$ not constant was shown in [2], from which one deduces that f_Ω in M must be $\neq 0$, if Ω is sufficiently close to Ω'.

We will conclude this paper with the question whether (and, if so, how) the Futaki character, whenever it is non-trivial, can be lifted to a complex multiplicative group character of the connected automorphism group $\mathfrak{H}_0(M)$. Here too it would appear that the answer could be facilitated by an existence of extremal Kähler metrics.

References

[1] Bando, S.: An obstruction for Chern class forms to be harmonic. Manuscript received, 1983
[2] Calabi, E.: Extremal Kähler Metrics. Seminars on Differential Geometry (S.T. Yau, ed.). Princeton Univ. Press & Univ. of Tokyo Press, Princeton, New York, 1982, pp. 259–290
[3] Futaki, A.: An obstruction to the existence of Einstein-Kähler metrics. Inv. Math. **73**, Fasc. 3 (1983), 437–443
[4] Lichnérowicz, A.: Sur les transformations analytiques des variétés kählériennes. C.R. Acad. Sci. Paris, 244 (1957), 3011–3014
[5] Matsushima, Y.: Sur la structure du groupe d'homéomorphismes analytiques d'une certaine variété Kählérienne. Nag. Math. J. 11 (1957), 145–150
[6] Levine, M.: A remark on extremal Kähler metrics, manuscript submitted.

Convex Affine Maximal Surfaces

Eugenio Calabi*

Result. Math. 13, No. 3-4, 199–223 (1988)

Summary

We consider a smooth surface in 3-space that is locally representable in suitable cartesian coordinates (x,y,z) as the graph of a strongly convex function

$$z = u(x,y)$$

such a surface is an extremal for the interior variation of the affinely invariant area

$$\iint w(x,y)\,dx\,dy \quad , \quad w = (u_{xx}u_{yy}-u_{xy}^2)^{\frac{1}{4}} \quad ,$$

if and only if it satisfies the following Euler-Lagrange equation

$$\frac{\partial}{\partial x}\left(\frac{u_{yy}w_x-u_{xy}w_y}{w^4}\right) + \frac{\partial}{\partial y}\left(\frac{-u_{xy}w_x+u_{xx}w_y}{w^4}\right) = 0$$

We prove that, under various types of technical assumptions concerning the asymptotic behavior of $u(x,y)$, the only solutions of this equation describe paraboloids. The problem is equivalent to a problem concerning the global behavior of entire holomorphic maps with a type of constraint not studied previously.

*Supported by NSF Grant #DMS84-04535

DOI: 10.1007/BF03323241

§1. **Introduction** We denote by $\mathbf{A}_3$ the 3-dimensional, real, unimodular affine space, by $T\mathbf{A}_3$ its associated (3-dimensional) tangent space, which can also be viewed as the translation group of $\mathbf{A}_3$ and by $(T\mathbf{A}_3)^*$ the dual vector space of $T\mathbf{A}_3$. A locally strongly convex (ℓ.st.c.) surface in $\mathbf{A}_3$ is an oriented, immersed surface of class at least C^4, denoted by $X: \Sigma \to \mathbf{A}_3$ or more briefly by X, with the following property: for any euclidean structure superimposed on and compatible with the affine structure of $\mathbf{A}_3$, the second fundamental form that X induces on Σ is positive definite everywhere.

For any point $p \in \Sigma$ there exists an open, connected neighborhood U of p in Σ, such that one can find in infinitely many ways a unimodular affine coordinate system $\{x,y,z\}$ in $\mathbf{A}$ with the property that $X(U)$ is described as the graph of a function

$$z = u(x,y)$$

that is strongly convex in an appropriate domain $U^1 \subset \mathbf{R}^2$; thus the quadratic form

$$(1.1) \qquad r\,dx^2 + 2s\,dx\,dy + t\,dy^2 \quad \left(r = \frac{\partial^2 u}{\partial x^2},\ s = \frac{\partial^2 u}{\partial x \partial y},\ t = \frac{\partial^2 u}{\partial y^2}\right)$$

is positive definite everywhere in U, and conformally equivalent to any of the possible second fundamental forms mentioned above. Thus the affine structure of $\mathbf{A}_3$ induces on each ℓ.st.c. surface X an orientation and a conformal structure representable by the common conformal equivalence class of (1.1): therefore Σ can be regarded naturally as a Riemann surface. We shall derive the structure equations of the affine differential geometry in terms of holomorphic local uniformizing parameters ζ and their conjugates $\bar{\zeta}$. It will be useful in the sequel to restrict our notion of immersed surfaces $X: \Sigma \to \mathbf{A}_3$ to simply connected, abstract surfaces Σ; this entails, of course, no loss of generality.

A differentiable immersion of a (simply connected) Riemann surface Σ into $\mathbf{A}_3$ is ℓ.st.c. and compatible with the complex structure of Σ, if and only if the immersion map X, expressed in terms of a local complex parameter ζ and its conjugate $\bar{\zeta}$, and re-written as $X(\zeta,\bar{\zeta})$ [1] satisfies the following conditions, where we denote by (A,B,C) the determinant of the triple of vectors $\{A,B,C\}$:

$$(X_\zeta, X_{\bar{\zeta}}, X_{\zeta^2}) = (X_\zeta, X_{\bar{\zeta}}, X_{\bar{\zeta}^2}) = 0 \ ; \ -i(X_\zeta, X_{\bar{\zeta}}, X_{\zeta\bar{\zeta}}) > 0$$

everywhere; here $X_\zeta, X_{\bar{\zeta}}, X_{\zeta^2}$, etc. denote the respective complex partial derivatives $\partial X/\partial\zeta$, $\partial X/\partial\bar{\zeta}$, $\partial^2 X/\partial\zeta^2$, etc.

For each holomorphic local parameter ζ in Σ, define the positive function

[1] The conjoint notation of $\zeta,\bar{\zeta}$ as independent variables is a typographical device to remind that the function expressed is not necessarily holomorphic.

$$F(\zeta,\bar{\zeta}) = \sqrt{-i(X_\zeta, X_{\bar{\zeta}}, X_{\zeta\bar{\zeta}})} \tag{1.2}$$

and note that the hermitian differential form

$$ds^2 = 2F(\zeta,\bar{\zeta})|d\zeta|^2 \tag{1.3}$$

is independent of the parameter: it defines on Σ the Riemannian metric known as the Berwald-Blaschke metric, or first fundamental form of affine differential geometry for the ℓ.st.c. surface X. We shall derive now the structure equations for ℓ.st.c. surfaces, expressed in terms of these complex parameters, because we feel that this is the most convenient way to handle calculations on any oriented Riemannian surface. The analogous form of these equations in the case of hyperbolic surfaces expresses them in terms of asymptotic parameters: it is this form of presentation that was wisely chosen by Blaschke in his classic text [2, pp. 131-141] for his "first run" exposition of the structure equations.

The vector product ("cross" product) will denote the alternating bilinear mapping that assigns to any two vectors in $\mathbf{TA}_3$ (respectively $(\mathbf{TA}_3)^*$) the vector in the dual space $(\mathbf{TA}_3)^*$ (respectively $\mathbf{TA}_3$) such that, for any third vector C in the original space,

$$(A\times B)\cdot C = (A,B,C) .$$

The same relations hold if $\mathbf{TA}_3$ and $(\mathbf{TA}_3)^*$ are replaced by their respective complexifications.

For any ℓ.st.c. surface $X: \Sigma \to \mathbf{A}_3$ and any local holomorphic parameter ζ in a subdomain $U \subset \Sigma$ we introduce the corresponding complex frame fields, first for $(TA_3)^* \otimes \mathbf{C}$, then for $TA_3 \otimes \mathbf{C}$ as follows: the frame for the dual space consists of the so-called affine co-normal Y^* and its first partial derivatives:

$$\{Y^*, Y^*_\zeta, Y^*_{\bar\zeta}\} ;$$

where

$$Y^* = -iF^{-1} X_\zeta \times X_{\bar\zeta} ; \qquad Y^*_\zeta = \frac{\partial Y^*}{\partial \zeta} ; \qquad Y^*_{\bar\zeta} = \frac{\partial Y^*}{\partial \bar\zeta} ;$$

the frame for TA_3 consist of the two generating tangent vectors of $X(U)$ and the affine normal Y :

$$\{X_\zeta, X_{\bar\zeta}, Y\} ; \; Y = F^{-1} X_{\zeta\bar\zeta} ;$$

the two frames satisfy each the determinant relation

$$(Y^*, Y^*_\zeta, Y^*_{\bar\zeta}) = (X_\zeta, X_{\bar\zeta}, Y) = iF$$

the mutual inner product relation of the two frames can be written in matrix form as follows:

$$(1.4) \quad \begin{Bmatrix} Y^* \\ Y^*_\zeta \\ Y^*_{\bar\zeta} \end{Bmatrix} \cdot \{X_\zeta, X_{\bar\zeta}, Y\} = \begin{Bmatrix} 0 & 0 & 1 \\ 0 & -F & 0 \\ -F & 0 & 0 \end{Bmatrix} .$$

With these notations, we can write the structure equations of X in terms of either one frame or the other. We use the complex tensor notation for surfaces introduced in [5, pp. 111-114]: each tensor space of the algebra is reduced to a direct sum of bigraded complex line bundles $E_{(r,s)}$ $(r,s \in \mathbf{Z})$, where $E_{(r,s)}$ is locally generated by $d\zeta^r \otimes d\bar{\zeta}^s$. Where $d\zeta^r$ and $d\bar{\zeta}^s$, for r or s negative, denote the contravariant tensors $(\frac{\partial}{\partial\zeta})^{-r}$ or $(\frac{\partial}{\partial\bar{\zeta}})^{-s}$ respectively; the tensor products are regarded arbitrarily as commutative unless otherwise specified; thus the metric coefficient F is the fibre coordinate of a cross section in $E_{(1,1)}$; the tangent bundle of Σ tensored with $\mathbf{C}$ splits into the direct sum of $E_{(-1,0)}$ and $E_{(0,-1)}$, generated locally by $\frac{\partial}{\partial\zeta}$ and $\frac{\partial}{\partial\bar{\zeta}}$ respectively.

The Levi-Civita covariant derivative operator ∇ on smooth sections in $E_{(r,s)}$ splits into $\nabla' + \nabla''$, where ∇' is of bidegree $(+1,0)$ and ∇'' of bidegree $(0,+1)$ as follows. We start with the axiomatic conditions:

i) if $\phi = \phi(\xi,\bar{\xi})$ and $\psi = \psi(\zeta,\bar{\zeta})$ are the local coefficients of smooth sections in $E_{(r,0)}$ and $E_{(0,s)}$ respectively, then the local coefficients of $\nabla''\phi$ in $E_{(r,1)}$ and of $\nabla'\psi$ in $E_{1,s}$ are respectively $\phi_{\bar{\zeta}} = \frac{\partial\phi}{\partial\bar{\zeta}}$ and $\psi_\zeta = \frac{\partial\psi}{\partial\zeta}$;

ii) for scalars w, $\nabla w = dw$;

iii) for the metric coefficient F, $\nabla'F = \nabla''F = 0$.

These axioms, combined with the usual ones for connections yield the formula for the general case, where f is a smooth section in $E_{(r,s)}$

$$\nabla' f = F^{r} \frac{\partial}{\partial \zeta} (F^{-r} f) = \frac{\partial f}{\partial \zeta} - r\, f\, \frac{\partial \log F}{\partial \zeta}$$

$$\nabla'' f = F^{s} \frac{\partial}{\partial \bar\zeta} (F^{-s} f) = \frac{\partial f}{\partial \bar\zeta} - s\, f\, \frac{\partial \log F}{\partial \bar\zeta} .$$

Computing the second covariant derivatives, one obtains the Ricci identity

$$(1.5) \qquad [\nabla',\nabla'']f = \nabla'\nabla'' f - \nabla''\nabla' f = (s-r)FKf ,$$

where $K = -F^{-1} \frac{\partial^2 \log F}{\partial\zeta\partial\bar\zeta}$ is the intrinsic Gaussian curvature of the Berwald-Blaschke metric.

The covariant differentials of the two frame fields are respectively represented by

$$(1.6) \qquad \nabla'\{X_\zeta, X_{\bar\zeta}, Y\} = \{F^{-1} A X_{\bar\zeta} , \; FY , \; -HX_\zeta - F^{-1} B X_{\bar\zeta}\} ,$$

$$\nabla''\{X_\zeta, X_{\bar\zeta}, Y\} = \{FY , \; F^{-1}\bar{A} X_\zeta , \; -F^{-1}\bar{B} X_\zeta - HX_{\bar\zeta}\} ,$$

$$\nabla'\{Y^*, Y^*_\zeta, Y^*_{\bar\zeta}\} = \{Y^*_\zeta , \; -BY^* - FA^{-1} Y^*_{\bar\zeta} , \; -HFY^*\} ,$$

$$\nabla''\{Y^*, Y^*_\zeta, Y^*_{\bar\zeta}\} = \{Y^*_{\bar\zeta} , \; -HFY^* , \; -\bar{B}Y^* - F\bar{A}^{-1} Y^*_\zeta\}$$

(each of the four lines above implies the remaining three, either by complex conjugation or by applying (1.4)).

Aside from the metric coefficient F, the coefficients A, B and H that appear in the above equations are expressible as determinants as follows:

$$A = -i(Y, X_\zeta, X_{\zeta^2}) = i(Y^*, Y^*_\zeta, Y^*_{\zeta^2}) \quad \text{with } A d\zeta^3 \text{ invariant} \tag{1.7}$$

$$B = i(Y, X_\zeta, Y_\zeta) = iF^{-1}(Y^*_\zeta, Y^*_{\bar\zeta}, Y^*_{\zeta^2}) \quad \text{with } B d\zeta^2 \text{ invariant}$$

$$H = iF^{-1}(Y, Y_\zeta, Y_{\bar\zeta}) = iF^{-1}(Y, X_\zeta, Y_{\bar\zeta}) = iF^{-2}(Y^*_\zeta, Y^*_{\bar\zeta}, Y^*_{\zeta\bar\zeta}) ;$$

the function H is real valued, independent of the choice of local parameter and is called the <u>affine mean curvature</u>. In the textbooks on affine differential geometry these coefficients are introduced as coefficents of the second and third fundamental forms of X, as follows. The first fundamental form is the Berwald-Blaschke metric; the second fundamental form is the cubic form

$$II = A d\zeta^3 + \bar{A} d\bar\zeta^3 \quad ;$$

it is apolar with respect to the metric, meaning that it consists solely of parts in $E_{(3,0)}$ and $E_{(0,3)}$; the third fundamental form is the quadratic form

$$III = B d\zeta^2 + 2HF d\zeta d\bar\zeta + \bar{B} d\bar\zeta^2 \quad .$$

The integrability conditions are obtained by applying (1.5) to any one of the four lines in (1.6); they consist of five constraints involving the coefficients F, A, B and H, in the form of two conjugate pairs of complex equations (Codazzi equations of first and second kind)

$$(1.8)\qquad \nabla''A = \frac{\partial A}{\partial\bar\zeta} = -FB\,, \qquad \frac{\partial\bar A}{\partial\zeta} = -F\bar B\,,$$

$$(1.9)\qquad \nabla''B = \frac{\partial B}{\partial\bar\zeta} = F\frac{\partial H}{\partial\zeta} + F^{-1}A\bar B\,; \qquad \frac{\partial\bar B}{\partial\zeta} = F\frac{\partial H}{\partial\bar\zeta} + F^{-1}B\bar A\,,$$

and the (self-conjugate) "affine theorema egregium",

$$(1.10)\qquad K = -F^{-1}\frac{\partial^2 \log F}{\partial\zeta\partial\bar\zeta} = F^{-3}A\bar A + H\,.$$

§2. Affine-Maximal Surfaces.

The ℓ.st.c. surfaces whose affine mean curvature H is identically zero are called affine-maximal surfaces, because the equation $H = 0$ is the Euler-Lagrange equation characterizing surfaces whose affine area is stationary (critical) with respect to smooth deformations with compact interior support. We prefer to designate them as maximal, rather than minimal (as in the case of euclidean geometry), because the second variation of the affine area for such surfaces is always negative [7].

Affine-maximal surfaces admit a representation in terms of holomorphic maps that is strangely reminiscent of the Weierstrass-Schwartz representation of minimal surfaces in the classical sense. The key to this representation is in the third and fourth equations of (1.6); the condition $H = 0$ is equivalent to

$$\frac{\partial^2 Y^*}{\partial\zeta\partial\bar{\zeta}} = -HFY^* = 0 \quad . \tag{2.1}$$

Lemma. A ℓ.st.c. surface is affine-maximal, if and only if its associated co-normal indicatrix map is a harmonic immersion of the surface with the Riemann surface structure induced by the Berwald-Blaschke metric.

In general, the co-normal indicatrix map $Y^*: \Sigma \to (T\mathbf{A}_3)^*$ is a differentiable immersion that is radially transversal, i.e., locally strongly starlike with respect to the origin. The inversion problem of finding an ℓ.st.c. surface, given its co-normal indicatrix is a non-linear p.d.e. problem of elliptic type, that is equivalent to the open Minkowski problem [6]. However, if the immersion Y^* is harmonic in terms of a preassigned conformal structure on Σ, then there exists a solution, unique up to a translation, of the corresponding Minkowski problem, that is ℓ.st.c., affine-maximal, and with Berwald metric conformally equivalent to Σ; furthermore, this solution is obtainable by means of a simple complex quadrature, we we shall now show.

To begin with, we recall our assumption that the abstract Riemann surface Σ is simply connected; then Y^* is harmonic, if and only if there exists a holomorphic immersion $Z = Z(\zeta): \Sigma \to (T\mathbf{A}_3)^* \otimes \mathbf{C} \sim \mathbf{C}^3$ (unique up to a real translation) such that Y^* is the projection of $-2Z$ to its imaginary part,

$$Y^* = i(Z - \bar{Z}) \quad . \tag{2.2}$$

The positivity relation (radial transversality and positive orientation) of Y^*, represented by the inequality

$$F = -i\left(Y^*\ ,\ \frac{\partial Y^*}{\partial \zeta}\ ,\ \frac{\partial Y^*}{\partial \bar{\zeta}}\right) > 0\ ,$$

is equivalent to the following condition in terms of Z :

$$(2.3) \qquad F = (Z-\bar{Z}\ ,\ Z'\ ,\ \bar{Z}') > 0 \qquad \left(Z' = \frac{dZ(\zeta)}{d\zeta}\right)\ .$$

This last inequality has the following interpretation of the immersed holomorphic curve $Z(\Sigma) \subset (T\mathbf{A}_3)^* \otimes \mathbf{C}$: one compactifies the complex "dual" vector space $(T\mathbf{A}_3)^*$ to a complex projective 3-space in the standard way by writing any complex vector $z \in (T\mathbf{A}_3)^* \otimes \mathbf{C}$ with an extra component equal to 1 as a column vector $\binom{1}{z}$ and regarding it as a homogeneous coordinate representation of a point in a complex projective 3-space. Then the Plücker coordinates of the complex projective tangent line to the curve at $Z(\zeta)$ are those of the bivector

$$\binom{1}{Z(\zeta)} \wedge \binom{0}{Z'(\zeta)}\ ;$$

its six components can be listed as those of the pair of vectors $\{Z', Z\times Z'\}$. If we introduce the complex conjugate line to this tangent line, similarly represented by $\{\bar{Z}'\ ,\ \bar{Z}\times\bar{Z}'\}$; the condition (2.3) is equivalent to

$$\det\begin{pmatrix} 1 & 0 & 1 & 0 \\ Z & Z' & \bar{Z} & \bar{Z}' \end{pmatrix} > 0$$

and expresses the fact that the projective tangent of each point of $Z(\Sigma)$ is skew with respect to its complex conjugate; in other words, it has no real points and lies in the "positive" one of the two components in the set of all lines in $P^3_{\mathbf{C}}$ that have no real points.

Given the holomorphic immersion $Z: \Sigma \to (T\mathbf{A}_3)^*$ with the required property, we recapture the immersion X of the affine-maximal surface by virtue of the following theorem, (we rename henceforth the ambient spaces $\mathbf{R}^3$ or $\mathbf{C}^3$ as the case may be).

Theorem 2.1. A holomorphic immersion $Z: \Sigma \to \mathbf{C}^3$ with Σ simply connected has the property that $i(Z - \overline{Z})$ is the affine co-normal indicatrix of a ℓ.st.c. affine-maximal surface $X: \Sigma \to \mathbf{R}^3$ with a compatible Berwald-Blaschke metric, if, and only if each of the complex tangent lines of $Z(\Sigma)$ is positively skew to its complex conjugate. Furthermore the affine-maximal surface X in question is unique up to a translation, and has the following expression in terms of Z :

$$X = -i(Z\times\overline{Z} + \int Z\times dZ - \int \overline{Z}\times d\overline{Z}) . \tag{2.4}$$

Proof. The necessity part of the first statement is proved in the preamble to the formal statement of the theorem; the sufficiency can be verified by direct computation from (2.4).

The uniqueness of X up to a real translation is shown as follows. The datum of $Y^* = i(Z-\overline{Z})$ determines the affine normal indicatrix Y uniquely, since Y can be uniquely characterized by the following three equations

$$Y\cdot Y^* = 1 \ , \quad Y\cdot Y^*_\zeta = Y\cdot Y^*_{\overline{\zeta}} = 0 \ ,$$

whence

$$Y = -iF^{-1}(Y^*_\zeta \times Y^*_{\overline{\zeta}}) \quad ;$$

we express this in terms of Z as follows:

$$(2.5) \qquad Y = -i(Z-\overline{Z}\ ,\ Z'\ ,\ \overline{Z}')^{-1}\ (Z'\times\overline{Z}') \quad .$$

Since $Y = F^{-1}\dfrac{\partial^2 X}{\partial\zeta\partial\overline{\zeta}}$, it follows that the real vector-valued map of Σ defined by

$$X + iZ\times\overline{Z}$$

is a harmonic map: there must exist therefore another holomorphic, 3-dimensional vector valued map $W\colon \Sigma \to \mathbf{C}^3$ such that

$$X = -i(Z\times\overline{Z} + W - \overline{W})$$

The conformal (Riemann surface) structure on Σ is characterized by the equations

$$\frac{\partial X}{\partial \zeta} \cdot Y^*_\zeta = 0 \quad , \quad \frac{\partial X}{\partial \bar\zeta} \cdot Y^*_{\bar\zeta} = 0 \quad , \quad F = -\frac{\partial X}{\partial \zeta} \cdot \frac{\partial Y^*}{\partial \bar\zeta} = -\frac{\partial X}{\partial \bar\zeta} \cdot \frac{\partial Y^*}{\partial \zeta} > 0 :$$

hence

$$(Z'\times\bar Z + W')\cdot Z' = 0 \;, \quad (Z\times\bar Z' - \bar W')\cdot\bar Z' = 0 \;,$$

and

$$F = (Z'\times\bar Z + W')\cdot\bar Z' = -(Z\times\bar Z' - \bar W')\cdot Z' > 0$$

These equations determine W' and its conjugate uniquely as follows

$$W' = Z\times Z' \quad , \quad \bar W' = \bar Z\times\bar Z' \quad .$$

This verifies the uniqueness of (2.4) up to a real translation as an expression for the immersion X, and thereby completes the proof of the theorem.

We list here for further reference the expressions for the structural forms $Ad\zeta^3$ and $Bd\zeta^2$ in terms of the holomorphic map Z; the corresponding expression for the metric coefficient F was calculated earlier in (2.3):

$$(2.6) \quad A = (Z-\bar Z, Z', Z'') \quad , \quad B = -F^{-1}\frac{\partial A}{\partial\bar\zeta} = (Z-\bar Z, Z', \bar Z')^{-1}(\bar Z', Z', Z'')$$

Corollary 1. <u>If</u> $X: \Sigma \to \mathbf{R}^3$ <u>is a</u> ℓ.st.c. <u>affine-maximal surface and</u> ζ <u>a local holomorphic parameter for the associated Riemann surface structure, then the vector-valued cubic form</u>

$$(2.7) \qquad \Xi = (AY + B\frac{\partial X}{\partial \zeta})\, d\zeta^3$$

and the scalar valued sextic form

$$(2.8) \qquad \Phi d\zeta^6 = (B\nabla' A - A\nabla' B)\, d\zeta^6$$

are holomorphic on Σ .

Proof. By combining equations (2.3), (2.4), (2.5) and (2.6) one readily verifies that Ξ and Φ have the following formulation in terms of Z

$$(2.9) \qquad \Xi = -iZ'\times Z''\, d\zeta^3 \quad , \quad \Phi = (Z', Z'', Z''')\, d\zeta^6 \quad ;$$

the assertion then is obvious.

Corollary 2. In an affine-maximal surface, either $Ad\zeta^3$ vanishes identically, or else A and B may vanish simultaneously at most in a discrete set of points.

The proof in view of (2.7), is obvious from Corollary 1, since Ξ is holomorphic.

One distinguishes within the class of all affine-maximal surfaces certain special subclasses: we shall list and name them in order of inclusion, starting with the narrowest one.

a) An affine-maximal surface satisfying $A = 0$ identically, also satisfies $B = 0$, $\Xi = 0$ and $\Phi = 0$; this condition is equivalent to the property that $Z(\Sigma)$ lies in a complex affine line and characterizes the case where $X(\Sigma)$ is locally an <u>elliptic paraboloid</u>.

b) An affine-maximal surface with A not necessarily zero but with $B = 0$ can be characterized equivalently by any one of the following properties: i) the affine normal Y is constant; ii) the co-normal indicatrix $Y^*(\Sigma)$ lies in a real plane; iii) the holomorphic curve $Z(\Sigma)$ lies in a complex affine plane in $\mathbf{C}^3$ that is parallel to but distinct from its conjugate plane; iv) the cubic form $A d\zeta^3$ is holomorphic; this last property implies that $\Phi = 0$. Surfaces with this property are called <u>improper affine spheres</u>.

c) Finally there are affine-maximal surfaces $X(\Sigma)$ for which B is not necessarily zero, but with Φ identically zero; this last condition means that the associated holomorphic curve $Z(\Sigma)$ lies in a complex affine plane in $\mathbf{C}^3$. This class of affine-maximal surfaces apparently has not been previously recognized in the literature; for lack of imagination we call these affine-maximal surfaces <u>free of torsion</u>.

§3. The Main Theorem.

A Bernstein theorem regarding affine-maximal surfaces is any proved statement whereby, if any such surface satisfies certain "completeness" conditions at infinity (to be formulated ad hoc), then it belongs necessarily to one of the proper subclasses described at the end of the last section. A theorem of this type was established in [ ]; it says that, if a ℓ.st.c. affine-maximal surface satisfies simultaneously the two following conditions, (a) it is complete as a Riemannian manifold with respect to the Berwald-Blaschke metric; (b) it is globally a convex surface without finite boundary points, then it is an elliptic paraboloid.

The unsatisfactory feature of the statement is that it is difficult to imagine the assumptions (a) and (b) as being independent of one another; on the other hand efforts to deduce either of the two assumptions as a consequence of the other have been unsuccessful to this date. However we shall give two successive versions of the theorem; the first one is a slight improvement of the Bernstein theorem just quoted, while the second version, in the next section, substantially weakens the Riemannian completeness assumption (a). The proofs of the two versions are quite different; we include both, because the first one (actually found more recently than the second) uses mainly real variable techniques and is probably more suggestive of methods of attacking analogous problems in higher dimensions; the second proof uses complex variable techniques and suggests some interesting possible ramifications in the theory of entire holomorphic functions.

Theorem 3.1. Let $X: \Sigma \to \mathbf{A}_3$ be a ℓ.st.c. surface that is affine-maximal, Σ being endowed with the convexity orientation and the conformal structure determined by the Berwald-Blaschke metric. If (a) the surface Σ is complete with respect to the Berwald-Blaschke metric, and if (b) the affine co-normal indicatrix $Y^*: \Sigma \to (T\mathbf{A}_3)^*$ has the property that the image $Z(\Sigma)$ of the associated holomorphic curve lies in a half-space of $\mathbf{C}^3$: then $X(\Sigma)$ lies in an elliptic paraboloid in $\mathbf{A}_3$.

§4. First proof of the main theorem.

We recall the integrabililty condition (1.10); in the special case of affine-maximal surfaces, the condition $H = 0$ implies, first, that the curvature of the Berwald-Blaschke metric is nonnegative, since it equals at each point the squared hermitian norm of the cubic form $A d\zeta^3$.

$$K = \| A \|^2 = F^{-3} A\bar{A} \quad .$$

We calculate the laplacian of K using the integrability conditions (1.11), (1.12):

$$\frac{1}{2}\,\Delta K = F^{-4}\,\nabla''\nabla'(A\bar{A}) = F^{-4}\,\nabla''((\nabla'A)\bar{A} + A(\overline{\nabla''A}))$$

$$= F^{-4}\{(\nabla'A)(\overline{\nabla'A})) + (\nabla'\nabla''A + 3FKA)\,\bar{A} + \nabla''(-AF\bar{B})\}$$

$$= \|\nabla'A\|^2 + 3\|A\|^4 + F^{-4}(-(\nabla'(FB))\bar{A} - F(-FB\bar{B} + A(\overline{\nabla'B})))$$

$$= \|\nabla'A\|^2 + 3\|A\|^4 + \|B\|^2 - 2\mathrm{Re}(F^{-3}\bar{A}\nabla'B)\ .$$

In order to majorize reasonably the last term appearing in the last expression, the only one that is not necessarily ≥ 0 , we use the definition of the holomorphic sextic form (2.8) and replace $\nabla'B = \frac{\partial B}{\partial\zeta} - 2B\,\frac{\partial \log F}{\partial\zeta}$ formally by the substitution

$$\bar{A}\nabla'B = \frac{\bar{A}}{A}\ (B\nabla'A - \Phi)$$

and observe that

$$\|\nabla'A\|^2 + \|B\|^2 - 2F^{-3}\ \mathrm{Re}(\frac{\bar{A}}{A}\ (B\nabla'A-\Phi)) \geq \|\nabla'A - \frac{FA\bar{B}}{\bar{A}}\|^2 - \|\Phi\|\ .$$

Therefore we have the inequality

$$\frac{1}{2}\,\Delta(\|A\|^2) \geq 3\|A\|^4 - \|\Phi\|\ . \tag{4.1}$$

Next, we show that the two assumptions a) and b) about the global properties of Σ and X imply that Φ is identically zero.

In fact, since Σ is complete with respect to the Berwald-Blaschke and since the curvature K is non-negative, it follows from the theorem of Blanc and Fiala [1] that Σ is conformally of parabolic type. Assumption (b) means that some nontrivial complex linear functional on $\mathbf{C}^3$, composed with Z. yields a holomorphic function $\psi: \Sigma \to \mathbf{C}$ with range contained in a half-plane; but Σ is of parabolic type, and therefore the function ψ is constant on Σ; in other words $Z(\Sigma)$ lies in a complex affine plane; hence, by virtue of (2.9)

$$\Phi = (Z', Z'', Z''')$$

is identically zero. Consequently the general inequality (4.1) for affine-maximal surfaces reduces to

$$\Delta K \geq 6K^2 \quad . \tag{4.2}$$

We quote now the following generalization of the Haviland-Osserman theorem [8,3,4]. Given any complete Riemannian manifold M with nonnegative Ricci curvature and a nonnegative, monotone increasing, differentiable function $f: \mathbf{R} \to \mathbf{R}$ with $\inf\{t: f(t) > 0\} = t_0$ and, for every $t_1 > t_0$,

$$\int_{t_1}^{\infty} \left(\int_{t_1}^{s} f(t)dt\right)^{-\frac{1}{2}} ds < \infty \quad ,$$

then there exists no function $u: M \to \mathbf{R}$ satisfying

$$\Delta u \geq f(u)$$

except if $u(x) < t_0$ everywhere in M .

Applying this to (4.2) with $f(t) = 6(\max\{t,0\})^2$ and $u = K \geq 0$, we see that $K = 0$ identically; hence $A = 0$, proving that $X(\Sigma)$ is contained in a paraboloid, as we set out to prove.

Corollary. Assume again that a ℓ.st.c. surface $X: \Sigma \to \mathbf{A}_3$ is affine-maximal and complete with respect to the Berwald-Blaschke metric; we replace the technical assumption about the associated holomorphic curve $Z: \Sigma \to \mathbf{C}^3$ (essentially, that X be torsion free) with the apparently weaker one that, for some positive constant $c < 6$ the sextic form $\Phi\, d\zeta^6$, $\Phi = B\nabla' A - A\nabla' B = (Z', Z'', Z''')$ be dominated by the cubic form $A\, d\zeta^3$ as follows

$$(4.4) \qquad \|\Phi\| = F^{-3}|\Phi| \leq c\|A\|^4 = cF^{-6}\, A^2\bar{A}^2 = cK^2 \quad .$$

Then $X(\Sigma)$ is again limited to being a paraboloid, meaning that A , B and Φ vanish identically.

Proof: Under the special assumption (4.4), it follows from (4.2) that $K \geq 0$ and

$$\Delta K \geq 2(6-c)K^2 \quad .$$

The rest of the argument leading to $K = A^2 = 0$ follows as in the proof of the main theorem. The assumption (4.4) can be interpreted descriptively as the absolute torsion of $Z(\ )$ being bounded by a constant $c < 6$.

It would be desirable to be able to obtain the same conclusion without any restriction on : one has to deal otherwise with the expression $F^{-3}\,\mathrm{Re}((\nabla'B)\bar{A})$ in the last term of (4.1). For this purpose one computes the laplacian of $\|B\|^2 = F^{-2}B\bar{B}$ or, better still, of its square root, $\|B\|$: the reason for calculating $\Delta\|B\|$ rather than $\Delta(\|B\|^2)$ is that under scale changes of X, the function $\|B\|$ has the same nonzero weight as $\|A\|^2$; therefore it seems natural to juxtapose $\|A\|^2$ with $\|B\|$ in a linear combination.

A straightforward calculation yields the following relation:

$$\Delta\|B\| = \|B\|^{-1}F^{-3}\nabla''\nabla'(B\bar{B}) - \frac{1}{2}\|B\|^{-3}F^{-5}(\nabla'(B\bar{B}))(\nabla''(B\bar{B})) =$$

$$= \|B\|^{-1}F^{-3}(5F^{-2}A\bar{A}\;B\bar{B} + 2F^{-1}\mathrm{Re}((\nabla'A)\bar{B}^2) + \nabla'B\;\overline{\nabla'B}) -$$

$$- \frac{1}{2}\|B\|^{-3}F^{-5}(B\bar{B}(\nabla'B)\overline{\nabla'B} + F^{-2}A\bar{A}\;B^2\bar{B}^2 + 2F^{-1}\mathrm{Re}(A(\nabla'B)\bar{B}^3)) =$$

$$= \frac{9}{2}\|A\|^2\|B\| + \frac{1}{2}\|B\|^{-1}\|\nabla'B\|^2 + 2\|B\|^{-1}F^3\mathrm{Re}((\nabla'A)\bar{B}^2) -$$

$$- F^{-6}\|B\|^{-3}\,\mathrm{Re}(A(\nabla'B)\bar{B}^3) \quad .$$

Looking at this identity together with (4.1), one verifies that there exists only one nontrivial linear combination of the two (up to a positive factor) such that the right-hand side is universally nonnegative; we write the result in several equivalent ways, as follows:

$$(4.5)\quad \Delta\left(\frac{1}{2}\|A\|^2 + \|B\|\right) =$$

$$= 3\|A\|^4 + \|\nabla'A + \|B\|^{-1}B^2\|^2 + \frac{1}{3}\|B\|^{-1}\|\nabla'B - 3\|B\|A\|^2 +$$

$$+ \frac{1}{6}\|B\|^{-1}\|\nabla'B - 3F^{-1}\bar{A}B^2\bar{B}^{-1}\|^2$$

$$= 3\|A\|^4 + \|\nabla'A + B^{\frac{3}{2}}\bar{B}^{-\frac{1}{2}}\| + \frac{1}{2}\|B\|^{-1}\|\nabla'B - 2\|B\|A -$$

$$- F^{-1}\bar{A}B^2\bar{B}^{-1}\|^2 + \|B\|^{-1}\|A\bar{B}^{\frac{3}{2}} - \bar{A}B^{\frac{3}{2}}\|^2$$

$$\geq 3\|A\|^4 + \|B\|^{-1}\|A\bar{B}^{\frac{3}{2}} - \bar{A}B^{\frac{3}{2}}\|^2 \quad .$$

We plan to study further the various terms of the above inequality in detail with a view to their geometrical interpretations and possible implications.

§5. Main Theorem, Second Version.

We now prove a slightly different form of Bernstein's theorem in which the metric completeness assumption is replaced by the weaker one that the Riemann surface Σ is of parabolic type; this means that the Riemann surface Σ is conformally equivalent to the complex line $\mathbf{C}$.

Theorem 5.1. _Let_ $Z: \mathbf{C} \to \mathbf{C}^3$ _be an entire holomorphic immersion satisfying_ $(Z-\overline{Z}, Z', \overline{Z}') > 0$ _(in other words, each complex tangent line to_ $Z(\mathbf{C})$ _and its complex conjugate line are mutually "positively" skew). Assume that_ $Z(\mathbf{C})$ _lies in a complex affine 2-plane_ P _in_ $\mathbf{C}^3$ _; then_ $Z(\mathbf{C})$ _actually lies in a complex line._

Proof. Observe that the complex affine plane P containing the curve $Z(\mathbf{C})$ lies could not possibly coincide with its complex conjugate, for such a plane would contain all complex tangent lines to $Z(\mathbf{C})$ as well as their conjugates, so no conjugate pairs of these lines could be mutually skew.

Consider the compactification $P^3_{\mathbf{C}}$ of $\mathbf{C}^3$ as a complex projective 3-space, obtained by adjoining a projective plane "at infinity", the projective completion Π of P in $P^3_{\mathbf{C}}$, and its conjugate $\overline{\Pi}$: Π and $\overline{\Pi}$ do not coincide, and therefore intersect in a complex projective line $L = \overline{L} = \Pi \cap \overline{\Pi}$. The line L contains a real projective line $L_{\mathbf{R}}$ (topologically, a circle), consisting of its real points; the other points of L lie in two connected components, each holomorphically equivalent to a disk, each disk being the conjugate of the other. The map from $\mathbf{C}$ to L which assigns to each $\zeta \in \mathbf{C}$ the point where the tangent line to $Z(\mathbf{C})$ at $Z(\zeta)$ intersects L (both lines lie in Π) is a holomorphic one; the image cannot be a real point, for the tangent lines to $Z(\mathbf{Z})$ contain no real point. Hence, by continuity, we have a holomorphic map from $\mathbf{C}$ to one of the components of $L \setminus L_{\mathbf{R}}$; but the latter is holomorphically equivalent to a disk; hence by Liouville's theorem the holomorphic map from **C** to L is a constant, proving that Z(**C**) is contained in a complex line. The theorem is thereby proved.

It would be interesting to determine to what extent the assumption that Z(**C**) lies in a plane can be weakened; in particular, it would be especially important to deduce the geometric "straightness" of Z(**C**) from the positivity of $(Z-\bar{Z}, Z', \bar{Z}') = F$ and the completeness of **C** with respect to the metric $ds^2 = 2F|d\zeta|^2$ or perhaps from some other, weaker assumption.

References

[1] Blanc, Ch. & Fiala, F., Le type d'une surface et sa courbure totale, Comm. Math. Helv. 14 (1941-42), 230-233.

[2] Blaschke, W., Vorlesungen über Differentialgeometrie; II, Affine Differentialgeometrie, Berlin, J. Springer, 1923.

[3] Calabi, E., An extension of E. Hopf's maximum principle with an application to Riemannian geometry, Duke Math. Jour. 25 (1958), 45-56.

[4] Calabi, E., Improper affine hyperspheres of convex type and a generalization of a theorem by K. Jörgens, Mich. Math. Jour., 5 (1958), 105-126.

[5] Calabi, E., Minimal immersions of surfaces in spheres, Jour. Diff. Geom., I (1967), 111-125.

[6] Calabi, E., Géométrie différentielle affine des hypersurfaces, Sém. Bourbaki 33rd year (1980-81), N° 575; Lect. Notes in Math. #901, pp. 189-204, Springer-Verlag, Berlin-Heidelberg, 1982.

[7] Calabi, E., Hypersurfaces with maximal affinely invariant area, Amer. Jour. of Math., 104 (1982), 91-126.

[8] Osserman, R., On the inequality $\Delta u > f(u)$, Pac. Jour. Math. 7 (1957), 1641-1647.

Affine Differential Geometry and Holomorphic Curves

Eugenio Calabi

Complex geometry and analysis, Proc. Int. Symp.
in Honour of E. Vesentini, held in Pisa, Italy, 1988, 15–21,
Lecture Notes in Math., 1422, Springer, Berlin (1990)

Consider a smooth, immersed, locally strongly convex surface in euclidean 3-space $\mathbb{R}^3$ oriented so that the second fundamental form II_e is positive definite everywhere, and denote by K_e and dA_e its Gaussian curvature ($K_e > 0$) and element of euclidean area respectively. The following objects are of interest in this note:

(a) the positive density

$$dA = K_e^{1/4}\, dA_e; \tag{0.1}$$

(b) the positive definite quadratic form

$$g = K_e^{-1/4}\mathrm{II}_e; \tag{0.2}$$

(c) the linear functional Y^* on the tangent space of $\mathbb{R}^3$ at each point of the surface, whose value at each vector $\vec{u}$ is

$$Y^*(\vec{u}) = K_e^{-1/4}\langle N_e, \vec{u}\rangle, \tag{0.3}$$

where N_e denotes the unit normal vector to the surface.

The importance of dA, g and Y^* is due to the fact that they are unchanged if one replaces the given euclidean structure on $\mathbb{R}^3$ by any affinely equivalent one inducing the same orientation and volume form; they are in fact the most elementary among the basic *unimodular affine invariants* of the surface. The quadratic form (0.2) is used as an affinely invariant Riemannian metric on the surface, and is called the *Berwald-Blaschke metric;* the positive density (0.1) expresses the corresponding affinely invariant element of area, and the linear functional Y^*, whose null-plane is the tangent plane at each point of the surface, is called the *affine co-normal vector* to the surface.

One of the more obvious questions in affine differential geometry is the one concerning locally strongly convex surfaces that are extremals for the affinely invariant area

$$\int dA = \int K_e^{1/4}\, dA_e \tag{0.4}$$

under interior deformations. The Euler-Lagrange equation for this variational problem is equivalent to the system of equations

$$\Delta Y^* = 0, \tag{0.5}$$

DOI: 10.1007/BFb0089401

where Δ is the Laplace-Beltrami operator associated to the Berwald-Blaschke metric, applied here independently to each component of the affine co-normal.

The techniques of complex analysis are useful in connection with the Riemannian geometry of surfaces, in the first place because they allow us to simplify many of the otherwise tedious calculations. Accordingly, we shall introduce local complex parameters ζ for the surface, as well as their complex conjugates $\overline{\zeta}$: they are characterized by the orientation of the surface and by the conformal class of the Berwald-Blaschke metric (or, equivalently, of the euclidean second fundamental form II_e). Let the immersion functions of the surface be represented locally as a differentiable mapping X of a parameter domain $\Omega \subset \mathbb{C}$ into $\mathbb{R}^3$ (without loss of generality we shall always assume Ω to be simply connected):

$$X = \begin{pmatrix} x_1(\zeta,\overline{\zeta}) \\ x_2(\zeta,\overline{\zeta}) \\ x_3(\zeta,\overline{\zeta}) \end{pmatrix}, \qquad (\zeta \in \Omega).$$

The relation of the complex structure of the surface, represented by the holomorphy class of the parameter ζ, and the affine invariants (0.1), (0.2), (0.3) is expressed as follows, where (A,B,C) denotes the determinant functional of any ordered triple of vectors A, B, C, and subscripts denote partial derivations with respect to the indicated parameters:

$$\begin{aligned} (X_\zeta, X_{\overline{\zeta}}, X_{\zeta\zeta}) &= 0; \\ (X_\zeta, X_{\overline{\zeta}}, X_{\overline{\zeta}\overline{\zeta}}) &= 0; \\ -i(X_\zeta, X_{\overline{\zeta}}, X_{\zeta\overline{\zeta}}) &> 0. \end{aligned}$$

The Berwald-Blaschke metric then is expressed by

$$ds^2 = 2F\,|d\zeta|^2,$$

where

$$F = \left[-i(X_\zeta, X_{\overline{\zeta}}, X_{\zeta\overline{\zeta}})\right]^{1/2};$$

the affine area element is given by

$$dA = iF\, d\zeta \wedge d\overline{\zeta},$$

and the affine co-normal by

$$Y^* = iF^{-1}\, X_\zeta \times X_{\overline{\zeta}}.$$

The Euler-Lagrange equations (0.5) can be expressed equivalently by the condition that $-\frac{1}{2}Y^*$ is the imaginary part of an immersed holomorphic curve $Z\colon \Omega \to \mathbb{C}^3$, uniquely determined by X up to a real translation:

$$Y^*(\zeta,\overline{\zeta}) = i\big(Z(\zeta) - \overline{Z(\zeta)}\big). \tag{0.6}$$

Conversely, Y^* determines X uniquely up to a real translation, as stated more precisely below (see [2]):

Proposition 0.1: *Given any immersed, holomorphic curve $Z:\Omega \to \mathbb{C}^3$, there exists a locally strongly convex surface $X:\Omega \to \mathbb{R}^3$ with $Y^* = i(Z - \overline{Z})$ as its co-normal indicatrix and a Berwald-Blaschke metric conformally related to ζ if and only if Z satisfies the following inequality, in which $Z' = dZ/d\zeta$:*

$$(Z - \overline{Z}, Z', \overline{Z'}) > 0. \tag{0.7}$$

Furthermore, X is uniquely determined by Z up to a translation, and can be expressed explicitely in terms of Z by

$$X = -i\left(Z \times \overline{Z} + \int Z \times dZ - \int \overline{Z} \times \overline{dZ}\right). \tag{0.8}$$

The left-hand member of (0.7) expresses the value of F in terms of Z and ζ.

The inequality (0.7), disregarding a possible change of sign due to a switch of orientation by interchanging the rôles of ζ and $\overline{\zeta}$, has the following geometric interpretation in terms of the holomorphic curve Z: *for each point $Z(\zeta)$ of the holomorphic curve Z, the complex tangent line to it and the line obtained from it by complex conjugation are mutually skew, i.e., not coplanar.*

A locally strongly convex surface X satisfying the Euler-Lagrange equation (0.5), or equivalently (0.6) and (0.8), is called an *affine maximal surface.* Certain important subclasses of affine maximal surfaces have corresponding equivalent descriptions in terms of their representation by holomorphic curves Z as follows:

Proposition 0.2: *A strongly locally convex surface X is a trivial solution of the equation (0.5) characterizing it as an affine maximal surface, i.e., an elliptic paraboloid, if and only if the corresponding holomorphic curve Z describes a complex affine line, necessarily skew to the complex conjugate line.*

Proposition 0.3: *An affine maximal surface X belongs to the special subclass of such surfaces called improper affine spheres, characterized by the property that their co-normal indicatrix traces a plane region, if and only if the corresponding holomorphic curve Z lies in a complex affine 2-plane in $\mathbb{C}^3$ that is parallel to, but distinct from, its complex conjugate.*

The representation of an affine maximal surface in terms of a class of holomorphic curves in $\mathbb{C}^3$ may be interpreted as an affine counterpart of the Weierstrass-Schwartz representation of minimal surfaces in $\mathbb{R}^3$ in terms of "minimal" holomorphic curves in $\mathbb{C}^3$.

The affine analogue of the Plateau problem (which has not been studied as yet) consists of determining an immersion of a closed disk in $\mathbb{R}^3$, with boundary mapped onto a prescribed closed smooth curve, and tangent planes at boundary points in smoothly varying given directions (all compatible with at least one locally strongly convex immersion of the whole disk) in such a way as to *maximize*, whenever possible, the affinely invariant area (0.4) among all the allowable immersions.

For a solution of this problem it will be necessary to find *a priori* estimates for the location of interior point images, as well as of the direction of the tangent plane and of other invariants depending on derivatives of up to third order. One of the conceivable methods that would implicitely furnish estimates for third derivatives in terms of lower order ones consists of solving a "Bernstein problem" for affine maximal surfaces. In its simplest form, this problem amounts to formulating reasonably weak asymptotic conditions at the boundary of the parameter domain, under which any affine maximal surface satisfying them is a trivial solution, i.e., an elliptic paraboloid. The crucial idea in this version of the Bernstein problem exploits the fact that the Gaussian curvature K of the Bernwald-Blaschke metric for an affine maximal surface is non-negative everywhere, and its vanishing identically characterizes the trivial solutions. In fact, for $F(\zeta,\overline{\zeta}) = (Z - \overline{Z}, Z', \overline{Z'})$ according to (0.7), we have

$$K = -F^{-1}\frac{\partial^2 \log F}{\partial\zeta\partial\overline{\zeta}} = F^{-3}A\overline{A},$$

where

$$A = A(\zeta,\overline{\zeta}) = (Z - \overline{Z}, Z', Z''),$$

and $A = 0$ easily implies that $Z' \times Z'' = 0$ or, equivalently, that Z traces holomorphically a complex affine line in $\mathbb{C}^3$.

Theorem 0.4: *Let Z be an immersed holomorphic curve in $\mathbb{C}^3$, parametrized by $\zeta \in \Omega \subset \mathbb{C}$, such that the real valued density*

$$F(\zeta,\overline{\zeta}) = (Z - \overline{Z}, Z', \overline{Z'})$$

is strictly positive for each $\zeta \in \Omega$. If Z satisfies the following two conditions:
(a) *Z is complete with respect to the induced metric*

$$ds^2 = 2F(\zeta,\overline{\zeta})\,|d\zeta|^2; \tag{0.9}$$

(b) *Z lies in a complex affine plane in $\mathbb{C}^3$,*
then Z traces a complex affine line in $\mathbb{C}^3$.

Proof: If the complex curve is complete with respect to the metric (0.9), and since its curvature K is known to be non-negative everywhere, it follows from the theorem of Blanc and Fiala that it is of parabolic type; in particular it implies that the global uniformizing parameter domain Ω is the full complex line $\mathbb{C}$.

If $Z(\mathbb{C})$ lies in a complex affine plane $\Pi \subset \mathbb{C}^3$, it follows that $\Pi \neq \overline{\Pi}$, for if Π coincided with $\overline{\Pi}$, then every tangent line to the curve would be coplanar with its conjugate, and hence $(Z - \overline{Z}, Z', \overline{Z'})$ would be identically zero. We now compactify $\mathbb{C}^3$ to the complex projective 3-space $\mathbb{P}^3(\mathbb{C})$ in the usual way, and denote by Π and $\overline{\Pi}$ the projective completion of the plane containing $Z(\mathbb{C})$ and its conjugate respectively, and consider the complex projective line $\Lambda = \Pi \cap \overline{\Pi} \subset \mathbb{P}^3(\mathbb{C})$. Since $\Lambda = \overline{\Lambda}$, Λ consists of a circle (real projective line) of real points, and two disjoint disks (real hyperbolic planes) H_+ and H_- of imaginary points. Consider the holomorphic map $\varphi: \mathbb{C} \to \Lambda$ which assigns to each $\zeta \in \mathbb{C}$ the point of intersection of the tangent line to $Z(\mathbb{C})$ at ζ with Λ. Since $\varphi(\zeta) \neq \varphi(\overline{\zeta})$ (the latter is the intersection of Λ with the conjugate to the tangent line), it follows that the image of $\mathbb{C}$ by φ lies entirely either in H_+ or H_-. It follows from Liouville's theorem that φ is necessarily a constant map, from which one deduces immediately that $Z(\mathbb{C})$ traces a complex affine line, proving the assertion, **q.e.d.**

A corollary of the theorem just proved is the "weak Bernstein theorem" previously announced in [1].

Corollary 0.5: *If X is an immersed, locally strongly convex, affine maximal surface in $\mathbb{R}^3$ which is*
(a) *complete with respect to the Bernwald-Blaschke metric, and*
(b) *globally convex as an (open) surface in $\mathbb{R}^3$,*
then X traces an elliptic paraboloid.

Proof: Since X is an open, globally convex surface in $\mathbb{R}^3$, there exists at least one real affine line L in $\mathbb{R}^3$ such that, for every point of the surface, the line parallel to L at that point intersects there the surface transversally; denoting by $\vec{v}$ a fixed, non-zero vector representing the direction of L, or its opposite, we see that, without loss of generality, $Y^*(\vec{v}) > 0$ everywhere on the surface, where Y^* is the affine co-normal. But the surface is of parabolic type, by the same argument as in Theorem 0.4, and $Y^*(\vec{v})$ is harmonic and everywhere positive: hence $Y^*(\vec{v})$ is constant. Since $i(Z - \overline{Z}) \cdot \vec{v}$ is constant, it follows that $Z \cdot \vec{v}$ is constant, so that Z lies in a complex affine plane, and the argument follows as in Theorem 0.4, **q.e.d.**

The unsatisfactory feature of the statement of this corollary as a "Bernstein theorem" is that it is hard to conceive of the two completeness assumptions, the metric one in terms of the Berwald-Blaschke metric and the affine one in terms of global convexity of the surface, as being both independently essential, especially since neither of the two assumptions seems to be exploited to its fullest extent. Howewer, we have not succeded so far in deducing either one of these two assumptions as a consequence of the other one. On the other hand we can prove another somewhat sharper version of the corollary, resulting by substituting another sort of a restrictive assumption, which still seems unnecessarily strong, as follows:

Theorem 0.6: *Let Z be an entire, immersed holomorphic curve, i.e., $Z: \mathbb{C} \to \mathbb{C}^3$ with $Z'(\zeta) \neq 0$ everywhere. Assume that, for any two points $\zeta_1, \zeta_2 \in \mathbb{C}$ the complex tangent line to the curve at ζ is skew with respect to the complex conjugate to the tangent line at ζ_2, i.e.,*

$$\left(Z(\zeta_1) - \overline{Z(\zeta_2)}, Z'(\zeta_1), \overline{Z'(\zeta_2)}\right) \neq 0. \tag{0.10}$$

Then $Z(\mathbb{C})$ lies in a complex affine line.

Proof: We first compose the immersion Z with the natural imbedding of $\mathbb{C}^3$ in $\mathbb{P}^3(\mathbb{C})$ by adjoining the constant 1 before the three components of Z, and treating the four resulting functions as homogeneous coordinates representing an immersion in $\mathbb{P}^3(\mathbb{C})$, denoted by Z_p. Then the six homogeneous Plücker coordinates of the tangent lines to the curve are represented, up to a scalar factor, by $Z_p \wedge Z'_p$ and can be described equivalently by the six components of the pair of vector functions $(Z', Z \times Z')$.

Lemma 0.7: *The linear span in $\mathbb{C}^6$ of the values of the six Plücker coordinates $Z_p \wedge Z'_p$ of the tangent lines of $Z(\mathbb{C})$ is at most three-dimensional over $\mathbb{C}$.*

Proof of the Lemma: Denote by k the complex dimension of the linear span of $Z_p \wedge Z'_p$ (a priori, $1 \leq k \leq 6$). Consider the inequality (0.10), at first for a fixed ζ_2, as ζ_1 varies over $\mathbb{C}$. This can be rewritten as the (4×4)-determinant equation

$$Z_p(\zeta_1) \wedge Z'_p(\zeta_1) \wedge \overline{Z_p(\zeta_2)} \wedge \overline{Z'_p(\zeta_2)} \neq 0;$$

for fixed ζ_2 this implies one homogeneous linear constraint on $Z_p \wedge Z'_p \colon \mathbb{C} \to \mathbb{C}^6$; in other words, the image of $\mathbb{C}$ in $\mathbb{P}^5(\mathbb{C})$ represented by treating $Z_p \wedge Z'_p$ as homogeneous coordinates is disjoint from the hyperplane dual to the point $\overline{Z_p \wedge Z'_p(\zeta_2)}$.

We now repeat the argument taking all values of ζ_2, at least in a set that is dense in some open subdomain of $\mathbb{C}$. This shows that the image of $\mathbb{C}$ in $\mathbb{P}^5(\mathbb{C})$ under $\zeta_1 \mapsto (Z_p \wedge Z'_p)(\zeta_1)$ avoids an infinite collection of hyperplanes, whose common intersection is a projective subspace of $\mathbb{P}^5(\mathbb{C})$, exactly $(5-k)$-dimensional. It follows from the Jensen-Nevanlinna theory, or even the most elementary generalization of Picard's theorem to higher dimensions, that $(Z_p \wedge Z'_p)(\mathbb{C})$ is contained in the join of one of its point with that $(5-k)$-dimensional projective subspace. But that number again is *at least* equal to $k-1$. This implies that $2k \leq 7$, and so it follows that $k \leq 3$, as claimed, **q.e.d.**

To complete the proof of Theorem 0.6, we observe that the image of $\mathbb{C}$ in $\mathbb{P}^5(\mathbb{C})$ represented by $Z_p \wedge Z'_p$ or equivalently by $Z' \oplus (Z \times Z')$ lies in the intersection of a complex projective 2-subplane with the nonsingular quadric described by the identity $Z' \cdot (Z \times Z') = 0$. Such an intersection is either:

(a) a projective 2-plane in the Grassmannian of all projective lines in $\mathbb{P}^3(\mathbb{C})$ (the quadric in $\mathbb{P}^5(\mathbb{C})$ is just that), consisting of all lines through a fixed point in $\mathbb{P}^3(\mathbb{C})$, and this is clearly absurd;

(b) a projective 2-plane in the same Grassmannian consisting of all lines in a 2-subplane of $\mathbb{P}^3(\mathbb{C})$; this would reduce $Z(\mathbb{C})$ to a plane curve in $\mathbb{C}^3$, where the conclusion follows by Theorem 0.4;

(c) a reducible quadric curve if the 2-plane is tangent to but not contained in the 4-dimensional quadric; in this case, since the image of $\mathbb{C}$ is irreducible, it would be a line in $\mathbb{P}^5(\mathbb{C})$, and this would again lead to an absurdity;

(d) finally, it could be a generic intersection. This would mean that the image of $\mathbb{C}$ in $\mathbb{P}^5(\mathbb{C})$ is relatively open in an irreducible conic section: the family of lines in $\mathbb{P}^3(\mathbb{C})$ represented by this curve in the Grassmannian quadric in $\mathbb{P}^5(\mathbb{C})$ consists of the rulings on an irreducible quadric surface in $\mathbb{P}^3(\mathbb{C})$, either singular or nonsingular. Since neither of these two cases can occur as the family of tangents to an immersed holomorphic curve, the theorem is completely proved, **q.e.d.**

A subsequent study on this topic should lead to a *quantitative* counterpart of the "Bernstein theorem" just established, or possibly of a sharper version of it. This means to seek a result of the following type. Consider a locally strongly convex, affine maximal surface in $\mathbb{R}^3$ that is not trivial, as measured quantitatively, for instance, by a positive value K_0 of the curvature of the Berwald-Blaschke metric at some interior point p_0; one would want to estimate an upper bound for K_0 in terms of the distance, defined geodesically or otherwise, of p_0 from the nearest singularity or "ideal boundary point"; the latter measure is to be interpreted as a quantitative expression of the extent to which the conditions for the Bernstein theorem are not satisfied.

References.

[1] E. Calabi: *Hypersurfaces with maximal affinely invariant area.* Amer. J. Math. **104** (1982), 91–126.

[2] E. Calabi: *On affine maximal surfaces.* Technical report, Technische Universität, Berlin, 1987 (to appear).

Simple Closed Geodesics on Convex Surfaces

Eugenio Calabi and Jianguo Cao

J. Differential Geom. 36, No. 3, 517–549 (1992)

Abstract

A geodesic is said to be simple if it does not have any self-intersection point. It will be shown that the shortest closed geodesic is simple on any smooth Riemannian 2-sphere of nonnegative curvature.

We will also derive various estimates for lengths of simple closed geodesics, in terms of the diameter D, total area A, and curvature K of a given surface M^2. In particular, if we let L be the length of the longest simple closed geodesic on a smooth Riemannian sphere of curvature $0 \leq K \leq 1$, then $2D \leq L \leq A/2$. Furthermore, equality $L = A/2$ holds if and only if M^2 is isometric to the unit sphere.

Finally, if M^2 is a Riemannian sphere with nonnegative curvature, then we find that the isoperimetric inequality $A \leq 8D^2/\pi$ is useful.

Introduction

The purpose of this note is to study simple closed geodesics on compact oriented convex surfaces. A geodesic γ is said to be simple if γ has no self-intersections. In what follows, all geodesics are assumed to be nontrivial. Hence, any point curve will not be counted as a closed geodesic. If a Riemannian surface M^2 is homeomorphic to the two-sphere S^2 and if M^2 has nonnegative sectional curvature, then M^2 is called a convex surface.

First, we would like to find out which closed geodesics are simple on a given surface. The following theorem gives a partial answer.

Theorem D. *If g is a C^3 smooth metric on a two-sphere S^2 with nonnegative curvature, then any nontrivial closed geodesic of the shortest length is simple.*

In Theorem D, we only consider the C^3 smooth metric g, since there are examples of nonsmooth metrics on a two-sphere S^2 in which the shortest geodesics are not simple. For instance, the bi-equilateral triangle (two

Received January 26, 1990 and, in revised form, October 23, 1990. The first author was supported in part by National Science Foundation grant DMS 87-02359, and the second author by National Science Foundation grant DMS-8610730 at the Institute for Advanced Study.

DOI: 10.4310/jdg/1214453180

equilateral triangles glued along the boundary) has both simple and nonsimple closed geodesics of shortest length (cf. [10, p. 20]). The convexity assumption is needed in Theorem D because when the curvature becomes negative, one can find some examples where the shortest closed geodesic is not simple (cf. [19, p. 31]).

Second, we would like to work on other aspects of closed geodesics. In particular, it might be an interesting problem to study simple closed geodesics more quantitatively. Hence, we are led to estimate lengths of simple closed geodesics, in terms of other geometric data: such as the diameter, total area, and upper bound of curvature of a given surface. The first main estimate in this note is:

Theorem A. *Let M^2 be a C^3 smooth Riemannian surface diffeomorphic to S^2, with curvature $0 \leq K \leq 1$ and area A. Then the length of any simple closed geodesic γ is less than or equal to $A/2$, i.e., $L(\gamma) \leq A/2$. Furthermore, equality holds if and only if M^2 is isometric to the unit sphere.*

Using Theorem A, one can easily derive the following fact.

Corollary A. *Let $\mathscr{M}_1(M)$ be the moduli space of all simple closed geodesics in a smooth Riemannian surface M. If M is a two-sphere with nonnegative curvature, then $\mathscr{M}_1(M)$ is a compact subspace of $C^0(S^1, M)$.*

The conclusion of Theorem A (resp. Corollary A) fails if the hypothesis $K \geq 0$ is weakened to $K \geq -\epsilon^2$ for any $\epsilon > 0$. For example, on a flat torus or a two-sphere of *bone* type, there is no upper bound for lengths of all simple closed geodesics. If M_n^2 is a hyperbolic surface of genus n, then $\text{Area}(M_n^2) = 4\pi(n+1)$. It is known that lengths of primitive simple closed geodesics are related to the moduli space of hyperbolic surfaces with a fixed genus n, thus, they can be arbitrary long (cf. [21]). Therefore, it is reasonable to limit our attention to two-spheres of nonnegative curvature.

There are also examples of nonsmooth metrics on a two-sphere in which the least upper bound for lengths of simple closed geodesics is infinity. For instance, a special tetrahedron (two rectangles glued along its boundary) has a family of simple closed geodesics whose lengths are unbounded.

Our next step is to give a lower bound for the length of longest simple closed geodesics. Hence, we introduce

$$L(M) = \sup\{L(\gamma) | \gamma \text{ is a simple closed geodesic on } M\}$$

During the summer of 1989, the first author found a nonsharp lower bound for $L(M)$ in terms of the diameter and curvature of M. Later on, Chris Croke kindly pointed out to us that there is a sharp estimate for $L(M)$, which only depends on the diameter. We are grateful to C. Croke for permitting us to quote his unpublished result.

Theorem B. *Let M^2 be a Riemannian manifold, diffeomorphic to S^2, of diameter D. Then there is a simple closed geodesic γ with $L(\gamma) \geq 2D$, i.e., $L(M^2) \geq 2D$. Moreover, if equality $L(M^2) = 2D$ holds, then there are infinitely many simple closed geodesics of length $2D$.*

Finally, we wish to point out that the following isoperimetric inequality is useful.

Theorem C. *Let M^2 be a Riemannian two-sphere with nonnegative curvature. Then the following inequality holds:*

$$\operatorname{Area}(M^2) \leq \frac{8}{\pi} D^2 .$$

The best estimate of A/D^2 for convex surfaces could be $\pi/2$, which is achieved by two flat discs glued together along the boundary. Our estimate $8/\pi$ is off the best conjectured constant only by a factor ≤ 1.62.

To illustrate the usefulness of our main theorems, we present some applications.

Corollary 0.1. *Let M^2 be a smooth Riemannian sphere with $0 \leq K \leq 1$, and let D and A be as above. Then the longest simple closed geodesic γ exists and the length of γ satisfies the following inequalities:*

$$\sqrt{A\pi/2} \leq L(\gamma) \leq A/2, \qquad 2D \leq L(\gamma) \leq 4D^2/\pi .$$

Furthermore, equality $L(\gamma) = A/2$ holds if and only if M^2 is isometric to the unit sphere.

Corollary 0.2. *Let M^2 be a smooth Riemannian sphere of nonnegative curvature, and let γ be the shortest simple closed geodesic on M^2. Then $L(\gamma) \leq 9D$.*

We would like to say a few words about the ratio $L(M)/D^2$, when D becomes infinite. In §3 of Part I, we will construct a one-family of smooth metrics g_i on S^2 whose curvature satisfies $0 \leq K \leq 1$, in which $L(M_i)/D_i^2 \geq (32\pi)^{-1}$ and $D_i \to \infty$ as $i \to \infty$. Hence, the upper bound for the rate of growth $L(M^2)$ has to be quadratic in terms of D in Corollary 0.1.

The least upper bound of $L(M)/D^2$ for all smooth convex surfaces of $K \leq 1$ is still not known. Comparing the unit two-sphere, our estimate is off the best estimate at most by a factor 2. Our estimates $2D \leq L(M) \leq A/2$ are sharp and optimal.

There are several known methods to find simple closed geodesics. On the two-sphere S^2 with an arbitrary smooth Riemannian metric g, there always exist three simple closed geodesics by the Lusternik-Schnirelmann theorem (cf. [2] or [11]). However, those three geodesics may not be shortest ones. Poincaré also suggested that one could find a simple geodesic on a convex surface M by minimizing the arclength functional over the set $\mathscr{A}$ of all simple smooth closed curves which separate M into two pieces of equal total curvature. His method was carried out correctly by C. Croke in [9].

The proof of Theorem D introduces a completely different approach. We will study the space of one-cycles (also called rimmed domains) instead of the ordinary loop space on a surface. The space of one-cycles has some nice properties. For example, it allows us to *cut and paste* any nonsimple closed geodesic along its intersection points to get a nearby family of one-cycles, in which the lengths of one-cycles become smaller (cf. §2 of Part II). Such perturbation is one of the basic techniques in Part II. A minimax argument for the space of one-cycles is also needed to prove Theorem D. In order to keep the proof short, we will quote some results of Almgren and Pitts, which are related to the geometric measure theory. For the reader who is not familiar with the geometric measure theory, we will also give an independent proof of the minimax principal on the space of one-cycles in the appendix.

In the proof of Theorem A, we will make use of an integral formula, which is due to L. Santaló (cf. §1 of Part I). The higher dimensional analogues of Theorem A are still not known to the authors. Some relevant results may be found in [3]. The proofs of Corollaries 0.1–0.2 will be given in §3 of Part I.

Conventions. When a metric g is continuous up to its third derivatives on a two-dimensional manifold M^2, (M^2, g) is called a C^3 smooth Riemannian surface. In what follows, we always let M^2 be an oriented two-sphere S^2 endowed with a C^3 smooth metric g unless otherwise stated. Given a piecewise smooth curve γ (not necessarily closed), by $L(\gamma)$ we will mean the length of γ. The injectivity radius of the Riemannian surface M^2 is denoted by $\mathrm{inj}(M^2)$. If Ω is an open domain, $A(\Omega)$ is defined to be the area of Ω. K stands for the curvature of (M^2, g). These notations will be used throughout this paper.

Acknowledgment

The second-named author is very indebted to Professor C. Croke, his teacher, for continuous support and encouragement, especially the help in writing §2 of Part I. Both authors would like to thank H. Gluck and W. Ziller for their helpful comments and interest in this work.

Finally, the authors want to thank the referee for pointing out some minor errors and misprints in an earlier version of this paper.

PART I
AREA, DIAMETER, AND LENGTH OF SIMPLE CLOSED GEODESICS

1. Total surface area and the length of simple closed geodesic

In this section, we will derive a sharp upper bound, in a smooth Riemannian sphere of nonnegative curvature, for the length of all simple closed geodesics in terms of the total area and the curvature k. By rescaling the metric, we may assume without loss of generality that $0 \le K \le 1$.

We will show that if M is a C^3 smooth Riemannian surface, diffeomorphic to S^2, with curvature $0 \le K \le 1$ and area A, then the length of any simple closed geodesic γ is less than or equal to $A/2$. This estimate is sharp, which is achieved by the unit sphere. Inequality $L(\gamma) \le A/2$ fails if the hypothesis $0 \le K \le 1$ is weakened to $-\epsilon^2 \le K \le 1$ for any $\epsilon > 0$. The estimate also fails for flat tori and surfaces of higher genus except for the projective plane $\mathbf{R}P^2$.

We begin with an elementary example to demonstrate the main idea of this section. Let Ω be a rectangle with width a and height b on the Euclidean plane E^2. Obviously, one sees that $A(\Omega) = ab$, the area of Ω, and $L(\partial\Omega) = 2(a+b)$., the perimeter of Ω. Our goal is to study the upper bound of the ratio $L(\partial\Omega)/A(\Omega)$. If $a = 1$ and $b \to 0$, then the ratio becomes infinite. However, if we denote the width of Ω by $W_0(\Omega) = \min\{a, b\}$, we can easily verify that $L(\partial\Omega) \le 2A(\Omega)/W_0(\Omega)$.

Let γ be a simple, closed geodesic on a Riemannian sphere M. Clearly, γ divides M into two simply connected, open components, ay Ω_1 and Ω_2. We will define the width, $W(\Omega_j)$, of each Ω_j, and then estimate $L(\gamma)$ from above in terms of the area $A(\Omega_j)$ and $W(\Omega_j)$ for either of two domains Ω_j. There is a classical formula due to L. Santaló which gives a sharp estimate of $L(\gamma)/A(\Omega_i)$ in terms of the "width" (cf. [20] or [8]). To set the stage for our application of the Santaló formula, we must introduce an analogous notion of $W(\Omega_j)$, the width of any given, open, connected surface Ω_j with boundary $\partial\Omega_j$, as follows.

The new width $W(\Omega_j)$ is defined to be the infimum for lengths of all geodesic chords within Ω_j. More precisely, we let N represent the inwardly pointing unit normal vector field of $\partial\Omega_j$, UM represent the unit circle bundle of M, and $U\Omega_j = UM|_{\Omega_j}$. The upper semicircle bundle is denoted by

$$U^+\partial\Omega_j = \{v | v \in UM|_{\partial\Omega_j}, \langle v, N_{\Pi(v)}\rangle > 0\}.$$

For any $v \in U^+\partial\Omega_j$ or $v \in U\Omega_j$, we let σ_v be a geodesic with $\sigma'(0) = v$ and $\sigma(0) = \Pi(v)$, where $\Pi\colon UM \to M$ is the canonical projection map. Naturally, we let $l(v)$ be the smallest value of $t > 0$ (possibly ∞) such that $\sigma_v(t) \in \partial\Omega_j$. Finally, one can define

$$W(\Omega_j) = \inf\{l(v) | v \in U^+\partial\Omega_j\}.$$

The Santaló formula (cf. [10, p. 421]) states that

$$(1.1)\qquad \begin{aligned} 2\pi A(\Omega_j) = \mathrm{Vol}(UM|_{\Omega_j}) &\geq \int_{U^+\partial\Omega_j} l(v)\langle v, N_{\Pi(v)}\rangle\, dv \\ &\geq W(\Omega_j)L(\partial\Omega_j)\int_0^\pi \sin\theta\, d\theta = 2W(\Omega_j)L(\gamma). \end{aligned}$$

Hence, we have

$$(1.2)\qquad L(\gamma) \leq \frac{\pi}{W(\Omega_j)} A(\Omega_j), \qquad j = 1, 2.$$

In what follows, we want to estimate $W(\Omega_j)$ from below in terms of curvature of a given surface. Our lower bound for $W(\Omega_j)$ will be independent of the choice of any simple closed geodesic γ and Ω_j.

Lemma 1.1. *Let M be a C^3 smooth Riemannian sphere whose curvature K satisfies $0 \leq K \leq 1$, and let γ, Ω_j, and $W(\Omega_j)$ be as above. Then $l(v) \geq \pi$ for all $v \in U^+(\partial\Omega_j)$, i.e.,*

$$(1.3)\qquad W(\Omega_j) \geq \pi.$$

Moreover, if $l(v) = \pi$ for all $v \in U^+\partial\Omega$, then $L(\partial\Omega_j) = 2\pi$.

The proof of Lemma 1.1 is quite involved, and we postpone it to the end of this section. At this moment, under the assumption that Lemma 1.1 holds, we would like to prove

Theorem A. *Let M be a C^3 smooth Riemannian surface diffeomorphic to S^2, with curvature $0 \leq K \leq 1$ and area A. Then the length of any simple closed geodesic γ is less than or equal to $A/2$, i.e., $L(\gamma) \leq A/2$. Furthermore, equality holds for some γ if and only if M is isometric to the unit sphere.*

Proof of Theorem A. Let Ω_1 be an open component of $M - \gamma$ with the property that $\mathrm{Area}(\Omega_1) \leq \frac{1}{2}\mathrm{Area}(M)$. It follows from Lemma 1.1 and

(1.2) that

$$L(\gamma) \le \frac{\pi}{2W(\Omega_1)} A(M) \le A/2.$$

Furthermore, equality holds if and only if $\text{Area}(\Omega) = \text{Area}(M)/2 = \text{Area}(M \setminus \overline{\Omega})$, $l(v) = \pi$ for all $v \in U^+\partial\Omega$. Using a result of Bangert, we now conclude that Ω is isometric to the unit semisphere (cf. [4]). One can also give an alternative proof as follows: Lemma 1.1 implies that $L(\partial\Omega) = 2\pi$. In this case, by (1.2) we get $A(M) = 4\pi$. It will be shown in Lemma 1.2 that $A(M) \ge 4\pi$, and equality holds if and only if M is the standard unit two-sphere. This completes the proof of Theorem A under the assumption that Lemmas 1.1 and 1.2 are true. q.e.d.

In order to carry out the proof of Lemma 1.1, we need to derive some preliminary facts.

Lemma 1.2. *Let* $M = (S^2, g)$ *with nonnegative curvature* K, $0 \le K \le 1$. *Then the injectivity radius of* M, $\text{inj}(M)$, *is greater than or equal to* π, *i.e.,*

(2.4) $$\text{inj}(M) \ge \pi.$$

Consequently, $\text{Area}(M) \le 4\pi$ *and* $A(M) = 4\pi$ *if and only if* M *is isometric to the standard unit two-sphere.*

Remark. Inequality (1.4) clearly fails on certain flat tori. We include a proof of Lemma 1.2 here for the sake of completeness.

Proof. It follows from the Gauss-Bonnet theorem that $\max K(x) > 0$. Hence, the set $G = \{x | x \in M, K(x) > 0\}$ is nonempty. By Corollary 5.7 in [7], we may assume that there is a smooth closed geodesic γ, parametrized by its arc-length, with $\gamma(0) = p$ through p and q such that $L(\gamma) = 2d(p, q) = 2\,\text{inj}(M)$. Let $\overline{G}$ be the closure of G. If $\gamma \cap \overline{G} = \varnothing$, we can move γ within the flat region $M \setminus G$ until γ hits $\overline{G}$. Therefore, we may assume that $\gamma \cap \overline{G} \ne \varnothing$ at the beginning. Let N be the unit normal vector field of γ which is pointing towards G and

$$h_s(t) = \exp_{\gamma(t)}[sN(t)].$$

When $0 < s < \epsilon_0 =$ the focal radius of γ, it follows from the Corollary of Rauch II that $L(h_s) \le L(\gamma)$; the equality holds if and only if the strip $\Delta_s = \{h_\alpha(t) | 0 \le \alpha \le s\}$ is flat (see Lemma 1.4 below, [12] or [7, p. 31]). In our case, $L(h_s) < L(\gamma)$ whenever $s > 0$, since $K(x) \ge 0$ and $K > 0$ somewhere in any strip Δ_s. Now, for the same reason as in [7, p. 99], we conclude that q is conjugate to p. Hence, $\text{inj}(M) = d(p, q) \ge \pi$.

Using the fact that $\text{inj}(M) \ge \pi$ and $K \le 1$, one can use the standard area comparison theorem or the Berger-Kazdan inequality to conclude that

$A(M) \geq 4\pi$ and the equality holds if and only if M is the standard unit 2-sphere (cf. [12], or [5], [14]). In fact, when $A(M) = 4\pi$ and $0 \leq K \leq 1$, the Gauss-Bonnet formula implies that $K(p) = 1$ for all $p \in M$. Hence, M is a round sphere of curvature 1. q.e.d.

As an application of Lemma 1.2, one has

Corollary 1.3. *Let M be as in Lemma 1.2, and let σ_1 and σ_2 be two distinct geodesic segments with the same endpoints. Then*

$$\max\{L(\sigma_1), L(\sigma_2)\} \geq \pi.$$

We also need a special sharp version of the second Rauch comparison theorem (Berger's Lemma) to prove Lemma 1.1. Since it was not clearly stated in any literature, we present it here with a simple proof.

Let M be a two-dimensional Riemannian manifold with nonnegative curvature $K \geq 0$, $\eta: [0, 1] \to M$ be a geodesic with $\|\eta'(0)\| = 1$, and $N(0)$ be a unit normal vector of η at $\eta(0)$.

Lemma 1.4. *Suppose J is a Jacobi field along η with $J(0) = a\eta'(0) + bN(0)$, $b \neq 0$, and $J'(0) = 0$. If η has no focal points of the geodesic σ defined by $\sigma(s) = \exp_{\eta(0)}[sN(0)]$ in $[-1, 1]$, then*

$$\|J(t)\| \leq \|J(0)\| \quad \textit{for all } t \in [0, 1].$$

Moreover, $\|J(s)\| = \|J(0)\|$ for some $s > 0$ if and only if $K(\eta(t)) = 0$ for all $t \in [0, s]$.

Proof. Let $N(t)$ be the unit normal vector field along η. Since $\dim M = 2$, one gets $N'(t)k = 0$. Since $J'(0) = 0$, it follows that J can be decomposed into

$$J(t)k = a\eta'(t) + f(t)N(t),$$

where $f(0) = b$ and $f'(0) = 0$ (cf. [7, p. 19]). Clearly, one sees

$$\|J(t)\|^2 = a^2 + f^2(t).$$

Since σ does not have any focal point on η, f does not change its sign on $[0, 1]$. By the assumption $K \geq 0$, we know that

$$f''(t) = -K(\eta(t))f(t),$$
$$f(t)f''(t) \leq -K(t)f^2(t) \leq 0.$$

If $f(0) = b > 0$, then $f''(t) \leq 0$ as long as $f(t) > 0$. This together with $f'(0) = 0$ implies that $f'(t) \leq 0$ and $0 < f(t) \leq f(0) = b$.

When $f(0) = b < 0$, one can also show that $-f(t) \leq -f(0) = -b$ and $f^2(t) \leq f^2(0)$. Hence,

$$\|J(t)\|62 = a^2 + f^2(t) \leq a^2 + b^2 = \|J(0)\|^2.$$

$\|J(s)\| = \|J(0)\|$ holds for some $s > 0$ if and only if $K(\eta(t)) = 0$ for all $t \in [0, s]$. q.e.d.

Now, we are ready to prove Lemma 1.1.

Proof of Lemma 1.1. Notice that Ω is an open domain. By the definition of $W(\Omega)$, we are interested in all geodesic chords which lie in Ω except for endpoints. In order to give a criterion for geodesics which are transversal to $\partial\Omega$, we need to introduce an intrinsic distance function d_γ of $\partial\Omega = \gamma$.

For any pair of points $\{p, q\}$ on a simple closed geodesic γ, p and q divide γ into two geodesic segments γ_1 and γ_2. We define $d_\gamma(p, q) = \min\{L(\gamma_1), L(\gamma_2)\}$.

Let σ be a geodesic segment σ with endpoints $\{p, q\} \subset \gamma$ and of length $L(\sigma) < \pi$. If σ is different from γ_1 and γ_2, then it follows from Corollary 1.3 that $d_\gamma(p, q) \geq \pi$. Conversely, if η is a geodesic with endpoints $p_0, q_0 \in \partial\Omega$, and if $L(\eta) < d_\gamma(p_0, q_0)$, then η is clearly transversal to $\partial\Omega$. This observation will be used later on.

We will use a contradiction method to finish the proof of Lemma 1.1. It takes several steps to get a contradiction.

Suppose to the contrary that Lemma 1.1 is false. Then there would be a family of geodesics $\sigma_i: (0, L_i) \to \Omega$ with endpoints $\{p_i, q_i\} \subset \gamma$ and of length $L_i < \pi - \epsilon_0$, where $\epsilon_0 = \frac{1}{2}[\pi - W(\Omega)] > 0$ and $L_i \to W(\Omega)$ as $i \to \infty$. Clearly, by the argument above, we have

$$(1.5) \qquad d_\gamma(p_i, q_i) \geq \pi \quad \text{for all } i.$$

Since $\overline{\Omega}$ is compact, we may choose a subsequence of $\{\sigma_i\}$ which is convergent to a normal geodesic $\sigma_0: [0, L(\sigma_0)] \to \overline{\Omega}$, with endpoints $\{p_0, q_0\} \subset \partial\Omega$, and of length

$$(1.6) \qquad L(\sigma_0) = W(\Omega) < \pi \leq d_\gamma(p_0, q_0).$$

Because of (1.6), we know that $p_0 \neq q_0$ and σ_0 is a nontrivial geodesic which lies in the interior of $\overline{\Omega}$ except for its endpoints.

Let $L_0 = L(\sigma_0)$. Lemma 1.2 tells us that $\text{inj}(M) \geq \pi > L_0$. By the definition of $W(\Omega)$ and the first variational formula, one can show that σ_0 is perpendicular to $\partial\Omega = \gamma$ at points p_0 and q_0. Let N be the unit normal vector field along σ_0 and $G = \{x | x \in \Omega, K(x) > 0\}$. Since Ω is diffeomorphic to a disc and $\partial\Omega$ forms a closed geodesic, one sees that $G \neq \varnothing$. If $\sigma_0 \cap \overline{G} = \varnothing$, we move σ_0 in direction N within the flat region $\Omega \setminus G$ until σ_0 hits $\overline{G}$ somewhere. Therefore, we may assume that $\sigma_0 \cap \overline{G} \neq \varnothing$ and N is pointing towards G. As we did in the proof of

Lemma 1.2, we introduce a family of curves

$$h_s(t) = \exp_{\sigma_0(t)}[sN(t)], \qquad t \in [0, L_0]. \tag{1.7}$$

If δ is the focal radius of σ_0 in Ω, we claim that

$$L(h_s) < L(\sigma_0), \quad \text{whenever } 0 < s < \delta. \tag{1.8}$$

This can be seen as follows. For any fixed s with $0 < s < \delta$, it follows from Lemma 1.4 that

$$L(h_s) \le L(\sigma_0), \tag{1.9}$$

and the equality holds in (1.9) if and only if the strip $\Delta_s = \{h_u(t) | 0 \le u \le s, 0 \le t \le L_0\}$ is flat. However, our strip is not flat as long as $s > 0$; thus, the strict inequality (1.8) holds.

Now, we fix and $s > 0$ and apply the Birkhoff curve shortening process to h_s with fixed endpoints. From this, we get a geodesic σ with endpoints $h_s(0), h_s(L_0) \in \partial\Omega$, and

$$L(\sigma) \le L(h_s) < L(\sigma_0) = W(\Omega). \tag{1.10}$$

It remains for us to show that $\sigma \subset \overline{\Omega}$. It is clear that there is an $\epsilon > 0$ such that for all $x, y \in \partial\Omega$, with $d(x, y) < \epsilon$, the minimizing geodesic τ from x to y satisfies $\tau \subset \overline{\Omega}$. In other words, $\partial\Omega$ is convex to Ω. Hence, it follows that $\sigma \subset \overline{\Omega}$ (cf. [10, pp. 3–7]). Furthermore, if s is sufficiently small, we have

$$d_\gamma(\sigma(0), \sigma(L_0)) = d_\gamma(h_s(0), h_s(L_0)) \ge d_\gamma(p_0, q_0) - 2s > \pi - \epsilon_0 > L(\sigma).$$

Therefore, σ is transversal to $\partial\Omega$ and lies in Ω except for its endpoints. This fact together with (1.10) gives a contradiction to the definition of $W(\Omega)$. This finishes the proof of the first part of Lemma 1.1:

$$W(\Omega) \ge \pi. \tag{1.11}$$

The second part of Lemma 1.1 follows immediately from Bangert's result (cf. [4]).

Remark. There is another interesting estimate for the length of any given simple closed geodesic in the two-sphere of nonnegative curvature. Let γ be a simple closed geodesic which bounds a domain Ω. We introduce the notion of $\rho(\Omega)$, the radius of domain Ω, by letting

$$\rho(\Omega) = \max\{d(p, \gamma) | p \in \Omega\}.$$

If $K \ge 0$, then we claim

$$A(\Omega) \ge \tfrac{1}{2} L(\gamma)\rho(\Omega). \tag{$*$}$$

This can be seen as follows.

Let $\sigma_s = \{q | d(q, \gamma) = s\}$. Using the co-area formula, one gets

$$A(\Omega) \geq \int_0^{\rho(\Omega)} L(\sigma_s)\, ds.$$

Let $f(s) = L(\sigma_s)$. It is clear that $f'(0) = 0$, since σ_0 is a closed geodesic. The function f is also semicontinuous. When f is smooth at $s = s_0$ and $K \geq 0$, then the second variational formula tells us that $f''(s_0) \leq 0$ (cf. [7, p. 20]). Using this fact together with $f(0) = L(\gamma)$ and $f(\rho) \geq 0$, we conclude that $f(s) \geq [1 - s/\rho(\Omega)]L(\gamma)$. This leads us to

$$A(\Omega) \geq \int_0^{\rho} f(s)\, ds \geq \tfrac{1}{2} L(\gamma)\rho(\Omega).$$

Since the assumption that $K \leq 1$ is not needed here, this estimate is independent of the one given by Theorem A.

2. Diameter of a surface and the length of simple closed geodesics

We shall discuss the relation between the length of a simple closed geodesic and the diameter of a Riemannian sphere M. In this section, the curvature K of M is not necessarily nonnegative. M is allowed to have an arbitrary metric.

Theorem B. *Let M be a Riemannian manifold, diffeomorphic to S^2 of diameter D. Then there is a simple closed geodesic γ of length $L(\gamma) \geq 2D$. Furthermore, if the longest simple closed geodesic γ exists and $L(\gamma) = 2D$ holds, then there are infinitely many simple closed geodesics of length $2D$.*

Recall that $L(M)$ is defined to be $\sup\{L(\gamma) | \gamma$ is a simple closed geodesic$\}$. It is well known that the energy functional E defined on the loop space satisfies the Palais-Smale condition. The critical points of E are closed geodesics. Hence, if $L(M) < \infty$, then the set of all simple closed geodesics is a compact subset of $C^0(S^1, M)$. In particular, the longest simple closed geodesic exists as long as $L(M) < \infty$.

Our new observation in this section is based on the proof of the theorem of Lusternik and Schnirelmann, which is due to Ballmann (cf. [2] or the appendix of [15]) and Grayson [11]. The Lusternik-Schnirelmann theorem asserts that, on any two-dimensional sphere with an arbitrary metric, there exist at least three simple closed geodesics. We denote these three geodesics by γ_1, γ_2, and γ_3 and assume that $L(\gamma_1) \leq L(\gamma_2) \leq L(\gamma_3)$. C. Croke observed that $L(\gamma_2) \geq 2D$. The following proof was suggested by him.

Proof of Theorem B. We will adopt some notation from the appendix of [15]. It was known that the second simple closed geodesic γ_2 was found by a minimax method among some special two-parameter families of closed curves. Denote the set of all these two-parameter closed curves by Γ, where Γ is a subset of $C^0([-1, 1] \times [0, 1]; C^0(S^1, S^2))$ and S^1 is the unit circle. The precise definition of Γ will be given later. At the moment, we want to explain the main idea in the proof of Theorem B. Our strategy is to show that

$$\max\{L(u_{t_1, t_2}) | -1 \leq t_1 \leq 1, \ 0 \leq t_2 \leq 1\} \geq 2D \tag{2.1}$$

holds for each $u \in \Gamma$. This assertion would imply

$$L(\gamma_2) = \inf_{u\in\Gamma} \max_{\substack{-1\leq t_1\leq 1 \\ 0\leq t_2\leq 1}} \{L(u_{t_1, t_2})\} \geq 2D.$$

Theorem B would follow immediately.

It will take several steps to verify (2.1). First, let us give the precise definition of Γ as follows.

Let $S^2 = \{(x, y, x) | x^2 + y^2 + z^2 = 1\}$. We first introduce u^0, the generator of Γ. All other elements of Γ will be homotopic to u^0 under some extra conditions (see (2.2)–(2.5) below). Our u^0 will consist of *all small* circles on S^2 that meet the yz-plane orthogonally. For any $(t_1, t_2) \in [-1, 1] \times [0, 1]$, we assign a small circle u_{t_1, t_2} of radius $|\cos \frac{\pi}{2} t_1|$ as follows:

$$\begin{aligned} u^0_{t_1, t_2}(s) &= u^0(t_1, t_2, s) \\ &= \left(\cos \frac{\pi}{2} t_1 \cos 2\pi s, \cos \frac{\pi}{2} t_1 \sin 2\pi s, \sin \frac{\pi}{2} t_1\right) \\ &\quad \cdot \begin{bmatrix} 1, & 0, & 0 \\ 0, & \cos \pi t_2, & \sin \pi t_2 \\ 0, & -\sin \pi t_2, & \cos \pi t_2 \end{bmatrix}. \end{aligned}$$

Notice that

(i) For each $t_2 \in [0, 1]$, $u^0_{\pm 1, t_2}$ is a point curve.

(ii) $u^0(t_1, 0, s) = u^0(-t_1, 1, -s)$.

(iii) For any pair of points $\{p, q\} \subset S^2$, there is a small circle in u^0 that passes through p and q.

Naturally, we require that $u \in \Gamma$ if and only if u satisfies:

(2.2) u induces a continuous map from $[-1, 1] \times [0, 1] \times S^1$ to S^2,

(2.3) For each $t_2 \in [0, 1]$, $u_{\pm 1, t_2}$ is a point curve,

(2.4) $u(t_1, 0, s) = u(-t_1, 1, -s)$ for all t_1 and s,

(2.5) there is a continuous homotopy h^λ between u^0 and u, $\lambda \in [0, 1]$, such that $h^0 = u^0$, $h^1 = u$, and each h^λ satisfies conditions (2.2)–(2.4).

It is sufficient to verify the following assertion:

Claim 1. *For each $u \in \Gamma$ and any given pair of points of $p, q \in S^2$, there is a closed curve u_{t_1, t_2} which passes through p and q for some $(t_1, t_2) \in [-1, 1] \times [0, 1]$.*

If one chooses p, q on the Riemannian sphere $M = (S^2, g)$ satisfying $d(p, q) = D$, then inequality (2.1) follows from Claim 1 immediately. The proof of Claim 1 can be carried out by an elementary topological argument below. (Claim 1 is obviously true for u^0.)

Denote $[-1, 1] \times [0, 1] \times S^1 \times S^1$ by T. Inspired by (2.4), we introduce the following equivalence relations

$$(t_1, 0, s_1, s_2) \sim (-t_1, 1, -s_1, s_2),$$
$$(t_1, t_2, s_1, s_2) \sim (t_1, t_2, s_2, s_1).$$

Using the equivalence relations above, we let $\widetilde{T}$ be the quotient space of T with the quotient topology and K^3 be the subspace defined by $\{\pm 1\} \times [0, 1] \times S^1 \times S^1 / \sim$. Clearly, $\widetilde{T}$ is a four-dimensional orbifold with boundary K^3. In fact, the four-dimensional Z_2-homology of $(\widetilde{T}, K^3)$ is nonzero, i.e.,

$$H_4(\widetilde{T}, K^3; Z_2) = Z_2 \neq 0,$$

where $Z_2 = 2/2Z$ and Z is the ring of all integers.

On $S^2 \times S^2$ we also introduce an equivalence relation by letting $(p, q) \sim (q, p)$ for any pair of points $p, q \in S^2$. Let $[p, q]$ be the equivalent class of (p, q), $[S^2 \times S^2]$ be the quotient space of $S^2 \times S^2$, and $\Delta = \{[p, p] | p \in S^2\}$. Obviously $[S^2 \times S^2]$ is a four-dimensional orbifold with boundary Δ. Clearly, one sees that $H_4([S^2 \times S^2], \Delta : Z_2) = Z_2$.

For each $u \in \Gamma$, we assign a continuous map F_u from $\widetilde{T}$ to $[S^2 \times S^2]$ as follows:

$$F_u(t_1, t_2, s_1, s_2) = [u(t_1, t_2, s_1), u(t_1, t_2, s_2)]. \tag{2.6}$$

It follows from (2.3) that F_u maps K^3 to Δ. Let $\deg_2(F_u)$ denote the Z_2-degree of the map $F_u: (\widetilde{T}, K^3) \to ([S^2 \times S^2], \Delta)$. In what follows, we will compute $\deg_2(F_u)$.

When $u = u^0$, F_{u^0} is a one-to-one map on the open set

$$G = \{(t_1, t_2, s_1, s_2)/\sim \mid -1 < t_1 < 1,\ 0 < t_2 < 1,\ 0 < s_1 < s_2 < 1\}.$$

Note that G is dense in $\widetilde{T}$. Hence, $\overline{G}$ represented a generator of $H_4(\widetilde{T}, K^3; Z_2)$, and so does its image. Therefore, we have demonstrated that F_{u^0} induces an isomorphism

$$(F_{u^0})_*: H_4(\widetilde{T}, K^3; Z_2) \to H_4([S^2 \times S^2], \Delta; Z_2).$$

In particular, the Z_2-degree of the map $F_{u^0}: (\widetilde{T}, K^3) \to ([S^2 \times S^2], \Delta)$ is nonzero, i.e., $\deg_2(F_{u^0}) = \pm 1 \neq 0 \pmod 2$. It follows from (2.5) that $\deg_2(F_u) \neq 0$ for all $u \in \Gamma$. Hence, Claim 1 has been verified. This completes the proof of (2.1).

When $L(M) = 2D$ holds, we know

$$2D \leq L(\gamma_2) \leq L(\gamma_3) \leq L(M) = 2D.$$

Hence, $L(\gamma_2) = L(\gamma_3)$ holds in the Lusternik-Schnirelmann theorem, which guarantees the existence of infinitely many simple closed geodesics of length $2D$ (cf. [15]).

3. An isoperimetric inequality and its applications

In this section, we will give a nonsharp isoperimetric inequality and derive various estimates for lengths of simple closed geodesics.

We begin with a very crude observation:

Theorem C. *Let M be a Riemannian surface, diffeomorphic to S^2, with nonnegative curvature $K \geq 0$. Then*

$$\text{Area}(M) \leq \frac{8}{\pi} D^2.$$

Proof. This is an easy consequence of the two eigenvalue estimates which are due to Hirsch and Zhong-Yang respectively. Hirsch's theorem tells us that, on any Riemannian sphere M, the first eigenvalue of M, λ_1, is less than or equal to $8\pi/A$, i.e.,

$$\lambda_1 \leq 8\pi/A. \tag{3.1}$$

(See [13] and [22].)

On the other hand, for any Riemannian manifold with nonnegative Ricci curvature, Zhong and Yang found a lower bound of λ_1 in terms of the diameter D (cf. [24]):

$$\lambda_1 \geq \pi^2/D^2 . \tag{3.2}$$

Combining (3.1) and (3.2), we get the desired result.

Remark. Estimate (3.2) is optimal for certain flat tori, not for round two-spheres (cf. [23, p. 114]).

Using Theorems A, B, and C, one can easily derive:

Corollary 0.1. *Let M be a smooth Riemannian sphere with $0 \leq K \leq 1$. Then the longest simple closed geodesic γ exists and*

$$\sqrt{A\pi}/\sqrt{2} \leq L(\gamma) \leq A/2 , \tag{0.1}$$

$$2D \leq L(\gamma) \leq 4D^2/\pi . \tag{0.2}$$

Furthermore, equality $L(\gamma) = A/2$ holds if and only if M is isometric to the unit sphere.

We would like to say a few words about the ratio L/D^2, when D becomes infinite. The following example shows that there is a family of smooth metrics g_i on S^2 whose curvature is bounded by $0 \leq K \leq 1$, in which diameter $D_i \to \infty$ and $L_i/D_i^2 \geq (40\pi)^{-1}$ as $i \to \infty$.

Example 3.1. Let us take two copies of a square rectangle of width $2\pi\kappa$, and then glue them along the boundary smoothly except for corner points. The resulting surface is a flat, nonsmooth two-sphere $\widehat{S}^2$ with four exceptional points. Let us call these four points p_1, p_2, p_3, p_4. Around each point p_j we circle out U_j, a neighborhood of p_j. Replace U_j by a quarter of the unit sphere and make the metric smooth under the curvature restriction $0 \leq K \leq 1$ within a metric ball of radius π centered at p_j. Thus, we get a smooth Riemannian sphere of M_k of diameter D_k satisfying $\kappa\pi < D_k \leq 2\pi\kappa\sqrt{5} + \delta$, where δ is independent of κ.

We claim that there is a simple closed geodesic γ_k of length $L(\gamma_k) \geq k\pi\sqrt{16+\kappa^2}/2$ within the flat region of M_k. In order to see this, one needs to view M_k in a different way. Let $\mathbf{Z}$ be the ring of all integers and let us take a lattice $\Gamma\kappa$ in $\mathbf{R}^2$ given by

$$\Gamma_k = \{(4\kappa\pi i_1 + 2\kappa\pi , 4\kappa\pi i_2 + 2\kappa\pi) | i_1 , i_2 \in \mathbf{Z}\} .$$

Clearly, $\mathbf{R}^2/\Gamma_k$ is a flat torus. Notice that the origin $(0, 0)$ is not an element of Γ_k, when κ is a fixed positive integer. Denote the group of $\{\pm \mathrm{Id}\}$ by $\mathbf{Z}_2$, where $-\mathrm{Id}$ is the antipodal map of $\mathbf{R}^2$. It is not hard to see that the orbit space of the group $\mathbf{Z}_2$ acting on $\mathbf{R}^2/\Gamma_k$, say $\mathbf{Z}_2 \backslash (\mathbf{R}^2/\Gamma_k) = \widehat{S}^2$, is homeomorphic to M_k. A fundamental domain of this orbifold can be chosen to be an open rectangle $Q = (0, 2\pi\kappa) \times (-2\pi\kappa, 2\pi\kappa)$. Let us identify Q with the birectangle given at the beginning. Denote the quotient map from $\mathbf{R}^2$ to $\widehat{S}^2$ by F_k. The preimage of singular points $\{p_j\}$ via F_k is exactly $2\pi\kappa(\mathbf{Z}\oplus\mathbf{Z})$. If $\mathfrak{B}$ denotes the union of all disks of radius π whose centers are elements of $2\pi\kappa(\mathbf{Z} \oplus \mathbf{Z})$, then F_k maps $\mathbf{R}^2 - \mathfrak{B}$ to the flat region of (M_k, g_k), which is locally an isometry. Clearly, there is a straight line $\tilde{\gamma}_k$ within $\mathbf{R}^2 - \mathfrak{B}$ which passes through the point $(0, 3k\pi)$ with slope $\kappa/4$. The projection of $\tilde{\gamma}_k$ is the desired closed geodesic γ_k of length $\kappa\pi\sqrt{16+\kappa^2}/2$. Hence, we conclude that $L(\gamma_k)/D_k^2 > (40\pi)^{-1}$ and $D_k > \kappa\pi$, when κ is sufficiently large.

Example 3.2. In Example 3.1, if we replace Γ_k by an affine lattice $\widehat{\Gamma}_k$ of angle 60° and mesh size $4\kappa\pi$, then it is easy to see that the resulting surface is a tetrahedron, which is homeomorphic to S^2 with four singular points. However, in this case, the fundamental domain for this quotient space $\mathbf{Z}_2 \backslash (\mathbf{R}^2/\widehat{\Gamma}_k)$ can also be chosen to be an equilateral triangle of length $4\pi\kappa$. Smoothing the metric along four vertices of the tetrahedron as in Example 3.1, we get a smooth Riemannian sphere $\widehat{M}_k$ of curvature $0 \le K \le 1$. One can easily show that $\kappa\pi < D_k \le 4\kappa\pi + \delta$, where δ is independent of κ. Obviously, there is an affine map from $\mathbf{R}^2$ to $\mathbf{R}^2$, which takes Γ_k to $\widehat{\Gamma}_k$. Playing the same game as before, one can find a simple closed geodesic γ_k in the flat part of $\widehat{M}_k$ of length $L(\gamma_k) = \kappa\pi\sqrt{\kappa^2 + 4k + 16}/2$. This fact implies that $L(\widehat{M})/D_k^2 > (32\pi)^{-1}$ and $D_k > k\pi$, when $\kappa \gg 1$.

Finally, we wish to point out that there is an estimate for the length of the shortest simple closed geodesic, which grows at most linearly in terms of D.

Corollary 0.2. *Let M be a smooth Riemannian sphere of nonnegative curvature. If γ is the shortest simple closed geodesic on M, then $L(\gamma) \le 9D$.*

Proof. It will be shown in Part II that the shortest closed geodesic is simple on a smooth Riemannian sphere of nonnegative curvature (cf. Theorem D). On the other hand, $L(\gamma) \le 9D$ holds for the shortest closed geodesic γ (cf. Theorem 4.1 in [10]).

PART II
THE SHORTEST CLOSED GEODESIC IS SIMPLE ON SMOOTH COMPACT CONVEX SURFACES

1. Birkhoff's ideas and rimmed domains

In order to find the shortest closed geodesic, we will work on a variational problem for the space of all one-cycles in a given closed, smooth Riemannian surface. In order to get started, we want to borrow two major ideas from Birkhoff which are related to the ordinary loop space (cf. [2], [10]). The first one we will use is his method of finding closed geodesics on spheres, which is called the minimax method.

Let M be the two-sphere S^2 endowed with a C^3 smooth metric g. Given a piecewise smooth curve γ (not necessarily closed), by $L(\gamma)$ we will mean the length of γ. We denote the injectivity radius of the Riemannian surface M by $\mathrm{inj}(M)$.

If f_t is a one-parameter family of closed curves starting and ending at a point curve in such a way that the induced map $f: S^2 \to S^2$ (see Figure 1) has nonzero degree, then Birkhoff's argument (or minimax argument) allows us to conclude that M has a nontrivial closed geodesic of length less than or equal to the length of the longest curves in this one-parameter family. We shall extend his method to the space of one-cycles to get an analogous minimax principle (cf. §2 of this part).

The second idea that we will use is the Birkhoff case shortening process, B.C.S.P. (which Birkhoff used in the above mentioned argument). The precise definition of B.C.S.P. is given in [15] and [10]. Since we need to derive some new properties of B.C.S.P., we recall it here.

In general, the B.C.S.P. β^N depends on an integer $N > 2$, where N is chosen so large that $L(\gamma)/N$ is smaller than the injectivity radius, $\mathrm{inj}(M)$, of the surface M. For any piecewise smooth closed curve γ, we will define a new curve $\beta^N(\gamma)$ as well as a homotopy γ_s, $s \in [0, 1]$, from $\gamma = \gamma_0$ to $\beta^N(\gamma) = \gamma_1$. The homotopy γ_s will be defined in such a way that $L(\gamma_{s_1}) \le L(\gamma_{s_2})$ whenever $s_2 \ge s_1$.

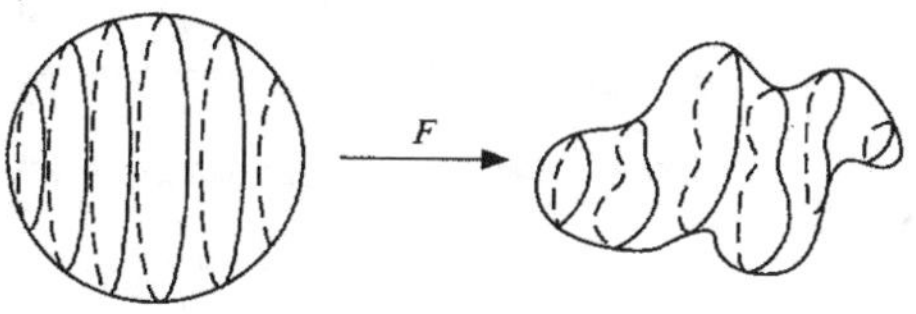

FIGURE 1

We may assume that $\gamma\colon [0, 1] \to M$ is a closed curve parametrized proportional to arc-length; if not, the first part of the homotopy reparametrizes γ so that it is. We define $\gamma_{1/2}$ to be the unique piecewise geodesic closed curve such that $\gamma_{1/2}(i/N) = \gamma(i/N)$ for all integers $i = 1, 2, \cdots, N$. For $s \in [0, \frac{1}{2}]$, γ_s will be given by

$$\gamma_s(i/N + t) = \tau_i^s(t), \qquad \text{when } 0 \le t \le 2s/N,$$
$$\gamma_s(i/N + t) = \gamma(i/N + t), \quad \text{when } 2s/N \le t \le 1/N,$$

where τ_i^s is the minimizing geodesic from $\gamma(i/N)$ to $\gamma(i/N + 2s/N)$ parametrized on the interval $[0, 2s/N]$ proportional to arc-length. Finally, γ_1 is defined as the unique closed broken geodesic with $\gamma_1(i/N + 1/(2N)) = \gamma_{1/2}(i/N + 1/(2N))$ which is parametrized proportional to arc-length on each interval $[i/N + 1/(2N), (i+1)/N + 1/(2N)]$. We then define γ_s for $s \in [\frac{1}{2}, 1]$ to be a homotopy between $\gamma_{1/2}$ and γ_1 in the same way that γ_s, $s \in [0, \frac{1}{2}]$, homotopes from γ_0 to $\gamma_{1/2}$.

We will apply B.C.S.P.. to some special closed curves which have nonnegative geodesic curvatures. More precisely, we let γ be a simple (no self-intersections) closed curve on M which divides M into two components. Let Ω (open) be one of these components.

Definition 1.0. A closed curve γ will be called convex to Ω if there is an $\epsilon > 0$ such that for all $x, y \in \gamma$, with $d(x, y) < \epsilon$, the minimizing geodesic τ from x to y satisfies $\tau \subset \overline{\Omega}$.

The following has been shown in [10].

Lemma 1.1. *Let γ be convex to Ω and have length L, and let $N > L/\operatorname{inj}(M)$ (also $N \gg 2$). Then if we apply B.C.S.P. with N breaks to γ the resulting curves γ_t satisfy*

(1) $\gamma_t \subset \overline{\Omega}$,

(2) γ_t *is simple and convex to* $\Omega_t = \Omega - \{x \in \gamma_s | 0 \le s \le t\}$.

Using Lemma 1.1, we have

Lemma 1.2. *Let γ be convex to Ω and have length L. Then there exists either a simple closed geodesic σ of length $L(\sigma) \le L$, or a homotopy σ_s, $s \in [0, 1]$, which satisfies the following conditions:*

(1.1) $\sigma_1 = \gamma$, $\sigma_0 =$ *a point curve, and* $L(\sigma_s) \le L$ *for all* s,

(1.2) $\{\sigma_s | 0 \le s \le 1\}$ *gives rise in a natural way to a map* F_σ *from the two-disk* D^2 *onto* Ω *with* γ *as the boundary, and* $_\sigma$ *has degree* ± 1.

Proof. Let Λ^L be the space of piecewise smooth closed curves $\alpha\colon [0, 1] \to M$ with length less than or equal to L, and where Λ^L has the C^0-

topology. It is known that the B.C.S.P. $\beta^N : \Lambda^L \to \Lambda^L$ is continuous when $N > (4(\mathrm{inj}(M))^{-1})L$ (cf. [15]). The nontrivial closed geodesics or point curves are the only fixed points of β^N. It can be shown that, for the curve γ given above, the sequence of simple curves $\{\gamma_i\}$ defined by $\gamma_0 = \gamma$ and $\gamma_i = \beta^N(\gamma_{i-1})$ converges to either a simple closed geodesic σ of length $L(\sigma) \le L$ or a point curve (cf. [10, pp. 4–7]). If the limit is a point curve, then the homotopies $\{\gamma_s\}_{0\le s<\infty}$ give rise in a natural way to a map from the two-disk D^2 into Ω with γ as the boundary. In fact, setting

$$D^2 = \{se^{t2\pi i} | (s, t) \in [0, 1] \times [0, 1)\}, \tag{1.3}$$

one can define

$$\sigma_s(t) = \gamma_{(1-s)/s}(t), \quad F_\sigma : D^2 \to \Omega, \quad \text{and} \quad F_\sigma(se^{2\pi i}) = \sigma_s(t).$$

Assertion (1.1) follows from the definition of σ_s. Using Lemma 1.1, we know that F_σ maps D^2 onto Ω with γ as boundary and F_σ has degree ± 1. q.e.d.

In the applications in this paper γ will be a piecewise geodesic curve. In this case, the definition that γ being convex to Ω reduces to the condition that all the angles of γ are convex to Ω.

Let us now consider an interesting example. Suppose $\eta: [0, 1] \to M$ is a nonsimple closed geodesic and M is a sphere. Obviously, η divides M into more than two components, say, $\Omega^1, \Omega^2, \cdots, \Omega^k$, $k \ge 3$. Now, we fix an orientation of M and let $\Omega^1, \Omega^2, \cdots, \Omega^k$ have the inherited orientation from M. For each Ω^i, $1 \le i \le k$, it is not hard to see that $\partial\Omega^i$, the boundary of Ω^i, consists of broken geodesics whose angles are convex to Ω^i. This leads us to make the following observation.

Corollary 1.3. *Let η, Ω^i, and $\partial\Omega^i$ be as above. Then there is either a simple closed geodesic σ of length $L(\sigma) < L(\partial\Omega^i)$ or a homotopy α_s, $s \in [0, 1]$, which satisfies*

(1.5) $\alpha_1 = \partial\Omega^i$, $\alpha_0 =$ *a point curve, and* $L(\alpha_s) \le L(\partial\Omega^i)$ *for all* s.

(1.6) $\{\alpha_s\}$ *gives rise in a natural way to a map* F_α *from the two-disk* D^2 *onto* Ω^i *with* po^i *as the boundary of degree* ± 1.

Proof. If $\partial\Omega^i$ has no self-intersection points, Corollary 1.3 follows from Lemma 1.2 immediately. In the case that $\partial\Omega^i$ has self-intersection points $q_1, \cdots, q_n$, $n \ge 1$, and if q_i is one of them, we notice that the angles of $\partial\Omega^i$ at q_i are strictly convex to Ω^i. Without loss of generality, for each q_i, we may assume that there are locally two curves α and β

such that $\max\{L(\alpha), L(\beta)\} < \text{inj}(M)$,

$$\alpha: [a, b] \to \partial\Omega^i, \qquad \beta: [a, b] \to \partial\Omega^i,$$

such that $\alpha(t) \neq \beta(t)$ for $t \neq (a+b)/2$ and $\alpha(a+b)/2 = \beta((a+b)/2) = q_i$. We move α away from β, by the following local homotopy:

$$(1.7)\quad \alpha_s(t)k = \alpha(t), \qquad \text{when } t \notin \left[\frac{a+b-s(b-a)}{2}, \frac{a+b+s(b-a)}{2}\right],$$

$$(1.8)\quad \alpha_s(t) = \tau^s(t), \qquad \text{when } t \in \left[\frac{a+b-s(b-a)}{2}, \frac{a+b+s(b-a)}{2}\right],$$

where τ^s is the minimizing geodesic from $\gamma((a+b-s(b-a))/2)$ to $\gamma((a+b+s(b-a))/2)$ parametrized proportional to the arc-length on the interval $[(a=b-s(b-a))/2, (a+b+s(b-a))/2]$.

Therefore, when $\partial\Omega^i$ has self-intersection points, we choose the first part of the deformation α_s, $s \in [0, \epsilon]$, locally given by (1.7)–(1.8). The resulting new domain Ω^i_ϵ is still connected and the boundary $\partial\Omega^i_\epsilon$ is the union of simple closed curves. Moreover, $\partial\Omega^i_\epsilon$ is convex to Ω^i_ϵ. Let σ be one of those closed curves, $\sigma \subset \partial\Omega^i_\epsilon$. In what follows we shall show that $\partial\Omega^i_\epsilon = \sigma$ if there is no shorter closed geodesics. In fact, Lemma 1.2 tells us that there is either a simple closed geodesic γ with the length $L(\gamma) \leq L(\partial\Omega^i_\epsilon) < L(\partial\Omega^i)$, or there is a homotopy σ_s given in Lemma 1.2. When the second case occurs, one can show that the set $G = \{x \in \sigma_s | 0 \leq s \leq 1\}$ is equal to Ω^i_ϵ by the Jordan curve theorem, since Ω^i_ϵ is path-connected. Therefore, one knows that $\partial\Omega^i_\epsilon = \sigma$. Hence, $\partial\Omega^i$ is connected (by letting $\epsilon \to 0$). This completes the proof of Corollary 1.3.

We remark that the orientation of α_1 in Corollary 1.3 does not necessarily coincide with the original orientation of η.

2. The space of one-cycles and a minimax principle

In this section we will treat nonsimple closed geodesics as one-cycles. In addition, we will discuss a minimax principle (the generalized Birkhoff argument) on the space of one-cycles with integer coefficients over a surface M.

There are many ways to define the topology on the space of one-cycles. At the moment, we use the weak topology which was first introduced by deRham. Roughly speaking, the space of one-cycles can be thought of as a subspace of the "dual" space of one-forms on M.

Let $\Lambda^1(M)$ be the space of all C^∞ smooth one-forms on M. For any oriented one-dimensional rectifiable set T, there is a corresponding linear function defined on $\Lambda^1(M)$ which is

$$T(\varphi) = \int_T \varphi \quad \text{for all } \varphi \in \Lambda^1(M). \tag{2.1}$$

In what follows, T will also be called a rectifiable current.

Furthermore, if a rectifiable current T satisfies

$$T(df) = \int_T df = 0 \tag{2.2}$$

for all smooth functions f, then T is said to be closed, or T is a one-cycle.

The weak topology of the space of one-cycles can be described in the usual way. We only have to define the limit of a sequence of currents $\{T_i\}$. We require that

$$\lim_{i\to\infty} T_i = T \tag{2.3}$$

holds if and only if

$$\lim_{i\to\infty} T_i(\varphi) = T(\varphi) \quad \text{for all } \varphi \in \Lambda^1(M). \tag{2.4}$$

This weak topology is closely related to the "mass" distance $\mathfrak{M}$. Recall that g is the metric defined on S^2 which induces a natural pointwise metric on $\Lambda^1(M)$. For any currents T, we introduce

$$\mathfrak{M}(T) = \sup\left\{ \int_T \varphi \,\middle|\, |\varphi(x)| \le 1,\ \varphi \in \Lambda^1(M) \text{ and } x \in M \right\}. \tag{2.5}$$

Suppose γ is a piecewise smooth curve and γ has only finitely many self-intersection points. Then one can verify that

$$\mathfrak{M}(\gamma) = L(\gamma) = \text{ the length of } \gamma. \tag{2.6}$$

From now on, we let Z be the ring of all integers and we will work on the set

$$\mathbf{Z}_1(S^2, Z) = \left\{ \sum a_i T_i \,\middle|\, a_i \in Z \text{ is an integer and } T_i \text{ is a rectifiable one-cycle} \right\}. \tag{2.7}$$

$\mathbf{Z}_1(S^2, Z)$ is called a one-dimensional cycle group, or the space of one-cycles with integer coefficients in Z. It has the weak topology mentioned above.

The space of one-cycles $\mathbf{Z}_1(S^2, Z)$ has been studied by Almgren and others. In particular, one of Almgren's results implies the following.

Lemma 2.1. *Let* $\Pi_1(\mathbf{Z}_1(S^2, Z), \{0\})$ *be the fundamental group of space* $\mathbf{Z}_i(S^2, Z)$ *with the topology above. Then there is a natural isomorphism*

$$\Psi: \Pi_1(\mathbf{Z}_1(S^2, Z), \{0\}) \to H_2(S^1, Z) = Z \tag{2.8}$$

between $\Pi_1(\mathbf{Z}_1(S^1, Z))$ *and* $H_2(S^2, Z)$, *the two-dimensional homology group of* S^2 *with integer coefficients in* Z.

Proof. This is a special case of the main theorem in [1].

Notice that the isomorphism Ψ above was intuitively taken by using the "evaluation map" F in Figure 1. There are some useful methods to find a one-parameter family of one-cycles σ_t, $t \in [0, 1]$, with the property $\Psi([\sigma]) \neq 0$. For instance, let $f: M \to [0, 1]$ be a Morse function. Then, if we take the sublevel set $D_t = f^{-1}([0, t])$ and the level set $\sigma_t = f^{-1}(t) = \partial D_t$ with the induced orientation from D_t, it is easy to see $\Psi([\sigma])$ is a generator of $H_2(M, Z)$. In some special cases, σ_s and $\psi(\sigma)$ can also be constructed by using broken geodesics. For example, see the following lemma.

Lemma 2.2. *Let* η *be a nonsimple closed geodesic of length* L. *Then there exist either a nontrivial closed geodesic* σ *of length* $L(\sigma) < L$, *or a one-parameter family of one-cycles* σ_t, $t \in [-1, 1]$, *satisfying the following conditions*:

$$L(\sigma_t) \leq L \quad \textit{for all } t \in [-1, 1], \tag{2.9}$$

(2.10) $\{\sigma_t\}$ *gives rise in a natural way to a map* $F_\sigma: S^2 \to S^2$ *of degree* ± 1.

Proof. Since η is a nonsimple closed geodesic, η divides M into k connected components $\Omega^1, \Omega^2, \cdots, \Omega^k$ with $k \geq 3$. Let Ω^i (open) be one of these components. We think of M as a k-legged star fish with legs $\Omega^1, \cdots, \Omega^k$ (see Figure 2).

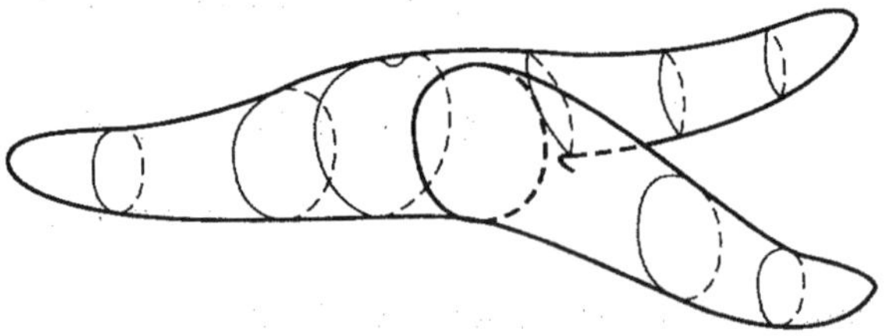

FIGURE 2. 3-LEGGED STAR FISH.

Applying Corollary 1.3 to $\partial\Omega^i$ and Ω^i, we get either a nontrivial closed geodesic σ of length $L(\sigma) < L$, or a homotopy α_s^i which satisfies (1.5)–(1.6). We may assume that there is no nontrivial closed geodesic σ of length $L(\sigma) < L$. The rest of the proof uses $\{\alpha^i\}$, $1 \le i \le k$, to construct σ_t, $t \in [0, 1]$, which satisfy (2.9)–(2.10).

Fix an orientation of M, let Ω^i have the inherited orientation from M, and let Z_2 be the group $Z/2Z$. Since the one-dimensional homology group of M with Z_2 coefficients vanishes, $[\eta] \in H_1(M, Z_2) = 0$, there exist domains $\Omega^{k_1}, \cdots, \Omega^{k_n}$ such that

$$\eta = \sum \partial\Omega^{k_i} \quad \text{in } \mathbf{Z}_1(M, Z_2). \tag{2.11}$$

Let $H_1(M, Z)$ be the first homology group with integral coefficients. By using (2.11), one can also show that

$$\eta = \sum n_i \partial\Omega^{k_i} \quad \text{in } Z_1(S^2, Z) \text{ and } n_i = \pm 1. \tag{2.12}$$

Finally, we choose σ_0 to be

$$\sigma_0 = \sum \partial\Omega^{k_i} \quad \text{in } \mathbf{Z}_1(S^2, Z). \tag{2.13}$$

Applying Corollary 1.3 to σ_0 and Ω^{k_i}, $i = 1, \cdots, n$, one gets half of the homotopy σ_s, $s \in [-1, 0]$. The other half can be derived from the rest of the domains by the same method. We remark that the orientation of σ_0 does not necessarily coincide with the orientation of η, since some of $\{n_i\}$ in (2.12) may be -1. q.e.d.

We are interested in the one-parameter family of one-cycles σ_t given by Lemma 2.2, because there is a minimax principle which works on such families. In general, we shall consider all one-parameter families of one-cycles $\{\sigma_t\}$ which are not homotopic to 0 in the space of one-cycles.

More precisely, let $\sigma\colon [0, 1] \to \mathbf{Z}_1(S^2, Z)$ be a continuous loop in $\mathbf{Z}_1(S^1, Z)$, $\sigma_t = \sigma(t)$, such that

(2.14) σ_t starts and ends at the sum of point curves, hence, $\sigma_0 = \sigma_1 = 0$;

(2.15) $[\sigma] \neq 0$, i.e., σ_t is a nonnull homotopy loop in $\Pi_1(\mathbf{Z}_1(S^2, Z), \{0\})$.

We define a critical value (minimax value) of the mass functional $\mathfrak{M}$ to be

$$c_0 = \inf_{[\sigma]\neq 0} \sup_{0\le t\le 1} \{\mathfrak{M}(\sigma_t)\} \simeq \inf_{[\sigma]\neq 0} \sup_{0\le t\le 1} \{L(\sigma_t)\}. \tag{2.16}$$

We would like to say a few words about the relations between the length of the nonsimple closed geodesic and the critical value c_0 given in (2.16). Lemma 2.2 tells us that, for any nonsimple closed geodesic γ of $L(\gamma)$, either γ is not the shortest, or $L(\gamma) \geq c_0$. The next lemma asserts that the equality $L(\gamma) = c_0$ never holds when γ has more than one self-intersection point (counting multiplicities).

Let γ be a closed geodesic. Then γ will be said to have $k-2$ self-intersection points if $M - \gamma$ has k connected components.

Lemma 2.3 (Almgren-Pitts). *Let γ be a nonsimple closed geodesic of length L which has more than one self-intersection. Then there exists either a nontrivial closed geodesic σ of length $L(\sigma) < L$, or a one-parameter family of one-cycles σ_t, $t \in [0, 1]$, satisfying* (2.14)–(2.15) *and* $\max_{0 \leq t \leq 1} L(\sigma_t) < L$. *Furthermore, σ_t can be chosen so that σ_t is made of at most two closed curves for each $t \in [0, 1]$.*

Proof (Pitts). If there is no closed geodesic σ of $L(\sigma) < L$, we will construct a one-family of one-cycles σ_t so that $L(\sigma_t) < L(\gamma)$ and σ_t satisfies (2.14)–(2.15). Such a one-parameter family σ_t could be constructed by the same argument as in the proof of Lemma 2.2 with some modifications.

Let Ω^i, $\partial\Omega^i$, and α^i be as in the proof of Lemma 2.2, $i = 1, \cdots, k$. Since γ has more than one self-intersection point, one knows that $k \geq 4$. Almgren and Pitts, in particular, found the desired σ_t for the case γ has two self-intersections p_1, p_2 and a four-legged star fish $k = 4$ (cf. [19, pp. 35–40]). For $k > 4$, the proof remains the same with little changes. We recall Pitts' argument here, since it will be used in the appendix.

To describe the path σ_t, $t \in [0, 1]$, we subdivide the interval $[0, 1]$ into five subintervals (cf. Figure 3). For $t \in [0, 1/5]$, we bring a cycle up to two of the legs of the star fish; $\sigma(1/5)$ is a figure eight, and so has one self-intersection. From $1/5$ to $2/5$, using (1.4)–(1.5), we "open up" the figure eight $\sigma(1/5)$ by a length-decreasing homotopy. When t is in $[2/5, 3/5)$, σ_t has two components: the curve $\sigma(2/5)$ plus a second cycle moving up the third leg of the star fish, which can be done by using α_s given in Corollary 1.3. At $t = 3/5$, $\sigma(3/5)$ has one self-intersection. From $3/5$ to $4/5$, we once again "open up" the figure eight $\sigma(3/5)$ by a mass-decreasing homotopy; from $t = 4/5$ to $t = 1$, we "slide" the cycle $\sigma(4/5)$ to a point by using Lemma 1.2. It is clear that $\sigma(3/5)$ is the cycle of longest length in the path $\{\sigma_t\}$ and $L(\sigma(3/5)) < L(\gamma) = L$. q.e.d.

The following one-dimensional minimax theorem is due to Almgren and Pitts.

Theorem 2.4 (Almgren-Pits). *Let c_0 be as in* (2.16). *Then there is a nontrivial closed geodesic γ_0 of length $L(\gamma_0) = c_0$. Furthermore, if γ_0 is also the shortest closed geodesic on M, then γ_0 is either simple, or a figure eight.*

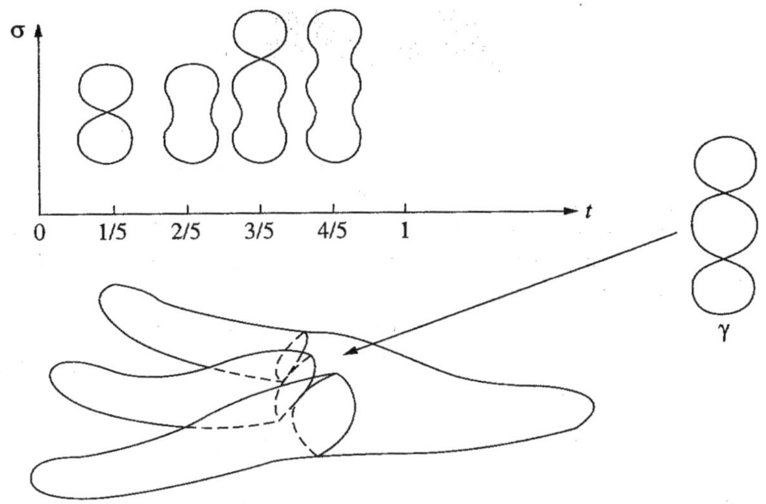

FIGURE 3. 4-LEGGED STAR FISH.

Proof. See [18] and [19, p. 162], or an independent proof in the appendix, which does not use the geometric measure theory.

3. The proof of Theorem D

In this section, we shall show that, on any smooth Riemannian two-sphere of nonnegative curvature, the shortest closed geodesic is simple. In this case, we will also demonstrate that the length of any shortest closed geodesic is equal to the minimax value c_0, given by (2.16).

Notice that, if the curvature is negative somewhere on a given two-sphere, then the length of the shortest closed geodesic is not necessarily equal to the minimax value c_0. Standing as the counterexample is the "dumbbell" (cf. [6]). The "dumbbell" manifold is homeomorphic to S^2, shown in Figure 4 (next page). The pipe connecting the two halves is to be thought of as having fixed length L_0 and varying radius $\epsilon \to 0$. Using the Clairaut's relation, one can verify that the minimax value $c_0 > 2\pi\epsilon$, where $2\pi\epsilon$ is the length of the shortest geodesic.

Now, we are ready to prove

Theorem D. *If g is a C^3-smooth metric on a two-sphere S^2 with nonnegative curvature, then any nontrivial closed geodesic of shortest length is simple.*

Proof. Let γ be a nontrivial closed geodesic of the shortest length. It follows from Lemma 2.3 and Theorem 2.4 that γ is either a figure eight

FIGURE 4. THE DUMBBELL.

or a simple closed geodesic. We will show that γ cannot be a figure eight by a contradiction method. According to the minimax principal (Theorem 2.4), it is sufficient to verify the following assertion:

Claim 1. *If γ were a figure eight, there would be either a shorter closed geodesic σ of $L(\sigma) < L(\gamma)$, or a one-parameter family of one-cycles σ_t, $t \in [-1, 1]$, satisfying*

(3.1) $$L(\sigma_t) < L(\gamma) \quad \textit{for all } t \in [-1, 1];$$

(3.2) *$\sigma_{-1} = \sigma_1 = 0$ are zeros as one-currents. Moreover, σ_{-1} and σ_1 are sums of point curves;*

(3.3) $$\sigma_t \textit{ is a nonnull homotopy in } \Pi_1(\mathbf{Z}_1(S^2, Z), \{0\}),$$

i.e., $[\sigma] \neq 0$.

The construction of σ_t is done in several steps by the same method as in the proof of Lemma 2.2, with some changes. The new ingredient in the construction is to use the second variational formula. First, we need to deform the closed geodesic γ to a new figure eight σ_0 which satisfies (3.1).

Step 1. *A new figure eight σ_0 of length $L(\sigma_0) < L(\gamma)$.* Clearly, the given γ violates restriction (3.1). Hence, we have to replace γ by a "nearby" figure eight with smaller length. The desired σ_0 can be found by using the exponential map along the parallel normal vector field $\vec{n}$ of γ.

By the assumption, the curvature K of the metric g is nonnegative and M is homeomorphic to S^2. It follows from the Gauss-Bonnet formula that there is a nonempty open set G in M on which the curvature is strictly positive, i.e., $G = \{x | K(x) > 0\} \neq \emptyset$. If the metric is flat in a neighborhood of γ, $K = 0$, then one can move γ by a translation to get a family of new closed geodesics of figure eight type. Such a translation will stop until a new geodesic hits $\overline{G}$, the closure of G. Therefore, we may assume that there is a point q on γ such that $q \in \overline{G}$.

Let $\vec{n}(s)$ be the smooth unit normal vector field along $\gamma(t)$ pointing inward to G. Now we define a small perturbation of γ by the exponential map:

(3.4) $$f_t(s) = \exp_{\gamma(s)}[t\vec{n}(s)].$$

For the same reason as in the proof of Lemma 1.2 in Part I, we see that there is an $\epsilon > 0$ with

(3.5) $$L(f_t) < L(\gamma) \quad \text{for all } t \text{ with } 0 < t < 2\epsilon.$$

For simplicity, we change the parametrization of f_ϵ so that $f_\epsilon : [-1, 1] \to S^2$ and $f_\epsilon(0) = f_\epsilon(-1) = f_\epsilon(1) = p_0$, which is the unique intersection point.

Step 2. *Cut and paste* f_ϵ *at its intersection point* p_0. In order to get a new closed curve σ_0 which is a limit of simple closed curves, we let

$$f^+ = f_\epsilon|_{[0,1]}, \quad f^- = f_\epsilon|_{[-1,0]}, \quad \sigma_0 = f^+ - f^-,$$
$$\eta^- = -f^-|_{[\epsilon-1,-\epsilon]}, \qquad \eta^+ = f^+|_{[\epsilon,1-\epsilon]},$$
$$p_1 = \sigma_0(\epsilon - 1), \quad p_2 = \sigma_0(\epsilon), \quad p_3 = \sigma_0(1-\epsilon), \quad p_4 = \sigma_0(-\epsilon).$$

For any pair of points $\{p, q\}$ with $d(p, q) < \frac{1}{2}\mathrm{inj}(M)$, we denote the minimal geodesic from p to q by $\overrightarrow{pq}$. Finally, we define

$$\sigma_{-\epsilon} = \eta^- + \overrightarrow{p_1p_2} + \eta^+ + \overrightarrow{p_3p_4}, \qquad \sigma_\epsilon = \eta^- + \overrightarrow{p_1p_4} + \eta^+ + \overrightarrow{p_3p_2}.$$

One can find the homotopy between $\sigma_{-\epsilon}$ and σ_ϵ, say σ_t, $t \in [-\epsilon, \epsilon]$, such that $L(\sigma_t) \leq L(\sigma_0) < L(\gamma)$ for $t \in [-\epsilon, \epsilon]$. This part of the construction of $\{\sigma_t\}$, $t \in [-\epsilon, \epsilon]$, is straightforward.

Our final step is to shrink σ_ϵ and $\sigma_{-\epsilon}$ to the point curves.

Step 3. *The deformation from* σ_ϵ *and* $\sigma_{-\epsilon}$ *to point curves.* We cannot apply Lemma 1.2 to σ_ϵ or $\sigma_{-\epsilon}$ directly, since $\sigma_{\pm\epsilon}$ may not be convex to the domain it encloses. This leads us to make the following observations on each part of $\sigma_{-\epsilon}$ and σ_ϵ. Let

$$\gamma^- = \gamma|_{[\epsilon-1,-\epsilon]}, \qquad \gamma^+ = \gamma|_{[\epsilon,1-\epsilon]},$$
$$q_1 = \gamma(\epsilon - 1), \quad q_2 = \gamma(\epsilon), \quad q_3 = \gamma(1-\epsilon), \quad q_4 = \gamma(-\epsilon),$$
$$\sigma_{2-\epsilon} = \gamma^{-1} + \overrightarrow{q_1q_2} + \gamma^+ + \overrightarrow{q_3q_4}, \qquad \sigma_{2\epsilon} = \gamma^- + \overrightarrow{q_1q_4} + \gamma^+ + \overrightarrow{q_3q_2}.$$

Obviously, there is a homotopy of σ_t, $t \in [\epsilon, 2\epsilon] \cup [-2\epsilon, -\epsilon]$, such that

(3.6) $$L(\sigma_t) \leq L(\gamma) - \delta < L(\gamma),$$

where δ is a positive constant which depends on

$$\min_{1 \leq i \leq 3}\{d(q_i, \sigma_0(0)) + d(q_{i+1}, \sigma_0(0)) - d(q_i, q_{i+1})\} > 0.$$

Because each component $\sigma_{2\epsilon}$ and $\sigma_{-2\epsilon}$ is a broken closed geodesic which is convex to the domain it encloses, one can use Lemma 1.2 to find either a shorter closed geodesic σ of $L(\sigma) < L(\gamma)$, or the rest of σ_t, $t \in [-1, -2\epsilon] \cup [2\epsilon, 1]$, such that

(3.1) $$L(\sigma_t) < L(\gamma) \quad \text{for all } t \in [-1, -2\epsilon] \cup [2\epsilon, 1].$$

(3.2) $\sigma_{-1} = \sigma_1 = 0$ are zeros as one-currents. Moreover, they are sums of point curves.

Combining Steps 1–3, we get the $\{\sigma_t\}$, $t \in [-1, 1]$, as claimed in (3.1)–(3.3).

This finishes the proof of Claim 1. Thus, Theorem D follows from Lemma 2.3 and Theorem 2.4 immediately.

We conclude this part by studying the length of the shortest closed geodesic.

Theorem 3.1. *Let g be the metric on the two-sphere S^2 with nonnegative curvature $K \geq 0$ and let γ be a nontrivial closed geodesic of shortest length L. Then L is equal to the minimax value c_0 given in* (2.16).

Proof. It follows from Theorem 2.4 that $L(\gamma) \leq c_0$. Hence, we only have to demonstrate $L(\gamma) \geq c_0$. For a closed geodesic γ of the shortest length, it suffices to construct a one-parameter family of closed curves f_t, $t \in [-1, 1]$, starting and ending at a point curve in such a way that the induced map $F: S^2 \to S^2$ (see Figure 1) has nonzero degree, $L(f_t) \leq L(\gamma)$ and $f_0 = \gamma$.

Theorem D tells us that γ is simple. Therefore, γ divides S^2 into two components Ω^- and Ω^+. We will shrink $\gamma = \partial\Omega^+$ to a point curve with the domain Ω^+ by f_t, $t \in [0, 1]$. One can play the same game on the other "half" of S^2, Ω^-.

The perturbation f_t of γ, for $t \geq 0$, will be carried out by the same method as in Step 1 in the proof of Theorem D. Suppose γ: $[0, 1] \to M$ is a parametrization proportional to the arc-length, and let $\vec{n}(s)$, $s \in [0, 1]$, be the unit normal vector field which points inward to Ω^+. Let $G = \{x | x \in \Omega^+ \text{ and } K(x) > 0\}$. For the same reason as above, we may assume that $\gamma \cap \overline{G} \neq \emptyset$ and $\vec{n}$ points inward to G. Let

(3.4) $$F(s \cdot t) = f_t(s) = \exp_{\gamma(s)}[t\vec{n}(s)], \qquad s \in [0, 1].$$

Using Lemma 1.4 of Part I and $K > 0$ on G, one can show that there is an $\epsilon > 0$ and

(3.7) $$L(f_t) < L(\gamma) \quad \text{for all } t \text{ with } 0 < t < \epsilon.$$

Now applying the Birkhoff curve shortening process to f_ϵ, we get a limiting curve f_1. Since γ is the shortest closed geodesic, f_1 has to be a point curve. Because $\partial\Omega^+$ is convex to Ω^+ (cf. §1), we know that $f_1 \in \Omega^+$. Using the same argument as in the proof of Lemma 1.2, we get a homotopy f_t, $t \in [\epsilon, 1]$, which satisfies $L(f_t) \le L(f_\epsilon) < L(\gamma)$. This gives the desired f_t for $0 \le t \le 1$. The remaining part of f_t, $-1 \le t \le 0$, can be produced by the same method. Hence, we get $\{f_t\}_{-1\le t\le 1}$ which induces a map $f: S^2 \to S^2$ of degree ± 1, $L(f_t) \le L(\gamma)$ and $f_0 = \gamma$. This completes the proof of Theorem 3.1.

Appendix

In this appendix, we will give an alternative proof of the minimax principal for the space of one-cycles (cf. Theorem 2.4) without using the geometric measure theory. All notation remain the same as in§2 of Part II unless otherwise specified.

The main tool that we will use is the variational method. However, there are some technical difficulties. Recall that we introduced the "mass" functional $\mathfrak{M}$ on the space of one-cycles (cf. (2.5)). The "mass" functional $\mathfrak{M}$ is only a semicontinuous function with respect to the weak topology on $\mathbf{Z}_1(S^2, Z)$. For example, let $\{(x, y)\}$ be the geodesic normal coordinates at a given point p, and let σ_ϵ be the ellipsoid given by $(x/c)^2 + (y/\epsilon)^2 = 1$ with counter-clockwise orientation, where $c < \frac{1}{2}\operatorname{inj}(M)$ is a constant. One can see that, in such a family of one-cycles, σ_ϵ,

(A.1) $\quad \sigma_\epsilon \to 0 \quad \text{in } \widetilde{Z}^1(S^2, Z) \qquad \text{and} \qquad \mathfrak{M}(\sigma_0) = 0 \quad \text{as } \epsilon \to 0;$

(A.2) $\quad \mathfrak{M}(\sigma_\epsilon) \ge 2c \ne 0 \quad \text{for all } \epsilon > 0.$

In order to get around the unpleasant situation above, we need to replace the mass functional $\mathfrak{M}$ by the length functional L on the space of all parametrized one-cycles. The length functional L has some nice properties. For instance, the sum of the lengths of two curves is independent of the choice of orientations on two curves. We also notice that, for any finite number of paths $\varphi_1, \varphi_2, \cdots, \varphi_n$, $\varphi_i: [0, 1] \to S^2$, with the constraint $\sum[\varphi_i(1) - \varphi_i(0)] = 0$, the sum of these paths gives us a one-cycle $\sum \varphi_i$. For the purpose of this appendix, Lemma 2.3 tells us that it is sufficient to consider all one-cycles which have at most two connected components. Therefore, we are going to work on the set of all pairs of paths $\{(\varphi_1, \varphi_2)\}$ with the constraint

(A.3) $\quad \varphi_1(1) + \varphi_2(1) - \varphi_1(0) - \varphi_2(0) = 0.$

For simplicity, we denote this set by Γ:

(A.4) $\Gamma = \{\Phi | \Phi = (\varphi_1, \varphi_2),\ \varphi_i: [0, 1] \to S^2$ is a piecewise smooth path and $\sum \varphi_i(1) - \sum \varphi_i(0) = 0$, i.e., Φ satisfies (A.3)$\}$.

Notice that there is a natural map which takes Γ to the space of one-cycles:

$$\text{(A.5)} \qquad \begin{aligned} \mathscr{P} : \Gamma &\to \mathbf{Z}_1(S^2, Z) \\ (\varphi_1, \varphi_2) &\to \varphi_1 + \varphi_2 . \end{aligned}$$

When $\Phi = (\varphi_1, \varphi_2)$, by $L(\Phi)$ we mean the sum $L(\varphi_1) + L(\varphi_2)$. We would like to define a strong topology on Γ so that $\mathscr{P}$ and L become continuous maps. One good choice is the modification of the Fréchet topology.

We shall now define the modified Fréchet distance $\hat{d}(\Phi, \tilde{\Phi})$ between two elements $\Phi = (\varphi_1, \varphi_2)$ and $\tilde{\Phi} = (\tilde{\varphi}_1, \tilde{\varphi}_2) \in \Gamma$ as follows. For simplicity, let $U_\epsilon = [0, \epsilon] \cup [1 - \epsilon, 1]$. We may also assume that the surface M is isometrically embedded in the Euclidean space R^n by Nash's theorem. The distance $\hat{d}$ consists of two parts. First, we consider the matched pair of arcs:

$$\text{(A.6)} \qquad \underline{d}_{\epsilon_i}(\Phi, \tilde{\Phi}) = \max_{\epsilon_i \le t \le 1-\epsilon_i} d(\varphi_i(t), \tilde{\varphi}_i(t)) + \int_{\epsilon_i}^{1-\epsilon_i} |\varphi_i'(t) - \tilde{\varphi}_i'(t)|^2 \, dt .$$

The residue part can be measured by

$$\text{(A.7)} \qquad \underset{\sim}{d}_{\epsilon_i}(\Phi, \tilde{\Phi}) = L(\varphi_i|_{U_{\epsilon_i}}) + L(\tilde{\varphi}_i|_{U_{\epsilon_i}}) .$$

Finally, we define $\hat{d}$ to be the greatest lower bound for every choice of $\epsilon_i \in [0, 1]$, $i = 1, 2$,

$$\text{(A.8)} \qquad \hat{d}(\Phi, \tilde{\Phi}) = \inf_{0 \le \epsilon_i \le 1} \left\{ \sum_i \underline{d}_{\epsilon_i}(\Phi, \tilde{\Phi}) + \sum_i \underset{\sim}{d}_{\epsilon_i}(\Phi, \tilde{\Phi}) \right\} .$$

From now on, we will simply denote the topological space $(\Gamma, \hat{d})$ by Γ. It follows from the definition of $\hat{d}$ that $\mathscr{P}$ is in fact a continuous map from Γ to $\mathbf{Z}_1(S^2, Z)$. Since Γ is relatively easy to deal with, we will transfer our geometric problem on $\mathbf{Z}_1(S^2, Z)$ to a variational problem on Γ via the map $\mathscr{P}$.

For any given path $\underline{\Phi}$: $[0, 1] \to \Gamma$, if $\mathscr{P}(\underline{\Phi})$ satisfies conditions (2.14)–(2.15), then $\underline{\Phi}$ is said to be admissible, or briefly $[\underline{\Phi}] \neq 0$.

For the set of all admissible paths in Γ, one can find the following fact about its minimax value.

Lemma A.1. *Let* c_0 *be the minimax value of the mass function* $\mathfrak{M}$ *on* $\mathbf{Z}_1(S^2, Z)$ *given in* (2.16). *Then*

$$c_0 = \inf_{[\Phi]\neq 0}\max_{0\leq s\leq 1} L(\Phi^s). \tag{A.9}$$

Proof. Using the same argument as in the proof of Lemma 2.3 of Part II, one can show that c_0 can be taken among all nontrivial paths $\{\sigma^s\}$, $s \in [0, 1]$, in the subspace $\mathscr{P}(\Gamma)$ of $\mathbf{Z}_1(S^2, Z)$, where $\mathscr{P}(\Gamma)$ is the set of all one-cycles which have at most two components.

Recall that, if Φ only has finitely many self-intersections, then, by (2.7), one knows $\mathfrak{M}(\mathscr{P}(\Phi)) = L(\Phi)$ .1 The set of all such Φ is a dense subset of Γ. Hence, (A.9) holds.

Proof of Theorem 2.4. In order to carry out the proof, we need to consider the variation of the length functional L on Γ. Let $\mathbf{X}(M)$ be the set of all smooth vector fields on M. For any $X \in \mathbf{X}(M)$, X generates a one-parameter group of diffeomorphism h_t (cf. [16, p. 10]). The derivation of L at $\Phi \in \Gamma$ in direction X is defined by

$$\delta L_\Phi(X) = \frac{d}{dt}(L(h_t \circ \Phi))|_{t=0}. \tag{A.10}$$

If $\delta L_\Phi(X) = 0$ for all $X \in \mathbf{X}(M)$, then Φ is called a critical point of L in Γ and $L(\Phi)$ is called a critical value of L.

Our next step is to show that c_0 is a critical value of L. This can be done by a finite-dimensional approximation of a subspace of $\Gamma^{c_1} = \{\Phi | \Phi \in \Gamma,\ L(\Phi) \leq c_1\}$, where $c_1 > c_0$ is a constant. Such an approximation can be described as follows. Choose an integer $N > 2$ so large that $c_1/N > \operatorname{inj}(M)$. For any $\Phi = (\varphi_1, \varphi_2) \in \Gamma^{c_1}$, applying the first half of B.C.S.P. to each φ_i, $i = 1, 2$, one can get a new element $\Phi^{1/2} \in \Gamma$, as well as a homotopy Φ^s, $s \in [0, 1/2]$, from $\Phi^0 = \Phi$ to $\Phi^{1/2}$ (cf. §1 of Part II). The homotopy was defined in such a way that $L(\Phi^s) \leq L(\Phi^\tau)$ whenever $s \geq \tau$. Hence, if $\mathscr{N}$ is the product of $2N$ copies of M, then Γ^{c_1} is homotopic to a compact quotient space of $\mathscr{N}$. Using the same argument as in [16, pp. 88–100], one can show that c_0 is a critical value of L.

Suppose that $\Phi = (\varphi_1, \varphi_2)$ is a critical point in Γ with $L(\Phi) = c_0$; we want to show that $\varphi_1 \cup \varphi_2$ forms a closed geodesic. When Φ is a critical point, φ_1 and φ_2 have to be geodesic paths. Since $\varphi_1 + \varphi_2$ is a closed one-cycle, endpoints of φ_1 and φ_2 have to meet at at most two distinct points. When the set $\{\varphi_1(0), \varphi_1(1), \varphi_2(0), \varphi_2(1)\}$ has exactly two distinct points, it is easy to verify that φ_1 and φ_2 form a closed geodesic or a union of closed geodesics (or possibly point curves). If endpoints φ_1 and φ_2 meet at a single point and both are nontrivial geodesic loops, then we let

$$e_1 = \varphi_1'(0)/\|\varphi_1'(0)\|, \qquad e_2 = \varphi_1'(1)/\|\varphi_1'(1)\|,$$
$$e_3 = \varphi_2'(0)/\|\varphi_2'(0)\|, \qquad e_4 = \varphi_2'(1)/\|\varphi_2'(1)\|.$$

Using the fact $\delta L_\Phi(X) = 0$ for all $X \in \mathbf{X}(M)$ and the first variational formula in [7, p. 24], one knows that

$$e_1 - e_2 + e_3 - e_4 = 0. \tag{A.11}$$

Using (A.11) and the fact that $\|e_i\| = 1$, $0 \le i \le 4$, one may assume that $e_1 = -e_3$ and $e_2 = -e_4$ after reindexing e_i, $i = 1, 2, 3, 4$, suitably. Therefore, $\varphi_1 \cup \varphi_2$ forms a closed geodesic after changing the orientation of φ_1 appropriately.

Moreover, if there exists the shortest closed geodesic γ_0 of length $L(\gamma_0) = c_0$ on (S^2, g), then using Lemma 2.3 and the argument above, we conclude that γ_0 has at most one self-intersection. This completes the proof of the minimax principle.

Remark. Suppose M is a closed smooth Riemannian surface of genus k and γ is a nontrivial closed geodesic of shortest length among its homology class $[\gamma]$. Then using the argument above, one can estimate the number of self-intersections of γ. In particular, if $[\gamma] = 0$ in $H_1(M, Z)$, then γ has at most $2k + 1$ self-intersections.

References

[1] F. J. Almgren, *The homotopy groups of the integral cycle groups*, Topology **1** (1960) 257–299.

[2] W. Ballmann, *Der Satz von Lusternik und Schnirelmann*, Bonner Math. Schriften **102** (1978), 1–25.

[3] W. Ballmann, G. Thorbergsson & W. Ziller, *Closed geodesics on positively curved manifolds*, Ann. of Math. (2) **116** (1982) 213–247.

[4] V. Bangert, *Manifolds with geodesic chords of constant length*, Math. Ann. **265** (1983) 273–281.

[5] M. Berger, *Blaschke's conjecture for spheres*, Manifolds, All of Whose Geodesics are Closed (A. Besse, ed.), Ergebnisse Math. u. i. Grenzgeb., Vol. 93, Appendices D and E, Springer, Berlin, 1978.

[6] J. Cheeger, *A low bound for the smallest eigenvalue of the Laplacian*, Problems in Analysis, A Symposium in Honor of Salomon Bochner (R. C. Gunning, ed.), Princeton University Press, Princeton, NJ, 1970.

[7] J. Cheeger & D. Ebin, *Comparison theorems in Riemannian geometry*, North-Holland Math. Library, North-Holland, Amsterdam, 1975.

[8] C. Croke, *Some isoperimetric inequality and eigenvalue estimates*, Ann. Sci. École Norm. Sup (4) **13** (1980) 419–435.

[9] ——, *Poincare's problem and the length of shortest closed geodesic on a convex hypersurface*, J. Differential Geometry **17** (1980) 595–634.

[10] ——, *Area and the length of shortest closed geodesic*, J. Differential Geometry **27** (1988) 1–22.
[11] M. Grayson, *Shortening embedded curves*, Ann. of Math. **129** (1989) 71–111.
[12] E. Heintze & H. Karcher, *A general composition theorem with applications to volume estimates for submanifolds*, Ann. Sci. École Norm. Sup. (4) **11** (1978) 451–470.
[13] J. Hersch, *Quarter proprietes isoperimetriques de membrances spheriques homogenes*, C. R. Acad. Sci. Paris Sér. A **270** (1970) 1645–1648.
[14] J. L. Kazdan, *An isoperimetric inequality and Wiedersehen manifolds*, Seminar on Differential Geometry (S. Y. Yau, ed.), Ann. of Math. Studies, No. 102, Princeton University Press, Princeton, NJ, 1982, 525–537.
[15] W. Klingenberg, *Lectures on closed geodesics*, Appendix, Springer, Berlin, 1978, 203–219.
[16] L. A. Lyusternik, *The topology of function spaces and the calculus of variations in the large*, Trudy Mat. Inst. Steklov. **19** (1949); English transl., Transl. Math. Monographs, Vol. 16, Amer. Math. Soc., Providence, RI, 1966.
[17] J. Milnor, *Morse theory*, Annals of Math. Studies, No. 51, Princeton University Press, Princeton, NJ, 1963.
[18] J. Pitts, *Regularity and singularity of one dimensional stational integral varifolds on manifolds arising from variational methods in the large*, Symposia Mathematics, Vol. XIV, Roma, Italy, 1974.
[19] ——, *Existence and regularity of minimal surfaces on Riemannian manifolds*, Math. Notes, No. 27, Princeton University Press, Princeton, NJ, 1981.
[20] L. A. Santalo, *Integral geometry in general spaces*, Proc. Internat. Congr. Math. (Cambridge 1950), Vol. I, Amer. Math. Soc., Providence, RI, 1952, 483–489.
[21] W. P.Thurston, *The topology and geometry of* 3*-manifolds*, Lecture Notes in Math., preprint, Princeton University, 1979.
[22] P. Yang & S. T. Yau, *Eigenvalues of the Laplacian of compact surfaces and minimal submanifolds*, Ann. Scuola Norm. Sup. Pisa Cl. Sci. (4) 7 (1980) 55–63.
[23] S. T. Yau, *Nonlinear analysis in geometry*, Enseignement Math. **33** (1987) 109–138.
[24] J. Q. Zhong & H. C. Yang, *On the estimate of first eigenvalue of a compact Riemannian manifold*, Sci. Sinica, Ser. A **27** (1984) 1252–1265.

UNIVERSITY OF PENNSYLVANIA
CORNELL UNIVERSITY

Sums of Generalized Harmonic Series and Volumes

Frits Beukers, Johan A. C. Kolk, and Eugenio Calabi

Nieuw Arch. Wisk. (4) 11, nr. 3, 217–224 (1993)

Introduction. It is well-known that the Riemann zeta-function (as defined by Euler):

$$\zeta(s) = \sum_{k=1}^{\infty} k^{-s} \qquad (s \in \mathbf{R},\ s > 1) \tag{1}$$

takes "elementary" values (i.e., values definable in a first-year calculus class) when s is a positive even integer; more precisely:

$$\zeta(2n) = (-1)^{n-1} B_{2n} \frac{1}{2(2n)!} (2\pi)^{2n} \qquad (n \in \mathbf{N}).$$

Here, B_{2n} denotes the $2n$-th Bernoulli number; we use the convention whereby the rational numbers B_n are defined from the Maclaurin series coefficients of the following function:

$$\frac{t}{e^t - 1} = \sum_{n=0}^{\infty} B_n \frac{t^n}{n!} \qquad (|t| < 2\pi). \tag{2}$$

Equivalently, the series (1) can be replaced by one with the same terms restricted to odd values of k, yielding:

$$(1 - 2^{-s})\zeta(s) = \sum_{k=0}^{\infty} (2k+1)^{-s},$$

whence:

$$\sum_{k=0}^{\infty} \frac{1}{(2k+1)^{2n}} = (-1)^{n-1} B_{2n} \frac{2^{2n} - 1}{2(2n)!} \pi^{2n} \qquad (n \in \mathbf{N}). \tag{3}$$

If we replace the series of either (1) (for $s = 2n$) or (3) with another of the same terms in absolute value but *alternating in sign*, then there is no similar elementary formula for the sum. However, for *odd* positive integer values of s there is the similar formula due to Euler:

$$\sum_{k=0}^{\infty} \frac{(-1)^k}{(2k+1)^{2n+1}} = E_n \frac{1}{2(2n)!} \left(\frac{\pi}{2}\right)^{2n+1}, \tag{4}$$

where the Euler numbers E_n, for $n \in \mathbf{N}_0$, are the natural numbers defined according to:

$$\sec t = \frac{1}{\cos t} = \sum_{n=0}^{\infty} E_n \frac{t^{2n}}{(2n)!} \qquad \left(|t| < \frac{\pi}{2}\right). \tag{5}$$

Among the conventional proofs of (3) and (4), most of them involve either residue theory of complex functions, or Mittag-Leffler expansions, or techniques from Fourier analysis. This means that the average undergraduate majoring in mathematics could rarely learn it before the senior year (unless she/he is unusually talented).

The aim of this paper is to present a simultaneous proof of (3) and (4) in the joint formulation:

$$\sum_{k=0}^{\infty} \frac{(-1)^{nk}}{(2k+1)^n} = \delta_n \left(\frac{\pi}{2}\right)^n \qquad (n \in \mathbf{N}). \tag{6}$$

Here each δ_n is a positive rational number, arising as the volume of a certain n-dimensional convex polytope Δ^n with rational vertices. By direct computation of $\mathrm{vol}_n\, \Delta^n$ we find:

$$\sum_{n=1}^{\infty} \delta_n\, t^{n-1} = \frac{1}{2}(\sec t + \tan t) \qquad \left(|t| < \frac{\pi}{2}\right). \tag{7}$$

On expanding the right side of (7) we obtain the expressions for the δ_n in terms of the Bernoulli and Euler numbers as indicated above.

The proof requires no special knowledge beyond what is included in a good second-year calculus class, and it is in four steps. Step I: using standard techniques on convergence we convert the sum of the series into an integral of a rational function over the unit cube in $\mathbf{R}^n$. Step II: by means of a nontrivial substitution of variables, and this is the heart of the matter, the integral is shown to be equal to $\delta_n(\frac{\pi}{2})^n$, with δ_n as above. Step III: by an analysis of some elementary inequalities, we dissect Δ^n into n congruent pyramids; and we express δ_n in terms of the $(n-1)$-dimensional volume of the basis of such a pyramid. Step IV: the volume of this basis is determined by means of a two-step recursion. This requires mathematical induction with respect to n.

From series to integral over a cube. We begin with the following

Lemma 1. *The infinite series (6) converges for each $n \in \mathbf{N}$, and its sum is represented by the integral:*

$$\sum_{k=0}^{\infty} \frac{(-1)^{nk}}{(2k+1)^n} = \int_{\Box^n} \frac{1}{1-(-1)^n (x_1 \cdots x_n)^2}\, dx. \tag{8}$$

Here $\Box^n$ is the open unit cube of $x \in \mathbf{R}^n$ satisfying $0 < x_i < 1$, for $1 \le i \le n$, while $\int_{\Box^n} \cdots dx$ denotes n-dimensional integration over $\Box^n$ with respect to variables $x = (x_1, \ldots, x_n)$.

Proof: First note that the definition of the integral needs some care in the case of n even, as the integrand becomes infinite at the corner $(1, 1, \ldots, 1)$ of the closure of $\Box^n$. Since the integrand is positive in the open set $\Box^n$, in order to prove convergence of the integral it suffices to show the existence of an increasing sequence of sets $K_\lambda \subset \Box^n$ such that $\cup_{0<\lambda<1} K_\lambda = \Box^n$ and that the $\int_{K_\lambda} \frac{1}{1-(-1)^n(x_1\cdots x_n)^2}\, dx$ exist and converge as $\lambda \uparrow 1$. We shall do this, thereby proving our lemma, by choosing the subsets K_λ to be the contracted hypercubes $\lambda\Box^n$ with $0 < \lambda < 1$. Note that:

$$\int_{\lambda\Box^n} \frac{1}{1-(-1)^n (x_1 \cdots x_n)^2}\, dx = \int_{\Box^n} \frac{\lambda^n}{1-(-\lambda^2)^n (x_1 \cdots x_n)^2}\, dx.$$

Expansion of the right side into a geometric series yields:

$$\int_{\Box^n} \sum_{k=0}^{\infty} \lambda^{n(2k+1)} (-1)^{nk} (x_1 \cdots x_n)^{2k}\, dx.$$

For fixed $0 < \lambda < 1$, the series converges uniformly for $x \in \Box^n$, allowing us to interchange integration and summation to obtain:

$$\sum_{k=0}^{\infty} \lambda^{n(2k+1)} (-1)^{nk} \int_{\Box^n} (x_1 \cdots x_n)^{2k} \, dx.$$

The integration is easy now in view of:

$$\int_{\Box^n} (x_1 \cdots x_n)^{2k} \, dx = \Big(\int_0^1 x_1^{2k} \, dx_1\Big) \cdots \Big(\int_0^1 x_n^{2k} \, dx_n\Big) = \frac{1}{(2k+1)^n}.$$

The final result is:

$$\int_{\lambda\Box^n} \frac{1}{1 - (-1)^n (x_1 \cdots x_n)^2} \, dx = \sum_{k=0}^{\infty} \lambda^{n(2k+1)} \frac{(-1)^{nk}}{(2k+1)^n}.$$

The series converges uniformly for $\lambda \in [0, 1]$; and hence we can take the limit for $\lambda \uparrow 1$, finding the series of the lemma. But this implies that the limit of the integrals exists for $\lambda \uparrow 1$, and that it is equal to the integral of the lemma. This yields the desired equality. □

From cube to polytope. If $n = 1$ the integral in (8) is immediately seen to be equal to $\frac{1}{2}\frac{\pi}{2}$, hence $\delta_1 = 1/2$. So we assume $n \geq 2$ from now on. In order to evaluate the integral (8) (this is by no means elementary), we make a surprising change of variables. In what follows we shall regard the indices i of the n coordinates of a point in $\mathbf{R}^n$ as integers modulo n, so we compute with them cyclically. The substitution of variables is given by the following equations with the conventions on indices just described:

$$x_i = \frac{\sin y_i}{\cos y_{i+1}} \qquad (i \in \mathbf{N} \mod n). \tag{9}$$

Lemma 2. *The change of variables represented by (9) gives a differentiably invertible one-to-one correspondence between the set $\Box^n$ for x and the set $\frac{\pi}{2}\Delta^n$ for y. Here Δ^n is the open n-dimensional convex polytope consisting of the $x \in \mathbf{R}^n$ satisfying:*

$$0 < y_i, \qquad y_i + y_{i+1} < 1 \qquad (i \in \mathbf{N} \mod n). \tag{10}$$

The Jacobian determinant $\frac{\partial x}{\partial y}$ of the correspondence equals $1 - (-1)^n (x_1 \cdots x_n)^2$.

Notice that the Jacobian is exactly the denominator of the integrand in (8) (this is of course the raison d'être of the substitution).

Proof: We denote the correspondence by $\Psi : \frac{\pi}{2}\Delta^n \to \Box^n$. For $y \in \frac{\pi}{2}\Delta^n$, we have $0 < y_i < \frac{\pi}{2} - y_{i+1} < \frac{\pi}{2}$ (cyclically); and thus:

$$0 < \sin y_i < \sin\big(\frac{\pi}{2} - y_{i+1}\big) = \cos y_{i+1}.$$

Hence $0 < x_i < 1$, and using this we get $\Psi(y) \in \Box^n$. Conversely, given $x \in \Box^n$, consider the linear function $\phi : \mathbf{R} \to \mathbf{R}$ given by:

$$\phi(t) = x_n^2(1 - x_1^2(1 - \cdots (1 - x_{n-1}^2(1 - t)) \cdots)).$$

Since $|\phi(t) - \phi(t')| = (x_1 x_2 \cdots x_n)^2 |t - t'|$, the mapping ϕ sends $]0, 1[$ into itself contractively, and hence ϕ has a unique fixed point $t_0 \in \Box$. Now select the unique element $y \in \frac{\pi}{2}\Delta^n$ satisfying:

$$\sin^2 y_n = t_0, \qquad \sin^2 y_i = x_i^2(1 - \sin^2 y_{i+1}) \quad (n - 1 \geq i \geq 1).$$

Then $\Psi(y) = x$ comes down to $\sin^2 y_n = x_n^2(1 - \sin^2 y_1)$, and this follows from $\phi(\sin^2 y_n) = \sin^2 y_n$. That is, the equation $\Psi(y) = x$ has a unique solution $y \in \frac{\pi}{2}\Delta^n$. This gives that Ψ is differentiable, one-one and onto.

The Jacobian matrix $D\Psi(y)$ has the following simple form, where we use ψ_i and ψ_i' as an abbreviation for $\sin y_i$ and $\cos y_i$ respectively:

$$\begin{pmatrix} \frac{\psi_1'}{\psi_2'} & \frac{\psi_1\psi_2}{\psi_2'^2} & 0 & \cdots & 0 \\ 0 & \frac{\psi_2'}{\psi_3'} & \frac{\psi_2\psi_3}{\psi_3'^2} & \cdots & 0 \\ \vdots & \vdots & \vdots & \ddots & \vdots \\ \frac{\psi_n\psi_1}{\psi_1'^2} & 0 & 0 & \cdots & \frac{\psi_n'}{\psi_1'} \end{pmatrix}.$$

Consequently the Jacobian $\frac{\partial x}{\partial y} = \det D\Psi(y)$ equals:

$$\prod_{i=1}^{n} \frac{\psi_i'}{\psi_{i+1}'} - (-1)^n \prod_{i=1}^{n} \frac{\psi_i\psi_{i+1}}{\psi_{i+1}'^2} = 1 - (-1)^n \prod_{i=1}^{n} \left(\frac{\psi_i}{\psi_{i+1}'}\right)^2 = 1 - (-1)^n (x_1 \cdots x_n)^2.$$

The differentiability of Ψ^{-1} follows from the inverse function theorem (see [MT], p.288) and the fact that the Jacobian is positive on $\frac{\pi}{2}\Delta^n$. □

Note that the equations (9) can be solved explicitly; one obtains:

$$y_i = \arcsin\left(x_i \sqrt{\frac{1 + \sum_{j=1}^{n-1} \prod_{k=i+1}^{i+j} (-x_k^2)}{1 - (-1)^n (x_1 \cdots x_n)^2}}\right) \qquad (i \in \mathbf{N} \mod n).$$

Now transform the integral (8) by the substitution (9), and deduce from Lemma 2:

$$\int_{\Box^n} \frac{1}{1 - (-1)^n (x_1 \cdots x_n)^2}\, dx = \left(\frac{\pi}{2}\right)^n \int_{\Delta^n} dy = \left(\frac{\pi}{2}\right)^n \mathrm{vol}_n(\Delta^n) = \left(\frac{\pi}{2}\right)^n \delta_n.$$

Combining this result with Lemma 1 we find the following

Theorem 3. *The formula (6) above holds with δ_n equal to the volume of the Δ^n of Lemma 2.*

Dissecting the polytope into congruent pyramids. The cyclic permutation of the n coordinates in $\mathbf{R}^n$ defines an orthogonal linear transformation C of $\mathbf{R}^n$. The transformation C preserves the inequalities (10) defining Δ^n, and therefore Δ^n is mapped isometrically onto itself by C.

Now denote by $\Gamma^n = \Gamma^n_n$ the collection of $x \in \Delta^n$ satisfying $x_n < x_i$ for $1 \le i \le n-1$. In addition, define the sets Γ^n_j, for $1 \le j < n$ similarly, but with the index n interchanged with j. Since $C^j \Gamma^n = \Gamma^n_j$, all these sets have equal volume and are pairwise disjunct, and the closure of Δ^n is the union of the closures of the Γ^n_j's. Hence (cf. Theorem 3):

$$\delta_n = \mathrm{vol}_n(\Delta^n) = n\,\mathrm{vol}_n(\Gamma^n). \tag{11}$$

Note that $x \in \Gamma^n$ is characterized by the inequalities:

$$0 < x_n < x_i \quad (1 \le i \le n-1), \qquad x_i + x_{i+1} < 1 \quad (1 \le i \le n-2). \tag{12}$$

Indeed, the two "missing" equations now are a consequence of the given ones, since $x_{n-1} + x_n < x_{n-1} + x_{n-2} < 1$ and $x_n + x_1 < x_2 + x_1 < 1$.

Let Θ^{n-1} be the intersection of the closure $\overline{\Gamma}^n$ with the coordinate hyperplane $x_n = 0$. Obviously $y \in \Theta^{n-1}$ if and only if:

$$0 < y_i \quad (1 \le i \le n-1), \qquad y_n = 0, \qquad y_i + y_{i+1} < 1 \quad (1 \le i \le n-2). \tag{13}$$

The next lemma shows that $\overline{\Gamma}^n$ is the pyramid in $\mathbf{R}^n$ with $a = (\frac{1}{2}, \frac{1}{2}, \ldots, \frac{1}{2})$ as its apex and Θ^{n-1} as its basis, i.e., $\overline{\Gamma}^n$ is the convex hull of a and Θ^{n-1}.

Lemma 4. *$\Psi : \Theta^{n-1} \times \,]0,1[\, \to \Gamma^n$ given by $x = \Psi(y,t) = (1-t)\,y + t\,a$ is a differentiably invertible one-to-one correspondence, and its Jacobian is given by $\dfrac{\partial x}{\partial(y,t)} = \dfrac{1}{2}(1-t)^{n-1}$.*

Proof: Suppose x belongs to the pyramid $\mathrm{im}(\Psi)$. Then $0 < x_n = t/2 < (1-t)y_i + t/2 = x_i$, since $0 < y_i$. Furthermore $y_i + y_{i+1} < 1$ implies $x_i + x_{i+1} = (1-t)(y_i + y_{i+1}) + t < 1 - t + t = 1$, for $1 \le i \le n-2$. According to (12) this means $x \in \Gamma^n$. Conversely, assume $x \in \Gamma^n$. Then $x = \Psi(y,t)$ is solved by $t = 2x_n$ and $y_i = (x_i - x_n)/(1 - 2x_n)$. First note that $0 < x_n < 1/2$, for every $x \in \Gamma^n$; hence $0 < t < 1$. The inequality $x_n < x_i$ implies $0 < y_i$, while $x_i + x_{i+1} < 1$ gives $y_i + y_{i+1} < 1$, for $1 \le i \le n-2$. In view of (13) we have $y \in \Theta^{n-1}$, and so $x \in \mathrm{im}(\Psi)$. The computation of the Jacobian is immediate. □

As a consequence of the lemma we find:

$$\mathrm{vol}_n(\Gamma^n) = \frac{1}{2}\int_{\Theta^{n-1}}\left(\int_0^1 (1-y_n)^{n-1}\,dy_n\right)dy = \frac{1}{2n}\,\mathrm{vol}_{n-1}(\Theta^{n-1}) =: \frac{1}{2n}\theta_{n-1}. \tag{14}$$

Combining (11) and (14) we see:

$$\delta_n = n\,\mathrm{vol}_n(\Gamma^n) = \frac{1}{2}\theta_{n-1}. \tag{15}$$

Two-step recursion for the volume of the basis of the pyramid. From (13) we see $x \in \Theta_n$ if and only if $0 < x_1 < 1, 0 < x_2 < 1 - x_1, \ldots, 0 < x_n < 1 - x_{n-1}$. Hence:

$$\theta_n = \int_0^1 \left(\int_0^{1-x_1} \cdots \left(\int_0^{1-x_{n-1}} dx_n \right) \cdots dx_2 \right) dx_1.$$

In order to compute this integral we introduce polynomial functions p_n on $\mathbf{R}$ by:

$$p_0(x) = 1, \qquad p_n(x) = \int_0^{1-x} p_{n-1}(t)\, dt; \qquad \text{then} \quad \theta_n = p_n(0).$$

Now we have, for $x \in \mathbf{R}$:

$$\begin{aligned} p_{n+2}(x) &= \int_0^1 p_{n+1}(t)\, dt - \int_{1-x}^1 p_{n+1}(t)\, dt = p_{n+2}(0) - \int_0^x p_{n+1}(1-t)\, dt = \\ &= p_{n+2}(0) - \int_0^x \Big(\int_0^t p_n(s)\, ds \Big) dt. \end{aligned}$$

Hence by mathematical induction on n we find, for $x \in \mathbf{R}$:

$$p_n(x) = \sum_{0 \le i \le [\frac{n}{2}]} p_{n-2i}(0) \frac{(-1)^i}{(2i)!} x^{2i} - s(n) \frac{(-1)^{[\frac{n}{2}]}}{n!} x^n,$$

where $s(n) = 0$ or 1 if n is even or odd, respectively. Since $p_n(1) = 0$, for $n \in \mathbf{N}$, we obtain:

$$\sum_{i=0}^{n} \theta_{2n-2i} \frac{(-1)^i}{(2i)!} = 0, \qquad \sum_{i=0}^{n} \theta_{2n+1-2i} \frac{(-1)^i}{(2i)!} = \frac{(-1)^n}{(2n+1)!} \quad (n \in \mathbf{N}). \tag{16}$$

Notice that (16) gives an algorithm for recursive computation of the θ_n, and hence also of the δ_n. The left hand sides in (16) occur as the coefficient of t^{2n}, and t^{2n+1} respectively, if one performs the multiplication in the left hand side of the following identity of formal power series; and this implies:

$$\Big(\sum_{n=0}^{\infty} \theta_n t^n \Big) \Big(\sum_{n=0}^{\infty} \frac{(-1)^n}{(2n)!} t^{2n} \Big) = 1 + \sum_{n=0}^{\infty} \frac{(-1)^n}{(2n+1)!} t^{2n+1}.$$

Recognizing the Maclaurin expansion for cos and sin we find in view of (15) the result of (7):

$$2 \sum_{n=1}^{\infty} \delta_n t^{n-1} = \sum_{n=0}^{\infty} \theta_n t^n = \frac{1}{\cos t} + \tan t = \sec t + \tan t \qquad (|t| < \frac{\pi}{2}).$$

Recall the expansion (5) for sec. The function tan is an odd function whose odd-order derivatives at 0 belong to $\mathbf{N}$, and it has the power series $\tan t = \sum_{n=1}^{\infty} T_n t^{2n-1}$, where the T_n are given by (see [C], p.423 for a short proof):

$$T_n = (-1)^{n-1} B_{2n} \frac{2^{2n} - 1}{(2n)!} 2^{2n}.$$

In [KB] one finds tables for the Euler and Bernoulli numbers. Inserting both these expansions in (7) we obtain the following equality of power series, which relates the volumina δ_n to the analytically defined numbers E_n of (5) and B_{2n} of (2):

$$\sum_{n=1}^{\infty} \delta_n t^{n-1} = \sum_{n=0}^{\infty} E_n \frac{1}{2(2n)!} t^{2n} + \sum_{n=1}^{\infty} (-1)^{n-1} B_{2n} \frac{2^{2n} - 1}{2(2n)!} 2^{2n} t^{2n-1} \qquad \Big(|t| < \frac{\pi}{2} \Big).$$

If we combine this result with (6) we obtain (3) and (4). With this we feel we have completed our short circular excursion through some analysis, some geometry, some combinatorics and back to the tried and familiar grounds of elementary subjects.

Acknowledgement. The authors are grateful to D. Zagier for acting as a trait d'union between them, and to T.A. Springer for his interest during the preparation of this manuscript.

References

[C] Courant, R. - *Differential and Integral Calculus. Volume I. Second Edition.* Wiley Classics Library Edition. Interscience Publishers: 1988

[KB] Knuth, D.E., Buckholtz, T.J. - Computation of tangent, Euler and Bernoulli numbers. Math. Comp. **21**, 663-688 (1967)

[MT] Marsden, J.E., Tromba, A.J. - *Vector Calculus. Third Edition.* W.H. Freeman and Company: New York 1988

What are the Best Almost-Complex Structures on the 6-Sphere?

Eugenio Calabi and Herman Gluck

Greene, Robert (ed.) et al.,
Differential Geometry,
Part 2: Geometry in Mathematical Physics and Related Topics,
Proceedings of a summer research institute, held at
University of California, Los Angeles, CA, USA, July 8–28, 1990,
99–106, Proc. Sympos. Pure Math., 54, Part 2,
Amer. Math. Soc., Providence, RI (1993)

ABSTRACT. The most familiar almost-complex structures on the 6-sphere S^6 are defined in terms of Cayley multiplication. We observe here, using a twistor type transformation and the method of calibrated geometries, that these structures are "best" in the sense that they have the smallest volume when pictured as cross-sections of the appropriate bundle $E(S^6) = O(8)/U(4)$ over S^6.

To carry this out, we use the Triality Theorem for $SO(8)$ to construct an explicit isometry of each component of $E(S^6)$ with the real Grassmann manifold G_2R^8 of oriented 2-planes in 8-space, which matches these smallest cross-sections with the subgrassmannians G_1R^7. These in turn are known from an earlier paper to be volume-minimizing in their homology class because they are calibrated by the Euler 6-form on $G_2R^8 \cong G_6R^8$.

A *complex structure* on a real vector space V is a linear map $J : V \to V$ such that $J^2 = -I$. An *almost-complex structure* on a smooth manifold M is a smoothly varying choice of complex structures on the tangent spaces to M at each of its points. If a complex structure on a real inner product space V preserves the inner product, then we call it a *hermitian structure.* Likewise, if an almost-complex structure on a Riemannian manifold M preserves the inner product on each tangent space, then we call it an *almost-hermitian structure.*

If (M^{2n}, g) is a Riemannian manifold, let $E(M)$ denote the set of all pairs (x, J_x), where x is a point of M and J_x is a hermitian structure on the tangent space to M at x. $E(M)$ is a bundle over M with fibre the set of all hermitian structures on R^{2n}. An almost-hermitian structure $\mathcal{J}$ on M is a smooth cross-section of this bundle, and its image is a smooth submanifold of $E(M)$. We define the *volume* of the almost-hermitian structure to be the volume of this submanifold, measured in the natural Riemannian

1991 *Mathematics Subject Classification.* Primary 53C15.
This paper is in final form and no version of it will be submitted for publication elsewhere.

metric induced on $E(M)$ by (M, g) and its Levi-Civita connection. By *best* almost-hermitian structures on M, we mean those of minimum volume.

Let S^n denote the unit n-sphere in R^{n+1}. Both S^2 and S^6 admit almost-complex structures; see below for details. In 1947, Hopf **[Ho 1]** and Ehresmann **[Eh 1]** each showed that S^4 does not admit an almost-complex structure. In the same year, Kirchhoff **[Ki]** showed that if S^n admits an almost-complex structure, then S^{n+1} is parallelizable; see also **[St, p. 216]**. The next year, Hopf **[Ho 2]** showed that S^8 does not admit an almost-complex structure. In 1950, Ehresmann **[Eh 2]** pointed out that Kirchhoff's result implies that S^{4n} does not admit an almost-complex structure. Finally, in 1953, Borel and Serre **[BS]** showed that S^2 and S^6 are the only spheres which do admit almost-complex structures.

$E(S^2)$ consists simply of two copies of the base S^2, corresponding to its two possible orientations, so there are exactly two almost-hermitian structures on S^2. By contrast, $E(S^6)$ is a 12-dimensional manifold, with two disjoint copies of the complex projective space CP^3 as fibres.

There are many almost-hermitian structures on S^6: the most familiar example $\mathcal{J}$ is obtained by thinking of R^7 as the purely imaginary Cayley numbers, and of S^6 as those of unit length. For each $x \in S^6$, let J_x denote, say, right Cayley multiplication by x, acting on the tangent space to S^6 at x. Note that this tangent space consists of all Cayley numbers orthogonal to both 1 and x, and hence is taken to itself by right multiplication by x. Since multiplication by a unit Cayley number is an orthogonal transformation, each J_x is a hermitian structure.

Simple modifications of this example are obtained by using O(7) to alter the orthogonal identification of S^6 with the unit imaginary Cayley numbers. We refer to all of these almost-hermitian structures on S^6 as *Cayley structures*, and will prove

THEOREM. *The almost-hermitian structures of minimum volume on S^6 are the Cayley structures, and no others.*

In 1951, Echmann and Fröhlicher **[EF]** showed that these Cayley structures on S^6 do not come from complex structures. Indeed, the question as to whether or not S^6 admits a complex structure seems still to be undecided.

We thank Lyndon Woodward of the University of Durham for suggesting this problem to us, and the National Science Foundation for its support.

1. Cayley structures on S^6

We observe in this section that the space $\mathrm{CaS}(S^6)$ of Cayley structures on S^6 is homeomorphic to the union of two disjoint copies of the 7-dimensional real projective space RP^7.

Let $\mathcal{J}$ be our original Cayley structure on S^6, in which $J_x = R_x =$ right Cayley multiplication by $x \in S^6$. Write $\mathcal{J} = \{(x, R_x) : x \in S^6\}$. The other

Cayley structures can be written as

$$\mathcal{J}^A = \{(A^{-1}(x), A^{-1}R_x A) : x \in S^6\},$$

where $A \in \mathrm{O}(7)$, viewed as the subgroup of $\mathrm{O}(8)$ fixing the Cayley number 1. It is easy to see that $\mathcal{J}^A = \mathcal{J}$ precisely when A is an automorphism of the Cayley algebra. Hence $\mathrm{CaS}(S^6)$ is a copy of $G_2\backslash\mathrm{O}(7)$, where G_2 is the 14-dimensional exceptional Lie group of automorphisms of the Cayley algebra. Each component of $\mathrm{CaS}(S^6)$ is a copy of $G_2\backslash\mathrm{SO}(7)$.

To help see the homeomorphism $G_2\backslash\mathrm{SO}(7) \cong RP^7$, we will exhibit $G_2\backslash\mathrm{SO}(7)$ as a bundle over S^4 with fibre RP^3.

Let $B \in \mathrm{SO}(7)$, and interpret B as acting on the Cayley numbers with $B(1) = 1$. Let S^4 denote the 4-sphere of unit Cayley numbers orthogonal to 1, i and j. Define a map $p : \mathrm{SO}(7) \to S^4$ by

$$p(B) = B^{-1}(B(i)B(j)).$$

Since $B(1) = 1$, we note that $B(i)$ and $B(j)$ are orthogonal purely imaginary Cayley numbers, and hence their product $B(i)B(j)$ is a purely imaginary Cayley number orthogonal to both of them. Hence $p(B) = B^{-1}(B(i)B(j))$ is a purely imaginary Cayley number orthogonal to i and j, and therefore an element of the 4-sphere defined above.

Now suppose $B = CD$, where $C \in G_2$ is an automorphism of the Cayley algebra. Then

$$\begin{aligned} p(CD) &= (CD)^{-1}(CD(i)CD(j)) \\ &= D^{-1}C^{-1}(C(D(i)D(j))) = D^{-1}(D(i)D(j)) \\ &= p(D). \end{aligned}$$

Hence p is constant on right cosets G_2D, and therefore descends to a map $p : G_2\backslash\mathrm{SO}(7) \to S^4$.

This map p is a fibration, and we want to examine the fibre $p^{-1}(k)$. That means we are looking at those $B \in \mathrm{SO}(7)$ which satisfy $B(k) = B(i)B(j)$. That is, the restriction of B to the subalgebra of quaternions is an isomorphism into the Cayley algebra. Let ε be a unit Cayley number orthogonal to the quaternions. There is a unique automorphism C of the Cayley algebra which agrees with B on the $1, i, j, k, \varepsilon$-plane. Again writing $B = CD$, we see that $D = C^{-1}B$ is the identity on this 5-plane. Let D' denote the restriction of D to the orthogonal 3-plane. The correspondence $G_2B \leftrightarrow D'$ matches the fibre $p^{-1}(k)$ with the $\mathrm{SO}(3)$ of orientation-preserving isometries of this 3-plane. Hence the fibre of p is homeomorphic to a real projective space RP^3.

Thus each component $G_2\backslash\mathrm{SO}(7)$ of the space of Cayley structures on S^6 is a bundle over S^4 with fibre RP^3. The space RP^7 also appears as a bundle over S^4 with fibre RP^3: just take the Hopf fibration of S^7 by great

3-spheres and divide out by the antipodal map. One checks easily that these two bundles are the same.

Indeed, the correspondence between $G_2\backslash SO(7)$ and $RP(7)$ can be confirmed at the end of the paper, when we have seen an explicit isometry between $E^+(S^6)$ and the Grassmann manifold G_2R^8, matching the individual right-handed Cayley structures on S^6 with the individual subgrassmannians G_1R^7. These subgrassmannians consist of all oriented 2-planes through the origin in R^8 which contain a fixed line. There is an RP^7's worth of such lines, hence an RP^7's worth of such subgrassmannians, and hence an RP^7's worth of right-handed Cayley structures.

2. Complex structures and Grassmann manifolds

Let x be a point of S^6 and J_x a hermitian structure on the tangent space T_xS^6. We translate T_xS^6 with its hermitian structure to the parallel 6-plane through the origin, and then extend J_x to a hermitian structure on $R^8 \cong \mathrm{Ca}$ by defining $J_x(1) = x$ and $J_x(x) = -1$. Vice versa, if we are given a hermitian structure J on $R^8 \cong \mathrm{Ca}$, let $x = J(1)$ and let J_x be the restriction of J to the orthogonal complement of the 2-plane spanned by 1 and x, that is (in effect), to T_xS^6. In this way, we have a one-to-one correspondence between points (x, J_x) of $E(S^6)$ and hermitian structures J on $R^8 \cong \mathrm{Ca}$. See **[St, pp. 215–216]**. The set of these hermitian structures on R^8 is just the homogeneous space $O(8)/U(4)$; hence

$$E(S^6) = O(8)/U(4).$$

This space has two isometric components; we focus on one of them:

$$E^+(S^6) = SO(8)/U(4).$$

In this section we describe a specific isometry between the above space and the Grassmann manifold

$$G_2R^8 = SO(8)/(SO(2) \times SO(6))$$

of oriented 2-planes through the origin in 8-space. In his 1950 paper **[Eh 2]**, Ehresmann remarked that $SO(8)/U(4)$ is the complex hyperquadric in CP^7, which in turn is well known to be a copy of G_2R^8. The isometry we give here is based on the

TRIALITY THEOREM FOR $SO(8)$. *Consider any triple of elements A, B and C in $SO(8)$ such that*

$$A(mu) = B(m)C(u), \quad \textit{for all } m, u \in \mathrm{Ca},$$

where Cayley multiplication is used on both sides of the equation. If any one of these three orthogonal transformations is preassigned, then the other two exist and are unique up to changing sign for both of them.

We will write $(A, B, C) \in \mathrm{Tr}$ to indicate that this triality relation holds.

For proofs and further discussion of this theorem, the reader is referred to Cartan [**Cr**, pp. 370 and 373] as well as to [**Fr**, pp. 14–16] and [**GWZ**, pp. 192–194].

Suppose now that B is a hermitian structure on R^8. Thus $B^2 = -I$. Let A and C be the elements of SO(8) promised by the Triality Theorem. Then

$$\begin{aligned} A^2(mu) &= A(A(mu)) = A(B(m)C(u)) \\ &= B^2(m)C^2(u) = -mC^2(u). \end{aligned}$$

By the uniqueness part of the Triality Theorem, we either have $A^2 = I$ and $C^2 = -I$ or else $A^2 = -I$ and $C^2 = I$. That is, either A is an involution and C is an hermitian structure, or vice versa. Which of these two alternatives occurs depends on whether the hermitian structure B is "right-handed" or "left-handed."

Let J be a hermitian structure on R^{2n}. Fix an orientation for R^{2n}, and suppose that $x_1, \dots, x_n$ are chosen there so that the vectors

$$x_1, J(x_1), x_2, J(x_2), \dots, x_n, J(x_n)$$

form an orthonormal basis. If this ordered basis agrees with the orientation, we say that the hermitian structure J is *right-handed*; otherwise, we say it is *left-handed*. This definition is independent of the choice of elements $x_1, \dots, x_n$.

Next we see how the handedness of the hermitian structure B determines which of A and C is the involution and which the hermitian structure.

Let x be a purely imaginary Cayley number of unit length. Let R_x and L_x denote right and left multiplication by x. Both are hermitian structures on R^8. Choose the orientation of R^8 so that R_x is right-handed. Then L_x will be left-handed.

LEMMA. $(R_x, R_x, -L_xR_x) \in \mathrm{Tr}$.

PROOF. To verify this, we must show that

$$(mu)x = -(mx)(xux).$$

To do this, we use the Moufang identity [**Mo**]

$$z(xyx) = ((zx)y)x,$$

one of three identities representing vestiges of associativity which hold for the Cayley numbers. Note that no parentheses appear in the expression xyx on the left-hand side of this identity, since associativity holds whenever only two letters are involved; in particular, the maps R_x and L_x commute with one another. In this Moufang identity, put $x = x$, $y = u$ and $z = mx$, getting (since $x^2 = -1$)

$$(mx)(xux) = ((mxx)u)x = -(mu)x,$$

as desired.

Denoting this "charter member" of Tr by (A, B, C), we observe that

$$\begin{aligned} C^2 &= (-L_x R_x)(-L_x R_x) = (L_x)^2 (R_x)^2 \\ &= (-I)(-I) = I, \end{aligned}$$

since R_x and L_x commute. Thus C is an involution. And clearly $A = R_x$ is a hermitian structure.

So in this particular example, in which the hermitian structure B is right-handed, it is C which turns out to be the involution and A the hermitian structure. We see next that the same alternative occurs for every right-handed hermitian structure B, which one naturally expects by continuity.

To do this, first notice the "multiplicative structure" of the triality relation. That is, suppose (A, B, C) and (A', B', C') satisfy the relation. Then so do (AA', BB', CC') and (A^{-1}, B^{-1}, C^{-1}). This is easily checked, and reflects the fact that the correspondences between orthogonal transformations in a triality triple can be expressed as automorphisms of either Spin(8) or PSO(8), though not, incidentally, of SO(8). Here, Spin(8) denotes the simply connected double cover of SO(8), while PSO(8) denotes the quotient of SO(8) by its center $\{I, -I\}$.

Suppose now that B is a right-handed hermitian structure on R^8. Since any two hermitian structures of the same handedness on R^8 are conjugate within SO(8), there is an element $G \in SO(8)$ such that $B = GR_x G^{-1}$. Suppose $(F, G, H) \in \mathrm{Tr}$. By the lemma, $(R_x, R_x, -L_x R_x) \in \mathrm{Tr}$. By the multiplicative structure of triality,

$$(A, B, C) = (FR_x F^{-1}, GR_x G^{-1}, H(-L_x R_x)H^{-1}) \in \mathrm{Tr}.$$

Hence C is the involution and A the hermitian structure, as claimed.

By contrast, if B is a left-handed hermitian structure on R^8 and $(A, B, C) \in \mathrm{Tr}$, then A is the involution and C the hermitian structure.

Return once more to the triple

$$A = R_x, \quad B = R_x, \quad C = -L_x R_x,$$

and focus on the involution C. Notice that

$$C(1) = -L_x R_x(1) = -x1x = 1,$$

and

$$C(x) = -L_x R_x(x) = -xxx = x,$$

since $x^2 = -1$. On the other hand, if y is orthogonal to both 1 and x, then

$$C(y) = -L_x R_x(y) = -xyx = xxy = -y,$$

since orthogonal pure imaginaries anticommute. Thus C restricts to the identity on the 2-plane P_x spanned by 1 and x, but to minus the identity on the orthogonal 6-plane. So to the hermitian structure $B = R_x$ on R^8, we intend to associate the 2-plane P_x.

Now suppose that B is any right-handed hermitian structure on R^8. As before, we have $B = GR_xG^{-1}$ for some $G \in SO(8)$. Let $(F, G, H) \in \mathrm{Tr}$. Then the involution $C = H(-L_xR_x)H^{-1}$ restricts to the identity on the 2-plane $H(P_x)$, and to minus the identity on the orthogonal 6-plane. So to the hermitian structure B we associate the 2-plane $H(P_x)$, and denote it by P_B. By uniqueness of H up to sign, this 2-plane is well-determined by B.

If the hermitian structure B on R^8 is right-handed, then so is $-B$. If, in the triality relation, B is replaced by $-B$, then exactly one of A and C will change sign. Hence the 2-plane $P_{-B} = P_B$. By uniqueness in the Triality Theorem, one can only have $P_B = P_{B'}$, if $B' = B$ or $-B$.

Thus we get a one-to-one correspondence between pairs $(B, -B)$ of hermitian structures on R^8 and unoriented 2-planes $P_B = P_{-B}$ in R^8. Passing to the simply connected double covers, we get a one-to-one correspondence between hermitian structures B and oriented 2-planes, still denoted by P_B. That is, we have a bijection between $E^+(S^6) = SO(8)/U(4)$ on the one hand, and the Grassmannian $G_2R^8 = SO(8)/(SO(2) \times SO(6))$ of oriented 2-planes in 8-space on the other.

It is easy to see that this bijection is an isometry between these two symmetric spaces, when each is given the Riemannian submersion metric induced from a common bi-invariant metric on $SO(8)$. The triality "correspondence" in $SO(8)$, which takes $B = R_x$ to $C = -L_xR_x$, descends to an automorphism

$$T : PSO(8) \to PSO(8),$$

which is an isometry by the uniqueness up to scale of bi-invariant metrics on simple Lie groups. Noting that $U(4)$ and $SO(2) \times SO(6)$ are the centralizers in $SO(8)$ of B and C, respectively, one checks easily that T takes $P(U(4))$ to $P(SO(2) \times SO(6))$. Thus T descends to an isometry taking

$$E^+(S^6) = PSO(8)/PU(4)$$

to

$$G_2R^8 = PSO(8)/P(SO(2) \times SO(6)).$$

Now suppose the right-handed almost-hermitian structure $\mathcal{J}$ on S^6 is our primitive Cayley structure: $\mathcal{J} = \{(x, R_x) : x \in S^6\}$. The image of J in G_2R^8 under the above isometry then consists of all the oriented 2-planes spanned by 1 and x, for $x \in S^6$. But this is a subgrassmannian of the type G_1R^7, and as mentioned in the introduction, is known to be calibrated by the Euler 6-form on G_2R^8, and is hence volume-minimizing in its homology class. The reader is referred to **[HL]** or **[GMZ]** for the definition of "calibrate," and to **[GMZ]** for the proof of this assertion.

Since all the other right-handed Cayley structures are images of this primitive one under the isometric action of $SO(8)$, they too are volume-minimizing in this homology class, and correspond to all the other realizations of G_1R^7 inside G_2R^8.

It is easy to see that all right-handed almost-hermitian structures on S^6 lie in the same homology class in $E^+(S^6)$. In fact, as Ehresmann **[Eh 2]** pointed out, $\pi_6 E^+(S^6)$ is infinite cyclic, and hence all the cross-sections of this bundle are homotopic.

Likewise, all the left-handed Cayley structures are volume-minimizing in their homology class, in the other component of $E(S^6)$.

Since it was proved in **[GMZ]** that the subgrassmannians G_1R^7 are the unique volume minimizers in their homology class in G_2R^8, and since the Cayley structures correspond to these subgrassmannians, it follows that the Cayley structures are the unique volume minimizers in their two homology classes, completing the proof of the theorem.

References

[BS] A. Borel and J.-P. Serre, *Groupes de Lie et puissances réduites de Steenrod*, Amer. J. Math. **75** (1953), 409–448.

[Ca] E. Calabi, *Construction and properties of some 6-dimensional almost complex manifolds*, Trans. Amer. Math. Soc. **87** (1958), 407–438.

[Cr] E. Cartan, *Le principe de dualité et la théorie des groupes simples et semi-simples*, Bull. Sci. Math. **49** (1925), 361–374.

[Ec] B. Eckmann, *Complex-analytic manifolds*, Proc. Internat. Congr. Math., Cambridge, Mass., 1950, pp. 420–427.

[EF] B. Eckmann and A. Fröhlicher, *Sur L'integrabilite des structures presque-complexes*, C. R. Acad. Sci. Paris **232** (1951), 2284–2286.

[Eh 1] C. Ehresmann, *Sur la théorie des espaces fibrés*, Topologie Algebrique—Paris 1949, Edition CNRS, Paris, 1949, pp. 3–15.

[Eh 2] ——, *Sur les variétés presque complexes*, Proc. Internat. Congr. Math., Cambridge, Mass., 1950, pp. 412–419.

[Fr] H. Freudenthal, *Oktaven, Ausnahmegruppen und Octaven geometrie*, Utrecht, 1951.

[GMZ] H. Gluck, F. Morgan and W. Ziller, *Calibrated geometries in Grassmann manifolds*, Comment. Math. Helv. **64** (1989), 256–268.

[GWZ] H. Gluck, F. Warner, and W. Ziller, *The geometry of the Hopf fibrations*, Enseign. Math. **32** (1986), 173–198.

[HL] R. Harvey and H. B. Lawson, Jr., *Calibrated geometries*, Acta Math. **147** (1981), 47–157.

[Ho 1] H. Hopf, *Sur les champs d'éléments de surface dans les variétés à 4 dimensions*, Topologie Algebrique—Paris 1947, Editions CNRS, Paris, 1949, pp. 55–59.

[Ho 2] ——, *Zur Topologie der komplexen Mannigfaltigkeiten*, Studies and Essays Presented to R. Courant on his 60th Birthday, Interscience 1948, pp. 167–185.

[Ki] A. Kirchhoff, *Sur l'existence de certains champs tensoriels sur les spheres*, C. R. Acad. Sci. Paris **225** (1947), 1258–1260.

[Mo] R. Moufang, *Zur Struktur von Alternativkörpern*, Math. Ann. **110** (1935), 416–430.

[St] N. Steenrod, *The topology of fibre bundles*, Princeton Univ. Press (1951).

University of Pennsylvania

Extremal Isosystolic Metrics for Compact Surfaces

Eugenio Calabi

Arthur L. Besse (ed.),
Actes de la table ronde de géométrie différentielle en l'honneur de Marcel Berger
(Luminy, France, 12–18 juillet, 1992), 167–204,
Sémin. Congr., 1, Soc. Math. France, Paris (1996)[1]

Abstract. Given a closed, orientable surface M of genus ≥ 2, one seeks an *extremal isosystolic metric* on M: this is a Riemannian metric that induces on M the smallest possible area, subject to the constraint that the corresponding *systole*, or shortest length of any non-contractible closed curve, is a fixed, positive number. The geometric problem is rendered into an analytic one by reducing it to solving a nonlinear, partial differential equation with free boundaries. Examples are shown, to illustrate some possible candidates for solutions of the problem in special cases.

Résumé. Sur une surface M compacte orientable de genre ≥ 2, on cherche une métrique isosystolique extrémale : c'est une métrique riemannienne d'aire la plus petite possible sous la contrainte que la systole, i.e. la courbe fermée lisse non contractible de longueur minimale, soit un nombre positif fixé. Le problème géométrique est transformé en un problème analytique en le réduisant à la résolution d'une équation aux dérivées partielles non-linéaire à frontière libre. Des exemples sont donnés pour illustrer des candidats possibles à être solution du problème dans des cas particuliers.

[1] The headers on pages 170–204 of the original scan were partially obscured. They have been replaced by retyped copies.

TABLE OF CONTENTS

1. INTRODUCTION

Given a compact Riemannian or Finslerian manifold (M, g), where g denotes the Riemannian (respectively, Finsler) metric, a base point $x_0 \in M$, and an element γ in the fundamental group $\pi_1(M, x_0)$, the *local systole* $Sys_\gamma(M, x_0, g)$ is defined to be the minimum length of any loop path through x_0 in the homotopy class γ. Denote by $\overline{\gamma}$ the conjugacy class of γ in $\pi_1(M, x_0)$; then the free local systole of (M, g) at $\overline{\gamma}$ is defined to be the minimum length of any closed path representing the free homotopy class $\overline{\gamma}$, and is denoted by $Sys_{\overline{\gamma}}(M, g) = Inf_{x_0 \in M}(Sys_\gamma(M, x_0, g))$. The systole (with no added qualifier) $Sys(M, g)$ is understood to be the least value of $Sys_{\overline{\gamma}}(M)$ as $\overline{\gamma}$ ranges over all non-trivial free homotopy classes.

In the terminology of M. Gromov [6], an n-dimensional, differentiable manifold M is called *essential*, if, for all Riemannian (respectively, Finsler) metrics g in M, the isosystolic ratio [1] $Vol(M, g)/(Sys(M, g))^n$ has a positive lower bound depending only on the topology of M. Gromov's compactness theorem asserts that, if M is essential, then for any positive constant c the function space of all metrics g in M, normalized by a positive factor so that $Sys(M, g) = 1$ and satisfying the volume inequality $Vol(M, g) \leq c$, is compact in the Fréchet-Hausdorff topology. In particular, all closed, 2-dimensional surfaces except for the 2-sphere are essential. With these facts in mind, it is natural to raise the question of estimating the minimum isosystolic ratio for any closed surface, orientable or not, in terms of its genus. Many variants of this question have been studied, some of them formulated to include more general spaces, such as manifolds with boundary, others dealing with restricted classes of metrics, such as Riemannian metrics with non-positive, or constant, negative curvature, or metrics in a given, conformal class, to name a few. While some statements in this paper apply to surfaces with boundary, we shall limit our consideration almost exclusively to Riemannian metrics in closed, orientable surfaces, leaving other cases for another occasion. The only types of closed surfaces for which one knows an explicit, extremal

[1] In Gromov's definition the isosystolic ratio is expressed by $Sys(M,g)/(Vol(M,g))^{1/n}$.

isosystolic metric, i.e. a Riemannian metric minimizing the isosystolic ratio, are the projective plane (P.M. Pu, [7]), the torus (C. Loewner, unpublished, cf. M. Berger, [3,4]) and the Klein Bottle (C. Bavard, [1,2]). For each of the other types of surfaces (i.e. for surfaces with negative Euler characteristic) there is a very wide gap between the best available estimates of upper and of lower bounds for the extremal isosystolic ratio. The main purpose of the paper is to reduce the problem of extremal isosystolic metric to a variational problem that may be studied by the methods of classical calculus of variations. At the end of this paper we shall exhibit for the record two explicit examples of metrics in an orientable surface of genus 3: both metrics attain locally minimum values of the isosystolic ratio, relative to small deformations of the metric in its function space, the second metric having an isosystolic ratio about 1.5% lower than the first ; it is believed that the value achieved by the second metric $((7\sqrt{3})/8 \approx 1,51554)$ is very close to, if not actually equal to the absolute minimum value for surfaces of genus 3. The two examples consist of piecewise flat metrics in the surface, each one constructed in terms of a corresponding, explicit, well known triangulation, with a large group of symmetries.

No similar construction has been found to yield an extremal isosystolic metric in surfaces of any genus $g = 2$, or ≥ 4, suggesting that the genera of surfaces whose extremal isosystolic metrics are piecewise flat may be quite sparse: it is this particular observation that has motivated the present study ; its ultimate goal is that of studying the general local properties of extremal isosystolic metrics, especially when they are not piecewise flat. Unfortunately the partial differential equations obtained have not yielded methods to construct any non-trivial, explicit solutions. However it is shown in Sections 6 and 7 that, merely by using the maximum principle, one can obtain some fairly close *a priori* estimates of the minimum isosystolic ratio in two examples, that illustrate also a useful generalization of the isosystolic problem. The first example consists of seeking a Riemannian metric in a 2-disk, admitting the group of symmetries of a regular hexagon, that minimizes the area subject to the condition that the least distance between each of the three pairs of opposite "sides" equals 2; the second example deals with the extremal isosystolic metrics in a torus with one open disc deleted: in this case the "systole" consists of two independent, positive, real numbers, representing, respectively, the "boundary systole" and the least length of any closed path representing a non-trivial homology class of cycles. Both of these examples illustrate some of the singularities that extremal isosystolic metrics may exhibit in general.

2. STRUCTURE OF k-REGULAR DOMAINS

Let M be a closed, orientable surface of genus $g \geq 2$ and consider the complete function space G of singular, generalized Riemannian (respectively, Finsler) metrics g on M, such that:

(i) g is bounded, locally, from above and below, by smooth Riemannian metrics ;

(ii) the g-length functional on the space of rectifiable arcs (the latter with the Fréchet topology) is lower semicontinuous.

This class of metrics is invariant under homeomorphisms of M of Lipschitz class; its definition ensures the compactness of any set of paths of bounded length, in any compact domain. In particular, the g-distance $d(x, y)$ between any two points $x, y \in M$ is achieved by a compact (non-empty) set of shortest paths. The function space G has the topology of uniform Lipschitz convergence of $d(x, y)$ in each compact subset of M: this topology ensures both the equivalence of the area functional $Vol(D) = Vol_g(D)$ [2] with the Lebesgue measure of any Borel set $D \subset M$ and its continuity with respect to the metric $g \in G$. Given any element $\overline{\gamma}$ in the set $\overline{\pi}_1^*(M)$ of non-trivial, homotopy classes of free, closed paths in M, the (free) local systole $Sys_{\overline{\gamma}}(M, g)$ is achieved by a compact family of oriented, closed paths of length $Sys_{\overline{\gamma}}(M, g)$, representing the class $\overline{\gamma}$: such closed paths will be referred to as systole-long paths ; for any given, positive real number A, the set $\Gamma_A \subset \overline{\pi}_1^*(M)$ consisting of all classes $\overline{\gamma}$ such that $Sys_{\overline{\gamma}}(M, g) \leq A$ is a finite set. The metrics in the class G may be discontinuous: for example, they may include isolated "short-cut" (or "fast-track") curves ; however it is a complete function space, to which an

[2] In the case of a Finsler metric g, the volume element form $dVol_g$ in terms of local parameters (u,v) is defined to be $\pi^{-1}\sigma(u,v)|du \wedge dv|$, where $\sigma(u,v)$ denotes the area of the unit g^*-disc in the cotangent bundle of M, with respect to the dual Finsler form g^* of g.

extremal isosystolic metric may be reasonably expected to belong, by a process of convergence of metrics whose isosystolic ratio approaches the lower bound, avoiding the full generality of the Fréchet-Hausdorff-Gromov topology.

Throughout the paper we shall tacitly assume that every metric $g \in G$ in M is normalized by the condition $Sys(M, g) = 1$, with possible exceptions explicitly stated.

Given any (M, g) with $g \in G$ and any real constant $A \geq 1$, consider the set $\Gamma_A \subset \overline{\pi}_1^*(M)$ consisting of all classes $\overline{\gamma}$ such that $Sys_{\overline{\gamma}}(M, g) \leq A$ and, for each $\overline{\gamma} \in \Gamma_A$, the union $K_{\overline{\gamma},A} \subset M$ of all oriented, systole-long paths representing the free homotopy class $\overline{\gamma}$. Let $\overline{\gamma}^{-1}$ denote the free homotopy class of the closed, oriented paths, whose reversal of orientation yields a path representing $\overline{\gamma}$: obviously $Sys_{\gamma^{-1}}(M, g) = Sys_{\gamma}(M, g)$, so that Γ_A is a finite, symmetric set, and $K_{\overline{\gamma}^{-1},A} = K_{\overline{\gamma},A}$. For any subset $S \subset \Gamma_A$ the subset $B_S = \cup_{\overline{\gamma} \in S}(K_{\overline{\gamma},A}) \subset M$ is compact, and therefore for any integer $k \geq 0$, the subset $U_{A,k} \subset M$ consisting of all points that are included in $K_{\overline{\gamma},A}$ for exactly $2k$ elements $\overline{\gamma} \in \overline{\pi}_1^*(M)$ (counting $\overline{\gamma}$ and $\overline{\gamma}^{-1}$ separately) is relatively open in the subset of points that are covered by at least $2k$ of the sets $K_{\overline{\gamma},A}$. For any $A \geq 1$, any non-empty, open subdomain $U \subset M$, contained in $U_{A,k}$ is called a <u>k-regular domain</u> in M.

Now assume that the metric g is an extremal isosystolic one ; we shall examine the possible open k-regular domains $U_{A,1} \subset M$ for small values of k.

Lemma 2.1. — *If g is an extremal isosystolic metric in M, then, for any constant $A \geq 1$, the subsets $U_{A,0}$ and $U_{A,1}$ of M are empty, and consequently $U_{A,2}$ is an open subdomain of M.*

Proof. The set $U_{A,0}$ is open. Hence, if it is not empty, there is a non-empty, open subdomain V such that its closure $\overline{V}$ is compact and $\subset U_{A,0}$. In addition there is a positive ϵ such that, for every point $x \in V$, the least length of any homotopically non-trivial loop based at x is $\geq A + \epsilon$. Replace the metric g by a conformally equivalent one $g' = g \cdot exp(-\delta\phi)$, where ϕ is a non-negative, non-zero function with support in V, and δ is a positive constant. Then the volume of M in terms of the metric g' is strictly smaller than the original one in terms of g ; at the same time, for δ sufficiently small, the systole $Sys(M, g')$ would remain identical with $Sys(M, g)$. This shows that

g could not be an extremal isosystolic metric, and consequently $U_{A,1}$ is an open subset of M.

Assume, as before, that there is a non-empty, simply connected, open set V such that $\overline{V} \subset U_{A,1}$ is compact, so that, for each point $x \in V$, there is at least one non-oriented, systole-long, closed curve passing through x ; any such curve with its two opposite orientations represents a unique, non-oriented, free homotopy class $\overline{\gamma}^{\pm 1} \in \overline{\pi}_1^*(M)$. While V is not necessarily foliated by its intersections with the covering family of systole-long paths, it is foliated by their orthogonal trajectories. From this foliation and a choice of orientation, say the one defined by $\overline{\gamma}$, one constructs a function u, that is constant along each orthogonal trajectory, and whose restriction to each of the systole-long paths provides its parametrization by its oriented arc length ; these properties determine u uniquely up to an added constant ; we shall refer to u as a potential function in V ; the formal definition follows this proof.

Given the potential function u in V, one chooses a second function v of Lipschitz class, such that (u, v) is a system of local parameters for M ; then the metric g may be represented almost everywhere as a quadratic form on the cotangent bundle, so that the norm $|\alpha|_g$ of a Pfaffian form $\alpha = \xi du + \eta dv$ is given by $|\alpha|_g^2 = \eta^2 + 2 \cdot f(u,v)dudv + g(u,v)\eta^2$ with $g(u,v) > (f(u,v))^2$ almost everywhere. Thus the metric is determined by the two functions $f(u,v)$, $g(u,v)$. The corresponding volume form is $dVol = (g(u,v) - (f(u,v))^2)^{-\frac{1}{2}}|du \wedge dv|$. As in the previous case, one could replace the metric g with another metric g', identical with g outside V and, inside V, defined by a quadratic form $|\xi du + \eta dv|_{g'}^2 = \xi^2 + 2 \cdot f(u,v)dudv + g'(u,v)\eta^2$ with $g'(u,v)$ slightly larger than $g(u,v)$ in a set of positive measure ; the resulting metric g' then would have an isosystolic ratio strictly smaller than that of g. This fact shows that, if g is an extremal isosystolic metric, then $U_{A,1}$ is empty as well as $U_{A,0}$. It follows that, for any extremal g and for any constant $A \geq 1$, $U_{A,2}$ is an open subdomain of M.

Definition 2.2. — *Given, in a surface (M, g), a family of oriented paths of shortest length, filling an open, simply connected domain $U \subset M$, and such that no two of the paths cross each other, a <u>(local) geodesic potential function</u> for the family of paths in U is a function $u : U \to \mathbf{R}$, that is constant along an orthogonal trajectory of the paths* *of the given family, and whose restriction to each path provides a parametrization of that path by its oriented arc length.*

It is well known that the choice of orthogonal trajectory in the above definition is immaterial, so that a local geodesic potential function for a given family of shortest paths is unique up to an additive constant. From a local viewpoint, a geodesic potential u is a function of Lipschitz class, with Lipschitz coefficient identically equal to 1 everywhere in V and no topological critical points (this second property must be added, since u is not necessarily differentiable) ; any function with these two properties has a gradient flow, whose orbits are shortest paths in V.

Lemma 2.3. — *If g is an extremal isosystolic metric in M, then the open subset $U_{A,2} \subset M$ is locally flat, and the systole-long paths belonging to the two distinct homotopy classes that meet at each point of $U_{A,2}$ intersect each other orthogonally almost everywhere.*

Proof. Let U be a simply connected subdomain of $U_{A,2}$, let $\overline{\gamma}_1, \overline{\gamma}_2$ be two distinct, non opposite homotopy classes of closed paths, among the four that are representable by systole-long paths of length $\leq A$, that meet U, and let u, v be potential functions in U, as in Definition 2.2, for the two respective families of systole-long paths in U. One deduces from Lemma 2.1 , after replacing U, if necessary, by a smaller domain, that the paths of the two families passing through any given point $x \in U$ cross each other transversally ; hence the pair of local potential functions (u, v) forms a system of local parameters. In any set where the angles between the paths are bounded away from zero, the Riemannian distance function, expressed in terms of (u, v) is of Lipschitz class ; it follows from Rademacher's theorem that (u, v) is differentiable almost everywhere in U. Therefore one may represent almost everywhere the Riemannian metric, as before, as a quadratic form on the cotangent bundle in U. The norm $|\alpha|_g$ of any given Pfaffian form $\alpha = \xi du + \eta dv$ in U is now defined by the quadratic form

$$|\alpha|_g^2 = |\xi du + \eta dv|_g^2 = \xi^2 + 2f(u,v)\xi\eta + \eta^2 \ , \tag{2.1}$$

where $f(u, v)$ is a measurable function satisfying $|f(u, v)| < 1$ almost everywhere, and the corresponding volume element is

$$dVol = (1 - (f(u,v))^2)^{-\frac{1}{2}} |du \wedge dv| \ \ ; \tag{2.2}$$

the statement of Lemma 2.2 is therefore reduced to showing that $f(u,v) = 0$ almost everywhere in U.

Suppose that $f(u,v)$ is not identically zero in a subset $V \subset U$ of positive measure. One could then alter the metric g in V, as in the proof of Lemma 2.1, by replacing $f(u,v)$ by its product with a smooth function slightly smaller than 1, so that the value of the systole of M with respect to the new metric would be unchanged ; at the same time the resulting total volume of M would be decreased. Therefore the given metric g could be an extremal isosystolic one in M, only if the local functions $f(u,v)$ defined in $U_{A,2}$ were identically zero almost everywhere. This concludes the proof of Lemma 2.2.

The lemma just proved does not exclude the possibility that $U_{A,2}$ may contain a set of measure zero of singular points.

The next lemma generalizes the last one in the case of domains $U_{A,k}$ with $k \geq 3$. In order to state it, we must recall the notion of generalized angle, adapted from A.D. Aleksandrov, between two paths of shortest length with a common point of origin x, when the metric may be singular and the paths may fail to be differentiable at x.

In the first place, even if the point x is not an isolated point of intersection, one may assume without loss of generality that the two paths do not cross each other (in the topological sense) anywhere else in a neighborhood of x. In fact, if they meet and cross at any point $y \neq x$, the segments between x and y along the two paths have obviously equal length ; if one then redefines the two paths by interchanging their traces along the segments between x and y, the new paths are again length minimizing and have "fewer" crossings, since they now meet at y without crossing each other. By applying Zorn's lemma, for any pair of shortest paths (or rays) issued from a common origin x, one may replace it with another pair of rays, respectively of equal lengths and jointly tracing the same continuum as the original pair, and not crossing each other anywhere in a neighborhood U of x ; then there exists a simply connected, compact neighborhood $V \subset U$ of x, such that the union Γ of the two rays splits V into two compact, pathwise connected subsets $V' \cup V$", with Γ as their common boundary.

Let $\delta_0 > 0$ be sufficiently small, so that the closed metric ball $B(x, \delta_0)$ of radius δ_0 and center x is contained in V, let V' be one of the two "halves" of V that are bounded by Γ and, for any $\delta (0 < \delta \leq \delta_0)$, let $E'(\delta) = V' \cap B(x, \delta)$. Then the angle spanned by V' is defined to be number $2 \cdot \lim_{\delta \to 0}(Vol(B(x, \delta))/\delta^2)$, if the limit exists, or, if the limit does not exist, the generalized angle is defined to be the set of accummulation values of $2 \cdot \lim_{\delta \to 0}(Vol(B(x, \delta))/\delta^2$. If the metric is continuous and non-degenerate at x, the angle just defined coincides with the elementary notion ; in the singular case, the sum of the two opposite angles spanned by two rays does not necessarily equal 2π. However, if x is an interior point of a path of least length, then the total angle spanned by the two resulting rays from x is necessarily $\geq 2\pi$.

Lemma 2.4. — *Let g be an extremal isosystolic metric in a closed, orientable surface M and let $U \subset M$ be any k-regular domain ($k \geq 2$). Introduce, for each point $x \in U$, a family of k unoriented, systole-long, closed curves passing through x, with the property that the $2k$ free homotopy classes, represented by each of the k curves with its two orientations, constitute a complete list of the $2k$ homotopy classes thus obtainable. Consider, locally at x, the corresponding family of $2k$ segments of these paths, originating at x (rays), ordered in their natural, counterclockwise cyclic order in terms of an orientation of M. Then, for almost all $x \in U$, the total of the angles at x from each of these $2k$ rays to the next equals 2π, the two angles formed by any two opposite pairs of paths are equal and, most importantly, the angle at x between any two consecutive paths is $\leq \pi/2$.*

Proof. In the first place we recall from the proof of Lemma 2.3 that, choosing as local coordinates the geodesic potential functions (u, v) corresponding to two of the families of systole-long paths in U, the coordinates are differentiable almost everywhere and the Riemannian metric form (2.1) is determined by the measurable function $f(u, v)$; f has the property that $|f(u, v)| < 1$ almost everywhere and $(1 - (f(u, v))^2)^{-\frac{1}{2}}$ is integrable. The function $f(u, v)$ represents almost everywhere in U the cosine of the interior angle between the oriented, systole-long paths chosen to define u and v: this proves the first two, more elementary, assertions. Suppose that the main conclusion failed: this would mean that, on choosing (u, v) corresponding to a consecutive pair among the $2k$ oriented, systole-long paths, the resulting function $f(u, v)$ would be strictly positive in a set $V \subset U$ of positive measure, and, in V, the remaining $k-2$ unoriented, systole-long paths would be representable as graphs of monotone decreasing functions, either expressing v in terms of u or vice-versa. The remainder of the proof would be similar to the corresponding arguments in Lemma 2.3. Choosing a bounded, positive function $\phi(u, v)$ with support in V, one could construct a metric g'_ϵ from g by replacing the function $f(u, v)$ by $f_\epsilon(u, v) = f(u, v) \cdot exp(-\epsilon \cdot \phi(u, v))$ for any constant $\epsilon > 0$. The change could only leave unchanged or decrease the norms of the geodesic potential functions of the original $2k$ families of systole long paths meeting V. As a result, the length of these paths could not decrease under the change of metric ; for ϵ sufficiently small, no additional systole-long paths, representing homotopy classes other than the $2k$ original ones could appear, and the total area of M would decrease. The combined effect would be that the substitute metric g'_ϵ would have an isosystolic ratio smaller than that of g, contrary to the assumption. This completes the proof of Lemma 2.4.

The proofs of the two last lemmas demonstrate the importance of the geodesic potential functions ; indeed they play an essential role in what follows. The first task is to extend the notion of these potential functions, so that they are defined, in some sense, globally, rather than just in the union of the respective systolic paths ; this is the topic of the next section.

3. SYSTOLIC BANDS AND POTENTIAL FUNCTIONS

Let M be a closed, oriented surface of genus $g \geq 2$, and let g be any Riemannian metric on M in the class G defined at the beginning of Section 2. Let $M^\sim$ denote the universal covering surface of M with $\Pi : M^\sim \to M$ denoting the covering map, choose an arbitrary point $x_0^\sim \in M^\sim$ as a base point, and let $x_0 = \Pi(x_0^\sim)$ be the corresponding base point for M. The group $\pi_1(M, x_0)$ operates freely by translation on $M^\sim$ and the covering map $\Pi : M^\sim \to M$ induces in $M^\sim$ the "pull-back" metric $g^\sim = \Pi^*(g)$, which is invariant under the action of $\pi_1(M, x_0)$. Thus, for any non-trivial element $\gamma \in \pi_1(M, x_0)$ and any $x^\sim \in M^\sim$ with $x = \Pi(x_0) \in M$, the Riemannian distance $d_{g^\sim}(x^\sim, \gamma(x^\sim))$ describes the least length of any x-based loop in M, representing the homotopy class $\psi^{-1}\varphi\psi$, where $\psi = \Pi(\psi^\sim)$, $\psi^\sim$ is a path in $M^\sim$ from $x_0^\sim$ to $x^\sim$, and φ is any loop in M based at x_0, representing the homotopy class γ. Then the free local systole, $Sys_{\overline{\gamma}}(M, g)$, is the value of $Inf_{x^\sim \in M^\sim}(d(x^\sim, \gamma(x^\sim)))$.

Given any non-trivial $\gamma \in \pi_1(M, x_0)$, choose any $x^\sim \in M^\sim$ that minimizes the distance $d(x^\sim, \gamma(x^\sim)) = Sys_{\overline{\gamma}}(M, g)$ and any path $\varphi^\sim$ from $x^\sim$ to $\gamma(x^\sim)$ that achieves that distance as its own length. It is not hard to verify that, since M is an orientable surface, for any integer n one has the identity $Sys_{\overline{\gamma}^n}(M, g) = |n| Sys_{\overline{\gamma}}(M, g)$; thus, if one takes the union of the following translations of that path, $U_{-\infty<n<\infty}(\gamma^n(\varphi^\sim))$, one obtains a complete path (i.e. a complete geodesic, in the smooth case), that achieves the minimum distance between any two of its points. To any non-trivial, cyclic subgroup $\langle\gamma\rangle \subset \pi_1(M, x_0)$ generated by γ, one associates the family Σ_γ of all complete, unbounded "geodesics" $U_{-\infty<n<\infty}(\gamma^n(\varphi^\sim))$ generated by all possible paths $\varphi^\sim$ of length $Sys_{\overline{\gamma}}(M, g)$, connecting any suitable $x^\sim$ with $\gamma(x^\sim)$. One calls the paths of the family Σ_γ the systolic band directed by γ, and its trace $B_{\gamma^\sim} \subset M^\sim$ is the union of all the paths belonging to Σ_γ.

Given a non-trivial element $\gamma \in \pi_1(M, x_0)$, and the corresponding systolic band Σ_γ of paths in $M^\sim$, one may define a (global) potential function u_γ axiomatically as follows.

Definition 3.1. — *Given a closed, orientable surface M of genus ≥ 2 with a Riemannian metric g, a non-trivial element $\gamma \in \pi_1(M, x_0)$ and the corresponding systolic band Σ_γ of complete, shortest-length paths in $M^\sim$ directed by γ, a global potential function $u_\gamma : M^\sim \to \mathbf{R}$ directed by γ is a function that satisfies the following axioms.*

(1) *For each $x^\sim \in M^\sim$, the function u_γ satisfies the relation*

$$u_\gamma(\gamma(x^\sim)) = u_\gamma(x^\sim) + Sys_{\overline{\gamma}}(M, g) .$$

(2) *The function* u_γ *is of Lipschitz class everywhere in* $M^\sim$, *its Lipschitz constant satisfies* $0 < Lip(u_\gamma, x^\sim) \leq 1$ *at each* $x^\sim \in M^\sim$; *in the complement of the trace* $B_{\gamma^\sim}$ *of* Σ_γ, *the function* u_γ *is of class* $C^{1,1}$ *and its differential* $du_\gamma(x^\sim)$ *has norm* ≤ 1 *with respect to* g, *while if* $x^\sim$ *lies in the trace* $B_{\gamma^\sim}$ *of* Σ_γ, *then* $Lip(u_\gamma, x^\sim) = 1$, *and the directional derivatives of* u_γ *take the value* 1 *precisely in the direction of any path in* Σ_γ *that passes through* $x^\sim$.

(3) *For any given constant* δ *with* $0 < \delta \leq 1$, *there exists a positive constant* C *such that, for each point* $x^\sim \in M^\sim$ *at a distance* $\geq C$ *from the nearest path in* Σ_γ *the function* u_γ *satisfies* $Lip(u_\gamma, x^\sim) \leq \delta$.

(4) *The function* u_γ *has no critical points, in the sense that each level set of* u_γ *is a rectifiable, connected, properly imbedded curve in* $M^\sim$, *and for each* $x^\sim \in M^\sim$, *the image in* M_γ *under* Π'_γ *of each path of steepest ascent (respectively, descent) of* u_γ *from* $x^\sim$ *with* $\gamma(x^\sim)$ *(respectively,* $\gamma^{-1}(x^\sim)$*) is contained in a compact set, invariant under the translation by* γ.

Since addition of constants to potential functions does not affect the properties that characterize them as such, one may include an additional requirement that they vanish at a designated base point $x_0^\sim \in M^\sim$. To any potential function u_γ in $M^\sim$ one associates the corresponding reduced potential function $\overline{u}_\gamma : M_\gamma \to \mathbf{R}/(Sys_{\overline{\gamma}}(M,g)\mathbf{Z})$,

$$\overline{u}_\gamma(\Pi'_\gamma(x^\sim)) = u_\gamma(x^\sim) mod \cdot (Sys_{\overline{\gamma}}(M,g)\mathbf{Z}) \ . \tag{3.1}$$

The proof of the existence of a potential function u_γ for each non-trivial $\gamma \in \pi_1(M, x_0)$ when the metric has the required full generality is too long and technical to be fully included in the present paper ; however the initial step of a construction of these functions is easy and achieves the purpose, if the metric of (M, g) is smooth, at least of class $C^{1,1}$, and if the geodesics in $M^\sim$ have no conjugate points. This part of the proof is included for heuristic reasons.

Let (M, g) be an arbitrary closed, oriented surface of genus $g \geq 2$ with a metric g of class G ; let $M^\sim$, $g^\sim$, $x_0^\sim$, and γ be as before, and consider the systolic band Σ_γ of paths of shortest length in $M^\sim$ directed by γ. For any given path $\varphi^\sim \in \Sigma_\gamma$, choose an auxiliary base point $y_0^\sim \in \varphi^\sim$ as its initial point and parametrize $\varphi^\sim$ by its oriented arc length s from $y_0^\sim$; thus, for any real number s, we denote by $y^\sim(s)$ the point of $\varphi^\sim$ at an oriented distance s from $y_0^\sim$. Denoting by $d(x^\sim, y^\sim)$ the Riemannian distance between $x^\sim$ and $y^\sim$, the Busemann functions v_+ and v_- determined by the data of γ, $x_0^\sim$, $\varphi^\sim$, and $y_0^\sim$, are the real valued functions on $M^\sim$ defined by

$$v_+(x^\sim) = \lim_{s\to+\infty} (d(x^\sim, y^\sim(s)) - s) \ , \tag{3.2}$$

$$v_-(x^\sim) = \lim_{s\to-\infty} (d(x^\sim, y^\sim(s)) + s) \ .$$

By means of these functions, one introduces the functions u_γ and $h_{\gamma,\varphi^\sim}$, described respectively as the preliminary potential and stream functions, defined as follows:

$$u_\gamma(x^\sim) = \frac{1}{2}(v_-(x^\sim) - v_-(x_0^\sim) - v_+(x^\sim) + v_+(x_0^\sim)) \ , \tag{3.3}$$

$$h_{\gamma,\varphi^\sim}(x^\sim) = \frac{1}{2}(v_-(x^\sim) + v_+(x^\sim)) \ . \tag{3.4}$$

The following list of properties of the functions just introduced are either elementary, so that their proofs may be omitted.

1. The Busemann functions $v_\pm$ satisfy the Lipschitz condition $Lip(v_\pm) = 1$ and the functional identities

$$v_+(\gamma(x^\sim)) = v_+(x^\sim) - Sys_{\overline{\gamma}}(M, g) \ ,$$

$$v_-(\gamma(x^\sim)) = v_-(x^\sim) + Sys_{\overline{\gamma}}(M, g) \ ;$$

furthermore, if the metric is smooth and if geodesics have no conjugate points, the Busemann functions are of class C^1, with $|\bigtriangledown V_\pm| = 1$ and $\bigtriangledown V_- \neq -\bigtriangledown V_+$ everywhere.

2. For any fixed γ and under different choices of $x_0^\sim \in M^\sim$, $\varphi^\sim \in \Sigma_\gamma$, and $y_0^\sim \in \varphi^\sim$, the resulting functions v_+, v_-, u_γ, and $h_{\gamma,\varphi^\sim}$ are modified by additive constants.

3. The preliminary stream function $h_{\gamma,\varphi^\sim}$ is invariant under the action of γ ; it vanishes identically on the path $\varphi^\sim$ and takes values ≥ 0 everywhere else in $M^\sim$; furthermore it satisfies everywhere the Lipschitz condition $Lip(h_{\gamma,\varphi^\sim}) \leq 1$. If the

metric is smooth and if there are no conjugate points, then $\bigtriangledown h_{\gamma,\varphi^\sim}$ exists everywhere, is continuous, and $| \bigtriangledown h_{\gamma,\varphi^\sim}| < 1$.

4. The preliminary potential function $u\gamma$ satisfies the Lipschitz condition $Lip(u_\gamma) \leq 1$ everywhere in $M^\sim$; in addition it satisfies the relation $u_\gamma(\gamma(x^\sim)) = u_\gamma(x^\sim) + Sys_{\overline{\gamma}}(M,g)$; it is normalized additively so that it vanishes at $x_0^\sim$, and depends only on γ, not on the particular choice of path $\varphi^\sim \in \Sigma_\gamma$. If the metric is smooth and if there are no conjugate points, then $\bigtriangledown u_\gamma$ exists everywhere, is continuous, and is nowhere zero.

If the metric g is sufficiently smooth and if the geodesics in $M^\sim$ have no conjugate points, then the gradient flow of the function u_γ constitutes a foliation of $M^\sim$, including each of the paths $\varphi^\sim \in \Sigma_\gamma$. The dynamical system of this flow is invariant under the translation group generated by γ and the preliminary stream function $h_{\gamma,\varphi}$ is constant along each orbit. In addition, since $\bigtriangledown h_{\gamma,\varphi^\sim} = 0$ only along complete geodesics of the systolic band Σ_γ, it follows that, under the special assumptions on g, the preliminary stream functions $h_{\gamma,\varphi^\sim}$ are actual stream functions, constant along, and locally separating, the orbits of the gradient flow of u_γ, so that each of these orbits is invariant under the translation of γ. This shows that, if the metric g is of class at least $C^{1,1}$ and if there are no conjugate points, the function u_γ satisfies all the four properties characterizing a global potential function directed by γ. In the general case, the potential functions can be obtained similarly from formulas (3.3), (3.4), in which the Busemann functions $V_\pm$ are replaced by corresponding functions with similar properties, but constructed by a process yielding functions better suited to our purposes. An important fact, applied in that construction is the uniform exponential growth property of $\pi_1(M, x_0)$ in terms of its generators.

Given a non-trivial $\gamma \in \pi_1(M, x_0)$, consider the covering surface M_γ of M the cyclic subgroup of $\pi_1(M, x_0)$ generated by γ with the metric induced by the covering map from g. The surface M_γ is homeomorphic to a cylinder, with $M^\sim$ as its universal covering surface. The family Σ_γ of paths of least length in $M^\sim$ directed by γ corresponds to the family of systole-long paths in M_γ, also denoted by Σ_γ. Since the genus g of M is assumed to be ≥ 2, the family Σ_γ is compact ; if it consists of more than one essential closed path in M_γ, there are two such paths, bounding a retract of M_γ, that contains all the systole-long paths in M_γ. This retract will be called the *systolic strip* of γ, denoted by Y_γ ; if γ is a "simple" element of $\pi_1(M, x_0)$, meaning that γ is representable by a simple loop in M, then the covering map Π_γ of M by M_γ is a one-to-one isometry of Y_γ onto its image in M ; the latter will likewise be denoted by Y_γ, despite the small risk of ambiguity. This is the case, in particular, if $\overline{\gamma}$ is a critical, free homotopy class for the systole of (M, g). We observe that, for general metrics g, the systolic strips Y_γ, whether considered in M_γ or in M, are not necessarily covered by the paths of the systolic band Σ_γ.

4. THE PRELIMINARY VARIATIONAL PROCESS

We shall consider now the problem of characterizing extremal isosystolic metrics in a closed, orientable surface M of genus $g \geq 2$ in terms of local properties such as, for instance, solutions of partial differential equations. The surface M is assumed to be *polarized*, meaning that the fundamental group $\pi_1(M, x_0)$, regarded as the group of homotopy classes of x_0-based loops, is identified with its standard presentation as an abstract group, by an explicit choice of $2g$ generators $p_i, q_i (1 \leq i \leq g)$, satisfying the relation

$$p_1 q_1 p_1^{-1} q_1^{-1} \cdots p_g q_g p_g^{-1} q_g^{-1} = e \ , \tag{4.1}$$

in the usual way, first by orienting M, then by assigning to each of the $2g$ abstract generators p_i, q_i a corresponding system of $2g$ x_0-based, simple, oriented loops, that are pairwise disjunct away from x_0, chosen in such a way that, if one cuts M along these $2g$ loops, one obtains a simply connected domain D. In addition, the loops are constructed, so that one may read the oriented boundary of D as a sequence of $4g$ oriented loops in M corresponding to the left-hand side of the relation (4.1).

The first question that arises in considering the extremal isosystolic problem is that of characterizing the subsets $S = S_g \subset \pi_1(M, x_0)$ that may possibly occur as the *critical* subsets for some extremal isosystolic metric g: the sets S_g consist of the elements $\gamma \in \pi_1(M, x_0)$, whose conjugacy classes $\overline{\gamma}$ are represented by its systole-long paths. The question is somewhat ambiguous as posed, because extremal isosystolic metrics are not necessarily continuous in M. In fact, consider a surface M with an extremal isosystolic metric g, normalized under the condition $Sys(M, g) = 1$: there may be (cf. the second example in Section 6) a sufficiently small constant $\delta > 0$ and a non-empty family of non-critical homotopy classes γ, whose corresponding local systoles $Sys_{\overline{\gamma}}(M, g)$ lie in the interval $1 < Sys_{\overline{\gamma}}(M, g) \leq 1 + \delta$, and are achieved by closed paths $\varphi_{\overline{\gamma}}$. Choosing one (or a finite set) of such paths $\varphi_{\overline{\gamma}}$, one may replace the metric g with a discontinuous metric g', identical with g outside the trace of the chosen path(s) $\varphi_{\overline{\gamma}}$, and uniformly smaller than g along each $\varphi_{\overline{\gamma}}$, that reduces the length of $\varphi_{\overline{\gamma}}$ to unity, thereby rendering it of systole length: if δ is sufficiently small, the change of metric will not affect the length-minimizing property of the paths that are systole-long in terms of g. The metric g' then is again an extremal isosystolic one, having the same systole and total area as g, but its critical set S', interpreted literally, would include S together with all the homotopy classes γ affected by the change. The following definition is proposed in order to clarify the ambiguity.

Definition 4.1. — *Let (M, g) be a closed surface of genus $g \geq 2$ with a Riemannian metric g that is extremal isosystolic. Then the* <u>essential critical set</u> *with respect to g is the subset $S = S_g \subset \pi_1(M, x_0)$ consisting of the homotopy classes γ, such that the conjugacy classes $\overline{\gamma}$ of each $\gamma \in S$ are represented each by a band of systole-long paths, whose trace in M has* <u>positive measure</u>. *A* <u>maximal critical set</u> *with respect to g is a set $S' = S'_g$, $S_g \subseteq S'_g \subset \pi_1(M, x_0)$, that is maximal with respect to the property of being representable by systole-long paths in terms of any metric g' such that $g' = g$ almost everywhere and each systolic band of g is also a systolic band of g', so that the values of both the area of M with respect to either metric coincide.*

One observes that any extremal isosystolic metric g in M may be represented as the limit, almost everywhere, of an increasing sequence (g_n) of smooth Riemannian metrics, such that the isosystolic critical set of free homotopy classes of each g_n is, for instance, a maximal critical set (or else, trivially, an essential one) of classes with respect to g. The following proposition describes some general properties of the subsets $S \subset \pi_1(M, x_0)$ that may occur as the set of critical classes (essential or maximal) with respect to an extremal isosystolic metric.

Proposition 4.2. — *If g is an extremal isosystolic metric in M, then the following assertions are valid.*

(i) *The essential critical subset $S = S_g \subset \pi_1(M, x_0)$ with respect to g is necessarily the union of a finite family of non-trivial conjugacy classes, symmetric with respect to inversion.*

(ii) *Each element of S is representable by a simple, closed path.*

(iii) *Any two elements of S may be represented by two simple, closed paths of least length, that either cross each (transversally) at exactly one point, or are disjunct, or may be approximated uniformly by two disjunct, simple, closed paths.*

(iv) *For each element of S there is at least one other element of S, such that any pair of simple, closed curves representing the two respective classes have a non-empty intersection.*

(v) *(conjectured) The canonical image of S in the homology group $H_1(M, \mathbf{Z})$ includes a family of $2g$ elements constituting an integral basis of $H_1(M, \mathbf{Z})$.*

Proof. The finiteness and symmetry properties (i) of S are trivial. The intersection properties (ii) and (iii) are well known, elementary properties, that are verifiable by suitable, smooth approximations of length minimizing paths in surfaces with path-length metrics (and by the geodesic paths themselves, if the metric is smooth). The symplectic property (iv) is a consequence of Lemma 2.1 and Property (iii).

Property (v), the conjectured property of homology fullness of the set S, is the only one that requires some comment. Even though it may seem intuitively obvious, in actual fact it is essentially equivalent to the strict monotonicity of the minimum isosystolic ratio as a function of the genus of the surface M. Current attempts to prove the strict monotonicity of the minimum isosystolic ratio depend on sharp numerical estimates of the "systole relative to the boundary" for surfaces with boundary, with an extremal isosystolic metric g_c $(0 < c < 3)$, in which the homology systole is unity, and c is the minimum length of any curve homotopic to the boundary. A numerical computation of these estimates is now in progress (cf. the first example in Section 6).

A consequence of Gromov's compactness theorem [5] is that, for any given value $g \geq 2$ of the genus of M, there can be at most finitely many subsets $S \subset \pi_1(M, x_0)$, pairwise inequivalent under the action of the automorphism group of $\pi_1(M, x_0)$, such that each is the set of critical homotopy classes for some extremal isosystolic metric in M. On the other hand the properties on the sets S stated in Lemma 4.1 do not appear to be even sufficient to deduce that there are finitely many sets S, up to equivalence, that verify these properties. The problem of characterizing *a priori* the sets of critical homotopy classes, either essential or maximal, of extremal isosystolic metrics for surfaces of genus ≥ 2 is open and probably very difficult. For the present purposes it is sufficient to remark that, given any set $S_0 \subset \pi_1(M, x_0)$ with the five properties listed in Proposition 4.2, there are at most finitely many sets S' with $S_0 \subseteq S' \subset \pi_1(M, x_0)$, verifying the same properties.

The "variational search" for an extremal isosystolic metric in M consists of two "processes", performed alternately infinitely many times, where each process replaces a given, non-extremal metric with another metric, exhibiting a lower isosystolic ratio. The present section outlines the first process, while Section 5 is devoted to describing the second one.

The first process consists by itself of an infinite iteration of two alternating steps, Steps 1 and 2, described below.

Step 1. Given any admissible metric g_0 in M (a smooth one, to begin with), normalized by the condition $Sys(M, g_0) = 1$, one introduces the infinite dimensional numerical torus $\mathbf{T} = \Pi_{\gamma \in \pi_1(M,x_0) \setminus \{e\}}(\mathbf{R}/\mathbf{Z})$, whose components are indexed by the non-trivial elements of the fundamental group of $\pi_1(M, x_0)$. One then considers the mapping $F : M^{\sim} \to \mathbf{T}$, where $M^{\sim}$ is the universal covering surface of M, defined by

$$(4.2) \quad y_\gamma(x^{\sim}) = \{(Sys_{\overline{\gamma}}(M, g_0))^{-1} u_\gamma(\Pi'_\gamma(x^{\sim}))\} mod \mathbf{Z} \ ; \ F = (y_\gamma)_{\gamma \in \pi_1(M,x_0) \setminus \{e\}} \ ,$$

In this equation, for each $\gamma \in \pi_1(M, x_0) \setminus \{e\}$, the function $u_\gamma : M_\gamma \to \mathbf{R}$ is a potential function directed by γ, as in definition (3.1). Clearly each component function y_γ of F is of Lipschitz class, with a Lipschitz ratio $\leq (Sys_{\overline{\gamma}}(M, g_0))^{-1} \leq 1$ at each point. The mapping F is equivariant under the diagonal action of $\pi_1(M, x_0)$ on $M^{\sim} \times \mathbf{T}$ as a transformation group, acting simultaneously by translation on $M^{\sim}$, and on $\mathbf{T}$

by permuting the components according to the adjoint action of the group on itself. Therefore (see the lemma that follows) there exists a unique Riemannian metric $g_1^\sim$ on $M^\sim$, determined at each point $x^\sim$ by the property of having the least possible element of volume, subject to the constraint that, for each γ, the weak differential [3] $dy_\gamma(x^\sim)$ has a norm ≤ 1 with respect to $g_1^\sim$; since this metric is obviously invariant under translations of $M^\sim$ by $\pi_1(M, x_0)$, it induces a metric g_1 in M with stystole ≥ 1 and area no larger than the original one with respect to g_0. It is precisely this step that would be carried out more naturally in the context of Finsler metrics, while the restriction to Riemannian metrics introduces some complications. The preference given here to Riemannian metrics is due to the greater familiarity with Riemannian geometry by most people.

The result of the change of metric just described may alter, in general, the set of isosystolic critical classes, by either addition or deletion ; furthermore the potential functions with respect to g_0 for each $\gamma \in \pi_1(M, x_0)\backslash\{e\}$ need not satisfy the axioms set in Definition 3.1 in terms of g_1: in particular the condition $|dy_\gamma(x^\sim)|_{g_1} = 1$ is no longer equivalent, in general, to the property of $x^\sim$ lying in the trace of the systolic band directed by an essential critical class γ.

Step 2. The second step of the process consists of replacing the family of potential functions u_γ with respect to of the original metric g_0 with a corresponding family in terms of g_1, thereby altering the mapping $F : M^\sim \to \mathbf{T}$.

One then repeats this two-step operation. It follows once more from the compactness theorem for the isosystolic problem, that, after a finite number of iterations, the operation first stabilizes the set of critical classes of paths, and leads to a sequence of metrics, converging to a metric g, such that neither of the two steps just outlined necessarily lead to any further change. Such a metric g is characterized by the three following properties:

[3] A Lipschitz-continuous function f is said to be non-critical at an interior point $x \in M$, if every neighborhood of x contains an open neighborhood U of x, within which both subsets $\{y{:}f(y)>f(x)\}$ and $\{y{:}f(y)<f(x)\}$ are non-empty and contractible. The differentials of two such functions f and g are said to be weakly linearly independent at x, if, for each pair of constants $(a,b)\neq(0,0)$, the function $af+bg$ is non-critical at x. The weak differential $df(x)$ of f at x is the closure of the set of all differentials at x of smooth functions g, such that $f-g$ is critical at x. It follows that the set $df(x)$ is non-empty and is tightly contained in the euclidean ball of radius $Lip(f,x)$.

(i) the metric g is uniquely determined by the mapping $F : M^{\sim} \to \mathbf{T}$ defined in (4.2), under the property of minimizing almost everywhere the element of area, subject to the constraint $|du_\gamma(\Pi'_\gamma(x^{\sim}))|_g \leq 1$ for every $x^{\sim} \in M^{\sim}$ and every γ ;

(ii) for any given $x^{\sim}$ and γ, $|du_\gamma(\Pi'_\gamma(x^{\sim}))|_g = 1$, if, and only if $x^{\sim}$ lies in the systolic band in $M^{\sim}$ directed by γ ;

(iii) The systolic strips cover M in such a way that almost every point $x \in M$ is a transversal intersection point of at least two systole-long paths, and the angles between the successive, oriented, systole-long paths meeting at x are each $\leq \pi/2$ intersection, with a total angle $\geq 2\pi$ (the proof essentially the same as in Lemmas 2.1, 2.2 and 2.4).

Recall now the definition of the systolic strips, given in the closing paragraph of Section 3: for each non-trivial homotopy class $\gamma \in \pi_1(M, x_0)$ the systolic strip Y_γ directed by γ is a compact retract of M_γ (the covering space of M corresponding to the cyclic subgroup $\langle\gamma\rangle \subset \pi_1(M, x_0)$), obtained by "filling in" the compact subdomain between the two outermost systole-long paths in M_γ. Restricting γ now to the critical isosystolic classes of M, the corresponding systolic strips may be identified with the corresponding images in M under the covering map, one sees that M is covered (more descriptively, "bandaged") by this finite family of critical systolic strips, so that each point is covered by at least one pair of mutually transversal strips, and every tangent vector makes an angle $\leq \pi/4$ with a gradient vector of the potential function at least one critical class. The total surface M is thus decomposed into a finite number of compact, convex, geodesic polygons, partially ordered by inclusion, each one determined by some interior point x, and defined as the component of x in the intersection of all the critical systolic strips containing x in their interior. Each of these geodesic polygons, determined by the intersection of k unoriented systolic strips (or $2k$ oriented ones), is completely described by the corresponding k pairs of mutually opposite potential functions, and is handled in the next section in a way that generalizes the treatment of k-regular domains in Section 2.

5. THE EULER-LAGRANGE EQUATIONS

Let M be a closed, orientable surface with a geodesic Riemannian metric g, such that the critical systolic strips cover M with the conditions described at the end of Section 4, and let D be a geodesic polygon in M, determined by a finite intersection of k unoriented, critical systolic strips, and such that there exists some interior point of D that is not interior to any additional, critical systolic strip. Let $\pm u_1, \pm u_2, \cdots, \pm u_k$ denote the corresponding potential functions and

$$\mathbf{u} = (u_1, u_2, \cdots, u_k) : D \to \mathbf{R}^k$$

the mapping that they define, where the choice of orientation of each component u_j is immaterial. The Riemannian metric in D, by assumption, is the one which, at almost every point $x \in D$, has the least area density form under the condition $|di_j(x)|_g \leq 1$ for each j $(1 \leq j \leq k)$; in particular the metric at x is controlled by the subset among the functions u_j such that $|du_j(x)|_g = 1$. If two of these functions (let us say u_1 and u_2) suffice to determine g in an open subset $U \subset D$, then du_1 and du_2 are mutually orthogonal (by Lemma 2.3) ; otherwise, for almost all x there is a neighborhood U of x, where three of them, say u_1, u_2 and u_3 determine g by themselves under the condition $|du_j(x)|_g = 1$ $(j = 1, 2, 3)$, and the six disjunct angles in terms of g formed by the six differentials $\pm du_j$ in the cotangent space of x are all $< \pi/2$ (cf. Lemma 2.4). In any case the mapping $\mathbf{u}$ of equation (5.1) is almost everywhere an immersion of D, inducing a metric, that is determined at each point by the direction of the tangent plane. This situation is precisely one considered by É. Cartan in his 1933 monograph [5] for the purpose of studying invariants attached to variational problems. What makes the isosystolic problem difficult from the viewpoint of Cartan's treatment is that the area functional is not of class C^2 in terms of the direction parameters of the tangent plane. For instance, if $k = 3$, suppose that the image $\mathbf{u}(D)$ in $\mathbb{R}^3$ is represented locally as the graph of a function, $u_3 = f(u_1, u_2)$, or, more briefly, $z = f(x, y)$, and denote by (u, v, w) the homogeneous direction numbers of any plane at the point (x, y, z) ; then the tangent plane of $\mathbf{u}(D)$ at (x, y, z) has direction numbers proportional to $(p, q, -1)$, where $p = \partial f/\partial x$ and $q = \partial f/\partial y$). It follows from the assumptions $|dx|_g = |dy|_g = 1$ that the area functional $dVol_g$ is expressed by (2.1), (2.2). In order to account for the alternative possibilities that the least-area metric be controlled by all three of the potential functions or just two of them, it is best to express the area density function in terms of the three homogeneous direction numbers $\xi = (\xi_1, \xi_2, \xi_3)$ of the tangent plane, by means of a function $L(\xi)$, defined to be positively homogeneous of degree one, such that $dVol_g = L(p, q, -1)$. Introduce the function (Heron's function)

$$\Delta(\xi_1, \xi_2, \xi_3) = Re\{(2(\xi_2^2\xi_3^2 + \xi_3^2\xi_1^2 + \xi_1^2\xi_2^2) - (\xi_1^4 + \xi_2^4 + \xi_3^4))^{\frac{1}{2}}\} ,$$

representing, when it is positive valued, twice the area of the euclidean triangle with sides of length $|\xi_1|$, $|\xi_2|$, $|\xi_3|$; then direct computation leads to the following expression for $L(\xi)$:

$$L(\xi) = \begin{cases} 2|\xi_1\xi_2\xi_3|/\Delta(\xi_1, \xi_2, \xi_3) , & \text{if } \mathrm{Max}\{\xi_1^2 , \xi_2^2, \xi_3^2\} < \frac{1}{2}(\xi_1^2 + \xi_2^2 + \xi_3^2) \\ \mathrm{Max}\{|\xi_1|, |\xi_2|, |\xi_3|\}, & \text{if } \mathrm{Max}\{\xi_1^2, \xi_2^2, \xi_3^2 \geq \frac{1}{2}(\xi_1^2 + \xi_2^2 + \xi_3^2) \end{cases}$$

The two alternative analytic expressions given above for $L(\xi)$ reflect the two respective possibilities, whether the metric is controlled by all three of the potential functions or by just two of them. The global function $L(\xi)$ represents geometrically the least diameter of any disk that contains a triangle with sides of length $|\xi_1|$, $|\xi_2|$, $|\xi_3|$, and is the lower convex envelope of the one expressed by the algebraic function in the top line.

Consider now a small variation of the immersion $\mathbf{u}$ in a compact subdomain of D, reflected by a corresponding variation of the functional relation between x, y, z in a compact subdomain of its domain of definition. Such a change induces a corresponding change in the metric in the global surface M, that generally leaves all critical local systoles unchanged, and, if the change is sufficiently small, does not introduce any new critical homotopy classes. Thus the isosystolic variational problem is reduced, locally, to that of finding functions $z = f(x, y)$ with prescribed boundary values in a compact domain $D \subset \mathbb{R}^2$, that minimize the integral $\int_D L((\partial f/\partial x, \partial f/\partial y, -1)|dx \wedge dy|$. A routine calculation yields readily the Euler-Lagrange equation for this problem, in non-parametric form:

$$\frac{\partial^2 L(p,q,-1)}{\partial p^2}\frac{\partial^2 f(x,y)}{\partial x^2} + 2\frac{\partial^2 L}{\partial p \partial q}\frac{\partial^2 f(x,y)}{\partial x \partial y} + \frac{\partial^2 L}{\partial q^2}\frac{\partial^2 f(x,y)}{\partial y^2} = 0 . \tag{5.3}$$

The convexity of the homogeneous function $L(\xi)$ guarantees that the local boundary value problem has a unique, weak solution of class $C^{1,1}$, that assumes the assigned boundary values, provided that the domain is strictly convex. On the other hand $L(\xi)$, in contrast with the classical analogue $L_0(\xi) = ((L\xi_1^2 + \xi_2^2 + \xi_3^2))^{\frac{1}{2}}$ arising from Plateau's problem, is not strictly convex in every non-homogeneous plane in $\mathbb{R}^3$. Indeed it is strictly convex, locally, only in non-homogeneous planes at points where the top-line formula of (5.2) holds, i.e. in the open domain $\Omega \subset \mathbb{R}^3$, where ξ_1^2, ξ_2^2 and ξ_3^2 satisfy the triangle inequality. For any solution $f(x,y)$ of the Euler-Lagrange equation (5.3), the solution is of elliptic type, and hence real analytic, at any point of D for which $\xi \in \Omega$.

As the point for which $\xi \in \Omega$ approaches a point (x,y,z) for which ξ lies on a regular part of the boundary of Ω (consisting of all lines through the origin, making an angle of 45° with exactly one of the three coordinate axes), the solution of (5.3) becomes a degenerate elliptic solution, approaching a parabolic point, where the bicharacteristics are tangent to the curve of parabolic points, Hence, if the analytic solution of the elliptic equation admits an analytic continuation across such a transition curve, it provides an example of solution of a non-linear differential equation of Tricomi type. However, the coefficients of the actual equation (5.3) and the discontinuity of the second derivatives of L at the boundary of Ω imply that, when ξ lies in the interior of the complement of Ω (consisting of six disjunct, convex, open, quadric cones), then the function $f(x,y)$ is not restricted by the Euler-Lagrange equation at all ; indeed two of the three potential functions constitute a local cartesian chart for a flat metric g, and the third one is only required to have a Lipschitz constant ≤ 1 in terms of g. We summarize the main results of preceding discussion in the statement of the following theorem.

Theorem 5.1. — *Let (M,g) be a closed surface with an extremal isosystolic metric g, and let $\overline{\gamma}_j^{\pm 1}$ $(j = 1,2,3)$ be six distinct, critical free homotopy classes of closed paths in M, admitting an open, simply connected region $D \subset M$ that does not meet any critical isosystolic strip other than the six strips directed by $\overline{\gamma}_j^{\pm 1}$. Let $u_j : D \to \mathbb{R}$ be local potential functions in D directed by each $\overline{\gamma}_j$: then the combined map (u_1,u_2,u_3): $D \to \mathbb{R}^3$ is an immersion, that may be described locally in non-parametric form as* *the graph of a function* $z = f(x, y)$ *as* (x, y, z) *traces the image* $(u_1, u_2, u_3)(D) \subset \mathbb{R}^3$. *In addition the function* f *satisfies the Euler-Lagrange equation (5.3). The latter equation is vacuous in the interior of any subdomain of* D*, whose image in* $\mathbb{R}^3$ *has the property that each of its tangent planes makes an euclidean angle* $< 45°$ *with any one of the three coordinate planes. In the interior of any region for which the tangent planes make angles* $> 45°$ *with each of the three coordinate planes, the equation (5.3) is elliptic.*

One distinguishes, among the elliptic solutions of the Euler-Lagrange equation (5.3), the trivial class, consisting of potential functions that are related by linear equations: the metric that they induce is obviously flat. Conversely, it is not hard to verify that, if an extremal isosystolic metric, determined by three potential functions related by an elliptic solution of (5.3), is locally flat, then the functions must be linearly related.

In order to complete the variational analysis, one must still consider the domains of M, where an extremal isosystolic metric may be controlled by four or more overlapping, unoriented isosystolic bands. Clearly, the functional relation between any three local potential functions (directed by pairwise transversal critical isosystolic bands) may be expressed locally by choosing two of them as local parameters, so that the third one is represented as a function of two variables, namely $z = f(x, y)$, in the discussion leading to Theorem 5.1. If $f(x, y)$ satisfies the Euler-Lagrange equation (5.3) in a domain D, then the local potential functions directed by any number of additional critical isosystolic classes could conceivably enter the local picture without disturbing the equilibrium condition implied by the Euler-Lagrange equation: in fact, given any smooth Riemannian metric in D, any smooth curve in D is locally an orthogonal trajectory of a uniquely determined, local geodesic foliation, that may generate such a potential function. One can prove that it is not possible to have two functions $f_1(x, y)$ and $f_2(x, y)$ that are simultaneously elliptic solutions of (5.3), inducing the same volume form (and hence the same metric), unless f_1 and f_2 are being either both linear functions of x and y, or else differ from each other by an additive constant, or have a constant value for their sum. It seems plausible to conjecture that the interior extremality condition in the domain D of intersection of four or more isosystolic strips should lead to a decomposition of D into compact sub-regions, in each of which only three of the potential functions would be related by (5.3). In addition, it still seems very hard to investigate the regularity (or non-regularity) of the interface curves between the resulting triple intersections of systolic bands of strips. The next section is the result of an attempt to determine the free boundary that arises between subdomains where a solution of (5.3) is of elliptic type on one side, while, on the opposite side, the extremal metric is controlled by just two potential functions.

6. A SPECIAL FREE BOUNDARY PROBLEM

The following boundary value problem related to the extremal isosystolic metric problem leads to a special, non-elementary, local solution of the Euler-Lagrange equation (5.3) ; there is some hope that this solution may be evaluated explicitly (in the sense of classical function theory), thus providing a useful comparison barrier for estimates of more general solutions.

Problem. — Given a closed, regular euclidean hexagon Ω, find a Riemannian metric in Ω that minimizes its total area, subject to the constraint that each point of each side of Ω has least Riemannian distance equal to 2 from the diametrically opposite side.

We postulate the following two additional properties of the metric sought in this problem, with the hope that at least one of them may be deduced as consequences of the other data.

(i) Opposite pairs of sides of the hexagon Ω have a one-to-one correspondence, so that each point of any one side of Ω is the only one that has distance $= 2$ from the corresponding point on the opposite side ;

(ii) the metric achieving the minimum area in the problem is unique, and hence admits the dihedral group D_6 of symmetries of the hexagon as its own group of isometries.

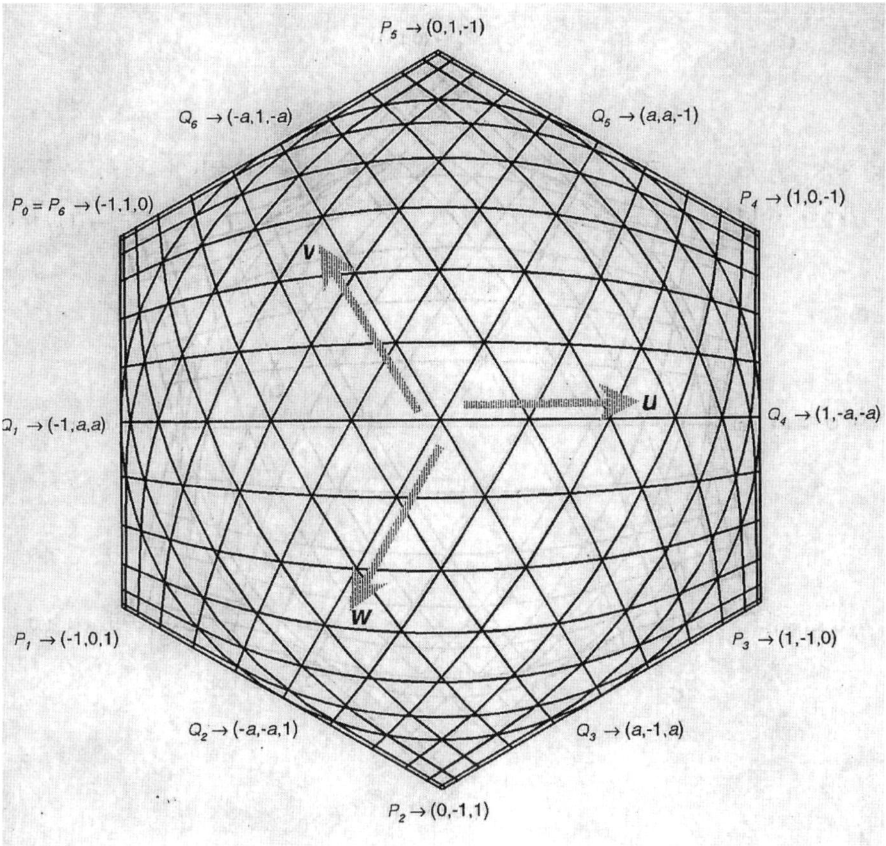
$P_5 \to (0,1,-1)$
$Q_6 \to (-a,1,-a)$
$Q_5 \to (a,a,-1)$
$P_0 = P_6 \to (-1,1,0)$
$P_4 \to (1,0,-1)$
v
u
$Q_1 \to (-1,a,a)$
$Q_4 \to (1,-a,-a)$
w
$P_1 \to (-1,0,1)$
$P_3 \to (1,-1,0)$
$Q_2 \to (-a,-a,1)$
$Q_3 \to (a,-1,a)$
$P_2 \to (0,-1,1)$

Figure 1.

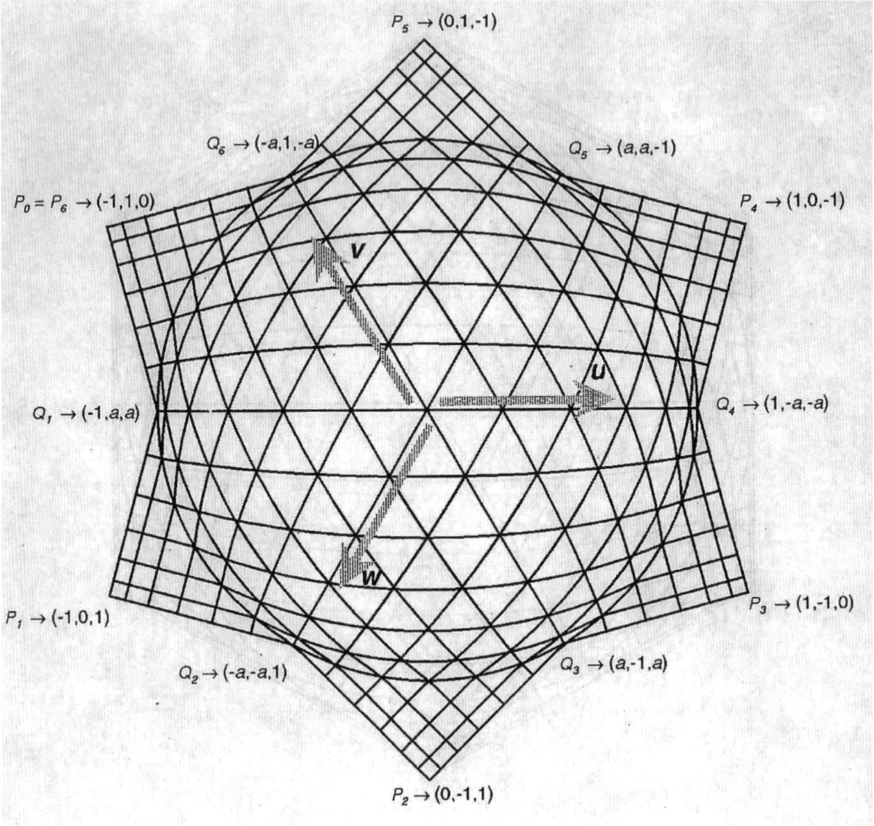
$P_5 \to (0,1,-1)$
$Q_6 \to (-a,1,-a)$
$Q_5 \to (a,a,-1)$
$P_0 = P_6 \to (-1,1,0)$
$P_4 \to (1,0,-1)$
V
U
$Q_1 \to (-1,a,a)$
$Q_4 \to (1,-a,-a)$
W
$P_1 \to (-1,0,1)$
$P_3 \to (1,-1,0)$
$Q_2 \to (-a,-a,1)$
$Q_3 \to (a,-1,a)$
$P_2 \to (0,-1,1)$

Figure 1a.

To begin with, we label and orient the six sides of Ω (see Figure 1) by orienting and tracing its boundary, starting and ending at a vertex $P_0 = P_6$, and denote each successive, oriented side by l_j ($j = 1, \cdots, 6$ *modulo* 6), in the order in which it is traced, so that l_j runs from P_{j-1} to P_j. According to the results of Section 4, the problem is equivalent to the following one. Consider any family $(u_j)_{j \in \mathbb{Z}/6\mathbb{Z}}$ of six real valued functions of Lipschitz class in Ω, such that $u_{j+3} = -u_j$, and satisfying the following four conditions.

(i) Each u_j ranges in the interval $[-1, 1]$, taking the values -1 and 1 respectively and exclusively on the sides l_j and l_{j+3}.

(ii) The functions u_1, u_3 and u_5, also renamed u, v and w, define an imbedding of class $C^{1,1}$

$$(u, v, w) : \Omega \to K^3 = \{(x, y, z) \in \mathbb{R}^3 : max\{|x|, |y|, |z|\} \leq 1\}$$

into the cube K^3 of edge length $= 2$.

(iii) The image surface is transversal to the direction of all three coordinate axes at each of its points.

(iv) The image surface $\Sigma = (u, v, w)(\Omega)$ is invariant under the group $(D_6 D_3)$ of rotations and reflections of $\mathbb{R}^3$ that map the cube K^3 and the line $\{x = y = z\}$ onto themselves.

Under these conditions, in particular, the image surface is the graph of a function $w = f(u, v)$ in a certain (as yet undetermined) subdomain $\Theta \subset \{(u, v) \in \mathbb{R}^2 : |u| \leq 1, |v| \leq 1\}$, with both first order partial derivatives $p = \frac{\partial f}{\partial u}$, $q = \frac{\partial f}{\partial v}$ strictly negative everywhere. The main problem is to find a function $f(u, v)$ that satisfies the partial differential equation (5.3), with $L(p, q-1)$ defined in (5.2), such that its graph in K^3 satisfies the symmetry and the boundary conditions described above.

In the next few paragraphs we refer directly to Figure 1 and 1a ; the latter is a schematic representation of the hexagon Ω, indicating next to each labeled point the local values taken by the triple values taken by the triple (u, v, w) of potential functions. The thick, shaded arrows indicate the gradient flow of the indicated functions u, v and w near the center of the respective systolic bands, while the thin lines suggest the web of shortest paths connecting opposite sides.

Consider first the possible configuration of the six shortest paths $(\varphi_j)_{j\in\mathbb{Z}/6\mathbb{Z}}$ in Ω where φ_j is the shortest path from the vertex P_{j-1} to the vertex P_{j+1}. Each of these six paths constitutes one of the two boundary paths of the corresponding "systolic band" of shortest paths from the side l_{j-1} to the side l_{j+2}, and is also an orbit of the gradient flow of u_{j-1}. One sees that the union of these six "boundary" shortest paths must cover the whole boundary of the hexagon Ω, because otherwise there would be open sets in Ω, meeting at most one systolic band, contradicting Lemma 2.1. Recalling the postulated hexagonal symmetry of the solution of the problem, it is easily seen that, for each $j mod.6$, an initial segment of the path φ_j must trace at least the initial "half" of the side l_j (see Fig.1, i.e. the segment of l_j from P_{j-1} to the midpoint Q_j. By symmetry, a terminal segment of the path φ_j traces the terminal half of l_{j+1}, from Q_{j+1} to P_{j+1}.

On the other hand the paths φ_j cannot lie totally in the boundary of Ω . In fact, suppose that one of the paths φ_j, and hence all six, were to be entirely contained in $\partial\Omega$; in other words, suppose that each φ_j were to just trace the two sides l_j and l_{j+1}: this would mean that since each φ_j has length 2 and describes a shortest path between any two of its points, and because of the symmetry of Ω, the function $u_1 + u_3 + u_5 = u + v + w$ would be identically zero on the whole boundary $\partial\Omega$. Note that $u + v + w$ satisfies the same linearized form of the second order equation (4.8) as w, in terms of the four variables u, v, p and q and the equation would have to be of elliptic type everywhere in the interior of Ω ; hence $u+v+w$ satisfies the conditions for the maximum-minimum principle and, since it vanishes identically of the boundary of Ω, it follows that $u+v+w = 0$ identically in Ω. Therefore the gradient flows of the functions u_j would all be straight lines, contrary to the assumption. This argument shows that each vertex P_j has a neighborhood U_j in Ω, bounded by a segment ψ_j in the middle of the path φ_j, such that U_j is 2-regular (in the sense of Section 2). By the same token, the interior Ω_0 of $\Omega\setminus(\cup_{1\leq j\leq 6} w(U_j))$, bounded by a part of $\cup_{1\leq j\leq 6}(\varphi_j)$, is 3-regular, and the functional relation $w = w(u, v)$ in Ω_0 makes $w(u, v)$ a non-trivial, elliptic solution of (4.8), with parabolic degeneration in the interior of each of the boundary arcs ψ_j. It is likely, but has not been proved, that the remainder of $\partial\Omega_0$, consisting of $\cup_j(\varphi_{j-1} \cap \varphi_j)$ reduces to the six points Q_j, as suggested in Fig. 1 and 1a.

The intrinsic geometry of the closed neighborhoods $\overline{U}_j$ of each P_j is locally euclidean, with a cartesian frame consisting of the functions (u_j, u_{j+1}), while u_{j-1} becomes essentially irrelevant, subject only to the conditions of symmetry and the inequality $Lip(u_{j-1}) \leq 1$. The shape of each $\overline{U}_j$ is better suggested by Figure 1a. In terms of this geometry, the boundary arc ψ_j is a concave curve ; on the other hand, ψ_j is a geodesic in the intrinsic geometry of $\overline{\Omega}_0$; therefore $\overline{\Omega}_0$ is a simply connected disk, bounded by a closed geodesic curve: denote by a the length of one twelfth of its circumference. The intrinsic geometry of the whole hexagon Ω then looks like the visible surface of a fried egg, with $\overline{\Omega}_0$ representing the yolk, and the albumen represented by six isosceles, right triangles with right angle sides of length $1 - a$, with a neighborhood of the hypotenuse hallowed out, leaving in its place a concave curve of length $2a$.

The length of each side of the hexagon Ω is $2 - 2a$. The geometry of the flat domains U_j implies that the positive constant a satisfies the inequalities

$$\sqrt{2} - 1 < a < \frac{1}{2} \ . \tag{6.1}$$

A preliminary calculation leads to a fairly accurate estimate of the area A of (Ω, g) of about 2.48, slightly less than the area $1.5 \cdot \sqrt{3}$ of the flat hexagon with apothem of unit length. These estimates will be applied in the next example.

7. OTHER EXAMPLES

We shall apply the example of the hexagon of the last section as a "building block" to estimate the case of an extremal isosystolic metric in a once punctured torus M. Even though this is somewhat outside of the announced scope of this paper, the complete solution of this case is important in comparing the extremal isosystolic ratios for closed surfaces of genera g and $g + 1$. At this point we shall outline the example only in the case of a torus with a "sufficiently large" hole. If (M, g) is a once punctured torus, M, including its boundary circle ∂M, with a Riemannian metric g, let $\beta(M, g)$ denote the least length of any closed path in M freely homotopic to ∂M, and introduce the parameter $c = \beta(M, g)/Sys(M, g)$ to indicate the "relative size" of the puncture. Using isoperimetric inequalities of Minkowski and Mahler, one sees that the area $(Vol(M, g)$ has a positive lower bound in the space of allowable metrics, if and only if $c < 3$: the limit case, $c = 3$, yields a degenerate metric g on M, attained as a limit of metrics on the flat torus with Loewner's metric (the quotient of $\mathbb{R}^2$ by the translation lattice of a regular hexagonal tiling) by deleting increasingly large, convex, open sets from the interior of each hexagonal tile. If c is in some open interval $]c_0, 3[$, any metric g in M with relative size of the puncture equal to c and sufficiently small isosystolic ratio is an approximation of the collapsed metric of the 1-skeleton of the regular hexagonal tiling of $\mathbb{R}^2$. One can identify therefore (M, g) as the union of six narrow corridors of comparable length, each corridor meeting both others at each end, so that all three corridors intersect in two "nodes" , whose closures are disjunct ; each node meets ∂M along three disjunct arcs. Without loss of generality, we may assume that each node minimizes the total length of its boundary, subject to the constraint of meeting ∂M in three components. Then each systole-long, closed path in M runs through two corridors and both nodes. If the metric g is an extremal isosystolic one, there must be exactly three pairs of mutually opposite homotopy classes of closed paths, within which systole-long paths may occur, so that each corridor is traversed by systolic bands belonging to pairs of opposite homotopy classes. From the results of Section 2, it follows that the interior of each corridor (unless it is collapsed to a segment) is locally flat. Therefore the gradient flows of the respective potential functions must impact the boundary of the corridor at angles 45°. In fact, the two gradients are mutually perpendicular at each interior point so that at least one is transversal to the boundary of the corridor, making an angle $\alpha \neq 0$ with the generalized normal (wherever such a normal exists). In order for a systole-long path to hit the boundary of M and continue beyond as a shortest length path, it follows from Snell's laws of geometrical optics that the metric g must be discontinuous on ∂M, so that the length of any small segment σ of ∂M, each of whose points is a point of impact by some systole-long path from the interior, with a mean incidence angle α, compared to the lower limit of arc length of smooth, interior paths approaching σ from the interior of M, has ratio approaching $\sin\alpha$. Therefore the two systolic bands in each corridor, being mutually orthogonal in the interior of M, hit ∂M at any point where there is a normal, making equal opposite angles of $45°$, as claimed.

It follows from the above description that the nodes, when the metric is locally an extremal isosystolic one, may be obtained from the metric of the hexagon Ω described in the last section, with the following modifications. First one deletes an open neighborhood of each of the vertices P_1, P_3, P_5, in the shape of an isosceles, right triangle, with short sides of length $2-4a$; next, one rescales the arc length of the hypotenuse (where the triangle has been cut away from the hexagon Ω) by dividing it uniformly by $\sqrt{2}$, so that it too then has length $2-4a$ and becomes part of the boundary of an enneagon (polygon with nine sides) ; the intrinsic, extremal area of the resulting "fried egg" is now $A-6(1-2a)^2$, or about 2.31 ; three of its sides have, as arranged, length $2-4a$, while the other six are segments of length $2a$ in the original sides of Ω. Then one attaches to each of these last six sides of the enneagon an isosceles, right triangle with hypotenuse of the same length $2a$, so that one of its shorter sides becomes a straight continuation of the side of the enneagon, previously shortened to length $2-4a$: this continuation too becomes shortened by the factor $\sqrt{2}$, so that its new length is a. The new perimeter now consists of six straight segments, alternating each one of three sides of length $2a\sqrt{2}$ (consistent with length of nearby paths in the interior) with a side of length $2-2a$, representing a "fast track", relative to the interior, by the uniform factor $1/\sqrt{2}$. The resulting configuration of the albumen surrounding the yolk of the fried egg is now connected in the interior. This describes each of the two nodes of the surface (M,g) with the extremal isosystolic metric. The total area of each node is now $A+6a(1-a)(3a-1)$.

Finally one builds the punctured torus (M,g) by connecting the two nodes with three equal corridors, each with the following description. The interior of each corridor is isometric to a euclidean rectangle of width $2a\sqrt{2}$ and length $b\sqrt{2}$ for some arbitrary constant b. One shortens first the two sides of length $b\sqrt{2}$ of each of the three rectangles, again uniformly by the factor $\sqrt{2}$, reducing their length to b ; then one attaches one of the remaining sides of each rectangle, of length $2a\sqrt{2}$, to each of the three sides of the same length, on one of the two nodes, and the opposite, free sides similarly to the other node, matching the sides attached in such a way that the resulting surface M is orientable, and has a connected, "fast track" boundary of total length $B(M,g) = 6b + 12(1-a)$, representing its boundary systole. As a result, the manifold (M,g) has total area $Vol(M,g)$, and systole $Sys(M,g)$ expressed by the following formulas:

$$\begin{aligned} Vol(M,g) &= 2A + 12(1-a)(3a-1), \\ Sys(M,g) &= 2b + 4\ . \end{aligned} \tag{7.1}$$

If the boundary systole $\beta(M,g)/Sys(M,g) = 3 - 6a/(b+2)$ is prescribed and equal to a constant $c \in \,]0,3[$, this determines b uniquely in terms of the numerical constant a estimated in (6.1), at the end of the last section. This leads to a fairly accurate estimate for the extremal isosystolic ratio for M in terms of the relative length c of the boundary, provided that c is sufficently close to 3 (roughly, $c > 2.5$). While the formulas for the estimate, eventually sharpened by numerical analysis, describe what is believed to be the extremal isosystolic ratio for the punctured torus, possibly as long as $b \geq 0$ the formula cannot remain valid as c decreases indefinitely ; there must indeed be a lower limit for c for the present model of the metric to remain extremal isosystolic ; to prescribe the relative boundary systole at smaller levels would bring the nodes relatively closer to one another, until a catastrophe occurs, and the two yolks get scrambled.

The next two examples, with which we will conclude this paper, are easier: they consist of two different metrics in a closed surface M of genus 3, that are locally flat, generated by linear solutions of the Euler-Lagrange equations (5.3) for extremal isosystolic metrics ; it seems likely that they attain, locally in the function space of metrics, relative minimum values for the isosystolic ratio. Despite a careful search, no other examples of piecewise flat, extremal isosystolic metrics have been found, for surfaces of any genus ≥ 2 ; indeed there may not exist any. In both examples, the metrics admit a large, finite group of isometries ; this makes them much easier to describe.

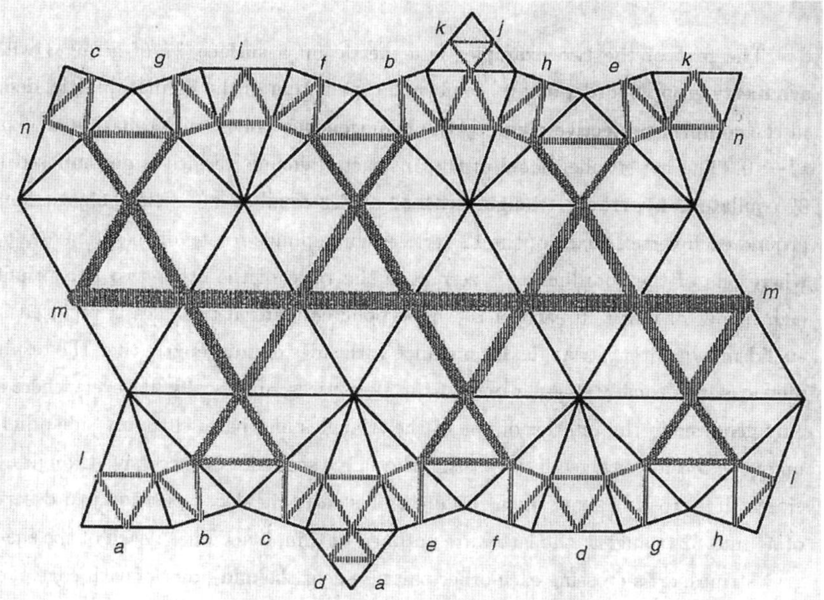

Figure 2.

Figure 3.

The first of the two examples is a metric on a surface M of genus 3 with a symmetry group of order 96 ; its conformal type is that of the Fermat quartic, defined in the complex projective plane by the homogeneous polynomial equation $x^4 + y^4 + z^4 = 0$. The hyperbolic model of the abstract Riemann surface is decomposed into 32 equilateral hyperbolic triangles with all angles equal to $\pi/4$, fitting eight triangles around each vertex ; each of the 12 vertices corresponds to one of the points (x, y, z), where one of the coordinates is zero, and the ratio of the other two is a primitive eighth root of unity. If each of the hyperbolic equilateral triangles is replaced by a euclidean equilateral triangle, for instance with sides of unit length, then the resulting metric is intrinsically singular only at the 12 vertices, and locally flat everywhere else. Any geodesic in the interior of one of the triangles and parallel to any side admits a unique extension as a geodesic, passing through a sequence of successive triangles, and closes after crossing six of them. Figure 2 illustrates the decomposition just described of M into 32 triangles: the latter are outlined in thin, black lines ; each of the shaded, grey stripes, criss-crossing each other pairwise near the mid-point of each edge of each triangle, represents the central portion of each systolic band, and the matching pairs of letters labelling the outer edges of the diagram indicate which pairs of those edges are to be identified (preserving, of course, the orientation). The surface with this metric has a systole equal to 3, total area equal to $8 \cdot \sqrt{3}$, and hence the systolic ratio is $(8\sqrt{3})/9 \sim 1.5396$. The surface M is completely covered (generically 3 times) by the traces of 16 unoriented systolic bands.

In addition there are 24 additional, isolated, unoriented, simple, closed curves of length $2\sqrt{3}$ (not shown in the figure), that represent exclusively the next smallest value attained by local systoles. Each of them consists of a cyclic sequence of four altitudes of adjoining triangles, alternately bisecting a side and the singular vertex angle (the total angle at each vertes measures 480°). One may modify the metric in M by shrinking these 24 curves uniformly by the factor $\frac{1}{2}\sqrt{3}$, giving it the length $3 = Sys(M, g)$, while leaving g unchanged elsewhere. The resulting metric, despite its 24 "fast tracks", does not alter the property of the original systolic bands achieving the minimum length within their homotopy classes. No other non contractible curve in M may be shrunk to one of length 3 without causing some of the original systole-long curves to cease being length-minimizing in their class. Therefore M , with the metric described above, and altered by shrinking the 24 curves down to length 3 has a maximal set of critical isosystolic classes consisting of 80 free homotopy classes, out of which 32 make up the essential critical classes, while the other 48 classes systolic bands consisting of just one curve each.

The second example is conformally equivalent to Felix Klein's sextic curve and has a symmetry group isomorphic to $PSL\,(2,\mathbb{Z}/7\mathbb{Z})$, of order 168 (it is the well known example achieving A. Hurwitz's upper bound for the order of the group of symmetries of such a surface). It corresponds, as a Riemann surface, to the compactification of the quotient of Poincaré's upper half-plane by the free action of the congruence subgroup

$$\Gamma_y - \mathrm{Ker}\{(PSL(2,\mathbb{Z}) \to PSLL(2,\mathbb{Z}/7\mathbb{Z})\} \subset PSL(2,\mathbb{Z})\ ;$$

the compactification involves adjoining 24 points, to form a compact Riemann surface of genus 3. The hyperbolic metric in the resulting, compact surface consists of 56 hyperbolic, equilateral triangles with angles measuring $2\pi/7$, whose vertices are the 24 points adjoined in the compactification. As before, each of these 56 triangles is replaced by a euclidean equilateral triangle of unit side length. The resulting surface, illustrated diagrammatically in Figure 3, if locally flat, like the one in the previous example, except for the 24 singular points. The systole is achieved by any geodesic path parallel to any side of any triangle ; its completion as a geodesic always closes after crossing 8 triangles (see Fig. 3). Thus the area of M with this metric is $14\cdot\sqrt{3}$, the systole has length 4, and therefore the systolic ratio is $(7\sqrt{3})/8 \sim 1.5155$, or about 1.5% lower than the one for the previous metric. It seems very likely that the systolic ratio $(7\sqrt{3})/8$ achieved by this metric is the absolute minimum value for a surface of genus 3, or else is probably very close to the minimum. With this metric, the surface is covered almost everywhere three times by 21 bands of unoriented systole-long paths. The next smallest value of a local systole is $3\sqrt{3}$, achieved by 28 bands of unoriented closed curves (not shown in the figure), also covering M, generically 3 times: The excess of this "second" systole over the principal value 4 is too great, to permit any "fast track" shrinking of curves in any other non-trivial, free homotopy class down to length 4, without destroying the minimal length property of the original systole-long curves.

It may be interesting to note that the conformal equivalence classes of the metrics described in the two above examples are quite far apart in the space of moduli of surfaces of genus 3. Assuming that both metrics represent relative minimizing points for the isosystolic ratio, this indicates that the isosystolic ratio, as a functional on the space of all metrics in a surface M, is not a convex variational functional.

It was mentioned in the introduction that it seems unlikely that, for surfaces of genus $g \geq 2$ and $\neq 3$, the extremal isosystolic metrics may be piecewise flat, as it is in the last two examples ; in the case $g = 2$ the non-existence of a piecewise flat, extremal isosystolic metric may be shown by a reduction to a symmetry argument. Some attempts, so far without success, have been made in the case $g = 4$, using metrics with a symmetry group of order 120.

In conclusion, I wish to recall that a great deal of what I originally learned about the isosystolic problem has been from lectures by Marcel Berger. For this and for many other fruitful ideas that conversations with him have suggested, I wish to express my appreciation, and thereby end this exposé.

BIBLIOGRAPHY

[1] C. Bavard, *Inégalité isosystolique pour la bouteille de Klein,* Math. Ann. **274** (1986), 439–441.

[2] C. Bavard, *Inégalités isosystoliques conformes pour la bouteille de Klein,* Geometriae Dedicata **27** (1988), 349–355.

[3] M. Berger, *Du côté de chez Pu,* Ann. Sci. Ecole Norm. Sup. Paris **5** (1972), 1–44.

[4] M. Berger, *A l'ombre de Loewner,* Ann. Sci. Ecole Norm. Sup. Paris **5** (1972), 241–260.

[5] E. Cartan, *Les espaces métriques fondés sur la notion d'aire,* Exposés de Géométrie, I, Paris, Hermann (1933).

[6] M. Gromov, *Filling Riemannian Manifolds,* J. Differential Geom. **18** (1983), 1–147.

[7] P.M. Pu, *Some inequalities in certain nonorientable Riemannian manifolds,* Pac. Jour. Math. **2** (1952), 55–71.

Affine Geometry, Curve Flows, and Invariant Numerical Approximations

Eugenio Calabi*, Peter J. Olver†, and Allen Tannenbaum‡

Adv. Math. 124, No. 1, 154–196 (1996)

A new geometric approach to the affine geometry of curves in the plane and to affine-invariant curve shortening is presented. We describe methods of approximating the affine curvature with discrete finite difference approximations, based on a general theory of approximating differential invariants of Lie group actions by joint invariants. Applications to computer vision are indicated. © 1996 Academic Press, Inc.

1. INTRODUCTION

This paper is concerned with a modern presentation of the basic theory of affine geometry in the plane and related questions of invariant approximations of affine differential invariants. Although affine geometry does not have as long or as distinguished a history as either Euclidean or

* Supported in part by NSF Grant DMS 92-03398. E-mail address: calabi@math.upenn.edu.
† Supported in part by NSF Grant DMS 95-00931. E-mail address: olver@ima.umn.edu.
‡ Supported in part by NSF Grant ECS-9122106, by the Air Force Office of Scientific Research Grant F49620-94-1-00S8DEF, by Army Research Office Grants DAAL03-91-G-0019, DAAH04-93-G-0332, and DAAH04-94-G-0054, and by Image Evolutions, Ltd. E-mail address: tannenba@ee.umn.edu.

DOI: 10.1006/aima.1996.0081

projective geometry, its recent importance in the rapidly developing area of computer vision warrants a modern reassessment of the basics. Affine geometry received its first comprehensive treatment in the seminal work of Blaschke [6], who was inspired by Klein's general Erlanger Programm, which provided the foundational link between groups and geometry, and Einstein's theory of relativity. (The latter motivation, though, is, to a modern thinker, more mysterious.) Affine geometry is based on the affine, or on the unimodular affine group. In the plane, affine geometry is the "geometry of area," just as Euclidean geometry is the geometry of distance. Besides the basic work of Blaschke, we refer the reader to [22] and to the more modern texts [14, 27] for a more comprehensive treatment of the subject.

Even though our primary focus is mathematical, a key motivation for pursuing this line of research comes from certain practical issues in computer vision. Indeed, certain visually based symmetry groups and their associated differential invariants have, in recent years, assumed great significance in the study of computer vision and image processing. One such problem is that of finding and recognizing a planar object (which may be occluded) whose shape has been transformed by a geometric viewing transformation (that is, an element of the projective group acting on the plane). This common type of shape recognition task naturally brings in the use of invariants under various groups of viewing transformations. Research in model-based shape analysis and recognition has already resulted in many useful products, such as optical character recognizers, handwriting recognition systems for computers, and printed-circuit board visual inspection systems. Space limitations preclude us from discussing direct applications of our results to computer vision, which shall be dealt with in subsequent papers.

In the practical application of invariant theory to computer vision, a robust and efficient numerical computation is crucial. We are interested in numerical approximations to differential invariants which are themselves invariant under the transformation group in question. This will enable us to compute the "differential invariant signatures" for plane curves in a manner which will be unaffected by group transformations. The ideal approximation will be geometric, in the sense that it can be computed by specifying a finite number of points, and hence its invariance means that it must be re-expressed in terms of the joint invariants of the group in question. Thus our general question is how to systematically utilize joint invariants to approximate differential invariants. One motivation comes from the results of Green [11], generalized in [17], that relate the number of differential invariants of curves to the number of joint invariants of the group action; the numerological implications of Green's results are thus to be given an analytical justification.

The construction of efficient and practical numerical approximations to differential invariants is a nontrivial problem in that the more important differential invariants, such as the affine and projective curvatures, depend on high order derivatives of the parametrizing functions of the curve. The theory of "noise resistant" differential invariants developed by Weiss [29] provides one approach to this problem. Weiss replaces the higher order differential invariants by lower order derivatives, but, in our view, this is only a partial resolution of the difficulty. In our approach, a fully noise-resistant finite-difference approximation to the affine (and Euclidean) curvatures is proposed. Another approach to invariant numerical schemes for solving partial differential equations having a prescribed symmetry group appears in the work of Dorodnitsyn [7, 8].

Our approach to approximating differential invariants and invariant differential equations is governed by the following philosophy. Consider a group G acting on a space E. We are particularly interested in how the geometry, in the sense of Klein, induced by the transformation group G applies to (smooth) curves[1] $\Gamma \subset E$. A differential invariant I of G is a real-valued function, depending on the curve and its derivatives, which is unaffected by the action of G. The simplest example is the Euclidean curvature of a plane curve, which is invariant under the Euclidean group consisting of translations and rotations. The theory of differential invariants dates back to the original work of Lie [16]; see [17] for further historical remarks and a modern exposition. In order to construct a numerical approximation to the differential invariant I, we employ a finite difference approach and use a mesh or discrete sequence of points $P_i \in \Gamma$, $i \in 0, 1, 2, \ldots$, to approximate the curve, and use appropriate combinations of the coordinates of the mesh points in our approximation scheme. The approximation will be invariant under the underlying group G, and hence its numerical values will not depend on the group transformations, provided it depends on the joint invariants of the mesh points. In general, a joint invariant of a group action on E is a real valued function $J(P_1, \ldots, P_n)$ depending on several points $P_i \in E$ which is unaffected by the simultaneous action of G on the points, so $J(g \cdot P_1, \ldots, g \cdot P_n) = J(P_1, \ldots, P_n)$. Again, the simplest example is provided by the Euclidean distance $\mathbf{d}_e(P, Q)$ between points in the plane, which depends on two points. Thus, *any G-invariant numerical approximation to a differential invariant must be governed by a function of the joint invariants of* G. For instance, any Euclidean invariant approximation to the curvature of a plane curve must be based on the distances between the mesh points. Such a formula is known—see Theorem 3.2 below. In this paper, we illustrate this general method by deriving

[1] More generally, we can develop the same theory for surfaces or arbitrary submanifolds of the space E. In this work, just for simplicity, we restrict our attention to curves.

a fully affine invariant finite-difference approximation to the affine curvature of a plane curve. The resulting Taylor series expansion leads us to a general conjecture on the approximation of group-invariant curvatures for arbitrary regular transformation groups in the plane. We will also indicate some methods for determining similar approximations to higher order invariants.

Motivated by such questions, in this paper we will give a detailed discussion of equiaffine geometry, which includes new geometric approaches to the equiaffine normal and curvature. We discuss finite difference approximations of Euclidean and affine differential invariants. Finally, we provide some new, remarkable solutions to the affine curvature flow. Even though this paper is essentially devoted to the derivation of a number of new results in the theory of affine invariants, we will also provide a number of background results to make this work accessible to the largest possible audience of mathematicians and researchers in computer vision, so that the paper will also have a tutorial flavor.

2. SOME FUNDAMENTAL CONCEPTS

When one treats Euclidean or affine geometry from the analytic standpoint, one must deal with two distinct spaces: the *space of points* (the Euclidean space proper), denoted E, and the finite-dimensional real[2] *vector space* TE consisting of translations (or displacements) of E. Within the space of points, there is one main operation—subtraction: Given two points $P, Q \in E$, the object $v = Q - P$ is the unique displacement vector in TE mapping E onto itself that takes the point P to the point Q. The group of transformations of E that preserve this structure is known as the *affine group*, denoted by $\mathbf{A}(n)$ or $\mathbf{A}(n, \mathbb{R})$, where n is the dimension of E. An element of $\mathbf{A}(n)$ consists of a linear transformation $A \in \mathbf{GL}(n)$, which operates on TE, coupled with a displacement vector $b \in TE$; the full action on the point space takes the form $P \mapsto AP + b$. Note that this induces the purely linear action $v \mapsto Av$ on the displacement vector space, and thus it underlies the necessity of distinguishing between E and TE.

An affine coordinate system on E is prescribed by an *affinely independent set* of points $(P_0, P_1, ..., P_n)$ in E, meaning that the displacement vectors $e_i = P_i - P_0$ form a basis of E. A displacement vector $v = \sum_k y^k e_k \in TE$ is identified with the coordinate n-tuple $(y^1, ..., y^n)$, while we associate points $P \in E$ with their relative displacement vectors $v_P = P - P_0 = \sum_i x^i e_i$. In this way, we identify the affine group $\mathbf{A}(n) \simeq \mathbf{GL}(n) \ltimes \mathbb{R}^n$ with the semidirect product of the general linear group with the displacement or translation subgroup.

[2] One can, of course, develop much of the general theory over the complex numbers or other fields. Again, for simplicity, we restrict our attention to real geometry throughout.

If TE has dimension n, then the space $\bigwedge^n TE$ of volume forms on E is a one-dimensional vector space. The affine transformations act on $\bigwedge^n TE$ according to the determinantal representation $(A, b) \mapsto \det A$. Given two sets of points $(P_0, P_1, ..., P_n)$, $(Q_0, Q_1, ..., Q_n)$, not necessarily distinct, such that $(P_0, P_1, ..., P_n)$ is an affinely independent set, there is a unique affine endomorphism of E that maps P_i onto Q_i for each $1 \leqslant i \leqslant n$. Its homogeneous linear part, i.e., the linear endomorphism of TE taking each $v_i = P_i - P_0$ to $w_i = Q_i - Q_0$, has a determinant that, if nonzero, expresses the ratio of (oriented) volumes of the n-parallelotope determined by the w's to that determined by the v's, or, equivalently, the ratio of volumes of the n-simplex spanned by the Q's to that spanned by the P's. Thus the full affine group $\mathbf{A}(n)$ preserves the *ratios* between volumes of subsets of E, or of TE, while volumes themselves are *relative invariants* of the group.

An *orientation* on TE is prescribed by the choice of one of the two connected components of $\bigwedge^n TE \backslash \{0\}$; the orientation-preserving affine transformations are those having positive determinant. The notion of volume on E is fixed by specifying what constitutes a "unit volume," which is represented by a fixed form $\Omega_0 = e_1 \wedge e_2 \wedge \cdots \wedge e_n \in \bigwedge^n TE$, where $\{e_1, ..., e_n\}$ form a basis of TE, and the volume of the n-paralleltope spanned by the e_i's is normalized to be 1. In this case, the oriented volume of the parallelotope determined by the displacement vectors $v_i = \sum_k y_i^k e_k$, $i = 1, ..., n$, is calculated by the fundamental determinantal bracket expression

$$[v_1, ..., v_n] = \det \begin{vmatrix} y_1^1 & y_1^2 & \cdots & y_1^n \\ y_2^1 & y_2^2 & \cdots & y_2^n \\ \vdots & \vdots & \ddots & \vdots \\ y_n^1 & y_n^2 & \cdots & y_n^n \end{vmatrix}. \tag{2.1}$$

Similarly, the volume of the n-simplex having vertices $P_0, P_1, ..., P_n$ in E, with P_i having coordinates $(x_i^1, ..., x_i^n)$ with respect to some affine coordinate system, is given by

$$\Delta(P_0, P_1, ..., P_n) = \frac{1}{2^n} [P_0, P_1, ..., P_n], \tag{2.2}$$

where

$$\begin{aligned} &[P_0, P_1, ..., P_n] \\ &\quad = [P_1 - P_0, P_2 - P_0, ..., P_n - P_0] \\ &\quad = \det \begin{vmatrix} x_1^1 - x_0^1 & x_1^2 - x_0^2 & \cdots & x_1^n - x_0^n \\ x_2^1 - x_0^1 & x_2^2 - x_0^2 & \cdots & x_2^n - x_0^n \\ \vdots & \vdots & \ddots & \vdots \\ x_n^1 - x_0^1 & x_n^2 - x_0^2 & \cdots & x_n^n - x_0^n \end{vmatrix} \end{aligned}$$

$$= \det \begin{vmatrix} x_0^1 & x_0^2 & \cdots & x_0^n & 1 \\ x_1^1 & x_1^2 & \cdots & x_1^n & 1 \\ x_2^1 & x_2^2 & \cdots & x_2^n & 1 \\ \vdots & \vdots & \ddots & \vdots & \vdots \\ x_n^1 & x_n^2 & \cdots & x_n^n & 1 \end{vmatrix}. \tag{2.3}$$

Note particularly that, in an n-dimensional affine space, the respective bracket expressions (2.1), (2.3) depend on n displacement vectors, but $n+1$ points. Restricting the group of affine transformations to those that preserve volume produces the so-called *equiaffine*, or *unimodular affine* transformation group, denoted by $\mathbf{SA}(n) \simeq \mathbf{SL}(n) \ltimes \mathbb{R}^n$, consisting of all pairs (A, b) where $\det A = 1$ and $b \in TE$. The associated equiaffine geometry in E and TE will form the principal subject of this paper.

In Euclidean geometry, one endows the displacement vector space TE with the additional structure, determined by a *norm* $v \mapsto |v|$. The geometric properties of the Euclidean norm come from the fact that it is characterized as the square root, $|v| = \sqrt{\langle v; v \rangle}$, of a positive definite quadratic form, associated to a symmetric, bilinear, scalar product $\langle v; w \rangle$. The norm on the displacement space TE induces the Euclidean distance[3] $\mathbf{d}_e(P, Q) = |Q - P|$ between pairs of points in E. The group of Euclidean motions is the set of all transformations of E that preserve the norm in TE. It has the form $\mathbf{E}(n) \simeq \mathbf{O}(n) \ltimes \mathbb{R}^n$, being a semidirect product between the orthogonal group, consisting of rotations and reflections, along with the translations. Choosing an orientation, which amounts to a choice of an orthonormal basis $\{e_1, \ldots, e_n\}$ of TE, restricts us to the proper (or unimodular) Euclidean motions of E, which excludes the reflections, and so is given by $\mathbf{SE}(n) \simeq \mathbf{SO}(n) \ltimes \mathbb{R}^n$.

In general, given a group G acting on a space M, by an *invariant* of G we mean a real-valued function $I: M \to \mathbb{R}$ which is unaffected by the group action: $I(g \cdot x) = I(x)$ for all $x \in M$, $g \in G$. For example, the norm $|v|$ defines an invariant for the Euclidean group action on the displacement space TE. On the other hand, since the action on space of points includes the translations, there are no (non-constant) invariants of the Euclidean group action on E itself. In this case, we must look at invariants depending on more than one point. In general, a *joint invariant* of a group action is an invariant of the product action of G on the m-fold Cartesian product $M \times \cdots \times M$. Thus $I(x^1, \ldots, x^m)$ is a joint invariant if and only if

[3] We shall consistently employ the subscript $_e$ for Euclidean invariant quantities, so as to distinguish them from the affine and equiaffine invariants that are the primary focus of this paper.

$I(g \cdot x^1, ..., g \cdot x^m) = I(x^1, ..., x^m)$ for all $g \in G$. The simplest joint invariant of the Euclidean group acting on E is the distance function $\mathbf{d}_e(P, Q)$. In fact, according to [30], every joint invariant of the Euclidean group can be written in terms of the distances between pairs of points. For example, the inner product $v \cdot w = (P - P') \cdot (Q - Q')$ between two displacement vectors can be re-expressed via the Law of Cosines: $v \cdot w = \frac{1}{2}\{|v - w|^2 - |v|^2 - |w|^2\}$. Further, since the Euclidean group is a subgroup of the affine group, any (equi-)affine invariant is automatically a Euclidean invariant, and hence can also be rewritten in terms of Euclidean distances. Thus, the volume $|\Pi(w_1, ..., w_n)|$ of the parallelotope $\Pi(w_1, ..., w_n)$ spanned by n displacement vectors $\{w_1, ..., w_n\} \in TE$ has its square rationally determined by the mutual scalar products:

$$|\Pi(w_1, ..., w_n)|^2 = \det(\langle w_i; w_j \rangle). \tag{2.4}$$

In the case of the unimodular affine group, there are no non-constant invariants on either E or TE. The simplest joint invariant associated with the equiaffine group action on TE is the fundamental bracket (2.1) governing the volume element. See Weyl [30] for a proof that the brackets constitute a complete set of joint affine invariants for displacement vectors, meaning that any equiaffine joint invariant can be written as a function of the various brackets between sets of n displacement vectors. We note that the brackets are *not* algebraically independent; their functional interrelationships are completely governed by the fundamental system of syzygies

$$\sum_{i=0}^{n} (-1)^k [v_0, ..., \widehat{v_k}, ..., v_n][v_k, w_1, ..., w_{n-1}] = 0, \tag{2.5}$$

valid for any set of displacement vectors $v_0, ..., v_n, w_1, ..., w_{n-1}$. Similarly, the fundamental joint invariants of the action of $\mathbf{SA}(n)$ on E itself are the simplex volumes (2.2) prescribed by $n+1$ points in E. Besides the syzygies induced by the displacement bracket syzygies (2.5), the point bracket expressions are subject to an additional linear syzygy

$$[P_0, P_1, ..., P_n] = \sum_{i=0}^{n} [P_0, ..., P_{k-1}, Q, P_{k+1}, ..., P_n], \tag{2.6}$$

valid for any $n+2$ points $P_0, P_1, ..., P_n, Q$. Finally, in the case of the full affine group $\mathbf{A}(n)$, relative ratios of brackets (or volumes) provide the required joint invariants.

3. EUCLIDEAN CURVATURE AND CURVE FLOWS

We now specialize to the Euclidean geometry of the plane, so that E denotes the two-dimensional Euclidean space, with displacement space TE.

If we introduce coordinates on E via the choice of an origin $O \in E$ and orthonormal basis e_1, e_2 of TE, then each point $A \in E$ can be identified with its coordinates $(x_A, y_A) \in \mathbb{R}^2$, such that $A - O = x_A e_1 + y_A e_2$. The basic equiaffine invariant geometric quantity is the area of a displacement parallelogram

$$[v, w] = v \wedge w = \det \begin{vmatrix} x_v & y_v \\ x_w & y_w \end{vmatrix}. \tag{3.1}$$

We note that, in accordance with the general theory, the affine-invariant area of the triangle having vertices A, B, C, which is

$$\Delta(A, B, C) = \tfrac{1}{2}[A, B, C] = \tfrac{1}{2}(B - A) \wedge (C - A) = \tfrac{1}{2} \det \begin{vmatrix} x_A & y_A & 1 \\ x_B & y_B & 1 \\ x_C & y_C & 1 \end{vmatrix}, \tag{3.2}$$

cf. Eqs. (2.2), (2.3), can be written in terms of their Euclidean distances $a = \mathbf{d}_e(A, B)$, $b = \mathbf{d}_e(B, C)$, $b = \mathbf{d}_e(C, A)$, via the well known semi-perimeter formula

$$[A, B, C] = \sqrt{s(s-a)(s-b)(s-c)}, \qquad \text{where} \quad s = \tfrac{1}{2}(a + b + c). \tag{3.3}$$

Consider a regular, smooth plane curve $\Gamma \subset E$ of class C^2. The Euclidean curvature of Γ at a point $B \in \Gamma$ is defined as the reciprocal $\kappa_e = 1/r$ of the radius of the osculating circle to Γ at B. Let us choose an affine coordinate system (x, y) on E and parametrize the curve by a pair of smooth functions $\mathbf{x}(r) = (x(r), y(r))$, where the parameter r ranges over an interval $I \subset \mathbb{R}$. In terms of the parametrization, then, the Euclidean curvature has the well-known formula

$$\kappa_e = \frac{\mathbf{x}_r \wedge \mathbf{x}_{rr}}{|\mathbf{x}_r|^3}, \tag{3.4}$$

in which subscripts denote derivatives. In particular, if we choose a coordinate system such that the part of Γ near B is represented by the graph of a function $y = u(x)$, then

$$\kappa_e = \frac{u_{xx}}{(1 + u_x^2)^{3/2}}. \tag{3.5}$$

In this form, κ_e describes the simplest differential invariant of the Euclidean group in the plane [17]. The Euclidean arc length parameter is defined as $ds_e = \sqrt{1 + u_x^2}\, dx$, the right-hand side representing the simplest invariant one-form for the Euclidean group. The arc length integral $\int_\Gamma ds$ determines the Euclidean distance traversed along the curve. Higher order differential invariants are provided by the successive derivatives of curvature with respect to arc length. In fact, the functions

$$\kappa_e, \frac{d\kappa_e}{ds_e}, \frac{d^2\kappa_e}{ds_e^2}, \frac{d^3\kappa_e}{ds_e^3}, \ldots, \tag{3.6}$$

provide a complete list of differential invariants for the Euclidean group, in the sense that any other differential invariant can be (locally) expressed as a function of the fundamental differential invariants (3.6).

As a first illustration of our general philosophy of approximating differential invariants by joint invariants, we describe how to use standard geometrical constructions to obtain a numerical approximation to the Euclidean curvature that is unaffected by rigid motions, so that any translated or rotated version of the curve will provide precisely the same numerical approximation for its curvature. We first approximate the parametrized curve $\mathbf{x}(r) = (x(r), y(r))$ by a sequence of mesh points $P_i = \mathbf{x}(r_i)$, not necessarily equally spaced. Our goal is to approximate the Euclidean curvature of Γ by a Euclidean invariant numerical approximation based on the mesh points. Clearly, because the curvature is a second order differential function, the simplest approximation will require three mesh points. (A deeper, but related, reason for this is because the joint invariants of the Euclidean group are the distances between two points, so that one can only produce numerical joint invariant approximations by comparing the joint invariants involving three or more points.)

With this in mind, we now derive the basic approximation formula for the Euclidean curvature. Let A, B, C be three successive points on the curve Γ such that the Euclidean distances are $a = \mathbf{d}_e(A. B)$, $b = \mathbf{d}_e(B, C)$, $c = \mathbf{d}_e(A, C)$, which are assumed to be small; see Fig. 1. The key idea is to use the circle passing through the points A, B, C as our approximation to the osculating circle to the curve at B. Therefore, the reciprocal of its radius $r = r(A, B, C)$ will serve as an approximation to the curvature of the curve at B. We can apply Heron's formula to compute the radius of the circle passing through the points A, B, C, leading to the exact formula

$$\tilde{\kappa}_e(A, B, C) = 4\,\frac{\Delta}{abc} = 4\,\frac{\sqrt{s(s-a)(s-b)(s-c)}}{abc}, \tag{3.7}$$

cf. (3.3), for its curvature. Since formula (3.7) only depends on the Euclidean distances between the three points, it provides us with a completely Euclidean invariant numerical approximation to the curvature of Γ at the middle point B. In other words, the approximation for two curves related by a Euclidean motion will be *identical.*

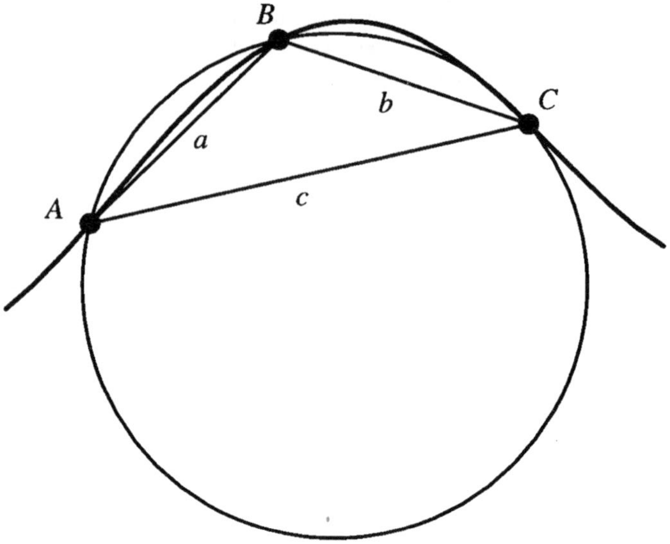

FIG. 1. Euclidean Curvature Approximation.

We now need to analyze how closely the numerical approximation $\tilde{\kappa}_e(A, B, C)$ is to the true curvature $\kappa_e(B)$ at the point B. Our analysis is based on the following series expansion of the distance c in terms of the other two distances a and b, which are assumed small.

THEOREM 3.1. *Let A, B, C be three successive points on the curve Γ, and let $a = \mathbf{d}_e(A, B)$, $b = \mathbf{d}_e(B, C)$, $c = \mathbf{d}_e(A, C)$ be their Euclidean distances. Let $\kappa_e = \kappa_e(B)$ denote the Euclidean curvature of Γ at the middle point B. Then the following expansion is valid*:

$$c^2 = (a+b)^2 - \frac{1}{4}\, ab(a+b)^2\, \kappa_e^2 + \frac{1}{6}\, ab(a+b)^2\, (a-b)\, \kappa_e \frac{d\kappa_e}{ds_e}$$
$$- \frac{1}{24}\, ab(a+b)(a^3+b^3)\, \kappa_e \frac{d^2\kappa_e}{ds_e^2} - \frac{1}{36}\, ab(a+b)^2\, (a-b)^2 \left(\frac{d\kappa_e}{ds_e}\right)^2$$
$$- \frac{1}{64}\, ab(a+b)^2\, (a-b)^2\, \kappa_e^4 + \cdots. \tag{3.8}$$

The omitted terms involve powers of the distances a, b of order $\geqslant 7$.

Proof. This is found by a direct, albeit complicated, Taylor series expansion. We represent the curve between A and C as the graph of $y = u(x)$, which, assuming the three points are sufficiently close, can always be arranged via a Euclidean motion. The points can be assumed to be $A=(h, u(h))$, $B=(0, 0=u(0))$, and $C=(k, u(k))$, with $h<0<k$ if B is the middle point. We then expand $c=\sqrt{(k-h)^2+(u(k)-u(h))^2}$ as a Taylor series in powers of h, k. Then we substitute for h and k their expansions in powers of a, b, obtained by inverting the Taylor series for $a=\sqrt{h^2+u(h)^2}$ and $b=\sqrt{k^2+u(k)^2}$. (The computations are quite complicated, and were done with the aid of the computer algebra system Mathematica.) Q.E.D.

Remark. Since a, b, and c are Euclidean invariants, *every* coefficient of the powers $a^m b^n$ in the full expansion of c must be a Euclidean differential invariant, and hence a function of κ_e and its arc length derivatives. The precise formulas for the coefficients were found by inspection—we do not know the general term in the expansion (3.8).

We now substitute the expansion (3.8) into Heron's formula (3.7) to obtain the following expansion for the numerical approximation.

THEOREM 3.2. *Let A, B, C be three successive points on the curve Γ, and let a, b, c be their Euclidean distances. Let $\kappa_e=\kappa_e(B)$ denote the Euclidean curvature at B. Let $\tilde{\kappa}_e=\tilde{\kappa}_e(A, B, C)$ denote the curvature of the circle passing through the three points. Then the following expansion is valid:*

$$\tilde{\kappa}_e=\kappa_e+\frac{1}{3}(b-a)\frac{d\kappa_e}{ds_e}+\frac{1}{12}(b^2-ab+a^2)\frac{d^2\kappa_e}{ds_e^2}+\cdots. \tag{3.9}$$

In particular, if we choose the points to be equally spaced, meaning that $a=b$ (not that their x coordinates are equally spaced), then the first error term in the approximation (3.9) is of second order.

Remark. The same general method can also be used to find Euclidean-invariant numerical approximations for computing the higher order differential invariants $d\kappa_e/ds_e$, etc., using more points and more distances, as needed.

In recent years, the analysis and geometrical and image processing applications of curve flows based on curvature have received a lot of attention. We consider a one-parameter family of curves $\mathbf{x}(\cdot, t)$ that satisfy a geometric evolution equation. Here t represents either the time, or, in computer vision applications, a scale parameter. The partial differential equation governing the time evolution of the curve family is assumed to be geometric, meaning that it does not depend on the precise mode of parametrizing the family of curves, but, rather, on purely intrinsic geometric quantities associated with the curve at a given time. The most fundamental of these geometric flows is the Euclidean curve shortening flow, in which one moves in the normal direction to the curve according to its Euclidean curvature:

$$\frac{d\mathbf{x}}{dt} = \kappa_e \mathbf{n}_e. \tag{3.10}$$

Here $\mathbf{n}_e$ denotes the Euclidean inward normal. When the curve is given as the graph of a function $y = u(x, t)$, the Euclidean curve flow takes the form:

$$u_t = \frac{u_{xx}}{1 + u_x^2}. \tag{3.11}$$

This flow has the effect of shrinking the Euclidean arc length of the curve as rapidly as possible, cf. [10]. The Euclidean curve shortening flow is of great interest in differential geometry, computer vision, and other fields, and has been studied by many authors. See [3] for applications to image enhancement, and [13] for applications to the theory of shape in computer vision. Clearly the flow (3.10) is invariant under the Euclidean group acting on the plane, and so a fully invariant numerical integration must rely on Euclidean joint invariants, meaning intermesh distances.

Two particular types of solutions are of immediate interest. First, if the initial curve is a circle, with constant curvature, then it remains circular, with its radius satisfying $r_t = 1/r$, so that the curve shrinks to a point in a finite time. The results of Gage and Hamilton [9] and Grayson [10] show that *any* smooth, embedded, closed curve converges to a round point when deforming according to the flow (3.10). This means that, first, if the initial curve is not convex, it becomes convex, and then the resulting convex curve shrinks to a point, asymptotically becoming circular before disappearing.

A second class of solutions are the "grim reapers" which are found by assuming that the curve has constant velocity. Taking the velocity to be in the vertical direction and using the graphical form (3.11) means that we assume that $u_t = c$ where c is a constant. The resulting Euclidean-invariant ordinary differential equation

$$\frac{u_{xx}}{1 + u_x^2} = c$$

can be readily integrated, leading to the general form

$$u(x, t) = -\frac{1}{c} \log[\cos c(x - x_0)] + c(t - t_0),$$

for constants x_0 and t_0, for the grim reaper. At this point, we conclude our brief survey of Euclidean curve flows, and turn to our main subject of interest.

4. THE EQUIAFFINE LENGTH INTEGRAL

We now turn to our primary focus: the affine geometry of curves in the plane. Affine invariants are not suited for the study of curves with inflection points; therefore we shall deal only with *strongly locally convex* curves. In this section, all curves will be assumed to have not only no inflection points, but to be continuously differentiable with respect to suitable parameters of order up to 5, although in the next few paragraphs derivatives of orders at most 3 will appear. Many of our constructions will refer to a sufficiently short piece of the convex curve, in the following precise sense.

DEFINITION 4.1. Let Γ be a smooth plane curve without inflection points. A compact arc $\Gamma(A, B) \subset \Gamma$, i.e., with both end points A, B included, will be called a *short arc* if no two tangent lines to $\Gamma(A, B)$ are mutually parallel, including the tangents at the end points.

This condition, in Euclidean geometry, is equivalent to the statement that the total turning angle of the tangential direction of $\Gamma(A, B)$ is less than half a revolution; in terms of purely affine invariants of Γ, the property means that the arc $\Gamma(A, B)$ may be inscribed in a *support triangle*, which is bounded by the segment joining the endpoints A, B, and by the tangent lines at the two endpoints.

DEFINITION 4.2. Let Γ be a strongly convex curve, and let $\Gamma(A, B)$ be a short arc of Γ with end points A and B. The *support point* of $\Gamma(A, B)$ is the point P where the two end-point tangent lines intersect. The *support triangle* of $\Gamma(A, B)$ is defined as the triangle $\mathbf{T}(A, B) = APB$; see Fig. 2.

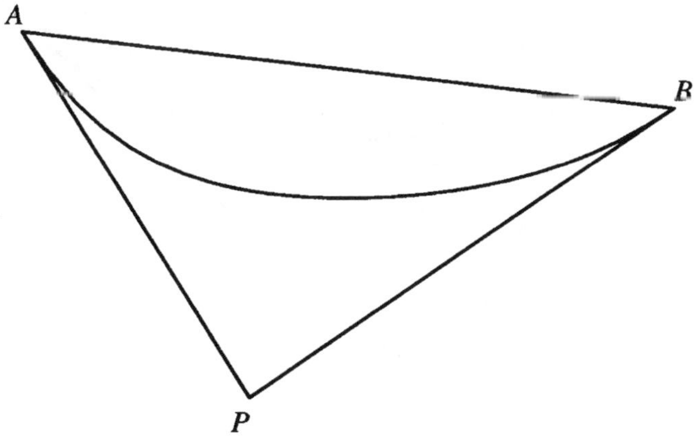

FIG. 2. Support point and support triangle.

Note that, by convexity, the support triangle circumscribes the short arc Γ. We regard the (positive) area of the support triangle,

$$\mathbf{A}(A, B) = |\Delta(A, P, B)| = \tfrac{1}{2}|[A, P, B]|,$$

cf. (3.2), as an equiaffinely invariant "indicator" of the distance between the (non-oriented) *tangent line elements* (A, AP) and (B, BP). More precisely, we want to introduce an equiaffinely invariant distance function between the two line segments, so that, if we break the arc $\Gamma(A, B)$ in two at any intermediate point C and compare the distance from the tangent line element at A to the one at C, with the distance from the tangent line element at C to the one at B, we obtain, asymptotically for very short smooth arcs, the original distance defined from $\mathbf{A}(A, B)$. It is obvious that the area $\mathbf{A}(A, B)$ itself does not have this asymptotic property. However, its cube root does, as the following theorem of Blaschke [6] shows.

THEOREM 4.3. *Let $\Gamma(A, B)$ be a short arc of a strongly convex curve Γ, joining a point A to a point B, and let $C \in \Gamma$ be another point, interior to the arc $\Gamma(A, B)$. Draw tangent lines to Γ at each of the three points, as well as the three chords joining them, as shown in Fig.* 3. *We let P denote the support point for the arc $\Gamma(A, B)$, and P_A, P_B the respective support points for the respective sub-arcs $\Gamma(A, C)$, $\Gamma(C, B)$; thus P_A, P_B are the points where the tangent line at C intersects the tangents AP and BP. Each of the three support triangles $\mathbf{T}(A, B) = APB$, $\mathbf{T}(A, C) = AP_AC$, $\mathbf{T}(C, B) = CP_BB$*

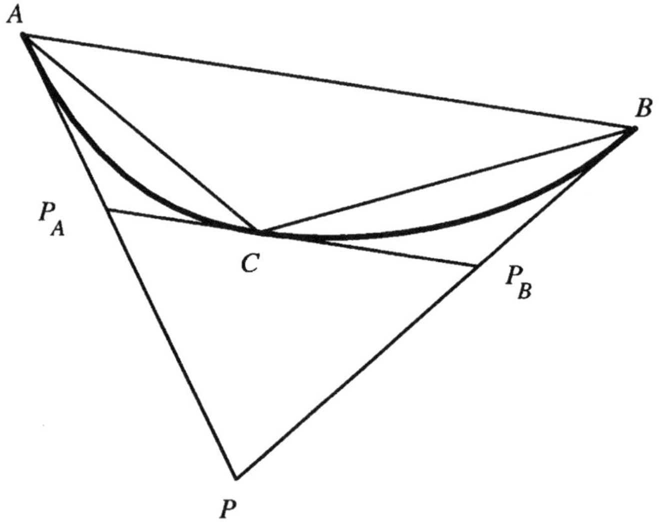

FIG. 3. The affine anti-triangle inequality.

circumscribes the corresponding arc of Γ. *Let* $\mathbf{d}(A, B) = 2\sqrt[3]{\mathbf{A}(A, B)}$, $\mathbf{d}(A, C) = 2\sqrt[3]{\mathbf{A}(A, C)}$, $\mathbf{d}(C, B) = 2\sqrt[3]{\mathbf{A}(C, B)}$ *denote twice the cube roots of their respective areas. (The factor of 2 is merely included for later convenience.) Then the following anti-triangle inequality is true*:

$$\mathbf{d}(A, B) \geqslant \mathbf{d}(A, C) + \mathbf{d}(C, B). \tag{4.1}$$

Equality is achieved if and only if the following affinely invariant length relations (length ratios among pairs of segments in the same line) hold:

$$\frac{AP_A}{AP} = \frac{PP_B}{PB} = \frac{P_A C}{P_A P_B}. \tag{4.2}$$

Furthermore, if one fixes the two boundary line elements (A, AP) *and* (B, BP), *then the set of line elements* (C, CP_B) *that satisfy* (4.2), *with* C *in the interior of the triangle* $\mathbf{T}(A, B)$, *constitute a one-parameter family of tangent line elements of the unique arc of the parabola having the prescribed tangent elements at the end points.*

Proof. Since any two (non-degenerate) triangles, with their vertices in a given order, are affinely equivalent in a unique way, we may fix the two boundary line elements (A, AP) and (B, BP) in such a way that the area of the resulting triangle $\mathbf{T}(A, B)$ equals unity. Then the set of line elements (C, CP_B) that may occur as tangent line elements of any strongly convex, short arc joining the given boundary elements is in a natural correspondence with the triple of real numbers (u, v, w) with $0 < u < 1$, $0 < v < 1$, $0 < w < 1$, according to the following recipe.

First, choose the point P_A on the line segment AP according to the vector relation $A - P_A = u(A - P)$. Then choose P_B on the segment PB so that $P_B - P = v(B - P)$, and, finally, choose C on the segment $P_A P_B$ so that $C - P_A = w(P_B - P_A)$. One readily verifies that

$$\mathbf{A}(A, C) = |\Delta(A, P_A, C)| = u\, |\Delta(A, P, C)|,$$

$$|\Delta(A, P, C)| = w\, |\Delta(A, P, P_B)|, \qquad |\Delta(A, P, P_B)| = v\, |\Delta(A, P, B)| = v,$$

whence $\mathbf{A}(A, C) = uvw$. Similarly,

$$\begin{aligned}\mathbf{A}(C, B) &= |\Delta(C, P_B, B)| = (1 - v)\, |\Delta(C, P, B)| \\ &= (1 - v)(1 - w)\, |\Delta(P_A, P, B)| = (1 - v)(1 - w)(1 - u).\end{aligned}$$

Thus the "distances" between the line elements in question satisfy the relations

$$\mathbf{d}(A, C) = \sqrt[3]{uvw}\, \mathbf{d}(A, B), \qquad \mathbf{d}(C, B) = \sqrt[3]{(1 - u)(1 - v)(1 - w)}\, \mathbf{d}(A, B).$$

It is well known that the geometric mean of any finite family of positive real numbers is strictly smaller than their arithmetic mean; applying this to the identities above, and adding, one sees that

$$\mathbf{d}(A, C) + \mathbf{d}(C, B) \leqslant \tfrac{1}{3}(u+v+w)\,\mathbf{d}(A, B) + \tfrac{1}{3}[(1-u)+(1-v)+(1-w)]\,\mathbf{d}(A, B) = \mathbf{d}(A, B),$$

with equality achieved only when $u=v=w$. This proves the first part of our assertion.

In order to constructively verify the second assertion, we take the circumscribed triangle $\mathbf{T}(A, B)$ as before and adapt an affine coordinate system (x, y) to it with origin at P, so that the points A and B have respective coordinates $(0, 1)$ and $(1, 0)$. Setting $u=v=w=r$, where $0<r<1$ is a parameter, the line $P_A P_B$ has equation $(1-r)\,x+ry=1$, and the point C on that line is defined parametrically by its coordinates

$$C = \mathbf{x}(r) = (x(r), y(r)) = (\tfrac{1}{2}r^2, \tfrac{1}{2}(1-r)^2). \tag{4.3}$$

This shows that the point $\mathbf{x}(r)$ traces the arc of the parabola $y = x + \frac{1}{2} - \sqrt{2x}$ bounded between the points $P_A = (0, \frac{1}{2})$ and $P_B = (\frac{1}{2}, 0)$, with the corresponding axes as tangents. This completes the proof of the theorem. Q.E.D.

The construction of the parametric equation (4.3) of the parabola and the statement of Theorem 4.3 show, in addition, that for any two values $r_1 < r_2$ of the parameter r, the area of the triangle circumscribed to the arc corresponding to $[r_1, r_2]$ equals $\frac{1}{8}(r_2 - r_1)^3$. We recall here that the usual, formal definition of the equiaffinely invariant arc length for locally convex smooth curves $\mathbf{x}(r)$ is expressed by the invariant integral

$$s = \int \sqrt[3]{\left|\left[\frac{d\mathbf{x}}{dr}, \frac{d^2\mathbf{x}}{dr^2}\right]\right|}\, dr, \tag{4.4}$$

where we are considering the derivatives of $\mathbf{x}(r)$ as displacement vectors in TE and using the notation of (3.1). In the case of the parametric representation (4.3) of a parabola, the parameter r describes the affine arc length.

More generally, let Γ be a convex curve of class C^2 traced by $\mathbf{x}(r)$ for r in a closed interval $I = [r_0, r_1]$. Subdivide I into a finite sequence of n small subintervals using a mesh $r_0 < r_1 < \cdots < r_{n-1} < r_n$, and let $P_k = \mathbf{x}(r_k)$ be the corresponding points on Γ. Inscribe each subarc $\gamma_k = \Gamma(P_{k-1}, P_k)$ in a corresponding support triangle $\mathbf{T}_k = \mathbf{T}(P_{k-1}, P_k)$. Let $\mathbf{d}_k$ equal twice the cube root of the area of $\mathbf{T}_k$. Then, on the one hand, the sum of the quantities $\mathbf{d}_k$ is nonincreasing under successive refinements of the subdivision, while, on the other hand, the sum converges downward to the value of the integral (4.4). With this observation, we make the following definition of the *pseudo-distance* between any two (non-oriented) line elements in general position in the equiaffine plane.

DEFINITION 4.4. Suppose the two line elements (A, AX) and (B, BY) are in general position, meaning that the lines AX and BY are not parallel, intersecting at a point P, and that the three points A, B, and P are distinct. Then the *distance* (or *pseudo-distance*) between (A, AX) and (B, BY) is defined to be twice the cube root of the area of the triangle $\mathbf{T} = APB$.

There are two easy, alternative geometric interpretations of the equiaffine arc length of a convex curve. One can replace the cube root of eight times the area of the small triangles by either the cube root of twelve times the area of the region between the small arcs of the subdivision of Γ and the corresponding chords, or that of 24 times the area between the small arcs and their endpoint tangents. Either of these two definitions is easier to adapt to the case of convex hypersurfaces in $\mathbb{R}^n$ than the one presented here; however, the approximation of the true affine length by subdivision is no longer monotone in either of the two modified cases.

The geometric interpretation of the equiaffine arc length just described admits two natural generalizations to higher dimensions. One generalization pertains to curves in n-dimensional space. Here the equiaffinely invariant arc length of an arc of class C^n parametrized by $\mathbf{x}(r)$ is formally defined by the integral

$$s = \int \left| \left[\frac{d\mathbf{x}}{dr}, \frac{d^2\mathbf{x}}{dr^2}, \ldots, \frac{d^n\mathbf{x}}{dr^n} \right] \right|^{2/n(n+1)} dr.$$

The other generalization deals with hypersurfaces (mainly in the strongly locally convex case) of class C^2 in $\mathbb{R}^n$. In this case, the easiest description of the formally equiaffinely invariant metric structure is in terms of a Euclidean structure on $\mathbb{R}^n$ defined by a positive definite quadratic form $\langle \cdot, \cdot \rangle$, inducing the familiar Euclidean invariants: the first fundamental form ds_e^2, the element of surface area dA_e, the unit normal vector N_e, the second fundamental form $\mathrm{II}_e = \langle N_e, d^2X \rangle$, assumed to be positive definite, and the Gaussian curvature $K_e > 0$. Then the positive definite quadratic form $ds^2 = K_e^{-1/(n+1)} \mathrm{II}_e$ and the corresponding $(n-1)$-dimensional area form $dA = K_e^{1/(n+1)}\, dA_e$ may easily be shown to be invariant under the equiaffine group. The intent of the geometrization of these formulas is to enable us to "see," from the shape of the surface, the length of paths and areas of subdomains.

5. THE EQUIAFFINE STRUCTURE EQUATIONS

Having introduced the element of equiaffine arc length for a smooth curve Γ without inflection points, the remaining equiaffine invariants are best described, analytically, in terms of the derivatives of the parametric representation of the curve, when the oriented, equiaffine arc length s itself is used as a parameter. However, one should observe that the existence and continuity of $d^k\mathbf{x}(s)/ds^k$ for any $k \geqslant 1$ require existence and continuity of the $(k+1)$st derivative of $\mathbf{x}$ with respect to a general parameter. The formal definition of s implies that the first two derivatives $\mathbf{x}_s = d\mathbf{x}(s)/ds$ and $\mathbf{x}_{ss} = d^2\mathbf{x}(s)/ds^2$ are linearly independent and, indeed, satisfy the identity

$$[\mathbf{x}_s, \mathbf{x}_{ss}] = \pm 1. \tag{5.1}$$

If necessary, one may replace the parameter s by $-s$ in order to reduce the right-hand side of (5.1) to $+1$. Either way, $\mathbf{x}_{ss}$ points toward the concave side of the curve Γ, while the positive sign in (5.1) indicates that, as s increases, the curve turns toward the left. For each point $\mathbf{x}(s_0)$ one defines what Cartan called the "moving frame" (repère mobile) of Γ, namely the affine coordinate system with origin at $\mathbf{x}(s_0)$, such that the coordinate pair (u, v) corresponds to the point $\mathbf{x}(s_0) + u\mathbf{x}_s(s_0) + v\mathbf{x}_{ss}(s_0)$. The two "unit" coordinate vectors $\mathbf{x}_s(s_0)$, $\mathbf{x}_{ss}(s_0)$ are then called the *(affine) unit tangent* and *unit normal* respectively, and are accordingly denoted by $\mathbf{t}(s_0)$ and $\mathbf{n}(s_0)$ respectively.

Differentiating both sides of (5.1) with respect to s, we see that $[\mathbf{x}_s, \mathbf{x}_{sss}] = 0$, implying that $d\mathbf{n}(s)/ds$ is a scalar multiple of $\mathbf{t}(s)$. One is thus led to the formal definition of the *(equi-)affine curvature* $\kappa(s)$ via the equation

$$\frac{d\mathbf{n}(s)}{ds} = -\kappa(s)\,\mathbf{t}(s). \tag{5.2}$$

The seemingly capricious choice of sign in the above equation is contrived so that, in the case of non-singular conic sections (in which case κ is constant), κ is positive, zero, or negative, according to whether the conic is, respectively, an ellipse, a parabola, or a hyperbola. See Theorem 6.4 below.

The data consisting of the equiaffine arc length parameter s and the affine curvature κ furnish the total generating system of equiaffine invariants of a curve Γ. In fact, the structure equations for Γ may be deduced from (5.1), (5.2), and can be written in Cartan's notation as the evolution of the moving frame $(\mathbf{x}(s), \mathbf{t}(s), \mathbf{n}(s))$ as follows:

$$d\begin{pmatrix} \mathbf{x}(s) \\ \mathbf{t}(s) \\ \mathbf{n}(s) \end{pmatrix} = \begin{pmatrix} 1 & 0 \\ 0 & 1 \\ -\kappa(s) & 0 \end{pmatrix} \begin{pmatrix} \mathbf{t}(s) \\ \mathbf{n}(s) \end{pmatrix} ds. \tag{5.3}$$

The initial conditions $(\mathbf{x}(s_0), \mathbf{t}(s_0), \mathbf{n}(s_0))$ consist of an arbitrary unimodular affine coordinate frame, and the solution $(\mathbf{x}(s), \mathbf{t}(s), \mathbf{n}(s))$ is unique, meaning that the frame corresponding to any s to which the solution of (5.3) may be extended is related to the initial frame by a unique equiaffine motion. However, since the system reduces to a scalar third order equation, namely

$$\frac{d^3\mathbf{x}(s)}{ds^3} + \kappa(s)\,\frac{d\mathbf{x}(s)}{ds} = 0, \tag{5.4}$$

it is not easy to estimate the geometric shape of the solution. For instance, when does a periodic curvature function $\kappa(s)$ produce a closed curve solution?

A suggested exercise at this point is to compute the equiaffine arc length, the moving frame, and the affine curvature for the closed, convex curve defined as follows:

$$\mathbf{x}(s) = (\cos t - \tfrac{1}{10}\cos 3t,\ \sin t + \tfrac{1}{10}\sin 3t).$$

6. LOCAL COORDINATES

Let Γ be a short, compact arc of a convex curve. One can choose, in many ways, an equiaffine coordinate system (x, y) such that Γ is the graph of a convex function $y = u(x)$, with x ranging over a compact interval $[x_0, x_1]$. We now rewrite the affine arc length, normal, and curvature in the given coordinate system. First, the element of equiaffine arc length of Γ is given by $ds = \sqrt[3]{u_{xx}}\,dx$, where the subscripts indicate successive differentiations with respect to x. It follows that the affine tangent and normal vectors at the point corresponding to x are

$$\mathbf{t} = (u_{xx})^{-1/3}\,(1, u_x), \qquad \mathbf{n} = \tfrac{1}{3}(u_{xx})^{-5/3}\,(-u_{xxx},\ 3u_{xx}^2 - u_x u_{xxx}). \tag{6.1}$$

In particular, we have

LEMMA 6.1. *The y-axis is parallel to the affine normal at a point $(x, u(x))$ if and only if $u_{xxx} = 0$.*

Finally, one deduces the formula

$$\kappa = \frac{3u_{xx}u_{xxxx} - 5u_{xxx}^2}{9(u_{xx})^{8/3}}. \tag{6.2}$$

from the structure equations (5.3). As in the Euclidean case, the element of equiaffine arc length

$$ds = \sqrt[3]{u_{xx}}\, dx \tag{6.3}$$

is the simplest invariant one-form, and the curvature κ is the simplest differential invariant for the equiaffine group in the plane. Every other differential invariant can be expressed as a function of κ and its successive derivatives with respect to arc length. Since the equiaffine curvature is a fourth order differential invariant, the following equiaffine version of the definition of the Euclidean curvature via an osculating circle is immediate.

DEFINITION 6.2. Let Γ be a smooth, convex curve, and let $A \in \Gamma$. The *osculating conic* to Γ at A is the unique conic passing through A having fourth order contact with Γ at A.

THEOREM 6.3. *Two smooth, convex curves passing through a common point A have the same equiaffine curvature at A if and only if they have fourth order contact at A. In particular, the curvature to a curve Γ at A equals the (constant) curvature of its osculating conic at A.*

In particular, we need to know the explicit formula for the curvature of a general conic.

THEOREM 6.4. *Consider a nondegenerate conic $\mathscr{C}$ defined by the quadratic equation*

$$Ax^2 + 2Bxy + Cy^2 + 2Dx + 2Ey + F = 0. \tag{6.4}$$

The equiaffine curvature of $\mathscr{C}$ is given by

$$\kappa = \frac{S}{T^{2/3}}, \tag{6.5}$$

where

$$S = AC - B^2 = \det \begin{vmatrix} A & B \\ B & C \end{vmatrix}, \qquad T = \det \begin{vmatrix} A & B & D \\ B & C & E \\ D & E & F \end{vmatrix}. \tag{6.6}$$

Both S and T are equi-affine invariants of the conic. The invariant S vanishes if and only if the five points lie on a parabola. The invariant T vanishes if and only if the conic degenerates to a pair of lines, and hence fails our convexity hypothesis.

COROLLARY 6.5. *The equiaffine curvature of an ellipse in the plane is given by $\kappa = (\pi/\mathbf{A})^{2/3}$, where $\mathbf{A}$ is the area of the ellipse.*

7. THE AFFINE NORMAL

We now begin our discussion of geometric approximations to the affine geometric quantities associated with a convex plane curve. Let Γ be, as before, a short arc, with end points A, B. Let M be the midpoint of the chord AB. Let the tangents to Γ at A and B intersect at a point P, so that Γ is inscribed in the support triangle $\mathbf{T} = APB$ (see Fig. 4).

THEOREM 7.1. *The direction of the median PM of the triangle $\mathbf{T}$ is a mean affine normal direction of Γ, in the sense that if Γ is of class C^3, then there exists at least one point of Γ where the affine normal is parallel to PM.*

Proof. Choose an equiaffine coordinate system (x, y) such that the y-axis includes the median PM in the direction indicated. Then Γ is the graph of a convex function $y = u(x)$ and, since M lies on the y-axis, u is defined over a symmetric interval $-a \leqslant x \leqslant a$ for some $a > 0$. At the same time, since P also lies on the y-axis, u satisfies the boundary condition $au_x(a) - u(a) = (-a)\, u_x(-a) - u(-a)$, which may be translated by integration by parts as follows:

$$0 = \int_{-a}^{a} d[xu_x - u] = \int_{-a}^{a} xu_{xx}\, dx = \int_{-a}^{a} \tfrac{1}{2}(a^2 - x^2)\, u_{xxx}\, dx.$$

Since the "weight" function $a^2 - x^2$ is positive in the interior of the interval, the third derivative u_{xxx} has a weighted mean value of zero. Lemma 6.1 completes the proof. Q.E.D.

There are several analogous statements, giving alternative geometric interpretations of some mean direction of the affine normal, but none is as

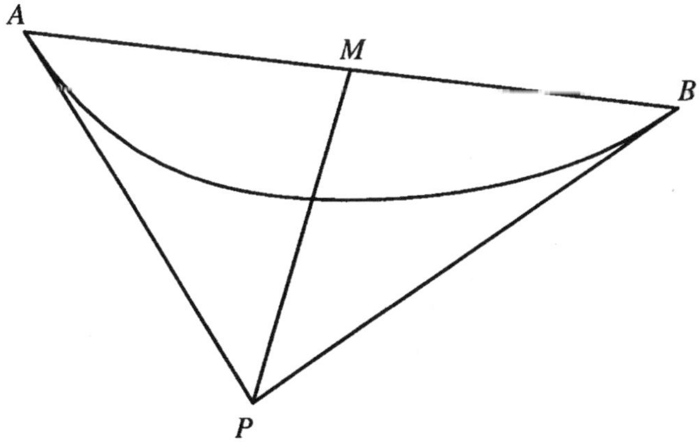

FIG. 4. Median of support triangle.

simple to state or prove as the one just shown. However, we shall present some of these alternatives, because they may be better suited for generalizations to locally convex hypersurfaces in $\mathbb{R}^n$. All of them deal with the support triangle APB, the midpoint M of the chord AB and various choices of an interior point $C \in \Gamma$, so that both CM and PC represent mean directions of the affine normal. We shall deal first with the case where C is the unique point of Γ where the tangent line is parallel to the line AB.

THEOREM 7.2. *Let Γ be a short, strongly convex arc, inscribed in the triangle $\mathbf{T} = APB$, where A, B are the end points of Γ and the corresponding tangent lines intersect at P. Let M be the midpoint of the chord AB, and $C \in \Gamma$ where the tangent line is parallel to the line AB. Then there exist* (a) *a point $C' \in \Gamma$ where the affine normal is in the same direction as the directed line PC, and* (b) *a point $C'' \in \Gamma$ where the affine normal is in the same direction as the directed line CM.*

Proof. We first prove the existence of C'. Let (x, y) be an affine coordinate system such that the y-axis contains the segment PC with the same orientation. Then Γ is the graph of a convex function $y = u(x)$ defined over a closed interval $a \leqslant x \leqslant b$ with $a < 0 < b$. The assumptions on PC correspond, in this coordinate system, to the following conditions: (a) the point $C = (0, u(0))$ satisfies $u_x(0) = (u(b) - u(a))/(b - a)$, corresponding to the boundary condition

$$bu_{xx}(0) - u(b) = au_x(0) - u(a); \tag{7.1}$$

(b) the point $P = (0, y_0)$ lying on the y-axis corresponds to the boundary condition

$$-y_0 = bu_x(b) - u(b) = au_x(a) - u(a). \tag{7.2}$$

The two boundary conditions (7.1), (7.2) can, in turn, be expressed in integral form, respectively, as follows:

$$0 = \int_a^b d[xu_x(0) - u(x)] = \int_a^b [u_x(0) - u_x(x)]\, dx = -\int_a^b \int_0^\xi u_{xx}(x)\, dx\, d\xi$$

$$= -\int_0^b (b - x)\, u_{xx}\, dx + \int_a^0 (x - a)\, u_{xx}\, dx,$$

$$0 = \int_a^b d[xu_x(x) - u(x)] = \int_a^b xu_{xx}(x)\, dx.$$

Adding the two equations and integrating by parts once more, we obtain:

$$0 = -\int_0^b (b-2x)\, u_{xx}(x)\, dx + \int_a^0 (2x-a)\, u_{xx}(x)\, dx$$

$$= -\int_0^b x(b-x)\, u_{xxx}(x)\, dx + \int_a^0 (-x)(x-a)\, u_{xxx}(x)\, dx.$$

The last expression expresses the vanishing of the integral of a continuous third derivative u_{xxx}, weighted by a positive function, over the interval $[a, b]$. Therefore, we deduce the existence of an interior point $C' = (x', u(x'))$ such that $u_{xxx}(x') = 0$, meaning that the affine normal at C' is vertical.

In order to prove assertion (b), we choose an affine coordinate system (x, y) such that the y-axis contains the segment CM. We prove the assertion, at first, under the additional assumption that Γ is the graph of a convex function $y = u(x)$. In this case, the assumptions translate into the following two statements: (i) since M lies on the y-axis, u is defined on a symmetric interval $[-a, a]$ for some $a > 0$; (ii) since the tangent to Γ at C is parallel to AB, $u_x(0) = (u(a) - u(-a))/(2a)$, leading to the following argument, similar to the previous ones:

$$0 = \int_{-a}^a d[x u_x(0) - u(x)] = \int_{-a}^a [u_x(0) - u_x(x)]\, dx = -\int_{-a}^a \int_0^\xi u_{xx}(x)\, dx\, d\xi$$

$$= -\int_0^a (x-a)\, u_{xx}\, dx + \int_{-a}^0 (a+x)\, u_{xx}\, dx = \int_{-a}^a \tfrac{1}{2}(a-|x|)^2\, u_{xxx}\, dx.$$

Once more, the vanishing of the last integral implies the existence of a point C'' where the affine normal is vertical—under the extra condition that Γ is a graph over $[-a, a]$.

Assume now that Γ is no longer a graph, and let A and B have coordinates $(-a, y_0)$ and (a, y_1) respectively. Then Γ includes either a point $A' = (-a, y_0')$ with $y_0' < y_0$, in which case the subarc from A' to B is the graph of a function, or else there exists a point $B' = (a, y_1')$ with $y_1' < y_1$, in which case the subarc from A to B' if the graph of a function (but not both, since Γ is a short arc). Without loss of generality, we assume the former case, as in Fig. 5. In this case, a vertical line drawn downward (i.e., in the same direction as MC) from A meets Γ at another point A', and a line from A' in the direction of AB (i.e., to the *right*) meets Γ at a point B'. It is clear that the midpoint M' of the segment $A'B'$ is to the *left* of the segment CM. Therefore, replacing Γ by the subarc Γ' from A' to B', one may apply the previous argument, whereby there is a point in Γ' where the affine normal is in the same direction as CM', that is to say, to the *left* of CM. On

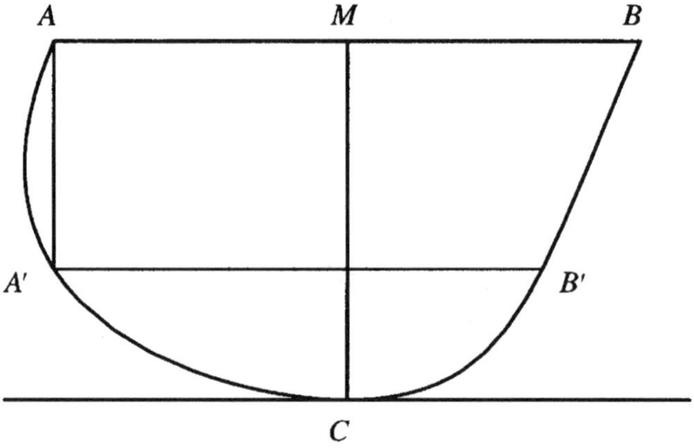

FIG. 5. Mean affine normal.

the other hand, the arc of Γ from A to A', by a similar argument, contains a point where the affine normal points to the right of the direction of $A'A$, or, equivalently, CM. By continuity, there exists a point C'' in the subarc of Γ from A to B' where the affine normal is in the same direction as CM. This concludes the proof of Theorem 7.2. Q.E.D.

A somewhat different geometrical construction of mean affine normals for a short convex arc is described by the following theorem.

THEOREM 7.3. *Let Γ be a short, smooth, convex arc, inscribed in the triangle $\mathbf{T} = APB$, as above, and let M be the midpoint of the chord AB. For any point $C \in \Gamma$, not an endpoint, denote by P_A and P_B the points of intersection of the tangent to Γ at C with the segments PA and PB, respectively. Then there exists at least one point $C \in \Gamma$ which is the midpoint of the associated segment $P_A P_B$. Furthermore, for any such point, there exist points $C', C'' \in \Gamma$ where the affine normal is in the same direction as PC or CM, respectively.*

Proof. To show the existence of the point C, consider the area of the triangle $P_A P P_B$ as C varies between A and B. The area is always positive, continuously dependent on C, and approaches zero as C approaches either A or B. The desired point C occurs when this area attains a (local) maximum value. Note that if Γ were to include a sub-arc Γ' of a hyperbola having PA and PB as asymptotes, then *each* point $C \in \Gamma'$ would have the desired property.

Assuming, then, that C is the midpoint of its associated segment $P_A P_B$, we proceed to prove, first, that the direction PC occurs as a direction of an affine normal to Γ. Let (x, y) be an affine coordinate system such that the

y-axis contains the segment PC with the same orientation. The assumptions imply, first of all, that Γ is the graph of a convex function $y = u(x)$ defined over a closed interval $a \leqslant x \leqslant b$ with $a < 0 < b$. The assumption that P lies on the y-axis is equivalent to the boundary condition (7.2). It is now convenient to choose the direction of the x-axis to be parallel to $P_A P_B$, which means that $u_x(a) = -u_x(b)$ and $u_x(0) = 0$. Our assumption that C is the midpoint of $P_A P_B$ means that

$$u_x(b) - 2u_x(0) + u_x(a) = 0. \tag{7.3}$$

We now introduce a Legendre transform of the function u: choose the strictly monotone function $u_x(x) - u_x(0)$ of x as the new independent variable $\hat{x}$ and the transformed function $\hat{y} = \hat{u}(\hat{x}) = xu_x(x) - u(x)$ as the new dependent variable. Then

$$d\hat{x} = u_{xx}\, dx, \qquad d\hat{y} = xu_{xx}\, dx = x\, d\hat{x}.$$

Therefore

$$\frac{d^2\hat{u}}{d\hat{x}^2} = \frac{1}{u_{xx}} > 0.$$

The boundary condition (7.3), in terms of the transformed function, sets the interval of definition of $\hat{u}$ to be $[-\hat{a}, \hat{a}]$, where $\hat{a} = u_x(b) - u_x(0) = u_x(0) - u_x(a)$, while (7.2) becomes $\hat{u}(-\hat{a}) = \hat{u}(\hat{a})$. In addition, we have $\hat{u}_{\hat{x}}(0) = 0$. These conditions on the transformed variables and the function $\hat{u}$ include the properties of the function u in the proof of assertion (b) of Theorem 7.2, namely the graph $\hat{\Gamma}$ of $\hat{u}$, with endpoints $\hat{A} = (-\hat{a}, \hat{u}(-\hat{a}))$, $\hat{B} = (\hat{a}, \hat{u}(\hat{a}))$, and the point $\hat{C} = (0, \hat{u}(0))$ such that $2\hat{a}\hat{u}(0) = \hat{u}(\hat{a}) - \hat{u}(-\hat{a})$ (both sides here being zero). Therefore there exists an intermediate value $\hat{x}'$ corresponding to $x' = \hat{u}_{\hat{x}}(\hat{x}') \in [a, b]$, for which $\hat{u}_{\hat{x}\hat{x}\hat{x}}(\hat{x}') = -u_{xxx}(x')/[u_{xx}(x')]^2 = 0$. This shows the existence of $C' = (x', u(x')) \in \Gamma$ where the affine normal is in the same direction as PC.[4]

[4] It is possible, of course, to give the same proof without using Legendre transforms; however the steps to deduce, by two integrations by parts, from the assumptions on u the corresponding identity

$$0 = \int_a^b du(x) = \frac{1}{2}\int_a^b [u_x(b) - u_x(0) - |u_x(x) - u_x(0)|]^2 \frac{u_{xxx}}{(u_{xx})^2}\, dx$$

would seem much more opaque than the proof presented here.

To show the existence of a point $C'' \in \Gamma$ whose affine normal is in the same direction as CM, we arrange the y axis of our equiaffine coordinate system to include the segment CM in the positive direction. Introduce the chords AC and CB, and let M_A and M_B be their respective midpoints, such that the corresponding affine normals are positive scalar multiples of the vectors $M_A - P_A$ and $M_B - P_B$ respectively, as in Fig. 6. On the other hand, taking into account the identities

$$M_A - P_A = \tfrac{1}{2}(A - P_A) + \tfrac{1}{4}(P_B - P_A),$$
$$M_B - P_B = \tfrac{1}{2}(B - P_B) + \tfrac{1}{4}(P_A - P_B),$$

we see that the vector

$$M - C = \tfrac{1}{2}(A + B) - \tfrac{1}{2}(P_A + P_B) = (M_A - P_A) + (M_B - P_B).$$

It follows that the direction of $M - C$ is intermediate between those of $M_A - P_A$ and $M_B - P_B$. Therefore, by continuity, there exists a point $C'' \in \Gamma$ between C''_A and C''_B where the affine normal is a positive scalar multiple of the vector $M - C$. This completes the proof of Theorem 7.3. Q.E.D.

Having shown various ways to represent an "average" direction of the affine normal, one naturally seeks a corresponding "average" normalization

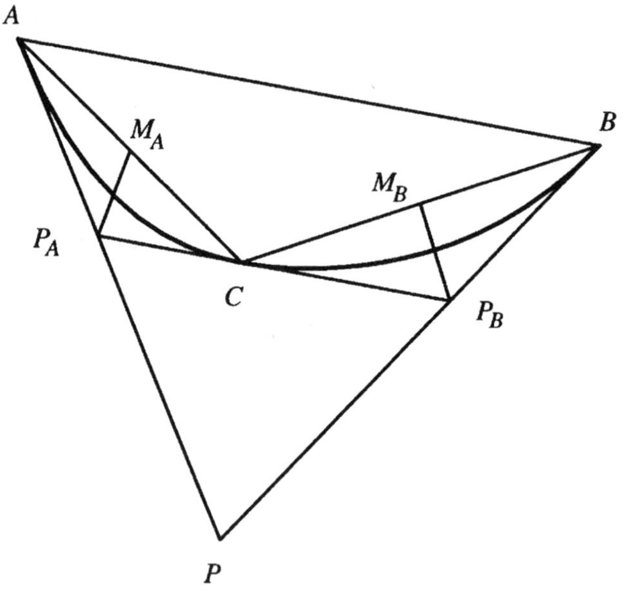

FIG. 6. Mean affine normal.

in agreement with the formal definition of the equiaffine invariants. Recalling the definition of the equiaffine length of a short arc $\Gamma(A, B)$, cf. Definition 4.4, we denote by $\mathbf{d}(A, B)$ the equiaffine distance between the line elements (A, AP) and (B, PB). Then the vector $B-A$ on the one hand is the "average" direction of the oriented tangents over Γ; on the other hand, it represents the integral $\int_A^B \mathbf{t}\, ds$, where $\mathbf{t}$ is the equiaffine tangent vector and ds is the element of equiaffine arc length. The estimation of the average value of $\mathbf{t}$ by $(B-A)/\mathbf{d}(A, B)$ is then obvious.

We choose now the average direction of the affine normal of the arc Γ according to Theorem 7.1, represented by the median vector $M-P$ of the support triangle APB. Observe that the alternating product $[B-A, M-P]$ is twice the area of the triangle APB, which is $\frac{1}{4}\mathbf{d}(A, B)^3$. Since $B-A$ is approximately equal to $\mathbf{d}(A, B)\,\mathbf{t}$, and, from the structure equations $[\mathbf{t}, \mathbf{n}]=1$, it follows that the equiaffine normal vector $\mathbf{n}$ should, in the mean, be represented by $4\mathbf{d}(A, B)^{-2}\,(M-P)$. We formally summarize these estimates in the following theorem.

THEOREM 7.4. *Let Γ be a short arc and let $\mathbf{d}(A, B)$ be the equiaffine distance between its endpoints A and B, i.e., twice the cube root of the area of its support triangle APB. Let M be the midpoint of the chord AB. Then a mean value for the equiaffine frame* $(\mathbf{t}, \mathbf{n})$, *consisting of the tangent and normal to Γ, is represented by*

$$\mathbf{t}_{\mathrm{av}} = \frac{B-A}{\mathbf{d}(A, B)}, \qquad \mathbf{n}_{\mathrm{av}} = 4\,\frac{M-P}{\mathbf{d}(A, B)^2}.$$

8. THE AFFINE CURVATURE

The structure equation (5.3) has two obvious consequences that serve to interpret it in geometric terms. In the first place, under infinitesimal displacements of a point on the curve, the equiaffine normal shifts parallel to the tangent. Second, the sign of the equiaffine curvature κ tells us which way the affine normal varies over small arcs of a convex curve. More precisely, if κ is everywhere positive in a short arc, then the affine normals at its endpoints, both pointing into the concave side of the curve, lean toward each other, like the Euclidean normals of a convex arc, while if $\kappa<0$ everywhere, then the affine normals lean away from each other. One can apply the results of the last section to make these statements more precise.

PROPOSITION 8.1. *Let Γ be a short arc of a smooth, convex curve, and let APB be its support triangle. Let $C \in \Gamma$ be the point whose tangent line is* *parallel to the chord AB, and let the tangent line at C intersect the segments PA and PB at P_A and P_B respectively. Let $t = t_{A,B}$ be the real number, $0 < t < 1$, defined by the equivalent vector relations $P_A - P = t(A - P)$ or $P_B - P = t(B - P)$. Then there exists a point on Γ where the equiaffine curvature κ is positive, negative, or zero according to whether t is, respectively, $> \frac{1}{2}$ or $< \frac{1}{2}$ or $= \frac{1}{2}$.*

Proof. We refer to Fig. 7. Draw the chords of Γ from A to C and from C to B, and let M_A and M_B be their respective midpoints. It follows that the vector $M_B - M_A$ equals $\frac{1}{2}(B - A)$. Since the tangent at C is parallel to the line AB, it follows that $P_B - P_A = t(B - A)$, where t is as in the statement of the proposition. From Theorem 7.1 we known that the directed half-lines $P_A M_A$ and $P_B M_B$ represent mean directions of the equiaffine normal in the respective portions of Γ. To compare these two directions, one immediately verifies that

$$(M_B - P_B) - (M_A - P_A) = (M_B - M_A) - (P_B - P_A) = (\tfrac{1}{2} - t)(B - A).$$

The required conclusion now follows. Q.E.D.

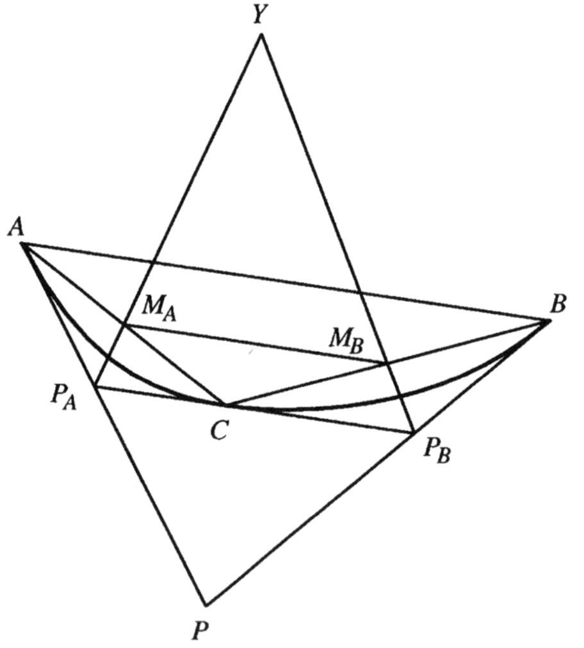

FIG. 7. Affine curvature construction.

COROLLARY 8.2. *Let Γ be a smooth, closed, convex curve without inflection points in the affine plane. Let* **B** *denote the convex body bounded by Γ. Let* $\mathbf{B}^*$ *denote the convex body neighborhood of* **B**, *obtained as the Minkowski sum* $\mathbf{B}^* = 2\mathbf{B} + (-\mathbf{B})$.[5] *Then, from every point on the boundary of* $\mathbf{B}^*$ (*and, a fortiori, from every exterior point of* $\mathbf{B}^*$) *one can "see" at least one point on Γ where the equiaffine curvature is positive.*

Remark. This statement is considerably stronger than one found in various textbooks, asserting the existence of a point with positive equiaffine curvature on any "half-oval," i.e., on any locally convex, smooth bounded arc whose tangents at the endpoints are parallel, and with no other pair of parallel tangents.

Proof. Let P be any point on the boundary of $\mathbf{B}^*$, and let PA and PB be the two tangent lines to Γ from P, where A and B are the respective points of contact with Γ. Let A' and B' be the midpoints of the respective segments PA and PB. It follows from the definition of $\mathbf{B}^*$ that the line $A'B'$ cannot meet **B**. Consequently, if one draws the tangent line to the short arc of Γ between A and B, i.e., the set of points of Γ that are visible from P, then the ratio $t_{A,B}$ defined in Proposition 8.1 is strictly greater than $\frac{1}{2}$. Q.E.D.

To conclude this section, we shall refine the last proposition to yield a numerical approximation to the actual value of the equiaffine curvature of a short arc.

THEOREM 8.3. *Let Γ be a short arc of a smooth, convex curve, with endpoints A, B, and the same construction as in Proposition* 8.1. *Let* $\mathbf{d}(A, B)$ *denote the equiaffine distance from A to B. In Fig.* 6, *prolong the lines* $P_A M_A$ *and* $P_B M_B$ *to their intersection point Y (if necessary, in the projective completion of the plane), as depicted in Fig.* 8. *The three points P, C, and Y lie on a common line. Let Q_C denote the intersection of that line with the chord AB, and consider the (negatively valued) cross ratio*

$$\rho(A, B) = [Q_C, P : Y, C] = \frac{(Q_C Y : PY)}{(Q_C C : PC)}.$$

Then a mean value of the equiaffine curvature κ over Γ is represented by

$$\overline{\kappa_\Gamma} = 8\,\frac{1 + \rho(A, B)}{\mathbf{d}(A, B)^2}. \tag{8.1}$$

[5] In other words, $\mathbf{B}^*$ is the set of points P for which one can find points $M, Q \in \mathbf{B}$ such that M is the midpoint of the segment PQ.

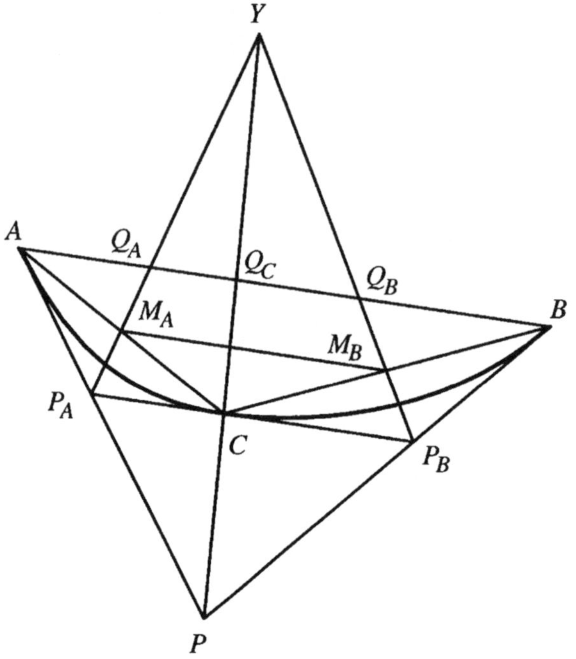

FIG. 8. Equiaffine curvature approximation.

Proof. The collinearity of the three points P, C, and Y follows from Desargues' Theorem. From the perspective point A, the four points P, C, Q_C, Y, defining the cross ratio ρ, may be projected to the corresponding points P_A, M_A, Q_A, Y in the line $P_A M_A$. (Equivalently, we can project from B to obtain P_B, M_B, Q_B, Y in the line $P_B M_B$.) Since M_A is the midpoint of $P_A Q_A$, the cross ratio ρ reduces to the scalar coefficient in the linear vector relation $Y - Q_A = -\rho(Y - P_A)$, whence

$$M_A - P_A = \tfrac{1}{2}(1+\rho)(Y - P_A). \tag{8.2}$$

According to Theorem 7.4,

$$M_A - P_A = \tfrac{1}{4}\mathbf{d}(A, C)^2\, \mathbf{n}(A, C), \tag{8.3}$$

where $\mathbf{d}(A, C)$ is twice the cube root of the area of the triangle $AP_A C$ (and asymptotically the affine length of the arc AC in Γ) and $\mathbf{n}(A, C)$ is a mean vector value of the equiaffine normal over the same arc. Furthermore, the point Y, marking the intersection of two neighboring affine normal lines, approximates, as the arc Γ is shortened to a point $X \in \Gamma$, the corresponding point $X + \kappa^{-1}\mathbf{n}$ of the affine evolute of Γ at X. Combining the approximate relations $Y - P_A \approx \kappa^{-1}\mathbf{n}(A, C)$ with (8.3), we see from

(8.2) that an approximate value $\bar{\kappa}$ of the equiaffine curvature is given by $\frac{1}{8}(1+\rho)\,\mathbf{d}(A, C)^{-2}$. Interchanging A and B, we have another approximation $\bar{\kappa} \approx \frac{1}{8}(1+\rho)\,\mathbf{d}(C, B)^{-2}$. Recalling Theorem 4.3, the equiaffine length of Γ is approximated by $\mathbf{d}(A, B)$, or, alternatively, by $2\mathbf{d}(A, C)$ or $2\mathbf{d}(C, B)$. Combining these formulas completes the proof of (8.1). Q.E.D.

Although (8.1) can in principle be used as a method for approximating the affine curvature, it has several numerical difficulties that preclude its direct use. First, the construction relies on the introduction of the tangent lines at the points A and B, and hence we need to introduce an additional numerical approximation. Moreover, the approximation needs to incorporate affine invariances, and so the standard difference quotient is not satisfactory for this purpose. More serious is the instability in the computation of the intersection point Y, which can be at infinity (and, indeed, is if the curve is a parabola), and is thus highly unstable from a numerical point of view. Presumably, one can overcome the latter difficulty by multiplying the numerator and denominator in the ratio (8.1) by an appropriate factor, although we have not thoroughly investigated this as of yet.

9. FINITE DIFFERENCE APPROXIMATIONS OF AFFINE INVARIANTS

In this section, we discuss a fully affine-invariant finite difference approximation to the affine curvature of a convex curve in the plane. The starting point is the result that one can approximate the (positive) affine curvature at a point of a plane curve by the affine curvature of the conic section passing through five nearby points. We will explicitly show how this may be used to produce an affine-invariant finite difference approximation to the affine curvature. The first item is to determine the formula for the affine curvature of a conic passing through five points.

Given a numbered set of points P_i, $i = 0, 1, 2, \ldots$, we let

$$[ijk] = [P_i, P_j, P_k] = (P_i - P_j) \wedge (P_i - P_k)$$

denote twice the (signed) area of the triangle with vertices P_i, P_j, P_k, cf. (3.2). See [23, 28], for a proof of the following elementary fact.

THEOREM 9.1. *Let $P_0, \ldots, P_4$ be five points in general position in the plane. There is then a unique conic section $\mathscr{C}$ passing through them whose quadratic equation has the affine-invariant form*

$$[013][024][\mathbf{x}12][\mathbf{x}34] = [012][034][\mathbf{x}13][\mathbf{x}24], \tag{9.1}$$

where $\mathbf{x} = (x, y)$ is an arbitrary point on $\mathscr{C}$.

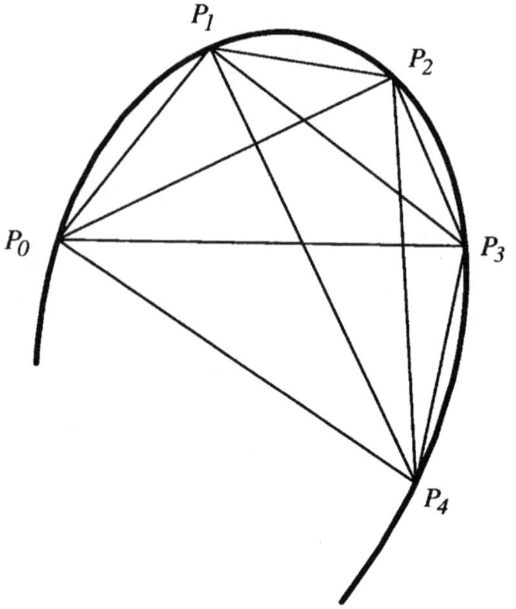

FIG. 9. The affine pentagram.

In order to compute the affine curvature of the conic (9.1), we use formula (6.5), and thus need to compute the two equiaffine invariants S, T, as given in (6.6), in equiaffine invariant form. In other words, the resulting formula should be written in terms of the areas of the $\binom{5}{3}=10$ triangles determined by the points taken three at a time; see Fig. 9. (Of course, only 5 of these areas are independent, due to the syzygies listed below.) Substituting the formulas for the coefficients, we find a particularly nice affine-invariant expression for our first affine invariant

$$4T = \prod_{0 \leqslant i<j<k \leqslant 4} [ijk]; \tag{9.2}$$

in other words, to compute T, multiply together all 10 triangular areas in the pentagram described by the 5 points. The fact that T has such a form is not so surprising, since T vanishes if and only if the conic degenerates to a pair of lines, which requires that three of the five points lie on a line, meaning that $[ijk]=0$ for some $i<j<k$.

One affine-invariant formula for S is found directly:

$$\begin{aligned} 4S = {} & [013]^2\,[024]^2\,[1234]^2 - 2[012][034][013][024] \\ & ([123][234] + [124][134]) + [012]^2\,[034]^2\,[1324]^2, \end{aligned} \tag{9.3}$$

where

$$[ijkl]=(P_i-P_k)\wedge(P_j-P_l)=[ijl]-[ijk].$$

Formula (9.3) is not nearly as pleasant as (9.2), particularly because the right-hand side appears to be unsymmetrical with respect to permutations of the five points. However, S must clearly be symmetrical with respect to these permutations. Of course, the explanation lies in the syzygies among the triangular areas: these are

$$[123]=[012]+[023]+[031], \tag{9.4}$$

$$[012][034]-[013][024]+[014][023]=0, \tag{9.5}$$

and the analogous formulas obtained by permutation of the symbols 0, ..., 4, cf. (2.5), (2.6). A judicious application of (9.4), (9.5) will suffice to demonstrate that (9.3) is symmetrical under permutation. A completely symmetrical formula for S can, of course, be obtained by symmetrizing (9.3), i.e., summing over all possible permutations of the set $\{0, 1, 2, 3, 4\}$ and dividing by $5!=120$, although the result is much more complicated than (9.3). We have been unable to find a simple yet symmetrical version of the formula for S.

As in the Euclidean case, we are interested in numerical approximations to the affine curvature of a strongly convex plane curve Γ which are invariant under the special affine group. As before, we approximate the parametrized curve $\mathbf{x}(r)=(x(r), y(r))$ by a sequence of mesh points $P_i=\mathbf{x}(r_i)$. Any affine-invariant numerical approximation to the affine curvature κ (as well as any other affine differential invariant $d^n\kappa/ds^n$) must be a function of the joint affine invariants of the mesh points, which means that it must be a function of the areas $[ijk]=[P_iP_jP_k]$ of the parallelograms (or triangles) described by the mesh points. Because the affine curvature is a fourth order differential function, the simplest approximation will require five mesh points, so that the approximation will depend on the ten triangular areas (or, more basically, the five independent areas) in the pentagram whose vertices are the five mesh points; see Fig. 9.

With this in mind, let us number the five successive mesh points as P_0, P_1, P_2, P_3, P_4. (This is just for simplicity of exposition; of course, in general, one should replace the indices 0, ..., 4 by $i, i+1, i+2, i+3, i+4$.) Since we are assuming that Γ is convex, the mesh points are in general position. Let $\mathscr{C}=\mathscr{C}(P_0, P_1, P_2, P_3, P_4)$ be the unique conic passing through the mesh points. Let $\tilde{\kappa}=\tilde{\kappa}(P_0, P_1, P_2, P_3, P_4)$ denote the affine curvature of the conic $\mathscr{C}$, which we evaluate via the basic formula (6.5), where the invariants S, T are computed in terms of the triangular areas according to (9.3), (9.2). We regard $\tilde{\kappa}$ as a numerical approximation to the affine curvature $\kappa = \kappa(P_2)$ of Γ at the middle point P_2. We now need to analyze how closely the numerical approximation $\tilde{\kappa}$ is to the true curvature κ at the point P_2. Assuming the points are close together (see the discussion below), we need to compute a Taylor series expansion of the distance $\tilde{\kappa}$. An extensive Mathematica computation produces the desired result.

THEOREM 9.2. *Let P_0, P_1, P_2, P_3, P_4 be five successive points on the convex curve Γ. Let κ be the affine curvature of Γ at P_2, and let $\tilde{\kappa}$ denote the affine curvature of the conic section passing through the five points. Let*

$$L_i = \int_{P_2}^{P_i} ds, \qquad i = 0, ..., 4, \tag{9.6}$$

denote the signed affine arc length of the conic form P_2 to P_i; in particular $L_2 = 0$. We assume that each L_i is small. Then the following expansion is valid:

$$\tilde{\kappa} = \kappa + \frac{1}{5}\left(\sum_{i=0}^{4} L_i\right)\frac{d\kappa}{ds} + \frac{1}{30}\left(\sum_{0 \leqslant i \leqslant j \leqslant 4} L_i L_j\right)\frac{d^2\kappa}{ds^2} + \cdots. \tag{9.7}$$

The higher order terms are cubic in the distances L_i.

Remark. The property of "being close" is therefore expressed in affine-invariant form as the statement that all the arc lengths $L_0, ..., L_4$ are small. In this way, we are able to introduce a fully affine-invariant notion of "distance," albeit one that requires knowledge of five, rather than two, points.

Proof. This is found by a direct Taylor series expansion of the affine-invariant expressions (9.3), (9.2), for the affine invariants S, T, and then substitution into the formula (6.5) for the curvature of the conic section. We represent the curve as the graph of $y = u(x)$, which, assuming that the five points are sufficiently close, can always be arranged. The points can be assumed to be $P_0 = (h, u(h))$, $P_1 = (i, u(i))$, $P_2 = (0, 0) = (0, u(0))$, $P_3 = (j, u(j))$, $P_4 = (k, u(k))$, where h, i, j, k are small. The areas are then given, for example, by

$$\begin{aligned}[013] &= (h-i)(u(h)-u(j)) - (h-j)(u(h)-u(i)) \\ &= hu(i) - iu(h) - hu(j) + ju(h),\end{aligned}$$

with elementary Taylor series expansion. The result is a Taylor series expansion for $\tilde{\kappa}$ in terms of h, i, j, k, with leading term κ, as given in (6.2), the derivatives being evaluated at 0. However, h, i, j, k, being the differences of the x coordinates of the mesh points, are not affine invariant, and hence the coefficients of the expansion are not affine differential invariants. To remedy this, we must introduce the affine arc lengths (9.6) as our basic affine-invariant parameters. Using the formula (6.3) for the affine-invariant arc length element, the expansion

$$L_0 = \int_0^h \sqrt[3]{u_{xx}}\, dx = h\, \sqrt[3]{u_{xx}} + \frac{1}{6} h^2 \frac{u_{xxx}}{(u_{xx})^{2/3}} + \cdots \tag{9.8}$$

can be inverted to produce a Taylor series expressing h in terms of L_0. Plugging this, and the analogous series for i, j, k into the previous Taylor series, produces the final result. Q.E.D.

Remark. An affine invariant finite difference approximation to the affine normal can also be found by computing the affine normal to the approximating conic $\mathscr{C}$ at the middle point P_2 and expressing this in terms of the triangular areas. The method can also produce invariant numerical methods for computing $d\kappa/ds$, etc., using more points.

10. A GENERAL CONJECTURE

The reader has probably already noted that the Euclidean and affine curvature approximation series (3.9), (9.7) bear a remarkable similarity to each other. This suggests a generalization which we indicate here, albeit as a conjecture without proof. We begin by surveying the general theory of differential invariants of finite-dimensional Lie groups of transformations in the plane; see [17] for a detailed presentation. Let G be an r-dimensional Lie group acting on $E = \mathbb{R}^2$, with coordinates x, y, and let $\mathfrak{g}$ denote its Lie algebra of infinitesimal generators, which are vector fields $\mathbf{v} = \xi(x, y)\, \partial_x + \eta(x, y)\, \partial_y$ on E. Curves in the plane are then (locally) represented as functions $y = u(x)$. Let $\mathbf{J}^n$ denote the nth jet space of E, which has coordinates $(x, u^{(n)}) = (x, u, u_x, u_{xx}, ..., u_n)$. There exists a G-invariant arc length element $ds_G = P(x, u^{(n)})\, dx$ represented by the simplest (lowest order) G-invariant one-form, and a G-invariant curvature κ_G, which is the simplest (lowest order) differential invariant. We also assume that G determines an "ordinary" action, meaning that it acts transitively and locally effectively on E, and, moreover, its prolonged actions $G^{(n)}$ are also locally transitive on a dense open subset of $\mathbf{J}^n$ for all $0 \leqslant n \leqslant r-2$, where r is the dimension of G. (In the language of [17], G admits no pseudo-stabilization of the prolonged orbit dimensions.) Indeed, Lie's complete classification of all finite-dimensional transformation groups on the plane, [15, 17], shows that, of the transitive groups, only the elementary similarity group $(x, u) \mapsto (\lambda x + c, \lambda u + d)$ and some minor variants thereof fail this hypothesis. Under these assumptions, the G-invariant arc length has order

$n \leqslant r-2$ and the G-invariant curvature $\kappa(x, u^{(r-1)})$ has order exactly $r-1$. The solutions to the ordinary differential equation

$$\kappa(x, u^{(r-1)}) = c, \tag{10.1}$$

for c constant, determine the curves of constant curvature for the group action. In fact, one does not need to integrate the ordinary differential equation (10.1), since these curves can be found directly from the group action.

PROPOSITION 10.1. *A curve $\Gamma \subset M$ has constant G-invariant curvature if and only if it is the orbit, $\Gamma = \exp(t\mathbf{v})\, P_0$, of some point $P_0 \in M$ under a one-parameter subgroup* $\exp(t\mathbf{v}) \subset G$ *determined by an infinitesimal generator* $\mathbf{v} \in \mathfrak{g}$.

Thus, for the Euclidean group, we recover the circles and straight lines as the constant curvature curves, while for the special affine group these are the conic sections. Since (10.1) has order $r-1$, given r points $P_1, ..., P_r \in E$ in "general position," there exists a unique constant curvature curve $\Gamma_0(P_1, ..., P_r)$ passing through them. Let $\kappa_0(P_1, ..., P_r)$ denote its curvature. Since (10.1) is a G-invariant ordinary differential equation, $\kappa_0(P_1, ..., P_r)$ is a joint invariant of the r points.

Let $\Gamma \subset M$ be an arbitrary curve in the plane. We are interested in constructing a G-invariant finite difference approximation to its G-invariant curvature $\kappa(P_1)$ at a given point $P_1 \in \Gamma$ in the curve. Choose $r-1$ nearby points $P_2, ..., P_r \in \Gamma$. Then the curvature $\kappa_0 = \kappa_0(P_1, ..., P_r)$ of the constant curvature curve $\Gamma_0 = \Gamma_0(P_1, ..., P_r)$ passing through the points determines our approximation to $\kappa(P_1)$.

Conjecture. The following series expansion holds:

$$\kappa_0 = \kappa + \frac{1}{r}\left(\sum_{i=1}^{r} L_i\right)\frac{d\kappa}{ds} + \frac{1}{r(r+1)}\left(\sum_{1 \leqslant i \leqslant j \leqslant r} L_i L_j\right)\frac{d^2\kappa}{ds^2} + \cdots, \tag{10.2}$$

where κ, $d\kappa/ds$, etc., are evaluated at P_1, and

$$L_i = \int_{P_1}^{P_i} ds \tag{10.3}$$

denotes the G-invariant "distance" from the point P_1 to P_j, measured as the G-invariant arc length along the constant curvature curve Γ_0. (In particular, $L_1 = 0$.) The expansion assumes that all the arc lengths L_i are small.

EXAMPLE 10.2. Consider the translation group $(x, u) \mapsto (x+c, u+d)$. In this case, $\kappa = du/dx$, and the constant curvature curves are the straight lines. Then $\kappa_0(p_1, p_2) = (u_2 - u_1)/(x_2 - x_1)$. Therefore, the expansion (10.2) is merely the Taylor series, and so is valid to general order! (Note that since dx is the translation-invariant arc length, the "length" of a straight line segment is $\int_{P_1}^{P_2} dx = x_2 - x_1$.)

Thus, the conjectured series expansion (10.2) is valid up to order 2 for the translation group, the Euclidean group, and the special affine group. Direct verification for other planar groups appears to be problematic because the formulas for the finite difference approximation κ_0 are not as easy to come by, because the constant curvature curves involve transcendental functions. Indeed, more detailed computations demonstrate that, for the Euclidean group, the natural generalization of (10.2) is not valid to order 3.

11. AFFINE CURVATURE FLOW

In this section, we will present several new solutions to an affine invariant nonlinear heat equation which arose out of certain problems in vision and image processing. Indeed, invariant theory has recently become a major topic of study in computer vision. Since the same object may be seen from a number of points of view, one is motivated to look for shape invariants under various transformations. A closely related topic that has been receiving much attention from the image analysis community is the theory of scale-spaces or multiscale representations of shapes for object recognition and representation; see [12] for an extensive list of references on the subject. Initially, most of the work was devoted to linear scale-spaces derived filtering using a Gaussian kernel or equivalently by running the shapes through the linear heat equation. Here the variance of the filter (or equivalently, the time of the heat equation) plays the role of a scale-space parameter. The greater the variance (or time), the more the given shape is smoothed. Of course, the diffusion being isotropic, the shape will be blurred as well. To remedy such problems, in the past few years, a number of nonlinear and geometric scale-spaces have been investigated as well. The idea is to introduce a nonlinear smoothing which preserves edges while it smoothes on either side of the edge. Such nonlinear smoothing methods may be found in [21, 4, and 13]. See also the references in [12]. The combination of invariant theory and geometric multiscale analysis was investigated in [24] and [25]. There, the authors introduced an affine invariant geometric scale-space. Part of this work was extended to other groups as well in [18, 19]; see also [4]. The shape representations which we derive allow us to compute invariant signatures at different scales and in a robust way. These flows are already being used with satisfactory results in various applications [12].

The affine curvature flow was introduced precisely to give an affine invariant multiscale representation of planar shape. For closed convex curves, the affine curvature flow is found by evolving the curve in the direction of its affine normal, proportional to the affine curvature, and so has the form

$$\mathbf{x}_t = \kappa \mathbf{n}, \tag{11.1}$$

where κ is the affine curvature and $\mathbf{n}$ is the affine normal. Interestingly, although the affine curvature depends on fourth order derivatives of the parametrized curve, we can replace (11.1) by a much simpler flow whose image curves are the same, differing only by an inessential reparametrization. Specifically, we rewrite the affine normal $\mathbf{n}$ in terms of the tangential and Euclidean normal directions to the curve, leading to an evolutionary flow of the form

$$\mathbf{x}_t = \sqrt[3]{\kappa_e}\, \mathbf{n}_e + F\mathbf{t}, \tag{11.2}$$

where κ_e is the *Euclidean* curvature, $\mathbf{n}_e$ is the *Euclidean* (inward) normal, and F is some function of the curvature and its derivatives whose precise form is irrelevant. Indeed, since the tangential component $F\mathbf{t}$ in (11.2) only affects the reparametrization of the image curve, this term can be safely omitted (or even replaced by any other convenient tangential term $\tilde{F}\mathbf{t}$ if desired). As a result, the flow (11.1) can be written in the equivalent (but non-affine-invariant) form

$$\mathbf{x}_t = \sqrt[3]{\kappa_e}\, \mathbf{n}_e. \tag{11.3}$$

Note that, in the form (11.3), the flow can be extended to nonconvex curves—inflection points have zero Euclidean curvature, but that does not cause any difficulties for either the Euclidean normal or curvature, even though the affine curvature is not well defined at such points. In particular, if the curve is given as the graph of a function $y = u(x)$, then the normal version (11.3) of the affine curvature flow has the particularly simple form

$$u_t = \sqrt[3]{u_{xx}}. \tag{11.4}$$

An alternative, useful formulation of the affine flow is obtained by taking the cross product of (11.2) with the tangent vector $\mathbf{x}_r$ to the parametrized curve $\mathbf{x}(r, t)$ at time t. Using the formula (3.4) for the Euclidean curvature of a parametrized curve $\mathbf{x}(r) = (x(r), y(r))$, and the fact that $\mathbf{x}_r \wedge \mathbf{n}_e = |\mathbf{x}_r|^2$, we find that the components $x(r)$, $y(r)$ of $\mathbf{x}$ must satisfy the underdetermined second order partial differential equation

$$(\mathbf{x}_r \wedge \mathbf{x}_t)^3 = \mathbf{x}_r \wedge \mathbf{x}_{rr} \qquad \text{or} \qquad (x_r y_t - x_t y_r)^3 = x_r y_{rr} - x_{rr} y_r. \tag{11.5}$$

Conversely, since $\mathbf{x}_r$ is parallel to the unit tangent vector $\mathbf{t}$, given a solution to (11.5), it must also satisfy (11.2) for some choice of tangential component $F\mathbf{t}$, and hence the individual image curves for each value of t will describe a solution to the affine curvature flow (11.1) (although they will not necessarily have the correct parametrization to satisfy the normal version (11.3)).

Very few nontrivial particular solutions to the affine curvature flow are known. First, the curves of constant affine curvature will evolve analogously to the Euclidean circles. Thus, an initial ellipse will remain elliptical, having the same eccentricity and orientation, while collapsing to a point. Moreover, recent results of Angenent *et al.* [5] demonstrate that a smooth closed embedded curve remains regular, eventually shrinking to a point.

There is an affine analogue of the grim reaper. If we choose the vertical axis to translate along, the solution is obtained by setting $u_t = c$, for constant c in (11.4). The result is a vertically translating parabolic solution

$$y = u(x, t) = \tfrac{1}{2} c^3 (x - x_0)^2 + c(t - t_0). \tag{11.6}$$

We have also constructed some examples of solutions having inflection points. The simplest example is found by assuming $u_t = cx$ in (11.4). Omitting an inessential integration constant, the resulting solution is

$$y = u(x, t) = \tfrac{1}{20} c^3 x^5 + ctx,$$

which, at each value of t, has a stationary inflection point at the origin. The successive image curves of the "quintic fan" are depicted in Fig. 10.

To produce explicit solutions with a nonstationary inflection point is more difficult. The following procedure allows us to describe another solution having a moving inflection point. In fact, the whole curve precesses under the one-parameter affine flow $(x, y) \mapsto (x + s, y + sx + \frac{1}{2}s^2 + cs)$, with infinitesimal generator $\partial_x + (x + c)\,\partial_y$, where c is a constant. With this in mind, we consider the parametrized family of curves

$$\mathbf{x}(r, t) = (w(r) + t, \tfrac{1}{2}(w(r) + t)^2 + br + ct), \tag{11.7}$$

where $w(r)$ is to be determined. Substituting into (11.5), we find that w must satisfy the reduced ordinary differential equation

$$bw_{rr} = bw_r^3 - (cw_r - b), \tag{11.8}$$

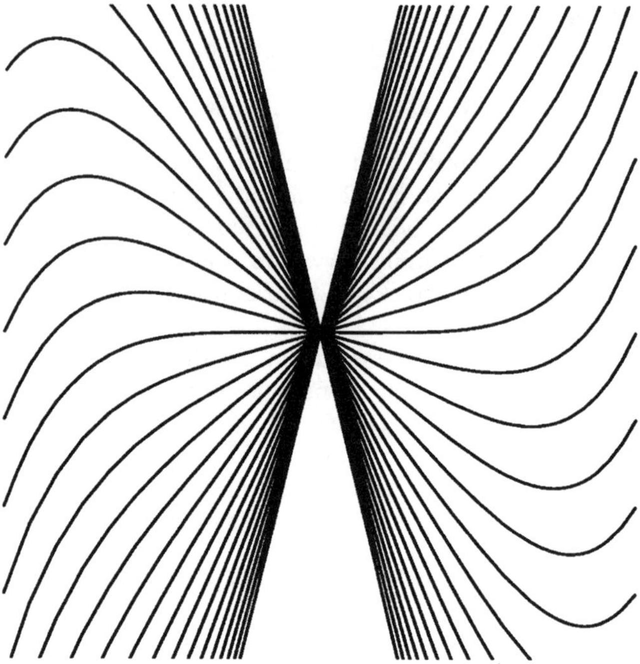

FIG. 10. Quintic fan.

which can clearly be integrated by quadrature. Let us specialize to the case $b = c = 1$, whereby (11.8) reduces to the first order differential equation

$$\frac{dz}{dr} = 1 - 3z + 3z^2 \tag{11.9}$$

in $z = w_r$. Solving (11.9) (and ignoring integration constants), we find

$$z(r) = \frac{1}{2} + \frac{1}{2\sqrt{3}} \tan \frac{\sqrt{3}}{2} r, \qquad w(r) = \frac{1}{2} r - \frac{1}{3} \log \cos \frac{\sqrt{3}}{2} r, \tag{11.10}$$

which, upon substitution into the defining equation

$$\mathbf{x}(r, t) = (w(r) + t, \tfrac{1}{2}(w(r) + t)^2 + r + t), \tag{11.11}$$

produces the desired solution. The curve looks like a version of the Euclidean grim reaper, but bent around a parabola. It has a triple inflection point on the interior half of the curve, close to the point of maximum curvature. Since $\mathbf{x}_r \wedge \mathbf{x}_{rr} = 0$ at the inflection point, it is explicitly given by

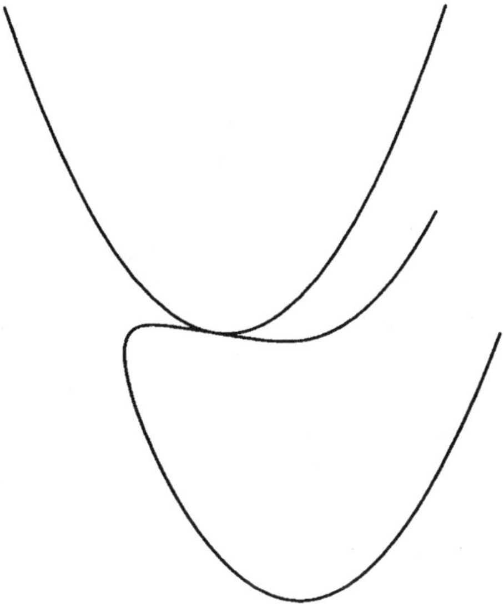

FIG. 11. Parabolic scooper with inflection point locus.

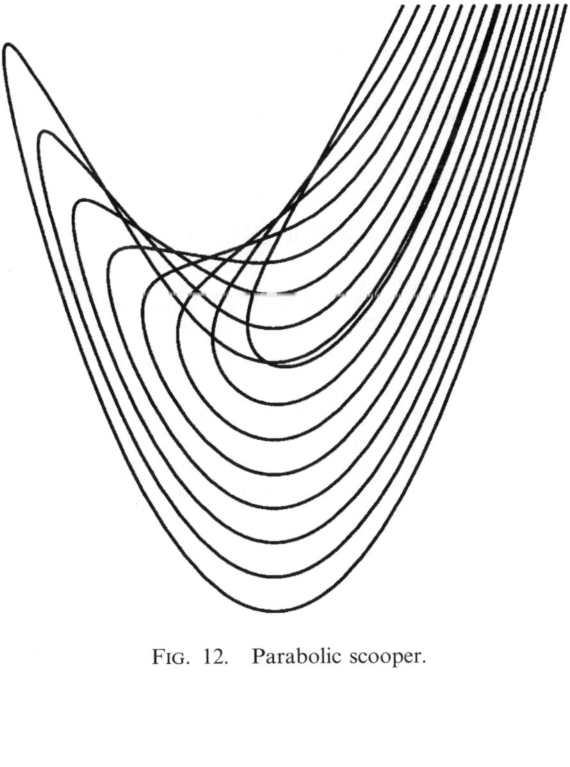

FIG. 12. Parabolic scooper.

solving $z = w_r = 1$, so that $r = 2\pi/3\sqrt{3}$. Thus, the inflection point traces out a parabola

$$y = \frac{1}{2}x^2 + x + \frac{\pi}{3\sqrt{3}} - \frac{1}{3}\log 2;$$

see Fig. 11 for a picture of the curve (at $t=0$) and the parabolic locus traced out by the inflection points. The actual curve evolution, which we propose to call the "parabolic scooper," is depicted in Fig. 12; note the inflection parabola reappearing as an envelope to the curve family.

An important point of interest of this example is that it describes the asymptotic behavior in the small of any (triple) inflection point that is precessing. Indeed, translating the inflection point for $t=0$ to the origin, one finds the following convenient reparametrization

$$x(r) = -\int \frac{dr}{1+r^3} = -r + \frac{1}{4}r^4 - \cdots,$$

$$y(r) = \frac{1}{2}x(r)^2 - \int \frac{r\,dr}{1+r^3} = -\frac{1}{20}r^5 + \frac{11}{112}r^8 - \cdots,$$

clearly demonstrating the nature of the inflection point.

Finally, for numerical implementations of the affine flow based on the Osher–Sethian level curve algorithm [20] as well as explicit applications to invariant planar curve representation, see [25]. Implementations based on the applications of our affine-invariant finite difference approximations are under development and will form the subject of a subsequent work.

REFERENCES

1. M. Ackerman and R. Hermann, "Sophus Lie's 1880 Transformation Group Paper," *Math. Sci. Press*, Brookline, MA, 1975.
2. M. Ackerman and R. Hermann, "Sophus Lie's 1884 Differential Invariant Paper," *Math. Sci. Press*, Brookline, MA, 1976.
3. L. Alvarez, P. L. Lions, and J. M. Morel, Image selective smoothing and edge detection by nonlinear diffusion, *SIAM J. Numer. Anal.* **29** (1992), 845–866.
4. L. Alvarez, F. Guichard, P. L. Lions, and J. M. Morel, Axioms and fundamental equations of image processing, *Arch. Rational Mech.* **123** (1993), 199–257.
5. S. Angenent, G. Sapiro, and A. Tannenbaum, On the affine invariant heat equation for nonconvex curves, preprint in preparation.
6. W. Blaschke, "Vorlesungen über Differentialgeometrie," Vol. II, Springer-Verlag, Berlin, 1923.
7. V. A. Dorodnitsyn, Transformation groups in net spaces, *J. Sov. Math.* **55** (1991), 1490–1517.

8. V. A. Dorodnitsyn, Symmetry of finite difference equations, *in* "CRC Handbook of Lie Group Analysis of Differential Equations," Vol. 1, (N. H. Ibragimov, Ed.), pp. 349–403, CRC Press, Boca Raton, FL, 1994.
9. M. Gage and R. S. Hamilton, The heat equation shrinking convex plane curves, *J. Differential Geom.* **23** (1986), 69–96.
10. M. Grayson, The heat equation shrinks embedded plane curves to round points, *J. Differential Geom.* **26** (1987), 285–314.
11. M. L. Green, The moving frame, differential invariants and rigidity theorems for curves in homogeneous spaces, *Duke Math. J.* **45** (1978), 735–779.
12. B. ter Haar Romeny (Ed.), "Geometry Driven Diffusion in Computer Vision," Kluwer, Dordrecht, 1994.
13. B. Kimia, A. Tannenbaum, and S. Zucker, On the evolution of curves via a function of curvature, *J. Math. Anal. Appl.* **163** (1992), 438–458.
14. A.-M. Li, U. Simon, and G. Zhao, "Global Affine Differential Geometry of Hypersurfaces," de Gruyter, New York, 1993.
15. S. Lie, Theorie der Transformationsgruppen I, *Math. Ann.* **16** (1880), 441–528; also "Gesammelte Abhandlungen," Vol. 6, pp. 1–94, Teubner, Leipzig, 1927; see [1] for an English translation.
16. S. Lie, Über Differentialinvarianten, *Math. Ann.* **24** (1884), 537–578; also "Gesammelte Abhandlungen," Vol. 6, pp. 95–138, Teubner, Leipzig, 1927; see [2] for an English translation.
17. P. J. Olver, "Equivalence, Invariants, and Symmetry," Cambridge Univ. Press, Cambridge, 1995.
18. P. J. Olver, G. Sapiro, and A. Tannenbaum, Classification and uniqueness of invariant geometric flows, *Compt. Rend. Acad. Sci. Paris, Sér. I* **319** (1994), 339–344.
19. P. J. Olver, G. Sapiro, and A. Tannenbaum, Differential invariant signatures and flows in computer vision: a symmetry group approach, *in* "Geometry-Driven Diffusion in Computer Vision" (B. M. Ter Haar Romeny, Ed.), pp. 205–306, Kluwer, Dordrecht, 1994.
20. S. Osher and J. Sethian, Front propagation with curvature dependent speed: Algorithms based on Hamilton-Jacobi formulations, *J. Comput. Phys.* **79** (1988), 12–49.
21. P. Perona and J. Malik, Scale-space and edge detection using anisotropic diffusion, *IEEE PAMI* **12** (1990), 629–639.
22. E. Salkowski, "Affine Differentialgeometrie," de Gruyter, Berlin, 1934.
23. G. Salmon, "A Treatise on the Higher Plane Curves," 3rd ed., Hodges, Foster, and Figgis, Dublin, 1879.
24. G. Sapiro and A. Tannenbaum, On affine plane curve evolution, *J. Funct. Anal.* **119** (1994), 79–120.
25. G. Sapiro and A. Tannenbaum, Affine invariant scale-space, *Int. J. Computer Vision* **11** (1993), 25–44.
26. G. Sapiro and A. Tannenbaum, On invariant curve evolution and image analysis, *Indiana J. Math.* **42** (1993), 985–1009.
27. A. P. Schirokow and P. A. Schirokow, "Affine Differentialgeometrie," Teubner, Leipzig, 1962 [German translation of Russian original].
28. B. Sturmfels, "Algorithms in Invariant Theory," Springer-Verlag, New York, 1993.
29. I. Weiss, Geometric invariants and object recognition, *Int. J. Comput. Vision* **10** (1993), 207–231.
30. H. Weyl, "Classical Groups," Princeton Univ. Press, Princeton, NJ, 1946.

Printed in Belgium

Where Does Geometry Go? A Research and Education Perspective

Jean-Pierre Bourguignon, Eugenio Calabi, James Eells,
Oscar Garcia-Prada, and Misha Gromov

Fernández, Marisa (ed.) et al.
Global Differential Geometry: The Mathematical Legacy of Alfred Gray,
held in Bilbao, Spain, September 18–23, 2000, 442–457,
Contemp. Math., 288, Amer. Math. Soc., Providence, RI (2001)

ABSTRACT. After brief statements by panel members, the topic of the round table has been discussed with the audience. The main points that emerged from the exchange are: the evolution of geometry in the past century and years to come, the extent to which geometers are doing enough efforts to keep informed of geometric works done by other scientists, the impact of computer tools on the future of geometry, the future ability of mathematicians, and geometers in particular, to attract good students.

Jean-Pierre Bourguignon (JPB): It was the organisers' idea to close the conference with a round table. They wanted to take advantage of the proclamation of the year 2000 World Mathematical Year to look back and also to give us an apportunity to try and project ourselves into the future.

For a mathematician it is very tempting to draw a perspective comparing the year 2000 to the year 1900, which is remembered through David Hilbert's lecture at the International Congress of Mathematicians held that year in Paris, in which he proposed his famous list of 23 problems. If one now looks back at Hilbert's list, it is a fact that geometry as a whole is not that much present. Indeed many more problems are devoted to algebra. It is interesting to contrast this with the fact that the 19th century has been extremely fruitful as far as geometry is concerned. Recall, for example, that in the early part of that century the fundamental barrier of having a unique model for geometry had been broken. This major event in the history of mathematics was followed up by many others. We can quote a few. the groundbreaking work of Bernhard Riemann establishing Riemannian geometry as a comprehensive subject in which curvature plays a central role; the Erlangen programme of Felix Klein, identifying a geometry with the set of properties invariant under a group action; and something perhaps less well known, the systematic use of geometric models to describe natural phenomenon by Hermann von Helmholtz.

1991 *Mathematics Subject classification.* Primary 53; Secondary 51, 52, 97.

The participants to the round table thank Marisa Fernandez and her co-organizers for the great efforts they put into many details to set up the conference which resulted in a wonderfully congenial atmosphere.

DOI: 10.1090/conm/288/04867

The century closed on a great variety of works, from which I would like to single out the work of Sophus Lie on continuous transformation groups.

At that time, that is 1900, nobody could really foresee that in the 20th century geometry would undergo a major evolution and play a central role in the development of mathematics. This is indeed the impression that comes out of Henri Poincaré's attempt to propose some avenues for the future of mathematics in the 20th century in his opening address to the Paris Congress. Nevertheless, it is a fact that this century has really been a great century for geometry, in particular in its relations with physics, to start with nothing less than the big boost given to these relations by Albert Einstein's theory of general relativity.

So, here we are today, trying to look into the future, and asked to point to some directions in which, according to our opinion, geometry will develop.

We know how daring this exercise is. Nevertheless, I personally find such an endeavour very appropriate at the end of a conference devoted to the memory of Alfred Gray. Alfred definitely contributed to many areas of geometry. At the end of his life he was actively engaged in convincing colleagues that one should really take advantage of the new tools provided by computers to make good propaganda for geometry in order to help form new insights through images.

The format proposed is to have the panelists make brief statements. The most important part of the morning should be what I hope will be a very lively and open discussion with people in the audience.

Eugenio Calabi (EC): Over the years, I came to the conclusion that one can only speak very subjectively about what constitutes geometry. I see it as a field in which the primary source of intuition is ultimately tied with our sensory perceptions of the world. Of course, as we reach more abstract areas, one has to interpret what sensory experience means. And I've tried to make that as visible as possible and to communicate this view. I must say that communication can be difficult in that respect, something that made me very much concerned with the process of education in this area... although I've never been known to be an effective lecturer at the undergraduate level.

The idea of communicating what is our source of intuition has been changed by the big impact the advent of computers had - note that it is only twenty years ago that I saw the first one in operation. And that affects only the last one fifth of the century. In the future it will have an impact that is now difficult to evaluate in full. For me, geometry is as essential as other areas of mathematics such as analysis, algebra, combinatorics, statistics, probability. It is an applied field because being very largely problem-solution oriented. We are bugged by different problems that we try to solve, and we try to master the tools that seem to bring us closer to the solutions or give us more insight on them until, being very optimistic, we really make a breakthrough. But I've gone through periods when I felt I had to study algebra and analysis, functional analysis and so on.

Aside from that, I would like to know how my colleagues sitting here feel about the same problem: what constitutes geometry? It's certainly very different from what I thought of as geometry when I was a high-school, or even a college, student.

Oscar Garcia-Prada (OGP): The point I want to emphasise is the interplay between geometry and physics, and in particular how physics is a source of very challenging and deep problems in geometry.

Here are a few instances of this interplay tailored to my taste and to my knowledge and limitations. They are coming most of the time from quantum physics, and in particular from the understanding of the geometry of gauge theories:

- the influence of general relativity on differential geometry;
- the influence of classical mechanics on symplectic geometry;
- the appearance of instantons in Yang-Mills' theories and their use by Simon Donaldson to actually define invariants of 4-manifolds, using the moduli spaces of instantons, as structuring objects.

In many of the examples, I see a path going from physics to differential equations and partial differential equations, continuing to differential geometry and algebraic geometry, and finally arriving at topology. In fact this connection stresses the deep relations existing between differential geometry and algebraic geometry. In particular in the example I just mentioned, the moduli of instantons has a reincarnation in algebraic geometry in the moduli space of algebraic bundles. And this link was the tool used by Donaldson to compute his invariants.

There are many instances where topological field theories have been used to define knot invariants such as Jones-Witten theory, leading to substantial applications to low dimensional geometry and topology. There are also important applications from physics to higher dimensional geometry such as the contribution of Yang Mills' theories to the study of Kähler groups via the theory of Higgs bundles initiated by Hitchin and Simpson, leading to the solution of problems that people like Gromov and others had studied.

The list is very long. Coming back to relativity theory I want to mention the development of twistor theory by Penrose some thirty years ago. For him this was a way of understanding space-time in a deeper way, trying to solve Einstein's equations by converting them into complex geometric terms, in a more fundamental space that he calls twistor space. This dictionary has become an important tool in the study of differential, in particular Riemannian, geometry, for the study of subfields such as quaternionic manifolds, harmonic maps and many other things. This lead to the explicit solution of instanton equations via the beautiful construction of Atiyah-Drinfeld-Hitchin-Manin.

These are also instances in which somehow an important aspect of the interaction is a general idea of symmetry, such as the recent and challenging idea of supersymmetry and the understanding of the geometry underpinning it. Many beautiful theories are coming from supersymmetric theories, like e.g. Seiberg-Witten theory and a deeper understanding of Morse theory. In the same list one finds the systematic study of Calabi-Yau manifolds by physicists and their discovery of a new type of symmetry, mirror symmetry, with a new emphasis on interesting objects such as special Lagrangian submanifolds.

The point I want to make is that today a working geometer should always be looking for new ideas coming from physics. Since we are supposed to discuss here also about education and how the teaching of geometry should be done in universities and schools, I would like to point out that physics must be part of the curriculum of the undergraduate degree in mathematics. And this is almost never the case in Spain.

James Eells (JE): When I got this invitation from Marisa, or I should say order from Marisa, to participate in this round table, of course I accepted. It's hard to refuse Marisa, as you know, but I certainly entered with a good deal of trepidation.

My real, serious question is: is this worthwhile at all? For instance, I found that I could take the list of topics and suggestions that Jean-Pierre [Bourguignon] had put down and, systematically going through it, I could replace "geometry" by any one of many other terms of significance. I mentioned at breakfast this morning that one could replace it by "beauty" as a concept and get just as interesting and in my view just as meaningful a formulation of a problem.

I was a little bit nervous about this round table because the one thing that I would find quite unfortunate and inappropriate would be that we should legislate too much, in our own minds at least, on what geometry is and where it's going, and what should be included, let's say at the research level. The basic viewpoint that I personally have is that whatever geometry is, in my study about whatever subject I deal with, it is very important to think geometrically. I can't do it otherwise. And certainly when I think of analysis, as I remarked to Jean-Pierre [Bourguignon], I certainly think in geometric terms. I had the very good fortune of having three very great influences in my formative years in mathematics. Coming from English Literature I went to Harvard and studied with Hassler H. Whitney. We never spoke a word of mathematics, I might say, in the forty years of our friendship, but of much else. And then, surprising for many, Jean Leray, who treated me like a son. I don't know how many people can say that, but I'm very proud to be able to say it from the heart. Then Victor Hurewicz. If any of you ever knew Victor Hurewicz you knew a man of just unbounded love for all human aspects of life. These mathematicians certainly taught me above all not to make sharp delimitations of what is geometry or what is algebra or what is what. If you look at the works of any of those three splendid teachers of mine, you'll see how much they influenced me in being absolutely unrestricted in their spheres of interest.

The most important single point about which I had concerns here was that this limitation should not be formulated in any way that is morally binding or even predicated in our thinking. And with that in mind I pass the word to Misha Gromov, who is certainly one of the most universal amongst us, and I dare say he will have very clearly defined views.

Misha Gromov (MG): I would like to continue what I was saying in my lecture *"Separating inside and outside views of mathematics"*. You have just heard the inside view. So let me propose an outside one. It will be a specific outside view on geometry and geometers given by one of the most powerful geometric minds of our time whom I met. You may not know him, his name is Manfred Eigen. He studied physics and invented a new type of spectrography. This is a remarkable technique, based on geometry. And then he worked on chemical evolution. But in some sense the kind of geometry he does, geometers don't do it and don't know it. When I first met him, I didn't quite understand. I asked him how he understood geometry so well, and then I realised, after somebody explained to me later, his major work in physics is based on how the geometry of high dimensional space gets projected to lower dimensions. He doesn't just analyse them. He found specific formulas and technology for how to measure relaxation times for interactions between molecules, and also the association/disociation of water.

His next contribution was to the theory of chemical evolution. He actually made some specific things with his hands, and gave a very elegant model of evolution specifically applicable for viruses and bacteria. It very well applies to the evolution of geometry and geometers. The model is a kind of random cloud of points moving in a landscape following a gradient exhibiting small perturbations. This determines some critical values, and depending on these a variety of different behaviours. And when a parameter reaches a specific value, ones goes to a deep well and stay there for a very long time. One may become extinct if there is no fluctuation there. If one gets out, successful development will occur, depending on a particular range of parameters, provided there are not too spread out in order that species preserve their identity, and they rarely make big jumps in order not to stagnate.

When I think of geometers, I see a major point in the jumps. We should give more attention to them. In my view, we are stagnating. We don't make jumps. We sit in a very narrow niche and we don't even know about great geometric developments happening next door.

This is my criticism of present day geometers. My suggestion is that we must make efforts to allow big jumps, and not idealise what we do and say how great we are. This is really why I feel we are in a still very primitive stage. We don't know the basic facts about space. Since Riemann we haven't progressed at all, and even spaces studied by other people like physicists, e.g., the huge evolution spaces studied by Eigen, we are not even aware of them. Really we are in very poor shape. That's my statement.

JPB: Thank you, Misha, for these provocative thoughts.

We should really start the discussion now. I would just add for myself two comments. The first one is that it's extremely hard to identify what we need to do and to make predictions. It's very important that we explore many predictions so that we don't have only a one-way street ahead of us.

Nevertheless, we still make decisions, e.g., we chose topics of courses and we design curricula. This is in fact our everyday responsibility as teachers. We actually train the next generation in a certain way, and this is not neutral. Of course training is not the end of the story because, as you know, the best students are the ones who, after some time, completely forget the details of the training and just keep its spirit. That was my first point.

The second point concerns boundaries. Certainly one of the worst things that could happen for the development of our discipline would be to dissect mathematics in a way that finally creates very strong separations between its subfields. In relation with this question, at the end of the 20th century, within geometry one witnessed a blurring of the boundaries between the various types of geometry studied today. One has for example realised that complex geometry is not that far away from real geometry, that symplectic geometry is not that far away from Riemannian geometry in some ways. Another idea is exemplified by the study of moduli spaces of structures. It's indeed important, when studying a specific structure, to look at neighbouring ones, something that physicists have been using consistently over the years with great success. Mathematicians have been slow in getting interested in this type of questions, and were more or less forced into it, because they were asked very specific questions about, say, dimensions of moduli spaces or more specific information about them. But it took time to them to really take it as an interesting object in itself. In some sense my second comment echoes what Misha [Gromov] said. It concerns the extreme emphasis that has been put for a long time on continuous structures and, correlatively, the great neglect in which in some countries, France is one of them, the discrete side of things and combinatorics in various disguises have been kept, completely separated from geometry. The result is that they remain largely ignored by many colleagues, and often considered as weak mathematics. This is one of the walls that, I think, has to be put down. We need to help our students form a better global idea of the whole field.

Let us now listen to questions, comments, statements, everything from the audience.

Luc Lemaire *(Université libre de Bruxelles)*: I have a specific question. Non commutative geometry has not been mentioned by anybody so far, and although it's difficult to make predictions about the future, do people in the panel feel that non commutative geometry is an important subject that will develop. Should we ask our students to make the rather heavy investment that studying this subject represents.

JPB: Two comments on this. The first one is that, as Misha [Gromov] said, since Riemann we haven't had any real new insight into the structure of space. It's true that non commutative geometry forces people who want to use it to change their approaches and their feelings about space.

The second comment is that non commutative geometry is a structure in which the continuum and the discrete are unified, precisely one of the features that I put forth in my preliminary point. When developing non commutative geometry, Alain Connes, and also others, were definitely inspired by physics. In the last three or four years, it has been amazing to see a number of physicists, coming from different viewpoints, who got interested in formulating their problem in the non commutative world. This applies to Ed Witten, originally skeptical about non commutative geometry, who realised that results he recently achieved could be obtained following a non commutative path.

Misha [Gromov] was asking us to look for such enlargements. In my opinion non commutative geometry is definitely a direction, but there should be steps in other directions too. I don't believe it would be a good thing to think only of a one dimensional space. One has to be ready to go in several different dimensions.

EC: This reminds me still of the other question about intergenerational communication - the form of teaching, I mean. Each generation grows up and is educated with one view of mathematics and experiences of the evolution of it, and then has to try and communicate to the next generation how one's own feelings are. I remember myself when I was in graduate school. I would be starting a course on which I thought I knew something to begin with, and when a slightly different viewpoint was being presented in the lecture room my private reaction to that was one of cold hostility. My first impression was always that the lecturer was trying to put something over on us. Almost invariably I was wrong. It was just that I had to modify my way of thinking, my way of looking at the same subject. The small amount of technical things I would learn was really over-ridden by these changes.

My next thought is that, human beings having roughly the same capacity for learning from one generation to another, we have to learn as teachers what is important and decide what is best understandable if not forgotten in our own teaching, even though we may retain some love for it. I still love the idea of geometrical constructions with ruler and compass, even though nobody ever teaches them nowadays. I would like to hear people express their views on what areas of geometry not only should be emphasised, but what areas can or may be omitted or at least areas we could take the risk of not communicating in detail.

Antonio Ros *(Granada University)* **:** I'd like to introduce an idea that is closer to the area I am interested in at the present date, which is geometry of surfaces in 3-space. This area is somewhat special inside the world of differential geometry because we have at our universities an undergraduate course called something like "classical differential geometry of curves and surfaces". In many countries and many institutions we exist as professional geometers because this course belongs to the curriculum, and if tomorrow this course disappears, then our universities and our government may not want to have differential geometers any more. In my opinion our future is linked to the future of this particular course.

On the other hand almost everywhere in the world a new geometry is emerging very quickly. I mean the kind of computational and discrete geometry that puts more emphasis on applications. For instance, every day one can hear that *"video games are a direct consequence of computational geometry"*, and this sort of things. Presidents of universities may start to wonder about the type of geometry a student must learn. And here is my question: what is the attitude we must take in order to preserve classical differential geometry and in order to convince people that this geometry is also a useful one? Basically, must we or even can we do something about this question?

JE: Had I spoken three minutes longer - and I won't speak three minutes now - I would have remarked about a concrete example or two, amongst many available at this meeting, to illustrate my point about the basic unity in mathematics as someone with a geometric taste. I would have chosen two spectacularly fine examples: One by Antonio Ros, who gave a lecture which had both a geometric spirit and an analytical one; And the other one by Eugenio Calabi (may I say "of course"), whose presentation was both analytical and geometrical, and peculiarly "Calabian". I say this with a good deal of affection.

But it does seem to me that a basic unity is at work, e.g. when one considers the basic influence geometry has when one does analysis, or probability theory, or algebra, or when one works in the direction of Jean Leray. This inter-relationship is vastly more important than course assignations. This is a point I wanted to make, and that Antonio [Ros] himself made very, very clear to me in his lecture. I told him so afterwards, so excuse the embarrassment of repeating it now, but I thought this was worth it.

Santiago Lopez de Medrano *(University Autonoma de Mexico)*: We must observe that we are reaching the end of the 20th century with a good balance between - there are different ways of saying this - geometry and analysis, pictures and theorems. But we must remember that this hasn't been the case all the time. Let me recall that Lagrange was very proud of his book on celestial mechanics, that didn't have a single picture in it. Of course, out of the work of Lagrange came a beautiful new geometry that should be called Lagrangian, but is now called symplectic.

In the middle of this century there was a big attack on geometry from which we are still suffering in schools. The emphasis was then put on anything that had to do with logic and rigour, and pictures and geometry and the thinking around geometric figures was practically banned. In research mathematics the stress was put more on algebraic and logical aspects than on geometric ones. So it is very good that we have arrived at a point where, in my opinion, there is a good balance, and of course Alfred Gray was one of the proponents of improving the balance between these aspects. But, as Professor Ros mentioned, there are new challenges, there are new dangers, maybe we are now being pushed to a situation where we will have beautiful pictures but no theorems. And in particular in teaching, there is a very strong emphasis on computers and calculators in school, and of course it is not necessarily the best aspects of calculators and computers that are taught. So we have to be very careful about this. We still have to fight for geometry. Everyone of us that had something to do with setting up programmes in high schools and universities has witnessed that geometry is always under attack.

JPB: Let me just try to pick up what you've just said. It's true that in some cases there are attempts to really wipe out geometry from programmes, but doesn't that mean also that we are just off the main stream and that we, geometers, have not made the necessary efforts to really incorporate into our teaching, or even into the tools that we are using fluently, some new tools that some people feel are really relevant for many things, including modelling. It is quite clear that doing modelling today without at some stage using a computer is just inappropriate. You know that the computer is just doing what you ask it to do. But for the moment, for example, the fact that the level of development of adapted software able to deal comfortably with images is not so satisfactory may also be the proof of our limited engagement. So there is always some balance between the pressure from outside, which most of the time is over-simplifying the situation, and that coming from inside. Incorporating computer tools into teaching is something which can in the end introduce into the minds of students the idea that after all you don't need to think geometrically since a computer is doing that for you. But, if you ease too much the understanding of geometric concepts, you may actually achieve the exact opposite.

So are there some people who have personal experience about this kind of thing or strong personal opinions? In my opinion, in the future it's something we will not be allowed to neglect. We will even have to make sustained efforts in this direction.

Karsten Grove *(University of Maryland)*: I would like to return to the beginning of the discussion of the panel, because I believe it is actually destructive to leave it where it was. From a personal point of view I consider it is very important in research to be always critical about where you are, what you do, but also to enjoy it. And the same thing goes for talking about geometry and mathematics in general. We heard from the panel what a great century this has been for geometry on the one hand, and then we hear from Misha [Gromov] what a terrible shape geometry is in. It would be too bad to leave it like that. I would like to have a much more balanced view. We should recognise and be happy about what we all know about geometry. But at the same time we should be critical as Misha is and try to do better.

On the other hand, it is always good to try and have some ideas about how we would do better. And I would like to hear more about the positive side of these views, namely not to say *"we're in terrible shape"* but *"let's see what we can really do"*. That would be very important.

Jair Koiller *(Laboratorio Nacional de Computacao Cientifica, Brazil)*: I'd like to go on with the line of questioning of Professor Grove, actually. Just to provoke Professor Gromov, because his talk was so inspiring about the views of in and out math, I think he's hiding something, because I keep hearing about the necessity of new ideas of space. We heard this twice today and once in his talk. Is it possible that he can give some hints of what the structure of space is that he has in mind, learning from his context with biologists? So that we will have a more optimistic view.

MG: You know there are two separate issues here. One is purely mathematical - what type of spaces you may envisage or expect to be there if you don't know much about them - and the other is what sorts of geometric problems are being considered not only in biology but also in everyday physics that we are not aware of. There is remarkable geometry going on, as I mentioned, in the work of Eigen. Recently there was also some very remarkable work about how one can apply geometry - and exactly the kind of geometry we are doing here - to kinetic equations. They were looked at from a geometric angle, namely via some projection on the space of solutions to the Boltzmann equation. It's very simple and clear but, when you write it down, you have huge, complicated equations, which happen to be pertinent. They describe correctly the kinetics of chemical reactions, and people doing that can improve the efficiency of these reactions and can control them better, giving some 20% better input. We don't know this geometry. If we were more involved in that, we would not complain that universities don't take us seriously. We could come to them and say *"you see, we can improve your chemistry"*, something we just do not do. I learned this recently and feel really ashamed that I never knew it before. One point for us generally is we have to learn for our research before we start thinking about how we teach our students. In my opinion we do not know enough geometry to say what geometry is anyway. These are just two instances that I have learned - I'm certain there are billions of them in science - of remarkable geometric structures, which are extremely useful and new, and we don't know about them. We have not so much to teach but to learn. That's one part. I have been doing this for a few years and found more and more things others use in elementary courses that we still don't know about. This is unpleasant to discover if you have lived all your life as a geometer.

Secondly, the geometry of space requires a more technical discussion. I can just make some suggestions on the situations where you expect the emergence of new spaces. And one of them is class field theory. There are indications of the existence of a very remarkable space in the little glimpses we have on the other side of the mirror. We need to find the proper lens. Definitely this must be part of a much bigger body of knowledge containing a mixture of symmetries. Objects of the same nature seems to appear in pure differential geometry, e.g. in Hodge theory. This determines much of the remarkable structure we observe, and we don't know what it is.

JPB: If you will allow me one word on this. Instead of claiming we already know what space is about, another approach is to look where challenges are, because very often mathematics has been changing because mathematicians were facing big challenges. Just to give an example, I was really struck by the lecture given on data analysis by David Leigh Donoho, professor of statistics at Stanford University, at the American Mathematical Society meeting *"Mathematical Challenges of the 21st century"* last August. He presented very challenging problems which in my opinion require developing some kind of a geometric vision. In the area of data analysis, because of the availability of new computer tools one can present the data more and more using pictures, and the question is then: can one have geometric structures emerge from these pictures? It is likely that an approach completely different from the one we have been developing so far will be required. Requests for such a treatment will be coming from many different fields, e.g. seismology, astronomy, molecular biology, in particular genomics, in fact from all areas which can generate huge amounts of data. There is a need to organise these data, in order to deal with them as a global object with an internal geometry that needs to be identified.

This will require on the one hand we be not too conservative and do not claim too easily we already have all the tools, and on the other hand we be not so destructive that we claim everything we learned has to be forgotten. For me such a mix is needed, but a good way to have our minds pressed to go deeper is definitely having to face some big challenges.

Joseph Wolf *(University of California at Berkeley)*: One of the things that I wanted to comment on is the use of geometry and special functions. A lot of special function theory can be put into the context of looking at a manifold of some sort, for instance a sphere, and looking at eigenfunctions of the Laplacian. Up to now, people have mostly looked at the spectrum of the Laplacian. They haven't actually looked at the eigenfunctions, except in very specific situations, for example spherical harmonics.

Now, in those specific situations often there is good classical theory which is extremely boring unless you are going to use it for something. But then it turns out to be related to something more geometric, some group action or something of that sort. Very often you have an interesting geometric situation which you discover by computing some explicit examples such as Riemannian symmetric spaces, and then turn out to be - with the right viewpoint - something more general in Riemannian geometry, for example product decomposition or de Rham theory.

So, what I suggest is that one looks to the possibility that you have a surface or a Riemannian manifold of some sort which is described in a very geometric way, and then you could hope to look for the Laplacian and try to find the eigenfunctions using the computer in some way. This is an area which simply has not been explored except, as I say, in a few special cases where there are auxiliary tools. Computer technology should provide more and more exciting ways of penetrating this open domain as it improves in the future.

EC: I just want to bring back this idea that we do have a mixed view of what geometry really is. On the one hand, we have the viewpoint that geometry evolves from axioms to theorems. On the other hand, there is the definition at the opposite end that geometry is a part of mathematics which draws its intuition ultimately from our perceptions of physical experiences with the real world. Now, you can be extreme in one way or the other. From the standpoint of the perception from the physical world, that empirical view of geometry could be regarded as science fiction, if you push it to extremes. Real numbers which were initially conceived as ratios of measurement of quantities became mathematically efficient only when you apply the Jonathan Swift principle that everything can be re-scaled up and down indefinitely often, which is in fact completely contrary to our current conception of the real world. So this ambiguous origin, namely being on the one hand very rigidly tied up with logical reasoning and rooted imaginatively in our perceptions of the world, is in part the charm of geometry, but more generally of mathematics.

MG: For me, the position of logic versus intuition is purely emotional. You cannot justify it in any way, and moreover this may be a negative phenomenon because we try to use words with connotations and conceptions that actually don't exist.

EC: I will agree with you that logic is a process of instinctive acceptance, according to rules that most of us agree upon. Otherwise, if everybody challenges every argument that's being presented ...

MG: No but there is a particular specific issue here. We believe there is a foundation of logic, however we fail to prove it, and the proof would be making an artificial proof system.

Joseph Wolf: One has to admit that there is some basic level on which we understand each other, or at least understand the words that other people are using. So you cannot say that our notion of logic is completely wrong. Perhaps it leaves something to be desired.

MG: Indeed it is not completely wrong, but it's not what helps us understand how mathematics work. For example in communications we get the vibration of air and the theory of sound. There again is a basic ingredient which probably makes sense of everything we know. First it is the absence - manifested by the absence of a computer program of this nature. People believe that logic responds to thinking, they try to come up with such a programme and fail. So when you're correcting it, it's like religious belief. But if you want to make a scientific theory, completely forget it and go from a different angle.

Simon Gindikin *(Rutgers University)*: For me, we have to face a very painful problem, about which our friends from Spain and Mexico already talked. What we can observe now is that mathematics has lower and lower priority in the world and also that less and less young people want to become mathematicians. This is at least the case in the USA where I am now. I have worked for 40 years in mathematics, and sometimes I think this was the best time. It may be that the best possibility for mathematics is in totalitarian countries, as it was in the Soviet Union, where the best young people went into mathematics. I believe we have a problem there. For example in the USA at my university nobody would think to teach physics as part of the education of professional mathematicians, and only 5 students every year take the differential geometry course.

There is one point we did not discuss here, namely the relationship between pure and applied mathematics. I understand perfectly that to sell mathematics and obtain some grants we need to insist that mathematics has a lot of applications. But let us be honest with ourselves: how did the whole thing start? In Plato's academy future kings studied geometry as first subject, not military arts or economics. But after they developed their minds, they could do many things. We now observe that, when good mathematicians after their PhD's cannot find a reasonable job, they go to Wall Street and apply not exactly what they studied during their PhD or graduate education but use the mind that was the result of their mathematical education. This helps even in such a strange environment as Wall Street.

So, in my opinion, it is very important to keep the traditional role of a mathematical education, not as direct and applied information, but as part of the general culture, for the sake of developing reasoning ability. And I must say I am very pessimistic at least in my environment in New Jersey.

JPB: I'll challenge what you just said. My starting point is that, in recent years, the call for mathematical knowledge in a very broad sense came from many different areas of industry and services with an unprecedented strength. Usually, when a mathematician is employed by a company, his training, in particular if he has gone as far as obtaining a PhD, is used more comprehensively than through his or her academic speciality. But at the same time the knowledge needed to deal with very different types of problems that one meets outside academia is extremely broad. Note that this also corresponds to the present situation in mathematics where ones has to borrow knowledge from many different sub-areas. So I don't feel appropriate to state that, if one is employed as a mathematician somewhere, one has to forget one's mathematics. It has to be used in a different way, but certainly for more than just the training to reasoning. One is actually forced to use various sorts of mathematics, some of which simple, some more advanced, and sometimes even some waiting to be developed.

This new situation has to do with resources presently at the top of the shop list for industry: more and more formal data, more and more things of a virtual nature, more and more complex systems... Mathematics is precisely the science which has been developed to deal with such data. So, in a sense, in terms of interactions with society in general and more specifically concerning jobs to be offered to students, I would be more worried about the future of my discipline if I was a physicist.

It is true though that we, as mathematicians, tend to see the choice made by some of our very best students to take jobs outside academia as a loss for the whole community. My viewpoint on this is that we have to be less definitive about it, because, if we have people in these positions who have a great respect for mathematics because they enjoyed doing it, they will be our friends, or even our ambassadors. They will help us develop mathematics possibly in different ways. One of the best ways in which mathematics has developed over the centuries has been by coping with challenges coming from many different areas. It is true that sciences, and natural sciences in particular, have been one of the absolute main sources of challenges for mathematicians, but there are many others. In his lecture, Misha [Gromov] has been stressing challenges coming from biology, but there will also be many challenges coming from the services, human activities that didn't exist at that level and with this extension in the past. As you probably know, the development of probability theory came originally from problems connected to census, insurance, and demography.

Therefore, I would be less pessimistic than Simon Gindikin. Again we, as teachers, must take a broader view on what training is to be offered to our students and what relations we can keep with them. Of course when they go and work in companies, we don't see them as often as if they were in research institutes, but we have to find mechanisms by which we still are in contact with them, so that they can come back to us when they see challenges they feel are deep enough to be addressed in themselves. I am sure many such challenges will be brought to us. I say this on the basis of my personal experience with students at L'École Polytechnique. Some of them having left completely the scientific world and working as engineers come up to me with really challenging questions. In fact they have still sufficient trust in mathematics to call me up and pose questions to me as a mathematician. This is one of the reasons why I am less pessimistic than Simon, but I may be wrong.

MG: I absolutely agree with what Jean-Pierre [Bourguignon] said but the main point is that it's up to us to do this, and to broaden our perspectives. And we will have to work hard to achieve this. In particular we will have to learn things we presently don't know.

Luc Lemaire: I also agree with the point of view just expressed about the way to attract people to mathematics. If we present the situation in the way we see it, we will be tempted to say that the good guys remain at the universities and somehow are paid to teach and given free time to do research. This leads to a very small mathematical family not really connected to the outside world, and in particular it makes us regret when mathematicians leave our small world. The risk is that we shall dry out somehow, and also we won't be able to attract young people. On the other hand, if more and more often good mathematicians go outside academia and obtain positions in which they can use their mathematics in very different ways, in ways visibly useful for society, a proof being that they are given good jobs, then you can expect to have more people starting to study mathematics. And so, even if we seem to lose some students who go away, we can win a much larger number of them because more will come and study mathematics.

Concerning the question raised of the disappearance of geometry in high-school programmes, the problem is not limited to geometry. In fact, in many countries, governments tend to press down the level of difficulty of high-school studies, as a form of demagogy to please parents. Children have to be able to pass and obtain good marks in exams without really working or thinking hard. And so everything that is too difficult is sort of pushed away from the programmes. And of course geometry is an example of something difficult because one has to think when practicing it. You don't have a crank that you just turn to obtain the answer. Some types of algebraic manipulations and computations can be easy, not all of them of course, and they are easier to teach. When students have made ten of them, they can make the eleventh one, and so everybody is happy because they can pass the exams. For me, the fight to maintain geometry in curricula should really be the fight to maintain opportunities to think and face some difficulties to solve problems. Whether they are problems in geometry or in probability or in combinatorics is really not so important for me. It is really the teaching of minimally difficult things which is presently put in jeopardy in high schools.

EC: We've been touching on our role in dealing, as Misha [Gromov] put it earlier, with mathematics and on the need for mathematicians to talk to the outside world. I have often been asked *"What good is what you're doing as a mathematician?"*, and I've tried to work out the most convincing answers. We live in a complicated world where nobody does one thing completely. I mean there is a tradesman, a builder and a carpenter. They need tools. Some people are attracted to working with their hands and doing carpentry and building, and others prefer being toolmakers. There are people who love making money out of oil, but they also need prospectors and wildcatters to find the oil. In some sense, we are at the wildcatter end of the industrial world. But we have to sell the fact that even an oil magnate would not get very far unless he did some prospecting for unexpected sources of oil. We are the wildcatters in the sense that we have to look for the new sources of information for science.

Santiago Lopez de Medrano: The point of Professor Gindikin was not that it was bad that some mathematicians leave universities and go and do different things. For me the important point is that something is happening in teaching the way we do that stops students from going into mathematics, into that real mathematics that will be useful in universities and outside.

The Chairman asked about experiences with teaching and computers. Let me go back to Alfred Gray, who, as you will all agree, was enthusiastic about everything. But one thing he wasn't enthusiastic about, when he visited us, was going back home to teach. No, he wasn't enthusiastic about that. He wasn't teaching differential geometry of curves and surfaces. He was teaching some low-level course with calculators, and he said *"I have to go back and teach students how to push buttons"*. That's exactly what he said.

Given the programmes, given the structure of the curriculum and the current general philosophy of education, many mathematicians will have to do teach some very mechanical aspects of mathematics. In that way we are not creating the geometers who will be very good for research and for applications and all that. This is the point.

JPB: You may be right in stating or formulating the question this way but I don't believe that this state of affairs is completely out of our hands. It depends a little bit on how we project ourselves and whether we make the personal investment into having some knowledge of what is around us, and how we can approach physics and other disciplines from that point of view. We need to be probably more open than we are in general. From that point of view the key issue can be what sort of documents are available to really get more knowledge about other things, but also how can we create events in which we really get some inside view on how mathematics is used elsewhere. If one really wants colleagues from the mathematical community absorb information on how mathematics gets used here or there, one really needs more than personal contacts, documents in particular. Documents of this nature are rarely available. As a community we probably have to create these opportunities, both in reading, in writing and in hearing and meeting the right people.

Joseph Wolf: When you come to a problem in applying mathematics or where you're asked questions about possible mathematical tools in some industrial or biophysical setting, usually the problem is very difficult to understand. The hard part is often not the mathematical tool but to understand well enough what the problem isto pick out the appropriate tool. Some of our students who do not find jobs in academia go into industry or analogous settings. We need to have them come back to us and talk to us. This is extremely important, because it helps build a bridge through which one can have the sufficient background in some non academic setting so that one can understand questions and perhaps have the possibility of contributing something to such discussions.

Helaman Ferguson *(Washington)*: I am a sculptor and in my sculpture I celebrate mathematics. So I would like to return to the title of this discussion: Where does geometry go? We are living in a golden age of mathematics. I don't think anyone would contest that. Not only is it a golden age because of the theorems and achievements but also because of the problems that we have in mathematics, and the ones that we are confronted with coming from outside. As a group of humans on the planet, we are communicating much better and much more clearly than we ever have at any time in the past. As mathematicians, we get to view problems in other disciplines that we would not ordinarily be able to do. This brings me back to the question of where does geometry go? In my work as a mathematical sculptor working in stone primarily, I endeavour to celebrate mathematics, celebrate the theorems and also celebrate the problems.

I have a patron who has acquired 200 acres of property in Maryland, and he has invited me to develop it as a mathematical theme sculpture park. And it is my particular goal to put in good settings in various places of this property, mathematical theorems - multi-tone granite theorems. And also multi-tone granite problems as well.

Geometry can go on this 200 acres somewhere, and so I am inviting you to suggest what you think are the most beautiful theorems and what are the most beautiful problems in mathematics and geometry specifically and communicate those to me by e-mail, or some of you may have some immediate thoughts right now. But I am inviting your participation in this mathematical garden of delights.

Joseph Wolf: Let me say something optimistic, although perhaps it doesn't fit in too well with some things that other people have said. I think that one of the things that led to the flowering of geometry was the Russian launch of the Sputnik. Indeed, as a result, it was no longer so bizarre to study mathematics, it wasn't so weird to learn something about science, and in the USA at least many more students in secondary schools and universities were exposed to much more science than they would have been otherwise. Suddenly, it became respectable to study, to learn some science, even to go to a library on a weekend. Of course I realise this is quite different in other countries, and in particular was very different in Russia, where the intellectual tradition was rather different.

But now what has happened is that respectability has shifted more towards applications, more towards things that you deal with directly with computers. I don't really think that in the long run the study of geometry is going downhill. It is perhaps an oscillation which will rebound, perhaps in view of applicability, in view of some technological developments that are difficult to foresee at the moment.

Luc Lemaire: If I can also make an optimistic statement, when I think of what I have simply observed in the last 30 years in geometry, I really feel I have seen some fantastic things. I could for example take the work of Donaldson on 4-dimensional manifolds, since he's not here. It is fantastic and completely unexpected. It was certainly not planned in advance, and it is not a panel which said 20 years before *"we should study such and such equations of physics because it will be relevant to understand the geometry of 4-dimensional manifolds"*. That is of course part of the excitement, that there are all those unexpected discoveries.

About the future I am very impatient to see what surprises will come, and what new results will come, and that cannot be foreseen at all now. If they could be foreseen, they would not be that interesting!

JPB: The topic we tried to focus on was definitely where does geometry go, for the good reason that, a priori, we are the experts who should be able to say something about this question. At least we should be able to come up with some vision of this. What we heard shows that we are definitely all conscious of the fact that it's very important to keep geometry open, and not to restrict it to things we have been doing. Each one of us has his or her own little garden that we like to cultivate, and having these little gardens is certainly extremely useful, because we need to have fun when we do mathematics, when we do our research. But we should know that our little garden lies on a much bigger planet, and we need to get to know the other side of the planet.

The other point made concerns the need for this openness in relation with the training of students. Eugene [Calabi] said it very well: at some point you really need to communicate to the students what is really most important, and what you really consider the heart of the matter, which is why you like things and the things you do. At some stage this has to be carried across to the students, and this is not so easy. Very often students don't immediately realise what is the weight going with this message. Sometimes they even get frustrated or irritated by the faces that each of us puts on this side or the other side of the questions we discuss. But quite often students, after some time, realise there is a message in what we say, which finally gets carried across.

Concerning the development with computers or other new sides of society, different views have been expressed, and in my opinion it is good that it remains this way: some people are really fearing that the technology will taking over so much that finally the heart of the matter, that Simon Gindikin called old fashioned geometry, will be lost in the battle. But even if you take technology in a strict sense, as Professor Lopez de Medrano remarked, to teach how to push buttons require knowing which button there is to push, and this already requires some thinking.

If one moves the thinking from one place to the other, it is still fine, but with the deluge of pictures, under which we will be living, we as geometers will not be able to be out of that. We will have to look at these pictures and be able to say something about them. What lies ahead of us is more challenges and, as I tried to suggest, many other new disciplines are going to challenge us as geometers.

There are definitely some disciplines which for the moment have not been so much inspiring or so much challenging for geometers. One of them is certainly medicine, but I'm pretty sure with all the imaging which goes into it, very soon there will be really very big challenges coming from there which will require the interest of a large number of mathematicians. At some stage we will have to use a number of things we know about pictures, illusions with pictures and so on, because for medical doctors this will be absolutely crucial in their practical work.

Altogether, I feel that we have to keep this openness, and also a sense of responsibility according to which, if some bad things happen, we accept a share of them. Therefore, if there are not just bad devils trying to kill geometry from outside, it is not true either that we are defending an old field a bit out of breath and trying to survive. In my view the correct picture is rather different: we have to look outward, identify new challenges, really face them and at the same time prepare our students to look forward to more work and more joy while doing it. This last point is extremely important for me because this is one face of passion.

The Space of Kähler Metrics II

Eugenio Calabi and Xiuxiong Chen

J. Differential Geom. 61, No. 2, 173–193 (2002)

Abstract

This paper, the second of a series, deals with the function space $\mathcal{H}$ of all smooth Kähler metrics in any given n-dimensional, closed complex manifold V, these metrics being restricted to a given, fixed, real cohomology class, called a polarization of V. This function space is equipped with a pre-Hilbert metric structure introduced by T. Mabuchi [10], who also showed that, formally, this metric has nonpositive curvature. In the first paper of this series [4], the second author showed that the same space is a path length space. He also proved that $\mathcal{H}$ is geodesically convex in the sense that, for any two points of $\mathcal{H}$, there is a unique geodesic path joining them, which is always length minimizing and of class $C^{1,1}$. This partially verifies two conjectures of Donaldson [8] on the subject. In the present paper, we show first of all, that the space is, as expected, a path length space of nonpositive curvature in the sense of A. D. Aleksandrov. A second result is related to the theory of extremal Kähler metrics, namely that the gradient flow in $\mathcal{H}$ of the "K energy" of V has the property that it strictly decreases the length of all paths in $\mathcal{H}$, except those induced by one parameter families of holomorphic automorphisms of M.

1. Introduction and main results

1.1 Riemannian metrics and nonpositively curved space.

Let (V, ω_0) be an n-dimensional, compact Kähler manifold without boundary. Consider the space of Kähler distortion potentials

$$\mathcal{H} = \{\varphi \subset C^\infty(V) : \omega_\varphi - \omega_0 + \partial\bar{\partial}\varphi > 0 \text{ on } V\}$$

Clearly, each fibre of the tangent space $T\mathcal{H}$ of $\mathcal{H}$ is $C^\infty(V)$, and each Kähler metric ω_φ defined by any potential $\varphi \in \mathcal{H}$ defines a measure

Received 03/03/00 and revised 05/21/01. The second author is supported partially by NSF postdoctoral fellowship.

DOI: 10.4310/jdg/1090351383

$d\,\mu_\varphi = \frac{1}{n!}\omega_\varphi^n$. The L_2 norm of functions in V determined by the measure $d\,\mu_\varphi$ and applied to $T\mathcal{H}$ defines a Weil-Peterson type metric on the function space $\mathcal{H}$ (see Section 2.1 for historical remarks on this metric).

In 1997, following a program of Donaldson [8], the second author proved that this space is convex with respect to geodesics of class $C^{1,1}$, and used this fact to prove that such geodesics (and only they) achieve the infimum length of all paths joining their end points. The strongest result on these geodesics established so far is that they are of class $C^{1,1}$. If one could prove that they are, as expected, of class at least C^4, then the formal calculations in [10] would imply that the curvature of $\mathcal{H}$ is nonpositive in the geometric sense of A. D. Aleksandrov. It is an interesting question whether one could prove that the geometric curvature of $\mathcal{H}$ is nonpositive regardless of the regularity of geodesics. The following theorem confirms that this is indeed the case.

Theorem 1.1. *The space of Kähler potentials in any given, polarized manifold V is a nonpositive curved space in the following sense. Let A, B, C be three points in the space of Kähler potentials and denote $d(\cdot,\cdot) : \mathcal{H} \times \mathcal{H} \to \mathbf{R}^+$ the distance (in terms of the metric aforementioned). For any λ, with $0 \le \lambda \le 1$, let P_λ be the point on a geodesic path joining B and C with $d(B, P_\lambda) = \lambda d(B,C)$ and $d(P_\lambda, C) = (1-\lambda)d(B,C)$. Then the following inequality holds:*

$$d(A,P_\lambda)^2 \le (1-\lambda)d(A,B)^2 + \lambda d(A,C)^2 - \lambda\cdot(1-\lambda)d(B,C)^2.$$

In [4], the second author proved that every geodesic segnent achieves the minimum length of all paths connecting its end points. However, the problem remained unsettled whether any length minimizing sequence of paths joining any two points of $\mathcal{H}$ must include a subsequence converging to the unique geodesic. That question is now settled, according to the following statement:

Theorem 1.2. *Let $\varphi_i \in \mathcal{H}(i = 1,2)$ be two Kähler distortion potentials, generating two corresponding Kähler metrics in V, and let $\{C_i\}$ be any sequence of paths in $\mathcal{H}$ from φ_0 to φ_1 and whose length converge to the least possible limit. Then $\{C_i\}$ converges, in the topology of Hausdorff distance, to the unique geodesic of class $C^{1,1}$ from φ_0 to φ_1.*

1.2 The gradient flow

The notion of extremal Kähler metrics was first introduced in [2] by the first author of the present article. In order to attack the problem of the existence of extremal Kähler metrics, he proposed a "steepest descent" process, consisting of the initial value problem corresponding to the following parabolic equation of order four:

$$\frac{\partial \varphi}{\partial s} = R(\varphi) - \underline{R}. \tag{1.1}$$

Here $R(\varphi)$ denotes the scalar curvature of the Kähler metric ω_φ and $\underline{R}$ is the average value over V — a constant depending only the polarized class. The flow in $\mathcal{H}$ represented by the solutions of (1.1) has sometimes been refered as the Calabi flow. In general, a solution to the initial value problem for (1.1) is known to exist for a short interval of the "time" paremeters. In the case of Riemann surfaces ($n = 1$), Chrusciel [7] proved, using the fact that metric of constant curvature are known *a prori* to exist, that the solution of (1.1) exists for all s and converges at infinity.[1] It is remarkable that, under the flow in $\mathcal{H}$ represented by solutions of (1.1), each smooth path in $\mathcal{H}$ has decreasing length, as stated more precisely below.

Theorem 1.3. *Let $\varphi_i \in \mathcal{H}(i = 0, 1)$ be two Kähler distortion potentials, generating two corresponding Kähler metrics in V, and let $\varphi(t)(0 \leq t \leq 1)$ be a smooth path in $\mathcal{H}$ from $\varphi(0) = \varphi_0$ to $\varphi(1) = \varphi_1$. Suppose that, for some constant $s_0 > 0$, there is a smooth family of Kähler distortion potentials $\phi(s,t)(0 \leq s \leq s_0, 0 \leq t \leq 1)$ such that, for each $t \in [0,1]$, $\phi(s,t)$ satisfies th initial value problem $\frac{\partial \phi}{\partial s} = R(\phi) - \underline{R}$, $\phi(0,t) = \varphi(t)$. Then the length $L(s)$ of the path $\{\phi(s,t) : 0 \leq t \leq 1\}$ in $\mathcal{H}$ is strictly decreasing function of s, except for the case where the complex vector field $g^{\lambda\overline{\mu}}(\phi(s,t))\frac{\partial^2 \phi(s,t)}{\partial t \partial \overline{z}_\mu} \frac{\partial}{\partial z^\lambda}$ in V is holomorphic for each (s,t) in a neighborhood of a segement $\{(s,t) : 0 \leq t \leq 1\}$.*

Thc proof of thio statoment consists of a formal calculation involving the second order differential operator on real or complex functions, defined below and named after André Lichnérowicz.

Definition 1.4. For any smooth function $f : V \to \mathbf{C}$ and any Kähler metric g in V, the Lichnérowicz operator D is defined in local terms by $D_g f = \sum_{\alpha,\beta=1}^{n} f_{,\alpha\beta} dz^\alpha \otimes dz^\beta$ where the local functions $f_{,\overline{\alpha}\overline{\beta}}$

[1]The second author gave a new geometric proof to Chrusciel's theorem [6]. Following the approach taken in [6], M. Struwe [13] gave a unified treatment of both Ricci flow and Calabi flow in Riemann surfaces. For higher dimensional Kähler manifolds, very little is known about the global existence of the flow.

are components of the second covariant derivatives of f in terms of g:

$$f_{,\bar{\alpha}\bar{\beta}} = \sum_{\lambda,\mu=1}^{n} g_{\lambda\bar{\alpha}} \frac{\partial}{\partial \bar{z}_{\beta}} \left(g^{\lambda\bar{\mu}} \frac{\partial f}{\partial \bar{z}_{\mu}} \right).$$

The length $L(s)$ of the path $\{\phi(s,t) : 0 \leq t \leq 1\}$ in $\mathcal{H}$ is, by definition,

$$L(s) = \int_0^1 \left(\int_V \left(\frac{\partial \phi(s,t)}{\partial t} \right)^2 d\,\mu_{\phi(s,t)} \right)^{\frac{1}{2}} d\,t.$$

If one differentiates both sides with respect to s, one readily obtains from the standard variational formulas the desired expression

$$\frac{d\,L}{d\,s} = - \int_0^1 \left(\int_V \left| D \frac{\partial \phi}{\partial t} \right|^2_{\phi(s,t)} d\,\mu_{\phi(s,t)} \right) \cdot \left(\int_V \left| \frac{\partial \phi}{\partial t} \right|^2 d\,\mu_{\phi(s,t)} \right)^{-\frac{1}{2}} d\,t.$$

Theorem 1.5. *The following two statements are immediate consequences of Theorem* 1.3.

1. *If the flow in $\mathcal{H}$ generated by* (1.1) *exists for all time for any smooth initial data, then the distance between any two Kähler metrics decreases under that flow.*

2. *If the K-energy*[2] *is a weakly convex function along geodesics,*[3] *and in particular if the first Chern class of V is nonpositive, then the same flow decreases the distance between metrics.*

The flowing question/conjecture is highly interesting:

Question/Conjecture 1.6. *Is the distance function in $\mathcal{H}$ strictly decreasing under the gradient flow* (1.1), *in particular, if the two metrics are not equivalent up to some holomorphic isometric transformation?*

Remark 1.7. The only conceivable paths in $\mathcal{H}$ that do not shrink under the flow under discussion are those induced by paths in the group

[2]The K-energy is defined by T. Mabuchi in 1987 [10] while the flow was introduced by the first author in 1982 [2]. And it is commonly known as the "Calabi flow" in the literature.

[3]A function $f(t)(0 \leq t \leq 1)$ is weakly convex if for any t, we have $f(t) \leq (1 - t)\,f(0) + t\,f(1)$. For this theorem, we additionally assume that f is differentiable at both end points, i.e., $f'(1) \geq f'(0)$.

of holomorphic transformations of V that are of *hyperbolic* type at each point. A path in that group is said to be of hyperbolic type with respect to a given Kähler metric if the holomorphic vector field in V represented by the tangent to a given Kähler metric is a multiple of a vector field generating a group of holomorphic isometries. A rigorous verification of this conjecture would require locally uniform estimates of the regularity of the flow (1.1), independent of second derivatives of the initial metrics.

Acknowledgements. The second author wishes to express his thanks to S. Donaldson for many enlightening discussions, to L. Simon and R. Schoen for their encouragement throughout the project, and Guofang Wang for a careful reading of an earlier version of this paper. Both of us would like to thank the referee for some interesting suggestions.

2. $\mathcal{H}$ is a nonpositively curved space

In this section, we want to show that $\mathcal{H}$ is a nonpositively curved space in the sense of Aleksandrov.

2.1 A Riemannian metric in the infinite dimensional space.

In 1987, T. Mabuchi introduced ([10]) the infinite dimensional Riemannian (or pre-Hilbert) structure, in the function space of Kähler metrics in V and showed that the space with this metric is, *formally*, a locally symmetric space with nonpositive curvature. The same metric was rediscovered, apparently unaware of Mabuchi's work, by S. Semmes [12] and Donaldson [8], each describing it from somewhat different viewpoints. We shall briefly review the definition and elemntary properties of this metric.

For each Kähler metric g in V, any smooth, real-valued function ψ in V represents a tangent vector to the space of all Kähler metrics.[4] Thus any "basis" $\{\psi_\lambda\}$ for the space of all smooth functions, attached to each Kähler distortion potential $\varphi_0 \in \mathcal{H}$, constitutes a global field of tangential frames. Since any two of these vector fields obviously

[4]Since constants, and only constants, represent trivial infinitesimal deformations of metric, it is convenient to replace the actual space of Kähler metrics with the product of $\mathcal{H}$ with the real line $\mathbf{R}$, so that, for each metric g, constants act by translation on $\mathbf{R}$ alone, while function φ orthogonal to constants are regarded as infinitesimal Kähler distortion potentials, acting trivially on $\mathbf{R}$.

commute, the total frame may be considered, formally, as a sort of a global, flat affine coordinate system for $\mathcal{H}$, in the following sense: the "point" $\varphi_0 \in \mathcal{H}$ being the origin, the numerical range of the coordinates consists of the set of arrays (y^λ) such that the function $\varphi_1 = \varphi_0 + \sum_\lambda y^\lambda \psi_\lambda$ satisfies the condition to be the Kähler distortion potential for a Kähler metric, that is then assigned to the array (y^λ).

Given any tangent vector on $\mathcal{H}$, represented by a smooth function $\psi : \mathcal{H} \times V \to \mathbf{R}$, the square length of ψ at the point $\varphi \in \mathcal{H}$ is defined to be

$$\|\psi\|_\varphi^2 = \int_V \psi^2 \, d\mu_\varphi,$$

where $d\mu_\varphi = \frac{1}{n!}\omega_\varphi^n$. The corresponding symmetric bilinear form for any two tangent "vectors" f_1, f_2 at the point $\varphi \in \mathcal{H}$, is written as

$$\langle f_1, f_2\rangle_\varphi = \int_V f_1 \cdot f_2 \, d\mu_\varphi.$$

When no confusion is arisen, we just write

$$\langle f_1, f_2\rangle = \int_V f_1 \cdot f_2 \, d\mu_\varphi.$$

The "tensor notation" for the metric, in terms of the global tangent frame $\{\psi_\lambda\}$ is then $\mathcal{G}_{\psi_\lambda\,\psi_\mu} = \langle \psi_\lambda, \psi_\mu\rangle$, $d\mathcal{S}^2 = \mathcal{G}_{\psi_\lambda\,\psi_\mu}\, dy^\lambda d\,y^\mu$.

One can calculate directly the Christoffel symbols of the first kind by differentiating formally $\}_{\psi_\lambda\,\psi_\mu}$ in the direction represented by any ψ_ν:

$$\partial_{\psi_\nu} \mathcal{G}_{\psi_\lambda\,\psi_\mu} = \int_V \psi_\lambda \psi_\mu \triangle_\varphi \psi_\nu \, d\mu_\varphi,$$

where

$$\begin{aligned}
\{\psi_\lambda, \psi_\mu; \psi_n u\} &= \frac{1}{2}\left(\partial_{\psi_\lambda}\mathcal{G}_{\psi_\nu\,\psi_\mu} + \partial_{\psi_\nu}\mathcal{G}_{\psi_\lambda\,\psi_\mu} - \partial_{\psi_\nu}\mathcal{G}_{\psi_\lambda\,\psi_\mu}\right)\\
&= \frac{1}{2}\int_V \left(\psi_\mu\psi_\nu\triangle_\varphi\psi_\lambda + \psi_\lambda\psi_\nu\triangle_\varphi\psi_\mu - \psi_\lambda\psi_\mu\triangle_\varphi\psi_\nu\right) d\mu_\varphi\\
&= -\frac{1}{2}\int_V \psi_\lambda g_\varphi^{\alpha\bar\beta}\left(\frac{\partial\psi_\mu}{\partial z_\alpha}\frac{\partial\psi_\nu}{\partial \bar z_\beta} + \frac{\partial\psi_\nu}{\partial z_\alpha}\frac{\partial\psi_\mu}{\partial \bar z_\beta}\right) d\mu_\varphi.
\end{aligned}$$

The last expression then yields the covariant derivative formula for tangent vector fields ψ_λ:

$$\nabla_{\psi_\mu}\psi_\lambda(\varphi) = \nabla_{\psi_\lambda}\psi_\mu(\varphi) = \frac{1}{2}g_\varphi^{\alpha\bar\beta}\left(\frac{\partial\psi_\mu}{\partial z_\alpha}\frac{\partial\psi_\lambda}{\partial\bar z_\beta} + \frac{\partial\psi_\lambda}{\partial z_\alpha}\frac{\partial\psi_\mu}{\partial\bar z_\beta}\right).$$

Similarly, the equation for a geodesic path $\varphi(t)$ in $\mathcal{H}$ in terms of a parameter $t(0 \le t \le 1)$ proportional to the arc length is

$$\frac{\partial^2\varphi}{\partial t^2} = g_\varphi^{\alpha\bar\beta}\frac{\partial^2\varphi}{\partial t\partial z_\alpha}(t)\frac{\partial^2\varphi}{\partial t\partial\bar z_\beta}(t) = \frac{1}{2}\left|\nabla\frac{\partial\varphi}{\partial t}\right|^2_{\varphi(t)}, \tag{2.1}$$

where the norm of the gradient in the right-hand side is taken in V with respect to the metric associated with the form $\omega_{\varphi(t)}$. One may derive the same equation, alternatively, as the Euler-Lagrange equation for the variational problem of minimizing the energy integral

$$\int_0^1 \left\langle\frac{\partial\varphi}{\partial t}(t), \frac{\partial\varphi}{\partial t}(t)\right\rangle^2_{\varphi(t)} dt = \int_0^1 \int_V \left(\frac{\partial\varphi}{\partial t}\right)^2 d\mu_{\varphi(t)}\, dt,$$

over the set of smooth paths in $\mathcal{H}$ connecting the fixed end points $\varphi(0)$ and $\varphi(1)$. A calculation with the derivatives of the Christoffel symbols, analogous to the familiar formalism of the Ricci calculus, enables us to verify, *formally*, that each 2-dimensional plane in the tangent bundle of $\mathcal{H}$ has nonpositive curvature. The details may be found in [10], [12] and [8].

In [8], S. K. Donaldson also formulated serval important conjectures about geometric properties and their possible relation with properties of the complex manifold V. In 1997, the second author proved some of his conjectures, whose contents are summarized by the following theorem:

Theorem B ([4]). *The following statements are true:*

1. *The space of Kähler potentials $\mathcal{H}$ is convex by $C^{1,1}$ geodesics. More specifically, if $\varphi_0, \varphi_1 \in \mathcal{H}$ and $\varphi(t)$ $(0 \le t \le 1)$ is a geodesic connecting these two points in $\mathcal{H}$, then the second order mixed covariant derivatives of $\varphi(t)$ are uniformly bounded from above.*

2. *$\mathcal{H}$ is a metric space.*[5] *In other words, the infimum of the lengths of all possible curves between any two different points in $\mathcal{H}$ is strictly positive.*

3. *If $C_1(V) < 0$, then the extremal Kähler metric is unique in each Kähler class.*

[5]This is a conjecture of Donaldson[8].

2.2 A lemma on approximate geodesics

In a local coordinate system of V, let $\omega_0 = \sum_{\alpha,\beta=1}^{n} g_{0_{\alpha\bar\beta}}\, dz_\alpha\, \overline{dz_\beta}$ and

$$\omega_\varphi = \sum_{\alpha,\beta=1}^{n} g_{\alpha\bar\beta}\, dz_\alpha\, \overline{dz_\beta}, \qquad \text{where} \qquad g_{\alpha\bar\beta} = g_{0_{\alpha\bar\beta}} + \frac{\partial^2 \varphi}{\partial z_\alpha \partial \overline{z_\beta}}.$$

In this subsection, $z_1, z_2, \ldots, z_n$ are local coordinates in V. For any path $\varphi(\cdot, t) : [0,1] \to \mathcal{H}$, we can view it as a function defined in the product manifold $V \times [0,1]$. Following an idea of S. Semmes, we introduce a dummy variable θ such that $V \times ([0,1] \times S^1)$ is a $(n+1)$-dimensional Kähler manifold and $t = \mathrm{re}(z_{n+1})$. Here S^1 is the unit circle, and we always use the following notations:

$$i, j, k = 1, 2, \ldots, n, n+1 \qquad \text{and} \qquad \alpha, \beta, \gamma = 1, 2, \ldots, n.$$

Consider the projection

$$\begin{array}{rcl} \pi : V \times ([0,1] \times S^1) & \to & V \\ (z, t, \theta) & \to & z. \end{array}$$

Consider the pullback metric $\pi^* g_0$. Note that $\pi^* \omega_0$ is a degenerate Kähler form of co-rank 1 in $V \times ([0,1] \times S^1)$.

Definition 2.1. A path $\varphi(t)(0 < t < 1)$ in $\mathcal{H}$ is a convex path if

$$\det \left(\pi^* g_{0_{i\bar j}} + \frac{\partial^2 \varphi}{\partial z_i \partial \overline{z_j}} \right)_{(n+1)(n+1)} > 0, \qquad \text{in } V \times (I \times S^1).$$

Definition 2.2. A convex path $\varphi(t)$ in the space of Kähler metrics is called an ϵ-approximate geodesic if the following holds:

(2.2)
$$\begin{aligned} \det \left(\pi^* g_{0_{i\bar j}} + \frac{\partial^2 \varphi}{\partial z_i \partial \overline{z_j}} \right)_{(n+1)(n+1)} &= \left(\frac{\partial^2 \varphi}{\partial t^2} - \frac{1}{2} \left| \nabla \frac{\partial \varphi}{\partial t} \right|^2_{g(t)} \right) \det g(t) \\ &= \epsilon \cdot \det g_0 \end{aligned}$$

where $g(t)_{\alpha\bar\beta} = g_{0_{\alpha\bar\beta}} + \frac{\partial^2 \varphi}{\partial z_\alpha \partial \bar z_\beta}$ $(1 \le \alpha, \beta \le n)$.

Also from [4], we have the following, which plays a crucial role in the proof in the next subsection.

Lemma 2.3 (Geodesic approximation lemma, [4]). *Suppose given two smooth curves $\phi_1(\cdot, s), \phi_2(\cdot, s) : [0,1] \to \mathcal{H}$. For ϵ_0 small enough, there exist two-parameter smooth families of curves*

$$\varphi(\cdot, t, s, \epsilon) : [0,1] \times [0,1] \times (0, \epsilon_0](0 \le t, s \le 1, \text{ and } 0 < \epsilon \le \epsilon_0) \to \mathcal{H}$$

(from $\phi_1(\cdot, s)$ to $\phi_2(\cdot, s)$) such that the following properties hold:

1. *For any fixed s and ϵ, $\varphi(\cdot, t, s, \epsilon)$ is an ϵ-approximate geodesic connecting $\varphi_1(\cdot, s)$ and $\varphi_2(\cdot, s)$. More precisely, $\varphi(\cdot, t, s, \epsilon)$ solves the corresponding Monge-Ampere equation*

   $$\det\left(\pi^* g_{0_{i\bar{j}}} + \frac{\partial^2 \varphi}{\partial z_i \partial \overline{z}_j}\right) = \epsilon \cdot \det(g_0), \text{ in } V \times \mathbf{R}; \tag{2.3}$$

   and

   $$\varphi(\cdot, 0, s, \epsilon) = \phi_1(\cdot, s), \quad \varphi(\cdot, 1, s, \epsilon) = \phi_2(\cdot, s).$$

   Here φ is independent of $\mathrm{Im}(z_{n+1})$.

2. *There exists a uniform constant C which depends only on $\phi_1(\cdot, s), \phi_2(\cdot, s)$ such that*

   $$|\varphi| + \left|\frac{\partial \varphi}{\partial s}\right| + \left|\frac{\partial \varphi}{\partial t}\right| < C; \qquad 0 < \frac{\partial^2 \varphi}{\partial t^2} < C, \qquad \frac{\partial^2 \varphi}{\partial s^2} < C.$$

3. *For fixed s, as $\epsilon \to 0$, the ϵ-approximating geodesic $\varphi(\cdot, t, s, \epsilon)$ converges to the unique geodesic between $\phi_1(\cdot, s)$ and $\phi_2(\cdot, s)$ in weak $C^{1,1}$ topology.*

4. *Define the energy element along $\varphi(\cdot, t, s, \epsilon)$ by*

   $$E(t, s, \epsilon) = \int_V \left|\frac{\partial \varphi}{\partial t}\right|^2 dg(t, s, \epsilon),$$

   where $g(t, s, \epsilon)$ is the corresponding Kähler metric defined by the Kähler potentials $\varphi(t, s, \epsilon)$. Then there exists a uniform constant C such that

   $$\max_{t,s}\left|\frac{\partial E}{\partial t}\right| \le \epsilon \cdot C.$$

   In other words, both the energy and length element converge to a constant along each convex curve if $\epsilon \to 0$.

2.3 Length of Jacobi vector field grows supre-linearly

In this subsection, we use the same notation as in Lemma 2.3. We want to prove that the Jacobi vector field along any geodesic grows supre-linearly.

Lemma 2.4. *Let $\varphi(\cdot, t, s, \epsilon)$ be the two-parameter families of approximating geodesics defined as in Lemma 2.3. Let $Y(\cdot, t, s, \epsilon) = \frac{\partial \varphi}{\partial s}$ be the deformation vector fields and $X(\cdot, t, s, \epsilon) = \frac{\partial \varphi}{\partial t}$ the tangential vector fields along the approximating geodesics. Then the second derivatives of Y along the approximating geodesics are positive:*

$$\nabla_X \nabla_X Y \geq 0.$$

Note that Y converges to a Jacobi vector field as $\epsilon \to 0$. Moreover, we have

$$\langle Y, \nabla_X Y \rangle \geq \langle Y, Y \rangle.$$

Proof. The equation for a family of ϵ-approximate geodesics is:

$$\left(\frac{\partial^2 \varphi}{\partial t^2} - \frac{1}{2} \left| \nabla \frac{\partial \varphi}{\partial t} \right|^2_{g(t)} \right) \det\, g(t) = \epsilon \cdot \det\, g_0, \quad 0 \leq\ s,\ t\ \leq 1.$$

Denote

$$X = \frac{\partial}{\partial t}, \qquad Y = \frac{\partial}{\partial s}, \qquad Y' = \nabla_X Y,$$

and

$$H = \frac{\det\, g_0}{\det\, g}, \qquad f = \frac{\partial^2 \varphi}{\partial t^2} - \frac{1}{2} \left| \nabla \frac{\partial \varphi}{\partial t} \right|^2_{g(t)} = \nabla_X X.$$

Then the approximating geodesic equation becomes

$$f = \nabla_X X = \epsilon \cdot H.$$

Note that any two-parameter family of smooth functions $F(s,t)$ can be viewed as a two parameter family of tangent vectors at $T_{\varphi(\cdot,t,s,\epsilon)}\mathcal{H}$. Then the Riemannian metric in $\mathcal{H}$ gives the following covariant derivatives:

$$\nabla_X F = \nabla_{\frac{\partial}{\partial t}} F(s,t) = \frac{\partial F}{\partial t} - \frac{1}{2} \nabla_g \frac{\partial \varphi}{\partial t} \cdot \nabla_g \frac{\partial F(s,t)}{\partial t}$$

and

$$\nabla_Y F = \nabla_{\frac{\partial}{\partial s}} F(s,t) = \frac{\partial F}{\partial s} - \frac{1}{2}\nabla_g \frac{\partial \varphi}{\partial s} \cdot \nabla_g \frac{\partial F(s,t)}{\partial s}.$$

Clearly, as $\epsilon \to 0$, Y is the Jacobi vector field along the geodesic. By definition, the length of Y at t is:

$$|Y|^2(t,s) = \int_V \left|\frac{\partial \varphi}{\partial s}\right|^2 \det g.$$

Then

$$\frac{1}{2}\frac{\partial}{\partial t}|Y|^2 = \langle \nabla_X Y, Y\rangle = \langle \nabla_Y X, Y\rangle.$$

Let $K(X,Y)$ denote the sectional curvature of the space of Kähler metrics at point $\varphi(\cdot, t, s, \epsilon)$. By a formal calulation (cf. [10], [12] and [8]), we have[6]

$$K(X,Y) = -|\{X,Y\}_\varphi|_g^2 \leq 0.$$

Therefore, we have

$$\begin{aligned}
\frac{1}{2}\frac{\partial^2}{\partial t^2}|Y|^2 &= \langle \nabla_Y X, \nabla_X Y\rangle + \langle \nabla_X \nabla_Y X, Y\rangle \\
&= |Y'|^2 - K(X,Y) + \langle \nabla_Y \nabla_X X, Y\rangle \\
&\geq |Y'|^2 + \int_V \epsilon \frac{\partial \varphi}{\partial s} \nabla_{\frac{\partial}{\partial s}} H \det g \\
&= |Y'|^2 + \int_V \epsilon \frac{\partial \varphi}{\partial s} \cdot \left(\frac{\partial H}{\partial s} - \frac{1}{2}\nabla \frac{\partial \varphi}{\partial s} \cdot \nabla H\right) \det g \\
&= |Y'|^2 + \frac{\epsilon}{2}\int_V \left|\nabla \frac{\partial \varphi}{\partial s}\right|^2 H \cdot \det g \geq |Y'|^2.
\end{aligned}$$

The last equality holds since

$$\frac{\partial H}{\partial s} = \frac{\partial}{\partial s}\left(\frac{\det g_0}{\det g}\right) = -\triangle_g \frac{\partial \varphi}{\partial s} \cdot H$$

and

$$\begin{aligned}
&-\int_V \frac{\partial \varphi}{\partial s} \triangle_g \frac{\partial \varphi}{\partial s} \cdot H \det g \\
&\qquad = \frac{1}{2}\int_V \left|\nabla \frac{\partial \varphi}{\partial s}\right|^2 H \cdot \det g + \frac{1}{2}\int_V \frac{\partial \varphi}{\partial s} \nabla \frac{\partial \varphi}{\partial s} \nabla H \det g.
\end{aligned}$$

[6] For any two functions f_1, f_2 and a Kähler form ω_φ, the term $\{f_1, f_2\}_\varphi$ is defined to be the Possion bracket of f_1 and f_2 with respect to the sympletic form ω_φ.

It follows that

$$\frac{\partial^2}{\partial t^2}|Y| \geq 0.$$

In other words, $|Y(t)|(0 \leq t \leq 1)$ is a convex function of t. Since $Y(0) = 0$, we have

$$\frac{\partial}{\partial t}|Y(t)|_{t=1} \geq \frac{|Y(1)|}{1}.$$

or at time $t = 1$

$$\langle Y, Y' \rangle \geq \langle Y, Y \rangle. \tag{2.4}$$

q.e.d.

2.4 Proofs of Theorems 1.1 and 1.2

In this subsection, we want to show that $\mathcal{H}$ is a nonpositively curved space. We follow again the notations in Lemma 2.3 and the preceding subsection.

Proof of Theorem 1.1. Consider a special case of Lemma 2.3 when $\phi_1(\cdot, s) = \phi_1$ is one point on $\mathcal{H}$ (instead of a curve). We denote this point as P. Let $Q = \phi_2(\cdot, 0) \in \mathcal{H}$ and $R = \phi_2(\cdot, 1) \in \mathcal{H}$. Furthermore, we assume that $\phi_2(\cdot, s)$ (denoted as by QR) is an ϵ-approximate geodesic connecting Q and R. In other words, it satisfies the following equation:

$$\nabla_Y Y \cdot \det\, g = \left(\frac{\partial^2 \varphi}{\partial t^2}_{ss} - \frac{1}{2}\left|\nabla \frac{\partial \varphi}{\partial t}_s\right|_g^2 \right) \det\, g = \epsilon \cdot \det\, g_0.$$

Let $Q(s)$ denote the point $\phi_2(\cdot, s)$ and denote by $E(s)$ the energy of the ϵ-approximate geodesic from P to $Q(s)$. As $\epsilon \to 0, E(s) \to$ a constant which, by our normalization, is the square of the geodesic distance from P to $Q(s)$. Thus it is enough to work with $E(s)$. Next

$$E(s) = \int_0^1 \langle X, X \rangle dt = \int_0^1 \int_V \left(\frac{\partial \varphi}{\partial t}\right)^2 \det\, g\, dt$$

and

$$E(QR) = \int_0^1 \langle Y, Y \rangle ds = \int_0^1 \int_V \left(\frac{\partial \varphi}{\partial s}\right)^2 \det\, g\, ds.$$

Thus

$$\begin{aligned}\frac{1}{2}\frac{d\,E(s)}{d\,s} &= \int_0^1 \langle \nabla_Y X, X\rangle dt = \int_0^1 (X\langle X,Y\rangle - \langle \nabla_X X, Y\rangle)dt \\ &= \langle X,Y\rangle_{t=1} - \int_0^1 \int_V \frac{\partial \varphi}{\partial s}\cdot \epsilon\, H \;\det\, g\; dt \\ &= \langle X,Y\rangle_{t=1} - \epsilon\cdot \int_0^1 \int_V \frac{\partial \varphi}{\partial s} \;\det\, g_0\; dt.\end{aligned}$$

Now the second derivatives:

$$\begin{aligned}\frac{1}{2}\frac{d^2\,E(s)}{d\,s^2} &= \frac{d}{d\,s}\langle X,Y\rangle_{t=1} - \epsilon\cdot \int_0^1 \int_V \frac{\partial^2 \varphi}{\partial s^2} \;\det\, g_0\; dt \\ &\geq \langle Y',Y\rangle_{t=1} + \langle X, \nabla_Y Y\rangle_{t=1} - C\,\epsilon \int_V \det\, g_0 \\ &\geq \langle Y,Y\rangle_{t=1} + \int_V \frac{\partial\varphi}{\partial t}\,\epsilon\cdot H\cdot \det g - C\,\epsilon \int_V \det\, g_0 \\ &\geq \langle Y,Y\rangle_{t=1} + \int_V \frac{\partial\varphi}{\partial t}\,\epsilon\cdot \det\, g_0 - C\,\epsilon \int_V \det\, g_0 \\ &\geq E(QR) - C\cdot\epsilon\cdot \mathrm{vol}(V).\end{aligned}$$

Here we have used the inequality (2.4) in the second inequality from the top. And $E(QR)$ denotes the energy of the path $\phi_2(\cdot, s)$. For the energy elements of curves, the following inequality holds:

$$E(s) \leq (1-s)E(0) + sE(1) - s(1-s)(E(QR) - C\cdot\epsilon\cdot \mathrm{vol}(V)).$$

Now fix s;[7] as $\epsilon \to 0$, each energy element of a path approaches the square of the length of that path. Thus the above inequality reduces to

$$|PQ(s)|^2 \leq (1-s)|PQ|^2 + s|PR|^2 - s(1-s)|QR|^2. \tag{2.5}$$

Thus the space of Kähler metrics satisfies the defining inequality for a nonpositively curved space and hence it is a nonpositive space. Here $|PQ(s)|$ represents the distance from P to $Q(s)$; $|PQ|$ represents the distance from P to Q; $|PR|$ represents the distance from P to R; and $|QR|$ represents the distance from Q to R. q.e.d.

Next we prove Theorem 1.2.

[7] Actually, using successive subdivision one sees that knowing the inequality (2.5) holds for $s = \frac{1}{2}$ is enough to prove it for all $0 < \lambda < 1$, cf. [9].

Proof of Theorem 1.2. Let φ_0 and φ_1 be two points in $\mathcal{H}$ with distance $l > 0$. Suppose that $\varphi(\cdot, t)(0 \leq t \leq 1)$ is a $C^{1,1}$ geodesic which connects these two points in $\mathcal{H}$. Let $\varphi_i(t)(0 \leq t \leq 1)$ be an arbitrary family $(i = 1, 2, \ldots, n, \ldots)$ of curves between φ_0 and φ_1 with length $l_i \geq l > 0$. Next we assume that this is a distance minimizing sequence of curves. In other words,

$$\lim_{i \to \infty} l_i = l.$$

Then, we need to show that $\varphi_i(\cdot, t)(0 \leq t \leq 1)$ converges to $\varphi(\cdot, t)(0 \leq t \leq 1)$ in some reasonable topology. For convenience, we assume that every curve involved has been parameterized proportionally to the arc-length. Then we only need to show that for each fixed $s > 0$, we have

$$\lim_{i \to \infty} d(\varphi_i(\cdot, s), \varphi(\cdot, s)) \to 0.$$

Since $\mathcal{H}$ is a nonpositively curved space, we have (comparing with the Euclidean space):

$$d(\varphi_i(\cdot, s), \varphi(\cdot, s)) \leq \sqrt{\frac{l_i^2 - l^2}{4}} \to 0.$$

Theorem 1.2 is then proved. q.e.d.

3. The gradient flow of the K-energy

In this section, we prove Theorem 1.3.

Proof of Theorem 1.3. We follow the notations in Theorem 1.3. Let $\phi(\cdot, t) : [0, 1] \to \mathcal{H}$ be any smooth curve in $\mathcal{H}$ and let $\phi(\cdot, t, s)$ be the image of this curve under the gradient flow after time s. Recall

$$\frac{\partial \phi}{\partial s} = R(\phi) - \underline{R}.$$

Denote by $g(s, t)$ the Kähler metric associated with the Kähler potentials $\phi(s, t)$. Use $\triangle$ to denote the complex Laplacian operator of metric $g(s, t)$. Following a calculation in [2], we have

$$\frac{\partial R}{\partial t} = -D^* D \, \frac{\partial \phi}{\partial t} + \sum_{\alpha=1}^{n} \left(\frac{\partial \phi}{\partial t} \right) \phi^{\alpha} R_{,\alpha}$$

and

$$\frac{\partial}{\partial t} \det\, g(s,t) = \triangle\, \frac{\partial \phi}{\partial t}\, \det\, g.$$

Recall that the energy of the path $\phi(\cdot, t, s)$ (at time s fixed) is:

$$E(s) = \int_0^1 \int_V \left(\frac{\partial \phi}{\partial t}\right)^2 \det g\, dt.$$

Under the gradient flow (1.1), we have

$$\begin{aligned}
\frac{dE}{d\,s} &= \int_0^1 \int_V 2\frac{\partial \phi}{\partial t}\frac{\partial^2 \phi}{\partial t \partial s} \det\, g\, dt + \int_0^1 \int_V \left(\frac{\partial \phi}{\partial t}\right)^2 \triangle \frac{\partial \phi}{\partial s} \det\, g\, dt \\
&= \int_0^1 \int_V 2\frac{\partial \phi}{\partial t}\frac{\partial R}{\partial t} \det\, g\, dt - \int_0^1 \int_V 2\frac{\partial \phi}{\partial t}\left(\frac{\partial \phi}{\partial t}\right)^{\alpha} \left(\frac{\partial \phi}{\partial s}\right)_{\alpha} \det\, g\, dt \\
&= \int_0^1 \int_V 2\frac{\partial \phi}{\partial t}\left(-D^* D\, \frac{\partial \phi}{\partial t} + \phi^{,\alpha} R_{,\alpha}\right) \det\, g\, dt \\
&\quad - \int_0^1 \int_V 2\frac{\partial \phi}{\partial t}\left(\frac{\partial \phi}{\partial t}\right)^{\alpha} R_{\alpha} \det\, g\, dt \\
&= - \int_0^1 \int_V \left|D\frac{\partial \phi}{\partial t}\right|_g^2 \det\, g\, dt.
\end{aligned}$$

It follows that

$$\frac{dL}{d\,s} = - \int_0^1 \left(\int_V \left|D\frac{\partial \phi}{\partial t}\right|^2_{\phi(s,t)} dg(s,t) \cdot \sqrt{\int_V \left|\frac{\partial \phi}{\partial t}\right|^2 dg(s,t)}^{\,-\frac{1}{2}} \right) dt,$$

where $L(s)$ is the length of the evolved curve at time $s > 0$. From this formula, if the length of a smooth curve is not decreasing, then

$$\int_0^1 \left(\int_V \left|D\frac{\partial \phi}{\partial t}\right|^2_{\phi(s,t)} \det\, g(s,t) \sqrt{\int_V \left|\frac{\partial \phi}{\partial t}\right|^2 dg(s,t)}^{\,-\frac{1}{2}} \right) dt = 0.$$

It follows that

$$\int_0^1 \int_V \left|D\frac{\partial \phi}{\partial t}\right|^2_{\phi(s,t)} = 0$$

or

$$\left(\frac{\partial \phi}{\partial t}\right)_{,\alpha\beta} \equiv 0, \qquad \forall\, \alpha, \beta = 1, 2, \ldots n;\ \forall\, t \in [0,1].$$

In other words, the curve $\phi(t)(0 \leq t \leq 1)$ is either trivial (depending only on t) or it represents a family of holomorphic transformations. Theorem 1.3 is then proved. q.e.d.

Next we give a proof of the first part of Theorem 1.5.

Proof of Theorem 1.5(1). For any $\varphi_0, \varphi_1 \in \mathcal{H}$, consider the space of all smooth curves which connect φ_0 with φ_1. We denote it by $\mathcal{L}(\varphi_0, \varphi_1)$. For any curve $c \in \mathcal{L}(\varphi_0, \varphi_1)$, we denote its length by $L(c)$. Then the distance between the two points φ_0 and φ_1 can be defined as

$$d(\varphi_0, \varphi_1) = \inf_{c \in \mathcal{L}(\varphi_0, \varphi_1)} L(c).$$

We also define a map in $\mathcal{H}$ via the gradient flow (1.1): for a fixed time s, and for any $\varphi \in \mathcal{H}$, we define that the image of φ under the map π_s is the image of φ along the gradient flow after time $s > 0$, provided the gradient flow initiated at φ does exist for time $s > 0$. It is clear that for any φ, the map is defined for small $s > 0$. However, for a fixed $s > 0$, π_s is not necessarily defined for all $\varphi \in H$ since we don't know the global existence of the gradient flow.

On the other hand, if the gradient flow exists for all the time for any smooth initial metric, then this induces a well-defined map from $\mathcal{L}(\varphi_0, \varphi_1)$ to $\mathcal{L}(\pi_s(\varphi_0), \pi_s(\varphi_1))$ for any $s > 0$. Since the length of any smooth curve in $\mathcal{H}$ is decreased under the gradient flow, we have

$$\inf_{c \in \mathcal{L}(\pi_s(\varphi_0), \pi_s(\varphi_1))} L(c) \leq \inf_{c \in \mathcal{L}(\varphi_0, \varphi_1)} L(c), \qquad \forall\, s > 0.$$

Thus,

$$d(\pi_s(\varphi_0), \pi_s(\varphi_1)) \leq d(\varphi_0, \varphi_1), \qquad \forall\, s > 0.$$

q.e.d.

Before proving the second part of Theorem 1.5, we need to use a theorem in [4] where an explicit formula for the first derivatives of the distance function in $\mathcal{H}$ is given. For the convenience of the readers, we will include this theorem here. The second part of Theorem 1.5 is essentially a corollary of this theorem.

Theorem 3.1 ([4]). *For any two Kähler potentials φ_0, φ_1, the distance function $d(\varphi_0, \varphi_1)$ is at least C^1. More specifically, if φ_0, φ_1 move along two curves $\varphi_0(s)$, and $\varphi_1(s)$ respectively, and if we denote the distance between $\varphi_1(s)$ and $\varphi_2(s)$ by $L(s)$, then*

$$\begin{aligned}\left.\frac{d\,L(s)}{d\,s}\right|_{s=0} &= \langle X, Y_1\rangle |X|^{-\frac{1}{2}}\Big|_{t=1} - \langle X, Y_0\rangle |X|^{-\frac{1}{2}}\Big|_{t=0}\\ &= \int_V \frac{\partial \varphi_1}{\partial s}\frac{\partial\,\varphi}{\partial\,t}\,dg(s)\cdot\left.\left\{\int_V \left|\frac{\partial\varphi}{\partial t}\right|^2 dg(s)\right\}^{-\frac{1}{2}}\right|_{t=1}\\ &\quad - \int_V \frac{\partial \varphi_0}{\partial s}\frac{\partial\,\varphi}{\partial\,t}\,dg(s)\cdot\left.\left\{\int_V \left|\frac{\partial\varphi}{\partial t}\right|^2 dg(s)\right\}^{-\frac{1}{2}}\right|_{t=0}.\end{aligned}$$

Here $\varphi(t)(0 \le t \le 1)$ denotes the $C^{1,1}$ geodesic connecting the two metrics φ_0 and φ_1; and $X = \frac{\partial\varphi}{\partial t} \in T_{\varphi(t)}\mathcal{H}$ and $Y_i = \frac{\partial\varphi_i}{\partial s} \in T_{\varphi_i}\mathcal{H}$ $(i = 0, 1)$.

Now we complete the proof of Theorem 1.5.

Proof of Theorem 1.5(2). If the gradient flow (1.1) exists for all the time, then it is straightforward to show that flow (1.1) decreases the distance between any two points in $\mathcal{H}$ unless they are connected by a holomorphic transformation. Thus, we only deal with the case when the K-energy is weakly convex. By definition, for any curve $\varphi(t) \in \mathcal{H}$, the K-energy is defined as

$$\frac{dM(\varphi(t))}{dt} = -\int_V \frac{\partial\varphi}{\partial t}\,(R - \underline{R})\,\det g.$$

Along a $C^{1,1}$ geodesic, the second derivative of the K-energy is convex in the weak sense that

$$\frac{d^2\,M(\varphi(t))}{d\,t^2} \ge 0.$$

In particular, we have

$$\left.\frac{dM(\varphi(t))}{dt}\right|_{t=1} \ge \left.\frac{dM(\varphi(t))}{dt}\right|_{t=0}. \tag{3.1}$$

Suppose that $\varphi(t)(0 \le t \le 1)$ is the unique $C^{1,1}$ geodesic which connects φ_1 and φ_2, and suppose it is parameterized proportionally to arc length. If we flow φ_1 and φ_2 by the gradient flow (1.1), we have

$$\frac{\partial \varphi_1}{\partial s} = R(\varphi_1(s)) - \underline{R} \qquad \text{and} \quad \frac{\partial \varphi_2}{\partial s} = R(\varphi_2(s)) - \underline{R}.$$

Plugging this into the corresponding formula in Theorem 3.1, we have

$$\begin{aligned}
\frac{d\,L(s)}{ds} &= \left\{ \int_V \left| \frac{\partial \varphi}{\partial t} \right|^2 d\,g(s) \right\}^{-\frac{1}{2}} \\
&\quad \cdot \left(\int_V (R(\varphi_2) - \underline{R})\ \frac{\partial \varphi}{\partial t}\ dg(s) \Big|_{t=1} \right. \\
&\qquad \left. - \int_V (R(\varphi_1) - \underline{R})\ \frac{\partial \varphi}{\partial t}\ dg(s) \Big|_{t=0} \right) \\
&= - \left\{ \int_V \left| \frac{\partial \varphi}{\partial t} \right|^2 d\,g(s) \right\}^{-\frac{1}{2}} \cdot \left(\frac{dM}{dt}\Big|_{t=1} - \frac{dM}{dt}\Big|_{t=0} \right) \\
&= - \left\{ \int_V \left| \frac{\partial \varphi}{\partial t} \right|^2 dg(s) \right\}^{-\frac{1}{2}} \cdot \int_{t=0}^{1} \frac{d^2 M}{dt^2}\ dt \le 0.
\end{aligned}$$

q.e.d.

4. Some further corollaries, remarks and the relationship with stability

Corollary 4.1 ([8]). *If all geodesics are smooth, then any extremal Kähler metric is unique up to a holomorphic automorphism.*

Proof. Suppose that there exist two extremal Kähler metrics in a fixed Kähler class. It was proved in [3] that any extremal Kähler metric must be symmetric with respect to a maximal compact subgroup. Without loss of generality, one may assume that both metrics are symmetric under the same maximal compact subgroup. Then every metric in the geodesic which connects this two extremal metrics must also have the same symmetry group (via the Maximum Principle). If the scalar curvature is constant, then an argument of Donaldson [8] on the convexity of the K energy implies that the extremal metric must be unique. If the scalar curvature is not a constant, then the gradient vector field of the scalar curvature is a holomorphic vector field and it is unique in each Kähler class once the maximal compact subgroup is fixed. In particular, the gradient flow (1.1) restricted to the two extremal metrics induces the same holomorphic transformation. Thus the distance of these two extremal Kähler metrics under the gradient flow is unchanged. Suppose φ_1, φ_2 are the two extremal metrics and $\varphi(\cdot, s)$ is the unique geodesic connecting them. Since the distance of φ_1 and φ_2 is not decreased under the gradient flow, by Theorem 1.3, the path $\varphi(\cdot, s)$ must either be totally trivial or represent a holomorphic transformation. q.e.d.

Remark 4.2. For the uniqueness of the extremal Kähler metrics, the known results are as follows:

1) In the 1950s, the first author showed the uniqueness of Kähler-Einstein metric if $C_1 \leq 0$.

2) In 1987, T. Mabuchi and S. Bando [1] showed the uniqueness of Kähler-Einstein metric up to a holomorphic transformation if the first Chern class is positive.

3) In [4], the second author proved that any metric with constant scalar curvature is unique in each Kähler class if $C_1(V) < 0$.

The problem for the general case is still open. However, the second author [5] had examples of nonuniqueness of some degenerated extremal Kähler metrics in S^2.

S.-T. Yau predicted in [14] that the existence of a Kähler-Einstein metric is related to the stability in the sense of Hilbert schemes and geometric invariant theory. His conjecture should be extended to include the case of extremal Kähler metrics. From Theorem 1.3, we observe some kind of link, perhaps still a bit mysterious, between the the existence of extremal metrics and "stability" of the infinite dimensional space $\mathcal{H}$ in some sense. At least formally, it fits nicely in the general picture that Yau's conjecture describes. The following paragraph is essentially speculative in trying to explain this point. If we are willing to put aside the regularity issue, then Theorem 1.3 implies that the gradient flow of the K energy is a distance contracting flow in $\mathcal{H}$. In this infinite dimensional path length space $\mathcal{H}$, we choose a large enough ball, which hopefully contains any possible candidates for extremal Kähler metrics. Now flow the entire ball by this gradient flow. If a global solution of the gradient flow always exists for all smooth initial metrics, then the contracting nature of the flow will shrink the size of the ball. In the limit, the ball will shrink to a point, which must be an extremal

Kähler metric we are looking for. However, this formal picture is not quite complete. A dichotomy can possibly take place: As the size of the ball shrinks, the ball may also drift away to infinity. In the first possibility when the ball stays in a finite domain, the infinite dimensional manifold is considered "stable" in some sense and we arrive at the unique extremal Kähler metric in the limit of the flow. In the second case when the ball drifts to infinity, then the infinite dimensional space is considered "unstable" in some sense, and the gradient flow converges to an extremal Kähler metric in a different Kähler manifold.

References

[1] S. Bando and T. Mabuchi, *Uniqueness of Einstein Kähler metrics modulo connected group actions*, Algebraic Geometry (Sendai, 1985), Adv. Stud. Pure Math., **10**, North-Holland, Amsterdam, 1987, 11–40, MR 89c:53029, Zbl 0641.53065.

[2] E. Calabi, *Extremal Kähler metrics*, Seminar on Differential Geometry, Ann. of Math. Stud., **102**, Princeton Univ. Press, Princeton, N.J., 1982, 259–290, MR 83i:53088, Zbl 0487.53057.

[3] E. Calabi, *Extremal Kähler metrics*, II, *Differential Geometry and Complex Analysis*, Springer, 1985, 96–114, MR 86h:53067, Zbl 0574.58006.

[4] X. X. Chen. *Space of Kähler metrics*, J. Differential Geom. **56** (2000), 189–234, MR 1 863 016.

[5] X. X. Chen, *Obstruction to the existence of metric whose curvature has umbilical Hessian in a K-surface*, Comm. Anal. Geom., **8**(**1**) 267–299, 2000, MR 2001k:53060, Zbl 0971.53029.

[6] X. X. Chen, *Calabi flow in Riemann surfaces revisited: a new point of view*, International Mathematics Research Notices (**6**) 2001, 276–297, MR 2002c:53111.

[7] P. T. Chrusciel, *Semi-global existence and convergence of solutions of the Robinson-Trautman (2-dimensional Calabi) equation*, Comm. Math. Phys., (**137**) 1991, 289–313, MR 92a:83007, Zbl 0729.53071.

[8] S.K. Donaldson, *Symmetric spaces, Kähler geometry and Hamiltonian dynamics*, Northern California Symplectic Geometry Seminar, Amer. Math. Soc. Transl. Ser. 2, **196**, Amer. Math. Soc., Providence, RI, 1999, 13–33, MR 2002b:58008, Zbl 0972.53025.

[9] N. J. Korevaar and R. Schoen, *Sobolev spaces and harmonic maps for metric space targets*, Comm. Anal. Geom., **1**(**4**) 1993, 1–99, MR 95b:58043, Zbl 0862.58004.

[10] T. Mabuchi, *Some sympletic geometry on compact Kähler manifolds* I, Osaka, J. Math., **24** 1987, 227–252, MR 88m:53126, Zbl 0645.53038.

[11] A. Petrunin, *Metric minimizing surfaces*, Electronic Research Announcements of the Amer. Math. Soc., **5** 1999, 47–54, MR 2000a:53061, Zbl 0916.53032.

[12] S. Semmes, *Complex Monge-Ampere and sympletic manifolds*, Amer. J. Math., **114** 1992, 495–550, MR 94h:32022, Zbl 0790.32017.

[13] M. S. Struwe, *Curvature flows on surfaces*, preprint, August 2000.

[14] S.-T. Yau, *Open problems in geometry*, Chern—a Great Geometer of the Twentieth Century, Internat. Press, Hong Kong, 1992, 275–319, MR 1 201 369

University of Pennsylvania, Philadelphia, PA 19104-6395
Princeton University, Princeton, NJ 08544-1000

Acknowledgements

The Editors of this volume and Springer express their gratitude to the copyright owners for granting permissions. The reprints in this book are published with the kind permission of:

1. **American Mathematical Society, Providence, RI**

[2] Calabi, E. & Dvoretzky, A. (1951). Convergence- and sum-factors for series of complex numbers
[3] Calabi, E. & Spencer, D. C. (1951). Completely integrable almost complex manifolds
[5] Calabi, E. & Rosenlicht, M. (1953). Complex analytic manifolds without countable base
[9] Calabi, E. (1954). The variation of Kähler metrics. I.
[10] Calabi, E. (1954). The variation of Kähler metrics. II.
[15] Calabi, E. (1958). Construction and properties of some 6-dimensional almost complex manifolds
[21] Calabi, E. (1961). On compact, Riemannian manifolds with constant curvature. I
[23] Calabi, E. (1964). Linear systems of real quadratic forms
[32] Calabi, E. (1970). Examples of Bernstein problems for some nonlinear equations
[35] Calabi, E. (1975). A construction of nonhomogeneous Einstein metrics
[40] Calabi, E. (1982). Linear systems of real quadratic forms. II
[48] Calabi, E. & Gluck, H. (1993). What are the best almost-complex structures on the 6-sphere?
[51] Bourguignon, J.-P., Calabi, E., Eells, J., García-Prada, O., Gromov, M. (2001). Where does geometry go? A research and education perspective

2. Annals of Mathematics, Princeton, NJ

[**6**] Calabi, E. & Eckmann, B. (1953). A class of compact, complex manifolds which are not algebraic
[**7**] Calabi, E. (1953). Isometric imbedding of complex manifolds
[**20**] Calabi, E. & Vesentini, E. (1960). On compact, locally symmetric Kähler manifolds
[**22**] Calabi, E. & Markus, L. (1962). Relativistic space forms

3. Odile Berger, Paris

Berger, M. (1996). Encounter with a geometer: Eugenio Calabi

4. Birkhäuser Verlag, Basel

[**44**] Calabi, E. (1988). Convex affine maximal surfaces

5. Jean-Pierre Bourguignon, Bures-sur-Yvette

Bourguignon, J.-P. (1996). Eugenio Calabi and Kähler metrics

6. Duke University Press, Durham, NC

[**17**] Calabi, E. (1958). An extension of E. Hopf's maximum principle with an application to Riemannian geometry
[**18**] Calabi, E. (1959). Errata: An extension of E. Hopf's maximum principle with an application to Riemannian geometry
[**26**] Calabi, E. (1967). On Ricci curvature and geodesics
[**31**] Calabi, E. & Hartman, P. (1970). On the smoothness of isometries

7. Elsevier (Academic Press), San Diego, CA

[**36**] Calabi, E. & Wilf, H. S. (1977). On the sequential and random selection of subspaces over a finite field
[**50**] Calabi, E., Olver, P. J., Tannenbaum, A. (1996). Affine geometry, curve flows, and invariant numerical approximations

8. International Press of Boston, Somerville, MA

[**25**] Calabi, E. (1967). Minimal immersions of surfaces in Euclidean spheres
[**46**] Calabi, E. & Cao, J. (1992). Simple closed geodesics on convex surfaces
[**52**] Calabi, E. & Chen, X. X. (2002). The space of Kähler metrics. II

9. Istituto Nazionale di Alta Matematica, Roma

[**34**] Calabi, E. (1972). Complete affine hyperspheres. I

10. Johns Hopkins University Press, Baltimore, MD

[**42**] Calabi, E. (1982). Hypersurfaces with maximal affinely invariant area

11. Koninlijk Wiskundig Genootschap (KWG), Amsterdam

[**8**] Calabi, E. (1954). The space of Kähler metrics

12. L'École normale supérieure, Paris

[**37**] Calabi, E. (1979). Métriques kählériennes et fibrés holomorphes

13. Mathematischen Forschungsinstituts, Oberwolfach

[**33**] Calabi, E. (1971). Über singuläre symplektische Strukturen

14. Michigan Mathematical Journal, Ann Arbor, MI

[**16**] Calabi, E. (1958). Improper affine hyperspheres of convex type and a generalization of a theorem by K. Jörgens

15. N. Bourbaki

[**39**] Calabi, E. (1981). Géométrie différentielle affine des hypersurfaces

16. Princeton University Press, Princeton, NJ

[**4**] Calabi, E. (1953). Metric Riemann surfaces
[**11**] Calabi, E. (1957). On Kähler manifolds with vanishing canonical class
[**30**] Calabi, E. (1970). On the group of automorphisms of a symplectic manifold
[**41**] Calabi, E. (1982). Extremal Kähler metrics

17. Société Mathématique de France, Paris

[**19**] Calabi, E. & Vesentini, E. (1959). Sur les variétés complexes compactes localement symétriques
[**49**] Calabi, E. (1996). Extremal isosystolic metrics for compact surfaces